October 21–25, 2012
Paphos, Cyprus

Association for
Computing Machinery

Advancing Computing as a Science & Profession

MSWiM'12

Proceedings of the 15th ACM International Conference on

Modeling, Analysis and Simulation of Wireless and Mobile Systems

Sponsored by:

ACM SIGSIM

Association for
Computing Machinery

Advancing Computing as a Science & Profession

The Association for Computing Machinery
2 Penn Plaza, Suite 701
New York, New York 10121-0701

ISBN: 978-1-4503-1628-6

Additional copies may be ordered prepaid from:

ACM Order Department
PO Box 30777
New York, NY 10087-0777, USA

Phone: 1-800-342-6626 (USA and Canada)
+1-212-626-0500 (Global)
Fax: +1-212-944-1318
E-mail: acmhelp@acm.org
Hours of Operation: 8:30 am – 4:30 pm ET

ACM Order Number: 617121

Printed in the USA

MSWIM 2012 Chairs' Welcome

On behalf of the Organizing Committee, it is our great pleasure to welcome you to the 15th ACM International Conference on Modeling, Analysis and Simulation of Wireless and Mobile Systems (MSWiM), held in 2012 in Paphos, Cyprus Island. MSWiM has a long tradition of introducing important, high quality and timely work, and we are very pleased that this year's event maintains the high standard set in previous years.

In 2012, the call for papers attracted 161 registered papers in all areas of mobile and wireless systems of which 115 were accepted into the review process. The submitted papers came from 37 countries, and the accepted papers from 27 countries, reflecting the truly international profile of MSWiM. It is also informative to note that the five most commonly listed topics for submissions to MSWiM this year were:

- Algorithms and protocols for energy efficiency and power control
- Analytical models
- Capacity, coverage and connectivity modeling and analysis
- Correctness, survivability and reliability evaluation
- Design methodologies

The submissions included a large number of papers of very high quality making the selection process difficult and competitive. In the end, we selected 41 regular papers, which is an acceptance rate of 25%, and an additional 9 short papers were recommended for the program for their high quality and contribution. The MSWiM 2012 technical program also comprises two distinguished keynotes addresses, a poster and demonstration session and tutorials. In addition, there are four one-day Symposia: MobiWac, PE-WASUN, DIVANet and Q2SWinet as well as three workshops: WMuNeP, PM^2HW^2N and HP-MOSys co-located with the main conference. Among the 42 regular papers, three have been selected as candidates for the best paper award at the conference and will be announced in the opening session. The papers selected are:

"A Rigorous Analysis of AODV and its Variants," P. Hofner, R. van Glabbeek, W. L. Tan, M. Portmann, A. McIver, A. Fehnker

"Influence of Spatial Statistics of Spectrum Use on the Performance of Cognitive Wireless Networks," J. Riihijärvi, J. Nasreddine, P. Mähönen

"Modeling and Connectivity Analysis in Obstructed Wireless Ad Hoc Networks," M. G. Almiron, O. Goussevskaia, A. A. F. Loureiro, A. C. Frery

The winner among these three papers will be announced at the conference banquet, and will be reported in the proceedings of MSWiM 2013. At this point, we take the opportunity to congratulate the winner of the best paper award of MSWiM 2011:

"Maximal utility rate allocation for energy harvesting wireless sensor networks," Bo Zhang, Robert Simon and Hakan Aydin, George Mason University, USA.

Our thanks go to Professor Azzedine Boukerche and the Advisory Board for their vision in guiding MSWiM from workshop to symposium and finally to conference status over the past 14 years. A special thank you goes to the Technical Program Committee as well as the external reviewers, for all the hard work in reviewing any judging the contributions of all the submitted papers. We thank our organizing committee, as putting together a conference such as MSWiM is a team effort. Thanks to tutorial co-chairs Oliver Holland and Yi Qian, workshop chair Joel Rodrigues, demo/tools co-chairs Gennaro Boggia and Ivan Martinovic, and poster chair Stefano Basagni, for organizing the value-added sessions MSWiM attendees have come to expect. Our thanks go to Mirela A. M. Notare and Salil Kanhere for advertising the conference. Thanks are also due to Robson De Grande, our registration chair and Constandinos Mavromoustakis, our local arrangements chair. Last but not least we want to thank Richard W. Pazzi for helping out with the web site, putting together the final proceedings as publications chair and for helping out with numerous tasks during the process of organising MSWIM 2012.

Finally, we thank our sponsors, ACM and ACM SIGSIM, for their continued support of these successful meetings.

In closing, we warmly welcome you to MSWiM 2012. We are confident that both the conference and your stay in Paphos will be informative, interesting, and above all enjoyable!

Bjorn Landfeldt
MSWIM'12 Program Co-Chair
Lund University, Sweden

Ravi Prakash
MSWIM'12 Program Co-Chair
University of Texas at Dallas, USA

Albert Zomaya
MSWIM'12 General Chair
University of Sydney, Australia

Vasos Vasiliou
MSWIM'12 Vice General Chair
University of Cyprus

Table of Contents

Session 8: Ad Hoc and Mesh Networks

Session 9: Scheduling

Session 10: MAC & Physical Layers I

Session 11: MAC and Physical Layers II

Session 12: Energy Efficiency and Management

Session 13: Cellular and Vehicular Networks

Session 14: Social Networks

Session 15: Wireless Sensor and Mesh Networks

MSWiM 2012 Conference Organization

General Chair: Albert Zomaya, *University of Sydney, Australia*

General Vice-Chair: Vasos Vasilliou, *University of Cyprus, Cyprus*

Program Chairs: Bjorn Landfeldt, *Lund University, Sweden*
Ravi Prakash, *University of Texas at Austin, USA*

Workshop Chair: Joel Rodrigues, *University of Beira Interior, Portugal*

Poster Chair: Stefano Basagni, *Northeastern University, USA*

Demo/Tools Co-chairs: Gennaro Boggia, *Politecnico di Bari, Italy*
Ivan Martinovic, *University of Oxford, UK*

Tutorials Co-Chairs: Oliver Holland, *Kings College London, UK*
Yi Qian, *University of Nebraska - Lincoln, USA*

Publications Chair: Richard W. Pazzi, *University of Ottawa, Canada*

Local Arrangements Chair: Constandinos Mavromoustakis, *University of Nicosia, Cyprus*

Publicity Chairs: Mirela A. M. Notare, *Barddal University, Brazil*
Salil Kanhere, *University of New South Wales, Australia*

Registration Chair: Robson De Grande, *University of Ottawa, Canada*

Steering Committee Chair: Azzedine Boukerche, *University of Ottawa, Canada*

Program Committee: Nael Abu-Ghazaleh, *State University of New York at Binghamton, USA*
Christer Ahlund, *Lulea University of Technology, Sweden*
Ozgur Akan, *Koc University, Turkey*
Brahim Bensaou, *The Hong Kong University of Science and Technology, Hong Kong*
Azzedine Boukerche, *University of Ottawa, Canada*
Roberto Beraldi, *Sapienza University of Rome, Italy*
Luciano Bononi, *University of Bologna, Italy*
Raffaele Bruno, *IIT-CNR, Italy*
Chi Chung Cheung, *The Hong Kong Polytechnic University, Hong Kong*
Carla-Fabiana Chiasserini, *Politecnico di Torino, Italy*
Kaushik Chowdhury, *Northeastern University, USA*

Program Committee (continued):

Marco Di Felice, *University of Bologna, Italy*
Marcelo Dias de Amorim, *UPMS Sorbonne Universites, France*
Lorenzo Donatiello, *University of Bologna, Italy*
Falko Dressler, *University of Erlangen, Germany*
Stephan Eidenbenz, *Los Alamos National Laboratory, USA*
Ehab Elmallah, *University of Alberta, Canada*
Silvia Giordano, *University of Applied Sciences SUPSI, Switzerland*
James Gross, *Aachen University, Germany*
Andrei Gurtov, *University of Oulu, Finland*
Zugmunt Haas, *Cornell University, USA*
Shengming Jiang, *South China University of Technology, PR China*
Guillaume Jourjon, *NICTA, Australia*
Merkourios Karaliopoulos, *National and Kapodistrian University of Athens, Greece*
Holger Karl, *University of Paderborn, Germany*
Peng-Yong Kong, *Institute for Infocomm Research, Singapore*
Jerzy Konorski, *Gdansk University of Technology, Poland*
Bjorn Landfeldt, *Lund University, Sweden*
Renato Lo Cigno, *University of Trento, Italy*
Francesco Lo Presti, *Universita' di Roma Tor Vergata, Italy*
Vincenzo Mancuso, *IMDEA, Spain*
Michela Meo, *Politecnico di Torino, Italy*
Enzo Mingozzi, *University of Pisa, Italy*
Hamid Nafaa, *University College Dublin, Ireland*
Andrea Passarella, *IIT-CNR, Italy*
Marius Portmann, *University of Queensland, Australia*
Tahiry Razafindralambo, *INRIA Lille, France*
Dirk Staehle, *University of Wuerzburg, Germany*
Violet Syrotiuk, *Arizona State University, USA*
Mineo Takai, *University of California, Los Angeles, USA*
Terence D. Todd, *McMaster University, Canada*
Yu-Chee Tseng, *National Chiao-Tung University, Taiwan*
Damla Turgut, *University of Central Florida, USA*
Emmanouel Varvarigos, *University of Patras, Greece*
Andreas Willig, *University of Canterbury, New Zealand*
Adam Wolisz, *Technical University of Berlin, Germany*
OHongyi Wu, *University of Louisiana at Lafayette, USA*
Yaoqing Yang, *University of Nebraska-Lincoln, USA*
Sherali Zeadally, *University of the District of Columbia, USA*
Hongwei Zhang, *Wayne State University, USA*
Jingyuan Zhang, *University of Alabama, USA*
Jun Zhang, *The Hong Kong Polytechnic University, Hong Kong*

Additional reviewers:

Karl Andersson
Zin Ao
Arash Asadi
Ketan Bahulkar
Luca Bedogni
Ignacio Bermudez
A.Ozan Bicen
Marthinus Booysen
Faycal Bouhafs
Huayang Cao
Xin Che
Yu Chen
Delia Ciullo
Ekaterina Dashkova
Christian Dombrowski
Juergen Eckert
Ozgur Ergul
Stefano Ferretti
Alessandro Finamore
Scott Fowler
Li Gao
Andres Garcia-Saavedra
Ali J. Ghandour
Vittorio Ghini
Quasheng Guan
Bingyi Guo
Yassine Hadjadj-Aoul
Abdulla Hammad
Hanan Hassanien
Jan-Hinrich Hauer
Oswald Jumira
Merkourios Karaliopuolos
Murat Kocaoglu
Evangelia Kokolaki
Pardeep Kumar
Dimitriy Kupstov
Frank Lehrieder
Chuan Li
Lavy Libman
Li Liu
Xiaohui Liu
Joong Soo Ma
Adnan Majeed
Derya Malak
Francesco Malandrino
Gustavo Marfia
Valentina Martina
Khalil Massri
Constandinos Mavromoustakis
Emiliano Miluzzo
Vicram Munishwar
Sean Murphy
Ghasem Naddafzadeh Shira
Suneth Namal
Enrico Natalizio
Ilya Nikolaevsky
Andreas Panayides
Donald Parruca
Paul Patras
Tatiana Polishchuk
Oscar Punal
Qilin Qi
Nandakishore Santhi
Olov Schelen
Michele Segata
Charalambos Sergiou
Ghalib Shah
Mohamed Shazly
Mohammad Sheikh
Dirk Staehle
Poitr Szczechowiak
Hwee-Xian Tan
Lin Tang
Bige Unluturk
Salvatore Vanini
Vasos Vassiliou
Yuehua Wang
Zhijun Wang
Hongyi Wu
Wedong Xiao
Ye Yan
Yu-Ting Yu
Ming-Tuo Zhou
Zinon Zinonos
Albert Zomaya
Dimitrios Zorbas

MSWiM 2012 Sponsor

Sponsor: SIGSIM

Modeling and Optimization of Vehicular Wireless Ad-Hoc Networks

Kin K. Leung
Electrical and Electronic Engineering Dept
Imperial College
London, UK
kkleung@ieee.org

Abstract:

In this talk, the speaker will give an overview of his current research work in the area of wireless ad-hoc, mesh and sensor networks at Imperial College. By focusing on vehicular networks, the speaker will present new stochastic traffic models for vehicular ad-hoc networks (VANETs) in urban environments and their applications to quantify communication connectivity and identify locations for placing road-side communication nodes to optimize connectivity. Results on distributed optimization for ad-hoc networks with power control will be discussed. The talk will then be concluded with a discussion on future work to extend the traffic models, optimization techniques and cross-layer protocol designs for VANETs.

Categories and Subject Descriptors

H.1.1 [Models and Principles]: Systems and Information Theory

Keywords

network design, network optimization, performance modeling, wireless ad-hoc networks, wireless networks

Bio:

Kin K. Leung received his B.S. degree from the Chinese University of Hong Kong in 1980, and his M.S. and Ph.D. degrees from University of California, Los Angeles, in 1982 and 1985, respectively.

He joined AT&T Bell Labs in New Jersey in 1986 and worked at its successor companies, AT&T Labs and Bell Labs of Lucent Technologies, until 2004. Since then, he has been the Tanaka Chair Professor in the Electrical and Electronic Engineering (EEE), and Computing Departments at Imperial College in London. He serves as the Head of Communications and Signal Processing Group and as the Deputy Director for the University Defense Research Center in Signal Processing in the EEE Department at Imperial College. His research interests focus on networking, protocols, optimization and modeling issues of wireless broadband, sensor and ad-hoc networks. He also works on multi-antenna and cross-layer designs for the physical layer of these networks.

He received the Distinguished Member of Technical Staff Award from AT&T Bell Labs in 1994, and was a co-recipient of the 1997 Lanchester Prize Honorable Mention Award. He was elected as an IEEE Fellow in 2001. He received the Royal Society Wolfson Research Merits Award from 2004 to 2009. He has actively served on many conference committees. He serves as a member (2009-11) and the chairman (2012-13) of the IEEE Fellow Evaluation Committee for Communications Society. He was a guest editor for the IEEE Journal on Selected Areas in Communications (JSAC), IEEE Wireless Communications and the MONET journal, and as an editor for the JSAC: Wireless Series, IEEE Transactions on Wireless Communications, and IEEE Transactions on Communications. Currently, he is an editor for the ACM Computing Survey and International Journal on Sensor Networks. Survey and International Journal on Sensor Networks.

MSWiM'12, October 21–25, 2012, Paphos, Cyprus.
ACM 978-1-4503-1628-6/12/10.

Sensing, Tracking and Contextualizing Entities in Ubiquitous Computing

Antonio A.F. Loureiro
Department of Computer Science
Universidade Federal de Minas Gerais
Belo Horizonte, Minas Gerais, Brazil
loureiro@dcc.ufmg.br

ABSTRACT

Information and Communications Technology is increasingly becoming a part of our daily activities. Nowadays, we are able to sense a broad range of entities in the world: from physical entities that comprise the Internet of Things (IoT) to social entities that include people in social networks. We can expect to have a ubiquitous sensing infrastructure in different environments, with varying needs and complexities. While these sensors/entities are mostly static, smart phones are taking center stage as the most widely adopted and ubiquitous computing device. Smart phones have enormous potential for the study of human social networks and human behavior "in vivo," in a natural context outside laboratories. Besides their computing power, smart phones are currently available with an increasing rich set of embedded sensors, such as GPS, accelerometer, microphone, camera, gyroscope and digital compass. Sensing vast areas becomes more feasible when people carrying their portable devices collect data and collaborate among themselves. Systems that enable sensed data in this way are named participatory sensing systems (PSSs). In those networks, the shared data is not limited to sensor readings passively generated by the device, but also includes proactive user observations. A challenge is how to obtain meaningful information from this large amount of data at different scales along the time. Some of the traditional techniques that have been used to process those distinct data sources are information fusion, data mining and machine learning. Notice that the consumers of such information can be, in one extreme, a large set of cooperating or non-cooperating entities, and, in the other, an individual or a "thing", for instance.

Two important attributes of the sensed data are the location and time where they are collected. Localization and tracking play a key role in ubiquitous computing due to the fact that a wide variety of envisioned applications that collect sensed data rely on their ability to detect, localize and track different entities. This adds a new dimension to this problem when obtaining meaningful information from data associated to different entities. An important problem here is target tracking, which is the capacity of detecting and continuously tracking the state of a target (entity), or a set of targets. This is another fundamental problem in ubiquitous computing.

Finally, the last dimension considered in this talk is context, which can be defined as any information used to characterize an entity. Context and context-awareness provide a ubiquitous computing environment with the capacity of adapting available services by autonomically deriving the entities' needs from the context they are.

In this talk, we will discuss each one of these basic building blocks, and discuss how the combination of sensing, tracking and contextualization give us information to make sound decisions in ubiquitous computing that can definitely lead to useful services in areas as diverse as economics, social network, ecology, security, environment and health.

Categories and Subject Descriptors

C.2.4 [**Distributed Systems**]: Distributed applications—*Ubiquitous computing*

Keywords

Sensing, tracking, context, ubiquitous computing

Short bio

Antonio A.F. Loureiro is a professor in the Computer Science Department of UFMG, Brazil. He holds a Ph.D. in Computer Science from the University of British Columbia, Canada, 1995. Back in 1998, he established the first mobile computing research group in Brazil, namely the Ubiquitous and Wireless Sensor Networks Lab. Besides, he coauthored the book "Introduction to Mobile Computing", in Portuguese), perhaps the first book on mobile computing published in Portuguese. Afterwards, in 2002, he coordinated the SensorNet project (www.sensornet.dcc.ufmg.br), a Brazilian pioneering project on sensor networks. He has published over 200 papers in referred publishing venues. His main research areas include wireless sensor networks, mobile computing, vehicular networks, and distributed systems.

MSWiM'12, October 21–25, 2012, Paphos, Cyprus.
ACM 978-1-4503-1628-6/12/10.

Influence of Spatial Statistics of Spectrum Use on the Performance of Cognitive Wireless Networks

Janne Riihijärvi, Jad Nasreddine and Petri Mähönen
Institute for Networked Systems, RWTH Aachen University
Kackertstrasse 9, D-52072 Aachen, Germany
email: {jar, jad, pma}@inets.rwth-aachen.de

ABSTRACT

We study the performance characteristics of cognitive wireless networks under different dynamic spectrum access scenarios. Our focus is especially on the influence of spatial structures of the primary and secondary user networks on the achievable performance of the secondary network. We adopt techniques from spatial statistics to develop stochastic models for the structure and interaction of these networks. The chosen models are based on Gaussian random fields and Gibbs point processes, and are firmly grounded on empirical data. We then apply extensive Monte Carlo simulations to study the behavior of these models and their related performance properties. The models and the applied techniques are applicable also to a wider variety of networking problems. Our results provide first quantitative assessments on the influence of the spatial structure, and especially correlation properties, of the involved networks on the expected performance and thus on the utility of dynamic spectrum access based systems.

Categories and Subject Descriptors

C.2 [**Computer-Communication Networks**]: Miscellaneous; I.6.4 [**Simulation and Modeling**]: Model Validation and Analysis; I.6.5 [**Simulation and Modeling**]: Model Development

Keywords

Performance evaluation, dynamics spectrum access, cognitive wireless networks, spatial statistics

1. INTRODUCTION

Cognitive Wireless Networks (CWNs) and in particular Dynamic Spectrum Access (DSA) based systems have become one of the most intensely studied research topics in wireless communications [1–3]. There has been significant progress during recent years in understanding theoretical foundations of the individual technologies involved, as well as studying different application scenarios for DSA (see, for example, [4–7]). However, the system-level performance properties of dynamic spectrum access networks remain relatively poorly understood, especially when the primary user is not a TV broadcasting network. This is in part due to the very complicated dynamics over space, time, and frequency domains inherent in such networks. These dynamics arise because the reuse of spectrum effectively couples the dynamics of the involved networks together in a manner that does not occur in more traditional wireless communication systems. From practical point of view, most significant results have probably been estimates on available spectrum for DSA use in the context of TV white spaces based on actual TV tower data sets [8–12]. These data sets have also been used to study the potential performance of Wi-Fi or cellular network deployments in the TV bands (see, for example, [13]). Some more general asymptotic capacity results have also been presented in the literature, but these are usually derived assuming either very large-scale multihop networks, or uniformly random deployment of nodes in both primary and secondary networks [14–17].

In this paper we provide detailed results on the influence of spatial structure of the primary and secondary networks on the achievable performance of the secondary system. We formulate a stochastic model for the structure and interaction of the primary and secondary user networks using techniques from spatial statistics [18] in an effort to strike a balance between mathematical tractability and realism of the model. In particular, we argue based on extensive spectrum use measurements and several empirical data sets on wireless network structure that a combination of *Gaussian random fields* [19] and *Gibbs point processes* [20] with careful selection of parameters yield good models. While our focus here is on the influence of the spatial structure of the networks involved on performance, the models can be extended towards the full space-time case as well in the manner of [21][1]. We believe that the modeling approach adopted is also of independent interest, and can be used as a foundation for treatment of wide variety of other types of networking problems, wireless or otherwise. The price to be paid for the added realism in this approach is the scarcity of closed-form analytical results to rely on, except in various asymptotic cases. Accordingly, in order to obtain results that are relevant for expected deployments, we focus instead on the use

[1]In this paper we generalize the approach used in [21], where we applied actual cellular network base station locations as a model of the primary network. In the present work much more general spatial models are used.

MSWiM'12, October 21–25, 2012, Paphos, Cyprus.

of Monte Carlo simulations in deriving our performance estimates.

The rest of this paper is structured as follows. In Section II we give a short overview of the DSA scenarios considered in the rest of the paper, and discuss the different design alternatives. We then give the precise model formulation of the arising networks based on the scenarios in Section III. In Section IV we present our results on the arising secondary network structures and their allowed transmit powers, and then extend these results towards capacity estimates in Section V. We finally draw our conclusions in Section VI.

2. DEFINITION OF THE CONSIDERED SCENARIOS

Before presenting the modeling framework in detail, we shall briefly recall some basic facts about DSA based networks. The defining property of such networks is that a *secondary network* can access *opportunistically* the licensed spectrum of a *primary network* that is willing to share its spectrum under some constraints. In the classical DSA scenario the primary network is an incumbent holding a license for a range of frequencies used, e.g., for broadcasting services, whereas the secondary user is a packet radio network seeking to use the same range of frequencies when and where the secondary activity does not degrade the primary quality of service more than given by a specified threshold. However, the range of potential applications for DSA techniques is much broader. For example, the primary and secondary systems might be operated by the same operator (e.g. the primary being a mobile network and the secondary an IEEE 802.11 hotspot deployed by the same operator). The secondary network could even be an integrated part of the primary network such as in the case of femtocell network that might need to use some of the allocated sub-channels to the macrocells due to high traffic loads. In our model we do not make specific assumptions about the technologies involved. In particular, primary networks will not be limited to broadcasting networks with very large coverage as is usually considered in the literature (e.g. IEEE 802.22).

The scenario in which DSA is being used defines the constraints the secondary must follow with respect to the access to the primary spectrum. These constraints can be either related to a reduction in the coverage area or to an acceptable interference probability [6, 17, 22, 23]. In this paper, we mainly use the interference probability as the primary constraint. However, it should be noted that a mapping between the two approaches can be derived when specific primary networks with known system characteristics are considered. By using the interference probability constraint, the primary network will ensure that all its users have some required level of Quality of Service (QoS), meaning that the outage probability is kept at a low level. This is reflected by the fact that the interference probability experienced by any primary network client should not exceed a threshold ε. Depending on the chosen application scenario this can either be achieved by on-line algorithms estimating the interference probability for a given secondary network configuration, or by defining through off-line means a fixed spectrum access policy for secondary use. The latter can be based on, for example, use of an energy detector and a limit on maximum allowed transmit power.

The interference probability with respect to primary user, p, is the probability that the experienced interference I by p due to secondary activity exceeds a predefined threshold $\iota_{\max}$

$$\mathbb{P}_{\text{int}} \equiv \mathbb{P}\{I \geq \iota_{\max}\}. \tag{1}$$

In general, one can define a spectrum hole as "the region of space–time–frequency in which a particular secondary use is possible" [6]. In this paper we focus on the spatial dimension, defining spectrum holes as regions in which either the secondary can transmit with a positive power while respecting the interference constraint, or in which the given energy detection threshold θ_{E} is not exceeded. These will be referred to as the *power control model* and the *energy detection model* in the following, respectively.

In order to relate the interference probability to the spatial structure of the involved networks, a model for interference and thus for radio propagation is needed. We assume that the interference I experienced by the primary due to the activity of a secondary user at distance d and transmitting with power P is given in logarithmic scale by

$$I(\text{dBm}) = P(\text{dBm}) - L(d) + \xi, \tag{2}$$

where $L(d)$ is the large-scale deterministic path loss and ξ is a normal random variable used to model shadowing. Unless mentioned otherwise, we assume ξ to have zero mean, and standard deviation of $\sigma_s = 7\,\text{dB}$. For path loss $L(d)$, we adopt the Xia-Bertoni propagation model [24] defined for a given frequency f (expressed in GHz) and distance d by

$$L(d) = K + A_1 \log_{10}(f) + A_2 \log_{10}(d), \tag{3}$$

where the constants K, A_1 and A_2 are taken to have values of 122.1 dB, 21 dB and 37.6 dB, respectively. With these assumptions and considering ε as the threshold of the interference probability the primary constraint can be written using (1) and (2) as

$$\frac{1}{2}\left[1 - \operatorname{erf}\left(\frac{\iota_{\max} + L(d) - P}{\sigma_s\sqrt{2}}\right)\right] = \varepsilon. \tag{4}$$

Let us now relate the primary constraint (4) to the considered scenarios. In the scenario based on energy detection, the size of the no-talk zone [6], in which the secondary user is not allowed to transmit, depends on the distribution of power received from the primary, and the energy detection threshold θ_E. The latter is in turn set by, for example, a regulator or the operator of the primary network based on the maximum allowed transmit power P, maximum interference $\iota_{\max}$, interference probability ε and the propagation model. Further, assumptions on primary and secondary network structure are needed to make the connection to the interference probability from these quantities. A typical example of this scenario is the IEEE 802.22 system [11,12]. For further discussion on detector based approaches, see [5,7,25].

In the scenario based on power control, the definition of the spectrum hole becomes more flexible. In this scenario, the no-talk zone is reduced to match the coverage area of the primary[2] and the allowed transmit power of the secondary user will depends on its relative position towards the primary coverage area, or the nearest client of the primary network within this coverage area [26–28]. Implementing such an approach is, of course, more challenging as the

[2]This is due to the fact that the receiver can be at any place inside the coverage area, and therefore no secondary can be allowed to be active there.

coverage area of the primary has to be either communicated explicitly to the secondary (resulting effectively into a geolocation-based scheme) or estimated on-line. The latter can be accomplished using collaborative measurements and techniques from spatial statistics and signal processing communities [29–31]. Once the distance d to the coverage area or the nearest primary client has been estimated, the secondary network can invert (4) to obtain the allowed transmit power, given by

$$P(\text{dBm}) = \iota_{\max} + L(d) + \sigma_s \sqrt{2}\,\text{erf}^{-1}(2\varepsilon - 1)\,. \quad (5)$$

3. MODELS OF NODE LOCATIONS AND SPECTRUM USE

We shall now introduce the models used for primary and secondary networks that will be used throughout the rest of the paper to study the performance characteristics of the two scenarios introduced above. We shall begin by introducing our model of the primary network, and then consider deployment models for the secondary system.

3.1 Random Field Models of Primary Spectrum Use

The influence of the primary network on the secondary one is usually studied by separately modeling the locations of primary transmitters, and combining these with models of transmitter activity and radio propagation to arrive at a value of the power spectral density (PSD) at each point in space. This results in models with large number of parameters that tend to be fixed to particular technologies and deployment scenarios. We instead aim at constructing a more parsimonious family of models, combining all of these effects into a single stochastic model that still can be fitted to empirical data. More precisely, we shall assume that the power spectral density of the primary transmitter network (measured in dBm) is a realization of a Gaussian random field $Z(\boldsymbol{x})$, with $\boldsymbol{x}$ taking values in some chosen domain $D \subset \mathbb{R}^2$.

The actual marginal PSD distributions for large number of transmitter usually have heavier upper tails than Gaussian random fields, closer to α-stable distributions [32]. However, for our applications the property that matters is the lower tail of the marginal distribution that is usually very well approximated by Gaussian random fields. This assumption is, of course, related to the common practice of approximating sums of certain log-normal random variables by a log-normal random variable [33], and seems to hold well in practice as illustrated in Figure 1. For more details on the measurement campaign based on which the figure was generated, see [34].

Assuming now second-order stationarity and isotropy such a model is defined by the mean $\mu_Z \equiv \mathbb{E}\{Z(\boldsymbol{x})\}$ and the covariance function

$$C(r) \equiv \text{Cov}\big(Z(\boldsymbol{x}), Z(\boldsymbol{x}')\big), \quad (6)$$

where $r \equiv \|\boldsymbol{x} - \boldsymbol{x}'\|$. We choose the covariance function considered here based on extensive measurement campaigns we have conducted on spectrum use over space, described in detail in [34,35]. Based on those results we shall assume that $C(r)$ is given by the *Matérn model*, defined by

$$C_{\text{Matérn}}(r) \equiv \sigma^2 \left(\frac{1}{2^{\kappa-1}\Gamma(\kappa)} \left(\frac{r}{\phi}\right)^{\kappa} K_\kappa\left(\frac{r}{\phi}\right) \right), \quad (7)$$

where $\kappa > 0$, $\phi > 0$, $\sigma^2 > 0$, Γ is the gamma function, and K_κ is the modified Bessel function of the second kind.

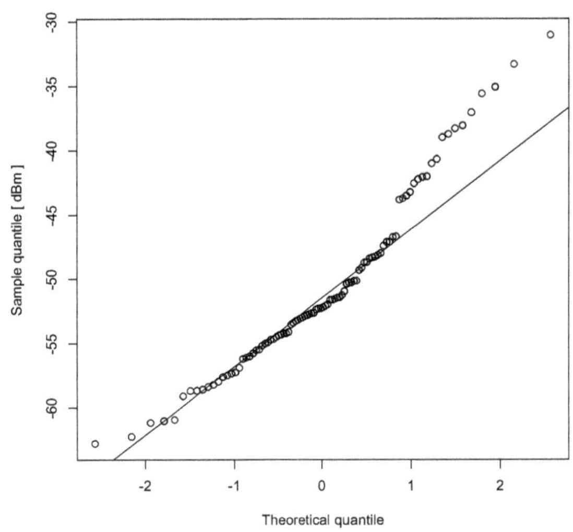

Figure 1: The normal quantile-quantile plot for the marginal PSD distribution for the DVB-T channel 37 measured at 96 locations in downtown Aachen, Germany [34]. The lower tail is clearly very well approximated by the normal distribution.

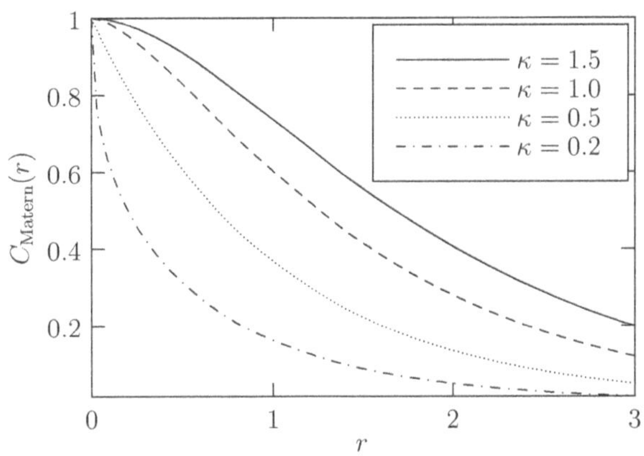

Figure 2: The Matérn covariance function for different values of the smoothness parameter κ (with $\sigma^2 = \phi = 1$).

Figure 2 illustrates the shape of the covariance function for different values of κ. The parameter σ^2 controls the variance of the marginal distribution of Z. The *effective correlation range* of the model is defined as the value of r at which $C(r) = 0.05$. It is influenced by both ϕ and κ, with the influence of the latter shown in Figure 3. Typical values of the effective correlation range used below will range from some hundreds of meters to few kilometers. The parameter κ is also related to the differentiability of realizations of $Z(\boldsymbol{x})$. Namely, if $\kappa \geq (2n+1)/2$ the outcomes of Z will be n times differentiable. Because of this property, we shall refer κ as the *smoothness* of the model.

Corresponding to the two approaches to dynamic spec-

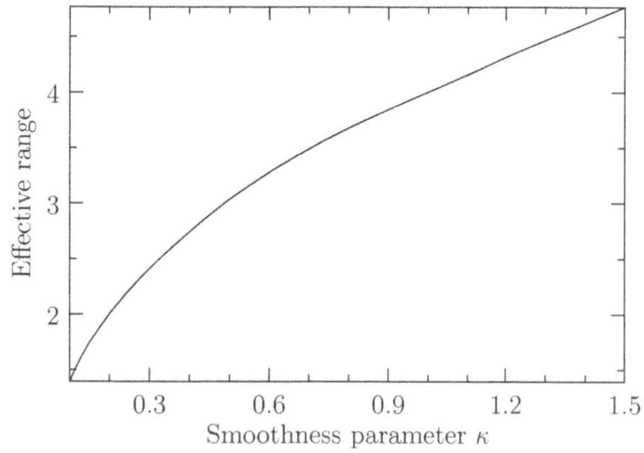

Figure 3: Influence of the smoothness parameter κ on the effective range of the Matérn covariance function (with $\sigma^2 = \phi = 1$).

trum access discussed in the previous section, we introduce two threshold parameters that together with Z define further random sets that will be central for evaluating the secondary network performance. The *energy detection threshold* θ_{E} defines together with Z the *no-talk zone*

$$D_{\text{NT}} \equiv \{ \boldsymbol{x} \in D \,|\, Z(\boldsymbol{x}) \geq \theta_{\text{E}} \}, \tag{8}$$

inside which secondary use based on energy detection is not allowed. For secondary use based on power control, we define the *primary network terminal sensitivity* θ_{S} yielding the *potential primary user service area*[3]

$$D_{\text{SA}} \equiv \{ \boldsymbol{x} \in D \,|\, Z(\boldsymbol{x}) \geq \theta_{\text{S}} \}. \tag{9}$$

The service area D_{SA} defines also another random field, given at each point $\boldsymbol{x} \in D$ as the value of P in (5) evaluated at distance $d = \min_{\boldsymbol{y} \in D_{\text{SA}}} \|\boldsymbol{x} - \boldsymbol{y}\|$. We shall call this random field of *allowed transmit powers*. Figure 4 shows an example realization of the different random sets and fields defined until now.

3.2 Models for Node Locations in the Secondary Network

For modeling the node locations in the secondary network we shall also apply a probabilistic approach. We assume that the locations are given by a *point process*. Intuitively, a point process is a random process that yields as realizations collections of node locations $\{X_i\}_{i=1}^{n}$. The individual locations X_i as well as the total number of locations n are random, but in general with complicated dependency structure. More formally, a point process is defined as a random counting measure N, assigning to each measurable set $A \subseteq D$ a random variable $N(A)$, the number of points in A [36]. To make the connection with the more intuitive definition, we can always write

$$N = \sum_{i=1}^{n} \varepsilon_{X_i}, \tag{10}$$

[3]Notice that for most actual networks D_{SA} would be slightly larger than the actual service area, with the difference being dependent on the inter-transmitter interference characteristics of the primary network.

where ε_x is a point mass at x (the measure equivalent to Dirac delta distribution), the X_i are D-valued random variables, and n is a random variable with values in the set of extended natural numbers (i.e., in $\mathbb{N} \cup \{+\infty\}$). Usually N is assumed to be *simple*, meaning that the points X_i are almost surely distinct. Due to its analytical tractability, N is often assumed to be a *homogeneous Poisson point process* in the related work in the literature. We shall instead choose N to yield better correspondence to the spatial structure of actually deployed wireless networks. In earlier work we have shown that the *Geyer saturation process* [37] provides a very good fit for a number of existing node location data sets [38]. Unfortunately obtaining closed-form analytical solutions within the Geyer model is extremely difficult, and we have to rely on Monte Carlo estimates instead to study the model outcomes. However, we feel this is well justified by the increased realism achieved.

The Geyer saturation model belongs to the family of *Gibbs point processes*, which are characterized by existence of a density function f with respect to the homogeneous Poisson point process of unit intensity. The Geyer saturation process itself is a generalization of the Strauss process [39], defined by the density

$$f(\boldsymbol{X}) = \alpha \beta^{\#(\boldsymbol{X})} \gamma^{s_r(\boldsymbol{X})}, \tag{11}$$

where $s_r(\boldsymbol{X})$ denotes the number of point pairs of $\boldsymbol{X}$ that are closer than distance r apart, $\beta > 0$ and $0 \leq \gamma \leq 1$. Notice that α is for normalization only, and is not an independent parameter. The Strauss model is powerful tool for modeling *regular* point processes, in which individual points tend to be separated by some minimum distance, but it cannot model *clustered* distributions of locations common in wireless networks. For the Geyer process an additional *saturation threshold* ζ is added, bounding the contribution of the exponent of γ and thereby enabling both clustered (for which $\gamma > 1$) and regular processes to be modeled. The case $\zeta \to \infty$ yields the Strauss process as a limit.

The parameters of the Geyer saturation model can be fitted into empirical data sets using maximum likelihood analysis, or pseudolikelihoods. The relevant techniques, including tests for goodness-of-fit, have been primarily developed by Baddeley and Turner, who have also provided an open source implementation of the relevant routines [40,41]. We applied these techniques in [38] on, among others, the access point locations of the Google Mountain View Wi-Fi network [42], obtaining an excellent fit with parameters $r = 150\,\text{m}$, $\zeta = 2$, $\beta \approx 3.329 \times 10^{-5}$ and $\gamma \approx 0.4112$. The corresponding point process forms our baseline location model, well suited for representing especially deployments of Wi-Fi like dynamic spectrum access networks.

3.3 Simulation Environment and Parameters

As discussed above, we shall in several instances rely on Monte Carlo estimates in studying the performance characteristics for the models introduced above. For all the simulations, we consider a square domain D of $10\,\text{km} \times 10\,\text{km}$, resulting in, on average, $N(D) = 1350$ secondary user access points being generated. Unless mentioned otherwise, the parameters of the correlation model defining Z are set to $\sigma^2 = 140\,\text{dBm}^2$, $\phi = 480\,\text{m}$, $\kappa = 1$ and $\mu_Z = \text{-}90\,\text{dBm}$, whereas the thresholds used are set to $\varepsilon = 0.05$, $\theta_{\text{E}} = \text{-}105\,\text{dBm}$ and $\theta_{\text{S}} = \text{-}85\,\text{dBm}$ (all given for fixed but arbitrary channel bandwidth). The interference threshold $\iota_{\max}$ of the primary

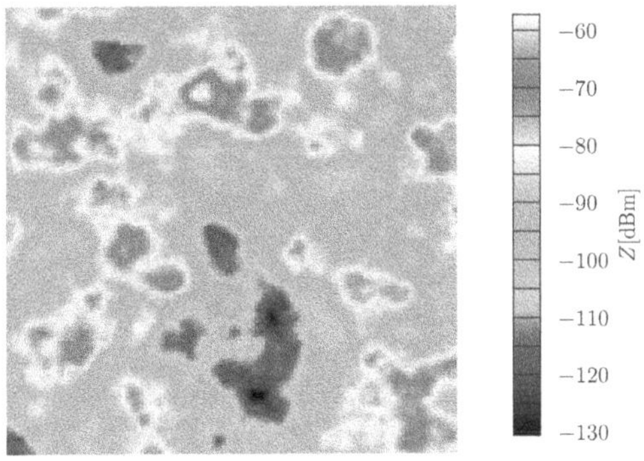

(a) Realization of Z with Matérn covariance.

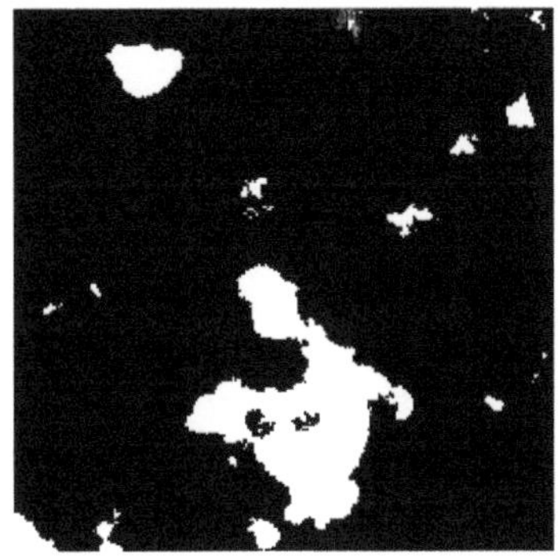

(b) Corresponding random set D_{NT}.

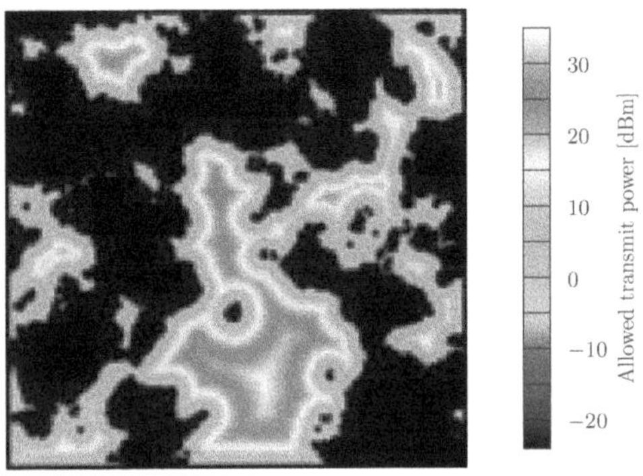

(c) Allowed transmit powers for power control model.

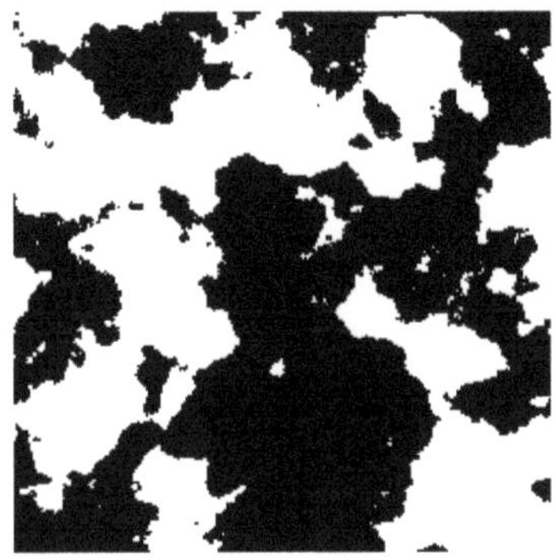

(d) Primary user service area D_{SA}.

Figure 4: Illustrations of the model outcomes with parameters $\kappa = 1$, $\phi = 360\,\mathrm{m}$, $\sigma^2 = 140\,\mathrm{dBm}^2$, $\mu_Z = -90\,\mathrm{dBm}$, $\theta_{\mathrm{E}} =$ -105 dBm and $\theta_{\mathrm{S}} =$ -90 dBm.

and the transmit power for the energy detection model are fixed at -100 dBm and 20 dBm, respectively. The effective correlation range of Z with these parameters is 1920 m. The generated realizations of Z consisted of 250×250 discrete pixels, each $40\,\mathrm{m} \times 40\,\mathrm{m}$ in size. All the simulations were carried out using the R environment [43], with the libraries spatstat [41], geoR [44] and RandomFields [45] being used for some of the computations needed. Formal error estimates were used in each case to ensure that simulations were repeated in sufficient numbers, and wherever possible, we report robust statistics of the obtained empirical distributions to ensure maximal confidence in obtained results.

4. IMPACT OF THE PRIMARY NETWORK STRUCTURE ON SECONDARY TRANSMIT OPPORTUNITIES

We shall first study the influence of the primary network structure on the occurrence of transmit opportunities for the secondary user based on the models introduced above. We are interested in the number of secondary user nodes that have transmit opportunities in the energy detection and power control models, as well as the quality of those opportunities measured in terms of the allowed transmit power. The related capacity estimates will be discussed in Section V.

The occurrence of transmit opportunities is dependent solely on the structure of the random sets D_{NT} and D_{SA}. While the shapes of those sets can be complicated, depending on correlation structure of Z, their *sizes* are very simple to characterize. In fact, since the marginal distribution of Z is assumed to be normal, the sizes of D_{NT} and D_{SA} and thereby the numbers of nodes of N in these sets depend only on the standardized differences $\Delta_{\mathrm{E}} \equiv (\mu_Z - \theta_{\mathrm{E}})/\sigma$ and $\Delta_{\mathrm{S}} \equiv (\mu_Z - \theta_{\mathrm{S}})/\sigma$.

Describing D_{NT} and D_{SA} beyond their sizes, and especially their influence on the *distribution* of active secondary nodes and their allowed transmit powers does not appear feasible through analytic means (although the theory of *excursion sets* [19] would provide some asymptotic results). Accordingly, the following results are obtained from extensive Monte Carlo simulations based on the models introduced above. We shall first consider the energy detection model, and the relation of the spatial structure of *active* nodes $N_{\mathrm{A}} \equiv N \cap (D - D_{\mathrm{NT}})$ to that of our Geyer saturation model N.

In the limit of the vanishing effective correlation range (effected by letting $\phi \to 0$) the local structure of the active secondary network would approach that of N subjected to independent thinning, with the retention probability dependent on Δ_{E}. However, for larger effective correlation ranges

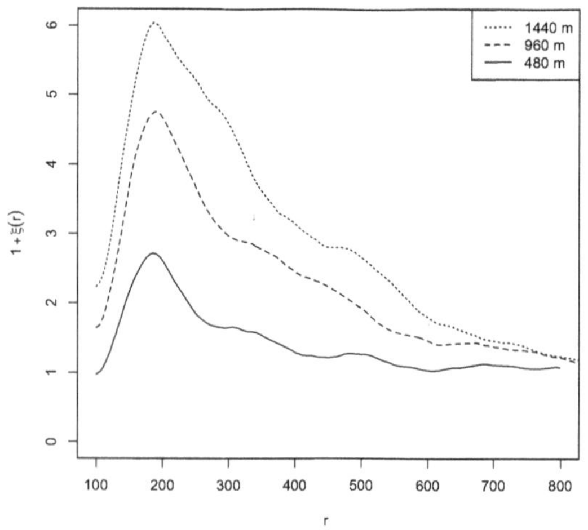

Figure 5: The influence of the effective correlation range of Z on the pair correlation function of active APs in the energy detection model.

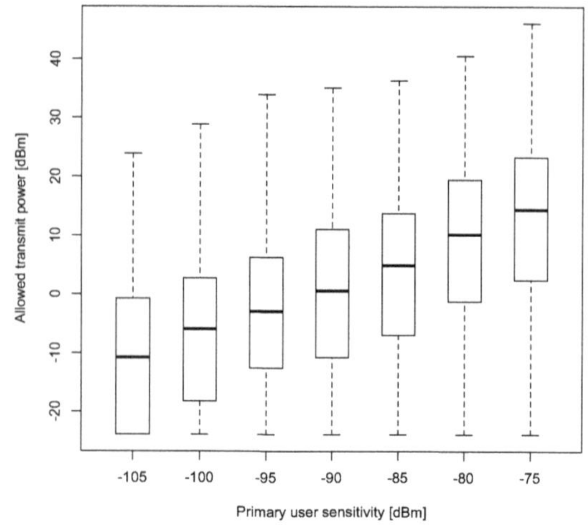

Figure 6: Distribution of the allowed transmit power for the power control model as a function of θ_S. The three horizontal lines of the box correspond to the median and the 25% and 75% quantiles of the data, with the whiskers giving the data point at most 1.5 times the interquartile range from the edge of the box.

the situation is more complex. The correlations in nearby values of Z will cause the set $D - D_{NT}$ to consist of typically several connected components, with distribution of component sizes being controlled by the correlation function $C(r)$ and the parameter Δ_E. It is intuitively clear that the resulting point process N_A should be clustered. A simple method for confirming this is to estimate the *pair correlation function* $\xi(r)$ of N_A. It is defined by the joint probability density

$$dP = \nu^2\big(1 + \xi(r)\big)\, dA_1\, dA_2 \qquad (12)$$

of finding one point in each of the two area elements dA_1 and dA_2 at distance r apart (ν is the *intensity* of the locations, giving the mean number of points per unit area). If the locations were uniformly random, we would obviously have $\xi(r) \equiv 0$. At length scales in which clustering is present, ξ obtains positive values proportional to the level of clustering of locations. Figure 5 shows the estimate of $\xi(r)$ for N_A for three different effective ranges, with the clustering induced by correlations in Z being clearly visible. The high peak at roughly $r = 200$ m is induced by the typical distance to the nearest neighboring node in the Geyer model, whereas the high values of ξ around that point are due to the correlations in Z.

These correlations are especially significant in scenarios in which N becomes very dense, since they will heavily influence the interference and contention statistics in the secondary network. For a recent interesting asymptotic analysis of this influence, see [46].

In the power control model the structure of the components of $D - D_{SA}$ influence mainly the distribution of the allowed transmit powers for the active nodes. We shall study both the influence of the sensitivity threshold θ_S as well as the correlation structure of Z on the results. Notice that since we assume N to be independent of Z, the marginal distribution of the random field of allowed transmit powers will also closely reflect the distribution of allowed transmit powers at the active nodes. Thus we shall only focus on the

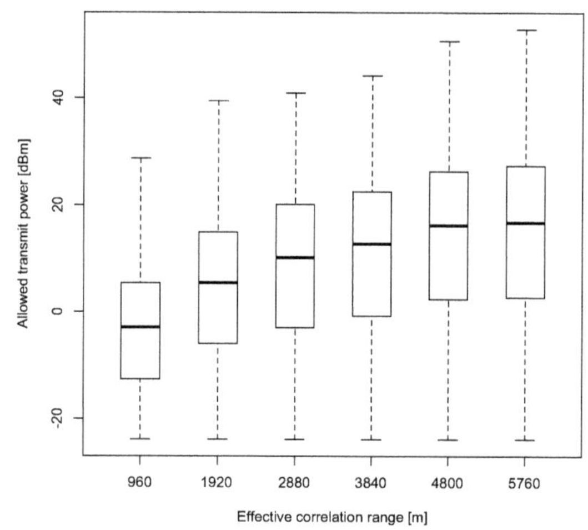

Figure 7: Distribution of the allowed transmit power for the power control model as a function of the effective correlation range of Z.

latter in the following. We shall also limit our focus on the distribution of allowed transmit powers of active nodes, as opposed to all the nodes of N.

Figure 6 shows the influence of the threshold θ_S on the distribution of allowed transmit power. The results clearly show that even with the relatively large effective correlation range of 1920 m threshold values below -90 dBm will result in very low allowed transmit powers. On the other hand

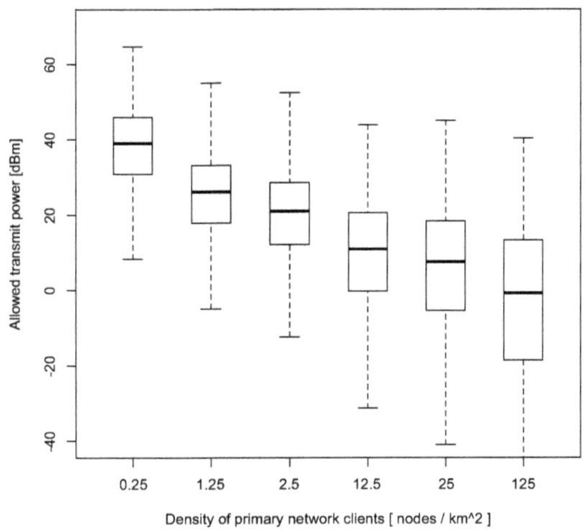

Figure 8: Distribution of the allowed transmit power for the power control model for the case of known primary user receiver positions.

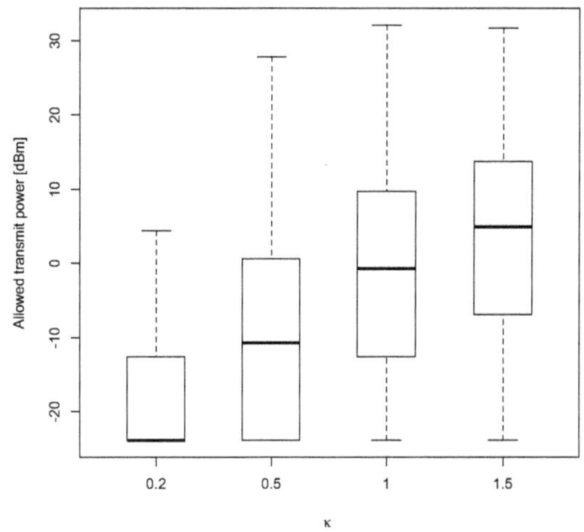

Figure 9: Distribution of the allowed transmit power for the power control model as a function of the smoothness parameter κ of Z.

increase in the effective correlation range rapidly increases the quality of transmit opportunities as shown in Figure 7. Especially for effective correlation ranges beyond four kilometers, the active secondary nodes would be able to offer service similar to present-day Wi-Fi hotspots. Of course, the low median powers are in part due to the extremely conservative way of estimating the allowed power. Any further information on expected distribution of the primary user receivers would also significantly increase the resulting transmit powers. A simple illustration of this is shown in Figure 8, corresponding to optimal power control where primary receiver locations are assumed to be known, obtained from a Poisson point process on D_{SA}. Probabilistic models of user distributions could also be used to achieve similar effect, albeit with reduced gains.

We shall conclude this section by studying the influence of the smoothness parameter κ on results. Many of the spatial correlation models used in the wireless communications community have assumed a simple parametric correlation model without the possibility to control the smoothness of the arising sample paths. An example of such a model is the exponential one proposed by Gudmundson as a model for correlated shadowing [47], roughly corresponding to the Matérn model with $\kappa = 1/2$. Because of this, there has been rather few studies on how details, such as the smoothness of the sample paths, influence the estimates obtained using these models. The results for our application, shown in Figure 9, are quite interesting in this regard. We see that even though the effective correlation range itself is kept constant at 1920 m, small values of κ result in very small transmit powers for the secondary network. This is because small values of κ allow Z to have occasional high variations in short scales, resulting in small "islands" in D_{SA} to occur, causing the degradation in allowed transmit powers. The conclusion to be drawn from this is that care should be applied to ensure that for any given application also the sample path behavior of the chosen stochastic model matches that of the phenomena being modeled, in addition to correlation range and variance.

5. ACHIEVABLE CAPACITIES

We shall now move our focus from allowed transmit powers to capacities. The configuration of the secondary user transmitters can be described in terms of an *activation vector* $\boldsymbol{a} \in \{0,1\}^{\#(\boldsymbol{X})}$ for which $a_i = 1$ if the node located at X_i active. This in turn depends on X_i and the operation of the selected MAC protocol or scheduling solution. We then assume that conditional on an activation vector $\boldsymbol{a}$, the capacity C_{ij} of the link from X_i to X_j depends on the *signal to interference and noise ratio* of the link, given by

$$\mathrm{SINR}_{ij} \equiv \frac{P_i/L_{ij}}{P_N + Z(X_j) + \sum_k a_k P_k / L_{kj}}, \tag{13}$$

where P_i is the transmit power of X_i, L_{ij} is the path loss from X_i to X_j, P_N is noise power, and the summation is over all the links except the one between X_i and X_j. Since we are interested in achievable capacity, we adopt the *generalized physical model* for which the capacity is estimated by Shannon's formula $C_{ij} = B \log_2(1 + \mathrm{SINR}_{ij})$, where B is the channel bandwidth, and interference is treated as noise. Alternative choices of capacity metrics are, of course, both possible and straightforward to introduce into the general framework used here. We report all capacity results in normalized form (bits per second per Hertz), since for any reasonable choice of B the system never became dominated by noise.

We focus throughout this section on the secondary network downlink. We assume that around each point of N there is a Poisson distributed number of clients, with mean of five clients per node, and their locations are sampled from a symmetric bivariate normal distribution centered at the points of N with standard deviation of $\sigma = 25$ m. This can be thought of as a variant of the Thomas point process oc-

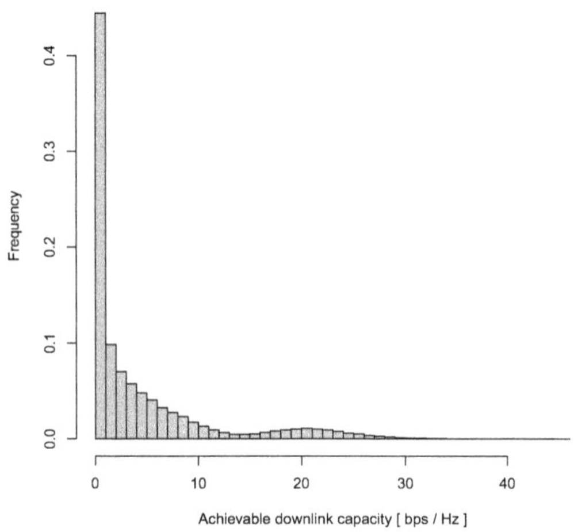

Figure 10: The distribution of achievable downlink capacity for the energy detection model.

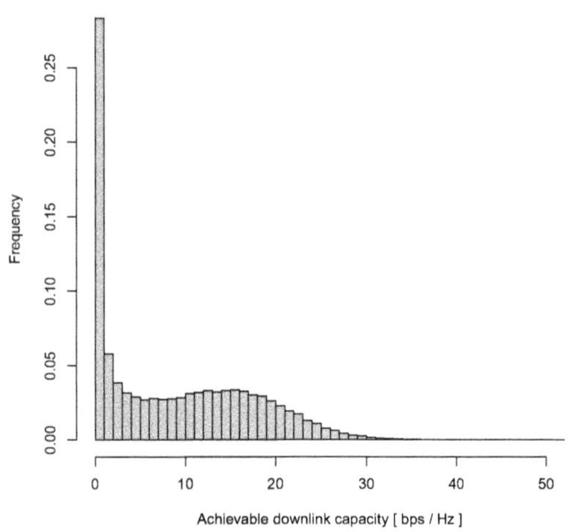

Figure 11: The distribution of achievable downlink capacity for the power control model.

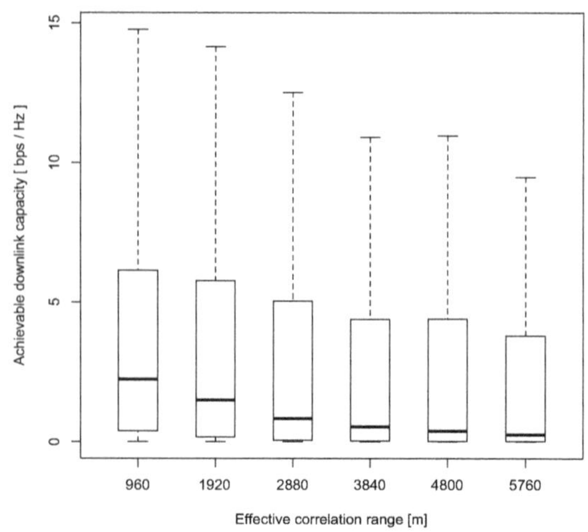

Figure 12: The influence of effective correlation range on achievable downlink capacities for the energy detection model.

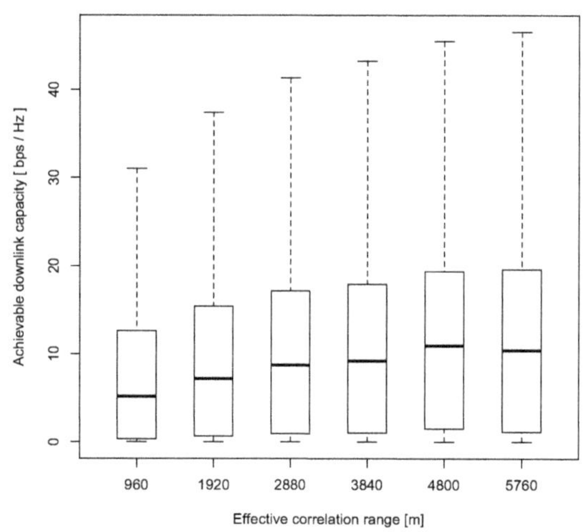

Figure 13: The influence of effective correlation range on achievable downlink capacities for the power control model.

casionally used in performance studies [32], but with more realistic distribution for cluster locations.

Figures 10 and 11 show the distribution of the achievable capacities, taken over the whole client population. We see that as expected, the power control model in general behaves better, resulting in fewer users being in near-outage, and enabling overall decent performance. Since the transmit power of the energy detection model is fixed at relatively high power (20 dBm), some of the clients will have very high SINR-values and correspondingly high capacity. However, the small number of active access points leads into poor overall performance. Figures 12 and 13 show the influence of the effective correlation range of Z on the results. Interestingly, increase in correlation range has opposite effects on the downlink capacity distributions for the energy detection and power control models. This is because with the power control model the secondary network can increase link capacities by increasing transmit powers as the correlation length increases (as was shown in Figure 7), and is also able to offer service in larger proportion of the overall domain D (since $\theta_E < \theta_S$).

The results discussed above are upper limits, since they assume only one active link in the secondary network. In a more realistic scenario multiple secondary transmitters will be active simultaneously, resulting in increased interference and thereby reduction in capacity. We have also studied this case in detail, and while the distribution of capacities does shift lower as more and more nodes become active, the performance did not degrade severely in any of our simulations, and even with all nodes of N transmitting simultaneously the median link capacity was reduced only by approximately 20% (corresponding figures are omitted for space reasons). This is because of the structure of N which is regular enough to ensure that the inter-transmitter distances are typically quite large. For a denser deployment, or in an ad hoc networking scenario, interference between secondary users would rapidly become an additional performance bottleneck to the limitations imposed by the primary.

6. CONCLUSIONS

In this paper we have studied in detail the influence of the spatial structure of the primary network on the performance characteristics of dynamic spectrum access based networks. We have also presented simple yet powerful stochastic modeling framework on which our work was based on. Our experiences indicate that Gaussian random fields and Gibbs point processes fitted to empirical data yield a solid foundation for studying in detail characteristics of DSA networks, and, more generally, cognitive wireless networks. The results of our study show clearly that details of the primary network structure and the level of interference the primary is ready to tolerate have a major impact on the expected secondary performance.

The presented work can of course be extended in several ways. Perhaps the simplest extension would be to consider multiple independent frequency channels instead of a single one. This would greatly reduce the outage probability especially for the energy detection model, and would increase the overall power budget and the resulting capacity of the secondary network. Another worthwhile extension would be to consider more realistic models for the physical and MAC layers. Understanding the upper performance limits is important, but should not be the exclusive goal. The most ambitious extension would finally be to develop a full space-time version of the presented models, and extend the treatment from one slice of time towards full dynamics of the primary and secondary networks. The major challenge in such a work is, in addition to greatly more complex analysis required, the lack of sufficient empirical data sets on which models could be constructed on. We have taken some early steps towards this direction in [21], albeit using a less general model of the primary network structure as done here.

Finally, we note that the methods applied here have numerous potential applications beyond the cognitive wireless networks. For example, many of the natural inhomogeneities in wireless systems, such as uneven distributions of users, correlated shadowing, or changes in parameters of the propagation environment are amenable to be treated as correlated random fields. Similarly, whenever models of user or node locations are needed that can be fitted to empirical location data, Gibbs point processes form a very powerful tool.

Acknowledgment

The authors would like to thank RWTH Aachen University and the German Research Foundation (Deutsche Forschungsgemeinschaft, DFG) for providing financial support through the UMIC research centre. We would also like to thank the European Union for providing partial funding of this work through the FARAMIR project.

7. REFERENCES

[1] J. Mitola III and G. Maguire Jr, "Cognitive radio: making software radios more personal," *IEEE personal communications*, vol. 6, no. 4, pp. 13–18, 1999.

[2] Q. Zhao and B. Sadler, "A survey of dynamic spectrum access," *IEEE Signal Processing Magazine*, vol. 24, no. 3, pp. 79–89, 2007.

[3] S. Haykin, "Cognitive radio: brain-empowered wireless communications," *IEEE journal on selected areas in communications*, vol. 23, no. 2, pp. 201–220, 2005.

[4] R. Tandra and A. Sahai, "SNR walls for signal detection," *IEEE Journal on Selected Topics in Signal Processing*, vol. 2, no. 1, pp. 4–17, 2008.

[5] T. Yucek and H. Arslan, "A survey of spectrum sensing algorithms for cognitive radio applications," *IEEE Communications Surveys & Tutorials*, vol. 11, no. 1, pp. 116–130, 2009.

[6] R. Tandra, M. Mishra, and A. Sahai, "What is a spectrum hole and what does it take to recognize one?" *Proceedings of the IEEE*, vol. 97, no. 5, pp. 824–848, 2009.

[7] A. Ghasemi and E. S. Sousa, "Spectrum sensing in cognitive radio networks: the cooperation-processing tradeoff," *Wireless Communications and Mobile Computing*, vol. 7, no. 9, pp. 1049–1060, 2007.

[8] A. Sahai, S. Mishra, R. Tandra, and K. Woyach, "Cognitive radios for spectrum sharing," *IEEE Signal Processing Magazine*, vol. 26, no. 1, pp. 140–145, 2009.

[9] K. Harrison, S. Mishra, and A. Sahai, "How much white-space capacity is there?" in *Proc. of IEEE DySPAN 2010*, 2010.

[10] J. van de Beek, J. Riihijärvi, A. Achtzehn, and P. Mähönen, "TV White Space in Europe," *IEEE Transactions on Mobile Computing*, vol. 11, no. 2, pp. 178–188, Feb. 2012.

[11] C. R. Stevenson, G. Chouinard, Z. Lei, W. Hu, S. J. Shellhammer, and W. Caldwell, "IEEE 802.22: The first cognitive radio wireless regional area network standard," *IEEE Communications Magazine*, vol. 47, no. 1, pp. 130 –138, january 2009.

[12] C. Cordeiro, K. Challapali, D. Birru, and S. Shankar, "IEEE 802.22: An introduction to the first wireless standard based on cognitive radios," *Journal of Communications*, vol. 1, no. 1, pp. 38–47, April 2006.

[13] A. Achtzehn, M. Petrova, and P. Mähönen, "On the Performance of Cellular Networks in TVWS," *In Proc. of IEEE ICC'12, Ottawa, Canada*, 2012.

[14] J. Lee, S. Lim, J. G. Andrews, , and D. Hong, "Achievable transmission capacity of secondary system in cognitive radio networks," *IEEE International Conference on Communications*, May 2010.

[15] M. Vu, N. Devroye, and V. Tarokh, "An overview of scaling laws in ad hoc and cognitive radio networks," *Wireless Personal Communications*, vol. 45, no. 3, pp. 343–354, 2008.

[16] S. Srinivasa and S. Jafar, "Cognitive Radio Networks: How Much Spectrum Sharing is Optimal?" in *IEEE Global Telecommunications Conference, 2007. GLOBECOM'07*, 2007.

[17] E. G. Larsson and M. Skoglund, "Cognitive radio in a frequency-planned environment: some basic limits," *IEEE Transactions on Wireless Communications*, vol. 7, no. 12, pp. 4800–4806, December 2008.

[18] N. Cressie, *Statistics for spatial data.* Wiley-Interscience, 1993.

[19] R. Adler and J. Taylor, *Random fields and geometry.* Springer Verlag, 2007.

[20] J. Møller and R. Waagepetersen, *Statistical inference and simulation for spatial point processes.* CRC Press, 2004.

[21] J. Riihijärvi, J. Nasreddine, and P. Mähönen, "Influence of primary network structure and dynamics on achievable performance of cognitive wireless networks," in *Proc. of IEEE DySPAN 2011.* IEEE, 2011.

[22] B. L. Mark and A. O. Nasif, "Estimation of maximum interference-free power level for opportunistic spectrum access," *IEEE Transactions on Wireless Communications*, vol. 8, no. 5, pp. 2505 -2513, May 2009.

[23] J. Nasreddine, O. Sallent, J. Pérez-Romero, and R. Agustí, "Positioning-based framework for secondary spectrum usage," *Physical Communication*, vol. 1, no. 2, pp. 121–133, 2008.

[24] L. Maciel, H. Bertoni, and H. Xia, "Unified approach to prediction of propagation over buildings for all ranges of base station antenna height," *IEEE Transactions on Vehicular Technology*, vol. 42, no. 1, pp. 41–45, 1993.

[25] B. Farhang-Boroujeny and R. Kempter, "Multicarrier communication techniques for spectrum sensing and communication in cognitive radios," *IEEE Communications Magazine*, vol. 46, no. 4, pp. 80–85, April 2008.

[26] M. Haddad, M. Debbah, and A. M. Hayar, "Distributed power allocation for cognitive radio," in *the 9th International Symposium on Signal Processing and Its Applications (ISSPA 2007)*, Feb. 2007, pp. 1–4.

[27] Y. Yu, H. Murata, K. Yamamoto, and S. Yoshida, "Interference information based power control for cognitive radio with multi-hop cooperative sensing," *IEICE Transactions on Communications*, vol. E91-B, no. 1, pp. 70–76, 2008.

[28] J. Nasreddine, J. Riihijärvi, and P. Mähönen, "Location-based adaptive detection threshold for dynamic spectrum access," in *Proc. of IEEE DySPAN 2010*, Singapore, April 2010.

[29] S. Kim, E. Dall'Anese, and G. Giannakis, "Cooperative Spectrum Sensing for Cognitive Radios Using Kriged Kalman Filtering," *IEEE J. Sel. Topics Sig. Proc*, 2011, to appear.

[30] J. Riihijärvi, P. Mähönen, M. Wellens, and M. Gordziel, "Characterization and Modelling of Spectrum for Dynamic Spectrum Access with Spatial Statistics and Random Fields," in *Proc. of 1st International Workshop on Cognitive Radios and Networks (CRNETS), in conjunction with IEEE PIMRC 2008*, Cannes, France, September 2008.

[31] B. Sayrac, S. Ben Jemaa, and E. Moulines, "Interference Cartography for Hierarchical Dynamic Spectrum Access," in *Proc. of IEEE DySPAN 2008*, 2008, pp. 1–5.

[32] M. Haenggi, J. G. Andrews, F. Baccelli, O. Dousse, and M. Franceschetti, "Stochastic Geometry and Random Graphs for the Analysis and Design of Wireless Networks," *IEEE Journal on Selected Areas in Communications*, Sep. 2009.

[33] N. Beaulieu and Q. Xie, "An optimal lognormal approximation to lognormal sum distributions," *IEEE Transactions on Vehicular Technology*, vol. 53, no. 2, pp. 479–489, 2004.

[34] A. Achtzehn *et al.*, "Improving Coverage Prediction for Primary Multi-Transmitter Networks Operating in the TV Whitespaces," *accepted to IEEE SECON'12, Seoul, Korea*, 2012.

[35] M. Wellens, J. Riihijärvi, and P. Mähönen, "Spatial statistics and models of spectrum use," *Computer Communications*, vol. 32, no. 18, pp. 1998–2011, 2009.

[36] A. F. Karr, *Point Processes and Their Statistical Inference*, 2nd ed. Marcel Dekker, 1991.

[37] C. Geyer, "Likelihood inference for spatial point processes," *Stochastic Geometry: Likelihood and Computation*, pp. 79–140, 1999.

[38] J. Riihijärvi and P. Mähönen, "Modeling Spatial Structure of Wireless Communication Networks," in *INFOCOM IEEE Conference on Computer Communications Workshops, 2010*, 2010, pp. 1–6.

[39] D. Strauss, "A model for clustering," *Biometrika*, vol. 62, no. 2, p. 467, 1975.

[40] A. Baddeley and R. Turner, "Practical Maximum Pseudolikelihood for Spatial Point Patterns (with Discussion)," *Australian & New Zealand Journal of Statistics*, vol. 42, no. 3, pp. 283–322, 2000.

[41] ——, "Spatstat: an R package for analyzing spatial point patterns," *Journal of Statistical Software*, vol. 12, no. 6, pp. 1–42, 2005, ISSN 1548-7660. [Online]. Available: www.jstatsoft.org

[42] J. Robinson, R. Swaminathan, and E. Knightly, "Assessment of urban-scale wireless networks with a small number of measurements," in *Proceedings of the 14th ACM international conference on Mobile computing and networking.* ACM New York, NY, USA, 2008, pp. 187–198.

[43] R Development Core Team, *R: A Language and Environment for Statistical Computing*, R Foundation for Statistical Computing, Vienna, Austria, 2007, ISBN 3-900051-07-0. [Online]. Available: http://www.R-project.org

[44] P. J. Ribeiro Jr and P. J. Diggle, "geoR: a package for geostatistical analysis," *R-NEWS*, vol. 1, no. 2, pp. 14–18, June 2001, iSSN 1609-3631. [Online]. Available: http://CRAN.R-project.org/doc/Rnews/

[45] M. Schlather, *RandomFields: Simulation and Analysis of Random Fields*, 2009, R package version 1.3.41.

[46] R. K. Ganti and J. G. Andrews, "A new method for computing the transmission capacity of non-poisson wireless networks," in *Proc. of ISIT*, 2010.

[47] M. Gudmundson, "Correlation model for shadow fading in mobile radio systems," *Electronics letters*, vol. 27, no. 23, pp. 2145–2146, 1991.

A Class of Backpressure Algorithms for Networks Embedded in Hyperbolic Space with Controllable Delay-Throughput Trade-off

Eleni Stai
Institute of Communications and Computer Systems
School of Electrical & Computer Engineering
National Technical University of Athens
Athens, Zografou, 15780, Greece
estai@netmode.ntua.gr

John S. Baras
Institute for Systems Research
Department of Electrical & Computer Engineering
University of Maryland
College Park, MD 20742, USA
baras@umd.edu

Symeon Papavassiliou
Institute of Communications and Computer Systems
School of Electrical & Computer Engineering
National Technical University of Athens
Athens, Zografou, 15780, Greece
papavass@mail.ntua.gr

ABSTRACT

Future communications consist of an increasing number of wireless parts, while simultaneously need to support the widespread multimedia applications imposed by social networks. These human-machine systems, driven by both real time social interactions and the challenges of the wireless networks' design, call for efficient and easy to implement, distributed cross-layer algorithms for their operation. Performance metrics such as throughput, delay, trust, energy consumption, need to be improved and optimized aiming at high quality communications. We investigate the coveted throughput-delay trade-off in static wireless multihop networks based on a "computer-aided" design of the backpressure scheduling/routing algorithm for networks embedded in hyperbolic space. Both routing and scheduling exploit the hyperbolic distances to orient the packets to the destination and prioritize the transmissions correspondingly. The proposed design provides us with the freedom of controlling its theoretical throughput optimality and of counterbalancing its practical performance through simulations, leading to significant improvements of the throughput-delay trade-off.

Categories and Subject Descriptors

C.2 [**Computer Systems Organization**]: Computer - Communication Networks; C.2.1 [**Computer - Communication Networks**]: Network Architecture and Design—*Wireless Communication, Network Communications, Distributed Networks*

General Terms

Performance, Algorithms, Design

Keywords

Wireless Networks, Greedy Embedding, Hyperbolic Geometry, Delay, Throughput

1. INTRODUCTION

The new era of networks integrates increasingly the wireless communications either as autonomous wireless network topologies or as parts of hybrid network infrastructures. In addition, we are witnessing the spread of multimedia applications imposing real-time traffic over wireless networks and calling for high throughput with low packet delays. Even more, the emerging development of social network applications over wireless ad hoc topologies strengthens the calling for efficient throughput-delay trade-offs.

The backpressure algorithm, introduced in its original form in [14], satisfies the requirement of throughput optimality. It performs routing and scheduling based on congestion gradients, by allowing transmission to the links that maximize the sum of differential queue backlogs in the network. However, by deploying routing without any information about the position or distance to the destination, the backpressure explores all possible source-destination paths leading to high delays. Several approaches have been developed towards the direction of reducing the delay imposed by the pure backpressure algorithm. To begin with, the authors in [15] combine backpressure and shortest path routing by imposing hop-count constraints on each flow, assuming that each node knows a-priori its hop distance from all others. Following another approach, the authors in [3] use shadow queues for the backpressure scheduling/routing, improving in this way the delay of the backpressure algorithm while simultaneously reducing the number of real queues need to be stored at each node. In the same spirit of "virtual" queues, the authors in [8] examine the problem of delay reduction in stochastic network optimization problems. After proving the relation of the backlog vector with the dual optimal solution (Lagrange multiplier) of the optimization problem, they use this finding to develop the FQLA algorithm that uses virtual place holder bits in order to reduce delays. Last but not least, in [9], a backpressure algorithm based on packet delays instead of queue backlogs is developed and shown to keep the average per flow delays well concentrated around their average values and to respond to the problem of finite flows.

In this paper, we propose a "computer-aided" backpressure algorithm which admits a static network embedded in hyperbolic space according to a greedy embedding. A greedy embedding in hyperbolic space is a correspondence between nodes and hyperbolic coordinates such that the greedy routing algorithm, employed in hyperbolic space, does not have local minima, i.e. every node can find at least one neighbor closer than itself to all possible destinations [4, 12]. We consider a greedy embedding of the network following [4] and we impose routing constraints on the backpressure algorithm, by determining as possible next-hop neighbors for a specific destination only "greedy" neighbors, i.e. those that strictly reduce the hyperbolic distance to the destination. Furthermore, we scale the differential backlogs on the links with a properly designed function of hyperbolic distances, in order to incorporate the hyperbolic distances in the scheduling procedure. As it is shown through simulations, this approach reduces the average per flow delay and increases the average per flow throughput for each examined network topology and social flow. The term "computer-aided" is used due to the derivation of the form of the scaling function through simulations. The basic intuition behind our idea lies in the occasion of specific network topologies such as the wireless multihop network with clique interference constraints, where it is delay-optimal to schedule the packets being closest to the destination [7]. We therefore believe that we should take into account the topological properties of the network together with the backlog information, in order to improve the delay performance.

The rest of the paper is organized as follows. In section 2, we describe the system model and basic knowledge regarding hyperbolic space and greedy embeddings. In section 3, we describe the proposed algorithm, prove its throughput optimality and discuss its complexity and distributivity. In section 4, we provide a numerical study for choosing the scaling function and illustrating the improvement in the throughput-delay trade-off, while possible directions for extensions to dynamic networks are briefly presented. Finally, section 5, concludes the paper.

2. SYSTEM MODEL AND HYPERBOLIC EMBEDDING

The system consists of a wireless multihop network with N nodes that function in a time-slotted fashion. Let us denote as $\mathbf{N}$ the set of the network nodes. At each time slot t, the scheduling/routing algorithm decides which set of non-interfering links is going to transmit and which flows will be served by each link, while new packets may arrive at the nodes. The number of packets that arrive at node i for destination d is a random variable $A_i^d(t)$, i.i.d. in all t, with expected value λ_i^d. We suppose that each node i stores a queue $q_i^d(t)$ for each destination d. We denote with $\mu_{ij}(t)$ the packets served by the link (i,j) and with the $N \times N$ matrix, $[\mu_{ij}(t)]$, the traffic over all the network, at time t. Also, we denote with $\mu_{ij}^d(t)$ the communication traffic on the link (i,j) for destination d at time t. The arrival and service rates are considered bounded and $\lambda_i^d \leq A_{\max}$. We use the term I_S to refer to the set of service rate matrices of all possible independent sets of the graph, i.e. maximal sets of links that do not interfere with each other. Also, we use the notation $\mathcal{N}(i)$ for the one-hop neighborhood of node i.

The whole infinite hyperbolic plane can be represented inside the finite unit disc $\mathcal{D} = \{z \in \mathcal{C} \,||z| < 1\}$ of the Euclidean space; the Poincaré Disc model. The greedy embedding used in this work is based on the Poincaré Disc model. The hyperbolic distance function $d_H(z_i, z_j)$, for two points z_i, z_j, in the Poincaré Disc model is given by [4, 1]:

$$\cosh d_H(z_i, z_j) = \frac{2|z_i - z_j|^2}{(1-|z_i|^2)(1-|z_j|^2)} + 1 \qquad (1)$$

The Euclidean circle $\partial\mathcal{D} = \{z \in \mathcal{C} \,||z| = 1\}$ is the boundary at infinity for the Poincaré Disc model. In addition, in this model, the shortest hyperbolic path between two nodes is either a part of a diameter of $\mathcal{D}$, or a part of a Euclidean circle in $\mathcal{D}$ perpendicular to $\partial\mathcal{D}$.

The greedy embedding is constructed by choosing a spanning tree of the graph of the initial network and then embedding the spanning tree into the hyperbolic space according to the distributed algorithm of [4]. If a spanning tree of the graph admits a greedy embedding in the hyperbolic space then the whole graph admits also a greedy embedding [12]. Let us denote as "greedy" paths, the paths consisting of nodes with strictly decreasing hyperbolic distances to their destinations. From definition, the greedy embedding ensures the existence of at least one greedy path between each source-destination pair in the case of static networks. Every pair of nodes i, j is connected through a unique path, let us denote it as $i,\ i_1,\ i_2, \ldots\ i_k,\ j$, lying on the spanning tree which is embedded in the hyperbolic space. Due to the particular embedding, i_1 is at least one greedy neighbor of i for j and i_k is a greedy neighbor of j for i. Let us denote as $dist_H(i,d)$ the hyperbolic distance between nodes i, d.

We adapt the capacity region of [6], so as to include the routing constraints of the proposed class of algorithms. Therefore, the capacity region should allow routing only through greedy paths. The capacity region Λ_G is the set of all input rate matrices (λ_i^d) with $\lambda_i^d \neq 0$ if $i \neq d$ and (i,d) is a source destination pair, such that there exists a rate matrix $[\mu_{ij}]$ satisfying the following constraints:

- Efficiency constraints: $\mu_{ij}^d \geq 0$, $\mu_{ii}^d = 0$, $\mu_{dj}^d = 0$, $\sum_d \mu_{ij}^d \leq \mu_{ij}$, $\forall\ i, d, j$.
- Flow constraints: $\lambda_i^d + \sum_l \mu_{li}^d \leq \sum_l \mu_{il}^d$, $\forall\ i, d:\ i \neq d$.
- Routing constraints: $\mu_{ij}^d = 0$ if i has at least one greedy neighbor for d and j is not one of the $i's$ greedy neighbors.

Finally, we define the notion of strong stability of the queues, which will be used in the proof that follows. According to the Definition 3.1 of [6], a queue, $q_i^d(t)$ is strongly stable if $\limsup_{t\to\infty} \frac{1}{t}\sum_{\tau=0}^{t-1} E(q_i^d(\tau)) < \infty$. If all the queues of the network are strongly stable, then the whole network is strongly stable.

3. GREEDY BACKPRESSURE WITH QUEUE DIFFERENCE SCALING

In this section, we describe the proposed algorithm, which modifies the classic backpressure algorithm by performing routing only through greedy paths, and scheduling considering, in addition to the queue backlogs, the hyperbolic distances to the final destination of the transmitting and receiving nodes. The scheduling process weights the queue difference for destination d of each link (i,j) by a function $f(dist_H(i,d), dist_H(j,d)) : \mathbf{N}^3 \rightarrow \Re$, $i \neq j, d$, thus assigning priorities to particular links for transmission, choosing as crucial links for delay reduction those being highly prioritized. Algorithm 1 describes in pseudo-code the proposed scheduling/routing algorithm.

Algorithm 1: Greedy Backpressure with Queue Difference Scaling (GBP QDS)

---

for *each directed link* (i,j) **do**
  for *each destination* d **do**
    %Greedy backpressure%
    if $dist_H(i,d) > dist_H(j,d)$ **then**
      %Queue Difference Scaling
      $P_{ij}^d(t) =$
      $f(dist_H(i,d), dist_H(j,d))(q_i^d(t) - q_j^d(t))$;
    else
      $P_{ij}^d(t) = -\infty$;
  %Define the weight $P_{ij}(t)$ *as follows* :
  $P_{ij}(t) = \max(\max_d P_{ij}^d(t), 0)$;
  $d^*(i,j) = arg \max_d P_{ij}^d(t)$;
Choose the rate matrix through the maximization :
$[\mu_{ij}(t)] = arg_{\mu' \in \mathbf{I_S}} \max \sum_{(i,j)} \mu'_{ij} P_{ij}(t)$
for *each directed link* (i,j) **do**
  if $\mu_{ij}(t) > 0$ **then**
    the link (ij) *serves* $d^*(i,j)$ *with* $\mu_{ij}^{d^*}(t) = \mu_{ij}(t)$;
  For $d \neq d^*$ *we set* $\mu_{ij}^d(t) = 0$

It is important to mention that the greedy embedding ensures the existence of a greedy path for a source destination pair. So the routing constraints of the Algorithm 1 are well defined and there are no holes that can stack the packets to a specific node. Therefore, the correctness of the routing in terms of leading the packet to the destination is ensured. The function $f(dist_H(i,d), dist_H(j,d)) = f(i,j,d)$ is not arbitrarily defined and should satisfy some properties. More precisely, it should be a positive function. Due to the finiteness of the network we can define minimum and maximum values for $f(i,j,d)$, $f_{\min} \leq f(i,j,d) \leq f_{\max}$, where the condition $f_{\min} > 0$ should apply. As we will see from simulations, different types of functions $f(i,j,d)$ are suitable for different network topologies or types of social flows. However, the idea of the "hyperbolic distance-based" scheduling remains common in all different cases of physical topologies and social flows. In the case of $f(i,j,d) = 1$, let us denote the GBP QDS algorithm simply as GBP.

The Queue Difference Scaling approach provides us with freedom in choosing how to prioritize transmissions based on the network topology. As will be seen from simulations, it appears to behave uniformly over different network topologies and social flows, in the sense that we can find a function $f(i,j,d)$ that improves the throughput-delay trade-off, in all the examined cases. As a trade-off, the stability of queues under this approach can be easily proved only for a subset of the capacity region Λ_G, which depends on the values $f_{\max}, f_{\min}$ and leads to throughput optimality (the whole Λ_G) when $f_{\max} \rightarrow f_{\min}$ (Theorem 1). Finally, in the Queue Difference Scaling Approach, only links with positive differential backlog $q_i^d(t) - q_j^d(t)$ can transmit as in the classic backpressure algorithm.

THEOREM 1. *Let us define* $\epsilon_0 = \frac{f_{\max} - f_{\min}}{f_{\min}} A_{\max}$. *If we assume that the average arrival rates* $\lambda_i^d + \epsilon, \epsilon > \epsilon_0$ *lie inside the capacity region* Λ_G, *then the queues of the network are strongly stable, under the Greedy Backpressure with Queue Difference Scaling algorithm (GBP QDS) for static networks.*

PROOF. We define two indicator functions:
$I_1 = \{dist_H(i,d) > dist_H(j,d) \bigwedge j \in \mathcal{N}(i)\}$,
$I_2 = \{dist_H(i,d) < dist_H(j,d) \bigwedge i \in \mathcal{N}(j)\}$,
We use $|I_1$ to denote that the I_1 condition is satisfied for nodes i,j. Similarly for I_2. The queue dynamics in the case of Algorithm 1 are

$$q_i^d(t+1) = \max\{q_i^d(t) - \sum_{j|I_1} \mu_{ij}^d(t), 0\} + \sum_{j|I_2} \mu_{ji}^d(t) + A_i^d(t). \quad (2)$$

We denote by $Q(t) = [q_i^d(t)]$, the column array of the queues of the network. From the Lemma in page 52 of [6], we have that

$$(q_i^d(t+1))^2 \leq (q_i^d(t))^2 + \left(\sum_{j|I_1} \mu_{ij}^f(t)\right)^2 + \left(\sum_{j|I_2} \mu_{ji}^d(t) + A_i^d(t)\right)^2$$
$$-2q_i^d(t)\left(\sum_{j|I_1} \mu_{ij}^d(t) - \sum_{j|I_2} \mu_{ji}^f(t) - A_i^d(t)\right). \quad (3)$$

As aforementioned, $f_{\min} \leq f(i,j,d) \leq f_{\max}$. The backpressure scheduling under the Queue Difference Scaling approach, is
$[\mu_{ij}(t)] = arg_{\mu' \in \mathbf{I_S}} \max \sum_{(i,j,d)} f(i,j,d) \mu_{ij}^{d\,\prime} \left(q_i^d(t) - q_j^d(t)\right)$.
Due to this maximization each summand of the form $f(i,j,d)\mu_{ij}^d(t)\left(q_i^d(t) - q_j^d(t)\right)$, $f(i,j,d) > 0$, with optimal $\mu_{ij}^d(t)$, is non negative or in other words, $\mu_{ij}^d(t) = 0$ whenever $q_i^d(t) - q_j^d(t) < 0$. Therefore, we can write that

$$\sum_{(i,j,d)} f_{\min} \mu_{ij}^d(t) \left(q_i^d(t) - q_j^d(t)\right),$$
$$\leq \sum_{(i,j,d)} f(i,j,d) \mu_{ij}^d(t) \left(q_i^d(t) - q_j^d(t)\right),$$
$$\leq \sum_{(i,j,d)} f_{\max} \mu_{ij}^d(t) \left(q_i^d(t) - q_j^d(t)\right), \quad (4)$$

where the service rates $\mu_{ij}^d(t)$ are chosen by the maximum weight matching (scheduling as in Algorithm 1). Also,

$$\max_{\mu' \in \mathbf{I_S}} \sum_{(i,j,d)} f_{\min} \mu_{ij}^{d\,\prime} \left(q_i^d(t) - q_j^d(t)\right),$$
$$\leq \max_{\mu' \in \mathbf{I_S}} \sum_{(i,j,d)} f(i,j,d) \mu_{ij}^{d\,\prime} \left(q_i^d(t) - q_j^d(t)\right),$$
$$\leq \max_{\mu' \in \mathbf{I_S}} \sum_{(i,j,d)} f_{\max} \mu_{ij}^{d\,\prime} \left(q_i^d(t) - q_j^d(t)\right). \quad (5)$$

We define the Lyapunov function $L(Q(t)) = \sum_{i,d} f_{\max} q_i^d(t)^2$

and we compute the Lyapunov Drift

$$E(L(Q(t+1)) - L(Q(t))|Q(t)) = E\left[\sum_{i,d} f_{\max} q_i^d(t+1)^2 \,\middle|\, Q(t)\right]$$
$$-E\left[\sum_{i,d} f_{\max} q_i^d(t)^2 \,\middle|\, Q(t)\right] \overset{Eq.(2,3)}{\leq} \sum_{i,d} f_{\max} q_i^d(t)^2 +$$
$$\sum_{i,d} f_{\max}\left\{E\left[\left(\sum_{j|I_1}\mu_{ij}^d(t)\right)^2 + \left(\sum_{j|I_2}\mu_{ji}^d(t) + A_i^d(t)\right)^2 \,\middle|\, Q(t)\right]\right\}$$
$$-\sum_{i,d} f_{\max}\left\{2q_i^d(t)E\left[\left(\sum_{j|I_1}\mu_{ij}^d(t) - \sum_{j|I_2}\mu_{ji}^d(t) - A_i^d(t)\right)\,\middle|\, Q(t)\right]\right\}$$
$$-\sum_{i,d} f_{\max} q_i^d(t)^2 \leq f_{\max}B + 2\sum_{i,d} f_{\max} q_i^d(t)\lambda_i^d$$
$$-2\sum_{i,d} f_{\max} q_i^d(t)E\left[\left(\sum_{j|I_1}\mu_{ij}^d(t) - \sum_{j|I_2}\mu_{ji}^d(t)\right)\,\middle|\, Q(t)\right], \quad (6)$$

where B is an upper bound of
$\sum_{i,d} E\left[\left(\sum_{j|I_1}\mu_{ij}^d(t)\right)^2 + \left(\sum_{j|I_2}\mu_{ji}^d(t) + A_i^d(t)\right)^2 \,\middle|\, Q(t)\right]$.
Therefore, we conclude to the following inequality

$$E(L(Q(t+1)) - L(Q(t))|Q(t)) \leq f_{\max}B + 2\sum_{i,d} f_{\max} q_i^d(t)\lambda_i^d$$
$$-2\sum_{i,d} f_{\max} q_i^d(t)E\left[\left(\sum_{j|I_1}\mu_{ij}^d(t) - \sum_{j|I_2}\mu_{ji}^d(t)\right)\,\middle|\, Q(t)\right],$$
$$= f_{\max}B + 2\sum_{i,d} f_{\max} q_i^d(t)\lambda_i^d$$
$$-2\sum_{d}\sum_{(i,j)|I_1} f_{\max}(q_i^d(t) - q_j^d(t))E\left[\mu_{ij}^d(t)\,\middle|\, Q(t)\right],$$
$$\overset{Eq.(4)}{\leq} f_{\max}B + 2\sum_{i,d} f_{\max} q_i^d(t)\lambda_i^d$$
$$-2\sum_{d}\sum_{(i,j)|I_1} f(i,j,d)(q_i^d(t) - q_j^d(t))E\left[\mu_{ij}^d(t)\,\middle|\, Q(t)\right]. \quad (7)$$

If $\lambda_i^d + \epsilon$, $\epsilon > \epsilon_0$ lie inside the capacity region, then from Corollary 3.9 in [6], there exist rates $\hat{\mu}_{ij}^d(t)$ determined according to the network topology and which, independently of the queue backlog, satisfy
$\lambda_i^d + \epsilon = E\left[\sum_{j|I_1}\hat{\mu}_{ij}^d(t) - \sum_{j|I_2}\hat{\mu}_{ji}^d(t)\right] \ \forall\, i, d.$
As a result, we have

$$\sum_{i,d} f_{\min} q_i^d(t)\lambda_i^d + \sum_{i,d} f_{\min} q_i^d(t)\epsilon$$
$$= \sum_{i,d} f_{\min} q_i^d(t)E\left[\sum_{j|I_1}\hat{\mu}_{ij}^d(t) - \sum_{j|I_2}\hat{\mu}_{ji}^d(t)\right],$$
$$= \sum_{d}\sum_{(i,j)|I_1} f_{\min}E\left[\hat{\mu}_{ij}^d(t)\right](q_i^d(t) - q_j^d(t)),$$
$$\leq \max_{\mu' \in \mathbf{I_S}} \sum_{d}\sum_{(i,j)|I_1} f_{\min}\mu_{ij}^{d\,\prime}(q_i^d(t) - q_j^d(t)),$$
$$\overset{(5)}{\leq} \max_{\mu' \in \mathbf{I_S}} \sum_{d}\sum_{(i,j)|I_1} f(i,j,d)\mu_{ij}^{d\,\prime}(q_i^d(t) - q_j^d(t)), \quad (8)$$

where $\hat{\mu}_{ij}^d(t)$ are the service rates chosen by the random scheduling algorithm. Therefore the Lyapunov drift (Eq. (7)) becomes

$$E[L(Q(t+1)) - L(Q(t))|Q(t)] \leq f_{\max}B$$
$$+2\sum_{d}\sum_{i} f_{\max}\lambda_i^d q_i^d(t) - 2\sum_{d}\sum_{i} f_{\min}(\lambda_i^d + \epsilon)q_i^d(t),$$
$$E[L(Q(t+1)) - L(Q(t))|Q(t)] \leq f_{\max}B$$
$$+2\sum_{d}\sum_{i}(f_{\max} - f_{\min})A_{\max}q_i^d(t) - 2\sum_{d}\sum_{i} f_{\min}q_i^d(t)\epsilon. \quad (9)$$

Therefore if the assumption $\epsilon > \epsilon_0 = \frac{f_{\max} - f_{\min}}{f_{\min}} A_{\max}$ is satisfied then from the Lemma 4.1 of [6], the network is strongly stable, as when $\sum_d \sum_i q_i^d(t) > \frac{f_{\max}B}{2(-(f_{\max} - f_{\min})A_{\max} + f_{\min}\epsilon)}$ the Lyapunov drift becomes negative. □

At this point, we should note that for the GBP algorithm, $f_{\min} = f_{\max} = 1$, thus Theorem 1 holds for $\epsilon > 0$, meaning that GBP is throughput optimal, ensuring stability for the entire capacity region Λ_G. Also, Theorem 1 is proved for a general positive function with finite value field, although in this paper we assume $f(i,j,d) = f(dist_H(i,d), dist_H(j,d))$.

3.1 Discussion on the Complexity and Distributivity of Algorithm 1

In this section, we briefly discuss the issues of the complexity and the distributivity of the Algorithm 1, mostly compared to those of the pure backpressure algorithm. The lack of centralized infrastructure in wireless multihop networks leads to the search of algorithms characterized by low-complexity to be implemented distributively. The proposed algorithm, has at most the same complexity as the original backpressure, due to the fact that each node computes fewer queue differences by checking at each neighboring link only the destinations for which this link is on a greedy path. Regarding the creation of the greedy embedding and the computation of the hyperbolic distances, they can be performed only once and stored at each node, and therefore they do not add any extra complexity at each step of the execution of the algorithm. Furthermore, the greedy embedding of [4] is based on a distributed algorithm where the hyperbolic coordinates are computed sequentially from the root to the leaves of the spanning tree. Additionally, with respect to the distributivity issue, the proposed algorithm if combined with a CSMA based backpressure scheme [10] achieves a completely distributed implementation, a fact that is also true for the classic backpressure algorithm. The method followed in [10] can be modified, so that the transmission aggressiveness of the CSMA algorithm is characterized by the product of the $f(i,j,d)$ and the queue difference instead of only the queue difference. Finally, several approximations of the maximum weight matching [5] which constitutes the caveat in the distributivity of the backpressure-style algorithms, can also be applied to Algorithm 1.

4. NUMERICAL EVALUATION

In this section, we study the delay-throughput trade-off performance in different topology and traffic scenarios. We examine two different physical layer topologies for the wireless multihop network, each one consisting of $N = 16$ nodes (Fig. 1). The first one is a 4x4 grid topology (Fig. 1(a)), while the second constitutes a Random Geometric Graph (RGG), where the nodes are randomly and uniformly distributed in a square region of side $L = 4m$ (Fig. 1(b)). Each link can transmit one packet during a time slot under the one-hop interference model. Throughput is expressed as the percentage of packets that reach their destination divided by those sent from the source for each flow and both throughput and delay are expressed as averages for all flows (source-destination pairs). Considering the social network of flows which is developed over our wireless network, three cases are simulated. The first one is denoted as the "All Pair Communication Model" where each node at each time slot chooses a random node and generates traffic for it with probability ranging from $\lambda = 0.01$ to $\lambda = 0.3$ with step increase 0.01. For each λ we run the algorithms for 5000 slots. The second examined social graph, is inspired by Kleinberg's model [11]. According to this graph each node communicates with all its one-hop physical neighbors and with one more node in further hop-distance, i.e. long-range social contact. At each time slot, each node chooses randomly one of its possible destinations and generates traffic for it with probability ranging from $\lambda = 0.01$ to $\lambda = 0.3$ with step increase 0.01. We call this model the "Kleinberg Model". Finally, the third type of social graph is denoted as the "Long Range Model", according to which each node communicates with 3 other long-distant (not one-hop neighbors) in physical layer nodes, chosen randomly. The traffic generation for the latter model is similar with the Kleinberg Model. According to previous research, the matching models for the social traffic over wireless networks are the Kleinberg Model or the Long Range Model [2]. In all simulations, we employ the Backpressure (BP), the GBP and the GBP QDS algorithms. Let us denote as $d_{\max}$ the largest hyperbolic distance among all node pairs in the network. Indeed, we need to note that each hyperbolic distance has a unique value.

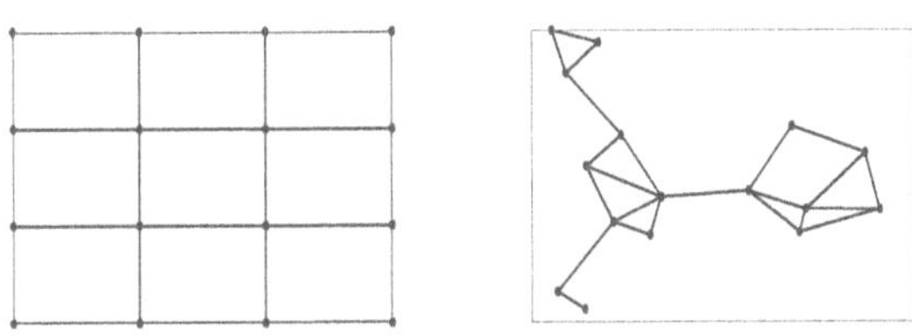

(a) Grid Topology (b) Random Geometric Graph Topology

Figure 1: Physical Layer Topologies.

The range of λ is chosen suitably from simulations in order to avoid excessive increase in the sum of queue lengths in the network under all physical and social topologies and algorithms, which would indicate the transgression of the capacity region. An example of the queue scaling for the All Pair Communication Model in the RGG topology is depicted in Fig. 2. We observe that after a certain value of λ, which concides with the value of λ at which the degradation in the throughput and delay begins (as shown in Fig. 6), the sum of queue backlogs in the network starts to increase, indicating that the network capabilities have been exceeded. The functions f_1, f_2, f_3 in Fig. 2 will be explained in the sequel of this section. Similar results are obtained for the other combinations of physical and social graphs and determine the suitable range of values for λ. For the Long Range Model and the grid physical topology, the sum of queue lengths increases much faster with respect to the increase of λ and therefore, we use a more limited range for λ (Fig. 5).

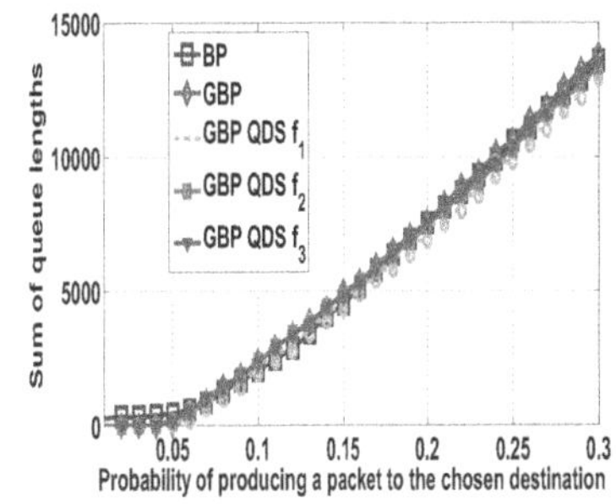

Figure 2: Queue Scaling for the RGG Topology - All Pair Communication Scenario.

The combination of the grid topology and the All Pair Communication Model is examined first. In Fig. 3, the delay and throughput performance of the algorithms BP, GBP and GBP QDS for different chosen functions is illustrated. The employed functions are the following, $f_1 = |dist_H(i,d) - dist_H(j,d)|$, which gives higher weight to links that transfer the packets for longer hyperbolic distance, $f_2 = (2 \cdot d_{\max} - dist_H(i,d) - dist_H(j,d)) \cdot |dist_H(i,d) - dist_H(j,d)|$, which matches higher weights with links both close to the destination and with a high improvement of hyperbolic distance and $f_3 = \frac{|dist_H(i,d)-dist_H(j,d)|}{dist_H(i,d)+dist_H(j,d)}$, that has approximately the same meaning with f_2. It is important to note that the necessity of Greedy Routing constraints is reflected on the choices for the $f(i,j,d)$ function, as they ensure the correctness of the meaning of the $f(i,j,d)$, i.e. high difference $|dist_H(i,d) - dist_H(j,d)|$ implies moving closer in hyperbolic distance to destination. More precisely, due to the greedy routing constraints, we can ignore the absolute value and simply multiply by $(dist_H(i,d) - dist_H(j,d))$. We can observe that GBP QDS for f_1, f_2, f_3 performs better than GBP and BP, leading to higher throughput and lower delay, for all values of λ except from a small region $\lambda \simeq 0.14 - 0.18$. We conclude that GBP QDS with f_3 has a satisfactory behavior for this combination of physical topology and social communication graph.

Similarly with Fig. 3, in Fig. 4 the delay and throughput performance for Kleinberg's social graph over the grid topology, is shown. In the GBP QDS scheduling/routing the f_1 and f_3 scaling functions are examined. Similar conclusions as in the All Pair Communication Model apply, with a very good delay-throughput trade-off driven by the f_1 function for all values of λ.

With respect to the combination of the grid topology and the Long Range Model, its performance is depicted in Fig. 5 where GBP QDS is applied with f_2, f_3 and $(f_3)^2$. Obviously, GBP QDS with $(f_3)^2$ leads to a significant improvement of the throughput-delay trade-off in the entire range of λ.

In the sequel, we examine the Random Geometric Graph physical topology, in order to pinpoint any differences in performance due to changes in the physical layer graph.

In Fig. 6, 7, 8, we depict the performance of the RGG physical topology and the three examined models of social flows. We employ the same functions as in the grid topology, with the exception of the Kleinberg Model, where the function $(f_3)^2$ achieves a better performance in terms of the

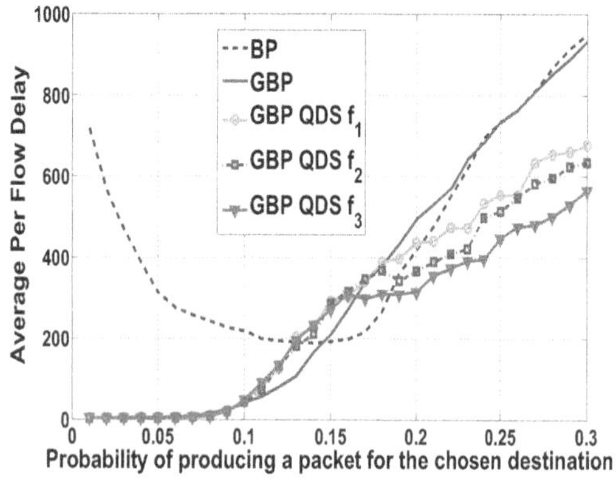

(a) Average Per Flow Delay

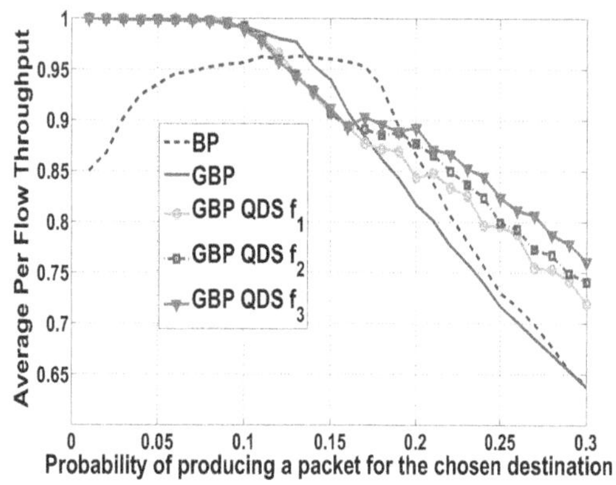

(b) Average Per Flow Throughput

Figure 3: GRID Topology - All Pair Communication Scenario.

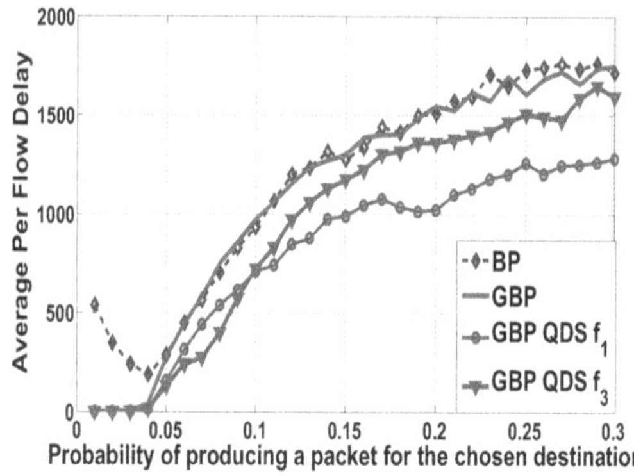

(a) Average Per Flow Delay

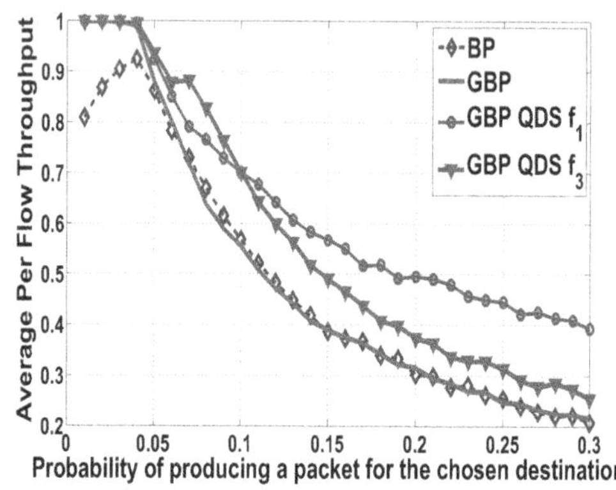

(b) Average Per Flow Throughput

Figure 4: GRID Topology - Kleinberg's Communication Scenario.

throughput-delay trade-off. It can be observed that similarly with the grid topology, for each combination of the RGG topology and the three social models, we can find a function $f(i, j, d)$ that improves the throughput-delay trade-off of the pure backpressure algorithm.

The GBP algorithm performs well in light traffic conditions due to routing the packets towards the direction of the destination and avoiding the long paths and cycles which characterize the classic backpressure algorithm. In higher traffic, it either presents the same or even sometimes worse performance than backpressure (depending on the social flows), as although orienting the packets towards the destination it restricts the number of paths creating more congested routes. However, the improvement in the throughput-delay trade-off is significant if topological properties in the hyperbolic space are taken into consideration in the scheduling part of the algorithm. Indeed, by combining greedy routing constraints with hyperbolic topology aware scheduling, we can improve the throughput-delay trade-off in a greater range of traffic values. In addition, from the simulations we can conclude that a good delay-throughput performance can be achieved by a function of the form

$$f(i,j,d) = \frac{(|dist_H(i,d) - dist_H(j,d)|)^{k_1}}{(dist_H(i,d) + dist_H(j,d))^{k_2}}, \; k_1, k_2 = 0, 1, 2, ... \qquad (10)$$

which can be used to initialize our searching for a suitable function in a different network topology and/or social flow. Specifically, we showed that this form of functions with $k_1 = k_2$ achieves a very good performance in our simulation scenarios. This class of functions, scales with $f_{\max} = 1$ the links adjacent to the destination and the rest of the links with diminishing values as we move further from the destination; depending also on the improvement that the link induces to the hyperbolic distance from the destination. This observation verifies our intuition to schedule with higher priority links closer to the destination.

4.1 Discussion on Extensions to Dynamic Networks

The Queue Difference Scaling approach of Algorithm 1 can be extended to dynamic networks, where the process of node churn is much slower than the routing/ scheduling process, by following the ideas proposed in our previous work [13]. In dynamic networks, new nodes join the network while existing nodes can resign from their network functionality. The greedy embedding algorithm of [4] allows for an easy and distributed integration in the greedy embedding of the newcomers. However, the deactivation of a node (supposing that the network still remains connected) can locally destroy the greedy embedding causing malfunctions to Algorithm 1. In this case, as proposed in [13] the GBP is replaced locally

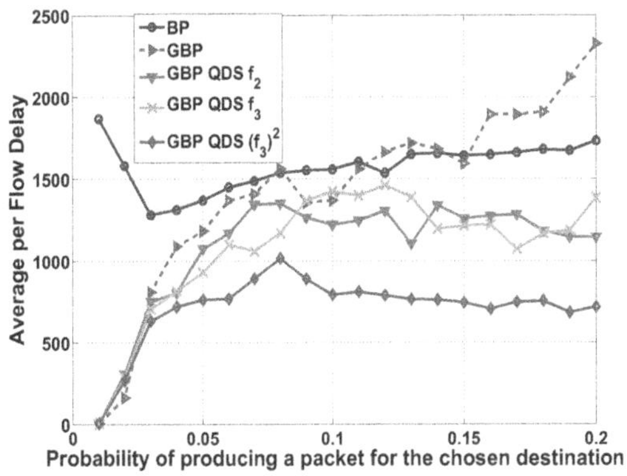

(a) Average Per Flow Delay

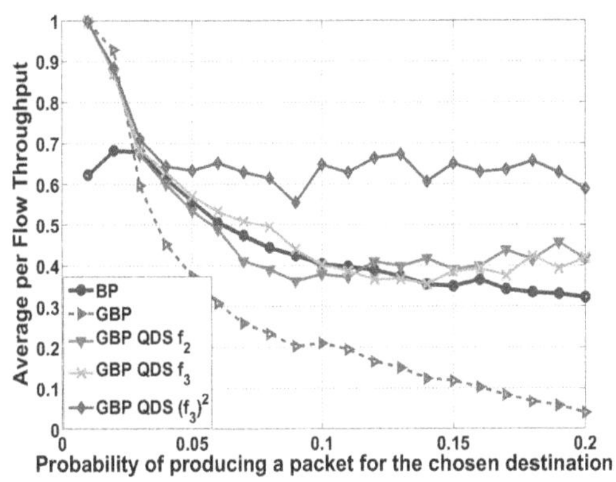

(b) Average Per Flow Throughput

Figure 5: GRID Topology - Long Range Communication Scenario.

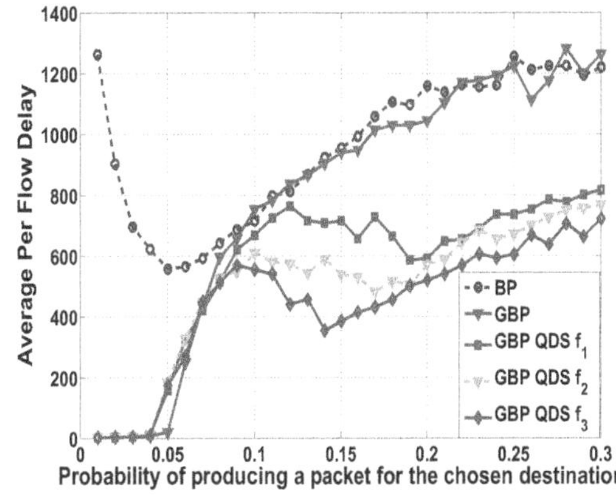

(a) Average Per Flow Delay

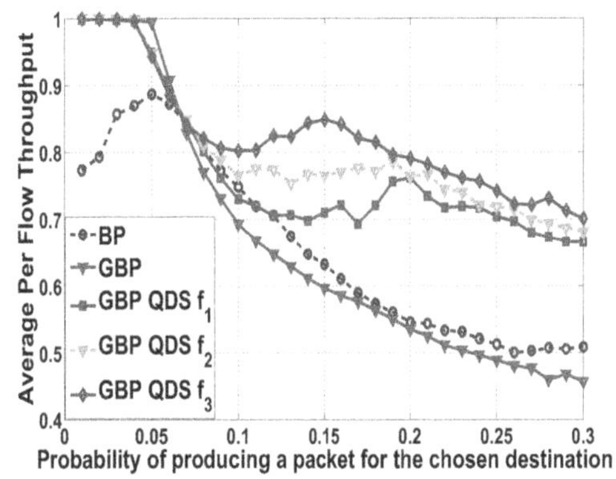

(b) Average Per Flow Throughput

Figure 6: RGG Topology - All Pair Communication Scenario.

by the classic backpressure algorithm. However in the GBP QDS the function $f(i,j,d)$ needs to be redefined when the greedy property is locally lost, because, as aforementioned, its form may not express a valuable meaning for scheduling, i.e. f_3 is proportional to $|dist_H(i,d) - dist_H(j,d)|$ which is meaningful only if j is a greedy neighbor of i for destination d. Therefore, the function $f(i,j,d)$ should be replaced by a time dependent function $f(i,j,d,t)$, possible forms of which, at the times t when the greedy property between i,d is lost, will be examined in our future work.

5. CONCLUSIONS

In this work, we focused on the throughput-delay trade-off in static wireless networks, and for the scheduling/routing algorithms analyzed. We proposed a design of the backpressure algorithm in the hyperbolic space that scales the queue differences by a suitable function. Through rigorous analysis, we proved the stability of the GBP QDS algorithm for a part of the capacity region which can be controlled through the min and max properties of the scaling function. Through extended simulations, we illustrated the improvement in the throughput-delay trade-off, which is possible with a sophisticated choice of the scaling function and introduced a class of topology-dependent functions that achieve a satisfactory performance in the examined physical and social topologies. The proposed algorithm if combined with a CSMA based backpressure scheme [10] can lead to high quality performance via a completely distributed implementation.

6. ACKNOWLEDGMENTS

The research of Eleni Stai and John Baras was partially supported by the US Air Force Office of Scientific Research under MURI grant FA9550-09-1-0538, and by the National Science Foundation (NSF) under grant CNS-1018346. Eleni Stai was also partially supported by the Foundation for Education and European Culture (IPEP).

7. REFERENCES

[1] J. W. Anderson. *Hyperbolic Geometry.* Springer 2nd Edition, 2007.

[2] B. Azimdoost, H. R. Sadjadpour, and J. J. Garcia-Luna-Aceves. Capacity of Composite Networks: Combining Social and Wireless Ad Hoc Networks. In *Proceedings of IEEE Wireless Communications and Networking Conference (WCNC)*, pages 464 – 468, March 2011.

[3] L. Bui, R. Srikant, and A. L. Stolyar. Novel Architectures and Algorithms for Delay Reduction in Back-pressure Scheduling and Routing. In *Proceedings of IEEE INFOCOM*, pages 2936 – 2940, April 2009.

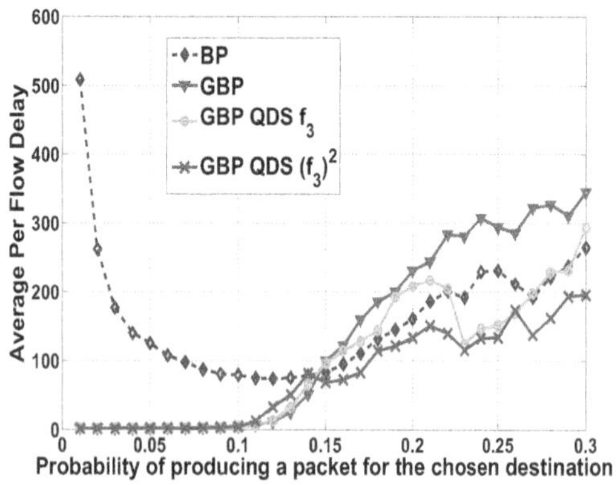

(a) Average Per Flow Delay

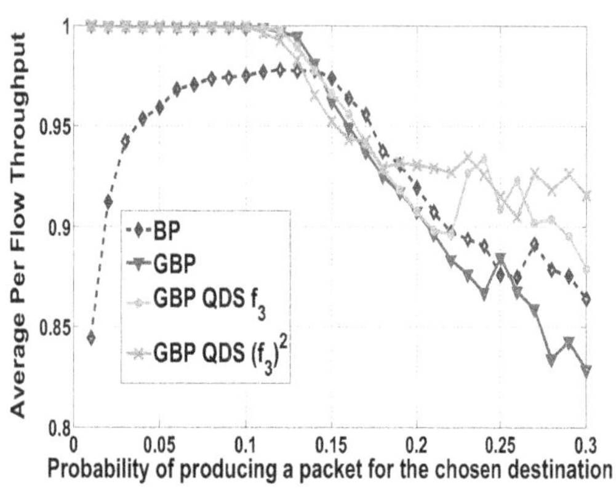

(b) Average Per Flow Throughput

Figure 7: RGG Topology - Kleinberg's Communication Scenario.

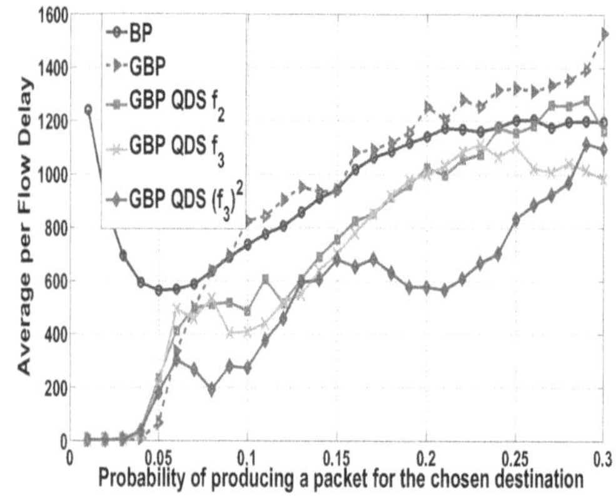

(a) Average Per Flow Delay

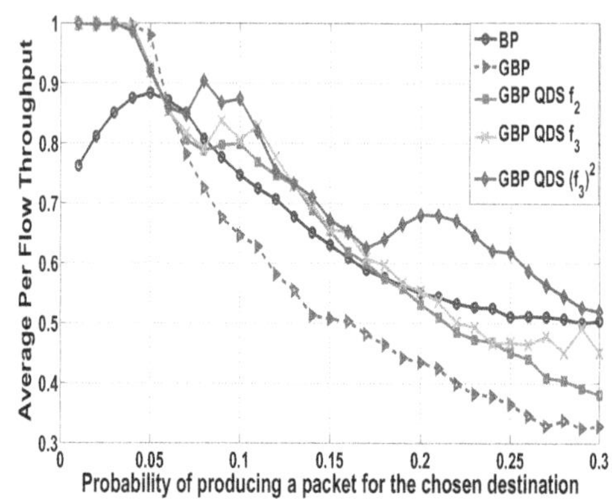

(b) Average Per Flow Throughput

Figure 8: RGG Topology - Long Range Communication Scenario.

[4] A. Cvetkovski and M. Crovella. Hyperbolic Embedding and Routing for Dynamic Graphs. In *Proceedings of IEEE INFOCOM*, pages 1647 – 1655, April 2009.

[5] A. Eryilmaz, A. Ozdaglar, D. Shah, and E. Modiano. Distributed Cross-layer Algorithms for the Optimal Control of Multi-hop Wireless Networks. *IEEE/ACM Transactions on Networking*, 18(2):638–651, April 2010.

[6] L. Georgiadis, M. J. Neely, and L. Tassiulas. *Resource Allocation and Cross-Layer Control in Wireless Networks*. Now Publications, 2006.

[7] G. R. Gupta and N. B. Shroff. Delay Analysis and Optimality of Scheduling Policies for Multihop Wireless Networks. *IEEE/ACM Transactions on Networking*, 19(1):129–141, February 2011.

[8] L. Huang and M. J. Neely. Delay Reduction via Lagrange Multipliers in Stochastic Network Optimization. *IEEE Transactions on Automatic Control*, 56(4):842–857, April 2012.

[9] B. Ji, C. Joo, and N. B. Shroff. Delay-based Back-pressure Scheduling in Multi-hop Wireless Networks. In *Proceedings of IEEE INFOCOM*, pages 2579 – 2587, April 2011.

[10] L. Jiang and J. Walrand. A Distributed CSMA Algorithm for Throughput and Utility Maximization in Wireless Networks. *IEEE/ACM Transactions on Networking*, 18(3):960–972, June 2009.

[11] J. Kleinberg. The Small-World Phenomenon: An Algorithmic Perspective. In *Proceedings of the 32nd ACM Symposium on Theory of Computing*, 2000.

[12] R. Kleinberg. Geographic Routing Using Hyperbolic Space. In *Proceedings. of IEEE INFOCOM*, pages 1902 – 1909, May 2007.

[13] E. Stai, J. S. Baras, and S. Papavassiliou. Throughput-Delay Trade-off in Wireless Multihop Networks via Greedy Hyperbolic Embedding. In *Proceedings of the 20th Int'l Symposium on Mathematical Theory of Networks and Systems (MTNS)*, July 2012.

[14] L. Tassiulas and A. Ephremides. Stability Properties of Constrained Queueing Systems and Scheduling Policies for Maximum Throughput in Multihop Radio Networks. *IEEE Transactions on Automatic Control*, 37(12):1936–1948, December 1992.

[15] L. Ying, S. Shakkottai, A. Reddy, and L. Shihuan. On Combining Shortest-path and Back-pressure Routing over Multi-hop Wireless Networks. *IEEE/ACM Transactions on Networking*, 19(3):841–854, June 2011.

Model-based Analysis of a Protocol for Reliable Communication in Railway Worksites

Leonardo Montecchi[1], Paolo Lollini[1], Boris Malinowsky[2], Jesper Grønbæk[2], Andrea Bondavalli[1]

[1]University of Florence
Viale Morgagni 65
I-50134 Firenze, Italy
{lmontecchi,lollini,bondavalli}@unifi.it

[2]Forschungszentrum Telekommunikation Wien
Donau-City-Straße 1
A-1220 Wien, Austria
{malinowsky,gronbaek}@ftw.at

ABSTRACT

In this paper we perform a model-based analysis of the Timed Reliable Communication (TRC) protocol, which is being used within the EU funded ALARP project for railway worksite communication. TRC is a group communication protocol based on IEEE 802.11 networks, targeting safety-critical applications with limited bandwidth requirements. The paper contains an in-depth analysis of the performance and reliability characteristics of the protocol using a Stochastic Activity Networks model. The results are first compared with available experimental measurements for the sake of model validation. The validated model is then used for a thorough analysis of a set of key metrics under different environment and network conditions. The obtained results allow: i) to assess that the protocol allows to satisfy the ALARP targeted performance and reliability requirements, and ii) to evaluate the existing tradeoffs and help in choosing parameter values for the final implementation.

Categories and Subject Descriptors

C.4 [**Performance of Systems**]: Modeling Techniques; Reliability, availability and serviceability. C.2.2 [**Network Protocols**]: Protocol verification.

Keywords

broadcast, wireless, 802.11, ALARP, reliable communication

1. INTRODUCTION

Wireless communication technologies are rapidly improving, and replacing wired communication links in several application scenarios. Moreover, wireless devices are now easily embedded in everyday appliances, like telephones, cars, televisions, printers, and cameras. Technologies based on the IEEE 802.11 standard [11] allow indoor, office and home connectivity at low costs, 802.15 [21] supports wireless broadband access (e.g., WiMax), while 802.16 [22] allows communication within a limited range and it is commonly used for personal devices (e.g., Bluetooth).

The spread of such technologies allows for applications that were not possible in the past. However, ensuring reliable and timely message transmission in wireless networks is still a challenge. Wireless communication links reliability depends on the actual distances between nodes, which varies in mobile settings, on the presence of obstacles, and on possible interference from other transmitting devices. Executing critical distributed services and achieving the required reliability, safety and timeliness of communication on wireless devices requires a careful design and analysis of communications.

MSWiM'12, October 21–25, 2012, Paphos, Cyprus.

In this paper, we analyze a message dissemination protocol that is being used in building an Automated Track Warning System (ATWS) for the safety of railway workers, based on wirelessly connected personal devices. The analysis is performed using a Stochastic Activity Networks model, which is evaluated using a numerical solver. The results are first compared with available experimental measurements for the sake of model validation. The validated model is then used for a thorough analysis of a set of key metrics in different scenarios.

The paper is organized as follows. Section 2 describes the context in which analyses are drawn and their motivations, while the analyzed protocol is described in Section 3. Related work is reviewed in Section 4; the modeling approach and the target measures of interest are described in Section 5, while the evaluation and results are discussed in Section 6. Finally, conclusions are drawn in Section 7.

2. THE ALARP SYSTEM

ALARP ("A railway automatic track warning system based on distributed personal mobile terminals" [1]) is a research project funded within the Seventh Framework Programme (FP7). Its objective is to design and develop an innovative Automatic Track Warning System (ATWS), to improve the safety of trackside workers.

2.1 System Description

The ALARP ATWS is able to inform the trackside workers about dangerous events within the worksite, e.g., approaching trains on the track, maintenance events on power lines and/or safety equipment that may put at risk workers' safety (e.g. being hit by a train or by an electric shock), emergencies on tracks and tunnels nearby the workers (e.g. fires in a tunnel, toxic smoke, etc.). Additionally, it keeps track of the status and position of the workers, in order to identify and localize those at risk, suggest escape routes, or provide information in case of rescue.

The ALARP architecture is COTS-based and involves the following main components: i) the track-side train presence alert device (TPAD), able to sense an approaching train on the interested track, ii) a set of wearable wireless Mobile Terminals (MTs), to inform the workers about possible approaching trains and other events that could put at risk their safety, and iii) infrastructure for wireless communication [2].

The MT has to communicate and interact through wireless connections with other MTs and the TPADs. Most importantly, the MTs need to be capable to receive information about approaching trains from the TPADs, and notify it to the workers. Railway regulations require a Controller Of Site Safety (COSS) to supervise the workers team and take care of safety concerns. The ALARP system supports the COSS by providing useful information about the location of workers and the state of their MTs.

The ALARP communication architecture joins two technologies in a two-hop setup. Worksite communication relies on off-the-shelf IEEE 802.11 [11] technology, having limited range but some properties that can be exploited for timely communication. A coordinator (i.e., an Access Point) provides access to the medium and serves as gateway for Mobile Terminals. TPADs communicate with the worksite via longer range wireless technology, delivering messages to the coordinator at the worksite for dissemination. In this paper we focus on the protocol that is used for communication at the worksite.

2.2 Relevant Requirements

The communication protocol adopted in ALARP shall allow the communication layer to reliably distribute messages. It should support broadcast (distribution to all nodes), multicast (one-to-many communication), and unicast (message exchange between two dedicated nodes) communication primitives. The communication layer defines three criticality levels: high for system safety-critical messages (LEVEL HIGH), medium for messages affecting system availability (LEVEL MEDIUM), and low for messages with no requirements (LEVEL LOW, best effort) [10].

According to ALARP requirements [2], the alert of an approaching train shall reach the workers within the time interval [T_0, T_1] before the train approaches the worksite, i.e. the workers should be alerted at time t, $T_1 > t > T_0$, where T_0 is the minimum allowed time to reach a safe position and T_1 is a programmable value that depends on national regulations.

Typically, the TPAD would be placed at distance such that the train will take about one minute to reach the worksite, and the minimum time T_0 for workers to reach the safety position would be around 45 seconds. Therefore, an informal derived requirement on worksite communication states that any broadcast should take a maximum of 10-12 seconds in the worst case.

3. THE TRC PROTOCOL

In order to satisfy the ALARP requirements, the work in [6] defined the ALARP Timed Reliable Communication (TRC) protocol, by properly adapting the existing Real-time Group Communication Protocol (RGCP) [8]. The latter was originally designed to support real-time group decisions between autonomous robots communicating through wireless networks.

RGCP relies on the IEEE 802.11 Point Coordination Function (PCF) [11] and the existence of a coordinator (Access Point) to realize a polling scheme of nodes, and allocating fixed time slots using a round-based approach. During the "Contention Free Period" (CFP), a central coordinator controls the access to the medium for a group of stations. Every station remains silent until it is polled by the access point, which then grants exclusive access to the medium by transmitting the polling message.

The RGCP protocol is based on the following assumptions [8]:

- Messages delivered during the CFP are delivered correctly within a fixed time-bound (t_m).
- Messages may be lost (omission faults).
- Message losses are asymmetric; i.e., some stations may receive a broadcast message and some may not.
- The number of consecutive message losses is bounded by the omission degree OD.
- Stations may crash or leave the reach of the AP.
- The AP is not subject to any kind of error.

The communication is structured into rounds, and each round into N slots, with N being the number of nodes. During each slot, three messages are exchanged: "poll", "request", and "broadcast". In each slot, the AP polls a station, which responds with a "request" message containing the message to be broadcast. The AP assigns a sequence number to that message and broadcasts it to the group of stations. The "request" message is also used to acknowledge each of the preceding broadcasts by piggy-backing a bit field on the header of the request message. Therefore, one round after sending a broadcast message, the access point is able to decide whether each group member has received the message or not. In the latter case, the access point will retransmit the affected message. Each message is retransmitted at most $OD+1$ times to ensure, according to assumptions, that it is correctly delivered to all nodes. If the AP does not receive the request message from a station for $OD+1$ consecutive times, it considers that station to have left the group and broadcasts a message indicating the change in group membership [8].

A variant of the protocol, called RGCP *with resiliency degree*, allows the user to specify the maximum number of retransmissions of the messages. This bound on message retransmissions (called resiliency degree *res(c)*) may vary for different message classes *c*. Choosing *res(c)* smaller than *OD* allows trading reliability of message transmission for shorter transmission delays. The RGCP with resiliency degree allows trading the probability of broadcast delivery for resource utilization, which is a useful feature for the ALARP system: less critical messages could be transmitted using a lower resiliency degree, thus reducing network utilization and leaving room for possibly arriving safety-critical messages.

However, RGCP with resiliency degree ensures many more properties than are really necessary for the ALARP system. More in detail, it guarantees ordering and agreement by means of a final decision message which is broadcast by the AP upon reception of all the acknowledgments. The ALARP system does not require neither ordering nor agreement; therefore these two properties can be relaxed. Practically, this simply consists in removing the last part of the protocol, in which the decision message is sent [6]. In TRC after at most *res(c)+1* rounds the AP will start processing the next application-level message. Based on the received acknowledgments by that time it is able to know if all the nodes have received the broadcast, and potentially raise a warning if it is not the case.

Additional modifications have been made with respect to the original protocol, in order to adapt to the ALARP requirements and architecture. In particular, three different message classes have been defined, mapping the three ALARP message classes: *res(LH)*, *res(LM)*, and *res(LL)*. Protocol implementation relies on off-the-shelf equipment, fully compatible with the 802.11 standard. Although developed for the ALARP system, the protocol is therefore applicable to a wider range of COTS-based critical applications and infrastructures.

4. RELATED WORK

Communications are a fundamental part of modern systems and several works have been devoted to the analysis of their performance and dependability properties. Concerning wireless communication, a large collection of works focuses on physical properties of the wireless medium. The authors of [12] describe the physics of signal propagation in different environments, considering both indoor and outdoor scenarios. Other works focus on defining good approximations of the physical channel through a channel model; a nice survey on this topic can be found in [15]. Another set of works focuses on the analysis of a specific radio technology or protocol: in [17] and [18] the transmission of multicast packets in wireless LANs is analyzed, while in [13] capacity estimation for downlink WCDMA networks is performed. Conversely, [14] focus on the uplink direction, also considering inter-to-intra cell interference. The authors of [16] evaluate and improve the performance of the MAC protocol in 802.11 networks, in a dedicated short-range communication environment. Sensor networks are taken into account in [19], where a secure routing protocol is designed and analyzed. In [5] stochastic models and measurements are combined to analyze the performance of a consensus algorithm.

The original RGCP protocol has already been analyzed in a number of different ways. In [9] the protocol has been analyzed combining a Stochastic Activity Networks model and measurements; the model has then be refined in [3] to relax the assumption of independent losses, taking into account for loss correlation among consecutive packet transmissions. A modified version of the model in [9], for the analysis of the same protocol, has been proposed also in [20], although some major flaws impair the model, which actually does not represent the same system.

In this paper we take an orthogonal approach to the analysis in [9] and [3], where the main focus was broadcast throughput under continuous operation. In this paper we focus on transmission delay and delivery probability of a single application-level broadcast message, which are the main concerns in the ALARP system. Moreover, in this paper we also consider degraded protocol execution, as opposed to a simple failure/success model as in previous evaluations of the RGCP protocol. Finally, wireless technology has considerably improved over the past 10 years, and therefore we base our analysis on more recent experimental measurements, performed within the ALARP project.

A parallel evaluation work of the ALARP time reliable communication has been performed also in [6], where an experimental worksite test setup has been established, and results then used to parameterize a stochastic simulation model. The experimental setup used in such work is aligned to the ALARP system communication requirements (low bandwidth, but strict bounds on time behavior) and investigates cross-traffic and interference scenarios using unlicensed bands. In particular, a scenario with contending nodes and saturated channel conditions representing a highly challenging IEEE 802.11 communication environment is considered, as encountered in dense urban worksite setups. After successful validation against the experimental results, the simulation model has been used to generalize the experimental results to common railway worksite settings.

The results in [6] provide an input to support our analytical approach, in which we perform an extended analysis of TRC performance and reliability. Further details on this aspect are provided in Section 6. The main advantage of adopting an analytical solution approach relies in the accuracy of the computed solution, and in the avoidance of the rare event problem (i.e., a large number of simulation runs need to be executed to be able observe events that rarely occur). This is especially relevant in safety analysis as performed in this paper, where the objective is to evaluate the occurrence of hazardous events, which need to be extremely rare for the system to fulfill its requirements.

5. MODELING APPROACH

The main objective that drives this analysis is to assess the effectiveness of the TRC protocol in guaranteeing timely and reliable delivery of the safety-critical messages exchanged by the components of the ALARP system. In our approach, we focus on the transmission of a single application-level message using the TRC protocol, and aim to evaluate its quantitative properties.

In the following we will only consider broadcast messages, however the analysis can be easily adapted to multicast and unicast messages as well. The main reason is twofold: i) the most critical message in the ALARP system is a broadcast message (the "RISK_EVENT" message informing the workers of approaching trains), and ii) broadcast messages constitute the worst case, both in terms of successful delivery probability, and in delivery delay.

5.1 Measures of interest

Following from the definition of the protocol, the transmission of a message terminates when any of the following occurs: i) the poll-request fails for $OD+1$ consecutive rounds, ii) the same message is broadcast for $res(c)+1$ consecutive rounds, or iii) the coordinator receives the acknowledgment from all the recipient nodes. When any of these events occur, the execution of the protocol for that message terminates, and the next application-level message is processed.

Upon termination, the result of message transmission may be categorized in four classes. To precisely identify them, we first define the following three predicates, which are evaluated at the end of message transmission:

- PR_FMAX is true if the poll-request has failed $OD+1$ consecutive times for that message.
- BC_FMAX is true if the message broadcast has failed $res(c)+1$ consecutive times, i.e., if the message has not been delivered to all the nodes within $res(c)+1$ rounds.
- ACK_OK is true if the coordinator has received the acknowledgements from all the recipient nodes.

We now identify four different degrees of success in disseminating the message to all nodes, "Complete Success", "Partial Success", "Dissemination Failure", and "Poll-Request Failure":

$$csucc = \neg PR\_FMAX \wedge \neg BC\_FMAX \wedge ACK\_OK$$
$$psucc = \neg PR\_FMAX \wedge \neg BC\_FMAX \wedge \neg ACK\_OK$$
$$disfail = \neg PR\_FMAX \wedge BC\_FMAX$$
$$prfail = PR\_FMAX$$

It is easy to verify that the four cases above are disjoint, and that they represent the universe with respect to the possible outcomes of a message transmission using TRC. Our analysis aims to evaluate the relative probability that sending a message results in each of the four cases. Distinguishing between them will help us to better understand how the system parameters affect the execution of the protocol, and to assess the improvements that TRC introduced for the ALARP scenario over the reference RGCP protocol.

It should be noted that, from a strict safety point of view, both the failures defined above represent a catastrophic failure (i.e., workers are not notified of an approaching train). However, for the purpose of the analysis, it is of interest to distinguish between them. Firstly, the two different failures depict two completely different scenarios. If a "Poll-Request failure" occurs, it means that the message has not even been able to reach the coordinator, and therefore nobody (including the COSS) will be alerted of the approaching train. Conversely, if the "Dissemination failure" occurs, a part of the MTs may still have received the message, and thus part of the workers may be aware of the approaching train. Although a "Dissemination failure" would still be a violation of ALARP safety requisites, means to mitigate its consequences would still be applicable, especially considering the "physical" interaction and communication between workers, and the supervising of the COSS on the worksite. Concerning the "Partial Success" outcome, it affects both resource utilization and false alarms, since the coordinator may wrongly believe that the broadcast has not been received by all the MTs.

Another measure which is of interest is the time required to execute the protocol. Our model-based analysis supports the evaluation of the average time required to execute the protocol and, most importantly, its probability distribution, which allows to accurately assess the ability of the protocol to adhere to the time bounds imposed by the ALARP system. To summarize the following measures of interest can be evaluated:

- P_{csucc}, P_{psucc}, P_{prfail}, $P_{disfail}$: Relative probability that each of the four cases occur in the transmission of a message using the TRC protocol ("Complete Success", "Partial Success", "Poll-Request Failure" and "Dissemination Failure", respectively).
- $T_{deliver}$: Time required the deliver the message to all the recipients, measured from the instant at which the coordinator polls the sender for the first time.
- $N_{deliver}$: Number of nodes that have received the message when the protocol ends.
- T_{end}: Duration of a single execution of the protocol. It measures the interval between the instant at which the coordinator polls the node for the first time, and the instant at which it polls it again asking for the next application-level message.

5.2 Modeling assumptions

Here we summarize the assumptions that have been made in the construction of the model, in addition to those introduced by the definition of the protocol (see Section 3).

- All the nodes are communicating using the TRC protocol only. This assumption completely holds in the ALARP scenario, since all the messages at the worksite (broadcast, multicast, and unicast) are transmitted using TRC, with different *res(c)* values based on the reliability class of the message.
- Any application-level message is transmitted within a single 802.11 packet. This assumption holds within the ALARP system, because of the limited amount of data to be transmitted.
- A single unicast packet may be lost with probability p_m. This probability holds both for the "poll" and "request" messages. The fault model assumed in the definition of the protocol considers only omission faults, i.e., messages may not be delivered.
- A broadcast (or multicast) packet may not be received by a given station with probability p_b.
- Homogeneous links. All the links are homogeneous, i.e., they are characterized by the same properties.
- Independent losses. The loss probability of a message sent by a station *i* is independent from losses of messages sent by other stations and previous messages sent by the same station.

The above is a reasonable set of assumptions that are often used in the modeling of wireless communication. Moreover, these assumptions are in part motivated also by the objective of: i) obtaining a model that can be solved analytically, and ii) using existing experimental measurements as input parameters.

5.3 Stochastic Activity Networks model

The model used for our analyses has been created using the Stochastic Activity Networks (SAN) formalism [7], using the Möbius modeling framework [4]. Stochastic Activity Networks are an extension of Stochastic Petri Nets (SPN), having a powerful set of primitives that allow specifying very complex stochastic processes. The firing time of activities ("transitions" in classic Petri nets terminology) can follow different kinds of probability distributions, although the use of non-exponential activities (with some exceptions) prevents applying analytical solution techniques.

As described in Section 3, the TRC protocol is a slot-based protocol, in which the time advances in deterministic steps. Since the duration of slots is fixed, it suffices to count the number of elapsed slots to evaluate time-related quantities. For this reason, we are only interested in the number of firing of activities, and not in the actual elapsed time. Moreover, no concurrency behavior is present in the model, and therefore time distribution of activities has no impact on the computed measures of interest. Since we target an analytical solution, we choose to have only exponentially distributed (and instantaneous) activities; moreover, the rate of all exponential activities is set to one. Actually, the model could be implemented as a Discrete-Time Markov Chains (DTMC) as well; however a different model would be needed for each combination of *OD*, *res(c)*, and *N*, thus making it cumbersome to analyze different scenarios.

The overall SAN model is shown in Fig. 1, and it basically consists of four parts: a) poll-request, b) broadcast, b) acknowledgments, and d) support for measures of interest.

5.3.1 Poll-Request

The first step in sending a message using the TRC protocol is performing the poll-request message exchange with the coordinator. The corresponding part of the model is portion a) of Fig. 1.

Initially there is a token in place "Round", which means that we are at the beginning of a round of the protocol. For simplicity we assume that the slot reserved to the sender node is the first slot in the round. The elapsing of such slot is represented by the activity "MySlotDelay"; when it fires, a token is added in place "SlotCount", which counts the number of slots that have elapsed

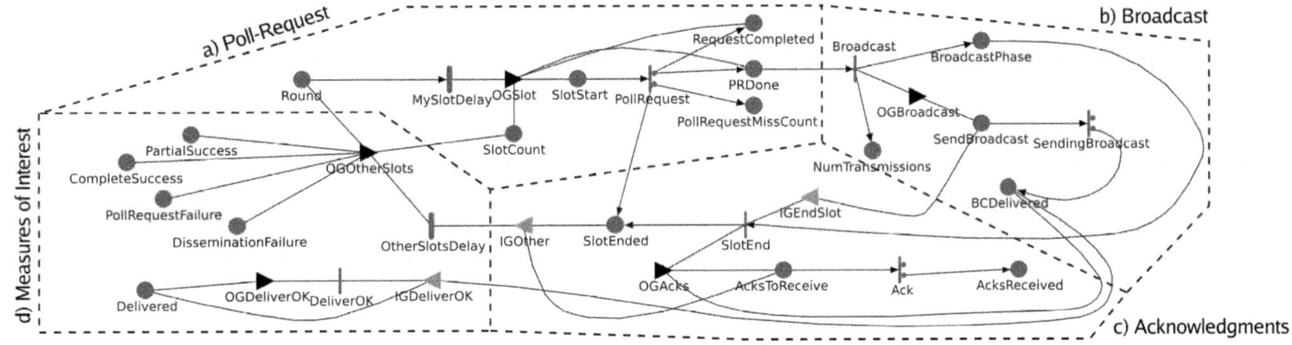

Figure 1. The SAN model used to evaluate the TRC protocol, and the four parts in which it is organized.

in the execution of the protocol. Then, it is checked if the poll-request message exchange has already been completed, by checking for the presence of a token in place "RequestCompleted".

If the poll-request part has not been completed yet, a token is added in place "SlotStart", causing the instantaneous activity "PollRequest" to become enabled and fire. The activity has two cases, representing the success and failure of the message exchange, respectively. The successful case occurs when both the "poll" and "request" messages are delivered correctly; the associated probability value is therefore $(1-p_m)(1-p_m)$. The failure case occurs with the complementary probability value of $1-(1-p_m)(1-p_m)$. If the poll-request succeeds, a token is added to "RequestCompleted" and "PRDone" places. Otherwise, a token is added to place "PollRequestMissCount", which counts the number of consecutive poll-request failures, and to place "SlotEnded", to model the end of the slot reserved for the node.

5.3.2 Broadcast

The part of the model that represents the actual broadcast is depicted in portion b) of Fig. 1. When place "PRDone" contains a token, the "Broadcast" activity is enabled and fires, removing the token from "PRDone", and adding a token in places "BroadcastPhase" and "TransmissionsCount", which counts the number of broadcast transmissions that have been performed for the current message. Place "BCDelivered" holds the number of nodes that have successfully received the broadcast message. When the activity "Broadcast" fires, the gate "OGBroadcast" checks if the broadcast has already been delivered to all nodes, by checking the number of tokens in "BCDelivered".

If the broadcast has not been delivered yet, the gate adds in place "SendBroadcast" a number of token equal to the number of nodes that still have to receive the broadcast. This number is exactly $(N-1)$ minus the number of tokens in "BCDelivered", where N is the number of nodes in the scenario. When a token is in place "SendBroacast" the activity "SendingBroadcast" is enabled and fires, removing the token from "SendBroadcast". The activity "SendingBroadcast" has two cases, corresponding to the successful reception of the broadcast by the node (occurring with probability p_b), and to its loss. If the broadcast succeeds, a token is added in place "BCDelivered".

Finally, when place "BroadcastPhase" contains a token and "SendBroadcast" is empty, the instantaneous activity "SlotEnd" fires, removing the token from place "BroadacastPhase" and adding a token in place "SlotEnded", modeling the end of the slot.

5.3.3 Acknowledgments

In the subsequent slots (each one reserved for the other nodes in the connected set), the coordinator will poll the other nodes, and possibly receive the acknowledgments piggy-backed to their "request" messages. Then, in the following round, within the slot reserved to the sender node, the coordinator will check if it has received all the $N-1$ acknowledgments. Otherwise, if the maximum number of retransmissions has not been reached, it will retransmit the broadcast.

The part of the model that represents this aspect of the protocol is shown in portion c) of Fig. 1. It is important to note that, for our measures of interest, we are not interested in the order in which the coordinator receives the acknowledgments, neither in *which* acknowledgments it has already received. Instead, we are interested in the number of nodes that have acknowledged the message; this number is given by the number of tokens in place "AcksReceived", which is initially empty.

When the activity "SlotEnd" fires, the output gate "OGAcks" adds in place "AcksToReceive" a number of tokens equal to the number of nodes that i) have received the broadcast, and ii) from which the coordinator has not yet received the acknowledgment, which is exactly the number of tokens in "BCDelivered" minus those in "AcksReceived". The activity "Ack" models the transmission of a single acknowledgment message, and fires once for each token in place "AsksToReceive". The coordinator receives the acknowledgment from a given node only if the poll-request succeeds with that node. Therefore, the probability that the coordinator receives the acknowledgment from a given node (within one round) is the probability that the poll-request succeeds, which equals to $(1-p_m)(1-p_m)$ as discussed above. If success occurs, a token is added in place "AcksReceived".

When "AcksToReceive" becomes empty, the "Ack" activity is disabled, while the activity "OtherSlotsDelay" becomes enabled and fires, representing the end of current round. The output gate "OGOtherSlots" adds $N-1$ tokens in place "SlotCount", to record that additional $N-1$ slots have elapsed (one for each of the other nodes). The gate also checks the stopping conditions for the protocol. If they are not satisfied, a token is added again in place "Round", and the next round is started; otherwise the transmission of the message ends.

5.3.4 Support for measures of interest

To support the evaluation of the measures of interest defined in Section 5.1 some additional elements have been added to the SAN model, forming portion d) of Fig. 1. The activity "DeliverOK" fires as soon as the broadcast is successfully delivered to all nodes (i.e., the number of tokens in place "BCDelivered" reaches $N-1$). The output gate "OGDeliverOK" adds to place "Delivered" the same number of tokens that are hold in place "SlotCount", thus also recording the number of slots that have elapsed. For each of the four outcomes defined in section 5.1 an additional place is added to the model, and they are used to record the outcome of protocol execution. While checking the stopping conditions for the protocol, the output gate "OGOtherSlots" also checks which of the four outcomes has occurred, and sets the marking of the corresponding place to the current marking of "SlotCount".

The relative probability that message transmission results in one of the defined outcomes (P_{csucc}, P_{psucc}, P_{prfail}, $P_{disfail}$) is evaluated by evaluating the probability that the number of tokens in the corresponding place is greater than zero at the end of the protocol. Similarly, the probability distribution of the time required to deliver the broadcast to all the nodes ($T_{deliver}$) is given by the probability distribution of the number of tokens in place "Delivered". The duration of the broadcast (T_{end}) can be evaluated by evaluating the number of tokens in place "SlotCount", while the number of nodes that have received the message ($N_{deliver}$) is evaluated by counting the number of tokens in place "BCDelivered".

The evaluation is performed by transient analysis, using a sufficiently large instant of time to ensure that the termination of the protocol is reached. Due to the nature of the protocol, termination is reached in at most $N(OD+res(c))+1$ slots; therefore, we can derive the maximum number k of consecutive exponential activities that should fire to reach termination. The selected instant of time t should then allow all the k activities to fire, i.e., the sum of the k exponentially distributed random variables should be smaller than t with a probability that approaches one. All the exponential variables in the model are independent and identically distributed; therefore, their sum is described by an Erlang distribution with shape k. It is then easy to verify if the selected value of t is really sufficiently large; in our analyses we have used $t=10000$, which corresponds to a probability lower than 10^{-50} that, in the worst case, some activity has still not fired.

6. EVALUATION AND RESULTS

6.1 Model parameters and settings

The described model is based on the set of parameters listed in Table 1. Some of them are part of the environment, and the system has no control on them, e.g., the message loss probabilities p_m and p_b, or the number of nodes N. Instead, other parameters are implementation-dependent and should be chosen with care. Although the number of parameters is limited, each of them has a great impact on the measures of interest.

Table 1 also shows the default values that will be used in the following evaluations, which have been carefully selected based on system requirements, typical working conditions, and existing experimental measurements. The number of nodes in the scenario (N) has been suggested by railway expertise within the ALARP project, stating that the "typical maximum" number of workers involved in a worksite is 20 [2]. Most worksites will probably feature a smaller amount of workers; in safety analysis it is however preferable to analyze worst-case scenarios. Concerning the duration of the timeslot we based on measurements and simulations described in [6], where a timeslot duration of 25 ms is identified as a lower bound to have sufficient time to deliver the "poll", "request" and "broadcast" messages in the same slot.

As discussed in Section 2.2, it has been agreed within ALARP that the maximum time to send the RISK_EVENT message should be around 10 seconds. After such amount of time all the nodes should have received the information, or know that they have missed it. The nodes that do not have received the message

Table 1. SAN model parameters and their default values.

Name	Description	Default
N	Number of nodes in the scenario (coordinator excluded)	20
t_{slot}	Duration of a timeslot	25 ms
OD	Omission degreee	10
res(c)	Resiliency degree of the message	10
p_m	Loss probabilitty of a single unicast packet	0.177
p_b	Probability that a station does not receive a broadcast (or multicast) packet	p_m

will be aware of that by receiving a message with a greater sequence number, or because they will be disconnected by the access point. The worst-case execution time is $N(OD+res(LH))+1$ protocol slots. By assuming $OD=res(LH)=10$, in a scenario comprising 20 nodes, and using a timeslot of 25 ms, this quantity is still within the acceptable bound (exactly 10.025 seconds).

The default value for the loss probability of unicast messages, p_m, has been set according to experimental results on the prototypal implementation of the protocol described in [6]. In particular, we are using results from the "cross-traffic" scenario, since interference is likely to occur, for example, in case the ALARP system is going to operate in an urban worksite. Finally, concerning the probability that a single station will not receive a broadcast packet (p_b), measurements have been performed in [9] on the original RGCP protocol. Reusing such measurements is problematic mainly because: i) wireless technology has greatly improved in the last ten years and ii) such results refer to a controlled office environment and good physical conditions, which are by far not the typical ALARP conditions. Using the same approach as in [9] we plan to perform up-to-date measurements of p_b, which are however not available at this time. Obviously some correlation exists between parameters p_m and p_b, since both are influenced by network conditions, e.g., interference or obstacles that may exist between the sender node and the receiver. As a conservative setting, in the following evaluations we set p_b to the same value as p_m, i.e., a broadcast packet is simply considered as N individual unicast packets. This is a conservative approximation, since in reality a broadcast message is transmitted using a single datagram.

6.2 Model validation with experimental data

In this section we compare the results obtained with the SAN model with results collected from the experimental runs conducted in [6]. In establishing an effective analytic framework, such experimental studies are an important aspect, since they allow to perform lower-layer measurements and assess the validity of the adopted assumptions. The scenario that is used for model validation is a simple scenario comprising two nodes and a dedicated access point, communicating with a prototype implementation of the TRC protocol. To keep the scenario as similar as possible to the experimental setup we have used the same parameters as in [6]: two nodes ($N=2$), an omission degree of 15, and a timeslot duration of 50 ms. The unicast message loss probability, p_m, is set as the experimentally measured packet loss rate.

The experimental test setup described in [6] allows to measure the following quantities: i) packet loss ratio, with packet denoting a single IP packet; ii) poll-request transmission time, describing the duration between the coordinator sending a poll message and the reception of the corresponding request; iii) broadcast reception

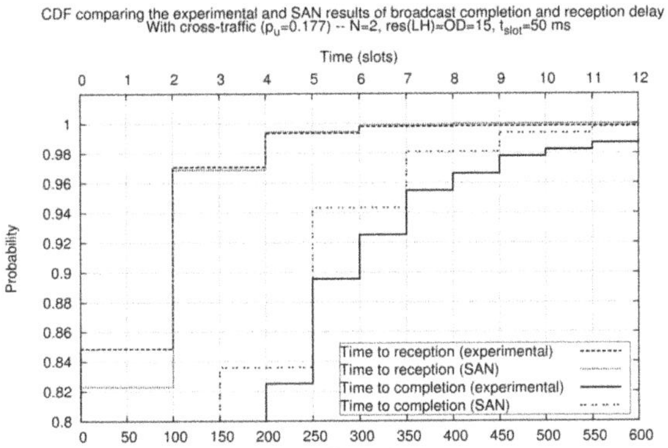

Figure 2. CDFs for validation with experimental data (experimental results taken from [6], for the case with cross-traffic.

delay, characterizing the delay between the coordinator receiving the broadcast message for dissemination, and the same broadcast message being successfully received by all nodes; iv) broadcast completion delay, characterizing the delay between the coordinator receiving the message for dissemination, and the instant in which it receives the last acknowledgement for that message.

Two scenarios were considered, one including cross-traffic and one without cross-traffic. The main coordinator is an off-the-shelf IEEE 802.11 Access Point running the TRC protocol. The scenario includes two additional nodes and another Access Point, contending the same channel. Channel contention is generated by constant transmission and saturated transmission buffers using a frame length of 1506 bytes. The remainder of this section discusses the experimental results with regard to their influence on TRC protocol configuration, and model parameterization.

Packet-loss ratio is estimated by counting the total number of executed protocol rounds and the number of successful poll-request packet sequences, thus providing an estimator p' of packet loss probability. This estimator is used to set the SAN model parameter p_m, thus allowing for a realistic setting regarding the probability of packet loss for unicast packets. To measure the *poll-request delay distribution* within a single slot, the transmission times are captured using the coordinator system clock. The first timestamp is captured immediately before sending a poll packet; the second timestamp is captured on receiving the request packet. Knowing this transmission characteristic is essential to appropriately configure the node slot parameter, t_{slot}. With a t_{slot} value set very high, bandwidth is allocated without being actually required for transmission, i.e., most poll-request-broadcast sequences finish in a fraction of the slot. This unnecessarily increases the overall protocol time bounds. Vice-versa, setting t_{slot} too low leads to an increase of necessary packet retransmissions, but having the benefit of shorter round times. The *completion delay* characterizes the TRC protocol performance, having influence on selecting the appropriate *OD* and *res(c)*. From a safety perspective in the ALARP system, the *reception delay* is crucial for disseminating information to workers. From a throughput perspective, the completion delay is more relevant, since it determines when the sender node will be able to issue a new broadcast request.

While the first two experimental metrics are used to set SAN model parameters, the other two can be used to validate the model against the experimental measurements. Fig. 2 provides a graphical comparison of the experimental CDF results and analytic SAN results, in the "cross-traffic" scenario. Both broadcast completion and broadcast reception delays have been taken into account. It should be

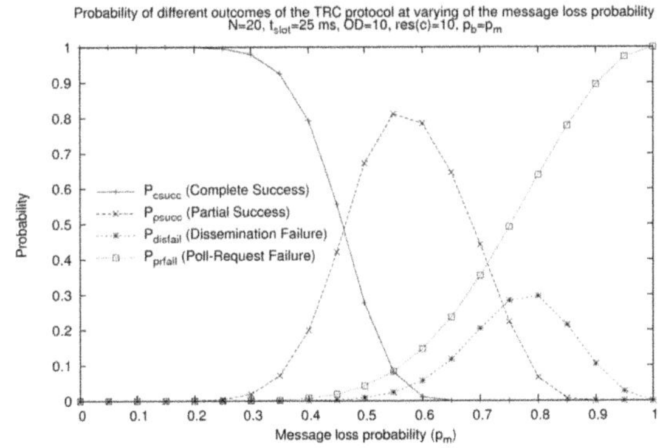

Figure 3. Probability of occurrence of the possible outcomes of the protocol, at varying of message loss probability.

noted that there is a minor difference in the way the metrics are obtained: in the experimental setup the delay was measured from when the coordinator receives the broadcast, while in the SAN model it is measured from when the node has the message ready to be sent (i.e., the beginning of the timeslot). Moreover, experimental measurements have a higher precision with respect to the SAN model, which evaluates the selected measures with a resolution of protocol rounds.

Results in Fig. 2 show however a good correlation between SAN results and experimental results, on both metrics. The higher resolution of measurements is however observable on the time to broadcast completion measure. A slight shift towards a higher required reception and completion time is observable in the experiments, compared to the SAN. This stems from regular, short send interruptions in the communication equipment for outgoing packets, therefore occasionally "missing" the slot, and has been observed and described in [6] as well. This behavior is particular to the COTS under test, and of course neglected in the model.

Similar results are obtained also in the case without cross-traffic, thus confirming that the stochastic model developed in this paper provides a good analytical match for the considered scenario. Link-layer network properties depend on several factors, including distance between nodes, weather conditions, interference sources. The validation performed in this paper demonstrates the ability of our model to match real-world data. More thorough measurement campaign could allow to accurately parameterize the model under different network conditions.

6.3 Evaluations

In this section we evaluate the target metrics using the SAN model described in Section 5.3. The model has been evaluated using the analytical solver provided by the Möbius framework [4].

6.3.1 Varying message loss probability

In Fig. 3 the relative probabilities of the four possible outcomes of message transmission are shown, at varying of the message loss probability p_m (and consequently p_b), with default parameters as listed in Table 1. Results shows that the probability of "Complete Success" drops rapidly to zero as p_m increases above 0.4, while the probability of "Partial Success" starts to increase: due to the higher loss probability the coordinator is not able to receive all the acknowledgments within the allowed retransmissions. For values of p_m above 0.7, the probability of the "Partial Success" outcome starts to decrease as well. Consequently, the probability of protocol failure starts to increase when p_m raises above 0.4, and continues to increase until it reaches 1 when p_m is equal to 1.

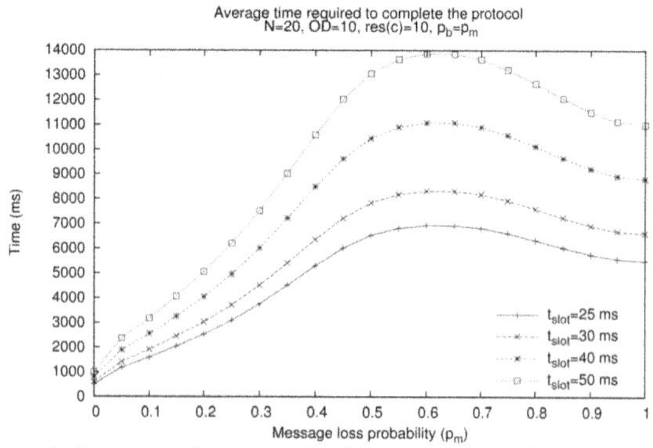

Figure 4. Average time required to complete the protocol, at varying of message loss probability, and timeslot duration.

Results in Fig. 3 highlight two interesting points. First, the changes introduced by TRC protocol with respect to "RGCP with resiliency degree" actually help in increasing the success probability in the ALARP system. More in detail, since in "RGCP with resiliency degree" the message is delivered to the application only if all the acknowledgments from all nodes are received, the "Partial Success" case would not have been possible. In TRC the "Partial Success" outcome provides a significant contribution to the success probability, especially when p_m increases above 50%. The second point is that, as the message loss probability increases, the failure of the protocol is almost always caused by a failure of the poll-request exchange. Failure to deliver the broadcast ("Dissemination Failure") occurs as well, but its occurrence probability does never exceed 0.3. Even if using a conservative setting, the most critical part of the protocol is the poll-request exchange. The probability of "Dissemination Failure" reaches its maximum when p_m is around 0.8, then it starts to decrease because the increasing failures of the poll-request exchange do not even allow the protocol to reach the "broadcast" stage.

Fig. 4 shows how the mean time required to complete the protocol varies at varying of the message loss probability p_m, for different timeslot durations. Starting from about $p_m=0.5$, the mean time required to complete the broadcast decreases with the decrease of p_m. The lower bound is reached when p_m is zero; under these conditions, the protocol always terminates successfully after the first round. With the default settings, comprising 20 nodes and timeslot duration of 25 ms, it takes exactly 500 ms to complete one round. Conversely, when the message loss probability is 1, the protocol always terminates after having failed the poll-request $OD+1$ consecutive times, which yields exactly 220 slots, corresponding to 5500 ms. On the average, the time required to complete the broadcast with default

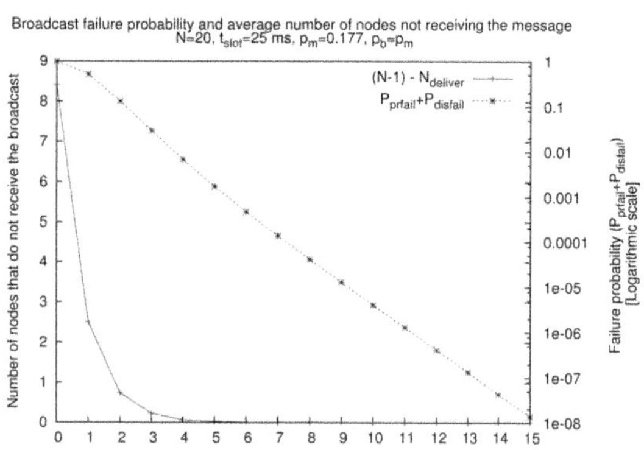

Figure 5. Failure probability, and number of nodes that do not receive the message, at varying of the resiliency degree.

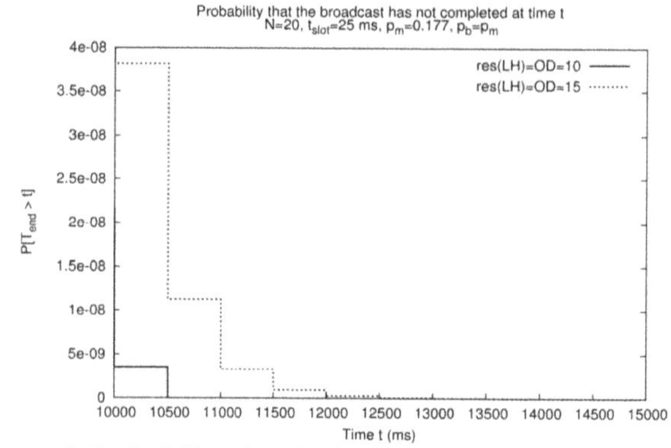

Figure 6. Probability that the broadcast has not completed at time t, at varying of the resiliency degree.

settings is below 7 seconds, leaving enough room even to process additional emergency messages.

It is interesting to note that the greatest amount of time is needed not when p_m approaches 1, but when it is around the value of 0.6. This behavior is due to the fact that when p_m approaches 1 it is more likely that the poll-request message exchange will fail $OD+1$ times, thus shortening the total time required for protocol execution, since the actual broadcast is not performed at all.

6.3.2 Varying the resiliency degree

In the following evaluations we hold the parameter p_m to its default value, and evaluate the impact of varying the *res(c)* parameter. The plot in Fig. 5 shows the impact of the resiliency degree of the message on two related measures of interest. The failure probability of the protocol ($P_{prfail} + P_{disfail}$) is plotted with respect to the y-axis on the right, in logarithmic scale. The other line depicts the number of nodes that do not have received the message when the protocol terminates (given by the quantity N-1-$N_{deliver}$) and refers to the left y-axis. As expected, both measures decrease as the resiliency degree parameter increases.

When using a resiliency degree of zero, meaning that no retransmissions are allowed, on the average more than 8 out of 19 nodes will not receive the broadcast. By increasing *res(c)* this quantity drops very rapidly to zero: by just increasing *res(c)* to 2 only one node, on the average, is not able to receive the message. The failure probability drops very rapidly as well: an increase of two points in resiliency degree yields a decrease of failure probability of about one order of magnitude. For example, in the default scenario ($OD=10$) the failure probability is between 10^{-5} and 10^{-6}; by increasing *res(c)* to 12 the failure probability drops below 10^{-6}.

Results in Fig. 5 show that the failure probability of the TRC protocol can be reduced, by increasing the omission degree parameter, and consequently the resiliency degree of the transmitted message. Such increase in success probability, however, comes at a price, which corresponds to an increase in protocol duration. Fig. 6 shows the probability that the protocol has not completed at a given instant of time, which is evaluated as 1–$F(t)$, where $F(t)$ is the cumulative distribution function of the time at which the protocol terminates. To emphasize the connection with ALARP requirements, only the portion after 10 seconds is shown. The figure clearly shows that increasing the resiliency degree from 10 to 15 increases such probability of almost an order of magnitude.

These kinds of tradeoffs are of greatest importance in the analysis of the ALARP system and results show that their careful analysis is mandatory, given the limited amount of time in which communication should take place. The model described in this paper has been proven useful for this kind of analysis, and will provide a valuable resource in the final assessment of the ALARP system.

7. CONCLUSIONS AND FUTURE WORK

In this paper we have performed a model-based analysis of the Timed Reliable Communication (TRC) protocol, a broadcast protocol based on the IEEE 802.11 wireless technology that is being used within the EU funded ALARP project for railway worksite communication. The model results have been compared with experimental measurements collected within the project, and good correspondence has been observed. A set of key metrics have then been defined and evaluated, showing that the protocol allows to satisfy the ALARP performance and reliability requirements under the specified assumptions. Finally, tradeoffs in choosing protocol parameters have been observed and discussed. Future work consists in performing accurate measurements in real worksite setups, to the extent to which it is possible, and analyzing the protocol behavior using parameters taken from such real scenario.

8. ACKNOWLEDGMENTS

This work has been partially supported by the European Project FP7-IST-234088 ALARP [1] and by the Italian Ministry for Education, University, and Research (MIUR) in the framework of the Project of National Research Interest (PRIN) "DOTS-LCCI": Dependable Off-The-Shelf based middleware systems for Large-scale Complex Critical Infrastructures.

9. REFERENCES

[1] ALARP – A railway automatic track warning system based on distributed personal mobile terminals – FP7-IST-2010-234088 http://www.alarp.eu/.

[2] ALARP D1.2, "Requirements Specifications", Project FP7-IST-2010-234088 ALARP, June 2010.

[3] A. Bondavalli, A. Coccoli, and F. D. Giandomenico. "QoS Analysis of Group Communication Protocols in Wireless Environment". Concurrency in Dependable Computing. Kluwer Academic Publishers. pp. 169-188. 2002.

[4] G. Clark, T. Courtney, D. Daly, D. Deavours, S. Derisavi, J. M. Doyle, W. H. Sanders, and P. Webster. "The Möbius Modeling Tool", Proc. of the 9th International Workshop on Petri Nets and Performance Models, pp. 241-250, 2001.

[5] A. Coccoli, P. Urban, A. Bondavalli, and A. Schiper, "Performance analysis of a consensus algorithm combining Stochastic Activity Networks and Measurements", Proc. International Conference on Dependable Systems and Networks (DSN'02). pp. 551-560, 2002

[6] B. Malinowsky, J. Grønbæk, H.P. Schwefel, A. Ceccarelli, A. Bondavalli, E. Nett, "Realization of Timed Broadcast via Off-the-Shelf WLAN DCF Technology for Safety-critical Systems", Proc. 9th European Dependable Computing Conference (EDCC'12), Sibiu, Romania, May 8-11, 2012.

[7] J. F. Meyer, A. Movaghar, and W. H. Sanders, "Stochastic activity networks: structure, behaviour and applications", Proc. International Workshop on Timed Petri Nets. IEEE, New York, pp. 106–115, 1985.

[8] M. Mock, E. Nett, and S. Schemmer, "Efficient reliable real-time group communication for wireless local area networks", Proc. 3rd European Dependable Computing Conference (EDCC'99), pp. 380-400, 1999.

[9] A. Coccoli, S. Schemmer, F. Di Giandomenico, M. Mock, and A. Bondavalli, "Analysis of Group Communication Protocols to Assess Quality of Service Properties". 5th IEEE High Assurance System Engineering Symposium (HASE00), 247-256, 2000.

[10] ALARP D2.2, "Preliminary wireless communication solution", Project FP7-IST-2010-234088 ALARP, July 2011.

[11] "IEEE Standard for Local and Metropolitan Area Networks, Part 11: Wireless LAN Medium Access Control (MAC) and Physical Layer (PHY) Specifications" IEEE Std 802.11-2007, June 12, 2007.

[12] J.B. Andersen, T.S. Rappaport, and S. Yoshida, "Propagation measurements and models for wireless communications channels" IEEE Communications Magazine, vol.33, no.1, pp.42-49, Jan 1995.

[13] K. Hiltunen and R. de Bernardi, "WCDMA downlink capacity estimation", IEEE 51st Vehicular Technology Conference Proceedings (VTC 2000), pp. 992-996, Tokyo, 2000.

[14] R. Owen, P. Jones, S. Dehgan, and D. Lister, "Uplink WCDMA capacity and range as a function of inter-to-intra cell interference: theory and practice" IEEE 51st Vehicular Technology Conference Proceedings, pp. 298-302, 2000.

[15] M. Zorzi and R.R. Rao, "On channel modeling for delay analysis of packet communications over wireless links", 36th Annual Allerton Conference on Communications, Control and Computing, Monticello, IL, Sept. 23-25, 1998.

[16] M.I. Hassan, H.L. Vu, and T. Sakurai, "Performance Analysis of the IEEE 802.11 MAC Protocol for DSRC Safety Applications" IEEE Transactions on Vehicular Technology, vol.60, no.8, pp. 3882-3896, Oct. 2011.

[17] C. Tang and P.K. McKinley, "Modeling Multicast Packet Losses in Wireless LANs", 6th ACM International Workshop on Modeling, Analysis and Simulation of Wireless and Mobile Systems (MSWIM'03), San Diego, USA, Sept. 2003.

[18] J.C.P. Wang, M. Abolhasan, D.R. Franklin, F. Safaei, "Characterising the Behaviour of IEEE 802.11 Broadcast Transmissions in Ad Hoc Wireless LANs", IEEE International Conference on Communications, (ICC '09). pp. 1-5, 2009.

[19] H. Cheng, C. Rong, and G. Yang, "Design and Analysis of a Secure Routing Protocol Algorithm for Wireless Sensor Networks" IEEE International Conference on Advanced Information Networking and Applications (AINA'11), pp. 475-480, March 22-25, 2011.

[20] M. Massink, D. Latella, and J.-P. Katoen, "Model checking dependability attributes of wireless group communication" International Conference on Dependable Systems and Networks (DSN'04) pp. 711-720, June 28th – July 1st, 2004.

[21] "IEEE Standard for Local and Metropolitan Area Networks, Part 15.1: Wireless Medium Access Control (MAC) and Physical Layer (PHY) Specifications for Wireless Personal Area Networks (WPANs)", IEEE Std 802.15.1-2005, 2005.

[22] "IEEE Standard for Local and metropolitan area networks, Part 16: Air Interface for Broadband Wireless Access Systems", IEEE Std 802.16m-2011, May 5, 2011.

Extending Graph-based Models of Wireless Network Structure with Dynamics

Elena Meshkova, Janne Riihijärvi, and Petri Mähönen
Institute for Networked Systems, RWTH Aachen University
{eme, jar, pma}@inets.rwth-aachen.de

ABSTRACT

Graphs are routinely used to approximate the structure of wireless networks especially when studying connectivity or various aspects of network performance. Although not as detailed as geometrical and spatial models dealing with individual node locations, graphs can nevertheless be used to capture many of critical interactions between the nodes. In this paper we explore the use of graphs to approximate *network dynamics* in addition to static network structure. We show that a simple linearization of network performance functionals results in natural correlation metrics for influence of parameter changes on performance. Further, based on extensive simulations we demonstrate that these correlation metrics have a high degree of robustness against perturbations in node locations. We also discuss the relationships between the arising graph approximations of network dynamics and the commonly applied connectivity and conflict graphs. Our results indicate that graphs formed by approximating dynamics combined with connectivity structures can shed considerable insight on "gray zone" behavior commonly encountered in wireless networks. They are promising candidates for autonomous context identification.

Categories and Subject Descriptors

C.2.1 [**Network Architecture and Design**]: Wireless communication

Keywords

Network dynamics, connectivity and conflict graphs, correlation, modeling, wireless, protocol parameters, optimization

1. INTRODUCTION

Graph-based models are widely applied to a number of networking problems, ranging from routing and load balancing in fixed networks to channel allocation and scheduling problems in wireless systems. While for fixed networks graph models are in a sense exact, for wireless networks they should be considered as an approximation. The connectivity structure of wireless networks is properly thought of as being *spatial* in nature [5, 14], and their performance fundamentally depends on dynamically changing distance and fading environment. However, spatial models can become quite complex and typically require computationally demanding processing to extract information from them. As an alternative, computationally effective *graph approximations* of wireless network structures have been developed, as it has been observed that often actual performance characteristics often depend only loosely on node locations.

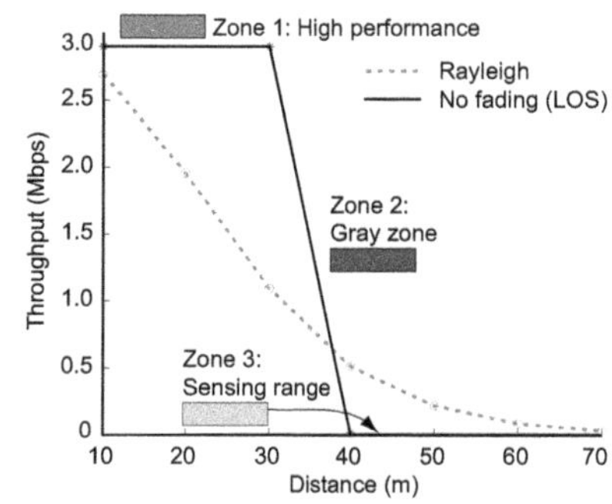

Figure 1: Dependency of performance on distance between two wireless nodes for different channel conditions. Three distinct performance behaviors for LOS channels are labeled.

The currently applied graph approximations are almost universally based on the concepts of connectivity and conflict graphs [8]. These can be thought of as encoding *constraints* how resources can be used in the network. For example, for CSMA-based networks contention relations expressed through a conflict graph dictate which configurations of transmitters can be active at the same time. The accuracy of such models relies on the observation that radio propagation in typical Wi-Fi deployments often feature strong large line-of-sight components, and the "gray" area, where performance is unstable and rapidly decreases with distance, is very narrow and can often be neglected. In this sense the contention relations are *robust* with regards to small changes in distances between the nodes. It is precisely the robustness property that enables graph-based approximation of the relationships between wireless nodes. This feature is further illustrated in Figure 1 that displays the performance of a point-to-point link as a function of the distance between the nodes for different wireless channel types. As mentioned above, we see that in the absence of fading, or when the line-of-sight (LOS) component is strong, the dependency of performance on the distance is weak, and it

MSWiM'12, October 21–25, 2012, Paphos, Cyprus.

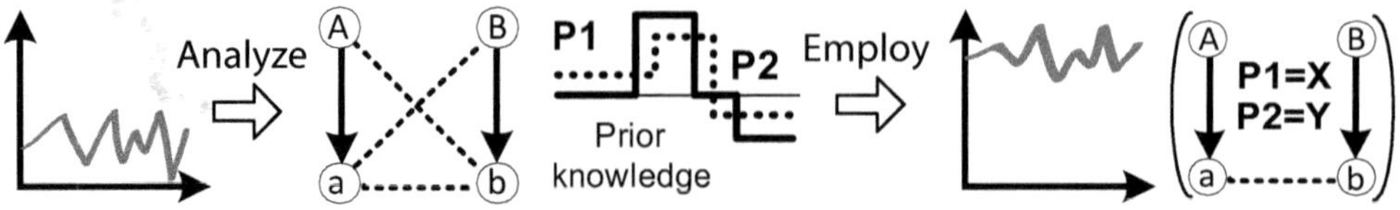

Figure 2: Illustration of the problem statement. In which context do we operate (e.g., with respect to current connectivity graph) and how it can be optimized (e.g., by tuning protocol parameter settings)?

can be quantized only in three states. However, in non-line-of-sight conditions this no longer completely holds, and the graph approximation with small number of edge types might become inaccurate. One should note that such situations can be prevalent for many present non Wi-Fi systems, and even in the case of MIMO enabled Wi-Fi we expect that small-scale fading will start to play more dominant role that needs to be taken into account in graph-based approaches.

In this paper we suggest a different approach to graph-based wireless network approximation that, among other features, enables robust quantization with smaller number of states for more complex operational conditions. We focus on *network dynamics* as opposed to simple structural relations between the nodes. Classical contention and conflict graphs are typically used in the literature to study the expected performance of a particular protocol or a protocol stack with *fixed* parameter settings. We focus instead on the impact different tunable parameters of the protocol stack for a given *context*, which might be defined by a number of variables such as node positions, propagation environment, application behavior, network optimization goals, etc. We aim to describe these in a similarly robust manner as has been done for contention relations in the literature. We argue based on classical dynamical systems theory that various linearizations of network dynamics should naturally yield robust metrics related to statistical correlations between performance and parameter values. Based on extensive simulations we show that this is indeed the case. We believe that such approximations of network dynamics have a number of interesting applications. As an example, they enable fast and robust context-aware network optimization based on observed properties of connectivity graphs.

2. NETWORK OPTIMIZATION AND MODELS OF DYNAMICS

We consider a classical utility-based optimization framework for network performance [3]. We denote by U the overall network utility, in practice defining the global optimization target as a statistical average from which random environmental contributions such as fading have been averaged out. In most scenarios U would be expressed as a sum of individual utility functions for different data flows or users, but this level of detail is not needed here. Now, in a given environment, and for given set of node locations, U is a function of a collection of *configurable* parameters $\boldsymbol{p}$, i.e., $U = U(p_1, \ldots, p_n)$. These parameters include, for example, transmit power, use of RTS/CTS mechanism, transport protocol settings, and so on. Then around any given operating point $\boldsymbol{p}_0$, the dynamics of U under changes in the parameter values can be expanded as Taylor series following standard arguments used in control theory and theory of dynamical systems. Formally this is done as follows. Let $\boldsymbol{\alpha} = (\alpha_1, \ldots, \alpha_n) \in \mathbb{Z}^n$ be a multi-index with standard definitions $|\boldsymbol{\alpha}| \equiv \sum_i \alpha_i$, $\boldsymbol{\alpha}! \equiv \alpha_1! \cdots \alpha_n!$, $\boldsymbol{p}^{\boldsymbol{\alpha}} \equiv p_1^{\alpha_1} \cdots p_n^{\alpha_n}$ and

$$D^{\boldsymbol{\alpha}} f(\boldsymbol{x}) \equiv \frac{\partial^{|\boldsymbol{\alpha}|} f}{\partial x_1^{\alpha_1} \cdots \partial x_n^{\alpha_n}}. \tag{1}$$

Then the Taylor series reads

$$U(\boldsymbol{p}) = \sum_{|\boldsymbol{\alpha}|=0}^{\infty} \frac{D^{\boldsymbol{\alpha}} U(\boldsymbol{p}_0)}{\boldsymbol{\alpha}!} (\boldsymbol{p} - \boldsymbol{p}_0)^{\boldsymbol{\alpha}}. \tag{2}$$

Usually the highest degree of robustness in optimization and control is achieved by using the linear terms as a basis for optimization decisions (i.e., $|\boldsymbol{\alpha}| = 1$). As the result (after normalization and appropriate choice in the range of parameter values considered) we obtain terms closely related to *statistical correlation coefficients* between U and p_i.

The application of the correlation coefficients can be understood as follows. Suppose the network is observed to perform poorly, measured as a low overall network utility. In order to improve this performance, complete optimization problem could be solved at network configuration, looking for the global optimum of $U(\boldsymbol{p})$. This can, however, be very costly due to the large number of parameter combinations and the typically complex global structure of the optimization problem. An intermediate choice is to respond to performance degradation by tuning only few parameters, perhaps even one at a time, based on the above-shown linear approximation and conflict/contention graphs identified for the given operational point. This process is illustrated in Figure 2. We determine the current operational context either through direct observation of selected Key Performance Indicators (KPIs) or by obtaining a connectivity graph [15]. For this context, as result of prior training or modeling, we have estimates of the correlation coefficients. Parameters with large correlation coefficients would be the ones to be adjusted first as they are likely to lead to the fastest performance improvement. Others may be tuned later if desired. The parameter adjustments are likely either to change the underlying connectivity graph, e.g., by avoiding hidden terminal problem by lowering transmit power, or changing packet dynamics by, for instance, enabling the RTS/CTS mechanism. The suggested models or rules are useful if the correlation coefficients are robust against small changes in network configuration, and the optimality gap between such a simplified optimization scheme and the global optimum is small enough. This turns out to often be the case, as we show below based on our simulations.

3. SIMULATION SETUP AND RESULTS

Our simulation scenario together with a more detailed graph-based view of the performance evaluation problem is illustrated in Figure 3. We consider two pairs of nodes. The distance between the pairs is increasing throughout the simulation time. The nodes communicate using IEEE 802.11a

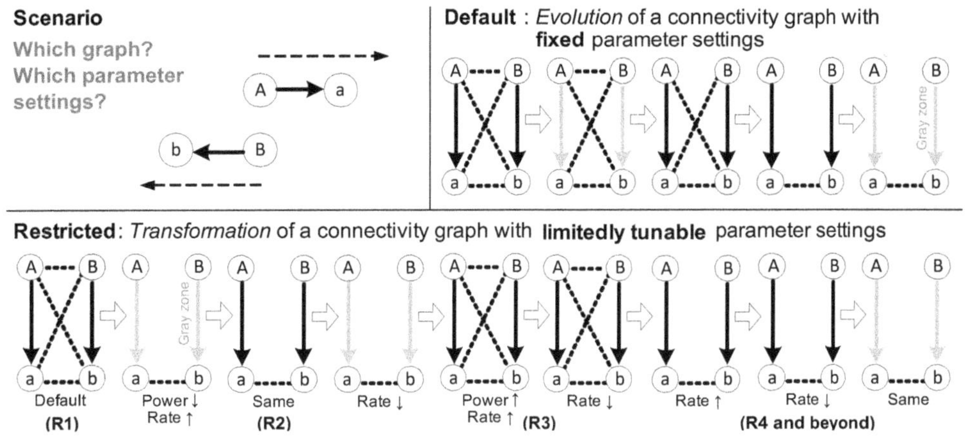

Figure 3: Scenario overview. Two transmitting nodes are horizontally moving away from each other. The connectivity graph evolution, which is typical for such movement with fixed parameter settings, is shown in the top panel. The lower panel displays the graph transformation if parameter setting can be limitedly tuned. Here the parameter adjustments allow avoiding hidden terminal (HT) problem and selected gray zones, for which poor performance is observed. (The R labels denote distinct operational regions that display distinct performance dynamics as shown in Fig. 4).

Wi-Fi. We systematically study the performance of the network by employing the QualNet commercial packet level network simulator [1]. The overall goodput as a function of all combinations of parameter and simulation settings is used as our main performance metric. We derive the correlation coefficients between parameter values and throughput as discussed above, as well as the performance gains that can be obtained by tuning different combinations of parameters from their default values. We consider the following changes in the simulation scenario, line-of-sight (LOS) propagation conditions and channels with the Rayleigh fading model, TCP and UDP traffic (FTP and CBR applications, respectively), internode distances between transmitters (between 2 m and 202 m), with receivers being separated by 50 m. The tunable parameters are transmit power, data rate, MAC retry limit, use of RTS/CTS mechanism, maximum contention window, and maximum fragment size.

Let us first look at the overall performance numbers. Figures 4a and 4b illustrate the impact of changes in parameter values on the UDP traffic in line-of-sight propagation conditions and the TCP traffic in the presence of Rayleigh fading (corresponding in a sense to a worst-case propagation environment for graph based models). From the results we can clearly see that the network performance is maximized with *very few* parameters. In vast majority of cases changes in the transmit power have the largest impact on achieved performance. In the case of UDP traffic the data rate change is another important aspect. This simple parameter optimization can result in high increase in the observed performance with respect to default settings that maximize network performance over the whole operational range. From these figures it is also clear that the results that can be achieved by parameter adjustment depend only loosely on the actual distances between the nodes, making graph approximation of these relationships feasible.

Given the strong dependency between performance and just few of the possible parameter values, the simple linear approximation procedure discussed above should work well unless there are significant non-linearities present in the dynamics of the network. Figure 4c shows the correlation coefficients for the influence of different parameters on throughput for the case of UDP with LOS channel taken over all parameter permutations. We see that the strongest correlations (either positive or negative) are with respect to either rate or power. Further, the correlation coefficients are seen to be quite robust against small changes in the distances between the nodes. Significant behavioral changes are rather observed for the regions that loosely correspond to a certain type of connectivity graphs and their possible transformation ($R1$ – $R4$). Therefore, very simple greedy optimization scheme would be able to keep the network performance at a high level, provided the correlation information would be available. Finally, in order to emphasize the distinctness of the regions $R1$ – $R4$ given in Figures 4a and 4b, Figure 4d shows the cumulative distribution functions for the throughput in the TCP case for the different operational regions. The performance properties are clearly highly distinct, again confirming that highly accurate graph approximation with only few zones can be obtained.

We conclude this section by discussing about the selected further results on the behavior of correlation coefficients between goodput and various parameters as functions of the distance between the node pairs. Figures 5 and 6 display correlation coefficients obtained for all individual parameter permutations color-coded on the basis of their behavioral similarity. The figures also display the correlation coefficients *clustered* into five classes. These values correspond to each of the colored groups, and can serve as additional labels for conflict graphs to indicate the likely network dynamics with respect to parameter changes. Figure 5 gives detailed results for the case of transmit power being changed. We see that for *all* of the parameter permutations the results are naturally divided into *few* different robustly defined zones, and strong correlations can be observer for majority of the cases. This indicates that for many cases even if we are not in full control of the network, e.g., there is an autonomous data rate control, we can choose based on correlation coefficients the behavioral strategy that would maximize the per-

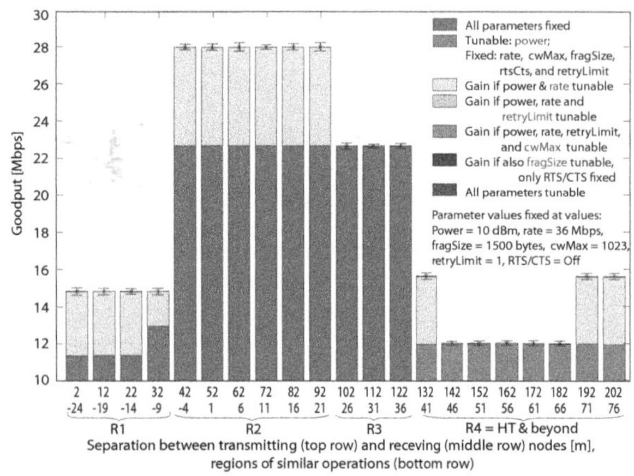

(a) The performance of UDP traffic in LOS propagation conditions, with no fading model applied.

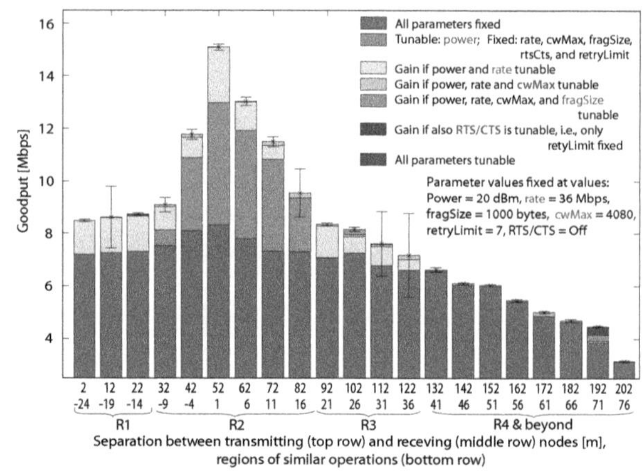

(b) The performance of TCP traffic with Rayleigh fading model applied.

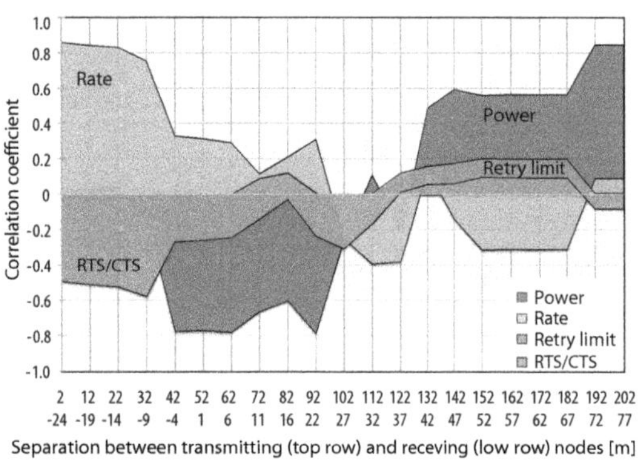

(c) The parameter value correlations for all possible permutations for UDP traffic with LOS propagation.

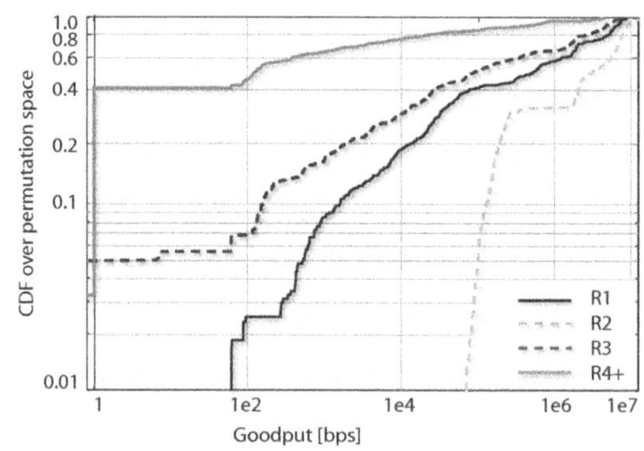

(d) Cumulative distribution functions for parameter permutation obtained for TCP traffic with Rayleigh fading for different regions R of the above graphs.

Figure 4: Performance of UDP and TCP traffic in scenarios without fading and Rayleigh fading. Figures show the overall achieved performance, as well as cumulative contributions from making MAC/PHY parameters tunable. The parameters are set flexible in order of their maximal contribution to the improvement in average (over all distances) performance. Sample overall correlation coefficient values for the scenarios, as well as the CDF for one of the scenarios is provided as well.

formance. Additionally, these zones roughly correspond to the possible range of connectivity graphs that can be formed at the current scenario point. Figure 6 on the other hand shows sample results, where parameters were found not to have a strong and similar performance impact in the majority of the scenarios. We see that the coefficient values either become insignificant (esp. in the cwMax case, Fig. 6c), depend much more heavily on the other parameters settings (Fig. 6a), or simply indicate that the given parameter should be chosen to have maximum value without additional tuning taking place (Fig. 6b). The other notable trend is that for such parameters the correlation coefficient behavior may differ between traffic profiles and channel conditions (Fig. 6a,b), i.e., they do not display such a robust dynamics as in case of highly influential settings (Fig. 5).

4. RELATED WORK

There is extensive prior work in the fixed networks domain utilizing random graphs, or graphs constructed as solution to specific performance optimization problems [4]. Graph approximations for wireless networks, and their subgraphs, were discussed in, e.g., [6,7,11,13], with only limited studies on multiple ways to define edges between nodes in [9,16]. The latter works mostly again solve specific types of problems, rather than positioning this model as a generalization of other types of network motifs. Passive estimation of conflict graphs was explored in [15]. Other works, e.g., [2,10,12], discuss on power and rate adaptation in order to practically address hidden/exposed terminal problems, as well as avoid gray zones. However, only low-level key-performance indications and raw SINR values are used for parameter adaptations. The latter two papers also provide example usage of conflict graphs with multiple channels or directional antennas. We generalize their optimization strategy based on the parameter correlation metrics.

5. CONCLUSIONS

One of the main modeling approaches is to capture dependencies between obtained performance and optimizable entities with respect to a range of operational conditions. Graph approximation based only connectivity and contention relations between nodes do not capture dynamics from changes in configurable network parameters. In this paper we have suggested using parameter-performance correlation coeffi-

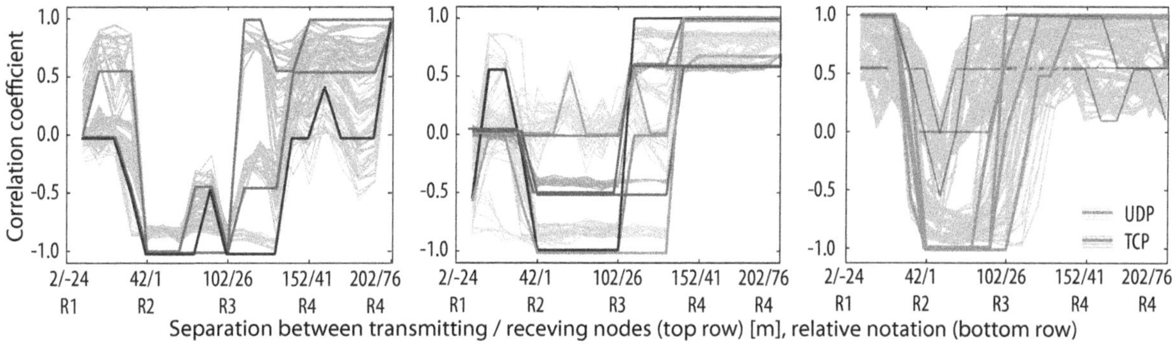

a) **Power**, UDP traffic, LOS. b) **Power**, TCP, LOS. c) **Power**, UDP & TCP, Rayleigh.

Figure 5: Correlations of the transmit power and obtained performance for all parameter permutations.

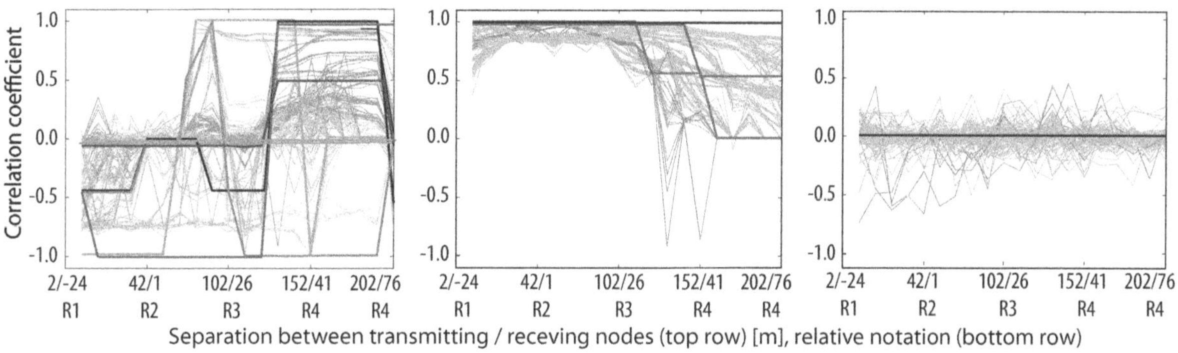

a) **Retry limit**, UDP traffic, LOS. b) **Retry limit**, TCP, Rayleigh. c) **cwMax**, TCP, Rayleigh.

Figure 6: Correlations of the retry limit and cwMax MAC parameters and obtained performance over all range of simulated inter-node distances and parameter permutations. Similar behaviors are color-clustered. Bolder lines denote the respective *clustered* correlation coefficients that later can be used for labeling the corresponding operational conditions (e.g., graph connectivity approximations).

cients to capture this dynamics. Our experiments have confirmed that this is, indeed, an efficient and robust metric, which can, along with graph-based approximations, lead to effective autonomous wireless network optimization.

Acknowledgments

We thank European Union for partial financial support from through the FARAMIR project. We acknowledge also support from DFG through UMIC Research Center. We also thank Md. Mohiuddin Khan for fruitful discussions.

6. REFERENCES

[1] QualNet. Scalable Network Technologies, http://www.scalable-networks.com [visited on 23.04.2011].

[2] N. Ahmed *et al.* Online estimation of RF interference. In *Proc. of ACM CoNEXT*, 2008.

[3] M. Chiang *et al.* Layering as Optimization Decomposition: A Mathematical Theory of Network Architectures. *Proceedings of the IEEE*, 95(1):255–312, 2007.

[4] J. Doyle *et al.* The "robust yet fragile" nature of the Internet. *Proceedings of the National Academy of Sciences*, 102(41):14497–14502, 2005.

[5] R. Ganti and M. Haenggi. Interference and outage in clustered wireless ad hoc networks. *IEEE Transactions on Information Theory*, 55(9):4067–4086, 2009.

[6] M. Garetto, T. Salonidis, and E. Knightly. Modeling per-flow throughput and capturing starvation in CSMA multi-hop wireless networks. *IEEE/ACM Transactions on Networking*, 16(4):864–877, 2008.

[7] C. Hua and R. Zheng. Starvation modeling and identification in dense 802.11 wireless community networks. In *In Proc. of IEEE INFOCOM*, 2008.

[8] K. Jain, J. Padhye, V. Padmanabhan, and L. Qiu. Impact of interference on multi-hop wireless network performance. *Wireless networks*, 11(4):471–487, 2005.

[9] H. Li *et al.* Multi-dimensional conflict graph based computing for optimal capacity in MR-MC wireless networks. In *Proc. of IEEE ICDCS*, pages 774–783, 2010.

[10] X. Liu *et al.* DIRC: increasing indoor wireless capacity using directional antennas. In *ACM SIGCOMM CCR*, volume 39, pages 171–182, 2009.

[11] E. Magistretti, O. Gurewitz, and E. Knightly. Inferring and mitigating a link's hindering transmissions in managed 802.11 wireless networks. In *Proc. of ACM MOBICOM*, pages 305–316, 2010.

[12] S. Rayanchu *et al.* FLUID: improving throughputs in enterprise wireless LANs through flexible channelization. In *Proc. of ACM Mobicom*, 2011.

[13] S. Razak, V. Kolar, and N. Abu-Ghazaleh. Modeling and analysis of two-flow interactions in wireless networks. *Ad Hoc Networks*, 8(6):564–581, 2010.

[14] J. Riihijärvi and P. Mähönen. Modeling Spatial Structure of Wireless Communication Networks. In *Proceedings of IEEE NetSciCom workshop*, pages 1–6, 2010.

[15] V. Shrivastava *et al.* PIE in the sky: online passive interference estimation for enterprise wlans. In *Proc. of ACM NSDI*, 2011.

[16] Y. Wu *et al.* Network planning in wireless ad hoc networks: a cross-layer approach. *IEEE JSAC*, 23(1):136–150, 2005.

Investigating the Interacting Two-Way TCP Connections over 3GPP LTE Networks

Yinsheng Xu†, Fengyuan Ren†, Shibo Xu†, Chuang Lin†, Sajal K. Das‡
†Tsinghua National Laboratory for Information Science and Technology
†Department of Computer Science and Technology, Tsinghua University, Beijing, China
†{xuysh08, renfy, xshbo, clin}@csnet1.cs.tsinghua.edu.cn
‡Department of Computer Science and Engineering, University of Texas at Arlington, TX, USA
‡das@uta.edu

ABSTRACT

This paper investigates the interactions between two-way TCP connections over 3GPP LTE networks. In the LTE network, the two-way TCP flows share buffers on a common bottleneck, i.e., the radio access links. The behaviors of TCPs significantly influence the others in the opposite direction. Specifically, the radio links of LTE are *asymmetric*, which may induce drastic interactions of TCPs and rapid draining of downlink buffer. The periodic idleness of downlink is a huge waste of the precious radio bandwidth and results in considerable performance degradation. In the viewpoint of *Coupled Queues*, we thoroughly understand the interacting TCPs and explain the reason for performance degradation. Based on a straightforward modeling procedure, we formalize the evolution of two-way TCPs and model the bottleneck queue size in every *slot*. The model indicates the queues are close coupled, which is verified with simulations on NS2. If the uplink (queue) is fully utilized, the downlink (queue) will always be underutilized even idle, and vice versa. Furthermore, an effective solution called *Preemptive ACK Queueing* (PAQ) is designed to decouple the queues, which improves the performance of two-way TCPs over LTE networks.

Categories and Subject Descriptors

C.2.2 [**Computer-Communication Networks**]: Network Protocols

General Terms

Performance

Keywords

Long Time Evolution (LTE), two-way TCP, asymmetric links, coupled queues

1. INTRODUCTION

The Third Generation Partnership Project (3GPP) Long Term Evolution (LTE) standardization is the next forward step in cellular network. LTE can achieve a high peak-data-rate that scales with scalable system bandwidths/capacity and coverage improvements, spectrum efficiency, latency reduction and packet optimized radio access. An overview of the LTE network is depicted in Figure 1. The functionality of LTE is mainly divided into three domains: User Equipment (UE), Evolved UMTS Terrestrial Radio Access Network (EUTRAN) and System Architecture Evolution (SAE) core, also known as Evolved Packet Core (EPC).

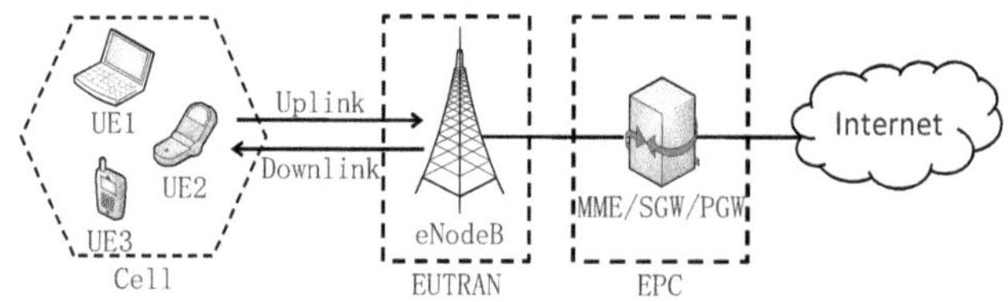

Figure 1: An overview of the 3GPP LTE network

Since the large-scale deployment of commercial LTE networks is being carried out, the performance of the most widely used Transmission Control Protocol (TCP) [6] over LTE should be investigated. TCP adjusts its congestion window (Cwnd) based on regularly receiving acknowledgement (ACK), which arrives at the source from the *backward* link. However, once the ACK packets are interfered on its way back (most probably on the bottleneck), the adjustment of *Cwnd* will be affected. In the LTE networks, the radio access link is commonly the bottleneck for TCP connections. The two-way TCP traffic share buffers on the common bottleneck, their behaviors significantly influence the TCPs in the opposite direction. In particular, the intermingling of ACK and TCP data in one buffer may result in deteriorating the reverse link utilization and performance of the reverse TCPs. For example, in the uplink buffer[1], the ACK packets of download TCP may be intermingled and queued with the upload TCP data packets. Once the upload traffic increases rapidly, the download ACKs will be seriously congested and their queueing delays are hence enlarged steeply. The improper arriving interval of ACKs will destroy the regular spacing of download TCP. If the interval is too small, the ACK packets will arrive at the source in a burst, which gives rise to rapid fluctuations of queue size and is defined as *ACK Compression* in [13]. On the contrary, the overlarge interval will slow down the TCP sending rate and thus the

[1]For simplifying expressions, we particularly refer buffers (and queues in those buffers) to the ones on the radio access bottleneck, unless specified otherwise.

MSWiM'12, October 21–25, 2012, Paphos, Cyprus.

downlink buffer is draining fast, which deteriorates the link utilization.

According to the 3GPP Release 8, an 20MHz channel can support peak rates of 100Mbps in the downlink and 50Mbps in the uplink [2]. Besides, the commercial LTE speed is previewed 28.7Mbps for download while 10.4Mbps for upload [1]. Thus we consider the downlink radio bandwidth/capacity is much larger than the uplink one in LTE, and the radio access links are *asymmetric bottleneck*. Due to the asymmetric bottleneck of LTE, the interaction of two-way TCP is much considerable, i.e., a long time idle link and throughput degradation. The situation will be worse especially with larger uplink buffer size. In the simulation experiments, we observe up to 20% of the whole period when the downlink is completely idle, which induces nearly 50% degradation of the total download goodput. In other words, the TCP performance over LTE will be seriously influenced on the presence of the opponents in the opposite direction. The two-way TCP problem have been considered in literatures since [13]. Most of them are based on the understanding of ACK Compression. A recent paper addresses the problem with new insight, namely *Data Pendulum* [5]. Nevertheless, the explanation for Data Pendulum is not completed. Hence, we believe a thorough understanding of the interactions of two-way TCPs will benefits improving the performance of LTE and other asymmetric access networks.

In this paper, we investigate the interactions between two-way TCP connections, which is a fundamental problem for LTE networks. First, we perform multi-flow simulations to confirm the interacting phenomenon exists ubiquitously and the performance degradation can be crucial. Second, we construct a straightforward model, which formalizes the TCP evolution and computes the queue size in every *slot*. The model is verified and presents a quantitative relations between the queue lengths. Based on the relations, the *coupled queues* is proposed as an appropriate explanation for interacting TCPs. In the LTE networks, the TCP Round Trip Time (RTT) is dominated by the sum of forward and backward queueing delays. If the uplink queue size increases, the download RTT will rise because of ACK waiting in the large queue. Thereby, the evolution of TCP *Cwnd* is decelerated and the downlink buffer will drain soon. Consequently, the precious downlink radio bandwidth of LTE is wasted. In other words, the queues are close coupled. Third, we obtain a lower bound $\mathcal{B}$ of the downlink queue size. As long as the LTE network is configured to keep $\mathcal{B} > 0$, the downlink buffer is prevented from idleness and the downlink bandwidth will not be wasted. Finally, an effective solution called *Preemptive ACK Queueing* (PAQ) is designed to decouple the queues and improve the performance of two-way TCPs over LTE networks.

The rest of the paper is organized as follows. Section 2 presents a quantitative view of performance degradation of two-way TCPs in LTE. Afterwards, the straightforward modeling and analysis of the coupled queues are provided in Section 3. The Preemptive ACK Queueing method is developed and evaluated in Section 4. We supplement discussions in Section 5 and report related works in Section 6, respectively. Finally, we draw conclusions in Section 7.

2. RELATED WORKS

The two-way TCP problem has been broadly considered and discussed ever since the observations of S.Shenker [11] and L. Zhang [13]. Conventional explanation for the problem is *ACK compression* [13], which affects TCP self clocking by clustering data segments and ACKs of both connections [12]. In a recent literature [5], Heusse et al. consider the problem and provide some new insights. They first argue ACK Compression, which only arises in a perfectly symmetric scenario including equal bandwidth and equal RTT. Nevertheless, either changing the network configurations slightly or activating the *Delayed ACK* mechanism will make it disappear. Afterwards, they propose a different explanation called *Data Pendulum* in the bottleneck buffers. Although the Data Pendulum is valid in the simulation, the two-way TCP problem is still not fully explained. Specifically, in the analysis section, the authors make an important assumption that the RTTs of two-way TCPs are *equal*. However, the Data Pendulum phenomenon exactly emerges in simulations only when RTTs are *unequal*. In other words, the understanding based on Data Pendulum is not completed, which can be considered as another description of the interacting phenomenon. In addition, only a simple single-flow topology is concerned throughout the simulation. In consequence, a thorough understanding of two-way TCP problem advantages not only explaining the performance degradation but also illumining a comprehensive solution.

In the existing literatures, some efficient techniques are proposed to address the two-way TCP challenge. RFC 3449 [3] recommends several up-stream link scheduling disciplines that effectively isolate the download ACKs from data segments, such as weighted fair queueing and class based queueing. ACK-first scheduling prioritizes ACK packets over data packets. However, it requires per-flow state, which may not be an appropriate solution for resource-constrained LTE networks. Connection-level bandwidth allocation [7] aims at providing a guaranteed minimum utilization for the downlink by data packets while providing the maximum possible throughput for the upload connection under this constraint. Nevertheless, it requires knowing that parameter to estimate the efficiency of download and upload connections. ACK Filtering (AF) [4] is a link-layer mechanism for reducing the number of ACKs sent to the upload direction. This approach is effective but can have subtle impacts on TCP throughput, depending on how the TCP version interprets the ACKs received. Based on the understanding of [5], Podlesny et al. develop a weighted queue approach called *Asymmetric Queue* to solve the problem [10]. However, the Asymmetric Queue is structured in packets rather than bytes, which is not reasonable for analyzing a buffer filled with data and negligible ACK of small size.

3. PERFORMANCE DEGRADATION OF INTERACTING TWO-WAY TCPS

In this section, we perform simulation experiments on NS2[2] to indicate the degradation of interacting two-way TCP connections with quantitative results.

3.1 Topology and Simulation Configuration

Throughout the paper, we consider the topology in Figure 2, including N_u upload TCPs and N_d download TCPs. In the simulation, a client-sever (CS) mode FTP is deployed as the application supported by TCP. The uplink buffer (B_u) and downlink buffer (B_d) reside in UE and eNodeB, respectively. The uplink bandwidth[3] is denoted by C_u while

[2]In this paper, all the simulations are performed on the Network Simulator 2 (NS2), version 2.35.

[3]In this paper, we consider the meaning of *bandwidth* is the same as *capacity*, which both denote the maximum transmission rate that a link can provide.

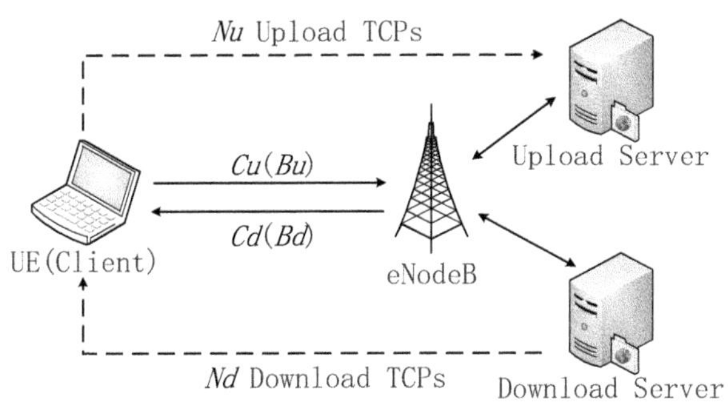

Figure 2: Topology

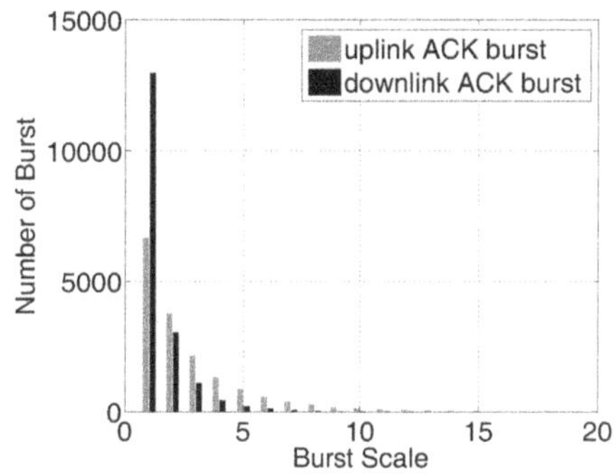

Figure 3: ACK bursts in multi-flow scenario (C_u=5Mbps, C_d=10Mbps)

Table 1: Summary of Nations

Notation	Meaning
N_u	number of upload TCP
N_d	number of download TCP
$Cwnd_u$	congestion window of upload TCP
$Cwnd_d$	congestion window of download TCP
C_u	uplink capacity between UE and eNodeB
C_d	downlink capacity between UE and eNodeB
B_u	uplink buffer size
B_d	downlink buffer size
Q_u	queue size in uplink buffer
Q_d	queue size in downlink buffer
L_u	packet size of upload TCP data
L_d	packet size of download TCP data
τ_u^i	the propagation delay of the ith upload TCP
τ_d^j	the propagation delay of the jth download TCP

Table 2: Simulation Configuration for Multi-Flow Scenario

Parameter	Value
N_u and N_d	10
TCP version	Sack
queue management	DropTail
activated period of upload TCPs	5s∼50s
activated period of download TCPs	0s∼50s
L_u and L_d	1000B
bottleneck buffer	structured in byte
B_u and B_d	30000B
τ_u^i	$i$ms
τ_d^j	$j$ms
C_u	5/1Mbps
C_d	10Mbps
bandwidth between eNodeB and server	100Mbps

the downlink is C_d. Main notations and the corresponding meanings are summarized in Table 1.

In this section, we simulate a heterogeneous multi-flow scenario, which represents a more general case. Specifically, the propagation delays of TCP connections are different flow by flow in both directions. The heterogeneous scenario avoids TCP synchronization and eliminates ACK Compression. Since the size of ACK is negligible compared to TCP data[4], it is reasonable to structure the buffer in byte instead of packet. As a result, in this paper we can strictly exclude the buffer overflow caused by ACK. Once the size of ACK packet is neglected, the bottleneck queues only consist of TCP data packets, which improves the accuracy when we calculate the queue size. In addition, the receive window (Rwnd) is set large enough to remove its effect, and all the other buffers can be considered infinite. The bottleneck bandwidth is asymmetric, we use 5/1Mbps for LTE uplink and 10Mbps for downlink. Note that C_u = 1Mbps is sufficient to accommodate the download ACK and upload data simultaneously. The detailed configuration parameters are summarized in Table 2.

3.2 A Word on ACK Compression

We start with a brief discussion on *ACK Compression*, which is defined as the large scale ACK burst. As illustrated in Figure 3, a quantitative view of ACK burst is presented to indicate the level of ACK Compression. Here, the ACK burst is calculated as the number of contiguous ACK packets between two separate TCP data packets. Obviously, either on the uplink or the downlink, the scale of ACK burst is almost less than 10. If we reduce the uplink bandwidth (i.e., C_u = 1Mbps), the downlink ACK burst will decrease rapidly while on the uplink it remains steady. In other words, the ACK Compression will not appear in most cases except a perfectly symmetric single-flow case shown in [5]. Moreover, the *Delayed ACK* mechanism can only provide limited improvement on two-way TCP connections.

3.3 Performance Degradation

The performance of two-way TCPs is evaluated with variable uplink bandwidth (i.e., C_u is 5Mbps and 1Mbps respectively). The reduction of C_u can be considered as introducing UDP traffic tp the uplink. We focus on the goodput (in Figures 4 and 5) and bottleneck queue size (in Figures 6 and 7). Once the upload TCPs are activated (after 5s), the downloads are seriously influenced. When C_u is reduced to 1Mbps, the total goodput of download TCPs degrades to about half of the downlink bandwidth (in Figure 5).

The corresponding evolutions of bottleneck queues are depicted in a local view. In the asymmetric links (Figure 6), the bottleneck queues always evolve towards the opposite direction and the buffers alternately drain/overflow. An earlier version of this phenomenon is observed and named *out-of-phase queue synchronization* in [13], but not thoroughly

[4]In general simulations of NS2, the size of ACK packet is 40 bytes while the TCP data packet is 1040 bytes by default.

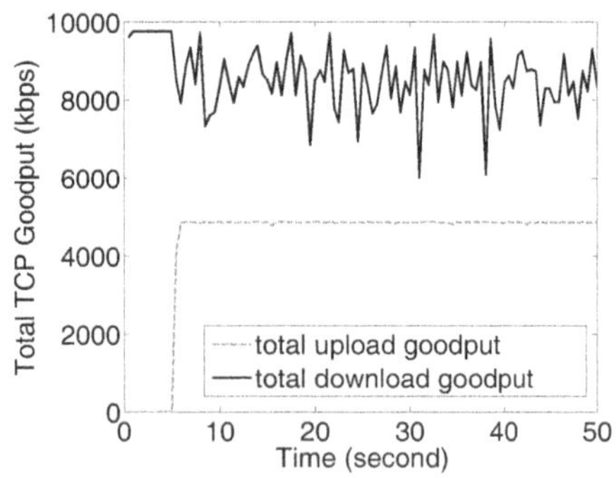

Figure 4: Total TCP goodput in multi-flow scenario (C_u=5Mbps, C_d=10Mbps)

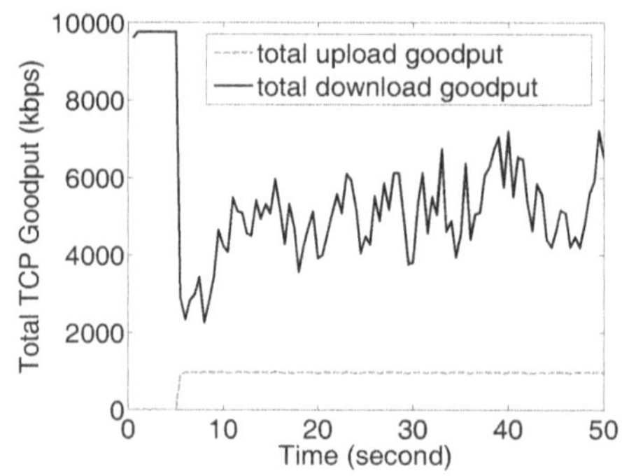

Figure 5: Total TCP goodput in multi-flow scenario (C_u=1Mbps, C_d=10Mbps)

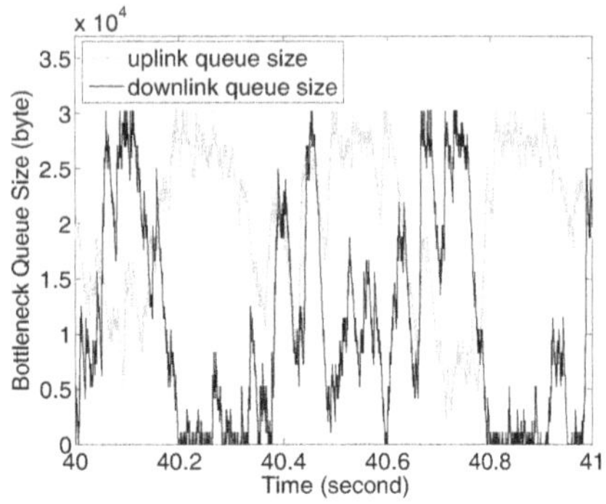

Figure 6: Queue size in multi-flow scenario (C_u=5Mbps, C_d=10Mbps)

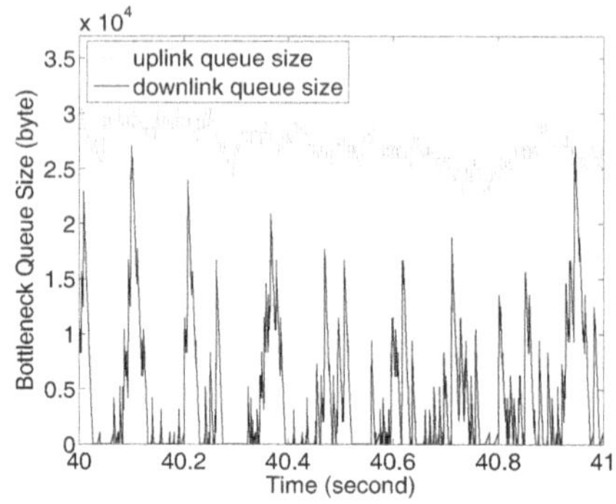

Figure 7: Queue size in multi-flow scenario (C_u=1Mbps, C_d=10Mbps)

explained yet. Since the ACKs of download TCPs are congested in the uplink buffer with large Q_u, the download *Cwnds* will be decelerated and the downlink is often underutilized. Especially in the highly asymmetric case (Figure 7), up to 20% time of the whole simulation, the downlink is completely idle, which deteriorates the downlink utilization. Besides, the *in-phase queue synchronization* [13] caused by ACK Compression never emerges in the above cases, which also proves the disappearance of ACK Compression.

The interacting phenomenon induces considerable degradation to two-way TCPs over LTE. In the subsequent section, we use a single-flow scenario to model and explain it.

4. MODELING AND ANALYSIS OF THE TWO-WAY TCPS

This section presents the understanding of the interacting two-way TCPs, which follows a straightforward modeling procedure. We aim at achieving a quantitative relation between queues to explain the phenomenon. For ease of understanding and explanation, we consider a single-flow asymmetric topology, where N_u=1 and N_d=1. The main difference between the single-flow case and the multi-flow case lies in the different propagation delay. Namely, we set $\tau_u > \tau_d$[5] to eliminate ACK burst. The result is depicted in Figure 8.

4.1 Formalization of TCP Evolution

The formalization of TCP evolution is tough but primary to analyze the two-way TCP problem. Most of the existing literatures, such as [8], utilize the methodology of differential equation to model TCP behaviors. However, in the two-way TCP case, the reverse TCP connection complicates the formalization and achieving an analytical solution. Therefore, we follow a round by round way to model TCP behavior like the methodology used in [9], which is more understandable and feasible in the two-way TCP problem.

We begin with observing the evolution of bottleneck queue size in the single-flow scenario. Figure 9 illustrates a local view of Q_u and Q_d, which both evolve in a period of about 1 second. The evolution regularity is more clear than the queues in multi-flow case, which benefits the modeling and analysis. In order to formalize the TCP behaviors, we combine the queue size with TCP outstanding data, shown in Figures 10 and 11. The outstanding data represents the corresponding TCP *Cwnd*. We observe the whole simulation and find out that, only a limited number of TCP timeouts (usually less than 4) collectively occur at the beginning of the simulation. Once the TCPs turn into stationary status, the timeout will no longer come out and TCP periodically evolve in one phase, namely the *congestion avoidance* phase after packet loss is detected, which is described in a complete period as follows.

When the uplink buffer overflows, $Cwnd_u$ will drop to its half and the upload TCP initiates the congestion avoidance phase. Then the upload sending rate is decelerated and Q_u will decrease to 0 rapidly (in Figure 10). At that moment, plenty of download ACKs will be released from the congested uplink buffer in $R1$. In consequence, the download TCP will be acknowledged in a little burst with $Cwnd_d$ increasing steeply (see the small peak at each periodic tail of $Cwnd_d$ in Figure 11). Subsequently, a short impact of

[5] In the single-flow scenario, τ_u^1 and τ_d^1 are replaced by τ_u and τ_d, respectively.

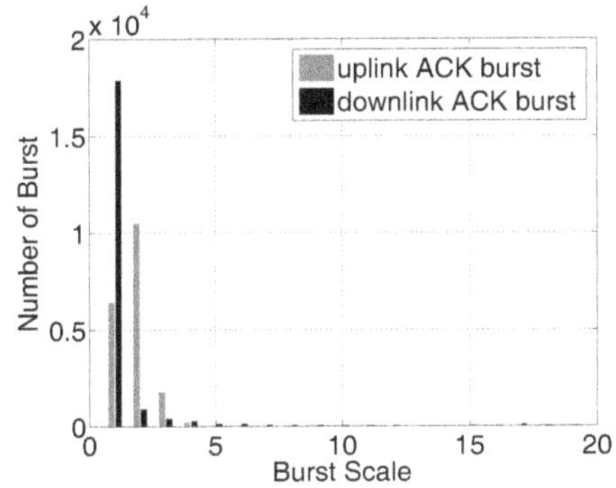

Figure 8: ACK bursts in single-flow scenario (C_u=5Mbps, C_d=10Mbps)

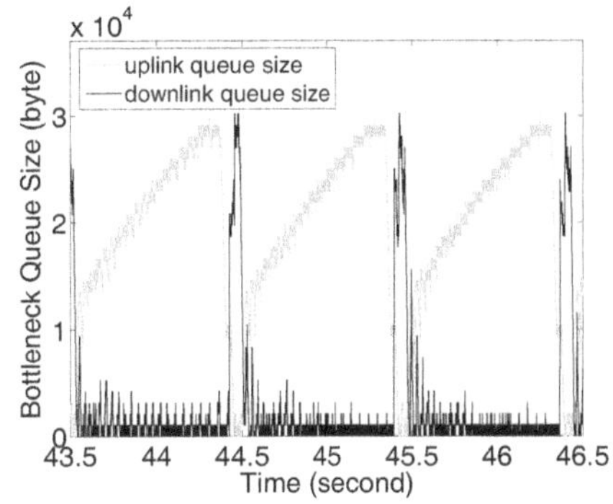

Figure 9: Bottleneck queue size in single-flow scenario (C_u=5Mbps, C_d=10Mbps)

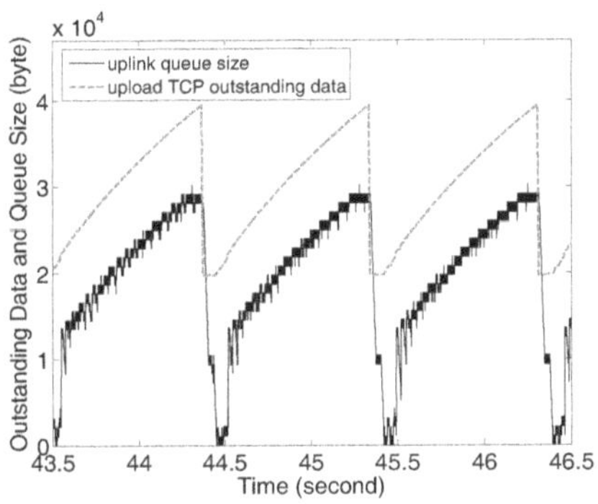

Figure 10: Upload outstanding data and uplink queue size (C_u=5Mbps, C_d=10Mbps)

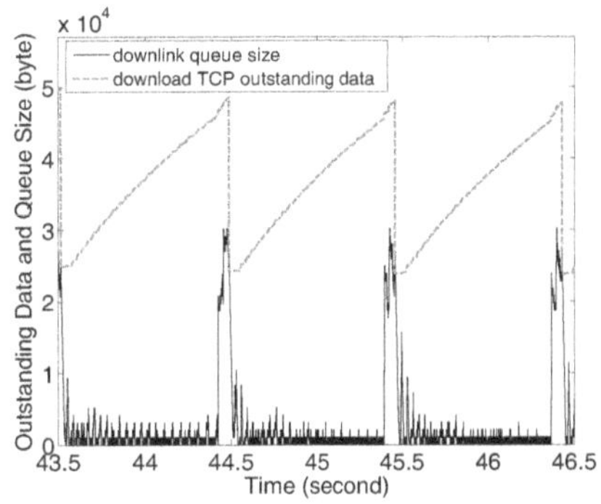

Figure 11: Download outstanding data and downlink queue size (C_u=5Mbps, C_d=10Mbps)

download data packets will arrive at downlink buffer in $R2$ and forms a large peak of Q_d. Likewise, the download TCP then experiences packet loss and starts a new phase of congestion avoidance. These periodic behaviors continuously repeat after a initiation phase at the beginning of the simulation (about 0s~7s). Accordingly, we mainly focus on the congestion avoidance phase and formalize it.

In the following modeling process, we compute congestion window and queue size in a slot. The *slot* is defined as the larger RTT of two-way TCP connections, which consists of two queueing delays and a constant propagation delay τ[6]. Without loss of generality, we assume $\tau_u > \tau_d$, thus $\tau = 2\tau_u$ is set as the constant part of a slot. Namely, the length of the kth slot ($S(k)$) is,

$$S(k) = \frac{Q_u(k-1)}{C_u} + \frac{Q_d(k-1)}{C_d} + \tau \tag{1}$$

where $Q_u(k-1)$ and $Q_d(k-1)$ represent the queue sizes at the end of slot $k-1$ and the first two terms on the right denote the forward/backward queueing delay.

In slot $k-1$, if the buffer overflows, the packet loss will be detected and $Cwnd(k)$ will drop by half. Otherwise, all the data in $Cwnd(k-1)$ are acknowledged and $Cwnd(k)$ will add 1. In other words, the $Cwnd$ is calculated as,

$$Cwnd_u(k) = \begin{cases} Cwnd_u(k-1)/2 & \text{if } Q_u(k-1) > B_u, \\ Cwnd_u(k-1)+1 & \text{otherwise.} \end{cases} \tag{2}$$

[6]In this case, the processing delay and the transmission delay are both neglected.

$$Cwnd_d(k) = \begin{cases} Cwnd_d(k-1)/2 & \text{if } Q_d(k-1) > B_d, \\ Cwnd_d(k-1)+1+\alpha & \text{otherwise.} \end{cases} \tag{3}$$

Note (3), since the download RTT is smaller than the upload one ($\tau_d < \tau_u$), through a whole slot, $Cwnd_d$ actually adds by more than 1. Hence a compensation coefficient α is introduced to enhance the increasing of $Cwnd_d$.

4.2 Computation of Bottleneck Queue Size

Given the formalization of $Cwnd$, it is straightforward to calculate the queue size via addition and subtraction with input and output, respectively. In each slot, the input (how many bytes arrive in a slot) is the product of $Cwnd$ and packet size while the output (how many bytes leave in a slot) is the product of slot length and bandwidth, i.e.,

$$Q_u(k) = [Q_u(k-1) + Cwnd_u(k-1)L_u - S(k)C_u]^+ \tag{4}$$

where notation $[x]^+$ denotes $max(0, x)$. We substitute $S(k)$ of (1) into (4) and obtain,

$$\begin{aligned} Q_u(k) =& [Q_u(k-1) + Cwnd_u(k-1)L_u - Q_u(k-1) \\ & - Q_d(k-1)\frac{C_u}{C_d} - \tau C_u]^+ \\ =& [Cwnd_u(k-1)L_u - Q_d(k-1)\frac{C_u}{C_d} - \tau C_u]^+ \end{aligned} \tag{5}$$

Similarly for the download TCP, we can also acquire,

$$Q_d(k) = [Cwnd_d(k-1)L_d - Q_u(k-1)\frac{C_d}{C_u} - \tau C_d]^+ \tag{6}$$

Besides, once the buffer overflows, Q_u or Q_d will be reset

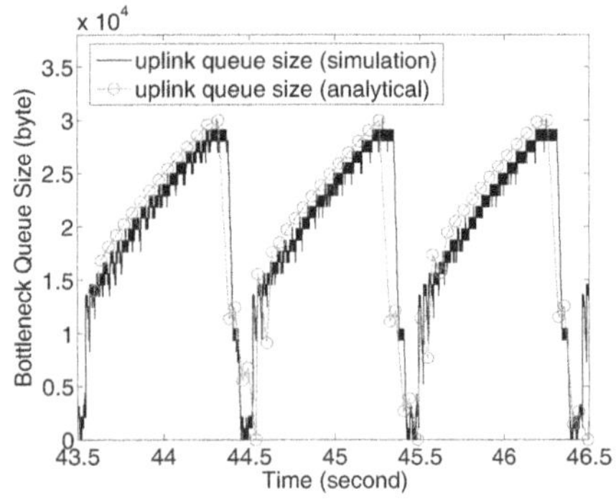

Figure 12: Analytical and simulation values of uplink queue size (C_u=5Mbps, C_d=10Mbps)

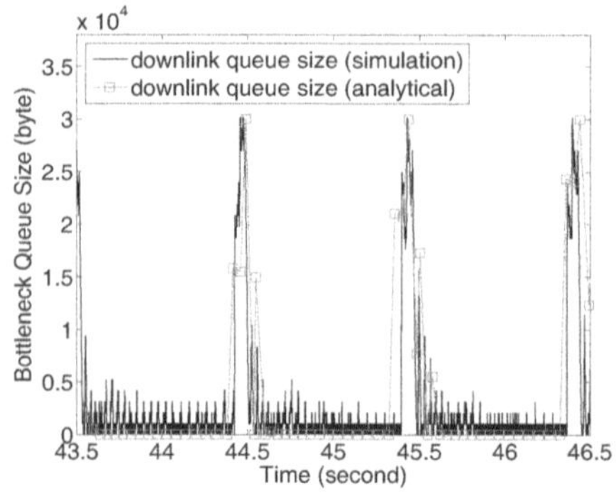

Figure 13: Analytical and simulation values of downlink queue size (C_u=5Mbps, C_d=10Mbps)

to B_u or B_d to ensure $0 \leq Q_u \leq B_u$ and $0 \leq Q_d \leq B_d$ always hold.

Combining equations (5) and (6), we achieve the quantitative relations between Q_u and Q_d. Suppose that $Q_u(k-1)$ increases, $Q_d(k)$ trends to decrease subsequently, and vice versa. In other words, the queues are confirmed to be close *coupled*, which challenges supporting high link utilization simultaneously in both directions. In fact, the coupling of queues directly indicates the interaction of two-way TCPs (the size of the reverse ACK is negligible compared to TCP data packet), which explains the phenomenon. Note that the coupling actually results from (1), which points out the ACKs have to experience a considerable queueing delay in the reverse buffer. Once the forward or the backward queueing delay is negligible, the interacting phenomenon will completely disappear. Moreover, the key parameters that influence the link utilization and TCP performance can be summarized, including link bandwidth (C_u and C_d), buffer size (B_u and B_d) and packet size of TCP data (L_u and L_d). In most cases, the propagation delay τ is not a configurable parameter.

The coupled queues can be also described as a pair of contending adversaries. Like a sine wave and a cosine wave (in Figure 6), the coupled queues always evolve towards the opposite direction, with an amplitude of half buffer size. If the uplink queue acts strong, its opponent on the downlink will be suppressed all the time (in Figure 7). The two adversaries can hardly achieve a win-win status. In other words, the proposed model exactly explains the *Data Pendulum* phenomenon observed in [5]. In consequence, the viewpoint of coupled queues provides an appropriate understanding and explanation for the two-way TCP problem, which is different from those in the existing literatures.

Using the parameters in Table 3, we conduct simulations and compare the values with analytical results of (5) and (6). The compensation coefficient α is heuristically set to 0.7. The comparative results are shown in Figures 12 and 13, indicating that the analytical results well matches the simulation values. So far, the model proposed in this paper is validated to be accurate and feasible for analyzing two-way TCP connections.

4.3 A Lower Bound of the Downlink Queue Size

Now that the proposed model is verified, we turn to figure out a lower bound of Q_d, which points out how to keep the downlink buffer from idleness. The bound is given in the following theorem.

Table 3: Simulation Configuration for Single-Flow Scenario

Parameter	Value
N_u and N_d	1
TCP version	Sack
queue management	DropTail
activated period of upload TCP	5s∼50s
activated period of download TCP	0s∼50s
L_u and L_d	1000B
bottleneck buffer	structured in byte
B_u and B_d	30000B
C_u	5Mbps
C_d	10Mbps
access bandwidth	100Mbps

THEOREM 1. *Given*

$$\mathcal{B} = C_d\left(\frac{B_d}{2C_d} - \frac{B_u}{C_u} - \frac{\tau}{2}\right) \quad (7)$$

$\mathcal{B}$ is a lower bound of Q_u.

PROOF. Before $\mathcal{B}$ is obtained, we calculate the minimum input bytes $[Cwnd_d(k-1)L_d]_{min}$ during a slot. Apparently, $[Cwnd_d(k-1)L_d]_{max}$ is precisely twice of $[Cwnd_d(k-1)L_d]_{min}$ provided that we focus on the stationary status of two-way TCPs that only evolve in congestion avoidance phase, i.e.,

$$\begin{aligned}[Cwnd_d(k-1)L_d]_{min} &= \frac{[Cwnd_d(k-1)L_d]_{max}}{2} \\ &= \frac{[Cwnd_d L_d]_{max}}{2} \\ &= \frac{(C_d\tau + B_d)}{2}\end{aligned} \quad (8)$$

where $[Cwnd_d L_d]_{max}$ represents the maximum outstanding data of download TCP (comprising the maximum bytes stored in the buffer and flying on the link during a slot).

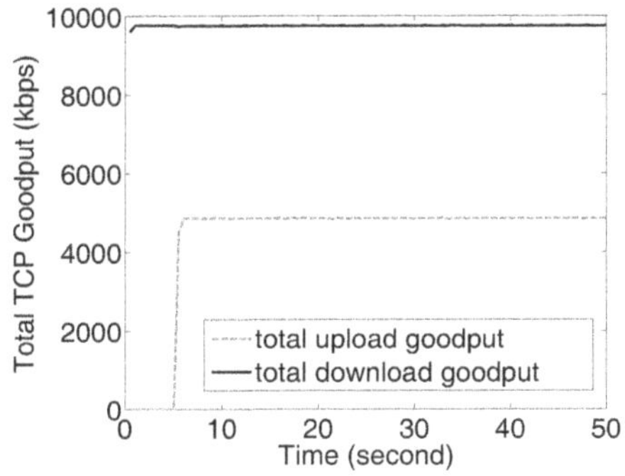

Figure 14: Goodput in multi-flow scenario with PAQ on uplink only (C_u=5Mbps, C_d=10Mbps)

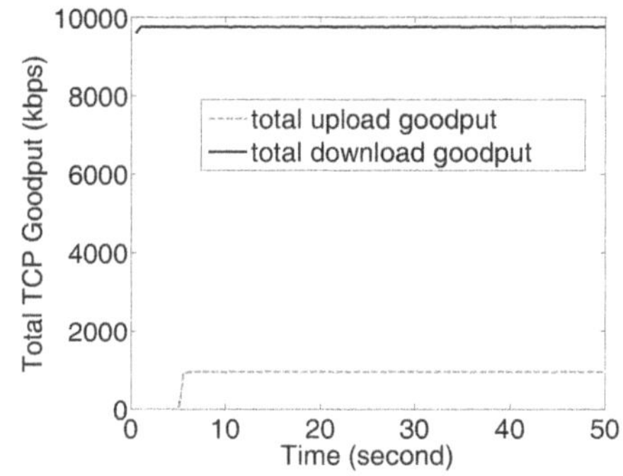

Figure 15: Goodput in multi-flow scenario with PAQ on uplink only (C_u=1Mbps, C_d=10Mbps)

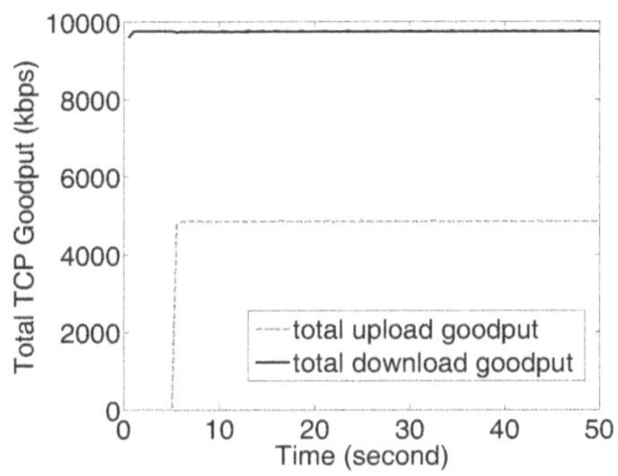

Figure 16: Goodput in multi-flow scenario, PAQ on both uplink and downlink (C_u=5Mbps, C_d=10Mbps)

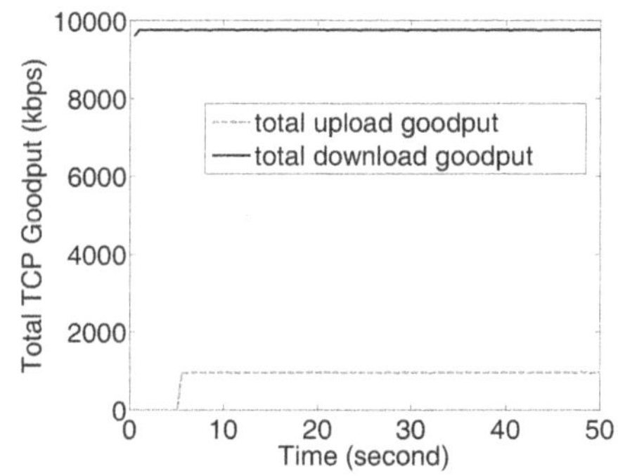

Figure 17: Goodput in multi-flow scenario, PAQ on both uplink and downlink (C_u=1Mbps, C_d=10Mbps)

Next, review equation (6) and we have,

$$\begin{aligned}\mathcal{B} =&[Q_d(k)]_{min}\\ =&[Cwnd_d(k-1)L_d - Q_u(k-1)\frac{C_d}{C_u} - \tau C_d]_{min}\\ =&[Cwnd_d(k-1)L_d]_{min} - [Q_u(k-1)]_{max}\frac{C_d}{C_u} - \tau C_d\\ =&\frac{C_d\tau + B_d}{2} - B_u\frac{C_d}{C_u} - \tau C_d\\ =&C_d(\frac{B_d}{2C_d} - \frac{B_u}{C_u} - \frac{\tau}{2})\end{aligned} \tag{9}$$

□

As long as the LTE network is configured to keep a positive $\mathcal{B}$, the downlink buffer will not drain and the downlink radio bandwidth is well utilized. Furthermore, B_u/C_u reflects the minimum draining time of uplink buffer. If the bottleneck asymmetry is exacerbated (reducing C_u or introducing more upload UDPs), the uplink congestion will become worse and degrades the download performance. In the meantime, the larger uplink buffer size will deteriorate the downlink utilization, which presents an unconventional insight. Nevertheless, larger B_d still benefits the download performance.

5. A DECOUPLING METHOD

Since the essential reason for interacting TCPs lies in the coupled queues, a straightforward idea to decouple them is to remove the reverse queueing delay of ACK. In other words, the decoupling mechanism keeps the ACK packets always being scheduled before data packets in the buffer. Accordingly, we exploit *Preemptive ACK Queueing* (PAQ) to make sure the ACKs no longer experience the queueing delay in the backward buffer and hereby eliminate the interactions between two-way TCPs. In NS2, the PAQ method is implemented in the queueing module. We perform comparative simulations in multi-flow scenario to evaluate the PAQ mechanism. The simulation parameters still comply with the configurations in Table 2.

The results are illustrated in Figures 14 ~17. Compared to Figures 4 and 5, the interactions are eliminated with PAQ only on uplink (Figures 14 and reffig::1m10m-goodput-up-priority) or with PAQ on downlink (Figures 16 and 17) at the same time. The total goodput nearly 100% approximates the link capacity, meaning PAQ is actually required on the uplink only. Moreover, PAQ hardly harms the goodput of the opponent upload TCPs.

The Cumulative Distribution Functions (CDFs) of queue size in the highly asymmetric case are presented at last. In Figure 18, the reason for about 10% duration of empty buffer with PAQ is that the upload TCPs are activated after 5s, which is exactly 10% period of the whole 50s simulation. Figure 19 implies the downlink utilization is considerably improved with PAQ. Particularly, PAQ prevents the idleness of downlink, which is about 20% of the whole simulation without PS. In other words, the PAQ method effectively decouples the interacting two-way TCPs and supports full utilization simultaneously in both directions.

6. DISCUSSION

Actually, the two-way TCP problem reminds us of paying attention to the *backward* link, where congestion may also occur. The congested backward link imposes significant impact on the *forward* link utilization. Almost all the existing

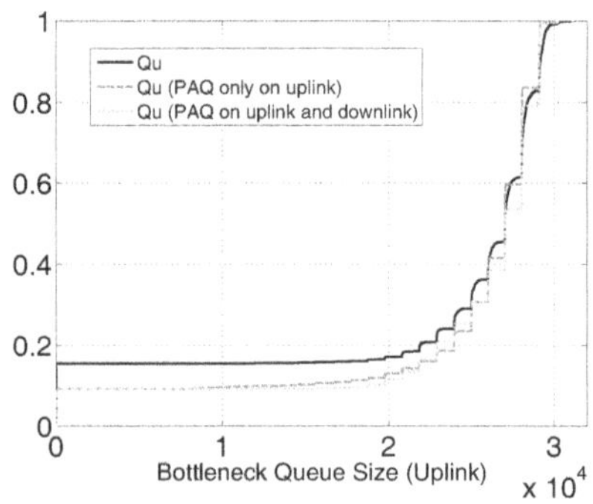

Figure 18: CDF of uplink queue size in multi-flow scenario (C_u=1Mbps, C_d=10Mbps)

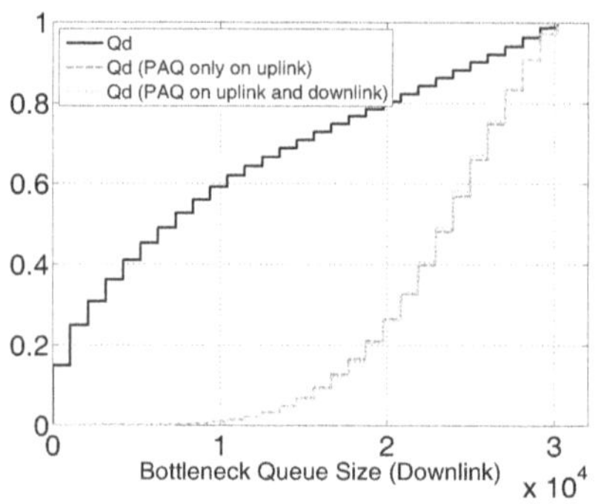

Figure 19: CDF of downlink queue size in multi-flow scenario (C_u=1Mbps, C_d=10Mbps)

research on TCP only concerns about the congested forward link. Indeed, in the backbone Internet, the backward influence may be not attractive even crucial. Nevertheless, in the asymmetric LTE network, the TCP performance will be badly impacted by its adversaries in the opposite direction (in Figures 4 and 5), which is worthy of understanding and improving.

The proposed model in this paper is based on an analytical procedure for single-flow scenario. As a matter of fact, the model and Theorem 1 can be extended to multi-flow case, which is much more general and left in the future work. In addition, the *slot* concept utilized in Section 3 is a tradeoff solution. Since the RTTs of upload and download are set different to eliminate ACK Compression, it is hard to make sure the kth Q_d is always consistent with the kth Q_u. We perform the calculation in a slot and supplement a compensation coefficient α to correct the inaccuracy. The tradeoff benefits simplifying the model and explanation, which is hence considered reasonable. Finally, the main purpose of this paper is not on designing a sophisticated solution. The PAQ method is acceptable as an heuristic attempt to decouple the queues. A comprehensive algorithm for thoroughly solving the two-way TCP problem is left in the future work.

7. CONCLUSION

In the viewpoint of *coupled queues*, we investigate the two-way TCP problem over LTE networks. The interaction between forward and backward TCPs exists ubiquitously and its influence can be crucial in the asymmetric bottleneck of LTE. Based on a straightforward model, we obtain the quantitative relations between the coupled queues and provide a new explanation different from those in the existing literatures. The lower bound of downlink queue size is figured out to indicate how to avoid the idleness of LTE downlink. Besides, the larger uplink buffer may result in further degradation, which presents an unconventional insight. The simulations conducted on NS2 verify the accuracy of the proposed model and evaluate the efficiency of the decoupling method.

8. ACKNOWLEDGMENTS

This work is supported in part by the National Grand Fundamental Research Program of China (973) under Grant No.2009CB320504, 2012CB315803 and 2010CB328105, and the National Natural Science Foundation of China (NSFC) under Grant No.60971102. and the National Science and Technology Major Project of China (NSTMP) under Grant No.2011ZX03002-002-02.

9. REFERENCES

[1] AT&T 4G LTE speed preview: 28.87Mbps downloads. http://www.slashgear.com/att-4g-lte-speed-preview-28-87mbps-downloads-20153265.

[2] LTE - An Introduction. Technical report, Ericsson, 2009.

[3] H. Balakrishnan, V. Padmanabhan, G. Fairhurst, and M. Sooriyabandara. RFC 3449 TCP Performance Implications of Network Path Asymmetry.

[4] H. Balakrishnan, V. N. Padmanabhan, and R. H. Katz. The effects of asymmetry on tcp performance. In *Proceedings of MobiCom '97*.

[5] M. Heusse, S. A. Merritt, T. X. Brown, and A. Duda. Two-way tcp connections: old problem, new insight. *SIGCOMM Comput. Commun. Rev.*, 41:5–15, Apr. 2011.

[6] V. Jacobson. Congestion avoidance and control. In *Proceedings of the ACM SIGCOMM '98*, pages 314–329, 1988.

[7] L. Kalampoukas, A. Varma, and K. K. Ramakrishnan. Two-way tcp traffic over rate controlled channels: effects and analysis. *IEEE/ACM Transactions on Networking*, 6(6):729–743, Dec. 1998.

[8] V. Misra, W.-B. Gong, and D. Towsley. Fluid-based analysis of a network of aqm routers supporting tcp flows with an application to red. In *Proceedings of the ACM SIGCOMM '00*, pages 151–160, 2000.

[9] J. Padhye, V. Firoiu, D. Towsley, and J. Kurose. Modeling tcp throughput: a simple model and its empirical validation. In *Proceedings of the ACM SIGCOMM '98*, pages 303–314, 1998.

[10] M. Podlesny and C. Williamson. Improving tcp performance in residential broadband networks: a simple and deployable approach. *SIGCOMM Comput. Commun. Rev.*, 42(1):61–68, Jan. 2012.

[11] S. Shenker, L. Zhang, and D. D. Clark. Some observations on the dynamics of a congestion control algorithm. *SIGCOMM Comput. Commun. Rev.*, 20:30–39, October 1990.

[12] R. Wilder, K. K. Ramakrishnan, and A. Mankin. Dynamics of congestion control and avoidance of two-way traffic in an osi testbed. *SIGCOMM Comput. Commun. Rev.*, 21(2):43–58, Apr. 1991.

[13] L. Zhang, S. Shenker, and D. D. Clark. Observations on the dynamics of a congestion control algorithm: the effects of two-way traffic. *SIGCOMM Comput. Commun. Rev.*, 21:133–147, August 1991.

Increasing the Performance of HSDPA with High-Speed Single Frequency Network

Oleksandr Puchko
Department of Mathematical Information Technology
University of Jyväskylä
Jyväskylä, Finland
oleksandr.puchko@jyu.fi

ABSTRACT

High Speed Single Frequency Network (HS-SFN) is a multi-cell transmission scheme for High-Speed Downlink Packet Access (HSDPA) that can assist user equipment (UEs) in the softer handover area by combining signals from two neighbouring cells and at the same time by reducing the inter-cell interference. This article analyzes performance of HS-SFN for macro cell scenario under different channels with different type of schedulers by means of extensive system level simulations. The results indicate that HS-SFN can increase throughput for UEs in softer handover area from 40% to 80% without decreasing the overall performance.

Categories and Subject Descriptors

C.2.1 [**Network Architecture and Design**]: Wireless communication

Keywords

HSDPA, HS-SFN, Phase adjustment, LMMSE equalizer, CoMP, Multipoint transmission

1. INTRODUCTION

The development of wireless communication systems has been traditionally associated with improving the peak rate of individual users and increasing the overall system throughput. At the same time, it has become more and more evident that the cell edge performance is equally important. Since not all of the mobile users actually move, they can stay at the cell edge area for quite long periods of time generating or downloading significant amount of data. The most typical example for such a scenario is the mobile subscription used with a wireless modem in the office or home environment. Failing to deploy techniques helping mobile users are the cell edge may lead to severe unfairness and/or failing to ensure the minimum QoS requirements as mandated by the subscription type. In this paper we address those issues at the cell edge, in particular, how to increase the peak data rate for UEs allocated at the cell edge.

One approach for solving the cell edge problem is to consider the multi-point transmission, which is often referred to as Coordinated Multipoint Transmission (CoMP) and which is currently actively researched in industrial and academical area. In the beginning this technique was researched under LTE-Advance technology since March 2008 as a study item (SI). In March 2010 it was specified in E-UTRA [1], after that the research was suspended until December 2010. Starting from 2011 a lot of research works were published on CoMP in LTE-A, such as [2], [3] and [4].

Similar techniques called "Multipoint transmission" have being researched under HSPA technology from December 2010. An official SI on Multipoint transmission schemes was opened in January 2011 [5], and in March 2011 two main schemes were proposed for further investigations. One of the schemes is HS-SFN where the same signal is transmitted on the same frequency and scrambling code from two cells. The main concept and first results were presented at [6], however results that were presented in the paper do not show maximum gain that can be achieved by using HS-SFN. Such things as "enhanced linear minimum mean square error (LMMSE) equalizer" and "Phase Adjustment" that will be described in this paper were not taken into account. Furthermore results at [6] show the performance of HS-SFN under only one propagation channel model. This paper will present performance of HS-SFN usage under different channel models, such as ITU Pedestrian A and B at 3 km/h and Vehicular A at 30 km/h.

The rest of the article is organized as follows. Section 2 present HS-SFN concepts that include description of the main concept of HS-SFN, signal model and C/I calculation for normal and HS-SFN transmission. The same section also describes additional techniques such as "Enhanced LMMSE equalizer" and "Phase adjustment". Overall system design aspects, specifically HS-SFN candidate selection and control channel considerations, are presented in section 3. Finally section 4 presents performance evaluation of HS-SFN with semi-static simulator and simulation results, and section 5 concludes the paper.

2. HS-SFN CONCEPTS

Unlike the Multiflow multi-point transmission scheme mentioned in [7], in HS-SFN, the *same* downlink (DL) data is transmitted on the High Speed-Physical Downlink Shared Channel (HS-PDSCH) from two cells by using the same carrier frequency and scrambling code in both cells. Data

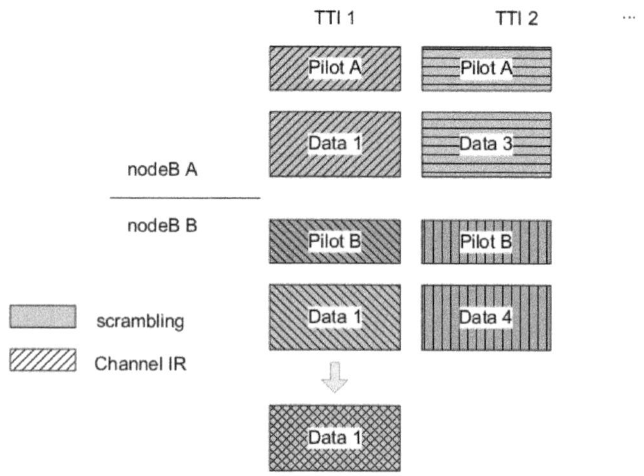

Figure 1: HS-SFN data signal formation.

transmission is Transmission Time Interval(TTI) aligned, so that the signals are combining over the air at the UE receiver, as presented in Fig. 1, [6]. Fig. 1 shows transmissions of two TTIs from cells A and B that belong to the same NodeB, where different colors represent different scrambling codes and crosshatching display different channel impulse response. The strict TTI alignment and a requirement to transmit exactly the same data poses a further constraint that the HS-SFN technique is most likely limited to the intra-site scenario, i.e. cells belonging to this cooperative transmission are under control of the same system module. It is however worth mentioning that recent advances in the core network architectures and, in particular, remote radio head deployment makes it possible to run HS-SFN in cases beyond the classical three-sector site scenarios.

The main gain of the HS-SFN usage comes from combining two strongest signals: the strongest serving one and the second strongest from the assisting cell. At the same time, it can reduce considerably the interference from the second cell as it starts to assist in data transmission. Equation (1) shows simplified version of enhanced C/I by HS-SFN usage:

$$C/I = \frac{0.7P_{own} + 0.7P_{2nd}}{\sum_{i=0}^{N} P_i - P_{own} - 0.7P_{2nd}}, \tag{1}$$

where i is the i-th cell and N the total number of cells. It is assumed that 70% of power is used for data channel (HS-PDSCH) and 30% is left for control (pilots and signaling) channel overhead.

Besides the TTI-level synchronization, the receiver should be able to deal with assisted transmission, which is another requirement that will be described in 2.1. Another challenge is that combining of two signals can have either positive or negative impact due to uncorrelated fast fading. This problem is solved if the data transmissions from the two cells are adjusted in such a manner that the signals would arrive in phase. This method is considered in section 2.4.

2.1 Signal model

According to [8], it is assumed that there is one Tx-antenna per each cell in NodeB and N_{Rx} Rx-antennas at UE. The received chip sequence equals to

$$r(m) = H^T s(m) + n(m), \tag{2}$$

where $s(m)$ is the transmitted chip stream (3) and $n(m)$ is the noise vector. $s(m)$ can be expressed as:

$$s(m) = [s(m+F+L-D), \dots, s(m), \dots, s(m-D)]^T, \tag{3}$$

$$0 \leq D \leq F. \tag{4}$$

In (3), L is the delay spread of the channel normalized to the chip interval and D is the delay parameter satisfying inequality 4 and F is the length of the linear filter w that the receiver uses in the filtering of r to obtain the estimate $\tilde{s}$ for the transmitted chip sequence s:

$$\tilde{s}(m) = w \cdot r(m). \tag{5}$$

Multiplication with the channel coefficient matrix H^T in (2) models the convolution with the impulse response of the channel. H is a $(F+L) \times (F \cdot N_{Rx})$ matrix given by the equation (6):

$$H = \begin{bmatrix} h & 0 & 0 & \dots & 0 \\ 0 & h & 0 & \dots & 0 \\ 0 & 0 & h & \dots & 0 \\ \vdots & \vdots & \vdots & \ddots & \vdots \\ 0 & 0 & 0 & \dots & h \end{bmatrix}^T, \tag{6}$$

where h is the $N_{Rx} \cdot (L+1)$ channel impulse response matrix defined by

$$h = \begin{bmatrix} h_1(0) & h_1(1) & \dots & h_1(L) \\ h_2(0) & h_2(1) & \dots & h_2(L) \\ \vdots & \vdots & \ddots & \vdots \\ h_{N_{Rx}}(0) & h_{N_{Rx}}(1) & \dots & h_{N_{Rx}}(L) \end{bmatrix}, \tag{7}$$

where $h_i(l)$ is the channel impulse of ith Rx-antenna at the delay l.

During the HS-SFN assisted transmission, only the data part of the signal uses the same scrambling, whereas both cells use their own scrambling for control channels which still need to be received by other legacy UEs. As a result, expression (2) is converted to:

$$r_c(m) = H_0^T s_0(m) + H_1^T s_1(m) + n(m). \tag{8}$$

In turn, in expression (8), $s(m)$ can be divided into control and data parts, so that:

$$r_c(m) = H_0^T(d(m)+c_0(m)) + H_1^T(d(m)+c_1(m)) + n(m) \tag{9}$$

where $H_{0,1}$ is the channel coefficient matrix for serving (H_0) and assisting (H_1) cells, $d(m)$ is the data part of the signal and $c_{0,1}(m)$ is the control part of the serving and assisting cell.

Since the data parts of the signals coming from the serving and assisting cells are the same, they can be combined according to the equation 10:

$$H_C^T = H_0^T + \sqrt{\frac{L_1 P_{1,data}}{L_0 P_{0,data}}} H_1^T, \tag{10}$$

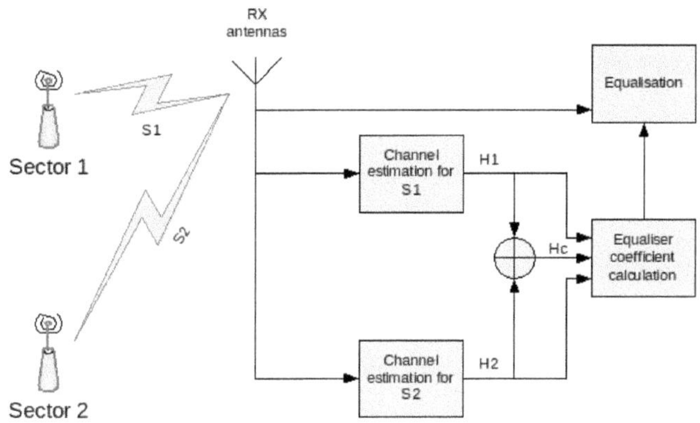

Figure 2: Receiver architecture.

where L_j is path loss from j^{th} cell to UE and $P_{j,data}$ is the transmit power for HS-PDSCH of the jth cell. The serving cell is referred to by $j = 0$, and the assisting cell by $j = 1$.

Using expressions (9) and (10), the received chip sequence can be found as follows:

$$r_c(m) = H_C^T d(m) + H_0^T c_0(m) + H_1^T c_1(m) + n(m). \quad (11)$$

2.2 Enhanced LMMSE Equalizer

During an HS-SFN transmission, UEs estimate the channel based on pilots transmitted in the control channel. As the pilots remain separated by scrambling also for an assisted transmission, the UE needs to estimate the two channels separately, combine the estimates, and then perform equalization of the signal, see Fig. 2.

This enhanced LMMSE equalizer is based on the LMMSE equalizer for standalone transmission (transmission from one cell) described in [9] and [10]. Comparison of LMMSE and enhanced LMMSE equalizers with HS-SFN can be found in [11].

To calculate the equalizer coefficient vector w of length $N_{Rx} \cdot F$, the following quadratic cost function should be minimized:

$$F(w) = E\left\{\left|w^T r(m) - s(m-\tau)\right|\right\}, \quad (12)$$

which is the distance between the received chip sequence $r(m)$ equalized and the $(m-\tau)$-th transmitted chip $s(m-\tau)$ for a given value of delay $\tau \in \{-F+D+1, -F+D+2, \ldots, D+L\}$.

Received chip sequence is defined in (11), and the transmitted $(m-\tau)$-th chip in (12) is expressed as

$$s(m-\tau) = d(m-\tau) + c(m-\tau). \quad (13)$$

Then, w that minimizes the cost function $F(w)$ can be found by deriving (12) with respect to w^T [12]:

$$\frac{\partial F}{\partial w^T} = \big(L_0 H_C^T R_{data} H_C^* + L_0 H_0^T R_{ctrl} H_0^* + L_1 H_1^T R_{ctrl} H_1^* + \\ + R_n\big) w^* - \big(P_{data}\sqrt{L_0} H_C^T \delta_\tau + P_{ctrl}\sqrt{L_0} H_0^T \delta_\tau\big) = 0, \quad (14)$$

where $R_{data,ctrl}$ is $(F+L-1) \times (F+L-1)$ the autocorrelation matrix of signal transmitted through data or control channel, that can be found as follows:

$$R_{data,ctrl} = \begin{bmatrix} P_{data,ctrl} & 0 & \ldots & 0 \\ 0 & P_{data,ctrl} & \ldots & 0 \\ \vdots & \vdots & \ddots & \vdots \\ 0 & 0 & \ldots & P_{data,ctrl} \end{bmatrix}. \quad (15)$$

Therefore, the enhanced LMMSE equalizer coefficient is equal to

$$w_e = \big(P_{data} L_0 H_C^H H_C + P_{ctrl} L_0 H_0^H H_0 + P_{ctrl} L_1 H_1^H H_1 + \\ + C_n\big)^{-1} \big(P_{data}\sqrt{L_0} H_C^H \delta_\tau + P_{ctrl}\sqrt{L_0} H_0^H \delta_\tau\big). \quad (16)$$

Note that only a single equalizer is required, leading to relative low complexity and consequently low power consumption at the UE.

2.3 C/I calculation

In order to account for inter-cell interference, intra-cell multipath interference, and equalization we may express the C/I according to [13] more accurately as

$$C/I = \frac{P_{data} \cdot \left|w^T H_c^T \delta\right|^2}{I_{own} + I_{col} + I_{thermal}}, \quad (17)$$

where total interference consists of three logically different interferences: own interference or so-called Multiple Access Interference (MAI) (18), colored interference (19), and thermal noise (20).

$$I_{own} = P_{data} w^T H_c^T \hat{\delta} H_c^* w^* + P_{ctrl} w^T H_0^T \hat{\delta} H_0^* w^*, \quad (18)$$

$$I_{col} = \sum_{j=1}^{N_{BS,all}} \frac{L_j P_j}{L_0 P_0} w^T H_j^T H_j^* w^*, \quad (19)$$

$$I_{thermal} = \frac{1}{L_0 P_0} C_{thermal}. \quad (20)$$

where:

H_c is combined channel impulse responce matrix (10);
P_{data} is power for HS-PDSCH;
P_{ctrl} is power for control and pilot channels;
w is equalizer weigth vector;
δ is delay of equalizer;
$C_{thermal}$ is thermal noise matrix;
L_j is pathloss from UE to jth nodeB;
P_j is transmit power of jth nodeB.

In (18) interference from the own cell was divided into the data and control parts due to the combining of data channel impulse response matrices. In (19), if $j = 1$, then only 30% of P_j should be considered since there is no interference from that cell on the HS-PDSCH channel.

2.4 Phase Adjustment

To improve further the HS-SFN performance gains, the transmitted signals from serving and assisting cells can be adjusted in such a way that the signals arrive to a UE in phase. Such an approach can definitely improve performance of the HS-SFN scheme since the similar solution has been used in the HSPA and LTE systems.

Similar to the existent HSPA single stream MIMO transmission[14], a set of transmitting antenna weights are defined. In the considered case, this set consists of two weights: the signal transmitted through HS-PDSCH from the serving cell is multiplied by the first weight, and the HS-PDSCH from the assisting cell by the second weight. The transmitting antenna weights are not applied to the pilot and control channels.

The first weight always remains fixed and equal to one:

$$\alpha_0 = 1. \tag{21}$$

The second weight provides a phase adjustment and is set to one of the following values according to the UE feedback:

$$\alpha_1 = \left\{ \frac{1+j}{\sqrt{2}}, \frac{1-j}{\sqrt{2}}, \frac{-1-j}{\sqrt{2}}, \frac{-1+j}{\sqrt{2}} \right\}, \tag{22}$$

where j is an imaginary unit. As one can notice, the weights applied to the assisting cell's HS-PDSCH differ by ninety degrees. Thus, there are four possible sets of transmitting antenna weights and one of these sets is chosen based on the UE feedback information. More information about phase adjustment and also comparison of HS-SFN with and without phase adjustment can be found in [15].

3. OVERALL SYSTEM DESIGN ASPECTS

3.1 HS-SFN candidate selection

A straightforward approach to select candidate UEs for the HS-SFN transmission is to rely upon the existing handover mechanisms and related measurements of link path strengths by the UE. Every UE maintains the active set of strongest cells and sends measurement reports whenever changes in the link strength are detected, so that the network can decide how the active set should be adjusted.

As a part of the active set update procedure, the radio network controller (RNC) can decide whether the HS-SFN scheme should be activated for a particular UE based on the cells quality reported in the measurement report. It should be noted that in the 3G/HSPA architecture, the measurement reports are sent towards the RNC, whereas the HS-SFN scheduling is performed by the NodeB based on the channel quality indicator (CQI) reports from the UE. Thus, the most anticipated approach is that RNC performs a so-called synchronous change configuration where both the UE and NodeB know that starting from a particular moment of time the UE will monitor the given HS-SFN assisting cell and send back related CQI reports. Similarly, it is up to the RNC to decide when the HS-SFN operation should be terminated or switched to a different assisting cell.

Once the serving cell starts to receive the CQI reports for the combined channel performance from the serving and assisting cells the site-wide scheduler can make a decision in which TTIs to enable the HS-SFN transmission by instructing the assisting cell to send exactly the same data as the serving one. This concept is similar to the existent HSPA multi-carrier joint scheduling approach with the difference that two cells belong to different sectors and are on the same frequency.

3.2 Control channel considerations

As shown in Fig. 1, while there is no interference on data channels in cell B, the pilot but also other control channels will remain on the native scrambling of cell B and will be subjected to higher interference by the use of scrambling A. As an example we may consider the High Speed Shared Control CHannel (HS-SCCH) which carries coding and modulation information for the UE that will be scheduled in the following TTI. The power of the HS-SCCH can be adjusted to guarantee a certain error rate at the designated UE receiver. Thus, while the interference on the HS-SCCH is raised by assisted transmissions in cell B, the C/I at the UE can be maintained by increasing the HS-SCCH power share (and decreasing that of the assisting data transmission).

Depending on the UE, the required C/I for the control channel may be -9dB down to -12dB [16]. When mapping the geometry distribution of all UEs in the network to HS-SCCH power requirements already 90% of all UEs will be served with a share of 22% of total transmit power on the HS-SCCH. With increased interference by HS-SFN assisted transmissions the 90%tile is achieved with an HS-SCCH power share of 30%. It should be noted that while the control channel power in cell B needs to be increased, it can be decreased in cell A.[6]

4. PERFORMANCE EVALUATION

In this section we present the system performance evaluation of HS-SFN done with a semi-static simulator. The network layout consists of 19 NodeBs with 1 km site to site distance. UEs are uniformly distributed over the network and transmitting according to the Next Generation Mobile Network (NGMN) traffic model. It means that a UE enters the network, transmits one file and disconnects. Depending on the load level, new UE comes after some time. Slow fading is modeled for each UE separately when a UE enters the network. Even though UEs are stationary, fast fading is modeled according to the International Telecommunication Union (ITU) channel model presented in the Table 1. The Signal to Interference plus Noise Ratio (SINR) is then mapped to packet error probabilities using the Actual Value Interface (AVI) approach. The main simulation parameters are introduced in Table 1.

The maximum active set size is set to 2 which has an impact on choosing HS-SFN candidate UE. Only the second strongest interfering cell is considered as the assisting cell with an additional requirement that this cell must belong to the same site as the main serving cell. According to the active set size, path loss model and NodeB antenna model from Table 1 the average number of the HS-SFN candidates is 4-5%. Since traffic load is uniformly distributed over the network there is no need to separate threshold value used for HS-SFN for different cells. However, for networks with unbalanced traffic load, the different threshold value might be used for each cell to achieve the best performance.

Two different types of scheduler were used in this paper. The first one is well know proportional fair scheduler [17], where the UE probability to be scheduled, depend on the proportional fair metric, which in turn depends on the achievable data rate and scheduler decision in previous TTIs. The second scheduler type is a so-called prioritizing scheduler. It means that HS-SFN candidates and UEs in the softer HO region will be scheduled first in both HS-SFN enabled and reference scenarios. This type of scheduler should bring the maximum gain that can be achieved from HS-SFN usage, because HS-SFN technique is able to help UEs only in the softer HO area. But still this gain would show the fair com-

Table 1: Simulation parameters.

Parameter	Value
Cell Layout	Hexagonal grid, 19 NodeBs, wrap-around
Number of sectors per NodeB	3
Inter-site distance	1000 m
Carrier Frequency	2000 MHz
Path Loss	L=128.1 + 37.6log10(R), R in kilometers
Log Normal Fading	
Standard Deviation	8dB
Inter-Node B Correlation	0.5
Intra-Node B Correlation	1.0
Correlation Distance	50m
Max BS Antenna Gain	14 dBi
Antenna pattern	
2D Pattern	$A(\theta) = -min(12(\theta/\theta_{3dB})^2, Am)$, $\theta_{3dB} = 35\ degrees$, $Am = 20dB$
Channel Model	PedA3, PedB3, VehA30
CPICH Ec/Io	-10 dB
Total Overhead power	30%
UE Antenna Gain	0 dBi
UE noise figure	9 dB
UE Receiver Type	enhanced LMMSE with RxD
Speading factor	16
Maximum Sector Transmit Power	43 dBm
Handover reporting range	3dB
Number of HARQ processes	8
Maximum active set size	2
Traffic	
Traffic model	NGMN Traffic Source Model
Offered load level	0.25, 0.5, 1 and 2 Mbits/sec
File size	Fixed at 1 Mbit
Inter-arrival time	Exponential, mean = 5 seconds
Flow control on Iub	Ideal and instantaneous
CQI	Ideal with 3 TTI delay
Scheduling	Common intra-NodeB scheduler, proportional fair algorithm

parison since UEs under similar geometry conditions were prioritized. In the worst case, the prioritizing scheduler can completely block UEs in non-softer HO area since they will not get scheduled during the presence of a softer HO UE. This can happen with the full buffer traffic model, however in NGMN traffic model, where a UE is removed from the network after it completed the packet transmission, probability of this issue is too small.

Regular independent scheduler for each cell is used for the reference scenario. And for HS-SFN scenario, common site-wide scheduler is enabled. It means that scheduler is able to form all possible scheduling combinations and choose the combination with the highest possible total data rate for the whole site while also taking into account proportional fair resource allocation.

4.1 Simulation Results

The simulation results will be presented in three type of statistics:

- Cumulative distribution function (CDF) versus burst rate
- Absolute mean burst rate versus offered load
- Mean burst rate gain versus offered load

All these results were gathered with 6000 finished calls, which means that for each simulation 6000 UEs received the packet that was enough to obtain reliable statistics.

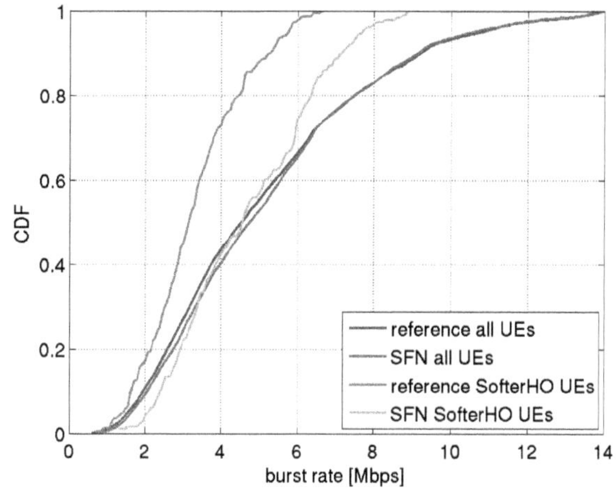

Figure 3: Burst rate CDF, 1Mbit/s, PedB channel, Proportional fair scheduler.

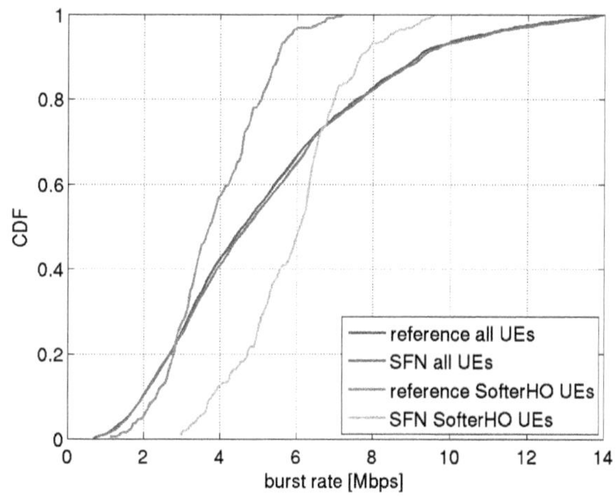

Figure 4: Burst rate CDF, 1Mbit/s, PedB channel, Prioritizing scheduler .

Fig. 3 and Fig. 4 show CDF for all UEs in the network and UEs in the softer handover (softerHO) area for the scenarios where HS-SFN is disabled (reference) and enabled (SFN). These figures show results for proportional fair and prioritizing schedulers accordingly. From these figures can be seen that HS-SFN significantly improved performance of UEs in the softer handover area. Furthermore these results also show that HS-SFN can marginally increase the performance of all UEs or at least do not decrease the overall performance. Compare Fig. 3 and Fig. 4 can be seen that the usage of prioritizing scheduler increased performance of UEs in the softer handover area and especially low burst rate UEs much more than with proportional fair scheduler.

Fig. 5, Fig. 6 and Fig. 7 show mean burst rate versus different load of the network for PedA, PebB and VehA channels correspondingly. The green lines display results for reference case and black lines display the results where HS-SFN is enabled; continuous and dash lines represent results for prioritizing and proportional fair schedulers accordingly.

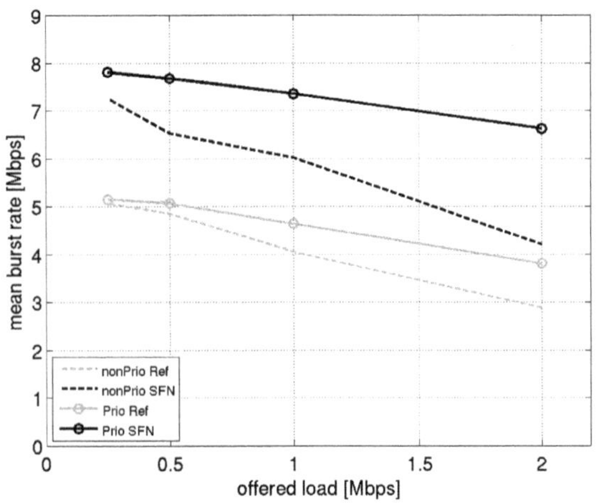

Figure 5: Mean burst rate for PedA channel .

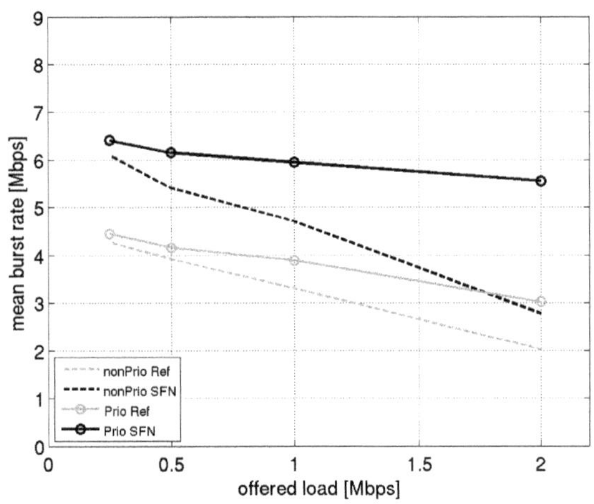

Figure 6: Mean burst rate for PedB channel .

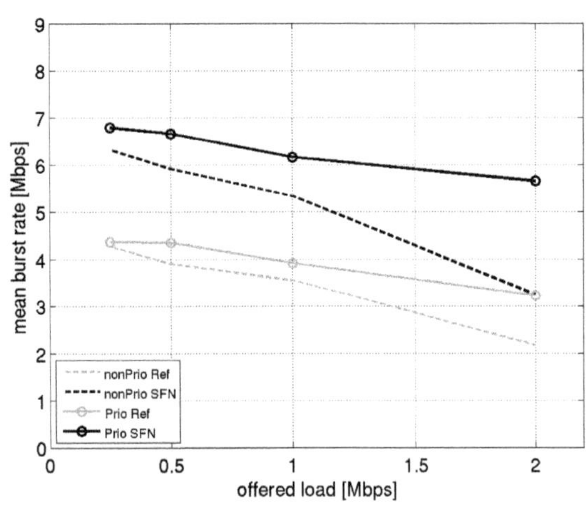

Figure 7: Mean burst rate for VehA channel .

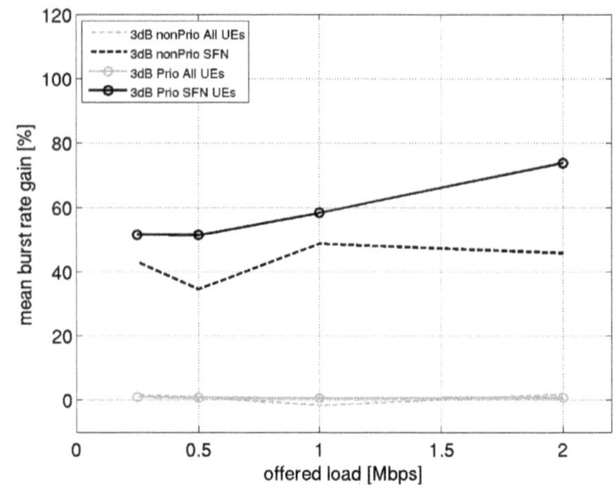

Figure 8: Mean burst rate gain for PedA channel .

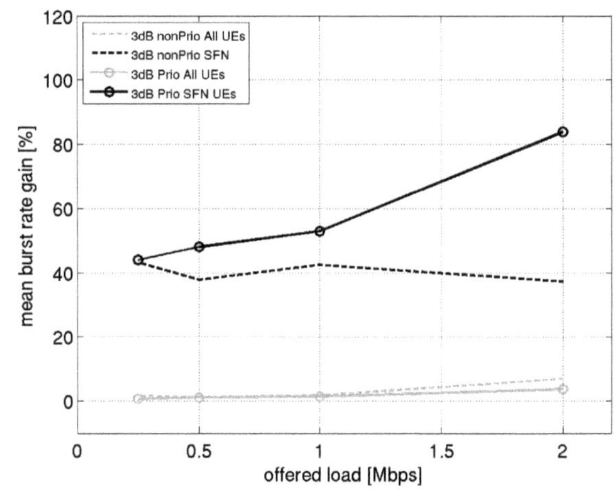

Figure 9: Mean burst rate gain for PedB channel .

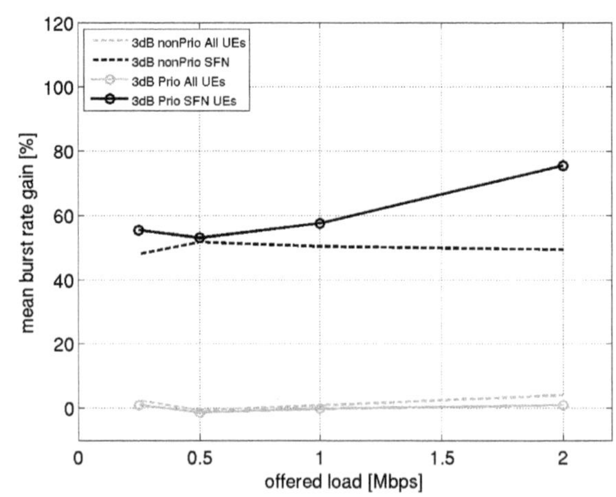

Figure 10: Mean burst rate gain for VehA channel .

Fig. 8, Fig. 9 and Fig. 10 show mean burst rate gain of HS-SFN over reference scenatios for PedA, PebB and VehA channels correspondingly. The green lines display results from all UE and black lines display the results from SFN UEs (or UEs in the softer HO area); continuous and dash lines represent results for prioritizing and proportional fair schedulers accordingly.

Basically figures for mean burst rate and throughput gains show the same results in different ways. The reason to show both of them is to show that even with different mean burst rate for different channels, HS-SFN still have roughly the same gain. It is worth mentioning that with proportional fair scheduler gain from HS-SFN usage vary from 40% to 50%. However with prioritizing scheduler the gain increasing with the increasing offered load. And can grow up to 80%.

5. CONCLUSIONS

In this paper, we have presented the main concepts of the HS-SFN multi-point transmission scheme and analyzed its performance under different channel models. HS-SFN can improve significantly performance of UEs in the softer handover area and also in the soft handover area if the network deployment relies upon the remote radio heads. To some extent, it can be thought as the DL soft combining approach, that the legacy HSPDA does not have. The simulation results have indicated gains from HS-SFN for the range of offered traffic load and for different channel types. It is worth mentioning that performance results can be improved by further increasing the amount of HS-SFN candidate UEs by adjusting the softer handover threshold or by using different network layout, such as 6 sectors per site. At the same time, the results can also decline if we consider imperfect channel estimation or legacy UEs incapable of utilizing HS-SFN. Also, UEs with receivers, which can handle efficiently the inter-cell interference, will benefit less from the HS-SFN technique. All these aspects would be good subjects for further considerations and research.

6. ACKNOWLEDGMENTS

The authors would like to thank Vesa Hytönen and Timo Hämäläinen from University of Jyväskylä, Alexander Sayenko, Thomas Höhne from Nokia Siemens Networks and Thomas Chapman from Roke Manor Research Ltd. for their ideas, comments and other help related to the development of HS-SFN concept and analyzing the results.

7. REFERENCES

[1] 3GPP TR 36.814 v9.0.0, "User equipment (UE) radio transmission and reception (FDD)," Mar 2010.

[2] M. Sawahashi, Y. Kishiyama, A. Morimoto, D. Nishikawa, and M. Tanno, "Coordinated multipoint transmission/reception techniques for LTE-advanced," *Wireless Communications, IEEE*, vol. 17, no. 3, pp. 26 –34, june 2010.

[3] W. Xiaoyi, K. Mingyan, and Q. Xin, "Downlink System Performance of Cooperative Multiple Points Transmission with Realistic RRM Structure," in *Computer Modeling and Simulation, 2010. ICCMS '10. Second International Conference on*, vol. 4, jan. 2010, pp. 467 –470.

[4] S. Brueck, L. Zhao, J. Giese, and M. Amin, "Centralized scheduling for joint transmission coordinated multi-point in LTE-Advanced," in *Smart Antennas (WSA), 2010 International ITG Workshop on*, feb. 2010, pp. 177 –184.

[5] 3GPP TSG-RAN, RP-101439, Nokia Siemens Networks, "Proposed study item on HSDPA multipoint transmission," Dec 2010.

[6] V. Hytönen, O. Puchko, T. Höhne, T. Chapman, "High-speed single-frequency network for HSDPA," in *Communication Technologies Workshop (Swe-CTW), 2011 IEEE Swedish*, Nov. 2011, pp. 1–6.

[7] V.Hytönen, O. Puchko, T. Höhne, T. Chapman, "Introduction of Multiflow for HSDPA," in *IEEE Fifth IFIP International Conferenceon New Technologies, Mobility and Security*, May 2012.

[8] T. Nihtilä, J. Kurjenniemi, M. Lampinen, T. Ristaniemi, "WCDMA HSDPA network performance with receive diversity and LMMSE chip equalization," in *IEEE 16th International Symposium on Personal, Indoor and Mobile Radio Communications, Vol. 2*, Sep. 2005.

[9] P. Darwood, P. Alexander, I. Oppermann, "LMMSE chip equalisation for 3GPP WCDMA downlink receivers with channel coding," *IEEE International Conference on Communications, 2001. ICC 2001*, vol. 5, 2001.

[10] T. Krauss, M. Zoltowski, G. Leus, "Simple MMSE equalizers for CDMA downlink to restore chip sequence: comparison to zero-forcing and RAKE," in *Proc. of 2000 IEEE International Conference on Acoustics, Speech, and Signal Processing*, Jun. 2000, p. 2865.

[11] O. Puchko, M. Zolotukhin, V. Hytönen, T. Höhne, T. Chapman, "Enhanced LMMSE equalizer for high-speed single frequency network in HSDPA," in *Communication Technologies Workshop (Swe-CTW), 2011 IEEE Swedish*, Nov. 2011, pp. 92–97.

[12] D. H. Brandwood, "A complex gradient operator and its application in adaptive array theory," in *Communications, Radar and Signal Processing, IEE Proceedings F*, Nov. 2008, p. 11.

[13] 3GPP TR 25.963 v9.0.0, "Feasibility study on interference cancellation for utra fdd user equipment (UE)," Dec 2009.

[14] J. Peisa, S. Wager, M. Sagfors, J. Torsner, B. Goransson, T. Fulghum, C. Cozzo, and S. Grant, "High speed packet access evolution - concept and technologies," in *Vehicular Technology Conference, 2007. VTC2007-Spring. IEEE 65th*, april 2007, pp. 819 –824.

[15] O. Puchko, M. Zolotukhin, V. Hytönen, T. Höhne, T. Chapman, "Phase adjustment in HS-SFN for HSDPA," in *IEEE Fifth IFIP International Conferenceon New Technologies, Mobility and Security*, May 2012.

[16] 3GPP TR 25.101 v10.0.1, "User equipment (UE) radio transmission and reception (FDD)," Jan 2011.

[17] A. Jalali, R. Padovani and R. Pankaj, "Data throughput of CDMA-HDR a high efficiency-high data rate personal communication wireless system," in *Vehicular Technology Conference Proceedings, 2000*, Tokyo, Japan, May 1997, pp. 1854–1858.

A Lightweight and Accurate Link Abstraction Model for the Simulation of LTE Networks in ns-3

Marco Mezzavilla[†], Marco Miozzo[§], Michele Rossi[†‡], Nicola Baldo[§], Michele Zorzi[†‡]
[†]Dept. of Information Engineering (DEI), Via Gradenigo 6B, 35131 Padova, Italy
[§]Centre Tecnològic de Telecomunicacions de Catalunya (CTTC),
Av. Carl Friedrich Gauss 7, 08860 Castelldefels, Barcelona, Spain
[‡]Consorzio Ferrara Ricerche (CFR), Via Saragat 1, 44122 Ferrara, Italy
{mezzavil,rossi,zorzi}@dei.unipd.it, {mmiozzo,nbaldo}@cttc.es

ABSTRACT

In this work we present a link abstraction model for the simulation of downlink data transmission in LTE networks. The purpose of this model is to provide an accurate link performance metric at a low computational cost by relying solely on the knowledge of the SINR and of the modulation and coding scheme. To this aim, the model combines Mutual Information-based multi-carrier compression metrics with Link-Level performance curves matching, to obtain lookup tables that express the dependency of the Block Error Rate on the SINR values and on the modulation and coding scheme being used. In addition, we propose a 3GPP-compliant Channel Quality Indicator evaluation procedure, based on the proposed Link Abstraction Model, to be used as part of the LTE Adaptive Modulation and Coding mechanisms. Finally, we discuss how these contributions have been tested, validated and integrated in the ns-3 simulator. The link abstraction model described in this paper has been included in the official ns-3 distribution since release 3.14.

Categories and Subject Descriptors

I.6.8 [**Simulation and Modeling**]: Discrete event; I.6.5 [**Model Development**]: Modeling methodologies; C.2.1 [**Computer Systems Organization**]: Computer - Communication Networks - Network Architecture and Design - Wireless communication

General Terms

Algorithms, Design, Performance

Keywords

LTE; OFDM; ns-3; Link Abstraction; BLER; MIESM; AMC.

1. INTRODUCTION

Long Term Evolution (LTE) [21] is today's most advanced cellular network technology, and is expected to be massively deployed in the upcoming years. It is a very versatile system, featuring many possible configurations able to achieve different performance tradeoffs. In the perspective of deploying and managing future LTE networks, operators and equipment vendors are strongly interested in identifying the configurations and solutions that can achieve the best possible performance in a variety of scenarios according to their needs. Ideally, this identification would be carried out experimentally by analyzing measurements obtained from an actually deployed network. However, doing so by either testbed experiments or field trials is a very time consuming and expensive process. Thus, simulation is an appealing alternative to perform some pre-tuning of the selected algorithms and protocols before they are deployed. In this respect, there is a tradeoff between choosing a very accurate simulation model, which typically has a high computational complexity and only allows for the simulation of a few network elements, and a more simplified model, which can scale to larger scenarios but which often has a limited accuracy.

The work that we present in this paper aims at reducing the gap between these two extremes. In detail, our aim is to provide an accurate and, at the same time, computationally lightweight Link Abstraction Model (LAM) for the LTE evolved Universal Terrestrial Radio Access (E-UTRA). This model shall allow the accurate prediction of transport block errors at the MAC layer taking into account channel fluctuations, multi-user interference as well as physical layer configurations (bandwidth assignment, modulation, coding, etc.). Thus, we integrate this model into the ns-3 Network-Level (NL) simulator of LTE, which accounts for multiple UEs and eNodeBs, and allows the simulation of a complete end-to-end LTE system, including architectural components. To this end, one might of course come up with a detailed implementation of the eUTRAN procedures [2] and especially of the LTE PHY (e.g., modeling its operations at the symbol level). However, this would lead to a very complex and computationally demanding model, which would not scale up to the medium and large network sizes typically considered for NL simulation. A more suitable approach, which is the one that we take here, is instead that of performing some offline preprocessing based on Link-Level (LL) simulations, so as to derive a simplified model of the influence of channel and system parameters on the PHY performance, represented by the Transport Block error rate. This pre-encoded mapping allows to retain a good amount of the accuracy of LL simulation when modeling phenomena such as multi-user interference, OFDMA bandwidth allocation and random channel realizations, while not retaining their complexity, thereby allowing for better scalability.

Evaluating the error distribution in OFDMA-based wireless systems is very challenging for a number of reasons. First of all, OFDM transmissions are typically used in scenarios affected by frequency selective fading, meaning that subcarriers may perceive very different channel gains. Besides, OFDMA further increases the system complexity as subcarriers are assigned to different users, whose signals are typically generated with different transmission powers and MCSs. This makes the task of predicting the error distribution per user rather complex, in terms of both collecting a reduced subset of parameters to describe performance trends, and generating a flexible error model in order to cover all possible scenarios. To address this problem, Link-to-System Mapping (LSM) has been previously proposed for use with generic multicarrier systems [11, 19, 23]. The first practical application of the LSM approach to modeling a real wireless technology was the evaluation of the Worldwide Interoperability for Microwave Access (WiMAX) radio technology by the IEEE 802.16 task force; in this context, several LSM techniques were applied and evaluated [24]. On this matter, two extensions of the well known network simulator 2 (ns-2) [5] called WINSE [8] and WiDe [17] had these solutions integrated; however, their code is not publicly available.

More recently, LSM has been extensively investigated for application to the LTE technology. Many papers have leveraged on LSM as part of simulation models aimed at the evaluation of interference management and allocation schemes [9, 16]. However, only a few of them [13, 14] made the simulation tool publicly available. [13] refers to a set of Matlab simulators that aim at providing a comprehensive framework for the simulation of link and MAC layer performance. The design choices of Matlab and the focus on lower layer aspects do not give to this tool the possibility of evaluating complex network scenarios, featuring mobility and traffic constraints. Some of these assumptions have been relaxed in [14], where c++ was adopted as the programming language and some networking functionalities were included. However, this simulator is designed primarily to evaluate MAC-level performance and does not properly model the Evolved Packet Core (EPC), in charge of handling, among other aspects, bearers, their Quality of Service (QoS), and mobility.

Recently, a new module called LTE-EPC Network Simulator (LENA) [4] has been developed for LTE as an extension of the ns-3 simulator [6]. LENA already includes EPC functionalities [10] and is designed in a product oriented fashion (i.e., it implements the Scheduling APIs defined by the Small Cell Forum [22], formerly known as Femto Forum). It is to be noted that LENA has all the advantages of a large open source project, including the support of a lively community for what concerns debugging, validation and maintenance.

The main contribution of this work, which is discussed in the following sections, is the design and implementation of a lightweight link abstraction model for the downlink transmission of LTE systems, and its integration with LENA. In addition, we show how this model can be exploited to design an algorithm for reporting channel quality indicator (CQI) feedback according to the 3GPP guidelines in order to test the online selection of the Modulation and Coding Scheme (MCS) for each user, subject to given BLock Error Rate (BLER) requirements. As a further introductory remark, we would like to note that a version of LENA including the code of the model presented in this paper has been part of the official ns-3 distribution since release 3.14.

2. THE PROPOSED ERROR MODEL

2.1 Overview

For an accurate evaluation of the user's performance, besides the MCS assigned by the LTE scheduler, it is important to track the residual errors, i.e., after link layer processing, that are due to channel phenomena such as fading, multiple-access interference, etc. However, as discussed above, a comprehensive simulation of link layer procedures would entail a high computational complexity, which is undesirable for multi-user scenarios. With BLER we refer to the residual error rate after all PHY layer procedures, i.e., affecting the code blocks at the output of the turbo decoder at the receiver side. The physical layer model of the LTE simulator returns SINR values $SINR_1, SINR_2, \ldots, SINR_N$ for each Resource Block (RB) $n \in \{1, 2, \ldots N\}$ for all users, calculated using an AWGN model and a Gaussian interference model. We recall that a RB corresponds to the allocation quantum in LTE, and is composed of 12 sub-carriers (15 kHz each) and 14 OFDM symbols, transmitted over a Time Transmission Interval (TTI) of 1 ms.

Following the Mutual Information Effective SINR Mapping (MIESM) method [12], for the transmission of each block we pick the instantaneous SINR vector $(SINR_1, SINR_2, \ldots, SINR_N)$ and map it onto a Mean Mutual Information per coded Bit (MMIB) metric. The obtained MMIB is a time-varying compressed representation of the channel quality as perceived by any given user at any given time.

In addition, we store offline calculated curves returning the BLER as a function of the SINR for each valid (MCS, CB_{size}) pair, where MCS is a modulation and coding scheme and CB_{size} represents the code block size. These curves have been obtained with the Vienna link level simulator [7, 18].

Finally, this offline calculated SINR to BLER mapping is utilized, together with the instantaneous MMIB information, to obtain the BLER traces for each user. This procedure is explained in greater detail in the following Section 2.2.

2.2 Effective SINR Mapping model

The link level simulations executed to build our abstraction model assume a frequency flat channel response at any given SINR (the so-called AWGN channel). Let us consider a given LTE user and let $SINR_n$ be the instantaneous SINR value associated with RB n, where $n = 1, 2, \ldots, N$ and N is the number of RBs allotted to this user.

Given this, let us assume that the simulator returns an instantaneous SINR sample for each RB, which means a vector $(SINR_1, SINR_2, \ldots, SINR_N)$. In order to obtain a lightweight and effective mapping from this vector to a single BLER metric we consider the *effective SINR mapping* (ESM) method, see [3]. Briefly, the instantaneous SINR vector is mapped onto a single scalar value as follows [19]:

$$eSINR = \alpha_1 I^{-1} \left(\frac{1}{N} \sum_{n=1}^{N} I \left(\frac{SINR_n}{\alpha_2} \right) \right) , \qquad (1)$$

where $I(\cdot)$ represents the *information measure* function, $I^{-1}(\cdot)$ is its inverse, whereas α_1 and α_2 are two scaling parameters that are tuned as a function of the selected MCS. These parameters are specified in Section 2.3 along with function $I(\cdot)$.

The results provided in [12] demonstrate that the MIESM method outperforms all the other mapping approaches in terms of approximation accuracy for the BLER curves. Thus, we adopted the Mutual Information (MI) metric for our implementation.

2.3 MIB Mapping

The MIB is defined in [3] as the mutual information between the bit input belonging to a specific constellation (MCS), and the corresponding log-likelihood ratio (LLR) output at the receiver.

As reported in [3], it can be approximated through the following function:[1]

$$J(t) = \begin{cases} 0, & t < 0.001 \\ a_1 t^3 + b_1 t^2 + c_1 t, & 0.001 \leq t < 1.6363 \\ 1 - e^{(a_2 t^3 + b_2 t^2 + c_2 t + d_2)}, & 1.6363 \leq t \leq 50 \\ 1, & t > 50 \end{cases}, \quad (2)$$

where the parameters have been obtained through numerical fitting and are reported in the following Table 1.

$a_1 = -0.04210661$	$a_2 = 0.00181492$
$b_1 = 0.209252$	$b_2 = -0.142675$
$c_1 = -0.00640081$	$c_2 = -0.0822054$
–	$d_2 = 0.0549608$

Table 1: J-function approximation parameters

Specifically, it has been demonstrated [3] that the MIB of any modulation m can be approximated as a mixture of $J(\cdot)$ functions as follows:

$$I_m(x) \simeq \sum_{k=1}^{K} \alpha_k J(\beta_k \sqrt{x}) \quad (3)$$

where $\sum_{k=1}^{K} \alpha_k = 1$ for some $K \geq 1$ and the argument x is the SINR associated with the transmission channel under study. Numerical fittings have been carried out (see again [3]) to obtain K, α_k and β_k for QPSK, 16-QAM and 64-QAM, as reported in the following Table 2.

Modulation m	**MIB function** $I_m(x)$
QPSK	$J(2\sqrt{x})$
16-QAM	$\frac{1}{2}J(0.8\sqrt{x}) + \frac{1}{4}J(2.17\sqrt{x}) + \frac{1}{4}J(0.965\sqrt{x})$
64-QAM	$\frac{1}{3}J(1.47\sqrt{x}) + \frac{1}{3}J(0.529\sqrt{x}) + \frac{1}{3}J(0.366\sqrt{x})$

Table 2: Numerical approximations for MIB mapping

As above, with $SINR_n$ we mean the instantaneous SINR value associated with RB n, where $n = 1, 2, \ldots, N$ and N is the number of RBs allotted to the user. According to the above discussion, the function $I_m(x)$ can be used to map $SINR_n$ onto the corresponding mutual information domain, where m is the adopted modulation scheme. Note that the argument x corresponds to $SINR_n$ and in LTE, for each sub-frame, the same modulation is picked for all RBs. Given all that, the Mean Mutual Information per coded Bit (MMIB) can be obtained as follows:

$$MMIB = \frac{1}{N} \sum_{n=1}^{N} I_m(SINR_n), \quad (4)$$

where N is the number of RBs assigned to a specific user and m is the modulation that this user is exploiting. To sum up, the model starts by evaluating the mutual information value for each RB from the corresponding SINR samples. Subsequently, the MMIB is computed by averaging (effective SINR mapping) the corresponding mutual information values as per Eq. (4).

2.4 BLER prediction

The data at the MAC layer of the LTE protocol stack (right above the LTE PHY) is arranged in Transport Blocks (TB), whose size depends on the specific configuration of the underlying PHY. TBs are split into a number of CBs which are independently encoded by the turbo encoder at the PHY layer. Each CB is then encoded and transmitted over the channel exploiting the N RBs allotted to the user. In this section we show how to efficiently compute the Transport BLock Error Rate (TBLER) from the results of Section 2.3.

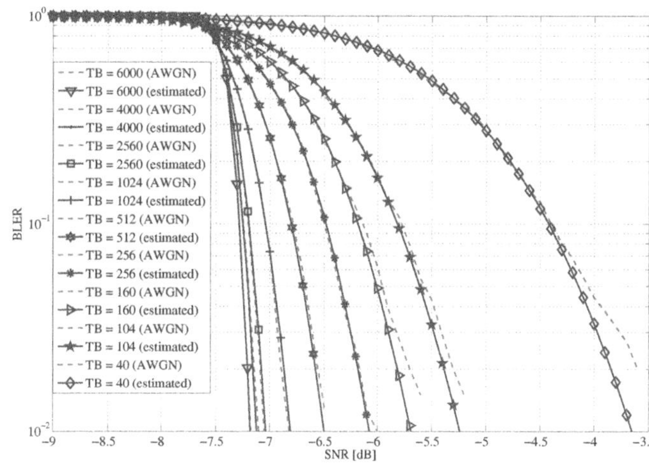

Figure 1: BLER vs SNR for MCS 1

For the moment, let us focus on the i-th CB of a given TB. As mentioned in Section 2.1, link-level simulations (whose results were obtained using the Vienna LL simulator) have been used to obtain the PHY layer performance in terms of BLER vs SINR over AWGN channels, accounting for the configuration of the PHY layer turbo encoder in terms of Code Block (CB) length and selected MCS. The 3GPP standard has been considered to assess the correct CB sizes in the simulations, according to [2]. As an example, the dotted lines in Fig. 1 show the BLER as a function of SINR for MCS 1. These curves have been calculated offline considering the LTE PHY layer procedures implemented in the LTE Downlink LL Vienna Simulator [7], as described in [15].

As can be seen from these plots, the CB size highly impacts the actual BLER performance for a given MCS.

As mentioned above, the selected CB i is transmitted over the channel using the N RBs that are assigned to the user. At the receiver side, a reference SINR value[2] is made available by the PHY layer model of the ns-3 simulator for each of these RBs, returning the SINR vector $(SINR_1, SINR_2, \ldots, SINR_N)$, as discussed in Section 2.1. From here, we obtain the MMIB metric using Eq. (4), as explained in Section 2.3. This MMIB corresponds to an equivalent SINR for the transmission of CB i over the allotted RBs. As a last step to obtain the residual error rate of CB i, we need to map its MMIB onto the corresponding BLER, which is referred to here as $CBLER_i$. This is done according to the following procedure.

In order to reduce the computational burden at simulation time as much as possible, an approximation based on the Gaussian cumulative model has been adopted. According to this, the estimated BLER curves as a function of MMIB are parameterized as follows:

$$CBLER_i(x) = \frac{1}{2}\left[1 - \text{erf}\left(\frac{x - b_{S,M}}{\sqrt{2}c_{S,M}}\right)\right], \quad (5)$$

where $b_{S,M}$ and $c_{S,M}$ are the mean and the standard deviation

[1]Note that, compared to [3], we have truncated the function to avoid that it takes values outside the [0,1] interval.

[2]We assume no frequency selectivity among the 12 sub-carriers composing the resource block.

of the Gaussian cumulative distribution, respectively, and x is the MMIB associated with CB i. S is the code block size and M is the MCS, which dictates the actual transmission rate. What we did at this point, was to find suitable pairs $(b_{S,M}, c_{S,M})$ for each MCS and code size. We did so through numerical fitting so that the curves from Eq. (5) would match those obtained from the Vienna LL simulator. The result of this procedure is shown in Fig. 1, where the solid curves represent the result of Eq. (5) where we have used the best fitting $(b_{S,M}, c_{S,M})$ pair for each MCS and code size. As can be seen from these plots, the approximated BLER from Eq. (5) (solid lines) closely match the BLER obtained through the numerical simulation of LTE PHY procedures (dotted lines). The overall Transport BLock Error Rate (TBLER) is thus found as:

$$TBLER = 1 - \prod_{i=1}^{C}(1 - CBLER_i), \qquad (6)$$

where C is the number of CBs contained in the TB.

Lookup tables: to limit the computational complexity and the memory space taken by the proposed link abstraction model, we only considered a subset of CB sizes, i.e., $\mathcal{S} = \{40, 104, 160, 256, 512, 1024, 2560, 4000, 6000\}$ bits. This choice is aligned with the typical performance of turbo codes, where large CB sizes do not strongly affect BLER performance. However, we note that for CB sizes smaller than 1000 bits, the BLER performance might significantly differ as we vary the block size (up to nearly 3 dB). Therefore, we accounted for an unbalanced quantization of CB sizes in order to get more accuracy in the critical zone (small code blocks). This is particularly evident from Fig. 1 that shows a similar BLER profile for large CB sizes (e.g., 2500, 4000 and 6000 bits), whereas the performance gap increases as the CB size gets smaller. Thus, $(b_{S,M}, c_{S,M})$ parameters have been tabulated for all valid combinations of MCS and block sizes in set $\mathcal{S}$. We remark that high MCS values with high order modulations and efficient coding rate schemes, such as 64-QAM with an Effective Coding Rate (ECR) of 0.92 (i.e., MCS 29), allow for a minimum CB size of 2560 bits. The latter is much larger than the minimum size at small MCS values, e.g., MCS 1, where the minimum size is 40 bits, see Fig. 1. This reflects the fact that turbo coding offers better performance as the code block size increases; thus, for high order modulations such as MCS 29, small code block lengths are inefficient as the resulting BLER performance is unacceptable.

3. LINK ADAPTATION IMPROVEMENT

In this section we propose an improved MCS assignment scheme supported by a new CQI evaluation scheme based on 3GPP guidelines. Note that this novel algorithm for CQI evaluation could not be tested in the previous ns-3 distribution as it is based on residual error estimates.

Link adaptation plays a fundamental role in modern wireless communications systems, which need to face issues such as strong interference from multiple users and their mobility, which makes the wireless channel frequency selective. These facts are coped with by LTE adaptive modulation and coding algorithms. Focusing on the downlink scenario, AMC has the role of tracking the perceived SINR and sending back to the base station (eNodeB) a so called CQI report. Hence, periodically, the UE reports to the eNodeB a single CQI value for all the RBs (the so called *wideband* CQI). This information is a "compressed" representation of the quality experienced by the UE in a specific sub-frame and is used at the base station side for the selection of the MCS. This process is continuously executed so as to adapt to channel and network dynamics.

Our proposed MCS assignment scheme relies on an SINR to CQI mapping approach based on the link error abstraction model presented in the previous section. As a competing approach we consider the algorithm that is currently implemented in the LENA ns-3 simulator, which is inspired by the *spectral efficiency* concept, see also [20].

Spectral efficiency-based approach: consider the generic RB n, and let $SINR_n$ be the corresponding SINR value, in linear units. We obtain the spectral efficiency η_n of RB n using the following equations:

$$\Gamma = -\frac{\ln(5\,\mathrm{BER})}{1.5}, \qquad (7)$$

$$\eta_n = \log_2\left(1 + \frac{SINR_n}{\Gamma}\right), \qquad (8)$$

where BER is the Bit Error Rate and Γ is the so called *SNR gap*, as it models the discrepancy between practical implementations and information-theoretic results.

Upon the calculation of η_n, which lies in the continuous interval $[0.15, 5.55]$, the procedure described in [1] is used to derive the corresponding CQI, which is a quantized version of η_n.

Error model-based approach: this model relies on the exploitation of our link abstraction model. Thanks to this approach, we can dynamically select the MCS that better complies with a given target transport block error rate for the connection, referred to as $TBLER_{\mathrm{th}}$. In the following, we describe our improved CQI evaluation procedure by abstracting away from the actual implementation details, i.e., on the actual representation of CQI values (at the receiver, e.g., number of CQI levels, etc.) and the subsequent mapping of these CQIs onto a suitable MCS (which is done at the eNodeB).

Our procedure works as follows. Periodically, each UE computes its received power spectrum profile, i.e., an SINR sample is acquired for each possible RB.[3] For any user $i = 1, 2, \ldots, N_{UE}$, it starts from MCS 29, which corresponds to the most aggressive transmission scheme, and evaluates the TBLER performance considering a transport block composed of all possible LTE RBs. The transport block error rate for user i, $TBLER_i$, is estimated through Eq. (5), taking as input the vector of SINRs for the selected user, and the MCS that we are currently evaluating. If $TBLER_i$ is larger than or equal to the target BLER defined by 3GPP (i.e., 0.1) [2], we keep on searching for a better (less aggressive) MCS; otherwise, the procedure stops. Thus, the corresponding CQI is obtained in order to satisfy the spectral efficiency constraints defined by the standard [2].

4. SIMULATION RESULTS

In the following we provide some technical results for selected LTE scenarios. Our main goals are: 1) to illustrate the usability of the proposed link abstraction model, and 2) to prove the efficiency of the link adaptation improvement proposed in Section 3. We would like to note that the validation of the error model proposed is simplified for the sake of readability; a more in depth validation can be found in the LENA documentation [4]. In Table 3 we report the considered system parameters.

First of all we would like to present some quantitative results on the computational complexity of the model. The simulator presented in this paper takes 23 seconds for simulating a 10 seconds scenario with one UE transmitting continuously at full bandwidth

[3]In this case, all RBs allowed by the selected LTE channel bandwidth are accounted for.

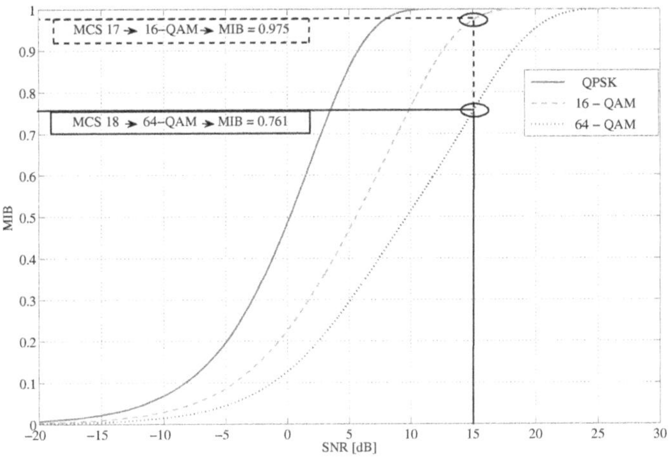

Figure 2: MI extraction

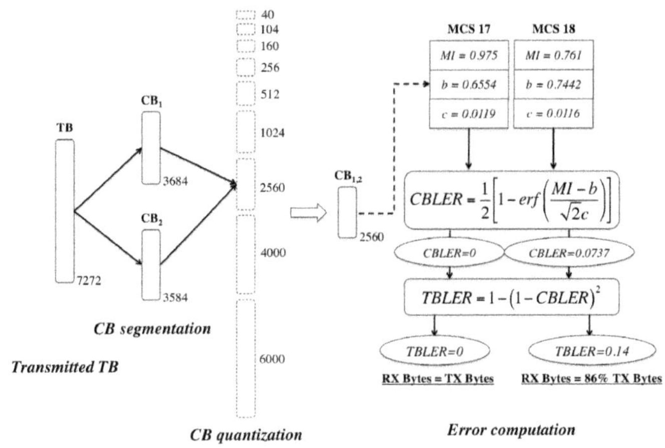

Figure 3: TBLER computation

PHY Configuration	
Parameter	**Value**
Frequency	2.1 GHz
Channel Bandwidth	5 MHz
Propagation Model	Friis free-space
Number of RBs	25
$RB_{bandwidth}$	180 kHz
$RB_{subcarriers}$	12
$RB_{OFDMsymbols}$	14
eNodeB TX power	30 dBm
Noise spectral density (N_O)	-174 dBm/Hz

Table 3: Main system parameters

to an eNB, while the same scenario takes 38 minutes when simulated with the LTE Vienna Link Layer simulator.[4]

1) Error model: we consider a downlink transmission from an eNodeB to a single static UE. For the wireless channel, we account for a Friis free-space propagation model, but note that the conclusions that we draw here are general and apply to more sophisticated models as well. The UE is placed 2150 meters away from the eNodeB and, according to the considered propagation loss model, it experiences an SINR of 15.1 dB for all its RBs.

We first evaluate the BLER performance resulting from the selection of a "safe" transmission scheme, MCS 17, which corresponds to 16-QAM (4 bits per OFDM symbol). As we now show, for this MCS the estimated error rate (through Eq. (5)) is below the standard TBLER threshold of $TBLER_{\rm th} = 0.1$. First of all, we extract the *mutual information* value associated with the experienced SNR, as shown in Fig. 2 (where we plot the approximation functions of Table 2). According to [2], from the selected MCS and the maximum number of assignable RBs, the TB size is 7272 bits (including the header). Following [2], the TB is split into two code blocks, CB_1 and CB_2, of size 3684 and 3584 bits, respectively. As shown in Fig. 3, these code sizes are mapped onto the closest CB size in set S. In fact, as per our discussion in Section 2.4, in our ns-3 simulator fitting parameters are only stored for a subset of all possible CB sizes. Thus, the resulting CB size that will be used for the prediction of the CBLER performance is 2560 bits. Now, using the $b_{S,M}$ and $c_{S,M}$ parameters associated with the latter code block size and the previously extracted MI with Eqs. (5) and (6), we obtain an estimated transport block error rate of $TBLER = 0$.

Next, we try to allocate a more aggressive modulation and coding scheme, MCS 18, for which the modulation order amounts to 6 bits per OFDM symbol (64-QAM). Thus, we repeat the procedure illustrated in the previous paragraph, obtaining the mutual information MI, and the fitting parameters $b_{S,M}$ and $c_{S,M}$. These quantities, together with Eqs. (5) and (6), return $TBLER = 0.14$, which means that MCS 18 is not compatible with the considered error requirements.

2) Link adaptation: we now consider a scenario with a single UE, and we vary its distance from the eNodeB. This leads to SNR values ranging from about 2 to 30 dB. Also, we consider the standard target transport block error rate of $TBLER = 0.1$. In Fig. 4, we show the effective spectral efficiency as a function of the SNR for the spectral efficiency-based (SE_MCS) and the error-based (EM_MCS) MCS selection schemes. The effective spectral efficiency metric reflects the actual bits per second per unit of frequency that are successfully transmitted from the eNodeB to the UE, by also accounting for the residual error after PHY layer processing. As can be seen from Fig. 4, EM_MCS outperforms the current approach, at all SNR levels. This indicates that SE_MCS tends to be too conservative, even though a more aggressive technique can be used while still adhering to the target error requirements.

5. CONCLUSIONS & FUTURE WORK

In this work we introduced a link abstraction model for the simulation of Transport Block errors in the downlink of an LTE system. Our objective here has been to provide a lightweight but still accurate procedure for the computation of the residual errors, after PHY layer processing, without having to go through the detailed simulation of LTE PHY procedures. Toward this end, we combined Mutual Information-based multi-carrier compression metrics with Link-Level performance curves matching. This allowed us to obtain pre-calculated lookup tables, which can be used in an online fashion to track residual bit errors after physical layer modulation and coding procedures. In addition, we have proposed a CQI evaluation procedure which can be used as part of the LTE Adaptive Modulation and Coding scheme, showing its superiority in terms of achievable spectral efficiency with respect to current ns-3 solutions. An implementation of both the link abstraction model and

[4]The reference hardware platform is an Intel Core2 Quad CPU Q8400 2.66GHz.

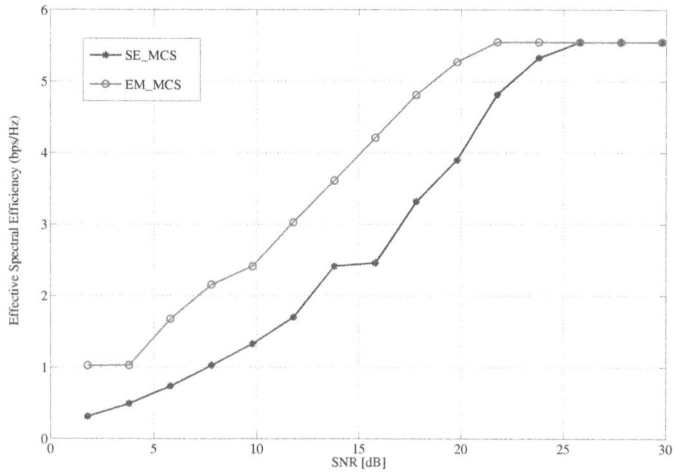

Figure 4: MCS assignment comparison

the CQI evaluation procedure is included with the official distribution of the ns-3 simulator since release 3.14, which is open source and publicly available. As future work we plan to investigate algorithms for resource allocation, especially targeting the transmission of streaming flows over LTE networks.

6. ACKNOWLEDGMENTS

The research leading to these results has received funding from the European Community Seventh Framework Programme (FP7-ICT-2009-5) under grant agreement n. 258053 (MEDIEVAL project). The work at CTTC was partially supported by Ubiquisys as part of the LENA project, by the Spanish Ministry of Science and Innovation under grant number TEC2011- 29700-C02-01 (project SYMBIOSIS), and by the Catalan Regional Government under grant 2009SGR-940.

7. REFERENCES

[1] 3GPP R1-081483, "Conveying MCS and TB size via PDCCH".

[2] 3GPP TS 36.213, "E-UTRA Physical layer procedures".

[3] IEEE 802.16 Broadband Wireless Access Working Group, Evaluation Methodology Document (EMD).

[4] The LTE-EPC Network Simulator (LENA) project, http://iptechwiki.cttc.es/LTE-EPC_Network_Simulator_(LENA).

[5] The Network Simulator - ns-2, http://www.isi.edu/nsnam/ns/.

[6] The Network Simulator - ns-3, http://www.nsnam.org/.

[7] The Vienna LTE Simulators, http://www.nt.tuwien.ac.at/about-us/staff/josep-colom-ikuno/lte-simulators/.

[8] A. Sayenko, O. Alanen, H. Martikainen, V. Tykhomyrov and O. Puchko. WINSE: WiMAX NS-2 Extension. In *ICST Simulation Tools and Techniques (SIMUTools)*, March 2009.

[9] B. Sadiq, R. Madan and A. Sampath. Downlink Scheduling for Multiclass Traffic in LTE. In *EURASIP Journal on Wireless Communications and Networking*, March 2009.

[10] N. Baldo, M. Requena-Esteso, J. Nin-Guerrero, and M. Miozzo. A new model for the simulation of the LTE-EPC data plane. In *ICST Workshop on ns-3 (WNS3)*, March 2012.

[11] Y. Blankenship, P. Sartori, B. Classon, V. Desai, and K. Baum. Link error prediction methods for multicarrier systems. In *IEEE Vehicular Technology Conference (VTC)*, September 2004.

[12] K. Brueninghaus, D. Astely, T. Salzer, S. Visuri, A. Alexiou, S. Karger, and G.-A. Seraji. Link performance models for system level simulations of broadband radio access systems. In *IEEE Personal, Indoor and Mobile Radio Communications (PIMRC)*, September 2005.

[13] C. Mehlfuhrer, J. Colom Ikuno, M. Simko, S. Schwarz, M. Wrulich and M. Rupp. The Vienna LTE simulators: Enabling reproducibility in wireless communications research. In *EURASIP Journal on Advances in Signal Processing*, July 2011.

[14] G. Piro, L.A. Grieco, G. Boggia, F. Capozzi and P. Camarda. Simulating LTE Cellular Systems: An Open-Source Framework. In *IEEE Transactions on Vehicular Technology*, pages 498–513, February 2011.

[15] J. Colom Ikuno, M. Wrulich and M. Rupp. Performance and Modeling of LTE-HARQ. In *ITG Workshop on Smart Antennas (WSA)*, February 2009.

[16] J. Fan, Q. Yin, G.Y. Li, B. Peng and X. Zhu. MCS Selection for Throughput Improvement in Downlink LTE Systems. In *International Conference on Computer Communications and Networks (ICCCN)*, August 2011.

[17] M. Miozzo and F. Bader. Accurate Modelling of OFDMA Transmission Technique using IEEE 802.16m Recommendations for WiMAX Network Simulator Design. In *ICST Conference on Mobile Networks and Management (MONAMI)*, September 2009.

[18] C. Mehlfuhrer, M. Wrulich, J. C. Ikuno, D. Bosanska, and M. Rupp. Simulating the Long Term Evolution Physical Layer. In *European Signal Processing Conference (EUSIPCO 2009)*, August 2009.

[19] M. Moisio and A. Oborina. Comparison of Effective SINR Mapping with Traditional AVI Approach for Modeling Packet Error Rate in Multi-State Channel. In *Next Generation Teletraffic and Wired/Wireless Advanced Networking (NEW2AN)*, June 2006.

[20] H. Seo and B. G. Lee. A proportional-fair power allocation scheme for fair and efficient multiuser OFDM systems. In *IEEE Global Telecommunications (GLOBECOM)*, December 2004.

[21] S. Sesia, I. Toufik, and M. Baker. *LTE: The UMTS Long Term Evolution - from Theory to Practice*. John Wiley & Sons, second edition, August 2011.

[22] SmallCell-Forum. LTE MAC Scheduler Interface Specification v1.11, October 2010.

[23] E. Tuomaala and H. Wang. Effective SINR approach of link to system mapping in OFDM/multi-carrier mobile network. In *Mobile Technology, Applications and Systems*, November 2005.

[24] WiMAX Forum. WiMAX System Evaluation Methodology Document. Version 1.0, January 2007.

An Analytical Model for Content Dissemination in Opportunistic Networks using Cognitive Heuristics*

Raffaele Bruno, Marco Conti, Matteo Mordacchini and Andrea Passarella
IIT–CNR, Pisa, Italy
{r.bruno,m.conti,m.mordacchini,a.passarella}@iit.cnr.it

ABSTRACT

When faced with large amounts of data, human brains are able to swiftly react to stimuli and assert relevance of discovered information, even under uncertainty and partial knowledge. These efficient decision-making abilities rely on so-called *cognitive heuristics*, which are rapid, adaptive, light-weight yet very effective schemes used by the brain to solve complex problems. In a content-centric future Internet where users generate and disseminate large amounts of content through opportunistic networking techniques, individual nodes should exhibit those properties to support a scalable content dissemination system. We therefore study whether such cognitive heuristics can also be used in such a networking environment. To this end, in this paper we develop an analytical model that describes a content dissemination mechanism for opportunistic networks based on one such heuristics, known as the recognition heuristic. Our model takes into account the different popularities of content types, and highlights the impact of the shared memory contributed by individual nodes to make the dissemination process more efficient. Furthermore, our model allows us to investigate the performance of the dissemination process for very large number of nodes, which might be very difficult to carry out through a simulation-based study.

Categories and Subject Descriptors

C.2.1 [**Network Architecture and Design**]: Wireless communication

Keywords

Opportunistic Networks, Cognitive heuristics, Data Dissemination, Analytical Model

1. INTRODUCTION

In the Future Internet scenario, the active participation of users in the production and diffusion of content, using mobile devices in connection with more traditional CDNs and P2P networks, will create a very large and dynamic information environment [15]. Opportunistic networking techniques will become a very important complement to infrastructure-based networks supporting mobile users (such as cellular and WiMAX) in order to efficiently disseminate content to interested users [16]. As mobile devices carried by users will have an active role in the data dissemination process, and considering the large volume of dynamic information that will be produced, devices will be faced with the very challenging task of determining the relevance of discovered content and selecting the most interesting data for the user. In order to be effective, this task should be performed rapidly. Furthermore, devices will have to make their decisions using only a partial knowledge of the environment and of the available content.

In this paper, we focus on a new (to the best of our knowledge) direction in the autonomic networking field, and we exploit results coming from the cognitive psychology area, by using models of how the human brain assesses the relevance of information under partial knowledge. In fact, the human brain is able to achieve effective decision-making results in front of partial information, and the presence of noisy data by exploiting cognitive processes known as *cognitive heuristics*. These heuristics are the key psychological tools that allow the brain to come up with a highly effective decision under conditions of uncertainty, limited time, knowledge, and computational capabilities. Moreover, they are adaptive tools that are able to refine their effectiveness through a continuous interaction with the environment. One of the most simple and effective heuristics is known as the *recognition heuristic*. The role of this heuristic is to assess the relevance of items when choosing among a pair (or a set) of objects. More specifically, it works by assuming that given a pair of objects, if one is recognized (i.e. the brain is able to recall that it already "heard" about that object) and the other is not, the recognized object has a higher value with respect to a given (possibly unknown) evaluation criterion. The simplicity of this heuristic, its effectiveness in decision making processes, and its adaptivity, make it a suitable candidate for embedding a human cognitive process into autonomic, self-aware ICT devices dealing with a content-centric Internet.

We have designed an algorithm for data dissemination in opportunistic networking environments that is inspired by the recognition heuristic. It allows us to simplify and limit the complexity of the data selection task in an opportunistic network, while maintaing a high effectiveness in the data dissemination process. The full specification

*This work is funded by the EC under the FET RECOGNITION (FP7-IST 257756) and EINS (FP7-FIRE 288021) projects.

of this algorithm and preliminary results based on simulations are shown in [3]. In this paper, we first summarize this algorithm. Then, in Section 5 we describe our analytical model of the content dissemination mechanism. This model takes into account the different popularities of content types, and highlights the impact of the shared memory contributed by individual nodes to make the dissemination process more efficient. In Section 6 we present a comparison between the model predictions and related simulation results, showing that the model is able to capture the transient and steady-state behaviors of the data dissemination process. Furthermore, our model allows us to explore network scenarios with large numbers of users and data items, which are generally difficult to study using simulation tools. Finally, Section 2, 3, and 7 complete the paper by summarizing related work, describing the main features of the recognition heuristic, and drawing the main conclusions of this work, respectively.

2. RELATED WORK

The first work that investigated the problem of content dissemination in opportunistic networks was developed in the PodNet Project [10]. Each node in the PodNet system is subscribed to a channel and devote part of its memory space for storing data items belonging to the channel it is subscribed to, and part for supporting a collaborative exchange of information. More advanced approaches exploit information about users social relationships to drive the content dissemination process [19, 4, 1]. Among them, the more important for our work is ContentPlace [1], where content dissemination is driven by the social structure of the network of users. Nodes evaluates which data items to store taking into account the social utility of the items, i.e. how they are likely of interest to users the node has social relationships with (and which, therefore, are expected to be in touch in the near future).

With respect to these approaches, in this paper we take a completely new direction, by borrowing models of human cognitive processes coming from the cognitive psychology domain. As this approach is still totally unexplored, in this paper we limit the set of contextual information that we use to the very minimum, and, for example, we do not exploit information about users social structures. This allows us to obtain initial exploratory results about the feasibility of this novel approach.

3. THE RECOGNITION HEURISTIC (RH)

Cognitive heuristics can be defined as simple rules used by the brain for facing situations in which people have to act quickly, relying on a partial knowledge of all the problem variables, the evaluation criterion of the different possible choices is not known, and the problem itself may be ill-defined in such a way that traditional logic and probability theory are prevented to find the optimal solution. One of the simplest and very popular cognitive heuristics is the *recognition heuristic* [7, 5], which is based on a very simple rule. When evaluating a couple of objects, and one is *recognized* (i.e. the brain is able to recall that it already "heard" about that object) and the other is not, this heuristic infers that the recognized object has a higher value with respect to a given evaluation criterion. The recognition heuristic is effective when the recognition of objects is highly correlated with the ideal evaluation criterion that should be used to select among the possible choices if complete information would be available.

To better understand how the recognition heuristic works, the foundational study on recognition heuristic [5] uses, as an example, the estimation of the university endowments. The evaluation criterion to be used in the example is the value of the endowment. This information is generally not publicly available. Nevertheless, it is argued that newspapers could act as mediators, since they periodically publish news related to the most important universities. Thus, the number of times a university appears on the newspapers could be a strong indicator that it has larger endowments than universities that do not or rarely appear on the media. When a person has to choose which university has the biggest endowment between a couple of institution names, he uses the recognition heuristic and chooses a recognized name against an unknown one. From this example it is straightforward to notice that the effectiveness of the recognition heuristic, i.e. the *recognition validity*, is continuously reinforced by the stimuli received from the environment.

The simplicity of the recognition heuristic has made it a powerful tool in support of decision-making processes. As such, it has been successfully used in various fields [12], like financial decision-making processes [13], the forecast of future purchase activities [8], the results of sport events [18], or political election outcomes [11].

4. AN RH-BASED DATA DISSEMINATION SCHEME

In this study we consider a network scenario where mobile nodes generate data items and other nodes can be interested in those items. More precisely, users are interested in data channels, i.e. high-level topics to which the data items belong. The goal is to bring all the data items of a given channel to all the nodes that are interested in it. To this end, each node contributes a limited amount of storage space to help the dissemination process, since *contacts* between users are the only way to disseminate data items. Ideally, upon encountering another node, a node should select which data items to fetch and keep in its storage space, such that the total *utility* of its local storage space for the overall dissemination process is maximized. Solving this problem exactly would clearly require global knowledge and central coordination, which is unfeasible in opportunistic networks. Therefore, the algorithm we have originally proposed in [3] (and summarized in this section) exploits the recognition heuristic to efficiently approximate the utility of storing the data items that can be exchanged during a pair-wise contact between two nodes. Based on these approximated utility values, each node stores only the most useful data items, selected among both those already available locally and those available on the encountered node, until its storage space is full.

Assuming that each individual node is interested in only one channel, nodes exchange only minimal information when meeting: the channel they are interested into and a summary of the data items they currently store. Using this information a node estimates how many users are interested in the channels of discovered items and how those items are spread in the network. At a very high level, nodes decide which items to store based on two simple assumptions: *i*) the more other nodes request a given channel, the more its data items are relevant, *ii*) the less a data item is replicated in the network, the more useful could be to store a copy of that item.

Since in our algorithm there are two types of decisions to be taken to estimate the utility of the data items to fetch (i.e., how relevant a channel is, and how relevant a data item is) we use two distinct recognition heuristics to separately recognize channels and data items. Intuitively, a node recognizes a channel as soon as it becomes "enough popular", i.e., as soon as that node encounters enough other nodes that are interested in the same channel. Hence, if a channel is recognized, it means that several users are interested in it and, thus, it is worth (for the overall community) to circulate its data items. Furthermore, a node recognizes a data item if it is "spread enough", i.e., as soon as it is encountered on at least a given number of other nodes. If a data item is not recognized, it means that only a few users have a copy of it in their memory, so it should be replicated more broadly to increase its diffusion. On the basis of these considerations, a node decides to fetch a data item from the encountered node if: *a*) it recognizes the channel the data item belongs to, and *b*) is does not recognize the data item itself. Finally, to recognize a channel (or a data item) each node maintains a counter that counts how many times it encountered a node that it is interested in that channel (stores that data item). Then a channel (data item) is recognized if this counter reaches a given recognition threshold. Furthermore, the recognition counters is decremented if the channel (data item) is not seen for a while. The exploitation of such "recognition thresholds" is consistent with the cognitive psychology research on how the recognition memory works in the brain (e.g. [17]).

As we have assumed that the shared storage space contributed by the nodes is limited, we have to define an algorithm that exploits the recognition level of channels and data items to select which items to keep (among the ones currently already available at the node or available on the encountered node) in case the storage space is insufficient to take all available data items. This algorithm thus works as a replacement algorithm for data items in the nodes storage space. To this end, we use a *modified Take-the-Best* algorithm [6], taken from the cognitive psychology literature. This algorithm ranks two (or more) options using several steps. On each step a heuristic is applied (e.g. the recognition heuristic), until it reaches the first (*best*) step where the heuristic can discriminate among the options. In our case, nodes decide which data items to store using a three-step version of the Take-the-Best algorithm. The first step applies the recognition heuristic to the channels, the second one uses the recognition of data items, while the last step ranks the items according to their recognition value, as explained below. The data items available on an encountered node (hereafter called "new" items) are evaluated together with the ones already stored in the node (hereafter called "old" items). We first apply the recognition heuristic to the channels, i.e. all the items whose channel is not recognized are discarded. If there is enough space to store all the remaining items, nothing else happens. Otherwise, we apply the recognition heuristic to data items, i.e. non-recognized data items are dropped. In case the number of remaining items still exceeds the size of the node cache, items are ranked according to their recognition level using the following two criteria. First of all, data items with lower recognition values (i.e., whose counter is lower) are more relevant that the ones with higher recognition levels, because the former ones are considered less widespread and, thus, more useful to fetch. Secondly, among the data items with the same recognition value, new ones are more relevant than old ones. Note that the heuristics applied in the subsequent steps are of increasing complexity. This is also a typical cognitive scheme, which tries to use the lower possible amount of resources to discriminate between possible choices.

A more detailed description of the recognition algorithms and of the modified Take-the-Best algorithm is given in [3].

4.1 Node Caches

To implement the above algorithm, each node makes use of several internal caches. We have two classes of caches: data caches and recognition caches.

Data Caches: (i) LI is the cache containing the *Local Items*, i.e. the items generated by the node itself; (ii)SC is the *Subscribed Channel* cache, i.e. the cache containing the items belonging to the channel the node is interested to and obtained by encounters with other peers (we assume that nodes are able to reserve enough space for the channel they are interested in, thus the SC cache is assumed unlimited); (iii) OC is the *Opportunistic Cache*, i.e. the cache containing the objects obtained through exchanges with other nodes and belonging to channels the node is not interested in. This cache is the part of the node's storage space contributed for the overall efficiency of the data dissemination process, beyond the particular interest of the individual node. Therefore, we assume that it has a limited size. Its content is selected using the values contained in the *Recognition* caches.

Recognition caches: (i) CC is the *Channel Cache*: whenever the node meets another peer subscribed to a given channel, the channel ID is put in this cache, along with a counter. The counter is incremented every time a node subscribed to the same channel is found. Note that the storage space required to maintain this information is very limited compared with the size of data items. Thus, the size of this cache is assumed unlimited; (ii) IC is the *Item Cache*: similarly to the previous cache, when a new data item is seen in an encountered node, its ID is put in this cache, along with a counter. The counter is incremented every time the object is seen again. We assume that also this cache has no space limits. Finally, we remind that the CC and IC caches make use of two different *recognition thresholds*, R_c and R, respectively. Items and channels whose counters exceed the value of the threshold are marked as recognized. This classification is used by the OC cache in order to handle its data, using the modified *Take-the-Best* algorithm described before.

5. ANALYTICAL MODEL

This section develops the analytical model describing the temporal evolution of the replication levels for data items belonging to a *tagged cannel c*. More precisely, in our model we use a set of embedded Markov chains to describe the status of the various caches deployed at the mobile nodes of the opportunistic network. Since the status of the caches of a node can change only when it exchanges data items upon encountering another node, the embedding points for our analysis are the time instants at which pairs of nodes meet.

Concerning the number of Markov chains needed in our model and their state space size, we use a distinct Markov chain for each channel to describe the evolution of the channel recognition level in the CC cache of a generic node. Similarly, we use a separate Markov chain for each channel to de-

Symbol	*Definition*
N	Number of nodes of the network
M	Number of items in the network
C	Number of channels in the network
$a.c$	Channel to which data item a belongs
$r_t(a)$	Fraction of nodes storing a copy of a
$Pop(c)$	Number of nodes subscribed to channel c
$\psi_a^t[1](\psi_a^t[0])$	Prob. that data item a is (is not) in SC at time t
$\nu_c^t[i]$	Prob. that channel c has a recognition level i in CC at time t
$v_a^t[i]$	Prob. that data item a has a recognition level i in IC at time t
$\varphi_a^t[i]$	Prob. that data item a is stored in OC at time t with a recognition level i
R_c	Threshold for channel recognition
R	Threshold for data item recognition

Table 1: List of mathematical notations

scribe whether a generic data item of that channel is stored in the SC cache. For the IC and OC caches, we consider a generic data item for each channel, and a Markov chain describes (i) the evolution of the recognition level of that data item in IC; and (ii) whether the data item is stored in OC and in which position, respectively. It is important to observe that in this study we consider a homogeneous network scenario, where each node meets the other nodes in the network with equal probability at an encounter event[1]. Therefore, it is sufficient to derive the steady state probabilities for the aforementioned Markov chains by considering a single *tagged node*, since the same average behavior is observed in all the other nodes. In conclusion, our model requires to track the evolution of four Markov chains per channel. As channels (data types) can be reasonably assumed to be much less than data items, the complexity of the model is reasonable.

The goal pursued by our analysis is to derive the fraction of nodes in the network that, at time t, have a copy of a data item a belonging to channel c either in SC or OC, hereafter denoted as $r_t(a)$. In fact, the knowledge of $r_t(a)$ is enough to fully characterize the dissemination process of the data items in the opportunistic network. In the following sections, we will describe how the state probabilities of the Markov chains describing the cache status are used to derive $r_t(a)$ for each channel. In Tab. 1 we list the main notations used in our analysis.

5.1 Subscribed Channel Cache

Let us consider a generic data item a belonging to the channel c to which the tagged node is subscribed. Then, in the Markov chain describing the status of the SC cache, state 1 is used to indicate that a copy of data item a is stored in SC, while state 0 indicates that SC does not hold a copy of a. Let us assume that at time $t+1$ the tagged node encounters another node, while the previous encounter event occurred at time t. Intuitively, the probability $p_{0,1}$ to move from state 0 to state 1 at time $t+1$, i.e., the probability that a copy of data item a is fetched by the tagged node from the node encountered at time $t+1$, is given by $p_{0,1} = r_t(a)$. On the other hand, the probability $p_{0,0}$ that a data item not present in SC at time t is also not present at time $t+1$ can be computed as $p_{0,0} = 1 - p_{0,1}$. Since we have assumed that the SC cache is large enough to contain all the data items of the channel the tagged node is subscribed to, a data item cannot be dropped by the SC cache. Formally, this implies that $p_{1,1} = 1$. Now, let $\psi_a^t = \{\psi_a^t[0], \psi_a^t[1]\}$ be the state vector of the Markov chain at time t, and let us denote with $P = \{p_{i,j}\}^2$, $i, j = 0, 1$, its transition matrix evaluated at time $t+1$. Then, we can compute the state vector at time $t+1$ as follows

$$\psi_a^{t+1} = \psi_a^t P \; . \tag{1}$$

The initialization distribution of the state vector is $\psi_a^0 = \{1, 0\}$, since the SC is empty at the beginning of the system.

5.2 Channel Recognition

As described in Sec. 4.1, CC stores the number of times each channel has been seen by the tagged node. Thus, the state i for the Markov chain used to describe the status of the CC for the tagged channel c represents its recognition level. The state 0 represents the condition of channel c not yet observed. It is important to note that the probability for a channel to be recognized (i.e. to reach the recognition threshold) depends only on its popularity $Pop(c)$. Then, the transition probabilities of this time homogenous Markov chain are given by

$$p_{i,j} = \begin{cases} Pop(c)/N & \text{if } j = i+1, i \in [0, R_c - 1] \\ \alpha^i(1 - Pop(c)/N) & \text{if } j = i-1, i \in [1, R_c] \\ 1 - (\frac{Pop(c)}{N}(1-\alpha^i) + \alpha^i) & \text{if } j = i, i \in [1, R_c - 1] \\ 1 - Pop(c)/N & \text{if } i, j = 0 \\ 1 - (1 - Pop(c)/N)\alpha^i & \text{if } i, j = R_c \\ 0 & \text{otherwise} \end{cases} , \tag{2}$$

where $Pop(c)/N$ is the probability of encountering a node subscribed to channel c, and α $(0 < \alpha < 1)$ is a discount parameter. More precisely, if a channel is not observed in the CC cache of the encountered node, then the tagged node decreases its recognition level i by one with a probability that decreases exponentially as i increases. This approach avoids that a tagged node considers relevant a channel that is not popular anymore in the network. On the other and, the probabilities of remaining in the same recognition state after the encounter event are simply the complements of the total probabilities of moving to other states. Special cases are state $i = 0$, where the recognition level cannot decrease, and state $i = R_c$, where the recognition level cannot increase.

Now let $\nu_c^t = \{\nu_c^t[0], \nu_c^t[1], \ldots, \nu_c^t[R_c]\}$ be the state vector at time t and $P = \{p_{i,j}\}$ the transition probability matrix. It holds that $\nu_c^{t+1} = \nu_c^t P$. Since at time t a node has no information on the channels, we have that the initial distribution is $\nu_c^0 = \{1, 0, \ldots, 0\}$.

5.3 Item Recognition

Conceptually, the recognition process of data items is similar to the channel recognition. In fact, IC maintains a list of the observed data items and their recognition levels, defined as the number of times they have been observed in encountered nodes. Thus, the state i of the Markov chain used to describe the status of the IC for the generic data item belonging to channel c represents its recognition level, while state 0 represents the condition of data item a not yet

[1]This condition holds for the stationary regime of several popular mobility model, such as the Random Waypoint model [14].

[2]Note that the $p_{i,j}$ values depend on the parameter t, but for the sake of notation simplicity we omit it.

observed. Considering that the probability that a tagged node finds a copy of data item a in the caches of another node encountered at time t+1 depends only on its replication level $r_t(a)$, the transition probability matrix of this Markov chain can be calculated as

$$p_{i,j} = \begin{cases} r_t(a) & \text{if } j=i+1, i \in [0, R-1] \\ \gamma^i(1-r_t(a)) & \text{if } j=i-1, i\in[1,R] \\ 1-[r_t(a)(1-\gamma^i)+\gamma^i] & \text{if } j=i, i\in[1,R-1] \\ 1-r_t(a) & \text{if } i,j=0 \\ 1-(1-r_t(a))\gamma^i & \text{if } i,j=R \\ 0 & \text{otherwise} \end{cases}, \quad (3)$$

where γ $(0<\gamma<1)$ is the discount factor for the item recognition level.

Now let $\upsilon_a^t = \{\upsilon_a^t[0], \upsilon_a^t[1], \ldots, \upsilon_a^t[R_c]\}$ the state vector at time t and $P=\{p_{i,j}\}$ the transition probability matrix. It holds that $\upsilon_a^{t+1} = \upsilon_a^t P$. Since at time t a node has not information on the data items, we have that the initial distribution is $\upsilon_a^0 = \{1, 0, \ldots, 0\}$.

5.4 Opportunistic Cache

Differently from the other caches, OC has a limited size B. As described in Sec. 4.1, only copies of data items belonging to recognized channels can be stored in the OC cache. However, given that OC can contain only a subset of available data items, a replacement policy is needed to drop less useful data items. To tackle the problem of modeling the evolution of OC, we represent this cache as a *queuing network*, as shown in Fig. 1. In this queuing network, a sub-queue i stores all data items that are in the OC and have the same recognition level i. The sub-queue 0 is a virtual queue that contains all the data items that are outside of the OC, independently of their recognition levels. Then, the ith element $(i>0)$ in the state vector $\varphi_a^t = \{\varphi_a^t[0], \varphi_a^t[1], \ldots, \varphi_a^t[R]\}$ represents the probability that the generic data item a is in the OC cache with recognition level i at time t. It is also important to note that each individual sub-queue has not a fixed size, but we must ensure that the sum of the numbers of data items stored in these sub-queues (excluding sub-queue 0) is lower or equal to B.

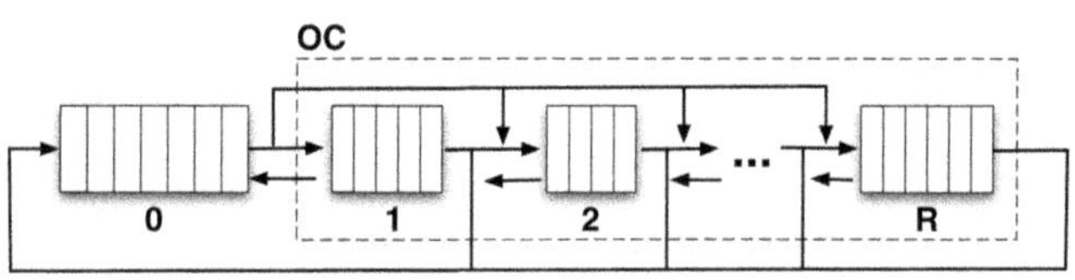

Figure 1: Queuing network modeling the OC cache.

It is intuitive to note that modeling the replacement policy used to manage the OC cache is a hard task. To facilitate the analysis we split the problem into two simpler sub-problems. First of all, we model the reordering of the data items stored in the OC cache due to changes in their recognition levels after an encounter event. For instance, a data item that was initially stored in sub-queue i should be moved to sub-queue $i+1$ if its recognition level is increased upon an encounter event. Since this process involves only an internal reordering of stored data items, no data items are dropped. The second step in the analysis takes into account that new data items fetched by the caches of the encountered node may enter the OC of the tagged node at a sub-queue that depends on their recognition level. Due to cache size constraints, some of these data items could not be allowed to enter the OC cache, or some data items already stored in OC could be removed to let new data items to enter the OC cache. In addition, in both steps we consider the possibility that data items in the OC are moved to sub-queue 0 – i.e. are dropped – if their channel is not recognized anymore after the encounter occurring at time $t+1$. In the following we separately describe these two modeling phases.

Step 1. Let us introduce an auxiliary Markov chain, whose state vector $\varphi'_a = \{\varphi'_a[0], \varphi'_a[1], \ldots, \varphi'_a[R]\}$ represents the probability that the generic data item a is in the OC cache with recognition level i *after* the encounter event, but *before* new data items are inserted in the OC. Then, we have that $\varphi'_a = \varphi_a^t P'$, where P' is the transition probability matrix of the auxiliary Markov chain modeling the data item reordering. Since no data items enter OC in this phase, we have that $p'_{0,0}=1$. However, a data item could be removed from OC if the channel it belongs to loses its recognition, which happens with probability $(1-\nu_a^{t+1}[R_c])$, or the recognition level of the data items becomes null. In other words

$$p'_{i,0} = 1-\nu_a^{t+1}[R_c], i\in[2,R]\ , \quad (4)$$

$$p'_{1,0} = (1-\nu_a^{t+1}[R_c]) + \nu_a^{t+1}[R_c]\gamma(1-r_t(a))\ . \quad (5)$$

Formula (5) can be explained by noting that a data item of a recognized channel can change its recognition level from one to zero only if it is not in the caches of the encountered node, and the tagged node applies the discount factor γ to the data-item recognition level stored in IC (see formula (3)). Following a similar line of reasoning we can compute the probability that a data item already in the OC moves to either backwards or forwards sub-queues. More formally we have that

$$p'_{i,i-1} = \nu_a^{t+1}[R_c]\gamma^i(1-r_t(a)), i\in[2,R]\ , \quad (6)$$

$$p'_{i,i+1} = \nu_a^{t+1}[R_c]r_t(a), i\in[1,R-1]\ . \quad (7)$$

On the other hand, the probability of remaining in the same sub-queue after the encounter event at time $t+1$ is simply given by the complement of the sum of the probabilities of moving backwards or forwards and to leave the OC cache:

$$p'_{1,1} = 1-(p_{1,0}+p_{1,2}) \quad (8)$$

$$p'_{i,i} = 1-(p_{i,0}+p_{i,i-1}+p_{i,i+1})\ , i\in[2,R-1] \quad (9)$$

$$p'_{R,R} = 1-(p_{R,0}+p_{R,R-1}) \quad (10)$$

We now exploit the knowledge of the state vector φ'_a to compute the new average number of data items in each sub-queue i, say B'_i, after the internal reordering, which is simply given by

$$B'_i = \sum_{a=1}^{M} \varphi'_a[i]\ . \quad (11)$$

Step 2: To derive the final status of the OC cache, i.e. $\varphi_a^{t+1}[i]$, in this step we first compute the average number $N_{0,i}$ of new data items that are eligible to enter OC at sub-queue i. Then, we compute the number F_i of available frees slot at each sub-queue i of the OC. The F_i value is the key parameter we need to compute the probability that a new data item is either discarded or cached, and the probability that an old data item stored in OC is removed.

To compute the $N_{0,i}$ quantity we should observe that a new data item not already stored in the OC of the tagged node is eligible for entering the OC cache in the sub queue i if and only if it is stored in the caches of the encountered node, if the channel it belongs to is recognized and its recognition level before the encounter event was $i-1$. Formally, this can be written as follows

$$N_{0,i} = \sum_{a=1}^{M} r_t(a)\nu_a^{t+1}[R_c]\upsilon_a^t[i-1]\varphi_a^t[0]\ . \tag{12}$$

It is important to remind that $N_{0,i}$ expresses the number of new data items that can be potentially copied in the OC cache. However, the actual number of new data items that are copied in OC will depend on the number of free slots. More precisely, let us denote with F_i the *maximum* number of free slots that new data items can occupy in the sub-queue i of the OC cache. It holds that

$$F_i = B - \sum_{j=1}^{i-1} B_j^{t+1}\ , \tag{13}$$

with $F_1 = B$. Indeed, data items (both new or old) with recognition level equal to one have the highest priority and they can use the entire OC. On the contrary, data items with recognition level equal to i ($i > 1$) can use only the part of OC not used by data items with lower recognition levels. It is also important to point out that in formula (13) we must use the B_i^{t+1} value because it provides the size of sub-queue i in OC *after* completing the internal reordering, the insertion of new items and the removal of old items. However, it is quite straightforward to observe that B_i^{t+1} is simply given by

$$B_i^{t+1} = \min(N_{0,i} + B_i', F_i)\ . \tag{14}$$

Formula (14) can be explained by noting that if $N_{0,i} + B_i' > F_i$, then there are enough free slots in OC for all the new items that should enter at level i and for the old items that are already at level i after the reordering. On the other case, some new items will be discarded and/or some old items will be dropped till only F_i slots are occupied[3]. Now, by using formula (13), (14) and the initial condition $F_1 = B$ we can iteratively compute all remaining B_i^{t+1} and F_i values.

Finally, to compute the state vector φ_a^{t+1} we introduce an auxiliary transition probability matrix P such that $\varphi_a^{t+1} = \varphi_a' P$. As observed in formula (12), a new data item is eligible for entering OC with probability $r_t(a)\nu_a^{t+1}[R_c]\upsilon_a^t[i-1]$. However, that new data item will certainly be fetched by the tagged node if $N_{0,i} \leq F_i$, otherwise only a fraction $F_i/N_{0,i}$ will be fetched. Thus, the probability $p_{0,i}$ that a new item effectively enters the OC cache can be expressed as

$$p_{0,i} = \begin{cases} r_t(a)\nu_a^{t+1}[R_c]\upsilon_a^t[i-1] & \text{if } N_{0,i} \leq F_i \\ r_t(a)\nu_a^{t+1}[R_c]\upsilon_a^t[i-1]\frac{F_i}{N_{0,i}} & \text{otherwise}\ , \end{cases} \tag{15}$$

On the other hand, an old data item that was already stored in the OC cache should be removed if the new data items have consumed all (or most of the) free slots. More precisely, we have that

$$p_{i,0} = \begin{cases} 0 & \text{if } N_{0,i} + B_i' \leq F_i \\ 1 & \text{if } N_{0,i} > F_i \\ 1 - \frac{F_i - N_{0,i}}{B_i'} & \text{otherwise}\ . \end{cases} \tag{16}$$

Formula (16) indicates that in case $N_{0,i} + B_i' \leq F_i$ there is no need of replacing data items because the free slots can accommodate both new and old data items. On the other hand if $N_{0,i} \leq F_i$ a fraction $1 - \frac{F_i - N_{0,i}}{B_i'}$ of old data items has to be removed from the OC cache.

5.5 Item Diffusion

At the beginning of network operations data items are not replicated, but they are stored in the LI caches of the nodes that have generated them. Since data items are assumed to be uniformly distributed over the network nodes, we have that at the initialization time $r_0(a) = 1/N$. After an encounter event occurring at time t the replication of each data item is given by

$$r_t(a) = r_0(a) + (1 - r_0(a))\left[\frac{Pop(a.c)}{N}\psi_a^t[1] + \left(1 - \frac{Pop(a.c)}{N}\right)\sum_{i=1}^{R}\varphi_a^t[i]\right] \tag{17}$$

where either the node has generated the data item with probability $r_0(a)$ or it is subscribed to the channel the data item belongs to and it has a copy of a in its SC cache, or it is not subscribed to the channel the data item belongs to and it has a copy of a in one of the sub-queues of its OC cache.

6. PERFORMANCE EVALUATION

6.1 Model Validation

In this section we validate the model accuracy comparing the analytical results with the results obtained using the simulator environment presented in [3]. More precisely, we consider a network composed of 45 nodes, 3 different channels with 99 data items each (297 data items in total). The channel popularity follows a zipfian distribution with parameter 1. In other words, giving a channel j, the probability of a node to be subscribed to that channel is given by $Pop(j) = N \cdot j^{\beta} / \sum_{i=1}^{M} i^{\beta}$, with $\beta = 1$. Thus, channel 1 is the most popular, while channel 3 the least popular. Finally, nodes are assumed to move according to a Random Waypoint model in a square area of side 1Km. The node speed is uniformly sampled in the range [1,1.86] m/s, which are typical pedestrian speeds [9], and we applied the techniques described in [14] to make sure that the model runs in stationary conditions. To better evaluate the impact of the OC cache on the data dissemination process, in the following simulations we assume that the content of SC is *not shared* during an encounter event. To take into account this assumption in the model, only formula (17) has to be changed to remove the $\psi_a^t[1]$ term. In the following graphs average values and 90% confidence intervals are computed by conducting 100 simulations of each scenario with different random seeds.

In Fig. 3 and 5 we plot the temporal evolution of the *hit ratio*, defined as the fraction of data items of the channel a node is subscribed to which are stored in SC, using $R_c = 5$ and $R = 5$ for all the channels, and a very small ($B = 3$) and relatively large ($B = 50$) OC size. As the curves indicate, the

[3]Following the same line of reasoning it is easy to compute the number $\hat{N}_{0,i}$ of new items that will effectively enter the OC cache at time $t+1$ as $\min(N_{0,i}, F_i)$.

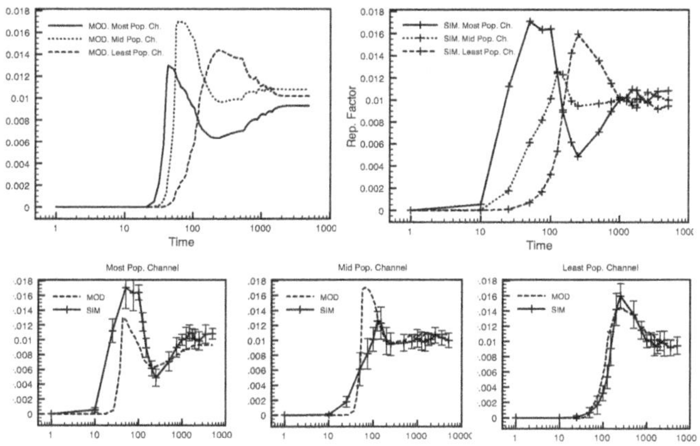

Figure 2: $r_t(a)$ for $R_c = 5$, $R = 5$ and $B = 3$.

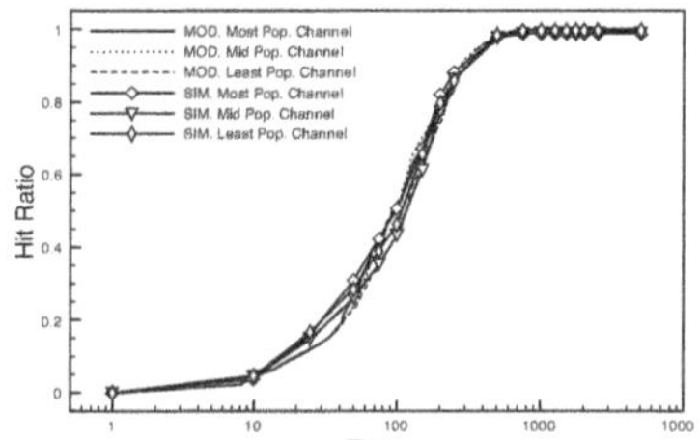

Figure 3: Hit ratios for $R_c = 5$, $R = 5$ and $B = 3$.

larger the OC the shorter the convergence time to hit ratios equal to one. For instance, all the channels reach a hit ratio equal to one around 800s when OC size is 3 slots, while this time is reduced to 200s if the size is 50 slots. Finally, the comparison between analytical and simulation results confirm that our model is sufficiently accurate to predict the temporal evolution of the hit ratios, independently of channel popularity or OC sizes. In Fig. 2, and 4 we show the temporal evolution of the data-item replication level $r_t(a)$ for the OC cache, using $R_c = 5$ and $R = 5$ for all the channels. For the sake of clarity, the sub-plots in the bottom rows directly compare the analytical and simulation results for the same channel, with the most popular channel in the rightmost sub-plot and the least popular one in the leftmost sub-plot. Important observations can be derived from the shown results. First of all, the replication levels show clear peaks. In addition, the peak recognition value is reached before for channels with high popularity than for channels with low popularity. This is due to the fact that the most popular channel and its data items get recognized before all the other channels. Thus, the data dissemination process for the most popular channel can start before than for other channels. However, after this peak the replication levels of all the channels decrease. Indeed, the more diffused a data item is (i.e., the more recognized), the less relevant it is for the dissemination process. This implies that less diffused items can increment their replication level more easily than data items of popular channels. This also explains why the hit ratio of the least popular channel is comparable to the hit ratio of the most popular channel. The second interesting observation is that the replication levels converge to steady values that are almost the same for all the channels. Typically, this happens when all the data items have reached the maximum recognition level, i.e. R. In this case, all data items are considered equivalent for the dissemination process and they are equally distributed in the OC caches. Finally, it is important to point out that our model is remarkably accurate in predicting this stationary behavior of the data-item replication levels. Furthermore, our model is able to predict the times at which the replication levels reach their maximum and minimum values.

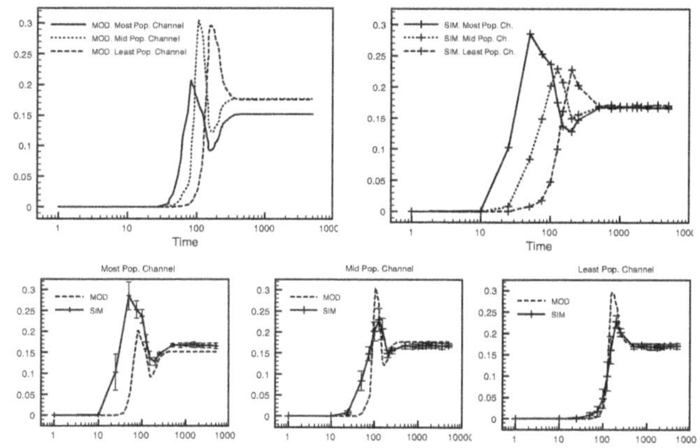

Figure 4: $r_t(a)$ for $R_c = 5$, $R = 5$ and $B = 50$.

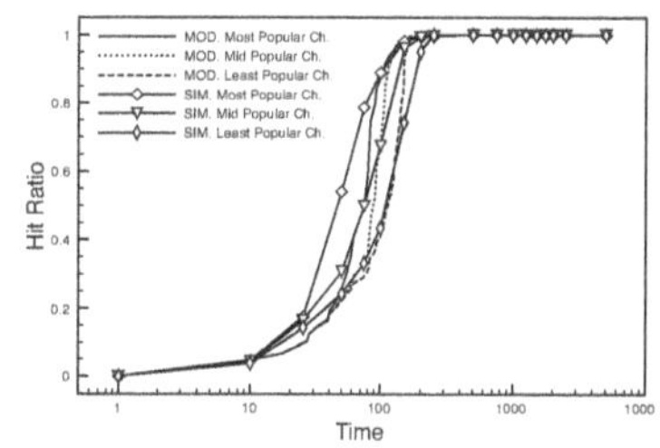

Figure 5: Hit ratios for $R_c = 5$, $R = 5$ and $B = 50$.

6.2 Scalability Analysis

In this section we aim at demonstrating the scalability of our model. More precisely, an accurate simulation of node mobility, communication protocols and cache management policies would be impossible to be executed in feasible time for large-scale opportunistic networks involving thousands of nodes and hundreds of channels, while analysis can easily handle such scales. To validate this statement, we have solved our model considering a network with $N = 1000$ nodes, $M = 100$ channels, and ten objects per-channel (1000 data items in total). This values have been chosen to better represent a "long-tail" scenario, where a relatively large number of small groups of users are subscribed to specific channels. Furthermore, channel popularity still follows a zipfian distribution of parameter 1. Fig. 6 and 7 show the

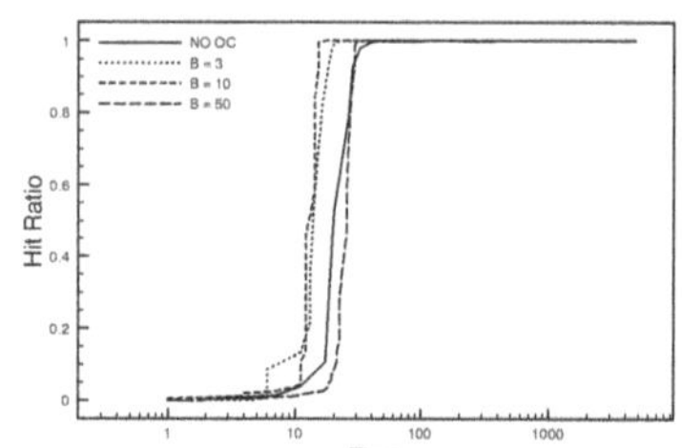

Figure 6: Hit ratio of the most popular channel for varying B values, $R_c = 5$ and $R = 5$.

hit ratios for the most and least popular channels, respectively, and for various sizes of the OC cache. As benchmark scheme we consider an epidemic-based data dissemination

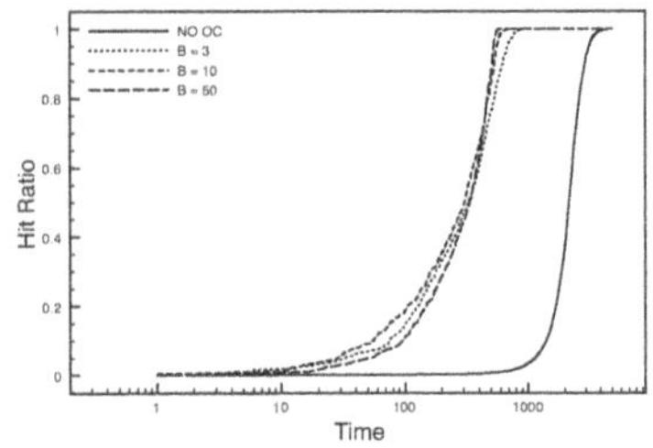

Figure 7: Hit ratio of the least popular channel for varying B values, $R_c = 5$ and $R = 5$.

protocol, where meeting nodes only share the data items for the channels they are both subscribed to. The first important observation is that the OC cache ensures a substantial performance gain over the epidemic scheme. In particular, for the least popular channel the time to reach a hit ratio equal to one is one order of magnitude shorter when using the OC cache. Nonetheless, it should to be noticed that, for the most popular channel, a greater OC does not necessarily imply a performance improvement over the epidemic scheme. This is due to the fact that, with a larger OC, the data items of the most popular channel quickly reach the maximum recognition threshold R, even before they have beed copied in all the nodes subscribed to their channel. However, when a data item is highly recognized, it is less relevant for the data dissemination process, and less diffused items have higher priority during item replacements in OC. The combination of these behaviors may negatively affect the effectiveness of the diffusion process for the high popular channels.

In the last set of results, shown in Fig. 8, we increase the number of nodes by ten times (10000 nodes in total). We can observe that when increasing the number of nodes the most popular channel does not experience significant improvements in the convergence time of the hit ratios. On the other hand, a quite significant improvement can be observed for the least popular channel because a larger pool of OC caches ensures that more copies of the most rare data items are circulating in the network.

Other results obtained by varying the values of the recognition thresholds are available in [2] and are omitted here due to space reasons.

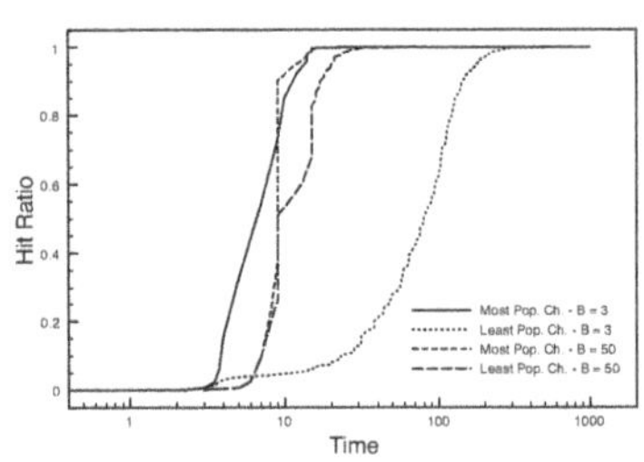

Figure 8: Hit ratios for the most and the least popular channels and varying B values, $R_c = 0$, $R = 5$ and $N = 10000$.

7. CONCLUSIONS

In this paper we have developed an analytical model to describe the performance of a content dissemination mechanism for opportunistic networks that relies on a cognitive scheme, known as the recognition heuristic. Our model takes into account the different popularities of content types, and highlights the impact of the shared memory contributed by individual nodes to make the dissemination process more efficient. In particular, we have shown that our model is sufficiently accurate to capture the transient and steady-state behaviors of key performance indexes of the data dissemination system, such as data replication levels and hit ratios. The low computational complexity of our model allowed us to explore network scenarios with large numbers of involved peers and objects, which are problematic to study using only a simulation-based approach. Future work involve the extension of our model to consider more heterogeneous environments. Moreover, we plan to study how to extend the recognition heuristic to take into account social-based information about the users.

8. REFERENCES

[1] C. Boldrini, M. Conti, and A. Passarella. Design and performance evaluation of ContentPlace, a social-aware data dissemination system for opportunistic networks. *Comput. Netw.*, 54:589–604, March 2010.

[2] R. Bruno, M. Conti, M. Mordacchini, and A. Passarella. Analysis of cognitive heuristics for content dissemination in opportunistic networks. Technical report, IIT-CNR `http://cnd.iit.cnr.it/andrea/docs/mswim12-tr.pdf`, 2012.

[3] Marco Conti, Matteo Mordacchini, and Andrea Passarella. Data dissemination in opportunistic networks using cognitive heuristics. In *Proc. of IEEE WOWMOM 2011*, 2011.

[4] P. Costa, C. Mascolo, M. Musolesi, and G.P. Picco. Socially-aware routing for publish-subscribe in delay-tolerant mobile ad hoc networks. *IEEE JSAC*, 26(5):748–760, 2008.

[5] G. Gigerenzer and D.G. Goldstein. Models of ecological rationality: The recognition heuristic. *Psychological Review*, 109(1):75–90, 2002.

[6] D.G. Goldstein and G. Gigerenzer. Reasoning the fast and frugal way: Models of bounded rationality. *Psychological Review*, 103(4):650–669, 1996.

[7] D.G. Goldstein and G. Gigerenzer. *The recognition heuristic: How ignorance makes us smart*, pages 37–58. Oxford University Press, 1999.

[8] D.G. Goldstein and G. Gigerenzer. Fast and frugal forecasting. *Int. Journal of Forecasting*, 25:760–772, 2009.

[9] D. Karamshuk, C. Boldrini, M. Conti, and A. Passarella. Human mobility models for opportunistic networks. *IEEE Comm. Mag.*, 46(12):157–165, Dec. 2011.

[10] V. Lenders, M. May, G. Karlsson, and C. Wacha. Wireless ad hoc podcasting. *ACM MC2R*, 12:65–67, Jan. 2008.

[11] J.N. Marewski, W. Gaissmaier, L.J. Schooler, D. Goldstein, and G. Gigerenzer. From recognition to decisions: Extending and testing recognition-based models for multialternative inference. *Psychonomic Bulletin & Review*, 17(3):287–309, 2010.

[12] Julian N. Marewski, Wolfgang Gaissmaier, and Gerd Gigerenzer. Good judgments do not require complex cognition. *Cogn. Process*, 11:103–121, 2010.

[13] M. Monti, L. Martignon, Gigerenzer. G., and N. Berg. The impact of simplicity on financial decision-making. In *Proc. of CogSci 2009*, 2009.

[14] W. Navidi and T. Camp. Stationary distributions for the random waypoint mobility model. *IEEE Trans. Mob. Comput.*, 3(1):99–108, 2004.

[15] A. Passarella. A survey on content-centric technologies for the current Internet: CDN and p2p solutions. *Comput. Comm.*, 35(1):1–32, January 2012.

[16] L. Pelusi, A. Passarella, and M. Conti. Opportunistic networking: data forwarding in disconnected mobile ad hoc networks. *IEEE Comm. Mag.*, 44(11):134 –141, Nov. 2006.

[17] L.J. Schooler and R. Hertwig. How forgetting aids heuristic inference. *Psychological Review*, 112(3):610, 2005.

[18] Sascha Serwe and Christian Frings. Who will win wimbledon? the recognition heuristic in predicting sports events. *J. Behav. Dec. Making*, 19(4):321–332, 2006.

[19] E. Yoneki, P. Hui, S.Y. Chan, and J. Crowcroft. A socio-aware overlay for publish/subscribe communication in delay tolerant networks. In *Proc. of ACM MSWIM'07*, pages 225–234, 2007.

CRISP: Collusion–Resistant Incentive–Compatible Routing and Forwarding in Opportunistic Networks

Umair Sadiq, Mohan Kumar, and Matthew Wright
University of Texas at Arlington
Arlington, Texas, USA
umair.sadiq@mavs.uta.edu, {mkumar,mwright}@uta.edu

ABSTRACT

In opportunistic environments, tasks such as content sharing and service execution among remote devices are facilitated by relays (devices with short–range wireless connectivity) that receive data, move around, and then forward the data. To achieve high throughput, it is important to secure forwarding and provide incentives for participation by relays. However, it is extremely challenging to monitor the behavior of relays in an opportunistic network due to sparse connectivity. Existing schemes do not work when selfish/malicious relays collude with each other to forge routing metrics, drop useful data, flood the network, or earn extra reward.

The credit scheme presented in this paper is the first in which routing as well as forwarding are incentive compatible. To design the scheme, the data transfer and loss in an opportunistic network are modeled as a non–linear generalized flow network. Then, optimality conditions for flow maximization describe the optimal behavior of a relay. This optimal behavior is made incentive compatible by requiring a relay to make a specific payment upon receiving the data and earn reward on forwarding the data. A cryptographic technique is used to make the scheme collusion resistant. Finally, a framework is proposed to implement CRISP in a completely distributed and opportunistic environment. Simulations on real and synthetic mobility traces validate a significant gain in throughput when compared with the existing credit schemes that are not incentive compatible.

Categories and Subject Descriptors

C.2.1 [**Computer Communication Networks**]: Network Architecture and Design—*store and forward networks*

Keywords

Credit Scheme, Incentive Scheme, Opportunistic Networks

1. INTRODUCTION

The primary challenge in opportunistic networks is to send data in a reliable and efficient way despite the absence of a connected path between the end points. Data transmission is critical to every fundamental service that an opportunistic network can provide, including content sharing , dissemination, collection, and collaboration [7,12,14,23]. The underlying mechanism used to forward data in these networks is to rely on any device in the area with short–range wireless connectivity to receive the data, expect the movements of the devices to make new connections possible, and to use these new connections to forward data onto other devices in the hope of eventually finding a path to the destination.

Numerous studies have investigated opportunistic routing in fully cooperative settings, in which each node participates in data forwarding to the best of its ability. Careful planning based on nodes' past and estimated future connection and movement behaviors can aid in making efficient forwarding in order to extract the most benefit out of each node's available resources [15, 19, 29]. In many realistic settings, however, we must anticipate selfish users who will not freely receive and send data at the cost of their devices' battery power and memory without some benefit to themselves. Additionally, we must be prepared to handle malicious users who seek to undermine the network's capabilities through manipulation of the routing system, e.g. by falsely claiming short routes to all destinations and dropping all data in a black hole attack, or by simply flooding the network.

Researchers have proposed a range of schemes to address these issues in opportunistic networks [17, 22, 37] as well as other network settings, such as ad hoc networks [31,32, 35,36]. The most promising approach to address these issues is to provide *credit* to nodes who participate in the forwarding process and allow only nodes who have credit to generate and send their own data packets[1]. Existing schemes, however, are not incentive compatible in routing or can be defeated by a small group of colluding attacker nodes [8,17,18,20,22,24,37]. Furthermore, existing schemes are designed to work in ad hoc networks in which the routing path is known [31, 35] or in mesh networks in which the routing and forwarding mechanisms are considered separately [10,33]. To date, no incentive–compatible schemes are designed to operate in opportunistic environments in a fully distributed way without - i) an online trusted authority and ii) *a priori* knowledge of routes.

Contributions: In this paper, we present a new credit scheme called CRISP (Collusion–Resistant Incentive–ComPatible routing and forwarding) that addresses aforementioned challenges. The basic CRISP mechanism requires

[1] we use the term data, packets, and data packets interchangeably.

, to pay credit to receive a data packet and later ob- credit for forwarding that packet. Nodes must also / credit to send new data packets of their own. Intu- ıvely, selfish nodes who fail to forward packets will not be able to send anything, and malicious nodes are prevented from flooding the network. Further, routing black holes are greatly limited due to having to pay for receiving the data.

Although these mechanisms are intuitive, ensuring both collusion resistance and incentive compatibility for routing as well as forwarding is challenging. In this work, we describe a principled method for setting the payment parameters. In particular, we first convert a contact graph of the network into an instance of network flow problem and model the loss of packets in the network as a queueing theory problem with *reneging*. With this model, we find the optimal operating point of individual nodes to maximize overall network flow. We then design our credit-based incentive scheme so as to maximize the profit of a node when it operates at this optimal operating point. Our scheme thus makes it the selfish choice to receive and forward data packets in the most efficient manner for the network as a whole.

To enforce our scheme's requirements, we present a simple and secure protocol that makes CRISP resistant to colluding attackers. In particular, payments are secured using a Merkle hash tree [37] over the transaction history, and the histories are monitored by other nodes in the network. We also present a framework for maintaining and tracking the credits in a distributed manner.

CRISP is thus the first credit scheme that is appropriate for opportunistic networks. It is fully distributed, requires neither a central bank nor *a priori* knowledge of routing paths. CRISP is robust to flooding and blackhole attacks and provides incentives for nodes to participate fully in routing and forwarding operations.

Organization: We now layout the organization of the rest of the paper. In Section 2, we provide background on opportunistic networks and credit schemes so as to best motivate our design choices. Section 3 describes a model of the opportunistic network contact graph as a flow network and shows how we further model each node as a M/M/1 queue with reneging. Based on this model, Section 4 shows how to find the optimal arrival rate under a given function of loss (the *attenuation*). We then describe our credit scheme in Section 5, with payments and rewards set so as to make nodes obtain their maximum profit when accepting and forwarding data at the optimal arrival rate. Section 6 shows the results of analysis and simulations that demonstrate the effectiveness of our approach. Finally, we cover related work in Section 7 and conclude in Section 9.

2. DESIGN CONSIDERATIONS

In an opportunistic network, a node acts as a: *source* – generates data; *relay* – receives, stores and forwards data; and *destination* – receives data. Therefore, the possible cheating actions in each role can be: source – generates dummy data (flooding attack); relay – drops data (blackhole attack)/ randomly forwards data (to earn credit); or destination – does not acknowledge received data. In addition, nodes can also collude with each other for such attacks. Payment for generating data and a reward for acknowledging received data are used in [34] to secure against cheating by a source and destination. However, no existing credit schemes or security frameworks secure the behavior of a relay in face of collusion (we discuss further in Section 7). To this end, CRISP ensures truthful forwarding and honest reporting of routing utility even when nodes collude. We now describe the key concepts used to design CRISP.

Routing and Forwarding: In an opportunistic network, nodes come into contact with each other at different times and locations. Upon contact with another node, a data packet is forwarded, only if the encountered node is *more* likely to deliver the packet. This likelihood of delivery is computed by a routing/forwarding scheme that assigns every node, a utility value for each destination where a node with better likelihood of delivery has a higher value of utility. For example, bubble rap [15], maxprop [5], and others [19, 27, 29] assign utility values for social based forwarding, vehicle based routing, and general opportunistic networks respectively. Thus, let U_v^x represent the utility value of node v for destination node x that is computed by a corresponding routing scheme (e.g., [5, 15, 19, 27, 29]). A packet for destination node x is forwarded from node v to another node w if node w is more likely to deliver the packet, i.e., if $U_w^x > U_v^x$. A packet also has an expiration time such that if a packet is not delivered by that time, it is dropped/lost at the node. In this paper, we consider the problem of truthful *reporting* of utility, with the assumption that the node *knows* its true utility (e.g., a node only updates its utility based on contact with a (trusted) node with high reputation value).

Incentive Compatibility: Incentive compatibility means that a node earns highest reward when it behaves honestly. There are two related concepts: strong incentive compatibility means that honest behavior is the best strategy irrespective of strategy of other nodes, whereas, weak incentive compatibility means that honest behavior is the best strategy with the requirement that other nodes behave honestly. It was proved in [35] that forwarding cannot be incentive compatible in the strong sense[2]. Thus, we use the term incentive compatibility to mean weak incentive compatibility. Typically, credit schemes are incentive compatible in forwarding, i.e., a relay earns highest reward when it forwards the data to another relay or the destination itself. However, none of the existing credit schemes for opportunistic networks is incentive–compatible in routing.

The distinction between the forwarding and the routing stages was first highlighted in [35] for ad hoc networks. Techniques from mechanism design [21], are adapted in [35] to make the routing *and* the forwarding stage incentive–compatible. However, these techniques are not applicable in opportunistic networks because the complete path from source to destination is not known. A related problem is that of opportunistic routing in mesh networks, as the path is determined on a per-hop basis. In [33], routing is made incentive–compatible by a carefully designed incentive scheme. However, the forwarding stage is not considered. Then, in [10], forwarding in mesh networks is made incentive compatible with the *assumption* that routing is incentive compatible. Thus, a relay makes maximum profit by honestly behaving in forwarding *only if* it is assumed that the relay behaves honestly in the routing stage. When considered together, routing and forwarding are not incentive compatible because a node can make higher profit by cheating in

[2]strong incentive compatibility is equivalent to a mechanism in which honest behavior is the dominant strategy

both routing *and* forwarding. Therefore, it is important to consider routing and forwarding together, without any assumption about the honest behavior in two separate stages.

Collusion: In ad hoc networks, nodes declare a cost of transfer, and it has been shown in [32] that no mechanism can be incentive compatible when nodes collude in declaring the cost. In order to bypass this restriction, we fix the cost of forwarding, i.e., every node gets the same minimum reward upon forwarding. Since every relay that takes part in the forwarding gets a reward, colluding nodes add fake relays in the path. As a result, even those relays that do not take part in the forwarding get a reward. One recent work [9] addresses this issue by decreasing the amount of reward based on the number of relays in forwarding path. But this approach requires a large percentage of payment to the central credit authority (e.g. 57%,73% in case of 3, 4 relays). Thus far, it is not clear how to redistribute this over payment back in the nodes such that the original scheme remains incentive compatible (e.g., redistribution of extra payment among relays that contribute allows nodes to increase reward by adding fake relays). In addition, [9] does not secure the routing behavior of nodes i.e. a relay can increase its reward and decrease network throughput by modifying its routing utility to receive and randomly forward those packets.

Design Principles: CRISP is designed based on following principles:

1. Data is always forwarded to a relay with higher routing utility than the one at which data was received,
2. Reward is based upon successful delivery of data, and the amount of reward is proportional to the improvement in utility made by the relay.
3. The ratio of payment (for receiving data) to reward (upon forwarding data) is adjusted so that reporting incorrect routing utility results in reduced profit.

These principles are explained below.

Principle (1) is used to secure against random forwarding. To enforce this requirement, a cryptographic technique of layered coins [20, 37] is used. Upon exchanging a packet, relays in contact sign (using their private key) the routing utility at which the packet is received (see [20,37] for details). As a result, e.g., if a node v receives a packet for destination x when its utility U_v^x is 0.2, it can only forward it to another node w if $U_w^x > 0.2$ (because $U_v^x = 0.2$ is stored in the layered coin in an un-forgeable way). This prevents a node from randomly forwarding a packet.

Principle (2) is used to secure against collusions that add fake relays. Thus, if a node v receives a packet at utility U_v^x and forwards it to a node w with utility U_w^x, then, the reward is proportional to $U_w^x - U_v^x$. As a node forwards more packets, its expected reward also increases. However, adding the fake relays in the path does not increase the total profit because U_w^x and U_v^x stay the same (U_w^x, U_v^x are stored in the layered coin).

Principle (3) is used to secure against collusions to modify declared routing utility. Suppose a node declares a routing utility $\hat{U}_v^x$ instead of its true utility U_v^x. If $\hat{U}_v^x > U_v^x$, the node receives more packets but it becomes less likely to forward (find neighboring nodes with a $U_w^x > \hat{U}_v^x$) the packets. If $\hat{U}_v^x < U_v^x$, the node receives less packets, but is more likely to forward the packets. If $\hat{U}_v^x = U_v^x$, the node receives packets at a certain rate. This is called the optimal arrival rate as it maximizes the throughput (the number of packets that are delivered to destination) of the network. Note that a node is required to make a payment to receive a packet, therefore, a node always tries to forward the received packets to get a reward. For analysis, we formulate this scenario as a network flow problem.

Network Flow and Optimization: Two important extensions of network flow are: i) generalized flow – a flow that is not conserved across an arc and ii) dynamic flow – a flow that takes time to travel across an arc. An opportunistic network is better modeled by a generalized dynamic flow, with multiple commodities and concave gains. The multiple commodities are for packets routed to different destinations and the concave gains are to model the packets lost/expired before delivery to the destination. Closely related problem of generalized flow with concave gains has been studied in [28, 30] and more recently hardness results for generalized dynamic flow are proved in [13]. To keep our analysis tractable, we drop the requirement of dynamic flow and model the opportunistic network as a generalized flow with multiple commodities and concave gains, and validate our derived results through simulations.

3. SYSTEM MODEL

This section describes: (i) contact graph of opportunistic network; (ii) flow network model for data transfer between nodes; and (iii) queue model for data loss at nodes (data is dropped from the queue when not serviced in time).

3.1 Contact Graph

In an opportunistic network, nodes come into contact with each other at different times and locations. Consider the network operation in a time interval I of duration T (to the order of few minutes). We construct a directed contact graph $G = (V, E)$, where each vertex $v, w \in V$ represents a node in an opportunistic network with $n = |V|$, and each arc $vw \in E$ represents a contact between node v and node w at any point in time during the interval I. Every node $v \in V$ in a network acts as a source, relay and destination. Let d_{vw} represent the amount of data that can be sent from node v to node w during any number of contacts in the interval I. Note that d_{vw} is the duration of contacts multiplied by the bandwidth available for communication. We consider the network to be symmetric, i.e., if $vw \in E$ then $wv \in E$ and $d_{vw} = d_{wv}$. Note that $d_{vw} = 0$ if $vw \notin E$, and $d_{vv} := 0$.

3.2 Flow Network

Graph G can be modeled as a multi–commodity flow network G_g. In this network, the *amount of flow* is used to represent the *rate of data transfer*. Thus, the terms arrival, loss and forwarding rate of packets are equivalent to the incoming, lost and outgoing flow respectively. These terms are used interchangeably in rest of the paper.

Let c_{vw} represent the flow capacity between nodes v and w. Then $c_{vw} := d_{vw}/T$, i.e., the constant *rate* of data transfer possible between node v and node w for the entire interval I. Network G_g has multiple commodities of flows $x \in V$ which correspond to data en–route for different destination nodes $x \in V$. Let g_{vw}^x be a flow of commodity x. The amount of flow g_{vw}^x is used to represent the rate of transfer of data (destined to node x) from node v to node w. Data of a particular commodity is only transferred in one direction, i.e., if $g_{vw}^x > 0$ then $g_{wv}^x = 0$. Let b_v^x be the amount of flow of commodity x generated at node v. Flow b_v^x is used to represent rate for the data generated during the interval I

or available at the start of interval I at node v. All flows are non–negative and the capacity constraint requires that the sum of flows of all commodities between nodes v and w is less than or equal to the capacity ($c_{vw} = c_{wv}$), i.e.,

$$\sum_{x \in V} g^x_{vw} + \sum_{x \in V} g^x_{wv} \leq c_{vw} \quad \forall v, w \in V, \tag{1}$$

Data packets have an expiration time, and a packet is lost at a node if it is not delivered before the expiration time. Let Q^x_v be the probability that a packet of commodity x is lost at node v. Then, if a node v receives packets for destination node x at the rate $\lambda^x_v = \sum_{w \in V} g^x_{wv}$, it is able to forward at the rate $\Gamma^x_v(\lambda^x_v) = \lambda^x_v(1 - Q^x_v)$.

Note that the data is lost at the node itself. This can be modeled in terms of a generalized flow network by splitting each vertex into two vertices joined by a new arc of flow λ^x_v, and then adding the attenuation function $\Gamma^x_v(\lambda^x_v)$ over the flow of this new arc. This process is similar to the conversion of a flow network with vertex capacities to a flow network with arc capacities. However, for notational simplicity, we avoid the standard notation, and describe the flow conservation as follows. For each node and commodity, the sum of outgoing flows is equal to the sum of locally generated flow and the attenuated incoming flow, i.e., for all $v, x \in V \quad \sum_{w \in V} g^x_{vw} = \Gamma^x_v(\lambda^x_v) + b^x_v$.

Notation: Let $\lambda_\mathbf{v} = (\lambda^x_v, \forall x \in V - \{v\})$ represent the column vector of arrival rate (incoming flow) of all commodities (except commodity destined to node v) at node v, such that, $\lambda_\mathbf{v} \in \mathbb{R}^{(n-1)\times 1}$. Let $\lambda = (\lambda_\mathbf{v}, \forall v \in V)$ represent the column vector of arrival rate (incoming flow) of all nodes and commodities (except commodities destined to each node), such that, $\lambda \in \mathbb{R}^{n(n-1)\times 1}$. Let $\hat{\lambda}_\mathbf{v}$ and $\hat{\lambda}$ represent the corresponding optimal arrival rate vectors. Let $\mathbf{h_v} = \lambda_\mathbf{v} - \hat{\lambda}_\mathbf{v}$ and $\mathbf{h} = \lambda - \hat{\lambda}$ represent the difference between a general and optimal rate vector where $h^x_v = \lambda^x_v - \hat{\lambda}^x_v$ is used to represent an individual component. Vector $\mathbf{S_v}$ and $\mathbf{S}$ are used to show sum of vector components, where $\mathbf{S_v} = (1, \forall x \in V - \{v\})$ and $\mathbf{S} = (1, \forall v \in V, x \in V - \{v\})$ are row vectors of ones. For example, $\mathbf{S_v}\lambda_\mathbf{v} = \sum_{x \in V - \{v\}} \lambda^x_v$ is the total incoming flow (excluding commodity v) at node v, similarly, $\mathbf{S}\lambda = \sum_{v \in V} \sum_{x \in V - \{v\}} \lambda^x_v$ is the total incoming flow at all nodes (excluding commodities destined to each node). The total outgoing flow (excluding locally generated flow) at all nodes is represented by $\Gamma(\lambda) := \sum_{v \in V} \Gamma_v(\lambda_\mathbf{v})$ where $\Gamma_v(\lambda_\mathbf{v}) := \sum_{x \in V - \{v\}} \Gamma^x_v(\lambda^x_v)$ is the outgoing flow at node v.

3.3 Queue Model

Packets received by a node have some expiration time, therefore, not all packets are forwarded. If a node v receives commodity x at rate λ^x_v, then, the rate *available* to forward the received commodity is μ^x_v, such that

$$\lambda^x_v + \mu^x_v = c^x_v. \tag{2}$$

The net capacity of relaying commodity x through node v is limited by capacity c^x_v, where c^x_v depends on the capacity c_{vw} available with each node w, the rate b^x_v consumed by forwarding the data generated at node v, and the schedule for splitting available capacity c_{vw} for each commodity x. In our solution, we look at the properties of the optimal arrival rate λ^x_v for a given capacity c^x_v for each node and commodity.

The rate available for forwarding μ^x_v is different from rate at which commodity x is actually forwarded $\Gamma^x_v(\lambda^x_v)$. For example, a node can receive data at the rate (in packets/min)

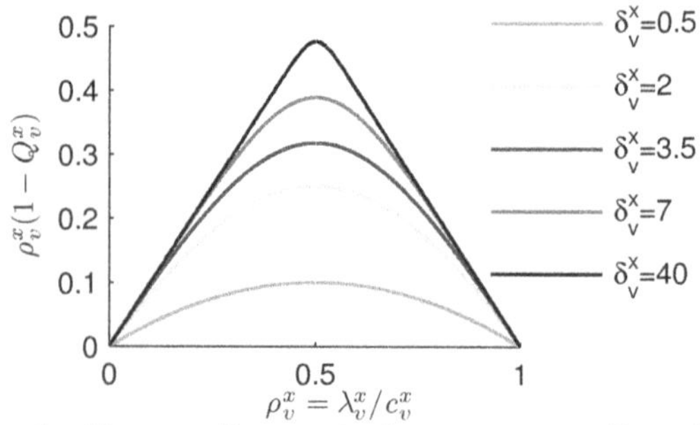

Figure 1: Forwarding rate is a concave function of packet arrival rate.

$\lambda^x_v = 5$ with the available rate of forwarding $\mu^x_v = 10$, but only able to forward at $\Gamma^x_v(\lambda^x_v) = 3$. $\Gamma^x_v(\lambda^x_v) \leq \lambda^x_v$ because a node cannot forward more packets than it receives, and some packets expire when there is no contact opportunity to forward the packet.

To derive the packet loss rate, as an approximation, the arrival rate λ^x_v and available forwarding rate μ^x_v are modeled as a Poisson Process. This scenario represents a $M/M/1$ queue with reneging [3]. The arrival process is Poisson with rate λ^x_v, the service process corresponds to forwarding process which is Poisson with rate μ^x_v, and each node v for a flow commodity x is a server. The holding time of a packet in queue is the time after which a packet is dropped if its service has not been completed i.e. if the packet has not been forwarded. We assume a constant holding time δ^x_v / c^x_v, where δ^x_v is a dimensionless constant that gives the order of holding time with respect to capacity of a node c^x_v.

4. ANALYSIS

The value of utility that a node declares determines the arrival rate of packets and consequently the packet loss rate. In this section we derive the relation for packet loss, and use it to find the optimal packet arrival rates that maximize the network flow. We call it the optimal packet arrival rate, as it corresponds to the optimal value of utility that a node should declare to maximize the network throughput.

4.1 Attenuation due to expired packets

We use the result (equation (29) in [3]) for the probability of a dropped packet Q^x_v in a $M/M/1$ queue with reneging. Substituting μ^x_v from (2) and replacing λ^x_v / c^x_v with ρ^x_v, the probability of packet loss is given by

$$Q^x_v = \frac{(1 - 2\rho^x_v)e^{\delta^x_v(2\rho^x_v - 1)}}{1 - \rho^x_v(1 - e^{\delta^x_v(2\rho^x_v - 1)})} \qquad \text{when} \quad \rho^x_v \neq 0.5,$$

where, $Q^x_v = 2/(2 + \delta^x_v)$ when $\rho^x_v = 0.5$. Though we were unable to prove, but from simulations we find that for a fixed value of δ^x_v, the function $\rho^x_v(1 - Q^x_v)$ is concave (first derivative decreases monotonically) in terms of ρ^x_v. Figure 1 shows the function $\rho^x_v(1 - Q^x_v)$ for different values of δ^x_v. Using a known property, that perspective of a concave function is also concave (replacing ρ^x_v by λ^x_v / c^x_v and multiplying by c^x_v) [4], we find that $\Gamma^x_v(\lambda^x_v) = \lambda^x_v(1 - Q^x_v)$ is concave in terms of λ^x_v.

4.2 Optimal flow with concave attenuation

In order to maximize the total flow delivered to the destination nodes, we minimize the total flow lost at individual relays given that all generated flow is pushed into the network. Using notation defined in Section 3.2, the total flow lost at all nodes is the sum of total flow lost at individual nodes for all commodities and is given by, $\mathbf{S}\lambda - \Gamma(\lambda)$.

Let the optimal incoming flow (arrival rate) at individual nodes be $\hat{\lambda}$ and the sum of optimal incoming flows be $L := \mathbf{S}\hat{\lambda}$. In order to find the properties of the optimal incoming flow that maximize the delivered flow, our objective is to minimize $L - \Gamma(\lambda)$ (total flow lost at relays) which is equivalent to maximizing $\Gamma(\lambda)$ such that $\mathbf{S}\lambda = L$. Also, flows cannot be negative, i.e., we have $\lambda \geq 0$, where $\geq$ is used for component wise inequality. Formally, this is described as,

$$\begin{aligned} \text{maximize} \quad & \Gamma(\lambda) \\ \text{subject to} \quad & \mathbf{S}\lambda = L, \qquad (3) \\ & \lambda \geq 0. \qquad (4) \end{aligned}$$

There is one additional capacity constraint on λ that we do not consider here. This is described in the form of individual flows g^x_{vw} in (1). One can solve the above optimization problem for the exact feasibility constraints. This problem is similar in structure to the problem of network utility maximization [16]. The concave attenuation function $\Gamma^x_v(\cdot)$ points to the law of diminishing returns, i.e., probability of successfully forwarding a packet decreases as the node declares a high utility value in routing to receive more packets. However, we relax the flow capacity constraint (1) to constraint (3) in order to analyze the optimal rate $\hat{\lambda}$. The reason is that, specifically in opportunistic network, the topology is not static over time, and as nodes move around the rate $\hat{\lambda}^x_v$ of one node (commodity) can be shifted to another node (commodity). In addition, simulation in a range of environments shows that honest behavior of nodes in routing and forwarding yields arrival rates that are very close to the optimal rate $\hat{\lambda}$ with constraints (3) and (4) only.

Based on analysis of attenuation in Section 4.1, assume that $\Gamma^x_v(\lambda^x_v)$ is concave and differentiable. Then, $-\Gamma^x_v(\lambda^x_v)$ is convex and $-\Gamma(\lambda)$ is convex because it is an increasing convex function of a convex function. Also, the constraints (3) and (4) are affine and convex. Solving the *Lagranian*, the optimal point is the global maximum and Karush–Kuhn–Tucker conditions for optimality [4] give us

$$\nabla(-\Gamma(\hat{\lambda})) + \xi\nabla(\mathbf{S}\hat{\lambda} - L) + \sum_{v\in V}\sum_{x\in V-\{v\}} \xi^x_v \nabla(-\hat{\lambda}^x_v) = 0,$$

$$\forall v \in V, x \in V - \{v\} \quad \frac{\partial\Gamma^x_v(\hat{\lambda}^x_v)}{\partial\lambda^x_v} = \xi - \xi^x_v, \quad (5)$$

$$\forall v \in V, x \in V - \{v\} \quad \xi^x_v\,\hat{\lambda}^x_v = 0, \quad (6)$$

where the Lagrange multipliers $\xi, \xi^x_v \geq 0$. Here $\xi \geq 0$ due to optimal value of L. It can be shown that $\Gamma(\lambda)$ for $\xi = 0$ is always greater than the case when $\xi < 0$ by changing constraint (3) to be $\mathbf{S}\lambda \leq L$. Equation (5) is the stationary condition, whereas (6) is the complementary slackness condition. These are in addition to the primal feasibility constraints (3) and (4). These optimality conditions can be further simplified. From (6), if $\hat{\lambda}^x_v > 0$, then $\xi^x_v = 0$ and $\partial\Gamma^x_v(\hat{\lambda}^x_v)/\partial\lambda^x_v = \xi$. So, our final result summarizes the optimality conditions of (3), (4), (5), and (6) as follows.

Main Result: Let $\hat{\lambda}$ be the arrival rate that satisfies primal feasibility constraints (3) and (4). Suppose there exists a constant $\xi \geq 0$ such that $\forall \hat{\lambda}^x_v > 0$,

$$\frac{\partial\Gamma^x_v(\hat{\lambda}^x_v)}{\partial\lambda^x_v} = \xi, \quad (7)$$

then $\hat{\lambda}$ is the optimal arrival rate, i.e., it maximizes $\Gamma(\hat{\lambda})$.

4.3 Optimal flow with general attenuation

In the previous section we derived the optimal arrival rate for continuous, differentiable and concave attenuation function. Our result on concavity of attenuation function is from an approximate $M/M/1$ queue model. Therefore, in this section, we consider general function $\Gamma^x_v : \mathbb{R}_+ \to \mathbb{R}_+$, without any requirement of continuity, differentiability and concavity. Following derivations use the insight about optimality conditions from the case of continuous concave function.

THEOREM 1. *Let $\hat{\lambda}$ be the arrival rate that satisfies feasibility constraints (3) and (4). Suppose there exists a constant $\xi \geq 0$ such that $\forall \hat{\lambda}^x_v > 0$,*

$$\frac{\Gamma^x_v(\lambda^x_v) - \Gamma^x_v(\hat{\lambda}^x_v)}{\lambda^x_v - \hat{\lambda}^x_v} \geq \xi, \quad \forall \lambda^x_v \in [0, \hat{\lambda}^x_v), \text{ and}$$

$$\frac{\Gamma^x_v(\lambda^x_v) - \Gamma^x_v(\hat{\lambda}^x_v)}{\lambda^x_v - \hat{\lambda}^x_v} < \xi, \quad \forall \lambda^x_v > \hat{\lambda}^x_v,$$

then $\hat{\lambda}$ is the optimal arrival rate, i.e., it maximizes $\Gamma(\hat{\lambda})$.

PROOF. In order to prove optimality, we show that $\Gamma(\hat{\lambda}) \geq \Gamma(\lambda)$ for any λ that satisfies the constraints (3) and (4). We can express any general rate $\lambda^x_v \geq 0$ in terms of the optimal $\hat{\lambda}^x_v$ as follows $\lambda^x_v = \hat{\lambda}^x_v + h^x_v$. Then, the conditions in the theorem can be re–written as

$$\Gamma^x_v(\hat{\lambda}^x_v) - \Gamma^x_v(\hat{\lambda}^x_v + h^x_v) \geq -\xi h^x_v, \quad \forall h^x_v \in [-\hat{\lambda}^x_v, 0),$$

$$\Gamma^x_v(\hat{\lambda}^x_v) - \Gamma^x_v(\hat{\lambda}^x_v + h^x_v) > -\xi h^x_v, \quad \forall h^x_v > 0,$$

Summing the above two relations over all nodes and commodities, we get

$$\Gamma(\hat{\lambda}) - \Gamma(\lambda) \geq -\xi\mathbf{S}\mathbf{h}. \quad (8)$$

From constraint (3), $\mathbf{S}\lambda = \mathbf{S}\hat{\lambda}$. Substituting $\lambda = \hat{\lambda} + \mathbf{h}$, $\Rightarrow \mathbf{S}(\hat{\lambda} + h) = \mathbf{S}\hat{\lambda} \Rightarrow \mathbf{S}\mathbf{h} = 0$. Substituting this result back in (8), we get $\Gamma(\hat{\lambda}) - \Gamma(\lambda) \geq 0 \Rightarrow \Gamma(\hat{\lambda}) \geq \Gamma(\lambda)$. □

Properties of Optimal Flow: The intuitive idea is simple. If *additional* packets are to be received at rate 10, the relay responsible for forwarding should be the one that can forward at rate 7 instead of the one that can only forward at rate 6. Therefore, with optimal allocation of flows, shifting individual flows from one node to another does not increase the overall forwarding rate. To see this property, let $M^x_v(\phi) = \frac{\Gamma^x_v(\lambda^x_v+\phi)-\Gamma^x_v(\lambda^x_v)}{\phi}$. Then $\phi M^x_v(\phi)$ is the increase in forwarding rate for increasing *optimal* arrival rate by ϕ. Similarly $\phi M^x_w(-\phi)$ is the decrease in forwarding rate for decreasing *optimal* arrival rate by ϕ. The optimal allocation of flows requires that the increase in forwarding rate $\phi M^x_v(\phi)$ at one node is always less than the decrease in forwarding rate $\phi M^x_w(-\phi)$ at another node, for the same amount of change ϕ in arrival rate, i.e., $\phi M^x_w(-\phi) > \phi M^x_v(\phi) \Rightarrow M^x_w(-\phi) > M^x_v(\phi)$. This condition is true because Theorem 1 requires $M^x_w(-\phi) \geq \xi$ and $M^x_v(\phi) < \xi$ at all values of $\phi \in (0, \hat{\lambda}^x_w]$ and $\forall v \in V, x \in V - v$. Also, the ratio of forwarding rate to the arrival rate is always greater than or equal to ξ (substituting $h^x_v = -\hat{\lambda}^x_v$ in Theorem 1 gives $\Gamma^x_v(\hat{\lambda}^x_v)/\hat{\lambda}^x_v \geq \xi$).

Consider the arrival (and corresponding forwarding) rates of a node in Table 1. Based on Theorem 1, for any value of $\xi \in (0.05, 0.85)$, the optimal arrival rate is $\lambda^x_v = 70$ with forwarding ratio 0.86. In a traditional credit scheme that rewards nodes based on the number of packets forwarded, a node maximizes its profit by maximizing its forwarding rate ($\Gamma^v_x = 62$) at a much lower forwarding ratio of 0.56. This is the key difference between CRISP and existing schemes and results in significant improvement in network throughput.

Table 1: CRISP operates at forwarding ratio 0.86 for any value of $\xi \in (0.05, 0.85)$.

Forwarding ratio	0.9	**0.86**	0.61	0.56	0.50
Forwarding rate Γ_v^x	9	**60**	61	62	60
Arrival rate λ_v^x	10	**70**	100	110	120

We call ξ the threshold constant and evaluate its effects further in Section 6.

5. CREDIT SCHEME, CRISP

In the previous section, we derived the optimal rate at which nodes should receive and forward the data. In this section, we design the credit scheme in which receiving and forwarding at the exact same rate maximizes a node's profit.

Payments and Rewards: We require a node to make a payment $\beta > 0$ to receive a data packet and earn a reward $\alpha > \beta$ for forwarding a data packet. The rate of profit P_v^x a node v makes on receiving commodity x at rate λ_v^x and forwarding at rate $\Gamma_v^x(\lambda_v^x)$ is given by $P_v^x = \alpha\Gamma_v^x(\lambda_v^x) - \beta\lambda_v^x$. Using notation from Section 3.2, the net rate of profit P_v at node v is given by $P_v = \sum_{x \in V-\{x\}} P_v^x = \alpha\Gamma_v(\lambda_\mathbf{v}) - \beta\mathbf{S_v}\lambda_\mathbf{v}$.

5.1 Incentive Compatibility

Based on the above payment mechanism, we show that setting the correct ratio of payment to the reward results in maximizing the profit at the optimal rate of arrival and hence, the credit scheme is incentive compatible.

THEOREM 2. *Suppose a node receives data at rate $\mathbf{S_v}\lambda_\mathbf{v}$, and forwards at rate $\Gamma_v(\lambda_\mathbf{v})$. The profit $P_v = \alpha\Gamma_v(\lambda_\mathbf{v}) - \beta\mathbf{S_v}\lambda_\mathbf{v}$ is maximum at the optimal rate of arrival $\hat{\lambda}_\mathbf{v}$ when β/α is set equal to ξ.*

PROOF. Let $\hat{\lambda}_\mathbf{v}$ be the vector of optimal rate of arrival as defined in Theorem 1, and the corresponding profit be $\hat{P}_v$. Next we show that $\hat{P}_v \geq P_v$. From the definition of profit, $\hat{P}_v - P_v = \alpha(\Gamma_v(\hat{\lambda}_\mathbf{v}) - \Gamma_v(\lambda_\mathbf{v})) - \beta\mathbf{S_v}(\hat{\lambda}_\mathbf{v} - \lambda_\mathbf{v})$. Using steps similar to the ones used to derive (8), we get

$$\hat{P}_v - P_v \geq -\alpha(\xi\mathbf{S_v}\mathbf{h_v}) - \beta\mathbf{S_v}(\hat{\lambda}_\mathbf{v} - \lambda_\mathbf{v})$$

$$\Rightarrow \hat{P}_v - P_v \geq (-\alpha\xi + \beta)\mathbf{S_v}\mathbf{h_v}$$

When $\beta/\alpha = \xi$, we get $\hat{P}_v - P_v \geq 0 \Rightarrow \hat{P}_v \geq P_v$. □

Credit Scheme: For $\beta > 0$ and $\alpha := \beta/\xi$, the exact payments to network and rewards from network for a node's role as source, relay and destination are as follows.
Relay: The payment to receive a packet is β and (expected) reward upon forwarding a packet is $\alpha := \beta/\xi$.
Destination: The payment to receive a packet is β and reward upon acknowledging the received packet is β.
Source: The payment to generate a packet is a fixed amount to cover net cost on network.

5.2 Collusion Resistance

Source, relays and destination can collude with each other to increase their profit, e.g., a source can give destination a behind–the–scene compensation to not acknowledge the received packet (as a result the source will not have to pay for the reward of relays). Possibilities of such collusions are analyzed in [34] which provides a number of conditions to ensure collusion resistance. In CRISP, the reward of relays when the packet is not delivered is zero (corresponds to $\gamma = 0$ in [34]). Based on analysis in [34], the only additional requirement to make CRISP collusion-resistant to such manipulations is that the source makes the same fixed payment to the network irrespective of whether the packet is delivered or not. When the packet is delivered, each node gets a reward based on its improvement, otherwise the payment stays in the network. However, analysis in [34] does not consider collusion among relays to add fake relays or to change routing utility. CRISP secures against these collusions [Section 2] as adding fake relays does not increase profit, and changing routing utility reduces profit.

5.3 Opportunistic Framework

Each relay pays β to the previous relay in the path in order to receive a packet. If the packet is delivered, the source makes the final payment to the relays and the destination; otherwise, the source pays to the network. To enforce this above payment scheme, any of the existing frameworks [20, 37] can be used. Here, we briefly discuss a completely decentralized framework, without any requirement of an offline bank. Relaxing the requirement of an offline bank is useful for maintaining credits on sensor nodes and devices that do not connect to a central server often enough to validate earned credits. Existing schemes have an online or an offline bank, which is a trusted authority. Each node submits all the transaction receipts or a summarized report of all transactions to the bank which can then check if a node is cheating and revoke its credits or its certificate. The problem becomes very challenging when such a trusted bank is not available to check all the receipts. A trivial solution is to keep track of all transaction receipts on each and every node, but it is prohibitive due a large overhead of maintaining these receipts on every node.

Forged Credits: To this end, we propose a novel solution based on a simple idea – *credit is neither created nor destroyed*; it is only transferred between nodes. Thus, each new node joining the network is initialized at some value, and then the only way for it to earn credit is by relaying traffic of other nodes. Thus, in the long run, a node provides the network some service proportional to the resources it consumes for itself. To identify collusion between two nodes to forge credit, each node also keeps track of the credit value at all other nodes. Then, if the value of total credit in the network does not stay constant, nodes can track back as to which node tried to reuse the credit.

Ensuring Payments: In case of a delivered packet, a source *has* to make a payment, because relays can identify the source of the packet, and upon not receiving the payment they can discontinue providing the service for non–paying source. To ensure a node's payment to the network when the packets are not delivered, a Merkel tree is used. We do not provide details here as a similar use for maintaining offline credit is proposed in [20, 37]. The difference is the leaf value of tree – in our case it is the hash of the packet a source generates. The root value of tree is flooded in the network by each source. To ensure that a source makes the payment, relays request the support of hash tree for all the packets that the relay forwarded. The overhead is just O(1) stored value per relay and verification of $O(\log(m))$ constant–size (e.g. 160–bit) hashes for m variable–sized packets that a relay forwards.

Consistency: To maintain and update credits of other nodes in a consistent manner, temporal distance [27, 29] is used. To secure the credits, each node signs the time–

stamped credit value with its private key. When two nodes come into contact, the node with smaller temporal distance with the destination, updates the other. Also, payments can be made either in a distributed manner through other nodes, or upon a direct contact with a relay. Payment upon a direct contact takes longer, but has low overhead in comparison to the other mechanism. In the next section, we evaluate our credit scheme. Due to limited space, we leave the detailed description and evaluation of payment framework for future work.

6. EVALUATION

A relay pays β credit to receive a packet and earns α credit upon forwarding a packet under the CRISP scheme. It has been shown in several studies that cheating in the routing and forwarding stage results in poor network performance [11]. Therefore, we consider a special type of cheating that is possible in the existing credit schemes. Overall, we consider three types of behavior:

- **Honest:** The relay truthfully follows the routing and forwarding protocol.
- **Selfish:** The relay modifies the routing utility to operate at the optimal rate [Theorem 1].
- **Cheating:** The relay modifies the routing utility to maximize the forwarding rate.

Existing schemes promote cheating (a node maximizes its profit by maximizing the forwarding rate) and therefore suffer from reduced network throughput. In our evaluation, we show that,

1. Cheating reduces the network throughput compared to honest behavior.
2. CRISP discourages cheating and promotes honest and selfish behavior.
3. Selfish behavior result in similar throughput compared to honest behavior.

CRISP promotes honest behavior and beneficial selfish behavior (by taking into account the packets received in addition to the packets forwarded) and thereby maintains good throughput. We show all of the above properties hold for a wide range of ξ values, packet expiration times, and packet generation rates in both real and synthetic mobility environments.

Simulation setup: Extensive simulations are run on real and synthetic mobility traces. The real trace was collected from people walking in a state fair [26], and synthetic trace is generated from a levy walk mobility model [25]. Messages are generated at each node for a random destination. The real mobility trace has been collected from participants that carry GPS receivers which log position at 30 second intervals. In order to make a comparison with suitable number of nodes, track logs from each day are considered to be a separate user. These logs have durations from around one to ten hours each day. We truncate all logs to 90 minutes during which 18 users record their location. Traces with levy walk are generated with the settings recommended in [25] with an area of $500 \times 500m^2$. All plots show the average of five simulation runs and the confidence interval is plotted on the first standard deviation. We use temporal distance [27, 29] as the routing scheme in simulation. All values of profit are normalized by the profit of selfish node and the value of throughput is normalized by the value of honest node.

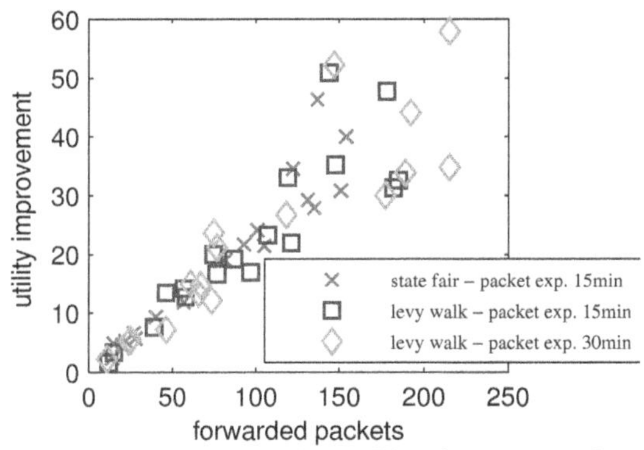

Figure 2: Improvement in utility is proportional to number of forwarded packets

6.1 Collusion Resistance

To secure against collusion to add fake relays, profit should be based on the total improvement a node makes in the path towards the destination as described in Section 2. Figure 2 shows that the expected reward based on improvement in utility is linear in the number of packets that a node forwards in different sets of environments. Thus, a reward based on improvement is equivalent to a reward based on number of packets (as described in our analysis) that a node forwards. Here, the value utility is an approximation for improvement that needs to be made to deliver the packet. We study the routing metric of temporal distance as is used in [19, 27, 29] and compute the utility by extending a relation derived in [27]. Thus, for a temporal distance t, the utility is computed as e^{-kt}, where k depends on the diffusion of the node (we take a value of 0.4/min, e.g., a 0.36 utility at a temporal distance of 2.5 mins). Thus, a node that receives a packet at $t = 10$ min and forwards it to a node at $t = 8$ min; receives an award of $e^{-0.32} - e^{-4}$ multiplied by the appropriate constant that makes the expected reward equal to α.

6.2 Value of threshold constant ξ

In order to find the actual value of ξ that maximizes throughput, we use the queue approximation result as described in Section 4.1. Let ψ be the average forwarding ratio among all nodes, i.e. the ratio of total packets forwarded to the total packets received over the whole network. From (7), the value of ξ is equal to $\frac{\partial \Gamma_v^x(\lambda_v^{x\star})}{\partial \lambda_v^x}$ where $\lambda_v^{x\star}$ is the solution for equation $\frac{\Gamma_v^x(\lambda_v^{x\star})}{\lambda_v^{x\star}} = \psi$. We only need to know value of δ_v^x in order to solve this equation[3]. In an actual network, the forwarding ratio of individual nodes is much different from the average value ψ. However, the arrival rates of all individual nodes still follow conditions of Theorem 1 for a constant value of ξ computed using the average value of ψ.

From simulation, we find ψ and the value of ξ that maximizes the throughput. Figure 3 compares the results from simulation and analysis. The lines show the mapping of forwarding ratio ψ to theoretically computed values of ξ at three different values of δ_v^x. The scatter points show the actual value ξ that maximizes throughput at the corresponding forwarding ratio ψ. Changing the generation rate or the

[3]The value of c_v^x is not required because ξ effectively depends on the ratio $\lambda_v^x / c_v^x = \rho_v^x$, i.e., $\xi = \frac{\partial(\rho_v^{x\star}(1-Q_v^x))}{\partial \rho_v^x}$ where $\rho_v^{x\star}$ is solution of equation $\frac{\rho_v^{x\star}(1-Q_v^x)}{\rho_v^{x\star}} = \psi$.

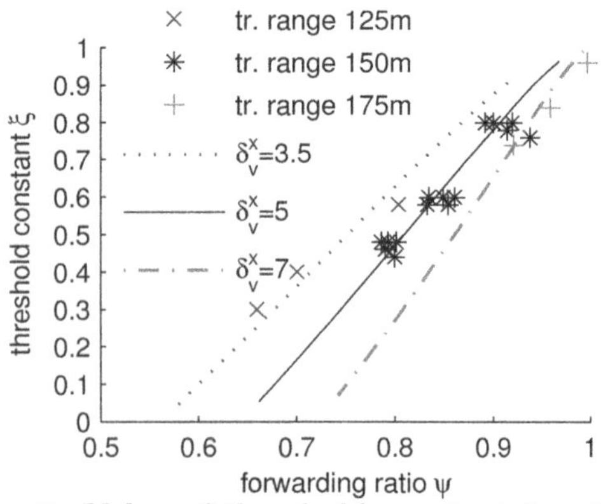

Figure 3: Value of threshold constant ξ – simulation and analysis

packet expiration time changes the forwarding ratio ψ at nodes. In a specific environment, the value of ξ with different forwarding ratios (generation rate or the packet expiration time) lies close to the line corresponding to a fixed value of δ_v^x. Changing the environment from being sparse (transmission range 125m, forwarding ratio 65%) to more connected (transmission range 175m, forwarding ratio 98%) corresponds to the value of δ_v^x from 3.5 to 5. Thus, at a given forwarding ratio ψ, using the values of $\delta_v^x \in [3.5, 5]$ one can estimate a range such that ξ that maximizes throughput belongs to that range. Later we show that CRISP is robust against inaccurate estimate of ξ.

6.3 Performance across nodes

In Fig. 4 and Fig. 5, each point corresponding to a node id is a complete simulation run in which that node's behavior is honest, selfish or cheating. Figure 4 shows that the network throughput when a node is honest or selfish is very close, but drops sharply when a node cheats. For example, the network throughput decreases by 15% when node #11 is cheating compared to the network throughput when node #11 is honest or selfish. The throughput degradation depends highly on the location and contacts of the node. For this reason, we show separate results for each of the cheating nodes. When two or more nodes cheat, then depending on their individual impact, they can drop a large percentage of data. For example, if node #1 and #2 cheat, they may only decrease the throughput by 5%, whereas if node #9 and #11 cheat, they can drop up to 25% of total data as can be seen in Fig. 4.

Figure 5 shows that the profit under CRISP of a selfish node is highest, a honest node is very close to it, and a cheating node is much lower (mostly negative). For example, when node #11 cheats, its profit is 500% less than when it is honest or selfish. The exception is when node #5 or #15 cheat, they make more profit than honest, but still less than selfish. This is not a problem because the throughput of selfish behavior is very close (same for node #5, slightly lower for node #15) to that of honest behavior as shown in Fig. 4.

6.4 Robustness and incentive compatibility

The value of ξ that maximizes throughput is 0.65 in Fig. 6 and 7 for the state fair trace. These figures show the results averaged over all nodes such that in one simulation run only a single node changes its behavior (honest, selfish or cheating). Figure 7 shows that profit under CRISP of honest (and selfish) behavior is greater than cheating for all

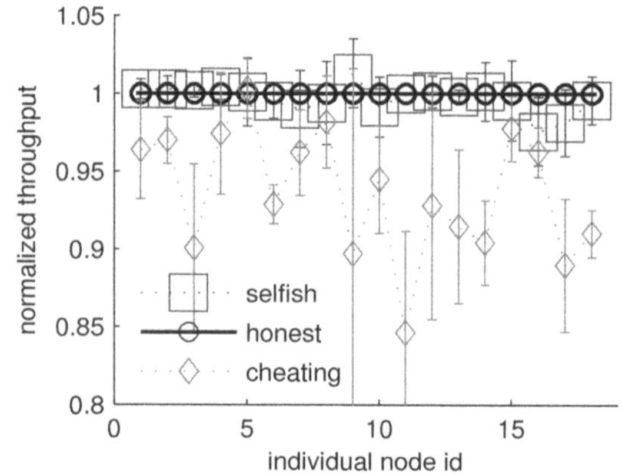

Figure 4: A single cheating node decreases network throughput by up to 15% (node #11).

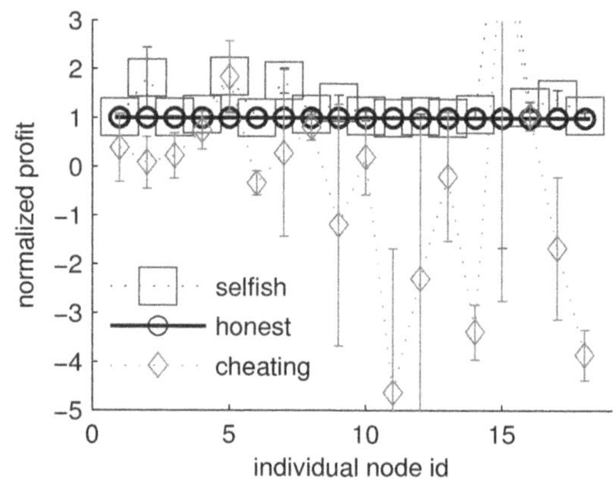

Figure 5: CRISP reduces the profit of cheating by up to 500% (node #11)

values of threshold constant $\xi > 0.4$. Also, for all values of ξ, Fig. 6 shows that throughput of selfish (and honest) behavior is greater than cheating. In real time, a node cannot figure out the exact utility it should declare for selfish or cheating behavior. In our simulations, we try out different (constant) shifts in the routing utility and plot the results only for the change in routing utility that maximizes forwarding rate (cheating) or profit under CRISP (selfish). Thus, CRISP is quite robust to inaccurate values of ξ and maintains incentive compatibility for a range (0.5 to 0.7) of ξ values because a node is guaranteed a good profit just by honestly routing and forwarding.

6.5 Performance across packet expiration times and packet generation rates

Figures 8 and 9, show mean throughput when packet expiration time and packet generation rates are varied in the state fair mobility trace. Each point corresponds to results averaged over 90 simulation runs (5 for each of 18 individual nodes when that node is honest, selfish or cheating). It is evident from Fig. 8 and 9 that the network throughput with selfish nodes is within 1% of that of the honest nodes whereas that of cheating nodes is around 5% lower. Note that the throughput is averaged over all nodes such that in one simulation only a single node changes its behavior (honest, selfish or cheating). Thus, the impact of multiple cheating nodes is much higher than 5%. Also, at a higher packet expiration time of 30 min, a cheating node attracts

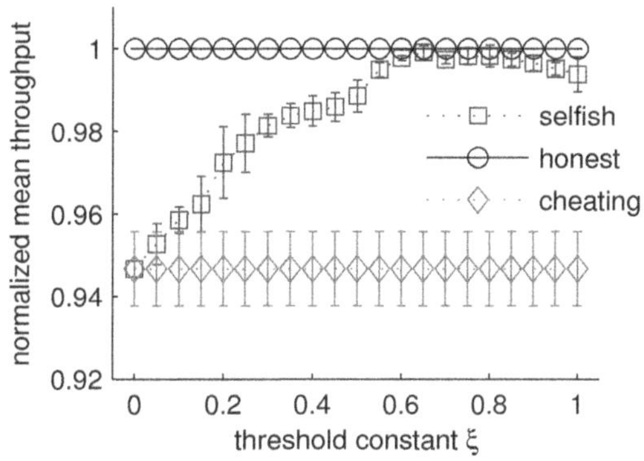

Figure 6: CRISP is robust over different values of threshold constant ξ – at all values of ξ throughput with selfish behavior is greater than that of cheating

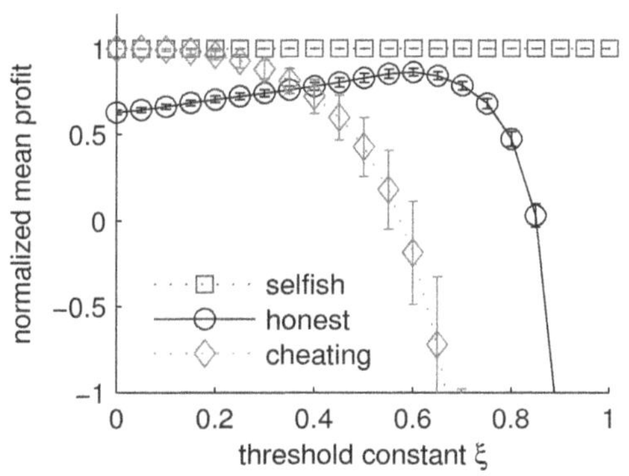

Figure 7: CRISP is incentive compatible over different values of threshold constant ξ – at values $\xi > 0.4$ profit of selfish (or honest) behavior is greater than cheating

more packets than it does at the expiration time of 15 min and less. Therefore, the throughput of the network decreases from 5% to 8% with high packet expiration times.

7. BACKGROUND AND RELATED WORK

In ad hoc networks, the earliest work proposes the use to a virtual currency 'nuglets' that are put with the generated packet and each relay on the forwarding path takes its share [6]. However, this scheme requires tamper–proof hardware. An elegant solution is proposed in [34] that provides credit to relays that forward a packet. The payment for source and relays is adjusted so that following the forwarding protocol is in their best interest. This solution, however, does not result in optimal selection of routing path. Therefore, a VCG mechanism based technique is proposed to select optimal routing paths as being truthful maximizes a nodes reward [1, 35]. [31] also considers the source with limited budget and provides a mechanism that makes routing/forwarding a Nash equilibrium strategy.

In opportunistic networks, [22] requires the relay to pay for receiving a packet, whereas it gets its reward by requesting a higher payment from the next relay. The exact amounts of payments are not discussed and the scheme requires a credit authority to be online in order to resolve payment conflicts. [20, 37] introduce the concept of layered coins for payments such that the credit authority never needs to be online at the time of transaction. The algorithm proposed in [8] secures information of each encounter in an un–forgeable

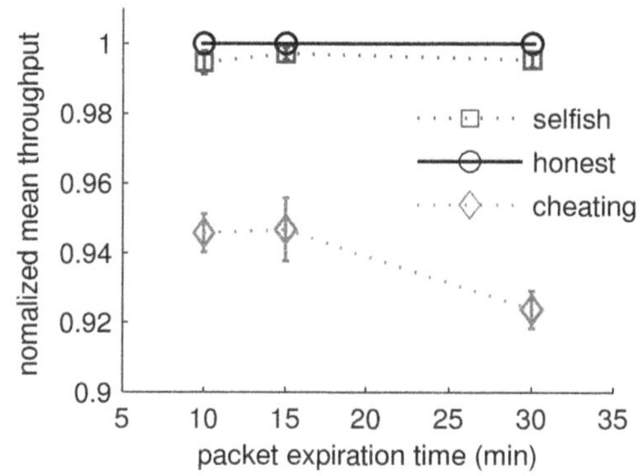

Figure 8: Throughput across different packet expiration times

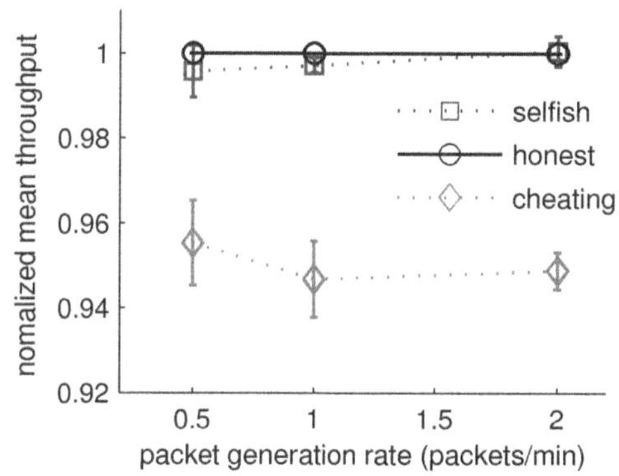

Figure 9: Throughput across different packet generation rates

ticket that is signed by private keys of encountering nodes. These tickets are then used to estimate delivery likelihood of nodes in [17]. While [17] secures the routing metric, a relay can still drop all the packets it receives using its true routing metric, and there is no check on that. [24] secures this limitation in [17], by recording the information about exchanged packets in encounter tickets, whereas, [18] address same limitation by using ACK packets and a more direct reputation–assisted forwarding mechanism.

Works in [1, 31, 35] still assume that nodes do not collude to increase their reward. [36] provides a mechanism to defend against collusion when colluding nodes do not completely trust each other and [32] shows that no mechanism can defend against collusion when colluding nodes can share each other's profit. Owing to this negative result on defense against collusion, [2] proposes a framework that monitors the network for special cases of collusion. However, such a framework cannot be used in opportunistic networks because observing network traffic is not possible due to disconnections and sparse connectivity. In opportunistic networks, all these techniques [8, 17, 18, 20, 22, 24, 37] fail when multiple nodes can collude with each other. In collusion, it can be assumed that nodes will share their private keys and earned rewards with each other. Therefore, some relays can add other relays in the path even when others do not take part in actual forwarding [8, 17, 18, 20, 22, 24, 37]. This way nodes can earn additional credit and use it to launch flooding or black–hole attacks. Similarly, colluding nodes can easily forge encounter tickets [8, 17], packet exchange records [24] or reputation values [18] by using their private keys and signing off fake receipts.

8. ACKNOWLEDGMENTS

This work was supported in part by the US National Science Foundation under grants ECCS-0824120, CSR 0834493, and CNS-1117866. Any opinions, findings, and conclusions or recommendations expressed in this paper are those of the author and do not necessarily reflect the views of the National Science Foundation.

9. CONCLUSIONS

In this paper, we proposed CRISP, a collusion-resistant credit scheme designed to ensure profit maximization when a node behaves honestly to receive and forward packets. Existing credit schemes reward upon forwarding packets which reduces the network throughput when a node tries to maximize its forwarding rate. In comparison, CRISP optimizes a specific packet arrival and forwarding rate by adjusting the ratio of reward to payment (for forwarding and receiving packets respectively). CRISP secures against collusion of nodes to change routing utility or add fake relays in the path and against both flooding and blackhole attacks. Extensive simulation results demonstrate that CRISP promotes honest (or selfish) behavior which maintains good throughput. CRISP is found to be robust with varying mobility characteristics, packet generation rates, packet expiration times and threshold constant. In future work, we plan to evaluate the opportunistic payment framework and extend CRISP for content sharing and service execution problems in opportunistic networks.

10. REFERENCES

[1] L. Anderegg and S. Eidenbenz. Ad hoc-vcg: A truthful and cost-efficient routing protocol for mobile ad hoc networks with selfish agents. In *ACM MobiCom*, 2003.

[2] M. Y. Arslan, K. Pelechrinis, I. Broustis, S. V. Krishnamurthy, P. Krishnamurthy, and P. Mohapatra. Detecting route attraction attacks in wireless networks. In *Proc. IEEE MASS*, 2011.

[3] D. Y. Barrer. Queuing with impatient customers and ordered service. *Operations Research*, 5(5):pp. 650–656, 1957.

[4] S. Boyd and L. Vandenberghe. *Convex Optimization.* Cambridge University, 2004.

[5] J. Burgess, B. Gallagher, D. Jensen, and B. N. Levine. Maxprop: Routing for vehicle-based disruption-tolerant networks. In *Proc. IEEE INFOCOM*, 2006.

[6] L. Buttyan and J. Hubaux. Stimulating cooperation in self-organizing mobile ad hoc networks. *ACM MoNet*, 8(5):579–592, 2003.

[7] M. C. C. Boldrini and A. Passarella. Contentplace: Social-aware data dissemination in opportunistic networks. In *Proc. MSWiM*, 2008.

[8] S. Capkun, L. Buttyan, and J. Hubaux. Sector: Secure tracking of node encounters in multi-hop wireless networks. In *Proc. ACM SASN*, 2003.

[9] B. B. Chen and M. C. Chan. Mobicent: a credit-based incentive system for disruption tolerant network. In *Proc. IEEE INFOCOM*, 2010.

[10] T. Chen and S. Zhong. Inpac: An enforceable incentive scheme for wireless networks using network coding. In *Proc. IEEE INFOCOM*, 2010.

[11] F. C. Choo, M. C. Chan, and E.-C. Chang. Robustness of dtn against routing attacks. In *Proc. COMSNETS*, 2010.

[12] M. Conti and M. Kumar. Opportunities in opportunistic computing. *Computer*, 43(1):42–50, 2010.

[13] M. Groβ and M. Skutella. Generalized maximum flows over time. In *Matheon Preprint 878*, 2012.

[14] A. Heinemann. *Collaboration in Opportunistic Networks.* VDM Verlag, 2007.

[15] P. Hui, J. Crowcroft, and E. Yoneki. Bubble rap: social-based forwarding in delay tolerant networks. In *Proc. ACM MobiHoc*, 2008.

[16] D. Lapsley and S. Low. An optimization approach to abr control. In *IEEE ICC*, 1998.

[17] F. Li, A. Srinivasan, and J. Wu. Thwarting blackhole attacks in disruption-tolerant networks using encounter tickets. In *Proc. IEEE INFOCOM*, 2009.

[18] N. Li and S. K. Das. Radon: reputation-assisted data forwarding in opportunistic networks. In *Proc. MobiOpp*, 2010.

[19] A. Lindgren, A. Doria, and O. Schelén. Probabilistic routing in intermittently connected networks. In *Proc. SAPIR*, 2004.

[20] R. Lu, X. Lin, H. Zhu, X. Shen, and B. Preiss. Pi: A practical incentive protocol for delay tolerant networks. *IEEE Trans. Wireless Commun.*, 9(4):1483–1493, 2010.

[21] N. Nisan and A. Ronen. Algorithmic mechanism design. In *Games and Economic Behavior*, pages 129–140, 1999.

[22] M. Onen, A. Shikfa, and R. Molva. Optimistic fair exchange in secure forwarding. In *Proc. MobiQuitous*, 2007.

[23] J. Ott, E. Hyytia, P. Lassila, T. Vaegs, and J. Kangasharju. Floating content: Information sharing in urban areas. In *Proc. IEEE PerCom*, 2011.

[24] Y. Ren, M. C. Chuah, J. Yang, and Y. Chen. Detecting blackhole attacks in disruption-tolerant networks through packet exchange recording. In *Proc. IEEE WoWMoM*, 2010.

[25] I. Rhee, M. Shin, S. Hong, K. Lee, and S. Chong. On the levy-walk nature of human mobility. In *Proc. IEEE INFOCOM*, 2008.

[26] I. Rhee, M. Shin, S. Hong, K. Lee, S. Kim, and S. Chong. CRAWDAD trace. `http://crawdad.cs.dartmouth.edu/ncsu/mobilitymodels/GPS/NC_State_Fair`, July 2009.

[27] U. Sadiq and M. Kumar. ProxiMol: Proximity and mobility estimation for efficient forwarding in opportunistic networks. In *Proc. IEEE MASS*, 2011.

[28] M. Shigeno. Maximum network flows with concave gains. *Mathematical Programming*, 107:439–459, 2006.

[29] T. Spyropoulos, K. Psounis, and C. S. Raghavendra. Efficient routing in intermittently connected mobile networks: the single-copy case. *IEEE/ACM Trans. Netw.*, 16(1):63–76, 2008.

[30] L. A. Végh. Concave generalized flows with applications to market equilibria. *CoRR*, 2011.

[31] W. Wang, S. Eidenbenz, Y. Wang, and X.-Y. Li. Ours: optimal unicast routing systems in non-cooperative wireless networks. In *Proc. ACM MobiCom*, 2006.

[32] W. Wang and X.-Y. Li. Low-cost routing in selfish and rational wireless ad hoc networks. *IEEE Trans. Mobile Comput.*, 5(5):596–607, 2006.

[33] F. Wu, T. Chen, S. Zhong, L. E. Li, and Y. R. Yang. Incentive-compatible opportunistic routing for wireless networks. In *Proc. ACM MobiCom*, pages 303–314, 2008.

[34] S. Zhong, J. Chen, and Y. Yang. Sprite: A simple, cheat-proof, credit-based system for mobile ad-hoc networks. In *Proc. IEEE INFOCOM*, 2003.

[35] S. Zhong, L. E. Li, Y. G. Liu, and Y. R. Yang. On designing incentive-compatible routing and forwarding protocols in wireless ad-hoc networks: an integrated approach using game theoretic and cryptographic techniques. *Wirel. Netw.*, 13:799–816, December 2007.

[36] S. Zhong and F. Wu. On designing collusion-resistant routing schemes for non-cooperative wireless ad hoc networks. In *Proc. ACM MobiCom*, 2007.

[37] H. Zhu, X. Lin, R. Lu, Y. Fan, and X. Shen. Smart: A secure multilayer credit-based incentive scheme for delay-tolerant networks. *IEEE Trans. Veh. Technol.*, 58(8), 2009.

A Social Node Model for Realising Information Dissemination Strategies in Delay Tolerant Networks

Saeed Bastani
University of Sydney, Australia
and NICTA, Australia
sbastani@it.usyd.edu.au

Bjorn Landfeldt
Lund University, Sweden
bjorn.landfeldt@eit.lth.se

Christian Rohner
Uppsala University, Sweden
christian.rohner@it.uu.se

Per Gunningberg
Uppsala University, Sweden
perg@it.uu.se

ABSTRACT

In the emerging Delay Tolerant Networks (DTNs) content dissemination platform, mobile nodes opportunistically exchange content as they meet, with the intent of sharing content among nodes with common interests. During a meeting, nodes can exchange both content of direct interest to themselves as well as content that is of interest to a larger set of nodes that may be encountered in the future. The utility of a DTN is governed by the content exchange opportunity (the amount of content that can be exchanged during a meeting) as well as the selection of content to be exchanged in order to maximise the interest nodes will have in information they are exposed to. Considering that there is a cost associated with the content exchange (e.g. battery usage, buffer occupancy or consumed transmission opportunity) the aim for nodes participating in content dissemination should be to maximise their payoff. In this paper, we contribute a generic framework for describing the characteristics of content exchange among participating nodes in a network. We incorporate a distributed information popularity measurement and the pairwise interaction of nodes modelled as a bargaining problem. The outcome of this process is the fair split up of transmission opportunity as a network resource and the selection of content objects to exchange in order to maximise the nodes' payoff. The framework is generally intended as a capstone for investigation of content dissemination properties and various content exchange strategies in a DTN. The paper further presents experiments conducted to validate the function and correctness of the proposed framework.

Categories and Subject Descriptors

C.4 [**Performance of Systems**]: Modeling techniques; C.2.4 [**Distributed Systems**]: Distributed applications

Keywords

Delay Tolerant Networks, Content Dissemination

1. INTRODUCTION

Delay Tolerant Networks (DTNs) consist of mobile devices communicating in a "store-carry-forward" fashion, without any assumption on the presence of infrastructure [14]. Sporadic connectivity and the lack of permanent end-to-end paths in these kind of networks hinder their applicability in delay sensitive applications; yet their potential utilization as a low cost network technology for delay tolerant content dissemination has attracted a significant attention from the research community [14, 13, 20, 2, 11]. Content dissemination in DTN is interpreted as a mechanism for carrying content to any node with interest in that content. Such a mechanism is typically realized by advertisement of nodes' interests/demands using the subscribe/publish paradigm [6] and selection of a group of nodes acting as forwarding brokers.

In a content dissemination network, there are few major players among others which determine the behaviour of the content dissemination process; for one, the dynamic physical network structure formed by meeting patterns of nodes governs the dissemination characteristics, albeit if the impacts of other players are ignored. Such a situation can be observed in existing works where the content dissemination by a node is governed by its membership of different node communities (e.g home community, exogenous or friend communities), where the membership of the node to a community is determined by relative meeting frequencies of nodes. The pairwise nodes interaction on meeting incidences is the second major player influencing the behaviour of the content dissemination process in terms of dissemination speed, dissemination preferences, etc. The nodes' interaction is driven by the current states of the participating nodes and the available network resource(s) to be used for content exchange. Moreover, depending on the selfish or cooperative interaction of nodes, various dissemination strategies emerge each having a different outcome. The third driving factor introduced in this work is the distribution of content popularity. Various dissemination strategies can be realized based on the popularity of different contents available/exposed in/to a node. Nevertheless, the realization of such strategies requires the information about content popularity to be accessible by the nodes either a priori or measured autonomously by nodes while they meet peers. In this paper, we propose a node model incorporating the fundamental characteristics of the latter two driving factors (i.e node interaction and content popularity). A game theoretic problem is proposed to describe nodes' interaction with the aim of fairly splitting the dwelling time (as the available content exchange opportunity) while taking into account the current state of the interacting nodes. Moreover, a distributed content popularity measurement mechanism is implemented in the nodes enabling a fresh perception of the popularity of the various contents in the network. This model enables a generic yet thorough investigation of the content dissemination process and potential content dissemination strategies.

2. RELATED WORK

Content dissemination has been the subject of extensive research in recent years. The efforts can be classified into two main directions: Routing in DTN and the application of various settings of the subscribe/publish paradigm.

Routing as a message forwarding mechanism is addressed in [17, 10, 9, 11] to mention but a few. These works attempt to deliver content opportunistically, selecting next-hop nodes as carriers mainly based on their mobility and collocation information. In [10], a probabilistic routing algorithm is proposed based on the period of collocation between nodes. [9], investigates methods for different target groups to disseminate content in an urban setting. [17] proposes a class of multi-copy protocols termed Spray routing to reach a trade-off between delay and transmission attempts. Other routing schemes propose different forms of flooding-control [21]. Routing based on knowledge oracles is proposed in [7]. The various knowledge oracles considered provide information about future contacts of nodes, available bandwidth, future traffic-demand of the nodes, and the queue sizes at each node. Other routing schemes propose different forms of flooding-control [21]. As a representative work, [18] proposed a controlled replication scheme termed *Spray and Wait* to reduce the number of copies of a given message, and hence the number of transmissions per message to L with the flexibility in tuning L in response to a delay requirement.

The Publish and Subscribe paradigm was originally applied to Internet-based scenarios. The authors in [5] introduced this paradigm to the context of mobile networks by establishing mobile publishers and subscribers and a set of backbone nodes responsible for content dissemination. In [15], a tree-based subscriber/publisher was proposed for relatively stable wireless environments. [20] proposed a broker based content dissemination scheme targeted for dynamic wireless environments. Brokers are elected from mobile nodes based on an election mechanism which relies on knowledge about the community structure of the underlying network. Thus, the structure of the network in terms of the existing communities is identified and a broker(s) is selected from each community to disseminate content. The *Haggle* project released implementations of data/content centric networks [13]. In Haggle, node descriptions form an individual interest vector of attributes and assigned weights that is matched to data attributes [1]. Content dissemination in Haggle is done on two layers: i) ranked searches, that is matching of nodes' individual interest vectors with the data in the cache. Ranked searches identify and prioritize data to be transmitted during a node contact, ii) traditional forwarding among nodes to identify delegate forwarders (nodes that are not interested in the data but are likely to deliver it). *Bloom filters* are used to avoid duplicate transmission of data that the other node already has; nodes exchange the Bloom filter instead of a long list with data in the caches. In [2], authors suggest each mobile node acts as a broker, arguing that building and maintaining of broker overlay is cost inefficient. In both [13] and [2], an autonomous community detection mechanism is used to identify communities and a weight is assigned to each cached object at the time of exchange to decide whether or not to fetch and forward that object, taking into account the current community of the node. In an attempt to minimise the computation and communication load imposed on intermediaries taking the burden of content relaying and in the meantime achieving a high delivery ratio, [11] proposes a mechanism termed Habit for content dissemination in MANETs.

Our work differs from the existing proposal in the following ways. First, most previous work consider a content as a black box without considering the subjects of the content and they also assume that there is a demand or a known popularity per content item, whereas in this work we consider the subject(s) of content and develop a distributed popularity measurement based on nodes' interests. Second, in contrast with some previous work oblivious to network resources when addressing node behaviour, we introduce nodes' contact time as an important resource of the communication network and propose a cooperative bargaining solution to determine and assign a share of the resource to each node. Furthermore, instead of forcing a certain set of nodes, like brokers or central nodes as in the existing proposals, to take the burden of content dissemination, we relax this reliance and let each node participate in content distribution where the degree of participation is governed by node's state and constraints. Third, in addition to the structural element of the content network, we take into account the impact of an important behavioural element realized in form of shift in individuals' information interests. Moreover, unlike the traditional interpretation of structural element as physical communities emerging based on the meeting patterns of nodes, the approach followed in this work is to employ the logical structure(s) in the network to devise realistic content dissemination strategies. A logical structure realized as a node community is a set of nodes with a certain meeting pattern and similar information interests. The last difference of this work with some of the previous proposals arises from the point of view of privacy. Unlike approaches using identities or positions of nodes to detect node communities, our approach does not use such attributes.

3. MODEL COMPONENT AND FUNCTION OVERVIEW

We model social node behaviour in a DTN as they meet and exchange information with their peers. A user model, as a component of the node model, captures the information interests of individual nodes and assign a weight to each information type the node is interested in. A second component termed social popularity model is responsible for collecting peers' interest vectors on meeting incidences and construct a pair of interest types and interest weights vectors representing the collective information interest of the network. We incorporate two fundamental features into the measurement mechanism; shift in individuals' information interests and the formation of local communities, the former is a behavioural element while the latter is a structural element. We apply the measured information popularity to assign forwarding priorities to the data objects currently in a node's cache. When it comes to data objects' exchange, the measured information popularity is coupled with the interaction model of meeting nodes, taking into account the state and constraints of nodes participating in information exchange sessions. Figure 1 illustrates the minimal components needed to define the social node behaviour. The internal functionalities associated with the components of the framework are described in the following sections.

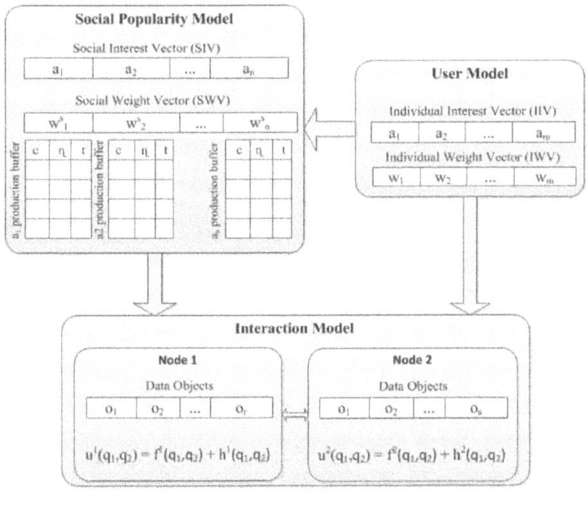

Figure 1: Node model in a delay tolerant network

3.1 User Model

We denote a consumer of information a 'user' regardless if the consumer is a person or a machine. The user model describes a social node's preferences or valuation of existing information attributes. It encompasses two major tasks; first, it uses a predefined internal process to identify the types of information the user is currently interested in. This may in practice be implemented using direct feedback from a user or by implementing a background process which monitors the user's activities and usage patterns. The second component is a representation of the measured valuation that is presented to the peers in the DTN for estimation of some information's social popularity.

In this work, we assume a background user model is present and generates a ranked list of information attributes, sorted in descending order of attribute ranks. In Figure 1 this implies that attributes a_1 and a_m have the highest and lowest ranks respectively. Such a ranked list determines the relative importance of information attributes and not a quantified absolute importance of a given attribute in the list. To quantify the absolute importance, a weight is assigned to each attribute representing, e.g. the fraction of time a user spends on consuming information with the given attribute. The exact meaning of importance can be defined for each application of the DTN.

We do not consider the implementation details of the actual user model and without loss of generality, we assign to each attribute a weight value generated from a distribution function. Among the candidate distribution functions, Zipf has proved to be a good representative of many real life complex systems. The application of Zipf has also been introduced to the context of Internet and social networks [4]. We apply the Zipf distribution to obtain attribute weights as follows:

$$f(k,\alpha,m) = \frac{\frac{1}{k^{\alpha}}}{\sum_{i=1}^{m}\left(\frac{1}{i^{\alpha}}\right)} \tag{1}$$

where m is the total number of attributes in the individual interest vector (IIV) of a node, k is the attribute rank, and α is an exponent value characterizing the distribution.

To this end, the proposed user model is represented by a ranked attribute list (i.e. IIV) and an individual weight vector (IWV). Given a list of ranked attributes in a node's IIV, the Zipf distribution determines the usage frequency for each attribute. The frequency values generated by Zipf distribution are then used as the weight elements in the IWV shown in the user model component in Figure 1.

3.2 Measurement of Information Social Popularity

A Social Interest Vector (SIV), captures the social popularity of information attributes currently advertised by nodes. The SIV comprises a subset $\{a_i | a_i \in \Omega\}$ of attributes where Ω is the set of all possible attributes in the network and has cardinality N. SIV is built and maintained continuously and incrementally by a node as it meets peers. More specifically, during a meeting incidence, nodes exchange IIV's and IWV's. New attributes from the collected IIV are added to the SIV and the attribute weights are adjusted accordingly and maintained in a vector denoted by SWV.

In the envisioned information popularity measurement model we incorporate two major social concepts, a behavioural shift component and a social structural component. The behavioural element reflects the transient nature of individual interests, i.e. the shift in user's information interest over time. This implies that the model should not rely on one-time collected interest vectors corresponding to peers encountered in the past. Instead, whenever a node is encountered, the SIV and SWV in a node are updated to reflect the possible changes in the interests of encountered nodes. This process guarantees that the model adapts to emerging events at all times during its life cycle. The social structural element involves the formation of *local communities*, among other structural entities representing the real world interactions of nodes. A local community from the standpoint of a given node is a set of nodes encountered frequently and recently compared to other peers in the network. It is not necessary for a node to have similar interests (i.e. IIV) to its local community. The notion of a local community implies that the SIV measured in a node will have a significant component induced by the local community compared to the component(s) induced by occasionally encountered peers. From a content exchange perspective, if the content in a node's cache is ordered with respect to the social interests measured in the node and represented by pair (SIV SWV), this leads to a state where a node with tight connection to its local community will prefer dissemination of the content that is of interest to its own community to the content of interest to the rest of the network. Moreover, nodes with balanced membership to several communities will act as *bridges*, muling data objects between those communities. To this end, we propose an adaptive information popularity measurement model, taking into account the properties described above. The popularity measurement process is illustrated by algorithm 1. When a node -termed 'target' for clarification- meets a peer node, it collects the peer's (IIV IWV) pair. For each newly detected attribute a from the peer's IIV, an entry is created in the target node's SIV and the corresponding weight value in peer's IWV denoted by w_a is considered as the current social weight as seen by the target node. Accordingly, a new entry in the target node's SWV is created which maintains the collected weight value. Obviously, the weight value reported by the peer is not guaranteed to be the only component of the social weight of attribute a measured in the target node since in future meetings other nodes will report possibly different weight values for the same attribute. This implies that a weight aggregation mechanism should be designed that produces a single value as the current designated social weight of the attribute. Such an aggregation mechanism requires the reported weight values to be maintained in the target node which in turn necessitates the allocation of buffer space to the weight values of the attribute. If not handled properly, the size of the buffer will therefore grow proportional to the number of encountered peers which in turn raises scalability issues. To tackle this, a limited buffer space is assigned to the weight values of the attribute and a weight clustering scheme is designed for aggregation purposes. Denote by b^p_a the buffer with a fixed size P assigned to attribute a to maintain the weight clusters corresponding to this attribute. We denote this buffer *production buffer*. As shown in Figure 1, three information entities are maintained in the b^p_a for each weight cluster: i) c^a_i represents the weight value of the cluster i in b^p_a , ii) t is the last time c^a_i was updated, and iii) η^a_i represents the frequency at which the cluster i was selected as the nearest neighbour of the new coming attribute weights. η^a_i is normalized over a time duration T to be independent from values far in the past. The nearest neighbour is determined based on value $|w_a - c^a_i|$ and a parameter δ. If $|w_a - c^a_i| \leq \delta$ and $|w_a - c^a_i| = min(|w_a - c^a_j|) \quad \forall j \in [1, P]$ then w_a is classified to c^a_i and the new aggregated weight is obtained and assigned to c^a_i. If no cluster exists that satisfies $|w_a - c^a_i| \leq \delta$ and there exists an unused cluster entry among the P available clusters , w_a is assigned as the initial weight of a new cluster, otherwise a forced classification and aggregation is applied. Finally, the update frequency η^a_i of the aggregated cluster c^a_i is modified to reflect this new aggregation.

The aggregation procedure of a chosen weight cluster denoted by c^a_{old} (to highlight the fact that the weight of this cluster is old and subject to change) and a recently collected attribute weight w_a is defined as follows (for simplicity, the index i is eliminated from the cluster name):

$$c^a_{new} = (1 - \alpha(\Delta t))c^a_{old} + \alpha(\Delta t)w_a \tag{2}$$

where c^a_{new} is the weight cluster after aggregation. Δt indicates the time difference between the collection time of w_a and the last time the chosen weight cluster was updated. $\alpha(.) \in [0, 1]$ is a monotonically increasing function of time difference Δt. We apply $\alpha(.)$ to assign a larger weight to the new attribute weight (i.e. w_a) in the aggregation scheme. This ensures that a recently collected attribute weight will have a larger component in the resultant weight cluster and thus alleviates the impact of stale attribute weights.

In the aftermath of an aggregation event, the new weight cluster may become the neighbour of an existing cluster. Thus, in step (19) of the algorithm, the entire production buffer is evaluated to find and apply potential aggregations. The aggregation of two weight clusters is slightly different from the aggregation of a weight cluster and a single attribute weight as described in (2). In the former case the update rates of the two weight clusters should also be aggregated to obtain a single rate. Denote by η_1 and η_2 the aggregation rates of weight clusters c_1 and c_2. We define the aggregation rate η_{12} of the two cluster weights as $\eta_{12} = \min(\eta_1 + \eta_2, 1)$. This implies that the update rate of the resulting weight cluster is the accumulated update rates of the two neighbour weight clusters. The aggregation scheme is expressed as:

$$c^a_g = \begin{cases} (1 - \alpha(\Delta t))c^a_1 + \alpha(\Delta t)c^a_2 & s.t. \Delta t = t_2 - t_1 \wedge t_2 \geq t_1 \\ (1 - \alpha(\Delta t))c^a_2 + \alpha(\Delta t)c^a_1 & s.t. \Delta t = t_1 - t_2 \wedge t_1 > t_2 \end{cases} \tag{3}$$

where c_g is the aggregate weight. t_1 and t_2 are the last update (or aggregation) times of c^a_1 and c^a_2, respectively.

The aggregation approaches described by (2) and (3) ensures that a weight cluster with a large η will have a large contribution in the ultimate social weight of the attribute from the standpoint of a network encountered by a node.

Algorithm 1 -Attribute Popularity Measurement

```
Input: (SIV, SWV), (IIV, IWV), δ, T
 1: i ← 1
 2: while i ≤ size(IIV) do
 3:   j ← indexOf(SIV, IIV(i))
 4:   if j ≤ 0 then
 5:     j ← createAttEntry(SIV)
 6:     createWeightEntry(SWV)
 7:     b_i^p ← assignBuffer(SWV(j))
 8:     appendAtt(SIV(j), IIV(i))
 9:     k ← appendWeight(b_i^p, IWV(i))
10:   else
11:     k ← findNearestNeighbour(b_i^p, IWV(i)
12:     if distance(b_i^p(k), IWV(i)) ≤ δ OR isFull(b_i^p)
        then
13:       b_i^p(k) ← aggregate(b_i^p(k), IWV(i))
14:     else
15:       k ← appendWeight(b_i^p, IWV(i))
16:     end if
17:   end if
18:   η_i(k) ← updateFrequncy(b_i^p(k), η_i(k), T)
19:   (b_i^p, η_i) ← aggregateClusters(b_i^p, η_i)
20:   SWV(j) ← updateSocialWeight(b_i^p, η_i)
21: end while
22: SWV ← normalize(SWV)
23: (SIV, SWV) ← sortDescend(SIV, SWV)
```

Two complementary steps are performed in a node prior to content exchange with a peer. First, a weight value is obtained for each attribute. Second, the social weight vector is normalized to indicate the relative popularity of different attributes. These steps are required to determine the ordering of several contents to be forwarded and to calculate the similarity between a node's view of information social popularity and the individual interest of an encountered peer (see section 3.3). The former step is indicated in line (20) of the algorithm and the attribute weight is calculated as the normalized weighted sum of the cluster weights associated with that attribute. i.e.:

$$w_a^s = \frac{\sum_{j=1}^{|c^a|} c_j^a \eta_j^a}{\sum_{j=1}^{|c^a|} \eta_j^a} \quad (4)$$

where $|c^a| \leq P$ is the effective size of cluster weight vector. The latter step is indicated in step (22) and the attribute weights are normalized with respect to the attribute with the largest weight updates. A ranked list of attributes is then created in step (23).

The information popularity measured in a node and represented by the pair (SIV, SWV) is used to evaluate the relevance of data objects received in a node with respect to the collective information interests of all peers in the network encountered so far. The evaluation process involves two steps: i) weights are assigned to attributes contained in the object. We assume weights are generated using Zipf distribution as discussed in section 3.1) a disjoint vector comparison (as described in section 3.4) is applied to evaluate the similarity of object attributes and social attributes.

On a meeting incidence with a limited dwelling time, the evaluated relevance of data objects, in addition to other factors, is used to choose a subset of available data objects in a node to exchange with a peer. The object exchange decision is also influenced by the interaction features of the meeting nodes, which we detail next.

3.3 Interaction during a node meeting

When two nodes carrying a number of data objects meet, each node tend to pursue a data exchange strategy producing the highest possible payoff, where payoff is the difference between the profit and the cost emerging from enforcement of an strategy. Considering the data exchange as the main subject of interaction, each node, as its strategy, aim at establishing a balance between the number of data objects it receives and transmits from/to the encountered party so that the data exchange will maximize the node's payoff . In a more general form, given a limited dwelling time d during which nodes are able to exchange data objects, nodes seek a balance between the fractions of time d consumed for reception and transmission. The time fractions identified by a node as its strategy of choice can be different from those identified by the other party and in some cases the mutual strategies may conflict due to the selfish behaviour nodes. In this sense, the behaviour of the meeting nodes can be captured using a two-player bargaining problem. The main step towards solving this problem is to design the utility functions describing the node's payoff. A generic utility function of a node i can be expressed as:

$$u^i(q_i, q_j, s^i) = f^i(q_j, s^i) + h^i(q_i, s^i) \quad s.t. \quad q_i + q_j \leq d \quad (5)$$

where f^i is the payoff accrued by node i if node j plays strategy q_j. h^i is the payoff accrued by player i if it chooses strategy q_i. s^i represents the current state of node i. h^i may be negative, hence become a net cost, since there may be a cost involved in data transmission e.g. fast depletion of energy resources. Conversely, a buffer discharge in a node with limited buffer space is a representative example where h^i is positive. q_i and q_j are the individual strategies chosen by nodes i and j respectively. The constraint in (5) ensures that the nodes' total transmission time do not exceed the dwelling time (d). We assume the meeting nodes have an identical estimate of the dwelling time. The utility function of node j denoted by u^j is symmetric to u^i.

To characterize the generic utility function (5), it is necessary to identify the utility parameters associated with nodes' interaction. The utility parameters impact the strategy played by a node and are classified into two categories. The first category includes parameters describing the state of a node at the time of a meeting. In order to keep the model parsimonious, we restrict these parameters to a minimal set including the fraction of buffer space occupied (b), the fraction of energy consumed(ξ), and the current satisfaction level of the node (ν), where $\nu \in [0, 1]$ and is determined by the meeting history of a node and increases accumulatively as a node receives data objects matching its interest vector. The model can readily be extended with further parameters for special cases. The second parameter category is not directly related to a node's state; these parameters rather provide complementary information about the encountered party, hence impacting the strategy selection of a node. The parameter r defined as the similarity between a social weigh vector (i.e. SIV) measured in a node and the individual weight vector (i.e. IIV) of the other node is an example of parameters belonging to this category. In our bargaining problem formulation we apply parameters r and ν to describe the willingness of a node to receive data objects from the other party.

To illustrate the effect of the above mentioned parameters on the players' strategies, without loss of generality, we instantiate the generic utility function (5) with a concrete function defined as:

$$u^i(q_i, q_j, s^i) = r_{ji}(1 - \nu_i)q_j + b_i(q_i - q_j) - \xi_i q_i$$
$$s.t. \quad q_i + q_j \leq d \quad (6)$$

Rearranging with respect to q_i and q_j leads to a more straightforward expression:

$$u^i(q_i, q_j, s^i) = (r_{ji}(1 - \nu_i) - b_i)\, q_j + (b_i - \xi_i)\, q_i$$
$$s.t. \quad q_i + q_j \leq d \quad (7)$$

where by analogy between (5) and (7), $f^i = (r_{ji}(1 - \nu_i) - b_i)\, q_j$ and $h^i = (b_i - \xi_i)\, q_i$.

The definitions of f^i and h^i in (7) are intuitive; a node i tends to choose a long fraction of time for transmission (i.e. q_i) if there are a significant number of packets currently occupying its buffer space (represented by b_i) and/or its consumed energy reserve is low enough to transmit more objects. In line with this argumentation, node i is willing to accept data objects from node j if node i: 1)has enough buffer space, 2)it realizes that the the other party has interesting data objects, and/or 3)node i is starving due to not being satisfied during the previous interactions with other nodes. We emphasize that the utility function in (7) is representative, hence can be redefined with respect to the application under investigation.

The generic game theoretic framework described above can be further customized based on the target technology in use and also the user preferences. In a 3G network with memory rich user devices, b_i is presumably large and thus the energy parameter (ξ) will act as the major driving factor influencing the strategy of a node. On the other hand, with technologies lacking an energy constraint, e.g. Vehicular Ad hoc Networks (VANETs), the available energy reserve is large most of the time and a node's transmission is only restricted to the available amount of data objects and the dwelling time d. To incorporate these flexibility into the model, we define threshold parameters b_{th} and ξ_{th} and redefine b_i and ξ_i as follows:

$$\hat{b_i} = \begin{cases} \frac{b_i}{b_{th}} & b_i < b_{th} \\ 1 & o.w \end{cases} \quad (8)$$

$$\hat{\xi_i} = \begin{cases} \frac{\xi_i}{\xi_{th}} & \xi_i < \xi_{th} \\ 1 & o.w \end{cases} \quad (9)$$

In devices with strict buffer or energy constraints, smaller values of b_{th} and ξ_{th} will be preferred by the user. This settings will also enable nodes to become generous or conservative in data dissemination process depending on their current state and the values of threshold parameters.

A further step towards full characterization of the proposed game theoretic framework described by (7) is to investigate scenarios emerging with respect to selfish and cooperative behaviour of nodes.

3.3.1 Non-Cooperative Game Scenario

To identify the Nash equilibria of the game defined by (7), we find the mutual best response strategies of the nodes. Fixing the strategy of node j at any q_j, the best response strategy q_i of node i is only dependent on h_i. Thus the best response strategy can be found by maximizing h_i while taking into account the constraint $q_i + q_j \leq d$. This yields:

$$q_i = \begin{cases} 0 & \beta_i < \alpha_i \\ d & o.w \end{cases} \quad (10)$$

where $\alpha_i = r_{ji}(1 - \nu_i) - \hat{b_i}$ and $\beta_i = \hat{b_i} - \hat{\xi_i}$. Applying the same argument to node j, the best response q_j is obtained as:

$$q_j = \begin{cases} 0 & \beta_j < \alpha_j \\ d & o.w \end{cases} \quad (11)$$

where $\alpha_j = r_{ij}(1 - \nu_j) - \hat{b_j}$ and $\beta_j = \hat{b_j} - \hat{\xi_j}$. Combining (10) and (11), the set of equilibria existing in the game are obtained: $(\mathbf{q_i^*}, \mathbf{q_j^*}) = \{(0,0), (0,d), (d,0), (d,d)\}$ Obviously, the last equilibrium (d,d) is not feasible since it violates the game constraint. This is in fact the sole case when an actual competition between nodes takes place. However, due to the infeasibility of the equilibrium, no node benefits from the competition. We tackle this situation in the context of a bargaining problem in next section.

3.3.2 Cooperative Bargaining Scenario

In cooperative games, players (in our case, meeting nodes) try reaching an agreement on the splitting of a resource that yields mutual advantage. In this case, the resource is the amount of time available for transmission, i.e. dwelling time d. A player i has its own utility function $u^i(q_i, q_j, s^i)$ that can be derived from the allocated resource and it also has a minimum desired utility $\left(u_0^i(q_i, q_j, s^i)\right)$, termed the disagreement point. The disagreement point is the minimum utility that each user expects to accrue by participating in the game without cooperation. Thus, it is safe to assume that the initial desired resource is at least guaranteed for each user in the cooperative game. Assume, $U = \left\{\left(u^i, u^j\right)\right\} \subset R^2$ is a feasible utility set which is convex, non-empty, closed and bounded. Also, assume $U_0 = \left\{\left(u_0^i, u_0^j\right)\right\} \subset R^2$ be the disagreement point. The pair (U, U_0) together describes the bargaining problem. We denote by B the subset of all rational and Pareto optimal pairs in region U. Pareto optimal points among players are points such that it is impossible to discover other points resulting in strictly larger advantage for the two players simultaneously.

Definition 1 (Pareto Optimality): In a two-player game with players i and j, a utility pair $\left(u^i, u^j\right) \subset R^2$ corresponding to a resource allocation pair (q_i, q_j) if for each $\left(\hat{u^i}, \hat{u^j}\right) \in U$, $\left(\hat{u^i}, \hat{u^j}\right) \geq \left(u^i, u^j\right)$ implies $\left(\hat{u^i}, \hat{u^j}\right) = \left(u^i, u^j\right)$.

Definition 1 implies that there may exist an infinite number of Pareto optimal points in the game. Hence, selection criteria are needed for the bargaining solution in order to identify a Pareto optimal point which is in the best interest of both players. Different bargaining solutions provides different criteria in terms of optimality and fairness for different bargaining problems. Nash Bargaining Solution (NBS)[12] and Kalai-Smorodinsky Bargaining Solution (KSBS)[8] are the most popular bargaining solutions used in the literature for different application domains. These solutions differ in several ways; KSBS preserves all the axioms of the NBS except the independence of irrelevant alternatives that is replaced by the axiom of individual monotonicity. This axiom can be used to solve application specific problems. For instance, it might be necessary to improve the quality of some selected players (e.g., players transmitting more important content) by allocating them additional resources. The KSBS does not impose restriction on convexity of feasible utility set, while the convexity is mandatory for a Nash bargaining solution to be applicable. Moreover, the KSBS provides different types of fairness as opposed to the Nash bargaining solution. As we avoid to be bounded to a specific utility function and utility set and also due to the possible fairness requirements, we opt for KSBS as our choice of bargaining solution throughout this paper.

As the first step towards developing the KSBS for the game problem defined by (7), we identify the Pareto frontier points forming the bargaining set B. A Pareto frontier point in this game is the pair $\left(u^i, u^j\right)$ of utilities corresponding to a feasible resource allocation (q_i, q_j) such that $q_i + q_j = d$. We solve (7) for q_i and q_j as functions of u^i and u^j and obtain the bargaining set B:

$$B = \left\{\left(u^i, u^j\right) \mid q_i\left(u^i, u^j\right) + q_j\left(u^i, u^j\right) = d \right.$$
$$\left. s.t \quad u^i \quad and \quad u^j > 0\right\} \quad (12)$$

Assuming that each node is aware of its desired utility, the KSBS solution must satisfy the following equation [16]:

$$\mathbf{u}^* = \mathbf{u_0} + k^* \left(\mathbf{u_{max}} - \mathbf{u_0}\right) \quad (13)$$

where $\mathbf{u}^* = \left((u^i)^*, (u^j)^*\right)$ is the Kalai-Smorodinsky solution, k^* is the maximum value of k such that $\mathbf{u}^* \in U$, and $\mathbf{u_{max}} = \left(u_{max}^i, u_{max}^j\right) \geq \mathbf{u_0}$ specifies the best alternative (or the desired utility pairs) in U for each player. From (13) and recalling that $\mathbf{u_0} = (0,0)$ it can easily be verified that $\frac{u_i}{u_{max}^i} = \frac{u_j}{u_{max}^j}$. Thus, the KSBS is the intersection of bargaining set B described by (12) and the line S defined as:

$$S = \left\{\left(u^i, u^j\right) \middle| \frac{u_i}{u_{max}^i} = \frac{u_j}{u_{max}^j} \quad s.t \quad u^i \quad and \quad u^j > 0\right\} \quad (14)$$

Setting $u^i = ku^i_{max}$, $u^j = ku^j_{max}$, and substituting in (12) determines k, and hence $\mathbf{u}^* = ((u^i)^*, (u^j)^*)$. Correspondingly, q_i and q_j are obtained by applying $(u^i)^*$ and $(u^j)^*$ to (7).
The fairness criterion for the meeting nodes should also be fulfilled during a data object exchange. Considering the stochastic nature of the communication channel which may cause disruption during data exchange, a scheduling mechanism similar to Weighted Fair Queuing (WFQ) is employed to maintain fairness at all time during data exchange.

3.4 Similarity of Individual and Social Information Interests

Parameter r_{ij} (and similarly r_{ji}) in (7) plays an important role in the proposed game theoretic model of node interaction, since it determines the degree of a node's willingness for cooperation. Recall that we defined r_{ji} (r_{ij}) as the amount of similarity between node j's (i's) SIV and node i's (j's) IIV. Also recall that data objects in a node are weighted and ranked for forwarding according to SIV and the corresponding SWV measured by the node. It follows that a high r_{ji} can be interpreted as a high probability that node j is carrying a number of objects in top positions of its buffer that are likely to be of interest to node i, thus giving node i enough motivation to receive those objects.
To calculate r_{ji}, we assume that node i shares its pair (IIV , IWV) with node j during the meeting incidence before the game starts. Then node j measures the similarity between node i's IIV ranked with respect to IWV and its own SIV ranked using SWV. Thus, the entire process can be reduced to the similarity measure of two ranked vectors. As per measure techniques [3, 19], the vectors subject to comparison here are generally disjoint and of non-uniform size. Furthermore, the similarity of the two vectors in positions with higher ranks is deemed to make a larger contribution to the overall measure. Such unique measurement requirements prevent direct application of the existing similarity measure techniques. Most relevant to our case is [19], where an indicator termed Ranked-Biased Overlapping (RBO) is calculated as the similarity measure of two infinite vectors. RBO of two vectors S and T is defined as:

$$RBO(S,T,p) = (1-p)\sum_{d=1}^{\infty} p^{d-1} A_d \quad (15)$$

where A_d is the degree of overlapping at depth d of the vectors and $0 < p < 1$ is a constant devised to assign an overlapping contribution proportional to the position in the vectors where overlapping occurs. In this sense, RBO is classified as a top-weighted technique. On the other hand, RBO is classified as an equal-size similarity measure, since the size of both vectors are assumed to be infinite. The latter characteristic of RBO hinders its direct applicability to our case. The main reason is that the overlap weights in (15) (i.e. $(1-p)p^{d-1}$) form a geometric series with their sum converging to 1 as the depth d approaches ∞. However, in our case, we deal with finite size vectors, thus the weighting scheme in (15) cannot be applied. Furthermore, in our case the vector sizes are not equal which implies a different similarity measure than RBO. Following this, we propose a new similarity measure which copes with the drawbacks of RBO while keeping its top-weighted feature intact.
Assume that the lengths of node j's SIV (and SWV) and node i's IIV (and IWV) are denoted by m and n, respectively. It follows that $m \geq n$. This is supported by the fact that the nodes exchange their IIVs on a meeting incidence and update their previous SIVs with respect to the received IIV from the other party. Thus, the updated SIV in a node is at least as long as the other party's IIV. We are also aware of the facts that IWV in a node i determines the relative importance of attributes contained in the node's IIV and the sum of weight elements in IWV is 1. Likewise, the relative importance of attributes in SIV are determined by weight elements in SWV and the sum of weight elements of SWV is 1. These features imply that the attribute weight elements maintained in node i's IWV and node j's SWV are good candidates for position-based weighting of the similarity measure. In other words, using the weight elements in IWV and SWV enables incorporation of the notion of top-weightedness in the similarity measure technique. Taking all these into consideration, we propose the following expression to calculate r_{ji}:

$$r_{ji} = \operatorname*{argmax}_k \left(p^k \sum_{d=1}^{n} \left(\frac{IWV_d + SWV_{k+d}}{2} \right) A\left(IIV_{1:d}, SIV_{k:k+d}\right) \right)$$
$$s.t.\ 0 \leq k \leq m-n \quad (16)$$

where k is the number of shifts applied to the smaller vector, IWV_d and SWV_{k+d} are the weight elements of IIV and SWV at attribute indexes d and $k+d$, respectively. $IIV_{0:d}$ and $SIV_{k:k+d}$ are the subset of attributes located at indexes 1 to d of IIV, and k to $k+d$ of SIV, respectively. We apply the arithmetic average of IWV and SWV as the overlapping weight. It is straightforward to show that $0 \leq \sum_{d=1}^{n} \left(\frac{IWV_d + SWV_{k+d}}{2}\right) \leq 1$, thus the similarity measure is normalized. Generally, any normalized combination of IWV and SWV which also preserves the top-weighted property can be applied to (16). Figure 2 depicts a physical interpretation of the proposed similarity measure with $p = 0.7$. In this figure the maximum similarity is achieved after one shift and $r_{ji} = 0.1204$.
We also apply (16) as a generic similarity measure of disjoint vectors to calculate the relevance between the data objects in a node's cache and the social information view of the node represented by (SIV SWV) pair.

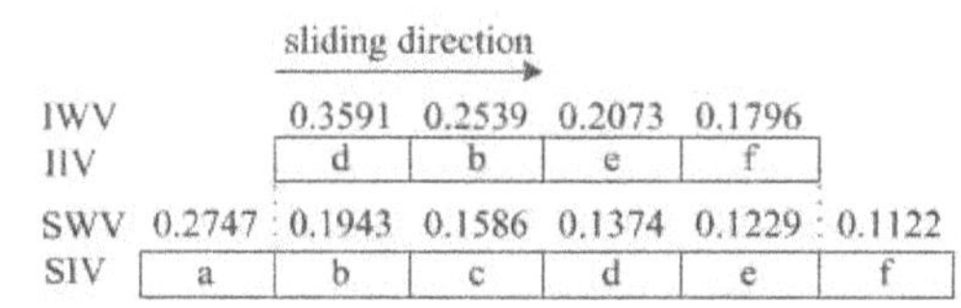

Figure 2: Similarity measure of individual and social interests. With $p = 0.7$ and $A = [0, 1/3, 2/4, 3/5]$, $r_{ji} = 0.1204$ is obtained at $k = 1$.

4. EXPERIMENTAL RESULTS

The model introduced in this paper represents a generic capstone architecture, allowing specific dissemination strategies to be realised by extending the generic model. Therefore, we carried out numerical studies to validate that the model behaves correctly according to basic elements of a content dissemination network, allowing such extensions to be

Table 1: Simulation Parameters

Ω	$d(sconds)$	p	$T(hours)$	δ	b_{th}	ξ_{th}
6	1.2	0.8	5	10^{-2}	0.7	0.2

realised. We focus on structural and behavioural elements to validate the proposed model. To validate the structural element, the impacts of the size of communities and the meeting pattern of nodes on the outcome of the content dissemination process were addressed. To perform validation with respect to the behavioural element, we addressed the impacts of strategies chosen by nodes as well as the shift in nodes' interests on the dissemination process. Performance aspects are left for the evaluation of such specific strategy extensions. For this reason, our studies are independent of performance considerations such as mobility rates and patterns, probabilities of successful transmissions etc.

We implemented a discrete event simulator in MATLAB to simulate meetings. Nodes were divided into separate groups where nodes in a group were assigned a subset of similar information attributes at the top of their interest vectors and the rest of a node's information attributes were chosen randomly and assigned to random positions in the interest vector. Different groups had different sets of attributes at the top of their nodes' interest vectors. The grouping scheme further allowed us to achieve high level of flexibility in representing various network conditions by defining custom meeting patterns inside and between nodes' groups. Similar to the node grouping, we also grouped the information objects with respect to the contained information types (i.e. attributes). The objects' attributes were assigned in a similar way to nodes' attributes. Each object group was targeted to a node group. Depending on the objectives of a simulation scenario, different number of groups were used. Throughout the simulations we assumed the total number of unique attributes (Ω) in the network to be 6. The object buffer in each node was set to a size of 1000 objects, and the attribute weight clusters used for popularity measurement in a node had a size of 4. Packet time and energy usage per packet were fixed to 0.3 second and 10^{-3}, respectively. Table 1 shows the remaining simulation parameters.

Figure 3 demonstrates the behaviour of the model in response to differences in node group sizes. The meeting rates of all nodes are identical. The x-axis represents the number of iterations and can be interpreted as the number of hours, days, or other unit used. The y-axis represents the penetration rate of objects of different types targeted to different nodes' groups. Figure 3(a) addresses a scenario where a weak majority exists in the network. Two groups of nodes are defined with sizes 51 and 49 nodes. The nodes in the former group are mainly interested in attributes 1 and 2 while the nodes in the latter group are mostly interested in attributes 3 and 4. Two groups of objects of equal size 50 are shared among two source nodes, each chosen from a group of nodes. A group of objects are of interest to a group of nodes; i.e. at the top of the nodes' attribute vectors, group 1 have attributes 1 and 2 and group 2 have attributes 3 and 4. As shown in Figure 3(a), the content of interest to the majority group (with size 51) is disseminated faster than the content targeted to the minority (with size 49). This scenario verifies that the popularity measurement and object forwarding model is capable of majority oriented dissemination even in presence of a weak majority. Figure 3(b) demonstrates a strong majority scenario. Three groups of nodes with sizes 30, 10, and 10 are defined and the interest vectors of the nodes in these three groups consist of attribute pairs (1,2), (3,4), and (5,6) respectively. 540 objects from three different types are initially shared by three sources selected from these groups. As expected, the group with a larger population dominates the community and receives its objects of interest faster. As the remaining two minority groups have identical sizes, they receive their objects of interest at identical rate as expected. In Figure 3(c) the three groups have equal sizes. Since the meeting rates of the three groups are also identical, no single group has an advantage and thus the content dissemination is performed uniformly. Figure 3(d) is an extension of the scenario demonstrated by 3(b). In this scenario, the group sizes are set to 30, 20, and 10 nodes. The result of this scenario leads us to conclude that in case of identical meeting rates, the dissemination priority of contents targeted to different node groups is identified based on the size of groups. It is worth mentioning that interpretation of network structure with respect to its physical and logical elements yield different number of communities in the above scenarios. The similar visit patterns of nodes implies for the presence of only a single community in all scenarios, whereas the content popularity distribution in the network implies for 2, 3, and 3 communities in scenarios (a), (b), and (c) respectively. The conclusion to be derived here is that given a uniform physical network structure, the dissemination behaviour is driven by the logical structure formed based on content popularity distribution and the network tends to serve majority communities with high priority.

Following this, we study scenarios where the meeting patterns of nodes play a key role in the outcome of content dissemination. We selected two scenarios of this kind as demonstrated in Figure 4. In these scenarios the groups were of identical sizes and we assigned a larger intra-group meeting rate to nodes in one of the groups whereas the meeting rates for the remaining groups and the inter-group meetings were identical. This setting only holds for a warm up period where nodes build their view of information social popularity. In the content dissemination phase we assigned identical visit rates to all groups. This approach enabled us to evaluate the function of the social popularity measurement component of the model. It is observed in Figure 4(a) that the content dissemination behaviour in this scenario is comparable to the scenario depicted in 3(b). This leads us to conclude that meeting patterns also contribute to the formation of a majority, thus forcing others to dedicate more resources to disseminate the content of interest to the majority. In the second scenario in this experiment, we introduced a degree of isolation in one of the two groups; i.e. a subset of nodes in a group never visits the others in the same group. As in the previous scenario, we defined this setting only for a warm up period. As shown in Figure 4(b), the content of interest to the isolated group is disseminated at a low rate. The intuition behind this phenomena is that the nodes in an isolated community will have a weaker belief in their social interests compared to the communities with strong bonds. This motivates the isolated nodes to join the non-isolated ones, forming a local community and contributing to the dissemination of the content they favour.

We investigate the impacts of social and individual oriented forwarding strategies on the outcome of content dissemina-

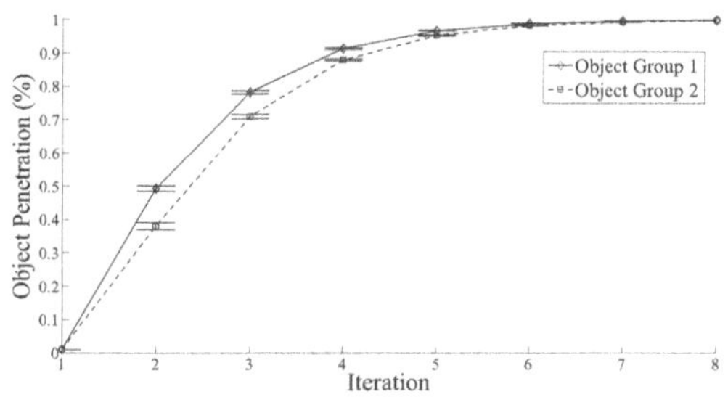

(a) group 2 is slightly less popular than group 1 (49 vs. 51 nodes)

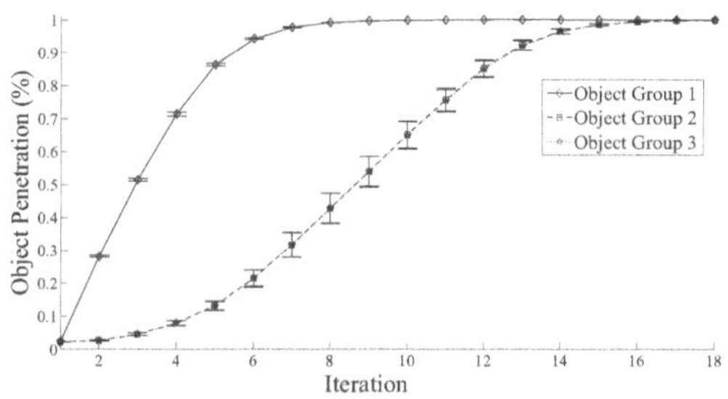

(b) group 1: high, group 2 & 3: medium popularity

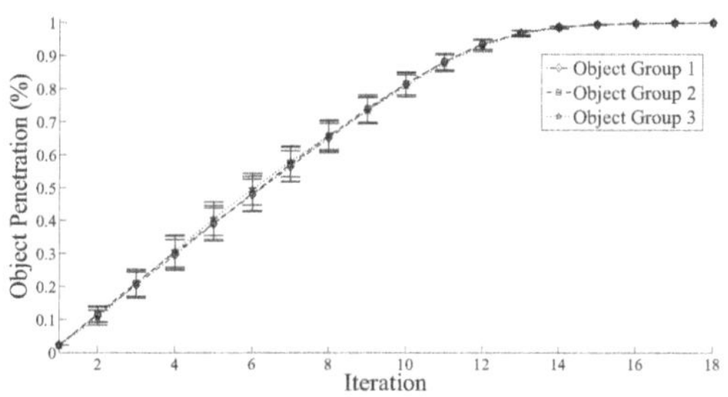

(c) groups with similar popularity

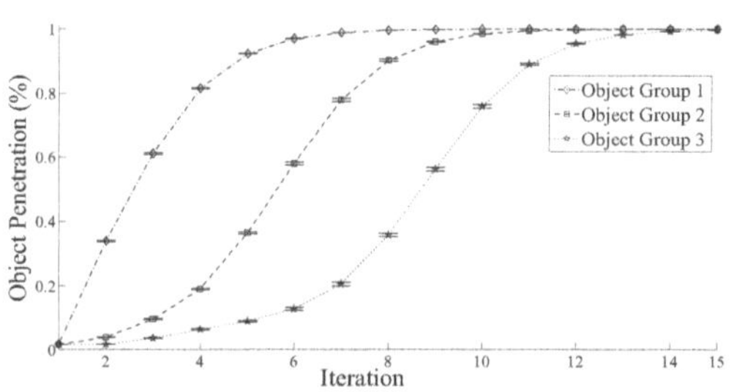

(d) group 1: high, group 2: medium, group 3: low popularity

Figure 3: Impacts of content popularity distribution on the content dissemination

tion in Figure 5. The solid curves show the penetration of object groups when the ordering of information objects are determined based on the individual interests of the encountered party and the dotted curves demonstrate the social oriented object forwarding. As in the scenario in Figure 3(d), three groups of nodes of sizes 30, 20, and 10 and identical meeting rates are configured. According to Figure 5, an individual oriented strategy causes the dominance of majorities on the content ordering to be reduced in favour of minorities and to be only proportional to the size of nodes with similar interests.

Finally, we address the impact of nodes' shifting interests on the content dissemination process. Two groups of nodes with sizes 12 and 10 and identical meeting rates were established. 2000 objects are initially shared by two sources selected from the two groups. After 10 iterations, 2 nodes from the larger group change their interests and become members of the smaller group. According to Figure 6, in the first 10 iterations, the first group becomes dominant and their targeted object group gains higher penetration. After the shift in interest, the second group becomes dominant and the dissemination priority changes as a result. The relatively higher fluctuations of the curves in 6 (compared to previous scenarios) are attributed to the elimination of the warm up period which in turn introduce some degree of randomness in the popularity measurement model.

As a general observation of all simulation scenarios, the pace of content dissemination decreases when the content density in the network increases. Such a finding is in agreement with dissemination objectives in DTN.

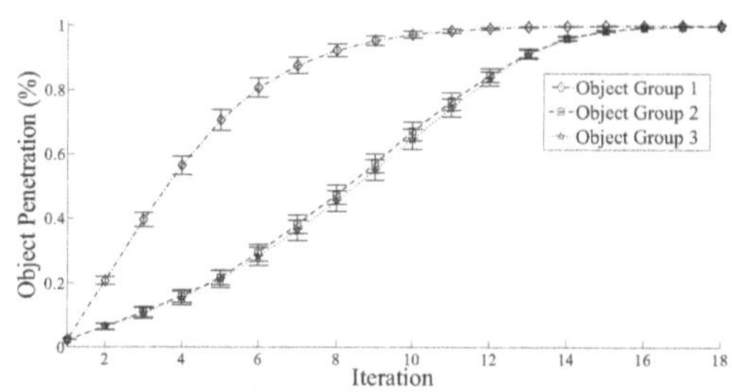

(a) group 1 and group 2 objects are interested by groups of nodes with high and low visit rates, respectively.

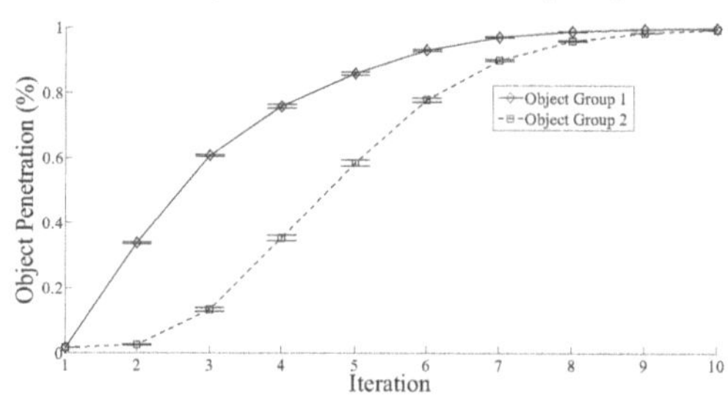

(b) group 1 and group 2 objects are interested by tight and scattered communities, respectively.

Figure 4: Impacts of non-uniform visit patterns on the content dissemination

5. CONCLUSION

We proposed a generic framework intended as a capstone architecture incorporating various content dissemination attributes. This framework, while being independent from the underlying network architecture and the mobility and meeting patterns among nodes, enables a typical node to dynamically build its view of the information interests of other participating nodes. Such capability coupled with the interaction model of meeting nodes facilitate the realization of various content dissemination strategies while fulfilling the individual nodes constraints in terms of buffer space, energy, etc. We numerically validated the model with respect

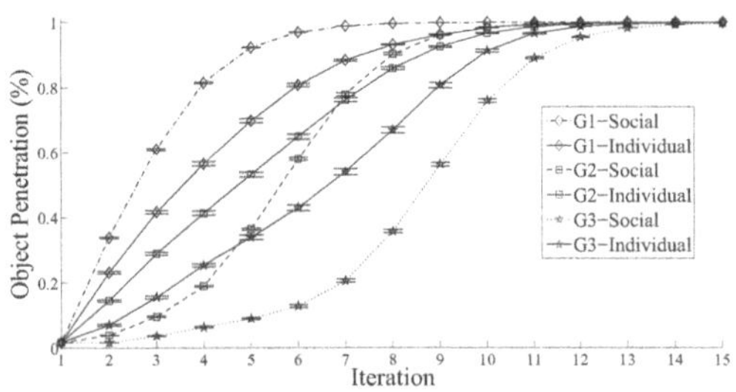

Figure 5: Comparison of social and individual oriented content dissemination strategies

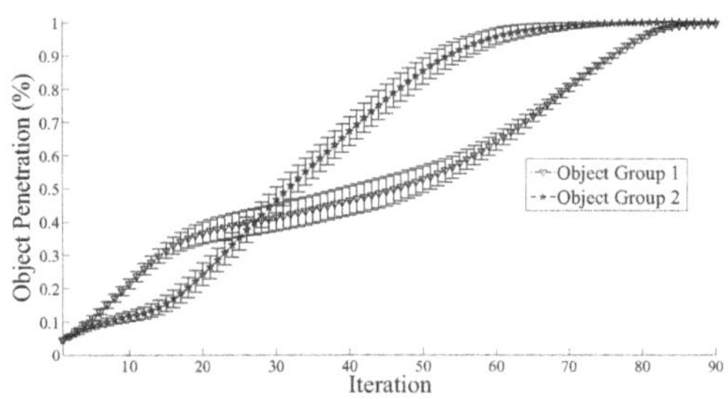

Figure 6: The impact of shift in nodes' interests

to structural and behavioural elements of content dissemination network, where we showed that the model well captures the essential network properties such as dynamic composition of nodes' communities and in the meantime reacts to the shift in information interests of nodes.

6. ACKNOWLEDGEMENTS

NICTA is funded by the Australian Government as represented by the Department of Broadband, Communications and the Digital Economy and the Australian Research Council through the ICT Centre of Excellence program.

7. REFERENCES

[1] F. Bjurefors, P. Gunningberg, E. Nordström, and C. Rohner. Interest dissemination in a searchable data-centric opportunistic network. In *Proc. of European Wireless Conference (EW)*, pages 889 –895, 2010.

[2] C. Boldrini, M. Conti, and A. Passarella. ContentPlace: social-aware data dissemination in opportunistic networks. In *Proc. of ACM MSWiM*, pages 203–210, USA, 2008.

[3] B. Carterette. On rank correlation and the distance between rankings. In *Proc. of ACM SIGIR*, pages 436–443, USA, 2009.

[4] C. Cattuto, V. Loreto, and L. Pietronero. Semiotic dynamics in online social communities. 46(2):33–37, 2006.

[5] G. Cugola, E. Di Nitto, and A. Fuggetta. The JEDI event-based infrastructure and its application to the development of the OPSS WFMS. *IEEE Transactions on Software Engineering*, 27(9):827 –850, September 2001.

[6] P. T. Eugster, P. A. Felber, R. Guerraoui, and A.-M. Kermarrec. The many faces of publish/subscribe. *ACM Computing Surveys*, 35(2):114–131, June 2003.

[7] S. Jain, K. Fall, and R. Patra. Routing in a Delay Tolerant Network. *SIGCOMM Computer Communications Review*, 34(4):145–158, 2004.

[8] E. Kalai and M. Smorodinsky. Other solutions to Nash's bargaining problem. *Econometrica*, 43(3):513–18, May 1975.

[9] J. Leguay, A. Lindgren, J. Scott, T. Friedman, and J. Crowcroft. Opportunistic content distribution in an urban setting. In *Proc. of ACM CHANTS*, pages 205–212, USA, 2006.

[10] A. Lindgren, A. Doria, and O. Schelén. Probabilistic routing in intermittently connected networks. *ACM SIGMOBILE Mobile Computing and Communications Review*, 7(3):19–20, 2003.

[11] A. Mashhadi, S. Ben Mokhtar, and L. Capra. Habit: leveraging human mobility and social network for efficient content dissemination in delay tolerant networks. In *Proc. of WoWMoM*, pages 1–6, 2009.

[12] J. Nash. The bargaining problem. *Econometrica*, 18(2):155–162, April 1950.

[13] E. Nordström, P. Gunningberg, and C. Rohner. Haggle: a data-centric network architecture for mobile devices. In *Proc. of ACM MobiHoc S3 Workshop*, pages 37–40, 2009.

[14] L. Pelusi, A. Passarella, and M. Conti. Opportunistic networking: data forwarding in disconnected mobile ad hoc networks. *IEEE Communications Magazine*, 44(11):134 –141, 2006.

[15] G. P. Picco, G. Cugola, and A. L. Murphy. Efficient content-based event dispatching in the presence of topological reconfiguration. In *Proc. of IEEE ICDCS*, USA, 2003.

[16] G. Rocheteau and C. Waller. Bargaining and the value of money. Technical report, 2005.

[17] T. Spyropoulos, K. Psounis, and C. Raghavendra. Efficient routing in intermittently connected mobile networks: The multiple-copy case. *IEEE/ACM Transactions on Networking*, 16(1):77 –90, 2008.

[18] T. Spyropoulos, K. Psounis, and C. S. Raghavendra. Spray and wait: an efficient routing scheme for intermittently connected mobile networks. In *Proc. of ACM SIGCOMM WDTN*, USA, 2005.

[19] W. Webber, A. Moffat, and J. Zobel. A similarity measure for indefinite rankings. *ACM Trans. on Inf. Syst.*, 28(4):20:1–20:38, 2010.

[20] E. Yoneki, P. Hui, S. Chan, and J. Crowcroft. A socio-aware overlay for publish/subscribe communication in delay tolerant networks. In *Proc. of ACM MSWiM*, pages 225–234, USA, 2007. ACM.

[21] Z. Zhang. Routing in intermittently connected mobile ad hoc networks and delay tolerant networks: overview and challenges. *Communications Surveys Tutorials, IEEE*, 8(1):24 –37, quarter 2006.

Lifetime Aware Deployment of k Base Stations in WSNs

Sabbir Mahmud
School of Computer Science and Engineering
The University of New South Wales
Australia
sabbirm@cse.unsw.edu.au

Hui Wu
School of Computer Science and Engineering
The University of New South Wales
Australia
huiw@cse.unsw.edu.au

ABSTRACT

In WSNs (wireless sensor networks), sensor nodes are typically battery powered. The number of base stations and their locations have significant impacts on the lifetime of a WSN. We study two multiple base station deployment problems. The first problem is to deploy k static base stations in a WSN such that the lifetime of the WSN is maximised. The second problem is to deploy k mobile base stations in a WSN such that the lifetime of the WSN is maximised. We propose a unified heuristic for both problems. In the special case of one static base station, we propose an optimal, polynomial time algorithm. In the special case of one mobile base station, we present an efficient heuristic. We have simulated our optimal algorithm, our heuristics, and the MinDiff-RE heuristic proposed by Azad and Chockalingam. The simulation results show that our heuristic for a single mobile base station prolongs the lifetime of a WSN by 253% on average compared to our optimal algorithm for a single static base station, and our heuristic for multiple mobile base stations performs 40% better than the MinDiff-RE heuristic on average.

Categories and Subject Descriptors

C.2 [**Computer-communication networks**]: Data communications; C.4 [**Performance of systems**]: Performance attributes

General Terms

Algorithms, performance

Keywords

Wireless sensor networks, base station deployment, network lifetime, wireless communication

1. INTRODUCTION

In wireless sensor networks, most of the energy of a sensor node is consumed by communications. The communication range of a sensor node is determined by its transmit power. The higher the transmit power, the longer the communication range [7]. sensor nodes are typically battery-powered. In order to save power, the transmit power of a sensor node needs to be low, leading to a short communication range. As a result, multi-hop communication between each sensor and its base station is typically used in large scale WSNs. In multi-hop communications, a sensor node may spend most of its energy on relaying transit data. Hence, it is important to shorten the hop distance between each source sensor node and its base station. The hop distance can be dramatically reduced by deploying multiple base stations. In order to deploy multiple base stations, we need to partition all the sensor nodes into multiple disjoint clusters, and place one base station in each cluster. Each sensor node sends its data only to its designated base station. The location of the base station of each cluster is very important. If a base station is deployed far from the data sources, many sensor nodes need to relay data packets. Therefore, the energy consumption of those sensor nodes will significantly increase. Hence, it is an important design issue to find the best location of a base station.

Given a cluster of sensor nodes and a base station at a fixed location, all the sensor nodes need to send their data to the base station via the neighbouring sensor nodes of the base station. Therefore, the sensor nodes near the base station consume much more energy than other sensor nodes, and die much sooner. A mobile base station is a solution to overcoming this problem. During the lifetime of a cluster, its base station is relocated to different locations to prolong the lifetime of the cluster.

In this paper, we study two multiple base station deployment problems. The first problem is to deploy k static base stations in a WSN such that the lifetime of the WSN is maximised. The second problem is to deploy k mobile base stations in a WSN such that the lifetime of the WSN is maximised. The lifetime of a WSN is defined as the shortest lifetime of all the sensor nodes in the WSN. We make the following major contributions.

1. We propose a unified, polynomial time heuristic for these two problems.

2. We present the first polynomial-time algorithm for optimally deploying a static base station in a cluster of sensor nodes such that the lifetime of the cluster is maximised.

3. We propose an efficient heuristic for deploying a mobile

MSWiM'12, October 21–25, 2012, Paphos, Cyprus.

base station in a cluster of sensor nodes such that the lifetime of the cluster is maximised.

4. We have simulated our optimal algorithm for deploying a static base station, our heuristic for deploying a mobile base station, our heuristic for deploying multiple mobile base stations, and the MinDiff-RE heuristic proposed by Azad and Chockalingam [2], on 36 instances of WSN with three different distributions. The simulation results show that our heuristic for deploying a single mobile base station prolongs the lifetime of a WSN by 253% on average compared to our optimal algorithm for deploying a single static base station, and our heuristic for deploying multiple mobile base stations performs 40% better than the MinDiff-RE heuristic on average.

This paper is organised as follows. Section 2 presents our network model and key definitions. Section 3 proposes a polynomial-time algorithm for optimally deploying a static base station and a heuristic for deploying a mobile base station in a cluster of sensor nodes. Section 4 describes a unified heuristic for deploying k static base stations and k mobile base stations. Section 5 discusses related work. Section 6 presents our simulation results, and Section 7 concludes this paper.

2. MODEL AND DEFINITIONS

A wireless sensor network consists of a set of n identical sensor nodes, each of which is located in a $2D$ plane. The location of each sensor node is known. All the sensor nodes have the same transmission range R. We assume that there are no communication obstacles between any two adjacent sensor nodes. Therefore, a sensor node v_i can directly communicate with a sensor node v_j if the Euclidean distance between v_i and v_j is not greater than R. There are k base stations to be deployed in a target WSN. As a result, all the sensor nodes need to be partitioned into k clusters with one base station in each cluster. A sensor node in each cluster sends its data to its designated base station only. If the Euclidean distance between a sensor node and its base station is greater than R, the data of the sensor node must be transmitted via other sensor nodes to the base station.

DEFINITION 1. *The connectivity graph of a wireless sensor network is an undirected graph* $G =< V, E >$, *where* $V = \{v_i : i = 1, 2, \cdots, n \text{ and } v_i \text{ is a sensor node}\}$, *and* $E = \{(v_i, v_j) : \text{if the Euclidean distance between } v_i \text{ and } v_j \text{ is not greater than } R\}$.

Without loss of generality, we assume that the connectivity graph G of a target wireless sensor network is connected. A connectivity graph is a unit disk graph [4].

DEFINITION 2. *Given two sensor nodes* v_i *and* v_j, *the shortest hop distance from* v_i *to* v_j *is the length of the shortest path from* v_i *to* v_j *in the connectivity graph.*

Let P be a set of n distinct points called sites, in a $2D$ plane. The Voronoi diagram [17] of P is the subdivision of the plane into n cells, one for each site. A point q lies in the cell of a site $p_i \in P$ iff the Euclidean distance between q and p_i is less than the Euclidean distance between q and p_j ($p_j \in P$ and $i \neq j$). The edges of the Voronoi diagram are all the points in the plane that are equidistant to the two nearest sites. The Voronoi diagram of n points in a $2D$ plane can be constructed in $O(n \log n)$ time [17].

DEFINITION 3. *A sensor node* v_i *is a neighbour of a sensor node* v_j *if the Voronoi cells of* v_i *and* v_j *share a Voronoi edge.*

DEFINITION 4. *Let* V *be a set of* n *sensor nodes in a* $2D$ *plane and* $C_i (i = 1, 2, \cdots, k)$ *be* k *disjoint clusters of* V. *A cluster* C_i *is a neighbour of a cluster* C_j *if there are two sensor nodes* $v_s \in C_i$ *and* $v_t \in C_j$ *such that* v_s *is a neighbour of* v_t.

DEFINITION 5. *Given a cluster* C_i *of sensor nodes and a sensor node* $v_j \notin C_i$, *the Euclidean distance from* v_j *to* C_i, *denoted by* $d(v_j, C_i)$, *is* $\min\{d(v_k, v_j) : v_k \in C_i \text{ and } d(v_k, v_j) \text{ is the Euclidean distance between } v_k \text{ and } v_j\}$.

The lifetime of a WSN (cluster) is defined as the shortest lifetime of all the sensor nodes of the WSN (cluster). We make the following assumptions about the target WSN. The initial energy of a sensor node v_i is e_i. Each sensor v_i generates f_i bits of data per unit time, where f_i is a constant. All the sensor nodes have the same transmit power and the same receive power. The energy consumed by each sensor node to send one bit of data is α_s, and the energy consumed by each sensor node to receive one bit of data is α_r. We only consider the energy consumption of wireless communications of each sensor node as it dominates the energy consumption of the whole sensor node. For ease of descriptions, we only consider the energy consumed by transmitting and receiving data.

Throughout this paper, the number of elements of a set S is denoted by $|S|$.

3. SINGLE BASE STATION

Deploying a single base station in a cluster of sensor nodes is a building block of our heuristic for optimally deploying k base stations. This problem has two versions. The first version considers a static base station. Its objective is to find the optimal location of the base station such that the lifetime of the cluster is maximised. The second version assumes a mobile base station. A mobile base station is relocated to different places during the lifetime of the cluster. The objective of the second version is to find a sequence of locations for the base station such that the lifetime of the cluster is maximised. We will propose an optimal algorithm for the static base station deployment problem and a heuristic for mobile base station deployment problem in the next two subsections.

3.1 Static Base Station Deployment

There are two key issues in the problem of optimally deploying a single static base station in a cluster. The first issue is to compute the maximum lifetime of the cluster with the static base station placed at a particular location. The second key issue is to find all the candidate locations of the base station such that one candidate location is the optimal location which maximises the lifetime of the cluster among all the candidate locations.

Given a particular location of the the static base station, the maximum lifetime of the cluster can be computed by using LP (linear programming) similar to the approach given in [16]. Let the base station be placed at a candidate location

c_i. Assume that T is the maximum lifetime of the cluster. Let $f_{ij}(f_{ij} \geq 0)$ be the amount of data in bits that v_i needs to send to v_j per unit time, and $N(v_i)$ be the set of all the neighbouring sensor nodes of v_i in the cluster. Hereinafter, we call f_{ij} the flow rate from v_i to v_j. For each neighbouring sensor node v_i of the base station located at the candidate location, $f_{ib}(f_{ib} \geq 0)$ denotes the amount of data in bits v_i needs to send to the base station. For each sensor node v_i which is at least two hops away from the base station, the following two constraints are satisfied.

$$\sum_{v_j \in N(v_i)} \alpha_s f_{ij} T + \sum_{v_j \in N(v_i)} \alpha_r f_{ji} T \leq e_i \quad (1)$$

$$\sum_{v_j \in N(v_i)} f_{ij} = f_i + \sum_{v_j \in N(v_i)} f_{ji} \quad (2)$$

Constraint 1 specifies that the total energy consumption of a sensor node v_i in sending and receiving data, does not exceed its initial energy. Constraint 2 implies that in a unit time, the total amount of data received plus the amount of data generated by v_i, is equal to the total amount of data v_i sends to its neighbouring sensor nodes. For each neighbouring sensor node v_i of the base station, the following two constraint are satisfied.

$$\alpha_s f_{ib} T + \sum_{v_j \in N(v_i)} \alpha_r f_{ji} T \leq e_i \quad (3)$$

$$f_{ib} = f_i + \sum_{v_j \in N(v_i)} f_{ji} \quad (4)$$

Constraint 3 specifies that the total energy consumed by v_i in sending and receiving data, does not exceed its initial energy. Constraint 4 implies that in a unit time, the total amount of data received plus the amount of data generated by v_i, is equal to the total amount of data v_i sends to the base station. Notice that constraints 1 and 3 are not linear. Now, we introduce a new variable t and set t to $\frac{1}{T}$. Divide both sides of the constraints 1 and 3 by T, we have the following constraints.

$$\sum_{v_j \in N(v_i)} \alpha_s f_{ij} + \sum_{v_j \in N(v_i)} \alpha_r f_{ji} \leq e_i t \quad (5)$$

$$\alpha_s f_{ib} + \sum_{v_j \in N(v_i)} \alpha_r f_{ji} \leq e_i t \quad (6)$$

Our optimisation objective function is as follows.

$$\min t \quad (7)$$

Therefore, given a candidate location of the base station, the maximum life time of a cluster can be computed by solving the above LP problem. Next, we show how to find all the candidate locations of the base station such that one of the candidate locations is an optimal location of the base station maximising the lifetime of the cluster. To find all the candidate locations, we consider each pair of sensor nodes v_i and v_j. If the Euclidean distance between v_i and v_j is greater than $2R$, where R is the maximum communication range of all the sensor nodes, we will ignore this pair. Otherwise, we find the candidate circles of v_i and v_j. A candidate circle of v_i and v_j is a circle that satisfies the following two constraints: 1) The radius of the circle is R. 2) v_i and v_j are on its circumference. The center of a candidate circle is a candidate location of the base station. Notice that for each pair of sensor nodes at most two candidate circles exist. If the Euclidean distance of a pair of sensor nodes is equal to $2R$, only one candidate circle of this pair exists. After finding all the candidate locations, our algorithm will search for the best candidate location of the base station sequentially.

Our algorithm for finding the optimal location of the static base station is shown in pseudo code as follows.

Algorithm 1: *Optimal single static base station deployment*

Input: *A cluster* $V = \{v_1, v_2, \cdots, v_m\}$ *of* m *sensor nodes in a 2D plane and a static base station.*
Output: *The optimal location of the static base station such that the lifetime of the cluster is maximised, and the lifetime of the cluster with the static base station placed at the optimal location .*

```
C = ∅;
for each pair of sensor nodes v_i and v_j (v_i, v_j ∈ V) do
  if the Euclidean distance between v_i and v_j ≤ 2R then
    Find the candidate circles C_1 and C_2 of v_i and v_j;
    Let c_1 and c_2 be the centers of C_1 and C_2;
    C = C ∪ {c_1} ∪ {c_2};
  end
end
for each candidate location c_i ∈ C do
  Place the base station at c_i;
  Construct the connectivity graph G(V ∪ {c_i})
  of all the sensor nodes and the base station;
  Compute the maximum lifetime t(c_i) of the cluster
  with the base station at c_i;
end
return max{t(c_j) : c_j is a candidate location in C};
```

THEOREM 1. *Algorithm 1 is guaranteed to find an optimal location of the base station.*

PROOF. Assume that the optimal location is c_{opt}. Let $S = \{v_1, v_2, \cdots, v_r\}$ be the set of sensor nodes that are one hop away from the base station at the optimal location c_{opt}. Draw a circle C_{opt} with the radius R and the center c_{opt}. According to the definition of the maximum communication range R, all the sensor nodes in S must be either in C_{opt} or on the circumference of C_{opt}. Next, we show that there is a candidate location c_k generated by our algorithm such that the set of sensor nodes that are one hop away from c_k is equal to S. Consider the following three possible cases.

1. There are two sensor nodes $v_i, v_j \in S$ such that v_i and v_j are on the circumference of C_{opt}. In this case, c_{opt} is one of our candidate locations.

2. Only one sensor node $v_i \in S$ is on the circumference of C_{opt}. Turn the circle C_{opt} clockwise around v_i until another sensor $v_j \in S$ is on the circumference of C_{opt}. Now all the sensor nodes in S are still in C_{opt}. Therefore, this case reduces to Case 1.

3. No sensor node is on the circumference of C_{opt}. Arbitrarily select a sensor node v_t, and move C_{opt} along the straight line $c_{opt}v_t$ until one sensor node in S is on the circumference of C_{opt}. Now all the sensors in S are still in C_{opt} or on the circumference of C_{opt}. Hence, this case reduces to Case 2.

Therefore, there exist a candidate location c_k such that all the sensor nodes in S can directly communicate with the

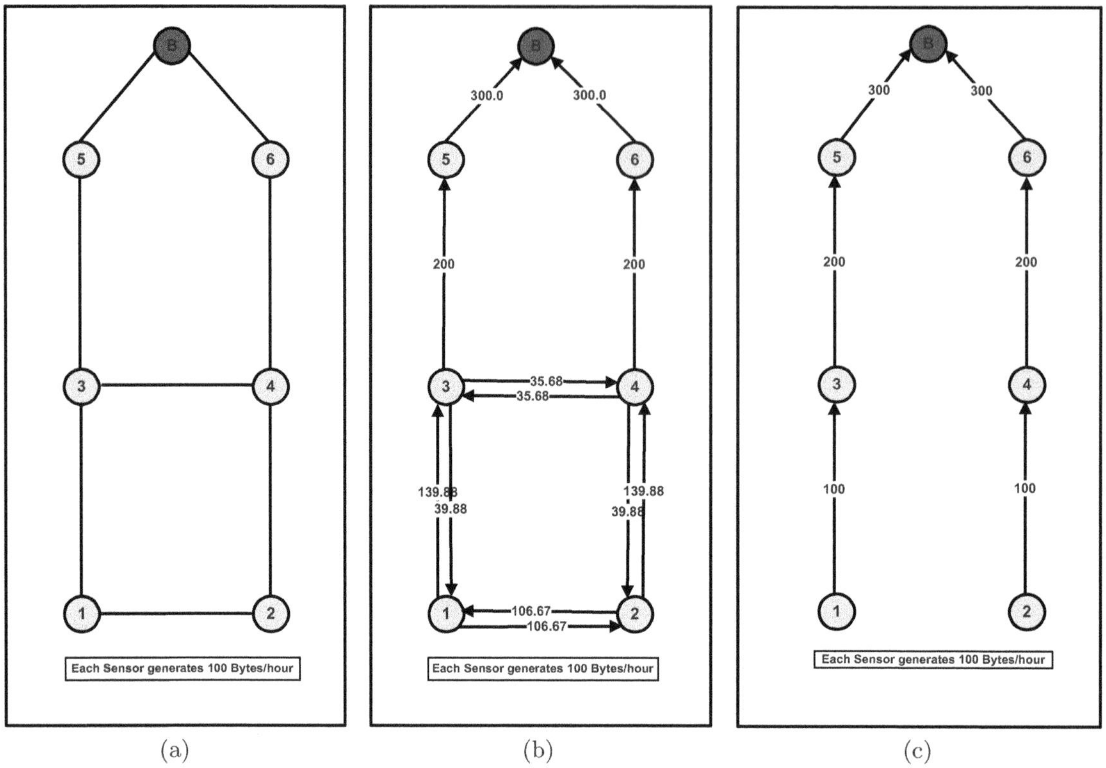

Figure 1: An example for the single static base station deployment problem

base station placed at c_k. According to our LP formulation for the optimal single static base station deployment problem, the candidate location at c_k and the optimal location of the base station leads to the same LP formulation. As a result, the maximum lifetime of the WSN with the base station placed at c_k is equal to that of the WSN with the base station placed at the optimal location. □

THEOREM 2. *Given a cluster of m sensor nodes, the time complexity of Algorithm 1 is $O(m^2 l)$, where l is the time complexity for solving an LP problem with $2m$ constraints and $2e$ variables.*

PROOF. There are $m(m-1)/2$ pairs of sensor nodes in a cluster with m sensor nodes. At most two candidate locations exist for each pair of sensor nodes. Therefore, the total number of candidate locations is $O(m^2)$. The connectivity graph $G(V \cup \{c_i\})$ of all the sensor nodes and the base station at the candidate location c_i, can be constructed as follows. Firstly, construct the connectivity graph $G(V)$. Secondly, for each neighbouring sensor node v_s of the base station at c_i, add an edge (v_s, c_i). Therefore, it takes $O(m)$ time to construct $G(V \cup \{c_i\})$ after $G(V)$ is constructed. After constructing the connectivity graph $G(V \cup \{c_i\})$, we construct an LP formulation for the problem of computing the maximum lifetime of the cluster with a base station placed at c_i, and find the maximum lifetime using an LP solver. For each candidate location c_i, the corresponding LP formulation has $2m$ constraints and $2e$ variables, where e is the number of edges in $G(V)$. Therefore, the time complexity of Algorithm 1 is $O(m^2 l)$, where l is the time complexity for solving an LP problem with $2m$ constraints and $2e$ variables. □

As shown in [11], an LP problem can be solved in polynomial time. Therefore, Algorithm 1 runs in polynomial time.

3.2 Mobile Base Station Deployment

For the mobile base station deployment problem, we relocate the base station to different places to make all the sensor nodes consume roughly the same amount of energy during the lifetime of the cluster.

There are two key issues in relocating a mobile base station. The first issue is when to relocate the base station. Relocating a base station incurs overheads, including informing all the sensor nodes of the new location of the base station and reconstructing a routing path from each sensor node to the base station. For the first issue, we use a fixed time interval, denoted by τ, between two successive relocations of the base station. Users may choose a proper value for τ. Therefore, the base station is moved to the next location after τ time has passed since the last relocation of the base station. The second issue is to find a good location for each relocation of the base station. One approach is to use our algorithm for deploying a single static base station to find the next location of the base station. The problem with this approach is that our algorithm for deploying a static base station does not minimise the total energy consumption of all the sensor nodes of the cluster.

Consider a cluster with 6 sensor nodes each of which generates 100 bytes of data per hour. Figure 1a shows the connectivity graph of the cluster with a base station at a particular location. Figure 1b shows the flow rates between sensor nodes and between sensor nodes and the base station generated by the LP formulation that maximises the lifetime of the cluster, where a directed edge (v_i, v_j) denotes that v_i

sends data to v_j and the edge weight on the edge (v_i, v_j) is the flow rate from v_i to v_j. As we can see, the flow rates shown in Figure 1b are not energy efficient for the sensor nodes $1-4$.

For the mobile base station deployment problem, the optimal location of the base station needs to minimise the total energy consumption of all the sensor nodes while achieving the maximum lifetime of the cluster. Next, we show that given a location c_k of the base station which maximises the lifetime of the cluster, how to find the optimal flow rates between sensor nodes and between each sensor node and the base station such that the total energy consumption of all the sensor nodes in the cluster is minimised. Let t_{max} be the maximum lifetime of the cluster with the base station located at c_k, $N(c_k)$ the set of all the sensor nodes that are one hop away from the base station located at c_k, and V the set of all the sensor nodes. Consider Constraints 2, 4, 5 and 6 for each sensor node v_i, where t and e_i are replaced by $1/t_{max}$ and the current energy of v_i, respectively. Let $w(c_k)$ be $\sum_{v_i \in N(c_k)}(\alpha_s f_{ib} + \sum_{v_j \in N(v_i)} \alpha_r f_{ji}) + \sum_{v_i \in V-N(c_k)}(\sum_{v_j \in N(v_i)} \alpha_s f_{ij} + \sum_{v_j \in N(v_i)} \alpha_r f_{ji})$, i.e., the total energy consumption per unit time of all the sensor nodes in the cluster with the base station. Our optimisation objective function is as follows.

$$\min w(c_k) \tag{8}$$

Clearly, this is an LP problem. Consider the example shown in Figure 1a. Figure 1c shows the flow rates between sensor nodes and between a sensor node and the base station that minimise the total energy consumption of all the sensor nodes while achieving the maximum lifetime of the cluster.

Our algorithm for a mobile base station relocates the mobile base station at a fixed time interval τ until one sensor runs out of energy. Each time when it relocates the base station, it performs the following tasks. Firstly, it computes the remaining maximum lifetime t_{max} by using the LP formulation given by constraints 2, 4, 5, 6 and optimisation objective function 7, where e_i is the current energy level of the sensor node v_i. Secondly, it finds a candidate location of the base station that results in the minimum total energy consumption of all the sensor nodes by using the LP formulation given by constraints 2, 4, 5, 6, and optimisation objective function 8, where t is $1/t_{max}$, and stores the candidate location in a queue. Lastly, it computes the accumulated maximum lifetime of the cluster and the residual energy of each sensor node. The details of our algorithm are shown as follows.

Algorithm 2: *Single mobile base station deployment*

Input: *A cluster* $V = \{v_1, v_2, \cdots, v_m\}$ *of* m *sensor nodes in a 2D plane and a base station.*
Output: *A queue* Q *storing all the locations of the mobile base station and the approximate maximum lifetime* t *of the cluster.*

```
C = ∅;
for each pair of sensor nodes v_i and v_j (v_i, v_j ∈ V) do
  if the Euclidean distance between v_i and v_j ≤ 2R then
    Find the candidate circles C_1 and C_2 of v_i and v_j;
    Let c_1 and c_2 be the centers of C_1 and C_2;
    C = C ∪ {c_1} ∪ {c_2};
  end
end
/* t is the maximum lifetime of the cluster */
t = 0;
Create an empty queue Q;
repeat
  for each candidate location c_i ∈ C do
    Place the base station at c_i;
    Construct the connectivity graph G(V ∪ {c_i})
    of all the sensor nodes and the base station;
    Compute the maximum lifetime t(c_i) of the
    cluster with the base station placed at c_i;
  end
  t_max = max{t(c_i) : c_i ∈ C};
  if t_max > τ then
    t = t + τ;
  else
    t = t + t_max;
  end
  A = {c_i : t(c_i) = t_max};
  for each candidate location c_i ∈ A do
    Find the minimum total energy consumption per unit time
    w(c_i) of all the sensor nodes with the base
    station placed at c_i;
  end
  Find the optimal candidate location c_j ∈ A
  satisfying w(c_j) = min{w(c_i) : c_i is a candidate
  location in A};
  Q.enqueue(c_j);
  for each each sensor node v_i ∈ V do
    Compute the residual energy e_i of v_i;
  end
until t_max ≤ τ
return Q and t;
```

THEOREM 3. *Given a cluster of m sensor nodes, the time complexity of Algorithm 2 is* $O(m^2 lt/\tau)$, *where t is the maximum lifetime of the cluster with a mobile base station, l is the time complexity for solving an LP problem with 2m constraints and 2e variables.*

PROOF. The loop body of the **repeat** loop has the same time complexity as Algorithm 1. The number of iterations of the loop is t/τ. Therefore, the theorem holds. □

4. MULTIPLE BASE STATIONS

Our unified heuristic for deploying multiple static base stations and multiple mobile base stations consists of two phases. In the first phase, it creates k initial disjoint clusters. Each cluster consists of a set of connected sensor nodes. In the second phase, it keeps moving sensor nodes from a cluster with a shorter maximum lifetime to a neighbouring cluster with a larger maximum lifetime until all the clusters are balanced in terms of their maximum lifetimes. In order to make the cluster balancing process faster, our heuristic makes use of both the maximum lifetime computed by using Algorithm 1, and the estimated maximum lifetime of each cluster during the cluster balancing process. The next two subsections describe these two phases in detail.

4.1 Creating k Initial Clusters

Our algorithm for creating k initial disjoint clusters consists of two major steps. Firstly, it constructs $s(s \geq k)$ disjoint clusters. At a time, it constructs a convex hull of all the sensor nodes that are not in any clusters, and creates a cluster that consists of the first $\lceil n/k \rceil$ sensor nodes in breadth-first search order starting with a sensor node on the convex hull of all the sensor nodes that are not yet in any cluster. Secondly, it repeatedly merges a cluster with the fewest sensor nodes with a neighbouring cluster with the fewest sensor nodes among all its neighbouring clusters until there are exactly k clusters. The details are shown in Algorithm 3.

THEOREM 4. *Given a WSN, the time complexity of Algorithm 3 is $O(n^2)$, where n is the number of sensor nodes.*

PROOF. The time complexity of Algorithm 3 can be broken down into the following components.

1. It takes $O(e)$ time to construct the connectivity graph of all the sensor nodes in V, where e is the number of edges in the connectivity graph of all the sensor nodes in the WSN.

2. It takes $O(n \log n)$ time to construct the convex hull for n sensor nodes [1]. After deleting a sensor node from the convex hull, it takes $O(\log^2 n)$ time to reconstruct the convex hull [14]. Therefore, the total time complexity of the convex hull construction and maintenance is $O(n \log n + \sum_{i=n:2} \log^2(i-1)) = O(n \log^2 n)$.

3. The time complexity of the first **while** loop is $O(e + n \log^2 n)$.

4. For the second **while** loop, a neighbour graph of all the clusters can be used. The neighbour graph is an undirected graph with each node representing a cluster and each edge denoting that the corresponding two clusters are neighbours. The neighbour graph of all the clusters can be constructed in $O(n^2)$ time. Each time when two neighbouring clusters are merged, the neighbour graph can be updated in $O(n)$ time. Therefore, the time complexity of the second **while** loop is $O(n^2)$.

As a result, the theorem holds. □

Algorithm 3: *Creating k initial clusters*

Input: *A set $V = \{v_1, v_2, \cdots, v_n\}$ of n sensor nodes in a 2D plane.*
Output: *k disjoint clusters of all the sensor nodes.*

```
S = V;
Create a convex hull G_h(S) for all the sensor nodes in S;
Create the connectivity graph G(V) of all the sensor nodes
in V;
i = 0;
while S ≠ ∅ do
    Arbitrarily select a sensor node v_s on G_h(S);
    Find the set A of the first ⌈n/k⌉ − 1 sensor nodes
    in breadth-first search order starting with v_s;
    i = i + 1;
    C_i = {v_s} ∪ A;
    S = S − C_i;
    Create a convex hull G_h(S) for all the sensor nodes in S;
end
while i > k do
    Find a cluster C_j with the fewest sensor nodes;
    Merge C_j with a neighbouring cluster with the fewest
    sensor nodes among all the neighbouring clusters ;
    i = i − 1;
end
```

4.2 Cluster Balancing

The k initial clusters created by Algorithm 3 are most likely not balanced in terms of the maximum lifetimes of all the clusters. Next, we propose an algorithm for balancing the k clusters such that they have almost the same lifetime.

Our cluster balancing algorithm starts with the initial k clusters and repeatedly moves a sensor node from a cluster with a shorter maximum lifetime to a neighbouring cluster with a longer maximum lifetime until all the clusters become balanced in terms of their lifetimes. When a node v_s is moved from a cluster C_i to a neighbouring cluster C_j, the maximum lifetimes of the clusters $C_i - \{v_s\}$ and $C_j \cup \{v_s\}$ need to be calculated. As the time complexity for computing the maximum lifetime of a cluster by using LP is high, we estimate the lifetimes of the cluster $C_i - \{v_s\}$ and the cluster $C_j \cup \{c_s\}$ as follows. Assume that the maximum lifetime of the cluster C_i is equal to the lifetime of a neighbouring sensor v_i of the base station of C_i [1]. Let $t(C_i)$ be the estimated maximum lifetime of the cluster C_i with a static base station at the optimal location, e_i the initial energy of v_i and w_i the energy consumed by v_i in sending and receiving data in a unit time. We have the following equation.

$$t(C_i) = e_i / w_i \tag{9}$$

Let $N_r(r = i, j)$ be the set of all the neighbouring sensor nodes of the base station of the cluster C_r that are reachable from v_s via non-neighbouring sensor nodes. Assume that the data generated by v_s can be equally distributed to all the sensors in N_i. We have Equations 10.

$$t(C_i - \{v_s\}) = \frac{e_i}{w_i - \frac{\alpha_s f_s + \alpha_r f_s}{|N_i|}} \tag{10}$$

From Equations 9 and 10, we have Equations 11.

$$t(C_i - \{v_s\}) = \frac{e_i t(C_i)|N_i|}{e_i|N_i| - (\alpha_s f_s + \alpha_r f_s) t(C_i)} \tag{11}$$

Similarly, we have the estimated maximum lifetime of the cluster $C_j \cup \{v_s\}$ as shown in Equation 12.

$$t(C_j \cup \{v_s\}) = \frac{e_j t(C_j)|N_j|}{e_j|N_j| + (\alpha_s f_s + \alpha_r f_s) t(C_j)} \tag{12}$$

In order to reduce the number of calculations of the maximum lifetime of a new cluster by using LP, we introduce a procedure named $fast - balancing(C)$, where C is the set of all the clusters. $fast - balancing(C)$ performs cluster balancing by using the estimated lifetime for each new cluster. Our algorithm for cluster balancing repeatedly calls $fast - balancing(C)$ until no sensor node can be moved to a neighbouring cluster, i.e., clusters are balanced in terms of the maximum lifetimes of all the clusters. Before calling $fast - balancing(C)$, our algorithm for cluster balancing calculates the maximum lifetime for each cluster by using LP. $fast - balancing(C)$ works as follows. In each iteration, a modifiable cluster C_i with the minimum estimated maximum lifetime among all the clusters in C is selected. A cluster C_i is modifiable if there exist a neighbouring cluster C_j with a longer estimated maximum lifetime and a sensor node $v_s \in C_i$ such that the following constraints are satisfied.

1. $t(C_j \cup \{v_s\}) > t(C_i)$.

2. The cluster $C_i - \{v_s\}$ is connected.

The first constraint ensures that after v_s is moved to C_j from C_i, the estimated maximum lifetime of C_j should be longer than that of C_i. The second constraint guarantees that removing v_s from C_i will not make C_i disconnected. If such a modifiable cluster does not exist, all the clusters

[1]Note that this assumption is not always true.

are balanced in terms of the estimated maximum lifetimes of all the clusters and $fast-balancing(C)$ returns. If such a modifiable cluster C_i exists, $fast-balancing(C)$ selects a neighbouring cluster C_j with the longest estimated maximum lifetime among all the neighbouring clusters of C_i and finds the set Q of sensor nodes in C_i which are the neighbouring sensor nodes of C_j and sorted in non-decreasing order of their Euclidean distances to C_j. Then, it keeps moving a sensor node in Q from C_i to C_j until no sensor node in Q can be moved. A sensor node $v_s \in Q$ is moved from C_i to C_j iff both of the above mentioned two constraints are satisfied.

After constructing k balanced clusters, our algorithm for cluster balancing computes the maximum lifetime of the WSN. The details are shown in Algorithm 4, and the procedure fast-balancing(C) is shown in Algorithm 5.

Algorithm 4: *Balancing clusters*

Input: *A set C of the k initial clusters $C_1, C_2, \cdots, C_k$.*
Output: *k balanced clusters $C_1, C_2, \cdots, C_k$, and the maximum lifetime of the WSN.*

```
/* A is the set of non-modifiable clusters at this moment */
/* B is the set of modifiable clusters at this moment */
balanced = false;
while balanced == false do
    A = {};
    B = C  /* C = A ∪ B holds all the time */
    for each cluster C_i ∈ C do
        Calculate the maximum lifetime t(C_i) of C_i
        by using Algorithm 1;
        modifiable(C_i) = true;
    end
    Node-Moved=false;
    fast-balancing(C);
    if Node-Moved==false then
        balanced = true;
    end
end
if this is a static base station deployment problem then
    for each cluster C_i do
        Compute the maximum lifetime t(C_i) of C_i
        by using Algorithm 1;
        /* t_max is the maximum lifetime of the WSN with k
        static base stations */
        t_max = min{t(C_i) : C_i ∈ C};
    end
else
    for each cluster C_i do
        Compute the maximum lifetime t(C_i) of C_i
        by using Algorithm 2;
        /* t_max is the maximum lifetime of the WSN with k
        mobile base stations */
        t_max = min{t(C_i) : C_i ∈ C};
    end
end
```

Given a WSN with n sensor nodes and e edges in the connectivity graph, the following theorem holds.

THEOREM 5. *The time complexity of Algorithm 4 is $O(rl + rkn^3)$, where r is the number of calls of fast-balancing(C), and l is the time complexity for solving an LP problem with $2n$ constraints and $2e$ variables.*

PROOF. For each call of fast-balancing(C), each sensor node is moved into a different cluster at most $k-1$ times. Hence, the total number of sensor node motions is bounded by $O(kn)$. Each sensor motion takes $O(n^2)$ time which is dominated by finding the sensor node v_s of C_i that has the shortest Euclidean distance to C_j by using exhaustive search. Therefore, the time complexity of fast-balancing(C) is $O(kn^3)$. In addition, for each iteration of the **while** loop of Algorithm 4, the total time complexity of computing the maximum lifetimes of all the clusters by using LP, is $O(l)$. Therefore, the theorem holds. □

r is quite small in practice. For all the 36 instances of WSN we used in our simulations, r is less than 10 for most of the instances. The largest value of r is 13 among all the 36 instances.

Algorithm 5: *fast-balancing(C)*

Input: *A set C of the k initial clusters $C_1, C_2, \cdots, C_k$.*
Output: *k balanced clusters $C_1, C_2, \cdots, C_k$ with respect to the estimated maximum lifetimes.*

```
while B ≠ ∅ do
    Select a cluster C_i with the shortest maximum lifetime
    and modifiable(C_i) = true from B;
    S = {C_s : C_s ∈ C and C_s is a neighbouring cluster of C_i};
    Find C_j ∈ S with the longest maximum lifetime;
    Q = {v_s : v_s ∈ C_i and v_s is a neighbouring sensor
    node of C_j};
    NodeMoved(C_i)=false;
    Create an array L of all the sensor nodes in Q;
    Sort L in non-decreasing order of their Euclidean
    distances to C_j;
    for s = 0, 1, ..., |Q| - 1 do
        if C_i - {L[s]} is connected then
            Compute N_i and N_j;
            temp = e_j t(C_j)|N_j| / (e_j|N_j| + (α_s f_s + α_r f_s) t(C_j));
            if temp > t(C_i) then
                C_j = C_j ∪ {L[s]};
                C_i = C_i - {L[s]};
                t(C_i - {L[s]}) = e_i t(C_i)|N_i| / (e_i|N_i| - (α_s f_s + α_r f_s) t(C_i));
                t(C_j ∪ {L[s]}) = temp;
                NodeMoved(C_i)= true;
                Node-Moved=true;
            else
                break;
            end
        end
    end
    if NodeMoved(C_i)= true then
        for each neighbouring cluster C_j of C_i do
            if modifiable(C_j) == false then
                modifiable(C_j) = true;
                A = A - {C_j};
                B = B ∪ {C_j};
            end
        end
    else
        modifiable(C_i) = false;
        A = A ∪ {C_i};
        B = B - {C_i};
    end
end
```

5. RELATED WORK

The problem of deploying multiple base stations in a large scale WSN has been extensively studied. Different objective functions have been used and many algorithms and heuristics have been proposed. [13] proposes a heuristic for deploying multiple base stations such that the maximum energy consumption of any cluster is minimised. Their heuristic uses a greedy approach to create k initial clusters, and repeatedly moves a sensor node from a cluster with higher

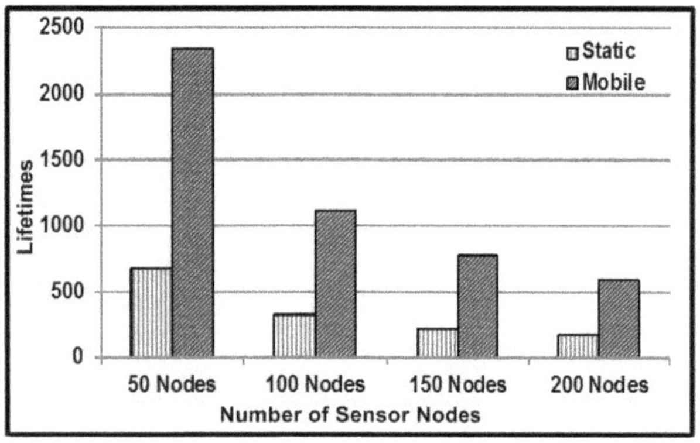

(a) Single BS with uniform distribution

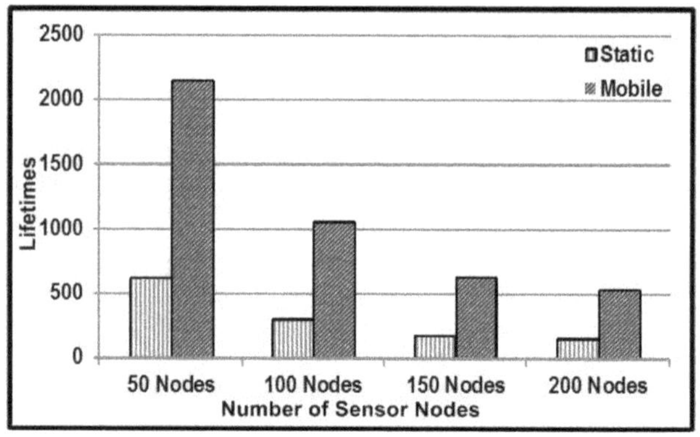

(b) Single BS with random distribution

Figure 2: Simulation results part I

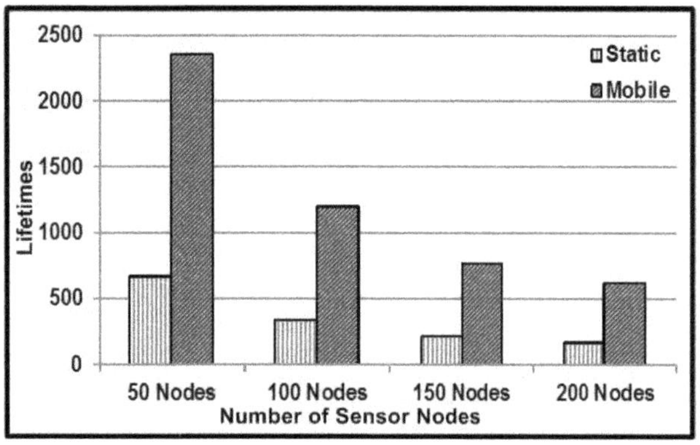

(a) Single BS with grid distribution

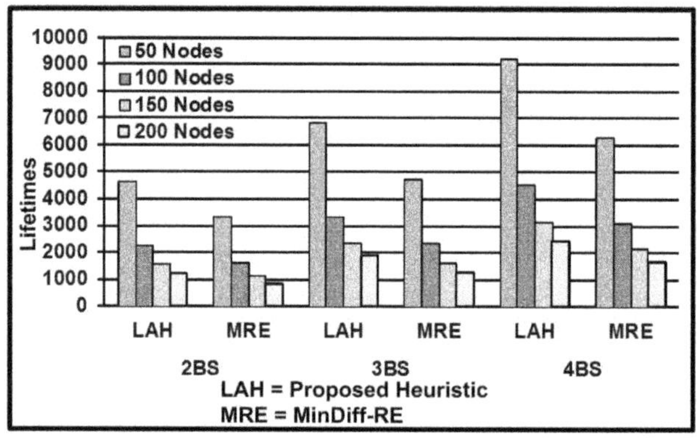

(b) Multiple mobile BS's with uniform distribution

Figure 3: Simulation results part II

total energy consumption to a cluster with lower total energy consumption until all clusters become balanced. [20] proposes a 2-approximation algorithm and a heuristic for the problem of deploying k base stations such that the maximum shortest hop distance between all the sensor nodes to their nearest base stations is minimised.

[18] proposes an algorithm and a heuristic for placing k base stations in an optimal way such that the average Euclidean distance between the sensor nodes and their base stations is minimised. The algorithm assumes that each base station knows the locations of all the sensor nodes, and the heuristic assumes that each base station only knows the locations of its neighbouring sensor nodes and other base stations. However, average Euclidean distance is not a good objective function. Consider the following example. All the sensor nodes of a WSN are uniformly distributed on a circle. The radius of the circle is greater than the maximum communication range of the sensor nodes. Assume that only one base station is deployed. Clearly, the center of the circle is the optimal location of the base station. If the base station is placed at the center, no sensor nodes can communicate with the base station. [21] studies the problem of placing k base stations in an optimal way such that the total latency of all the sensor nodes to their gateways is minimised. It proposes two heuristics for the problem using genetic algorithms.

[16] proposes an approximation algorithm that can guarantee $(1-\epsilon)$-optimal network lifetime performance for a single static base station placement problem with any desired error bound $\epsilon > 0$. The proposed $(1-\epsilon)$-optimal approximation algorithm is based on two novel techniques. The first technique is to discretize cost parameter with performance guarantee. Subsequently, the continuous search space can be broken up into a finite number of subareas. The second technique is to exploit the cost property of each subarea and represent it by a novel notion called fictitious cost point, each with guaranteed cost bounds. [16] also proposes an exponential time approximation algorithm for deploying multiple base stations. However, the approximation algorithm is not applicable to a large scale WSN due to its exponential time complexity.

[12] proposes a heuristic for statically selecting multiple sensor nodes as sinks in a WSN such that the lifetime of the WSN is maximised. The heuristic consists of a mixed ILP (Integer Linear Programming) formulation. Since the mixed ILP problem is NP-complete, the approach is not applicable to large scale WSNs [12]. [6] proposes a heuristic for deploying multiple mobile base stations to maximise the lifetime of the sensor network. The total lifetime of the network is divided into equal period of time known as rounds, and all mobile base stations change their locations at the beginning of every round. An ILP formulation is proposed to find the locations of base stations such that the maximum energy spent by each node in a round is minimised. [2] studies the problem of deploying multiple mobile base stations near the

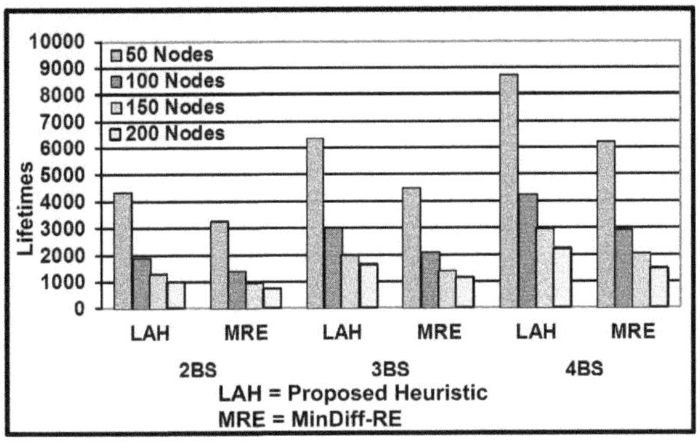

(a) Multiple mobile BS's with random distribution

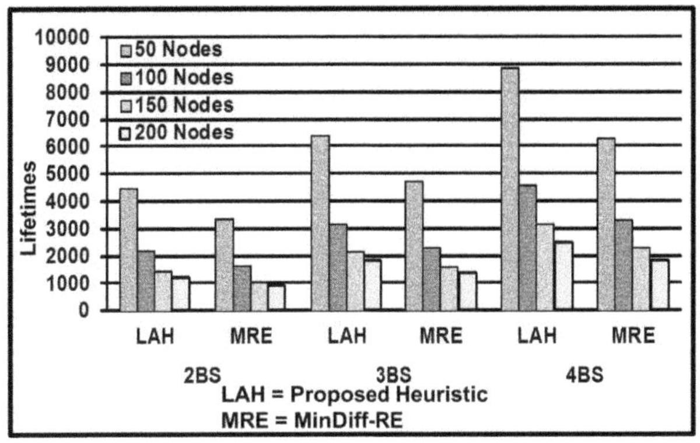

(b) Multiple mobile BS's with grid distribution

Figure 4: Simulation results part III

boundary of a WSN to prolong the lifetime of the WSN. It proposes three energy efficient low-complexity heuristics to determine the locations of the base stations near the boundary of a WSN. They are the top-Kmax heuristic, the heuristic for maximising the minimum residual energy, and the MinDiff-RE heuristic. The simulation results show that the proposed heuristics provide increased network lifetimes and amount of data delivered during the network lifetime compared to the heuristic proposed in [6].

6. SIMULATION RESULTS

We have implemented our heuristic for a single mobile base station, our unified heuristic for multiple static base stations and multiple mobile base stations, and our optimal algorithm for deploying a single static base station. When calculating the maximum lifetime of a cluster with a mobile base station, the value of τ is set to $t_{max}/20$, where t_{max} is the maximum lifetime of the cluster with a static base station. We have also implemented the MinDiff-RE heuristic proposed by Azad and Chockalingam [2] for comparison purpose. The reason for choosing MinDiff-RE heuristic is that it has the best performance among all the three heuristics proposed in [2] and outperforms the ILP approach proposed in [6] as demonstrated in [2].

We use QualNet Network Simulator [15] to generate 36 different network instances with three different distributions, namely, grid, uniform and random distributions. We use four different numbers of base stations, i.e., 1, 2, 3 and 4. For each distribution, we vary the number of sensor nodes from 50 to 200 with an increment of 50 nodes, and generate 3 network instances for each number of sensor nodes with the 4 different numbers of base stations, leading to a total of 12 network instances. Altogether, we generate 36 network instances for all the three distributions. Each sensor node is assigned an initial energy of e_i=0.05 J. The transmission range of all sensor nodes is fixed to 50 meters. The value of α_r is 50 nanoJ, and the value of α_s is 250 nanoJ. We assume that all the sensor nodes generate the same amount of data within a unit time. The data rate f_i of each sensor node v_i is 40 bits/hour. The hardware platform for our simulations is Intel Core 2 Duo processor with a clock frequency of 3 GHz and 4 GB RAM. The LP solver is the one provided by MATLab(version 7.10.0).

The simulation results for a single static base station and a single mobile base station with 36 network instances are shown in Figure 2a, Figure 2b and Figure 3a for the uniform, random and grid distributions, respectively. In each figure, the horizontal axis represents the number of sensor nodes, and the vertical axis denotes the average maximum lifetime of a WSN. For all the 36 network instances, the average, maximum and minimum lifetime improvements of our heuristic for a single mobile base station over our optimal algorithm for a single static base station are 253%, 274% and 241%, respectively.

For multiple base stations, we have simulated only multiple mobile base stations as a mobile base station significantly outperforms a static base station. The simulation results of our unified heuristic, denoted by LAH, and MinDiff-RE heuristic, denoted by MRE, are shown in Figures 3b, 4a and 4b. For all the 36 network instances, the average, maximum and minimum lifetime improvements of our heuristic over the MinDiff-RE heuristic are 40%, 48%, and 31%, respectively.

Moreover, we have computed the running times of our heuristic for all the instances. As expected, our unified heuristic runs much slower than the MinDiff-RE heuristic. For 200 sensor nodes, the average running time of our heuristic with 1, 2, 3 and 4 base stations is 45 minutes. The maximum running time is 65 minutes.

There are three key reasons why our unified heuristic outperforms the MinDiff-RE heuristic. Firstly, our unified heuristic is based on an optimal algorithm for a single static base station. Secondly, our unified heuristic selects the locations of multiple base stations from a set of candidate locations which include the optimal locations. On the contrary, the MinDiff-RE heuristic selects the locations of multiple base stations from a set of predefined feasible locations near the boundary of the WSN. Lastly, our unified heuristic constructs balanced clusters.

7. CONCLUSION

We study the problems of deploying k static base stations and k mobile base stations in a WSN such that the lifetime of the WSN is maximised. We propose a unified heuristic for both problems. In the special case of a single static base station, we present the first optimal, polynomial-time algorithm. In the special case of a single mobile base station, we propose an efficient heuristic. We have simulated our optimal algorithm, our heuristics and the MinDiff-RE heuristic proposed Azad and Chockalingam [2] on 36 instances of WSN with three different distributions, namely, grid, uniform and random distributions. The simulation results show that our heuristic for deploying a single mobile base station prolongs the lifetime of a WSN by 253% on average compared to our optimal algorithm for a single static base station, and our heuristic for multiple mobile base stations performs 40% better than the MinDiff-RE heuristic on average.

There are two open problems in the research on the lifetime aware deployment of multiple base stations. The first one is to deploy multiple base stations in a 3D environment. In such an environment, communication obstacles may exist between adjacent sensor nodes. The communication obstacles bring additional challenges to the lifetime aware deployment of multiple base stations. The second one is to deploy multiple base stations in a WSN where the transmit power of each sensor node is adjustable. In this problem, whether two sensor nodes can communicate directly depends on their transmit powers. The exponential time heuristic proposed in [16] is not applicable to large scale WSNs. An efficient polynomial time heuristic is desirable.

8. REFERENCES

[1] A. M. Andrew. Another efficient algorithm for convex hulls in two dimensions. *Information Processing Letter*, 9(5):216–219, 1979.

[2] A. Azad and A. Chockalingam. Mobile base stations placement and energy aware routing in wireless sensor networks. In *IEEE Wireless Communications and Networking Conference*, volume 1, pages 264–269. IEEE, 2006.

[3] R. Chandra, L. Qiu, K. Jain, and M. Mahdian. Optimizing the placement of integration points in multi-hop wireless networks. In *IEEE International Conference on Network Protocols*, pages 271–282. IEEE, 2004.

[4] B. N. Clark, C. J. Colbourn, and D. S. Johnson. Unit disk graphs. *Discrete mathematics*, 86(1):165–177, 1990.

[5] A. Efrat, S. Har-Peled, and J. S. B. Mitchell. Approximation algorithms for two optimal location problems in sensor networks. In *The 2nd International Conference on Broadband Networks*, pages 714–723. IEEE, 2005.

[6] S. R. Gandham, M. Dawande, R. Prakash, and S. Venkatesan. Energy efficient schemes for wireless sensor networks with multiple mobile base stations. In *IEEE Global Telecommunications Conference*, volume 1, pages 377–381. IEEE, 2003.

[7] Q. Gao, K. J. Blow, D. J. Holding, I. W. Marshall, and X. H. Peng. Radio range adjustment for energy efficient wireless sensor networks. *Ad Hoc Networks*, 4(1):75–82, 2006.

[8] G. Gupta and M. Younis. Load-balanced clustering of wireless sensor networks. In *IEEE International Conference on Communications*, volume 3, pages 1848–1852. IEEE, 2003.

[9] G. Gupta and M. Younis. Performance evaluation of load-balanced clustering of wireless sensor networks. In *The 10th International Conference on Telecommunications, 2003.*, volume 2, pages 1577–1583. IEEE, 2003.

[10] Y. T. Hou, Y. Shi, H. D. Sherali, and S. F. Midkiff. On energy provisioning and relay node placement for wireless sensor networks. *IEEE Transactions on Wireless Communications*, 4(5):2579–2590, 2005.

[11] N. Karmarkar. A new polynomial-time algorithm for linear programming. In *Proceedings of the sixteenth annual ACM symposium on Theory of computing*, pages 302–311. ACM, 1984.

[12] H. Kim, Y. Seok, N. Choi, Y. Choi, and T. Kwon. Optimal multi-sink positioning and energy-efficient routing in wireless sensor networks. *Information Networking: Convergence in Broadband and Mobile Networking*, pages 264–274, 2005.

[13] S. Mahmud, H. Wu, and J. Xue. Efficient energy balancing aware multiple base station deployment for wsns. In *The 8th European Conference on Wireless Sensor Networks*, pages 179–194. Springer, 2011.

[14] M. H. Overmars and J. Van Leeuwen. Maintenance of configurations in the plane. *Journal of computer and System Sciences*, 23(2):166–204, 1981.

[15] Qualnet. Scalable network technologies. `http://www.scalable-networks.com/content/products/qualnet`.

[16] Y. Shi and Y. T. Hou. Optimal base station placement in wireless sensor networks. *ACM Transactions on Sensor Networks (TOSN)*, 5(4):32:1–32:24, 2009.

[17] J. Urrutia. *Handbook of computational geometry*. North Holland, 1999.

[18] Z. Vincze, R. Vida, and A. Vidacs. Deploying multiple sinks in multi-hop wireless sensor networks. In *IEEE International Conference on Pervasive Services*, pages 55–63. IEEE, 2007.

[19] Q. Wang, M. Hempstead, and W. Yang. A realistic power consumption model for wireless sensor network devices. In *The 3rd Annual IEEE Communications Society on Sensor and Ad Hoc Communications and Networks*, volume 1, pages 286–295. IEEE, 2006.

[20] H. Wu and S. Mahmud. A 2-approximation algorithm for optimal deployment of k base stations in wsns. In *IFIP International Conference on Networking*, pages 378–391. Springer, 2012.

[21] W. A. Youssef and M. F. Younis. Intelligent gateways placement for reduced data latency in wireless sensor networks. In *IEEE International Conference on Communications*, pages 3805–3810. IEEE, 2007.

Interference-free Scheduling with Bounded Delay in Cluster-Tree Wireless Sensor Networks

Mario Di Francesco
Dept. of Comp. Sci. and Eng.
Aalto University
Espoo, Finland
mario.di.francesco@aalto.fi

Cristina M. Pinotti
Dept. of Comp. Sci. and Math.
University of Perugia
Perugia, Italy
pinotti@unipg.it

Sajal K. Das
Dept. of Comp. Sci. and Eng.
University of Texas at Arlington
Arlington, Texas, USA
das@uta.edu

ABSTRACT

Convergecast is a typical form of data collection in wireless sensor networks (WSNs), wherein nodes sample data from the environment and send them to a common destination. In order to prolong the network lifetime, a duty-cycle mechanism is usually coupled with a routing tree structure, in which nodes are organized in clusters. Each cluster aggregates data and sends them towards the root of the tree. However, clusters can interfere each other if their active time is not properly chosen. Furthermore, scheduling can lead to a long data delivery delay when a duty-cycle mechanism is used. In this article, we introduce a receiver-oriented scheduling algorithm for cluster-tree WSNs which provides a bounded latency for convergecast data collection. In contrast with most of the existing works in the literature, where two nodes are assumed to interfere if they are at most 2 hops away, we address the more general and realistic case where interfering nodes can be up to t hops away from each other, where $t \geq 2$. We first show that the minimum-latency convergecast problem is NP-hard for cluster-based WSNs with arbitrary topologies. We then focus on tree-based WSNs and derive bounds on the latency for convergecast data collection. We also propose a heuristic to obtain a t-interference-free scheduling in $O(nt)$ time, where n is the number of clusters in the WSN. We finally validate our findings by simulation on both synthetic topologies and routing trees obtained from WSN deployments.

Categories and Subject Descriptors

C.4 [**Performance of Systems**]: Performance attributes; C.2.1 [**Computer-Communication Networks**]: Network Architecture and Design—*Network communications*

Keywords

Wireless sensor networks, data gathering tree, convergecast, interference-free scheduling, bounded delay, clusters.

MSWiM'12, October 21–25, 2012, Paphos, Cyprus.

1. INTRODUCTION

In a typical wireless sensor network (WSN), sensor nodes are densely deployed over a sensing area to monitor a phenomenon of interest by sampling environmental data and relaying them to a data collection point, referred to as *sink*. Such many-to-one data collection paradigm is also known as *convergecast*, and is used in several application scenarios. Some of them, which include environmental monitoring as well as industrial automation and control, have long network lifetime and low latency as their major requirements [18].

One of the most efficient energy-conservation techniques consists in exploiting a duty-cycle [2], which can be coupled with a tree-based routing structure for convergecast data collection [13]. A characteristic of tree-based data collection is that data can be easily combined as they are routed towards the sink. This is usually accomplished by organizing nodes in *clusters*. One node within each cluster is designated as *cluster head* to collect data from other nodes, aggregate them, and send the result towards the sink. Cluster heads also regulate channel access and define the active time (i.e., through a duty-cycle mechanism) within each cluster. The way active times are scheduled determines the latency of the convergecast data collection [14]. This is particularly critical when data aggregation is performed in the network, since nodes in an intermediate level of the tree have to receive all the data coming from their children before forwarding them to the sink through multi-hop paths [16].

Despite the presence of cluster-heads, inter-cluster interference is still possible. This happens since multiple nodes belonging to different clusters can be potentially active at the same time. A solution to this problem is to find an interference-free schedule for the active periods of different clusters. This can be accomplished by a receiver-oriented approach which considers cluster-heads instead of individual nodes. Most of the solutions for interference-free scheduling in WSNs assumed that interfering nodes could be up to 2-hop away [13]. In practice, using 2-distance information may not be enough for an actual interference-free slot assignment. In fact, the radio signal propagation is affected by several different factors in real scenarios, where interferences can be unpredictable due to multi-path and shadowing effects [1]. Finally, the data collection tree significantly affects scheduling, especially when schemes different from the shortest path tree are used, as it usually happens in WSNs for energy conservation or coverage purposes (e.g., minimum spanning tree).

In this article, we investigate the problem of minimum-latency scheduling in cluster-tree WSNs with periodic traffic. In contrast with most of the literature, we address aggregated data and a cluster-based architecture. To this end, we take a receiver-oriented approach, and also consider the more general and realistic case where interfering nodes can be up to t hops away from each other. We formally define the interference-free minimum-latency convergecast problem, and show that it is NP-hard for cluster-based WSNs with arbitrary topologies. We then provide bounds on the latency when the problem is restricted to tree-based WSNs. Furthermore, we present a heuristic for low-latency data collection in cluster-tree WSNs that runs in $O(nt)$ time, where n is the number of clusters in the WSN. Finally, we evaluate our solution by simulation on both synthetic topologies and routing trees derived from WSN deployments. The results show that the latency bounds are accurate.

The rest of the article is organized as follows. Section 2 overviews the related work. Section 3 presents the t-interference-free Minimum-latency Convergecast (briefly, the t-MC) problem in cluster-based WSNs with arbitrary topologies. Section 4 focuses on cluster-tree WSNs, and provides bounds on the convergecast latency as well as a heuristic for interference-free scheduling. Section 5 evaluates the performance of our approach by simulation. Finally, Section 6 concludes the article.

2. RELATED WORK

Approaches for communication scheduling in WSNs have been extensively studied in the existing literature, with focus on energy conservation [2]. Specifically, scheduling for data collection in tree-based WSNs was surveyed in [13]. Many works actually focused on minimizing the length of the schedule. For instance, that problem was addressed in [6] and was shown to be NP-complete for generic WSN topologies, when interferences are described by an arbitrary graph, by mapping to the graph coloring problem. That work also specifically addressed tree-based WSNs, and proposed a heuristic for scheduling the tree based on nodes' level. However, minimizing the schedule length does not imply that the convergecast latency is also minimized. In contrast, we specifically consider the problem of minimum latency convergecast, and also provide delay bounds for tree-based data collection.

The problem of minimum-latency scheduling in WSNs was addressed in several works. In [8], a solution for optimal scheduling in WSNs was first obtained by considering linear topologies, and then extended to multi-line, tree, and general networks. Such a solution obtains a closed formula for the latency, given the number of messages that each source node has to transmit to the sink. A similar approach was taken in [4, 9], again by extending solutions for multi-line trees to generic trees. All the works mentioned above share some common assumptions: raw-data convergecast is considered, wherein all data generated by sensor nodes are relayed to the sink; and slots are assigned to individual sensor nodes. In contrast, in this article we consider aggregated traffic, hence specifically address cluster-based networks where data are reported to a cluster-head and then processed before being relayed towards the sink. Accordingly, we take a receiver-oriented approach where slots are assigned to clusters, rather than to individual nodes. This results in a radically different problem formulation and approach to scheduling.

The Minimum Delay Beacon Scheduling (MDBS) problem was introduced in [17] in order to find an interference-free slot assignment which minimizes the convergecast latency in tree-based ZigBee WSNs with aggregated traffic. The MDBS problem was shown to be NP-hard when interferences are described by an arbitrary graph, and a heuristic for a generic network modeled through the corresponding interference graph was also introduced. Optimal and sub-optimal solutions to the MDBS problem were presented for linear and ring topologies, respectively. To a certain extent, our approach is similar to [17]. In that work, however, only two-hop interferences were considered for tree topologies, while here we address the more general and realistic case where nodes can interfere up to t hops away. Furthermore, our heuristic solution runs in $O(nt)$ time, where n is the number of clusters in the WSN, as opposed to the one in [17] which requires $O(n^2)$ time. Finally, the approach presented in this article is more general, since it is not specifically targeted to ZigBee WSNs.

Some works specifically addressed delay-bound data collection in tree-based WSNs. For instance, bounds for convergecast data collection were provided in [7], and heuristics with a constant approximation factor were derived in [9]. As already mentioned, these results were obtained for raw-data convergecast where slots were assigned to individual nodes, hence cannot be applied in the scenario considered here. An algorithm for scheduling periodic traffic was presented in [15] for cluster-tree WSNs based on the ZigBee standard. However, the focus of the work is on the feasibility of a periodic schedule, with the goal of fair bandwidth allocation rather than a low latency. A a soft real-time scheduling algorithm with specific focus on energy conservation was considered in [11] for a similar scenario. The adopted approach relies on some features specific to the ZigBee standard, thus cannot be used in general cluster-tree scenarios.

3. MINIMUM-LATENCY CONVERGECAST

In this section, we will first introduce the system model, then we will formally define the minimum-latency convergecast problem in a cluster-tree WSN.

3.1 System model

We consider a multi-hop static WSN where nodes periodically report data to the sink, according to the convergecast communication paradigm. We assume that nodes are organized in a logical routing tree rooted at the sink, and that a duty-cycle mechanism is used for energy conservation [5]. Furthermore, we consider aggregated traffic, where parent nodes wait for all data generated by their own children before applying suitable operators on the data [16]. In the following, we will refer to a parent node as well as its children as a *cluster* (Figure 1a). We assume that nodes in a cluster use a certain medium access control (MAC) protocol – for instance, the IEEE 802.15.4 standard [12] – during a given time period denoted as *slot*, then they go to sleep to save energy. The slots of the different clusters in the network have the same duration, and repeat periodically within a *frame* (Figure 1b). In the following, we assume that the frame can accommodate up to k slots. Note that nodes at the intermediate levels of the tree belong to two different clusters, hence they are active for two different slots in each frame (Figure 1c).

We model a cluster-based WSN by the *connectivity graph* $G = (V, E)$, where V contains all parent nodes (simply re-

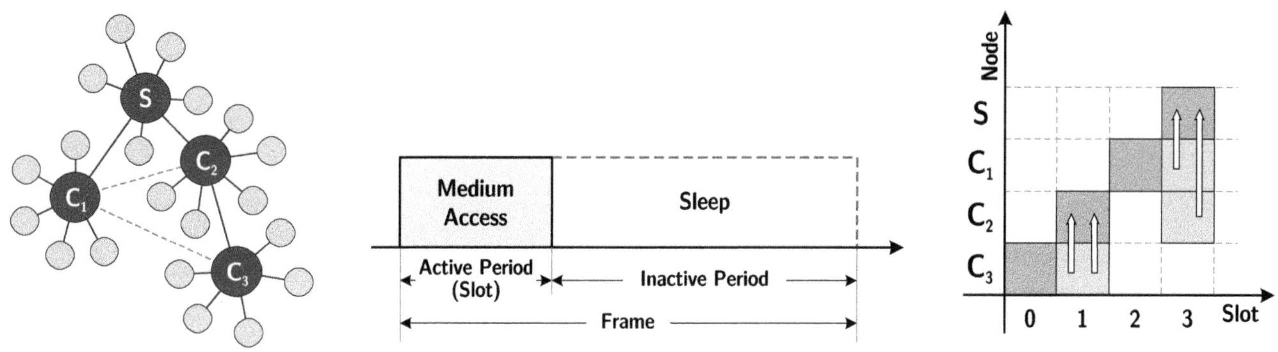

(a) (b) (c)

Figure 1: (a) Sample cluster-tree network topology: solid lines represent connectivity, while dashed lines interference. (b) Duty-cycled channel access in cluster-based WSNs. (c) Interference-free schedule for the WSN shown in (a).

ferred to as *nodes* in the rest of the article for brevity) and E contains all the symmetric communication links between the nodes in V. We do not make any assumption on how the connectivity graph is derived. For instance, it can be empirically obtained by the nodes after the deployment. In a cluster-based WSNs operating with a duty-cycle, nodes *interfere* if they are active at the same time and belong to different clusters. In this article, we explicitly address the scenario where interfering nodes can be at most t hops away from each other in the connectivity graph, where t is a constant not smaller than 2. In this case, interferences can be accounted for through the *interference graph* G^t defined as the t-th power of the connectivity graph G, that is, $G^t = (V, E^t)$. The interference graph has the same set of vertices V as G but the set of edges $E^t = \{(i,j) : d(i,j) \leq t\}$, where $d(i,j)$ is the shortest path between the vertices i and j in G.

3.2 Problem definition

The minimum-latency convergecast problem is to find a periodic sequence of slots in each frame so as to minimize the latency of communications between the sensor nodes and the sink, as well as to guarantee interference-free scheduling in G. Latency is measured as the number of slots between: the instant when the message is sent by the source node; and the instant when the message is received by the sink.

The first aspect to consider is related to the number of slots required to avoid interference up to the constant distance $t \geq 2$ in G. A *t-Interference-Free Schedule* (t-IFS) is a slot assignment $s : V \to \mathbb{N}$ which is guaranteed to be interference-free up to the distance t. Formally, given a connectivity graph $G = (V, E)$ and the associated interference graph $G^t = (V, E^t)$, a t-IFS scheduling assigns the slot $s(i)$ to router $i \in V$ so that $s(i) \neq s(j)$ if $(i,j) \in E^t$. Let the *size* S_t of the t-IFS be the largest slot assigned to the nodes in V, i.e, $S_t = \max_{i \in V} s(i)$. The *Minimum t-Interference-Free Schedule* (t-MIFS) problem is to find the t-IFS scheduling of minimum size, denoted by $S_t^*(G)$, over all possible t-IFS scheduling choices. It is easy to see that the t-MIFS problem is NP-complete when the WSN is represented by an arbitrary graph. In fact, the t-MIFS problem can be reduced (by mapping slots to colors) to a classical node coloring problem on the interference graph G^t. Thus, the size of the largest clique in G^t is a lower bound on the minimum size $S_t^*(G)$ of any t-IFS scheduling, but finding such a size is a hard problem. The value of $S_t^*(G)$ clearly affects the minimum bound of the convergecast latency, while the actual latency depends on how the slots are interleaved in the paths from source nodes to the sink. Note that slots assigned in decreasing order on a path are affected by a delay which depends not only by their relative offset, but also on the length k of the frame [14]. As a consequence, slots should be better assigned in non-decreasing order on the paths from the leaves to the sink.

In order to introduce the minimum-latency convergecast problem, let us define first the communication delay. Given a t-IFS slot assignment of size k for G, the *report latency* of two adjacent nodes i and j, such that $(i,j) \in E$, is the number of slots, denoted by w_{ij}, that node i has to wait to relay its collected sensory data to node j. If $s(j) \geq s(i)$ then $w_{ij} = s(j) - s(i)$, otherwise it is $w_{ij} = s(j) - s(i) + k$; in other words, $w_{ij} = (s(j) - s(i)) \bmod k$. The *cumulative report latency*, for each node $i \in V$ connected to the sink, is the sum of the report latencies on the edges of the shortest path from i to the sink in G. Clearly, if in the graph G there is no path from node i to the sink, the cumulative report latency is set to $+\infty$. For instance, consider the scheduling illustrated in Figure 1c. The cumulative report latency between the sink S and the cluster-head C_3 is equal to 3 slots. In the following, we say that an *inversion* occurs on the path from node i towards the sink if there is one edge (u,v) such that $s(u) > s(v)$. Note that the message delivery from a source node i to the sink lasts as many frames as the number of inversions on the shortest path from i to the sink. In detail, the cumulative report latency L on a path with $c \geq 1$ inversions yields $(c-1)k < L \leq ck$.

Finally, the *convergecast latency*, denoted as $L(G)$, is the maximum of the cumulative report latencies of individual nodes. Thus, given a number of available slots which is enough to achieve a t-IFS, the *t-interference-free Minimum-latency Convergecast* problem can be defined as follows:

DEFINITION 1. *Given $G = (V, E)$ and the constant interference distance t, the* t-interference-free Minimum-latency Convergecast *(t-MC) problem is to find a t-IFS mapping $s : V \to [0, k-1]$ in $G^t = (V, E^t)$ such that the convergecast latency $L(G)$ is minimized.*

THEOREM 1. *The t-MC problem is NP-hard.*

PROOF. The 3-CNF-SAT problem can be reduced in polynomial time to the decision version of the 2-MC problem[1]. As a consequence, the 2-MC problem is NP-hard, implying that the t-MC problem is NP-hard for $t \geq 2$. □

4. SCHEDULING IN CLUSTER-TREE WSNS

Before proceeding further, let us introduce some preliminary concepts. A graph $\mathcal{T} = (V, E)$ is a *free tree* when it is connected and it has exactly $|V| - 1$ edges. A *rooted tree* is a free tree in which a vertex r is identified as the *root* and all other vertices are ordered by levels, where the level $l(v)$ of a vertex v is equal to the distance $d(r, v)$. Thus, all the vertices adjacent to v are partitioned into its *parent* at level $l(v) - 1$, and into its children at level $l(v) + 1$. The *height* h of a tree $\mathcal{T}$ is the maximum level of its vertices. An *ordered tree* is a rooted tree in which an ordering is imposed among all the children of every node v.

In the following, for the sake of simplicity, an ordered tree will be called *tree*. Given a tree $\mathcal{T}$ of height h and an integer $\ell \leq h$, the *induced subtree* $\mathcal{T}_\ell$ consists of all the vertices v with $l(v) \leq \ell$ as well as all the original edges among them. For each vertex v of $\mathcal{T}$, let $A_i(v)$ denote the *ancestor* of v at distance i from v. Clearly, $A_i(v)$ is at level $l(v) - i$. Besides, let $D_i(v)$ denote the set of the *descendants* of v at distance i from v; thus the descendants belong to level $l(v)+i$. Moreover, let $|D_i(v)|$ denote the cardinality of $D_i(v)$.

Let us now revisit how to solve the t-MIFS problem on trees in order to provide a lower bound on the number k of slots required to solve the t-MC problem. To this end, we recall the simplicial property. A vertex x of $\mathcal{T}$ is *t-simplicial* when, for every pair of vertices u and v such that $d(x, u) \leq t$ and $d(x, v) \leq t$, it is also true that $d(u, v) \leq t$. A vertex x is called *strongly-simplicial* when x is t-simplicial for any value of t [3]. The existence of a strongly-simplicial vertex x implies that, for any integer t, all the nodes at distance at most t from x – denoted as the *t-simplicial neighborhood* $N_t(x)$ – need to be assigned to different slots to avoid interference.

From the observation above, it follows that the t-MIFS problem defined on a tree can be solved by visiting the tree in breadth-first-search order. Since each newly processed vertex v is a t-simplicial vertex at level $l(v)$ of the induced subtree $\mathcal{T}_{l(v)}$, for $1 \leq v \leq n$, it has to be assigned to a slot different from those slots assigned to the vertices in the neighborhood $N_t(v)$. Obviously, the size of the largest $N_t(v)$ scanned during the visit of $\mathcal{T}$ is the minimum number $S_t^*(\mathcal{T})$ of slots required for scheduling the communications without interference up to a distance t in $\mathcal{T}$. In [3], it has been shown how to compute $N_t(x)$ for each vertex $x \in \mathcal{T}$, and hence $S_t^*(\mathcal{T}) = \max\{N_t(x) : \forall x \in \mathcal{T}\}$, in $O(nt)$ time, where n is the total number of vertices in $\mathcal{T}$.

In the following, we derive a lower bound on the convergecast latency in a tree-based WSN. Recall that, since the root is the sink where data are accumulated, the shortest paths in $\mathcal{T}$ traverse the edges from the children towards the parents. Let us consider the tree logically partitioned into super-levels, each of $\lfloor \frac{t}{2} \rfloor + 1$ consecutive levels. In detail, given a fixed offset i from the root, the (i, j)-th super-level consists of the $\mathcal{T}$ levels

$$\left\{(j-1)\left(\left\lfloor \frac{t}{2} \right\rfloor + 1\right) + i, \ldots, j\left(\left\lfloor \frac{t}{2} \right\rfloor + 1\right) + i - 1\right\}$$

where $0 \leq i \leq \lfloor \frac{t}{2} \rfloor$ and $1 \leq j \leq \left\lfloor \frac{h+1-i}{\lfloor \frac{t}{2} \rfloor + 1} \right\rfloor$.

Let us now denote the subtree of height $\lfloor \frac{t}{2} \rfloor$ rooted at vertex x as $\mathcal{T}^{\lfloor \frac{t}{2} \rfloor}(x)$, and the vertices that belong to the first level of the (i, j)-th super-level as $\mathcal{S}(i, j) = \{x : l(x) = (j-1)(\lfloor \frac{t}{2} \rfloor + 1) + i\}$. Then, observe that $\cup_{x \in \mathcal{S}(i,j)} \mathcal{T}^{\lfloor \frac{t}{2} \rfloor}(x)$ is equal to all the vertices of the (i, j)-th super-level.

THEOREM 2. *Given a tree $\mathcal{T}$ of height h and the interference distance t, it is:*

$$L(\mathcal{T}) \geq \max \left\{ h, \max_{0 \leq i \leq \lfloor \frac{t}{2} \rfloor} \sum_{j=1}^{\left\lfloor \frac{h+1-i}{\lfloor \frac{t}{2} \rfloor + 1} \right\rfloor} \min_{x \in \mathcal{S}(i,j)} |\mathcal{T}^{\lfloor \frac{t}{2} \rfloor}(x)| - 1 \right\}$$

PROOF. In a subtree of height $\lfloor \frac{t}{2} \rfloor$, all the vertices are at pairwise distance at most t, and thus all the vertices must be assigned to all different slots. Thus, the convergecast latency in such a tree is no less than the size of the tree minus one.

Consider now the super-level (i, j) of $\mathcal{T}$. It consists of $|\mathcal{S}(i, j)|$ trees of height $\lfloor \frac{t}{2} \rfloor$, each rooted at a different vertex in $\mathcal{S}(i, j)$. The cumulative report latency for traversing such a super-level is at least equal to $\min_{x \in \mathcal{S}(i,j)} \left(|\mathcal{T}^{\lfloor \frac{t}{2} \rfloor}(x)| - 1\right)$. Hence, for a fixed offset i, the convergecast latency in $\mathcal{T}$ is contributed by: the sum of the cumulative report latency on each super-level; i slots for traversing the first i levels; one slot between any two consecutive super-levels. Therefore,

$$L(\mathcal{T}) \geq \sum_{j=1}^{\left\lfloor \frac{h+1-i}{\lfloor \frac{t}{2} \rfloor + 1} \right\rfloor} \min_{x \in \mathcal{S}(i,j)} \left(|\mathcal{T}^{\lfloor \frac{t}{2} \rfloor}(x)| - 1\right) + i + \left\lfloor \frac{h+1-i}{\lfloor \frac{t}{2} \rfloor + 1} \right\rfloor - 1$$

$$\geq \sum_{j=1}^{\left\lfloor \frac{h+1-i}{\lfloor \frac{t}{2} \rfloor + 1} \right\rfloor} \min_{x \in \mathcal{S}(i,j)} |\mathcal{T}^{\lfloor \frac{t}{2} \rfloor}(x)| + i - 1$$

Since the same reasoning can be repeated for each value of i, the convergecast latency is at least

$$L(\mathcal{T}) \geq \max_{0 \leq i \leq \lfloor \frac{t}{2} \rfloor} \sum_{j=1}^{\left\lfloor \frac{h+1-i}{\lfloor \frac{t}{2} \rfloor + 1} \right\rfloor} \min_{x \in \mathcal{S}(i,j)} |\mathcal{T}^{\lfloor \frac{t}{2} \rfloor}(x)| - 1$$

Finally, the claim holds since the latency cannot be smaller than the height of the tree. □

4.1 The SCHEDULING-LEVEL-BY-LEVEL Algorithm

In the following, an $O(nt)$ time algorithm is proposed to find a sub-optimal solution of the t-MC problem which uses $k \geq S_t^*$ slots.

Our algorithm exploits two facts:

1. the same slot can be assigned to any two vertices in the same level of the tree when their distance is at least equal to $t + 1$; and
2. the same slot can be assigned to any two vertices belonging to two levels at least $t + 1$ apart.

With reference to Fact 1), the vertices of every level of $\mathcal{T}$, visited from left to right, can be partitioned into disjoint subsets, each consisting of consecutive vertices at reciprocal distance not greater than t. The vertices in such subsets share

[1]The detailed proof is omitted here due to space limitations.

Algorithm 1: SCHEDULING-LEVEL-BY-LEVEL $(T, t, \{|\Gamma_i|\})$

output: the slot $s(v)$ for each vertex $v \in \mathcal{T}$
1 $k \leftarrow \sum_{h=0}^{t} |\Gamma_h|$; $\mu_{-1} \leftarrow k$;
2 **for** $0 \leq i \leq t$ **do** $\mu_i \leftarrow \mu_{i-1} - |\Gamma_i|$;
3 $r \leftarrow$ `root`(T); $s(r) \leftarrow \mu_0$; $d(r) \leftarrow 0$; $prec \leftarrow r$;
4 $Q \leftarrow \varnothing$; **if** $\lfloor \frac{t}{2} \rfloor > 1$ **then** `enqueue` (r, Q); `insert-queue` $(D_{\lfloor \frac{t}{2} \rfloor}(r), d(r) + \lfloor \frac{t}{2} \rfloor, Q_D)$;
5 **while** $Q \neq \varnothing$ **do**
6   $u \leftarrow$ `dequeue`(Q);
7   **foreach** $v \in$ `children`(u) **do**
8     $d(v) \leftarrow d(u) + 1$; $i \leftarrow d(v)$;
9   **if** $d(prec) \neq d(v)$ **then** $temp \leftarrow \mu_i$;
10     $s(v) \leftarrow temp$; $temp \leftarrow temp + 1$; $prec \leftarrow v$;
11   **if** $d(v) < \lfloor \frac{t}{2} \rfloor - 1$ **then** `enqueue`(v, Q);
12     $l_D \leftarrow d(v) + \lfloor \frac{t}{2} \rfloor$;
13   `insert-queue`$(D_{\lfloor \frac{t}{2} \rfloor}(v), l_D, Q_D)$;
14 **while** $Q_D \neq \varnothing$ **do**
15   `extract-queue`$(S, level, Q_D)$;
16   $u \leftarrow$ `first`(S); $d(u) \leftarrow level$; $i \leftarrow level \bmod (t+1)$;
17   $temp \leftarrow \mu_i$; $l_D \leftarrow d(u) + \lfloor \frac{t}{2} \rfloor$;
18   **foreach** $u \in S$ **do**
19     $s(u) \leftarrow temp$; $temp \leftarrow temp + 1$;
20   `insert-queue`$(D_{\lfloor \frac{t}{2} \rfloor}(u), l_D, Q_D)$;

a common ancestor at distance $\lfloor \frac{t}{2} \rfloor$. Indeed, each of such subsets at level ℓ corresponds to the descendants $D_{\lfloor \frac{t}{2} \rfloor}(x)$ at distance $\lfloor \frac{t}{2} \rfloor$ of some node x which belongs at level $\ell - \lfloor \frac{t}{2} \rfloor$. Consecutive subsets at level ℓ are then descendants of consecutive vertices at level $\ell - \lfloor \frac{t}{2} \rfloor$, thus the distance between vertices in such consecutive subsets is greater than t because their lowest common ancestor belongs to a level smaller than $\ell - \lfloor \frac{t}{2} \rfloor$. Therefore, in order to guarantee an interference-free assignment for each level ℓ, it is sufficient to assign as many slots as:

$$\gamma_\ell = \max\left\{ |D_{\lfloor \frac{t}{2} \rfloor}(x)| : x \in \mathcal{T} \text{ and } l(x) = \ell - \left\lfloor \frac{t}{2} \right\rfloor \right\} \quad (1)$$

By Fact 2), the set of slots assigned to level i can be reused for a family of levels $\mathcal{F}_i = \{\ell : \ell \equiv i \bmod (t+1)\}$ by still guaranteeing a t-IFS for $\mathcal{T}$. Specifically, the size of the same group Γ_i of slots assigned to the levels $\mathcal{F}_i$ is $|\Gamma_i| = \max\{\gamma_\ell : \ell \in \mathcal{F}_i\} = \max\{|D_{\lfloor \frac{t}{2} \rfloor}(x)| : x \in \mathcal{T} \text{ and } l(x) = \ell - \lfloor \frac{t}{2} \rfloor \text{ and } \ell \in \mathcal{F}_i\}$, with $0 \leq i \leq t$.

In the rest of this section, based on the observations above, we propose a t-IFS scheduling, i.e., SCHEDULING-LEVEL-BY-LEVEL, which uses $k = \sum_{i=0}^{t} |\Gamma_i|$ slots and provides a convergecast latency of $L(\mathcal{T}) < \lceil \frac{h+1}{(t+1)} \rceil k$.

The SCHEDULING-LEVEL-BY-LEVEL algorithm, illustrated in Algorithm 1, assumes that a preprocessing is performed to compute, for each vertex x of T, the set $D_{\lfloor \frac{t}{2} \rfloor}(x)$ and the size $|D_{\lfloor \frac{t}{2} \rfloor}(x)|$ of the descendants at distance $\lfloor \frac{t}{2} \rfloor$ from x. Such a computation can be performed in $O(nt)$ time by visiting the tree in postorder and by storing the descendants in linked lists such that the t unions of lists required for each node runs in $O(t)$ time. The corresponding recursive algorithm DESCENDANTS is invoked starting from the root r of $\mathcal{T}$ (Algorithm 2). Also the number $|\Gamma_i|$ of slots to be assigned at levels in $\mathcal{F}_i$ is computed by a preprocessing step with the algorithm COUNT-SLOTS, which performs a breadth-first search of $\mathcal{T}$ (Algorithm 3). Specifically, it uses a queue Q to store the vertices of $\mathcal{T}$ (line 2), in the same order in which they are visited, and computes for each vertex v its level $d(v)$ (line 6). Moreover, while traversing the uppermost $\lfloor \frac{t}{2} \rfloor$ levels, since all the vertices in such levels are at pairwise distance not greater than t, $|\Gamma_i|$ is set equal to the number of vertices at level i, $0 \leq i \leq \lfloor \frac{t}{2} \rfloor - 1$ (line 7). Then, while visiting the vertices at level $0 \leq d(v) \leq h - \lfloor \frac{t}{2} \rfloor$, the COUNT-SLOTS algorithm finds the size of the largest set of descendants at distance $\lfloor \frac{t}{2} \rfloor$ from the vertices at level $d(v)$ and stores it as $|\Gamma_i|$ where $i = (d(v) + \lfloor \frac{t}{2} \rfloor) \bmod (t+1)$ (lines 8-9). When the visit of $\mathcal{T}$ is completed, $|\Gamma_i|$ stores the size of the largest set of pairwise vertices at distance t at the levels in $\ell \in \mathcal{F}_i$ (line 11).

Algorithm 2: DESCENDANTS(x, T, t)

output: the sets $D_0(x), \ldots, D_t(x)$ and their sizes
1 $D_0(x) \leftarrow \{x\}$; $|D_0(x)| \leftarrow 1$
2 **for** $1 \leq i \leq t$ **do** $D_i(x) \leftarrow \varnothing$; $|D_i(x)| \leftarrow 0$;
3 **if** x **is not** leaf **then**
4   **foreach** $v \in$ `children`(x) **do**
5   DESCENDANTS(v, T, t);
6   **for** $i \leftarrow 1$ **to** t **do**
7       $D_i(x) \leftarrow D_i(x) \cup D_{i-1}(v)$;
8       $|D_i(x)| \leftarrow |D_i(x)| + |D_{i-1}(v)|$;

Algorithm 3: COUNT-SLOTS(T, t)

output: $|\Gamma_i|$ for level $\ell \in \mathcal{F}_i$, with $0 \leq i \leq t$
1 $d(r) \leftarrow 0$; $|\Gamma_0| \leftarrow 1$; **for** $1 \leq i \leq t$ **do** $|\Gamma_i| \leftarrow 0$;
2 $Q \leftarrow \varnothing$; $r \leftarrow$ `root`(T); `enqueue`(r, Q);
3 **while** $Q \neq \varnothing$ **do**
4   $u \leftarrow$ `dequeue`(Q);
5   **foreach** $v \in$ `children`(u) **do**
6     $d(v) \leftarrow d(u) + 1$;
7   **if** $d(v) \leq \lfloor \frac{t}{2} \rfloor$ **then** $|\Gamma_{d(v)}| \leftarrow |\Gamma_{d(v)}| + 1$;
8     $i \leftarrow (d(v) + \lfloor \frac{t}{2} \rfloor) \bmod (t+1)$;
9     $|\Gamma_i| \leftarrow \max(|\Gamma_i|, |D_{\lfloor \frac{t}{2} \rfloor}(v)|)$;
10   **if** $d(v) \leq h - \lfloor \frac{t}{2} \rfloor$ **or** $d(v) \leq \lfloor \frac{t}{2} \rfloor$ **then**
11   `enqueue`(v, Q);

Given the number of slots $|\Gamma_0|, \ldots, |\Gamma_t|$ to be devoted to each level of $\mathcal{T}$ and the sets $D_{\lfloor \frac{t}{2} \rfloor}(x)$ for each $x \in \mathcal{T}$, the algorithm SCHEDULING-LEVEL-BY-LEVEL finds the actual vertex-slot assignment in $O(n)$ time. In detail, the set of slots $\Gamma_i = \{\mu_i, \mu_i + 1, \ldots, \mu_{i-1} - 1\}$ dedicated to the levels $\ell \in \mathcal{F}_i$ of $\mathcal{T}$ are implicitly obtained by computing the first slot μ_i dedicated to the levels in $\mathcal{F}_i$, with $0 \leq i \leq t$. In fact, the algorithm assigns the slots $k - |\Gamma_0|, \ldots, k-1$ to the levels in $\mathcal{F}_0$, and sets $\mu_0 = k - |\Gamma_0|$. Similarly, it devotes the slots $\Gamma_1 = \{\mu_1, \ldots, \mu_0 - 1\}$, where $\mu_1 = \mu_0 - |\Gamma_1|$, to each level $\ell \in \mathcal{F}_1$, and so on (lines 1–3). Then, the tree is visited level-by-level by using two queues: Q and Q_D. In detail, Q stores one-by-one the vertices of the uppermost $\lfloor \frac{t}{2} \rfloor - 1$ levels. Note

$$N_t(x) = \begin{cases} \bigcup_{i=0}^{\lceil \frac{t}{2} \rceil - 1} [D_i(A_{t-i}(x)) \cup D_i(A_{t-1-i}(x))] \cup D_{\frac{t}{2}}(A_{\frac{t}{2}}(x)) & \text{if } t \text{ is even} \\ \bigcup_{i=0}^{\lceil \frac{t}{2} \rceil - 1} [D_i(A_{t-i}(x)) \cup D_i(A_{t-1-i}(x))] & \text{if } t \text{ is odd} \end{cases} \tag{2}$$

$$N_t(x) \subset \begin{cases} \bigcup_{i=0}^{\lceil \frac{t}{2} \rceil - 1} \left[D_{\lfloor \frac{t}{2} \rfloor}(A_{t-2\cdot i+\lfloor \frac{t}{2} \rfloor}(x)) \cup D_{\lfloor \frac{t}{2} \rfloor}(A_{t-2\cdot i+\lfloor \frac{t}{2} \rfloor+1}(x))\right] \cup D_{\frac{t}{2}}(A_{\frac{t}{2}}(x)) & \text{if } t \text{ is even} \\ \bigcup_{i=0}^{\lceil \frac{t}{2} \rceil - 1} \left[D_{\lfloor \frac{t}{2} \rfloor}(A_{t-2\cdot i+\lfloor \frac{t}{2} \rfloor}(x)) \cup D_{\lfloor \frac{t}{2} \rfloor}(A_{t-2\cdot i+\lfloor \frac{t}{2} \rfloor+1}(x))\right] & \text{if } t \text{ is odd} \end{cases} \tag{3}$$

Figure 2: The t-simplicial neighborhood $N(x)$ of a given node x.

that the vertices at level i, with $0 \le i \le \lfloor \frac{t}{2} \rfloor - 1$, are the descendants at distance i of the root, and are assigned, from left to right, to consecutive slots in Γ_i starting from slot μ_i (lines 8 - 10). All the vertices in the remaining levels of $\mathcal{T}$ are then stored in the queue Q_D of the descendants so as to preserve (in each level) the subsets of consecutive vertices which share the same ancestor $\lfloor \frac{t}{2} \rfloor$ levels above. Each subset at level $\ell + \lfloor \frac{t}{2} \rfloor$ represents the descendants at distance $\lfloor \frac{t}{2} \rfloor$ of some vertex at level ℓ. Specifically, for each vertex v at level ℓ, the vertices of the subset $D_{\lfloor \frac{t}{2} \rfloor}(v)$ are inserted together into D_Q when v is visited and extracted all together from Q_D while visiting level $\ell+\lfloor \frac{t}{2} \rfloor$. The vertices of $D_{\lfloor \frac{t}{2} \rfloor}(v)$ are then assigned from left to right to the slots in Γ_i where $i = (\ell+\lfloor \frac{t}{2} \rfloor) \bmod (t+1)$, starting from slot μ_i (lines 14 - 18).

THEOREM 3. *The* SCHEDULING-LEVEL-BY-LEVEL *algorithm uses k slots and finds in $O(n)$ time a t-IFS scheduling whose convergecast latency is $L(\mathcal{T}) < \lceil \frac{h+1}{t+1} \rceil k$. When $h \ge 3\lfloor \frac{t}{2} \rfloor + 1$, the algorithm provides an approximation $\rho < \frac{2}{c}\left(\lfloor \frac{t}{2} \rfloor + 1\right)\frac{\overline{\gamma}}{\widehat{\gamma}}$ where $\frac{1}{2} \le c < 1$, $\widehat{\gamma} = \min_{x\in\mathcal{T}} |D_{\lfloor \frac{t}{2} \rfloor}(x)|$, and $\overline{\gamma} = \max_{x\in\mathcal{T}} |D_{\lfloor \frac{t}{2} \rfloor}(x)|$.*

PROOF. Let us first show how the SCHEDULING-LEVEL-BY-LEVEL algorithm guarantees that, for each node $x \in \mathcal{T}$, there are no interferences in its t-simplicial neighborhood $N_t(x)$ in the induced subtree $\mathcal{T}_{l(x)}$ (Figure 2).

By inspection, it is easy to see that the t-simplicial neighborhood can be decomposed as in Equation (2). Observe that each subset of descendants in $N_t(x)$ at level $l(x)-t+z$, with $0 \le z \le t$, is a subset of the descendants at distance $\lfloor \frac{t}{2} \rfloor$ of some node y at level $l(x) - t + z - \lfloor \frac{t}{2} \rfloor$.

Namely, the set of vertices $D_i(A_{t-i}(x))$ belongs to level $l(x) - t + 2 \cdot i$ and it is a subset of $D_{\lfloor \frac{t}{2} \rfloor}(A_{t-2\cdot i+\lfloor \frac{t}{2} \rfloor}(x))$, for each $0 \le i \le \lceil \frac{t}{2} \rceil - 1$. Similarly, the set of vertices $D_i(A_{t-i-1}(x))$ belongs to level $l(x) - t + 2 \cdot i + 1$ and it is a subset of $D_{\lfloor \frac{t}{2} \rfloor}(A_{t-2\cdot i+\lfloor \frac{t}{2} \rfloor-1}(x))$, for each $0 \le i \le \lceil \frac{t}{2} \rceil - 1$. Moreover, when t is even, $D_{\lfloor \frac{t}{2} \rfloor}(A_{\lfloor \frac{t}{2} \rfloor}(x))$ is the same as $D_{\lfloor \frac{t}{2} \rfloor}(A_{t-2\cdot\frac{t}{2}+\lfloor \frac{t}{2} \rfloor}(x))$. As a result, Equation (3) holds. Since the algorithm SCHEDULING-LEVEL-BY-LEVEL assigns, for any x in $t+1$ consecutive levels of $\mathcal{T}$, the vertices in $D_{\lfloor \frac{t}{2} \rfloor}(x)$ to different slots, the proposed schedule is interference-free.

With reference to the convergecast latency, let us consider an arbitrary subpath of length $t+1$ starting at level $\ell \equiv t \bmod (t+1)$ and decreasing down to level $\ell-t$. The slots assigned on such a subpath form a sequence sorted in increasing order because the vertices at level $\ell \equiv i \bmod (t+1)$ are assigned to the slots in group Γ_i, with $t \ge i \ge 0$. Thus, the cumulative report latency along such a subpath is at most $k-1$, that is the difference between the largest and the smallest slot in Γ_t and Γ_0. Therefore, the cumulative report latency in $\mathcal{T}$ is $L(\mathcal{T}) \le \sum_{i=1}^{\lceil \frac{h+1}{t+1} \rceil}(k-1) + \sum_{i=1}^{\lceil \frac{h+1}{t+1} \rceil - 1} 1 < \lceil \frac{h+1}{t+1} \rceil k$ because there is at least a unitary report latency between two subpaths of length $t+1$.

With respect to the time complexity, the SCHEDULING-LEVEL-BY-LEVEL algorithm mainly performs a breadth-first-search of $\mathcal{T}$, therefore it requires $O(n)$ time.

The approximation ratio between the convergecast latency $L(\mathcal{T})$ provided by the SCHEDULING-LEVEL-BY-LEVEL algorithm and the optimal convergecast latency can be bounded from above by replacing the value of the optimal convergecast latency with its lower bound computed in Theorem 2. We express such a ratio as a function of the following parameters: $\widehat{\gamma} = \min_{x\in\mathcal{T}} |D_{\lfloor \frac{t}{2} \rfloor}(x)|$ and $\overline{\gamma} = \max_{x\in\mathcal{T}} |D_{\lfloor \frac{t}{2} \rfloor}(x)|$, which represent, respectively, the size of the minimum and the size of the maximum set of descendants at distance $\lfloor \frac{t}{2} \rfloor$ among all the vertices in $\mathcal{T}$.

Indeed, by Equation (1) and by recalling that $\mathcal{F}_i$ is the set of the ℓ levels in T such that $\ell \equiv i \bmod (t+1)$, the largest number of vertices that can be assigned to a level of $\mathcal{T}$ is indeed $\overline{\gamma}$, as $\max_{0\le i\le t} |\Gamma_i| = \max_{0\le i\le t}\{\max_{\ell\in\mathcal{F}_i} \gamma_\ell\} = \max_{x\in\mathcal{T}} |D_{\lfloor \frac{t}{2} \rfloor}(x)| = \overline{\gamma}$. Since $|\mathcal{T}^{\lfloor \frac{t}{2} \rfloor}(x)| \ge |D_{\lfloor \frac{t}{2} \rfloor}(x)|$, the lower bound in Theorem 2 can be rewritten as:

$$\max_{0\le i\le \lfloor \frac{t}{2} \rfloor} \sum_{j=1}^{\left\lfloor \frac{h+1-i}{\lfloor \frac{t}{2} \rfloor+1} \right\rfloor} \min_{x\in\mathcal{S}(i,j)} |\mathcal{T}^{\lfloor \frac{t}{2} \rfloor}(x)| - 1 >$$

$$\max_{0\le i\le \lfloor \frac{t}{2} \rfloor} \sum_{j=1}^{\left\lfloor \frac{h+1-i}{\lfloor \frac{t}{2} \rfloor+1} \right\rfloor} \min_{x\in\mathcal{S}(i,j)} |D_{\lfloor \frac{t}{2} \rfloor}(x)| - 1 \ge$$

$$\frac{h+1-\lfloor \frac{t}{2} \rfloor}{\lfloor \frac{t}{2} \rfloor + 1}\widehat{\gamma} - 1 \ge c\frac{h+1-\lfloor \frac{t}{2} \rfloor}{\lfloor \frac{t}{2} \rfloor + 1}\widehat{\gamma}$$

where $\frac{1}{2} \le c < 1$ since $\widehat{\gamma} \ge 1$ and $h \ge 3\lfloor \frac{t}{2} \rfloor + 1$. On the other hand, since $k \le (t+1)|\overline{\gamma}|$, $L(\mathcal{T}) < (h+1)\overline{\gamma}$ holds. Thus:

$$\rho < \frac{(h+1)\overline{\gamma}}{c\frac{h+1-\lfloor \frac{t}{2} \rfloor}{\lfloor \frac{t}{2} \rfloor+1}\widehat{\gamma}} < \frac{(h+1)\overline{\gamma}}{c\frac{h+1}{2(\lfloor \frac{t}{2} \rfloor+1)}\widehat{\gamma}} < \frac{2}{c}\left(\left\lfloor \frac{t}{2} \right\rfloor + 1\right)\frac{\overline{\gamma}}{\widehat{\gamma}}$$

□

Note that the approximation above becomes loose when

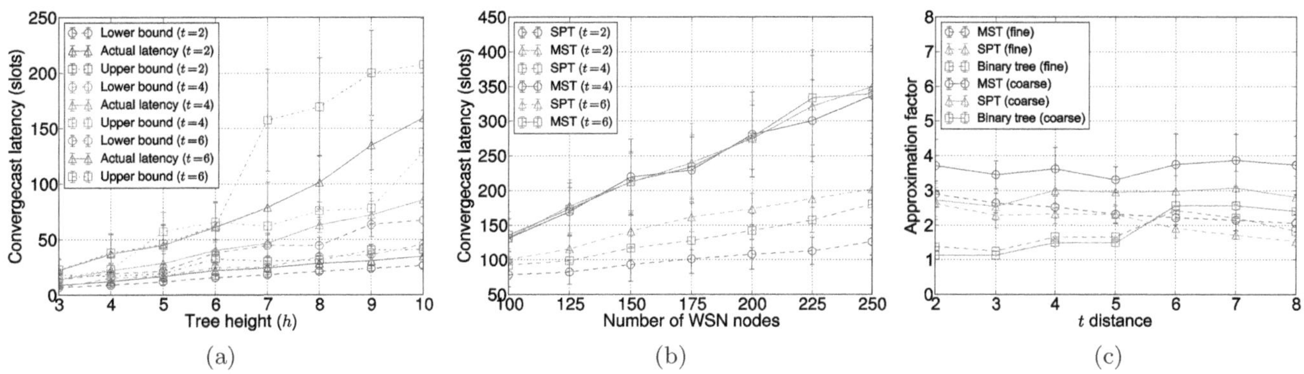

Figure 3: Convergecast latency in: (a) randomly generated tree topologies as a function of the height of the tree; (b) random WSN topologies as a function of the number of nodes. (c) Approximation factor of the SCHEDULING-LEVEL-BY-LEVEL algorithm as a function of the t distance.

the ratio $\frac{\overline{\gamma}}{\gamma}$ is high. However, the expression for ρ has the advantage of explicitly showing the dependency on the t-distance, and of requiring only the neighborhood information about the tree.

Finally, by accounting for the time complexity of the DESCENDANTS and COUNT-SLOTS algorithms, the t-MC problem is solved in $O(nt)$ time.

5. EVALUATION

In order to evaluate our solution, we performed several experiments by simulation. Specifically, we used a custom simulator written in python, and generated synthetic tree topologies as well as routing trees obtained by applying different routing protocols over a random WSN deployment. On the one side, the tree topologies were obtained starting with the sink, and adding to each node a number of children uniformly distributed between 1 and 3. On the other side, the WSN topologies were obtained by deploying nodes at random in a 200 m by 200 m sensing area. The radio transmission range was set to 30 m, and the connectivity between nodes was obtained by using the protocol model [10]. The convergecast tree was obtained as either the shortest path tree (SPT) or the minimum spanning tree (MST). The node closest to the center of the sensing area was selected as the sink in both cases.

We performed 25 independent replications of each experiment. In the plots, we reported the average values as well as the related standard deviations, shown as error bars. As for the considered metrics, in the following we will consider both the convergecast latency (as defined in Section 3.2) and the approximation factor.

5.1 Experimental Results

Figure 3a shows the convergecast latency in randomly generated trees as a function of their height. The values in the plot, which also shows the upper and the lower bounds, were derived for different values of the t distance. The figure shows that the actual convergecast latency (i.e., the one obtained by the SCHEDULING-LEVEL-BY-LEVEL algorithm) increases with the height h of the tree almost linearly, while the upper bound of the latency has a sharper behavior as it increases by k every $t+1$ levels. Clearly, the convergecast latency increases with the t distance as well, even though the major difference only appears for $t = 6$, while the latency is similar (and lower than 100 slots) for $t = 2$ and $t = 4$. Overall, the convergecast latency almost corresponds to the average value of the upper and the lower bounds, except for the region where $h \leq \lfloor \frac{t}{2} \rfloor$. This happens because slots cannot be reassigned in the first $\lfloor \frac{t}{2} \rfloor$ levels, and the convergecast latency is affected by the size of the entire tree rather than how the vertices are distributed at a given distance. This also explains why the upper bound and the actual latency are also close.

Figure 3b shows the convergecast latency in randomly generated WSN topologies as a function of the number of nodes. In the figure, the latency was obtained for different values of the t distance, as well as for the two different routing schemes (namely, SPT and MST). The results first show that the convergecast latency with the MST is higher than with the SPT, i.e., almost two times. This happens since the height of the MST is higher than that of the SPT, i.e., up to two times higher (on the average) in the experiments. Interestingly, the latency is not much affected by the t distance for the MST, while it increases with t for the SPT. This can be related to the relative impact of the height of the tree and the number of slots k on the latency, for the two cases. In detail, the height of the MST is almost independent of the number of nodes (i.e., about 20 in the experiments), since the deployment with 100 nodes is already dense enough to cover the sensing area. The height of the SPT, instead, decreases with the number of nodes, since the node density increases, hence, the length of the shortest paths decreases accordingly. In both cases, the number of slots k increases with the number of nodes, since the t-simplicial neighborhood of a given node increases in size accordingly (i.e., the number of nodes at distance t increases in denser trees). For the MST, the t-simplicial neighborhood increases almost linearly with t, and since $L(\mathcal{T}) \approx \frac{h}{t} \cdot k$, the convergecast latency is independent of t, which is not the case for the SPT.

Finally, Figure 3c shows the approximation factor as a function of the t distance, for different scenarios. In the figure, we reported both the accurate approximation factor, as the ratio between the actual latency and the lower bound obtained by considering a complete knowledge of the tree, and the coarser-grained one wherein the actual latency was replaced by the value in Theorem 3. Note that the coarse-grained approximation factor does not require to derive the actual schedule, but only needs the neighborhood information about the routing tree. We considered different routing trees: the SPT and the MST obtained from a random WSN deployment of 256 nodes; a balanced binary tree with height $h = 8$ as a reference. From the figure, we can see that the coarse-grained approximation factor always exceeds the fine-grained approximation factor obtained by the actual scheduling of Algorithm 1. Specifically, the two different approximation factors are close to each other (and also to 1) for the balanced binary tree when $t < 6$. This happens because of the regularity of the tree (equivalently, of the t-simplicial neighborhoods of any node), which makes the estimate of the bound more precise. The approximation factor is higher for the MST and the SPT, and the finer bound is roughly equal to 2.

In conclusion, our solution obtains a latency which is always bound to the optimal scheduling by a small constant factor, as it can be seen in Figure 3c. Specifically, the trend of the approximation ratio for $t = 2$ reflects the linear increase of t as the ratio $\frac{\overline{\gamma}}{\gamma}$ tends to 1. This confirms the validity and accuracy of the bounds obtained in Section 4.1.

6. CONCLUSION

In this article, we addressed latency-bound data collection in cluster-tree wireless sensor networks (WSNs) with reference to convergecast and aggregated traffic. We characterized the problem of minimum-latency data collection in tree-based WSNs where nodes are organized in clusters and use a duty-cycled mechanism for energy conservation. We considered interferences between clusters up to a distance $t \geq 2$ and provided bounds for latency-sensitive applications. We also designed a heuristic that runs in $O(nt)$ time, where n is the number of clusters in the WSN, to schedule the active time of clusters while reducing latency. Simulation results showed that the obtained bounds are accurate. In our future work, we will seek to refine the bounds introduced here as well as to improve the performance of the scheduling algorithm.

7. ACKNOWLEDGMENT

This work has been partially supported by the US National Science Foundation (NSF) under grant CNS-1049652 and by the Fondazione Cassa di Risparmio di Perugia (Italy) under grant "Ricerca di base 2010". Any opinions, findings and conclusions or recommendations expressed in this paper are those of the authors and do not necessarily reflect the views of the NSF.

Bibliography

[1] G. Anastasi, E. Borgia, M. Conti, E. Gregori, and A. Passarella. Understanding the real behavior of mote and 802.11 ad hoc networks: an experimental approach. *Pervasive Mob. Comput.*, 1(2):237–256, July 2005.

[2] G. Anastasi, M. Conti, M. Di Francesco, and A. Passarella. Energy conservation in wireless sensor networks: A survey. *Ad Hoc Networks*, 7(3):537–568, May 2009.

[3] A. Bertossi, C. M. Pinotti, and R. Rizzi. Channel assignment on strongly-simplicial graphs. In *Proc. of IPDPS 2003*, April 2003.

[4] H. Choi, J. Wang, and E. A. Hughes. Scheduling for information gathering on sensor network. *Wirel. Netw.*, 15(1):127–140, January 2009.

[5] M. Di Francesco, G. Anastasi, M. Conti, S. K. Das, and V. Neri. Reliability and energy-efficiency in IEEE 802.15.4/ZigBee sensor networks: A cross-layer approach. *Selected Areas in Communications, IEEE Journal on*, 29(8):1508–1524, September 2011.

[6] S. C. Ergen and P. Varaiya. TDMA scheduling algorithms for wireless sensor networks. *Wirel. Netw.*, 16(4):985–997, May 2010.

[7] C. Florens, M. Franceschetti, and R. McEliece. Lower bounds on data collection time in sensory networks. *Selected Areas in Communications, IEEE Journal on*, 22(6):1110–1120, Aug. 2004.

[8] C. Florens and R. McEliece. Packets distribution algorithms for sensor networks. In *Proc. of INFOCOM 2003*, pages 1063–1072, March 2003.

[9] S. Gandham, Y. Zhang, and Q. Huang. Distributed time-optimal scheduling for convergecast in wireless sensor networks. *Computer Networks*, 52(3):610–629, February 2008.

[10] P. Gupta and P. Kumar. The capacity of wireless networks. *Information Theory, IEEE Transactions on*, 46(2):388–404, March 2000.

[11] Z. Hanzálek and P. Jurčík. Energy Efficient Scheduling for Cluster-Tree Wireless Sensor Networks With Time-Bounded Data Flows: Application to IEEE 802.15.4/ZigBee. *Industrial Informatics, IEEE Trans. on*, 6(3):438–450, August 2010.

[12] IEEE 802.15.4, Part 15.4: Wireless Medium Access Control (MAC) and Physical Layer (PHY) Specifications for Low-Rate Wireless Personal Area Networks (LR-WPANs), September 2006.

[13] O. D. Incel, A. Ghosh, and B. Krishnamachari. *Scheduling algorithms for tree-based data collection in wireless sensor networks*, volume Theoretical aspects of distributed computing in sensor networks, pages 407–445. Springer, 2011.

[14] A. Keshavarzian, H. Lee, and L. Venkatraman. Wakeup scheduling in wireless sensor networks. In *Proc. of MobiHoc 2006*, pages 322–333, May 2006.

[15] A. Koubâa, A. Cunha, M. Alves, and E. Tovar. TDBS: a time division beacon scheduling mechanism for ZigBee cluster-tree wireless sensor networks. *Real-Time Syst.*, 40:321–354, Dec. 2008.

[16] S. R. Madden, M. J. Franklin, J. M. Hellerstein, and W. Hong. TinyDB: an acquisitional query processing system for sensor networks. *ACM Trans. Database Syst.*, 30:122–173, March 2005.

[17] M.-S. Pan and Y.-C. Tseng. Quick convergecast in ZigBee beacon-enabled tree-based wireless sensor networks. *Computer Communications*, 31(5):999–1011, March 2008.

[18] A. Willig. Recent and emerging topics in wireless industrial communications: A selection. *Industrial Informatics, IEEE Trans. on*, 4(2):102 –124, May 2008.

Coverage Management for Mobile Targets in Visual Sensor Networks

Vikram P. Munishwar
Computer Science
State University of New York at Binghamton
vmunish1@cs.binghamton.edu

Sameer Tilak
Calit2
University of California, San Diego
stilak@ucsd.edu

Nael B. Abu-Ghazaleh
Computer Science
State University of New York at Binghamton
nael@cs.binghamton.edu

ABSTRACT

Visual sensor networks (VSNs) are becoming increasingly popular in a number of application domains. A critical ability of such networks is to self-configure to minimize the need for operator control, improve scalability, and reduce cost. One of the areas of self-configuration is camera coverage control: how cameras adjust their field-of-view to allow automatic tracking of maximum number of targets. The problem is shown to be NP-hard for stationary targets, and efficient centralized, distributed and semi-centralized heuristics exist that perform close to optimal. For stationary targets camera configuration is a one-time activity and can happen offline and before the actual deployment. In contrast, if the targets are mobile, as the targets move away from their recorded positions, the cameras need to configure dynamically and in real-time to ensure coverage accuracy. In this paper, we propose several policies for automatic control of the cameras with a goal of coverage maximization for mobile targets. We study these policies using important performance metrics such as coverage gain, adaptability, scalability, and energy consumption. Our results indicate that factors such as target mobility models, target and camera scales/densities, and target velocities have significant impact on the performance of a given policy. For most of the scenarios, we found that the protocols that take into account non-local information (e.g. neighborhood information) and have self-adapting parameters (e.g. frequency of camera configurations) outperform the protocols that are either purely local or purely global and have non-adaptive parameters.

Categories and Subject Descriptors

C.2.1 [**Computer-Communication Networks**]: Network Architecture and Design—*Wireless communication, Centralized networks, Distributed networks*; C.4 [**Performance of Systems**]: *Design studies*

General Terms

Design, Experimentation, Performance

Keywords

mobile targets, coverage, camera configuration

1. INTRODUCTION

Visual sensor networks have a wide range of applications in areas such as monitoring and surveillance, traffic management and health care [3]. The existing multi-camera sensor systems need human operators to monitor and configure the network, who are prone to errors due to fatigue and lapses in attention. In fact, in the experiment conducted by Sandia National Laboratories for the U.S. Department of Energy demonstrates that even a dedicated and well-intensional individual has at most 20 minutes of attention span of acceptable levels [13].

To that end, recently smart camera networks have received considerable attention [1,30], which are able to self-configure to best match the application requirements. When networked, these smart cameras can coordinate to adapt their parameters in response to application-level metrics (e.g., coverage, image quality etc.) and system-level metrics (e.g., available network bandwidth/energy). This self-configuration reduces or eliminates the need for operator involvement both in monitoring and controlling the video stream and makes these system more vigilant and scalable (e.g., by the ratio of cameras to operators needed) than traditional surveillance networks [14].

Coverage management in sensor networks is an important problem and it has been studied thoroughly in omni-directional sensor networks [18,34]. The problem becomes more challenging for directional or visual sensor networks, since the coverage quality can change significantly depending on the current orientation of a sensor. One of the primary areas of self-configuration for visual sensor networks is the control of camera Field-of-Views (FoVs) to optimize cameras' ability to collectively monitor targets, events, or areas of interest. Many approaches in literature focus on achieving the goal of maximizing the coverage of stationary events, such as area or targets [2,19,25]. However, very little attention has been given to maximizing coverage when the events/targets to be monitored are mobile.

Majority of the camera network applications involve mobile targets, for which the coverage obtained for stationary targets would deteriorate as the targets move away from their recorded positions. While, the policies for stationary targets answer the question of *how* to configure cameras to cover maximum targets, we pose the further question of *when* or *how often* to reconfigure the cameras to track maximum number of mobile targets. To achieve stationary target coverage maximization, centralized, distributed, and semi-centralized camera configuration policies are proposed in the literature [2,25]. However, answering the latter question is non-trivial: in addition to the target mobility and/or velocity, it also needs to consider the wireless communication overhead that can potentially restrict the coverage policy to realize its full potential.

To address this problem, we propose camera reconfiguration policies that take into account the target mobility and network congestion. Specifically, our contributions can be summarized as below:
1) We show that it is hard to set the right reconfiguration interval for cameras due to the target mobility and network congestion. To address this problem, we propose two reconfiguration policies: Adaptive policy and Ondemand policy, and study their coverage accuracy and messaging overhead (Section 3, Section 5.1). The Adaptive and Ondemand policies provide excellent coverage for small scale networks. However, they require significant communication overhead, making them unsuitable for large scale networks. To address this problem, we investigate their variations in the form of hybrid policies that combine the reconfiguration policies with the distributed or semi-centralized camera configuration policies (Section 4).
2) We compared two target mobility patterns–depending on whether the targets are moving on pre-defined paths or not. We found that for a camera network deployed to monitor people, bikers, cars, etc., on streets, a simple and purely local greedy policy (each camera autonomously selects the pan covering the most targets) can provide close-to-optimal coverage with zero communication cost (Section 5.3).
3) We evaluate the proposed policies for different camera scale/density, target density, target velocities, and target mobility patterns, and show that these factors have a huge impact on coverage. For almost all of the cases, we found that the combination of semi-centralized camera configuration policy with the Adaptive reconfiguration policy outperforms all the other policies, and provides close-to-optimal coverage with low communication overhead (Section 5.4, Section 5.6, Section 5.5, Section 5.7).

2. CAMERA COVERAGE OVERVIEW

In this section, we present a brief overview of camera's FoV, followed by our assumptions and the coverage test for assessing whether a camera covers a target. Finally, we summarize the existing camera configuration approaches, and different target mobility models that are used in this work.

2.1 Field-of-view (FoV) of a camera

FoV of a camera represents the extent of its visible field. Typically, FoV is represented by its *angle-of-view* (AoV), α, and *depth-of-field*, R. AoV is represented in terms of the horizontal AoV, α_h, and vertical AoV, α_v. The depth-of-field represents the area of the visual scene that is acceptably sharp. Thus, for a given target, and a given level of acceptable sharpness, the depth-of-field of a camera spans from R_{min} to R_{max}. In order to simplify and abstract away the details, we consider the FoV of a camera as a spherical sector (as in the case of directional sensors), which can be described by its angle-of-view (α), and depth-of-field (R_{max}). Note that we use horizontal AoV (α_h) of the camera as α, since we consider pan-only cameras in this work. The FoV

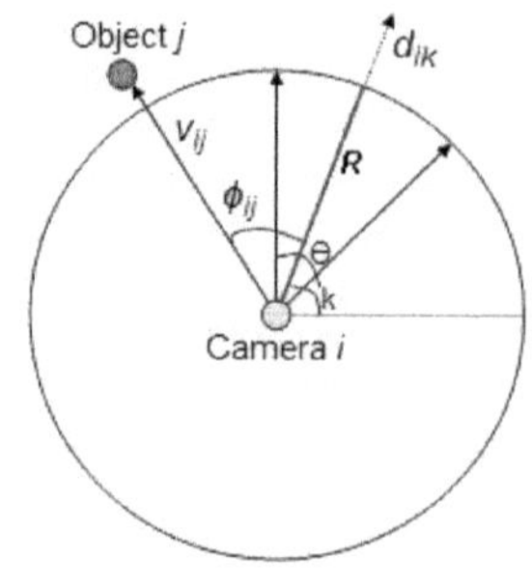

Figure 1: Camera coverage parameters.

adjustment can be performed by controlling camera's *pan*, which allows rotating the camera's FoV horizontally.

2.2 Coverage test and assumptions

Our work assumes that the cameras are already placed using some placement algorithm [17, 32], and focuses on configuring cameras' FoVs to maximize coverage. We assume that target locations can be obtained through an external localization mechanism, such as using camera calibration [5], a low resolution fisheye camera [24], stereoscopic localization [12], or LAser Detection And Ranging device (LADAR)-based localization [33]. This assumption is consistent with those made by other works [21] where PTZ cameras send their FoV and nearby targets' positions to a central node/base-station for computing optimal FoV parameters for all cameras. For target coverage, we assume that targets are covered if a selected point on them is covered; a target is either completely covered or not covered [27]. This point can be the centroid of the target, for example.

We consider a set of N static cameras trying to monitor M static targets (or objects) present in an area free of obstacles. Each camera is capable of panning to one of a discrete set of pans, P. The output of the coverage maximization policies is the list of $camera \in N$, $pan \in P$ pairs that collectively maximize the number of $targets \in M$ covered.

We define $\vec{v_{ij}}$ to be a vector (Euclidean vector) from camera i to target j. Vector $\vec{d_{ik}}$ denotes the pan direction k of camera i. ϕ_{ij} is a angle between $\vec{vij}$ and $\vec{d_{ik}}$. α represents the angle-of-view of a camera, defining the width of the FoV in the pan direction. In our formulation we assumed that the cameras are homogeneous, but this assumption can be directly relaxed by allowing the camera parameters (α, R_{max} and P) to be defined per camera.

The angle between camera and target vectors, ϕ_{ij}, is calculated as:

$$\phi_{ij} = \cos^{-1}\left(\frac{\vec{d_{ik}}.\vec{v_{ij}}}{|\vec{d_{ik}}||\vec{v_{ij}}|}\right). \tag{1}$$

A target is covered by a camera's FoV when the angular span of its FoV contains the target and the target is present within the coverable range of the camera. The former condition requires the ϕ_{ij} to be less than half of the AoV of camera i, and it can be represented as:

$$\phi_{ij} \leq \frac{\alpha}{2}. \tag{2}$$

The coverage test can be directly adapted to check for tilt constraints by comparing the vertical component of ϕ_{ij} with

α_v. The latter condition requires that the distance between the camera and the target is at most R_{max}, and can be represented as:

$$|\vec{v_{ij}}| \leq R_{max}. \tag{3}$$

We call the targets that obey the Equation 3 as the *coverable targets* of camera i. From the coverable targets, the targets that obey the Equation 2 are termed as *covered targets*.

2.3 Camera configuration policies

Camera configuration policies for covering maximum stationary targets can be broadly categorized in to three types: centralized, distributed, and semi-centralized policies.

Centralized policy relies on the global information. Each camera reports its coverable targets to a central location, also called as the base-station. Upon receiving packets from all the cameras, the base-station generates *optimal* camera-pan pairs by solving a coverage maximization problem [25] that maximizes the total unique targets covered. Once the optimal camera-pan pairs are generated, the base-station disseminates this information to the cameras.

Distributed policy relies on local or neighborhood specific information. Two cameras are considered as neighbors if their sensing regions (or FoVs) can overlap. In the *Distributed Greedy Approach (DGA)* [2], all cameras assign themselves a unique priority value. Each camera selects the pan that covers the maximum number of targets and informs it to its sensing neighbors (neighbors within $2 * R_{max}$ distance). Overlaps in the target coverage are avoided by accepting the decision of higher priority camera. The algorithm converges in iterative fashion, stabilizing from the highest priority camera in the network.

In the *pure Greedy policy (Greedy)*, each camera autonomously assigns a pan that maximizes the number of targets covered from its coverable targets.

Semi-centralized policy: Hierarchical [25]: The camera network is partitioned into multiple clusters such that there is *minimum* coverage dependency across clusters. Once such clusters are formed, the optimal problem can be solved for each cluster in a centralized manner.

2.4 Mobility models

In real-world, targets can follow different mobility patterns while moving, and their mobility is often dependent on the availability of physical paths. For instance, to move around on a university campus, students may use walk-ways, bike-lanes, or streets. They can walk or use either bike or car as a mode of transportation. To represent such mobility, we use the pedestrian mobility model provided in QualNet simulator [29] that simulates target mobility along streets. On the other hand, the mobility pattern used in parks or airports is not constrained by the path-ways, and thus can be represented by using the random-waypoint model [6,15,23]. Thus, we consider Pedestrian model and Random Waypoint model to evaluate the performance of the camera reconfiguration policies for different target mobility patterns.

3. CAMERA RECONFIGURATION POLICIES

In this section, we present camera reconfiguration policies that determine how often the cameras should be configured to track maximum number of targets.

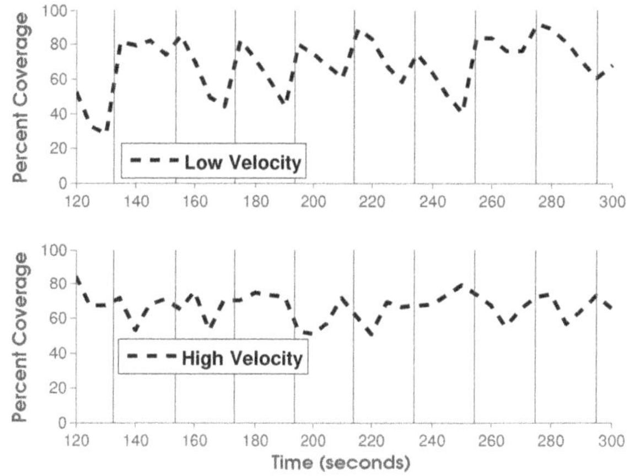

Figure 2: Observed coverage when targets are moving with different velocities with periodic interval set to 20 seconds. The vertical lines represent the reconfiguration times. The figure shows that the coverage can not be assumed to be monotonically decreasing between successive reconfigurations.

3.1 Periodic reconfiguration

Periodic policy is application-independent in nature. In this policy, cameras report the locations of targets coverable in different pans to the base station at regular intervals. Once the base station receives this information, the base station computes the optimal pan values for all the cameras and sends this information out to the network [1]. The periodic interval can be set at the deployment time; all cameras use the same reporting interval, and transmit at the same time.

Periodic policy is easy to implement and it performs well when the mobility pattern and the velocity of the targets to be monitored are known and do not vary over time. Unfortunately, setting the correct value for the reporting interval is non-trivial and depends on the target mobility/velocity and the network congestion state as shown below.

Impact of target mobility: Figure 2 plots coverage accuracy measured in terms of **percent coverage** for low-velocity targets (pedestrians) and high-velocity targets (cars) over time. The percent coverage represents the percentage of the total covered targets out of the total coverable (targets within the coverage range) targets. The y-axis represents the average of percent coverage taken at 5 seconds intervals.

The assumption that the coverage decreases between two reconfigurations may hold true for low velocity targets, but not for high velocity targets. As shown in the figure, the low-velocity targets exhibit coverage loss of up to 52% between successive reconfigurations. Thus, the reconfiguration interval can be decreased to obtain a desirable coverage. However, for the high velocity targets, coverage increases automatically when the targets move into the coverage region of some other camera. Thus, cameras need to be configured more frequently to accommodate sudden changes in the coverage.

However, if the target mobility is unpredictable, or varying over time, the periodic policy can suffer heavily in terms

[1]The amount of time required to compute the optimal camera-pan configurations for 100 cameras was found to be within tens of milliseconds.

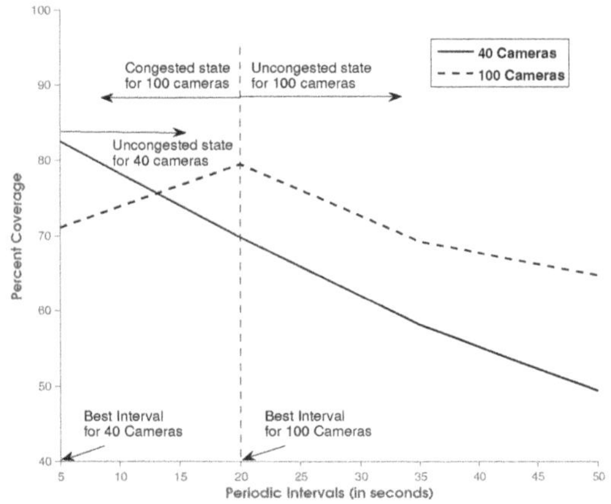

Figure 3: Impact of congestion on coverage. The coverage may not always decrease monotonically with the increase in the periodic interval. In fact, for a large scale network (of 100 cameras), the coverage first increases with the increase in the interval when the interval is set within the congestion limit. Thus, finding the right interval is non-trivial.

of coverage loss or the networking overhead. Specifically, a conservative periodic interval when there is a high target mobility can lead to low coverage. Similarly, an aggressive periodic interval when there is a low target mobility can lead to significant network overhead for a small improvement in coverage.

Impact of network congestion: Although, the coverage-specific adaptation may demand the interval to be as small as possible, the network may not be able to support such frequent reconfiguration operations. In such cases, the network enters the congestion state and further increases the reconfiguration operation's latency, causing more loss in coverage.

Figure 3 plots the percent coverage for 40 and 100 cameras for different periodic intervals. We use pedestrian mobility for targets. For a small-scale 40 camera network, the coverage drops monotonically with the increasing periodic interval. However, for a 100 camera network, aggressive periodic interval (e.g. 5 seconds) results in more than 10% less coverage due to the network congestion. Thus, the best periodic intervals differ based on the camera scale/density.

To address these challenges, we propose two policies that can adapt their reconfiguration frequency based on an application level parameter: target mobility/velocity, and a system level parameter: network congestion state.

3.2 Ondemand reconfiguration

In Ondemand policy, each camera continues to track the targets in its optimal FoV, after a reconfiguration operation. When a target leaves the FoV region or a new target enters the FoV region, the camera notifies about it to the base-station. When the base-station finds that the number of total lost targets is greater than a certain threshold, it sends a request to initiate reconfiguration operation to each camera. Upon receiving the request, the camera initiates the standard procedure by transmitting its coverable targets locations to the base-station, and setting the optimal FoVs received from the base-station.

The base-station keeps track of the lost targets, and computes the loss factor, L as:

$$L = \frac{|M_\ell|}{|M|} \quad (4)$$

where, M_ℓ represents the targets lost after the preceding reconfiguration, and M represents the total coverable targets obtained from the preceding reconfiguration data.

The Ondemand policy solves the problems associated with the periodic policy by reducing the reconfiguration rate in case of low mobility, and vice versa. However, it also has the intrinsic overhead of sending requests for initiating reconfiguration to all cameras. Note that it is critical to send these requests to cameras in order to obtain the locations of currently coverable targets for computing the optimal camera configurations.

3.3 Adaptive reconfiguration

Adaptive reconfiguration is a push-based policy, in which cameras send their coverable targets information to the base-station after certain interval. However, unlike the Periodic policy, this interval is set by the base-station, depending on target mobility/velocity and network congestion state.

Handling target mobility: In order to compute the next reconfiguration interval at the base-station, the cameras transmit their current pan value in addition to the set of coverable targets to the base-station. Base-station uses this information to obtain the instantaneous coverage, IC, of the network. Next, using the same information, it computes the optimal pan settings for all the cameras, and obtains the optimal coverage, OC, for the same target positions. Thus, the difference in coverage is obtained as $\delta = OC - IC$. Let τ_A represents the coverage loss-tolerance threshold.

To adapt the reconfiguration interval, we employ *Additive Increase Additive Decrease (AIAD)* [8] policy. Since, the drop in the coverage is not monotonic as shown in Figure 3, since more targets may appear in the FoV region than the lost targets at the time of next reconfiguration, we employ a *sliding window* mechanism to determine the appropriate adaptive interval. Specifically, we keep track of the previous k differences in the coverage, and adjust the adaptive interval AI as:

$$\tau_A - \delta_i \begin{cases} < 0 & \text{AI}-- \\ > 0 & \text{AI}++ \end{cases} \quad \forall 1 \le i \le k$$

If all k differences are neither positive nor negative, we do not change the value of AI. The applications that can tolerate certain coverage loss can set the appropriate value for the τ_A to obtain the desired coverage with minimum network overhead.

Handling network congestion: To ensure that the network does not enter the congestion state, we set a lower-bound on the AI, which is computed as the worst-case latency of the previous reconfiguration operation. Essentially, the worst-case latency is the difference between the times when the first camera sent its information to the base-station, and when the last camera received its optimal pan value from the base-station.

4. HYBRID RECONFIGURATION POLICIES

The reconfiguration policies discussed so far work well for small-scale networks due to their low reconfiguration laten-

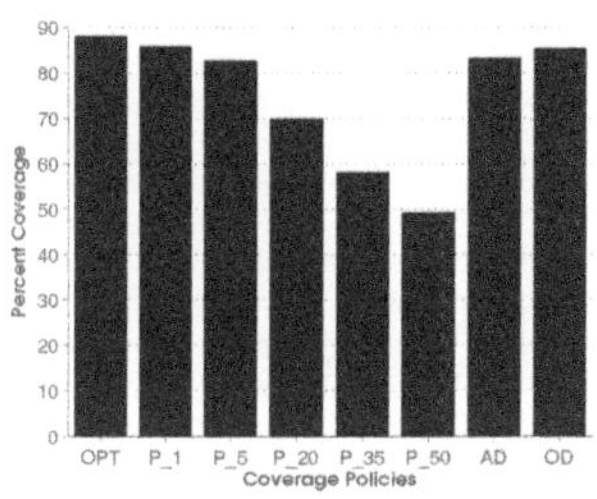

(a)

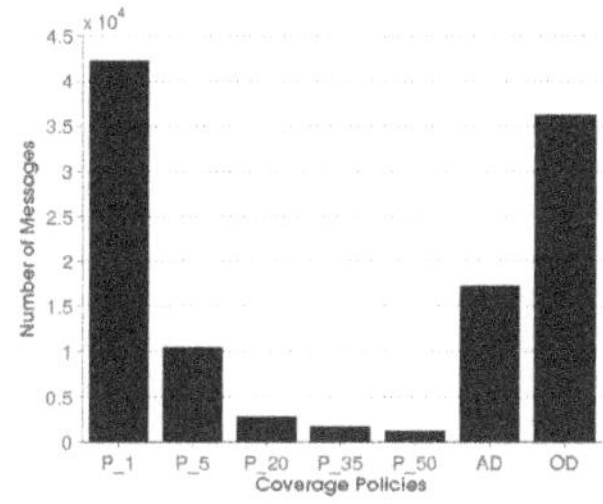

(b)

Figure 4: As we reduce the Periodic intervals, the improvement in coverage tends to flatten out, while the messaging overhead increases drastically. On the other hand, the proposed approaches, especially Adaptive protocol converges to the best interval, causing high coverage accuracy with low overhead.

cies. However, as the network scale increases, the reconfiguration latency increases (around 20 seconds for 100 cameras), causing increased loss in the coverage. To address this problem, we propose to combine the proposed reconfiguration policies with the existing distributed and semi-centralized coverage maximization policies [25], which we describe in the remainder of this section.

Table 2 summarizes the feasible combinations of coverage maximization approaches with different reconfiguration policies. Specifically, all the reconfiguration policies can be used in the centralized manner. We do not form more combinations of the Periodic approach, since Periodic approach is unable to adapt its reconfiguration frequency. In addition, Ondemand can not be combined with the DGA or the Greedy. Ondemand works by tracking the lost targets count. However, for DGA or Greedy, since FoVs are changed based on the local information, it is impossible to track the lost targets. Remaining feasible combinations work as below:

Adaptive with DGA (DGA_AD): In Adaptive with DGA, after centralized reconfiguration, cameras set their optimal pan, and continue to cover the targets under their FoVs. If a target moves out of a camera's FoV, it initiates the DGA procedure. Base-station adapts the interval using the usual procedure of the Adaptive approach.

Adaptive with Greedy (GR_AD): GR_AD works in the similar way as DGA_AD, except the Greedy procedure is invoked when a target moves out of a camera's FoV.

Adaptive, Ondemand with Hierarchical (AD_H, OD_H): Hierarchical works by invoking reconfigurations within clusters, as explained in Section 2.3. Thus, Adaptive with Hierarchical (AD_H) approach uses Adaptive as a reconfiguration policy within each cluster. Ondemand with Hierarchical (OD_H) can also be defined in the similar way.

We also form two additional combinations, DGA + Adaptive in cluster (DGA_AD_H) and Greedy + Adaptive in cluster (Greedy_AD_H), which are not shown in the table.

5. PERFORMANCE EVALUATION

In order to evaluate the coverage approaches, we built a coverage simulator by modeling the camera FoV specific parameters to match with that of the Axis 213 cameras [4]. Next, to enable studying the joint coverage and networking problem, we integrated the coverage simulator with the QualNet simulator [29]. Each camera is set with a predefined maximum length of the FoV $R_{max} = 60 meters$ and FoV angle $\alpha = 45°$. The camera can choose from 8 discrete pan values as its orientation, however the proposed approaches can use higher number of pans, which could be overlapping.

The goal of the evaluation is to study the performance of the proposed approaches for realistic settings where targets may use different mobility patterns, velocities of targets may change over time, and there could be a mixture of targets moving at different speeds. These scenarios represent a typical university campus environment [35], where students could be walking over streets, inside buildings (e.g. library), or using bikes or cars to get around the campus.

Evaluation Setup and Methodology: We use a fixed terrain of dimensions 500×500 sq. meters, where we place cameras and targets. Cameras are placed uniformly on the terrain, while different distributions are used for targets placement. The base-station is placed randomly within the terrain area. Cameras communicate using 802.11g protocol. Since the cameras' coverage range and their communication range are independent, we place additional routers to maintain reachability in the network.

We run each experiment 20 times with different seeds to avoid generating scenario specific results. At the start of the experiment, each node opens a TCP connection with the base-station, and also with the neighboring nodes (nodes with $2 \times$ coverage range of a camera), if required by the protocol. The nodes use Dynamic Source Routing (DSR) as a routing protocol. We now present evaluation of the proposed approaches using this setup.

5.1 Coverage-Energy trade-offs

We first study the behavior of the proposed approaches in terms of their coverage and communication-specific energy consumption. We deploy 40 cameras [35] in uniform manner, and use the *Pedestrian* mobility model provided by QualNet. We compare the proposed approaches with Optimal (OPT), which is obtained by solving the Integer Linear Programming based version of the stationary coverage maximization problem [25] on a snapshot of cameras and targets locations taken at every 5 seconds. We consider different intervals for the Periodic policy; for instance, P_5 represents the Periodic policy with reconfiguration interval set to 5 seconds. We also consider Adaptive (AD) and Ondemand (OD) policies. For this study, we use communication overhead to represent the energy expenditure of a camera.

We measure the networking overhead in terms of number of messages communicated within the network. The messaging overhead is captured at the MAC layer to consider MAC-level retransmissions, as well. We use TCP to ensure 100% reliability for the communication of control packets; however, we discovered that TCP introduces high delays under large scenarios due to its conservative congestion management protocol. Since, we are sending small control packets after certain intervals, we disable the exponential backoff mechanism of TCP that severely hampers the reconfiguration latency. Note that the TCP encounters packet losses primarily due to the channel contention.

As shown in Figure 4, as we reduce the interval, small increase in coverage results in significant increase in the messaging overhead. Specifically, moving from the interval of 5 seconds to 1 second increases the coverage accuracy only by 4%, while the messaging overhead by around 300%. On the

Reconfiguration Policies	Centralized	Distributed		Hierarchical
		DGA	Greedy	
Periodic	Section 3.1	NO	NO	NO
Adaptive	Section 3.3	Centralized Adaptive + DGA on target loss (DGA_AD)	Centralized Adaptive + Greedy on target loss (GR_AD)	Adaptive in cluster (AD_H)
Ondemand	Section 3.2	NO	NO	Ondemand in cluster (OD_H)

Table 1: Hybrid policies can be used to achieve coverage maximization for large-scale networks. The table shows the feasible hybrid approaches formed by combining reconfiguration policies with the distributed and semi-centralized coverage maximization policies.

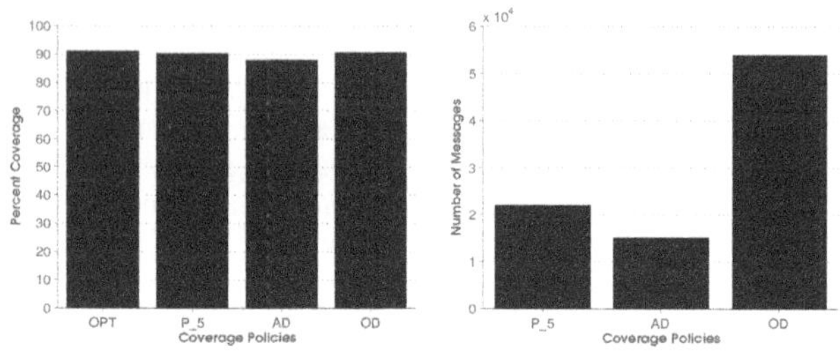

(a) (b)

Figure 5: Adaptive performs well in spite of the changes in mobility parameters, such as change in target velocity over time (common in real-world scenarios such as a campus-wide surveillance systems)

other hand, the proposed approaches, especially Adaptive protocol converges to the best interval, causing high coverage accuracy (within 3% of P_1) with low overhead (60% less overhead than P_1).

5.2 Adaptability to varying mobility

To study the impact of changing velocity of targets, we consider a scenario where students may be walking to the lecture hall, sit in the lecture hall for the class duration, and walk back to their dorms. As shown in Figure 5, Adaptive can effectively increase the reconfiguration interval, when the students are stationary, without much affecting the overall coverage accuracy. Thus, while as shown in Figure 4(b), under constant pedestrian mobility the adaptive policy has a 64% more messaging overhead than the P_5 policy, under varying mobility, the adaptive policy has 30% less messaging overhead than the P_5 policy.

5.3 Impact of mobility models on coverage

Performance of coverage maximization approaches depends on two factors: (1) overlaps in cameras' coverage–if there is no coverage overlap among cameras, purely local policy such as Greedy can lead to the optimal solution; and (2) target density – high target density leads to many feasible FoV choices, making it hard to find the FoV that leads to a globally optimal solution. In other words, if there is a single target for each camera to cover, then the Greedy can provide the optimal solution.

In this study, we varied the camera and target density. As shown in Figure 6 and Figure 7, for very low camera and target density, the Greedy performs quite close to the Optimal, and the difference in Optimal and Greedy increases substantially as the camera and target density increases.

Mobility	OPT	AD	OD	AD_H	OD_H	Greedy
Pedestrian	91.44	78.19	66.66	85.35	88.40	83.03
Rand WP	94.11	87.16	80.84	85.94	86.82	75.17
Group	91.70	79.35	72.62	82.14	82.94	72.77

Table 2: Study of different mobility patterns. The hierarchical approaches are least susceptible to the different mobility patterns.

The mobility pattern has a significant impact on the coverage maximization approaches. Specifically, the difference of Optimal and Greedy is much smaller for the pedestrian model than that for the random-waypoint model. This is because, for the pedestrian model, even if the cameras are not configured with optimal FoVs, as long as they are focused on the streets, the moving targets themselves will enter the FoV region, eventually. Thus, irrespective of camera and target density, Greedy can be a policy of choice for tracking targets moving on streets, since it provides near optimal coverage based on purely local decision (i.e. 0 messaging overhead). In contrast, in case of random waypoint model, targets can be present in *all possible* FoVs of a camera, complicating the process of finding the optimal FoV for each camera.

Table 2 shows the comparison of the proposed approaches for 60 cameras and 60 targets, for which the Greedy has relatively the worst performance. We also consider *Group mobility* of targets, where pedestrians are moving in groups in a random-waypoint pattern. For all the mobility patterns, we set pedestrian velocity as the velocity of targets. We observe that, while the performance of Greedy, Adaptive, and On-demand policies varies greatly with the different mobility models, the hierarchical policies are least susceptible to them, and outperform almost all of the other approaches.

5.4 Impact of target velocity

To evaluate the impact of target velocity on coverage, we consider three velocities: pedestrians (1-3 m/s), bikers (5-7 m/s), and cars (10-12 m/s). As shown in Figure 8 as the targets' velocity increases, all the reconfiguration policies tend to deviate from the optimal. However, the deviation for the hierarchical policies is minimum due to their low reconfiguration latency.

We also consider more realistic scenarios, where targets with mixed velocities as for pedestrians, bikers and cars, are to be tracked together by a camera network. As shown in

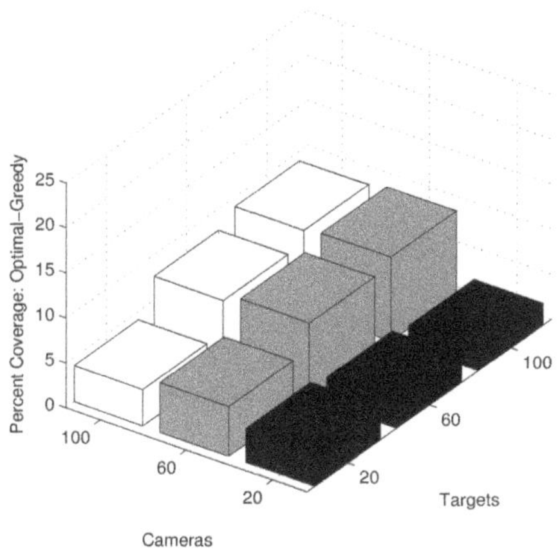

Figure 6: Optimal-Greedy for Pedestrian mobility, where targets walk on the streets. Greedy deviates from optimal only by a small margin (6%), because when cameras are focused *any where* on the street, the moving targets will eventually enter the FoV region, leading to high coverage accuracy.

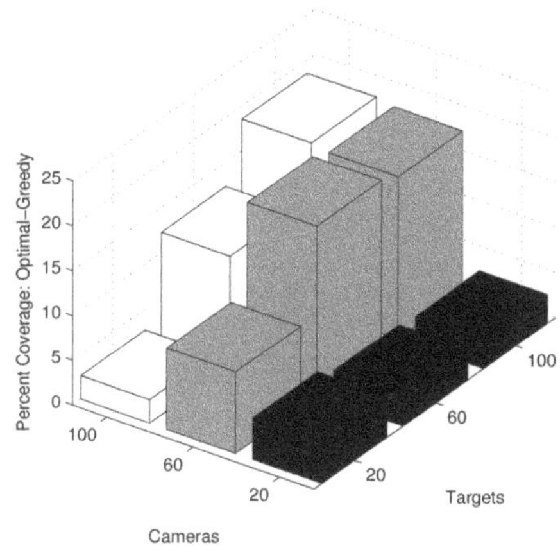

Figure 7: Optimal-Greedy for Random-waypoint mobility, which represents targets moving in a park or an airport. Cameras in such settings may have many FoV choices covering some unique target combinations, causing Greedy to significantly underperform than the optimal.

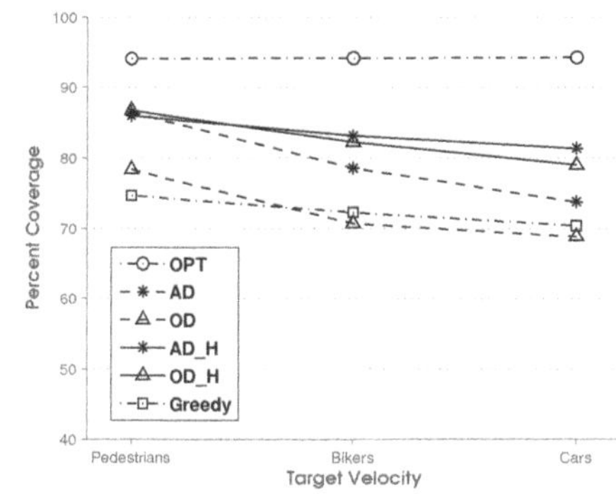

Figure 8: Impact of different target velocities on coverage. The hierarchical approaches have the least deviation; they can handle high target velocity well due to their low reconfiguration latency.

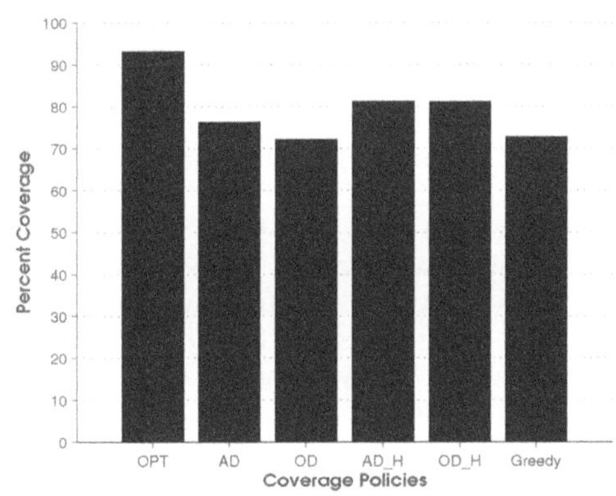

Figure 9: Coverage for mixed mobility. The proposed coverage approaches can effectively handle realistic scenarios such as covering pedestrians, bikers, and cars in a campus-wide surveillance network.

Targets	OPT	AD	OD	AD_H	OD_H	Greedy
20	99.31	95.65	93.08	95.22	95.77	90.24
60	94.11	87.16	80.84	85.94	86.82	75.17
100	84.91	78.3	69.82	77.03	78.15	66.81

Table 3: Impact of high, medium, and low target density/scale for 60 cameras. Greedy policy has high variations in the coverage, since the target density directly impacts the number of available FoV choices each camera gets. Furthermore, for the medium-scale network, the hierarchical approaches provide almost equal performance as the centralized Adaptive protocol.

Figure 9, the Hierarchical approaches outperform the other approaches because they can tune the reconfiguration interval separately for each cluster, depending on the velocities of the targets that they are covering.

5.5 Impact of target density on coverage

To study the impact of target density on coverage, we varied the number of targets (20, 60, and 100) while keeping the number of cameras constant (60). We would like to point out that (ref. Table 3) the coverage decreases as the target density is increased. This happens because the number of coverable targets increases at faster rate than the actual covered targets (an FoV with 45° AoV covers only $1/8^{th}$ of the coverable circle). As shown in Table 3, the hierarchical policies perform very close to the centralized adaptive policy, which performs well for small-scale networks.

5.6 Impact of camera density on coverage

Since, the Greedy policy performs considerably poor for random-waypoint mobility model, we evaluate the proposed policies for varying camera density. We vary the number of cameras (20, 60, and 100) while keeping the number of targets constant (60). As shown in Table 4, the coverage improves significantly (toward 100%) as we increase the camera density (100 cameras). However, hierarchical approaches provide high coverage and outperform all the other approaches for more dense networks, since the reconfiguration operations can be performed faster within the small-scale clusters.

5.7 Performance of distributed hybrid approaches for the large scale network

As shown in Table 4, since the centralized policies (AD and OD) fail to provide high coverage accuracy for large scale (100 cameras) network, we study the performance of hybrid distributed policies for large-scale networks. We use Greedy and DGA as the distributed approaches, where the DGA is the state-of-the-art approach for coverage maximiza-

Cameras	OPT	AD	OD	AD_H	OD_H	Greedy
20	55.41	54.17	55.17	53.2	54.15	51.91
60	94.11	87.16	80.84	85.94	86.82	75.17
100	99.96	90.84	88.14	96.03	97.07	87.74

Table 4: Impact of high, medium, and low camera density/scale for 60 targets. The hierarchical approaches outperform all the other policies, especially for more dense networks.

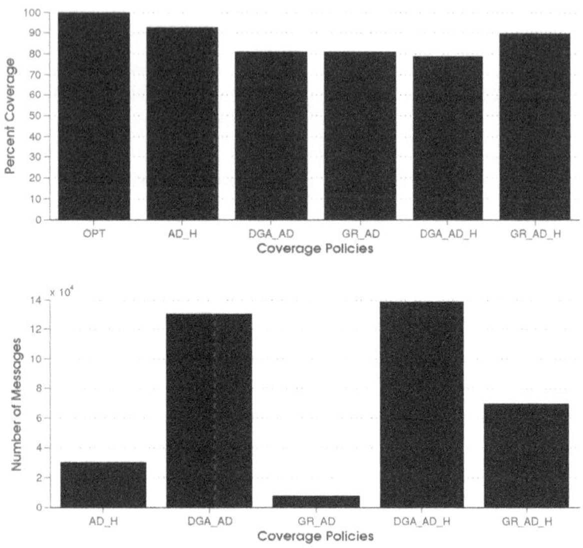

Figure 10: For a large-scale network with 100 cameras trying to monitor 100 targets, hierarchical with centralized policy within clusters (AD_H) provides better coverage with significantly low overhead compared to the other hybrid approaches. Combinations with DGA suffer heavily due to the inherent iterative nature of DGA's convergence.

tion of stationary targets [2]. We set both the number of cameras and number of targets to 100. We use AD_H from the Table 4, since it outperforms all the other approaches.

As shown in Figure 10, merging the simple Greedy approach with the reconfiguration policies yields equal or better coverage with significantly lower messaging overhead, in comparison with merging of the more complex DGA approach. Combinations with DGA suffer heavily due to the inherent iterative nature of DGA's convergence, which converges from the highest priority camera to the lowest priority camera in the network. On the other hand, the hierarchical approach with centralized policy within clusters (AD_H) provides better coverage with significantly low overhead compared to the other hybrid approaches.

6. RELATED WORK

In general, usefulness of coverage by directional sensors can be improved by either placing sensors carefully or configuring them (orientation and zoom, if available) so that minimum number of sensors can be used to meet the application specific coverage requirement. Thus, in this section, we present related work on sensor placement and configuration to maximize overall covered area or targets.

Area coverage maximization: For a given placement of cameras, Kansal et al [19] focus on configuring pan, tilt, and zoom parameters of the cameras to maximize the overall area covered by all the cameras, and propose a distributed mechanism to configure the cameras. Erdem et al. [10] consider the problem of determining automatic camera deployment layout to satisfy task-specific and floorplan-specific coverage requirements. Hoffmann et al [16] propose a distributed approach to maximize the total area coverage, where each camera tries to minimize coverage overlaps with its neighboring cameras. Ma et al [22] propose a 3D model for camera coverage, where cameras can adjust their pan and tilt values using a simulated annealing based heuristic to orient cameras such that the overall area covered gets maximized. Although, these works address the problem of maximizing coverage by directional sensors, they focus mainly on maximizing the coverage of a given area, which is different from our goal of maximizing target-oriented coverage.

Camera scheduling: When it is essential to cover all the coverable targets and when there are not enough cameras to satisfy that requirement, cameras can be scheduled to cover targets over time. Our work differs from the works of Krahnstoever et al [20] and Qureshi et al [28] in terms of the objective: we focus on applications where it is important to cover as many targets as possible continuously.

Target coverage maximization: Some of the recent works in the context of *directional sensor networks* (DSNs) have provided optimization based solutions to address a basic instance of the problem of covering maximum targets [2, 7]. Ai et al [2] formulate the coverage optimization problem as a Maximum Coverage with Minimum Sensors (MCMS) problem, and show that MCMS is NP-complete. However, since the optimization-based approach is not computationally scalable, they provide greedy based centralized (CGA) and distributed (DGA) policies for coverage maximization for DSNs. Cai et al [7] focus on the problem of maximizing the lifetime of DSNs. They compute Multiple Directional Cover Sets (MDCSs) for a given set of targets, where a Directional Cover Set (DCS) is a subset of directions of the sensors, which can cover all the targets present in the area under observation. The authors show that the problem of computing DCS and MDCS is NP-complete and propose heuristics to solve MDCS. Fusco et al [11] focus on achieving k-coverage of static targets and propose centralized and distributed heuristics with performance bounds. However, all these approaches focus on covering stationary targets.

Only one recent work focuses on the problem of maximizing the number of unique mobile targets covered by a camera network [26]. However, this work considers very few cameras (up to 4 cameras) in their evaluation, and thus ignores the challenges associated with camera scale and wireless networking issues.

Note that there is a significant research on tracking a single or multiple targets [9, 31], however, this research is not targeted towards maximizing the number of targets under observation (the coverage maximization problem).

7. CONCLUDING REMARKS

In this paper, we considered the coverage maximization problem for mobile targets in visual sensor networks. Our approach included combining the existing approaches that answer *how* to maximize coverage for static targets with the proposed reconfiguration policies that answer *when* or *how often* to reconfigure them with the optimal pans. Specifically, we proposed Adaptive and Ondemand policies that can use the right reconfiguration interval depending on the target mobility and network congestion state. We found that for the target mobility that uses pre-defined paths (such as streets), a simple and purely local greedy policy can yield close-to-optimal coverage. For general cases, we found that the combination of semi-centralized coverage maximization approach (Hierarchical) with the Adaptive reconfiguration policy outperforms all the other approaches, and provides close-to-optimal coverage with low communication overhead

In future, we plan to extend the coverage algorithms for pan, tilt, and zoom (PTZ) cameras. We also plan to use real-world mobility traces for different types of targets to study the performance of different coverage approaches.

8. REFERENCES

[1] H. Aghajan and A. Cavallaro. *Multi-Camera Networks.* Academic Press, 2009.

[2] J. Ai and A. Abouzeid. Coverage by directional sensors in randomly deployed wireless sensor networks. *Journal of Combinatorial Optimization*, 2006.

[3] I. Akyildiz, T. Melodia, and K. Chowdhury. A survey on wireless multimedia sensor networks. *Computer Networks*, 51(4):921–960, 2007.

[4] Axis 213 ptz network camera. Available at `http://www.axis.com/products/cam_213/`.

[5] A. Barton-Sweeney, D. Lymberopoulos, and A. Sawides. Sensor localization and camera calibration in distributed camera sensor networks. In *Broadband Communications, Networks and Systems*, 2006.

[6] J. Broch, D. Maltz, D. Johnson, Y. Hu, and J. Jetcheva. A performance comparison of multi-hop wireless ad hoc network routing protocols. In *ACM/IEEE international conference on Mobile computing and networking*, 1998.

[7] Y. Cai, W. Lou, M. Li, and X. Li. Target-Oriented Scheduling in Directional Sensor Networks. *IEEE Infocom*, 2007.

[8] L. Cui, S. Koh, X. Cui, and Y. Kim. Adaptive increase and adaptive decrease algorithm for wireless tcp. In *International Conference on Natural Computation.* IEEE, 2007.

[9] A. Ercan, A. El Gamal, and L. Guibas. Object tracking in the presence of occlusions via a camera network. *Proc. of the 6th ACM international conference on Information processing in sensor networks*, 2007.

[10] U. Erdem and S. Sclaroff. Automated camera layout to satisfy task-specific and floor plan-specific coverage requirements. *Computer Vision and Image Understanding*, 103(3):156–169, 2006.

[11] G. Fusco and H. Gupta. Selection and orientation of directional sensors for coverage maximization. In *Sensor, Mesh and Ad Hoc Communications and Networks, 2009.* IEEE.

[12] M. Garcia and A. Solanas. 3D simultaneous localization and modeling from stereo vision. In *IEEE International Conference on Robotics and Automation*, volume 1, 2004.

[13] M. Green. The appropriate and effective use of security technologies in us schools: A guide for schools and law enforcement agencies series: Research report. *NCJ*, 1999.

[14] A. Hampapur, L. Brown, J. Connell, A. Ekin, N. Haas, M. Lu, H. Merkl, S. Pankanti, I. Center, and N. Hawthorne. Smart video surveillance: exploring the concept of multiscale spatiotemporal tracking. *Signal Processing Magazine, IEEE*, 22(2):38–51, 2005.

[15] G. He and J. Hou. Tracking targets with quality in wireless sensor networks. In *International Conference on Network Protocols.* IEEE, 2005.

[16] M. Hoffmann, M. Wittke, J. Hahner, and C. Muller-Schloer. Spatial Partitioning in Self-Organizing Smart Camera Systems. *IEEE Journal of Selected Topics in Signal Processing*, 2008.

[17] E. Hörster and R. Lienhart. On the optimal placement of multiple visual sensors. In *ACM international workshop on Video surveillance and sensor networks*, 2006.

[18] C. Huang and Y. Tseng. The coverage problem in a wireless sensor network. *Mobile Networks and Applications Journal*, 10(4):519–528, Aug. 2005.

[19] A. Kansal, W. Kaiser, G. Pottie, M. Srivastava, and G. Sukhatme. Reconfiguration methods for mobile sensor networks. *ACM Transactions on Sensor Networks (TOSN)*, 2007.

[20] N. Krahnstoever, T. Yu, S. Lim, K. Patwardhan, and P. Tu. Collaborative Real-Time Control of Active Cameras in Large Scale Surveillance Systems. In *Workshop on Multi-camera and Multi-modal Sensor Fusion Algorithms and Applications*, 2008.

[21] P. Kulkarni, D. Ganesan, P. Shenoy, and Q. Lu. SensEye: a multi-tier camera sensor network. In *Proc. of the 13th annual ACM international conference on Multimedia*, 2005.

[22] H. Ma, X. Zhang, and A. Ming. A coverage-enhancing method for 3d directional sensor networks. In *Proc. INFOCOM Mini-conference*, 2009.

[23] J. Martínez Navarro. Evaluation of the impact of mobility models on handover in wlan indoor environments. 2009.

[24] K. Miyamoto. Fish eye lens. *JOSA*, 54(8):1060–1061, 1964.

[25] V. Munishwar and N. Abu-Ghazaleh. Scalable Target Coverage in Smart Camera Networks. In *ACM/IEEE International Conference on Distributed Smart Cameras*, 2010.

[26] P. Natarajan, T. Hoang, K. Low, and M. Kankanhalli. Decision-theoretic approach to maximizing observation of multiple targets in multi-camera surveillance. *International Conference on Autonomous Agents and Multiagent Systems*, 2012.

[27] C. Papageorgiou, M. Oren, and T. Poggio. A general framework for object detection. 1998.

[28] F. Qureshi and D. Terzopoulos. Planning Ahead for PTZ Camera Assignment and Handoff. In *ACM/IEEE International Conference on Distributed Smart Cameras*, 2009.

[29] Q. Simulator. Scalable Network Technologies. *Inc.[Online]. Available: www. qualnet. com.*

[30] C. Soto, B. Song, and A. Roy-Chowdhury. Distributed multi-target tracking in a self-configuring camera network. In *IEEE Conference on Computer Vision and Pattern Recognition*, 2009.

[31] N. Ukita, T. Nagao, and T. Matsuyama. Versatile Cooperative Multiple-Object Tracking by Active Vision Agents. In *IAPR Workshop on Machine Vision Applications.*, 2000.

[32] J. Urrutia. Art gallery and illumination problems. *Handbook of Computational Geometry*, pages 973–1027, 2000.

[33] C. Wang, C. Thorpe, and A. Suppe. Ladar-based detection and tracking of moving objects from a ground vehicle at high speeds. In *Proc. of the IEEE Intelligent Vehicles Symposium*, 2003.

[34] X. Wang, G. Xing, Y. Zhang, C. Lu, R. Pless, and C. Gill. Integrated coverage and connectivity configuration in wireless sensor networks. In *Proc. First International Conference on Embedded Networked Sensor Systems (SenSys)*, 2003.

[35] J. Xu, Z. Ni, C. De Leo, T. Kuo, and B. Manjunath. Spatial-temporal understanding of urban scenes through large camera network. In *ACM international workshop on Multimodal pervasive video analysis*, 2010.

A Lifetime Enhancing Node Deployment Strategy using Heterogeneous Nodes in WSNs for Coal Mine Monitoring

Subir Halder
Department of CSE
Dr. B. C. Roy Engineering College
Durgapur, WB 713206
subir_ece@rediffmail.com

Sipra DasBit
Deptartment of CST
Bengal Engg. and Science University
Shibpur, WB 711103
sdasbit@yahoo.co.in

ABSTRACT
Continuous monitoring of the environment in coal mines is an important application area in wireless sensor networks. It is very essential to know the conditions that exist in coal mines for ensuring safe working inside the mines. Based on the characteristics of room-and-pillar mining technique and the requirements for mine safety monitoring, we propose a heterogeneous node deployment strategy. To obtain real time and comprehensive monitoring, nodes are deployed in some strategic locations of the underground coal mine. We claim that the strategic locations are chosen in such a way that the proposed deployment scheme ensures 1-coverage and 3-connectivity. The claim has been justified by supporting lemma and proof. The present work achieves enhanced network lifetime by ensuring efficient energy usage amongst the nodes thereby prolonging network lifetime. Further, exhaustive simulation is performed considering a realistic routing protocol to substantiate our claim of energy efficiency and subsequent enhancement of network lifetime. Finally the scheme is compared with two existing deployment schemes. The results confirm our scheme's supremacy over both the existing schemes in terms of the performance metrics such as energy efficiency and network lifetime.

Categories and Subject Descriptors
C.2.1 [**Computer-Communication Networks**]: Network Architecture and Design—*Distributed networks; Network communications*;

General Terms
Algorithms, Performance, Design.

Keywords
Wireless sensor network, Room-and-pillar mining, Node deployment, Coverage and Connectivity, Network lifetime.

1. INTRODUCTION
Wireless Sensor Networks (WSNs) have drawn the attention of research community during the last decade. They are largely attributed for enabling new applications in large-scale networks consisting of small devices [1] [2] [3]. These devices are capable of gathering information from physical environment, processing on extracted data and transmitting data to remote locations.

In WSNs, one of the deployment methods is to deploy nodes in pre-determined locations. The applications include populating an area with highly precise seismic nodes, underwater WSN applications, underground coal mine monitoring etc. In this work we have considered underground coal mine monitoring as our application domain.

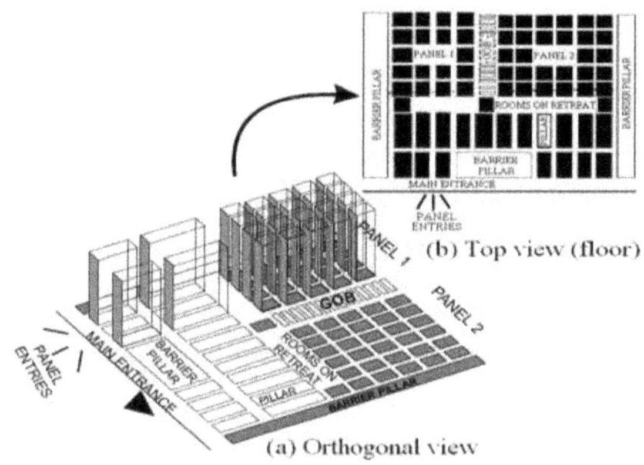

Figure 1: Room-and pillar mining

The structure of underground mining area is established by mining methods [2], which are determined by the shape and position of the ore body. One such mining method is known as room-and-pillar. Figure 1 shows a typical room-and-pillar mine. In room-and-pillar technique, mining is performed in two stages. In the first phase, rooms are created by keeping pillars while in the second phase the pillars are mined. A group of such pillars is known as panel and panels are separated from one another by solid barriers. The vacant area formed after coal extraction is known as the GOB. In normal practice after completion of extraction in one panel, the rooms on that panel are sealed in order to keep them isolated while the extraction process continues on other panels.

As the rooms in each panel are sealed, it is difficult to know the conditions that exist inside the sealed rooms. There may be high possibility of pillars collapsing mainly due to fire, gases and many other factors that can form inside the sealed rooms. This can sometimes become dangerous for miners as the extraction process goes on near the sealed rooms of the panel. Therefore, it is necessary to monitor the sealed rooms of a panel continuously.

MSWiM'12, October 21–25, 2012, Paphos, Cyprus.

Monitoring is essential at the sealed rooms of each panel as well as the panel entries and the main entrances. The utilization of a WSN for implementing the monitoring system benefits from flexible deployment and the multi-hop transmitting method provides more scalability for system building. This motivates us for undertaking such an application of underground coal mine monitoring using wireless sensor network.

Many works reported so far deal with the monitoring within underground coal mines by deploying sensor nodes. In one such work, Niu et al. [3] have proposed a distributed heterogeneous hierarchical mine safety monitoring prototype system for monitoring methane concentration, and locating miners in the underground mine. The proposed system is comprised of two parts. The first part is the lightweight overhearing-based data collection algorithm and the second part is the adaptive data sampling-rate adjusting mechanism. They have shown that the proposed system can monitor methane concentration and predict methane overproof timely under different conditions with less bandwidth and energy consumption than existing wireless solutions. Wang et al. [4] have proposed an unsupervised learning-based technique for detecting anomalies in gas monitoring within underground coal mines using Bayesian network. Authors claim that the Bayesian network model can learn spatio-temporal data patterns in the underground and thereby reduces false alarm. Li and Liu [5] have designed a topology maintenance protocol known as Structure-Aware Self-Adaptive (SASA) protocol. SASA is rapidly able to detect the structure variation during underground collapse. With an appropriate arrangement of sensor nodes and a collaborative mechanism, SASA is able to accurately report locations of collapses, for detecting and reconfiguring displaced nodes, thus maintaining the system integrity. R. Liu et al. [6] have proposed a relay node placement strategy for monitoring underground tunnel with an objective to minimize the deployment cost by reducing the number of relay nodes while maintaining connectivity. This scheme gives a general guideline for node deployment in wireless sensor network for underground tunnel application which gives guaranteed levels of radio link performance. Chen et al. [7] have proposed a cluster based node deployment algorithm for wireless underground mine sensor network where number of nodes in each cluster is uniform and cluster heads are chosen dynamically. The algorithm is designed based on this analysis and the cluster heads are placed non-uniformly such that coverage of the network is maintained with minimum number of cluster heads. Li et al. [8] have developed a coal mine surveillance system for detecting events in a 3D environment. They have proposed event feature patterns to specify complex events and developed a pattern-based event detection method on the computed 3D gradient data map. Events are detected by matching a time series of data map values against the known properties of certain events, e.g. water seepage.

In most of the above existing works on WSN based monitoring system, the deployment strategies have guaranteed the monitoring of events in the network. However, most of these strategies are silent about the maximization of network lifetime using minimum number of nodes while maintaing coverage and connectivity of the network area. Therefore, we have set the objective of the present work to achieve enhanced network lifetime by placing the least number of nodes in strategic locations of the mines while maintaining coverage and connectivity. Based on analysis made in [9] heterogeneous node deployment has been found as one of the efficient techniques for ensuring balanced energy consumption amongst the nodes thereby prolonging network lifetime. So in this work we have deployed heterogeneous nodes to fulfill the said objective.

The rest of the paper is organized as follows. The network architecture considered for the present work is presented in section 2. Description of the proposed energy efficient node distribution scheme is done in section 3. In section 4, the performance of the scheme is evaluated by providing simulation results. Finally the paper is concluded with some mention about the future scope of the work in section 5.

2. NETWORK ARCHITECTURE

In this section, we describe the network architecture along with the necessary assumptions. This section also presents communication and sensing model, energy model and the concept of network lifetime considered for this work.

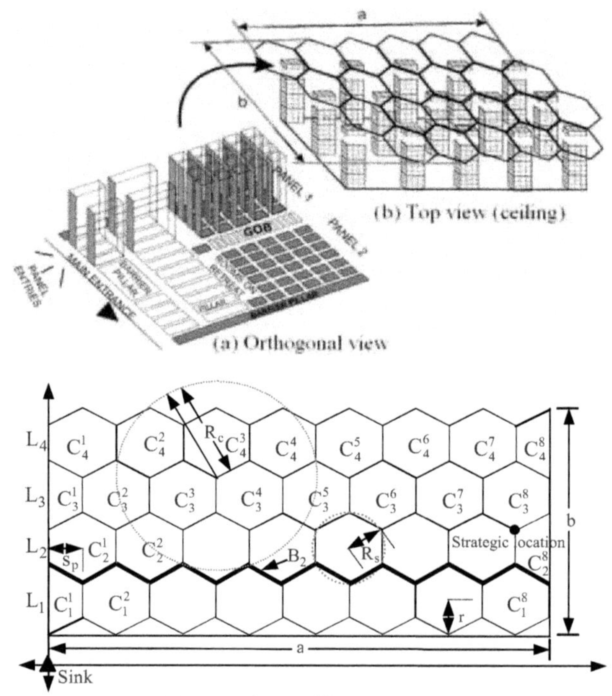

(c) Top view ceiling (zoomed)

Figure 2: Network architecture

2.1 Architecture

Figure 2 (a) shows the orthogonal view of network architecture where ceiling (Fig. (2b)) of the mine is considered as deployment area. We further consider that the deployment area is logically covered by regular hexagonal cell (RHC) [10]. The network coverage area a×b is divided into N layers containing regular hexagonal cells. A layer is identified as L_i where i=1,2,…,N. A layer is called odd layer when i is odd and is called even when i is even. For example, $L_2, L_4, L_6, \cdots$ (Fig. 2(c)) are the even layers whereas $L_1, L_3, L_5, \cdots$ are the odd layers. The radius of each regular hexagonal cell or cell is r. The sink is located at one corner of the network. Here i=1 indicates the layer nearest to and i=N indicates the layer farthest from the sink, whereas j=1 indicates the cell nearest to and j=M indicates the cell farthest from the sink within a layer. A cell is identified by C_i^j which indicates the jth cell in layer-i. For example, the cell C_3^7 identifies the 7th cell in layer-3. Further, we define B_i as the boundary between two consecutive layers-i & layer-(i-1). So the cells in layer-i & layer-(i-1) share a common boundary B_i. For example, B_2 is the boundary between layer-2 and layer-1. Each boundary has a number of vertices and for a boundary B_i there are 2M such vertices.

2.2 Assumptions

We consider static, heterogeneous nodes are deployed in the under ground coal mine. The nodes are heterogeneous in terms of energy. Here by energy heterogeneity [9] we mean some nodes are resource-constrained (SNs) and others nodes are resource-rich (RNs). A node (SN or RN) is considered dead if its energy is exhausted completely. Further, we consider the nodes in the underground coal mine are deployed at single-depth. The sink which is part of the monitoring station is located above the ground. We assume periodic data gathering applications where each sensor node uniformly generates n bits data and sends the data to the sink at fixed time-interval q(t). An ideal MAC layer with no collisions and retransmissions is assumed for our work.

2.3 Network Operation

We consider the network operation where a SN of a layer senses the environment, generates data, and periodically transmits the data to neighbouring RN. A RN forwards the data received both from the neighbouring SNs and neighbouring RNs which are farther away from the sink. In this way sensory data from a SN reaches the sink through intermediate RNs. This type of RNs are now-a-days commercially available [11].

In order to achieve enhanced network lifetime we have to deploy the number of SNs and RNs in such a way that overall energy consumption rate of SNs and RNs are minimized. In [12] [13] authors have observed that overall energy consumption by the SNs and RNs can be significantly reduced by deploying slightly more number of RNs than the required minimum number of RNs. Therefore, in our present strategy we have deployed minimum number of SNs to maintain 1-coverage and slightly more number of RNs than the required minimum number of RNs to maintain 1-connectivity. As a result it ensures 1-coverage and at least 3-connectivity.

2.4 Sensing and Communication Model

2.4.1 Communication Model

We consider a network as connected if any active node can communicate with any other active node, possibly using intermediate nodes as relays, so that the information collected by sensor nodes can be relayed back to the sink [14]. We assume two nodes can directly exchange messages if their Euclidean distance is not larger than the communication range R_C. We have considered the relationship between r and R_C must satisfy the condition $R_C \geq \sqrt{7}\,r$ for ensuring 3-connectivity in the network area (Fig. 2(c)). As we have considered heterogeneous sensor network, so there exists two kinds of communication links, one where the SNs send the sensed data to an RN or the sink within distance R_C, while the RNs can forward received packets to an RN or the sink within distance R_C.

Lemma 1: For a given network area a×b, in order to maintain connectivity of the network, the number of layers (N) should follow the relationship $N \geq \frac{\sqrt{7}}{3R_C}\left(2b - \frac{\sqrt{7}}{R_C}\right)$.

Proof: Let us consider that the radius of each cell is r. Now the network area is divided into number of layers and each layer consists of a number of cells. Let us consider that b (Fig. 2(c)) is divided into N number of layers (L_N), therefore using geometry we have

$$\left(\tfrac{3}{2}r \times N\right) + \tfrac{r}{2} = b \;\; or, \;\; N \geq \frac{\sqrt{7}}{3R_C}\left(2b - \frac{\sqrt{7}}{R_C}\right).$$

Here N represents the farthest or maximum layer number, i.e., layer number i=1,2,...N. Therefore, for a known network area and node's communication range using Lemma 1, we can compute the number of layers in the network while connectivity is preserved.

Corollary 1: For a given network area a×b, in order to maintain connectivity of the network, the number of cells (M) in a layer should follow the relationship $M \geq \frac{\sqrt{7}}{\sqrt{3}R_C}\left[a + \frac{\sqrt{3}}{2\sqrt{7}}R_C\right]$.

Proof: According to our network architecture, each layer consists of M number of cells. From geometry (similar to the Lemma 1), we derive the following:

$$\left[\sqrt{3}\,r \times (M-1)\right] + \tfrac{\sqrt{3}}{2}r = a \; or, M \geq \frac{\sqrt{7}}{\sqrt{3}R_C}\left[a + \frac{\sqrt{3}}{2\sqrt{7}}R_C\right].$$

Using Corollary 1, for a given network area and given communication range (R_C), we can compute the number of cells in a layer while ensuring coverage of the network area.

2.4.2 Sensing Model

A unit area is said to be covered [10], if every point in that area is within the sensing range of at least one active node. The nodes perform observation at an angle of 360°. The maximal circular area centered around a node v that can be covered by it is defined as its sensing area S(v). The radius of S(v) is called the v's sensing range R_S. We assume that the relationship between r and R_S must satisfy the condition $r \leq R_S$ for 1-coverage of the network area (Fig. 2(c)). Also, the coverage area C(X) of a set of nodes X is the union of the sensing areas covered by each node in X i.e. $C(X) = \cup_{\forall v \in X} S(v)$ [15].

Corollary 2: For a given network area a×b, in order to preserve network coverage, the number of layers (N) must follow the relationship $N \geq \frac{1}{3R_S}(2b - R_S)$.

Proof: From Lemma 1, the relationship between b and N is evaluated as,

$$N = \frac{1}{3r}(2b - r) \; or, \; N \geq \frac{1}{3R_S}(2b - R_S).$$

For a known network area and given sensing range (R_S), using Corollary 2, we can compute the number of layers in the network while coverage is preserved.

Corollary 3: For a given network area a×b, in order to preserve network coverage, the number of cells (M) must follow the relationship $M \geq \frac{1}{\sqrt{3}R_S}\left[a + \frac{\sqrt{3}}{2}R_S\right]$.

Proof: From Corollary 1, the relationship between a and M is evaluated as,

$$M = \frac{1}{\sqrt{3}r}\left[a + \frac{\sqrt{3}}{2}r\right] or, M \geq \frac{1}{\sqrt{3}R_S}\left[a + \frac{\sqrt{3}}{2}R_S\right].$$

Both the Lemma and Corollaries are important for designing the network. For a given network area, while Lemma 1 and Corollary

2 give guidelines about fixing the number of layers, Corollary 1 and Corollary 3 help in deciding the number of cells in each layer.

2.5 Energy Model

We have considered two-ray radio model [16] for our proposed strategy. In the proposed scheme no significant processing has been done, therefore, we have ignored processing energy. The energy consumption formulas considered here are as follows:

Energy consumption for transmitting (e_{tx}) n-bit data over a distance R_c is $e_{tx}(n, R_c) = (\beta_1 + \beta_2 R_c^4) n = e_t n$ where e_t is energy required to transmit one bit of data and β_1 & β_2 are both system parameters whose values depend on the receiver sensitivity and the receiver noise figure [16].

Energy consumption for receiving (e_{rx}) n-bit data from a distance R_c is $e_{rx}(n) = e_r n$, where e_r is energy required for receiving one bit of data. Energy consumption for sensing (e_{sx}) n-bit data around a distance R_c is $e_{sx}(n) = e_s n$, where e_s is energy required for sensing one bit of data.

3. DEPLOYMENT SCHEME

In the proposed deployment strategy nodes are deployed in two phases. In first phase, SNs are deployed at the centre of each cell of each layer for ensuring 1-coverage of the network. In second phase, RNs are placed on some strategic locations of the boundary of each layer in the network area for ensuring 3-connectivity of the network. One strategic location is shown in Fig. 2(c). In order to achieve 1-coverage and 3-connectivity equal number of SNs and RNs are deployed in the network area. The main objective of the present deployment scheme is achieving balanced as well as minimized overall energy consumption among SNs and RNs so that the network lifetime is enhanced with the least number of heterogeneous nodes.

3.1 Location of Node Deployment

The deployment scheme is location wise pre-determined in nature. As mentioned earlier, the sink is located on the ground surface at left bottom corner of the network. However, the sink can be located at any one of the four corners in the network. The co-ordinate where the sink is located is considered as (0, 0). We propose placing of SNs and RNs on the ceiling of the sealed room of each panel, panel entries and main entries. The location of SNs and RNs are calculated with respect to the sink's location and therefore, both SNs and RNs are located at same depth. SNs and RNs are deployed at the centre of each cell and on boundaries of different layers respectively. The location (x, y) where a SN or a RN is to be placed can be computed by controlling the different parameters in the following equations-

$$x = s_p + \sqrt{3}\, r \times (j-1) \qquad (1a)$$

$$y = r\left[(2-Q) + \frac{3}{2}(i-1)\right] \qquad (1b)$$

where i is the layer number, j is the cell number within the layer and r is cell radius. The other variables used in the above expressions are as follows:

$$s_p = \begin{cases} \dfrac{\sqrt{3}\,r}{2} & \text{for even layer} \\ 0 & \text{for odd layer} \end{cases} \qquad (2a)$$

Here, s_p (Fig. 2) signifies the shifting of centre location of a cell. For example, if i = 5, the value of s_p is 0 and it is $\frac{\sqrt{3}r}{2}$ for i=6 .

$$Q = \begin{cases} 0 & \text{for SNs} \\ 1 & \text{for RNs} \end{cases} \qquad (2b)$$

Here, Q signifies whether SNs or RNs are to be deployed. For example, at centre of each cell as SNs are deployed, the value of Q is 0. Similarly on the boundary as RNs are deployed, the value of Q is 1.

3.2 Sensor Node Deployment

In the first phase SNs are deployed at the centre of each cell which ensures network coverage. The location (x, y) where the SNs are placed is computed using Eqs. (1a) and (1b).

Let us consider a 4-layer architecture where the radius of each cell is 4 (r=4) and each layer has 8 cells as shown in Fig. 3. Therefore, in first phase for every layer, 8 SNs are required that are placed at the centre of each cell. Hence, total 8×4=32 SNs are required. As an example, for computing the centre location of cell C_2^5, putting j=5, r=4 and $s_p = 2\sqrt{3}$ in Eq. (1a) we get the value of x co-ordinate as $18\sqrt{3}$. For y co-ordinate, putting i=2, r=4 and Q=0 in Eq. (1b) we get the value of y co-ordinate as 14. So, (x, y) co-ordinate of the centre of C_2^5 cell is $(18\sqrt{3}, 14)$.

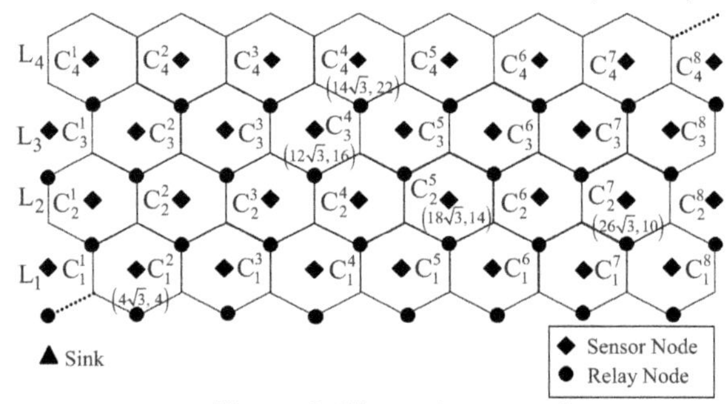

Figure 3: Illustrative example

3.3 Relay Node Deployment

In the second phase of distribution, RNs are distributed ensuring network connectivity. The number of RNs is same as the required number of SNs. These RNs are now deployed at specified vertices (strategic locations as shown in Fig. 2 (c)) of each cell located on the boundary of each layer.

Let us consider again the same 4-layer architecture (3.2). In all the layers there are 8 cells, so 8 RNs are deployed on each of the boundaries B_1, B_2, B_3 and B_4. Hence, total (4×8)=32 RNs are required for deployment in the network area. Here the predetermined places or vertices are located on the boundary between the adjacent layers. Now the co-ordinate of each vertex on boundary is computed by putting appropriate values of different parameters (i, j, r, s_p and Q). For example, location of a RN deployed at boundary B_2 associated with C_2^7 cell is computed as $(26\sqrt{3}, 10)$ by putting the parameter values as i=2, j=7, r=4, $s_p = 2\sqrt{3}$ and Q=1 in Eqs. (1a) and (1b). Similarly, location of a RN deployed at boundary B_3 associated with C_3^4 cell is computed as $(12\sqrt{3}, 16)$ by putting the parameter values of i=3, j=4, r=4, $s_p = 0$ and Q=1 in Eqs. (1a) and (1b). Similarly all the vertices on each boundary B_2, B_3 and B_4 where nodes are to be deployed are calculated by putting appropriate values of different parameters.

3.4 Algorithm for Node Deployment

Initially for a given network area a×b using Lemma 1 followed by Corollaries 1, 2 and 3 the value of N and M is calculated. The algorithm for the Node Deployment under Coal Mine (NDCM) described in sections 3.1, 3.2 and 3.3 is given below:

```
Begin
for i = 1, i ≤ N, i++
    for j = 1, j ≤ M, j++
        compute s_p ;                              // using Eq. (2a)
        if deployment-at-centre= =true then        // SNs deployment
            Q:=0;                                  // using Eq. (2b)
        else
            Q:=1;                                  // RNs deployment
        end if
        compute x;                                 // using Eq. (1a)
        compute y;                                 // using Eq. (1b)
    end for
end for
End
```

3.5 Routing Protocol

Once the nodes (SNs/RNs) are deployed, network set-up phase starts. In this phase, all the nodes learn about their one-hop neighbour RNs by exchanging messages among themselves. Here, by one-hop neighbour RNs we mean RNs which are within the communication range of a node (SN/RN). After the network set-up, routing starts. We have employed a routing protocol similar to one in [17] where a SN chooses a RN (among the one-hop neighbours) which is towards the sink and has the highest residual energy to send its data. Next, the forwardee RN employs same procedure to choose the next forwardee RN for sending its data. This process repeats till the data arrives to the sink.

For example in Fig. 4(a), the SN 'A' has six one-hop neighbour RNs viz. B, C, D, F, G and J. Now among these one-hop neighbour RNs, A may choose B, C, D and F as forwardee nodes as they are towards the sink. Similarly C, D, F, G and J are one-hop neighbour RNs of a RN B where B may choose C, D, F as forwardee nodes. During routing, A chooses one RN out of B, C, D, F that has the highest residual energy to send its data. Assuming B has been selected as a forwardee node by A, now B chooses a forwardee RN amongst C, D and F with the highest residual energy. Once the B selects forwardee RN, it forwards the data of its own as well as those received from the nodes (both SN and RN) of adjacent layers farther away from the sink. The selected RN repeats this process till the data arrives at RN in layer-1, after which the data is sent to the sink.

4. PERFORMANCE EVALUATION

Performance of the present node deployment strategy, reported in the earlier section is measured based on two parameters-coverage-connectivity and network lifetime. Both qualitative and quantitative analyses are presented here.

4.1 Qualitative Analysis

In this sub-section we have analyzed the performance of the proposed node deployment strategy qualitatively

4.1.1 Coverage-connectivity

Analysis on coverage and connectivity of the scheme is performed through lemma/claim and the corresponding proofs.

Lemma 2: For a given network area a×b, to cover the network by minimum number of SNs or 1-coverage, the required number of SNs is (M×N) where N is the number of layers and M is the number of cells in each layer.

Proof: Given a network area (with RHC architecture) to be covered by a set of sensing nodes, our objective is to determine the number of sensor nodes and their placements at the strategic points such that network area is covered by minimum number of sensing nodes. The area of each cell is given as:

$$A_{RHC} = \frac{3\sqrt{3}}{2} r^2 \approx 2.6 r^2.$$

The maximal circular area covered by each SN is given as $S(v) = \pi R_s^2 \approx 3.14 R_s^2$.

Further, we have partitioned the network area (a×b) into N number of layers and each layer consists of M number of cells. Therefore, the network area in terms of number of cells is represented as-

$$(M \times N) \times \frac{3\sqrt{3}}{2} r^2 \approx 2.6(M \times N) \times r^2.$$

Hence the number of sensing nodes needed to cover a network area (a×b) is given by-

$$\frac{2.6(M \times N) \times r^2}{3.14 R_s^2} = 0.8(M \times N).$$

In order to have an integer number of sensing nodes the upper ceil function ($\lceil . \rceil$) is employed, so the required minimum number of sensing nodes is (M×N).

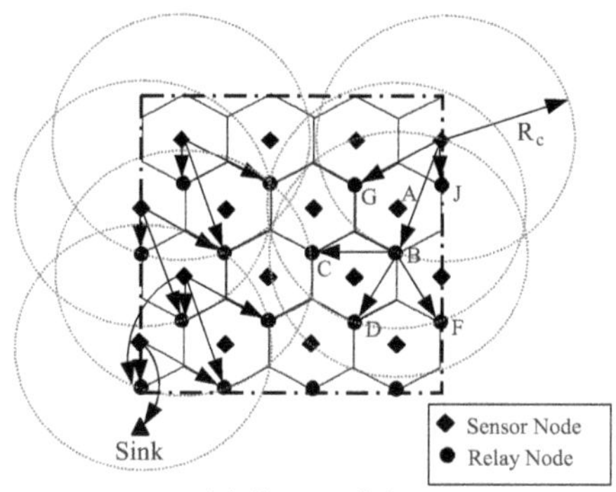

(a) Connectivity

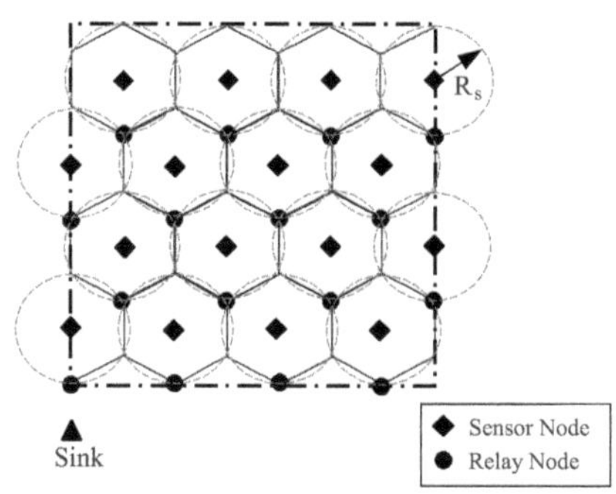

(b) Coverage

Figure 4: Coverage and Connectivity of the NDCM

Claim: If M×N number of SNs and RNs are deployed in a given network area a×b using the proposed deployment strategy 1-coverage and atleast 3-connectivity are achieved.

Proof: From Lemma 2, we can ensure that our node deployment strategy achieves 1-coverage. Let us consider a 4-layer network area where each layer consists of 4 cells shown in Figs. 4(a)-(b). In the first phase SNs are deployed in the centre of each cell. In Fig. 4 (b) we see that area of each cell is covered by one SN, therefore it justifies our claim that proposed deployment strategy achieves 1-coverage. In the second phase RNs are deployed in the strategic location of each cell. So there exist atleast three RNs in the communication range of a SN or a RN shown in Fig. 4(a), therefore justifying our claim that proposed deployment strategy achieves 3-connectivity.

For example, according to Lemma 2, for a 4-layer network area with 4 cells (Figs. 4(a)-(b)) in each layer, 4×4 number of SNs are required (Fig. 4(b)) in order to maintain 1-coverage. Similarly as per the above claim for the same network area, 4×4 number of RNs are required (Fig. 4(a)) to achieve at least 3-connectivity.

4.1.2 Network Lifetime

In this sub-section we have analyzed network lifetime of the proposed node deployment strategy for establishing the fact that judicious deployment of nodes can ensure enhancement of network lifetime.

Network lifetime has different definitions based on the desired functionality [18]. In presence of several existing state-of-the-art definition of network lifetime, we consider the following definition.

Network Lifetime: The network lifetime is defined as the time interval from the beginning of the network operation till a proportion of dead node exceeds a certain threshold, which may result in loss of coverage or connectivity of a certain region, and/or network partitioning.

From earlier discussion it is established that the set of one-hop neighbour RNs of each node (SN/RN) through which the node may transmit data towards sink contains at least three RNs. Hence network lifetime ends when a SN is dead and/or all the RNs from such a neighbor RNs of a node (SN/RN) are dead and subsequently loss of coverage and/or connectivity occurs.

Let us consider that initial energy of each SN and RN is E_{SN} and E_{RN} respectively. All the SNs uniformly generate and transmit data packet of size n bits each within a predefined time q(t) and therefore average energy consumption by each SN located at j[th] cell in layer-i is

$$e_{i,j} = n(e_s + e_t) \quad \text{for i=1,2,...,N and j=1,2,...,M.} \tag{3a}$$

where e_s and e_t denote energy required for sensing and transmitting one bit of data respectively.

Let us consider that in the network area V_k^{SN} numbers of set of one-hop neighbours SNs and V_h^{RN} numbers of set of one-hop neighbours RNs are formed where k=1,2,...,K and h=1,2,...,H. Each RN consumes energy for receiving data from one-hop neighbours SNs amongst V_k^{SN} and transmitting the same. In addition to this, it consumes energy for receiving data from one-hop neighbours RNs amongst V_h^{RN} and transmitting the received data. Hence, average energy consumption per q(t) time-interval by each RN located at j[th] cell in layer-i is

$$E_{i,j} = \left[V_k^{SN} + V_h^{RN}\right] n(e_r + e_t). \tag{3b}$$

The lifetime of a SN located at j[th] cell in layer-i is computed using Eq. (3a) and it is given as

$$LT_{i,j}^{SN} = \frac{E_{SN}}{e_{i,j}} = \frac{E_{SN}}{n(e_s + e_t)}.$$

Similarly, using Eq. (3b), the lifetime of a RN located at j[th] cell in layer-i is given as

$$LT_{i,j}^{RN} = \frac{E_{RN}}{E_{i,j}} = \frac{E_{SN}}{\left[V_k^{SN} + V_h^{RN}\right] n(e_r + e_t)}.$$

Therefore, the network lifetime is determined by the shortest lifetime of a SN or shortest lifetime of a set of one-hop neighbours RNs amongst V_h^{RN} that causes loss of coverage or connectivity of a certain region, and/or network partitioning and is expressed as

$$\min\left\{\min\left\{LT_{1,1}^{SN}, LT_{1,2}^{SN}, \dots, LT_{i,j}^{SN}\right\}, \min\left\{V_1^{RN}, V_2^{RN}, \dots, V_H^{RN}\right\}\right\}$$

$$\min\left\{\min\left\{LT_{1,1}^{SN}, LT_{1,2}^{SN}, \dots, LT_{i,j}^{SN}\right\}, \min\left\{\begin{matrix}\left(LT_{1,1}^{RN}, LT_{1,2}^{RN}, \text{Sink}\right), \\ \left(LT_{1,1}^{RN}, LT_{2,1}^{RN}, LT_{1,2}^{RN} LT_{2,2}^{RN}\right), \dots\end{matrix}\right\}\right\}.$$

4.2 Quantitative Analysis

The effectiveness of the proposed node deployment scheme is evaluated through simulation.

4.2.1 Simulation environment

The simulation is performed using MATLAB (version 7.1). We assume perfect MAC layer issues while performing the simulation. Simulation results of Node Deployment under Coal Mine (NDCM) are compared with two existing schemes, Relay Node Placement in Tunnel (RNPT) [6] and Chain-type Sensor Deployment Strategy (CSDS) [7]. Extensive simulations have been performed and average results of 100 independent runs has been taken while plotting the simulation graphs.

During implementation of the RNPT we have considered cluster-based architecture where relay nodes act as cluster heads and sensor nodes act as member nodes of one of the clusters. Here sensor nodes are deployed uniformly in predetermined locations whereas minimum numbers of relay nodes are deployed non-uniformly in such a way that connectivity in the network is ensured. Initially RNs are deployed non-uniformly so that distribution density of RNs near the sink is higher than those farther away from the sink. After deployment of RNs, SNs are equally deployed under each RN. Similar to RNPT, CSDS is also implemented as cluster-based architecture. Also during implementation of both the RNPT and CSDS, static cluster head has been considered.

Further, for all the above schemes including the proposed one we consider q(t) (section 2.2) as 5 sec and a routing protocol described in section 3.5. Table I provides the number of nodes (SNs and RNs) deployed during simulation of all the schemes (NDCM, RNPT and CSDS).

4.2.2 Simulation metrics

To evaluate the performance of NDCM, network lifetime and energy efficiency have been considered as performance matrices. We have chosen network coverage and connectivity as the yardsticks for measuring network lifetime defined in section 4.1.2. As mentioned in the previous sub-section, in our deployment scheme loss of coverage of a certain region occurs

Table I: Node deployed for various network sizes

Deployment scheme	Network size	Sensing node	Relay node
NDCM	4-layer	32	32
	10-layer	140	140
RNPT	4-layer	80	32
	10-layer	350	140
CSDS	4-layer	80	32
	10-layer	350	140

whenever one SN is dead or loss of connectivity occurs when all the RNs in a set of one-hop neighbour RNs of a SN/RN are dead. So similar to qualitative analysis, in simulation also we have obtained network lifetime by measuring the minimum value of the following parameters- minimum operative lifetime of a SN and minimum operative lifetime of a set of one-hop neighbor RNs. On the contrary, for RNPT and CSDS, SNs of the cluster collects information of the monitored region and feed RNs with local-view data. Therefore, if a RN runs out of energy, connectivity of a certain region is lost, and/or network partitioned, even though some SNs of the cluster may still be alive and functioning. Therefore, in simulation we have measured network lifetime by minimum operative lifetime of a RN.

Two more parameters are defined below for evaluating the extent of energy efficiency in the network:

Average energy consumption rate per RN (avg. ECR per RN): It is defined as energy consumption by a relay node in each q(t) time-interval. It is evaluated as

$$\text{Avg. ECR per RN}=\frac{\text{Total Energy consumption by RN in } q(t) \text{ time-interval in a layer}}{\text{Number of RNs in the layer}}$$

Average energy consumption rate per SN (avg. ECR per SN): It is defined as energy consumption by a sensor node in each q(t) time-interval. It is evaluated as

$$\text{Avg. ECR per SN}=\frac{\text{Total Energy consumption by SN in } q(t) \text{ time-interval in a layer}}{\text{Number of SNs in the layer}}$$

Two sets of experiments are conducted for evaluating the performance of the present scheme and two other competent schemes. One set of experiment measures energy efficiency and the other set verifies the enhancement of network lifetime. For each set, experiments have been conducted for two different network sizes viz. network with 4 and 10 layers having 8 and 14 cells in each of the layers respectively. We have assumed RNs with unlimited energy, however, in simulation we have considered RNs with initial energy 200J which is significantly greater than the SNs' initial energy of 50J. All the other parameters and their corresponding values used for simulation are listed in Table II.

4.2.3 Energy Efficiency

In this sub-section energy efficiency of the scheme is evaluated in terms of avg. ECR per SN/RN.

4.2.3.1 Average energy consumption rate per SN (avg. ECR per SN)

Figures 5(a) and 5(b) show avg. ECR per SN for 4-layered and 10-layered network respectively. The most important observation from Fig. 5 is, the plot for NDCM is relatively steady for both the network sizes and this indicates balanced energy consumption among all the layers. This has a direct effect on network lifetime which is plotted in Fig. 7.

Table II: Simulation parameters

Parameters	Value
β_1	$45\times10^{-9}\,\text{J/bit}$
β_2	$0.001\times10^{-12}\,\text{J/bit/m}^4$
e_r	$135\times10^{-9}\,\text{J/bit}$
e_s	$60\times10^{-9}\,\text{J/bit}$
Communication range (R_c)	61m
Sensing range (R_s)	40m
Packet size (n)	480 bits
Network area	500m×250m to 1000m×600m

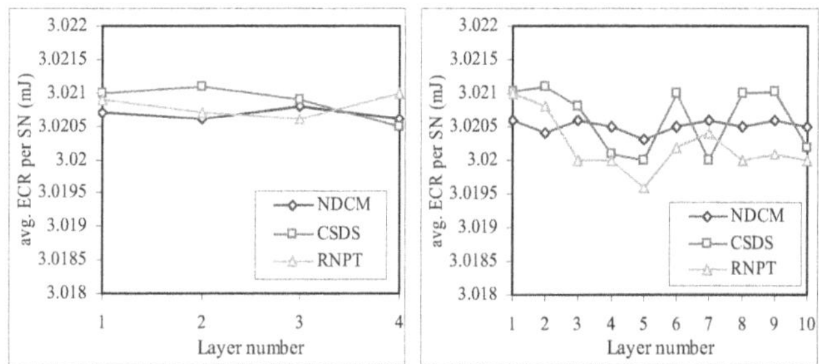

(a) 4-layer network (b) 10-layer network

Figure 5: Average energy consumption rate per SN

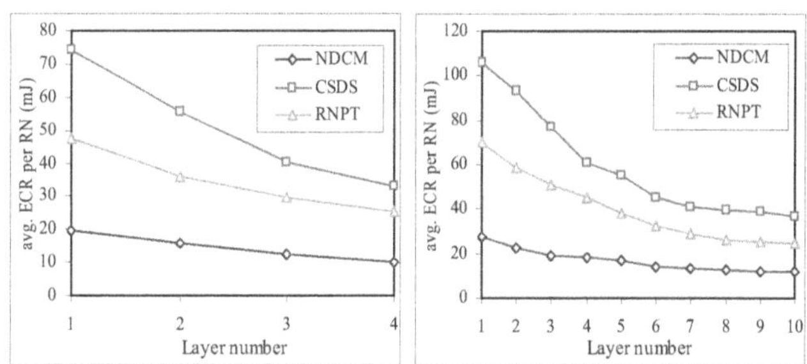

(a) 4-layer network (b) 10-layer network

Figure 6: Average energy consumption rate per RN

4.2.3.2 Average energy consumption rate per RN (avg. ECR per RN)

Figures 6(a) and 6(b) show avg. ECR per RN for 4-layered and 10-layered network respectively. For both the networks, in CSDS and RNPT, layer-1 has the highest avg. ECR per RN and the farthest layer has the lowest avg. ECR per RN. This is due to an imbalance in sharing of relay load, which according to us is the primary reason for uneven energy consumption among the RNs of different layers. Therefore, RNs deployed in layers nearer to the sink drain off their energy much more quickly in comparison to RNs deployed in layers farther away from the sink. It is also observed that for all three schemes, the mean of avg. ECR per RN is increasing in nature with increasing network sizes. For example, for network with 4 and 10 layers the mean of avg. ECR per RN is 14.51mJ and 16.95mJ for NDCM, 50.5mJ and 59.24mJ for CSDS, 34.65mJ and 40.05mJ for RNPT respectively. This result is expected as the network size increases more data need to be relayed by a RN.

4.2.4 *Network Lifetime*

In this sub-section network lifetime is evaluated for various network sizes.

The graph illustrated in Fig. 7 represents the network lifetime for two different network sizes. It is observed that the network lifetime of NDCM is 121% and 78% higher than that of CSDS and RNPT respectively for 4-layer network. For 10-layer network these values are 118% and 79.78% respectively. Further, similar to energy consumption in SN, here also we observe that the result plotted in NDCM is relatively steady along the layers and that indicates balanced energy consumption thereby affecting network lifetime positively. To be more specific, the plot of NDCM is almost flat signifying balanced consumption in all the layers and therefore it is more energy-efficient compared to CSDS and RNPT.

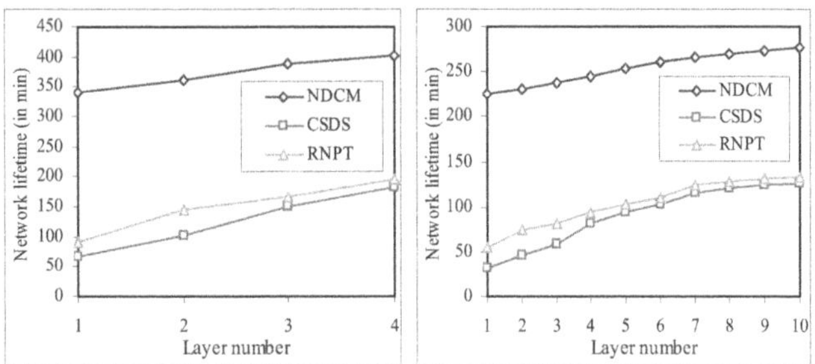

(a) 4-layer network (b) 10-layer network

Figure 7: Network lifetime for various network area

5. CONCLUSION AND FUTURE WORK

In the present work, we have proposed a node deployment strategy in wireless sensor network for monitoring of the environment in underground coal mines. The deployment strategy is heterogeneous in nature where heterogeneity has been introduced by using two types of nodes viz. sensing nodes and relay nodes. Nodes are deployed in two phases. In the first phase, sensing nodes are deployed at the center of each cell for ensuring 1-covergae. In the second phase, to make the deployment scheme more realistic relay nodes are placed at some strategic locations ensuring 3-connectivity. We have simulated our scheme to establish its energy efficiency leading to increased network lifetime. Finally the simulation results are compared with two of the existing strategies taking energy efficiency and network lifetime as performance metrics. The comparison results establish the present scheme's superiority over the existing schemes.

As a future extension of our work, the deployment strategy may be made more realistic by considering obstacles in the network area. Moreover, the scheme may be analyzed for further improvement considering QoS parameters.

REFERENCES

[1] I. F. Akyildiz, and E. P. Stuntebeck. Wireless Underground Sensor Networks: Research Challenges. *Ad Hoc Networks*, 4(6): 669–686, 2006.

[2] Z. Sun, and I. F. Akyildiz. Channel Modeling and Analysis for Wireless Networks in Underground Mines and Road Tunnels. *IEEE Trans. Communications*, 58(6): 1758–1768, 2010.

[3] X. Niu, X. Huang, Z. Zhao, Y. Zhang, C. Huang, and L. Cui. The Design and Evaluation of a Wireless Sensor Network for Mine Safety Monitoring. In *Proc. IEEE Int'l. Conf. GLOBECOM*, pages 1291–1295, 2007.

[4] X. R. Wang, J. T. Lizier, O. Obst, M. Prokopenko, and P. Wang. Spatiotemporal Anomaly Detection in Gas Monitoring Sensor Networks. In *Proc. Int'l. Conf. EWSN*, LNCS 4913, pages 90–105, 2008.

[5] M. Li, and Y. Liu. Underground Coal Mine Monitoring with Wireless Sensor Networks. *ACM Transactions on Sensor Networks*, 5(2): article 10, 29 pages, 2009.

[6] R. Liu, I. J. Wassell, and K. Soga. Relay Node Placement for Wireless Sensor Networks Deployed in Tunnels. In *Proc. IEEE Int'l. Conf. WiMob*, pages 144–150, 2010.

[7] G. Chen , Z. Zhu, G. Zhou, C. Shen, and Y. Sun. Sensor Deployment Strategy for Chain-type Wireless Underground Mine Sensor Network. *J. of China University of Mining and Technology*, 18(4): 561–566, 2008.

[8] M. Li, Y. Liu, and L. Chen. Nonthreshold-Based Event Detection for 3D Environment Monitoring in Sensor Networks. *IEEE Trans. on Knowledge and Data Engineering*, 20(12): 1699–1711, 2008.

[9] M. Yarvis, N. Kushalnagar, H. Singh, A. Rangarajan, Y. Liu, and S. Singh. Exploiting Heterogeneity in Sensor Networks. In *Proc. IEEE INFOCOM*, 2, pages 878–890, 2005.

[10] S. Halder, A. Ghosal, and S. DasBit. A Pre-determined Node Deployment Strategy to Prolong Network Lifetime in Wireless Sensor Network. *Computer Communications*, 34(11): 1294–1306, 2011.

[11] Crossbow Mica2 motes and Stargate-Xscale, http://www.xbow.com.

[12] Y. Zhu, and Q. Han. The More Relay Nodes, The More Energy Efficient?. In *Proc. Int'l Conf. Computational Science and Engineering*, 2, pages 87–92, 2009.

[13] A. J. Perez, M. A. Labrador, and P. M. Wightman. A Multiobjective Approach to the Relay Placement Problem in WSNs. In *Proc. IEEE Int'l Conf. WCNC*, pages 475–480, 2011.

[14] M. Cardei, and J. Wu. Energy-efficient Coverage Problems in Wireless Ad-hoc Sensor Networks. *Computer Communication*, 29(4): 413–420, 2006.

[15] D. Tian, and N. D. Georganas. Connectivity Maintenance and Coverage Preservation in Wireless Sensor Networks. *Ad Hoc Networks*, 3(6): 744–761, 2005.

[16] W. Heinzelman. Application-Specific Protocol Architectures for Wireless Networks. Ph.D. Thesis, Massachusetts Institute of Technology, 2000.

[17] X. Wu, G. Chen, and S. K. Das. Avoiding Energy Holes in Wireless Sensor Networks with Nonuniform Node Distribution. *IEEE Trans. on Parallel and Distributed Systems*, 19(5): 710–720, 2008.

[18] I. Dietrich, and F. Dressler. On the Lifetime of Wireless Sensor Networks. *ACM Trans. on Sensor Networks*, 5(1): 1–39, 2009.

Migrating Operator Placement for Compositional Stream Graphs

Vasvi Kakkad
School of IT
The University of Sydney
NSW 2006, Australia
vasvikak@it.usyd.edu.au

Andrew E. Santosa
School of IT
The University of Sydney
NSW 2006, Australia
santosa@it.usyd.edu.au

Bernhard Scholz
School of IT
The University of Sydney
NSW 2006, Australia
scholz@it.usyd.edu.au

ABSTRACT

Wireless sensor networks (WSN) and mobile clouds are composed of sensor nodes that have limited energy resources. For wireless sensor networks, query processing is the state-of-the-art for data gathering and processing applications to avoid low-level programming. The stream programming model has been widely used to represent queries as an information flow from the sensor nodes to the base station. The model describes queries as stream graphs consisting of operators that process data and channels that connect operators. Operators are deployed in the network to reduce the communication overhead and hence energy. The modification of WSN queries at runtime is of key importance due to changes in the environment and the network energy levels, resulting in the migration of operators between the network nodes.

In this work, we introduce the *migrating operator placement problem (MOPP)* that places operators of stream graphs on sensor nodes, such that energy costs are minimized. The placement takes changes of queries and migration of operators into account. The general MOPP is NP hard, and, therefore, we develop a dynamic program for a *compositional* subset of the stream graphs with polynomially-bounded running time. To improve the performance of our algorithm, we introduce a heuristic that reduces the search space to the proximity of the base station. We conduct various experiments using a simulator for wireless sensor networks with different sizes.

Categories and Subject Descriptors

C.2.1 [**Computer-Communication Networks**]: Network Architecture and Design—*distributed networks, network topology, wireless communication*; I.2.8 [**Artificial Intelligence**]: Problem Solving, Control Methods, and Search—*dynamic programming*

General Terms

Algorithms, Languages, Performance

Keywords

stream programming, operator placement, dynamic programming

1. INTRODUCTION

The technological advances in wireless sensor network (WSN) and mobile cloud enables the development of complex applications including health monitoring, environment sampling, and disaster area monitoring [2, 16, 24, 10]. Such applications deploy battery-powered sensing devices at remote locations with limited resources. The development of energy-efficient and complex wireless sensing applications requires in-depth embedded systems programming skills which makes it challenging, in particular for the domain experts. To overcome this challenge, the use of the database paradigm and query languages has emerged in the last few years as a viable solution [18, 25]. As siginificant part of energy consumption is due to communication, to save energy, some query processing schemes push the computations to the sensing devices allowing *in-network* processing instead of *data forwarding*, where the raw sensed data are sent directly to the base station [21, 1, 25]. However, in sensing applications, queries may need to be dynamically replaced on the nodes due to the changes in energy level of the nodes or in the network configurations. In this work, we propose a method for energy-efficient query execution considering operator migration.

We introduce the *migrating operator placement problem* (*MOPP*) that arises in applications such as mobile cloud computing [2, 17] and wireless sensor networks [19, 21, 13]. We consider the representation of the queries as stream graph using stream programming model similar to MaD-WiSe [1] and query processing [21]. The stream programming model (surveyed in [11]) expresses computations by independent computational units called *operators* and communication *channels* between them. In our context, the stream graph consists of operators that sense the environment and operators that process and forward the information to a base station. To minimize the energy cost of the system, we want to place operators on nodes such that the associated operator placement costs become minimal. The costs are comprised of (1) placement costs of an operator to a specific sensor node, (2) communication costs imposed by sending data between two connected operators, and (3) transition costs that are incurred by migrating operators from their current nodes to new nodes. The operator placement can be conducted either when the stream graph representing queries changes, the distributed system changes, and/or the energy levels of the nodes change after a fixed time interval.

As an example, let us consider a volcanic activity detection application that deploys several sensors in the proximity of the volcano and measures temperature, gas, and seismic activities. In an initial step, the application executes two queries: The first query collects the temperature reading from the sensors placed at distant locations and computes the average temperature. The second query collects the reading from seismometer, computes the high- and low-gain moving averages, and reports an event if the ratio of the two av-

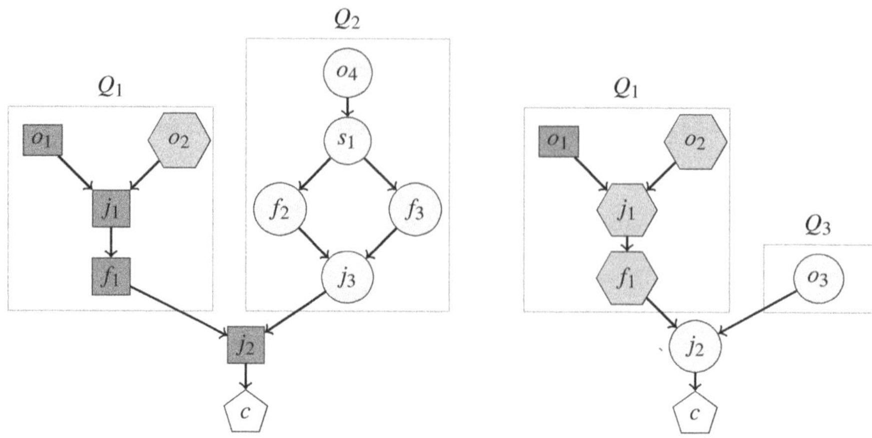

(a) $G^{pre} = (V^{pre}, E^{pre})$

(b) $G = (V, E)$

Figure 1: Running Example. (a) Stream graph G^{pre} for queries Q_1 and Q_2 with placement p^{pre}; (b) Stream graph G for query Q_1, newly added query Q_3, deleted query Q_2, and new placement p.

Op\Node	n_1	n_2	n_3	b
o_1	0	∞	∞	∞
o_2	∞	0	∞	∞
j_1	10	6	13	20
f_1	7	4	15	17
o_3	∞	∞	∞	∞
j_2	9	12	11	13
c	∞	∞	∞	0

(a) Placement Cost for Q_1 and Q_3

Op\Op	o_1	o_2	j_1	f_1	o_3	j_2	c
o_1	0	0	5	0	0	0	0
o_2		0	13	0	0	0	0
j_1			0	21	0	0	0
f_1				0	0	7	0
o_3					0	10	0
j_2						0	11

(b) Communication Cost for Q_1 and Q_3

Op\Node	n_1	n_2	n_3	b
j_1	0	10	16	21
f_1	0	11	13	19
j_2	0	8	5	10

(c) Transition Cost for Q_1

Figure 2: Incurred Costs in Operator Placement

erages is beyond threshold. The results from the two queries are merged to detect the volcanic activity as illustrated in Fig. 1 (a), which shows the stream graph of the queries. Subgraph Q_1 represents the temperature reading query with two sense operations o_1 and o_2, a join operation j_1 to merge the data from the two sensors, and a filter f_1 to compute the average temperature. The second subgraph Q_2 represents the seismic query with a sense operation o_4 that reads the data from a seismometer, a split operator s_1 that duplicates the incoming data to compute the high- and low-gain exponentially weighted moving average (EWMA) [24] using filters f_2 and f_3, and a joiner j_3 that merges the data from two filters to compute the ratio. Finally, a joiner j_2 combines the data from both the queries Q_1 and Q_2 before forwarding it to the sink c. In Fig. 1 (a), the placement of the operators is shown by the shapes of the graph nodes, i.e., rectangle, hexagon, circle, and pentagon denote the placement of operators to sensor nodes n_1, n_2, n_4 and base station b, respectively.

In our volcanic activity detection application, sensor nodes might be destroyed or become temporarily unavailable. Hence, the migration of operators from inoperable sensor nodes to operable ones is of paramount importance. In addition, the application may require to alter queries: for example, the observation from seismic event detection to the measurement of expelled gases. For this example, the altered stream graph is shown in Fig. 1 (b) that contains new query Q_3, and query Q_2 of the original graph was removed. For the operator placement of the altered stream graph, we take placement cost, communication cost, and transition costs into account that are tabularized in Fig. 2. Note that the placement of sense operators o_1, o_2, o_3, and the sink c are not arbitrary, and are fixed to nodes n_1, n_2, n_3, and base station b, respectively. The placement of the other operators is chosen such that energy costs of the placement are minimised. While recomputing the placement for modified stream graph, the difference of placement between the old and modified graphs incurs *transition* cost. In our example, query Q_1 carries transition cost for the operators j_1, f_1, and j_2, since they are placed on different sensor nodes in the old and new stream graphs (i.e, Fig. 1 (a) and Fig. 1 (b) respectively).

The overall contributions of this paper are as follows.

- We introduce the NP-hard *migrating operator placement problem* (*MOPP*) that computes a placement that minimizes computation, communication, and transition costs.
- We devise a dynamic program that computes an optimal operator placement for *compositional* streams which is a subclass of general stream graphs in polynomial time.
- To improve the performance of our approach, we introduce a locality heuristic that, at any step in the algorithm, only considers the placement onto a subset of the network close to the base station.
- We present an experimental evaluation confirming the efficiency of our approach and its effectiveness in reducing energy usage.

The structure of the paper is as follows. Section 2 explains the problem statement as a discrete optimization problem, and argues that the general problem is NP-hard. Section 3 details the graph composites. Section 4 gives the detail about the dynamic program for compositional streams and its correctness and complexity proofs. We present an experimental evaluation that confirm the efficacy of our approach, together with a locality heuristics that improve its efficiency in Section 5. In Section 6 we survey the related work and conclude this paper in Section 7.

2. MIGRATING OPERATOR PLACEMENT PROBLEM

In this section we introduce the MOPP. At any one time in the execution of the program, there are a number of active queries, such that the whole program can be represented as a stream graph $G = (V, E)$, with $V = V_{se} \cup V_{si} \cup V_{fi}$, with V_{se} a set of *sense* operators, V_{si} a singleton of a *sink* operator, and V_{fi} a set of *filters*, that performs computation on its input and sends the result to the output. V_{se}, V_{si}, and V_{fi} are pairwise disjoint, and $E \subseteq V^2$. We are also given a set N of sensor nodes with a distinguished element b called the *base station*. We introduce a *placement* $x : V \rightarrow N$ where x_u is a variable denoting the sensor on which $u \in V$ is placed. We write X for an unordered sequence of such variables, and X_G for a restriction of the sequence to placements of all nodes in V [1]. The placement x is constrained such that $x_u \neq x_v$ when $u \neq v$ and $u, v \in V_{se}$ (assuming $|N|$ is sufficiently large), and $x_u = b$ for $u \in V_{si}$.

We assume there is a change in the graph such as addition and removal of queries, such as we had a *pre-existing* graph $G^{pre} = (V^{pre}, E^{pre})$ that transformed into $G = (V, E)$. We denote the *pre-existing* placement $x^{pre} : V^{pre} \rightarrow N$, with variables x_u^{pre} for $u \in V^{pre}$ denoting $x^{pre}(u)$. The problem of MOPP is to find the placement x for the new graph $G = (V, E)$ with the following costs minimized.

Placement cost: The placement cost is the local cost incurred for placing an operator on a given sensor node. The cost depends on the available energy level on sensor node and computational load for executing the operator. For an operator $u \in V$ and the placement x, the placement cost is $w_P(u, x_u) \in \mathbb{R}^+$.

Communication cost: The communication cost is associated to each channel. For a channel $(u, v) \in E$, this is a product of two values: $w_C(u, v) \in \mathbb{R}^+$ denoting the bandwidth requirement of the channel, and $w_D(x_u, x_v) \in \mathbb{R}^+$ denoting the *distance factor* between the allocated nodes. By knowing the maximum distance factor it is possible to incorporate it into $w_C(u, v)$ and allowing $w_D(x_u, x_v) = 1$ when $x_u \neq x_v$ and 0 otherwise, as is assumed in Fig. 2 (b).

Transition cost: The transition cost is the cost incurred while moving an operator to a new sensor node. In other words, if an operator u is to be relocated to other sensor node, then the transition cost is $w_T(u, x_u^{pre}, x_u) \in \mathbb{R}^+$.

The following function encapsulates both the placement and transition costs.

$$w_{TP}(u, x_u) = \begin{cases} w_P(u, x_u) + w_T(u, x_u^{pre}, x_u) & \text{if } u \in V^{pre} \cap V, \\ w_P(u, x_u) & \text{otherwise.} \end{cases}$$

[1] A variable x_u for some u can appear more than once in the sequence, therefore for sequences X_1 and X_2, $X_1 = X_2$ modulo reordering and repetition. Moreover, $X_1 \setminus X_2$ removes all occurrences of x_u from the placement sequence X_1 for any $x_u \in X_2$. $X_1 \cup X_2$ concatenates both sequences.

Here we assume that the costs for removing or installing queries onto the network to be insignificant, where the nodes may already include the program or library functions to execute the operators. Although it may seem at first that it is less expensive to remove the queries entirely from the network and reinstall them rather than transitioning the operators, here we assume that there is a need to preserve the existing computation such that we are willing to pay the associated costs. The transition costs come from transferring computation along network nodes form the source to destination. In wireless sensors, this movement is costly [15].

The cost $f_G(X_G)$ of MOPP to be minimized, given a stream graph $G = (V, E)$ is the sum of computation and transition costs for each operator, and the communication costs:

$$f_G(X_G) = \sum_{u \in V} w_{TP}(u, x_u) + \sum_{(u,v) \in E} w_C(u, v) \cdot w_D(x_u, x_v) \quad (1)$$

The following holds (proof is in the appendix Section A.1).

THEOREM 1. *MOPP is NP-hard.*

3. GRAPH COMPOSITES

Here we introduce a class of stream graphs called *compositional* streams that are constructed from building blocks called *composites* (cf. StreamIt [23]). Although MOPP is NP-hard in general, compositional streams can be solved in polynomial time.

There are two types of composites: *stream* composites, which are graphs $G = (V, E, g, h)$ with $g = s(G)$ and $h = t(G)$ are respectively *input* and *output* operators, and *source* composites, graphs $G = (V, E, h)$, with $h = t(G) \in V$ as its output operator. A source composite G has no input operator, that is, $s(G) \notin V$. The templates for the composites are as follows.

(a) Stream Composite (b) Source Composite

Fig. 3 displays the types of stream and source composites. A stream composite is either (a) a filter or sink operator, (b) a *pipeline* that joins the output of a stream composite with the input of another stream composite, and (c) a *split-join* composite that splits an input into the inputs of two stream composites and combines the outputs of the streams into one. A source composite is either (d) a sense operator, (e) a pipeline joining the output of a source composite with the input of a stream composite, (f) a *join* composite joining the outputs of two source composites into one, or (g) a program, which is a pipeline that joins the output of a graph with a base station. The split-join and the join composites are constructed using special operators called *splitters* and *joiners* that respectively splits the tokens of a communication channel and joins them.

We define stream composite class $\mathbb{S}$ as follows.

(*R1*) If $V = V_{fi} = \{u\}$ or $V = V_{si} = \{u\}$ then $G_u = (\{u\}, \emptyset, u, u) \in \mathbb{S}$.

(*R2*) If $G_1 = (V_1, E_1, g_1, h_1) \in \mathbb{S}$ and $G_2 = (V_2, E_2, g_2, h_2) \in \mathbb{S}$, and $V_1 \cap V_2 = \emptyset$, then the pipeline $G_1 *^s G_2 = (V_1 \cup V_2, E_1 \cup E_2 \cup \{(h_1, g_2)\}, g_1, h_2) \in \mathbb{S}$.

(*R3*) If $G_1 = (V_1, E_1, g_1, h_1) \in \mathbb{S}$ and $G_2 = (V_2, E_2, g_l, h_l) \in \mathbb{S}$, and $V_1 \cap V_2 = \emptyset$, then the split-join composite $G_1 ||^s G_2 = (V_1 \cup V_2 \cup \{s, j\}, E_1 \cup E_2 \cup \{(s, g_1), (s, g_2), (h_1, j), (h_2, j)\}, s, j) \in \mathbb{S}$, where $s, j \notin V_1 \cup V_2$ are splitter and joiner respectively.

(*R4*) Nothing else is in $\mathbb{S}$.

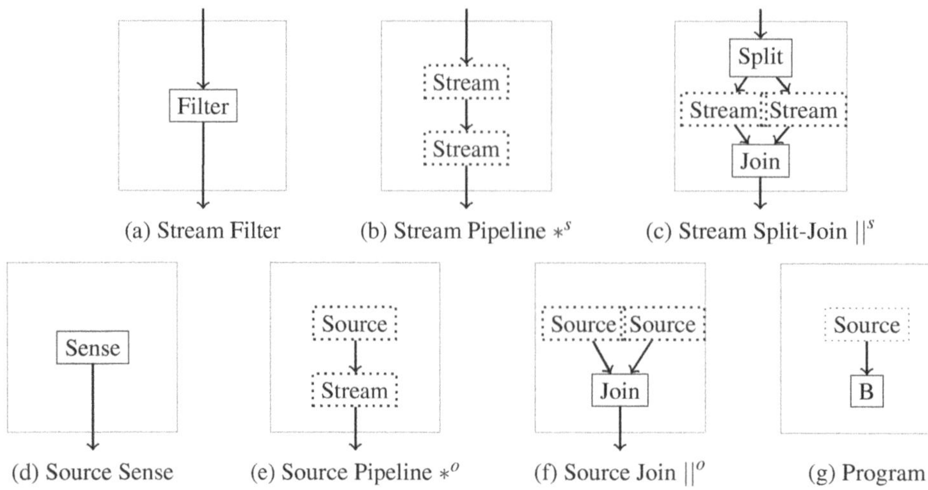

Figure 3: Stream and Source Composites

We define source composite class $\mathbb{O}$ as follows.

(*R5*) When $V = V_{se} = \{u\}$, then $G_u = (\{u\}, \emptyset, u) \in \mathbb{O}$.

(*R6*) If $G_1 = (V_1, E_1, h_1) \in \mathbb{O}$ and $G_2 = (V_2, E_2, g_2, h_2) \in \mathbb{S}$, and $V_1 \cap V_2 = \emptyset$ the pipeline $G_1 *^o G_2 = (V_1 \cup V_2, E_1 \cup E_2 \cup \{(h_1, g_2)\}, h_2) \in \mathbb{O}$.

(*R7*) If $G_1 = (V_1, E_1, h_1)$ and $G_2 = (V_2, E_2, h_2)$ are in $\mathbb{O}$, and $V_1 \cap V_2 = \emptyset$, then $G_1 ||^o G_2 = (V_1 \cup V_2 \cup \{j\}, E_1 \cup E_2 \cup \{(h_1, j), (h_2, j)\}, j) \in \mathbb{O}$, where $j \notin V_1 \cup V_2$ denotes a joiner.

(*R8*) Nothing else is in $\mathbb{O}$.

We note that a *program* is a composite satisfying (*R6*) but with $G_2 = G_u$ and $u \in V_{si}$. We now define the function *sub* that returns the set of immediate children of a composite G, called its *substreams*, as follows.

$$sub(G) = \begin{cases} \{G_1, G_2\} & \text{if } G = G_1 *^o G_2 \text{ or } G = G_1 *^s G_2, \\ \{G_j, G_1, G_2\} & \text{if } G = G_1 ||^o G_2 = (V, E, j), \\ \{G_s, G_j, G_1, G_2\} & \text{if } G = G_1 ||^s G_2 = (V, E, s, j), \text{ and} \\ \text{undefined} & \text{otherwise.} \end{cases}$$

4. DYNAMIC PROGRAMMING FOR MOPP

4.1 Dynamic Program

We define a dynamic program that computes the cost of allocating query operators to available sensors in the network. The program is defined using two functions $\hat{f}_G^{\mathbb{S}} : N^2 \to \mathbb{Q}^+$ and $\hat{f}_G^{\mathbb{O}} : N \to \mathbb{Q}^+$ depending on whether G is a stream or source composite. The arguments of $\hat{f}_G^{\mathbb{S}}$ are the sensors to which the input and output vertices of G are respectively placed, whereas the argument of $\hat{f}_G^{\mathbb{O}}$ is the placement of the output vertex of G.

DEFINITION 1 (DYNAMIC PROGRAM $\hat{f}_G^{\mathbb{S}}$ AND $\hat{f}_G^{\mathbb{O}}$).

1. $G \in \mathbb{S}$.

- *If* $G = G_u$ *with* $V = V_{fi} = \{u\}$ *then* $s(G) = t(G) = u$ *and*

$$\hat{f}_G^{\mathbb{S}}(x_{s(G)}, x_{t(G)}) = \begin{cases} w_{TP}(u, x_u) & \text{if } x_{s(G)} = x_{t(G)}, \\ \infty & \text{otherwise.} \end{cases}$$

- *If* $G = G_u$ *with* $V = V_{si} = \{u\}$, *then* $s(G) = t(G) = u$ *and*

$$\hat{f}_G^{\mathbb{S}}(x_{s(G)}, x_{t(G)}) = \begin{cases} w_{TP}(u, x_u) & \text{if } x_{s(G)} = x_{t(G)} = b, \\ \infty & \text{otherwise.} \end{cases}$$

- *If* $G = G_1 *^s G_2$ *then*

$$\begin{aligned} & \hat{f}_G^{\mathbb{S}}(x_{s(G)}, x_{t(G)}) \\ = \; & \min_{x_{t(G_1)}, x_{s(G_2)}} \left\{ \begin{array}{l} \hat{f}_{G_1}^{\mathbb{S}}(x_{s(G)}, x_{t(G_1)}) + \hat{f}_{G_2}^{\mathbb{S}}(x_{s(G_2)}, x_{t(G)}) + \\ w_C(t(G_1), s(G_2)) \cdot w_D(x_{t(G_1)}, x_{s(G_2)}) \end{array} \right\} \end{aligned}$$

Intuitively, the cost is the sum of the communication cost between the output operator $t(G_1)$ *of* G_1 *and the input operator* $s(G_2)$ *of* G_2 *and the costs for the substreams* G_1 *and* G_2.

- *If* $G = G_1 ||^s G_2$ *then*

$$\begin{aligned} & \hat{f}_G^{\mathbb{S}}(x_{s(G)}, x_{t(G)}) \\ = \; & \min_{x_{s(G_1)}, x_{t(G_1)}, x_{s(G_2)}, x_{t(G_2)}} \left\{ \begin{array}{l} \hat{f}_{G_s}^{\mathbb{S}}(x_s, x_s) + \hat{f}_{G_j}^{\mathbb{S}}(x_j, x_j) + \\ \hat{f}_{G_1}^{\mathbb{S}}(x_{s(G_1)}, x_{t(G_1)}) + \hat{f}_{G_2}^{\mathbb{S}}(x_{s(G_2)}, x_{t(G_2)}) + \\ w_C(s, s(G_1)) \cdot w_D(x_s, x_{s(G_1)}) + \\ w_C(t(G_1), j) \cdot w_D(x_{t(G_1)}, x_j) + \\ w_C(s, s(G_2)) \cdot w_D(x_s, x_{s(G_2)}) + \\ w_C(t(G_2), j) \cdot w_D(x_{t(G_2)}, x_j) \end{array} \right\} \end{aligned}$$

Here, given two substreams $sub(G) = \{G_1, G_2\}$ *(see also Fig. 3 (c)), joiner* j*, and splitter* s*, the cost is the sum of the communication costs for channels* $(s, s(G_1))$, $(t(G_1), j)$, $(s, s(G_2))$, *and* $(t(G_2), j)$*, the costs for the substreams* G_1 *and* G_2*, and the costs for the splitter* s *and joiner* j.

2. $G \in \mathbb{O}$.

- *If* $G = G_u$ *with* $V = V_{se} = \{u\}$ *then*

$$\hat{f}_G^{\mathbb{O}}(x_{t(G)}) = w_{TP}(u, x_u).$$

- *If $G = G_1 *^o G_2$ then*

$$\begin{aligned} & \hat{f}^{\mathbb{O}}_G(x_{t(G)}) \\ = \ & \min_{x_{t(G_1)}, x_{s(G_2)}} \\ & \left\{ \begin{array}{l} \hat{f}^{\mathbb{O}}_{G_1}(x_{t(G_1)}) + \hat{f}^{\mathbb{S}}_{G_2}(x_{s(G_2)}, x_{t(G)}) + \\ w_C(t(G_1), s(G_2)) \cdot w_D(x_{t(G_1)}, x_{s(G_2)}) \end{array} \right\} \end{aligned}$$

  Here the cost is the sum of the communication cost of $(t(G_1), s(G_2))$ and the costs of substreams G_1 and G_2.

- *If $G = G_1 ||^o G_2$ then*

$$\begin{aligned} & \hat{f}^{\mathbb{O}}_G(x_{t(G)}) \\ = \ & \min_{x_{t(G_1)}, x_{t(G_2)}} \\ & \left\{ \begin{array}{l} \hat{f}^{\mathbb{S}}_{G_j}(x_j, x_j) + \hat{f}^{\mathbb{O}}_{G_1}(x_{t(G_1)}) + \hat{f}^{\mathbb{O}}_{G_2}(x_{t(G_2)}) + \\ w_C(t(G_1), j) \cdot w_D(x_{t(G_1)}, x_j) + \\ w_C(t(G_2), j) \cdot w_D(x_{t(G_2)}, x_j) \end{array} \right\} \end{aligned}$$

  The cost here is the sum of communication costs on the channels $(t(G_1), j)$ and $(t(G_2), j)$ and the costs for substreams G_1 and G_2, and the joiner j.

The following theorem states the correctness of the dynamic program (proof in Section A.2).

THEOREM 2. *For any program $G \in \mathbb{O}$,*

$$\min_{x_{t(G)}} \hat{f}^{\mathbb{O}}_G(x_{t(G)}) = \min_{X_G} f_G(X_G).$$

4.2 Memo Table

For every call to $\hat{f}^{\mathbb{S}}_G$ and $\hat{f}^{\mathbb{O}}_G$ the algorithm allocates a table containing the possible return values of the call. Each table is a matrix $T(G)$, where for $G \in \mathbb{S}$, $T(G)_{x,y} = \hat{f}^{\mathbb{S}}_G(x,y)$, and when $G \in \mathbb{O}$, $T(G)$ is a row matrix with $T(G)_x = \hat{f}^{\mathbb{O}}_G(x)$. For a call to $\hat{f}^{\mathbb{S}}_G$, the rows and columns of the table correspond to the possible placements of $s(G)$ and $t(G)$. For a call to $\hat{f}^{\mathbb{O}}_G$, the table only has a single row, and the columns similarly correspond to the placements of $t(G)$.

We illustrate the tabling mechanism using the stream of Fig. 1 (b). The algorithm starts with the tables for the operators, as shown by the first three tables of Fig. 4. The tables reflect the computation cost of an operator on a specific sensor node. Note that the rows and columns of the tables are labeled with the sensor nodes and the base station. The entries of the tables are computed from the placement and transition costs of operators displayed in Fig. 2 (a) and (c). For example, $T(G_{j_1})_{n_2,n_2} = \hat{f}^{\mathbb{S}}_{G_{j_1}}(n_2,n_2) = 16$ reflects the fact that when the operator j_1 is placed on sensor node n_2, the sum of the placement and transition costs for the operator is 16, as the placement cost for j_1 on n_2 is 6 and the transition cost from n_1 to n_2 is 10. Fig. 4 (d) shows a table for one sense operator o_1 which has fixed placement on node n_1. Likewise, the tables are generated for other sense operators based on Fig. 2 (a).

After generating the tables for each operator, the algorithm builds the tables for the larger composites. Let us now consider how the elements of $T((G_{o_1}||^o G_{o_2}) *^o G_{f_1})$ (Fig. 4 (f)) is computed. In particular, when $x_{t(G)} = n_2$. To obtain this element, according to Definition 1 for $\hat{f}^{\mathbb{O}}_G$ with $G_1 *^o G_2$, we enumerate possible placements for $t(G_{o_1}||^o G_{o_2}) = j_1$ and for $s(G_{f_1}) = f_1$ such that $\hat{f}^{\mathbb{O}}_G(x_{t(G)})$ is minimal, for $x_{t(G)} = n_2$. The element is therefore the minimal value among the followings.

$$\begin{array}{lr} T(G_{o_1}||^o G_{o_2})_{n_1} + T(G_{f_1})_{n_2,n_2} + w_C(j_1, f_1), & \\ \quad \text{where } x_{j_1} = n_1 \text{ and } x_{f_1} = n_2 & (1) \\ T(G_{o_1}||^o G_{o_2})_{n_2} + T(G_{f_1})_{n_2,n_2}, & \\ \quad \text{where } x_{j_1} = n_2 \text{ and } x_{f_1} = n_2 & (2) \\ T(G_{o_1}||^o G_{o_2})_{n_3} + T(G_{f_1})_{n_2,n_2} + w_C(j_1, f_1), & \\ \quad \text{where } x_{j_1} = n_3 \text{ and } x_{f_1} = n_2 & (3) \\ T(G_{o_1}||^o G_{o_2})_{b} + T(G_{f_1})_{n_2,n_2} + w_C(j_1, f_1), & \\ \quad \text{where } x_{j_1} = b \text{ and } x_{f_1} = n_2 & (4) \end{array}$$

Here $t(G) = t(G_{f_1})$ such that $\hat{f}^{\mathbb{S}}_{G_{f_1}}(x_{s(G_{f_1})}, x_{t(G)}) < \infty$ only when $x_{s(G_{f_1})} = x_{t(G_{f_1})} = x_{t(G)} = n_2$ (see Fig. 4 (c)), and therefore in the above we do not consider the values computed from $T(G_{f_1})_{z,n_2}$ with $z \neq n_2$ as they cannot be minimal. From Fig. 4 (b) we know that the value of $w_C(j_1, f_1)$, is 21 and is added when $x_{j_1} \neq x_{f_1}$ as we assume that $w_D(x_{j_1}, x_{f_1}) = 1$ when $x_{j_1} \neq x_{f_1}$ and 0 otherwise. The minimal value of 36 is given by (2).

The tables also contain the information on optimal assignments. After constructing all tables, we reconstruct the solution recursively from the tables in the following way. Given a composite G in the graph, with an element of interest in the table $T(G)$ (this value is the optimal value when G is the top-level program), we can compute the placements for the input/output operators of the immediate children from the definition of the dynamic program. For example, when $G = G_1 *^o G_2$, the following holds from the dynamic program (Definition 1):

$$\begin{aligned} & T(G)_{x_{t(G)}} \\ = \ & \min_{x_{t(G_1)}, x_{s(G_2)}} \left\{ \begin{array}{l} T(G_1)_{x_{t(G_1)}} + T(G_2)_{x_{s(G_2)}, x_{t(G)}} + \\ w_C(t(G_1), s(G_2)) \cdot w_D(x_{t(G_1)}, x_{s(G_2)}) \end{array} \right\} \end{aligned}$$

Therefore, we compute the placement $x_{t(G_1)}$, $x_{s(G_2)}$, by enumerating their possible value pairs (for which there are $|N|^2$) and for each value pair, see if the rhs sub-expression $T(G_1)_{x_{t(G_1)}} + T(G_2)_{x_{s(G_2)}, x_{t(G)}} + w_C(t(G_1), s(G_2)) \cdot w_D(x_{t(G_1)}, x_{s(G_2)})$ equals $T(G)_{x_{t(G)}}$. We further explain this method via example. In Fig. 2, the optimal cost of 70 for the whole program is given in the table $T((((G_{o_1}||^o G_{o_2}) *^o G_{f_1})||^o G_{o_3}) *^o G_b)$. We indicate the assignment with a shaded element. This element is computed using the shaded elements in the table for $T(((G_{o_1}||^o G_{o_2}) *^o G_{f_1}) *^o G_{o_3})$ and $T(G_b)$ (not shown), and this value is further computed from the shaded cell in the table $T((G_{o_1}||^o G_{o_2}) *^o G_{f_1})$, and so on. After marking the values, the optimal placement can thus be inspected from the tables, which is $o_1 \mapsto n_1$, $o_2 \mapsto n_2$, $o_3 \mapsto n_3$, $j_1 \mapsto n_2$, $f_1 \mapsto n_2$, $j_2 \mapsto n_3$, and $c \mapsto b$. The placement is partially shown by the shaded elements of Fig. 4, and is indicated in Fig. 1 (b), with nodes enclosed in rectangle, hexagon, circle, and pentagon respectively indicate the placement of the operators to node n_1, n_2, n_3, and b.

4.3 Complexity Bound

We now discuss the complexity bounds of the algorithm based on the dynamic program. In the following discussions, $m = |V|$ is the number of operators of the composite under consideration, and $n = |N|$ is the number of sensor nodes. There are two important procedures in the algorithm:

1. construction of tables for each composite, and
2. reconstruction of optimal solution from the tables.

The following theorem holds.

THEOREM 3. *Given a compositional stream graph G and with operators V to be placed on nodes N, there is an algorithm to compute optimal placement with time complexity in $O(mn^4)$.*

$x_{s(G)} \backslash x_{t(G)}$	n_1	n_2	n_3	b
n_1	10	∞	∞	∞
n_2	∞	16	∞	∞
n_3	∞	∞	26	∞
b	∞	∞	∞	41

(a) $T(G_{j_1})$

$x_{s(G)} \backslash x_{t(G)}$	n_1	n_2	n_3	b
n_1	9	∞	∞	∞
n_2	∞	20	∞	∞
n_3	∞	∞	16	∞
b	∞	∞	∞	23

(b) $T(G_{j_2})$

$x_{s(G)} \backslash x_{t(G)}$	n_1	n_2	n_3	b
n_1	7	∞	∞	∞
n_2	∞	15	∞	∞
n_3	∞	∞	28	∞
b	∞	∞	∞	36

(c) $T(G_{f_1})$

$x_{t(G)}$	n_1	n_2	n_3	b
	0	∞	∞	∞

(d) $T(G_{o_1})$

$x_{t(G)}$	n_1	n_2	n_3	b
	23	21	44	59

(e) $T(G_{o_1} ||^o G_{o_2})$

$x_{t(G)}$	n_1	n_2	n_3	b
	49	36	70	78

(f) $T((G_{o_1} ||^o G_{o_2}) *^o G_{f_1})$

$x_{t(G)}$	n_1	n_2	n_3	b
	62	66	59	76

(g) $T(((G_{o_1} ||^o G_{o_2}) *^o G_{f_1}) ||^o G_{o_3})$

$x_{t(G)}$	n_1	n_2	n_3	b
	∞	∞	∞	70

(h) $T((((G_{o_1} ||^o G_{o_2}) *^o G_{f_1}) ||^o G_{o_3}) *^o G_b)$

Figure 4: Memo Tables for Operators and Composites (see Fig. 1(b))

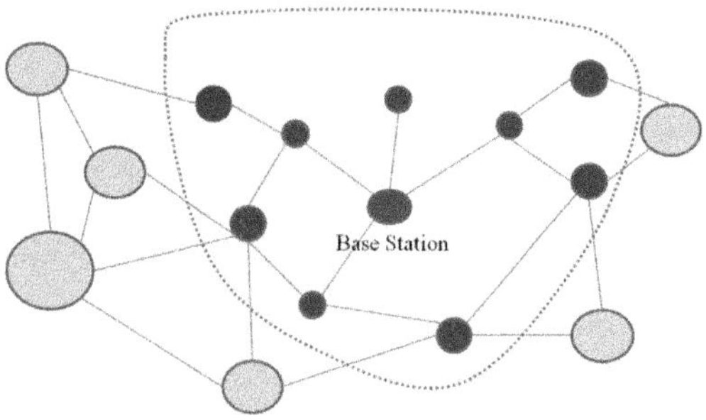

Figure 5: Sub-Network Two Hops Away from Base Station

The detailed proof is presented in the appendix Section A.3. Here we present the outline. The more expensive of both of the above steps is Step 1, where the computation cost is due to: a) generating $O(m)$ tables, one for each composite, and, b) computing the elements of each table. For each table, there are at most n^2 elements to be computed, and the computation requires examining the tables of the substreams, a process in $O(n^2)$. Therefore, the whole process is executed in $O(m \cdot n^2 \cdot n^2) = O(mn^4)$.

4.4 Mitigating Complexity by Locality

The time complexity of the dynamic program increases quartically in the number of sensors, which makes its running time prohibitively long for large networks. To mitigate this problem, we restrict its search space by creating a sub-network of the sensor network that contains sensor nodes executing the sense operators, the base station, and nodes that are in its proximity. The size of the sub-network is smaller and hence improves the running time of the dynamic program, however, the placement of the operators becomes restricted thus affecting its energy efficiency.

For creating the sub-network, we use a heuristic that selects nodes that are in proximity of the base station. Fig. 5 illustrates an example network with 16 nodes including a base station. In the figure, the size of the node indicates its distance from the base station, i.e., bigger nodes are further away from the base station. Our simple heuristic chooses a sub-network whose nodes are at most two hops away from the base station, indicated in Fig. 5 by nine nodes enclosed within a dotted shape. We increase or decrease the size of the sub-network by setting the distance parameter, i.e., number of hops from the base station.

5. EXPERIMENTAL EVALUATION

We evaluate our techniques for the Migrating Operator Placement problem (MOPP) as follows:

- Is dynamic programming more efficient than Integer Linear Programming (ILP) for MOPP ?
- What is the energy efficiency of the computed optimal placement in comparison with data forwarding?
- How does a restricted local search space affect speed and quality of the placement?

We employed simulation to answer the questions above. We implemented the simulator in Python, that generated sensor networks and stream queries, and ran the simulation in a fixed time step mode. In each time step the following events may be triggered: adding a query, migrating a query, deleting a query, adding a sensor node, and removing a sensor node. Activities including communication, execution, and migration of operators consume energy at the sensor nodes. Therefore, in each time step, the simulator adjusts the energy level of each sensor node depending on the activities in the time step and the running overheads of the sensor.

For our experiments we use instances of wireless sensor networks that are generated from the Erdös-Rényi graph model $G(n, p)$ [6], where n indicates the number of sensor or mobile nodes; and p is the probability with which each edge is included in the graph. This model is simple and widely used for mathematical analysis to generate network graphs with varying degree of connectivity [5, 9]. In our experiments we have generated the network graphs with $p = 0.5$.

User queries are introduced and deleted throughout the simulation. A query is generated by a probabilistic scheme: We assign a composite (cf. Fig. 3) a probability (i.e. 25% for sense and filter, and 10% for other composites). The composition tree of a query is recursively constructed. The composite of a node in the composition tree is chosen randomly according to its probability and the construction ends as soon as all leaves in the composition tree are sense operators (cf. Fig. 3(d)).

We run our simulator and the algorithm for MOPP on a Core2 Duo 2.99 GHz Intel machine with 4 GB RAM. The simulator and placement algorithm are implemented in Python.

5.1 ILP vs. Dynamic Programming

We formulated the MOPP as an AMPL program [7], and compared the running time of the dynamic program (cf. Section 4)

Operators	Nodes	ILP (s)	DP (s)
4	1	0.1	0.029
8	2	0.1	0.054
11	3	3.3	0.112
14	4	176.2	0.371
17	5	*	1.165
20	6	*	3.173
23	7	*	7.659
26	8	*	16.291
29	9	*	32.088
32	10	*	60.234

Table 1: GLPK and Dynamic Program Running Times

Operators	Nodes	cost(DP)	cost(Fwd)	DP (s)
2	2	39	39	0
6	2	95	121	0
8	3	208	296	0.015
12	5	428	225	0.031
17	3	395	584	0.016
18	8	215	517	3.5
20	6	295	579	1.64
26	13	470	844	72.782
38	10	669	1397	53.218
56	10	849	1782	60.891
57	12	815	1790	86.672

Table 2: Optimal In-Network vs. Data Forwarding

with the running time of the integer linear programming solver GLPK [8] in solving our AMPL program. Table 1 shows the results, where **Operators** are the total number of operators in a given set of queries, **Sensors** are the numbers of sensor nodes, **ILP** are the running times for GLPK runs in seconds (* represents a run that took more than two hours), and **DP** are the running times for the dynamic program in seconds. The time taken to execute the dynamic program was substantially less than the time to execute GLPK. As shown in the table, the difference became more pronounced with larger problem sizes.

5.2 Optimal In-Network vs. Data Forwarding

Our algorithm (cf. Section 4) cost-optimally places operators on sensor nodes. The resulting placement is that of an *in-network* processing scheme, i.e., the algorithm attempts to reduce communication overheads by largely placing operators within the network. An alternative scheme is that of *data forwarding*, which places all operators except the sense operations on the base station, assuming that the base station's energy resources are unconstrained.

In Table 2, we compare the energy costs of the in-network and the data forwarding schemes. The column **Operators** gives the number of operators in a query, and column **Nodes** gives the numbers of the sensor nodes in the network. The energy costs for the in-network processing scheme are shown in column **cost(DP)** and the data forwarding costs are shown in column **cost(Fwd)**. With smaller network and query sizes, the cost incurred by the in-network placement was similar to data forwarding. The data forwarding scheme became less efficient with bigger queries and sensor networks. In our experiment, the in-network processing scheme was up to 54% more energy-efficient (last row of Table 2) and therefore well suited for energy-constrained query processing.

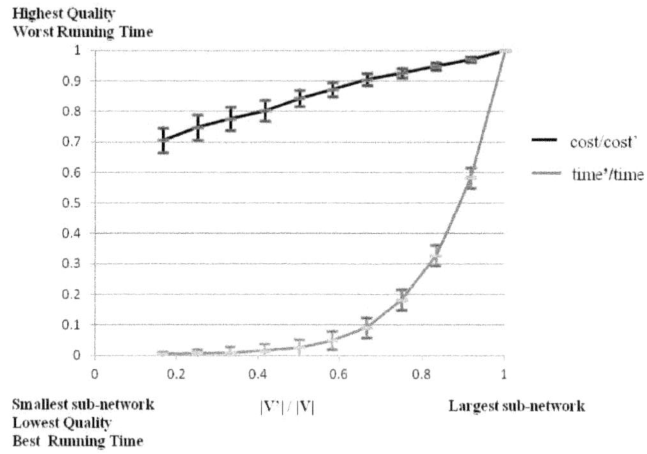

Figure 6: Locality Heuristic vs. Optimal In-Network

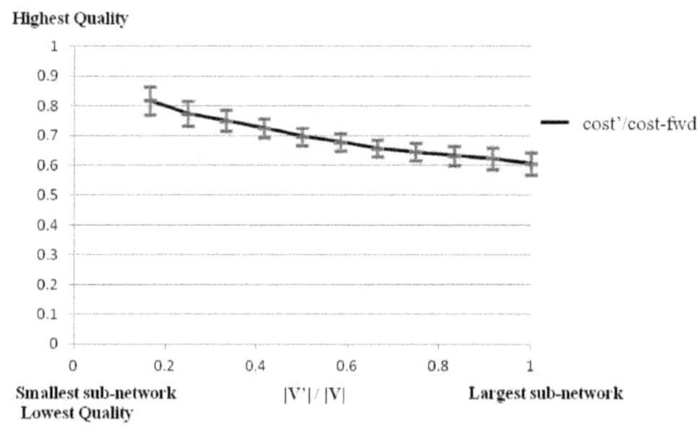

Figure 7: Locality Heuristic vs. Data Forwarding

5.3 Locality of Placement

As shown in Table 2, the running time of the dynamic program increases rapidly with the number of sensor nodes. For large sensor networks, the algorithm's running time becomes prohibitive. To mitigate this issue, we propose a locality strategy (cf. Section 4.4) which restricts the placement to a sub-network. The restriction impacts the placement quality resulting in less than optimal energy efficiency. To evaluate this approach, we computed the running times and in-network processing costs for the various sizes of the sub-network and compare them with the optimal value, i.e., the result when considering all nodes in the network.

Fig. 6 shows the effect of locality with 95% confidence intervals. The data were collected from experiments using 35 pairs of randomly-generated query and network topology (see Section 4.4), with network size of 12 sensor nodes. In the figure, we represent the fraction of the size (in nodes) of the sub-network to the whole network as $|V'|/|V|$. $cost/cost'$ values are the comparison of the computed optimal energy cost to the energy cost incurred by the sub-network, as a measure of quality of the locality heuristic. The in-network cost moves near to optimal as the size of the sub-network increases towards the actual size of the network. $time'/time$ values compare the execution time for the dynamic program on sub-network to that of the whole network. The result allows us to select the size of the sub-network using which we can compute the nearly optimal placement of the query operators in relatively short time. For our experimental data, a sub-network with the size of 66% of the actual network computes the placement in less than 10% of the running time and has a good placement quality that is already about 90% of the optimal placement.

Fig. 7 shows the comparison between the cost of our in-network processing with locality heuristic and the cost incurred due to data forwarding. We observe that the worst results using locality were still significantly better than that of data forwarding. We therefore conclude that optimized in-network processing is better than data forwarding even with the locality heuristic.

6. RELATED WORK

Implementing cloud computing on mobile devices stems from researches in *wireless sensor network* (*WSN*). Our research concerns *macro programming* abstraction, typically based on a declarative programming approach where the user submits high-level queries to the network. Existing macro-programming systems are targeted at resource-constrained embedded devices, and are therefore designed to preserve resources and handle faults. Their typically employ in-network processing, i.e. data aggregation/computing on the sensor nodes to minimize data transfers. This design is adopted in query-based WSN programming systems such as TinyDB [18], Cougar [25], Corona [20, 13], MaD-WiSe [1], MONET [26], [19, 21, 27], and our previous work on Cascade system [12].

TinyDB, Cougar, and Corona enable the user to indicate where, when and how often data should be physically sampled and delivered to save energy. TinyDB automatically reorders operators such that operators with higher data transfer rate are executed earlier and on the same node reducing data transfer over the network. It also batches events instead of executing query on every arrival of an event. Cougar provides a combination of a declarative query system with in-network computation. Corona allows the user to specify a *freshness* constraint on queries, which state the time limit of old sensor data to be reusable for the current query. This improves cache hit and minimizes the energy cost. These works, however, do not consider query optimization.

Similar to our approach, MaD-WiSe, MONET, [19, 21, 27], and Cascade are based on query plan optimization. MaD-WiSe is a declarative query system that uses an algebraic optimization approach that is based on transformation of a query execution plan into a semantically equivalent plan with lower cost. Their optimizer applies a greedy approach to choose the operator ordering based on various criteria that gives the best performance. In MONET, communications over multiple single-hop wireless channels are first modeled as additional linear constraints of an Integer Linear Programming (ILP) problem which is solved by a heuristic. In [19], the authors provide a model for the task-mapping problem for both energy balance and total energy spent. They suggest Mixed Integer Programming (MIP) formulations which gives optimal results with long run-time, coupled with a greedy heuristic. Srivastava et al. [21] consider stream query operator placement problem of tree-like network consisting of sequences of filters and joiners (links). They provide an algorithm with execution time in $O((m+n)\log(m+n))$ for the optimal operator placement, with m the number of tasks (filters) and n the number of nodes. Zhang et al. [27] present a cubic time algorithm to optimize *utilization* (data rate) under power level constraints, taking advantage of the tree structure of the network. Compared to the works of Srivastava et al. and Zhang et al., ours considers more complex graphs involving split-join structures together with operator migration while providing a polynomial time algorithm. We are not aware of any other work that provides a better complexity bound for the class of problem that we consider. Our work extends the approach taken by the Cascade system [12] by considering operator migration and using polynomial-time algorithm instead of an approximation heuristic applied to the general graph.

The *MOPP* introduced in paper is related to Stone's problem [22] that allocates a set of jobs on two nodes by reducing the problem to an s-t min-cut. The metric labeling problem [14] can be seen as a generalization of Stone's problem to an arbitrary number of nodes. The general metric labeling is NP-hard and good approximation algorithms are provided which rely on a series of s-t min-cuts to solve the problem [3]. This is the approximation heuristic used in the Cascade system [12].

7. CONCLUSION

In this paper we present the migrating operator placement problem (MOPP) on wireless sensor networks and mobile clouds. The model is applicable to stream graphs, where the system consists of computational entities called operators and communication channels between them. The objective is to minimize the energy costs of computation, communication, and transition due to the changes in the stream graph or the migration of operators. We devise an algorithm based on dynamic programming that computes optimal operator placement on the network with running time polynomially bounded to the input graph size for an important subclass of the stream graphs. To improve our algorithm, we consider a heuristic that only considers nodes with close proximity to the base station. We present an experimental evaluation that confirms the efficiency of our approach by comparing it with integer linear programming. Using a simulator, we demonstrate the effectiveness of the optimal placement by comparing the computed energy usage with that of data forwarding. The results show that our approach is 54% more energy-efficient. Our heuristic significantly improves the running time of the dynamic programming algorithm and computes nearly optimal placement in less than 10% of the running time.

Acknowledgments

This work is supported by Australian Research Council (ARC) grant *Computing with Nearly Consistent Data* no. DP 102-96-445. We also thank Thomas Martin Gawlitza for his comments.

8. REFERENCES

[1] G. Amato, S. Chessa, and C. Vairo. MaD-WiSe: a distributed stream management system for wireless sensor networks. *Softw. Pract. Exper.*, 40(5):431–451, Apr. 2010.

[2] P. Angin, B. K. Bhargava, and S. Helal. A mobile-cloud collaborative traffic lights detector for blind navigation. In T. Hara, C. S. Jensen, V. Kumar, S. Madria, and D. Zeinalipour-Yazti, editors, *11th MDM*, pages 396–401. IEEE Computer Society, 2010.

[3] Y. Boykov, O. Veksler, and R. Zabih. Fast approximate energy minimization via graph cuts. *IEEE Trans. Pattern Anal. Mach. Intell.*, 23(11):1222–1239, 2001.

[4] E. Dahlhaus, D. S. Johnson, C. H. Papadimitriou, P. D. Seymour, and M. Yannakakis. The complexity of multiterminal cuts. *SIAM J. Comput.*, 23(4):864–894, 1994.

[5] W. Du, J. Deng, Y. S. Han, P. K. Varshney, J. Katz, and A. Khalili. A pairwise key predistribution scheme for wireless sensor networks. *ACM Trans. Inf. Syst. Secur.*, 8(2):228–258, May 2005.

[6] P. Erdös and A. Rényi. On the evolution of random graphs. In *Publication of the Mathematical Institute of the Hungarian Academy of Sciences*, pages 17–61, 1960.

[7] R. Fourer, D. M. Gay, and B. W. Kernighan. *AMPL: A Modeling Language for Mathematical Programming*. Brooks/Cole, 2003. 2nd Edition.

[8] GLPK (GNU linear programming kit). http://www.gnu.org/s/glpk/, 2008.

[9] A. Hamlili. Adaptive schemes for estimating random graph parameters in mobile wireless ad hoc networks' modeling. In *3rd Wireless Days*, pages 1–5. IEEE, 2010.

[10] F. Ingelrest, G. Barrenetxea, G. Schaefer, M. Vetterli, O. Couach, and M. Parlange. Sensorscope: Application-specific sensor network for environmental monitoring. *TOSN*, 6(2), 2010.

[11] W. M. Johnston, J. R. P. Hanna, and R. J. Millar. Advances in dataflow programming languages. *ACM Comput. Surv.*, 36:1–34, March 2004.

[12] V. Kakkad, A. E. Santosa, and B. Scholz. Cascade: A stream programming environment for wireless sensor networks. Under submission, 2012.

[13] R. Khoury, T. Dawborn, B. Gafurov, G. Pink, E. Tse, Q. Tse, K. Almi'Ani, M. M. Gaber, U. Röhm, and B. Scholz. Corona: Energy-efficient multi-query processing in wireless sensor networks. In *DASFAA (2)*, volume 5982 of *LNCS*, pages 416–419. Springer, 2010.

[14] J. Kleinberg and E. Tardos. Approximation algorithms for classification problems with pairwise relationships: metric labeling and markov random fields. *J. ACM*, 49:616–639, September 2002.

[15] T. D. Le, N. Ahmed, and S. Jha. Location-free fault repair in hybrid sensor networks. In I. Chlamtac, editor, *1st InterSense*, volume 138 of *ACM International Conference Proceeding Series*, page 23. ACM, 2006.

[16] Y. Lee, S. S. Iyengar, C. Min, Y. Ju, S. Kang, T. Park, J. Lee, Y. Rhee, and J. Song. Mobicon: a mobile context-monitoring platform. *Commun. ACM*, 55(3):54–65, Mar. 2012.

[17] X. Li, H. Zhang, and Y. Zhang. Deploying mobile computation in cloud service. In M. G. Jaatun, G. Zhao, and C. Rong, editors, *1st CloudCom*, volume 5931 of *LNCS*, pages 301–311. Springer, 2009.

[18] S. Madden, M. Franklin, J. Hellerstein, and W. Hong. TinyDB: An acquisitional query processing system for sensor networks. *ACM Trans. Database Systems*, 30(1):122–173, 2005.

[19] A. Pathak and V. K. Prasanna. Energy-efficient task mapping for data-driven sensor network macroprogramming. *IEEE Trans. Computers*, 59(7):955–968, 2010.

[20] B. Scholz, M. M. Gaber, T. Dawborn, R. Khoury, and E. Tse. Efficient time triggered query processing in wireless sensor networks. In *ICESS*, volume 4523 of *LNCS*, pages 391–402. Springer, 2007.

[21] U. Srivastava, K. Munagala, and J. Widom. Operator placement for in-network stream query processing. In *24th PODS*, pages 250–258. ACM, 2005.

[22] H. S. Stone. Multiprocessor scheduling with the aid of network flow algorithms. *IEEE Trans. Software Eng.*, 3(1):85–93, 1977.

[23] W. Thies, M. Karczmarek, and S. P. Amarasinghe. StreamIt: A language for streaming applications. In *CC*, volume 2304 of *LNCS*, pages 179–196. Springer, 2002.

[24] G. Werner-Allen, K. Lorincz, M. Welsh, O. Marcillo, J. Johnson, M. Ruiz, and J. Lees. Deploying a wireless sensor network on an active volcano. *IEEE Internet Computing*, 10(2):18–25, 2006.

[25] Y. Yao and J. Gehrke. The Cougar approach to in-network query processing in sensor networks. *ACM SIGMOD Record*, 31(3):9–18, 2002.

[26] Y. Yu and V. K. Prasanna. Energy-balanced task allocation for collaborative processing in wireless sensor networks. *MONET*, 10(1-2):115–131, 2005.

[27] B. Zhang, R. Simon, and H. Aydin. Maximum utility rate allocation for energy harvesting wireless sensor networks. In A. Helmy, B. Landfeldt, and L. Bononi, editors, *14th MSWiM*, pages 7–16. ACM, 2011.

APPENDIX

A. PROOFS

A.1 Proof of Theorem 1

We show the NP-hardness by reducing the multi-way cut problem (see [4]) to MOPP. The input of a multi-way cut problem is an undirected graph $G^{MC} = (V^{MC}, E^{MC})$ and a terminal set $T = \{t_1, \ldots, t_l\} \subseteq V^{MC}$. A multi-way cut is a partition of V^{MC} into disjoint sets $R_1, \ldots, R_l$ $\left(\bigcup_i R_i = V^{MC}\right)$ such that terminal $t_i \in T$ is contained in R_i for all i. We denote by $\delta(R_i) = |(E^{MC} \cap R_i) \times (V^{MC} \setminus R_i)|$ the number of cut-edges that separate the disjoint set R_i from nodes outside the set. The goal of the multi-way cut is to minimize the number of cut-edges $\frac{1}{2}\sum_i \delta(R_i)$.

We map multi-way cut to an instance of MOPP such that the vertex set V^{MC} becomes set V, the edges E^{MC} become the channels E, and the terminals T correspond to the set of sensor nodes N, i.e., $N = V_{si} \cup V_{se}$. In the instance of MOPP, we assume in addition that the placement of the operators are unchanged such that transition costs are zero, and w_{TP} represents placement cost as follows.

$$w_{TP}(u, t_i) = \begin{cases} \infty, & \text{if } u \in T \wedge t_i \neq u \\ 0, & \text{otherwise,} \end{cases}$$

Note that the mapping $p : V \to N$ partitions the domain V, i.e., $R_i = \{u | x_u = t_i\}$ for all i, and $\bigcup_i R_i = V$. We also assume that for any $u, v \in V$, $w_C(u, v) = 1$. Further, $w_D(x_u, x_v) = 1$ when $x_u \neq x_v$ and 0 otherwise. By rewriting the objective of the *MOPP*, we obtain:

$$\begin{aligned} f_G(X_G) &= \sum_{u \in V} w_{TP}(u, x_u) + \sum_{(u,v) \in E} w_C(u, v) \cdot w_D(x_u, x_v) \\ &= \sum_i \begin{cases} \infty, & \text{if } x_{t_i} \neq t_i \\ 0, & \text{otherwise} \end{cases} + \sum_{(u,v) \in E} \begin{cases} 1, & \text{if } x_u \neq x_v \\ 0, & \text{otherwise} \end{cases} \end{aligned}$$

The first summand is a dualized constraint $x_{t_i} = t_i$, i.e., the function $f_G(X_G)$ is less than ∞ if $x_{t_i} = t_i$ for all i. Since, $x_{t_i} = t_i$, the terminal t_i is contained in R_i, and

$$\begin{aligned} \sum_{(u,v) \in E} \begin{cases} 1, & \text{if } x_u \neq x_v \\ 0, & \text{otherwise} \end{cases} &= \sum_{(u,v) \in E} (\exists i : (u, v) \in R_i \times (V \setminus R_i)) \\ &= \sum_i |(E \cap R_i) \times (V \setminus R_i)| = \sum_i \delta_i(R_i). \end{aligned}$$

A.2 Proof of Theorem 2

Theorem 2 is immediate from the following lemma.

LEMMA 1. *$\hat{f}^{\mathbb{S}}_G(x_{s(G)}, x_{t(G)}) = \min_{X_G \setminus \{x_{s(G)}, x_{t(G)}\}} f_G(X_G)$ when $G \in \mathbb{S}$ and $\hat{f}^{\mathbb{O}}_G(x_{t(G)}) = \min_{X_G \setminus \{x_{t(G)}\}} f_G(X_G)$ when $G \in \mathbb{O}$.*

PROOF. We prove by induction. Here we consider one base case when $G = G_u \in \mathbb{S}$ and $u \in V_{si}$, and one inductive case for $G = G_1||^o G_2$ with joiner j. Other cases can be proven similarly.

When $G = G_u \in \mathbb{S}$ and $u \in V_{si}$, then necessarily $x_{s(G)} = x_{t(G)} = x_u = b$. Now, $X_G \setminus \{x_{s(G)}, x_{t(G)}\}$ is empty, and therefore, $\hat{f}^{\mathbb{S}}_G(x_{s(G)}, x_{t(G)}) = w_{TP}(u, x_u) = f_{G_u}(x_u) = \min_{X_G \setminus \{x_{s(G)}, x_{t(G)}\}} f_G(X_G)$.

When $G = G_1||^o G_2 = (V, E, j)$, we consider that $G_1 = (V_1, E_1, t(G_1))$, $G_2 = (V_2, E_2, t(G_2))$, $G_j = (\{j\}, \emptyset, j, j)$, and $G = (\{j\} \cup V_1 \cup V_2, \{(t(G_1), j), (t(G_2), j)\} \cup E_1 \cup E_2, j)$. We assume inductively that

$$\hat{f}^{\mathbb{S}}_{G_j}(x_j, x_j) = \min_{X_{G_j} \setminus \{x_j\}} f_{G_j}(X_{G_j}) = f_{G_j}(x_j),$$
$$\hat{f}^{\mathbb{O}}_{G_1}(x_{t(G_1)}) = \min_{X_{G_1} \setminus \{x_{t(G_1)}\}} f_{G_1}(X_{G_1}), \text{ and}$$
$$\hat{f}^{\mathbb{O}}_{G_2}(x_{t(G_2)}) = \min_{X_{G_2} \setminus \{x_{t(G_2)}\}} f_{G_2}(X_{G_2})$$

Substituting the rhs of the above equations for the lhs in Definition 1 we get

$$\begin{aligned}
\hat{f}_G(x_{s(G)}, x_{t(G)}) = \; & \min_{X_{G_1} \cup X_{G_2}} f_{G_j}(x_j) + f_{G_1}(X_{G_1}) + f_{G_2}(X_{G_2}) + \\
& w_C(t(G_1), j) \cdot w_D(x_{t(G_1)}, x_j) + \\
& w_C(t(G_2), j) \cdot w_D(x_{t(G_2)}, x_j). \\
= \; & \min_{X_G \setminus \{x_j\}} \textstyle\sum_{u \in V} w_{TP}(u, x_u) + \\
& \textstyle\sum_{(u,v) \in \{(t(G_1), j), (t(G_2), j)\} \cup E_1 \cup E_2} w_C(u, v) \cdot w_D(x_u, x_v) \\
= \; & \min_{X_G \setminus \{x_{t(G)}\}} f_G(X_G)
\end{aligned}$$

□

A.3 Proof of Theorem 3

There are three important procedures in the algorithm, executed according to the following sequence:

1. construction of tables for each composite, and
2. reconstruction of optimal solution from the tables.

We first discuss several lemmas that specify the complexity bounds of each of the above procedures. Here $|V| = m$ and $|N| = n$.

To reason about the complexity bounds of steps 1 and 2, we first state a bound on the number of tables constructed by the algorithm.

LEMMA 2. *The number of tables constructed is in $O(m)$.*

PROOF. Here we first prove that for every composite $G = (V, E, g, h)$ or $G = (V, E, h)$, the number of generated tables is $|V| + b$, with $0 \leq b < |V|$.

In the base case where $|V| = 1$, the number of table is 1, which obviously satisfies the invariant for $b = 0$.

Next we show the proofs for when $G = G_1 *^o G_2$ and $G = G_1||^s G_2$ only, as other cases can be proven similarly.

In case $G = G_1 *^o G_2$, with $G_1 = (V_1, E_1, h_1)$ and $G_2 = (V_2, E_2, g_2, h_2)$, we assume the number of tables for G_1 and G_2 are respectively $|V_1| + b_1$ and $|V_2| + b_2$ where $0 \leq b_1 < |V_1|$ and $0 \leq b_2 < |V_2|$. We add another table for the composite such that the total number of table is $|V_1| + |V_2| + b_1 + b_2 + 1 = |V| + b_1 + b_2 + 1$ since $|V| = |V_1| + |V_2|$. Here, $0 \leq b_1 + b_2 + 1 < |V|$.

In case $G = G_1||^s G_2$, with $G_1 = (V_1, E_1, g_1, h_1)$ and $G_2 = (V_2, E_2, g_2, h_2)$, we assume the number of tables for G_1 and G_2 are respectively $|V_1| + b_1$ and $|V_2| + b_2$ where $0 \leq b_1 < |V_1|$ and $0 \leq b_2 < |V_2|$. Here we introduce two more actors: the splitter and the joiner such that $|V| = |V_1| + |V_2| + 2$. We also add three more tables, one table each for the splitter and the joiner, and another table for the whole composite, and therefore the number of tables is $|V_1| + |V_2| + 2 + b_1 + b_2 + 1 = |V| + b_1 + b_2 + 1$. Here, $0 \leq b_1 + b_2 + 1 < |V|$.

Since for any composite $G = (V, E, g, h)$ or $G = (V, E, h)$ the number of tables is $|V| + b$ for some $0 \leq b < |V|$, the number of tables is bounded by $2|V|$, hence in $O(m)$. □

The construction of each table itself is in $O(n^4)$, and therefore the following lemma holds.

LEMMA 3. *In placing a program G with operators V on nodes N, the minimal cost can be computed in $O(mn^4)$.*

PROOF. We first demonstrate that at each table construction step, the computation time required is in $O(n^4)$, with $n = |N|$. Here we consider two cases of composites.

- If G is an operator, the table construction involves considering just the table size, which is n^2.
- In case $G = G_1 *^s G_2$, to compute the value of a table entry $T(G)_{x_{s(G)}, x_{t(G)}}$, we need to consider all possible combinations for $x_{t(G_1)}$ and $x_{s(G_2)}$. Their number is in $O(n^2)$, and since there are n^2 entries in the table to be constructed, the table construction has a complexity in $O(n^4)$.

Although the proof for all cases are not shown here, the case when $G = G_1 *^s G_2$ has the worst complexity bound, and therefore the table construction has a complexity bound in $O(n^4)$. Since from Lemma 2 the number of tables constructed is in $O(m)$ with m the number of operators of a program, the complexity of table construction is in $O(mn^4)$. □

Lastly, the reconstruction of the optimal solution from the tables (discussed in Section 4.2) is in $O(mn^2)$.

LEMMA 4. *Given a program G with operators V to be placed on N, given minimal placement cost, the placement can be computed in $O(mn^2)$.*

PROOF. We first reason about the complexity bound for recovering a solution from the substream tables, given the optimal value in the table of a composite G. For this, we assume that G is an arbitrary composite, but for lack of space we only consider two of them. First, in case $G = G_u = (\{u\}, \emptyset, u, u)$, then we already have placements $x_{s(G)}$ and $x_{t(G)}$, and therefore the recovery terminates in $O(1)$. In case $G = G_1 *^o G_2$, then we need to choose the values for $x_{t(G_1)}$ and $x_{s(G_2)}$ such that $T(G)_{x_{t(G)}} = \min_{x_{t(G_1)}, x_{s(G_2)}} T(G_1)_{x_{t(G_1)}} + T(G_2)_{x_{s(G_2)}, x_{t(G)}} + w_C(t(G_1), s(G_2)) \cdot w_D(x_{t(G_1)}, x_{s(G_2)})$ (see Definition 1). This process examines the possible placement pairs for $t(G_1)$ and $s(G_2)$ for which there are n^2, and therefore the execution time is in $O(n^2)$, which is also the worst possible running time bound for any composite. Since from Lemma 2 the number of tables is in $O(m)$, the complexity bound for recovering the solution from the memo tables is in $O(mn^2)$. □

From Lemmas 3 and 4, the most complex of the three procedures mentioned above is table construction, hence Theorem 3.

A Novel Sensor Node Relocation Approach to Maintain Connectivity with a Center of Interest

Inès El Korbi
National School of Computer Science, University of Manouba, Tunisia,
ines.korbi@ensi.rnu.tn

Chayma Zidi
National School of Computer Science, University of Manouba, Tunisia,
chayma.zidi@gmail.com

Sherali Zeadally
Department of Computer Science and Information Technology, University of the District of Columbia, Washington DC, USA
szeadally@udc.edu

Leila Azouz Saidane
National School of Computer Science, University of Manouba, Tunisia,
leila.saidane@ensi.rnu.tn

ABSTRACT

We propose a new technique to relocate a set of redundant sensor nodes from their initial deployment area called ROI (region of interest) to a new location outside the ROI called COI (center of interest) where a new event happened (e.g., to measure the soil contamination level in the case of a chemical spill disaster). Relocating mobile sensor nodes from their initial deployment area ROI requires covering that area at a given center of interest while maintaining connectivity with the initial region of interest. Our proposed technique to relocate redundant mobile sensors nodes towards a new target position is a four-phase solution that includes the clustering phase and the cell head election, the evaluation of the number of nodes to be relocated, the propagation of the COI position, and finally the election of the redundant nodes and their relocation. A performance evaluation of our novel relocation approach shows that the consumed energy and the relocation time increase as the distance separating the initial ROI from the COI increases regardless of the size or the density of the initial deployment area.

Categories and Subject Descriptors

C.2.1 [**Network Architecture and Design**]: Wireless communication

General Terms

Algorithms, Performance

Keywords

Energy, networking, performance, sensors, wireless.

1. INTRODUCTION

A wireless sensor network consists of a large number of small, low-cost, low-power and mobile sensor nodes that are able to observe their environment. Due to these attractive characteristics, sensor networks have been deployed in many applications including military applications, smart homes, surveillance, and remote environment control. These sensor nodes are massively deployed in a region of interest (ROI) to collect data and process it [1]. Many applications such as monitoring remote harsh fields or toxic regions where human intervention is unsafe or burdensome require the mobility and self-management of sensor nodes. The motion capability is also required if an event outside the initial deployment area happened and have to be covered. The sensor nodes can then move to ensure the coverage of that new center of interest (COI). For example, if a chemical spill disaster occurs in a chemical plant, the noxious substance may propagate outside of the initial area of the chemical plant already controlled by a wireless sensor network. Hence, if we need to measure the soil contamination degree at a given point outside that initial zone, we would have relocate some sensor nodes to that center of interest to report such an information to the sink node.

The method used to relocate sensor nodes is a critical issue because it affects both the cost and the detection capability of a wireless sensor network. In this paper we investigate the problem of sensor relocation to respond to an event that needs a sensor node to be moved to its location to capture information and send this information back to the sink node. Additional nodes have to be relocated to ensure communication with the initial ROI in order to report information to the sink node. The sensor relocation problem differs slightly from the traditional sensor node deployment problem. In the case of the traditional sensor deployment problem, the initially deployed sensor nodes are moved within the coverage area either to avoid coverage holes, or to minimize the whole network energy consumption or to ensure both coverage and connectivity within the network [14], [19], [4]. In contrast, with the sensor relocation problem, we assume that both the coverage and the connectivity are already satisfied in the sensor network and the challenge is to satisfy the requirements associated with node relocation. The first requirement is the choice of the sensor nodes to be moved because the prospective relocated nodes may be used by some application and their movements may damage the sensing or communication tasks they are performing. The second major requirement aims to minimize the network energy consumption caused by movement because the latter is more expensive in terms of energy than communication and computation.

This rest of the paper is organized as follows. In section 2, we review related works on sensor node relocation. In section 3, we present the fundamental criteria that have to be satisfied with our approach especially the coverage-connectivity criterion [11] and we explain the different assumptions used by our proposed solution. Section 4 presents our proposed relocation approach to cover an event happening outside the ROI. The proposed solution needs to ensure that connectivity is still maintained between the

ROI and the COI where the new event occurred. In section 5, we evaluate the performance of our proposed approach using performance metrics such as the number of exchanged messages, energy consumption, and the relocation time. Finally, we make some concluding remarks in section 6.

2. RELATED WORKS

Several works have considered the relocation of wireless sensor nodes within a wireless sensor network. But most of them have considered sensor nodes relocation either to ensure fault tolerance, to remove coverage holes or for self organization purposes. Benhamed et al. [5] proposed a new technique of self organization in wireless sensor networks to ensure network robustness. This method aims to find the network articulation points that exist in order to ensure coverage and connectivity in the network. Their proposed approach eliminates critical positions by placing additional nodes around the articulation points to replace them if they fail. In [8], Chiasserini et al. considered the relocation of mobile sensor nodes based on virtual forces. Indeed, in the sensor field, each sensor behaves as a "source of force" for all other sensors. If two sensors are placed too close to each other, they exert negative forces on each other. On the other hand, if a pair of sensors is too far apart from each, they exert positive forces on each other. This ensures a global uniform sensor placement in the network [19]. When an event takes place, a potential force is generated by such an event (for example, upon the occurrence of a fire outbreak). Then the sensor nodes detecting such an event use virtual forces to move towards the event while maintaining a security distance that precludes them from destruction.

A grid-quorum based solution was presented in [18] to ensure fault tolerance in wireless sensor networks. Indeed, Wang et al. in [18] split the sensing field into m × m square cells in order to have a grid based topology. Each cell then elects a cell head node and the other nodes within that cell are considered as redundant. Taking advantage of the square sized grid topology, Wang et al. proposed a fault tolerance solution based on relocation of redundant nodes where supplier cells are along columns and demanding cells are along rows. Each cell head broadcasts a message with position information of the redundant nodes to all cell heads in the same column. After this step, all the cells' heads are aware of the redundant nodes within their column. If a cell head needs a redundant node to replace a faulty node within its cell, it first checks if there is an available redundant node along its column. If nothing is available, it then propagates its query to its neighboring cell head on the same row. The receiver performs the same redundant node research procedure as the sender. If it cannot satisfy the sender's query, it propagates the message to its neighbor on the same row until the message reaches a cell head node that can satisfy the initial sender query. Hence, the redundant node found can be relocated to replace the failed node.

In the same way Li et al. [12] proposed an information-mesh structure to ensure fault tolerance in the sensor network. They proposed a Mesh Based Sensor Relocation Protocol (MSRP) to relocate a redundant sensor node to a failed location. Indeed, nodes are initially divided into active nodes (A-nodes) and redundant nodes (R-nodes). Each R-node is associated with a proxy which is the nearest A-node in its vicinity. When the neighboring nodes detect an A-node failure, they cooperate to find the replacement R-node using the proxies' redundant nodes information. The target R-node is then relocated to its final position.

In [10], the authors proposed an optimized algorithm to detect node redundancy within the network by finding if a sensor node is completely covered by its neighborhood using a recursive geometric-based procedure. They proposed a cluster-based redundant node relocation algorithm that reduces the overall network energy consumption by choosing the nearest failed redundant sensor node as a replacement.

All the works presented above have considered sensor node relocation techniques that ensure fault tolerance or for self organization, network robustness and event coverage purposes by making use of an initial deployment area. None of the above proposed solutions consider sensor nodes relocation outside of the initial deployment area. Indeed, when relocating redundant nodes from their initial deployment area, we still need to maintain the continued operation of the initial network while ensuring the coverage and the communication with the initial ROI for the relocated nodes. Hence, we will propose a novel node relocation approach in the next section.

3. ASSUMPTIONS AND SOLUTION REQUIREMENTS

In this paper, we consider the issue of relocating a set of mobile sensor nodes from their initial deployment area (ROI) to cover a target event happening outside of the initial covered zone (COI). For instance our relocation approach aims to maintain both coverage and connectivity within the ROI when nodes relocate from their original positions towards the COI. Indeed, maintaining coverage within the ROI means that each point within that zone is covered at least by one sensor node. Hence only "redundant" nodes can be relocated from the ROI. Moreover, relocating "redundant" nodes from their initial zone to the COI must ensure nodes communications (connectivity) within the ROI and between the ROI and the COI. Coverage and connectivity are closely related because they are strongly related to sensing and communication ranges. According to the literature, as in [9], the relation $R_C \le 2R_S$ (where R_C is the communication range and R_S is the sensing range) is necessary and sufficient for coverage implies connectivity in a sensor network. Therefore, the sensors node relocation mechanism we propose in this work considers the coverage-connectivity criterion first and foremost requirement that must be satisfied.

In the following, we present a few assumptions related to the initial network deployment, sensor nodes characteristics and sensor nodes movement have to be considered.

3.1 Initial sensors deployment

Generally wireless sensor nodes are deployed randomly to cover events in a certain region of interest (ROI). Initial deployment may not ensure the total coverage of the ROI. Therefore, different mechanisms can be used to ensure covering the whole region and avoiding coverage holes [11]. A full discussion of the operations of the mechanisms that ensure the coverage-connectivity criterion is outside of the scope of this work. In this paper, we make the following assumptions:

- **Assumption 1:** The region of interest (ROI) is a square area of side a.
- **Assumption 2:** The wireless sensor network initially deployed satisfies the coverage-connectivity criterion. Indeed, this criterion can be satisfied using self deployment algorithms if the sensor nodes are initially deployed randomly [14], [16].
- **Assumption 3:** The number of sensor nodes within the network is large enough to cover both the ROI and to ensure connectivity with the new COI (center of interest).

The above assumptions can be justified by the fact that the event that has to be covered is in the near vicinity of the initial deployment zone. For example, in the case of a chemical spill disaster, we would like to detect the soil contamination in the surrounding zone (some hundred meters) away from the initial ROI. This could be done by relocating some redundant nodes from the initial zone.

Based on the assumptions above several grid-based techniques using various geometric shapes [1], [18], [12] have been proposed to ensure covering the initial ROI. The grid covering techniques are simple and enable easy coverage measurements. Grid-based techniques employ triangular shaped, rectangular shaped or hexagonal shaped grids [1]. Indeed, the hexagonal grids are optimal in terms of coverage calculations since they increase the maximum area covered by each sensor and therefore increase the number of calculated redundant nodes within each cell. However, in this work we choose a square-based grid because it is much more appropriate to the shape of the ROI given by assumption 1 and such a coverage technique will limit the number of exchanged messages which can be transmitted either on rows or on columns [18].

Moreover, we also consider a side size of each cell in the grid in such a way so as to satisfy both the coverage and the connectivity (as given by assumption 2). We then consider a side of each cell b equal to $\frac{2R_s}{\sqrt{5}}$ (with R_s is the sensing range) which is deduced by considering the fact that the furthest nodes in two adjacent cells can communicate together directly [15]. Therefore, the distance that separates them is at most equal to the communication range R_c (with $R_c = 2R_s$). Moreover, in order for assumption 3 to hold, we consider that initially more than one sensor is deployed within each cell.

3.2 Sensor nodes and sink node characteristics

In a wireless sensor network, sensor nodes detect events in a covered area and transmit them directly or using multi-hop routing to a particular node called a sink which transfers the collected information to the application. Therefore, we assume that:

- **Assumption 4:** The nodes are identical in terms of communication range, sensing range, and initial energy resources E_0. Each node is identified by a unique identifier. This is simplifying assumption and is frequently used in the literature [2], [10].
- **Assumption 5:** Each node *i* can detect its position at a given time using a localization system from [17], [3] and knows the limit of the ROI. It can therefore determine the row and column numbers (r_i, c_i) to which it belongs.
- **Assumption 6:** The sink node *p* is placed at one of the corners of the ROI area at coordinates (x_p, y_p) and is alerted by the application of the COI coordinates (x_{COI}, y_{COI}) when a new event occurred. It will notify the sensor nodes of the position of such an event. The sink node belonging to the cell is identified by row and column numbers (1,1). Generally the sink node is placed at the front of the sensor network to keep connecting with the wired structure.

4. PROPOSED SENSOR NODE RELOCATION APPROACH

In this section, we present our novel approach that relocates wireless sensor nodes from their initial ROI to cover a particular event happening at a COI outside the initial coverage zone and to maintain communication with the initial region of interest in such a way so as to report the captured information to the sink node. Our proposed approach uses four steps: clustering, evaluation of the number of nodes to be relocated, propagation of the COI position and finally redundant nodes election and relocation take place.

4.1 Clustering

When the nodes are initially deployed, all the nodes within the sensor network are in active mode. The clustering phase determines the nodes among the active nodes that will remain active to detect the events within the ROI. All the other nodes within the ROI switch to the passive state and are considered as redundant. Therefore, clustering elects a particular node (called cell head) within each cell of the square grid to cover the area of such a cell. All the other nodes belonging to the same cell switch to the passive state. The cell head also needs to maintain the list of its neighboring nodes belonging to the same cell and their positions. Clustering is one of the classical algorithms in hierarchical-based solutions [13]. The election of the cell head in each cell of the square grid can be made on the basis of multiple criteria (node identifier, energy, communication-coverage range). In our approach we elect the node having the highest remaining energy within a cell as a cell head. Moreover, if two nodes within the same cell have the same energy, the node with the smallest identifier is elected cell head.

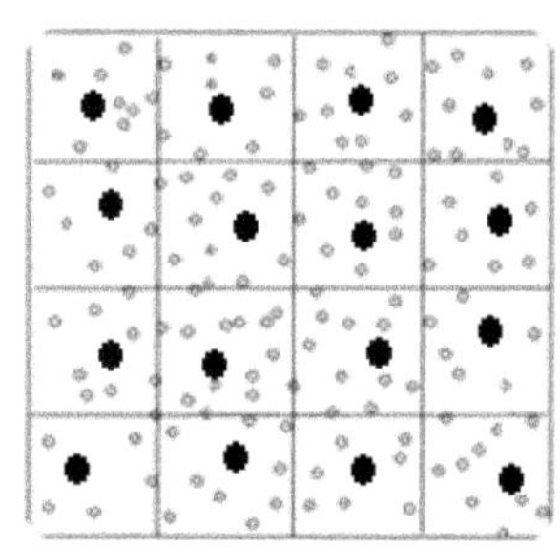

Figure 1: The clustering phase

Therefore, initially each node initializes itself as a *Cell_head* and initializes E_{max} with the maximum value of its remaining energy E_0. Then, each node periodically sends a *HELLO* packet that contains information such as its position, its remaining energy and its identifier. When each node *i* receives a *HELLO* packet from a sender *s*, it checks if the sender *s* is within the same cell (according to assumption 5, each node is aware of the cell row and column to which it belongs). If yes, it compares E_s (the energy received in the *HELLO* packet) with its local value E_{max}. If $(E_s > E_{max})$ or if $(E_s = E_{max})$ and $(s < i)$, then the sender *s* is saved within the node *i* as the *Cell_head*.

After a time α (corresponding to the time required to exchange *HELLO* packets), each node *i*, compares its identifier with the *Cell_head* value: if the *Cell_head* value is equal to *i*, then the node *i* considers that it is a cell head and broadcast a packet *CELL_HEAD_ADV* including its identifier and its position. Each node receiving the *CELL_HEAD_ADV* packet and is within the same cell of the sender acknowledges that packet with a *CELL_HEAD_ACK* including its position and its identifier. Upon reception of *CELL_HEAD_ACK* packets, the cell head node *i* updates its neighboring list with the characteristics of each neighboring node.

As the *HELLO* packets are transmitted periodically by all the nodes, a new cell head election procedure is triggered upon reception of a packet with a remaining energy greater than the one of the cell head node's remaining energy. Then, *CELL_HEAD_ADV* and *CELL_HEAD_ACK* packets are exchanged to update the new cell head's identifier.

4.2 Relocated node number evaluation

When a new event happens outside of the ROI, the application notifies the sink node of the position of such an event. The sink node will then trigger the algorithm to evaluate which nodes have to be relocated towards the COI while minimizing the overall network energy consumption. As mentioned before, since the mobility function is more expensive in terms of energy than communication and computation, the sink has to first determine the nearest cell: the cell whose position is closest to the center of Interest. Then, the sink node determines which cells surrounding the nearest cell may contain other redundant nodes to ensure connectivity between the ROI and the COI. We first proceed to the nearest cell computation.

4.2.1 The COI nearest cell computation

Determining the cell which has the closest position to the COI depends on both the COI and the sink node coordinates (x_p, y_p) and (x_{COI}, y_{COI}). Indeed, according to assumption 6, the sink node can be placed at one of the corners of the square shaped ROI. Different cases have to be discussed to determine r_N and c_N, the row and the column numbers of the COI nearest cell:

- **First case :** The sink node is located at the lower left corner of the grid, then:

$$r_N = \min\left(\max\left(\left\lceil \frac{x_{COI} - x_p}{b} \right\rceil, 1\right), \frac{a}{b}\right) \quad (1)$$

$$c_N = \min\left(\max\left(\left\lceil \frac{y_{COI} - y_p}{b} \right\rceil, 1\right), \frac{a}{b}\right) \quad (2)$$

Where a/b is the number of cells on one side of the ROI area.

- **Second case :** The sink node is located at the upper left corner of the grid, then r_N is given by equation (1) and:

$$c_N = \min\left(\max\left(\left\lceil \frac{y_p - y_{COI}}{b} \right\rceil, 1\right), \frac{a}{b}\right) \quad (3)$$

- **Third case :** The sink node is located at the lower right corner of the grid, then c_N is given by equation (2) and:

$$r_N = \min\left(\max\left(\left\lceil \frac{x_p - x_{COI}}{b} \right\rceil, 1\right), \frac{a}{b}\right) \quad (4)$$

- **Fourth case :** The sink node is located at the upper right corner of the grid, then:

$$r_N = \min\left(\max\left(\left\lceil \frac{x_p - x_{COI}}{b} \right\rceil, 1\right), \frac{a}{b}\right) \quad (5)$$

$$c_N = \min\left(\max\left(\left\lceil \frac{y_p - y_{COI}}{b} \right\rceil, 1\right), \frac{a}{b}\right) \quad (6)$$

Therefore, calculating the nearest cell row and column numbers helps in determining the redundant node within this cell that will be used to cover the event happening in the center of interest. In the next paragraph, we determine the other redundant nodes to be relocated to ensure connectivity between the initial ROI and the COI.

4.2.2 The Virtual sub grid computation

Based on the nearest cell research algorithm presented above and the COI coordinates, the sink node can determine the number of redundant nodes to be relocated to the COI to ensure connectivity with the initial ROI area.

The choice of the number of redundant nodes to be relocated is based on the following conditions:

1) Limiting the energy consumption due to mobility by moving the nodes whose positions are nearer to the COI.
2) Ensuring that the connectivity is maintained when the redundant nodes are relocated.

To satisfy the above conditions, we can choose to select the redundant nodes from a virtual sub-grid belonging to the initial ROI which, when being moved towards the COI, will ensure the coverage and the connectivity with the initial ROI (as shown is figure 2). Indeed, the virtual sub grid is chosen such a way that:

1) It propagates the square grid topology outside the ROI. This is beneficial because it ensures connectivity with the ROI since the distance that separates two nodes in two adjacent cells is always lower or equal to the communication range. Moreover, this can useful if another event happens outside the ROI and is in the same region of the COI.
2) The upper and the lower boundary of the virtual sub grid are computed based on the distance between the COI and the nearest cell.

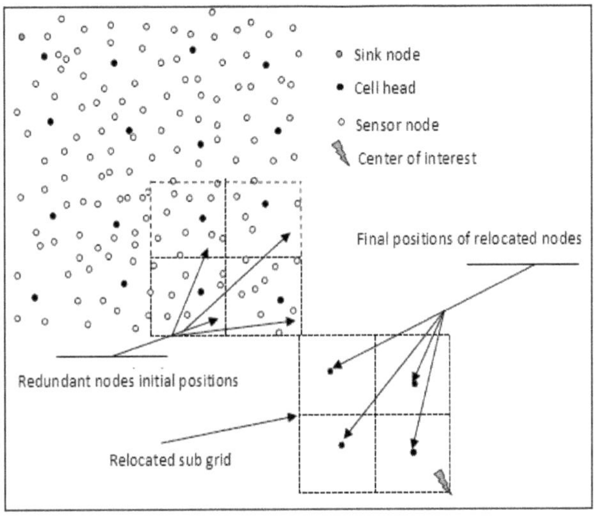

Figure 2: The virtual sub grid to be relocated to the COI

After detecting the virtual sub grid boundaries, the sink node will propagate the position information of the nearest cell, the sub grid boundaries, and the COI coordinates. When the nearest cell cell-head receives the virtual sub grid boundaries (propagated by the sink node), it will trigger the procedure that will relocate redundant sensors belonging to the virtual sub-grid zone to their final positions outside the ROI to capture the event information and to keep connectivity with the center of interest (COI). The virtual sub-grid boundaries propagation and the redundant sensor nodes relocation procedures are presented in the next section.

4.3 Propagation of the COI position

Once the sink node has determined the boundaries of the virtual sub grid that needs to be relocated outside the ROI, it sends a *NEAREST_CELL_ADV* message containing the coordinates of the COI and the coordinates (row and column numbers) of the upper left and the lower right cells of the virtual rectangular sub grid to be moved towards the COI. One of the coordinates sent in the *NEAREST_CELL_ADV* message correspond to the row and column numbers of the COI nearest cell.

As we are in presence of a square grid within the ROI, messages from the sink node can be transmitted along rows and columns. Therefore, the sink node propagates the *NEAREST_CELL_ADV* message to its next neighbor on the same row. When receiving the *NEAREST_CELL_ADV* message, the node *i* verifies if it is the neighbor of the message emitter belonging to the same row. If so, the node *i* then compares r_N, the COI nearest cell row number with r_i, the node *i* line number. If $r_N < r_i$, then the node *i* propagates the *NEAREST_CELL_ADV* message to its next neighbor along the same row until the message reaches a node *j* whose row number is equal to r_N. At that time, the node *j* forwards the *NEAREST_CELL_ADV* message to its neighbor on the same column until the message reaches the COI nearest cell.

4.4 Redundant nodes election and relocation

Once, the COI nearest cell (identified by row/column (r_N, c_N)) cell head receives the sink node message containing the COI position and the size of the virtual sub-grid to be moved to the COI, it computes the $Nearest\_node(r_N, c_N)$ coordinates. Indeed, the $Nearest\_node(r_N, c_N)$ is the sleeping node within the cell (r_N, c_N) which has the closest position to the COI among all its neighbors in the same cell. The coordinates $(x_N^{\min}, y_N^{\min})$ of the $Nearest\_node(r_N, c_N)$ satisfy the following equation:

$$d_{\min} = \sqrt{\left(x_N^{\min} - x_{COI}\right)^2 + \left(y_N^{\min} - y_{COI}\right)^2}$$
$$= \min\left\{d_i = \sqrt{\left(x_N^i - x_{COI}\right)^2 + \left(y_N^i - y_{COI}\right)^2}\ \forall i \in cell(l_N, C_N)\right\} \quad (7)$$

Then the $Cell\_Head(r_N, c_N)$ wakes up the node $Nearest\_node(r_N, c_N)$, creates and broadcasts two new messages *FORWARD_ROW* and *FORWARD_COLUMN* containing the same information as in the sink node's *NEAREST_CELL_ADV* message and are identified by a given sequence number.

Upon receipt of a *FORWARD_ROW* packet by the cell head k of the cell (r_k, c_k), (where r_k and c_k are row and column numbers of the cell which cell head is node *k*), the cell head *k* performs the following actions:

1) It verifies if it has already processed such a *FORWARD_ROW* message with the same sequence number.
2) If the *FORWARD_ROW* message has not been processed, the cell head compares its row number with the row number of the sender.
3) If they have the same row number, the receiver then verifies if the sender is its direct neighbor, i.e., the sender belongs to a cell whose column number is (c_k +1) or (c_k -1), depending on the COI's nearest cell position. Indeed, since node *k* column number is c_k, its neighboring cells column numbers can be either (c_k +1) or (c_k -1).
4) If condition 3) is satisfied, the receiver node k verifies if its coordinates are inside the virtual sub grid coordinates to be moved towards the COI.
5) If so, the node *k* generates a *FORWARD_ROW* packet and a *FORWARD_COLUMN* packet and then computes the $Nearest\_node(r_k, c_k)$: the node within the cell (r_k, c_k) which is the nearest node to the COI among

its neighbors in the same cell. The cell head k sends a message to wake up the node Nearest_node(r_k, c_k).

In the same way, upon receipt of a *FORWARD_COLUMN* packet by the cell head m of the cell (r_m, c_m), the node m performs the following actions:

1) It verifies if it has already processed such a *FORWARD_COLUMN* message with the same sequence number.
2) If the *FORWARD_COLUMN* message has not been processed, it compares its column number with the column number of the sender.
3) It they have the same column number, the receiver then verifies if the sender is its direct neighbor, i.e., the sender belongs to a cell whose row number is (r_m+1) or (r_m-1), depending on the COI's nearest cell position.
4) If condition (3) is satisfied, the receiver node m verifies if its coordinates are inside the virtual sub grid coordinates to be moved towards the COI.
5) If so, it generates a *FORWARD_COLUMN* packet and then computes the *Nearest_node*(r_m, c_m): the node within the cell (r_m, c_m) which is the nearest node to the COI. The cell head m sends a message to wake up the *Nearest_node*(r_m, c_m) sensor.

The above procedure is repeated until the *FORWARD_ROW* and the *FORWARD_COLUMN* packets reach a cell whose row and column numbers are outside of the virtual sub grid that should be moved towards the COI. After sending all the *FORWARD_ROW* and *FORWARD_COLUMN* packets and after waking up all redundant nodes, each redundant node in the virtual sub grid computes its trajectory based on the COI coordinates and the nearest cell coordinates transmitted in the *NEAREST_CELL_ADV*, *FORWARD_ROW* and *FORWARD_COLUMN* messages.

Since the trajectory is the same for all the redundant nodes, they move in parallel towards the COI and they all stop at the same time T_s : the time where the COI coordinates are within the *Nearest_node*(r_N, c_N) coverage region. In the next section, we present a performance evaluation of our proposed solution.

5. PERFORMANCE EVALUATION OF PROPOSED APPROACH

To evaluate the performance of our proposed sensor node relocation strategy described earlier, we have implemented our approach under the WSNet [7] simulator, an event driven simulator dedicated to wireless sensor networks. WSNet defines models written in C for the different network layers.

Using the WSNet-replay tool, we depict in figure 3 for a sample scenario, the initial square sized initial deployment zone and the relocated blue colored nodes covering the COI event and maintaining connectivity with the COI.

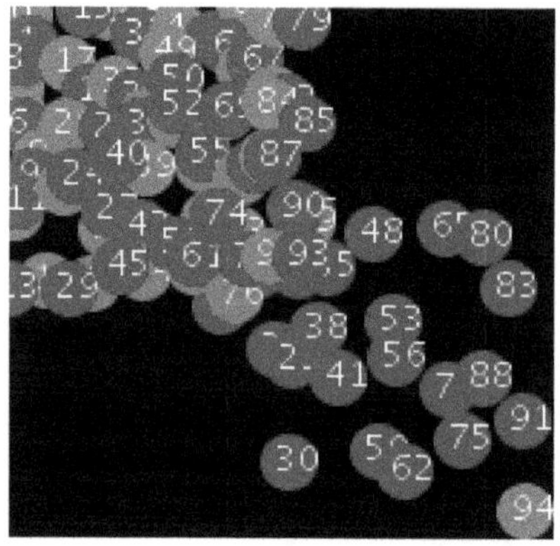

Figure 3: Wsnet-replay screen for the relocation approach

In this section, we evaluate the performance of our relocation approach in terms of the number of relocated nodes, the total energy consumption, and the relocation time. The total network energy consumption is directly affected by the number of relocated nodes. Another important evaluation criterion is the relocation time. In fact, this parameter will indicate how fast is the network reaction when being aware of a hazardous event happening outside the ROI. Hence, the shorter the relocation time is, the faster the network reacts in covering such an event.

Then, we consider three scenarios given by table 1, where we modify for each scenario the side size of the ROI, the number of sensor nodes initially deployed and the position of the COI. For each these configuration scenarios, the sink node is positioned at the upper left corner of the ROI at coordinates (0, 0).

Table 1: Simulation scenarios parameters

	ROI Area	Number of sensors	COI position
Scenario 1	100*100m	95	(180, 180)
Scenario 2	150*150m	200	(200, 190)
Scenario 3	200*200m	353	(220,250)

In figures 4 and 5, we depict the number of relocated sensors and the total energy consumption for the three scenarios above.

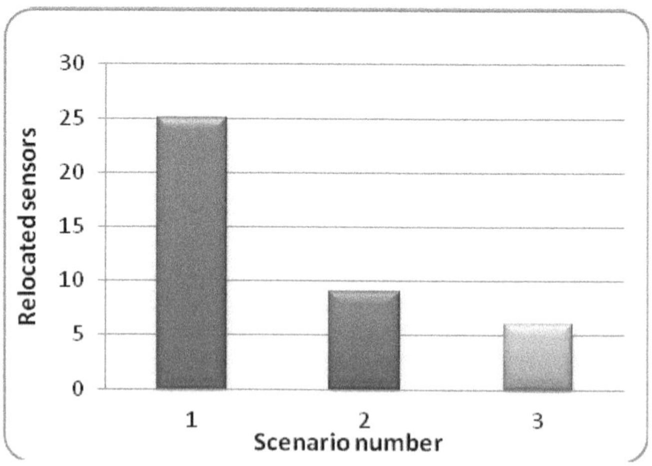

Figure 4: Number of relocated sensors

The figures show that the total consumed energy is proportional to the number of relocated sensors. This shows that mobility is the component that consumes the most energy in wireless sensor networks. Figures 4 and 5 also depict that the results obtained for the third scenario are the lowest in terms of the number of relocated sensors and the network energy consumption. This is because of the fact that the COI position is the closest one to the ROI compared to the two scenarios 1 and 2.

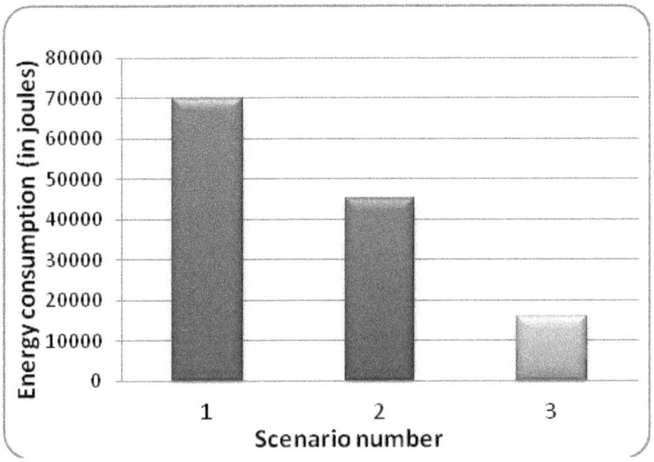

Figure 5: Total energy consumption for scenarios 1, 2, 3.

The close proximity of the COI to the ROI is further illustrated in figure 6 which illustrates the relocation time for the three different scenarios.

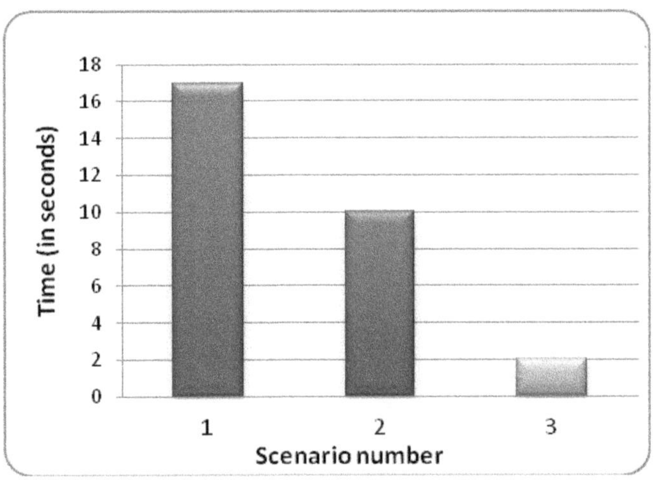

Figure 6: Relocation Time

The figure 6 shows that the relocation time has the highest value for the first scenario. This is due to the fact that for the first scenario, the distance that has to be crossed by the relocated sensors is the longest one among the three scenarios and since the redundant nodes are moving in parallel to their final locations, the relocation time obtained is the highest for that first scenario.

6. CONCLUSION

In this paper, we have proposed a novel approach to relocate wireless sensor nodes towards to center of interest when a new event has happened outside of the initial deployment zone. The relocated sensors have to cover the event happening at the COI while maintaining connectivity with their initial region of interest. To achieve this goal, we proposed a four phase based solution that includes cell heads election, evaluation of the number of nodes to be relocated, propagation of the COI position and redundant nodes election and relocation. We evaluated the performance of our proposed approach using the WSNet simulator. The obtained simulation results showed that the consumed energy and the relocation time increase as the distance separating the initial ROI from the COI increases no matter the size or the density of the initial deployment area are. As part of our future work, we will extend our approach to several centers of interests.

7. REFERENCES

[1] Ab. Aziz, N.A., Ab. Aziz, K. and Zakiah, W., "Coverage Strategies for Wireless Sensor Networks", World Academy of Science, Engineering and Technology, Issue 50, year 2009.

[2] Ahlswede, R., Cai, N., Li, S.-Y. R, and Yeung, R. W., "Network Information Flow", IEEE transactions oninformation theory, vol. 46, no. 4, year 2000.

[3] Albowitz, J., Chen, A. and Zhang, L., "Recursive position estimation in sensor networks," ICNP'01, 2001.

[4] Bartolini, N., Calamoneri, T., Fusco, E.G., Massini, A. and Silvestri, S., "Autonomous deployment of self-organizing mobile sensors for a complete coverage", In proc. of the third International Workshop on Self Organizing Systems IWSOS 2008, Vienna, Austria, December, 2008.

[5] Benahamed, K., Haffaf., H., Merabti, M. and Llewellyn-Jones, D., Khelifa, B., "Monitoring connectivity in wireless sensor networks", International Journal of Future Generation Communication and Networking, vol. 2, no. 2, year 2009.

[6] Biagioni, E. S. and Bridges, K. W., "The application of remote sensor technology to assist the recovery of rare and endangered species", The International Journal of High Performance Computing Applications, vol. 16, pp. 112-121, year 2002.

[7] Chelius, G., Fraboulet, A. and Ben Hamida, E., "The WSnet Network Simulator", http://wsnet.gforge.inria.fr, ver. 9.07, July 2009.

[8] Chiasserini, C.-F., Leonardi, E., Garetto, M. and Gribaudo, M., "A distributed sensor relocation scheme for environmental control", In proc of the 4th IEEE International Conference on Mobile Ad-hoc and Sensor Systems, MASS 2007, Pisa, Italy, October 2007.

[9] Hou, J. C. and Zhang, H., "Maintaining sensing coverage and connectivity in large sensor networks", Ad Hoc & Sensor Wireless Networks journal, vol.1, pp. 89-124, year 2005.

[10] Khedr, M. A., and Ramadan, H., "Effective Sensor Relocation Technique in Mobile Sensor Networks", International Journal of Computer Networks & Communications (IJCNC), vol.3, no.1, year 2011.

[11] Li, J., Lachlan, L.H., Andrew, Foh, C. H., Zukerman, M. and Chen, H.-H., "Connectivity, Coverage and Placement in Wireless Sensor Networks", Sensors Journal 2009, vol. 9, pp. 7664-7693.

[12] Li, X., Santoro, N. and Stojmenovic, I., "Mesh-Based Sensor Relocation for Coverage Maintenance in Mobile Sensor Networks," In proc. of the 4th Int. Conf. on Ubiquitous Intelligence and Computing, UIC 2007, Hong Kong, China , July 2007.

[13] Liu, C.-M., Lee, C.-H., "Distributed Algorithms for Energy-Efficient Cluster-head Election in Wireless Mobile Sensor Networks", In proc. of the 2005 International Conference on Wireless Networks, ICWN 2005, Las Vegas, USA, 2005.

[14] Li, W. and Cassandras, C.G., "A minimum-power wireless sensor network self-deployment scheme", the 2005 IEEE conference on Wireless Communications and Networking Conference, WCNC, Neaw Orleans, USA, march 2005.

[15] Roychowdhury, S. and Patra, C. "Geographic Adaptive Fidelity and Geographic Energy Aware Routing in Ad Hoc Routing", Special Issue of International Journal of Computer & Communication Technology (IJCCT) for International Conference on Computer Technology, (ICCT-2010), vol. 2 no. 2,3,4, year 2010.

[16] Sarazin, B., Rizvi, S., "A Self-Deployment Obstacle Avoidance (SOA) Algorithm for Mobile Sensor Networks", International Journal of Computer Science and Security, year vol. 4, issue 3, year 2010.

[17] Savvides, A., Han, C.-C. and Strivastava., M. B., "Dynamic fine-grained localization in ad-hoc networks of sensors", In proc. of the ACM/IEEE International conference on Mobile Computing and Networking, MobiCom'01, Rome, Italy, July 2001.

[18] Wang, G., Cao, G., Porta, T. and Zhang, W., "Sensor Relocation in Mobile Sensor Networks," In proc. of the 24th Annual IEEE International Conference on Computer Communications, INFOCOM 2005, Miami, USA, March 2005.

[19] Zou, Y. and Chakrabarty, K., "Sensor Deployment and Target Localization Based on Virtual Forces", In proc. the 22nd Annual Joint Conference of the IEEE Computer and Communication Societies, INFOCOM 2003, San Francisco, USA, April 2003.

A 2-Dimensional Approach to QoS Provisioning in Adversarial Mobile Ad Hoc Network Environments

Peter J. J. McNerney
School of Computer Science
University of Manchester, UK
mcnerney@cs.man.ac.uk

Ning Zhang
School of Computer Science
University of Manchester, UK
nzhang@cs.man.ac.uk

ABSTRACT

In an untrustworthy mobile ad hoc network (MANET), security and quality of service (QoS) should be considered in an integrated manner. Supporting QoS without considering security threats and their implications may not lead to an optimal QoS solution. However, applying too much security will almost certainly lead to a high level of resource consumption, which may deplete QoS. This paper attempts to address these two issues together by proposing a 2-Dimensional Adaptation ARChitecture (2-DAARC) for achieving QoS in MANETs containing blackhole attackers. The architecture supports two forms of adaptation: single-path adaptation (SPA) and multi-path adaptation (MPA). SPA uses adaptive bandwidth reservations along a single path; it is intended for supporting QoS when the threat-level in the network is low. MPA uses a duplicated, best-effort data packet delivery service along multiple paths; it is intended for supporting QoS when the threat-level is high. MPA employs a novel Priority-based Multi-path Type Selection (PMTS) algorithm for path selection. 2-DAARC switches between SPA and MPA based on the underlying network conditions measured by packet delivery ratios at the destination nodes. The architecture is evaluated against the INSIGNIA QoS framework, which uses a single-path bandwidth adaptation approach. The simulation results show that 2-DAARC performs better in terms of packet delivery ratios and end-to-end delays in lightly loaded networks.

Categories and Subject Descriptors

C.2.2 [**Computer-Communication Networks**]: Network Protocols

General Terms

Design, Performance, Security

Keywords

Mobile ad hoc networks, Security, Quality of Service

1. INTRODUCTION

A MANET is a collection of autonomous nodes which communicate with one another using wireless links [19]. Some MANET characteristics, such as the distribution of routing operations amongst peer devices, create a number of security vulnerabilities. These vulnerabilities can easily be exploited by malicious nodes to interrupt network operations. One such exploit is the *blackhole* attack: a blackhole attacker drops all data packets that it is asked to forward. This affects the QoS that a network can provide to a packet stream, and it makes QoS provisioning a challenging task.

QoS and security are two MANET research issues which are generally considered separately. However, the two issues are related. On one hand, it is difficult to support QoS in adversarial MANET environments without security provision: adaptive bandwidth reservations along a single path struggle to support QoS in the presence of attacks on data packets [13]; and blackhole attacks can prevent a QoS mechanism from adapting to the network conditions [14]. On the other hand, too much security provision will consume resources, and this may deplete the QoS which can be achieved. Thus it is necessary to consider both security and QoS requirements in an integrated manner. We argue that integrating security and QoS may be the most effective way of providing QoS in adversarial network environments, and that the integrated solution should adapt to network conditions.

For proof of concept, this paper presents and evaluates a novel QoS proposal, the 2-Dimensional Adaptation ARChitecture (2-DAARC). 2-DAARC is designed to support QoS for priority traffic in an adversarial MANET environment. It combines (1) adaptive bandwidth reservations and packet priority assignments over a single (primary) path, and (2) redundant data packet forwarding over multiple (i.e., primary and secondary) paths. The former is used when network conditions are favourable; the latter when conditions deteriorate and single-path bandwidth reservation is no longer sufficient to support QoS. A novel Priority-based Multi-path Type Selection (PMTS) algorithm is proposed for secondary path selection. PMTS attempts to select a path which is maximally disjoint with the primary path by calculating the paths' disjointedness. We use simulations to study the behaviour of 2-DAARC, and compare its performance with a single-path adaptation solution, INSIGNIA [18].

The rest of this paper is organised as follows. Section 2 reviews related work on QoS, security, and multi-path routing. Section 3 introduces 2-DAARC. In Section 4 we describe, analyse, and discuss the simulation results. Section 5 presents our conclusion and directions for future work.

2. RELATED WORK

This related work review is focused on multi-path routing to support QoS and security requirements. We first look at multi-path-based approaches to QoS provisioning. Next, we examine how multi-path routing can be used to support security requirements. We then turn our attention to integrated approaches to security and QoS provisioning.

Based on the literature, multi-path routing has been used to increase reliability in QoS provisioning. It has been used (1) to reserve bandwidth over multiple paths, and (2) to route packets over multiple paths in adaptation to network conditions. The QoS Multi-path Source Routing (QoS-MSR) protocol [24] uses the first approach. It uses a Multi-path Bandwidth Splitting Reservation (MBSR) mechanism to harness multiple paths' bandwidth to satisfy a single reservation request. MBSR splits a request into several smaller requests, each wanting to reserve bandwidth on a single path. Data packets are allocated to one of the paths using weighted round-robin scheduling. The work shows how the redundancy of multiple paths may be exploited to support QoS.

Multi-Path Dynamic Source Routing (MP-DSR) [10] uses the second approach. MP-DSR estimates the link availabilities of all the links along a path during the route discovery process, and uses these link availabilities to estimate the end-to-end reliability of the path. Based on the estimated end-to-end reliability of each path to a destination and the *end-to-end reliability* requirement specified by an application, MP-DSR selects the number of paths which should be used for data packet forwarding. Thus path selection is made in response to network and path conditions. MP-DSR does not prescribe a particular packet allocation scheme.

In the Segment Adaptive Multi-path Routing (SAMR) protocol [27], to increase delivery reliability intermediate nodes construct multiple paths to a destination node and duplicate data packets along them. To reduce potential congestion, intermediate nodes suppress duplicate data packets that they have previously forwarded. SAMR's authors remark that the extra bandwidth consumed by packet duplication is a fair sacrifice for a better QoS for real-time services.

The above solutions assume that all the nodes are trustworthy and that they always execute routing and QoS operations correctly. However, in an open MANET environment this assumption may not always be true. Some nodes in the network may be malicious and/or selfish. To overcome these problems a number of solutions have been proposed. These solutions aim to support data packet deliveries in adversarial MANETs. The essence of these solutions is largely on combining security provision with multipath routing.

The Secure Message Transmission (SMT) protocol [17] is one such solution. It uses a technique called *data dispersion* to cope with attacks on data packet forwarding. Data dispersion splits each data packet into a number of smaller packets, each containing extra bits. The extra bits are calculated in such a way that the original packet can be reconstructed given a subset of these packets. The packets are then transmitted across multiple paths. If an insufficient number of the smaller packets arrive at the destination node, the source node can retransmit them.

The work in [25] extends the SMT protocol with a misbehaving node detection mechanism. It detects misbehaviour using acknowledgements (ACKs) and binary search probing. Destination nodes send ACKs for each received data packet. If the number of missing ACKs exceeds a threshold the probing phase is triggered. Probes are piggy-backed on the data packets. Each node receiving a probe must send an ACK to the source node. If an attacker drops the probe/data packet, the downstream node will not receive the probe and will therefore not send an ACK. The source node can then identify the problem node. Trust levels are maintained for each link, and these are updated during probing. Links with a low trust level are avoided during path selection.

The work in [11] focuses on selecting paths which are the most likely to survive in the presence of attackers. It uses a fuzzy logic-based scheme to estimate a *path survivability level* (PSL) for each path, and selects paths which have the highest PSLs. This path selection process can be summarised as follows. Control packets are periodically transmitted to collect nodes' reputations, energy levels, and certificate expiration time (as well as other criteria) along paths between a source node and a destination node. These data are fed into a fuzzy inference system which produces a PSL. As new control data come in the PSLs may change, and if they do a new set of paths with the highest PSLs will be selected. In this way paths are selected dynamically in adaptation to the underlying network conditions.

Whilst the above solutions aim to support packet deliveries in the presence of attacker nodes, they do not directly consider the issue of QoS. There is little research which addresses the issue of QoS in the presence of attacks. Of what little has been done, the majority of the work focuses on the integration of security and QoS for infrastructure-based networks [23, 26]. Much of the related work for MANETs focuses on examining the time-costs of supporting confidentiality, integrity, and/or authentication, and the effects of these costs on the end-to-end packet delays and throughput [5, 22]. Other related work in this domain is to use cryptographic techniques to protect QoS signalling information [12, 21]. The focus of our work, however, is on achieving QoS in MANETs containing blackhole attackers. Our hypothesis is that if there are no or very few attackers in the network, then QoS measures, i.e., bandwidth reservation and packet priority assignment, may suffice to support priority packet deliveries. However, if the number of attackers cannot be ignored, then transmission redundancies (packet duplication over multiple paths) may be able to help to improve the chances of successful deliveries of priority traffic.

3. 2-DAARC

The 2-Dimensional Adaptation ARChitecture (2-DAARC), shown in Fig. 1, integrates two dimensions (forms) of adaptation: single-path adaptation (SPA), which uses adaptive, single-path bandwidth reservations and packet priority assignments; and multi-path adaptation (MPA), which duplicates data packets along multiple paths. SPA and MPA are used to support the delivery of priority traffic in a range of network conditions. SPA is used when network conditions are more favourable. MPA is used when network conditions deteriorate and SPA can no longer support the required QoS level. Switching between SPA and MPA (from SPA to MPA and vice versa) is performed at the source node and is governed by feedback data received from the destination node or a timeout in the absence of feedback (the lack of feedback is interpreted as an indicator of adverse conditions and should be a trigger to initiate adaptive actions [14]). The feedback data contain QoS statistics about the conditions of the in-use path(s). Message authentication of the feedback

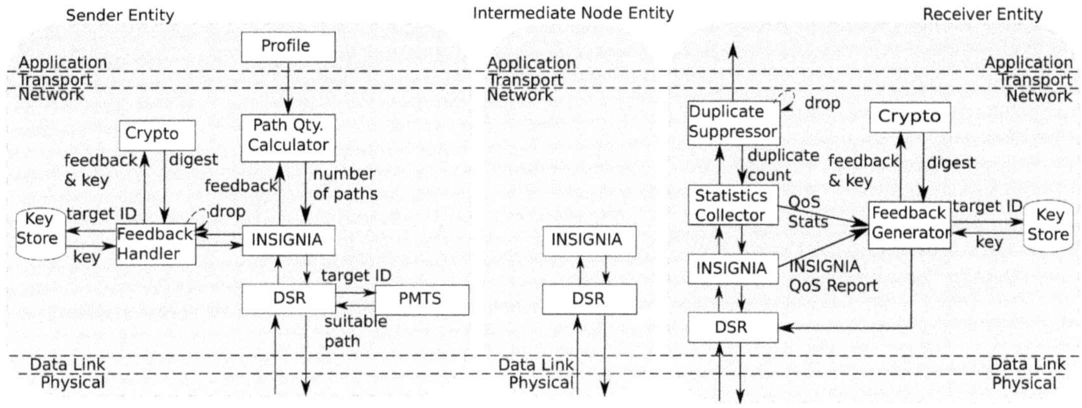

Figure 1: The 2-Dimensional Adaptation ARChitecture (2-DAARC)

Option	Service Mode	Payload Type	Bandwidth Indicator	Bandwidth Request	Generation Rate	Packet Type	Packet Status	Padding
Value	RES/BE	BQ/EQ	MAX/MIN	MAX/MIN	Packet Generation Rate	SPA/MPA	OR/DUP	-
Width	1 bit	1 bit	1 bit	16 bits	8 bits	1 bit	1 bit	3 bits

Figure 2: The 2-DAARC IP Options Header

assures the recipient that the data has not been altered in transit and that it has originated from the claimed node. We assume that the source and the destination nodes trust each other, and that the intermediate nodes are not trustworthy.

The SPA component is built on INSIGNIA [18], a QoS framework for MANETs. INSIGNIA uses adaptive bandwidth reservations and packet priority assignments to support QoS in the presence of node mobility. It has an 'in-band' QoS signalling system which makes use of the IP Options header of each data packet to convey QoS information. We extend this header to include options to support dynamic adaptation in 2-DAARC. The 2-DAARC value setting in the IP Options header is shown in Fig. 2. The first four options are original INSIGNIA header fields [18]. We have added three extra options. These are used to convey source node parameter values required by the destination node to calculate QoS statistics (to be described later). The Generation Rate option holds the source node's data packet generation rate. The Packet Type option specifies if the packet originated during SPA or MPA. The Packet Status option indicates if the packet is an original (OR) or duplicated (DUP) data packet. Padding ensures that the IP Options field ends on the 32-bit word boundary [3]. A path is determined using DSR's shortest hop-count path selection mechanism (this is also used to select primary paths during MPA).

The MPA component is built on the Dynamic Source Routing (DSR) protocol [6]. DSR is a best-effort, reactive, single-path routing protocol. However, it discovers and caches multiple paths to a destination node as part of its Route Discovery process. The additional paths can be utilized for multi-path routing. We have extended DSR to support our purpose in two ways: (1) we have devised a Priority-based Multi-path Type Selection (PMTS) algorithm for secondary path selection; and (2) we duplicate priority data packets along the multiple (primary and secondary) paths. PMTS calculates the *disjointedness* (a measure of redundancy) between an in-use primary path and each of the additional paths cached in the source node's Route Cache. The path with the greatest disjointedness will be selected as a secondary path. There are three levels of disjointedness: the highest level is *node-disjoint*, where paths do not share any nodes or links (except the source and the destination nodes); the second level is *link-disjoint*, where paths share nodes, but not links; and the lowest level is *non-disjoint*, where paths share nodes and links. Node-disjoint paths are most preferable, as they offer the greatest redundancy, and non-disjoint paths are least preferable as they offer the lowest redundancy. We use the terms 'primary path' and 'secondary path' only to distinguish the order of path selection, i.e., they are not used to differentiate between a main path and a back-up path, as both paths are used simultaneously to transmit duplicated data packets during MPA.

A number of design decisions have been made to minimize the bandwidth and performance costs introduced by the MPA component. As MPA is built on DSR, PMTS calculates the disjointedness of the paths cached in the source node's Route Cache (which are already discovered by DSR's Route Discovery mechanism during primary path discovery). In other words, PMTS does not inject control packets into the network to acquire secondary paths unless the Route Cache is empty, in which case a Route Discovery is required. Additionally, letting the source node calculate path disjointedness and select the secondary path avoids the needs to establish trust on intermediate nodes. It also allows the source node to switch between SPA and MPA with minimal delay.

The pseudocode for the PMTS algorithm is given in Algorithm 1. It calculates the disjointedness of the primary path with each of the paths maintained in the Route Cache. We refer to the paths in the Route Cache as *candidate paths.* Each node in a candidate path is compared with each node in the primary path. If a node is common to both paths, the algorithm checks whether the immediate previous node is also shared: if so, this hop is non-disjoint, and a counter for the number of non-disjoint hops is incremented; if not, this hop is link-disjoint, and a counter for the number of link-disjoint hops is incremented. In both cases, a counter for the total number of shared nodes and links is incremented. These counter values are used to select a link-/non-disjoint path with the least shared nodes/links as a secondary path if a node-disjoint path is not found. When the destination node is reached, the penultimate node in both paths is checked to determine whether the final hop is node-disjoint or non-disjoint. The final hop cannot be link-disjoint, as either the same node and therefore same link are used, or different nodes and different links are used. (A link-disjoint hop requires the same node but a different link; this is not possible in the final hop.) After comparing all the nodes in the candidate path with those in the primary path, the link-disjoint and non-disjoint counters are checked to determine the candidate path's disjointedness. Once the disjointedness of all the candidate paths is calculated 2-DAARC selects the candidate path with the highest disjointedness and the shortest hop-count as the secondary path. 2-DAARC can also be configured to use paths with a particular disjointedness for the duration of a session, e.g., only node-disjoint paths.

Having discovered a secondary path to the destination node, data packets are duplicated and transmitted across both the primary and the secondary paths. The purpose of this is to provide resilience to blackhole attacks. Whole data packet duplication is used so that the destination node only needs only one copy for use at the application layer. If both copies of the packet arrive at the destination node, the latter received copy is suppressed. A duplicate data packet detection and suppression mechanism is therefore used at the destination node. Suppressing duplicates at the network-layer minimises any effects (including performance and storage) they may have on higher layers. For example, receiving duplicated data packets when using the Real-time Transmission Protocol (RTP) can lead to negative values for the *fraction lost* and *cumulative number of packets lost* fields [20]. The duplicate detection and suppression mechanism is based on the anti-replay sliding-window mechanism used in IPsec (Internet Protocol security). It uses a window-size of 64 packets. The mechanism discards packets with a recently seen sequence number, or a sequence number that is less than the sequence number in the window's lower edge.

Every data packet arriving at the destination node is logged so that contextual information in the form of QoS statistics can be calculated. Five QoS statistics are calculated, three for SPA and two for MPA. For SPA, the *reserved packet delivery ratio*, the *degraded packet delivery ratio*, and the *SPA packet loss ratio* are calculated.[1] The packet delivery ratios (PDR) measure the *service quality* [14] of the in-use path. They are calculated as $(P/\tau)/G$, where P is the number of reserved or degraded packets received at the destination node during a past time-interval τ. The value of τ is set to 5 seconds. G is the source node's data packet generation rate, which is conveyed to the destination node in the IP Options header of a data packet. The *SPA packet loss ratio* is calculated as $1 - [((P_{reserved} + P_{degraded})/\tau)/G]$. For MPA, the *original packet loss ratio* (OPLR) and the *duplicated packet loss ratio* (DuPLR) are calculated. These describe the performance of the two paths used during MPA. OPLR is calculated using $1 - [(P_O/\tau)/G]$, where P_O is the number of original packets received during time-interval τ, and G is the packet generation rate. DuPLR is calculated using the same equation, but replacing P_O by P_D, where P_D is the number of duplicated packets received during τ.

These QoS statistics are periodically calculated and transmitted after an interval τ to the source node where they are used to govern adaptation. The time interval ($\tau = 5$ seconds) is consistent with the reporting interval recommended for the Real-Time Control Protocol [20]. The QoS statistics are communicated to the source node in a feedback (control) packet which is protected with a message authentication service to ensure its integrity and origin authentication. This service is provided using a keyed-hash function: SHA-224 [15] is used with a 112-bit symmetric key. This key-length is used because (1) a short key reduces the storage and processing costs, and (2) 112 bits is the minimum length currently recommended [1]. On receipt of a feedback packet, the source node computes the keyed-hash of the received feedback data and compares it with the received hash. If the two hash values are identical, then (1) the packet has not been tampered with during transit, and (2) the packet was generated by the claimed destination node with which the source node shares a symmetric key. The feedback data are used if and only if the hash values match, otherwise the packet is discarded. The source node then uses the received statistics to decide if the routing operation should continue to use SPA or switch to MPA or vice versa, or adaptation within SPA or MPA should take place. If the feedback packet is discarded, a new path is selected. If during SPA the source node does not receive a feedback packet before the expiration of a timeout, it will infer that there is a problem with the current path and it will switch to a different path. If no feedback is received during MPA, it is clear that both paths are not functioning correctly, so two new paths are selected.

The source node adapts within a dimension or between dimensions in response to received statistics. Adaptation decisions are based on whether the reported packet loss ratios (PLR) meet or exceed a packet loss threshold ρ. The value of ρ is 10%. (This value was determined via simulation.) Within the SPA dimension, a new path is used if the reported PLR is $\rho * 0.85 \leq PLR < \rho$. (The value of 0.85 was determined via simulation.) Within the MPA dimension, a

Algorithm 1: Pseudocode for the PMTS Algorithm

```
Input: Candidate path from the Route Cache
Output: Disjointedness of the candidate path
for node i ∈ primaryPath do
    for node j ∈ candidatePath do
        if primaryPath(i) == candidatePath(j) and j ≠ sourceNode then
            if primaryPath(i-1) == candidatePath(j-1) then
                nonDisjoint++;
                noOfSharedNodesAndLinks++;
            else
                if i ≠ primaryPathLength and j ≠ candidatePathLength then
                    linkDisjoint++;
                    noOfSharedNodesAndLinks++;
if nonDisjoint ≠ 0 then
    currentPathDisjointedness ← NON_DISJOINT;
else if nonDisjoint == 0 and linkDisjoint == 0 then
    currentPathDisjointedness ← NODE_DISJOINT;
else currentPathDisjointedness ← LINK_DISJOINT;
```

[1]A *degraded* packet is a priority packet which receives best-effort service after failing admission control during SPA.

new primary or secondary path is used if the original (primary path) and/or duplicated (secondary path) PLRs are higher than ρ. During SPA, if the PLR is larger or equal to ρ, MPA is enabled. During MPA, if both the OPLR is less than ρ and the DuPLR is less than ρ SPA is enabled.

4. PERFORMANCE EVALUATION

4.1 Simulation Model

Simulations are performed using NS-2 (v2.1b3) [16] on Fedora 14. NS-2 is extended with the CMU Monarch extensions (v1.1.2) [2], a HMAC-SHA2 implementation [4], and INSIGNIA [9]. NS-2 v2.1b3 is chosen for compatibility with the INSIGNIA extension. A random waypoint mobility model [7] is used to simulate node movement in a flat 1000m x 1000m space. 50 nodes move with minimum and maximum speeds of 1m/s and 19m/s, respectively, and a mean speed of 10m/s. Simulation duration is 900 seconds. A priority queuing mechanism is used with a buffer size of 50 packets. Other parameters include the IEEE 802.11 MAC protocol, a 2Mbps bandwidth, an omni directional antenna, a two-ray ground propagation model, and a 250 metre nominal range. Results are averaged over 10 simulation runs.

Constant bitrate traffic is sent by 10 and 20 source nodes at a rate of 4 packets/second (a common setting). The following parameters are set to default INSIGNIA values [8, 9]. 30% of source nodes send traffic demanding priority service; the remaining 70% send best effort traffic to provide a background load. Of the traffic demanding priority service during SPA, a 1:1 ratio of Base QoS (min. bandwidth requirement) and Enhanced QoS (max. bandwidth requirement) is demanded. Two other parameters used in INSIGNIA are an Adaptation Granularity of 4 and an Adaptation Parameter of 3. DSR [6] is INSIGNIA's underlying routing protocol.

Blackhole attacks are performed by 0%–10% of the nodes in 2% increments. They are simulated by dropping all the data packets received by an attacker (intermediate) node.

The following four assumptions are used in the simulation study. First, blackhole attackers only drop data packets, i.e., they forward control and feedback packets. The justifications for this are (1) attackers want to be included in paths, so they must participate in route discoveries; and (2) our focus is on the effects of blackhole attacks on data packets. Second, attackers do not collude. Third, we consider all information such as hop-counts, source routes, feedback data, etc., to be a faithful representation of the network and state information. The justifications for this are (1) source nodes make path selection and adaptation decisions based on correct data; and (2) source nodes do not adapt to the expensive MPA dimension based on false feedback data from the destination node. Finally, each pair of source-destination nodes communicating priority traffic has a security association and posses a pre-shared symmetric key.

4.2 Simulation Results

The purpose of this simulation study is two-fold: (1) to investigate the effectiveness of 2-DAARC's approach; is it more effective than the one used in INSIGNIA in adversarial conditions? And (2) to investigate which path selection mechanism, priority-based or node-disjoint-only, offers a better QoS. We first compare 2-DAARC with INSIGNIA under a range of network conditions including light and heavy loads, static and mobile nodes, and a range of attacker ratios. We then compare 2-DAARC using the PMTS path selection mechanism (marked 'PMTS' in the figures) with 2-DAARC using a node-disjoint-only path selection mechanism (marked 'node-disjoint') under the same conditions.

Fig. 3 plots the PDRs, normalised routing loads, and end-to-end delays (hereafter referred to as 'delay') of the three schemes under the static network condition (i.e., when the pause time is set to 900 seconds). There are 10 source nodes. Three of these transmit traffic demanding priority service. The other seven nodes transmit best-effort background traffic. From the results it can be seen that, across the 0–10% attacker ratios, 2-DAARC generally performs better than INSIGNIA in terms of PDRs and packet delays. The largest improvement in PDR that 2-DAARC brings over INSIGNIA is 10%. This improvement occurs when 2-DAARC only uses node-disjoint paths as secondary paths and at the 8% attacker ratio. Similarly, for most of the attacker ratio settings, INSIGNIA produces a higher delay than the 2-DAARC schemes. This is because 2-DAARC offers an alternative path to deliver priority traffic in an event when PDR is mostly affected by the blackhole attacks. This benefit comes at a cost. As shown in Fig. 3(b), 2-DAARC imposes more routing overhead than INSIGNIA. This is due to (1) the periodic transmission of feedback packets from the destination node to the source node; and (2) using multiple paths for data packet transmissions, as more control packets are required to support their delivery. In addition, among the two 2-DAARC schemes investigated, the scheme using only node-disjoint paths is more expensive that the other one. This is because only using node disjoint paths as secondary paths requires more route discoveries, as those paths which are not node-disjoint paths will be not be used, thus leading to a higher level of routing overhead.

Fig. 4 plots the same three metrics for the same number of nodes, but all nodes are now mobile (0 second pause time). It can be seen that 2-DAARC again generally performs better than INSIGNIA in terms of PDRs (Fig. 4(a)) and delays (Fig. 4(c)) across all the attacker ratios. Mobility has eroded the largest improvement in PDR that 2-DAARC brings over INSIGNIA to 4.5% for both 2-DAARC approaches. This reduction in the benefits offered by 2-DAARC is due to a combination of mobility-induced path breaks and nodes sometimes being unable to find a path to the destination. As shown in Fig. 3(b), the benefits of higher PDR and lower delays again come at the cost of higher routing overhead.

Fig. 5 shows the PDRs, normalised routing loads, and delays when the network load is increased to 6 priority traffic sources and 14 best-effort background sources. All nodes are mobile (0 second pause time). Under these conditions, INSIGNIA performs better than both 2-DAARC approaches in terms of PDR (Fig. 5(a)), normalised routing load (Fig. 5(c)), and delay (Fig. 5(e)). The greatest benefit in PDR that INSIGNIA brings over 2-DAARC is 31% at the 2% attacker ratio. The lower PDRs of 2-DAARC are a consequence of network congestion. Congestion leads to overflowing packet queues at intermediate nodes. The overflows occur due to the high volume of priority data packets transmitted during MPA. Fig. 5(c) shows that 2-DAARC's routing overhead is significantly greater than INSIGNIA's. As a significant percentage of packets are failing to be delivered (hence the lower PDR), a large number of ROUTE ERROR control packets are injected into the network. These are sent from an intermediate node to the source node when a data packet cannot

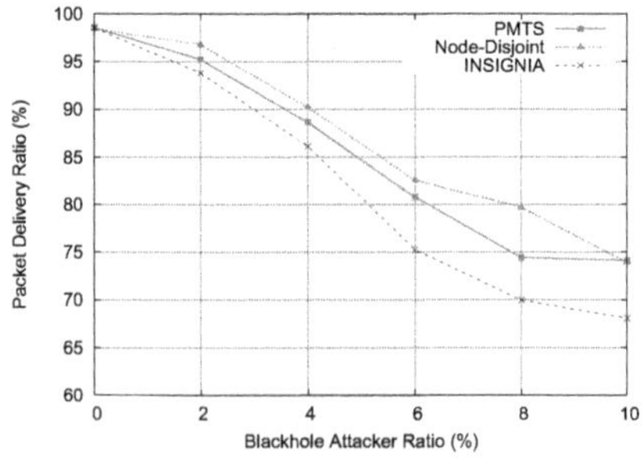

(a) Packet Delivery Ratio

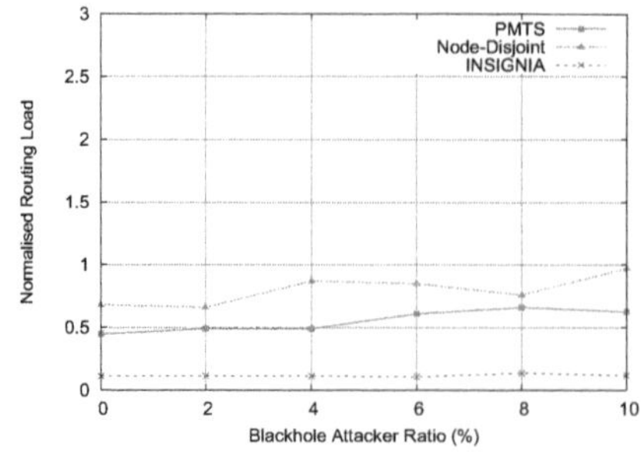

(b) Normalised Routing Load

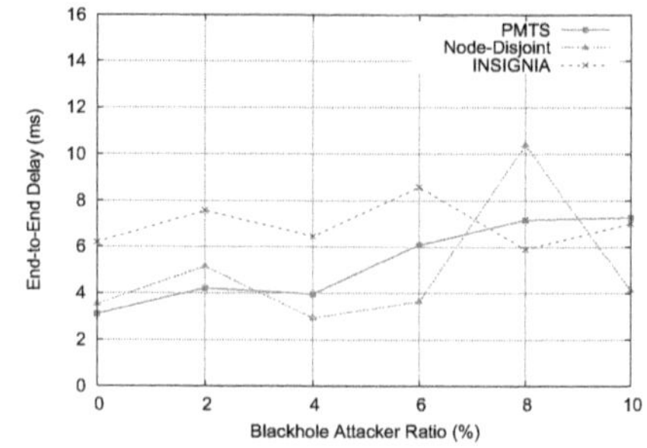

(c) End-to-End Delay

Figure 3: Packet Delivery Ratio, Normalised Routing Load, and End-to-End Delay for a 900 Second Pause Time and 3 QoS Sources (with 7 Best-Effort, Background Sources)

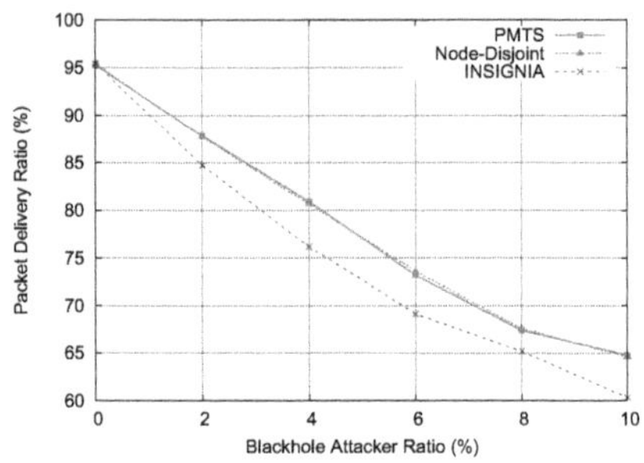

(a) Packet Delivery Ratio

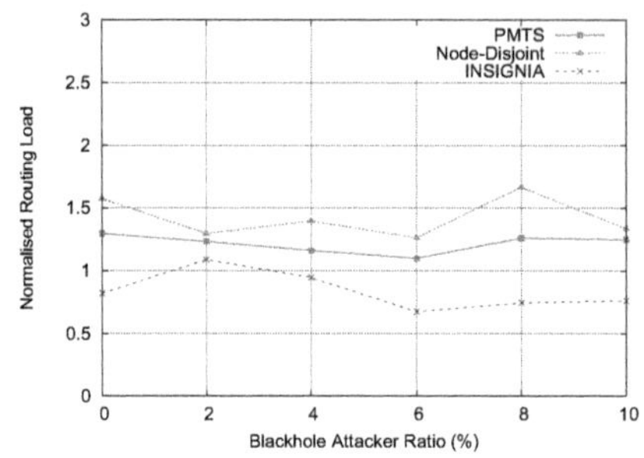

(b) Normalised Routing Load

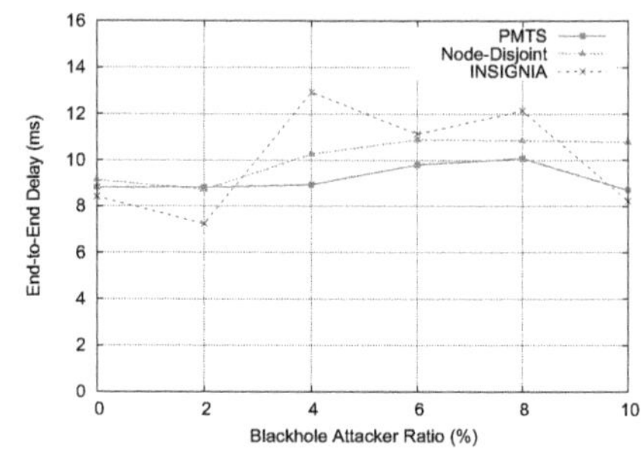

(c) End-to-End Delay

Figure 4: Packet Delivery Ratio, Normalised Routing Load, and End-to-End Delay for a 0 Second Pause Time and 3 QoS Sources (with 7 Best-Effort, Background Sources)

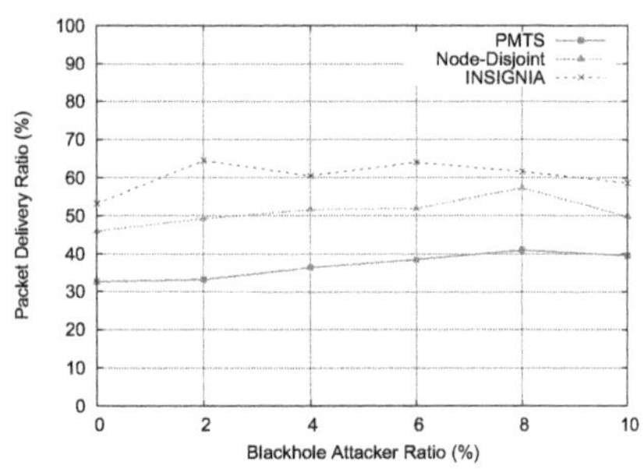

(a) PDR: Packet Salvaging On

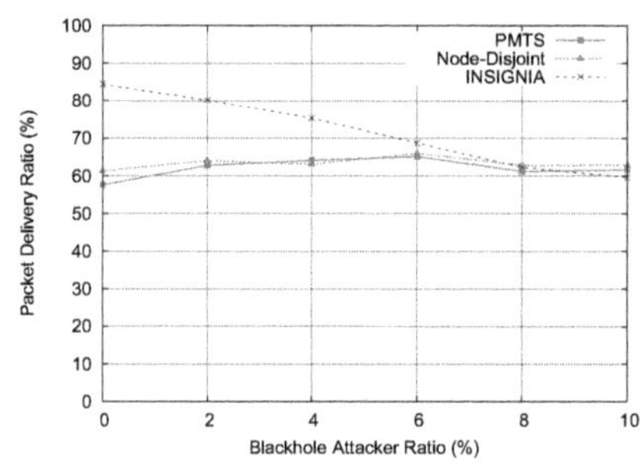

(b) PDR: Packet Salvaging Off

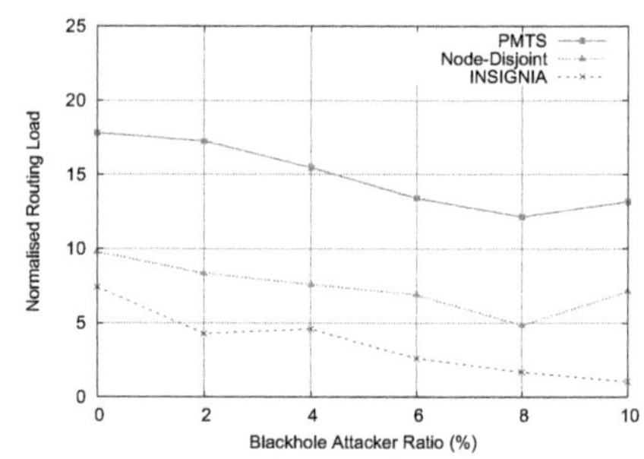

(c) Normalised Routing Load: Packet Salvaging On

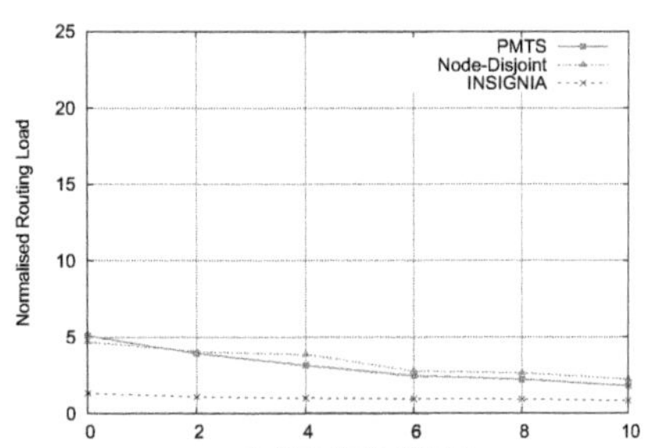

(d) Normalised Routing Load: Packet Salvaging Off

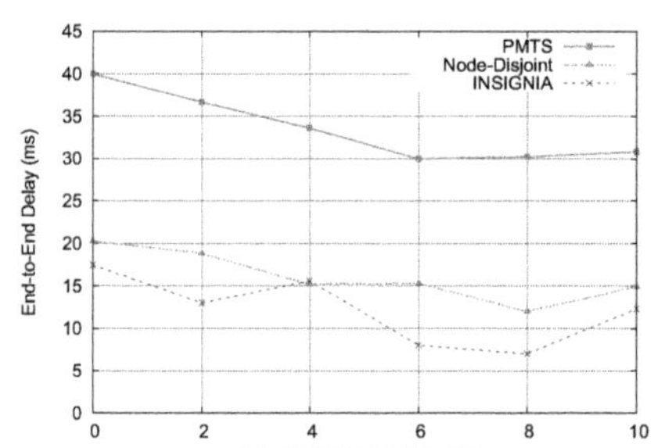

(e) End-to-End Delay: Packet Salvaging On

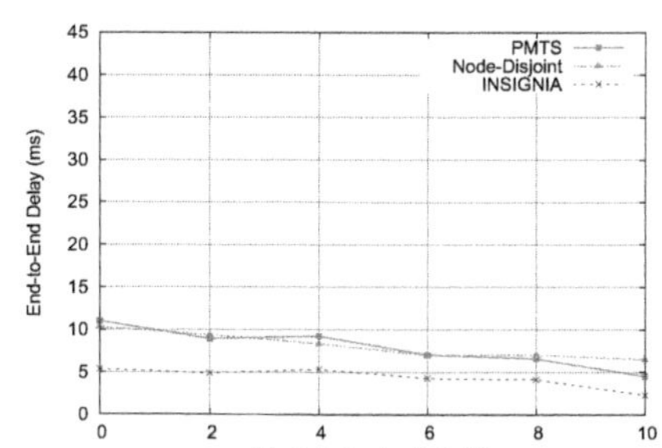

(f) End-to-End Delay: Packet Salvaging Off

Figure 5: Packet Delivery Ratio with Packet Salvaging Enabled and Disabled, Normalised Routing Load with Packet Salvaging Enabled and Disabled, and End-to-End Delay with Packet Salvaging Enabled and Disabled for a 0 Second Pause Time and 6 QoS Sources (with 14 Best-Effort, Background Sources)

be delivered to the next-hop node. This contributes to the higher control overhead. As shown in Fig. 5(e), the greatest difference in delay between INSIGNIA and 2-DAARC is 23 ms at the 8% attacker ratio. Congestion means that packet queue occupancy at intermediate nodes is high. The packets which are eventually delivered to the destination nodes experience significant delays due to the time it takes for them to be serviced by the congested intermediate nodes.

Another factor contributing to the low PDRs and high delays of 2-DAARC seen in Fig. 5 is DSR's packet salvaging mechanism. Packet salvaging is the process by which intermediate nodes attempt to retransmit data packets along an alternative path to the destination node when the next link in the current path has failed. However, intermediate nodes attempt to salvage both priority and non-priority data packets which are undeliverable due to congestion. Whilst salvaging aims to address failed packet transmissions, using it in a heavily loaded network exacerbates congestion. Disabling packet salvaging may therefore help to improve the QoS that 2-DAARC offers, as undeliverable packets are not retransmitted by intermediate nodes. Disabling salvaging, and keeping all other parameter vales unchanged, reveals a marked improvement in the PDR (Fig. 5(b)), control overhead (Fig. 5(d)), and delay (Fig. 5(f)) for 2-DAARC. These improvements are because packets are not being retransmitted when they have failed to be delivered to the next-hop nodes. This reduces congestion. However, the PDR is still less than or at best equal to the PDR obtained using INSIGNIA with packet salvaging disabled (DSR is INSIGNIA's underlying routing protocol). Moreover, the PDR is similar to the PDR achieved by INSIGNIA with packet salvaging enabled (Fig. 5(a)). 2-DAARC with packet salvaging disabled therefore does not improve on either the PDR or end-to-end delays of INSIGNIA in more heavily loaded networks. It is thus unsuitable for QoS provisioning in such conditions.

Having compared 2-DAARC with INSIGNIA, we now compare PMTS with a node-disjoint-only (NDO) path selection mechanism. We aim to determine whether PMTS can offer a better QoS than NDO so that we can configure 2-DAARC to use the path selection mechanism offering the highest QoS.

Fig. 3 shows the results for the lightly loaded network (3 priority traffic sources) with static nodes (900 second pause time). NDO supports higher PDR than PMTS at all nonzero attacker ratios. At the 8% attacker ratio, the PDR of PMTS paths is 74%, which is 6% less than NDO paths. This lower PDR is due to attackers appearing in the primary and the secondary paths. 40% of PMTS paths are non-disjoint and 1% are link-disjoint. Overall, 62% of the PMTS secondary paths contain attackers compared with 37% of secondary paths for NDO. For PMTS, 48% of both the in-use primary and secondary paths contain an attacker. Moreover, 65% of these paths contain the same attacker node. In contrast, only 16% of the pairs of NDO primary and secondary paths contain an attacker (the attacker in each path cannot be the same node as node-disjoint paths do not share any intermediate nodes). Thus the lower PDR of PMTS compared with NDO is due to a larger presence of attackers on primary and secondary paths. As seen in Fig. 3(b), NDO has a higher control overhead than PMTS. This is due to the volume of route discoveries being performed to find node-disjoint paths, rather than using link- and non-disjoint paths already available in the Route Cache (as is done with PMTS). The delay of PMTS, shown in Fig. 3(c), is 0.3 ms higher on average than for NDO. The slightly longer delay is due to a higher load on the link- and non-disjoint paths: the packet buffers on PMTS paths exceed 50% occupancy 2.1 times more often than NDO paths. This is a consequence of nodes being shared by the primary and the secondary paths. In contrast, the NDO paths are more lightly loaded as data packets are distributed across paths which do not share intermediate nodes, thus offering greater redundancy.

Fig. 4 shows the same load and the same number of nodes as before, but all nodes are mobile (0 second pause time). The PDR is similar for both approaches at all attacker ratios (Fig. 4(a)), but PMTS is clearly the better of the two approaches when the control overhead (Fig. 4(b)) and the end-to-end delay (Fig. 4(c)) are also considered: both are lower for PMTS than for NDO. The similarities in PDR are due to node mobility causing path breaks which, in turn, render some packets undeliverable. PMTS' lower control overhead is due to it using the changing contents of the Route Cache more effectively than the NDO mechanism. The shorter delay of PMTS paths is because intermediate nodes' packet buffers are more lightly loaded than those on NDO paths.

Fig. 5 shows the results when the network load is increased to 6 priority traffic sources and 14 best-effort background sources, and all other parameter values stay the same. It can be seen in Fig. 5(a) that NDO supports greater PDR than PMTS. This is because paths selected using PMTS suffer higher congestion that those selected using NDO: at the 10% attacker ratio, overflowing packet queues are responsible for 66% of the data packet loss for PMTS paths compared with 36% for NDO paths. As shown in Fig. 5(c), PMTS has a higher control overhead than NDO under the given settings. This is because 1.4 times as many ROUTE ERROR control packets are transmitted when using PMTS than with NDO, as PMTS suffers greater congestion and more path breaks than NDO. The delay on PMTS paths is on average double that of the NDO paths, and is shown in Fig. 5(e). This is a consequence of higher congestion on PMTS paths resulting from the link- and non-disjoint paths having shared nodes. Disabling packet salvaging leads to both path selection approaches performing similarly in terms of PDR (Fig. 5(b)), control overhead (Fig. 5(d)), and delay (Fig. 5(f)). However, the QoS from both approaches is still less than what is achievable using INSIGNIA with packet salvaging disabled.

5. CONCLUSION AND FUTURE WORK

In this paper we presented the 2-Dimensional Adaptation ARChitecture (2-DAARC) for integrated security and QoS support, and have used simulation to study its effectiveness in MANETs containing blackhole attackers. The study shows that 2-DAARC supports greater packet delivery ratios (PDR) and lower end-to-end delays than INSIGNIA in lightly loaded networks. Selecting paths with the highest available disjointedness using PMTS yields up to a 6% increase in PDR over INSIGNIA. Enforcing node-disjointedness during path selection yields up to a 10% increase in PDR. However, the latter path selection mechanism has a higher control overhead than the former. PMTS offers higher QoS than node-disjoint paths when nodes are mobile; the opposite is true for static nodes. As our future work we aim to support a better QoS by addressing the shortcomings identified in the 2-DAARC approach. 2-DAARC needs to make better informed adaptation decisions to improve QoS. It is necessary to differentiate between blackhole attacks and network congestion as the reason for priority data packet loss, and to adapt accordingly. This may be possible by combining knowledge of the number of received ROUTE ERROR packets with the received feedback data. The challenge here is to identify the reason for packet loss without establishing trust on intermediate nodes.

6. REFERENCES

[1] E. Barker and A. Roginsky. NIST Special Publication 800-131A, Transitions: Recommendation for Transitioning the Use of Cryptographic Algorithms and Key Sizes. `http://csrc.nist.gov/publications/nistpubs/800-131A/sp800-131A.pdf`, January 2011.

[2] Monarch Project: Wireless and Mobility Extensions to ns. `http://www.monarch.cs.rice.edu/cmu-ns.html`, Nov. 2000. Retrieved: 23/07/2010.

[3] J. P. (Ed.). Internet Protocol, DARPA Internet Program Protocol Specification. RFC 791 `http://www.ietf.org/rfc/rfc791.txt`, Sept. 1981.

[4] O. Gay. HMAC-SHA2. `http://www.ouah.org/ogay/hmac/`. Retrieved: 31/10/2011.

[5] W. He and K. Nahrstedt. An Integrated Solution to Delay and Security Support in Wireless Networks. In *IEEE Wireless Communications and Networking Conference (WCNC 2006)*, pages 2211–2215, 2006.

[6] D. Johnson, Y. Hu, and D. Maltz. The Dynamic Source Routing Protocol (DSR) for Mobile Ad Hoc Networks for IPv4. RFC 4728 (Experimental) `http://www.ietf.org/rfc/rfc4728.txt`, Feb. 2007.

[7] J. Y. Le Boudec and M. Vojnovic. Perfect Simulation and Stationarity of a Class of Mobility Models. In *INFOCOM 2005. 24th Annual Joint Conference of the IEEE Computer and Communications Societies. Proceedings IEEE*, volume 4, pages 2743–2754, 2005.

[8] S. B. Lee, G. S. Ahn, and A. T. Campbell. Improving UDP and TCP Performance in Mobile Ad Hoc Networks with INSIGNIA. *IEEE Communications Magazine*, 39(6):156–165, June 2001.

[9] S.-B. Lee, G.-S. Ahn, X. Zhang, A. T. Campbell, S.-Y. Nham, and C.-J. Yoo. INSIGNIA NS-2 Source Code. `http://comet.columbia.edu/insignia/ns_source_code.html`. Retrieved: 23/07/2010.

[10] R. Leung, R. L. Jilei, E. Poon, A.-L. C. Chan, and B. Li. MP-DSR: A QoS-aware Multi-path Dynamic Source Routing Protocol for Wireless Ad-Hoc Networks. In *In IEEE LCN'01*, pages 132–141, 2001.

[11] M. Lima, H. da Silva, A. dos Santos, and G. Pujolle. Survival Multipath Routing for MANETs. In *Network Operations and Management Symposium, 2008. NOMS 2008. IEEE*, pages 425–432, April 2008.

[12] B. Lu and U. W. Pooch. Security in QoS Signaling Systems for Mobile Ad Hoc Networks. In *Wireless And Mobile Computing, Networking And Communications, 2005. (WiMob'2005), IEEE International Conference on*, volume 3, pages 213–220, October 2005.

[13] P. J. J. McNerney and N. Zhang. Towards an Integration of Security and Quality of Service in IP-Based Mobile Ad Hoc Networks. In *Proceedings of the Global Communications Conference (GLOBECOM), IEEE*, Texas, USA, Dec. 2011.

[14] P. J. J. McNerney and N. Zhang. A Study on Reservation-Based Adaptation for QoS in Adversarial MANET Environments. In *Wireless Communications and Mobile Computing Conference, 2012. IWCMC '12. International*, Limassol, Cyprus, August 2012.

[15] National Institute of Standards and Technology (NIST). Federal Information Processing Standard (FIPS) 180-3, Secure Hash Standard (SHS). `http://csrc.nist.gov/publications/fips/fips180-3/fips180-3_final.pdf`, October 2008.

[16] The Network Simulator - ns-2. `http://nsnam.isi.edu/nsnam/index.php/Main_Page`. Retrieved:28/01/2010.

[17] P. Papadimitratos and Z. J. Haas. Secure Data Communication in Mobile Ad Hoc Networks. *IEEE Journal on Selected Areas in Communications*, 24(2):343–356, 2006.

[18] S.-B. Lee and G.-S. Ahn and X. Zhang and A. T. Campbell. INSIGNIA: An IP-Based Quality of Service Framework for Mobile Ad Hoc Networks. *Journal of Parallel and Distributed Computing*, 60:374–406, 2000.

[19] S. Corson and J. Macker. Mobile Ad Hoc Networking (MANET): Routing Protocol Performance Issues and Evaluation Considerations. RFC 2501 (Informational) `http://www.ietf.org/rfc/rfc2501.txt`, Jan. 1999.

[20] H. Schulzrinne, S. Casner, R. Frederick, and V. Jacobson. RTP: A Transport Protocol for Real-Time Applications. RFC 2550, July 2003.

[21] S. Sedaghat, F. Adibniya, and V. Derhami. A Secure Mechanism for QoS Routing in Mobile Ad Hoc Networks with QoS Requirements Consideration. In *Computational Intelligence and Communication Networks (CICN), 2010 International Conference on*, pages 320–324. IEEE, November 2010.

[22] Z. Shen and J. P. Thomas. Security and QoS Self-Optimization in Mobile Ad Hoc Networks. *Mobile Computing, IEEE Transactions on*, 7(9):1138–1151, July 2008.

[23] Y. Sheng, H. Cruickshank, A. Pragad, P. Pangalos, and A. Aghvami. An Integrated QoS, Security and Mobility Framework for Delivering Ubiquitous Services Across All IP-based Networks. In *IEEE 19th International Symposium on Personal, Indoor and Mobile Radio Communications*, pages 1–5, 2008.

[24] Y. T. Shu, G. H. Wang, L. Wang, O. W. W. Yang, and Y. J. Fan. Provisioning QoS Guarantee by Multipath Routing and Reservation in Ad Hoc Networks. *J. Comput. Sci. Technol.*, 19:128–137, Mar. 2004.

[25] R. Sivakami and G. Nawaz. Reliable Communication for MANETS in Military Through Identification and Removal of Byzantine Faults. In *Electronics Computer Technology (ICECT), 2011 3rd International Conference on*, volume 5, pages 377–381, April 2011.

[26] T. Taleb, Y. Aoul, and A. Benslimane. Integrating Security with QoS in Next Generation Networks. In *Global Telecommunications Conference (GLOBECOM 2010), 2010 IEEE*, December 2010.

[27] Y. Zhen and M.-Q. Wu and D.-P. Wu and Q.-J. Zhang and C.-X. Xu. Toward Path Reliability by Using Adaptive Multi-Path Routing Mechanism for Multimedia Service in Mobile Ad-Hoc Network. *The Journal of China Universities of Posts and Telecommunications*, 17(1):93–100, 2010.

Evaluating Resistance to Jamming and Casual Interception in Mobile Wireless Networks*

Lucia Gallina, Gian-Luca Dei Rossi, Andrea Marin, Sabina Rossi
DAIS, Università Ca' Foscari Venezia, Italy
{lgallina,deirossi,marin,srossi}@dais.unive.it

ABSTRACT

Mobile ad-hoc and sensor networks play an important role in several application fields. The usage of wireless links and the node mobility make the networks prone to security attacks; among these, jamming attacks are insidious and they consist of one or more nodes continuously transmitting dummy packets to keep some wireless links busy. The goal is to destroy the network connectivity or highly reduce its throughput. In this paper we propose a probabilistic formal method, based on a process algebraic approach, targeted at the analysis of connectivity and the evaluation of interference in mobile networks. We show our framework at work on the analysis of an indoor wireless communication scenario.

Categories and Subject Descriptors

C.4 [**Computer Systems Organization**]: Performance of systems—*modeling techniques*; C.2.1 [**Computer-Communication Networks**]: Network architecture and design—*Wireless Communication*

General Terms

Theory, reliability

Keywords

Process algebra, manets, interference, jamming

1. INTRODUCTION

Mobile ad-hoc networks (MANETs) consist of a set of wireless communicating devices that independently move in a given space. The mobility of the devices is one of the salient features of MANETs because it implies the impossibility of applying well-known protocols for networks based on static infrastructures. The design of MANETs must face a set of research challenges regarding functional and non-functional aspects. Among the functional requirements we mention *connectivity* that requires the development of reliable protocols on a dynamic infrastructure and unreliable wireless links. Network connectivity is a key-factor also when analysing MANETs' protocol robustness to *jamming* attacks. These are interferences generated by malicious users that constantly occupy some frequencies with dummy transmissions in the area surrounding their locations. Their goal is to destroy the connectivity of the network or highly reduce its throughput. Although the only possible counter-measure to a jamming attack is the physical removal of the malicious nodes from the network, different protocols may exhibit different robustness. Non-functional requirements are usually expressed in terms of indices for measuring the Quality of Service (QoS), e.g., the system's throughput, latency and overall energy consumption. In this paper we resort to the calculus introduced in [10] that allows for the formal specification of MANETs in terms of modular composition of its nodes. More technically, our calculus is a probabilistic and non-deterministic process algebra whose semantics is inspired by Segala's automata [13]. Node mobility is modelled probabilistically; the region where the MANET works is discretised into arbitrary small parts, and the mobility behaviour of the nodes is modelled by a Discrete Time Markov Chain (DTMC). The state of wireless links and the behaviour of the protocols are non-deterministic. The analysis of this model allows one to derive several functional properties of the MANETs, e.g., the observation equivalence of two protocols for a reliable communication over an unreliable channel can be proved [4]. In order to obtain quantitative indices of the network (e.g., throughput, response time, utilisation, energy consumption) the model must be transformed into a purely probabilistic one. This is done by means of schedulers in a similar fashion to [5]. The main contribution of this paper is the development of a methodology for the analysis of the resilience of MANETs' routing protocols to jamming attacks. Roughly speaking, we say that a routing protocol that governs a network N is resilient to a jamming attack if the functional behaviour of N without the jammer is equivalent - according to some observation semantics - to the functional behaviour of N when the jammer is introduced. It is worth of notice that the peculiarity of this approach is the analysis of a protocol in a scenario, i.e., the outcomes of the connectivity analysis depend on the protocol definition *and* the mobility properties of the nodes. For instance, one may decide to analyse the connectivity of a network in which highly mobile nodes have the roles of routing packets that could be destroyed

*Work partially supported by the MIUR Project IPODS "Interacting Processes in Open-ended Distributed Systems".

by jammers. The mobility of these nodes could ensure the resilience up to a given number of jammers. We apply our methodology to study the resilience of two MANETs' routing protocols to jamming attacks. In particular, we study if under different scenarios and spatial distribution of the jammers, the protocols are able to ensure the network connectivity among the nodes that are not directly disturbed by the attackers. We consider two different routing strategies for the Hybrid Wireless Mesh Protocol (HWMP), the first based on the reactive Ad-hoc On-Demand Distance Vector (AODV) protocol and the second using a proactive tree-based approach.

It is worth of notice that our calculus can be encoded into the probabilistic calculi used by automatic tools, such as the PRISM model checker [8]. As a consequence, more complicated scenarios than that considered here could be studied.

Related Works. Probabilistic and stochastic models are widely used in the design and verification of complex systems. Palamidessi et al. in [5] define an extension of the applied pi-calculus with non-deterministic and probabilistic choice operators. In the context of performance evaluation, in [7] Hillston introduces the Performance Evaluation Process Algebra (PEPA) that is used for modelling systems consisting of concurrently active components which co-operate and share work. Bernardo et al. introduce EMPA_{gr} [1], an extended Markovian process algebra including probabilities, priority and exponentially distributed durations. All these calculi abstract out the interferences, since they consider only atomic actions, and they do not allow multiple devices to transmit at the same time. The problem of interference is considered by Sangiorgi et al. [9] who propose a calculus to detect collisions due to the simultaneous transmissions of two or more devices. Differently from our work, their calculus does not support mobility of nodes. In the literature two types of jamming strategies are distinguished: *proactive jammers* keep the channel permanently occupied so that no transmissions are possible whereas *reactive jammers* only jam once an ongoing transmission has been detected. In [15], Xu et al. provide a detailed classification, and show how to distinguish interferences due to network collisions from the jamming caused by malicious attackers. Strasser et al. [14] show that it is extremely difficult to distinguish reactive jamming from simple network congestions.

Plan of the paper. Section 2 introduces our calculus. Section 3 defines the methodology for evaluating resilience to a jamming attack. Section 4 introduces a case study. Section 5 develops an effective proof technique and in Section 6 we apply it to the case study. Section 7 concludes the paper.

2. THE CALCULUS

The framework we propose is based on the probabilistic calculus presented in [10]. We introduce here a slightly simplified version of this calculus by focusing on its main peculiarity, the non-atomicity of the output and input actions to capture the presence of interferences caused by the simultaneous transmissions of two (or more) nodes.

Syntax. We use letters c for *channels*, n for *nodes*, l for *locations*, r for *transmission radii*, x and y for *variables*. *Closed values* contain nodes, locations, transmission radii and any basic value (booleans, integers, ...). *Values* include also variables. We use u and v for closed values and w for (open) values. We write $\tilde{v}$, $\tilde{w}$ for tuples of values. $\mathcal{N}$ denotes the set of all networks, and **Loc** the set of all locations.

Networks are collections of nodes, running in parallel and using channels to communicate messages.
- $\mathbf{0}$ denotes the empty network;
- $M_1|M_2$ is the parallel composition of two networks;
- $n[P]_l$ is a network node named n located at the physical location l, and executing the process P;
- $(\nu c)M$ is a network where the channel c is private with scope M. Note that given the structure of the syntactic productions, channels may not be dynamically created and thus $(\nu c)M$ simply plays the role of a CCS-style hiding operator. Of course, since channels represent radio frequencies, they may not be hidden in practice. Indeed, the use of the hiding operator is only meant to specialize the verification method to some specific class of contexts as we will see later.
Processes are sequential and live within the nodes:
- $\mathbf{0}$ is the inactive process;
- $\mathtt{in}(c, \tilde{x}).P$ is ready to listen to a transmission; here the variables in $\tilde{x}$ are bound with scope in P;
- $\mathtt{out}\langle c_r, \tilde{w}\rangle.P$ is ready to transmit; here tag r represents the transmission radius of the sender: the choice of specific transmission ranges may depend on various parameters, and is left to the process running inside the transmitter node.
- $[w_1 = w_2]P, Q$ behaves as P if $w_1 = w_2$, as Q otherwise.
- $A\langle\tilde{w}\rangle$ is the process defined via a (possibly recursive) definition $A(\tilde{x}) \stackrel{\text{def}}{=} P$, with $|\tilde{x}| = |\tilde{w}|$ where $\tilde{x}$ contains all channels and variables that are free in P.
In the dynamic of the calculus processes that are ready to send or receive evolve into active senders and receivers:
- $c(\tilde{x}).P$ is actively receiving a tuple $\tilde{w}$ of (closed) values via channel c and continues as $P\{\tilde{w}/\tilde{x}\}$, i.e., as P with $\tilde{w}$ substituted for $\tilde{x}$ (where $|\tilde{x}| = |\tilde{w}|$);
- $\bar{c}_r\langle\tilde{w}\rangle.P$ is transmitting a tuple of values $\tilde{w}$ via channel c and then continues as P.
We say that a process P is *active* if it is in prefix form, with the prefix denoting an active input or output action. Predicate $\mathtt{Active}(P)$ is $\mathtt{true}$ when P is active, and $\mathtt{A}(M)$ denotes the network composed of all the active nodes in M, i.e., all nodes $n[P]_l$ in M with P active.
Each node n is associated with a pair $\langle r_n, \mathbf{J}^n\rangle$, where r_n is a non negative real number denoting the maximum transmission radius that n can use to transmit, while $\mathbf{J}^n$ is the transition matrix of a discrete time Markov chain: each entry $\mathbf{J}^n_{lk}$ denotes the probability that the node n located at l may move to the location k. Hence, $\sum_{k\in\mathbf{Loc}} \mathbf{J}^n_{lk} = 1$ for all locations $l \in \mathbf{Loc}$. Static nodes are associated with the identity Markov chain, i.e., the identity matrix $\mathbf{J}^n_{ll} = 1$ for $l \in \mathbf{Loc}$ and $\mathbf{J}^n_{lk} = 0$ for $l \neq k$. We note by μ^n_l the probability distribution associated with node n located at l, that is, the function over **Loc** such that $\mu^n_l(k) = \mathbf{J}^n_{lk}$, for all $k \in \mathbf{Loc}$.
Let n be a node of a network M and l its location. We denote by $M\{n : l'/l\}$ the network obtained by substituting l by l' in n and by $[\![M]\!]_{\mu^n_l}$ the probability distribution over networks induced by μ^n_l and defined by: for all networks M',

$$[\![M]\!]_{\mu^n_l}(M') = \begin{cases} \mu^n_l(l') & \text{if } M' = M\{n : l'/l\} \\ 0 & \text{otherwise.} \end{cases}$$

Intuitively, $[\![M]\!]_{\mu^n_l}(M')$ is the probability that network M evolves to M' due to the movement of its node n located at l. We say that M' is in the support of $[\![M]\!]_{\mu^n_l}$ if $[\![M]\!]_{\mu^n_l}(M') \neq 0$. We write $[\![M]\!]_\Delta$ for the Dirac distribution on network M, namely the probability distribution defined as: $[\![M]\!]_\Delta(M) = 1$ and $[\![M]\!]_\Delta(M') = 0$ for all M' such that $M' \neq M$. Finally,

$$\text{(R-Bgn-Bcast)}\ \frac{\forall i \in I.d(l,l_i) > r_i \quad \forall i \in I\ \forall j \in J.d(l_i,l_j) > r_i \quad \forall h \in (J \cup K).d(l,l_h) \le r}{n[\mathtt{out}\langle c_r, \tilde{v}\rangle.P]_l \mid M \rightarrow [\![n[\bar{c}_r\langle\tilde{v}\rangle.P]_l \mid M']\!]_\Delta}$$

$$\text{where } M \equiv \prod_{i \in I} n_i[\bar{c}_{r_i}\langle\tilde{v}_i\rangle.P_i]_{l_i} \mid \prod_{j \in J} n_j[\mathtt{in}(c,\tilde{x}_j).P_j]_{l_j} \mid \prod_{k \in K} n_k[c(\tilde{x}_k).P_k]_{l_k},$$
$$M' \equiv \prod_{i \in I} n_i[\bar{c}_{r_i}\langle\tilde{v}_i\rangle.P_i]_{l_i} \mid \prod_{j \in J} n_j[c(\tilde{x}_j).P_j]_{l_j} \mid \prod_{k \in K} n_k[P_k\{\bot/\tilde{x}_i\}]_{l_k}$$

$$\text{(R-End-Bcast)}\ \frac{\forall j \in J.d(l,l_j) \le r}{n[\bar{c}_r\langle\tilde{v}\rangle.P]_l \mid \prod_{j \in J} n_j[c(\tilde{x}_j).P_j]_{l_j} \rightarrow [\![n[P]_l \mid \prod_{j \in J} n_j[P_j\{\tilde{v}/\tilde{x}_j\}]_{l_j}]\!]_\Delta}$$

$$\text{(R-Res)}\ \frac{M \rightarrow [\![M']\!]_\theta}{(\nu c)M \rightarrow [\![(\nu c)M']\!]_\theta} \qquad \text{(R-Move)}\ \frac{\mathtt{Active}(P) = \mathtt{false}}{n[P]_l \rightarrow [\![n[P]_l]\!]_{\mu_l^n}}$$

$$\text{(R-Par)}\ \frac{M \rightarrow [\![M']\!]_\theta}{M|N \rightarrow [\![M'|N]\!]_\theta} \qquad \text{(R-Struct)}\ \frac{N \equiv M \quad M \rightarrow [\![M']\!]_\theta \quad M' \equiv N'}{N \rightarrow [\![N']\!]_\theta}$$

Table 1: Reduction Semantics

we let θ range over $\{\mu_l^n \mid n \text{ is a node and } l \in \mathbf{Loc}\} \cup \{\Delta\}$. A message broadcast by a node is received only by the nodes that lie in the area delimited by the transmission radius of the sender. We presuppose a function $d(\cdot,\cdot)$ which takes two locations and returns the distance separating them.

Reduction semantics. The dynamics of the calculus is specified by the *probabilistic reduction relation* $\rightarrow$ in Table 1: $M \rightarrow [\![M']\!]_\theta$ denotes a transition that leaves from M and leads to a probability distribution $[\![M']\!]_\theta$. As usual, reduction relies on a structural congruence $\equiv$, such that, e.g., $M|N \equiv N|M$, $(M|N)|M' \equiv M|(N|M')$ and $M|\mathbf{0} \equiv M$.
The synchronization over a wireless channel is described by the rules (R-Bgn-Bcast) and (R-End-Bcast). (R-Bgn-Bcast) models the start of a transmission, with node n transiting from ready to active state to transmit message $\tilde{v}$ on channel c with radius r. The state change in n may cause a collision, which the rule captures as follows. We abuse the notation and write $n_h \in H$ to note nodes n_h with $h \in H$, for any index set H. The premise of the rule describes a situation in which nodes $n_i \in I$ and $n_k \in K$ are actively involved in a synchronization, while node n and the $n_j \in J$ are in (output and input, respectively) ready state. Given that all the active transmitters are out of n's range (because $d(l,l_i) > r_i$), n transits into active state: this awakes the $n_j \in J$, as they are now in range of an active transmitter, and at the same time causes a collision at the $n_k \in K$, which also are in range and were already active on input: as a result the $n_k \in K$ exit their active state, receiving the error signal $\bot$. All the remaining active receivers that do not sense a collision, and are in the range of an active sender may conclude the synchronisation, as described by the (R-End-Bcast) rule.
According to rule (R-Move), a node n located at l and executing a move action will reach a location with probability described by the distribution μ_l^n that depends on the Markov chain $\mathbf{J}^n$ statically associated with n. We assume that a node can move only if it is not actively involved in any synchronization. This is a fairly common assumption in wireless networks in which wireless synchronization may be assumed to be orders of magnitude faster than node mobility.
All the remaining rules are standard. However, notice that an application of the (R-Par) rule may cause messages to be lost by active receivers located within the range of an active sender, even when there is no interference. Similarly, an application of (R-Par) may exclude any set of active senders and/or receivers from a synchronization: in both cases, the network is left in an inconsistent state, with active senders (dually receivers) and no receiver (sender) in range. In order to disregard these situations, we will resort to the notion of "admissible scheduler" (to be discussed shortly) to guide the dynamics of networks through "well-formed" executions.
Given a network M, we write $M \rightarrow_\theta N$ if $M \rightarrow [\![M']\!]_\theta$ and N is in the support of $[\![M']\!]_\theta$. An execution for M is a (possibly infinite) sequence $M \rightarrow_{\theta_1} M_1 \rightarrow_{\theta_2} M_2 \ldots$. We write $Exec_M$ for the set of all possible executions starting from M, $last(e)$ for the final state of a *finite* execution e, e^j for the prefix execution $M \rightarrow_{\theta_1} M_1 \ldots \rightarrow_{\theta_j} M_j$ of length j of the execution $e = M \rightarrow_{\theta_1} M_1 \cdots \rightarrow_{\theta_j} M_j \rightarrow_{\theta_{j+1}} M_{j+1} \cdots$, and $e\!\uparrow$ for the set of e' such that $e \le_{prefix} e'$. We write $M \rightarrow^* M'$ if there exists a finite execution $e \in Exec_M$ with $last(e) = M'$. We denote by $behave(M)$ the set $\{[\![M']\!]_\theta \mid M \rightarrow [\![M']\!]_\theta\}$.
In order to solve the nondeterminism in a network execution, we consider each possible transition $M \rightarrow [\![M']\!]_\theta$ as arising from a *scheduler* (see [13]). A *scheduler* is a total function F assigning to a finite execution e a distribution $[\![N]\!]_\theta \in behave(last(e))$. We define the set of executions starting from a network M and driven by a scheduler F as:

$$Exec_M^F = \{e = M \rightarrow_{\theta_1} M_1 \rightarrow_{\theta_2} M_2 \ldots \mid \forall j,\ M_{j-1} \rightarrow [\![M_j']\!]_{\theta_j},\ [\![M_j']\!]_{\theta_j} = F(e^{j-1}) \text{ and } M_j \text{ is in the support of } [\![M_j']\!]_{\theta_j}\}.$$

Given a finite execution $e = M \rightarrow_{\theta_1} M_1 \ldots \rightarrow_{\theta_k} M_k$ starting from M and driven by a scheduler F we define

$$P_M^F(e) = [\![M_1']\!]_{\theta_1}(M_1) \cdot \ldots \cdot [\![M_k']\!]_{\theta_k}(M_k)$$

where $\forall j \le k$, $[\![M_j']\!]_{\theta_j} = F(e^{j-1})$. Given a scheduler F, $\sigma Field_M^F$ is the smallest sigma field on $Exec_M^F$ that contains the basic cylinders $e\!\uparrow$, where $e \in Exec_M^F$. The probability measure $Prob_M^F$ is the unique measure on $\sigma Field_M^F$ such that $Prob_M^F(e\!\uparrow) = P_M^F(e)$. Given a measurable set of networks H, we note by $Exec_M^F(H)$ the set of executions starting from M and crossing a state in H, i.e., $Exec_M^F(H) = \{e \in Exec_M^F \mid last(e^j) \in H \text{ for some } j\}$. We denote the probability for a network M to evolve into a network in H according to F as $Prob_M^F(H) = Prob_M^F(Exec_M^F(H))$.

We consider the class of well-formed networks (resp. executions) where, (1) a transmitter, before transiting in active state checks that, locally, the communication channel is not presently busy with other transmissions, and (2) each active receiver in the network is in the transmission cell of exactly one transmitter. Formally, we restrict the set of all schedulers to the specific set of *admissible* schedulers. We say that channel c is at the top level of a network M, denoted $c \in \mathtt{Top}(M)$, if $M \equiv (\nu\tilde{d})(n[P]_l \mid N)$ and P is of the form $\mathtt{in}(c,\tilde{x}).Q$; $c(\tilde{x}).Q$; $\mathtt{out}\langle c_{L,r},\tilde{w}\rangle.Q$; or $\bar{c}_{L,r}\langle\tilde{w}\rangle.Q$.
A network M is *well-formed* if either $\mathtt{A}(M) \equiv \mathbf{0}$ or $\mathtt{A}(M) \equiv (\nu\tilde{d})\big(\prod_{i\in I} n_i[\bar{c}_{L_i,r_i}\langle\tilde{v}_i\rangle.P_i]_{l_i} \mid \prod_{j\in J} n_j[c(\tilde{x}_j).P_j]_{l_j} \mid \mathtt{A}(N)\big)$ for some N and (1) $\forall i,i' \in I.d(l_i,l_{i'}) > max(r_i,r_{i'})$, (2) $\forall j \in J.\exists! i \in I$ such that $d(l_i,l_j) \leq r_i$, (3) $c \notin \mathtt{Top}(\mathtt{A}(N))$, and N is well-formed.
A scheduler F is *admissible* if for all executions e and for all networks M in the support of $F(e)$, M is well-formed. We shall denote the set of admissible schedulers by *Sched*.

3. RESISTANCE TO JAMMING

We present a methodology, based on observation equivalences, for evaluating resistance to jammings. Roughly speaking, we say that a network N, governed by a specific routing protocol, is resilient to a jamming attack if the behaviour of N in a context without the jammer is observational equivalent to the behaviour of N in the presence of the jammer.
We formalize the observational semantics of our calculus in terms of a notion *barb* [11]. We first introduce a notation for *strong barb*: for a network M, we write $M{\downarrow_{c@K}}$ when $M \equiv (\nu\tilde{d})(n[\bar{c}_r\langle\tilde{v}\rangle.P]_l \mid M')$, with $c \notin \tilde{d}$, $K \subseteq \mathbf{Loc}$ and for all $k \in K$, $d(l,k) \leq r$. We say that a well-formed network M has a *barb* with probability p on a channel c at locations in K according to the scheduler F, written $M{\Downarrow_p^F} c@K$, if $Prob_M^F(H) = p$ with $H = \{M' \mid M \rightarrow^* M'{\downarrow_{c@K}}\}$.
Schedulers constitute an essential feature for modeling communication protocols as they provide freedom in modeling implementation and incomplete knowledge of the system. However, many schedulers could be unrealistic or useless. Consider, e.g., schedulers giving priority to movements over communications which possibly prevent any node transmission, or schedulers giving priority to end broadcasting over begin broadcasting which will prevent any interference.
In this paper we specialize the observational semantics given in [10] in order to compare the behaviour of networks relative to a restricted set of schedulers. Since our semantics is contextual, we need to ensure that the set of schedulers we consider allows the specific networks we analyze to interact with any possible context. Hence, for a set $\mathcal{F}$ of schedulers and a set $\mathcal{M} = \{M_1, M_2, ...\}$ of networks, we define the *contextual superset* $\mathcal{F}_\mathcal{M}$ of $\mathcal{F}$ relative to $\mathcal{M}$, as the largest set of schedulers allowing M_1, M_2, ... to interact with any possible context even when driven by $\mathcal{F}$ (see [2] for a formal definition). Hereafter a context is a network term with a hole $[\cdot]$ defined by the grammar: $\mathcal{C}[\cdot] ::= [\cdot] \mid [\cdot]|M \mid M|[\cdot] \mid (\nu c)[\cdot]$.

Our probabilistic observational congruence relative to a specific set of schedulers is defined as follows.

Definition 1. *Let $\mathcal{M}$ be a set of networks, $\mathcal{F} \subseteq Sched$, and $\mathcal{R}$ be a relation over networks.*
- $\mathcal{R}$ *is* barb preserving *relative to $\mathcal{F}_\mathcal{M}$ if $M\mathcal{R}N$ and $M{\Downarrow_p^F} c@K$ for some $F \in \mathcal{F}_\mathcal{M}$ implies that there exists $F' \in \mathcal{F}_\mathcal{M}$ such that $N{\Downarrow_p^{F'}} c@K$.*
- $\mathcal{R}$ *is* reduction closed *relative to $\mathcal{F}_\mathcal{M}$ if $M\mathcal{R}N$ implies that for all $F \in \mathcal{F}_\mathcal{M}$, there exists $F' \in \mathcal{F}_\mathcal{M}$ such that for all classes $\mathcal{C} \in \mathcal{N}/\mathcal{R}$, $Prob_M^F(\mathcal{C}) = Prob_N^{F'}(\mathcal{C})$.*
- $\mathcal{R}$ *is* contextual *if $M\mathcal{R}N$ implies that for every context $\mathcal{C}[\cdot]$, it holds that $\mathcal{C}[M]\,\mathcal{R}\,\mathcal{C}[N]$.*
- Probabilistic observational congruence relative to $\mathcal{F}_\mathcal{M}$, *written $\cong_p^{\mathcal{F}_\mathcal{M}}$, is the largest symmetric relation over networks which is reduction closed and barb preserving relative to $\mathcal{F}_\mathcal{M}$, and contextual.*

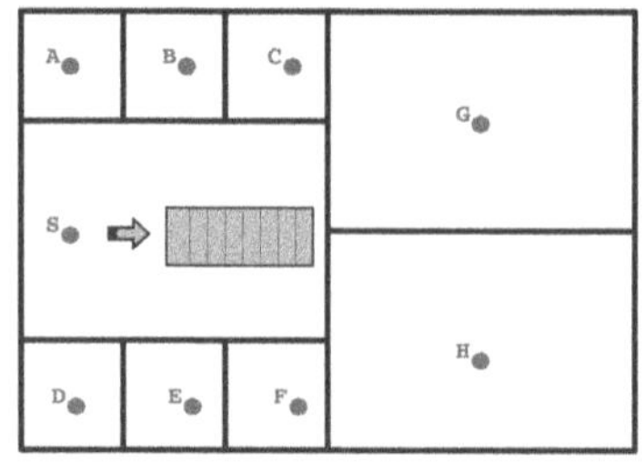

Figure 1: Topology of a building floor

Two networks are related by $\cong_p^{\mathcal{F}_\mathcal{M}}$ if they exhibit the same probabilistic behaviour relative to $\mathcal{F}_\mathcal{M}$.
The resistance to jamming of a network N governed by a specific routing protocol is evaluated by comparing the observational behaviour of N with the behaviour of the same network in the jamming context.

Definition 2 (Jamming-resistance). *Let N be a network and $C[\cdot]$ be a context consisting of a malicious jamming node. Let $\mathcal{F}$ be a set of schedulers and $\tilde{c}$ be a set of channels. We say that N is* jamming-resistant *in the context $C[\cdot]$, relative to the set of schedulers $\mathcal{F}$ and independently from communications over channels $\tilde{c}$, if*

$$(\nu\tilde{c})N \cong_p^{\mathcal{F}} (\nu\tilde{c})C[N].$$

The hiding operator $(\nu\tilde{c})$ is used here to specialize the verification method to some specific class of communications.

4. CASE STUDY

Let us consider an in-door MANET operating in a building of 25 × 25 meters with three floors whose height is 3 meters. For the sake of simplicity, we assume each floor to have the same topology: eight rooms connected by a central corridor as shown in Figure 1. A location is denoted by a pair $\langle f, r\rangle$ where $f \in \{1,2,3\}$ indicates the floor, while $r \in \{A,B,C,D,E,F,G,H,S\}$ indicates the room. Rooms A,B,C,D,E,F have dimension 5×5 meters, while G,H,S have dimension 10×15 meters. The MANET consists of seven devices: three stationary nodes (n_1, n_2, n_3), located respectively in the first, second and third floor, which represent the servers inside the building, and four mobile nodes (n_4, n_5, n_6, n_7). Nodes can transmit with three transmission radii: $r_1 = 5m$, $r_2 = 10m$ and $r_3 = 15m$. We assume the network uses omnidirectional antennas, and adopt the euclidean function to evaluate the distance between two locations. Hence, a node sending a message with radius r_1 can reach only those nodes lying in the adjacent rooms, or in the same room of an adjacent floor, while using radius r_2 the successfulness of a reception depends on the kind of rooms the sender and the receivers are occupying. We assume that each transmission always begins exactly from the centre of a room. In practice, if n_1 is located at $l = \langle 1, A\rangle$ and n_2 is located at $k = \langle 2, A\rangle$ then their distance is $d(l,k) = 3m$. Hence, they are able to mutually communicate by using any of the transmission radii in $\{r_1, r_2, r_3\}$. If n_1 is located at $l' = \langle 1, C\rangle$, then the Euclidean distance between l' and k is $d(l',k) = \sqrt{109}$. Since $10 < \sqrt{109} < 15$, n_1 and n_2 can communicate only using radius r_3.

The Hybrid Wireless Mesh Protocol (HWMP) is the routing protocol used by IEEE 802.11s [6]. It can be configured to operate in two modes: on-demand reactive mode and tree-based proactive mode. The reactive mode is based on the Ad-hoc On-demand Distance Vector Routing protocol (AODV) [3, 12]. It uses three types of control packets whose forwarding allows each node to update its information about the best path to reach a specific destination. Each node maintains information about its neighbours in its *route table*, where each entry contains the following data associated with a given destination d:

- $seq\#_d$: the sequence number of d: it is incremented before sending a request, or (only if the node is the destination) before broadcasting the response;
- $\mathtt{nexthop}_d$: the next intermediate node towards d;
- $\mathtt{hopcount}_d$: the number of hops from the source to d;
- `lifetime`: the life-time of the record in the route table;
- $\mathtt{List\ of\ precursors}_d$: the list of nodes using the actual node as `nexthop` to reach d.

The protocol exchanges three types of control packets:

- **RREQ** (Route Request): when a node needs to find a path, it broadcasts a RREQ message and then, when it receives the response, it chooses the cheapest path (in terms of energy costs, delays, number of hops, etc.);
- **RREP** (Route Response): When a node receives a RREQ message, it controls if it is the destination. In this case it immediately sends back the response, otherwise it searches a valid path in its *route table* to send back. If there are no valid paths, it propagates the RREQ packet;
- **RERR** (Route Error): the error message informing the network of a link breakage.

The tree-based proactive mode used by the HWMP protocol is based on a proactive protocol in which the network topology is built statically, forming a tree rooted in a chosen node. As in the previous mode, packet types, apart from root announcements, are RREQ, RREP and RERR. The difference lays in the fact that network topology is built statically from the start, and each node selects as the next hop the neighbour node that is nearest to the root according to its hop count, but it also maintains a table that contains the hop count to the root of its neighbours. The final result of these operations is that a spanning tree is created and used for the subsequent message forwarding. Whenever a node detects that the link with the upstream node, i.e., the node to which it would normally forward packets, is broken, it selects a new upstream node from its neighbours using the table of the hop counts, updates its own count and then broadcasts a RRER message to the downstream nodes. Then, these can select their own alternative paths if the hop count of the upstream node is no longer the shortest one. If a new path is chosen, the RERR message is propagated to the downstream nodes. The effect of this action is to build a new spanning subtree. In what follows we assume that the root node, e.g., a gateway between a wired network and the wireless one, is chosen statically and that it never moves.

Modelling the system. We model our case study into the framework of Section 2. The set of feasible locations is: $\mathbf{Loc} = \{\langle f, r\rangle : f \in \{1,2,3\} \wedge r \in \{A,B,C,D,E,F,G,H,S\}\}$. Each node n_i is statically characterised by a pair $\langle r_{n_i}, \mathbf{J}^{n_i}\rangle$, where r_n is the maximum transmission radius, i.e., in our scenario $r_{n_i} = r_3$ for all n_i, with $i = 1, \ldots, 7$, and $\mathbf{J}^{n_i}$ is the transition matrix of a discrete time Markov chain denoting the probability of movements. In its simplest definition, the routing table maintained by each node has the form:

$$\langle d, seq\#_d, \mathtt{nexthop}_d, \mathtt{hopcount}_d, LP_d\rangle$$

where d is the primary key of the table and identifies the node name of the destination associated with the entry. The packets used by HWMP are:

- $(\mathtt{rreq}, \mathtt{Bid}_{s,d}, (s, seq\#_s), (d, seq\#_d), \mathtt{hopcount}_{s,d})$ for the request of a path from s to d;
- $(\mathtt{rrep}, (s, seq\#_s), (d, seq\#_d), \mathtt{hopcount}_{s,d})$ for the response to a path request from s to d;
- $(\mathtt{rerr}, s, d, seq\#_d)$ denotes an error message;
- $(\mathtt{ack}, d, seq\#_d)$ to confirm the reception of a route response.

We consider the behaviour of the network assuming that a node needing a path to a given destination simply obtains it by calling the procedure $\mathtt{find\_path}(s,d)$ where s and d are the source and destination nodes. The effect of this procedure is to update the node's routing table.

We consider the following network N, where $\forall i \in [1-7], l_i \in \mathbf{Loc}$; here nodes n_4 and n_6 want to communicate:

$$N = \prod_{i=1}^{7} (n_i[P_i]_{l_i})$$

$$P_4 = \mathtt{find\_path}(n_4, n_6).\mathtt{out}\langle c_{r_2}, (msg, \mathtt{nexthop}_{n_6}, n_6)\rangle. \\ \mathtt{in}(c, (x_1, x_2, x_3)).[x_3 = n_4]\mathtt{out}\langle \mathtt{ok}_{r_2}, OK\rangle.P_4,$$

$$P_6 = \mathtt{in}(c, (x_1, x_2, x_3)).[x_2 = n_6](\mathtt{find\_path}(n_6, n_4). \\ \mathtt{out}\langle c_{r_2}, (ack, \mathtt{nexthop}_{n_4}, n_4)\rangle).P_6$$

and $\forall i \in \{1, 2, 3, 5, 7\}$:

$$P_i = \mathtt{in}(c, (x_1, x_2, x_3)).[x_2 = n_i](\\ [x_3 = n_4](\mathtt{find\_path}(n_i, n_4). \\ \mathtt{out}\langle c_{r_2}, (x_1, \mathtt{nexthop}_{n_4}, n_4)\rangle).P_i, \\ [x_3 = n_6](\mathtt{find\_path}(n_i, n_6). \\ \mathtt{out}\langle c_{r_2}, (x_1, \mathtt{nexthop}_{n_6}, n_6)\rangle).P_i).$$

Node n_4 uses the function `find_path` to discover the path to n_6, and then forwards the packet. When it receives the acknowledgment it fires the message OK through the channel `ok`. Node n_6 waits for the packet from n_4 and, when it receives it, it sends back the acknowledgment. Each intermediate node executes a simple process forwarding both the packet sent by n_4 and the ack of n_6.

These processes will be used to study both the *AODV* and the *tree-based* protocols, the only difference is the way the function `find_path` behaves: while in the *AODV* the path is discovered after the RREQ, RREP and RERR packages exchange, in the *tree-based* protocol, in order to find the best path, the process simply follows the predetermined routing spanning tree, built at the moment of the initial network setup through the algorithm informally described above.

$$\text{(Beg-Out)}\ \frac{-}{\mathtt{out}\langle c_r,\tilde{v}\rangle.P \xrightarrow{\bar{c}_r} \bar{c}_r\langle\tilde{v}\rangle.P} \qquad \text{(End-Out)}\ \frac{-}{\bar{c}_r\langle\tilde{v}\rangle.P \xrightarrow{\bar{c}_r\tilde{v}} P}$$

$$\text{(Beg-In)}\ \frac{-}{\mathtt{in}(c,\tilde{x}).P \xrightarrow{c} c(\tilde{x}).P} \qquad \text{(End-In)}\ \frac{-}{c(\tilde{x}).P \xrightarrow{c\vartheta} P\{\vartheta/\tilde{x}\}}$$

$$\text{(Then)}\ \frac{P \xrightarrow{\eta} P'}{[v=v]P,Q \xrightarrow{\eta} P'} \qquad \text{(Else)}\ \frac{Q \xrightarrow{\eta} Q' \quad v_1 \neq v_2}{[v_1=v_2]P,Q \xrightarrow{\eta} Q'} \qquad \text{(Rec)}\ \frac{P\{\tilde{v}/\tilde{x}\} \xrightarrow{\eta} P' \quad A(\tilde{x}) \stackrel{\text{def}}{=} P}{A\langle\tilde{v}\rangle \xrightarrow{\eta} P'}$$

Table 2: LTS rules for Processes

We consider different jamming attacks. We focus our attention on proactive jammers, executing the following process: $P = \mathtt{out}\langle c_{r_2}, JAM\rangle.P$ which continuously broadcasts the dummy message JAM with radius r_2 on channel c. We consider two malicious nodes: a static jamming attacker m_1 identified by $\langle r_2, \mathbf{I}\rangle$, with $\mathbf{I}$ being the identity matrix, located at $k = \langle 1, H\rangle$, and a mobile node m_2 identified by $\langle r_2, \mathbf{J}^{m_2}\rangle$ whose initial location is $k' = \langle 3, G\rangle$. Note that the jammer m_1 is a node blocking the activity of each node lying within $\mathbf{Loc}_{m_1} = \{\langle 1,H\rangle, \langle 2,H\rangle, \langle 3,H\rangle, \langle 1,F\rangle, \langle 2,F\rangle, \langle 3,F\rangle\}$.
We study the behaviour of the protocol in a context C_1 consisting of node m_1, and in a context C_2 consisting of both m_1 and m_2. We consider the set of schedulers $\mathcal{F}$ such that: (1) during the forwarding of the packets (x_1, x_2, x_3) where $x_1 \in \{msg, ack\}$, movements of nodes n_i with $i \in [4-7]$ occurs at each collision or JAM-message reception, (2) the beginning of output actions have priority on their ending. Let $C_1[\cdot] = \cdot \mid m_1[P]_k$. The robustness of the network using the HWMP protocol against the malicious node m_1 can be verified by checking if the observational behaviour of N is independent of the presence of the jammer inside the building. Formally, we have to prove that:

$$(\nu c)N \cong_p^{\mathcal{F}} (\nu c)C_1[N].$$

The restriction operator νc is used to limit the observation to the communications between nodes n_4 and n_6. Indeed, since c is hidden, the only observable action in $(\nu c)N$ and $(\nu c)C_1[N]$ is $\xrightarrow{\text{ok}!OK@K}$ for some $K \subseteq \mathbf{Loc}$. Moreover, we consider the dynamic context $C_2[\cdot] = \cdot \mid m_1[P]_k \mid m_2[P]_{k'}$ where m_2 is a mobile node, meaning that the *jamming area* may change in time. In this case, we have to prove that:

$$(\nu c)N \cong_p^{\mathcal{F}} (\nu c)C_2[N].$$

5. PROOF TECHNIQUE

We present a bisimulation-based proof technique for $\cong_p^{\mathcal{F}_{\mathcal{M}}}$ which is built upon a LTS semantics. Table 2 presents the LTS rules for processes. Transitions are of the form $P \xrightarrow{\eta} P'$, where η ranges over input and output actions, defined by:

$$\eta ::= c \mid c\vartheta \mid \bar{c}_r \mid \bar{c}_r\tilde{v} \quad \text{with} \quad \vartheta ::= \tilde{v} \mid \bot.$$

Rule (End-In) models either the correct reception of a message or the reception of a $\bot$ due to a collision.
Table 3 presents the LTS rules for networks. The transitions are of the form $M \xrightarrow{\gamma} [\![M']\!]_\theta$, where M is a network, $[\![M']\!]_\theta$ is a distribution over networks, and γ ranges over the labels:

$$\gamma ::= c?@l \mid c?\vartheta@l \mid c![l,r] \mid c!\tilde{v}[l,r] \mid c!\tilde{v}@K \triangleleft R \mid \tau.$$

We denote by $\mathtt{A}^s_M(c,l)$ the set of active senders of M on channel c reaching l. Rule (End-Rcv) models both the successful reception of a message or the reception of a failure message ($\bot$) due to an interference. Rule (Obs) models the observability of a transmission: every transmission may be detected (and hence *observed*) by any recipient located within the transmission cell of one sender and outside the interference area. The label $c!\tilde{v}@K \triangleleft R$ represents the transmission of the tuple $\tilde{v}$ of messages via c to the subset K of locations inside the reachable locations R within the transmission cell of the sender. Note that collisions are not observable. Rules (Lose1) and (Lose2) model both message loss and local activities. In Rule (Res) $\mathtt{Chan}(\gamma) = c$ if γ is of the form $c?@l$; $c?\vartheta@l$; $c![l,r]$; $c!\tilde{v}[l,r]$; or $c!\tilde{v}@K \triangleleft R$, and $\mathtt{Chan}(\tau) = \bot$.

Based on the LTS semantics, we define a notion of probabilistic labelled bisimulation relative to a set of schedulers $\mathcal{F}_{\mathcal{M}}$. It is built upon the following actions:

$$\alpha ::= c?@l \mid c?\vartheta@l \mid c!\tilde{v}@K \triangleleft R \mid \tau.$$

We write $M \xrightarrow{\alpha}_\theta N$ if $M \xrightarrow{\alpha} [\![M']\!]_\theta$ and N is in the support of $[\![M']\!]_\theta$. We write $M \xrightarrow{\alpha} N$ if $M \xrightarrow{\alpha}_\theta N$ for some θ. A labelled *execution* e of a network M is a finite (or infinite) sequence of steps: $M \xrightarrow{\alpha_1}_{\theta_1} M_1 \xrightarrow{\alpha_2}_{\theta_2} M_2 \dots \xrightarrow{\alpha_k}_{\theta_k} M_k$. With abuse of notation, we define $Exec_M$, $last(e)$, e^j and $e\!\uparrow$ as for unlabeled executions. We denote by $lbehave(M)$ the set $\{(\alpha, [\![M']\!]_\theta) \mid M \xrightarrow{\alpha} [\![M']\!]_\theta\}$. A scheduler[1] for the labelled semantics is a function F assigning a pair $(\alpha, [\![M]\!]_\theta) \in lbehave(last(e))$ with a finite labelled execution e. We note by $LSched$ the set of (admissible) schedulers for the LTS semantics and by $Exec^F_M$ the set of all labelled executions starting from M and driven by F. Since we are interested in weak equivalences, that abstract over τ-actions, we introduce the notion of *weak action* as: $\Longrightarrow$ is the transitive and reflexive closure of $\xrightarrow{\tau}$, $\xRightarrow{c?@k}$ denotes $\Longrightarrow\xrightarrow{c?@k}\Longrightarrow$, $\xRightarrow{c?\vartheta@k}$ denotes $\Longrightarrow\xrightarrow{c?\vartheta@k}\Longrightarrow$, $\xRightarrow{c!\tilde{v}@K\triangleleft R}$ denotes $\Longrightarrow\xrightarrow{c!\tilde{v}@K\triangleleft R}\Longrightarrow$, $\xRightarrow{\hat{\alpha}}$ denotes $\Longrightarrow$ when $\alpha = \tau$ and $\xRightarrow{\hat{\alpha}}$ when $\alpha \neq \tau$.
We denote by $Exec^F_M(\xRightarrow{\alpha}, H)$ the set of executions that, starting from M, according to the scheduler F, lead to a network in the set H by performing $\xRightarrow{\alpha}$. We define the probability of reaching a network in H from M through $\xRightarrow{\alpha}$, according to F as $Prob^F_M(\xRightarrow{\alpha}, H) = Prob^F_M(Exec^F_M(\xRightarrow{\alpha}, H))$. We are interested in a notion of bisimilarity which is a complete characterisation of our observational congruence. Since the latter is relative to a set of schedulers $\mathcal{F}_{\mathcal{M}}$, we need to define the corresponding set $\hat{\mathcal{F}}_{\mathcal{M}}$ for the LTS semantics [2].

[1] We still use F to denote a scheduler for the LTS semantics.

$$\text{(Beg-Snd)}\ \frac{P \xrightarrow{\bar{c}_r} P'}{n[P]_l \xrightarrow{c![l,r]} [\![n[P']_l]\!]_\Delta} \qquad \text{(End-Snd)}\ \frac{P \xrightarrow{\bar{c}_r\tilde{v}} P'}{n[P]_l \xrightarrow{c!\tilde{v}[l,r]} [\![n[P']_l]\!]_\Delta}$$

$$\text{(Beg-Rcv)}\ \frac{P \xrightarrow{c} P'}{n[P]_l \xrightarrow{c?@l} [\![n[P']_l]\!]_\Delta} \qquad \text{(End-Rcv)}\ \frac{P \xrightarrow{c\vartheta} P'}{n[P]_l \xrightarrow{c?\vartheta@l} [\![n[P']_l]\!]_\Delta}$$

$$\text{(Beg-Bcast)}\ \frac{M \xrightarrow{c![l,r]} [\![M']\!]_\Delta \quad N \xrightarrow{c?@l'} [\![N']\!]_\Delta \quad d(l,l') \le r \ \wedge\ \mathtt{A}^s_N(c,l) = \mathtt{A}^s_N(c,l') = \emptyset}{M|N \xrightarrow{c![l,r]} [\![M'|N']\!]_\Delta}$$

$$\text{(Coll-Bcast)}\ \frac{M \xrightarrow{c![l,r]} [\![M']\!]_\Delta \quad N \xrightarrow{c?\bot@l'} [\![N']\!]_\Delta \quad d(l,l') \le r \ \wedge\ \mathtt{A}^s_N(c,l) = \emptyset}{M|N \xrightarrow{c![l,r]} [\![M'|N']\!]_\Delta}$$

$$\text{(End-Bcast)}\ \frac{M \xrightarrow{c!\tilde{v}[l,r]} [\![M']\!]_\Delta \quad N \xrightarrow{c?\tilde{v}@l'} [\![N']\!]_\Delta \quad d(l,l') \le r}{M|N \xrightarrow{c!\tilde{v}[l,r]} [\![M'|N']\!]_\Delta}$$

$$\text{(Obs)}\ \frac{M \xrightarrow{c!\tilde{v}[l,r]} [\![M']\!]_\Delta \quad R = \{l' : d(l,l') \le r \wedge \mid \mathtt{A}^s_M(c,l') \mid = 1\} \quad K \subseteq R}{M \xrightarrow{c!\tilde{v}@K \triangleleft R} [\![M']\!]_\Delta}$$

$$\text{(Lose1)}\ \frac{M \xrightarrow{c![l,r]} [\![M']\!]_\Delta}{M \xrightarrow{\tau} [\![M']\!]_\Delta} \qquad \text{(Lose2)}\ \frac{M \xrightarrow{c!\tilde{v}[l,r]} [\![M']\!]_\Delta}{M \xrightarrow{\tau} [\![M']\!]_\Delta}$$

$$\text{(Move)}\ \frac{\mathtt{Active}(P) = \mathtt{false}}{n[P]_l \xrightarrow{\tau} [\![n[P]_l]\!]_{\mu^n_l}} \qquad \text{(Res)}\ \frac{M \xrightarrow{\gamma} [\![M']\!]_\theta \quad \mathtt{Chan}(\gamma) \neq c}{(\nu c)M \xrightarrow{\gamma} [\![(\nu c)M']\!]_\theta} \qquad \text{(Par)}\ \frac{M \xrightarrow{\gamma} [\![M']\!]_\theta}{M|N \xrightarrow{\gamma} [\![M'|N]\!]_\theta}$$

Table 3: LTS rules for Networks

DEFINITION 3. *Let M and N be two networks. An equivalence relation $\mathcal{R}$ over networks is a* probabilistic labelled bisimulation *relative to $\hat{\mathcal{F}}_{\mathcal{M}}$ if $M\mathcal{R}N$ implies: for all scheduler $F \in \hat{\mathcal{F}}_{\mathcal{M}}$ there exists a scheduler $F' \in \hat{\mathcal{F}}_{\mathcal{M}}$ such that for all α and for all classes $\mathcal{C}$ in $\mathcal{N}/\mathcal{R}$ it holds:*

- if $\alpha=\tau$ or $\alpha=c!\tilde{v}@K \triangleleft R$: $Prob^F_M(\xrightarrow{\alpha},\mathcal{C}) = Prob^{F'}_N(\stackrel{\hat{\alpha}}{\Longrightarrow}\mathcal{C})$;
- if $\alpha = c?@l$ or $\alpha = c?\vartheta@l$ then either $Prob^F_M(\xrightarrow{\alpha},\mathcal{C}) = Prob^{F'}_N(\stackrel{\alpha}{\Longrightarrow},\mathcal{C})$ or $Prob^F_M(\xrightarrow{\alpha},\mathcal{C}) = Prob^{F'}_N(\Longrightarrow,\mathcal{C})$.

Probabilistic labelled bisimilarity *relative to $\hat{\mathcal{F}}_{\mathcal{M}}$, written $\approx_p^{\hat{\mathcal{F}}_{\mathcal{M}}}$, is the largest symmetric probabilistic labelled bisimulation relative to $\hat{\mathcal{F}}_{\mathcal{M}}$ over networks.*

Probabilistic labelled bisimilarity is a complete characterisation our probabilistic observational congruence.

THEOREM 1. *Let $\mathcal{F}$ be a set of schedulers and $\mathcal{M}$ be a set of networks. Let M and N be two well-formed networks. $M \approx_p^{\hat{\mathcal{F}}_{\mathcal{M}}} N$ if and only if $M \cong_p^{\mathcal{F}_{\mathcal{M}}} N$.*

6. PROVING RESILIENCE TO JAMMING

In order to verify resilience of our network case study with respect to the jamming context C_1, we check whether

$$(\nu c)N \cong_p^{\mathcal{F}} (\nu c)C_1[N]. \qquad (1)$$

By using the proof technique presented in Section 5, it is sufficient to find a probabilistic bisimulation containing the pair $((\nu c)N, (\nu c)C_1[N])$. Let us consider the relation

$$\mathcal{R} = \{((\nu c)\bar{N}, (\nu c)\bar{C}_1[\bar{N}]) : (\nu c)\bar{C}_1[\bar{N}] \in \bar{\mathcal{M}}\}$$

where $\bar{\mathcal{M}} = \{(\nu c)\bar{C}_1[\bar{N}] : (\nu c)C_1[N] \rightarrow^* (\nu c)\bar{C}_1[\bar{N}]\}$. In order to prove (1), we show that $\mathcal{R}$ is a bisimulation relative to $\hat{\mathcal{F}}_{\mathcal{M}}$ with $\mathcal{M} = \{(\nu c)N, (\nu c)C_1[N])\}$. Since in $\bar{\mathcal{M}}$ channel c is hidden, the only actions that $\bar{N}$ can do are τ actions, or output through the channel ok, while context $\bar{C}_1[\cdot]$ can only make τ actions. Let $(\nu c)\bar{N} \xrightarrow{\tau} [\![(\nu c)\bar{N}']\!]_\theta$ driven by $F \in \hat{\mathcal{F}}_{\mathcal{M}}$. Then, $\forall \mathcal{C} \in \mathcal{N}/\mathcal{R}$, $Prob^F_{(\nu c)\bar{N}}(\xrightarrow{\tau},\mathcal{C}) = \sum_{(\nu c)\bar{N}'' \in \mathcal{C}}$ in the support of $[\![(\nu c)\bar{N}']\!]_\theta$ $[\![(\nu c)\bar{N}']\!]_\theta((\nu c)\bar{N}'')$.
If the τ action is due to the application of the rule (Move), we conclude by applying rule (Par) to m_1.
If $(\nu c)\bar{N} \xrightarrow{\tau} (\nu c)[\![\bar{N}']\!]_\Delta$, because of an application of rule (Beg-Bcast) then $\bar{N} \xrightarrow{c![l,r_2]} [\![\bar{N}']\!]_\Delta$ for some location l. If $\bar{C}_1[\mathbf{0}] \equiv C_1[\mathbf{0}]$, or $d(l,k) > r_2$ we are done, because it is enough to apply rule (Par), since both $\bar{C}_1[\bar{N}]$ and $\bar{C}_1[\bar{N}']$ are well-formed networks, and $\exists F' \in \hat{\mathcal{F}}_{\mathcal{M}}$ such that $Prob^F_{(\nu c)\bar{N}}(\xrightarrow{\tau}, \mathcal{C}) = Prob^{F'}_{(\nu c)\bar{C}_1[\bar{N}]}(\Longrightarrow, \mathcal{C})$, as required. If $\bar{C}_1[\mathbf{0}] \equiv C'_1[\mathbf{0}]$ and $d(l,k) \le r_2$, we can not apply rule (Par) because $C'_1[\bar{N}']$ is not well-formed. Now, we have to find in $\hat{\mathcal{F}}_{\mathcal{M}}$ a scheduler allowing $C'_1[\bar{N}]$ to reach $\bar{C}_1[\bar{N}']$ with probability 1, where $\bar{C}_1[\cdot] \in \{C_1[\cdot], C'_1[\cdot]\}$. In this case, there exists F' allowing m_1 to finish its communication, i.e., $C_1[\bar{N}] \xrightarrow{\tau} [\![C_1[\bar{N}']]\!]_\Delta$.
The most interesting case is when $d(k,l_h), d(l,l_h) \le r_2$ for some $h \in [1-7]$, where $\bar{N} \equiv \prod_{i=1}^7 n_i[Q_i]_{l_i}$ and $Q_h = c(\tilde{x}).Q'_h$. This means that n_h is in the jammer's transmission area, and it receives the correct packet in $\bar{N}$, while it receives a collision in $C'_1[\bar{N}]$ avoiding it to immediately reach $\bar{C}_1[\bar{N}']$ with probability 1. If the node receiving the collision ($\bot$)

is an intermediate node in the path from n_4 to n_6, we are done, since, when n_4 does not receive the acknowledge, it searches another path to reach n_6 and it sends again the message. If the node receiving the collision is the source or the destination of the communication (n_4, n_6), the bisimulation depends on the transition matrix modelling its mobility: if the node, with a finite number h of steps, goes far away from the jammer with probability 1, i.e., the probability to end up in an ergodic set in which all states represents locations inside the jammed area is 0, then the bisimulation is proved, otherwise the probability for $C_1'[\bar{N}]$ to reach a state in $\mathcal{C}$ is $1-p$, where $p = \sum_{l' \in \mathbf{Loc}_{jam}} \mathbf{J}_{ll'}^{n_i\,(h)}$. Moreover, when the protocol uses the tree-based strategy and the root nodes are static (e.g., n_1, n_2 and n_3), if the node receiving the collision is the root itself, it can never be possible to detect an alternative route, thus the bisimulation does not hold.

Consider now the context: $C_2[\cdot] = \cdot \mid m_1[P]_k \mid m_2[P]_{k'}$. To prove that $(\nu c)N \cong_p^{\mathcal{F}} (\nu c)C_2[N]$ we check whether

$$\mathcal{S} = \{((\nu c)\bar{N}, (\nu c)\bar{C}_2[\bar{N}]) : (\nu c)\bar{C}_2[\bar{N}] \in \mathcal{M}\},$$

where $\mathcal{M} = \{(\nu c)\bar{C}_2[\bar{N}] : (\nu c)C_2[N] \to^* (\nu c)\bar{C}_2[\bar{N}]\}$, is a bisimulation. The proof proceeds as above, but here the second jammer is mobile and thus we have to consider all its possible locations. Let the Markov chain $\mathbf{J}^{n_i m_2}$ be the joint process between $\mathbf{J}^{n_i}$ and $\mathbf{J}^{m_2}$, in which states are pairs (l, k) of states of the chains $\mathbf{J}^{n_i}$ and $\mathbf{J}^{m_2}$, which are associated with nodes n_i and m_2, respectively. Hence, each pair (l, k) of $\mathbf{J}^{n_i m_2}$ means that n_i is located at l and m_2 is located at k. We consider a subset $\mathbf{S}^{n_i m_2}$ of $\mathbf{J}^{n_i m_2}$, such that $(l, k) \in \mathbf{S}^{n_i m_2} \Leftrightarrow d(l, k) \geq r_{m_2}$, i.e., all combinations of states in which the jammer can interfere with the node n_i. n_i can always communicate successfully (then the bisimulation holds) only if the probability to end up in an ergodic set of states members of $\mathbf{S}^{n_i m_2}$ is 0. When using the tree-based mode, if the root nodes are static, even if n_4 and n_6, when jammed, are able to reach a safe location with probability 1, the success of the communication depends on the root nodes, and if the root nodes are jammed, n_4 and n_6 may not be able to find a valid path to complete their communication.

7. CONCLUSION

In the previous section, we applied our methodology for evaluating the resilience to jamming attacks of a wireless network governed by two different modes of protocol HWMP. Even if both relations $\mathcal{R}$ and $\mathcal{S}$ defined above are proved to be bisimulations only under particular conditions depending on the mobility of nodes, with our proof technique we have been able to show that the reactive approach used by the HWMP protocol is more robust against jamming than the proactive one, since, when using the tree-based routing, the bisimulations are proved under more restricted conditions which depend on the mobility behaviours of the root nodes.

8. REFERENCES

[1] M. Bernardo and M. Bravetti. Performance measure sensitive congruences for Markovian process algebras. *Theoretical Computer Science*, 290(1):117–160, 2003.

[2] M. Bugliesi, L. Gallina, S. Hamadou, A. Marin, and S. Rossi. Interference-sensitive preorders for manets. Technical Report DAIS-2011-10, University Ca' Foscari Venice, 2011.

[3] I. D. Chakeres and E. M. Belding-Royer. AODV routing protocol implementation design. In *Proc. of 24th International Conference on Distributed Computing Systems Workshops - W7: EC (ICDCSW'04)*, volume 7, pages 698–703. IEEE, 2004.

[4] L. Gallina, S. Hamadou, A. Marin, and S. Rossi. A probabilistic energy-aware model for mobile ad-hoc networks. In *Proc. of the 18th International Conference on Analytical and Stochastic Modelling Techniques and Applications (ASMTA'11)*, volume 6751 of *LNCS*, pages 316–330. Springer-Verlag, 2011.

[5] J. Goubault-Larrecq, C. Palamidessi, and A. Troina. A probabilistic applied pi-calculus. In *Proc. of the 5th Asian Symposium on Programming Languages and Systems (APLAS'07)*, volume 4807 of *LNCS*, pages 175–190. Springer-Verlag, 2007.

[6] G. Hiertz, D. Denteneer, S. Max, R. Taori, J. Cardona, L. Berlemann, and B. Walke. IEEE 802.11s: the wlan mesh standard. *Wireless Communications, IEEE*, 17(1):104–111, 2010.

[7] J. Hillston. *A Compositional Approach to Performance Modelling*. Cambridge University Press, 1996.

[8] M. Kwiatkowska and D. Parker. Advances in probabilistic model checking. In *Proc. 2011 Marktoberdorf Summer School: Tools for Analysis and Verification of Software Safety and Security*. IOS Press, 2012. To appear.

[9] I. Lanese and D. Sangiorgi. An operational semantics for a calculus for wireless systems. *Theoretical Computer Science*, 411(19):1928–1948, 2010.

[10] M.Bugliesi, L.Gallina, S. Hamaodu, A. Marin, and S. Rossi. Interference-sensitive preorders for manets. In *Proc. 9th International Conference on Quantitative Evaluation of SysTems (QEST'12)*. IEEE, 2012.

[11] R. Milner and D. Sangiorgi. Barbed bisimulation. In *Proc. of International Colloquium on Automata, Languages and Programming (ICALP'92)*, volume 623 of *LNCS*, pages 685–695. Springer-Verlag, 1992.

[12] E. M. Royer and C. E. Perkins. Multicast operation of the ad-hoc on-demand distance vector routing protocol. In *Proc. of the 5th Annual ACM/IEEE International Conference on Mobile Computing and Networking*, pages 207–218. ACM, 1999.

[13] R. Segala and N. Lynch. Probabilistic simulations for probabilistic processes. In *Proc. of the 5th International Conference on Concurrency Theory (CONCUR'94)*, volume 836 of *LNCS*, pages 481–496. Springer-Verlag, 1994.

[14] M. Strasser, B. Danev, and S. Čapkun. Detection of reactive jamming in sensor networks. *ACM Transactions on Sensor Networks*, 7:16:1–16:29, 2010.

[15] W. Xu, W. Trappe, Y. Zhang, and T. Wood. The feasibility of launching and detecting jamming attacks in wireless networks. In *Proc. of the 6th ACM International Symposium on Mobile ad hoc Networking and Computing*, pages 46–57. ACM, 2005.

Evaluation of Collaborative Selfish Node Detection in MANETs and DTNs

Enrique Hernández-Orallo, Manuel D. Serrat Olmos, Juan-Carlos Cano, Carlos T. Calafate, Pietro Manzoni
Departamento de Informática de Sistemas y Computadores
Universidad Politécnica de Valencia, Valencia, Spain
ehernandez@disca.upv.es, mdserrat@upvnet.upv.es, jucano@disca.upv.es, calafate@disca.upv.es, pmanzoni@disca.upv.es

ABSTRACT

Mobile ad-hoc Networks (MANETs) and Delay Tolerant Networks (DTN) rely on network cooperation schemes to work properly. Nevertheless, if nodes have a selfish behaviour and are unwilling to cooperate, the overall network performance could be seriously affected. The use of watchdogs is a well-known mechanism to detect selfish nodes. Nevertheless, the detection process performed by watchdogs can fail, generating false positives and false negatives that can induce a wrong behaviour.

In this paper we propose a collaborative watchdog approach based on the diffusion of selfish nodes awareness, that reduces the impact of false positives and false negatives. In order to evaluate the efficiency of our approach, we introduce an analytical model to evaluate the time of detection and the induced overhead of our collaborative watchdog. The results confirm the efficiency of our approach since the detection time of selfish nodes is reduced, the overall overhead is very low, and the impact of false positives and false negatives is minimised.

Categories and Subject Descriptors

C.2.1 [**Network Architecture and Design**]: Wireless communication; I.6 [**Simulation and modelling**]: Model validation and Analysis

Keywords

Wireless network, MANET, DTNs, Selfish Nodes.

1. INTRODUCTION

A Mobile ad-hoc network (MANET) is a network of mobile nodes connected by wireless links without using any pre-existent infrastructure. Nodes are free to move independently in any direction and can directly communicate with each other if a contact occurs (that is, if they are within communication range). Opportunistic and Delay Tolerant Networks (DTNs) constitute an emerging subclass of MANETs where only intermittent connectivity and opportunistic contacts take place. Opportunistic nodes collectively create dynamic networks that are built from short unpredictable contact times as nodes move in and out of connectivity. Applications of such networks include vehicular ad hoc networks (VANETs), and mobile social networks.

In these networks, for a proper functionality, nodes must forward traffic unrelated to their own use. That is, these networks rely on network cooperation schemes to work properly. Nevertheless, in the real world, nodes could have a selfish behaviour, being unwilling to forward packets for others. Selfishness means that some nodes refuse to forward other nodes' packets to save their own resources (mainly energy).

Several works studied node selfishness in MANETs and DTNs. A first study about misbehaving nodes and how watchdogs can be used to detect them was introduced in [14]. The authors proposed a Watchdog and Pathrater over the DSR protocol to detect non-forwarding nodes, maintaining a rating for every node. In [16] another scheme for detecting selfish nodes based on context aware information was proposed. The CONFINDENT protocol was proposed in [1], which combines a watchdog, reputation systems and bayesian filters from the node and its neighbours to securely detect misbehaving nodes. A Mobile Intrusion Detection System is described in [12] as an advanced watchdog. In [7] an analytical selfish model (which is tied specifically to a routing protocol) is proposed. Recent papers have focused on DTNs. In [11], the author introduces a model for DTN data relaying schemes under the impact of node selfishness. A similar approach is presented in [13] that shows the effect of socially selfish behaviour.

The impact of node selfishness on MANETs has been studied in [18]. When no selfishiness prevention mechanism is installed, the packet delivery rates become seriously degraded, from a rate of 80% when the selfish node ratio is 0, to 30% when the selfish node ratio is 50%. A recent survey [17] shows similar results: the number of packet losses rises 500% when the selfish node ratio increases from 0% to 40%. In DTNs the presence of selfish nodes can seriously degrade the performance of packet transmission. For example, in two-hop relay schemes, if a packet is transmitted to a selfish node the packet is not retransmitted, and so the packet is lost. Thus, if a node knows who are the selfish nodes, it will try to avoid them in order to boost performance.

Therefore, detecting such nodes quickly and accurately

MSWiM'12, October 21–25, 2012, Paphos, Cyprus.

is essential for the overall network performance. Previous works have demonstrated that watchdogs are appropriate mechanisms to detect misbehaving and selfish nodes. Essentially, watchdog systems overhear wireless traffic and analyse it to decide whether neighbour nodes are behaving in a selfish manner [9]. When the watchdog detects a selfish node it is marked as a *positive* (or a *negative* if it is detected as a non selfish node). Nevertheless, watchdogs can fail on this detection, generating *false positives* and *false negatives* that can seriously degrade the behaviour of the system.

This paper introduces a *collaborative watchdog* approach based on contact dissemination. If one node has previously detected a selfish node it can transmit this information to other nodes when a contact occurs. This way, nodes have second hand information about the selfish nodes in the network. The goal of our approach is to reduce the detection time and to improve the precision by reducing the effect of both false negatives and false positives. Although some of the aforementioned papers (such as [1,16]) introduced some degree of collaboration on their watchdog schemes, the diffusion is very costly since they are based on periodic message dissemination.

In order to evaluate the efficiency of our collaborative watchdog we introduce an analytical performance model. Assuming that the occurrence of contacts between two mobile nodes follows a Poisson distribution, we model the network as a Continuous Time Markov chain (CTMC) and derive expressions for obtaining the time and overhead (cost) of detection of selfish nodes. In a preliminary work [6] we introduced a basic collaborative approach based on the diffusion of the positives only. This model is now extended to the distribution of positives and negatives in order to reduce the side effect of false positives and negatives. The problem of false positives and negatives is that they can also be propagated in the network when a collaborative contact occurs, so it is important to reduce this impact. Regarding the false positives and negatives, as far as we know, this is the first work to study their effect on the detection of selfish nodes.

In general, our evaluation shows a significant reduction of the detection time of selfish nodes with a reduced overhead when comparing our collaborative watchdog against a traditional watchdog. From our experiments we conclude that, if only positives are transmitted, the false negatives have little impact on the performance, but the effect of false positives is magnified (due to the sole diffusion of positives). In the other case, if positives and negatives are transmitted, false negatives have a strong impact on performance, and the impact of false positives is reduced. Thus, a mix approach is proposed where the positives are always transmitted, and only a portion of the negatives are transmitted. This way, the effect of false negatives and false positives is minimised, improving the global precision of the watchdog.

2. ARCHITECTURE OVERVIEW

A node's watchdog consists on overhearing the packets transmitted and received by its neighbours in order to detect anomalies, such as the ratio between *packets received* to *packets being retransmitted* [8]. Initially, no node has information about the selfish node. When a node detects a selfish node using its watchdog, it is marked as *positive*, and if it is detected as a non selfish node, it is marked as *negative*. Later on, when this node contacts another node, it *can* transmit this information to it; so, from that moment on, both nodes store information about this positive (or negative). Therefore, a node can become aware about selfish nodes directly (using its watchdog) or indirectly through the collaborative transmission of information that is provided by other nodes.

Figure 1 shows the functional structure of the collaborative watchdog. It has three main components: the watchdog, the diffusion module and the network information:

- The *watchdog* has two functions: the detection of selfish nodes and the detection of new contacts. The detection of selfish nodes can generate the following events about neighbour nodes: `PosEvt` (*positive event*) when the watchdog detects a selfish node, `NegEvt` (*negative event*) when the watchdog believes that a node is not selfish, and `NoInfEvt` (*no info event*) when the watchdog does not have enough information about a node (for example if the contact time is very low or it does not overhear enough messages). The detection of new contacts is based on *neighbourhood* packet overhearing; thus, when the watchdog starts receiving packets from a new node it is assumed to be a new contact, and so it generates an event to the network information module.
- The *diffusion* module has two functions: the transmission and the reception of positives (and negatives). Although positives are always transmitted, sending the negatives can be troublesome, producing excessive messaging or fast diffusion of false negatives. Thus, we introduce a *negative diffusion factor* γ, that is the ratio of negatives that are actually transmitted. This value ranges from 0 (no negatives are transmitted) to 1 (all negatives are transmitted). The significance and importance of the γ factor will be detailed in the evaluation section. Finally, when the diffusion module receives a new contact event from the watchdog, it transmits a message including this information to the new neighbour node. When the neighbour node receives a message, it generates an event to the network information module with the list of these positives (and negatives).
- Network information: A node can have the following internal information about other nodes i: `NoInfo`, `Positive` and `Negative`. `NoInfo` means that it has no information about node i, `Positive` means it believes that node i is selfish and `Negative` means it believes that node i is not selfish. The updating of this information is based on the state transition diagram of figure 2. The network information about the nodes has an expiration time, so after some time without contacts it is deleted.

3. SYSTEM MODEL

The network is modelled as a set of N wireless mobile nodes, with C collaborative nodes and one selfish node ($N = C + 1$). Our goal is to obtain the time and overhead that a set of $D \leq C$ nodes need to detect who is the selfish node in the network. The overhead is the number of information messages transmitted up to the detection time.

First, we are going to characterize the inter-contact times. Then, we model the watchdog and the diffusion modules including the effect of false positives and false negatives.

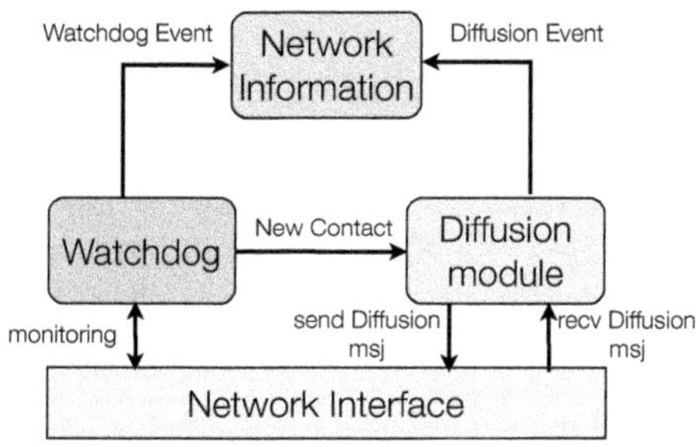

Figure 1: Arquitecture of the collaborative watchdog

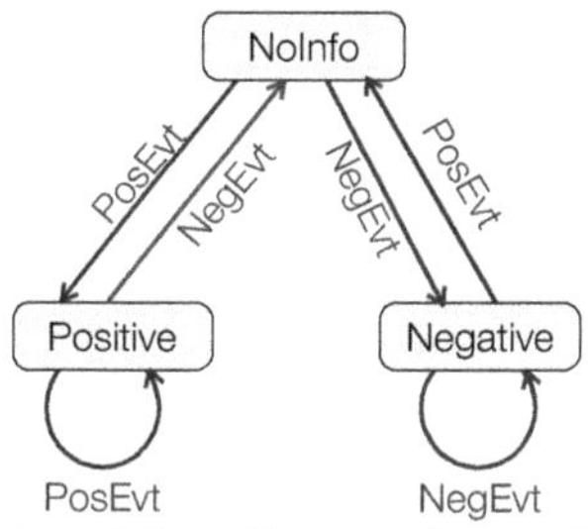

Figure 2: State transition diagram for updating the information about network nodes.

Then, we introduce a model for evaluating the detection of a selfish node, taking into account the effect of false negatives. Since this model evaluates the detection of a selfish positive (that is, a *true* positive), the effect of false positives can not be evaluated through it. Thus, a second model is introduced that evaluates the impact of false positives.

3.1 Characterizing Inter-contact Times

Characterizing inter-contact times (or inter-meeting times) between pairs of nodes is essential for analyzing the performance of contact based protocols. The inter-contact times distribution is obtained by *aggregating* the *individual pair distribution* of all combinations of pairs of nodes in the network. The *individual pair distribution* is defined as the distribution of the time elapsed between two consecutive contacts between the same pair of nodes [15].

The assumption that the aggregated inter-contact time follows an exponential distribution with rate λ has been shown to hold in several mobility scenarios of both human and vehicles [5,13,19]. Empirical results have shown that the aggregated inter-contact time distribution follows a power-law and has a long tail [3], meaning that there are some pairs of nodes that barely experience contact. In [2] it is shown that in a bounded domain (such as the one selected along this paper) the inter-contact distribution is exponential, but in an unbounded domain the distribution is power-law. The dichotomy of this distribution is described in [10]. The work in [4] analyzed some popular mobility traces and found that over 85% of the *individual pair distributions* fit an exponential distribution.

Our performance model assumes an exponential distributed inter-contact rate between nodes and therefore it is suited for both MANETs and DTNs. The main difference is the rate of contact, which is higher in MANETs.

3.2 Modelling system modules

The watchdog is modelled using three parameters: the probability of detection p_d, the ratio of false positives p_{fp}, and the ratio of false negatives p_{fn}. The first parameter, the probability of detection (p_d), reflects the probability that, when a node contacts another node, the watchdog has enough information to generate a `PosEvt` (or `NegEvt`) event. In other words (1-p_d) is the probability of the `NoInfEvt` event. This value depends on the effectiveness of the watchdog, the traffic load, and the mobility pattern of nodes. Furthermore, the watchdog can generate false positives and false negatives. A false positive is when the watchdog generates a positive for a node that is not a selfish node. A false negative is generated when a selfish node is marked as a negative. In order to measure the performance of a watchdog, these values can be expressed as a ratio or probability: p_{fp} is the ratio (or probability) of false positives generated when a node contacts a non-selfish node, and p_{fn} is the ratio (or probability) of false negatives generated when a node contacts a selfish node.

A contact does not always imply collaboration, so we model this probability of collaboration as p_c. The degree of collaboration is a global parameter, and it is used to reflect that either a message with the information about the selfish node is lost, or that a node temporally does not collaborate (for example, due to a failure or simply because it is switched off). In real networks, full collaboration ($p_c = 1$) is almost impossible.

Using the previous parameters we can model the probability of the `PosEvt` and `NegEvt` events when a contact occurs:

- `PosEvt` event: there are two possibilities: i) the node contacts with the selfish node and the watchdog detects it, with probability $p_d(1-p_{fn})$; and ii), the node contacts another node that has a `Positive` state about the selfish node with probability p_c. Note that a false positive can also be generated with probability $p_d \cdot p_{fp}$.

- `NegEvt` event: there are two possibilities: i) the node contacts with a non-selfish node with probability $p_d(1-p_{fp})$, and ii) the node contacts another node that has a `Negative`, being the probability $\gamma \cdot p_c$. A false negative can also be generated when it contacts with the selfish node with probability $p_d \cdot p_{fn}$.

Our model assumes that all nodes are selfish or collaborative. Security concerns, such as malicious nodes that spread false information about selfish nodes, are outside the scope of this paper.

3.3 A model for the detection of selfish nodes

In this subsection we introduce an analytical model for evaluating the performance of our collaborative watchdog approach. The goal is to obtain the detection time (and overhead) of a selfish node in a network. This model takes into account the effect of false negatives. False positives do not affect the detection time of the selfish node, so p_{fp} is not introduced in this model. The effect of false positives will be studied in subsection 3.4. For an easier exposition, we first introduce a model for $D = C$, and then this model is extended to the generic case of $D \leq C$.

Using λ we can model the network using a 2D Continuous Time Markov chain (2D-CTMC) with states $(c_p(t), c_n(t))_{t\geq 0}$, where $c_p(t)$ represents the number of collaborative nodes that have a `Positive` state about the selfish node at time t, and $c_n(t)$ represents the number of *collaborative* nodes that have a `Negative` state for the selfish node (note that, in this case, a `Negative` is a false negative). At the beginning all nodes have no information (`NoInfo` state). Then, when a contact occurs, $c_p(t)$ and $c_n(t)$ can be increased by one. As each node can be only in one state, then $c_p(t) + c_n(t) \leq C$. The final (absorbing) state is when $c_p(t) = C$. Our 2D-CTMC model has an initial state $(0, 0)$, a final state $(C, 0)$ and the transient states are all possible permutations that sum C: $\{(0,1), \ldots (0,C), (1,0), \ldots (1,C-1), (2,0) \ldots (2,C-2) \ldots (C-1,1)\}$. It is easy the derive that the number of permutations that sums C is $\mathrm{P^S}(C) = 0.5(C+1)(C+2)$. We define τ as the number of transient states ($\tau = \mathrm{P^S}(C)$) and υ as the number of absorbing states ($\upsilon = 1$). This model can be expressed using the following transition matrix $\mathbf{P}$ in canonical form:

$$\mathbf{P} = \begin{pmatrix} \mathbf{Q} & \mathbf{R} \\ \mathbf{0} & \mathbf{I} \end{pmatrix} \tag{1}$$

where $\mathbf{I}$ is a $\upsilon \times \upsilon$ identity matrix, $\mathbf{0}$ is a $\upsilon \times \tau$ zero matrix, $\mathbf{Q}$ is a $\tau \times \tau$ matrix with elements p_{ij} denoting the transition rate from transient state s_i to transient state s_j, and $\mathbf{R}$ is a $\tau \times \upsilon$ matrix with elements p_{ij} denoting the transition rate from transient state s_i to the absorbing state s_j.

Now, we derive the transition rates p_{ij}. Given the state $s_i = (c_p, c_n)$[1], the following transitions can occur:

- (c_p, c_n) to $(c_p + 1, c_n)$: A new collaborative node has a `Positive` state. The transition probability is $t_P = \lambda(p_d(1-p_{fn}) + p_c c_p)(C - c_p - c_n)$. The term $p_d(1-p_{fn})$ represents the probability of a `PosEvt` event from the watchdog, and $p_c c_p$ the probability of a `PosEvt` event from the diffusion module (it depends on c_p, so this probability is higher if more nodes have a `Positive` state). Finally, the factor $(C - c_p - c_n)$ represents the number of pending collaborative nodes. If there are no pending nodes, this value is 0.
- (c_p, c_n) to $(c_p, c_n + 1)$: A new collaborative node has a `Negative` state (a *false negative*). The transition probability is $t_N = \lambda(p_d p_{fn} + \gamma p_c c_n)(C - c_p - c_n, 0)$.
- $(c_p + 1, c_n)$ to (c_p, c_n): A collaborative node that has a `Positive` state changes to `NoInfo`. This occurs when the watchdog or diffusion module generates a `NegEvt` event. So, the transition probability is similar to t_N: $t_{P'} = \lambda(p_d p_{fn} + \gamma p_c c_n) c_p$.
- $(c_p, c_n + 1)$ to (c_p, c_n): A collaborative node that has a `Negative` changes to `NoInfo`. This occurs when the node detect or receives a `PosEvt`. So, the transition probability is similar to t_P: $t_{N'} = \lambda(p_d(1 - p_{fn}) + p_c c_p) c_n$.
- (c_p, c_n) to (c_p, c_n): This is the probability of no changes and is calculated as $t_0 = 1 - t_P - t_N - t_{P'} - t_{N'}$.

For example, for $N = 3$, we have $C = 2$, so $\tau = 5$ and $\upsilon = 1$, the transition matrix is:

$s_i \to s_j$	0,0	0,1	0,2	1,0	1,1	2,0
0,0	t_0	t_N	0	t_P	0	0
0,1	$t_{N'}$	t_0	t_N	0	t_P	0
0,2	0	$t_{N'}$	t_0	0	0	0
1,0	$t_{P'}$	0	0	t_0	t_N	t_P
1,1	0	$t_{P'}$	0	$t_{N'}$	t_0	0
2,0	0	0	0	0	0	1

[1]For simplicity, we omit the time in the states (that is $(c_p, c_n) = (c_p(t), c_n(t))$)

Using the transition matrix $\mathbf{P}$ we can derive two different expressions: one for the detection time T_d and another for the overall overhead (or cost) O_d. We start with the detection time. From the 2D-CTMC we can obtain how long it will take for the process to be absorbed. Using the fundamental matrix $\mathbf{N} = (\mathbf{I} - \mathbf{Q})^{-1}$, we can obtain a vector $\mathbf{t}$ of the expected time to absorption as $\mathbf{t} = \mathbf{N}\mathbf{v}$, where $\mathbf{v}$ is a column vector of ones ($\mathbf{v} = [1, 1, \ldots, 1]^T$). Each entry t_i of $\mathbf{t}$ represents the expected time to absorption from state s_i. Since we only need the expected time from state $s_1 = (0, 0)$ to absorption (that is, the expected time for all nodes to have a `Positive`), the detection time T_d, is:

$$T_d = E[T] = \mathbf{v_1}\mathbf{N}\mathbf{v} \tag{2}$$

where T is a random variable denoting the detection time for all nodes and $\mathbf{v_1} = [1, 0, \ldots, 0]$.

Concerning the overhead we need to obtain the number of transmitted messages for each state s_i. First, the duration of each state s_i can be obtained using the fundamental matrix $\mathbf{N}$. By definition, the elements of the first row of $\mathbf{N}$ are the expected times in each state starting from state 0. Then, the duration of state s_i is $f_i = \mathbf{N}(1, i)$.

Now, we calculate the expected number of messages m_i. The number of messages depends on the diffusion model. For an easier exposition, we start with $\gamma = 0$, that is, only the positives are transmitted. From state $s_1 = (0, 0)$ to s_{C+1}=(0,C) no node has a `Positive` state, so no messages are transmitted and $m_1 = 0$. From states $s_{C+2} = (1, 0)$ to $s_{2C+2} = (1, C - 1)$, one node has a `Positive` state. In these cases, the `Positive` can be transmitted to all nodes (except itself) for the duration of each state i ($\mathbf{N}(1, i)$) with a rate λ and probability p_c. Then, the expected number of messages can be obtained as $m_i = \mathbf{N}(1, i)\lambda(C - 1)p_c$. From states $s_{2C+3} = (2, 0)$ to $s_{3C+2} = (2, C - 2)$, we have two possible senders and $m_i = 2\mathbf{N}(1, i)\lambda(C - 1)p_c$. Summing up, the overhead of transmission (or the expected number of messages) is:

$$O_d = E[M] = \lambda(C - 1)p_c \sum_{i=1}^{\tau} \Phi(s_i)(1, i) \tag{3}$$

where $\Phi(s_i) = c_p$ is the number of nodes with a `Positive` for state s_i. Finally, for $\gamma > 0$, the ratio of nodes c_n that will transmit the negative is precisely γ, so $\Phi(s_i) = c_p + \gamma c_n$.

The previous model can be extended to the case of $D \leq C$. In this generic model the collaborative nodes are divided into two sets: a set with D *detecting* nodes, and a set of $M = C - D$ *middle* (or *non-detecting*) nodes. Note that the *detecting* and *middle* nodes have the same behaviour (both are collaborative nodes). The only purpose of this division is to analytically obtain the time and the overhead required for the subset of *detecting* nodes to detect the selfish node. We therefore use a 4D Continuous Time Markov chain (4D-CTMC) with states $(d_p(t), d_n(t), m_p(t), m_n(t))$, where $m_p(t)$ represents the number of *middle* nodes that have a `Positive`, $m_n(t)$ the *middle* nodes with a `Negative`, d_p the *detecting* nodes with a `Positive` and d_n the *detecting* nodes with a `Negative`. In this case the states must verify the following conditions $d_p(t) + d_n(t) \leq D$ and $m_p(t) + m_n(t) \leq M$. The final (absorbing) states is when $d_p(t) = D$. The number of transient and absorbing states is $\tau = (\mathrm{P^S}(D) - 1)\mathrm{P^S}(M)$ and $\upsilon = \mathrm{P^S}(M)$ respectively. We can derive the transition rates (p_{ij}) of the transition matrix $\mathbf{P}$ in a way that is similar to the previous model:

$$p_{ij} = \begin{cases} \lambda(p_d(1-p_{fn}) + p_c(m_p+d_p)) \cdot \mathcal{M}() & m_p+ \\ \lambda(p_d p_{fn} + \gamma p_c(m_n+d_n)) \cdot \mathcal{M}() & m_n+ \\ \lambda(p_d p_{fn} + \gamma p_c(m_n+d_n)) \cdot m_p & m_p- \\ \lambda(p_d(1-p_{fn}) + p_c(m_p+d_p)) \cdot m_n & m_n- \\ \lambda(p_d(1-p_{fn}) + p_c(m_p+d_p)) \cdot \mathcal{D}() & d_p+ \\ \lambda(p_d p_{fn} + \gamma p_c(m_n+d_n)) \cdot \mathcal{D}() & d_n+ \\ \lambda(p_d p_{fn} + \gamma p_c(m_n+d_n)) \cdot d_p & d_p- \\ \lambda(p_d(1-p_{fn}) + p_c(m_p+d_p)) \cdot d_n & d_n- \end{cases} \quad (4)$$

where $\mathcal{M}() = (M - m_p - m_n$, $\mathcal{D}() = D - d_p - d_n$, $x+$ represents a transition from state $(\cdots, x, \cdots)$ to $(\cdots, x+1, \cdots)$, and $x-$ represents a transition from state $(\cdots, x+1, \cdots)$ to $(\cdots, x, \cdots)$. Finally, p_{ii} is $1 - \sum_{j \neq i} p_{ij}$

Using transition matrix $\mathbf{P}$ we can obtain the detection time using equation 2 and the overhead from equation 3, where $\Phi(s_i) = d_p + m_p + \gamma(d_n + m_n)$.

3.4 A model for false positives

In this subsection we develop a model for evaluating the effect of the false positives. When a node has a false positive the problem is that, due to the diffusion of positives, this false positive can be quickly distributed in the network. A way to evaluate this diffusion is to obtain the time (and cost) that a set of D nodes have a false positive about a given node. Following the same process that in the model for the false negatives, we have (for $D = C$) a 2D-CMTC with the same states (c_p, c_n), but in this case c_p represents the number of nodes with false positives, and c_n the number of nodes with a negative. The transition rates (p_{ij}) of the transition matrix $\mathbf{P}$ are:

$$p_{ij} = \begin{cases} \lambda(p_d p_{fp} + p_c c_p) \cdot \mathcal{C}() & c_p+ \\ \lambda(p_d(1-p_{fp}) + \gamma p_c c_n) \cdot \mathcal{C}() & c_n+ \\ \lambda(p_d(1-p_{fp}) + \gamma p_c c_n) \cdot c_p & c_p- \\ \lambda(p_d p_{fp} + p_c c_p) \cdot c_n & c_n- \end{cases} \quad (5)$$

where $\mathcal{C}() = C - c_p - c_n$. We can see that the transition rates are the same than in the false negative model if we replace $p_{fp} = 1 - p_{fn}$. Therefore, we can use the previous model for obtaining the detection time T_d and the overhead O_d.

4. EVALUATION

This section is devoted to evaluating the performance of our collaborative watchdog approach. First, we study the global performance depending on the degree of collaboration and the number of nodes. Then, we focus our study on the impact of false negatives and false positives. Finally, we compare our approach to the classic periodic diffusion model. Note that, since λ is a multiplying factor of the transition matrix $\mathbf{P}$, the concluding results of this section are valid for any value of λ. Thus, for the evaluations that follows, we consider a λ value of 0.01 contacts/s, which has been shown to be a valid value in vehicular scenarios [19].

4.1 Impact of collaboration and nodes

The first evaluation shows the impact that the degree of collaboration (p_c) has over the efficiency of the collaborative watchdog for $\gamma = 0$ (that is, only positives are transmitted). Figure 3a shows the detection time and overhead for one detecting node ($D = 1$) in a network with 25 nodes ($N = 25$) with different probabilities of detection and false negatives

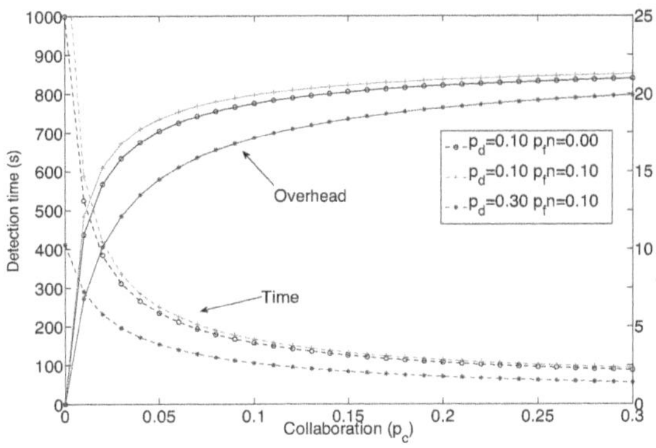

(a)

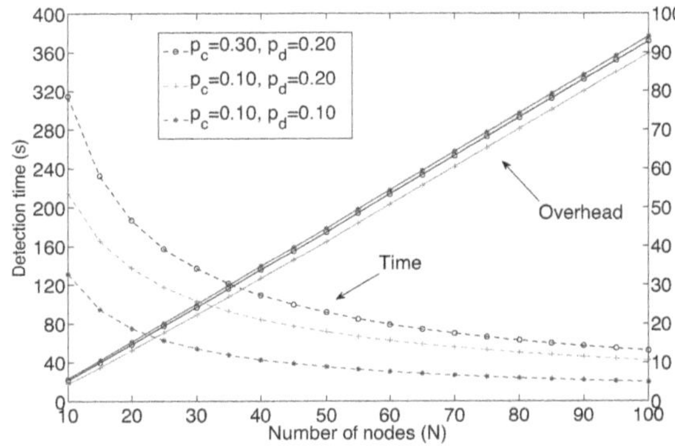

(b)

Figure 3: Impact of collaboration and nodes for only positives diffusion ($\gamma = 0$). a) depending on collaboration in a network of $N = 25$, b) depending on the number of nodes.

(p_d,p_{fn}). We observe that when increasing the degree of collaboration from 0 to 0.2 the detection time is reduced exponentially. The effect of p_d is the expected: for greater values of p_d, the detection time is reduced. The effect of false negatives produces a small increase on the detection time when p_{fn} increases, but this will be studied later.

When repeating the previous experiment to evaluate the detection time (and overhead) required by all nodes to detect the selfish node (that is, $D = N - 1 = 24$) we obtained a similar pattern. For example, for $p_d = 0.1$ and $p_{fn} = 0.1$, the detection time with no collaboration ($p_c = 0$) is $12,993s$. This value can be greatly reduced by using our collaborative watchdog. Thus, if all nodes implement the collaborative approach, and even for a low collaboration rate ($p_c = 0.2$), the detection time for all nodes is reduced to $191s$ with an overhead of just 85 messages.

We now evaluate the impact of the number of nodes, ranging from 10 to 100 (see figure 3b). Three different sets of values for p_c and p_d were used. In all the cases the value of p_{fn} is 0.1. We observe that, in general, the greater the number of nodes, the smaller the detection time and the greater the number of messages. The main reason is that, when the number of nodes is greater, the number of contacts is increased and so the information about the positive detection is disseminated more quickly. Reduced values for the collaboration and detection probabilities imply greater detection

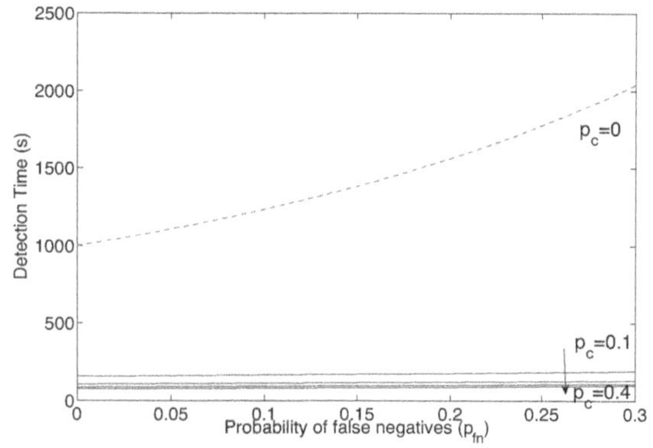

Figure 4: Evaluation of the impact of false negatives for $p_d = 0.1$ with $\gamma = 0$ for several values of p_c

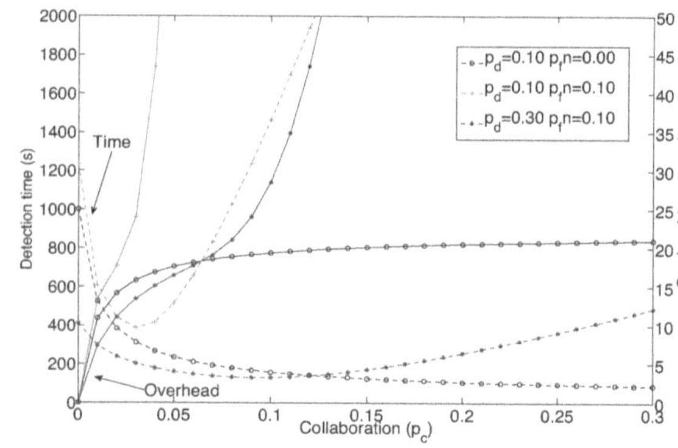

Figure 5: Detection time and overhead depending on collaboration for full transmission of negatives ($\gamma = 1$).

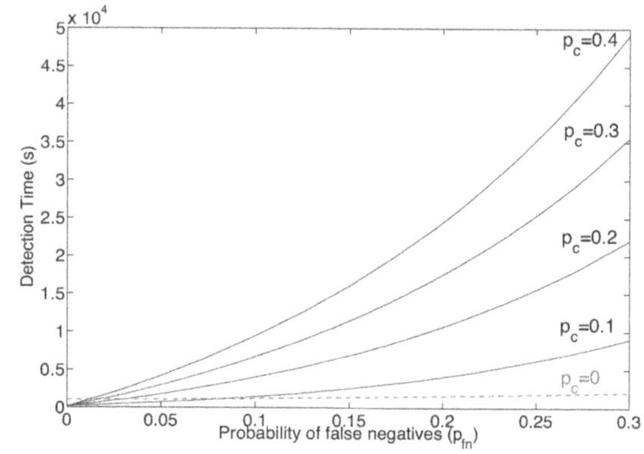

Figure 6: Evaluation of the impact of false negatives with $\gamma = 1$ for several values of p_c

times (as expected). Nevertheless, the cost depends mainly on N.

4.2 Impact of false negatives

The goal of the following experiments is to evaluate the impact of false negatives. Figure 4 shows the detection time for one node ($D = 1$) in a network of 25 nodes ($N = 25$) depending on false negatives for several values of p_c We can see that the detection time is greatly reduced when p_c is greater than zero, and that false negatives do not affect this detection time. Regarding the overhead, the experiment showed little influence on the number of messages, that is always around 20 messages. The results show that false negatives have a small influence on the detection time.

Figure 5 shows the results for $\gamma = 1$ (that is, full transmission of the negatives) with the same network parameters ($N = 25$, $S = 1$, $D = 1$). The results when p_{fn} is zero are very similar to the only positives diffusion case ($\gamma = 0$). However, when p_{fn} is not zero we can observe that for low degrees of collaboration the detection time decreases and the overhead increases in a similar way to the only positives case. Nevertheless, when p_c increases, the detection time increases again, and the overhead increases exponentially. It seems that the collaboration amplifies the impact of false negatives. This effect is confirmed in figure 6, where we can see that the curves for greater values of p_c have a greater exponential slope. This is particularly evident in the ($p_d = 0.1$,$p_{fn} = 0.1$) curve. Regarding the overhead, the results showed a similar behaviour (in general, a greater detection time implies a greater overhead).

Summing up, if only positives are transmitted, the detection time is greatly reduced and the impact of false negatives is also reduced; however, when all known negatives are transmitted, collaboration amplifies the effect of false negatives.

4.3 Impact of false positives

In this subsection we evaluate the influence of false positives using the model developed in section 3.4. In this case, we expect that the diffusion of negatives reduces the influence of false positives and that when γ is zero, the influence of false positives will be amplified. Figure 7a shows the detection time for $\gamma = 0$. We observe that for the curves where $p_c > 0$ the effect of false positives is amplified, leading to a drastic reduction on the detection time, meaning that these false positives are spread on the network rather quickly. For example, for $p_{fp} = 0.1$, the detection time is $1 \times 10^5 s$ for $p_c = 0$ and $1394s$ for $p_c = 0.1$. This detection time is equivalent to a value of $p_{fp} = 0.82$ when there is no collaboration ($p_c = 0$). Thus, the undesired effect is that the false positives rate is increased. Consequently, we need to transmit the negatives in order to compensate for these false positives. Figure 7b shows the results for $\gamma = 1$. In this case, we can see that the detection time is highly increased when the collaboration increases and so the effect of false positives is reduced.

Therefore, we have the inverse effect that in the false negatives case. If only positives are transmitted the effect of false positives is magnified and so the transmission of negatives is necessary in order to reduce the impact of false positives. This effect can be regulated using the γ factor. We evaluated the same scenario selected in figures 6 and 7a for $\gamma = 0.25$. First, we can see in figure 8a that the detection time is reduced, even if the ratio of false negatives is high. Second, figure 8b shows that the detection time is increased when the collaboration increases, effectively reducing the effect of false positives. For example, for $p_{fp} = 0.1$, the equivalent false positive rate for collaboration of $p_c = 0.3$ is reduced to 0.05.

The conclusions is that the γ value must be tuned up in order to achieve the desired behaviour. A γ value near to zero greatly reduces the detection time of selfish nodes, but increases the diffusion of false positives. A value near to one

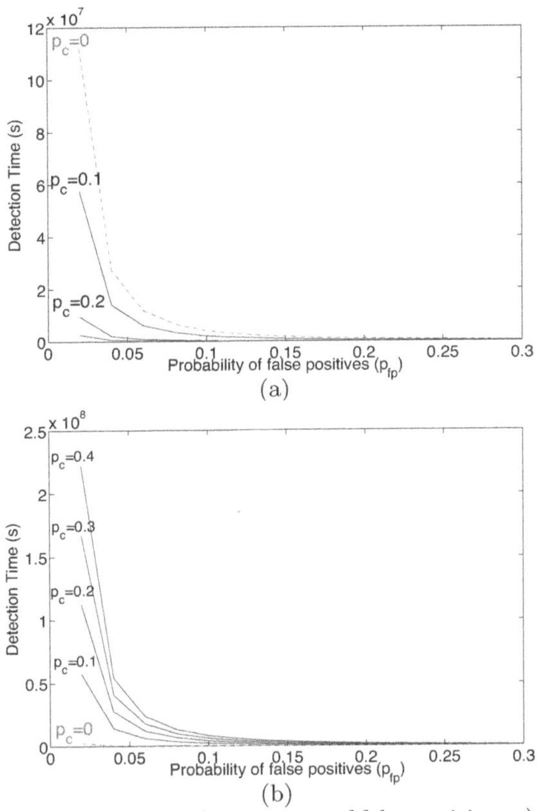

Figure 7: Evaluation of the impact of false positives a) when $\gamma = 0$, b) when $\gamma = 1$.

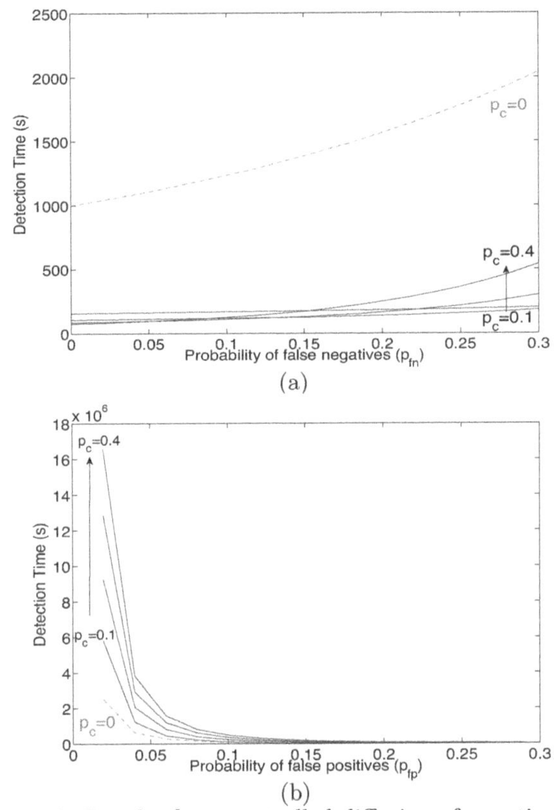

Figure 8: Results for a controlled diffusion of negatives $\gamma = 0.25$ a) impact of false negatives, b) impact of false positives.

increases the detection time (due to the the effect of the false negatives), but reduces the diffusion of false positives.

4.4 Comparison with other approaches

We now proceed by comparing our collaborative watchdog approach with previous cooperative approaches that use periodic messages for the diffusion of information about selfish node detections (such as the ones presented in [1,16]). Note that this comparison focuses only on the diffusion protocol. If a node has information about a positive (or negative), it will periodically broadcast a message with a given period P. This message will be received by all nodes that are within the communication range of the sender. The performance of this protocol clearly depends on the period P. A short period will reduce the detection time, but the number of messages transmitted (the overhead) will be high. A large period will increase the detection time by reducing the overhead.

The comparison of both protocols was based on simulations. We implemented the periodic diffusion protocol, as described in the previous paragraph. In this periodic approach only the positives are sent. Regarding our collaborative approach, the watchdog parameters are ($p_{fp} = 0.17$, $p_{fn} = 0.08$,$p_d = 0.11$), that were obtained based on a set of real test bed experiments from [9]. The rest of parameters are $p_c = 0.2$ and $\gamma = 0.25$. By using the ns-2 *setdest* command we generate mobility scenarios that are used to simulate both approaches.

Figure 9 shows the detection time and overhead for the periodic diffusion protocol when the its period P ranges from 1 to 30s with three different number of nodes ($N = 30, 40, 50$). The results confirm that increasing the period P implies a higher detection time while the overhead is reduced. We compare these results with the detection time and overhead values for our collaborative watchdog (that are in the legend of the plot for each value of N). For example, for $N = 50$, the periodic diffusion for periods below 4s has a shorter detection time than our model, but with a higher overhead. For example, for $P = 2s$, the detection time is 963s (a reduction of 9%) and the overhead is 5212 messages (an increment of 4738%) with respect to our collaborative watchdog approach. For $P = 4s$, the detection time is similar to our approach, and the overhead is 3210 messages (2972% higher). Regarding the false positives, in the periodic model the diffusion time of false positives is reduced for low values of P. For example, for $N = 40$ the detection time of false positives is reduced from $15,024s$ when there is no diffusion of positives to $900s$ when $P = 1$. This is equivalent to a false positives rate of 0.72, that is an unreasonable value.

Summing up, although using periodic diffusion can reduce the detection time slightly, this implies a large overhead and the impact of false positives is very high, so is not a viable strategy for low period values.

5. CONCLUSIONS

This paper proposes a *collaborative watchdog* to improve the detection time and efficiency of selfish nodes, reducing the effect of false positives of negatives (that is, improving the global precision of the detection process). The col-

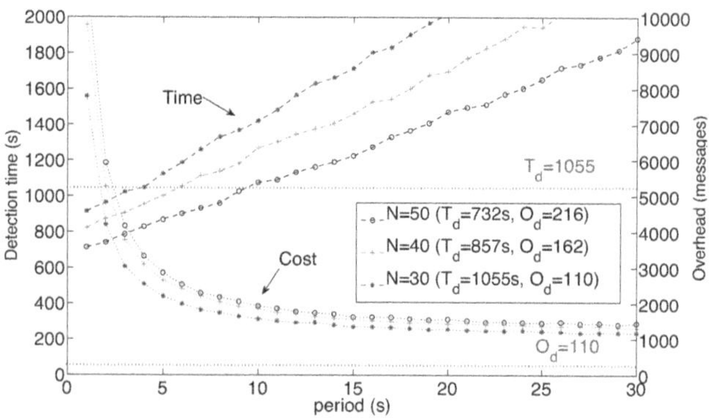

Figure 9: Detection time and overhead depending on period P for the periodic approach. The main parameters for the mobility model are mean-speed = 5m/s, side-area = 1000 m, pause-interval = 1s, range = 100m

laborative watchdog is based on the diffusion of the known positives and negatives. When a contact occurs between two collaborative nodes, the diffusion module transmits and processes the positives (and negatives).

Numerical results show that our collaborative watchdog can reduce the overall detection time with respect to the original detection time with no collaboration scheme with a reduced overhead (message cost). This reduction is very significant, with a percentage of reduction ranging from 20% for very low degree of collaboration to 99% for higher degrees of collaboration. Regarding the overall precision we show how by selecting a factor for the diffusion of negatives the harmful impact of both false negative and false positive is diminished. Summing up, the controlled effect of collaboration of our approach can reduce the detection time while increasing the global accuracy using a moderate local precision watchdog.

As future work, we plan to extend the model to introduce a reputation scheme in order to give more or less weight to the information received from other nodes.

6. ACKNOWLEDGMENTS

This work was partially supported by the *Ministerio de Ciencia e Innovación*, Spain, under Grant TIN2011-27543-C03-01.

7. REFERENCES

[1] S. Buchegger and J.-Y. Le Boudee. Self-policing mobile ad hoc networks by reputation systems. *Communications Magazine, IEEE*, 43(7):101 – 107, jul. 2005.

[2] H. Cai and D. Y. Eun. Crossing over the bounded domain: From exponential to power-law intermeeting time in mobile ad hoc networks. *Networking, IEEE/ACM Transactions on*, 17(5):1578 –1591, oct. 2009.

[3] A. Chaintreau, P. Hui, J. Crowcroft, C. Diot, R. Gass, and J. Scott. Impact of human mobility on opportunistic forwarding algorithms. *IEEE Transactions on Mobile Computing*, 6:606–620, Jun 2007.

[4] W. Gao, Q. Li, B. Zhao, and G. Cao. Multicasting in delay tolerant networks: a social network perspective. In *Proceedings of MobiHoc '09*, pages 299–308, 2009.

[5] R. Groenevelt, P. Nain, and G. Koole. The message delay in mobile ad hoc networks. *Performance Evaluation*, 62:210–228, October 2005.

[6] E. Hernández-Orallo, M. D. Serrat, J.-C. Cano, C. M. T. Calafate, and P. Manzoni. Improving selfish node detection in manets using a collaborative watchdog. *IEEE Communications Letters*, 16(5):642–645, 2012.

[7] M. Hollick, J. Schmitt, C. Seipl, and R. Steinmetz. On the effect of node misbehavior in ad hoc networks. In *Proceedings of IEEE International Conference on Communications, ICC'04*, pages 3759–3763, 2004.

[8] J. Hortelano, J.-C. Cano, C. T. Calafate, M. de Leoni, P. Manzoni, and M. Mecella. Black hole attacks in p2p mobile networks discovered through bayesian filters. In *In P2P Collaborative Distributed Virtual Environments (P2P CDVE 2010)*, 2010.

[9] J. Hortelano, J. C. Ruiz, and P. Manzoni. Evaluating the uselfusness of watchdogs for intrusion detection in vanets. In *In ICC'10 Workshop on Vehicular Networking and Applications*, 2010.

[10] T. Karagiannis, J.-Y. Le Boudec, and M. Vojnović. Power law and exponential decay of inter contact times between mobile devices. In *Proceedings of MobiCom '07*, pages 183–194. ACM, 2007.

[11] M. Karaliopoulos. Assessing the vulnerability of dtn data relaying schemes to node selfishness. *Communications Letters, IEEE*, 13(12):923 –925, december 2009.

[12] F. Kargl, A. Klenk, S. Schlott, and M. Weber. Advanced detection of selfish or malicious nodes in ad hoc networks. In *In Proceedings of the 1st European on Security in Ad-Hoc and Sensor Networks*, pages 152–165. Springer Verlag, 2004.

[13] Y. Li, G. Su, D. Wu, D. Jin, L. Su, and L. Zeng. The impact of node selfishness on multicasting in delay tolerant networks. *Vehicular Technology, IEEE Transactions on*, 60(5):2224 –2238, jun 2011.

[14] S. Marti, T. J. Giuli, K. Lai, and M. Baker. Mitigating routing misbehavior in mobile ad hoc networks. In *Proceedings of MobiCom '00*, pages 255–265, 2000.

[15] A. Passarella and M. Conti. Characterising aggregate inter-contact times in heterogeneous opportunistic networks. In *Proceedings of the 10th international conference on Networking*, pages 301–313, 2011.

[16] K. Paul and D. Westhoff. Context aware detection of selfish nodes in dsr based ad-hoc networks. In *In Proceedings of IEEE Globecom*, 2002.

[17] C. Toh, D. Kim, S. Oh, and H. Yoo. The controversy of selfish nodes in ad hoc networks. In *Proceeding of Advanced Communication Technology (ICACT), 2010*, volume 2, pages 1087 –1092, feb. 2010.

[18] Y. Yoo, S. Ahn, and D. Agrawal. A credit-payment scheme for packet forwarding fairness in mobile ad hoc networks. In *Proceedings of IEEE ICC*, volume 5, pages 3005 – 3009 Vol. 5, may 2005.

[19] H. Zhu, L. Fu, G. Xue, Y. Zhu, M. Li, and L. M. Ni. Recognizing exponential inter-contact time in vanets. In *Proceedings of INFOCOM'10*, pages 101–105, 2010.

Security Evaluation Methodology for Intrusion Recovery Protocols in Wireless Sensor Networks

Eliana Stavrou, Andreas Pitsillides
Department of Computer Science
University of Cyprus
Nicosia - Cyprus

{cs98is1,Andreas.Pitsillides}@cs.ucy.ac.cy

ABSTRACT

Researchers have designed intrusion recovery protocols in Wireless Sensor Networks (WSN) with the aim of restoring compromised services and promoting a secure and reliable environment in WSNs. Currently, evaluation frameworks in WSNs have been proposed in a generic security context and do not constitute a representative assessment of intrusion recovery protocols. This paper proposes an evaluation methodology to aid the evaluation and comparison of intrusion recovery protocols in WSNs. The methodology defines the intrusion recovery protocol aspects that should be evaluated using a number of evaluation criteria and guides researchers in identifying the evaluation direction they should follow.

Categories and Subject Descriptors

C.4 [**Performance of systems**]: Design studies, Performance attributes, Reliability, Availability and Serviceability.

Keywords

Intrusion recovery, WSN, performance evaluation, security.

1. INTRODUCTION

Intrusion recovery protocols in Wireless Sensor Networks (WSNs), i.e. [1,2], are developed in order to restore compromised services and allow the network to continue its operation in a secure manner, and offer reliable decision-making. The design phase of intrusion recovery protocols is only one step of the protocol development process. One has to evaluate the performance of the protocol towards its objectives, the restoration level that is achieved and also assess if the associated cost is acceptable to recover from security attacks. This is an urgent task as there are a number of threats [3,4] in WSNs that can compromise their operation. These threats need to be addressed by intrusion recovery protocols in a way that the cost of implementing restoration mechanisms is justified. The evaluation process can help researchers assess if their protocol can balance the trade-off between recovery, threats, compromisation and cost.

Evaluating intrusion recovery features is not a trivial procedure as it is often difficult for a developer to identify which are the most appropriate aspects that need to be considered for the assessment of these protocols. The aspects of intrusion recovery protocols that need to be evaluated in terms of security requirements [4] are generally not investigated consistently. This is a challenging task since each protocol is developed for specific applications [3] with different needs and requirements and also different deployment decisions. Furthermore, by reviewing the literature on the area of intrusion recovery protocols in WSNs, it has been observed that researchers use different sets of criteria to evaluate the performance of their protocol in terms of security requirements, and that these are almost universally not applicable to use in comparison with other protocols, even superficially.

Currently, evaluation frameworks [5-8] in WSNs have been proposed in a generic security context and cannot constitute a representative assessment of intrusion recovery protocols. Therefore, the need arises to establish such a common evaluation methodology in order to a) define the security evaluation aspects that should be considered to assess the security and performance of intrusion recovery protocols in WSNs, b) support researchers into evaluating and fine-tuning their protocol design and c) promote intrusion recovery protocols comparison. The paper is organized as follows: section 2 defines the intrusion recovery evaluation methodology. Section 3 presents a case study demonstrating the comparison of existing intrusion recovery protocols using the proposed methodology and section 4 constitutes conclusions.

2. EVALUATION METHODOLOGY

This section defines the proposed evaluation methodology that drives the evaluation process of intrusion recovery protocols in WSNs.

2.1 Definition of intrusion recovery evaluation aspects

In order to assess the effectiveness of a recovery protocol, the application's requirements in terms of intrusion recovery aspects must be considered. The main tasks [3] of sensors include sensing the environment, communicating with other sensors and reporting the observations made to the control center/sink. If any of these tasks is compromised by an adversary, an application's operation may be jeopardized. Table 1 indicates the WSN services that can be recovered when a specific security requirement is supported by intrusion recovery protocols.

In terms of sensing and processing, a sensor monitors its environment if it has enough energy. Since sensors often use batteries [3] as their main source of energy, batteries depletion leads to failure in sensing the environment. Energy depletion also leads in communication and reporting failure. Intrusion recovery protocols need to be assessed as to their **survivability** capabilities. Survivability refers to the ability of a node to apply intrusion

MSWiM' 12, October 21–25, 2012, Paphos, Cyprus.

recovery measures, remain alive after a security attack is launched and continue supporting the fundamental WSN services.

Assessment should also take into consideration the **availability** requirement. If sensors are prohibited from communicating, then reporting of observations cannot be achieved and decision-making can be compromised. This means that a recovery countermeasure needs to be evaluated as to its ability to restore the blocked nodes/communication link. Moreover, reliable decision-making is depended on the ability of the WSN to report the observations to the decision entity/control center. **Reliability** requirement needs to be supported, meaning that the ability of the network for packet delivery has to be restored in case of compromisation.

Restoring compromised services does not mean that the adversary will stop attacking the network. It is crucial for the recovery measures to resist to new attacks that aim to interrupt the recovered WSN services. Evaluation should consider the **resilience** level that can be achieved by the deployed intrusion recovery countermeasures when the network continuous to be under attack. The results could be used to improve the intrusion recovery countermeasure under evaluation.

Intrusion recovery would be very effective if sensors are equipped with the logic of applying recovery according to the situation. The **self-healingness** capability of a recovery solution should be assessed to indicate if it can address different situations.

The fact that many WSNs applications are time sensitive [3] challenges the design of intrusion recovery countermeasures. Recovery solutions should try to minimize the delays caused by security attacks and also restore the network to a stable state at the minimum time possible. The evaluation should consider the **responsiveness** capability of the recovery measures.

Table 1. Recovered WSN services based on intrusion recovery requirements

Intrusion recovery requirements	Operational WSN services		
	Sensing	Communicating	Reporting
Availability		x	x
Survivability	x	x	x
Reliability			x
Responsiveness		x	x
Resilience		x	x
Self-healingness		x	x

2.2 Definition of evaluation criteria

This section identifies the evaluation criteria that should be considered in a potential experiment/simulation scenario.

- Number of blocked nodes: The severity of an attack should be measured by calculating the number of nodes that a malicious node can compromise. The assessment will indicate how effective the recovery solution is to restore the availability of sensor nodes, therefore the availability of the communication link.

- Node density: The proposed recovery solutions should be assessed as to whether they can retain an acceptable node density level to retain the network communication.

- Energy consumption: The energy consumption will showcase how the survivability of sensors is affected by the operation of the recovery protocol, in the presence of a persistent adversary.

- Routing overhead: Control packets are usually exchanged to support the objectives of the routing phases and intrusion recovery mechanisms. It is important to assess the volume of control packets produced by the recovery protocol and observe how the survivability of nodes is affected.

- Retransmissions: Retransmissions lead to energy consumption and packet delivery delays, therefore they should be minimized. Evaluation will indicate if the proposed countermeasures can minimize packet retransmissions in the face of security attacks.

- Path length: Assessment should consider if recovery solutions lead to longer path lengths and decide if it is acceptable in terms of energy consumption and delays.

- Packet delivery: The packet delivery is defined as the number of data packets received by the destination node divided by the number of data packets transmitted by the source nodes. This criterion allows assessment of the protocols effectiveness to restore packet delivery.

- Eavesdropped packets: The recovery solutions ability to isolate malicious nodes and minimize eavesdrop occurrence needs to be evaluated. This will promote a resilient network operation against the eavesdrop attack, prohibiting attacks such as traffic analysis.

- Packet loss: It is vital to evaluate the packet loss that occurs at the network during an attack and after a recovery solution is deployed. This will allow assessment of the robustness of the network, and thus the resilience level that can be obtained.

- End to end packet delivery delay: It is measured as the difference between the start time of the packet transmission at the source and the end time of the packet being received at the sink. Lost packets are considered having "infinite" delay and therefore are not considered in the calculations. Recovery should be evaluated as to whether it incurs acceptable delivery delays.

Table 3 maps the evaluation criteria to specific intrusion recovery security requirements. The following section discusses how the proposed mapping can be used to select evaluation criteria in order to evaluate specific intrusion recovery aspects of a protocol.

2.3 Evaluation selection procedure

In order to proceed with the evaluation or comparison of intrusion recovery protocols, it is essential to decide what aspects will be assessed according to the application's objectives. This section guides the user to decide which security aspects he needs to consider for the evaluation. The following three steps apply:

2.3.1 Evaluation aspects selection

The evaluation will assess the intrusion recovery protocols' capability to address security attacks, recover compromised network components/services and prohibit attack occurrence. The following list defines a number of statements that are related to a specific evaluation aspect. One should choose the statements that are valid from Table 2, taking into consideration the purpose of the evaluation and the application's objectives. The selected statements will indicate the evaluation aspects that need to be considered and which will drive the evaluation process.

Table 2. Selection of evaluation aspects

Assess Recovery Protocols' Capability to:	Evaluation Aspect
Recover from service interruptions. Restore blocked nodes and communication links.	Availability
Minimize energy consumption. Incur reasonable energy consumption due to recovery measures.	Survivability
Enable packets to reach destination. Support reliable decision making.	Reliability
Minimize packet delivery delays. Incur acceptable packet delivery delays due to recovery measures.	Responsiveness
Prohibit adversary to compromise network operation, after recovery is applied. Minimize eavesdropping on communication.	Resilience
Address dynamic attack strategies and respond dynamically according to case. Achieve graceful degradation of performance as attacks get more severe.	Self-healingness

2.3.2 Evaluation criteria selection

Table 3 should be referenced in order to choose the evaluation criteria based on the evaluation aspects selected from the previous step. The selected evaluation criteria should be used in the simulations to conclude on the effectiveness of the protocols to address attacks and recover compromised services.

Table 3. Selection of evaluation criteria

Evaluation criteria / Evaluation aspects	**Availability**	**Survivability**	**Reliability**	**Resilience**		**Responsiveness**	**Self-healingness**
				Eavesdrop	**Other**		
Blocked nodes	x				x		x
Node density	x						
Energy consumption		x					
Routing overhead		x					x
Retransmissions		x					x
Path length		x					
Packet delivery			x		x		x
Eavesdropped packets				x			
Packet loss					x		
Packet delivery delay						x	

2.3.3 Evaluation modes definition

The following three evaluation modes should be utilized to conclude on the effectiveness of a recovery protocol: 1) Normal. The network should be assessed under normal network conditions. This will constitute the benchmark case where all other modes will be compared against. 2) Security attack. In the case where malicious activity is considered, the attack type must be defined along with the percentage of nodes in the network that act maliciously. 3) Intrusion recovery. Consider the recovery countermeasures that are applied to address security attacks.

3. CASE STUDY

This section aims to demonstrate the adoption of the proposed evaluation methodology in order to compare the effectiveness of proposed intrusion recovery protocols to address specific attacks.

3.1 Intrusion recovery protocols overview

This section briefly describes the intrusion recovery protocols to be compared. The first protocol [1] utilizes multiple directional antennas to control the communication links established between nodes. Sensors dynamically enable or disable their antennas in order to prohibit communication from/to the malicious nodes. In this way, the network can isolate malicious nodes effectively and restrict them from continuing to affect the sensors, after the intrusion recovery is applied. The aforementioned protocol is compared against a common practice recovery countermeasure that is proposed in the context of omni-directional networks. Once the attack is detected, sensors blacklist [2] malicious nodes, they stop accepting or forwarding packets from/to blacklisted nodes and they update the active route paths to exclude malicious nodes.

3.2 Definition of evaluation approach

The intrusion recovery countermeasure will be utilized in healthcare applications [3]. The evaluation purpose is to assess which of the intrusion recovery protocols [1] or [2] is more suitable to support the application's objectives. Healthcare applications perform critical operations and the decision-making is strongly depended on the ability of the network to deliver observed events to the control center. Moreover, due to privacy issues, eavesdropping should be minimized. Taking into consideration these requirements, the evaluation aspects that are selected from table 2 concern the reliability and resilience on eavesdropping. Based on Table 3 packet delivery and eavesdropped packets need to be considered in the performance evaluation.

Evaluation figures present the performance of intrusion recovery protocol 1 (called DIR) against the equivalent common practice recovery protocol 2 proposed in the context of omni-directional network (called OMNI). Comparisons are made for the following simulation scenarios, each given a reference ID (x-axis on figure 1) as follows: 1. Normal network conditions. 2. Blackhole attack. 3. Intrusion recovery and 4. Eavesdropping and DoS.

3.3 Configurations

Simulation scenarios are configured in an IEEE 802.15.4 network with the AODV as the underlying routing protocol and evaluated using the ns2 simulator. The network consists of 100 nodes randomly placed with a uniform distribution in a 1050 x 1050 m and 750 x 750 m areas. Simulation results are averaged over 30 random topologies with 30 random sets of 10% malicious nodes. Source nodes generate two packets per second of CBR traffic. The packet size is 70 bytes. Sensor nodes are assumed to be equipped with 4 patch directional antennas with 8dBi gain in the DIR protocol. The receiver's sensitivity is considered to be -90 dBm in both DIR and OMNI networks. Moreover, the transmission range is configured to be 168m in both networks in order to provide a fair comparison. The simulations run for 270 seconds.

3.4 Simulation results

3.4.1 Normal network state

As it can be observed in Fig. 1, DIR case achieves 98% and 90% packet delivery when considering a sparse (1050x1050) and dense (750x750) topology respectively under normal network state. A dense topology yields more collisions and packet drops due to higher node density which affects the packet delivery. The DIR showcases 13% and 8% more packet delivery when compared to the OMNI sparse and dense topologies respectively. This is justified as the usage of directional antennas achieves spatial reuse and decreases packet collisions during normal network operation, and in this way enhances the network's performance.

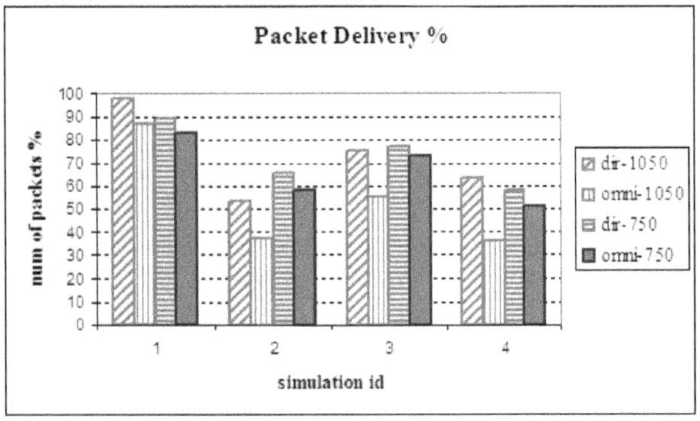

Figure 1. Packet delivery.

3.4.2 Blackhole attack

When the malicious nodes launch the blackhole attack, packet delivery (Fig. 1) is affected. The success of this attack depends on the location of the malicious nodes towards the active packet flow and the density of the network. Malicious nodes must be located in one of the active route paths in order to drop packets. When the network is sparse sensor nodes have fewer neighbors to select from as the next hop to the sink. As the network gets denser, nodes have more neighbors to choose from and therefore malicious nodes have fewer chances to be selected as forwarding points. The DIR presents a reduction of 45% packet delivery in the sparse network while the OMNI case shows a reduction of 57% when compared to their respective normal simulation scenario. In the case of the dense topology, DIR packet delivery is reduced by 26% while OMNI by 30%. This is due to the fact that there are more active malicious nodes in the sparse topology and therefore more active routing paths are affected by the attack. Overall, the DIR network retains a higher packet delivery in both dense and sparse topologies.

3.4.3 Intrusion recovery

As soon as the attack is detected, both networks apply their intrusion recovery countermeasures and successfully restore the network's packet delivery capability as depicted in Fig. 1. DIR restores around 77% packet delivery in both dense and sparse topologies, which is 36% and 5% more when compared to the OMNI sparse and dense networks respectively. As presented, both networks can effectively address the blackhole attack with the DIR network outperforming the OMNI case. The ability of the DIR case to reduce collisions and establish more stable routing paths aids the network to present a higher packet delivery.

3.4.4 Eavesdrop and DoS attacks

A DoS attack is launched by malicious nodes if they can eavesdrop on the network communication, after recovery is applied. When the network is dense, malicious nodes can eavesdrop on more neighbor nodes. Table 4 lists the total received packets by malicious nodes on eavesdrop attack. DIR reduces eavesdrop by 71% and 80% when considering the sparse and dense topologies respectively and compared to the OMNI cases. This significant reduction helps the network to prohibit traffic analysis on captured packets and potentially reveal sensitive data, i.e. related to security mechanisms and healthcare information. Due to this, packet delivery (Fig. 1) is greatly affected in the OMNI network when compared to the previously recovered results. DIR is affected as well, however with a much lower reduction rate. Overall, the DIR retains higher packet delivery than the OMNI as it isolates malicious nodes in a more effective manner and prohibits them from launching new attacks.

Table 4. Total received packets on eavesdrop attack

DIR-1050	OMNI-1050	DIR-750	OMNI-750
1030	3502	1405	7049

3.5 Intrusion recovery protocol selection

Evaluation has indicated that DIR and OMNI protocols can successfully address the blackhole attack. However, OMNI cannot suppress reactive adversaries that overhear communication and launch other attacks in order to continue affecting the WSN's services. Overall, DIR supports better the reliability and resilience requirements of healthcare applications. Eavesdropping is greatly minimized by DIR when compared to the OMNI case. Based on the results and taking into consideration the selected evaluation aspects, the DIR protocol is recommended to be utilized in the healthcare WSN application.

4. CONCLUSIONS

This research work proposes an evaluation methodology to promote evaluation and comparison of intrusion recovery protocols in WSNs. The methodology defines the aspects that should be assessed in the context of intrusion recovery protocols, maps appropriate evaluation metrics to each aspect and aids researchers deciding the evaluation approach they should utilize. A case study is presented demonstrating the usefulness of the proposed methodology in driving the evaluation procedure and facilitating the protocols' comparison. Future work will extent the evaluation methodology with more evaluation criteria under each intrusion recovery evaluation aspect.

5. REFERENCES

[1] Stavrou, E., and Pitsillides, A. Combating persistent adversaries in WSNs using directional antennas, ICT 2011.

[2] Crosby, G. V., Pissinou, N., and Makki, K. Location-aware, Trust-based Detection and Isolation of Compromised Nodes in WSN, Int. Journal of Network Security, 2006.

[3] Akyildiz, I.F., Su, W., Sankarasubramaniam, Y., and Cayirci, E. Wireless Sensor Networks: A Survey, Computer Networks (Elsevier), vol. 38, no. 4, pp. 393-422, March 2002.

[4] Stavrou, E., and Pitsillides, A. A Survey on Secure Multipath Routing Protocols in WSNs, Computer Networks, 2010.

[5] Zhang, Y., Huang, Y., and Lee, W. An extensible environment for evaluating secure MANET, SecureComm, pp. 339–352, 2005.

[6] Ashraf, A., et al. Design and Analysis of the Security Assessment Framework for achieving Discrete Security Values in WSN. IEEE CCECE, 855-860, 2008.

[7] Ransom, S., Pfisterer, D., and Fischer, S. Comprehensible Security Synthesis for WSN, MidSens'08, 2008.

[8] Kalpana Sharma, M K Ghose, Kuldeep, Complete Security Framework for WSN, IJCSIS, 3, 1, 2009.

RVIP: Bridging Live Networks and Software Virtual Networks for Large Scale Network Simulation at Real Time

Tingzhen Li[†], Jiejun Kong[*], Ping Li[†], Peng Gong[†]

[†]Beijing Institute of Technology, No.5 Zhongguancun South Street, Beijing, 100081, China

[*]Contact author, University of Florida, P.O.Box 116130, Gainesville, Florida 32611, USA

{tingzhen.li, jiejun.kong}@gmail.com, {liping85,penggong}@bit.edu.cn

ABSTRACT

In this work we investigate *real-virtual interface pair* (RVIP), a new interface system proposed for hybrid network simulation. In the hybrid system, multiple live networks (LNs) and multiple software virtual networks (SVNs) are connected together via standard IP protocols in an arbitrary topology and at real time. RVIP seeks to implement a new *Turing-indistinguishable mode* so that an LN and its counterpart SVN are indistinguishable in regards to a third-party live node. To realize the new mode, three necessary conditions must be satisfied: (1) All needed changes incurred by introducing an SVN into a live network scenario are put on the simulation's side, RVIP requires that no change is made on any live node; (2) An SVN does *not* exchange simulation events with LNs, that is, only standard IP protocol interactions between SVN and LN are allowed. (3) Any LN can be dynamically plugged into the hybrid scenario at real time, just like being plugged into an equivalent purely live network.

RVIP realizes the mapping between a live IP protocol stack and a virtual IP protocol stack (VIPS) by presenting a progressive series of connection patterns, namely *mapping single virtual node to live node*, *combining LN with SVN* and *connecting multi-SVNs*. Then we can construct more complex hybrid scenarios by combining these patterns. Compared to existing hybrid simulation efforts on NS-3, QualNet and OPNET, RVIP can support hybrid scenarios with multiple SVNs and multiple LNs connected by an arbitrary network topology. Our performance studies show that RVIP provides more efficient support in terms of common metrics such as larger throughput limit and smaller extra latency.

Categories and Subject Descriptors

C.2.1 [**Computer-Communication Networks**]: Network Architecture and Design—*Network communications*

General Terms

Design, Experimentation, Measurement, Performance

MSWiM'12, October 21–25, 2012, Paphos, Cyprus.

Keywords

Hybrid network simulation; Virtual protocol stack; Turing indistinguishable mode; No topological constraint; Zero live node change

1. INTRODUCTION

Existing study and test of a larger-scale network, for example, an IP network that is comprised of thousands of nodes, can be attributed to two major approaches, namely physical test and computational simulation. Either approach has its advantages and disadvantages. In physical test, the network objects under test run in a physical testbed while their performance is evaluated over numerous testing rounds. By using live networks and actual environment, this approach features the highest accuracy. However, when scalability becomes a critical concern in the test, this approach requires large amount of hardware units and considerable resources, thus incurs prohibitively expensive cost. On the other hand, computational simulation is a more cost-effective alternative and widely used approach that supports flexible and controlled experimentation of arbitrary network scenarios. However, the simulator is always implemented based on some degree of abstraction and simplification of the network protocols and live applications, thus may miss important system characteristics [18]. Therefore, it is appealing to construct a hybrid system which can combine the advantages of both approaches and provide a balance solution to accomplish the needed tests [10][23][22][4].

In the hybrid system, there are at least one software-implemented virtual network (SVN) and at least one live network (LN) interating with each other. The most widely adopted method to implement an SVN is using an off-the-shelf network simulator, such as NS-3, OPNET and QualNet. Then an interface system is devised to connect the LNs with the SVN(s). While the live applications run on the live hosts at real time, the simulator is responsible for modelling the discrete events in the link layer, the physical layer and the environmental scenario. Live packets generated by the live applications are injected into and extracted from the SVN(s) at real time. As proposed in [2], the emulation facility can operate in two modes, according to whether the live packet is transparent to SVN or not.

- *Opaque mode:* The simulator treats a live packet as an uninterpreted packet. Live packets sent by a live node flow through the SVN and arrive at other live nodes as if the SVN is a tunnel.
- *Protocol mode:* The simulator is able to interpret and/or generate live network traffic containing arbitrary field assignments. Live nodes can interact with virtual nodes on network protocols.

In the opaque mode, when a live packet enters an SVN, the simulator prepends and/or appends additional information, which will be used to process and forward the live packet by the virtual nodes. In the SVN, the packet may be dropped, delayed, reordered, or duplicated. However, since no protocol processing is performed in the opaque mode, protocol-specific traffic operations are impossible in the SVN. When the packet arrives at the virtual destination node, which acts as the proxy of the live destination node, the simulator removes the additional information from the packet and sends it out to the live node. Thus this mode can only be used in the "LN–SVN–LN" scenarios.

A distinct feature of the protocol mode is that a node in SVN can interpret any protocol field of an injected live packet and accordingly actualize protocol-specific behaviors. Live nodes can interact with virtual nodes on the protocols such as ARP, routing, ICMP, etc, gaining indistinguishability between LNs and SVNs at the granularity of network protocols. Unlike the opaque mode which can only let the live applications interact with the SVNs, in the ideal case the protocol mode can seamlessly integrate LN's and SVN's protocol stacks together by replacing any LN with an SVN, and vice versa. We name the ideal form of the protocol mode as the *Turing-indistinguishable mode*[1].

In the protocol mode emulation, there are two types of protocol entities: live IP protocol stack (LIPS) which runs on a live device in the physical world and virtual IP protocol stack (VIPS) emulated by a network simulator.

- LIPS and VIPS must be consistent with each other in protocol operations. The protocol implementation of LIPS on live system and the simulated protocol modeling of VIPS in network simulator must be strictly in accordance with the protocol standard.
- LIPS and VIPS must share the same time reference. This implies that the network simulator should have real-time (including faster than real-time) operation ability. Though it is impossible to achieve perfect timing synchronization between two network clocks [21], the clocks of LIPS and VIPS can be synchronized into a very small difference, for example, at the scale of 100ms [13].
- The packets between two end-to-end VIPS are always optimized by the simulation, making them different from live packets between two end-to-end LIPS. To realize the fusion of LNs and SVNs, an interface system is needed to convert the packets sent from one side to the other.

To realize the ideal Turing-indistinguishable mode, three necessary conditions must be satisfied:

1. All needed changes incurred by introducing an SVN into a live network scenario are put on the simulation's side. It is imperative that no change is made on any live node for the purpose of interacting with the SVN.
2. An SVN does *not* exchange simulation events with LNs. In other words, only standard IP protocol interactions between SVN and LN are allowed. All discrete simulation events for protocol stack modeling must be stopped at the SVN-LN boundary.
3. Any LN can be dynamically plugged into the hybrid scenario at real time, just like being plugged into an equivalent purely live network.

To realize the design goal, we present the RVIP interface, which is a highly-efficient, low-complexity, Turing-indistinguishable coupling of LNs and SVNs. RVIP realizes a one-to-one correspondence between a VIPS and its counterpart LIPS. It provides translation functions between live packets and virtual packets for multi-layer protocols such as ARP, IP, ICMP, IGMP, OLSR, OSPF, RIP, AODV, TCP and UDP, implements the interaction between live nodes and virtual nodes, and makes live nodes and virtual nodes indistinguishable in terms of network behavior. In addition, all the functions of RVIP are implemented on the simulation side, so that there is no protocol stack difference on a live node for the node to connect with an SVN or an LN. Live nodes in our system include broad range of heterogeneous IP nodes, from light-weight virtual machines such as LXC [9] to any fixed box which the simulator has no modifying permission. This greatly enhances the scalability and flexibility of our hybrid system.

The paper is organized as follows: Section 2 describes related work. In Section 3 we design and implement the RVIP interface system. Then Section 4 illustrates the advantages of using RVIP to construct a hybrid simulation. In Section 5, experiments are conducted to evaluate the efficiency of RVIP. Finally Section 6 concludes the paper.

2. RELATED WORK

Based on the NS-2 simulator, [10] improved the emulation facility by allowing live applications to be connected to specific nodes in the NS-2's SVN. It supports two working modes, the single host extension and the distributed clients extension. The former allows live applications run on a single live host only. The latter introduces a distribution layer between the live nodes and the NS-2 SVN, allowing live applications to run on several physical hosts. [5] evaluated the efficiency of this distribution layer. NS-2's successor, NS-3, also supports the single host emulation capability. Live nodes, some of them can be virtualized machines, are connected through a TUN/TAP device of the Linux kernel to proxy nodes in the SVN. [1] developed a distributed client extension for NS-3 adopting the same approach used in NS-2. Just like the existing systems, RVIP supports both the single host mode where a (virtualized) host's TAP device is mapped to a virtual node, and the distributed mode where a set of live nodes are connected to the SVN via LAN. However, RVIP does *not* require any change in the live nodes (such as installing or configuring an extra software), greatly simplifies the operational setup and aims at the Turing indistinguishability.

Based on the OMNeT++ simulator, [16] proposed an emulation framework "VirtualMesh", which requires live nodes create virtual wireless interfaces with parameters configured just like real interfaces. The virtual interface intercepts live traffic and forwards it to a simulation model. The live nodes and the simulator are connected by a UDP session, through which the live node can configure, manage, and communicate with the simulator. The same method is also adopted by [20]. These efforts only support the opaque mode in hybrid simulation, and require installing some new modules on live nodes (which may not grant the needed administrative permission).

There are previous efforts to improve the simulation scalability. [19] tried to increase the simulation network's scale by using "virtual time". Basically, the clocks on the network nodes are slowed down for the network applications, so that the applications work slower than what should be in the real scenario. This allows more discrete events be processed by the simulation server in unit time. Unfortunately, this approach does not apply to real-time simulation systems with real human and live devices in the loop.

[6] proposed the concept of "network immersion" so that the virtual network is indistinguishable from a physical test-bed in terms of network behavior, but it focuses on VPN rather than the protocol (e.g., AODV, ARP, etc.) interaction between LNs and SVNs.

[1]We borrowed the name from the AI field, where an AI entity passes the interactive *Turing Test* if it is indistinguishable from the real entity (human brain) in regards to black-boxed interactions.

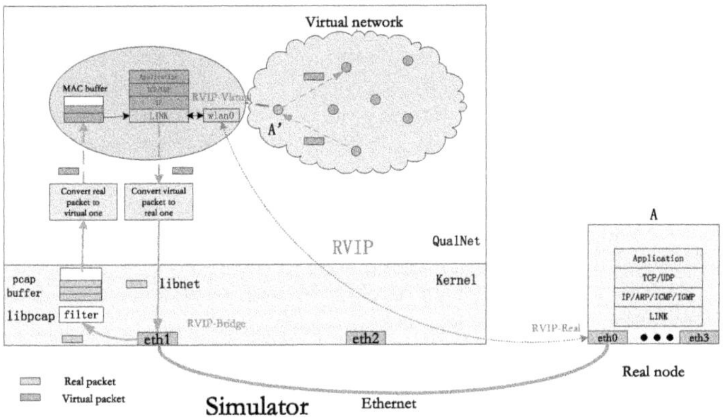

Figure 1: The structure of real-virtual interface pair

[8] constructed a mapping between a virtual interface in SVN and a TAP device in client machine, and the connection between them is realized by VPN, supporting geographically distributed applications to dynamically connect to the real-time network simulator. Compared with RVIP, this approach needs to install VPN clients on live nodes and the thoughput of the VPN connection is significantly reduced. In order to produce accurate result, the VPN emulation infrastructure requires a tight coupling (i.e. high-band width low-latency connection) between the live hosts and the real-time simulator. Similar to other existing emulators, it needs administrative privileges to establish the emulation infrastructure (e.g., setting up VPN on the client machines).

In the commercial market, OPNET corporation [4] has developed a module called system-in-the-loop (SITL), which transfers packets between LN and SVN to achieve interaction between live node and virtual node. By SITL, developer can construct two types of connections, namely *LN-SVN-LN* and *SVN-LN-SVN*. QualNet provides the interface of IPNE [14] and its successor EXata. IPNE supports NAT-YES, NAT-NO and TrueEmulation working modes. The first two modes are opaque, while TrueEmulation is the protocol mode. The above interfaces can accomplish the interaction between LN and SVN with limited real-time scalability and to a certain degree measured by the opaque mode or the protocol mode. The TrueEmulation mode of IPNE is difficult to configure, in particular for a large number of changes made on the live nodes; OPNET only supports a chain connection of LNs and SVNs. All the interfaces above cannot realize a simulation where LNs and SVNs are arbitrarily mixed. Additionally, to our best knowledge, neither OPNET nor QualNet/EXata has published performance analysis and evaluation of their LN-SVN interfaces.

3. RVIP: REAL-VIRTUAL INTERFACE PAIR

3.1 Design and Implementation

We design an interface of RVIP which realizes a one-to-one mapping between a real IP protocol stack (e.g., on a real IP network interface card) and a virtual IP protocol stack (VIPS). As demonstrated in Figure 1, a live node A has a real interface eth0, and its shadow node A' has a virtual interface wlan0, then eth0 and wlan0 form a RVIP, where eth0 is called "*RVIP-Real*", wlan0 is called "*RVIP-Virtual*", and they have the same IP address and hardware MAC address. RVIP seeks to implement the same behavior on RVIP-Real and RVIP-Virtual. In other words, when RVIP-Real sends out a live packet, the RVIP-Virtual sends out an equivalent packet in the SVN. Likewise, when RVIP-Virtual receives a virtual packet from other virtual nodes in the SVN, the RVIP-Real receives an equivalent packet. Thus RVIP-Real and RVIP-Virtual act like a single IP protocol stack with double faces, one in the actual world and the other in the virtual world.

RVIP has three basic modules: *packet capture*, *packet translation* and *packet construction*. After the *packet capture* module captures a live packet sent by RVIP-Real in real time, the *packet translation* module will analyze and convert the live packet to its equivalent virtual form. Then the virtual packet is inserted into the SVN and processed by the simulator. Similarly, when a virtual packet arrives at RVIP-Virtual, RVIP-Virtual does not deliver it to upper-layer protocols, instead, it translates it into a live form and send it to the RVIP-Real.

As Figure 1 shows, there is a live node connecting to the simulation server via an Ethernet NIC eth1 called "*RVIP-Bridge*" which serves as a bridge between a live node and the SVN. Several RVIPs can share a same RVIP-Bridge. At the beginning of the simulation, the simulator side does the following tasks:

- Disable the IP and upper-layer protocols on the virtual node, and keep its MAC and physical layers alive;
- Register an entry in the RVIP chain table for every RVIP;
- Assign a *libnet* handle [15] and a *pcap* handle [11] on RVIP-Bridge for every RVIP;
- Set the *pcap* filter for every RVIP, making it only receives packets sent by RVIP-Real.

After that, every RVIP has a *libnet* handle, a *pcap* handle and a *pcap* buffer. *libnet* handle is used by RVIP-Virtual to construct live packets, and *pcap* handle is used to read packets buffered in the *pcap* buffer.

A large-scale discrete-event simulation platform with high fidelity and faster than real time guarantees is needed to simulate a large number of network nodes in a near-realistic software virtual environment. Currrently in the commercial market, only QualNet and OPNET are approved by the DoD because they follow both civilian and military industrial standards in the simulation implementation. We choose QualNet due to scalability concerns. QualNet can simulate computer networks of more than 1000 nodes in an off-the-shelf multi-core or multi-processor server at real time. It is the largest and fastest simulation platform currently available, typically at least three times faster and with three times more network nodes than its competitors including OPNET. Therefore, we choose QualNet as the underlying simulator for RVIP.

The simulator runs in real time and keeps pace with RVIP-Bridge in the process of simulation. All packets sent by RVIP-Real get through the *pcap* filter and are buffered in the *pcap* buffer. A timer is set in the simulator and its time-out interval is T (currently set to 1ms). Every T, the timer times out, triggering the simulation engine to poll the RVIP successively. If the *pcap* buffer of an RVIP has packets, all of them are taken out by this RVIP's *pcap* handle. When a live packet is captured, RVIP will translate it to a virtual one and insert it into the MAC buffer of the RVIP-Virtual. Then the control is handed over to the simulator engine, which schedules and processes the virtual packet. Currently, we have provided the translation functions for the following protocols: OSPF, OSLR, AODV, UDP, TCP, ICMP, IGMP, ARP, RIP, and IP. The function is implemented in a modular framework to ease the efforts to add new protocol translation unit. The *packet construction* module is implemented at each RVIP-Virtual's MAC layer. When RVIP-Virtual receives a virtual packet, it calls its *libnet* handle to construct a new live packet and sends it out.

Essentially RVIP-Real is the master who simply runs its equipped IP protocol stack without any change required. RVIP-Virtual acts as the slave proxy of the RVIP-Real in SVN and only enables its MAC layer and physical layer functions. RVIP creates an integrated protocol stack, in which the IP and above layers are the live ones on live nodes and the MAC and physical layer are the virtual ones emulated by the simulator. By this way live nodes join into the SVN and interact with virtual nodes on network protocols.

3.2 Construct Hybrid Simulations based on RVIP

3.2.1 Mapping Single Virtual Node to Live Node

RVIP supports both distributed connections and single host connections when establishing the mapping between a live node and a virtual node. In the distributed mode, any device supporting TCP/IP and having a LAN card can be connected to the SVN to act as RVIP-Real. This mode distributes the workload to multi-hosts. In the single host mode, live nodes realized by virtualization technology, such as VMware and LXC, run on the same host with SVN and their TAP devices are connected to SVN as RVIP-Real via the Linux kernel bridge. In summary, applications and protocols above the link layer on the live nodes run as usual without any modification, while the lower layers below MAC are connected via the SVN. Packets generated by live applications can be injected into and routed through the SVN. Then the effect of network delays and protocol processing can be imposed on live network traffic.

3.2.2 Combining LN and SVN

In Figure 2, a live node called "*RVIP-Gateway*" connects a multi-node LN and an SVN together. The gateway can be an Internet gateway when LN is the Internet. The gateway has two network interfaces, one directly connecting with the LN and the other directly connecting with an SVN. On both interfaces, RVIP-Gateway runs the same routing protocol of the virtual nodes in SVN and forwards packets from one side to another, so that the topology information of one network can be introduced into the other. Now we could *ping*, *traceroute* each node in LNs or SVNs. When inspecting a live node's routing table, there are routing entries to those virtual nodes after the live node has interacted with virtual nodes on the network protocol level. As a result, the SVNs improve the flexibility and scalability of the target network modeling, and the LNs enable the actual users and the real devices to participate in the simulation at real time.

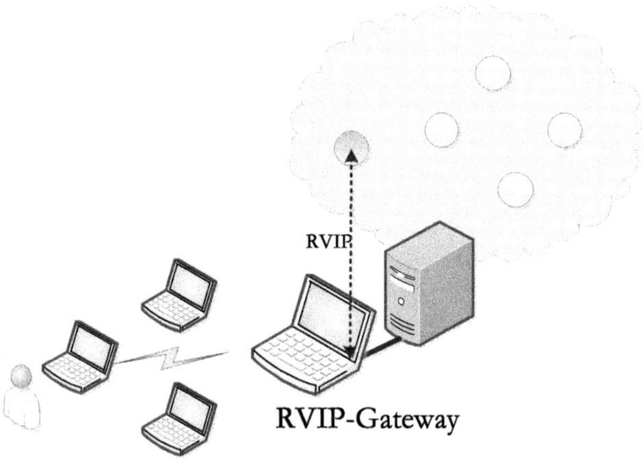

Figure 2: *RVIP-Gateway* combines LN with SVN

3.2.3 Connecting multi-SVNs

In Figure 3, a live node called "*RVIP-Router*" has multiple network interfaces, each of them connecting to an SVN and forming an RVIP. On every network interface of the RVIP-Router, the same routing protocols used by the corresponding SVN run on the interface and interact with the virtual nodes. The RVIP-Router exchanges topological information with every directly-connected SVN and distributes them into other SVNs. This way, virtual nodes in one SVN can communicate with the ones in other SVNs. As a result, we can use multiple simulation instances running at real time and at different simulation servers to enlarge the scale of network simulation. Each simulation server only takes charge of one SVN in the simulation and the RVIP-Routers connects all of them together.

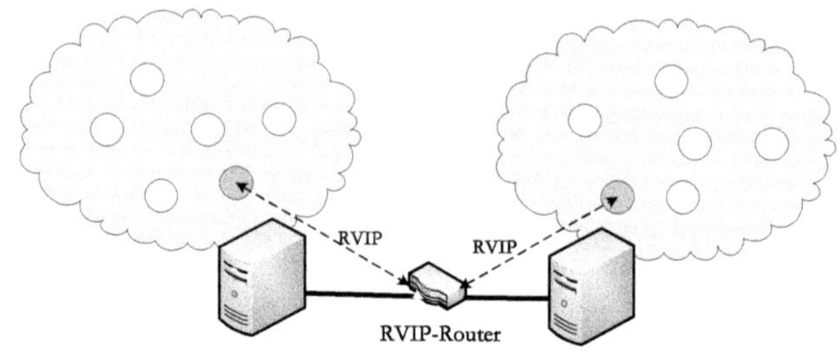

Figure 3: *RVIP-Router* connects multi-SVNs

Based on the above-mentioned basic patterns, we can construct more complex hybrid simulation scenarios where LNs and SVNs arbitrarily mixed together without topological constraints. This realizes a more scalable network simulation capability than the underlying simulation engine. We will show a case study in the next section.

4. CASE STUDY

This section presents a case study to demonstrate how RVIP can be used to integrate multi-SVNs and real world together, facilitating larger scale network emulation.

We carry out a simply prototype experiment. As Figure 4 shows, there are three SVNs, each has three virtual nodes connected by wire links with 20ms propagation delay. Live nodes, D and E, both have two NICs embedded into SVNs as RVIP-Reals. Each live node connects two SVNs together. We then instantiate two virtual IP protocol stacks (VIPS), F and G, in server B and C using LXC virtual machine. F and G also have two NICs; one is a TAP device functioning as an RVIP-Real towards an SVN, and the other as an interface to the real world. All nodes (simulated nodes in SVNs as well as D, E, F, G) run RIPv2 routing protocol during the experiment. The virtual nodes run QualNet's RIPv2 protocol while the live nodes run RIPv2 protocol from the Quagga [12] package. We measure the round trip time (RTT) between D and G using *ping*, and record the route between B and G using *traceroute* at the same time. Before the experiment, the interfaces to real world, X and Y, are connected together via an Ethernet cable. During the experiment, we disconnect X and Y for a while and then reconnect them. In the procedure, we find that, before the disconnection, the route path between D and G is D-SVN2-F-G, and after the disconnection, it becomes D-SVN1-E-SVN3-G. The RTT measurement between D and G shows a corresponding change in the experiment (Figure 5). We can draw the conclusion that RVIP connects SVNs and LNs together following the depicted random topology, and IP networking is accomplished because of fair contributions from routing protocols running on both live nodes and virtual nodes.

A large-scale network may have tens of thousands of nodes, making it difficult to simulate in real time. When network scale

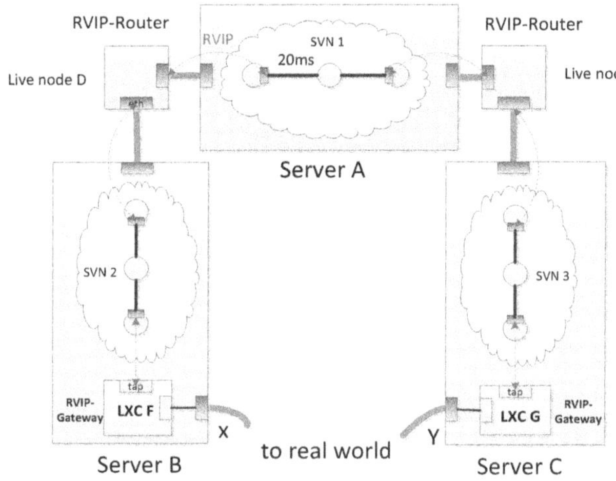

Figure 4: Scenario of the case study

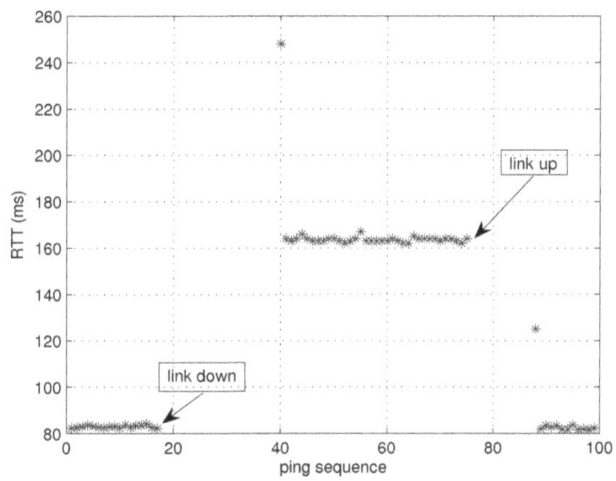

Figure 5: RTT measurement in the experiment

increases, a single simulator is unable to accomplish the real-time simulation task. An existing solution is using parallel computing to address the scalability concern and real-time concern. For example, PDNS [3] relies on computing clusters and supercomputers to simulate hundreds of thousands stations, with expensive hardware cost and decreased cost efficiency.

RVIP presents another solution with low cost. In RVIP, multiple simulation instances can run independently with each other in real time. Each simulation instance only takes charge of a single SVN in the entire simulation. The connection between different SVNs is realized by RVIP-Routers, which are live devices playing the role of hybrid router between different network parts. RVIP-Routers and SVNs can run the same set of routing protocols and other network protocols in the IP protocol stack, so that all SVNs and live nodes are seamlessly integrated as a monolith.

RVIP is in general a distributed solution. Like RVIP, existing solutions like *PDNS* and *PRIME* [7] also adopt a distributed framework. However, unlike RVIP, they use a set of parallel synchronization algorithms, instead of IP networks, to synchronize different simulation instances. For example, when packets are forwarded to a router simulated by the remote simulator, the local simulator must send a simulation event to that simulator. What they have realized are synchronization and interaction between different simulation instances at the granularity of discrete simulation event.

In contrast, RVIP is a pure networking solution which does *not* rely on the simulation events fired amongst different simulation instances. Unlike all previous work, RVIP seeks to eliminate the difference between an SVN and an LN in regards to a third-party live device/network. In a hybrid network scenario with multiple LNs and multiple SVNs, RVIP allows any LN be replaced by an equivalent SVN, and vice versa. In a nutshell, if the entire network topology is treated as a set of connected network tiles, each tile can be instantiated as an SVN or an equivalent LN without extra constraints in the SVN/LN tiling pattern.

In addition, we can dynamically connect any (IP-based) LN into an existing live hybrid scenario, and consequently attract more live devices, novel radios and actual users via the LN. This feature of *dynamic LN plug-in* is a new real-time simulation capability which improves scalability and flexibility of our solution. The Internet itself can be connected into an RVIP scenario as a dynamic LN plug-in.

Wireless Network Discussion: In wired networks, RVIP can easily address the scalability and flexibility concerns. Given the real-time capability of the underlying simulation engine (QualNet or equivalence), we can replace any LN (any connected subgraph of a reasonable size in the entire network topology) with an equivalent SVN running on the simulation engine at real time, or vice versa.

However, for wireless networks, due to node mobility and wireless signal propagation, two neighboring SVNs may have to exchange *environmental* discrete simulation events (in particular, node mobility and wireless signal propagation events) to deliver correct simulation results. Currently RVIP can only be used in wireless networks with FDMA or TDMA radios. All wireless nodes using the same frequency or TDMA schedule form an SVN. We are working on the following tasks to enable RVIP for other types of wireless networks in the real world.

- *Definition of geographical theater*: The entire wireless network area is divided into hexagon theaters, similar to cellular network's cells. Each theater is a wireless SVN (WSVN) which runs on a single simulation engine. The diameter of the theater is greater than the wireless interference radius.
- *Cross-WSVN mobile node*: A WSVN monitors the mobility pattern of its internal mobile nodes. If a mobile node steps into a neighboring WSVN, the previous WSVN and the next WSVN must exchange discrete events prior to the crossover.
- *Cross-WSVN signal propagation*: A WSVN monitors each wireless transmission event and notifies all neighboring W-SVNs about the event.

However, because QualNet's physical layer and signal propagation layer are not open to the public, currently we are unable to realize the WSVN features in our QUP-based implementation.

5. EVALUATION

5.1 Accuracy Test

The major overhead of RVIP comes from packet capturing, interpretation and construction when packets are transferred across the SVN-LN boundary. The overhead incurs inaccuracy in a simulated network as compared with its purely live equivalence. To ensure accuracy in hybrid simulation, RVIP should minimize the overhead. We conduct experiments to evaluate the accuracy of RVIP, mainly focusing on overhead in regards to network bandwidth and latency.

As shown in Figure 6, we construct a virtual wireless network in QualNet. In the SVN, there are 4 wireless nodes following a chain topology. Each node is configured with 802.11b MAC protocol

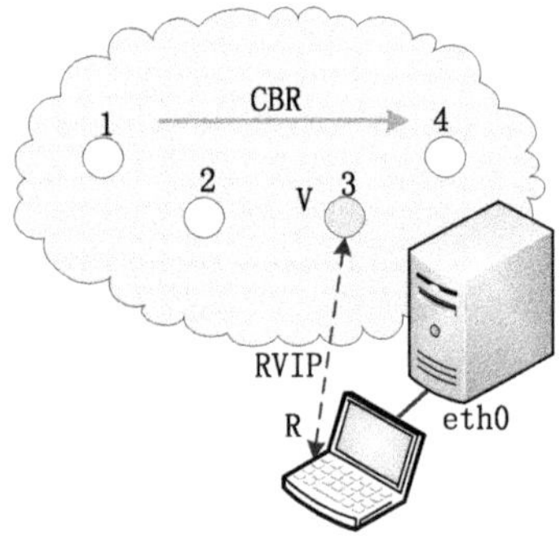

Figure 6: The accuracy test scenario

(without RTS/CTS), 2Mbps transmission rate and approximately 70m transmission radius. The distance between two neighbors is 50m. To focus on data traffic measurement, static routes are used in the network to remove routing traffic. Node 1 sends CBR packets to Node 4 with the following CBR parameters: CBR payload is 512Byte, packet sending rate is set to 1/s, 5/s, 10/s, 50/s, 100/s, 500/s and 1000/s progressively.

We conduct two comparative tests. Test 1 runs pure-software simulation without enabling RVIP, and Test 2 runs a hybrid simulation where node 3 was mapped to a live laptop. As Figure 6 shows, the RVIP-Virtual V interface and RVIP-Real interface R is connected by the RVIP-Bridge eth0. R/eth0 is a 1000Mbps Ethernet NIC. RVIP's overhead is highlighted by comparing the performance measured in the two tests.

Table 1: Delay at different packet processing phases

Phase	Task	Average time consumption
t1-t0	construct a live packett according to the virtual packet	6μs
t2-t1	a live packet is transmited by RVIP-Bridge and propagated	max≈1514Byte *8/1000Mbps=12μs
t3-t2	processed by the live node	varies per operation
t4-t3	same as *t2-t1*	same as *t2-t1*
t5-t4	waiting in the *pcap* buffer	*pcap* clears the buffer every ≤1ms
t6-t5	construct a virtual packet according to the live packet	2μs

Let's inspect the packet processing workflow of the hybrid simulation. A virtual packet sent by Node 2 arrives at V at *t0*; then the RVIP interprets the virtual packet and constructs a corresponding live packet, which takes *t1-t0*. Then the live packet is sent out by the RVIP-Bridge at *t1*, and arrives at R at *t2*. *t2-t1* is the transmission and propagation time of the live packet. From *t2* to *t3*, the laptop processes the live packet, such as doing route table lookup. Then the processed packet is sent out by R at *t3*, and arrives at RVIP-Bridge at *t4*. *t4-t3* is nearly identical to *t2-t1*. After received by RVIP-Bridge, the live packet is buffered in the *pcap* buffer until *t5*, when RVIP queries the buffer to fetch live packets. Then RVIP constructs a corresponding virtual packet and insert it into V's MAC buffer, which takes *t6-t5*. We have conducted experiments and recorded all above-mentioned timing points. The results are shown in Table 1, from which we know that all the phases result in additional delay compared with the purely simulated scenario. Amongst all phases, *t5-t4*, the waiting time in the *pcap* buffer, is the most significant one. This profiling result tells us to optimize the packet capture mechanism in our future work.

Other performance results are shown in Table 2. We learn that the deviation of network throughput caused by RVIP is insignificant (within 5% deviation). When the traffic load is low (packet sending rate≤50/s), in the pure simulation scenario, there is no packets loss and the average end-to-end delay is about 11ms; in hybrid simulation scenario, the extra delay caused by RVIP is about 0.5ms. As the traffic load increases, packet loss appears and the average end-to-end delay rapidly increases. The deviation caused by RVIP stays as insignificant compared to the measured results.

Table 2: Differential measurement to estimate RVIP overhead

Packet Rate	Performance	Pure Sim	Hybrid Sim	Difference	Deviation
1/s	*Latency(ms)*	11.30	11.77	0.47	4.2%
	Throughput(bps)	34.4K	34.4	0K	0%
5/s	*Latency(ms)*	11.11	11.53	0.42	3.8%
	Throughput(bps)	165.6K	165.6	0K	0%
10/s	*Latency(ms)*	11.07	11.51	0.44	4.0%
	Throughput(bps)	329.6K	329.6K	0K	0%
50/s	*Latency(ms)*	11.04	11.53	0.49	4.4%
	Throughput(bps)	1640K	1640K	0K	0%
100/s	*Latency(ms)*	1480	1430	-50	-3.4%
	Throughput(bps)	3072K	3072K	0K	0%
500/s	*Latency(ms)*	2860	2800	-60	-2.1%
	Throughput(bps)	3072K	3064K	-8K	-0.2%
1000/s	*Latency(ms)*	2880	2790	-90	-3.1%
	Throughput(bps)	2632K	2640K	-8K	0.3%

5.2 Stress Test

5.2.1 Distributed Mode

The distributed mode allows live nodes to be realized on hosts other than the simulation server, thus brings appealing features such as dynamic LN plug-in and multi-SVN connections. We conduct the following stress experiments under the distributed mode. In the scenario, live data streams originate from a live node, go through an SVN, and arrive at another live node. By comparing the metrics of the data stream at both live terminals, we can assess the overhead of RVIP compared to an equivalent live network. This method is also adopted by [5].

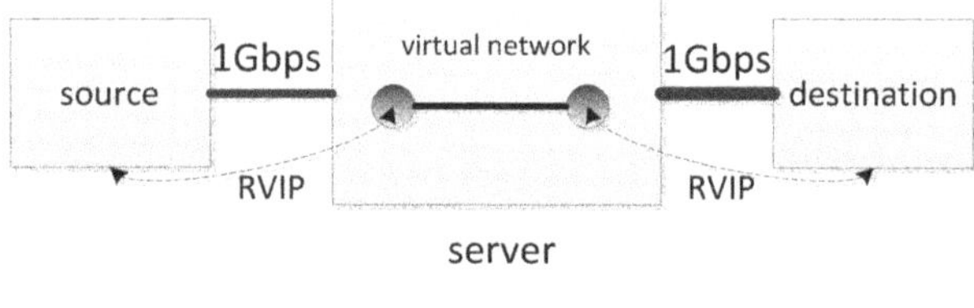

Figure 7: The stress test scenario

As Figure 7 shows, we construct a QualNet-based SVN where two nodes are connected by an 802.3 link with 10Gbps bandwidth and 1μs propagation delay. Each virtual node is mapped to a live laptop by RVIP. One laptop serves as the traffic source and the other as the traffic destination. The simulation server has an Intel Core 2 Duo 2.9GHz processor and 2GB RAM; each laptop has an Intel Core i5 2.4GHz processor and 3GB RAM. All hosts run Ubuntu Linux with a 2.6.35 kernel, and all NICs perform at 1Gbps.

We use a simple SVN so that the simulation server's CPU cycles are mostly consumed by processing events related to packet capturing and transferring from/to the live network. The packet operations invoke system calls to copy full packets between the operating system's kernel and user spaces, are thus much more expensive than processing the relatively simple simulation events.

In the test, the source sends CBR traffic to destination using the Traffic Generator (TG) tool [17], with packet size (P) set to 64-byte, 512-byte and 1024-byte progressively. During the experiment time, we also progressively change the packet sending rate (W) from 1K packet-per-second(pps) to 60K pps with 5Kpps increment per step. We monitor the sending and reception rates, also measure the route trip time (RTT) via *ping*.

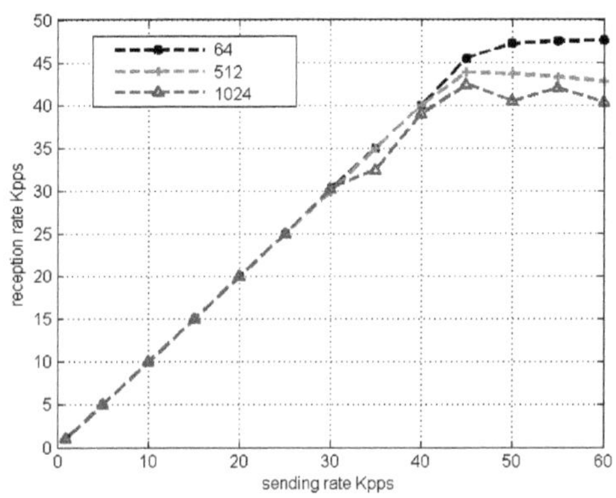

Figure 8: Relationship between sending rate and reception rate

The relationship between sending rate and reception rate is demonstrated in Figure 8. There is a clear point of saturation for all P and the saturation point is about 40Kpps when all additional traffic injected is dropped by the server. Considering that the live and virtual links are still far beyond their capability limit, the saturation point can be attributed to the exhaustion of the server's computational CPU power. We observe that the server's CPU is too busy according to the system monitor, thus the simulation server fails to fetch packets from its *pcap* buffer (PB) timely, results in the buffer overflow and packet drops. In RVIP's implementation, there are two buffers, PB and RVIP-Virtual's MAC buffer (MB). Packets sent by RVIP-Real are buffered in the PB and the simulator will clear N packets and insert them to MB every T (1ms under the ideal condition). N is determined by the free space of MB and P. We further conduct experiments to study the relationship between N, T and P when W is 50Kpps. Results are illustrated in Table 3. If P is smaller, the simulator can fetch more N every time, and the bigger N is the more time it will take the simulator to handle them, resulting in a larger T. N/T is the fetching rate of the simulator and when sending rate is larger than N/T, there comes the PB overflow and packet drops. N/T is mostly limited by the simulation platform's computational power and the complexity of SVN.

Table 3: Relationship amongst P, N and T

P(byte)	N (packet)	T (ms)	N/T (Kpps)
64	1453	32.0	45.4
512	277	6.7	41.3
1024	144	3.6	40.0

Figure 9 shows the measurement of RTT with an increasing W. Since our virtual link is configured with a negligible delay, the RTT is mostly attributed to the timing overhead incurred by RVIP. From

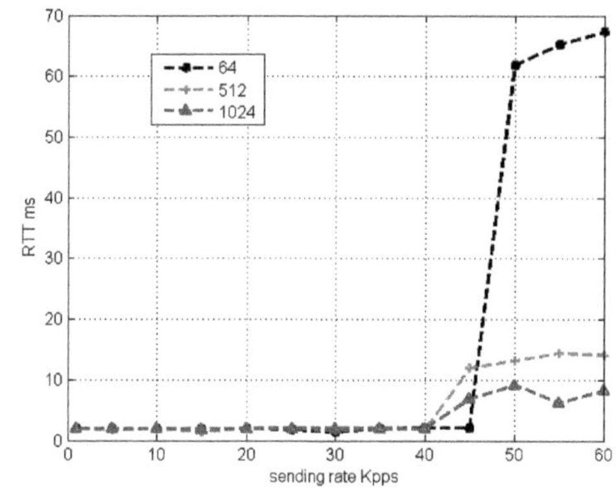

Figure 9: Relationship between sending rate and RTT

the figure we notice that RTT is a constant before reaching the simulation server's saturation point, then RTT increases significantly as W exceeds the saturation point. The most significant RTT increment happens at the saturation point when P is 64-byte. The underlying reason is that, as there are an increasing number of packets (N) to be inserted into the SVN, the average PB queuing time of the *ping* packets increases, then causes the observed jump in RTT measurement. For a smaller packet size, the queue size is larger and the expected waiting time is longer.

5.2.2 Single Host Mode

RVIP can also be used in single host mode scenarios for a comparison with NS-3's TapBridge model. We did the following tests to compare the performance between the two systems on the same equipment.

- Test1 (purely live network) is done with 2 laptops (used in Section 5.2.1) directly connected together via 1Gbps Ethernet link;
- Test2 (TapBridge with NS-3) is done on an NS-3 based SVN where two nodes are connected by a CSMA[2] channel with 1Gbps bandwidth and 1μs propagation delay. Each virtual node corresponds to an LXC virtual machine via NS-3's TapBridge UseBridge mode. NS-3 and both LXCs are running on the single simulation server (used in Section 5.2.1);
- Test3 (RVIP with QualNet) is done on a QualNet based SVN where two nodes are connected by an 802.3 link with 1Gbps bandwidth and 1μs propagation delay. Each virtual node corresponds to an LXC VIPS via RVIP. QualNet and both LXCs are running on the single simulation server (used in Section 5.2.1).

Table 4: Results of the three tests

	RTT (ms) before download	Throughput (Mbps)	RTT (ms)
Test1	0.36	344	0.4
Test2	0.4	28	460
Test3	1.2	160	4.8

In each test, we use *scp* on one host (live laptop or LXC) to download a massive file from the other one and measure the RTT by *ping*. Before the download, the average RTT of Test1 and Test2 are nearly the same, while Test3's RTT is about 1.2ms which is a

[2]NS-3's CSMA model is designed to approximate the 802.3 Ethernet model. For more details please refer to Chapter 9 of NS-3 Model Library.

constant as analyzed before. During the download, the throughput in Test1 can reach 344Mbps with 0.4ms RTT. This heavy load exceeds both NS-3 and RVIP's saturation points. RVIP (Test3) can achieve 160Mbps throughput with 4.8ms RTT, while NS-3 (Test2) can only achieve 28Mbps throughput with much larger RTT (about 10^2 times more).

6. CONCLUSION

Network emulation has been proposed to overcome the disadvantages of pure-software simulation and physical test for large-scale network research. The emulation facility can be subdivided into the *opaque mode* and the *protocol mode* according to whether the live packet is transparent to SVN or not. In comparison, the protocol mode has more advantages because it can realize protocol-specific interactions between LNs and SVNs. In this paper, we further enhance the protocol mode towards its ideal case, namely the *Turing-indistinguishable mode*. In this new ideal mode, any SVN and its counterpart LN would be indistinguishable from the angle of any third-party live node/network. This requires that live nodes with standard IP protocol stack must strictly follow standard IP protocols in interacting with the SVNs, and there is no simulator-specific changes required on any live node.

We design the RVIP interface to realize the new ideal mode. In a hybrid simulation scenario, multiple LNs and multiple SVNs are arbitrarily mixed together with no modification required on any live nodes. RVIP realizes the mapping between a live protocol stack and a virtual protocol stack (VIPS) by presenting a progressive series of connection patterns, that is, *mapping single virtual node to live node*, *combining LN with SVN* and *connecting multi-SVNs*. Then we can construct more complex hybrid scenarios by combining these patterns. We constructed real-time hybrid scenarios with multiple LNs and multiple SVNs, with each SVN instantiated on a single simulation server host running at real time. When simulating wired networks, there is no discrete event exchanged between two SVNs. When simulating wireless networks, there is no IP protocol discrete event exchanged between any two SVNs, only environmental discrete events (node mobility and wireless signal propagation events) are exchanged between two neighboring SVNs.

We present case studies to show RIVP's advantages in large-scale network simulation. In order to find out the accuracy of the hybrid simulation, experiments are carried out to evaluate the overhead incurred by RVIP on latency and throughput compared to an equivalent live network. The result shows that RVIP's performance strongly depends on the computational power of the underlying simulation platform. Before the computational power is exhausted at a saturation point, RVIP's overhead on throughput is insignificant and the extra delay is only about 0.5ms. We also did experiments to compare RVIP with NS-3's hybrid simulation. We draw the conclusion that RVIP bridges the gap between LN and SVN with little impact on the simulation accuracy and achieves relatively better performance than the existing solutions.

7. REFERENCES

[1] A. Alvarez, R. Orea, S. Cabrero, and et al. Limitations of network emulation with single-machine and distributed ns-3. In *Proceedings of the 3rd International ICST Conference on Simulation Tools and Techniques*, page 67. ICST, 2010.

[2] K. Fall. Network emulation in the vint/ns simulator. In *IEEE International Symposium on Computers and Communications*, pages 244–250, 1999.

[3] R. Fujimoto, K. Perumalla, A. Park, H. Wu, M. Ammar, and G. Riley. Large-scale network simulation: how big? how fast? In *IEEE MASCOTS*, pages 116–123, 2003.

[4] O. Inc. System-in-the-Loop (SITL) module. `http://www.opnet.com/solutions/network_rd/system_in_the_loop.htmll`.

[5] S. Kristiansen and T. Plagemann. Accuracy and scalability of ns-2's distributed emulation extension. *Simulation*, 87(1-2):45, 2011.

[6] J. Liu. Immersive real-time large-scale network simulation: a research summary. In *IEEE IPDPS*, pages 1–5, 2008.

[7] J. Liu, Y. Li, and Y. He. A large-scale real-time network simulation study using prime. In *WSC*, pages 797–806. IEEE, 2009.

[8] J. Liu, Y. Li, N. Vorst, and et al. A real-time network simulation infrastructure based on openvpn. *Journal of Systems and Software*, 82(3):473–485, 2009.

[9] LXC: Linux Containers. `http://lxc.sourceforge.net/`.

[10] D. Mahrenholz and S. Ivanov. Real-time network emulation with ns-2. In *IEEE DS-RT*, pages 29–36, 2004.

[11] S. McCanne, C. Leres, and V. Jacobson. Libpcap, 1989.

[12] Quagga Routing Software Suite. `http://www.nongnu.org/quagga/`.

[13] Scalable Network Technologies, Inc. Qualnet. `http://www.qualnet.com/`.

[14] Scalable Network Technologies, Inc. QualNet 4.5 Network Emulation Interface Model Library. `http://www.qualnet.com/`.

[15] M. Schiffman. The libnet packet construction library, 2005.

[16] T. Staub, R. Gantenbein, and T. Braun. Virtualmesh: an emulation framework for wireless mesh networks in omnet++. In *Proceedings of the 2nd International Conference on Simulation Tools and Techniques*, page 64. ICST, 2009.

[17] Traffic generator. `http://www.postel.org/tg/`.

[18] A. Vahdat, K. Yocum, K. Walsh, and et al. Scalability and accuracy in a large-scale network emulator. *ACM SIGOPS Operating Systems Review*, 36(SI):271–284, 2002.

[19] E. Weingärtner, F. Schmidt, T. Heer, and K. Wehrle. Synchronized network emulation: Matching prototypes with complex simulations. *ACM SIGMETRICS Performance Evaluation Review*, 36(2):58–63, 2008.

[20] E. Weingärtner, H. vom Lehn, and K. Wehrle. Device-driver enabled wireless network emulation. *Proc. SIMUTools*, 2011.

[21] L. Zhang. VirtualClock: A new traffic control algorithm for Packet switching networks. In *ACM SIGCOMM*, pages 19–29, Sept. 1990.

[22] J. Zhou, Z. Ji, and R. Bagrodia. TWINE: A Hybrid Emulation Testbed for Wireless Networks and Applications. In *IEEE INFOCOM*, pages 1–13, 2006.

[23] J. Zhou, Z. Ji, M. Varshney, and et al. WHYNET: a Hybrid Testbed for Large-scale, Heterogeneous and Adaptive Wireless Networks. In *Proceedings of the 1st international workshop on Wireless network testbeds, experimental evaluation & characterization*, pages 111–112. ACM, 2006.

Modeling of Correlated Large Scale Channel Parameters for System Level Simulations of Mobile Radio Networks

Denis Bareev, Alexey Trushanin, Mikhail Shashanov, Roman Maslennikov
Lobachevsky State University of Nizhny Novgorod
Gagarina, 23,
603950, Nizhny Novgorod, Russia
+7831 4656153
d.bareev@gmail.com, {alexey.trushanin, mikhail.shashanov, roman.maslennikov}@wcc.unn.ru

ABSTRACT

Modern evaluation methodologies for system level simulations of mobile radio networks (e.g., [1]) support modeling systems with multiple simultaneously operating base stations and mobile devices where accurate simulation of multi-user MIMO and other space-time signal processing techniques should be provided. To be able to predict realistic performance, such models introduce a number of channel parameters with complex auto- and cross-correlation properties. Particularly, such models include a set of large-scale parameters (tap delay spread, angular spreads of rays at the base station and mobile station sides, shadow fading, etc.) being cross-correlated as well as auto-correlated in the spatial domain (across the deployment area) with different correlation functions. This paper is devoted to a method for accurate generation of correlated channel parameters during a simulation based on filtering of the parameters defined in the nodes of a spatial grid. The proposed method combines formal mathematical exactness at least in the asymptotic case and a much lower computational complexity in comparison with existing solutions [2]. Method description, analytical derivation, implementation details and verification by numerical simulations for an exemplary deployment scenario are provided in the paper. The described method may be recommended for application in system level simulators for 3G and 4G mobile radio access networks such as WCDMA HSPA, IEEE 802.16m, LTE-A.

Categories and Subject Descriptors

H.4.3 [**Information Systems Applications**]: Communications Applications

General Terms

Algorithms, Performance, Verification

Keywords

Spatial correlation; cross-correlation; system level simulation; large-scale channel parameters; deployment model; shadow fading; angular spread; delay spread.

1. INTRODUCTION

Nowadays numerical simulations are the most widespread approach for wireless system performance evaluation. Such a trend may be explained by high accuracy of numerical methods, possibility to take into account detailed features of the considered technologies and significantly lower efforts in comparison with field measurements of deployed systems.

The main software tool used for numerical evaluation of modern mobile Radio Access Networks (RANs) is a System Level Simulator (SLS). The scope of SLS applications includes mobile systems performance evaluation in international standardization committees, by mobile operators, equipment manufacturers and scientists working in the field of the communication theory.

An SLS allows simulating the operation of a RAN consisting of multiple base stations and users working in parallel. Typically, the algorithms of the physical (PHY) and Media Access Control (MAC) layers are modeled by an SLS taking into account interaction between different nodes of the system as a whole. An SLS allows performing accurate simulation of power control, scheduling, radio resource management algorithms as well as physical layer algorithms for signal and data processing including coding, modulation, MIMO processing, equalization and others.

However, accuracy of the system level simulations and credibility of the simulation results are also very much dependent on the properties of the deployment model. To emulate realistic conditions of the system operation, the deployment models have to support the necessary level of details and correspond to characteristics of actual deployments. Typical deployment models include a layout model defining the geometry for placement of base stations and users, antenna and antenna pattern models at both base station and user sides, and channel models defining signal propagation conditions between different stations. Usually, such channel models comprise path loss, shadowing, multipath propagation, and user mobility models.

At present, there are a number of widely used deployment models developed by different standardization committees and other organizations [1]–[4]. The most advanced deployment models supporting simulation of multiple antenna signal processing techniques (such as multi-user MIMO) are rather complicated and introduce a number of spatially distributed channel parameters with complex auto- and cross-correlation properties. Such models are described, for example, in the evaluation guidelines for the ITU-R IMT-Advanced standardization process [1].

Particularly, such deployment models include a set of large-scale parameters (tap delay spread, angular spreads of rays at the base station and mobile station sides, shadow fading, etc.) being cross-correlated as well as auto-correlated in the spatial domain (across the deployment area) with different correlation functions. The topic of the present work is the algorithm of generation of the large-scale channel parameters in system level simulations so that to satisfy complex requirements for auto- and cross-correlation properties in the spatial domain.

Initial recommendations for generation of the large scale parameters with the discussed properties are provided in document WINNER D5.4 "Final Report on Link Level and System Level Channel Models" [2]. The recommendations suggest generation of correlation matrices including correlation coefficients between each pair of users simulated in the system. Such matrices contain all correlation coefficients between different parameters and different users and depend on the distance between the users. Such matrices can be considered as a sample of a multi-dimensional correlation function depending on the distance. After that well known algorithms can be applied for generation of a vector of correlated numbers with the specified correlation matrix from a vector of random uncorrelated numbers. This approach is formally accurate and conceptually very simple; however, its main drawback is vast computational complexity because of the requirement to process huge matrices having the size of each dimension equal approximately to the number of users in the system. For high user densities of about 30 users per cell, such number of users in the whole system may reach 570 for typical hexagonal layouts of 19 cells making complex processing of such matrices practically unacceptable.

Document [2] also notes a computationally simpler approach based on an introduction of a regular spatial grid of uncorrelated parameters independent on the user coordinates. Then, two-dimensional linear filtering of the parameter values in the nodes of the grid may be performed to generate a set of large-scale channel parameters with the required correlation properties in the spatial domain. However, a description of any specific method using that grid-based approach is not provided in the document. As such, there is a field for development of different exact and approximate methods for large-scale parameters generation. Such problems have been already solved during the development of existing SLS tools supporting complex deployment scenarios. But no complete descriptions of formally exact methods based on the mentioned grid-based approach for the channel models with complex auto- and cross-correlation properties of the parameters have been found in the literature.

The main goal of the current work is development of a novel method for generation of large-scale channel parameters having complex auto- and cross-correlation properties in the spatial domain based on filtering of the parameters defined in the nodes of a spatial grid. The proposed method combines formal mathematical exactness at least in the asymptotic case (for a small grid step) and a much lower computational complexity in comparison with the formal solution based on usage of correlation matrices for all users in the system.

Method description, mathematical derivation, implementation issues and results of numerical verification are provided in the paper.

2. DESCRIPTION OF PROPOSED METHOD

2.1 Formal Mathematical Description of Problem

The considered deployment models, for example described in the ITU-R IMT-Advanced evaluation guidelines [1], usually impose the following requirements on the correlation properties of large-scale channel parameters:

- A two-dimensional autocorrelation function is specified in the spatial domain $R_n(\Delta x)$ for each large-scale channel parameter s_n;
- A set of cross-correlation coefficients $K_{n,m}$ is specified between each pair of large-scale channel parameters s_n, s_m in the same point of space.

Autocorrelation functions $R_n(\Delta x)$ of the considered channel modeling methodologies are radially symmetric and have an exponential dependence on the separation distance (1). A correlation distance d_n is the distance of the correlation function decay in e times. Different large-scale parameters typically have different correlation distances.

$$R_n(r) = \exp\left[-\frac{|r|}{d_n}\right] \quad (1)$$

Thus, the formal requirement on the correlation matrix of large-scale parameters for two arbitrary selected users (points in the space) may be formulated as follows:

$$\mathbf{C}(x^{(1)}, y^{(1)}, x^{(2)}, y^{(2)}) = \left\langle \mathbf{s}(x^{(1)}, y^{(1)})\mathbf{s}(x^{(2)}, y^{(2)})^T \right\rangle =$$

$$= \mathbf{C}(r) = \begin{bmatrix} \exp(-r/d_1) & K_{12}(r) & \dots & K_{1N}(r) \\ K_{12}(r) & \exp(-r/d_2) & \dots & K_{2N}(r) \\ \dots & \dots & \dots & \dots \\ K_{1N}(r) & K_{2N}(r) & \dots & \exp(-r/d_N) \end{bmatrix}, \quad (2)$$

where $\mathbf{s}$ is a vector of large-scale parameter values, $x^{(1)}$, $y^{(1)}$ – coordinates of the first user, $x^{(2)}$, $y^{(2)}$ – coordinates of the second user, N – a number of correlated large-scale parameters, ^{T} denotes transposition, r is the distance between the users defined by the following formula:

$$r = \sqrt{(x^{(1)} - x^{(2)})^2 + (y^{(1)} - y^{(2)})^2}, \quad (3)$$

and $K_{n,m}(r)$ are arbitrary functions of r satisfying the following conditions:

$$K_{n,m}(0) = K_{n,m} \quad (4)$$

The values of $K_{n,m}(r)$ for non-zero separation distances are not specified by the scenario and, hence, there is freedom in their definition.

In the case when the same auto-correlation function is required for all large-scale parameters (i.e., all the correlation distances are equal) the problem of the generation of large-scale parameters satisfying formula (2) can be divided into two separate problems. Indeed, the correlation matrix in formula (2) in that case can be expressed as a product of an exponential correlation function and a constant matrix:

$$\mathbf{C}(x^{(1)}, y^{(1)}, x^{(2)}, y^{(2)}) = = \exp(-r/d) \cdot \begin{bmatrix} 1 & K_{12} & \dots & K_{1N} \\ K_{12} & 1 & \dots & K_{2N} \\ \dots & \dots & \dots & \dots \\ K_{1N} & K_{2N} & \dots & 1 \end{bmatrix} \quad (5)$$

Thus, as may be shown, uncorrelated large-scale parameters in each node can be first independently generated to satisfy the exponential autocorrelation for each large-scale parameter. Then the correlation between different parameters may be introduced for each user to satisfy the cross-correlation properties and that second step will not destroy the spatial correlation characteristics introduced at the first step. Or, in the opposite order, the cross-correlation in each node of the grid can be applied at first and then spatial-domain filtering may be performed for each parameter independently.

Both mentioned problems of filtering of a grid of uncorrelated scalars to provide the required autocorrelation function and a linear transform of a vector of uncorrelated parameters to provide the required correlation matrix have well known solutions.

But in the case of different parameters having different correlation distances, taking place in real scenarios, such approach with consecutive application of the autocorrelation and cross-correlation procedures to a set of uncorrelated values can be treated only as an approximate one because correlation matrix (2) in the general case cannot be expressed as product (5). It means that if the spatial auto-correlation procedure is applied first to provide the required autocorrelation properties for all the parameters, application of the cross-correlation procedure will destroy the auto-correlation properties. And, vice versa, if the cross-correlation procedure is applied first, the consecutive spatial-domain filtering will destroy the initial cross-correlation properties.

A novel algorithm is proposed in this paper to provide a formally accurate solution for the described problem of introducing joint auto- and cross-correlation properties to the initial set of uncorrelated parameters. The developed procedure is based on spatial-domain two-dimensional matrix filtering. A description of the method is provided in the next section.

2.2 Derivation of Proposed Method

The current section provides a general method description and its mathematical derivation.

A two-dimensional spatial grid is considered with each node k including a vector of N uncorrelated random Gaussian numbers $\mathbf{s}_k$ with zero mean and unit variance.

Then, an extension of a usual two-dimensional linear convolution is used to obtain the required correlated large-scale channel parameters. This extended convolution utilizes matrix and vector operations instead of usual scalar operations. So, the following formula (procedure) is proposed for calculation of a vector of correlated large-scale parameters in the point of the considered user location:

$$\mathbf{s}(x,y) = \sum_{k \in G} \mathbf{H}_k(x - x_k, y - y_k)\mathbf{s}_k(x_k, y_k), \quad (6)$$

where $\mathbf{s}$ is a vector of large-scale parameters for the considered user, x, y – coordinates of the considered user, k – a node index, G – the set of all grid nodes, x_k, y_k – coordinates of the k-th node, $\mathbf{H}_k$ – a set of matrices to be defined below to provide the required auto- and cross-correlation properties for $\mathbf{s}$.

After substitution of (6) into (2) and taking into account zero cross-correlation between vectors $\mathbf{s}_k$ and between their elements in the grid nodes, the equation for determination of $\mathbf{H}_k$ may be expressed as follows:

$$\mathbf{C}(\sqrt{(x^{(1)} - x^{(2)})^2 + (y^{(1)} - y^{(2)})^2}) = = \sum_{k \in G} \mathbf{H}_k(x^{(1)} - x_k, y^{(1)} - y_k)\mathbf{H}_k^T(x^{(2)} - x_k, y^{(2)} - y_k) \quad (7)$$

If the correlation properties are not dependent on the users absolute position (but only on their relative locations) then, without loss of generality, the coordinates of the first user can be selected as equal to zero. That leads to:

$$\mathbf{C}(\sqrt{\Delta x^2 + \Delta y^2}) = \sum_{k \in G} \mathbf{H}_k(x_k, y_k)\mathbf{H}_k^T(x_k - \Delta x, y_k - \Delta y), \quad (8)$$

where Δx, Δy are the separation distance of users by the x and y axes.

To solve equation (8), the following general form (solution) for $\mathbf{H}_k$ is proposed in this paper:

$$\mathbf{H}_k(\xi, \zeta) = = \begin{bmatrix} A_{11} \cdot h_1(\xi,\zeta) & \dots & A_{1N} \cdot h_1(\xi,\zeta) \\ \dots & \dots & \dots \\ A_{N1} \cdot h_N(\xi,\zeta) & \dots & A_{NN} \cdot h_N(\xi,\zeta) \end{bmatrix}, \quad (9)$$

where A_{nm} are constants, $h_n(\xi,\zeta)$ – spatial-domain functions, and ξ and ζ – spatial coordinates. A_{nm} can be combined into a matrix $\mathbf{A}$ (10):

$$\mathbf{A} = \begin{bmatrix} A_{11} & \dots & A_{1N} \\ \dots & \dots & \dots \\ A_{N1} & \dots & A_{NN} \end{bmatrix}, \quad (10)$$

Substitution of (9) into (8) provides the following equations for diagonal elements of $\mathbf{C}$:

$$\sum_{k \in G} h_n(x_k, y_k)h_n(x_k - \Delta x, y_k - \Delta y)\sum_{m=1}^{N} A_{nm}{}^2 = = \exp\left[-\frac{\sqrt{\Delta x^2 + \Delta y^2}}{d_n}\right], \quad n = \overline{1,N} \quad (11)$$

and for non-diagonal elements of $\mathbf{C}$:

$$\sum_{k \in G} h_n(x_k, y_k)h_m(x_k - \Delta x, y_k - \Delta y)\sum_{l=1}^{N} A_{nl}A_{ml} = = K_{nm}(\sqrt{\Delta x^2 + \Delta y^2}), \quad n,m = \overline{1,N} \quad (12)$$

To be able to calculate the sought parameters, following equations (13) and (14) are introduced to represent A_{nm} and $h_n(\xi,\zeta)$ in a different form so that if (13) and (14) are met then equations (11) and (12) can be met as well:

$$\mathbf{A}\mathbf{A}^T = \begin{bmatrix} 1 & K_{12} & \dots & K_{1N} \\ K_{12} & 1 & \dots & K_{2N} \\ \dots & \dots & \dots & \dots \\ K_{1N} & K_{2N} & \dots & 1 \end{bmatrix} \qquad (13)$$

$$\sum_{k \in G} h_n(x_k, y_k) h_n(x_k - \Delta x, y_k - \Delta y) = $$
$$= \exp\left[-\frac{\sqrt{\Delta x^2 + \Delta y^2}}{d_n}\right], \quad n = \overline{1,N} \qquad (14)$$

Thus, A_{nm} and $h_n(\xi,\zeta)$ calculated according to formulas (13) and (14) can be used for generating matrices $\mathbf{H}_k$. In that case calculation of $\mathbf{H}_k$ can be done in two separate steps: calculation of the matrix $\mathbf{A}$ using (13) and determination of the filtering functions $h_n(\xi,\zeta)$. Those steps are similar to the two step procedure considered above for successive auto- and cross-correlation, applied in the simplest case where the autocorrelation functions are identical for all large scale channel parameters.

The matrix $\mathbf{A}$ can be calculated from the cross-correlation matrix (13) using, for example, the well-known Cholesky factorization algorithm [5] or as a matrix square root from the correlation matrix.

An additional task is to determine functions $h_n(\xi,\zeta)$. As may be shown, the solution of equation (14) cannot be expressed using elementary functions. But within the proposed method, these dependences can be obtained numerically in the form of Look-Up Tables (LUTs) with a further linear interpolation between the closest points. Such an approach allows reaching sufficient accuracy of the method and a reasonable complexity of the implementation. The following procedure based on the usage of a two-dimensional Fast Fourier Transform (2D FFT) and a two-dimensional Inverse FFT (2D IFFT) is taken to solve (14) and to calculate the sought LUTs. Application of a 2D FFT to both left and right parts of equation (14) allows replacing the two-dimensional discrete convolution by a multiplication of the FFT-images. Then, a square root and an IFFT can be applied to both left and right parts of (14) to derive (15):

$$h_n(\xi,\zeta) = \mathrm{FFT}_{2D}^{-1}\left\{\left(\mathrm{FFT}_{2D}\left\{\exp\left[-\frac{\sqrt{\xi^2+\zeta^2}}{d_n}\right]\right\}\right)^{1/2}\right\}, \quad (15)$$

It should be mentioned that both matrix $\mathbf{A}$ and the LUTs of $h_n(\xi,\zeta)$ do not depend on user coordinates and depend only on the cross-correlation coefficients and correlation distances. Thus, the set of matrices $\mathbf{A}$ and the LUTs of $h_n(\xi,\zeta)$ for all the required deployment scenarios can be calculated offline and be available before the simulation.

2.3 Algorithm for Generation of Correlated Channel Parameters

This section describes the proposed algorithm for generation of spatially auto- and cross-correlated large-scale channel parameters. A two-dimensional spatial grid of uncorrelated parameters has to be generated once per simulation for all users in the system. The step of the grid should be selected such as to satisfy the requirements for accuracy and computational complexity. Then, generation of channel parameters with the required correlation properties for a selected user is performed according to the following procedure:

- Selecting a subset of nodes of the grid located inside the specified circle around the user's coordinates (taking into account fast decay of filtering functions and computational complexity). The radius of the circle should be selected for each considered deployment model so that to minimize the amount of computations while keeping the required accuracy of modeling;
- Building $\mathbf{H}_k(\xi,\zeta)$ matrices for each node selected at the previous step using formula (9) and $\mathbf{A}$ and $h_n(\xi,\zeta)$ preliminarily calculated using formulas (13) and (15);
- Performing matrix filtering by formula (6) to get a vector of the required large-scale channel parameters cross-correlated as well as auto-correlated in the spatial domain.

3. IMPLEMENTATION DETAILS

The described method for generation of auto- and cross-correlated large-scale parameters in the spatial domain has been implemented for the ITU-R IMT-Advanced deployment scenarios using the C++ programming language as a separate class called `Grid`. This class may be used by any SLS tools supporting the ITU-R IMT-Advanced scenarios.

`Grid` contains a two-dimensional spatial grid of parameters as well as functions of the algorithm described in the previous section. The main interface function of the class is intended for delivery of the generated large-scale parameters: Delay Spread (DS), Angular Spread of Departure (ASD), Angular Spread of Arrival (ASA), Shadow Fading (SF) and also the Ricean factor (for line-of-sight (LoS) channels only) satisfying the required correlation properties. The parameters are returned for the specified base station index and user coordinates.

According to the described method, the grid nodes store vectors of random parameters uncorrelated between each other and between different nodes. Such set of random parameters is generated once per users drop (i.e., once per independent system simulation). Then, the correlated large-scale parameters are generated for each link between each base station and each user in the simulated system. The member functions of `Grid` class are applied that perform filtering of the uncorrelated parameters in the grid nodes (as described in the previous section).

However, for the described procedure, storing of all five (in the general case of supporting LoS channels) parameters in each node of the grid would be extremely wasteful from the memory consumption point of view. To solve this problem, the following approach based on random generator seeds management has been implemented. For each grid node the set of five parameters (5 double numbers occupying 40 bytes of memory) is replaced by a pair of a Boolean flag (1 byte) and a random generator seed (typically, an unsigned integer number occupying 4 bytes of memory). Thus, the memory economy of 8 times is potentially available in that case. Utilization of the flag variable is explained below.

The following procedure is assumed for generation of the grid nodes parameters:

- Initialize all flags in the grid as false and keep all random seeds uninitialized;
- For each call from class member functions to get the parameters in the current node check the flag:
  - If the flag is false, store the current random generator seed into the corresponding field of the current node, change the flag state to true and randomly generate and return five Gaussian double parameters;
  - If the flag is true, store the current random generator seed into a temporary variable, set the seed stored for the current node into the random generator, randomly generate and return five Gaussian double parameters and set back the random generator seed stored in the temporary variable;

The described procedure ensures the consistency of parameter values for each node of the grid being generated starting from the same random generator's seed. In addition to the significant memory saving, the procedure allows avoiding redundant seed generations (calls of the random generator), in comparison with, for example, setting a random seed for each node at the stage of grid initialization. So, in the described algorithm, due to the presence of special flag fields for each node, the seeds are specified only for active nodes taking part in calculation of large-scale parameters. That provides significantly faster grid initialization especially for low user densities.

The memory saving gain of about 8 times as well as the initialization speed improvement are very important for the SLS tool performance because millions of nodes are typically taken for widely-used 19-cell hexagonal layouts. In that case, the array of the node parameters requires a significant amount of memory and may occupy even the most part of the total SLS program memory.

In addition to the array of seeds and flags for the grid nodes and the set of class member functions, a set of preliminarily calculated data is required for `Grid` class operation. Such set of data includes the required filtering functions $h_n(\xi,\zeta)$ and constants A_{nm} for each deployment scenario supported by the deployment model (Indoor Hotspot, Urban Micro, Urban Macro, etc.) [1] and each channel type (LoS/NLoS).

This data can be included directly into the C++ code as constant arrays (as have been done in the current implementation), or can be acquired during the SLS initialization by either reading from the input files or calculating it directly.

The next section provides the results of verification of the implemented method.

4. NUMERICAL VERIFICATION

The results of the numerical verification of the proposed method for calculation of correlated large-scale channel parameters and its implementation are provided below.

The IMT-Advanced Indoor Hotspot (InH) scenario and non-Line-of-Sight (NLoS) channels [1] are assumed for verification purposes. For the described case, a set of 4 large-scale channel parameters is used: DS, ASD, ASA and SF. The required cross-correlation coefficients between the parameters are provided in Table 1.

Table 1. Required cross-correlation coefficients between the large-scale channel parameters

	DS	ASD	ASA	SF
DS	1	0.4	0	-0.5
ASD	0.4	1	0	0
ASA	0	0	1	-0.4
SF	-0.5	0	-0.4	1

The autocorrelation function for each parameter in the spatial domain is an exponent and the correlation distances are given in Table 2.

Table 2. Required autocorrelation distances for different large-scale parameters

Parameter	Correlation distance, m
DS	5
ASD	3
ASA	3
SF	6

The selected grid parameters are provided in Table 3.

Table 3. Grid parameters

Parameter	Value
Grid step, m	0.5
Circle radius for filtering, m	6
Grid size, m^2	120x120

To verify the developed procedure, the users were placed in a regular rectangular grid with the step of 0.7 m (not overlapping with the reference grid of uncorrelated parameters). Then channel parameters initialization for each user has been performed according to the described methodology. Finally, the autocorrelation functions versus the separation distance and the cross-correlation coefficients (for zero separation distance) were measured and compared with the required ones.

Figure 1 provides an example of the generated spatial distribution for the DS parameter having Gaussian probability density function with zero mean and unit variance. As demonstrated in Figure 1, the generated DS changes smoothly in the spatial domain with the typical scale approximately equal to the correlation distance (5 m).

A comparison of the measured cross-correlation coefficients versus the required ones is provided in Table 4. The results of the comparison demonstrate that the measured values are close to the required cross-correlation coefficients. The average deviation is about 0.03 and may be explained by limited statistics for averaging and by restrictions of the proposed method, for example caused by the usage of a grid with a finite step size and by limitation of number of nodes taking part in convolution (6).

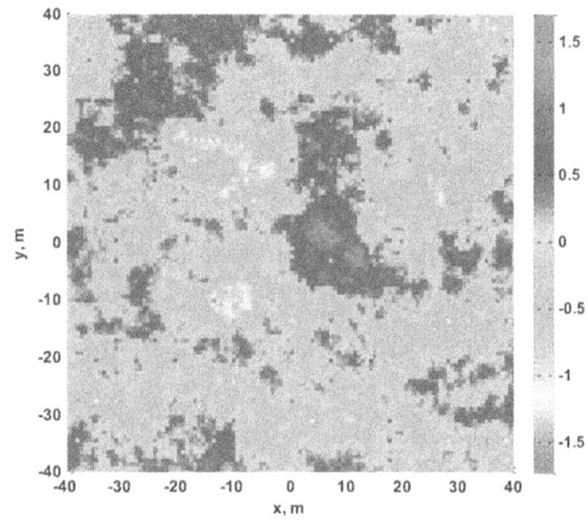

Figure 1. Example of DS spatial distribution

Table 4. Comparison of the measured cross-correlation coefficients versus the required coefficients

Parameters	Required cross-correlation coefficients	Measured cross-correlation coefficients	Deviation
DS vs. ASD	0.4	0.39	0.01
DS vs. ASA	0	0.02	0.02
DS vs. SF	-0.5	-0.50	0
ASD vs. ASA	0	-0.04	0.04
ASD vs. SF	0	0.07	0.07
ASA vs. SF	-0.4	-0.42	0.02

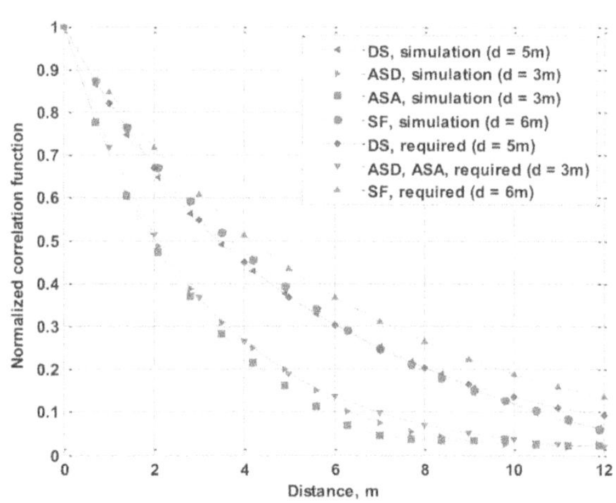

Figure 2. Comparison of the measured autocorrelation functions versus the required functions

A comparison of the measured autocorrelation functions versus the required ones is provided in Figure 2. As demonstrated in the figure, the numerical results are also close to the required exponential dependences, and small deviations can be explained by the same reasons of the limited simulation statistics and the method restrictions.

Thus, the applicability of the proposed method for the calculation of auto- and cross-correlated large scale channel parameters in the spatial domain and its correct implementation have been demonstrated.

5. SUMMARY

This paper is devoted to a method for accurate calculation of the large-scale channel parameters cross-correlated as well as auto-correlated in the spatial domain with different correlation functions for system-level simulations using modern system-level channel models, included, for example, in the ITU-R IMT-Advanced evaluation methodology [1].

The proposed method is based on filtering of parameters defined in the nodes of a spatial grid. A novel filtering approach is proposed for the method that extends the scalar two-dimensional filtering to the matrix case. This extension allows generating a set of parameters that are jointly auto-correlated with different correlation distances and cross-correlated with arbitrary correlation coefficients. The method combines formal mathematical exactness at least in the asymptotic case in contrast to other solutions exploiting the grid filtering concept and allows reaching much lower computational complexity in comparison with the existing formal solutions.

The method description, analytical derivation, implementation details and verification by numerical simulations for an exemplary deployment scenario are provided in the paper. The verification results confirm the applicability of the proposed method for generation of large-scale channel parameters during system level simulations.

The described method may be recommended for application in system level simulators for 3G and 4G mobile radio access networks such as 3GPP WCDMA HSPA, IEEE 802.16m, 3GPP LTE-Advanced.

6. REFERENCES

[1] Guidelines for evaluation of radio interface technologies for IMT-Advanced. 2010. Rep. ITU-R M.2135-1.

[2] Daniel S. Baum, Hassan El-Sallabi, Tommi Jämsä, *et al.* 2005. Final Report on Link Level and System Level Channel Models. WINNER D5.4.

[3] Spatial channel model for Multiple Input Multiple Output (MIMO) simulations. 2011. 3GPP TR 25.996.

[4] Guidelines for evaluation of radio transmission technologies for IMT-2000. 1997. Rep. ITU-R M.1225.

[5] Lloyd N. Trefethen, David Bau III. 1997. *Numerical Linear Algebra.* Philadelphia: Society for Industrial and Applied Mathematics.

TimeSync: Enabling Scalable, High-Fidelity Hybrid Network Emulation

Florin Sultan
Applied Communication Sciences (ACS)
1 Ericsson Dr, RRC-1D226
Piscataway, NJ 08854, USA
fsultan@appcomsci.com

Alex Poylisher
ACS
1 Ericsson Dr, RRC-1B313
Piscataway, NJ 08854, USA
apoylisher@appcomsci.com

John Lee
ACS
331 Newman Springs Rd, Bldg 2, NVC-2Z439
Red Bank, NJ 07701, USA
jlee@appcomsci.com

Constantin Serban
ACS
1 Ericsson Dr, RRC-1G318
Piscataway, NJ 08854, USA
cserban@appcomsci.com

C. Jason Chiang
ACS
1 Ericsson Dr, RRC-1N371
Piscataway, NJ 08854, USA
jchiang@appcomsci.com

Ritu Chadha
ACS
1 Ericsson Dr, RRC-1M310
Piscataway, NJ 08854, USA
rchadha@appcomsci.com

ABSTRACT

In this paper, we discuss a time synchronization approach to the time divergence problem in hybrid network emulation. Such emulation comprises primarily a discrete-event simulated network and virtual machines that send and receive traffic through the simulated network. For slower than real-time network simulations, the rate of time advance on virtual machines (real time) is faster than that of the discrete-event time. Consequently, packet transmission latency and other metrics in such hybrid network emulations will be distorted. As a result, e.g., TCP sessions between virtual machines may unduly time out. To address this problem, we have developed TimeSync, which tracks discrete-event simulation time to control time advance on virtual machines for slower than real time simulations so that time perception in the hybrid network emulation system is synchronized. We describe how TimeSync works and present our experimental evaluation and analysis.

Categories and Subject Descriptors

C.2 [**Computer Communication Networks**]: Miscellaneous

Keywords

Hybrid network emulation, virtual time

1. INTRODUCTION

Hybrid network emulation [7, 11, 9] (HNE) comprises primarily a discrete-event simulated network and virtual machines that send and receive traffic through the simulated network. It allows testing network applications rather than their models on simulated target networks, particularly mobile wireless networks. In some HNE approaches, e.g., [13, 3, 10], applications can run on top of their native operating systems without any code modification, so the same executable binary can be used in both HNE and real networks.

Fig. 1 shows a sample setup of a virtualized HNE testbed (e.g., in [10], using Xen [2] for host virtualization), where the network simulation runs on a dedicated machine and end-host VMs deployed on testbed machines run the unmodified OS, protocol stacks and applications. Some VMs have a corresponding shadow node inside the simulated network, and VMs communicate by injecting traffic into and receiving traffic from their corresponding shadow nodes, via VLAN or other encapsulation mechanisms.

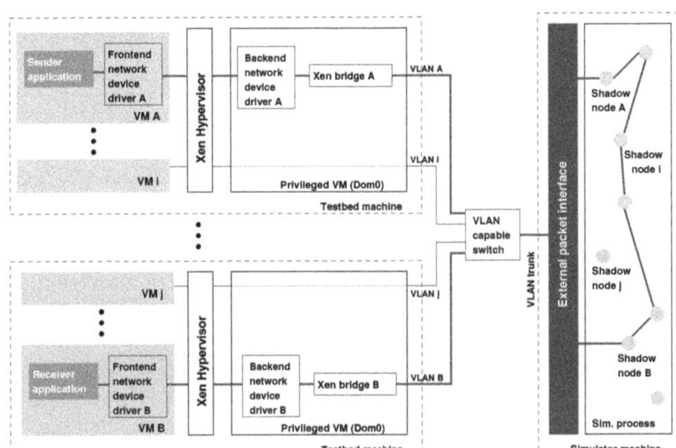

Figure 1: Hybrid network emulation (HNE): lower-layer network details in the simulator; application traffic originates and terminates at end-host VMs.

HNE can potentially address both feasibility and scalability concerns associated with testing applications over target networks. With respect to feasibility, as testing applications over hybrid emulated network only requires the models of network elements, the availability of network element hardware (e.g., next generation radio hardware) will not be an issue, and simulation can allow for testing over various different network topologies and configurations. With respect to scalability, as simulation is used to enable HNE, theoretically the scale of the target network is constrained only by a simulator's capabilities and hardware resource availability.

While the feasibility argument stands valid, the scalability of HNE is actually hindered by the *time divergence problem*: for complex, large-scale simulations, discrete-event simulation time advances slower than real time (typically in a non-uniform way), thus distorting packet propagation characteristics. For example, in a hybrid emulated network where the simulation time advances constantly two times slower than real time, the packet propagation latency perceived by applications running on virtual machines will be twice the expected value dictated by the simulation. To minimize and bound the possible loss of fidelity in HNE, time synchronization between simulation and the external OS domains becomes a necessity, particularly for large scale models where the loss of fidelity can be substantial.

To address the time divergence problem, we developed a system that uses the discrete-event simulation time to control and synchronize time advance on virtual machines for large-scale HNE. Our objectives are: (1) tight constraint on simulation time to advance no faster than real time, (2) tight synchronization of the VM system time with simulation time, (3) tight synchronization of the rate of flow of VM time (as perceived by software running inside a VM) with that of simulation time, (4) small footprint/low overhead.

This paper presents TimeSync, our solution to the time divergence problem, and its integration with an operational Xen-based HNE testbed [10]. The core idea of TimeSync is to create a simulator-driven virtual timeline in the VMs participating in emulation (as opposed to the 'real' timeline created by the hardware platform that VMs normally follow). It consists of two high-level components: *(i)* a simulator state introspection and control module, running alongside the simulator, that enables extraction of time information from the simulation and control over simulation time speed-up, and *(ii)* a dynamic time control and synchronization mechanism, implemented in the Xen hypervisor on the testbed machines, that enables tight control of VM time using time information from the simulation.

The remainder of the paper is organized as follows. In Section 2, we present the TimeSync architecture. We give implementation details in Section 3. We present an evaluation of our prototype in Section 4, discuss related work in Section 5, and conclude in Section 6.

2. SYSTEM ARCHITECTURE

The key idea in TimeSync is to make the time flow inside a VM closely *track* simulation time by dynamically controlling the Xen-VM time interface. Ideally, this should happen continuously, at infinitely small time granularity. In practice, however, extracting time information from the simulator and control of VM time can only be done at finite granularity. Our solution is to track the value of simulation time in small discrete steps, along with an approximation of its average rate of change between consecutive steps. Following simulation time dynamics in both discrete value and rate of progress is essential for good accuracy.

To this end, TimeSync uses two mechanisms: *(i)* simulator-side introspection, to extract time information as the simulation is running, and *(ii)* dynamic time virtualization in Xen, to apply this information dynamically to the VMs via the narrow hypervisor-VM interface. The time information consists of the value of simulation time at a given instant and its projected rate of progress relative to the real time.

To illustrate the time divergence problem, Fig. 2 depicts the simulation and real timelines during a slowdown period in the simulation. The simulation time advance is *discrete*, driven by processed events (shown as dots on the simulation timeline). Processing an event may take a variable amount of time, thus the advance of simulation time versus real time is also *non-linear*. As shown, if we sample the simulation time every Δ units of real time, the sampled values (the top line) yield intervals $d < \Delta$, likely of variable length [1].

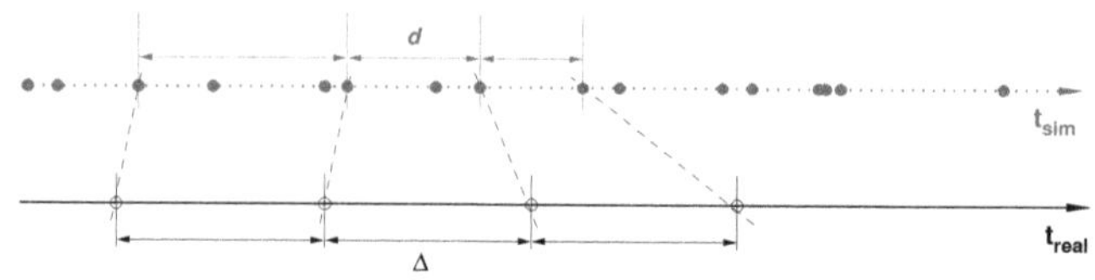

Figure 2: The flow of simulation time under slowdown (top), compared with real time (bottom).

Let ST(t) be the simulation time as a function of the real time t. Let also VT(t) be the virtual time (as perceived by a VM) as a function of real time t. Ideally, we would like VT to track ST, i.e., VT(t) = ST(t) for any t. Since in a real system this cannot be done continuously we propose a piece-wise linear approximation of ST(t) achieved as follows:

Introspection: every interval of constant length Δ in *real time*, we sample ST = ST(t) and predict a *slowdown factor* $SF \geq 1$ of the simulation time in the next interval.

Control: we constrain the simulator to run no faster than real time, which enforces that SF is never less than 1.

Dynamic Time Virtualization: we make VT(t_i) = ST at the beginning t_i of an interval and approximate VT(t) inside the interval as a linear function of t, VT(t) = $ST + (t - t_i) \;/\; SF$.

Fig. 3 shows the TimeSync system architecture. On the simulator side, a simulator introspection and control module (SICM) samples ST and computes SF every sampling period Δ, then sends them to the VM clock control module (CCM), running as a privileged process on all testbed machines. The CCM injects the (ST, SF) tuple into the Xen hypervisor. There, our time virtualization mechanism (details in Section 3) uses it to control all aspects of time perceived by VMs involved in the emulation: system time, its rate of progress, and timers. VMs run freely under the control of the Xen scheduler, but their time is virtualized: ideally, system time is set to ST at the beginning of an interval and flows at a rate of $1/SF$ until the next update.

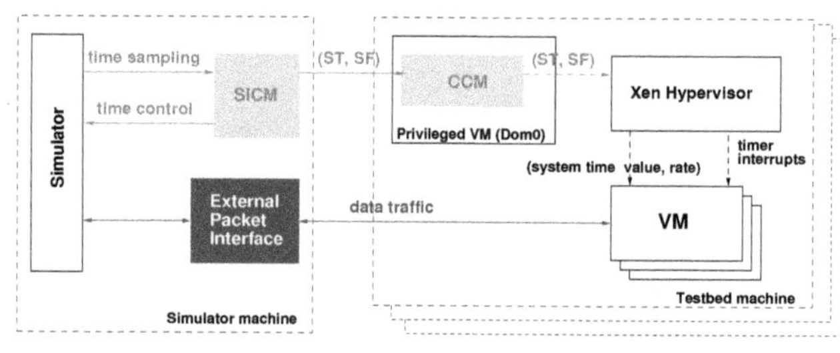

Figure 3: TimeSync system architecture.

Given the real and simulation time values at the beginning (rt_{prev} and st_{prev}) and at the end (rt and st) of the

[1]Simulation slowdown can be due to any such factors as heavy computational load for some event, large number of events, and transient lack of resources.

last sampling interval, SICM computes the actual slowdown factor SF for the interval as $SF = (rt - rt_{prev})/(st - st_{prev})$.

If the simulator did not advance at all in the last interval (so the denominator in the formula above is 0), we set SF to a predefined maximum slowdown factor SF_{max}.

Ideally, when applying SF to a VM, we would like to use the future SF (in the *next* interval). Since we cannot know it, we could rely instead on a prediction based on the history of slowdown across a window of recent intervals. In the current implementation, the SICM predicts SF by assuming that simulation time advances exactly as much in the next interval as in the last one, i.e., uses the actual SF value from the last interval as the projected SF value in the next.

Fig. 4 illustrates the impact of this dynamic time control mechanism on the time progression in a VM. The x axis shows the discrete moments in real time at which a (ST, SF) update is injected into Xen. The blue curve depicts the progression of simulation time used as a reference for sampling ST and for computing each SF value. The red curve shows the progression of VM system time.

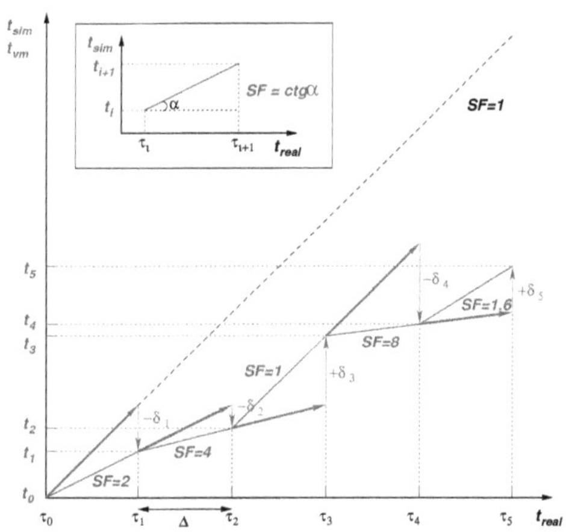

Figure 4: Progression of simulation (blue, thin solid line) and VM time (red, thick solid line with arrows) with dynamic SF control.

Every time CCM injects an (ST, SF) update (every sampling period), our Xen time virtualization mechanism forces the VM system time up or down by a difference δ_i, in an attempt to set it to the exact simulation time. In addition, an update adjusts VM timers and changes the progression rate of VM time according to the predicted SF value in the next interval. As a result, in each interval between two updates the red curve grows linearly with the same slope that the blue curve had exhibited in the previous interval (the inverse of the predicted SF). This causes the divergence between the two curves seen in the figure over each interval, marked by δ_i at the end of an interval. However, as the vertical arrows show, at the end of each interval this divergence is corrected by a new incoming update that forces the VM system time to the most recent simulation time sample ST received in the update. The error induced by SF prediction is bounded by Δ (reached only if the predicted SF was 1 but the real SF was infinity). Because the interval between updates is small (Δ is on the order of milliseconds), the instantaneous divergence between the two curves is small.

3. IMPLEMENTATION

This section details the implementation of TimeSync. We first give some background on the Xen time interface, and then describe the components of our implementation: (1) Xen-based dynamic time virtualization, (2) simulation-driven time synchronization, and (3) constrained advancement of simulation time.

3.1 Background: The Xen Time Interface

This section describes the notions of time maintained in the VMs (by guest OSs), and the mechanisms Xen uses to provide time-related information to the guest OSs. The details are limited to x86 CPUs, Xen 3.3.2, Unix-like guest OSs (Linux, NetBSD), and paravirtualized (PV) guests.

3.1.1 Time Keeping in a Guest OS

To support time services, a guest OS maintains two variables: *(i)* the system time (as nanoseconds elapsed since boot time), and *(ii)* a counter of interrupts (ticks) generated by a periodic hardware timer at a fixed rate (e.g., denoted by HZ in Linux and BSD systems, with typical values of 100 or 1000 ticks/s). Alternatively, to reduce interrupt overhead, some guest OSs (including Linux) can eliminate the periodic timer and run in the so-called tickless mode, in which the OS programs a single-shot timer to fire at the precise future moment when it needs to process an event. As such an event is usually expected much later than the next periodic timer interrupt would have regularly occurred, the single-shot timer eliminates useless interrupt overhead.

3.1.2 Guest-Xen Time Interface

Xen has two interfaces for communicating time-related information with a guest.

Xen-to-Guest: the Shared Info Page Xen passes information to a guest via a shared memory region called the shared info page. A guest kernel reads the shared info page to retrieve, among other things, time information dynamically updated by Xen as the system runs.

The shared info page holds an array of *per-VCPU* structures, each of which contains a `vcpu_time_t` structure. This, along with other fields in the shared info page that hold the wall-clock value at guest boot time, is used by Xen to implement time keeping on behalf of guests.

In addition, Xen provides to every guest VCPU a virtual periodic timer (with a default period of 10 ms) that the guest can arm to within a 1 ms period, and an optional single-shot timer. Depending on configuration, Linux guests may use the periodic timer for getting periodic virtual interrupts from Xen[2].

Guest-to-Xen: the Hypercall Interface A guest can make hypercalls into Xen to set the platform wall-clock time, and to schedule periodic and single-shot timers. The hypercalls of interest to TimeSync are the timer hypercalls (rather than the wall-clock hypercalls), since only a privileged guest (Dom0) can set the platform wall-clock time.

The hypercall interface provides primitives that manipulate, for each VCPU: *(i)* the periodic timer (start/stop); *(ii)* the single-shot timer (start). The periodic timer delivers virtual interrupts to the VCPU with the desired period. The single-shot timer delivers an interrupt to the VCPU at a target guest system time specified as an argument to the hypercall. A VCPU programs the single-shot prior to relinquishing the CPU to schedule a timer interrupt at the time it needs to process a known event (e.g., the next expiring guest timer).

[2]Other guests (e.g., NetBSD) do not rely on the periodic timer at all, using instead the single-shot timer, which they arm on every timer interrupt.

3.1.3 *Time Keeping in Xen*

In Xen, system time is an abstraction of the time in ns elapsed since the system was booted. This assumes an ideal, global notion of time, uniformly and instantly available to all CPUs, e.g., in an SMP system.

The TSC time is the number of CPU cycles that have elapsed since an arbitrary point in the past (provided by the 64-bit x86 Time-Stamp Counter CPU register). Xen uses the TSC time elapsed since a reference point to compute the current system time by adding the difference between two TSC samples (current, and the reference), multiplied by a TSC-to-ns conversion factor (denoted by mf), to the system time at the reference point.

In practice, to implement the system time abstraction efficiently on multi-CPU systems, Xen employs several approximations and optimizations. First, it maintains local (per-CPU) time variables independently on each physical CPU to track the last "good" known value of system time, along with the mf with respect to the TSC of that CPU. Second, on any given CPU, Xen does not continually maintain and update the CPU-local system time variable. Instead, it *computes* system time as follows: *(i)* periodically records a reference value of the system time, *(ii)* computes the TSC time elapsed since the last system time reference value, *(iii)* converts the TSC time elapsed to ns of real time using the CPU-local mf factor, and *(iv)* adds the value in ns to the reference system time to obtain the current system time.

Time Calibration in the Hypervisor: Since the rate of TSC change may vary over time (e.g., due to fluctuations in clock frequency), Xen performs time calibration every second, with two goals. First, it retrieves a "good" reference system time from a reliable time source and distributes it to all active CPUs. Second, it recomputes a new mf factor on each CPU, to be used until the next calibration event. These values are stored on each CPU in two CPU-local time variables.

Maintaining Time for Guest Domains: To provide time information for a guest domain, Xen: *(i)* pushes updates of system time to the guest, and *(ii)* manages periodic and single-shot timers on behalf of the guest. These take place independently and individually for each guest VCPU.

Xen uses three (logical) fields in the `vcpu_time_t` structure in the shared info page to pass CPU-local time information to a VCPU: *(i)* st: local reference system time at the last calibration on the CPU, *(ii)* ts: local TSC stamp at the time of the last calibration, and *(iii)* mf: the multiplication factor used by the VCPU when converting TSC time intervals into real time for its own computation of the current system time.

Note that the $T = (st, ts, mf)$ triple provided to a guest VCPU is exactly the same one that Xen itself uses to compute internally *its* system time on a given CPU, so it is specific to the current CPU that executes the VCPU time update. Moreover, guest kernels derive an *estimate* of the system time from T using a TSC-based scheme similar to that of Xen. Since guest accesses to the TSC are not virtualized this creates a dependency on the physical platform.

Xen updates the time information of a guest VCPU in three instances: *(i)* When the VCPU is scheduled for execution, but only if a time calibration has taken place while the VCPU was not running. *(ii)* When the VCPU is rescheduled on a different CPU than the one on which it has last run. *(iii)* When time calibration occurs on the underlying CPU.

In principle, a guest VCPU could read its time triple T from the shared info page and use it directly to compute the system time using the formula $st_{xen} = st + (TSC - ts) * mf$.

However, the guest kernel neither uses st_{xen} directly, nor does it count the timer interrupts received. Instead, it maintains its own view of system time in a system-wide variable (`processed_system_time`, or PST, in ns) that it advances in full increments of ticks at HZ frequency. On every timer interrupt received by a VCPU, the guest kernel: *(i)* Computes st_{xen} using the formula above. *(ii)* Compares st_{xen} with its PST and checks if at least a tick (at its own HZ rate) has passed since the last PST update; if so, it updates PST by a whole number of elapsed ticks, in ns. *(iii)* Increments its tick counter (jiffies, ticks, etc.) by this number of ticks.

This mechanism shields the guest OS from vagaries of virtual interrupt delivery by Xen, the most conspicuous of which is the loss of timer interrupts while guest VCPUs are not running. If virtual timer interrupts are lost or delayed, the guest will always advance its view of system time on the first interrupt it gets, and will make up for the lost/delayed interrupts strictly based on its own timer period (HZ) and the TSC time elapsed since its last PST update.

Also, because the guest does use the computed st_{xen} as a reference for comparison with its own PST (in step 2 above), it will *resist* changes in the st value that push st_{xen} back with respect to PST, and will *catch up* with st_{xen} if this jumps forward with respect to PST. This creates an asymmetry in the way the guest reacts to time updates from Xen.

Note that because the guest maintains its own notion of system time, any dynamic time virtualization scheme *cannot* know it and will have to make assumptions about its value at a given instant. Specifically, it is reasonable to assume that the guest computes st_{xen} using the formula above, and does this immediately when the triple T is provided by Xen.

3.2 Xen-based Dynamic Time Virtualization

Dynamic Time Virtualization: To keep the progression of guest time in sync with an external time source (such as the time progression generated by a network simulator), TimeSync exploits the Xen-guest time interface to virtualize both the absolute system time and the rate of time progression as perceived by the guest.

TimeSync introduces a thin layer of virtualization through which an external Dom0 process can dynamically control the st and mf parameters in the Xen-VCPU time interface. This enables: *(i)* fine-grained dynamic corrections to st, and *(ii)* specifying the rate at which time elapses in the guest.

Given an external time source that provides a pair of a system time value ST, along with a desired slowdown factor SF of the rate of time progression, our solution implements:

1. A **xenctl** call **toggle_slowdown()** that allows a Dom0 process to dynamically turn on and off time virtualization for several VMs.
2. A **xenctl** call **set_slowdown()** that allows a Dom0 process to specify the (ST, SF) pair to a list of VMs.
3. A mechanism for propagating changes in (ST, SF) to the target VMs. Currently our implementation only supports dynamic slowdown for single-VCPU VMs.
4. A mechanism for scaling the VCPU periodic and single-shot timers according to the currently effective SF and for dynamically updating the active VCPU timers on a change in SF, so that they would expire correctly in the new timeframe of the guest.

For each VCPU, we control the fields of the time triple T in its `vcpu_time_t` structure to supply to the VCPU a virtualized time triple $T_v = (st_v, ts_v, mf_v)$, where

$$st_v = \begin{cases} ST & \text{if ST is available,} \\ st_{est} & \text{otherwise.} \end{cases}$$

ts_v = TSC stamp at a T_v update or CPU switch, and
$mf_v = mf \ / \ SF$.

Here, st_{est} is a running estimate of the guest time that we maintain dynamically inside Xen, as a function of the sequence of all past SF seen since time virtualization has been turned on. To allow for fractional $SF > 1$ values, since Xen performs only integer arithmetic, we multiply the input SF by a fixed precision factor (e.g., 10^4 for 4-decimal precision) to obtain an integer, scale mf by the same factor and perform integer division in Xen, rounding the remainder.

Note that, because of the intrinsic dependency of both Xen and guest timekeeping on CPU-local parameters such as the TSC conversion factor mf, we cannot fully virtualize all parameters by just updating T_v when (ST, SF) changes. For example, besides propagating any change in SF via mf_v, we follow Xen in propagating changes to the CPU-specific conversion factor mf due to calibration or the VCPU being scheduled on a different CPU. This dependency may change in more advanced versions of Xen.

We propagate all changes to time-related parameters (ST, SF) in a controlled fashion, *lazily*, only upon scheduling a target VCPU for execution. This is important as the `set_slowdown()` call most often executes on a different CPU from the ones running target VCPUs. This call must propagate T_v values consistently from the CPU on which the target VCPU is executing, perform timer updates based on the (unknown) state of its timers, and avoid racing with the scheduler on the target CPU. For these reasons, we defer T_v propagation until the target VCPU is about to be (re)scheduled and the scheduler can execute the propagation code on the same CPU as the target VCPU.

Dynamic Timer Scaling: When a timer is first programmed, we scale the timeout value requested by the guest by the current SF in effect. This is easy for the periodic timer because it has a fixed period and is controlled only by Xen: it is started when a VCPU is about to be scheduled or whenever the timer fires, and stopped when a guest VCPU blocks and yields the CPU. Also, since the periodic timer timeout is *relative* to Xen system time, we simply multiply it by the effective SF to get a linearly slowed down timer.

Manipulation of the single-shot timer is more complex: *(i)* it is only started by the guest kernel, and can be programmed with unpredictable timeouts based on the guest needs (e.g., to fire when the next guest timer is due); *(ii)* it is programmed in terms of an absolute target timeout; *(iii)* the timeout is relative to the guest timeframe and not the Xen timeframe, i.e., the guest computes it based on its PST.

As described in Section 3.1.3, the guest PST does *not* follow Xen system time. We have verified that this discrepancy between guest and Xen system time is indeed present in native Xen. Because of it, when timeouts are small enough, the hypercall the guest uses to start the single-shot timer may start a Xen timer with a timeout into the past (if guest time lags behind Xen system time). The net effect of this lag is an imprecision in delivering the single-shot timer interrupt to the guest, i.e., the timer will fire sooner than expected by the guest, which will force the guest to reprogram it. The outcome is that the guest gets multiple interrupts for scheduling a single (desired) timer event.

All the above three factors need to be taken into account when scaling the single-shot timer. This requires converting a timeout from absolute guest timeframe into the Xen timeframe. First, we compute an estimate of the current guest time (at the time of the hypercall) based on the time elapsed in Xen since the last T_v change, the estimated guest time st_{est} at that instant, and the current SF in effect.

Second, we compute the timeout as a relative offset from our estimate of the current guest time and scale it based on the current SF into a relative Xen timeout that we use to program the single-shot timer inside Xen. During this whole process, as in the native Xen system, the unknown is the current guest system time, the lack of which we attempt to compensate by keeping our running estimate st_{est}.

Dynamic Timer Updates: If the SF changes while VCPU timers are running, we must update their timeouts. When an SF change from SF_{old} to SF_{new} takes effect (lazily, at the time a VCPU is scheduled), we stop the timer, compute the time until the timer is due to expire in the SF_{old} timeframe, re-scale it in the SF_{new} timeframe, and restart the timer with the new timeout.

The single-shot timer again poses a subtle problem at an SF change. If the VCPU was blocked and it is using the single-shot timer to schedule a wakeup, it would have programmed it in the SF_{old} timeframe, and will remain blocked (ineligible for execution) until the timer fires. A very large value of SF_{old} will have scheduled a wakeup timer interrupt far into the future. If now we get an $SF_{new} < SF_{old}$, the scheduler will not be invoked (unless some event needs to wake up the blocked VCPU) and the new SF will not take effect until the timer has fired in the old timeframe. We solve this problem by forcing a schedule event for the VCPU to get it out of the blocked state. This allows the guest to receive its new T_v (and thus SF value) from Xen, run, and block again, but not before programming its single-shot timer which will now be correctly scaled in the new timeframe.

Wall-clock Time Virtualization: To synchronize the wall-clock time on the end-host VMs at the beginning of a simulation, we configure them in independent wall-clock mode (i.e., they do not receive updates of wall-clock time from Xen). Inside each VM, we then perform a one-time `settimeofday()` call with a common value of the wall-clock time, multicast to all testbed machines from a reference machine before the start of the simulation. At the end of the simulation, we bring the VM wall-clock time up to date.

3.3 Simulation-driven Time Virtualization

The Xen-based time virtualization mechanism described above is generic, so it can be driven by *any* external time source that provides dynamic updates of ST and SF predictions on small time scales. In our emulation testbed implementation, the (ST, SF) pair is provided by the simulator introspection module (SICM).

We implement the introspection functionality in a separate *sampler process* that wakes up periodically every sampling period Δ in real time ($\Delta = 3$ ms in the current implementation). We use a process and not a thread in order to isolate it from interactions with unknown/unavailable simulator code, and to be able to tightly control it. E.g., we make it a real-time process and control its CPU affinity in order to isolate it from the scheduler and ensure it runs accurately on sampling period boundaries.

The sampler process communicates with the main simulator process via shared memory: on each invocation, it samples the last processed simulation time st (in shared memory) and records it along with the current real time rt. It then computes SF in the last sampling interval and uses this value as the projected SF value in the next interval, as described in Section 2. Since we cannot handle an infinite SF value (possible if simulator time does not advance in a sampling interval), we cap SF at a maximum SF_{max} (100 in the current implementation). SICM sends the tuple (ST, SF) via IP multicast to the VM clock control module (CCM) running on all physical machines in the testbed (see Fig. 3). This message may be sent over a dedicated network, to ensure isolation from other traffic.

CCM injects the (ST, SF) tuple it receives periodically from SICM into Xen using the `set_slowdown()` call (Section 3.2) . At the start of the simulation, CCM calls `toggle_slowdown()` to selectively enable time control for VMs used in the emulation; at the end, CCM calls it again to disable it. The effect of the latter call is to revert the VM timeframe of the target VMs to the default "normal" one as provided by Xen: it resets the VM system time to that of the host machine (as maintained by Xen), and stops scaling the rate of time progression and the VM timers.

3.4 Constraining Simulation Time Speedup

The simulator control functionality of SCIM, implemented as part of the simulation process, prevents speedup of simulation time. It advances the simulation in intervals no larger than a small number of simulation time units (100 μs in the current implementation). The module continuously samples the last processed simulation time and the real time at which this was recorded. Prior to advancing the simulation, SCIM checks if the last processed simulation time is ahead of the last real time, i.e., the simulator is attempting a speedup. If so, it then postpones processing of the next event until the real time has caught up with the simulation time.

This mechanism is currently implemented with a periodic simulation event. This may change in the future to take advantage of high-priority, hard deadline events, or by directly modifying the simulator scheduler, where possible.

4. EVALUATION

In this Section, we present the goals, methodology, and results of an evaluation of the TimeSync implementation integrated in the VAN [10] testbed.

4.1 Goals

The goal is to evaluate the impact of TimeSync on the fidelity of the emulation in a large-scale simulation. Intuitively, fidelity refers to how "close" the values of relevant network metrics are when measured in the end hosts, compared to the values for the same metrics as dictated by the simulation. We are thus interested in the distortion that the emulated end-to-end network introduces in these metrics.

It is *not* our goal to validate the accuracy of a given simulated network model at a certain scale. Since all our experiments are large-scale, we could not conduct similar experiments on real hardware at such scales to validate against (this is exactly the problem that TimeSync solves). In all experiments, we assume the simulated network is an accurate approximation of a similar real network.

Since our end goal is to evaluate a system that provides time virtualization in hybrid network emulation (TimeSync), we argue that the primary figure of interest in such a system is the *measured* end-to-end packet latency through the emulated network. Because measuring one-way latency requires taking local time samples at different hosts (the sender and the receiver of a packet), this provides good insight into three distinct areas: *(i)* how distorted the perception of time inside an end-host VM is in the base system (without TimeSync) due to simulation artifacts (e.g., slowdown induced by excessive load and/or number of simulated nodes); *(ii)* how effective is TimeSync at reducing or eliminating this distortion; *(iii)* what other discrepancies can be exposed once TimeSync eliminates the dominating components caused by simulator slowdown.

4.2 Experiment Setup

The testbed consists of one simulator machine running QualNet [9] under Linux and 7 testbed machines. All machines are Dell PowerEdge 1950 dual-processor Intel Core 2 Quad Xeon, having a total of 8 cores at 2.66 GHz and 16 GB of RAM each and two Intel e1000e 1Gbps network interfaces. Testbed machines run Xen 3.3.2 with TimeSync support, and use the default credit scheduler. Guest VMs run Linux kernel 2.6.18-164.11.1.el5xen in Dom0 and 2.6.21-2950.fc8xen in DomU, respectively. Each testbed machine runs seven end-host VMs, each with 1 GB of memory and 1 VCPU; a Dom0 has 8 VCPUs and 1 GB of memory. The data traffic between the simulator and testbed machines goes over a 1 Gbps switched Ethernet network. A separate 1Gbps Ethernet network is used to send the clock signal via multicast from the simulator to the testbed machines.

We simulate a network of 200 stationary nodes, structured into seven 802.11a subnets in ad hoc mode, connected by a core FastEthernet subnet (Fig. 5). Each 802.11 subnet consists of 25 nodes except for one subnet of 50 nodes. The core subnet consists of seven gateway nodes, one from each subnet. Inter-subnet traffic is forwarded through the core Ethernet. The main reason for choosing this setup is that wireless models are in general more complex and can achieve significant slowdown with a smaller number of nodes. This setup is also typical of the real wireless/mobile networks for which the VAN testbed is commonly used.

For unicast routing, we run OLSR with 2 seconds HELLO interval within the 802.11 subnets and OSPF within the core. For routing multicast traffic, we use PIM-DM with an add-on that supports ad hoc networks. The data rate of the 802.11 subnets is 18 Mbps and the radio range is 350 m, with free range propagation. The inter-gateway delay in the core subnet is 1 ms.

We assign one VM to each one of 49 nodes spread across all seven subnets as shown in Fig. 5. The rest of the nodes participate in the simulation but do not generate and do not receive application-layer traffic. The VMs generate application traffic and participate in IGMP and ARP, whereas all other network layer and MAC control traffic is generated by the nodes in the simulated network.

In all experiments, one VM (located in subnet 1) sends UDP packets of fixed size (460 bytes) at a constant rate to a single destination multicast group. All the other VMs in the network run a receiver application that joins the group and logs a trace of packets received. We specifically choose multicast traffic because it allows us to timestamp packets at a single source in the network (the unique sender), which makes comparison and aggregation of latency results over multiple hosts meaningful. We use `mgen` [12] as the sender/receiver application, and associated tools such as `trpr` to compute one-way end-to-end latency based on send/receive timestamps placed in the packets as they travel through the emulated network from the sender to receivers.

4.3 Methodology

To compute the latency distortion, we instrument the packet path to record two pairs of timestamps: *(i)* at the two

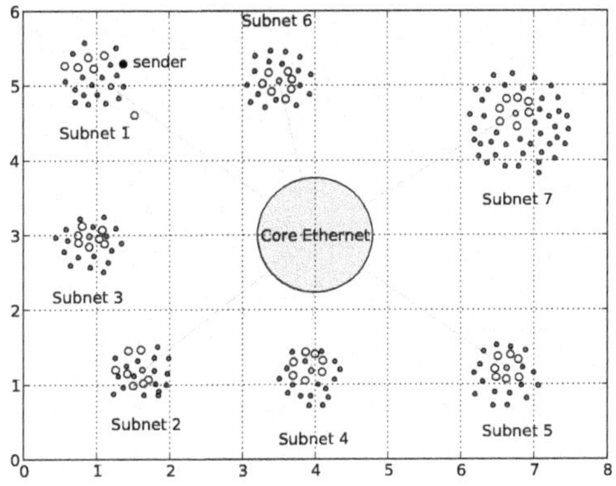

Figure 5: Layout (at scale, grid in km) of the simulated 200-node network, structured in 7 802.11 subnets. Receiver nodes (in white) and the sender each map to a VM in the testbed (49 VMs).

endpoints (sender/receiver), to obtain the "user-perceived" end-to-end latency in VM time, and *(ii)* at the ingress/egress points to the simulator, to obtain the simulated (reference) packet latency in simulation time. Both the sender and each individual receiver timestamp a packet locally, before sending and after receiving it, respectively, using `gettimeofday()` system calls: the difference δT between the timestamps gives us the one-way end-to-end latency. The simulator timestamps the packet at ingress/egress using *simulation time*: the difference δt between these two timestamps represents the simulated packet latency through the simulated network. Finally, we compute the *latency distortion metric* as $dist = \delta T - \delta t$.

We run experiments to evaluate the *dist* metric for two systems: *(i)* the "base system" (referred to as *Base*), and *(ii) Base* augmented with TimeSync (referred to as *TS*).

In our *Base* system, we modify the external packet interface of the QualNet simulator (IPNE/eXata). This interface has two properties: *(i)* it allows the simulation to run faster than real time when possible; *(ii)* it delays ingress packet injection into the simulation until simulation time reaches the packet arrival time (sampled in real time).

Since in *TS* there is no way to arbitrarily speed up VM execution, for a fair comparison we prevent the simulator from running faster than real time in *Base* as well. We achieve this by executing a periodic time-control simulation event with high frequency, as described in Section 3.4.

We vary the sender rate to study the impact of increasing simulation load on latency distortion. For each receiver host, we collect traces of packets received. We discard the first packet in a trace in order to eliminate the impact of ARP on latency (this effect is more prominent in *TS*, where the first packet taking the hit of ARP resolution would always get the maximum latency). Depending on its location, not every receiver may get the same number of packets. However, since the loss rates we observe are fairly small (under 1%), we can compute meaningful statistics over *any* of the receiver traces and compare them across different receivers. For each receiver, we compute statistics of the per-packet *dist* metric.

In order to factor out transients and noise caused by simulator startup initialization and transient regimes of routing protocols in the simulated network, traffic injection in all experiments starts consistently at 130 s in *simulation* time. Traffic injection lasts 500 s measured in the *end-host VM* time frame. In *Base*, this corresponds to 500 s of real time, while in *TS* it represents 500 s of VM time as controlled by TimeSync (approximating 500 s of simulation time). As a result, losses aside, the total number of packets in the trace for a given rate is the same for both *Base* and *TS*, which makes comparison meaningful.

4.4 Experiments with Base

These experiments illustrate the impact of simulation slowdown on the *Base* system, and its dependency on load. We inject traffic (see Section 4.3) at a rate of 1, 5, 10, and 20 pps (packets per second), respectively, and measure the latency distortion across the whole duration of each experiment.

Fig. 6 shows the average per-packet *dist* (a) and the average simulated latency (b), at 1, 5, 10, and 20 pps, across traces collected for one receiver located in subnet 2; results are similar for all the other 47 receivers. Averages were taken across multiple runs of each experiment at a given rate, to achieve confidence intervals of 95%. We plot each mean value inside a circle, along with its confidence interval. As can be seen, the distortion in latency goes up with the load, validating the intuition that higher loads on the simulator have a bigger impact on fidelity. The mean simulated latency (which is 4 orders of magnitude smaller than the distortion) is statistically invariant across all rates.

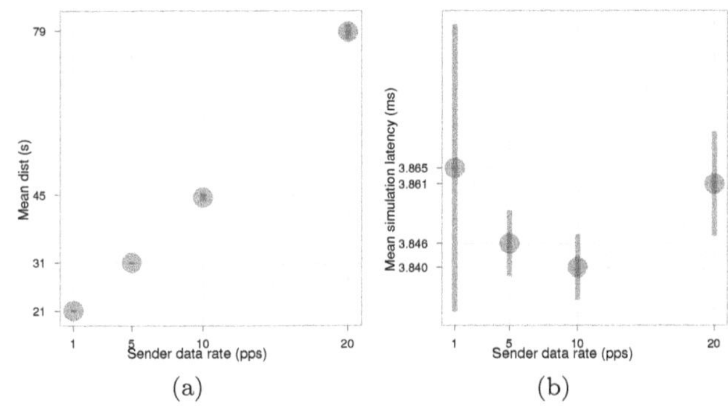

Figure 6: *Base*: Mean *dist* (a) and mean simulated latency (b) with 95% confidence, at 4 traffic loads, over one receiver trace (results are similar for all receivers). Note the different scales on the y axis (seconds vs. milliseconds).

To illustrate how the increase in load is reflected in the slowdown of the simulation time, Fig. 7 plots the progress of simulation time against real time at different traffic loads.

Two notable effects are visible in this graph. First, the subunit slope that every curve exhibits *before* traffic injection starts at 130 s in simulation time (marked by the horizontal line; the 130 s abscissa is also drawn for reference) indicates that the simulation time is already slowing down even in the absence of application traffic. The simulator cannot keep up with real time even with an idle network (note however that this "idle" network does require processing of link and network layer control traffic for 200 nodes). To confirm this effect, Fig. 8 plots the per-packet *dist* metric for a packet trace (at rate 10 pps), showing that the first packet in the trace already suffers about 8 seconds of latency inflation due to the slowdown cumulated prior to its injection. After that, the distortion keeps increasing due to the sustained divergence of simulation time (as seen in Fig. 7).

Second, the effect of dynamic increase in load is visible as the inflexion point in Fig. 7 at the start of traffic injection (130 s in simulation time). Subsequently, the decrease in the slope indicates a slower progression rate of the simula-

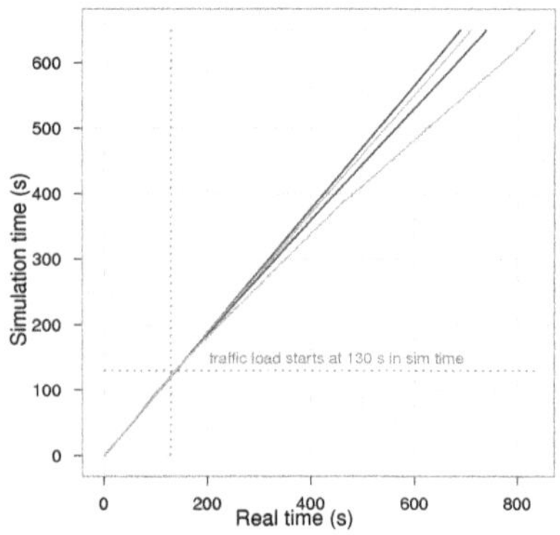

Figure 7: *Base*: Simulation time dynamics at various traffic loads.

tion time due to load. The higher the rate, the larger this divergence of simulation time with respect to the real time will be after the inflexion point.

In summary, these results show that in *Base* emulation fidelity is completely lost, as illustrated by the inflation in end-to-end latency by four orders of magnitude compared to the simulated latency (e.g., from 3.8 ms to 21 s at 1 pps). This makes the system practically unusable for accurate emulation in this setup (i.e., with this simulator and this simulated network), In our setup, the limiting factor is the network size: even without external load, the *Base* system cannot preserve fidelity. Applying traffic load further exacerbates this effect, worsening the distortion. We next evaluate how good TimeSync is at alleviating this problem.

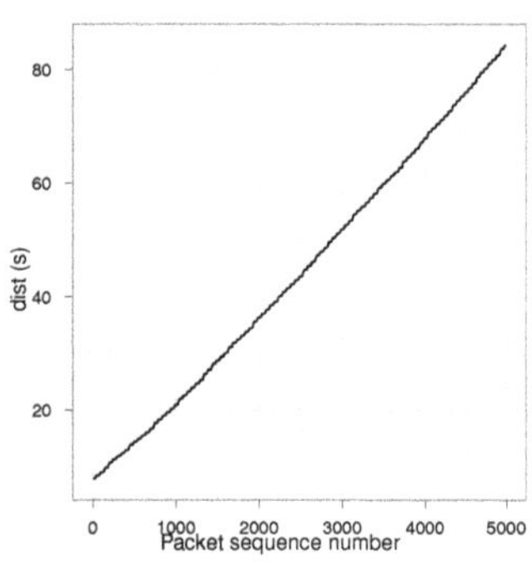

Figure 8: *Base*: Latency distortion across a whole packet trace at a traffic load of 10 pps, for one receiver (similar for all receivers).

4.5 Experiments with *TS*

First, we repeat the above load-varying experiments with the *TS* system. Fig. 10 shows the average per-packet *dist* (a) and the average simulated latency (b), at 1, 5, 10, and 20 pps, respectively, over traces collected for the same receiver as in *Base*. We note that: *(i) TS* achieves a mean *dist* between 0.47 and 0.54 ms; the relative error of the measured end-to-end latency over the simulated latency is 12 - 14%; *(ii)* the mean *dist* is fairly stable across all rates: it is statistically similar at 1, 5, and 10 pps, and only about 23 μs off (lower) at 20 pps; *(iii) TS* has no statistically discernible impact on the *simulated* latency (cf. the second graph in Figs. 10 and 6).

Next, we measure *dist* at a fixed traffic load (20 pps) and vary the simulated network size (and thus the slowdown due to model complexity). This is achieved by adding or removing nodes in the 802.11 subnets of the 200-node topology of Fig. 5, while keeping the original sender/receiver nodes. As shown in Fig. 9, mean *dist* is (a) well under 1 ms, and (b) is generally higher for lower slowdown and lower for higher slowdown. (b) can be explained by lower contention for testbed resources with lower load per unit of real time, and presents an interesting tradeoff of *dist* and running time.

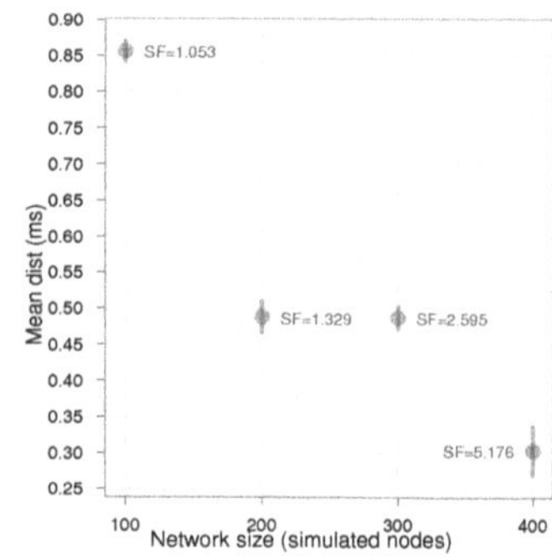

Figure 9: *TS*: Mean *dist* with 95% confidence, at the load of 20 pps and various network sizes, over one receiver trace (results are similar for all receivers). Mean slowdown (SF) is also shown.

These results show that TimeSync achieves useful fidelity of the emulation for layers 3 and above, as reflected by the dramatic reduction in latency distortion in *TS* vs. *Base*. It does so under scalable traffic load, with a complex network model and large network sizes which are otherwise not amenable in the *Base* system. Furthermore, TimeSync does not affect the accuracy of the simulation[3].

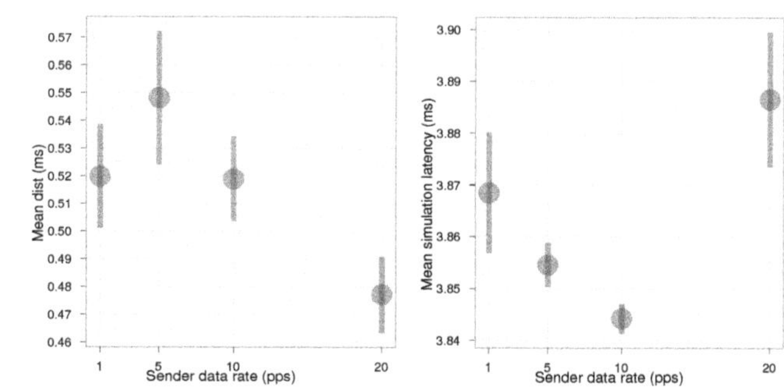

(a) (b)

Figure 10: *TS*: Mean *dist* (a) and mean simulated latency (b) with 95% confidence, at various traffic loads, over one receiver trace (results are similar for all receivers).

To illustrate how TimeSync controls VM time, Fig. 11 plots simulation time against VM time in *TS* at 1, 5, 10, and 20 pps, respectively. We collect two traces of time values during a run: the sender VM's timestamp on each packet sent (x axis), and the simulation time at which the packet arrives at the ingress interface to the simulation (y axis). The four curves are practically overlapping lines with a slope of one, which shows that the VM time successfully tracks simulation time at all rates, as visible at this scale. Note that, since trace collection is packet-driven, we do not have

[3]By design, TimeSync does not affect simulation correctness as it requires no change to the underlying models.

data points before traffic starts at 130 s of simulation time. However, we can infer that, prior to that, TimeSync had correctly synchronized VM and simulation time, since they are equal at the first data point.

4.6 Discussion

To explain these results, we note that *dist* has two components: $dist = sync\_err + extra\_err$

where *sync_err* is the error due to the disparity between simulation time and VM time, and *extra_err* is the error due to the extra latency introduced by various hw/sw components that interpose between the VMs and the simulator.

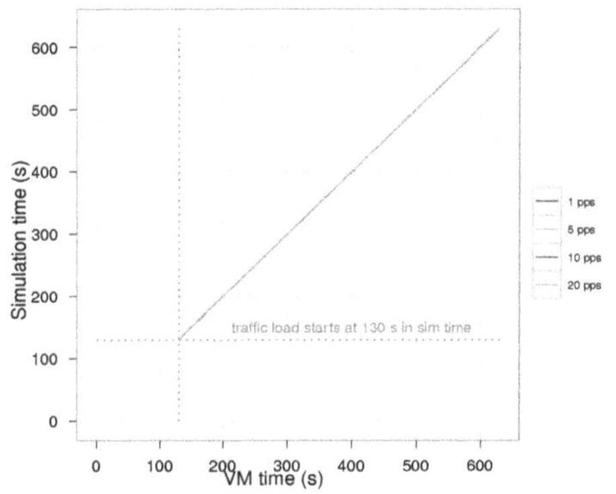

Figure 11: *TS*: Sim. time vs. VM time at various traffic loads.

In *Base*, *sync_err* (always positive) reflects the large divergence of VM time against the slower simulation time that causes the observed loss of fidelity. In *TS*, *sync_err* reflects the residual error intrinsic to TimeSync, caused by inaccuracies in tracking simulation time, e.g., due to discrete time sampling, wrong slowdown prediction, applying the slowdown at slightly different instants across VMs.

In both systems, *extra_err* reflects the delays that a packet incurs due to propagation (to/from the simulator machine) or processing in the software stacks it traverses (on the testbed and simulator machines). Note that when measuring *dist* in the *TS* experiments these delays are appropriately scaled due to time control by TimeSync. We call the sources of such delays *real-time components* because they operate in the real-world part of the system and introduce delays in real time (as opposed to the simulated delays in the simulator). For example, in Xen, such delays on the inbound path of a packet to a paravirtualized VM include hardware interrupt processing, virtual physical interrupt mapping and processing, Dom0 backend driver processing and demultiplexing, and Xen inter-domain signaling. Moreover, because of the Xen I/O model in which multiple VMs share a driver domain, the magnitude of these effects is unpredictable and not strictly related to the I/O volume of a VM.

In an hypothetical ideal system, if *extra_err* could be zero, then *dist* would estimate the error introduced by TimeSync. Conversely, if *sync_err* could be zero, *dist* would estimate the error introduced by the real-time components of the system. In a real system, both *sync_err* and *extra_err* are non-zero and their contributions to *dist* cannot be easily separated without heavy instrumentation.

Measurements on an idle system show a lower bound of about 200 μs for *extra_err* that includes traversal of Xen and simulator network stacks. We can conjecture that in the *Base* system *extra_err* is practically obscured by the much larger *sync_err* component caused by time divergence (tens of seconds in our experiments). Therefore, a naive user of *Base* may not even be aware of the existence of *extra_err*. Thus, an interesting side-effect of TimeSync is that it exposes *extra_err* by effectively removing from *sync_err* the dominant fraction due to simulation slowdown.

The interplay between synchronization and real-time errors in the *dist* metric has one other interesting implication. Because it exposes the real-time components of the error *extra_err*, TimeSync is also useful as a debugging and tuning aid, a feature on which we relied heavily to fine-tune *both* systems used in our experiments. In fact, our configuration reflects our experience while using *TS* and continuously monitoring the *dist* metric. We would routinely run the simulation with TimeSync enabled and identify various components not related to the simulation that were inflating the end-to-end latency. We would then take steps to reduce or eliminate their interference (one such example is the use of a real-time kernel on the simulator machine, which guarantees that the simulator process cannot suffer unbounded preemption due to other activities in the system).

This iterative process should stop when the distortion relative to simulated latency is reduced to a level acceptable for a given experiment. We believe that in the current implementation and experimental setup we have significantly reduced the impact of these factors, and the residual distortion *extra_err* is due to minimal delays on the packet path. Further instrumentation and research would help pinpoint and reduce it. For example, our system is limited by lack of real-time (RT) support and advanced I/O in the Xen version we use. Recent research in Xen RT support and I/O performance isolation [8] would benefit our system by reducing interference on the Xen packet processing path.

5. RELATED WORK

The idea of controlling the perception of time in an OS via virtualization has been used in other contexts [6, 5]. TimeWarp [6] slows down time statically by an integer "time dilation factor" (TDF) in order to present a Xen VM with the illusion of a faster (higher capacity) network. TimeWarp differs from TimeSync both in goals and approach: *(i)* it uses time dilation and existing hardware to study the effects of projected scaling of network hardware performance in physical (as opposed to simulated) environments; *(ii)* it sets TDF only once, and keeps it constant throughout the experimental study. In contrast, we address the problem of accurate network emulation in which the details of the network (of which performance scaling is only one of the many possible parameters) are simulated *externally*. We use *dynamic* virtualization of both system time and its rate of change to make the VM perception of time follow that of the simulator.

DieCast [5] is a framework for building a scaled testbed that reproduces a distributed system by running its nodes as Xen VMs while faithfully emulating their hardware capabilities. It uses time dilation by a pre-configured, *static* TDF to reduce a node's CPU requirements (which allows packing more VMs onto a physical machine) and scales down accordingly the perceived network and disk performance. DieCast is orthogonal and complementary to our work by providing a platform for accurate emulation of node capabilities, with constant TDFs. In contrast, time virtualization in TimeSync is *dynamic* and tracks fine-grained variations in the simulation timeline.

In [15] time virtualization is used in a single-OS network emulation system based on [1]. The simulator runs inside a VE (virtualization environment) alongside end-host VEs, on top of a shared Linux OS. The tight coupling enables [15] to control timing of I/O events and VE scheduling, such that VE time tracks simulation time (at the granularity of scheduler time slice in the worst case). This approach correctly slows down VE time under load, but also allows it to run uncontrollably faster than real time when possible. It also modifies the underlying OS, e.g., to control VE scheduling. In contrast, TimeSync addresses time divergence in a hybrid system consisting of a standalone simulator and end-host Xen VMs spread across multiple machines. It can support multiple VM OSs, and uses only a narrow interface for time virtualization. The loose coupling with the simulator enables TimeSync to easily scale out to multiple physical machines, as opposed to a single-machine system of [15].

[4] employs the TDF idea of [6] to cope with CPU overload that may bias the results of a network emulation. The system uses a Xen VM to host multiple OpenVZ VEs (the virtual end-hosts), and to enable aggregate CPU load monitoring and TDF control. It changes TDF across coarse-grained epochs, in response to variations in CPU load measured for the VM. The thresholds at which TDF changes, the corresponding TDF values and other parameters require extensive tuning and configuration. We address a similar fidelity problem, but in the different context of a fully-virtualized hybrid emulation testbed. It is interesting to note that in TimeSync the logical simulation time acts as the sole (implicit) load indicator. Tracking it at small time scales in the VMs has a natural feedback effect, reducing load on the simulator without complex system level load monitoring and intricate mappings from load to TDF.

SliceTime [14] is most closely related to our work. It targets a similar setup (distributed hybrid network emulation) and also tries to establish a common timeline for the VMs and the simulator. It however follows the different approach of fine-grained lock-step synchronization: at each step, all VMs *and* the simulator run for a fixed time slice in their own time frame, up to a common barrier, under the command of an external (centralized) controller. Strict real-time control of the running time of a VM requires a modified sEDF Xen scheduler. Within a time-slice, VM time advances at full speed, therefore it may arbitrarily drift from the simulation time. In order to hide this divergence, it is crucial for SliceTime to tightly control the time slice down to microsecond granularity. This forces a tradeoff between the drift magnitude and the scheduler overhead needed to bound it.

In [14], the limits of the approach with respect to multiple machines are not discussed, and the system is only evaluated with VMs collocated on a single machine. One problem of SliceTime in a multi-machine setup is that in an heterogeneous system (i.e., with uneven distribution of VCPUs to CPUs across different machines), an oversubscribed machine will slow down the whole system *in real time* since every VM it hosts has to execute its slice before the simulation can proceed. Note that this delay is different from simulation slowdown (e.g., due to overload): it is simply an artifact of the lock-step scheme stalling the simulator at every barrier. In addition, in [14] the signaling overhead is proportional to the number of VMs and inversely proportional to the slice size. Since [14] does not evaluate overhead, it is unclear how it would scale with large numbers of VMs running on multiple machines.

6. CONCLUSION AND FUTURE WORK

We have described TimeSync, a solution to the time divergence problem in virtualized hybrid emulation testbeds. TimeSync exploits the Xen-VM interface to make the time dynamics in a VM follow the simulation timeline. TimeSync *tracks* the simulation time to within tight error margins, acceptable for testing of applications and network protocols where timeouts are on the order of milliseconds or larger. Due to its small footprint (no coordination between the simulator and the VMs, no interference with simulator or VM execution, no interference with VM scheduling, no change to the VM OS), as we have shown, TimeSync is able to scale to handle large networks and complex simulation models.

We believe the TimeSync idea and mechanisms can be extended to hardware virtual machines (HVMs). Future implementations can take advantage of new features in Xen 4, including improvements in split network driver performance and TSC register virtualization. Another research direction is towards improvement in slowdown prediction, possibly based on longer history and/or internal simulator state (e.g., event queue sizes, processing requirements per event type, etc.). Finally, with the increasing demand for parallel/distributed simulation to help emulate even larger scale networks, an interesting problem is whether and how TimeSync can be integrated with such architectures.

7. ADDITIONAL AUTHORS

Additional authors: Keith Whittaker (U.S. Army CERDEC, email: keith.d.whittaker.civ@mail.mil), Chris Scilla (U.S. Army CERDEC, email: christopher.a.scilla.civ@mail.mil) and Syeed Ali (U.S. Army CERDEC, email: syeed.e.ali.civ@mail.mil).

8. REFERENCES

[1] OpenVZ: Container-based Virtualization for Linux, 2012.

[2] P. Barham, B. Dragovic, K. Fraser, S. Hand, T. Harris, A. Ho, R. Neugebauer, I. Pratt, and A. Warfield. Xen and the art of virtualization. In *Proc. of the 19th ACM symposium on OS principles*, 2003.

[3] P. Biswas, C. Serban, A. Poylisher, J. Lee, S.-C. Mau, R. Chadha, C.-Y. J. Chiang, R. Orlando, and K. Jakubowski. An Integrated Testbed for Virtual Ad Hoc Networks. In *Proc. of TRIDENTCOM*, 2009.

[4] A. Grau, S. Maier, K. Herrmann, and K. Rothermel. Time Jails: A Hybrid Approach to Scalable Network Emulation. In *Proc. of the 22nd PADS Workshop*, 2008.

[5] D. Gupta, K. V. Vishwanath, and A. Vahdat. DieCast: Testing Distributed Systems with an Accurate Scale Model. In *Proc. of NSDI*, 2008.

[6] D. Gupta, K. Yocum, M. McNett, A. C. Snoeren, A. Vahdat, and G. M. Voelker. To Infinity and Beyond: Time-warped Network Emulation. In *Proc. of NSDI*, 2006.

[7] T. H. J. Ahrenholz, C. Danilov and J. Kim. CORE: A Real-time Network Emulator. In *Proc. of MILCOM*, 2008.

[8] J. R. Santos, Y. Turner, G. Janakiraman, and I. Pratt. Bridging the Gap Between Software and Hardware Techniques for I/O Virtualization. In *Proc. of USENIX ATC*, 2008.

[9] Scalable Network Technologies. QualNet, 2012.

[10] C. Serban, A. Poylisher, and C.-Y. J. Chiang. Virtual Ad hoc Network Testbeds for Network-aware Applications. In *Proc. of NOMS*, 2010.

[11] U.S. Naval Research Laboratory. Extendable Mobile Ad hoc Network Emulator, 2012.

[12] U.S. Naval Research Laboratory. Multi-Generator (MGEN), 2012.

[13] S. Wang and Y. Huang. NCTUns Distributed Network Emulator. *Internet Journal*, 4(2):61–94, 2012.

[14] E. Weingärtner, F. Schmidt, H. V. Lehn, T. Heer, and K. Wehrle. SliceTime: A Platform for Scalable and Accurate Network Emulation. In *Proc. of NSDI*, 2011.

[15] Y. Zheng and D. M. Nicol. A Virtual Time System for OpenVZ-Based Network Emulations. In *Workshop on Parallel and Distributed Simulation*, 2011.

Modeling and Connectivity Analysis in Obstructed Wireless Ad Hoc Networks

Marcelo G. Almiron
Dept. of Computer Science
UFMG, Brazil
malmiron@dcc.ufmg.br

Olga Goussevskaia
Dept. of Computer Science
UFMG, Brazil
olga@dcc.ufmg.br

Alejandro C. Frery
Institute of Computing
UFAL, Brazil
acfrery@pq.cnpq.br

Antonio A.F. Loureiro
Dept. of Computer Science
UFMG, Brazil
loureiro@dcc.ufmg.br

ABSTRACT

Connectivity properties of wireless networks in open space are typically modeled using geometric random graphs and have been analyzed in depth in different studies. Such scenarios, however, do not often represent situations encountered in practice, like urban environments or indoor spaces, which are deeply affected by obstacles. In this work, we present a model for obstructed wireless ad hoc networks consisting of a set of n nodes, deployed at random in a lattice square of size $g \times g$, with a common transmission range r. For positioning the nodes in the field, all segments are considered as one-dimensional, but for communication purposes, we add a parameter ϵ to model the segments' width. Our model can be used to study the structure of obstructed networks analytically, as well as to simulate and evaluate a variety of node deployment strategies and the resulting network topologies. We derive analytical forms for the probability of existing crossing links between parallel and perpendicular segments sharing an intersection, toward a first topological characterization of our model. Moreover, we compute a lower bound for the probability of connectivity at intersections between segments, and apply percolation theory to derivate the Critical Transmission Range for connectivity in the overall network, i.e., the minimum transmission range that generates communication graphs that are connected with high probability.

Categories and Subject Descriptors

C.2.1 [**Computer-Communication Networks**]: Network Architecture and Design—*Wireless communication*

Keywords

Wireless ad hoc networks, obstructed networks, percolation theory, critical transmission range

1. INTRODUCTION

A popular model to analyze wireless ad hoc networks is the geometric random graph [1]. In this model, a set of n nodes is distributed at random (with a uniform distribution and independently) in space, and any pair of nodes within a certain distance r can communicate with each other. A rich body of results has been generated based on this model. Penrose [2], for instance, analyzes how connectivity in such networks is influenced by the parameters n and r. Gupta and Kumar [3] analyze how the capacity of the network scales as the number of nodes grows in the protocol and the physical interference models. Santi and Blough [4] analyze the Critical Transmission Range (CTR) for connectivity in sparse networks. This model, despite providing useful analytical tools, makes a limiting assumption: that a node can freely communicate with all nodes within the disk of radius r around it. Such an open space scenario often does not represent situations encountered in practice, such as urban environments, indoor spaces, or surfaces covered by dense vegetation, in which communication is deeply affected by obstacles.

One of the few studies addressing the modeling of networks in obstructed spaces is presented by Frieze et al. [5]. This model captures environments with regularly distributed obstacles by means of an $n \times n$ grid, in which columns and rows could represent street intersections and street blocks with unitary size. Each one of the n^2 street intersection hosts a node with probability p, and all nodes have the same communication range of w blocks. Among other results, Frieze et al. provide asymptotic bounds on k-connectivity of such networks.

In this work we propose an alternative model for obstructed wireless environments. We consider a network of n nodes positioned on a $g \times g$ grid of streets, where street blocks are comprised of uni-dimensional segments with an expected number μ of uniformly distributed nodes. Moreover, we introduce the street width parameter ϵ, which influences the probability of connectivity at street intersections. Figure 1a shows a schematic representation of the model, and Figure 1c illustrates an example of a network instance with $g = 10$, $n = 900$ (that leads to an expected density of $\mu = 5$) and $\epsilon = 0.3$. This model provides a more realistic representation of environments with regularly spaced arrays of obstacles than, for instance, the model in [5], because usually having only one node at each street intersection and no nodes along the street block is not enough to establish

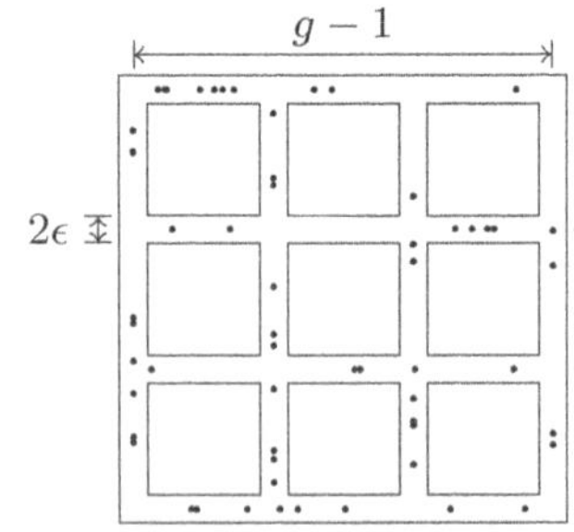

(a) Deployment of nodes in a scenario with $g = 4$

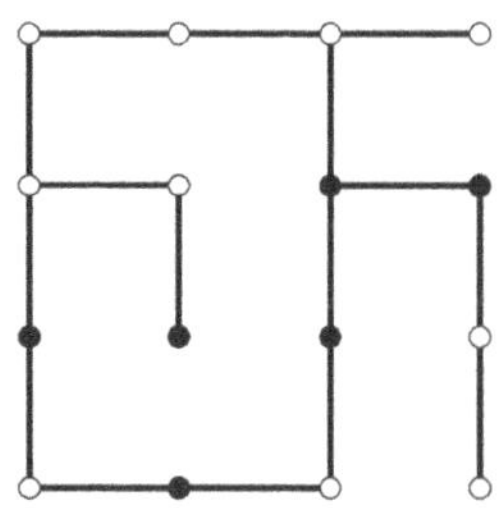

(b) Discrete percolation model

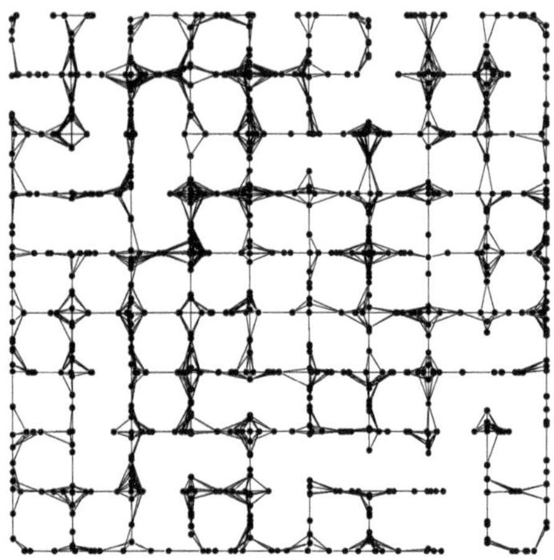

(c) An instance of a network with $n = 900$ and $g = 10$. The transmission range corresponds to the CTR

Figure 1: Modeling an obstructed wireless ad hoc network.

connectivity, especially if low-power/short-range radios are used. Our model is fundamentally different from the model in [5] because it does not allow all nodes to connect to neighbors in both horizontal and vertical directions, but rather is built from one-dimensional communication blocks and two-dimensional street intersections, which asks for new analytical tools.

Our model can be used to study the structure of obstructed networks analytically, as well as to simulate and evaluate a variety of node deployment strategies and the resulting network topologies. It can be potentially useful in simulating and analyzing several types of network scenarios, in particular, those that are comprised by segments of one-dimensional or "semi one-dimensional" arrays of nodes and regularly distributed arrays of obstacles. Examples of such networks are dense urban environments, smart grid communication networks, indoor environments with corridors and staircases connecting different floors, tunnels and subways, among others. Having such a large pool of application scenarios that include one-dimensional components, it becomes interesting to evaluate connectivity properties in such networks and determine their fundamental differences from non-obstructed environments.

We analyze the CTR for connectivity in this model, i.e., we compute the minimum transmission range that generates communication graphs that are connected with high probability (w.h.p.). Determining the CTR is a fundamental building block in the process of designing a communication infra-structure. The CTR provides insights of several important design questions, such as network planning, routing, and medium access control, while minimizing the transmitting power and, consequently, the energy consumption of the communication nodes.

As opposed to many existing theoretical results, such as [2, 5], which can be only applied to dense networks, our result can be applied to networks with any node density. More specifically, we prove that the CTR in our model is $r = \Theta((\ln g + \ln \mu)/\mu)$, where μ is the expected number of nodes on a segment and g is the granularity of the grid. This result can be compared to that of Santi and Blough [4], that states that, in non-obstructed random geometric graphs with field size ℓ in d dimensions, the CTR can be bounded as $r = O(\ell(\ln \ell/n)^{1/d})$. Note that the CTR in our model has good scaling properties, as it grows with the logarithm of the local neighborhood size plus the granularity of the grid.

The contributions of our work can be summarized as follows:

- We propose a framework to model the topology of random obstructed wireless communication networks. This framework can be used to analyze connectivity properties of different application scenarios and node deployment strategies, such as uniform and non-uniform distributions;

- We show how to apply discrete percolation theory to characterize and solve connectivity issues in wireless ad hoc networks. In particular, we characterize the CTR for connectivity in a regular obstructed environment;

- We derive analytical expressions for the probability of crossing links between parallel and perpendicular segments sharing an intersection, and these results are used for deriving a lower bound for the probability of connectivity at intersection;

- Combining results of discrete percolation theory and the probability of connectivity at intersections, we compute the CTR for connectivity in the overall network.

The rest of the paper is organized as follows. Section 2 presents our model. In Section 3 we compute the probability of crossing links between parallel and perpendicular segments sharing an intersection, and these results are used to derive a lower bound for the probability of connectitivy at intersections. Then, in Section 4 we combine the results of Section 3 with discrete percolation theory to compute the CTR for connectivity in the overall network. In Section 5 we validate our analytical results through simulations, and analyze the topology of the network in terms of different parameters. In Section 6 we discuss the related work. Finally, in Section 7 we present the concluding remarks.

2. MODEL

Our network model consists of a set of n nodes deployed uniformly at random in a lattice square of size $g \times g$. We refer to the parameter g as the *granularity* of the model. This lattice can be conceived as a Manhattan-style street map (see Figure 1a). In this sense, streets in the map are represented by segments in the lattice, and streets' intersections correspond to intersections between segments. Note that, without loss of generality (w.l.o.g.), the length of each street segment has a normalized length of 1, and the expected number of nodes per segment is $\mu = n/(2g(g-1))$. We consider that all segments have the same width 2ϵ. However, for the purpose of positioning the nodes, each segment individually is considered as a one-dimensional space, and we position all nodes over imaginary lines located at the center of the segments. The segment width defines the geometry of the space, which is used to determine the communication links at segments' intersections. Additionally, we consider that all nodes in the network have a common transmission range r, and that two nodes u and v, positioned at (x_u, y_u) and (x_v, y_v), respectively, are able to communicate with each other if the following two conditions are satisfied: (i) the minimum norm $\min\{\|x_u - x_v\|, \|y_u - y_v\|\}$ does not exceed ϵ, and (ii) the maximum norm $\max\{\|x_u - x_v\|, \|y_u - y_v\|\}$ does not exceed the common transmission range r (see Figure 2). Note that this communication rule for street intersections (L_∞ metric) is an oversimplification. However, we use it as a starting point because it simplifies the analysis to a manageable level.

To derive our analytical results, we use a powerful mathematical tool for solving connectivity issues in random environments; the mixed (discrete) percolation model [6]. In this model, we have a set $\mathbb{V}^2 = \{1, \ldots, g\}^2$ of g^2 vertices. Each vertex (x, y) defines its own position in the square lattice by means of its indices. Additionally, we have a set of edges $\mathbb{E}^2$, containing all pair of vertices (u, v) such that $\|x_u - x_v\| + \|y_u - y_v\| \leq 1$. Each vertex can be *open* or *closed* with probabilities p_s and $1-p_s$, and each edge can be open or closed with probabilities p_b and $1-p_b$, respectively. Then, the random grid is defined as the graph $\mathbb{G}_g = (\mathbb{V}^2, \mathbb{E}^2)$, and we consider that $\mathbb{G}_g$ *percolates*[†] if all vertices belong to the same connected component formed by open vertices and edges. Figure 1b presents an example of a random grid $\mathbb{G}_4$ that does not percolate (filled bullets represent open vertices, and the empty bullets represent closed vertices).

In order to solve the CTR problem, we describe the original network through the mixed percolation model defined above, as follows. We use the probability of connectivity at intersections, namely $\Pr(I_{con})$, as input value for the parameter p_s, which corresponds to the probability of open vertices, and also the probability of connectivity at segments, namely $\Pr(S_{con})$, as input value for the parameter p_b, which corresponds to the probability of open edges. Then we determine the critical value p_b^c, which grants (bond) percolation, whenever $p_s = 1$. The implications of this process are the following:

- The inverse of the expression that determines p_b^c leads to the CTR, whenever $p_s = 1$, and;
- p_b^c and p_s determine the critical segment width value ϵ_c, such that $\forall \epsilon \geq \epsilon_c$ the CTR obtained from p_b^c is valid.

3. PROBABILITY OF CONNECTIVITY

In this section we compute the probability of connectivity at segment intersections and apply a result from [7] to compute the probability of connectivity at (one-dimensional) segments.

Formally, we define connectivity at intersections as follows. Let $\mathcal{X}$, $\mathcal{Y}$, $\mathcal{W}$ and $\mathcal{Z}$ be four sets of μ nodes positioned in four adjacent segments sharing an intersection[‡]. This intersection is connected if and only if there exists a tree connecting at least one node from each of these four sets. Notice that, since the existence of a tree with four nodes, one per set, is sufficient to grant connectivity at the intersection, we only consider crossing links from one segment to another. We distinguish between two different types of crossing links: (i) links between parallel adjacent segments, and (ii) links between perpendicular adjacent segments. The first step toward the characterization of the probability of connectivity at segment intersections is computing the probabilities of the existence of these links. Lemma 1 presents the probability of existence of a crossing link between parallel adjacent segments, namely $p_\parallel(\mu)$.

LEMMA 1. *Let $\{X_i\}_{1\leq i\leq\mu}$ and $\{Y_j\}_{1\leq j\leq\mu}$ be two families of independent random variables, such that $X_i \sim \mathcal{U}(0,1)$[§] and $Y_j \sim \mathcal{U}(0,1)$, denoting the positions of μ nodes in each one of two parallel segments sharing an intersection. Let also r denote the common transmission range of all nodes, where $0 \leq r \leq 1$. The probability $p_\parallel(\mu)$ of existence of at least one link between two nodes in parallel adjacent segments is*

$$p_\parallel(\mu) \approx \frac{r}{6}\left(\mu - \mu(1-r)^\mu - 4\mu\left(1-\frac{r}{2}\right)^{\mu-1}\left(\left(1-\frac{r}{2}\right)^\mu - 1\right)\right).$$

PROOF. Sorting the realizations of $X_1, X_2, \ldots, X_\mu$ and $Y_1, Y_2, \ldots, Y_\mu$ in increasing order, we obtain the order statistics[¶] $X_{(1)}, X_{(2)}, \ldots, X_{(\mu)}$ and $Y_{(1)}, Y_{(2)}, \ldots, Y_{(\mu)}$, respectively. W.l.o.g., we consider that positions of nodes reflect the Euclidean distances from each node to the intersection point between segments. At least one crossing link between two nodes positioned in different parallel segments exists if and only if $X_{(1)} + Y_{(1)} \leq r$. Then, we can compute $p_\parallel(\mu)$ by

$$p_\parallel(\mu) = \iint_D f_{X_{(1)}Y_{(1)}}(x,y)\,\mathrm{d}y\,\mathrm{d}x, \tag{1}$$

where $D = \{(x,y) \mid 0 \leq x \leq r \wedge 0 \leq y \leq r-x\}$ and $f_{X_{(1)}Y_{(1)}}(x,y)$ is the joint distribution function of $X_{(1)}$ and $Y_{(1)}$. It is well known that the k-th order statistic of a family of μ independent standard uniform random variables is a Beta random variable $U_{(k)} \sim \mathcal{B}(k, \mu+1-k)$. Since all random variables are independent, we have that the joint distribution function of $X_{(1)}$ and $Y_{(1)}$ is $f_{X_{(1)}Y_{(1)}} = \mu^2(1-x)^{\mu-1}(1-y)^{\mu-1}$. Then, from (1) we have

$$\begin{aligned} p_\parallel(\mu) &= \int_0^r\int_0^{r-x} \mu^2(1-x)^{\mu-1}(1-y)^{\mu-1}\,\mathrm{d}y\,\mathrm{d}x \\ &= \int_0^r \mu\left(-(1-x)^{\mu-1}\right)\left((-r+x+1)^\mu - 1\right)\mathrm{d}x. \end{aligned} \tag{2}$$

[†]In the finite random grid, percolation means full connectivity between vertices.

[‡]Note that in total we consider 4μ nodes.

[§]I.e., distributed uniformly and at random in the interval [0, 1].

[¶]The k-th order statistic of a statistical sample is equal to its kth-smallest value.

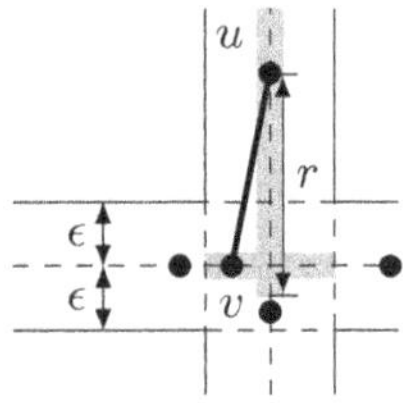

Figure 2: Determining the communication links present at street intersections. The gray region represents the coverage of node u.

Applying Simpson's rule with two intervals in (2), we have

$$p_{\|}(\mu) \approx \frac{r}{6}\left(\mu - \mu(1-r)^{\mu} - 4\mu\left(1-\frac{r}{2}\right)^{\mu-1}\left(\left(1-\frac{r}{2}\right)^{\mu}-1\right)\right).$$

As we will see in Section 5, this is an accurate expression for $p_{\|}(\mu)$. □

The exact probability of existing at least one crossing link between perpendicular segments, namely $p_{\perp}(\mu)$, is given in Lemma 2. Figure 2 can be useful to the reader in order to follow the proof.

LEMMA 2. *Let $\{X_i\}_{1\le i\le\mu}$, such that $X_i \sim \mathcal{U}(0,1)$, and $\{Y_j\}_{1\le j\le\mu}$, such that $Y_j \sim \mathcal{U}(0,1)$, be two families of independent random variables denoting the position of μ nodes in each one of two perpendicular segments sharing an intersection. Let also r denote the common transmission range of all nodes, where $0 \le r \le 1$. Additionally, let 2ϵ be the segments' width, where $\epsilon \le r$. The probability $p_{\perp}(\mu)$ of existence of at least one link between two nodes in different perpendicular segments is*

$$p_{\perp}(\mu) = 2\left((1-(1-r)^{\mu})(1-(1-\epsilon)^{\mu})\right) - (1-(1-\epsilon)^{\mu})^2. \tag{3}$$

PROOF. Let $X_{(1)}, X_{(2)}, \ldots, X_{(\mu)}$ and $Y_{(1)}, Y_{(2)}, \ldots, Y_{(\mu)}$ be the order statistics of $X_1, X_2, \ldots, X_\mu$ and $Y_1, Y_2, \ldots, Y_\mu$, respectively. W.l.o.g., we consider that positions of nodes reflect the Euclidean distances from each node to the intersection point between segments. There exists at least one link between two nodes in different perpendicular segments if and only if there is a link between $X_{(1)}$ and $Y_{(1)}$. Considering $\epsilon \le r$, we can compute $p_{\perp}(\mu)$ by

$$p_{\perp}(\mu) = 2\iint_{R_1} f_{X_{(1)}Y_{(1)}}(x,y)\,\mathrm{d}y\,\mathrm{d}x - \iint_{R_2} f_{X_{(1)}Y_{(1)}}(x,y)\,\mathrm{d}y\,\mathrm{d}x, \tag{4}$$

where $R_1 = \{(x,y) \mid 0 \le x \le r \wedge 0 \le y \le \epsilon\}$, $R_2 = \{(x,y) \mid 0 \le x \le \epsilon \wedge 0 \le y \le \epsilon\}$ and $f_{X_{(1)}Y_{(1)}}(x,y)$ is the joint distribution function of $X_{(1)}$ and $Y_{(1)}$. Since $X_{(1)}, Y_{(1)} \sim \mathcal{B}(1,\mu)$ are independent, the joint distribution function is $f_{X_{(1)}Y_{(1)}} = \mu^2(1-x)^{\mu-1}(1-y)^{\mu-1}$. Then, from (4), we have

$$\begin{aligned} p_{\perp}(\mu) &= 2\int_0^r\int_0^\epsilon \mu^2(1-x)^{\mu-1}(1-y)^{\mu-1}\,\mathrm{d}y\,\mathrm{d}x - \int_0^\epsilon\int_0^\epsilon \mu^2(1-x)^{\mu-1}(1-y)^{\mu-1}\,\mathrm{d}y\,\mathrm{d}x \\ &= 2\left((1-(1-r)^{\mu})(1-(1-\epsilon)^{\mu})\right) - (1-(1-\epsilon)^{\mu})^2. \end{aligned}$$

□

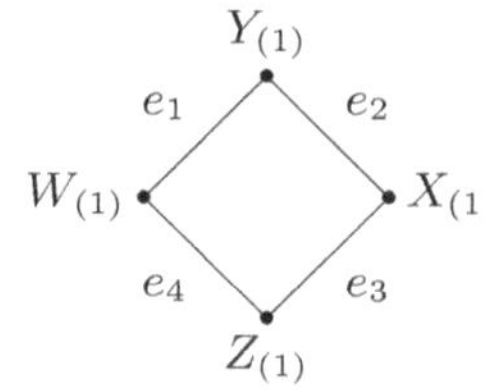

Figure 3: Computing a lower bound for the probability of connectivity at intersections. Each edge is an event, and $\Pr(e_1) = \Pr(e_2) = \Pr(e_3) = \Pr(e_4) = p_{\perp}(\mu)$.

At this stage, we have already computed the probability of existence of at least one crossing link between parallel ($p_{\|}(\mu)$) and perpendicular ($p_{\perp}(\mu)$) segments, respectively. The next step is to compute the probability of connectivity at intersections, namely $\Pr(I_{con})$. This probability is a (non-trivial) function of both $p_{\|}(\mu)$ and $p_{\perp}(\mu)$, which takes into account the geometric correlation between the existence of simultaneous parallel and perpendicular links. Nevertheless, we perceive that the influence of $p_{\|}(\mu)$ on $\Pr(I_{con})$ is very small in most scenarios, i.e., the probability of connectivity at an intersection is almost exclusively determined by the probability of perpendicular links. An exception to this behavior is when μ or ϵ are very small values, which are scenarios not so relevant in practice, given that the value of the CTR in these cases are too large. For the reason above and for the sake of simplification, in Theorem 1, we compute a lower bound on $\Pr(I_{con})$ using only $p_{\perp}(\mu)$, and show through simulations (in Section 5) that this results in a good approximation.

THEOREM 1. *Let $\{X_i\}_{1\le i\le\mu}$, $\{Y_j\}_{1\le j\le\mu}$, $\{W_k\}_{1\le k\le\mu}$ and $\{Z_m\}_{1\le m\le\mu}$ be four families of independent random variables, such that $X_i \sim \mathcal{U}(0,1)$, $Y_j \sim \mathcal{U}(0,1)$, $W_k \sim \mathcal{U}(0,1)$ and $Z_m \sim \mathcal{U}(0,1)$, denoting the positions of μ nodes in four adjacent segments sharing an intersection. Let also r denote the common transmission range of all nodes, with $0 \le r \le 1$, and 2ϵ be the segments' width, where $\epsilon \le r$. The probability of existence of a path between at least one node in a segment connecting at least one node in each one of the three other segments, denoted by $\Pr(I_{con})$, is*

$$\Pr(I_{con}) \ge 2\xi^2 - \frac{\xi^4}{p^2} + p^2, \tag{5}$$

where $\xi = ((1-\epsilon)^{\mu} - 1)((1-r)^{\mu} - (1-\epsilon)^{\mu})$ and $p = p_{\perp}(\mu)$.

PROOF. W.l.o.g., we consider that positions of nodes reflect the Euclidean distances from each node to the intersection point between segments. Consider the order statistics $X_{(1)}$, $Y_{(1)}$, $W_{(1)}$ and $Z_{(1)}$, from the families $\{X_i\}_{1\le i\le\mu}$, $\{Y_j\}_{1\le j\le\mu}$, $\{W_k\}_{1\le k\le\mu}$ and $\{Z_m\}_{1\le m\le\mu}$, respectively. There exists a path between at least one node in a segment connecting at least one node in each one of the three other segments if and only if there exists a path between $X_{(1)}$, $Y_{(1)}$, $W_{(1)}$ and $Z_{(1)}$.

Considering only perpendicular links between nodes, we can obtain a lower bound as follows. Let e_1 be the event that indicates the presence of a link between the realizations of $Y_{(1)}$ and $W_{(1)}$ (see Figure 3). Additionally, let e_2, e_3 and e_4 represent the existence of a link between $X_{(1)}$ and $Y_{(1)}$, $X_{(1)}$ and $Z_{(1)}$ and finally $W_{(1)}$ and $Z_{(1)}$, respectively.

Then, we have $e_1 \perp\!\!\!\perp e_3$ and $e_2 \perp\!\!\!\perp e_4$, where $\perp\!\!\!\perp$ indicates the stochastic independence between events. Consequently, we can compute the aforementioned lower bound by

$$\Pr(I_{con}) \geq \Pr(I_{con} \mid e_1 \cap e_3)p^2 + 2\Pr(I_{con} \mid e_1 \cap \overline{e_3})p(1-p), \quad (6)$$

where $p = p_{\perp}(\mu)$. Furthermore, we have

$$\Pr(I_{con} \mid e_1 \cap e_3) = \Pr(e_2 \mid e_1 \cap e_3)^2 + 2\Pr(e_2 \mid e_1 \cap e_3)\left(1 - \Pr(e_2 \mid e_1 \cap e_3)\right), \quad (7)$$

and

$$\begin{aligned}\Pr(e_2 \mid e_1 \cap e_3) &= 1 - \Pr(\epsilon \leq X_{(1)} \leq r \cap \epsilon \leq Z_{(1)} \leq r \mid e_1 \cap e_3) \\ &= 1 - \frac{\left(\int_\epsilon^r \int_0^\epsilon f_{X_{(1)}Z_{(1)}}(x,z)\,\mathrm{d}z\,\mathrm{d}x\right)^2}{p^2} \\ &= 1 - \frac{\xi^2}{p^2}, \end{aligned} \quad (8)$$

where $\xi = ((1-\epsilon)^\mu - 1)((1-r)^\mu - (1-\epsilon)^\mu)$.

On the other hand, we can compute $\Pr(I_{con} \mid e_1 \cap \overline{e_3})$ by

$$\Pr(I_{con} \mid e_1 \cap \overline{e_3}) = \Pr(A \mid e_1 \cap \overline{e_3}) \quad (9)$$

where $A = \{0 \leq Y_{(1)} \leq \epsilon \cap 0 \leq W_{(1)} \leq \epsilon \cap \epsilon \leq X_{(1)} \leq r \cap \epsilon \leq Z_{(1)} \leq r\}$. Since $Y_{(1)} \perp\!\!\!\perp \overline{e_3}$, $W_{(1)} \perp\!\!\!\perp \overline{e_3}$, $X_{(1)} \perp\!\!\!\perp e_1$ and $Z_{(1)} \perp\!\!\!\perp e_1$, from (9) we have

$$\Pr(I_{con} \mid e_1 \cap \overline{e_3}) = \left(\int_0^\epsilon \int_0^\epsilon f_{Y_{(1)}W_{(1)}}(y,w)\,\mathrm{d}w\,\mathrm{d}y \int_\epsilon^r \int_\epsilon^r f_{X_{(1)}Z_{(1)}}(x,z)\,\mathrm{d}z\,\mathrm{d}x\right) \frac{1}{p(1-p)}. \quad (10)$$

Solving the definite integrals in (10), we obtain

$$\Pr(I_{con} \mid e_1 \cap \overline{e_3}) = \frac{\xi^2}{p(1-p)}. \quad (11)$$

Finally, combining (7), (8) and (11) within Expression (6), we conclude that

$$\Pr(I_{con}) \geq 2\xi^2 - \frac{\xi^4}{p^2} + p^2.$$

□

We conclude this section with Theorem 2, from [7], which computes the probability of connectivity at segments.

THEOREM 2. *Let $\{X_i\}_{1\leq i\leq\mu}$ be a family of independent random variables, such that $X_i \sim \mathcal{U}(0,1)$, denoting the positions of μ nodes in a segment. Let also r denote the common transmission range of all nodes, where $0 \leq r \leq 1$. The probability of connectivity between all nodes is*

$$\Pr(S_{con}) = (1 - (1-r)^\mu)^{\mu-1}.$$

PROOF. See [7]. □

Theorems 1 and 2 are used in Section 4 for computing the CTR for connectivity in the overall network.

4. CRITICAL TRANSMISSION RANGE

In this section we compute the CTR for connectivity in the proposed model. The CTR for connectivity can be defined as follows.

DEFINITION 1. *Suppose n nodes are distributed in a region R. The Critical Transmission Range (CTR) is the minimum transmission range, r_c, which induces a communication graph with a unique connected component, including all the n nodes of the network.*

Theorem 3 presents the computed CTR for our model, namely r_c, and the critical value ϵ_c, such that $\forall \epsilon \geq \epsilon_c$ we have a scenario where r_c is the correct solution to the CTR problem. That is, r_c from Theorem 3 is the CTR when ϵ is large enough, and the lower bound for ϵ is determined by ϵ_c.

THEOREM 3. *Let $\mathcal{N}$ be a set of n nodes deployed uniformly at random in a lattice square of granularity g in the area $[0, g-1]^2$, with segments' width 2ϵ. The Critical Transmission Range for Connectivity, denoted by r_c is*

$$r_c = \frac{2(g^2-g)\ln\left(\frac{n}{2(g-1)} - g\right)}{n}$$

whenever $\epsilon \geq \epsilon_c$, for

$$\epsilon_c = F^{-1}(q; \mu, r_c), \quad (12)$$

where $q \approx 1$, $\mu = n/(2(g^2-g))$ and $F^{-1}(q; \mu, r_c)$ is the inverse, for the parameter ϵ, of the probability of connectivity at intersections, that is, the inverse of (5).

PROOF. Let us abstract the connectivity problem using a bond percolation model within a random grid of $g \times g$ vertices. Here, each vertex corresponds to an intersection in the lattice square and each edge corresponds to a segment. In the percolation model, each edge occurs with probability p_b.

We know, from Theorem 2, the exact probability of connectivity at segments, which in the abstraction corresponds to p_b. Additionally, from Theorem 1, we know the probability of connectivity at intersections, which in the abstraction corresponds to p_s, which must be close to one. Franceschetti and Meester [8] already studied the rate of convergence of p_b (that depends on g), and found that it must scale slightly faster than $\sqrt{g}$ in order for the random grid to be fully connected w.h.p., which means that $p_b = 1 - C_g/\sqrt{g}$ guarantees connectivity w.h.p. if and only if the sequence $C_g \to 0$ when $g \to \infty$. Then, from Theorem 2 we have

$$p_b = (1-(1-r)^\mu)^{\mu-1}$$

$$\frac{\ln\left(1 - \frac{C_g}{\sqrt{g}}\right)}{\mu - 1} = \ln(1 - (1-r_c)^\mu). \quad (13)$$

Applying Maclaurin series on both sides of (13), we obtain

$$-\frac{C_g}{\sqrt{g}(\mu-1)} + \frac{c_1}{\mu-1} = -(1-r_c)^\mu + c_2,$$

where c_1 and c_2 are the errors of the Maclaurin approximation taking the first term. Considering that c_1 and c_2 are small constants, let us write

$$-\frac{C_g}{\sqrt{g}(\mu-1)} = -(1-r_c)^\mu.$$

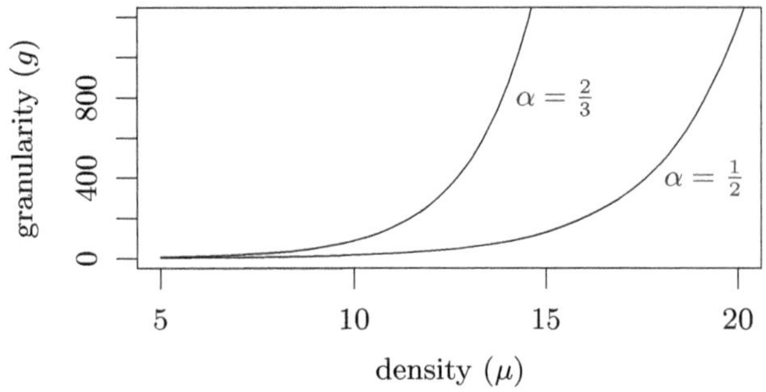

Figure 4: Relation between density and the upper bound for the granularity for two values of $r_c = \alpha$

Applying Maclaurin series in the right side of the expression above and considering $C_g = 1/\sqrt{g}$, we obtain

$$\frac{\ln(g) + \ln(\mu - 1)}{\mu} = r_c + c_3, \tag{14}$$

where c_3 is the error of the Maclaurin approximation. Since μ is the expected quantity of nodes per segment in the lattice square, we have that $\mu = n/(2(g^2 - g))$, then replacing μ in (14) and considering that c_3 is a negligible constant, we have

$$\frac{2(g^2 - g)\ln\left(\frac{n}{2(g-1)} - g\right)}{n} = r_c. \tag{15}$$

Note that Expression (15) corresponds to the CTR whenever connectivity at intersections is guaranteed, i.e., when $\Pr(I_{con}) \approx 1$.

We proceed to determine the critical value ϵ_c, which is critical in the sense that all other $\epsilon \geq \epsilon_c$ produce scenarios that, for the transmission range r_c, lead to connectivity at intersections with probability $q \approx 1$. That is, $F(\epsilon; \mu, r_c) = q$, where $q \approx 1$, $\mu = n/(2(g^2 - g))$ and $F(.) = \Pr(I_{con})$, is the probability of connectivity at intersections, derived in Theorem 1. Then, we have that $\epsilon_c = \min\{\epsilon \mid F(\epsilon; \mu, r_c) \geq q \wedge q \approx 1\}$, or equivalently $\epsilon_c = F^{-1}(q; \mu, r_c)$, where $q \approx 1$. We can think about $F^{-1}(q)$ as a q-quantile function for $F(.)$ with respect to ϵ. □

Analyzing the CTR derived in Theorem 3, we can discover some properties in the model. For instance, considering a desired CTR $r_c = \alpha$, then the granularity is limited by $g \leq e^{\alpha\mu}/(\ln(e^{\mu}) - 1)$. That is, supposing $\alpha = 1/2$ and $\mu = 10$, then the granularity cannot be greater than 16. On the other hand, considering $\mu = 15$, then the granularity can scale up to 129. Figure 4 shows the relation between density and granularity for two values of α.

Additionally, we have the bound ϵ_c, that decreases when r_c and μ grow, being μ the most influencing factor, as we show in Section 5.

In the following section, we present experimental results for validating Lemmas 1 and 2, as well as Theorems 1 and 3.

5. EXPERIMENTAL RESULTS

In this section we validate the results from Sections 3 and 4 through Monte Carlo simulations. In all cases we used the R system* in its version 2.13.1. In each figure, curves and lines represent analytical results, and empirical data is represented by bullets.

*Available in `http://www.R-project.org/`

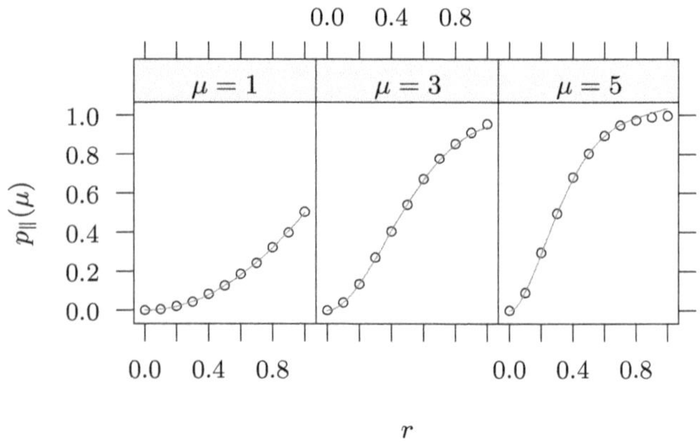

Figure 5: Probability of crossing links between parallel segments sharing an intersection

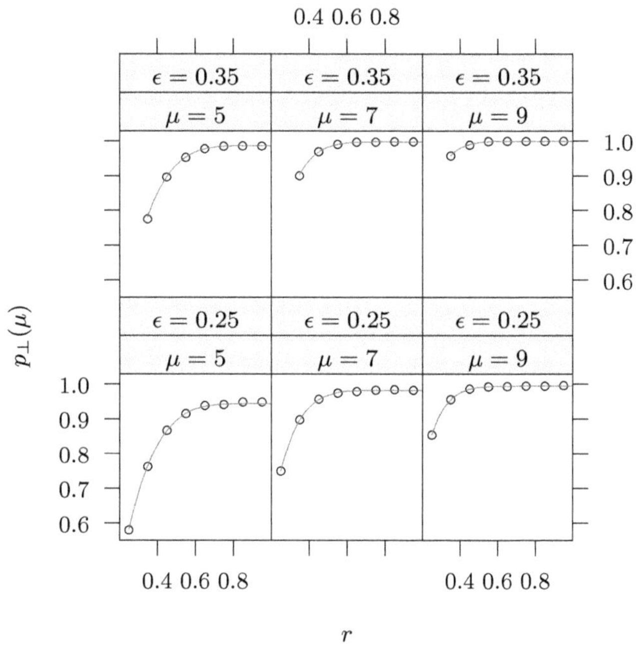

Figure 6: Probability of crossing links between perpendicular segments sharing an intersection

Figure 5 presents the probability of crossing links between parallel segments sharing an intersection in three scenarios, determined by the parameter μ. Here we perceive that the approximation we obtained, using Simpson's rule with two intervals in (2), is accurate.

For the case of crossing links between perpendicular segments, we have selected six scenarios, varying the segment width parameter, ϵ, and the expected number of nodes per segment, μ. Figure 6 presents these results considering the assumption $r \geq \epsilon$ from Lemma 2. It can be seen that the expression for $p_{\perp}(\mu)$ obtained in (3) is correct.

Figure 7 shows the empirical probability of connectivity at intersections against the analytical lower bound derived in Theorem 1. Here, we use the same scenarios as in Figure 6, and we observe that the higher the density of the

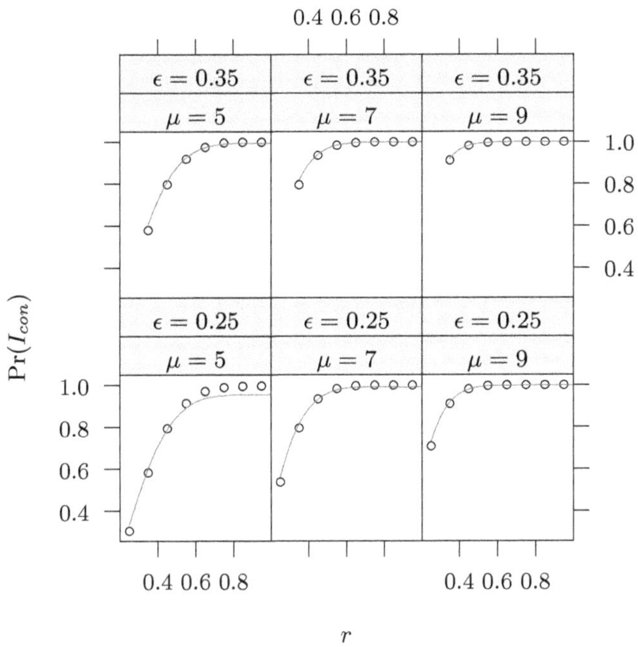

Figure 7: Probability of connectivity at intersection

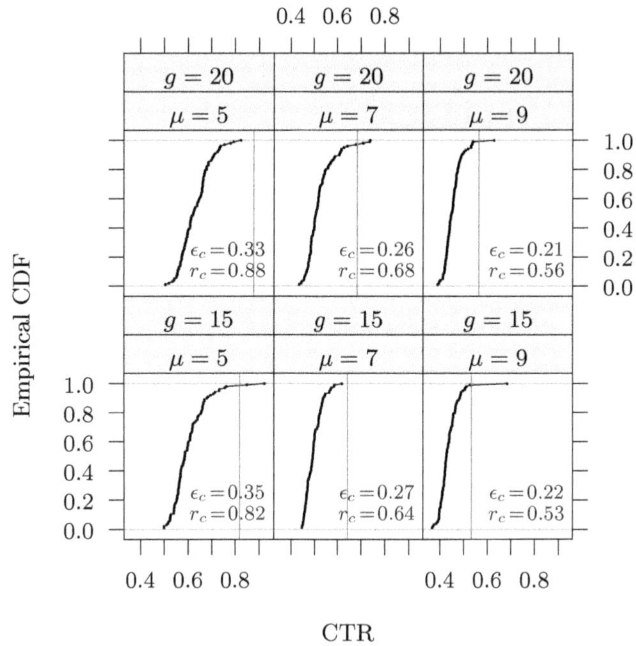

Figure 8: ECDFs and analytical CTRs

network and the greater the segments' width are, the closer the lower bound is to the proper value of $\Pr(I_{con})$. This happens because in these scenarios, crossing links between parallel segments are no longer needed to grant connectivity. On the other hand, when nodes are far from the intersection due to low density (small μ), or when the visibility is restricted (because of a small ϵ), the observed connectivity can be constructed, in a very small proportion, with crossing links between parallel segments.

We conclude this section by presenting experimental results for Theorem 3. Figure 8 shows Empirical Cumulative Distribution Functions (ECDFs) for the CTR considering different levels of granularity and density. Vertical lines in panels correspond to analytical CTRs. In each scenario, we used the critical value ϵ_c, computed numerically a priori, considering the analytical CTR in each case, and these values are reported in each panel. We observe in this figure that the denser the network and the lower the granularity are, the faster the phase transition is. In all cases, we perceive that the derived CTR corresponds indeed to the minimum transmission range that grants connectivity w.h.p., fact that can be confirmed observing intersections between vertical lines, that indicate analytical CTRs, and curves for the ECDFs.

6. RELATED WORK

The analysis performed in this work is closely related to the theory of random geometric graphs [1, 8], which studies properties of non-obstructed networks formed by a set of n points distributed at random in a d-dimensional region, capturing many essential characteristics of homogeneous and dense wireless ad hoc networks. Penrose [9], for instance, uses the fact that the longest edge of the Euclidean minimum spanning tree on n points has the same length as the CTR for connectivity, to analyze connectivity properties in such networks. The exact probability of connectivity in one-dimensional setting was derived in [7].

Several two-dimensional models based on continuum percolation were analyzed in [10]. Dousse et al. [11] analyze hybrid non-obstructed networks comprised by random uniform one-dimensional, two-dimensional, and a fixed-width strip combined with an overlaying connected backbone of base stations. In particular, they show that percolation (formation of a connected component of infinite size) never (for any node density) occurs on a line or a strip, and that gains in terms of connectivity from the hybrid structure are only significant when the node spatial distribution is close to one-dimensional.

Shakkottai et al. [12] analyze connectivity properties in a setting where nodes are regularly placed on a unit square region and a probability $p(n)$ that a node is active (or failed) is associated to each node. The authors argue that the randomness induced by node failure leads to a similar scaling in CTR as randomness due to uniform node placement. Santi and Blough [4] analyze the CTR for connectivity in sparse non-obstructed networks. They show that the communication graph in a region $R = [0, \ell]^d$ is connected w.h.p. if $r^d n = k\ell^d \ln \ell$, for some constant $k > 0$.

Connectivity properties of wireless networks with obstacles have received significantly less attention so far. In [5], Frieze et al. propose a model to represent wireless environments with regularly distributed obstacles. In their model, nodes are placed randomly on an $n \times n$ grid with probability $p(w)$, and each node has a communication range of w unitary blocks. Among other results, Frieze et al. show that the critical node placement probability $p(w)^* = O(\ln n/w)$, i.e., in the interval $O(1/w)$, the network goes from being k-connected with arbitrarily low probability to being k-connected with probability close to 1. In [13], Bollobas et al. extend this result by showing that $\lim_{w\to\infty} wp(w)^* = \log 3/2$. Another example of work on obstructed wireless networks can be seen in [14], which analyzes how the shape of obstacles might impair connectivity of wireless networks. This last result does not directly apply to regularly obstructed networks.

7. CONCLUSIONS

In this work we proposed new analytical tools to study connectivity properties of wireless ad hoc networks impaired by regularly distributed obstacles. We proposed a simple model that can be used to represent a range of network topologies encountered in practice, such as dense urban and indoor environments. In contrast to previously proposed models, this model uses one-dimensional communication segments and two-dimensional segment intersections as building blocks of a network. One important aspect of the model is its flexibility to be applied to different node deployment strategies and the possibility of working with different communication rules, such as L_1, L_2, or L_∞ metrics.

As a first topological characterization, we used a simplified connectivity model (L_∞ metric) to detemine the existence of links at grid intersections. Furthermore, we derived a lower bound for the probability of connectivity at intersections. Finally, we used this lower bound to compute the CTR for connectivity in the overall network. Our result can be compared to that of Santi and Blough [4] for non-obstructed random networks, where they show that the communication graph in a region $R = [0, \ell]^2$ is connected w.h.p. if the CTR $r = O(\ell\sqrt{\ln \ell/n})$. In our model, if the segment width is above a certain critical threshold ϵ_c, the CTR for connectivity must be $r = O((\ln g + \ln \mu)/\mu)$, where g is the granularity of the network (a global parameter, analogous to the parameter ℓ in [4]) and μ is the expected number of nodes in each (one-dimensional) segment (a local property). The critical segment width value ϵ_c is a fundamental property of our model, as it determines when the probability of connectivity at intersection is dominated (approaches to 1) by the probability of connectivity on one-dimensional segments. The lower the granularity of the network, g, and the higher the expected number of nodes at a segment, μ, the smaller is the critical segment width ϵ_c.

This work opens a number threads for future work. Among them, one might consider the calculation of other connectivity properties, such as almost-connectivity, as well as important network properties, such as degree distribution of nodes. Such characterizations can be used in the design of efficient network algorithms aware of topology. Another interesting extension would be to analyze the model in the Euclidean space with obstacles (L_2 metric), and to look at other kinds of regular lattices.

8. ACKNOWLEDGEMENTS

This work was partially supported by CNPq, FAPEAL and FAPEMIG.

9. REFERENCES

[1] M. Penrose, *Random geometric graphs*, ser. Oxford studies in probability. Oxford University Press, 2003.

[2] M. D. Penrose, "On k-connectivity for a geometric random graph," *Random Structures & Algorithms*, vol. 15, no. 2, pp. 145–164, Sep 1999.

[3] P. Gupta and P. R. Kumar, "The capacity of wireless networks," *IEEE Transactions on Information Theory*, vol. 46, no. 2, pp. 388–404, 2000.

[4] P. Santi and D. Blough, "The critical transmitting range for connectivity in sparse wireless ad hoc networks," *IEEE Transactions on Mobile Computing*, vol. 2, no. 1, pp. 25–39, 2003.

[5] A. M. Frieze, J. M. Kleinberg, R. Ravi, and W. Debany, "Line-of-sight networks," in *Proc. SODA*, New Orleans, Jan. 2007, pp. 968–977.

[6] G. Grimmett, *Percolation*, 2nd ed. Springer-Verlag, 1999.

[7] A. Ghasemi and S. Nader-Esfahani, "Exact probability of connectivity in one-dimensional ad hoc wireless networks," *IEEE Communications Letters*, vol. 10, no. 4, pp. 251–253, 2006.

[8] M. Franceschetti and R. Meester, *Random Networks for Communication: From Statistical Physics to Information Systems*, ser. Statistical and Probabilistic Mathematics. Cambridge University Press, 2007.

[9] M. D. Penrose, "The longest edge of the random minimal spanning tree," *The Annals of Applied Probability*, vol. 7, no. 2, pp. 340–361, 1997.

[10] I. Glauche, W. Krause, R. Sollacher, and M. Greiner, "Continuum percolation of wireless ad hoc communication networks," *Physica A: Statistical Mechanics and its Applications*, vol. 325, no. 3–4, pp. 577–600, 2003.

[11] O. Dousse, P. Thiran, and M. Hasler, "Connectivity in ad-hoc and hybrid networks," in *Proc. INFOCOM*, vol. 2, New York, June 2002, pp. 1079–1088.

[12] S. Shakkottai, R. Srikant, and N. Shroff, "Unreliable sensor grids: coverage, connectivity and diameter," in *Proc. INFOCOM*, vol. 2, San Francisco, Mar.–April 2003, pp. 1073–1083.

[13] B. Bollobás, S. Janson, and O. Riordan, "Line-of-sight percolation," *Combinatorics, Probability & Computing*, vol. 18, no. 1–2, pp. 83–106, 2009.

[14] M. Nekoui and H. Pishro-Nik, "Geometrical analysis of obstructed wireless networks." in *IEEE Information Theory Workshop*, 2009.

A Rigorous Analysis of AODV and its Variants

Peter Höfner
NICTA, Australia
University of New South Wales, Australia
Peter.Hoefner@nicta.com.au

Wee Lum Tan
NICTA, Australia
University of Queensland, Australia
WeeLum.Tan@nicta.com.au

Annabelle McIver
Macquarie University, Australia
NICTA, Australia
annabelle.mciver@mq.edu.au

Rob van Glabbeek
NICTA, Australia
University of New South Wales, Australia
rvg@cs.stanford.edu

Marius Portmann
NICTA, Australia
University of Queensland, Australia
marius@itee.uq.edu.au

Ansgar Fehnker
NICTA, Australia
University of New South Wales, Australia
Ansgar.Fehnker@nicta.com.au

ABSTRACT

In this paper we present a rigorous analysis of the Ad hoc On-Demand Distance Vector (AODV) routing protocol using a formal specification in AWN (Algebra for Wireless Networks), a process algebra which has been specifically tailored for the modelling of Mobile Ad Hoc Networks and Wireless Mesh Network protocols. Our formalisation models the exact details of the core functionality of AODV, such as route discovery, route maintenance and error handling. We demonstrate how AWN can be used to reason about critical protocol correctness properties by providing a detailed proof of loop freedom. In contrast to evaluations using simulation or other formal methods such as model checking, our proof is generic and holds for any possible network scenario in terms of network topology, node mobility, traffic pattern, etc. A key contribution of this paper is the demonstration of how the reasoning and proofs can relatively easily be adapted to protocol variants.

Categories and Subject Descriptors

C.2.2 [**Network Protocols**]: Routing protocols; Protocol verification; F.3.1 [**Specifying and Verifying and Reasoning about Programs**]: Invariants

General Terms

Reliability; Theory; Verification

Keywords

AODV; loop freedom; process algebra; routing protocols; wireless mesh networks

MSWiM'12, October 21–25, 2012, Paphos, Cyprus.

1. INTRODUCTION

Routing protocols are crucial to the dissemination of data packets between nodes in Wireless Mesh Networks (WMNs) and Mobile Ad Hoc Networks (MANETs). One of the most popular routing protocols that is widely used in WMNs and MANETs is the Ad hoc On-Demand Distance Vector (AODV) routing protocol [13]. AODV is one of the four protocols currently standardised by the IETF MANET working group, and it also forms the basis of new WMN routing protocols, including HWMP in the upcoming IEEE 802.11s wireless mesh network standard [10]. The details of the AODV protocol are laid out in the RFC 3561 [13]. However, due to the use of English prose, this specification contains ambiguities and contradictions. This can lead to significantly different implementations of the AODV routing protocol, depending on the developer's understanding and reading of the AODV RFC. In the worst case scenario, an AODV implementation may contain serious flaws, such as routing loops.

Traditional approaches to the analysis of AODV and many other AODV-based protocols [5, 10, 16, 18, 15] are simulation and test-bed experiments. While these are important and valid methods for protocol evaluation, in particular for quantitative performance evaluation, there are limitations in regards to the evaluation of basic protocol correctness properties. Experimental evaluation is resource intensive and time-consuming, and, even after a very long time of evaluation, only a finite set of network scenarios can be considered—no general guarantee can be given about correct protocol behaviour for a wide range of unpredictable deployment scenarios [3]. This problem is illustrated by recent discoveries of limitations in AODV-like protocols that have been under intense scrutiny over many years [12]. We believe that formal methods can help in this regard; they complement simulation and test-bed experiments as methods for protocol evaluation and verification, and provide stronger and more general assurances about protocol properties and behaviour.

This paper is based on a complete and accurate formal specification of the core functionality of the AODV routing protocol using the specification language AWN (Algebra of Wireless Networks) [7]. AWN provides the right level of abstraction to model key features such as unicast and broadcast, while abstracting from implementation-related details. As its semantics is completely unambiguous, specifying a protocol in such a framework enforces total precision and the removal of any ambiguities. A key contribution is to demonstrate how AWN can be used to support reasoning about protocol behaviour and to provide a rigorous proof of key protocol properties, using the example of loop freedom. In contrast to what can be achieved, e.g., by model checking, our proofs apply to all conceivable dynamic network topologies.

We analyse different readings of the AODV RFC, and show which interpretations do satisfy the loop freedom criterion, and which do not. We also discuss two limitations of the AODV protocol and propose solutions to them. We show how our formal specification can be used to analyse the proposed modifications and show that these AODV variants are loop free.

The rigorous protocol analysis discussed in this paper has the potential to save a significant amount of time in the development and evaluation of new network protocols, can provide increased levels of assurance of protocol correctness, and complements simulation and other experimental protocol evaluation approaches.

The remainder of this paper is organised as follows. We briefly describe AWN in Section 2, and use it to formally specify AODV in Section 3. We discuss one of several ambiguities and contradictions in the AODV RFC, and propose potential resolutions in Section 4. We then summarise the key points of a detailed proof of loop freedom of AODV in Section 5, and demonstrate how the reasoning and proof can relatively easily be adapted to variants of the AODV protocol in Section 6. We discuss related work in Section 7, and summarise our work in Section 8.

2. AWN—AN ALGEBRA FOR WIRELESS NETWORKS

Process algebras are standard tools to describe interactions, communications and synchronisations between a collection of independent agents, processes or network nodes. They provide algebraic laws that allow formal reasoning. For the specification of and for formal reasoning about AODV, we use AWN [7, 8], a process algebra specifically tailored for WMNs. AWN allows us to write a protocol specification in a simple language, which makes it easy to read and to use. Its key operators are *conditional unicast*—allowing error handling in response to failed communications while abstracting from link layer implementations of the communication handling—and *local broadcast*—allowing a node to send messages to all its immediate neighbours as implemented by the physical and data link layer.

In this section we only give an overview of the main operations (Table 1) and illustrate the use of AWN with a simple example. Additional explanations and a full description can be found in [7, 8].

The example considers a network of two nodes on which the same process is running. One node broadcasts an integer value. A received message will be delivered to the application layer if its value is 1. Otherwise the node decrements the value and broadcasts the new value. The behaviour of each node can be modelled by:

$$\texttt{X(n)} \stackrel{def}{=} \textbf{broadcast}(\texttt{n}).\texttt{Y()}$$
$$\texttt{Y()} \stackrel{def}{=} \textbf{receive}(\texttt{m}).([\texttt{m}{=}1]\ \textbf{deliver}(\texttt{m}).\texttt{Y()} + [\texttt{m}{\neq}1]\ \texttt{X}(\texttt{m}-1))$$

If a node is in a state X(n) it will broadcast n and continue

$X(exp_1, \ldots, exp_n)$	process name with arguments
$P + Q$	choice between processes P and Q
$[\varphi]P$	conditional process; execute P only if condition φ holds
$[\![\texttt{var} := exp]\!]P$	assignment followed by process P
broadcast$(ms).P$	broadcast message ms followed by P
groupcast$(dests, ms).P$	iterative unicast to all destinations $dests$ (if broadcast is inappropriate)
unicast$(dest, ms).P \blacktriangleright Q$	unicast ms to $dest$; if successful proceed with P; otherwise with Q
deliver$(data).P$	deliver data to application layer
receive$(\texttt{msg}).P$	receive a message
$P \Vert Q$	parallel composition of nodes

Table 1: Process expressions

in state Y( ). If a node is in state Y( ), and it receives m, it has two ways to continue. Process [m=1] **deliver**(m).Y( ) is enabled if m=1. In that case m will be delivered to the application layer, and the process returns to Y( ). Alternatively, if m≠1, the process continues as X(m−1). Note that calls to processes use expressions as parameters, in this case m−1.

Assume that the nodes A and B are within communication range of each other; node A in state X(2), and node B in Y( ). Then, node A broadcasts 2 and continues as Y( ). Node B receives 2, and continues as X(1). Next B broadcasts 1, and continues as Y( ), while node A receives 1, and, since the condition m=1 is satisfied, **delivers** 1 and continues as Y( ). This gives rise to transitions from one state to the other:

$$\texttt{X(2)} \Vert \texttt{Y()} \xrightarrow{A:\textbf{broadcast}(2)} \texttt{Y()} \Vert \texttt{X(1)} \xrightarrow{B:\textbf{broadcast}(1)} \xrightarrow{A:\textbf{deliver}(1)} \texttt{Y()} \Vert \texttt{Y()}$$

In state Y( ) ‖ Y( ) no further activity is possible; the network has reached a *deadlock*.

3. A FORMAL SPECIFICATION OF AODV

AODV is a reactive protocol, which means that routes are only established on demand. If a node S wants to send a data packet to node D, but currently does not know a route, it buffers the packet and initiates a route discovery process by broadcasting a route request (RREQ) message in the network. An intermediate node A that receives this message creates a routing table entry for a route towards S, referred to as a *reverse route*, and re-broadcasts the RREQ. This is repeated until the RREQ reaches the destination D, or alternatively a node with a route to D. In both cases, the node replies by unicasting a route reply (RREP) back to the source S, via the previously established reverse route. When forwarding RREP messages, a node creates a routing table entry for node D, called the *forward route*. When the RREP reaches S, a route between S and D is established and data packets can start to flow. In the event of link and route breaks, AODV uses route error messages (RERR) to notify the affected nodes. AODV uses sequence numbers to indicate the freshness of routes and to avoid routing loops. Full details are given in [13].

3.1 Modelling AODV

We present a model of AODV using AWN. The formalisation is a faithful representation of the core functionality of AODV as defined in [13]. We currently do not model optional features such as local route repair, expanding ring search, gratuitous route reply and multicast. We also abstract from all timing issues, since AWN currently does not

Process 1 The basic routine

```
AODV(ip,sn,rt,rreqs,store) ≝
 1.  receive(msg) .
 2.  /* depending on the message, different processes are called */
 3.  (
 4.    [msg = newpkt(data,dip)] /* new DATA packet */
 5.      PKT(data,dip,ip , ip,sn,rt,rreqs,store)
 6.    +[msg = pkt(data,dip,oip)] /* incoming DATA packet */
 7.      PKT(data,dip,oip , ip,sn,rt,rreqs,store)
 8.    +[msg = rreq(hops,rreqid,dip,dsn,oip,osn,sip)] /*RREQ*/
 9.      /* update the route to sip in rt */
10.      [[rt := update(rt,(sip, 0, val, 1, sip, ∅))]]
11.      RREQ(hops,rreqid,dip,dsn,oip,osn,sip , ip,sn,rt,rreqs,store)
12.    +[msg = rrep(hops,dip,dsn,oip,sip)] /* RREP */
13.      /* update the route to sip in rt */
14.      [[rt := update(rt,(sip, 0, val, 1, sip, ∅))]]
15.      RREP(hops,dip,dsn,oip,sip , ip,sn,rt,rreqs,store)
16.    +[msg = rerr(dests,sip)] /* RERR */
17.      /* update the route to sip in rt */
18.      [[rt := update(rt,(sip, 0, val, 1, sip, ∅))]]
19.      RERR(dests,sip , ip,rt,sn,rreqs,store)
20.  )
21. +[Let dip ∈ qD(store) ∩ vD(rt)] /* send a queued data packet */
22.  . . .
```

support time. In concrete terms, this means that the AODV timing parameters `ACTIVE_ROUTE_TIMEOUT`, `DELETE_PERIOD` and `PATH_DISCOVERY_TIME` are set to infinity.

In addition to modelling the complete set of core functionalities of the AODV protocol, our model also covers the interface to higher protocol layers via the injection and delivery of application layer data, as well as the forwarding of data packets at intermediate nodes. Although this is not part of the AODV protocol specification, it is necessary for a practical model of any reactive routing protocol, where protocol activity is triggered via the sending and forwarding of data packets.

Our AODV model consists of the following six processes:

- `AODV`, the main process, reads a message from the message queue (Line 1 of Process 1) and calls the appropriate process `PKT`, `RREQ`, `RREP`, or `RERR` to handle it (Lines 4–19). The process also handles the forwarding of any queued data packets if a valid route to their destination is known (Lines 21 ff.).
- `PKT` deals with received data packets, including forwarding if a route exists, and sending an error message if the route is broken. If the data packet originates at the local node and no route to the destination exists, the process buffers the data packet and initiates a new route discovery process.
- `RREQ` deals with received RREQ messages, and will be discussed in detail below.
- `RREP` deals with received RREP messages, including the updating of routing tables and handling of errors.
- `RERR` models the processing of AODV error messages.
- `QMSG` describes the general handling of incoming AODV messages: whenever a message is received, it is first stored in a FIFO queue. As soon as the corresponding node is able to handle a message it retrieves the oldest message from the queue and handles it.

Each node in an AODV network maintains a *routing table* to keep track of the node's routing information collected so far. A routing table consists of sets of entries of the form (*dip*, *dsn*, *flag*, *hops*, *nhip*, *pre*), with *dip* being the node identifier (typically IP address) of the ultimate destination node, and *dsn* the destination sequence number, which represents the "freshness" of this routing table entry. The *flag* parameter indicates whether an entry is valid or invalid, and *hops* represents the distance to the destination node *dip* in number of hops. *nhip* identifies the next hop node along the route to node *dip*, and *pre* is the set of *precursors*—nodes that "rely" on this routing table entry for their own routes. Following [13], a routing table entry would also contain a *sequence-number-status flag*. In the present paper we abstract from this flag, since (a) the main results are independent of the existence of the flag, and (b) none of the common implementations (AODV-UU [2], Kernel-AODV [1], AODV-UIUC [11], AODV-UCSB [6], AODV-ns2[1]) maintains this flag.[2] Hence the specification here follows the implementations available.

In a routing table *rt* there is at most one entry for each destination *dip*; $\mathtt{sqn}(rt,dip)$ denotes the sequence number of that entry and $\mathtt{flag}(rt,dip)$, $\mathtt{dhops}(rt,dip)$ and $\mathtt{nhop}(rt,dip)$ its validity, hop count and next hop. Furthermore the sets $\mathtt{kD}(rt)$ and $\mathtt{vD}(rt)$ of destinations contain all entries of *rt* for which there is an (arbitrary) entry or a valid entry, resp. The function `update` updates a routing table *rt* with an entry *r*, which is one of the major activities of `AODV`:

$$\mathtt{update}(rt,r) := \begin{cases} rt \cup \{r\} & \text{if } \pi_1(r) \notin \mathtt{kD}(rt) \quad \textit{//r is new} \\ nrt \cup \{nr\} & \text{if } \mathtt{sqn}(rt,\pi_1(r)) < \pi_2(r) \quad \textit{//fresher} \\ nrt \cup \{nr\} & \text{if } \mathtt{sqn}(rt,\pi_1(r)) = \pi_2(r) \\ & \quad \wedge\, \mathtt{dhops}(rt,\pi_1(r)) > \pi_4(r) \quad \textit{//shorter} \\ nrt \cup \{nr\} & \text{if } \mathtt{sqn}(rt,\pi_1(r)) = \pi_2(r) \quad \textit{//replaces} \\ & \quad \wedge\, \mathtt{flag}(rt,\pi_1(r)) = \mathtt{inv} \quad \textit{invalid} \\ nrt \cup \{nr'\} & \text{if } \pi_2(r) = 0 \quad \textit{//unk. sqn} \\ nrt \cup \{ns\} & \text{otherwise ,} \end{cases}$$

where the projections π_1, π_2 and π_4 select the respective component from an entry, namely the destination, the destination sequence number and the hop count. *s* is the current entry in *rt* for destination $\pi_1(r)$ (if it exists); and $nrt := rt - s$ removes *s* from *rt*. The entry *nr* is identical to *r* except that the precursors from the corresponding routing table entry are added and *ns* is generated from *s* by adding the precursors of *r*. The entry nr' is identical to *nr* except that the sequence number is replaced by the one from the routing table (route *s*).

If a route is not valid any longer, instead of deleting it, AODV sets its validity flag to invalid. This way, the stored information on the route, such as the sequence number and hop count, remains accessible. We model route invalidation by a function `invalidate` whose arguments are a routing table and a set *dests* of pairs (*rip*, *rsn*) of a destination *rip* to be invalidated, and the sequence number of the invalidated routing table entry. Normally, *rsn* is obtained by incrementing the last known sequence number of the route.

In our formalisation, a route request message has the form $\mathtt{rreq}(hops, rreqid, dip, dsn, oip, osn, sip)$, where *hops* is the number of hops the RREQ has already travelled from its origin *oip*, and *rreqid* (in combination with *oip*) is a unique identifier of the message. *dip* is the destination node identifier (IP address) of the route request and *dsn* the last known corresponding sequence number. The parameter *oip* is the

[1] `www.auto-nomos.de/ns2doku/aodv_8cc-source.html`

[2] Kernel-AODV implements the flag, but does not use it.

address of the originator of the route request and *osn* is its sequence number. Finally, *sip* represents the sender IP address, i.e., the address of the intermediate node from which the request was received. Any node forwarding such a message updates *sip* with its own address, increments *hops*, and retains all other parameters. A reply to such a message has the form **rrep**(*hops*,*dip*,*dsn*,*oip*,*sip*), where *dip* and *oip* are copied from the corresponding RREQ message and *hops* is the distance from *dip* to *sip*. The processes **RREQ** and **RREP** that handle incoming RREQ and RREP messages maintain variables `dip`, `oip`, etc. to store the values of the parameters of these messages, as summarised below.

Variables	Used for
ip	address of current node
dip	destination address
oip	originator of a route request or data packet
rip	destination of invalid route
sip	sender of AODV control message
nhip	next hop towards some destination

The process **AODV**, specified by Process 1, deals with the message handling of the node. It stores its own address in the variable `ip`, its own sequence number in `sn`, manages its routing table `rt`, records all route requests seen so far in `rreqs` and maintains in `store` data packets to be sent. Initially, `rt`, `rreqs` and `store` are set to empty, and `sn` to 1.

3.2 Route Request Handling

In this paper, we discuss only the model of the RREQ process; see [8] for a complete model of all AODV processes.[3]

A route discovery in AODV is initiated by a source node broadcasting a RREQ message. Process 2 shows our process algebra specification of the handling of a RREQ message received by a node *ip*.

If the RREQ with the same *oip* and *rreqid* has been seen previously by the node, it is ignored, and we go back to the main AODV process (Lines 1–2). If the RREQ is new (Line 3), we update the routing table by adding a "reverse route" entry to *oip*, the originator of the RREQ, via node *sip*, with distance *hops*+1 (Line 5). If there already is a route to *oip* in the node's routing table *rt*, it is only updated with the new route, if the new route is "better", i.e., fresher and/or shorter and/or replacing an invalid route (cf. Section 3.1). The process also adds the message to the list of known RREQs (Line 7).

Lines 9–20 deal with the case where the node receiving the RREQ is the intended destination, i.e., *dip*=*ip* (Line 9). In this case, a RREP message needs to be sent to the originating node *oip*. According to the AODV RFC, the node's sequence number is set to the maximum of the node's current sequence number and the destination sequence number (*dsn*) in the RREQ message (Line 10).

The RREP message is initialised as follows: hop count (*hops*) is set to 0, the destination (*dip*) and originator (*oip*) are copied from the corresponding RREQ message and the destination's sequence number is the node's sequence number `sn`. Of course, the sender's IP address (*sip*) is set to the node's ip (Line 12). The RREP message is unicast to the next hop along the reverse route back to the originator of the corresponding RREQ message, and if this is successful, the process goes back to the AODV process (Line 13).

[3]There, the sequence-number-status flag is modelled as well.

Process 2 RREQ handling

```
RREQ(hops,rreqid,dip,dsn,oip,osn,sip,ip,sn,rt,rreqs,store) ≝
1.  [(oip,rreqid) ∈ rreqs] /* the RREQ has been handled before */
2.    AODV(ip,sn,rt,rreqs,store) /* silently ignore RREQ */
3.  +[(oip,rreqid) ∉ rreqs] /* the RREQ is new to this node */
4.    /* update the route to oip in rt */
5.    ⟦rt := update(rt,(oip,osn,val,hops + 1,sip,∅))⟧
6.    /* update rreqs by adding (oip,rreqid) */
7.    ⟦rreqs := rreqs ∪ {(oip,rreqid)}⟧
8.    (
9.      [dip = ip] /* this node is the destination node */
10.       ⟦sn := max(sn,dsn)⟧ /* update the sqn of ip */
11.       /* unicast a RREP towards oip of the RREQ */
12.       unicast(nhop(rt,oip),rrep(0,dip,sn,oip,ip)) .
13.         AODV(ip,sn,rt,rreqs,store)
14.       ► /* if transmission fails, a RERR is generated */
15.         ⟦dests := {(rip,inc(sqn(rt,rip)))|rip ∈ vD(rt) ∧
                    nhop(rt,rip) = nhop(rt,oip)}⟧
16.         ⟦rt := invalidate(rt,dests)⟧
17.         ⟦pre := ⋃{precs(rt,rip) | (rip,*) ∈ dests}⟧
18.         ⟦dests := {(rip,rsn) | (rip,rsn) ∈ dests ∧
                    precs(rt,rip) ≠ ∅}⟧
19.         groupcast(pre,rerr(dests,ip)) .
20.         AODV(ip,sn,rt,rreqs,store)
21.     +[dip ≠ ip] /* this node is not the destination node */
22.       (
23.         /* valid route to dip that is fresh enough */
24.         [dip ∈ vD(rt) ∧ dsn ≤ sqn(rt,dip) ∧ sqn(rt,dip) ≠ 0]
25.           /* update rt by adding precursors */
26.           ⟦rt := addpreRT(rt,dip,{sip})⟧
27.           ⟦rt := addpreRT(rt,oip,{nhop(rt,dip)})⟧
28.           /* unicast a RREP towards the oip of the RREQ */
29.           unicast(nhop(rt,oip),
                 rrep(dhops(rt,dip),dip,sqn(rt,dip),oip,ip)) .
30.             AODV(ip,sn,rt,rreqs,store)
31.           ► /* if transmission fails, a RERR is generated */
32.             ⟦dests := {(rip,inc(sqn(rt,rip))) | rip ∈ vD(rt) ∧
                        nhop(rt,rip) = nhop(rt,oip)}⟧
33.             ⟦rt := invalidate(rt,dests)⟧
34.             ⟦pre := ⋃{precs(rt,rip) | (rip,*) ∈ dests}⟧
35.             ⟦dests := {(rip,rsn) | (rip,rsn) ∈ dests ∧
                        precs(rt,rip) ≠ ∅}⟧
36.             groupcast(pre,rerr(dests,ip)) .
37.             AODV(ip,sn,rt,rreqs,store)
38.         /* no valid route that is fresh enough */
39.         +[dip ∉ vD(rt) ∨ sqn(rt,dip) < dsn ∨ sqn(rt,dip) = 0]
40.           /* no further update of rt */
41.           broadcast(rreq(hops + 1,rreqid,dip,
                 max(sqn(rt,dip),dsn),oip,osn,ip)) .
42.           AODV(ip,sn,rt,rreqs,store)
43.       )
44.   )
```

If the unicast of the RREP fails, we proceed with Lines 14–20, in which a route error (RERR) message is generated and sent. This conditional unicast is implemented in our model with the AWN construct **unicast**(*dest*, *ms*).*P* ► *Q* (Lines 12ff.) We assume that, as is the case for relevant wireless technologies such as IEEE 802.11, unicast messages are acknowledged, and we therefore can determine whether the transmission was unsuccessful and the link to the next node towards *oip* is broken. In this case, the node sends a RERR message to all nodes that rely on the broken link for one of their routes. For this, we first determine which destination nodes are affected by the broken link, i.e., the nodes that have this unreachable node listed as a next hop in the routing table (Line 15). Here, the operator `inc` increments the sequence numbers of those entries. Then, we invalidate any affected routing table entries (Line 16), and determine the list of *precursors*, which are the neighbouring nodes that have a route to one of the affected destination nodes via the broken link (Line 17). Finally, using the AWN **groupcast** primitive, a RERR message is sent via unicast to all these precursors (Line 19), listing only those invalidated destinations with a non-empty set of precursors (Line 18).

Lines 21–42 deal with the case where the node receiving the RREQ is not the destination, i.e., $dip \neq ip$ (Line 21). The node can respond to the RREQ with a corresponding RREP on behalf of the destination node *dip*, if its route to *dip* is "fresh enough" (Line 24). This means that (a) the node has a valid route to *dip*, (b) the destination sequence number in the node's current routing table entry ($\mathtt{sqn}(rt,dip)$) is greater than or equal to the requested sequence number to *dip* in the RREQ message, and (c) the sequence number is valid, i.e., it is not unknown ($\mathtt{sqn}(rt,dip) \neq 0$). If these three conditions are met (Line 24), the node generates a RREP message and unicasts it back to the originator node *oip* via the reverse route. To this end, it copies the sequence number for the destination *dip* from the routing table *rt* into the destination sequence number field of the RREP message and it places its distance in hops from the destination ($\mathtt{dhops}(rt, dip)$) in the corresponding field of the new reply (Line 29). As usual, the unicast might fail, which causes the same error handling (Lines 32–35). Just before unicasting the RREP message, the intermediate node updates the forward routing table entry to *dip* by placing the last hop node (*sip*) into the precursor list for that entry (Line 26). Likewise, it updates the reverse routing table entry to *oip* by placing the first hop $\mathtt{nhop}(rt, dip)$ towards *dip* in the precursor list for that entry (Line 27).

If the node is not the destination and there is either no route to the destination *dip* inside the routing table or the route is not fresh enough, the route request received has to be forwarded. This happens in Line 41. The information inside the forwarded request is mostly copied from the request received. Only the hop count is increased by 1 and the destination sequence number is set to the maximum of destination sequence number in the RREQ packet and the the current sequence number for *dip* in the routing table. In case *dip* is an unknown destination, $\mathtt{sqn}(rt, dip)$ returns the unknown sequence number 0.

4. AMBIGUITIES IN THE RFC

The formal specification of AODV, outlined above and given in full detail in [8], closely follows the RFC 3561 [13], the official specification of the protocol. However, the RFC contains several ambiguities and contradictions; an inventory is presented in [8], and for each ambiguity or contradiction a number of ways to resolve them is listed. An *interpretation* of the RFC is given by the allocation of a resolution to each of the ambiguities and contradictions. Each reading, implementation, or formal analysis of AODV must pertain to one of its interpretations. The formal specification of AODV in [8] constitutes one interpretation; the inventory of ambiguities and contradictions is formalised by specifying each resolution of each of the ambiguities and contradictions as a modification of this formal specification, typically involving a rewrite of a few lines of code only.

A crucial contradiction in the RFC concerns the question of what would happen if a node has a valid routing table entry for a destination D, with destination sequence number n, and an error message is received from the next hop towards D, saying that the route to D is broken, and stating for this route a destination sequence number m, which may be smaller than n. Section 6.11 of the RFC unambiguously states that in such a case the node updates its routing table entry to D by marking the route as invalid, and *copying* the destination sequence number from the incoming route error message. However, Section 6.1 of the RFC states that if $m < n$, any information related to D in the AODV message must be discarded.

One can show [8] that in case no node will ever store a routing table entry to itself (a self-entry), the above situation will never occur. However, the RFC does not explicitly exclude self-entries[4], and they can in fact occur [8] in response to the standard handling of RREP messages.

The following ways to resolve this contradiction have been listed in [8]:

(a) Follow Section 6.11 of the RFC, in defiance of 6.1, i.e., *always* invalidate the routing table entry, and copy the destination sequence number from the error message to the corresponding entry in the routing table.[5]

(b) Follow Section 6.11 only where it does not contradict 6.1, i.e., invalidate the routing table entry and copy the destination sequence number *only* if $m \geq n$.

(c) Always invalidate the routing table entry, but update the destination sequence number to $\max(m, n)$.

(d) Always invalidate the routing table entry, but update the destination sequence number to $\max(m, n{+}1)$.

(e) Invalidate the routing table entry and update the destination sequence number to $\max(m, n{+}1)$ only if $m \geq n$.[6]

(f) Invalidate the routing table entry only if $m > n$.[7]

(g) Forbid self-entries; if an incoming RREP message would create a self-entry, discard that message.

(h) Forbid self-entries; if an incoming RREP message would create a self-entry, forward that message without updating the node's routing table.

It should be noted that only resolutions (a) and (b) are compliant with the RFC. However, in [9] we have shown that any interpretation based on resolutions (a) or (b) gives rise to routing loops, so in order to arrive at a loop-free version of AODV, one has to deviate from the RFC. Here, as in [8], we do so by choosing resolution (f).

The above is only one of many ambiguities; another one is presented in Section 6.1.

5. LOOP FREEDOM

We now formalise loop freedom and sketch a proof that our detailed specification of AODV cannot create routing loops. We also show how such a formal proof can form a baseline for evaluating variants of AODV—some of them will be loop free, others can yield loops.

First we formalise what it means for the routing tables established by AODV (our specification) to be free of loops. Let **IP** be the set of network nodes and $dip \in \mathbf{IP}$ a particular destination; let N be a state of the network, encompassing the current values of all variables maintained by all nodes.

[4]The Kernel-AODV, AODV-UIUC, AODV-UCSB and AODV-ns2 implementations allow self-entries to occur.

[5]It could be argued that this is not a reasonable interpretation of the RFC, since Section 6.1 should have priority over 6.11. However, this priority is not explicitly stated.

[6]The case $\max(m, n)$ if $m \geq n$ need no separate consideration, since it is equivalent to (b).

[7]Here, it does not matter whether we copy, take $\max(m, n)$ or $\max(m, n{+}1)$; they are all equivalent.

The *routing graph* $\mathcal{R}_N(dip)$ for destination *dip* in state N is the directed graph $(\mathbf{IP}, E)$ with set of vertices **IP** and set of edges $E \subseteq \mathbf{IP}\times\mathbf{IP}$ consisting of the pairs (ip, ip') such that $ip \neq dip$ and $(dip, *, \texttt{val}, *, ip', *)$ occurs in the routing table of *ip* in state N. Thus, there is an edge (ip, ip') if node *ip* is not the destination *dip*, but has a valid entry for *dip*, and *ip'* is the next hop according to that entry. Loops in directed graphs are defined to be paths following edges which return to a vertex.

A (network) state N is *loop free* if the routing graphs $\mathcal{R}_N(dip)$ are loop free for all $dip \in \mathbf{IP}$. The specification of AODV is *loop free* iff all reachable states are loop free.

Let us now turn towards a proof of loop freedom for AODV. It relies on a number of *invariants*—statements that hold for all reachable states of our model. An invariant is usually verified by showing that it holds for all possible initial states, and that, for any transition $N \xrightarrow{\ell} N'$ derived by our operational semantics [8], if it holds for state N then it also holds for state N', reached after performing some action ℓ. These transitions can be traced back to the line numbers in our process declarations `AODV`, `PKT`, `RREQ`, `RREP`, `RERR` and `QMSG`.

A proof of AODV's loop freedom using invariants has first been proposed in [4].[8] The main invariant of [4] states that *if node ip has a routing table entry for destination dip with next hop nhip, then also node nhip has a routing table entry for dip, and the latter has a larger destination sequence number, or an equal one with a strictly smaller hop count.*

This invariant is claimed to hold regardless whether the routing table entries for *dip* at *ip* and *nhip* are marked as valid or invalid. Nevertheless, the following example shows that it does not hold for the current version of AODV.

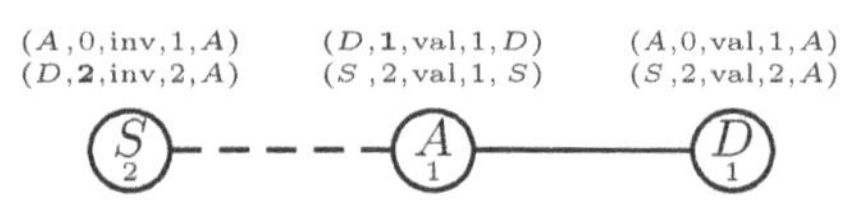

Figure 1: Sequence numbers can go down[9]

The network state depicted in Figure 1 can occur when, after a standard RREQ-RREP cycle, a link break is detected. We assume a simple linear topology of 3 nodes. Below the node names, the circles list the nodes' own sequence numbers, which we assume to be 1 at the initial state. The example starts with empty routing tables and node S searching for a route to node D. Before broadcasting a RREQ message, node S increments its sequence number by 1. Due to the successful exchange of RREQ-RREP messages, all routing tables are updated. After that, the link between nodes S and A goes down, S detects the link break and updates its routing table: it sets all entries in its routing table with next hop A to *invalid* and increases the destination sequence number of the route to D to 2; the destination sequence number in the route to A is unknown (0) and hence not incremented. Now the destination sequence number on the route to node D is 2 in the routing table of S and 1 inside A's routing table, hence this number can go down.

[8] In fact, the same idea occurs already in [14], but without the formalisation in terms of invariants. However, that proof fails to consider some cases that do occur in AODV and might yield routing loops [9].

[9] We omit the precursor set; the routing table entries in each node are represented as (*dip, dsn, flag, hops, nhip*), described in Section 3.1.

One way to avoid this problem is to claim the invariant only for the case that the routing table entries at *ip* and *nhip* are both marked as valid. This is what we do in Theorem 5.4 below, and it suffices to obtain loop freedom of AODV. However, in order to prove Theorem 5.4, we need an invariant that also takes invalid routing table entries into account (cf. Proposition 5.3), so it is not possible to avoid the above problem altogether.

To compensate for the increase of a sequence number in case of route invalidation, we introduce the concept of a *net sequence number* of a route to *dip* according to the routing table of node *ip* in state N, which combines "freshness" and validity:

$$\texttt{nsqn}_N^{ip}(dip) := \begin{cases} \texttt{sqn}_N^{ip}(dip) & \text{if } \texttt{flag}_N^{ip}(dip) = \texttt{val} \vee \texttt{sqn}_N^{ip}(dip) = 0 \\ \texttt{sqn}_N^{ip}(dip) - 1 & \text{otherwise .} \end{cases}$$

Here, we write $\texttt{sqn}_N^{ip}(dip)$ for $\texttt{sqn}(rt,\texttt{dip})$ in case *rt* happens to be the routing table maintained by a node with IP address *ip* in state N of the network. Likewise $\texttt{flag}_N^{ip}(dip)$ denotes the validity of the route from *ip* to *dip* according to the routing table of *ip* in state N, $\texttt{dhops}_N^{ip}(dip)$ its hop count, and $\texttt{nhop}_N^{ip}(dip)$ its next hop. Furthermore $\texttt{kD}_N^{ip}$, abbreviating $\texttt{kD}(rt)$, is the set of destinations for which there is a valid entry in the routing table of *ip*.

In this section we state the key theorems and sketch some of the proofs; all details can be found in [8]. In particular, we show only proofs w.r.t. Process 2, the RREQ handling, and the displayed portion of Process 1.

Proposition 5.1. *If a route request is sent (forwarded) by a node ip_c different from the originator of the request then the content of ip_c's routing table must be fresher or at least as good as the information inside the message.*

$$\begin{aligned} &N \xrightarrow{R:\textit{broadcast}(\texttt{rreq}(hops_c,*,*,*,oip_c,osn_c,ip_c))} N' \wedge ip_c \neq oip_c \\ &\Rightarrow oip_c \in \texttt{kD}_N^{ip_c} \wedge \big(\texttt{sqn}_N^{ip_c}(oip_c) > osn_c \vee (\texttt{sqn}_N^{ip_c}(oip_c) = osn_c \\ &\quad \wedge \texttt{dhops}_N^{ip_c}(oip_c) \leq hops_c \wedge \texttt{flag}_N^{ip_c}(oip_c) = \texttt{val})\big) \end{aligned}$$

The next proposition says that in the routing table of a node *ip*, an entry to a destination *dip* will never be deleted, and the net sequence number of the entry will never go down.

Proposition 5.2. *Let $ip, dip \in \mathbf{IP}$, and assume $N \xrightarrow{\ell} N'$, i.e. the network proceeds from state N to state N' by the occurrence of some action ℓ. If $dip \in \texttt{kD}_N^{ip}$ then $dip \in \texttt{kD}_{N'}^{ip}$, and $\texttt{nsqn}_N^{ip}(dip) \leq \texttt{nsqn}_{N'}^{ip}(dip)$.*

Proof. In our model of AODV, the only way to change a routing table is by means of the operations `update`, `addpreRT` and `invalidate`. None of them ever deletes an entry altogether, or decreases its net sequence number. □

In this proof it is essential that we use resolution (f) of the ambiguity in the RFC presented in Section 4. Proposition 5.2 would not hold under resolutions (a), (b) or (c).

Proposition 5.3. *If, in a reachable network state N, a node $ip \in \mathbf{IP}$ has a routing table entry to dip, then also the next hop nhip towards dip, if not dip itself, has a routing table entry to dip, and the net sequence number of the latter entry is at least as large as that of the former.*

$$\begin{aligned} &dip \in \texttt{kD}_N^{ip} \;\wedge\; nhip \neq dip \\ &\Rightarrow dip \in \texttt{kD}_N^{nhip} \;\wedge\; \texttt{nsqn}_N^{ip}(dip) \leq \texttt{nsqn}_N^{nhip}(dip) \;, \end{aligned}$$

where $nhip := \texttt{nhop}_N^{ip}(dip)$ is the IP address of the next hop.

PROOF. In the initial state, the invariant holds since all routing tables are empty (cf. Section 3.1). Next we assume that the property holds and check each line in Pro. 1 and Pro. 2 which could invalidate it.

A modification of the routing table of *nhip* is harmless, as it can only increase $\texttt{kD}_N^{nhip}$ as well as $\texttt{nsqn}_N^{nhip}(dip)$ (cf. Proposition 5.2).

Adding precursors to routes of *ip* does not harm since the invariant does not depend on precursors. It remains to examine all calls of `update` and `invalidate` to the routing table of *ip*. Without loss of generality we restrict attention to those applications of `update` or `invalidate` that actually modify the entry for *dip*, beyond its precursors; if `update` only adds some precursors in the routing table, the invariant—which is assumed to hold before—is maintained.

Pro. 1, Lines 10, 14, 18: The entry $(\texttt{sip}, 0, \texttt{val}, 1, \texttt{sip}, \emptyset)$ is used for the update; its destination is $dip := \texttt{sip}$. We assume this entry is actually inserted in the routing table of *ip*. Since $dip = \texttt{sip} = \texttt{nhop}_N^{ip}(dip) = nhip$, the antecedent of the invariant to be proven is not satisfied.

Pro. 2, Line 5: The entry $(\texttt{oip}, \texttt{osn}, \texttt{val}, \texttt{hops}{+}1, \texttt{sip}, *)$ is used for the update; again we assume it is inserted into the routing table of node *ip*. So $dip := \texttt{oip}$, $nhip := \texttt{sip}$, $\texttt{nsqn}_N^{ip}(dip) := \texttt{osn}$ and $\texttt{dhops}_N^{ip}(dip) := \texttt{hops}{+}1$. This information is distilled from a received route request message (cf. Lines 1 and 8 of Pro. 1). By Proposition 7.1 of [8], this message was sent before, say in state N'; by Proposition 7.8 of [8], the sender of this message has identified itself correctly, and is `sip`.

By Proposition 5.1, with $ip_c{:=}\texttt{sip}{=}nhip$, $oip_c{:=}\texttt{oip}{=}dip$, $osn_c := \texttt{osn}$ and $hops_c := \texttt{hops}$, and using that $ip_c = nhip \neq dip = oip_c$, we get that $dip \in \texttt{kD}_{N'}^{nhip}$ and

$$\texttt{sqn}_{N'}^{nhip}(dip) = \texttt{sqn}_{N'}^{ip_c}(oip_c) > osn_c = \texttt{osn} \text{ , or}$$
$$\texttt{sqn}_{N'}^{nhip}(dip) = \texttt{osn} \wedge \texttt{flag}_{N'}^{nhip}(dip) = \texttt{val} \text{ .}$$

We first assume that the first line holds. Then, by Proposition 5.2,

$$\texttt{nsqn}_N^{nhip}(dip) \geq \texttt{nsqn}_{N'}^{nhip}(dip) \geq \texttt{sqn}_{N'}^{nhip}(dip){-}1 \geq \texttt{osn} = \texttt{nsqn}_N^{ip}(dip) \text{ .}$$

We now assume the second line to be valid. From this we conclude

$$\texttt{nsqn}_N^{nhip}(dip) \geq \texttt{nsqn}_{N'}^{nhip}(dip) = \texttt{sqn}_{N'}^{nhip}(dip) = \texttt{osn} = \texttt{nsqn}_N^{ip}(dip) \text{ .}$$

Pro. 2, Lines 16, 33: In these applications of `invalidate`, the next hop *nhip* is not changed. Since the invariant has to hold before the execution, it follows that $dip \in \texttt{kD}_N^{nhip}$ also holds after execution. Furthermore, in view of Lines 15 and 32, the route is invalidated while the destination sequence number is incremented. For this reason the net sequence number stays the same, and the invariant is maintained. □

THEOREM 5.4. *If, in a state N, a node ip∈***IP** *has a valid entry to dip, and the next hop is not dip and has a valid dip-entry as well, then the latter entry has a larger destination sequence number or an equal one with a smaller hop count.*

$$dip \in \texttt{vD}_N^{ip} \cap \texttt{vD}_N^{nhip} \;\wedge\; nhip \neq dip$$
$$\Rightarrow \texttt{sqn}_N^{nhip}(dip) > \texttt{sqn}_N^{ip}(dip) \vee \big(\texttt{sqn}_N^{nhip}(dip) = \texttt{sqn}_N^{ip}(dip) \wedge \texttt{dhops}_N^{nhip}(dip) < \texttt{dhops}_N^{ip}(dip)\big) \text{ ,}$$

where $nhip := \texttt{nhop}_N^{ip}(dip)$*, the next hop in the routing table entry at ip for the route to dip.*

The proof [8] is similar to the previous one, but makes use of Proposition 5.3—including the case where $dip \notin \texttt{vD}_N^{nhip}$—in an essential way.

From Theorem 5.4, we can conclude

THEOREM 5.5. *AODV is loop free.*

PROOF. If there were a loop in a routing graph $\mathcal{R}_N(dip)$, then for any edge $(ip, nhip)$ on that loop one would have $\texttt{sqn}_N^{ip}(dip) \leq \texttt{sqn}_N^{nhip}(dip)$, by Theorem 5.4. Hence the value of $\texttt{sqn}_N^{ip}(dip)$ is the same for all nodes *ip* on the loop. Thus, by Theorem 5.4, the sequence numbers keep decreasing when travelling around the loop, which is impossible. □

6. ANALYSING VARIANTS OF AODV

In this section, we use AWN to model interpretations and variants of the AODV protocol. Interpretations are just different readings of the RFC, variants are the result of modifications to address existing limitations. Thanks to the use of process algebra, we can easily adapt the proofs of established correctness properties of the protocol, such as loop freedom. This is in contrast to the analysis of variants of existing protocols via simulations and testbed experiments, where all the work typically has to be redone from scratch (and even then cannot provide the same level of assurance).

6.1 Interpretations

In this section we briefly discuss two ambiguities of the RFC together with possible interpretations. More can be found in [8]. Of course each interpretation may possibly create routing loops and hence has to be examined separately. The specification and the proofs formalised in AWN can easily be adapted.

6.1.1 Invalidating Routing Table Entries

We have already presented one contradiction of the AODV RFC in Section 4. It was based on the question of what would happen if a node that has a valid routing table entry for a destination *D* receives an error message and invalidates the corresponding routing table entry. In the same section we also list eight possible resolutions. The first two, the only interpretations compliant with the RFC, violate Theorem 5.4, and yield routing loops [9]. The same holds for Interpretation (c). As stated before, to guarantee loop freedom one has to create an interpretation of AODV that is (literally speaking) not compliant with the RFC. Interpretations (d) and (e) are loop free—the proof is identical to the one of (f), given in Section 5. The remaining two resolutions can be proven to be loop free as well [8].

6.1.2 Updating with the Unknown Sequence Number

The AODV RFC [13] states that whenever a node receives a forwarded AODV control message from a neighbour (i.e., the neighbour is not the originator of the message), it creates a new or updates an existing routing table entry to that neighbour. In the presented specification, this update is modelled in Lines 10, 14 and 18 of Process 1. In the event a new routing table entry is created, the sequence-number-status flag is set to false to signify that the sequence number corresponding to the neighbour is unknown. This interpretation is modelled in [8] and is compliant with the RFC.

However, in most implementations of AODV (e.g. [2, 6]), an unknown sequence number is simply represented by the value 0, rather than by setting a flag. In the specification of Section 3 we follow this approach of using the value 0.

Since the RFC does not make the update mechanism clear, different interpretations arise when an existing valid routing table entry for the neighbour has to be updated. While it is clear that expiry values for timers associated with the routing table entry will be updated, it is not clear if a valid sequence number with value n (>0) will remain unmodified as it is, or be updated to the value 0. In order to verify which interpretations are reasonable, we check which ones satisfy the invariants specified before (e.g., Theorem 5.4).

If we assume that an entry $(dip, 0, \texttt{val}, hops', *, *)$ replaces an entry $(dip, dsn, \texttt{val}, hops, *, *)$ (where $dsn>0$) in a routing table, it is easy to see that Theorem 5.4 is violated. In fact, it is not hard to exploit this to create a routing loop, since it allows the possibility of decreasing destination sequence numbers [8, Sect. 9.1]. This is the interpretation that is implemented by AODV-UIUC and AODV-UCSB. Our interpretation follows AODV-UU: an entry $(dip, dsn, \texttt{val}, hops, nhip, *)$ is replaced by $(dip, dsn, \texttt{val}, 1, sip, *)$, i.e., in the existing entry only the next hop and the hop count is updated—the sequence number stays the same. This interpretation is not in line with the RFC—the RFC never merges information of two routes. However, it is loop free (cf. Section 5). AODV-ns2 does not perform an update if a routing table entry already exists, i.e., it uses a version of `update` as in Section 3.1, but without the fifth clause. This interpretation is also loop free; the proof is identical to the one presented in the previous section.

There are more interpretations possible if the sequence-number-status flag (the flag indicating whether a sequence number is known or unknown) is modelled.

(a) Set the flag to unknown and the sequence number to 0. Using the same argument as before, this can yield routing loops.

(b) Set the flag to unknown, but keep the destination sequence number stored in the routing table. This interpretation is indeed loop free and is most likely the intention of the AODV RFC.

Detailed proofs can be found in [8].

6.2 Variants

Let us now turn to variants of AODV and look at (known) shortcomings of the AODV protocol, present possible improvements and then use AWN to verify that the modified AODV is still loop free.

6.2.1 *Non-Optimal Route Selection*

In AODV's route discovery process, a destination node (or an intermediate node with an active route to the destination node) will generate a RREP message in response to a received RREQ message. The RREQ message is then discarded and not forwarded. This termination of the route discovery process at the destination can lead to other nodes inadvertently creating non-optimal routes to the source node [12], where route optimality is defined in terms of a metric, for example hop count. In [12] it is shown that during the route discovery process in AODV, the only nodes that generally discover optimal routes to the source and destination nodes are those lying on the selected route between the source node and the destination node (or the intermediate node) generating the reply. All other network nodes receiving the RREQ message (in particular those located "downstream" of the destination node) may inadvertently be creating non-optimal routes to the source node due to the unavoidable receipt of RREQ messages over other routes.

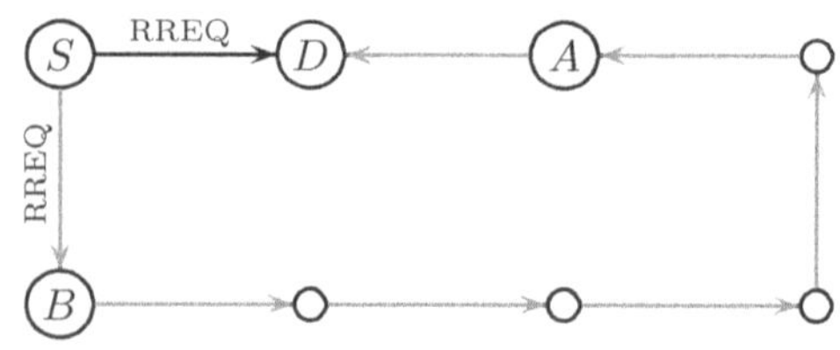

Figure 2: Non-optimal route selection

We illustrate this by the example in Figure 2. There, node S wants to find a route to node D. It generates and broadcasts a RREQ message that is received by its neighbours D and B. Since D is the destination, it responds with a RREP message. The received RREQ message is discarded and not forwarded. On the other hand, B continues to forward its received RREQ message, which eventually arrives at A. At node A, a routing table entry is created for the source S, with a hop count of six. This is clearly not optimal, as A is only two hops away from S. Due to the discarding of the RREQ message at D, A is prevented from discovering its optimal route to S.

A possible modification to AODV to solve this problem is to allow the destination node to continue to forward the RREQ message. This will then enable A in Figure 2 to discover its optimal route to S. In addition, the forwarded RREQ message from the destination node is modified to include a flag that indicates a RREP message has already been generated and sent in response to the former message. This is to prevent other nodes (with active routes to the destination) from sending a RREP message in response to their reception of the forwarded RREQ message.

The entire specification of this variant (in [8]) differs only in five lines from the original—all of which are contained in the process **RREQ**; the other processes remain unchanged. The changes introduce the new flag and a case distinction based on that, as well as three new broadcasts. For example, after initiating a route reply at the destination (Process 2, Line 12), the route request message is forwarded:

broadcast(`rreq(hops+1,rreqid,dip,dsn,oip,osn,ip,true)`) ,

where the last component of the RREQ message is the newly introduced flag. The proofs of important properties (e.g., loop freedom) are still valid. The proofs of the invariants proceed by examining lines in our processes where the invariant might be invalidated: The proof of Proposition 5.1, which can be found in [8], checks all occurrences of sending a RREQ message, and the proofs of Proposition 5.3 and Theorem 5.4 check all occurrences of `update` and `invalidate`. For the former, three new **broadcast**-commands have to be examined; however these cases are similar to the **broadcast** already implemented in the original process **RREQ** (Line 41). For the latter, no extra effort is needed, as the modification does not involve occurrences of `update` and `invalidate`.

6.2.2 *Failure of Route Discovery Process*

In AODV's route discovery process, a RREP message from the destination node is unicast back along the selected route

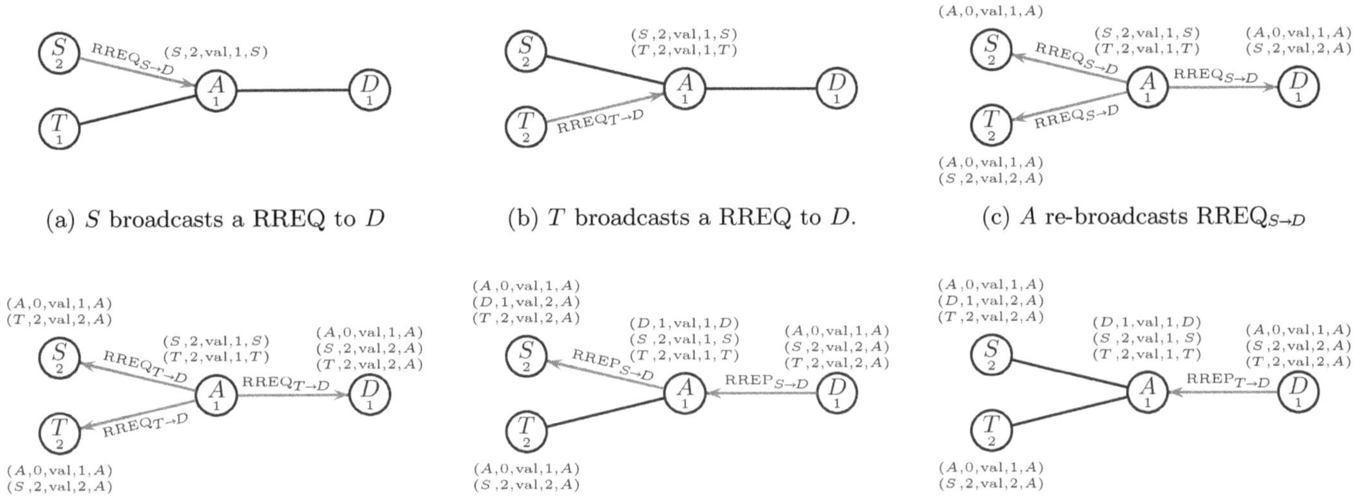

(a) S broadcasts a RREQ to D

(b) T broadcasts a RREQ to D.

(c) A re-broadcasts RREQ$_{S\to D}$

(d) A re-broadcasts RREQ$_{T\to D}$

(e) D unicasts RREP$_{S\to D}$ to A; A forwards to S

(f) D unicasts RREP$_{T\to D}$ to A; A drops RREP$_{T\to D}$

Figure 3: Failure of route discovery process

towards the originator of the RREQ message. Every intermediate node on the selected route will process the RREP message and, in most cases, forwards it towards the originator node. However, there is a possibility that the RREP message is discarded at an intermediate node and hence results in the originator node not receiving a reply. The discarding of the RREP message is due to the RFC specification of AODV [13] stating that an intermediate node only forwards the RREP message if it is not the originator node *and* it has created or updated a routing table entry to the destination node described in the RREP message. The latter requirement means that if a valid routing table entry to the destination node already exists, and is not updated when processing the RREP message, then the intermediate node will not forward the message. We illustrate this problem with an example.[10]

Figure 3 shows a four-node topology. In Figures 3(a) and 3(b), source nodes S and T, respectively, initiate a route discovery process to search for a route to D. When generating a RREQ message, the source node increments its sequence number before inserting it into the message. In processing the RREQ messages from S and T, node A creates routing table entries in its routing table.

In Figures 3(c) and 3(d), node A re-broadcasts the RREQ messages that it received previously. The destination node D receives the RREQ messages, and creates corresponding entries in its routing table. In Figure 3(e), D replies with a RREP message in response to the RREQ message from node S. Since the RREQ message from S does not contain any information on the destination sequence number for D, node D inserts its sequence number of 1 into the RREP message. This message is then processed by A (a routing table entry to D is created) and forwarded to S.

Similarly, in Figure 3(f), node D replies with a RREP message in response to the RREQ message from node T. Since the RREQ message from node T does not contain any information on the destination sequence number for node D, node D again inserts its sequence number of 1 into its RREP message. When the intermediate node A receives the RREP message, it processes the message. However, the existing routing table entry that node A already has for destination node D contains the same information (same destination sequence number and same hop count) as in the received RREP message. Therefore, node A does not update its routing table entry for node D and thus, according to the RFC specification, will not forward the RREP message to the source node T. This then leads to an unsuccessful route discovery process for node T.

A solution to this problem is to require intermediate nodes to forward *all* RREP messages that they receive using the newest available information on the route to the destination node: if the node's routing table contains an entry for the destination node that is valid and "fresher" than that in the RREP message, the intermediate node updates the contents of the RREP message to reflect this. The intermediate node A will then forward a RREP message containing up-to-date information on the destination node D.

As in Section 6.2.1, this solution does not violate any of the invariants; again this follows by adapting the original proofs [8].

7. RELATED WORK

Previous attempts to prove loop-freedom of AODV have been reported in [14, 4, 20], but none of these proofs are complete and valid for the current version of AODV [13]. [14] fails to consider the effect of RERR messages, which can result in routing loops; [4] analyses an earlier draft of AODV and uses an invariant that does not hold when following the AODV RFC; and [20] only considers a restricted version of AODV, not covering the important case of route replies by intermediate nodes. Details of the limitations of these proof attempts are provided in [9].

Graph Transformation Systems were used in [17] to model DYMO v10, a protocol derived from AODV. The paper provides a semi-algorithm, based on graph rewriting, which was

[10] A slightly different example was given on the MANET mailing list `http://www.ietf.org/mail-archive/web/manet/current/msg05702.html`

used to verify loop-freedom for DYMO. Model checking is also used to verify properties of routing protocols for WMNs. For example, [19] shows loop freedom of the ad-hoc protocol LUNAR, for fixed topologies or set changes in the topology. Model checking in general lacks the ability to verify protocols for an arbitrary and changing topology. It is used to check specific scenarios only.

8. CONCLUSIONS

In this paper, we have presented a complete and accurate model of the core functionality of AODV using the process algebra AWN, which has been tailored specifically for the formal modelling of wireless mesh networks and MANETs. The unique set of features and primitives of AWN allows the creation of accurate, concise and readable models of relatively complex and practically relevant network protocols, which we have demonstrated with AODV. This is in contrast to some prior related work, which either modelled only very simple protocols, or modelled only a subset of the functionality of relevant WMN or MANET routing protocols.

The currently predominant practice of informally specifying WMN and MANET protocols via English prose has a potential for ambiguity and inconsistent interpretation. The ability to provide a formal and unambiguous specification of such protocols via AWN is a significant benefit in its own right. We have demonstrated how AWN can be used as a basis for reasoning about critical protocol correctness properties, illustrated with the example of loop freedom. We have further shown how relevant proofs can relatively easily be adapted to protocol variants. In contrast to protocol evaluation using simulation, test-bed experiments or model checking, where only a finite number of specific network scenarios can be considered, our reasoning with AWN is generic and the the proofs hold for any possible network scenario in terms of topology and traffic pattern. None of the experimental protocol evaluation approaches can deliver this high degree of assurance about protocol behaviour.

9. REFERENCES

[1] Kernel AODV (ver. 2.2.2), NIST. `http://www.antd.nist.gov/wctg/aodv_kernel/`.

[2] AODV-UU: An implementation of the AODV routing protocol (IETF RFC 3561). `http://sourceforge.net/projects/aodvuu/`.

[3] K. Bhargavan, C. A. Gunter, M. Kim, I. Lee, D. Obradovic, O. Sokolsky, and M. Viswanathan. Verisim: Formal analysis of network simulations. *IEEE Transactions on Software Engineering*, 28(2):129–145, 2002.

[4] K. Bhargavan, D. Obradovic, and C. A. Gunter. Formal verification of standards for distance vector routing protocols. *J. ACM*, 49(4):538–576, 2002.

[5] I. Chakeres and C. Perkins. Dynamic MANET on-demand (AODVv2) routing. Internet Draft (Standards Track), draft-ietf-manet-dymo-22, 2012. `tools.ietf.org/html/draft-ietf-manet-dymo-22`.

[6] I. D. Chakeres and E. M. Belding-Royer. AODV routing protocol implementation design. In *Workshop on Wireless Ad Hoc Networking (WWAN'04)*, 2004.

[7] A. Fehnker, R. J. van Glabbeek, P. Höfner, A. McIver, M. Portmann, and W. L. Tan. A process algebra for wireless mesh networks. In H. Seidl, editor, *European Symposium on Programming (ESOP'12)*, volume 7211 of *LNCS*, pages 295–315. Springer, 2012.

[8] A. Fehnker, R. J. van Glabbeek, P. Höfner, A. McIver, M. Portmann, and W. L. Tan. A process algebra for wireless mesh networks used for modelling, verifying and analysing AODV. Technical Report 5513, NICTA, 2012. `http://www.nicta.com.au/pub?id=5513`.

[9] R. J. van Glabbeek, P. Höfner, W. L. Tan, and M. Portmann. Sequence numbers do not guarantee loop freedom—AODV can yield routing loops, 2012. `http://rvg.web.cse.unsw.edu.au/pub/AODVloop.pdf`

[10] IEEE P802.11s. IEEE draft standard for information technology—telecommunications and information exchange between systems—local and metropolitan area networks—specific requirements—part 11: Wireless LAN Medium Access Control (MAC) and physical layer (PHY) specifications-amendment 10: Mesh networking, July 2010.

[11] V. Kawadia, Y. Zhang, and B. Gupta. System services for ad-hoc routing: Architecture, implementation and experiences. In *Mobile Systems, Applications and Services (MobiSys'03)*, pages 99–112. ACM, 2003.

[12] S. Miskovic and E. W. Knightly. Routing primitives for wireless mesh networks: Design, analysis and experiments. In *Information Communications (INFOCOM'10)*, pages 2793–2801. IEEE, 2010.

[13] C. Perkins, E. Belding-Royer, and S. Das. Ad hoc on-demand distance vector (AODV) routing, RFC 3561 (experimental), 2003. `http://www.ietf.org/rfc/rfc3561.txt`.

[14] C. Perkins and E. Royer. Ad-hoc On-Demand Distance Vector Routing. In *Mobile Computing Systems and Applications (WMCSA'99)*, pages 90–100, 1999.

[15] A. A. Pirzada, M. Portmann, and J. Indulska. Performance analysis of multi-radio AODV in hybrid wireless mesh networks. *Computer Communications*, 31(5):885–895, 2008.

[16] K. Ramachandran, M. M. Buddhikot, G. Chandranmenon, S. Miller, E. Belding-Royer, and K. Almeroth. On the design and implementation of infrastructure mesh networks. In *Workshop on Wireless Mesh Networks (WiMesh'05)*. IEEE, 2005.

[17] M. Saksena, O. Wibling, and B. Jonsson. Graph grammar modeling and verification of ad hoc routing protocols. In C. R. Ramakrishnan and J. Rehof, editors, *Tools and Algorithms for the Construction and Analysis of Systems (TACAS'08)*, volume 4963 of *LNCS*, pages 18–32. Springer, 2008.

[18] A. Subramanian, M. Buddhikot, and S. Miller. Interference aware routing in multi-radio wireless mesh networks. In *IEEE Workshop on Wireless Mesh Networks (WiMesh'06)*. IEEE, 2006.

[19] O. Wibling, J. Parrow, and A. N. Pears. Automatized verification of ad hoc routing protocols. In D. de Frutos-Escrig and M. Núñez, editors, *Formal Techniques for Networked and Distributed Systems (FORTE '04)*, volume 3235 of *LNCS*, pages 343–358. Springer, 2004.

[20] M. Zhou, H. Yang, X. Zhang, and J. Wang. The proof of AODV loop freedom. In *Wireless Communications & Signal Processing (WCSP'09)*. IEEE, 2009.

VCM: The Vector-Based Coloring Method for Grid Wireless Ad Hoc and Sensor Networks

Cedric Adjih
Inria Paris Rocquencourt
78153 Le Chesnay cedex, France
cedric.adjih@inria.fr

Ichrak Amdouni
Inria Paris Rocquencourt
78153 Le Chesnay cedex, France
ichrak.amdouni@inria.fr

Pascale Minet
Inria Paris Rocquencourt
78153 Le Chesnay cedex, France
pascale.minet@inria.fr

ABSTRACT

Graph coloring is used in wireless ad hoc and sensor networks to optimize network resources: bandwidth and energy. Nodes access the medium according to their color. It is the responsibility of the coloring algorithm to ensure that interfering nodes do not have the same color. In this paper, we focus on wireless ad hoc and sensor networks with grid topologies. How does a coloring algorithm take advantage of the regularity of grid topology to provide an optimal periodic coloring, that is a coloring with the minimum number of colors? We propose the *Vector-Based Coloring Method*, denoted *VCM*, a new method that is able to provide an optimal periodic coloring for any radio transmission range and for any h-hop coloring, $h \geq 1$. In h-hop coloring, no nodes that are p-hop away, with $1 \leq p \leq h$ use the same color. This method consists in determining where a color can be reproduced in the grid without creating interferences while minimizing the number of colors used. We compare the number of colors provided by *VCM* with the number of colors obtained by a distributed coloring algorithm with line and column priority assignments. Finally, we discuss the applicability of this method to a real wireless network.

Categories and Subject Descriptors

C.2.1 [**Computer - Communication Networks**]: [Network Architecture and Design - Wireless communication]

General Terms

Algorithms, Theory

Keywords

Graph coloring, wireless sensor networks, grid, lattice, periodic coloring, Vector-based Method, valid coloring, optimal coloring, chromatic number bounds

1. CONTEXT AND MOTIVATION

In graph theory, vertex or node coloring consists in assigning colors to vertices such that no two interfering nodes have the same color. It has been proved that the 1-hop coloring problem in [1] and more generally the h-hop node coloring problem with $h \in \mathbb{N}^*$ is NP-complete [2, 3]. That is why, approximation algorithms or heuristics are adopted. A whole class of results expresses properties related to the *worst-case* performance of approximation algorithms: they typically prove that, for any input graph of a given family, the coloring obtained by a given algorithm uses at most α times the optimal number of colors. Such an algorithm is denoted an α-approximation algorithm. The approximation ratio of coloring algorithms that is defined as the ratio of the number of colors obtained by the algorithm to the optimal number is of crucial interest. The reader can refer to [7] for approximation ratio of coloring algorithms in grids, triangular lattices and hexagonal graphs. Various approximation ratio are established depending on the priority, graph type and class of algorithms [8]. A classic general heuristic is the FirstFit [9] coloring algorithm that assigns the first available color to the uncolored node with the highest priority.

Many coloring algorithms have been designed to be used in wireless networks to make the medium access more efficient. Coloring is used in radio cellular networks and more generally in wireless networks to allocate frequencies to cells or nodes in order to reduce the interferences and provide bandwidth spatial reuse (see for instance [10, 11, 12]). Indeed, a color corresponds to a frequency and two interfering nodes do not share the same color.

In wireless ad hoc and sensor networks, coloring is applied in time slot assignment. Each time slot is assigned to a color. In vertex coloring, any node has a color that is not shared with any of its interfering nodes. The color benefits are multiple. First, the avoidance of the interference between nodes, since interfering nodes do not have the same color. Second, compared to classical TDMA-based medium access schemes, the coloring enhances the bandwidth spatial reuse by allowing nodes with the same color to transmit during the same time slot. Third, coloring reduces the energy consumption by: (1) allowing nodes to sleep in the time slots during which they are neither transmitters nor receivers, (2) and avoiding collisions. Many works have addressed this problem. Among them, we can cite TRAMA [14], FLAMA [15], ZMAC [16] which is based on the DRAND [17] coloring algorithm, TDMA-ASAP [18], FlexiTP [19] and SERENA [23]. In general, the primary concern of such protocols is to reduce the number of colors. Other criteria are also considered. Schneider et al. [13] for instance aim at reducing the running time of their coloring method. They propose a randomized coloring algorithm that runs in $O(\Delta+\log(\Delta)\log(n))$ given an estimate of n and Δ, and $O(\Delta+log^2 n)$ without knowledge of Δ. n and Δ are respectively the number of nodes and the maximum nodes degree. Their vertex coloring requires $(\Delta+1)$ colors.

In this paper, we focus on vertex coloring applied to wireless ad hoc and sensor networks with regular topologies which are grids.

MSWiM'12, October 21–25, 2012, Paphos, Cyprus.

Wireless sensors organized according to a grid topology are a natural choice and exist in many real applications. Often, the grid organization is easy to deploy, and is efficient in terms of coverage, connectivity and management. For instance, a grid topology is one of the best methods to ensure sensor coverage for surveillance [4, 5]. It also helps to collect measurements with a uniform spatial sampling such as for instance in precision agriculture and irrigation as in [6].
Given such grid topologies, the question (and our problem statement), is: what is the method to color grids with the minimum number of colors? Is it possible to take advantage of the regularity of grids to design a h-hop coloring algorithm able to find the optimal number of colors for any grid with any transmission range value? Moreover, does a periodic color pattern that can tile the whole topology for a given radio range, exist? What is the optimal number of colors?

In this paper, we answer these questions by proposing the Vector-Based Coloring Method, denoted *VCM*, for assigning colors to nodes organized as a grid and which is asymptotically optimal when radio range increases. *VCM* computes the color of a node directly from its coordinates.

The paper is organized as follows. In Section 2, we present the problem statement and an overview of the *VCM* method. Sections 4, 5 and 6 describe the different components of *VCM*. In Section 7, we evaluate the complexity of *VCM*. Sections 8 and 10 show how to apply *VCM* in real wireless ad hoc and sensor networks. Section 9 compares the coloring results obtained by *VCM* with results obtained by different heuristics.

2. PROBLEM STATEMENT AND OVERVIEW

2.1 Notations, Definitions and Assumptions

In the remaining of this paper, we use the following notations:

- $\vec{UV}$ denotes a vector of extremities the nodes U and V.
- For any point W and any vector w we use the terminology: W is a point of coordinate w where $\vec{OW} = w$ and O is the origin of plan.
- $|\vec{UV}|$ denotes the norm, or length of the vector $\vec{UV}$, and $d(U,V)$ the euclidean distance between nodes U and V.
- $det(\vec{UV_1}, \vec{UV_2})$ is the determinant of the two vectors.
- $\vec{UV_1} \cdot \vec{UV_2}$ denotes the scalar product of the vectors.
- $\Lambda(\vec{UV_1}, \vec{UV_2})$ denotes the lattice having as a basis the vectors $(\vec{UV_1}, \vec{UV_2})$. We call $\vec{UV_1}$ and $\vec{UV_2}$ the basis (or generator) vectors. $\mathcal{P}(\vec{UV_1}, \vec{UV_2})$ is the fundamental parallelotope associated to $\Lambda(UV_1, UV_2)$. In 2 dimensions, $\mathcal{P}$ is a parallelogram. Moreover, the number of nodes in $\mathcal{P}(\vec{UV_1}, \vec{UV_2})$ called *lattice determinant* is given by $det(\vec{UV_1}, \vec{UV_2})$. These notations and definitions are adopted from [21]. We also denote $\mathcal{P}_U(\vec{UV_1}, \vec{UV_2})$ the fundamental parallelotope translated at node U.
- For x, y in $\mathbb{Z}^2$, with $y \neq 0$, we denote "x modulo y" the integer z in $\{0, 1, \dots |y| - 1\}$ such that we have $x \equiv z(mod|y|)$.

Moreover, the following definitions and assumptions are used:

- **Nodes:** The nodes are disposed in a grid. Without loss of generality we assume that the grid step is 1, hence the set of nodes is identified with $\mathbb{Z}^2$.
- **Neighborhood:** R denotes the radio range and is a real number ≥ 1. The underlying assumption is a unit disk model, assuming that nodes U and V can communicate directly in both directions if and only if their euclidean distance is lower than R. Hence, the set of neighbors of a node U is: $\mathcal{N}(U) \triangleq \{V \in \mathbb{Z}^2 \mid 0 < d(U,V) \leq R\}$.
- **Coloring:** A coloring ϕ is a mapping from the nodes $\mathbb{Z}^2$ to a set of colors, identified with the set of integers $\{1, \dots k\}$, with k a positive integer.
- **Valid Coloring:** a coloring is said to be a *valid h-hop coloring*, with h an integer ≥ 1, when all nodes that are at less than or equal to h-hop away are assigned different colors.
- **Periodic Coloring:** A coloring is denoted "periodic" if the set of nodes with the same given color, is the lattice $\Lambda(v_1, v_2)$ for some fixed vectors v_1, v_2 of $\mathbb{Z}^2$, after it is translated by some amount. In this paper, we consider a "strict" definition of periodicity because colors might otherwise be periodically repeated with a pattern generating more than a single lattice for instance.
- **Reduced vectors:** We call reduced vectors of a lattice, the vectors obtained by applying the Gauss lattice reduction on these vectors. Indeed, for any pair of vectors u_1, u_2 generating a lattice $\Lambda(u_1, u_2)$, the Gauss lattice reduction algorithm provides two *reduced* vectors v_1, v_2 generating exactly the same lattice and verifying the System (1)

$$\begin{cases} |v_1| \leq |v_2| \\ 2|v_1 \cdot v_2| \leq |v_1|^2 \end{cases} \tag{1}$$

See for instance [22] for more details.

2.2 Problem Statement

Because the set of colorings of $\mathbb{Z}^2$ is infinite, in this paper we restrict ourselves to colorings exhibiting a periodic color pattern (following the definition of periodic coloring of section 2.1).

Problem Statement: Find one of the valid periodic h-hop colorings with the minimum number of colors for $h \geq 1$.

2.3 Overview

In this paper, we propose the *periodic Vector-Based Coloring Method*, denoted **VCM** that answers the problem statement in polynomial time. This method consists in producing a periodic coloring obtained by tiling a color pattern. This color pattern is generated by two vectors verifying some conditions that will be detailed in the paper.

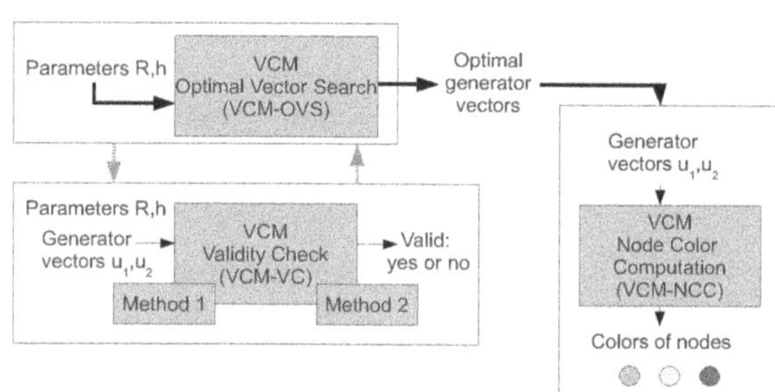

Figure 1: Components of VCM

As illustrated in Figure 1, *VCM* is composed of the following components:
C1. VCM-OVS An algorithm to search for optimal generator vectors, that is, yielding the minimum number of colors. Indeed, we limit the set of candidate vectors to find the vectors providing the optimal number of colors.
C2. VCM-NCC A periodic coloring based on two generator vectors. This component determines how any node selects its color.
C3. VCM-VC An algorithm to check the validity of the periodic coloring: the VCM incorporates two methods to verify the validity of the coloring for a given couple of generator vectors.

In the Sections 4 to 6 we detail the components of *VCM*.

3. PERIODIC COLORING

Figure 2 presents an example of a periodic 3-hop coloring of a 10×10 grid with a transmission range $R = 1$, and two generator vectors $\vec{UV_1}$ and $\vec{UV_2}$ (see [20] for more examples). The principles of the periodic coloring are:
P1. (Generator vectors) The two vectors $\vec{UV_1}$ and $\vec{UV_2}$, if linearly independent, generate the parallelogram $\mathcal{P}(\vec{UV_1}, \vec{UV_2})$ of the color pattern.
P2. (Parallelotope color unicity) Inside $\mathcal{P}(\vec{UV_1}, \vec{UV_2})$, there is no color reuse.

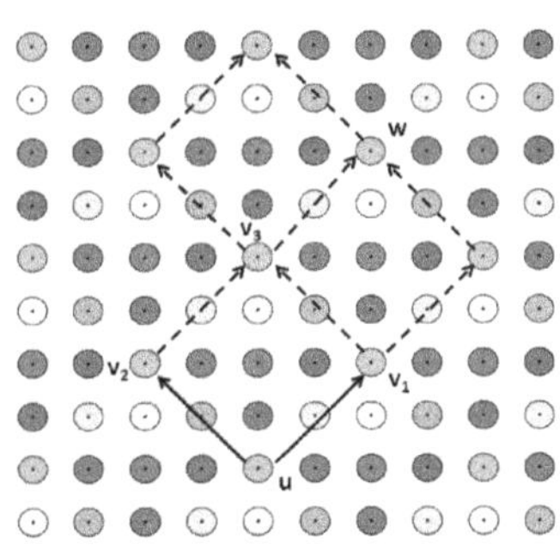

Figure 2: Example of periodic 3-hop coloring (R=1)

P3. (Lattice color repetition) Because the periodic coloring is obtained by repeating the color pattern, the nodes W such that $\vec{UW}$ is a linear combination of $\vec{UV_1}$ and $\vec{UV_2}$ have the same color as U. Precisely, the set of nodes W having the same color as U forms a *lattice* of $\mathbb{Z}^2$ with generator vectors $\vec{UV_1}$ and $\vec{UV_2}$: the vector $\vec{UW}$ can be written as $\vec{UW} = \alpha\vec{UV_1} + \beta\vec{UV_2}$ with α and β in $\mathbb{Z}^2$.
P4. (Coordinate-based color computation) The grid can be seen as a tiling with the pattern $\mathcal{P}_{\mathcal{U}}(\vec{UV_1}, \vec{UV_2})$. Thus, each node W belongs to its own parallelogram, and has coordinates relative to this parallelogram. Consequently, W has the same color as any node having the same coordinates in its own parallelogram.

To be applied to a wireless ad hoc and sensor network, these principles have to be enhanced. For instance, since nodes having the same color can simultaneously access the wireless medium, validity of the coloring must be verified. Hence, we search **valid** vectors (if they produce valid coloring). Moreover, to ensure an efficient usage of the bandwidth, the number of colors used should be minimized. These criteria are taken into account in our work and progressively introduced in the paper.
In the following we detail the components of *VCM*. As previously said, we only consider grid colorings that periodically reproduce a color pattern.

4. OPTIMAL VECTOR SEARCH (OVS)

To achieve an optimal spatial reuse, the coloring algorithm should minimize the number of colors used. For *VCM*, this number is entirely defined by the two generator vectors u_1 and u_2 describing the periodic tiling (see property **P3**). Hence, and our aim is to judiciously choose these vectors in order to reduce the number of colors used to color a grid.
However, by default, the infinite lattice $\mathbb{Z}^2$ is a possible set for candidate vectors. So, to find the optimal vectors in a small set, our approach is taking advantage of the following:
1. We determine the upper and lower bounds on the number of colors needed in a h-hop coloring of the grid, for $h \geq 1$.
2. Because for any couple of initial generator vectors, reduced vectors always exist, it is sufficient to search for some optimal vectors in the space of reduced vectors defined by System (1).
This is used to restrict the search space. We will show in Section 4.2 how to bound the set of candidate vectors and use properties of *lattice reduction* to decrease the complexity of the search for the optimal vectors.

4.1 Bounds on the number of colors in periodic colorings

In this section, we prove that the number of colors of optimal coloring when $R \to \infty$ is asymptotically $\frac{\sqrt{3}}{2}h^2R^2 + O(R)$. We first start with two examples of construction of valid generator vectors, that are used to determine the bounds on the number of colors, and to limit the vectors search space.

4.1.1 Examples of valid generator vectors

• Near hexagonal generator vectors
Here, we detail how to construct two valid vectors $\vec{UV_1}$ and $\vec{UV_2}$ which provide an approximation of an hexagonal lattice (when R is large). Figure 3 illustrates how some points V_1 and V_2 are constructed starting from some reference point U.

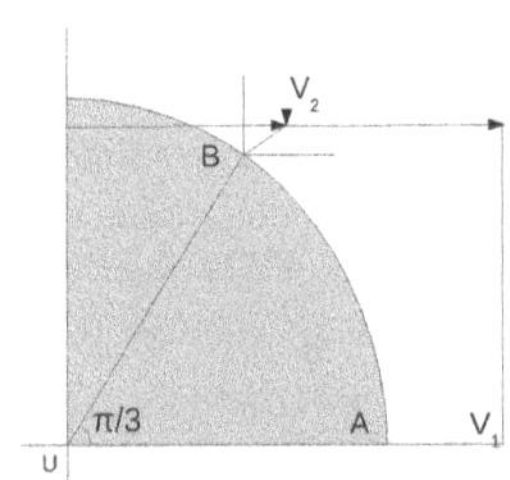

Figure 3: Selecting vectors for a near-hexagonal lattice

- Starting from the point U, the line with an angle $\frac{\pi}{3}$ with the horizontal line is considered, and its intersection with the circle of radius hR yields the point B.
- Next, let V_2 be the closest point of B on the grid having larger coordinates x_2 and y_2 than B.
- Then V_1 with coordinates (x_1, y_1) is selected with $(x_1, y_1) = (2x_2, 0)$.

Notice that by construction x_1 is greater than hR, and we have a valid choice of vectors $\vec{UV_1}$ and $\vec{UV_2}$.

• **Square generator vectors**

An alternate, simpler, choice of vectors is to compute the integer $\lambda = \lfloor hR \rfloor + 1$, (this symbol denotes the integer part of hR) and select the vectors with coordinates $u_1' = (\lambda, 0)$ and $u_2' = (0, \lambda)$. They generate a square lattice.

4.1.2 Lower bound

For the lower bound on the number of colors, we have the following theorem that is valid for any coloring (not just periodic coloring):

THEOREM 1. *The number of colors required to color an infinite grid with $R > \sqrt{2}$ is at least $\frac{\sqrt{3}}{2}h^2(R-\sqrt{2})^2$.*

PROOF. Consider any h-hop coloring of the grid. Consider a fixed color c, and now let S_c be the set of nodes having this color. We first establish a lower bound of the distance of nodes in S_c. Let us define $\rho = (R-\sqrt{2})h$. Consider two nodes A, B of S_c. By contradiction: if their distance verifies $d(A,B) \leq \rho$, they would be at most h-hop away, contradicting the definition of a h-hop coloring (see Lemma 4). Therefore, all nodes of S_c are at a distance at least ρ from each other.

Now consider the set of circles C of radius $\frac{1}{2}\rho$ and whose centers are the nodes of S_c. The fact that any two nodes of S_c are distant of more than ρ, implies that none of the circles in C overlap. Hence C is a *circle packing* by definition. From the Thue-Tóth theorem [24, 25] establishing that the hexagonal circle packing is the densest packing, with a density of $\frac{\pi}{\sqrt{12}}$, we deduce that C must have a lower or equal packing density. This implies an upper bound of the density of set S_c of centers of the disks of $\frac{1}{(\rho/2)^2\sqrt{12}}$.

Because each color yields a set of nodes with at most this density, it follows a lower bound of the number of colors that is the inverse of this quantity, hence the theorem. □

4.1.3 Upper bound

For an upper bound, we have the following theorem:

THEOREM 2. *The number of colors required to color an infinite grid is at most $\frac{\sqrt{3}}{2}h^2R^2 + 2hR + (2+hR)\sqrt{2}$.*

PROOF. This number is obtained from the near hexagonal vectors constructed in section 4.1.1, which thus yields an upper bound.

Denote (γ, δ) the coordinates of $\vec{BV_2}$ from section 4.1.1. By construction: $0 \leq \gamma \leq 1$ and $0 \leq \delta \leq 1$, hence: $|\vec{BV_2}| \leq \sqrt{2}$. Moreover,

$$\begin{aligned} |\vec{AV_1}| &= 2x_2 - hR \\ &= 2(hR\cos(\frac{\pi}{3}) + \gamma) - hR, \text{ with } \gamma \leq 1 \\ &\leq 2 \end{aligned}$$

Consequently, we can write n_c, the number of colors in the associated coloring as:

$$\begin{aligned} n_c &= det(\vec{UV_1}, \vec{UV_2}) \\ &= det(\vec{UA}, \vec{UB}) + det(\vec{AV_1}, \vec{UB}) + det(\vec{UV_1}, \vec{BV_2}) \\ &\leq det(\vec{UA}, \vec{UB}) + |\vec{AV_1}||\vec{UB}| + |\vec{UV_1}||\vec{BV_2}| \\ &\leq \frac{\sqrt{3}}{2}h^2R^2 + 2hR + (hR+2)\sqrt{2}. \end{aligned}$$

Thus the theorem. □

REMARK 1. *The number of colors of the construction of square generator vectors, proposed in previous section, is higher than for the near hexagonal vectors. It is $h^2R^2(1+O(\frac{1}{R}))$, hence the vectors cannot yield the tight bound, but the vectors are used later for an alternate upper bound.*

4.1.4 Asymptotic Number of Colors

THEOREM 3. *The number of colors $n_c(R)$ of an optimal periodic h-hop coloring for a fixed h verifies:*

$$n_c(R) = \frac{\sqrt{3}}{2}h^2R^2(1+O(\frac{1}{R}))$$

when $R \to \infty$.

PROOF. Combining the lower bound and the upper bound of the two theorems yields the result. □

COROLLARY 1. *Theorem 3 is true even considering periodic and non periodic coloring.*

PROOF. A periodic coloring is a special case of general coloring (including periodic and non periodic). Hence, the optimal number of colors in general coloring is less than or equal to n_c. □

COROLLARY 2. VCM *is asymptotically optimal even considering all possible valid colorings. In other terms* VCM *is an $(1+g(R))$-approximation of the optimal coloring of the grid with some g verifying $g(R) \to 0$ when $R \to \infty$.*

PROOF. *VCM* will find vectors with better or equal performance than those in Theorem 2. Indeed, in the worst case, the generator vectors for the near-hexagonal lattice will be selected by *VCM*. □

4.2 Finding the optimal vectors

In this section, we show how to limit the set of candidate vectors. Let u_1, u_2 be two candidate vectors and θ be the angle between them. We search l_{1min}, l_{1max} (respectively l_{2min}, l_{2max}) the lower and upper bounds of the length of u_1 (respectively u_2).

• Considering a h-hop coloring, the vectors u_1 and u_2 must be valid. According to Lemma 4, we have:

$$|u_1| > (R-\sqrt{2})\,h \text{ and } |u_2| > (R-\sqrt{2})\,h. \quad (2)$$

• As previously said, in order to reduce the set of candidate generator vectors, we reduce the size of these vectors by using the lattice reduction algorithm of Gauss. A consequence of Gauss property 1 is that: $|\cos\theta| \leq \frac{1}{2}$, and hence:

$$|\sin\theta| \geq \frac{\sqrt{3}}{2}. \quad (3)$$

• As shown in Theorem 1,

$$det(u_1, u_2) \leq S_h = \frac{\sqrt{3}}{2}h^2R^2 + 2hR + (2+hR)\sqrt{2}.$$

It results: $|u_1||u_2||\sin\theta| \leq S_h$. Using 3, we get:

$$\frac{\sqrt{3}}{2}|u_1||u_2| \leq |u_1||u_2||\sin\theta| \leq S_h \quad (4)$$

And as $|u_1| \leq |u_2|$, from 4 we have:

$$\frac{\sqrt{3}}{2}|u_1|^2 \leq S_h. \quad (5)$$

We now separate the cases where $R > \sqrt{2}$ and $R \leq \sqrt{2}$

4.2.1 Case $R > \sqrt{2}$

Using 2, from 5 we get:

$$\frac{\sqrt{3}}{2}|u_1|(R-\sqrt{2})h < \frac{\sqrt{3}}{2}|u_1|^2 \leq S_h. \quad (6)$$

And using 2 in 4:

$$\frac{\sqrt{3}}{2}|u_2|(R-\sqrt{2})h < S_h. \quad (7)$$

To summarize, the two generator vectors should verify for $R > \sqrt{2}$:

$$l_{1min} < |u_1| \leq l_{1max} \text{ and } l_{2min} < |u_2| < l_{2max} \quad (8)$$

with:

$$\begin{cases} l_{1min} = h(R-\sqrt{2}) \\ l_{1max} = \sqrt{\frac{2}{\sqrt{3}}S_h} \\ l_{2min} = h(R-\sqrt{2}) \\ l_{2max} = \frac{2}{\sqrt{3}}\frac{S_h}{h(R-\sqrt{2})} \end{cases} \quad (9)$$

In practice, to compute the two generator vectors, we determine the upper and the lower bounds of the coordinates of u_1, and u_2 using the System (9). Notice that we can search the vectors in the half plane $(y \geq 0)$, because if u_1 and u_2 are generator vectors, then their symmetric vectors with respect to $(y = 0)$ axis are also generator vectors. Consequently, we have:

$$\begin{cases} -l_{1max} \leq x_1 \leq l_{1max} \\ 0 \leq y_1 \leq l_{1max} \end{cases} \quad (10)$$

and

$$\begin{cases} -l_{2max} \leq x_2 \leq l_{2max} \\ 0 \leq y_2 \leq l_{2max} \end{cases} \quad (11)$$

4.2.2 Case $R \leq \sqrt{2}$

We use $l_{1min} = 0$. In addition, we can use the bound implied by the square generator vectors as proposed in Remark 1. Let $S_s = (hR+1)^2$, which replaces S_h for the computation of l_{1max}. Then, instead of a fixed bounds for l_{1max} and l_{2max}, we propose a bound depending on u_1, by using (4) we have: $l_{2max}(u_1) = \frac{2\sqrt{3}S_s}{3|u_1|}$, and because $|u_2| \geq |u_1|$ we have: $l_{1min}(u_1) = |u_1|$. Hence:

$$\begin{cases} 0 < |u_1| \leq hR+1 \\ |u_1| \leq |u_2| \leq \frac{2\sqrt{3}(hR+1)^2}{3|u_1|} \end{cases} \quad (12)$$

To find the optimal vectors, we define the Method OVS.

Method OVS:
1. The first step is to search S_1 the initial set of generator vectors u_1 and u_2. S_1 is the set of vectors having as coordinates the integers (x_1, y_1) and (x_2, y_2) verifying System (10) if $R > \sqrt{2}$ or otherwise System (12).
2. Now, the set S_1 should be filtered to keep only reduced and valid vectors. Indeed, for each couple of vectors (u_1, u_2) in S_1, we should verify:
2.1. (u_1, u_2) are reduced, that is they verify System (1).
2.2. to check the validity of the coloring, we use one of the methods that will be explained in Section 6. Indeed, two cases are possible:
2.2.1 if $R > \sqrt{2}$ apply VC2 (see Section6).
2.2.2. otherwise, apply VC1 (see Section6).
3. After the step 2, we obtain the set of valid reduced vectors. Now, the optimal vectors are then the vectors having the smallest absolute value of their determinant.
Notice that the search of the optimal vectors can be performed by a central unit that disseminates two vectors to all nodes. It is also possible that each node in the grid computes the two generator vectors.

5. NODE COLOR COMPUTATION (NCC)

5.1 Assigning Colors to Nodes

The node color computation (NCC) component of VCM method takes as parameters two generator vectors u_1,u_2 (computed as in Section 4). Let (x_1, y_1) and (x_2, y_2) be their coordinates and let $d = \det(u_1, u_2)$.

Method VCM-NCC: *VCM* assigns the color of a node $W \in \mathbb{Z}^2$ based on its coordinates (x, y) by computing first the integer quantities $c_1(W), c_2(W)$ as in (13),

$$\begin{cases} c_1(W) = (xy_2 - yx_2) \text{ modulo } d \\ c_2(W) = (-xy_1 + yx_1) \text{ modulo } d \end{cases} \quad (13)$$

and then using a bijective mapping between the couple (c_1, c_2) and a color $\in \{0, 1, \ldots |d|-1\}$.

PROPERTY 1. *With the previous coloring VCC-NCC, it is indeed possible to define a bijection from $(c_1(W), c_2(W))$ to $\{0, 1, \ldots |d|-1\}$. Moreover, the coloring verifies principles* **P3** *and* **P4** *defined in section 3.*

PROOF. We assume that $d > 0$ without loss of generality. We need to prove that the set $\{(c_1(w), c_2(w)) \mid w \in \mathbb{Z}^2\}$ has cardinality d and that the set of nodes with the same color is exactly the lattice $\Lambda(u_1, u_2)$ translated at w.

Let W be a grid point of coordinates $w = (x, y)$. Performing a change of vector basis in $\mathbb{R}^2$ from $\{(1,0),(0,1)\}$ to $\{u_1, u_2\}$, the new coordinates $(\alpha, \beta) \in \mathbb{R}^2$ of W in $\Lambda(u_1, u_2)$ verify $w = \alpha u_1 + \beta u_2$ and

$$\begin{cases} \alpha = \frac{\det(w, u_2)}{d} \\ \beta = \frac{\det(u_1, w)}{d} \end{cases} \quad (14)$$

with $d = det(u_1, u_2)$.

Let α' and β' be the integer parts of α, β, i.e., $\alpha' = \lfloor \alpha \rfloor$ and $\beta' = \lfloor \beta \rfloor$. For arbitrary nonzero integers λ, μ, with also $\mu > 0$, we have the identity: $\frac{\lambda}{\mu} = \lfloor \frac{\lambda}{\mu} \rfloor + \frac{\lambda \text{ modulo } \mu}{\mu}$. Thus 14 becomes:

$$\begin{cases} \alpha = \alpha' + \frac{\det(w, u_2) \text{ modulo } d}{d} = \alpha' + \frac{c_1(w)}{d} \\ \beta = \beta' + \frac{\det(u_1, w) \text{ modulo } d}{d} = \beta' + \frac{c_2(w)}{d} \end{cases} \quad (15)$$

Let W' the point with coordinates $w' = (\alpha', \beta')$. W' is on the lattice since α', β' are integers, and observe that W is in fact inside the parallelogram of the lattice $\Lambda(u_1, u_2)$ placed at node W' (i.e. inside the parallelogram defined by the 4 points of the lattice: w', $w' + u_1$, $w' + u_2$, $w' + u_1 + u_2$). Then (15) means simply that $(\frac{c_1(w)}{d}, \frac{c_2(w)}{d})$ are the coordinates of W relative to this parallelogram (with the basis vectors u_1, u_2).

Since there is a bijection between the set of coordinates of nodes in a parallelogram of $\Lambda(u_1, u_2)$ and the nodes themselves; and since $(\frac{c_1(w)}{d}, \frac{c_2(w)}{d})$ are these coordinates, we have the two properties: 1) there are exactly d possible values of (c_1, c_2) (because there are exactly d nodes in the parallelogram), and 2) no two nodes inside the parallelogram have the same values c_1, c_2 since these are their coordinates, relative to one vertex of the parallelogram. □

LEMMA 1. *With the color computation given by the System 13, the color of the node U is repeated at the nodes W such that $w = \alpha u_1 + \beta u_2$, $\forall$ α, and $\beta \in \mathbb{Z}^2$, where w is the vector with extremities the origin and W.*

PROOF. Actually, by construction (see end of section 5.2), the color of a node is given by its relative coordinates inside the parallelogram it belongs to. Hence, the color of any node U is reused at nodes w, such that $\vec{UW} = \alpha\vec{UV_1} + \beta\vec{UV_2}\ \forall\ \alpha,\beta$ which have the same relative coordinates. □

5.2 Computing the Number of Colors

The number of colors used in a periodic h-hop coloring is given by the next Property.

PROPERTY 2. *For any node U, the color pattern defined by the two generator vectors u_1, and u_2 meeting the aforementioned principles contains exactly $|x_1y_2 - x_2y_1|$ colors where (x_1,y_1) and (x_2,y_2) are the coordinates of u_1 and u_2.*

PROOF. By definition, no two nodes within the parallelogram defined by u_1 and u_2 use the same color. Hence the number of colors is equal to the number of nodes in this parallelogram. Moreover, as we said, the number of nodes in $\mathcal{P}(u_1,u_2)$, called *lattice determinant*, is equal to the absolute value of $\det(u_1,u_2)$ [21]. Hence the property. □

5.3 Example of bijection for NCC

One of the steps of VCM-NCC, is that a bijection needs to be established between the set of values $\{(c_1(w),c_2(w)) \mid w \in \mathbb{Z}^2\}$ and the set of colors $\{0,1,\ldots,d-1\}$: an example of bijection is provided in this section.

A bijection can be constructed by computing the values of $(c_1(w),c_2(w))$ for any node in $\mathcal{P}(u_1,u_2)$ in a list, sorting the list by lexicographical order, and, finally, setting the color associated with a couple $(c_1(w),c_2(w))$ to be its index in the sorted list minus 1.

Example: if $(0,0)$ appears as the 1st item of the sorted list [as it can be proved it will], the color assigned to that couple is 0. Then, for instance, for the point W of coordinates $w = (x_1,y_1)$, in other terms $w = u_1$, we have $(c_1(w),c_2(w)) = (0,0)$ and therefore the color assigned to W is 0.

Note that, from a pure implementation point of view, it may be difficult to enumerate exactly the points of $\mathcal{P}(u_1,u_2)$, but then, instead, it is sufficient to enumerate all the nodes in a superset, the *bounding box* of $\mathcal{P}(u_1,u_2)$, itself computed from its four vertices $O = (0,0)$, $O+u_1$, $O+u_2$ and $O+u_1+u_2$. Computing the set of values $(c_1(w),c_2(w))$ for the points in the bounding box, will yield all possible values for any $w \in \mathbb{Z}^2$.

An alternate method for coloring nodes is VCM-NCC2, detailed in [26]: it is based on identical ideas, but removes the need for establishing explicitly a bijection at the expense of a slightly more complicated color computation (and more complex proof). With VCM-NCC2, the color of a node is directly computed from its coordinates (x,y) with a few integer arithmetic operations.

6. VALIDITY CHECK (VC)

As defined previously, a h-hop coloring algorithm is valid if and only if no two nodes that are at less than or equal to h-hop from each other use the same color. Based on the color computation method **VCM-NCC** (that is explained in Section 5) and the generator vectors u_1 and u_2, **VCM-VC** checks beforehand if the coloring generated by the vectors is valid. However, only a subset of the grid node is tested. We distinguish two methods.

6.1 Method VC1: Verification around Origin

Method VC1: For each node W in the h-hop neighborhood of the origin node O, we compute the color of this node based on the given generator vectors u_1 and u_2 and the coloring method VCM-NCC. If W has the same color as O, then we conclude that the vectors u_1 and u_2 do not provide a valid coloring. Otherwise, if the color of O is not repeated at any point W in its h-hop neighborhood, then the coloring is valid.

The idea of Method VC1 is based on the Lemma 2.

LEMMA 2. *If two nodes V_1 and V_2 with coordinates v_1, v_2 in $\mathbb{Z}^2$ have the same color, then the color of the origin node is repeated at the node W of coordinates $v_2 - v_1$.*

PROOF. The functions c_1,c_2 computed from System (13) are actually linear modulo d. That is, if W is the node with coordinates $v_1 - v_2$, and w is the vector of nodes extremities the origin and W, we get: $c_1(W) = c_1(V_1) - c_1(V_2)$ modulo d. Hence, if $c_1(V_1) = c_1(V_2)$ we have $c_1(W) = c_1(O)$. This is true also for c_2, hence the lemma. □

THEOREM 4. *If none of the nodes inside the h-hop neighborhood of the origin node $O = (0,0)$ has the same color as O itself, then the coloring is valid.*

PROOF. By contradiction: assume that the coloring is invalid because two nodes V_1,V_2 at less or equal to h-hop have the same color. From Lemma 2, the node W such as $\vec{OW} = \vec{V_1V_2}$ has the same color as O. Notice that the distance in terms of hops number between O and W is the same as the distance between V_1 and V_2. Hence we have found a conflict of color between O and a node W which is at less than h hops from O. Hence the theorem. □

6.2 Method VC2: Verification in a Few Points

Method VC1 requires $\Theta(R^2)$ verifications when $R \to \infty$. In the following, we propose Method VC2 requiring only a bounded number of verifications. Method VC2 performs a check on a few nodes on the lattice $\Lambda(u_1,u_2)$ to guarantee that the h-hop coloring associated to u_1,u_2 is valid. This method is based on Gauss lattice reduction [22]: u_1 and u_2 should be reduced.

Method VC2: For the grid nodes that will have the same color as the origin (which are in fact, the lattice nodes), this method verifies that these nodes are at least $(h+1)$-away from O, in which case the coloring is valid. However, not all grid nodes need to be checked. It is sufficient to check only nodes W with coordinates α,β in $\Lambda(u_1,u_2)$, such that $|\alpha|$ and $|\beta| < \mu(R)$, with $\mu(R) = \frac{2\sqrt{3}R}{3(R-\sqrt{2})}$. The coloring is valid if and only if these nodes are strictly more than h hops from the origin node.

The theorem 5 illustrates this method.

THEOREM 5. *For $R > \sqrt{2}$, the coloring provided by two reduced vectors u_1,u_2 is valid iff:*
for all α,β integers verifying $|\alpha| < \mu(R)$, and $|\beta| < \mu(R)$, the node with coordinates α, β is at strictly more than h hops from the origin node O, where $\mu(R) = \frac{2\sqrt{3}R}{3(R-\sqrt{2})}$.

PROOF. The property comes from the fact that the points on the lattice $\Lambda(u_1,u_2)$ are "far" from the origin node, because the vectors u_1,u_2 are reduced. Indeed, let V_1 and V_2 the nodes such that $\vec{OV_1} = u_1$ and $\vec{OV_2} = u_2$. Lemma 7 (see Annex) means that any node V_3 on the lattice such that $\vec{OV_3} = \alpha u_1 + \beta u_2$ with $|\alpha|$ or $|\beta| \geq \mu(R)$ can reuse the color of the origin node O because they are at strictly more that h-hop from O (provided that the points V_1 or V_2 are themselves strictly more than h-hop away from O). Hence, to check the validity of the coloring provided by *VCM*, it is necessary and sufficient to check that for all $|\alpha|$ and $|\beta| < \mu(R)$, nodes of coordinates α, β in the lattice $\Lambda(u_1, u_2)$ are strictly more than h hops away from the origin of the lattice. This check includes checking the validity of u_1 and u_2 themselves (cases $(\alpha, \beta) = (1, 0)$ and $(\alpha, \beta) = (0, 1)$) □

Notice that for dense grids ($R \to \infty$), $\mu \to 1.15\ldots$. This small bound reduces the set of nodes to be checked in order to verify the validity of the coloring for given vector candidates. In fact for $R > \frac{3\sqrt{2}}{3-\sqrt{3}}$, that is for $R > 3.3461$, we have $\mu < 2$, hence only 4 points need to be checked (considering symmetries): u_1 (with $\alpha = 1, \beta = 0$), u_2 (with $\alpha = 0, \beta = 1$), $u_1 + u_2$ (with $\alpha = 1, \beta = 1$), $u_1 - u_2$ (with $\alpha = 1, \beta = -1$).

On the other hand, Method VC1 is applicable for any radio range R, whereas Method VC2 requires ($R > \sqrt{2}$).

7. COMPLEXITY OF VCM

We can now evaluate the complexity of *VCM* that lies in the generator vectors computation and in their validity check.

THEOREM 6. *Depending on* VC *method,* VCM *complexity is in* $\Theta(R^6)$ *for Method VC1 and* $\Theta(R^4)$ *for Method VC2.*

PROOF. The vector search space is in $\Theta(R^2)$ for each vector. The time complexity of the validity check is in $\Theta(R^2)$ for Method VC1 and $\Theta(1)$ for Method VC2. The theorem follows. □

Since method VC2 can be used as soon as $R > \sqrt{2}$, we deduce:

COROLLARY 3. *VCM complexity is* $\Theta(R^4)$

8. SUMMARY: HOW TO APPLY VCM IN PRACTICE

In practice, to apply *VCM*, we start from a set of sensor nodes arranged as a two-dimensional lattice (identified by their integer coordinates). In an actual wireless network, the set of neighbors may not be exactly given by the set of nodes within a fixed range R. However, notice that a valid h-hop coloring for a given R, is also a valid h-hop coloring for $R' < R$ (although likely non-optimal). Hence, we start by selecting the value of R, a radio range such as two nodes at a distance greater than R are never neighbors (maybe using measurements or neighborhood detection), and a value h. Then, each node proceeds as follows:

1. Find the optimal valid vectors u_1 and u_2 using the Method OVS.
2. Each node computes its color using for instance the bijection defined in Section 5.3.
2.1. It computes the values (c_1, c_2) for each grid node in $\mathcal{P}(u_1, u_2)$ using System (13). Sorting the set of the values (c_1, c_2) yields an example of mapping between every possible value of (c_1, c_2) and a unique integer in $[0, |d| - 1]$.
2.2. Knowing its coordinates in the grid, each node deduces its two components c_1 and c_2 according to System (13) and then its color from the previous mapping.

We can notice that *VCM* allows each node to know its color in a single round.

9. COLORING RESULTS WITH VCM

9.1 Examples of vectors

Table 1: VCM vectors generating the optimal number of colors.

Radio range	2-hop coloring v_1	2-hop coloring v_2	2-hop coloring colors	3-hop coloring v_1	3-hop coloring v_2	3-hop coloring colors
1	(2,1)	(-1,2)	5*	(2,2)	(-2,2)	8*
1.5	(-3,0)	(0,3)	9*	(4,0)	(0,4)	16*
2	(3,2)	(-2,3)	13*	(4,3)	(-3,4)	25*
2.5	(4,3)	(-1,5)	23*	(5,5)	(-7,2)	45*
3	(5,3)	(-1,6)	33*	(7,5)	(-8,4)	68*
3.5	(5,4)	(-6,3)	39*	(8,5)	(-8,5)	80*
4	(7,3)	(-6,5)	53*	(8,8)	(-11,3)	112*
4.5	(9,2)	(-6,7)	75*	(13,3)	(-9,10)	157*
5	(9,4)	(-1,10)	94*	(14,4)	(3,15)	198*
5.5	(9,6)	(-1,11)	105*	(16,0)	(8,14)	224*
6	(11,4)	(-9,8)	124*	(17,4)	(-12,13)	269*
6.5	(13,1)	(-7,11)	150*	(-19,0)	(9,17)	323*
7	(10,9)	(-4,13)	166*	(15,13)	(-19, 7)	352*

Table 1 gives for different radio ranges two vectors v_1 and v_2 generating the optimal periodic pattern as well as the minimal number of colors obtained by a periodic pattern, for both a 2-hop coloring and a 3-hop coloring. The '*' symbol highlights the optimality of the number of colors used.

We observe that 2-hop coloring of the grid with radio range $R = 3$ is not equivalent to 3-hop coloring of a grid with $R = 2$ in terms of the number of colors (33 vs. 25). We conclude that the optimal number of colors is not determined only by the product $h * R$, but also the values of h and R taken separately.

9.2 Comparison with other methods

Table 2 depicts the simulation results obtained with *VCM* for various grids. The method is compared to a distributed coloring algorithm using line/column as priority assignment heuristics. Results are given for 3-hop coloring. For high radio range values, the number of nodes should be high enough to allow the reproduction of the color pattern.

Table 2: Number of colors obtained for 3-hop coloring.

Radio range	Grid size	Colors VCM	Colors line/column
1	10x10	8*	8
	20x20	8*	8
	30x30	8*	8
1.5	10x10	16*	16*
	20x20	16*	16*
	30x30	16*	16*
2	10x10	25*	30
	20x20	25*	33
	30x30	25*	33
3	20x20	68*	80
	30x30	68*	83
3.5	20x20	80*	91
	30x30	80*	91

We observe that *VCM* provides an optimal 3-hop coloring, for any radio range. This is not true for any other priority assignment tested. Moreover, the number of colors does not depend on the grid size, provided that the grid size is high enough to allow the repetition of the color pattern defined by the generator vectors.

10. VCM IN REAL WIRELESS NETWORKS

In this paper, *VCM* has been described for grid topology since this topology is used by real applications briefly presented in Section 1. Notice however, that wireless communication may differ from what is expected by the theory that often uses simplified models: radio links may be asymmetric, a radio link may exist even if the remote node is at a distance higher than the transmission range R or conversely not exist even if the remote node is in the theoretical radio range. That is why, the first step in *VCM* is to select R such that two nodes that are at a distance greater than R are not neighbors.

Now, another real aspect in wireless ad hoc and sensor networks is the late arrival (mobility or in case of late start-up) and disappearance of nodes (a node is out of battery for instance). What is the impact of such impairments on *VCM*? We distinguish two categories:

1. Radio links disappearance: in this case, *VCM* always provides a valid coloring. The periodic coloring may still be optimal, as long as the percentage of missing radio links is below a given threshold L_1.
2. Radio links appearance: in this case, nodes that should not be neighbors (or heard nodes in case of asymmetric links) are. As a consequence, nodes having the same color may interfere because of these additional radio links. The periodic coloring provided by *VCM* may still be perfectly acceptable by the application as long as the percentage of additional radio links is below a given threshold L_2.

As a further work, we will evaluate the thresholds L_1 and L_2 and study how random topologies can be mapped in grids.

11. CONCLUSION

In this paper, we have presented *VCM*, the Vector-Based Coloring Method, able to provide an optimal periodic h-hop coloring of any grid, with h an integer ≥ 1, for any radio range R. This method is easy to use: a single round is needed. It suffices to compute the two generator vectors, as shown in this paper. Knowing its coordinates within the grid, each node deduces its color from a simple computation. We have shown that this h-hop node coloring is optimal in terms of colors and rounds. We presented also an upper and lower bound for the number of colors needed to color an infinite grid. *VCM* provides the optimal number of colors compared to all possible coloring including non periodic ones. Finally, we discussed how to apply *VCM* in real wireless networks.

12. REFERENCES

[1] Garey, M.; Johnson, D., *Computers and intractability: a guide to theory of NP-completeness*, W.H. Freeman, San Francisco, California, 1979.

[2] McCormick, S. T., *Optimal approximation of sparse Hessians and its equivalence to a graph coloring problem*, Math. Programming, v26, pp. 153-171, 1983.

[3] Amdouni, I.; Minet, P.; Adjih, C., *Node coloring for dense wireless sensor networks*, INRIA Research Report 7588, Paris-Rocquencourt, France, March 2011.

[4] Baumgartner, T.; Fekete, S. P.; Kamphans T.; Kröller A., Pagel, M. *Hallway Monitoring: Distributed Data Processing with Wireless Sensor Networks*, REALWSN, December 2010.

[5] Chakrabarty, K.; Iyengar S. S.; Qi H. R. and Cho E., *Grid Coverage for Surveillance and Target Location in Distributed Sensor Networks*, IEEE Transactions on Computers, Vol. 51, No. 12, 2002, pp. 1448-1453.

[6] McCulloch, J.; Mccarthy, P.; Siddeswara M. Guru; Wei, P.; Hugo, D.; Terhorst, A.; *Wireless sensor network deployment for water use efficiency in irrigation*, Proc. of the workshop on Real-world wireless sensor networks, March 2008.

[7] J.C. Bermond, F. Havet, F. Huc, C. Linhares-Sales, *Improper colouring of weighted grid and hexagonal graphs*, Discrete Mathematics, Algorithms and Applications, 2(3):395-411, 2010.

[8] Cargiannis, I.; Fishkin A.; Kaklamanis C.; Papaioannou E., A tight bound for online coloring of disk graphs, *Theoretical Computer Science*, 384, 2007.

[9] Kierstead H. A., Qin J., *Coloring interval graphs with First-Fit*, Discrete Math, Vol. 144, pp. 47-57, Sept. 1995.

[10] Riihijarvi J., Petrova M., Mahonen P., *Frequency Allocation for WLANs Using Colouring Techniques*, Journal of Ad Hoc and Sensor Wireless Networks, Vol. 3, pp. 121-139, 2007.

[11] Even G., Lotker Z., Ron D., Smorodinsky S., *Conflict-Free Colorings of Simple Geometric Regions with Applications to Frequency Assignment in Cellular Networks*, SIAM J. Comput, Vol. 33, No. 1, pp 94-136, 2004.

[12] Shakhar Smorodinsky, *Conflict-Free Coloring and its Applications*, CoRR abs/1005.3616, 2010.

[13] Schneider, J. and Wattenhofer R., Coloring unstructured wireless multi-hop networks, The 28th ACM symposium on Principles of distributed computing, PODC'09, USA, 2009.

[14] Rajendran, V.; Obraczka, K.; Garcia-Luna-Aceves, J.-J., *Energy-efficient, collision-free medium access control for wireless sensor networks*, Sensys'03, Los Angeles, California, November 2003.

[15] Rajendran, V.; Garcia-Luna-Aceves, J.J.; Obraczka, K., *Energy-efficient, application-aware medium access for sensor networks*, IEEE MASS 2005, Washington, November 2005.

[16] Rhee, I.; Warrier, A.; Aia, M.; Min, J., *Z-MAC: a hybrid MAC for wireless sensor networks*, SenSys'05, San Diego, California, November 2005.

[17] Rhee, I.; Warrier, A.; Xu, L., *Randomized dining philosophers to TDMA scheduling in wireless sensor networks*, Technical Report TR-2005-21, Dept of Computer Science, North Carolina State University, April 2005.

[18] Gobriel, S.; Mosse, D.; Cleric, R., *TDMA-ASAP: sensor network TDMA scheduling with adaptive slot stealing and parallelism*, ICDCS 2009, Montreal, Canada, June 2009.

[19] Lee, W. L.; Datta A.; Cardell-Oliver, R., FlexiTP: a flexible-schedule-based TDMA protocol for fault-tolerant and energy-efficient wireless sensor networks, *IEEE Transactions on Parallel and Distributed Systems*, vol. 19, 6, June 2008.

[20] Vector-Based Coloring Method page, `http://hipercom.inria.fr/SensorNet/VCM/`, INRIA, 2011.

[21] WuÌLbben, D.; Seethaler, D.; JaldeÌAn, J.; Matz, G., Lattice Reduction, A survey with applications in wireless communications, *IEEE Signal Processing Magazine*, vol. 29, May 2011.

[22] Vallée, B.; Vera, A.; *Lattice reduction in two dimensions: analyses under realistic probabilistic models*, Proceedings of AofAâĂŹ07,

[23] Minet, P.; Mahfoudh, S., *SERENA: SchEduling RoutEr Nodes Activity in wireless ad hoc and sensor networks*, IWCMC 2008, IEEE International Wireless Communications and Mobile Computing Conference, Crete Island, Greece, August 2008.

[24] Thue, A., *Über die dichteste Zusammenstellung von kongruenten Kreisen in einer Ebene*, Norske Vid. Selsk. Skr. No.1 (1910), 1-9.

[25] Tóth, L.F.; *Über die dichteste Kugellagerung*, Math. Z. 48 (1943), 676-684.

[26] Amdouni, I.; Adjih, C.; Minet, P.; *On the Coloring of Grid Wireless Sensor Networks: the Vector-Based Coloring Method*, Inria Research Report, RR-7756, September 2011.

13. ANNEX

LEMMA 3. *For any point V of $\mathbb{R}^2$, there exists a node V' of the grid $\mathbb{Z}^2$ such that $d(V,V') \leq \sqrt{2}/2$.*

PROOF. In the worst case, the node V occupies the center of a grid cell. It is at equal distance of two grid nodes that are diagonally opposed. Hence, its distance to one of them is equal to $\sqrt{2}/2$. □

LEMMA 4. *For any transmission range $R > \sqrt{2}$, for any grid node U, any node V that meets $d(U,V) \leq (R-\sqrt{2})h$ is at most h-hop away from U.*

PROOF. We consider the $h-1$ points of $\mathbb{R}^2$ that allow us to divide the segment $[U,V]$ in h equal parts.
Let W_i be these nodes, with $i \in [1,h-1]$.
For any $i \in [1,h-1]$, let W_i' the grid point the closest to W_i. For simplicity reason, we denote $W_0' = U$ and $W_h' = V$. We have $d(U,V) \leq \sum_{i=0}^{h-1} d(W_i', W_{i+1}')$.
We have $d(W_i',W_{i+1}') \leq d(W_i',W_i) + d(W_i,W_{i+1}) + d(W_{i+1},W_{i+1}')$.
According to Lemma 3, we have $d(W_i,W_i') \leq \sqrt{2}/2$. Hence, we get $d(W_i',W_{i+1}') \leq \sqrt{2} + d(W_i,W_{i+1})$. By construction, $d(W_i,W_{i+1}) = d(U,V)/h$.
If $d(U,V) \leq (R-\sqrt{2})h$, then $d(W_i',W_{i+1}') \leq R$. Hence, nodes W_i for $i \in \{1,2,\ldots,h-1\}$ constitute a h-hop path between U and V. □

LEMMA 5. *For any transmission range R, for any two grid nodes U and V, in h-hop coloring, if $d(U,V) > hR$ then U and V are at least $(h+1)$-hop away.*

PROOF. By contradiction assume that, U and V are h-hop away or less. Let W_i be the $k-1$ nodes constituting the k-hop path between U and V, with $k \leq h$. Let $W_1 = U$, and $W_h = V$. Since nodes W_i are 1-hop neighbors, we have:

$$|\vec{UV}| = |\sum_{i=1}^{h} \vec{W_iW_{i+1}}| \leq \sum_{i=1}^{h} d(W_i,W_{i+1}) \leq hR.$$

Hence the contradiction. □

LEMMA 6. *If u_1 and u_2 are reduced generator vectors of a lattice $\Lambda(u_1,u_2)$, with $|u_1| \leq |u_2|$, then for any vector w such that $w = \alpha u_1 + \beta u_2$, and $\alpha, \beta \in \mathbb{Z}^2$, we have $|w| \geq \frac{3}{4}\alpha^2|u_1|^2$, and $|w| \geq \frac{3}{4}\beta^2|u_1|^2$.*

PROOF. Let W the node of coordinates (α,β). We have:

$$\begin{aligned} |\vec{UW}|^2 &= \alpha^2|u_1|^2 + \beta^2|u_2|^2 + 2\alpha\beta\,(u_1 \cdot u_2) \\ &\geq \alpha^2|u_1|^2 + \beta^2|u_2|^2 - 2|\alpha||\beta|\,|u_1 \cdot u_2|. \end{aligned}$$

Since u_1 and u_2 are reduced vectors, we can use the property given in the System (1), we get:

$$\begin{aligned} |\vec{UW}|^2 &\geq \alpha^2|u_1|^2 + \beta^2|u_1|^2 + |\alpha||\beta||u_1|^2. \text{ Hence,} \\ |\vec{UW}|^2 &\geq ((|\alpha|-|\beta|)^2 + |\alpha||\beta|)|u_1|^2. \end{aligned}$$

Notice that this quantity does not change if we change the sign of α or of β. Thus, we assume $(\alpha \geq 0)$, $(\beta \geq 0)$, and let $f(\alpha,\beta) = (\alpha-\beta)^2 + \alpha\beta$.
By a change of variable $\beta = \frac{\alpha}{2} + \lambda$, we get:
$f(\alpha,\beta) = \frac{3}{4}\alpha^2 + \lambda^2 \geq \frac{3}{4}\alpha^2$. Similarly, we have $f(\alpha,\beta) \geq \frac{3}{4}\beta^2$.
Hence the lemma. □

LEMMA 7. *Consider any transmission range $R > \sqrt{2}$, two reduced generator vectors u_1 and u_2 of the lattice $\Lambda(u_1,u_2)$, and a node W with $\vec{UW} = \alpha u_1 + \beta u_2$ for some α and β in $\mathbb{Z}$. Assume also that the point V_1 such that $\vec{UV_1} = u_1$ is at strictly more than h hops from U. Then:*
if $|\alpha| \geq \mu(R)$ or $|\beta| \geq \mu(R)$ where $\mu(R) = \frac{2\sqrt{3}R}{3(R-\sqrt{2})}$, we have: W is strictly more than h hops away from U.

PROOF. Since u_1 and u_2 are reduced, we can apply Lemma 6 and obtain:

$$\begin{aligned} |\vec{UW}|^2 &\geq f(\alpha,\beta)|u_1|^2 \\ &\geq \frac{3}{4}\alpha^2|u_1|^2, \text{ and as } \alpha \geq \mu(R) \\ &\geq \frac{3}{4}\left(\frac{2\sqrt{3}R}{3(R-\sqrt{2})}\right)^2|u_1|^2 \\ &\geq \left(\frac{R}{R-\sqrt{2}}\right)^2|u_1|^2 \end{aligned}$$

Since the point V_1 is strictly more than h-hop away from U, the lemma 4 implies by contradiction that $|u_1| = |\vec{UV_1}| > (R-\sqrt{2})h$. It follows that:

$$|\vec{UW}|^2 > R^2h^2$$

Applying Lemma 5, we obtain the result. □

Relay-Aided High-Throughput Path Selection in Multi-rate Wireless Mesh Networks

Bipul Sen, Xin Zhao, Lavy Libman, Sanjay Jha
School of Computer Science and Engineering
University of New South Wales
Sydney, NSW 2052, Australia
{bsen,xinzhao,llibman,sjha}@cse.unsw.edu.au

ABSTRACT

This paper proposes an *opportunistic rate-adaptive expected transmission time (ORETT)* metric, which exploits cooperative (relay-aided) retransmissions to achieve high-throughput routing in multi-rate wireless mesh networks. This metric captures the combined effects of MAC-layer cooperative retransmission by neighbor nodes with transmission rate diversity (and rate-dependent link quality). In our approach, a relay node is selected among one-hop neighbors to assist with packet retransmissions and minimize the expected retransmission time. The paper describes the design and implementation of the ORETT routing metric using the DSR routing protocol. Our extensive simulation on the Qualnet platform confirms that multi-rate routing using ORETT significantly reduces the overall transmission time while yielding higher packet delivery ratio compared to single rate unicast or non-cooperative ETT based routing.

Categories and Subject Descriptors

C.2.1 [**Network Architecture and design**]: Wireless Communication; C.2.2 [**Network Protocols**]: Metrics—*Routing Protocols, performance measures*

Keywords

Cooperative routing, routing metric, wireless mesh.

1. INTRODUCTION

The IEEE 802.11 standards support multi-rate capability [8], allowing the physical layer to adapt to varying channel conditions using different modulation techniques and send data at different transmission rates. The method to select rates in multi-rate radio is not specified by the 802.11 standards. As a result, several rate adaptation protocols have been proposed with the aim of selecting the rate that optimizes the throughput for given channel conditions. Most prevailing rate adaptation methods, e.g. [3,10,11], are purely sender-side schemes where the sender selects the best rate based on the number of consecutive successes or losses of MAC-layer acknowledgment (ACK) frames. As an alternative, the Receiver Based Auto Rate (RBAR) scheme [7] allows the receiving node to select the rate using the signal-to-noise ratio (SNR) of the request-to-send (RTS) packet. The Opportunistic Auto Rate (OAR) protocol [15] applies the same receiver-based approach, but sends back-to-back data frames when channel conditions are good. However, all these schemes only consider the data rate of the direct link between the sender and the receiver, without taking into account the possibility of opportunistic overhearing by neighboring relay nodes. In some cases, it may be advantageous to retransmit packets using cooperative relay nodes that have better links to the receiver than the direct one.

Cooperative communication schemes at the MAC layer have been the subject of several studies in recent years, the majority of which are focused on improvements within the IEEE 802.11 MAC context by exploiting the benefit of cooperative communications [4, 13, 20, 22, 23]. In particular, *r*DCF [22], RAMA [23] and CoopMAC [13] are three similar MAC-layer approaches that take into account the physical-layer transmission rate diversity and volatility, and adaptively respond to variations in the rate that a link can support. Their idea is to temporarily reroute around a slow (low-rate) direct link using a two-hop local "detour" via a neighboring relay. Conversely, COBRA-MAC [1] and coop80211 [14] are MAC schemes based on *opportunistic* cooperation, where relays overhear the transaction on the direct link and only step in to make a retransmission if the direct transmission has failed. From the routing perspective, a few studies have attempted to design a cross-layer metric that reflects the MAC-layer cooperation gain in the route choice; among these, most closely related to our work are the ETTC [21] and CETT [19] metrics, whose designs are based on the CoopMAC [13] and COBRA-MAC [1] cooperative schemes respectively. Despite the extensive research on the benefits of cooperative communications at the physical and MAC layers for single-hop communications, the question of exploiting it for high-throughput routing in multi-hop wireless networks remains largely an open problem; in this context, our proposed ORETT metric extends these prior proposals, combining both the opportunistic overhearing and rate diversity aspects for high-throughput route choice.

The main contributions of this paper are as follows. We propose a new metric, namely the opportunistic rate-adaptive expected transmission time (ORETT), to capture the combined effects of MAC-layer cooperative retransmission and per-link estimates of packet transmission time for achiev-

MSWiM'12, October 21–25, 2012, Paphos, Cyprus.

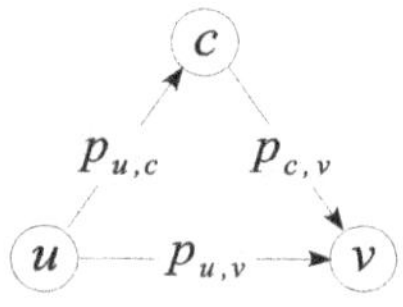

Figure 1: One-hop transmission from node u to node v using node c for cooperative retransmission.

Table 1: PDR at different transmission rates

	$r_1 = 5.5$ Mbps	$r_2 = 11$ Mbps
v	0.7	0.2
c	0.95	0.8

ing high throughput in multi-rate wireless mesh networks (WMN). Using an implementation on the Qualnet platform [16] of a minimal, backward-compatible modification of the 802.11 MAC incorporating a cooperative retransmission mechanism proposed in our earlier work (which we call Δ-MAC [17]), we provide detailed simulation results of the DSR routing protocol with the ORETT metric that demonstrates its effectiveness in multi-rate networks.

The rest of this paper is organized as follows. Section 2 presents a motivating example, followed by our system model in Section 3. Section 4 discusses the ORETT metric and the optimal rate and relay node selection. Section 5 introduces Δ-MAC and our implementation of the DSR protocol. Simulation experiments and results are provided in Section 6. Finally, Section 7 concludes the paper.

2. MOTIVATING EXAMPLE

In wireless networks, link layer protocols employ automatic repeat request (ARQ) mechanisms to combat packet losses. Because of the broadcast nature of wireless transmission, any packet can be opportunistically overheard by neighboring nodes. If a retransmission is necessary, it may often be advantageous to make it from one of the overhearing nodes rather than the original packet source, if this reduces the expected time until the packet is delivered to the receiver; hence, relay-aided cooperative retransmission may reduce the network bandwidth consumption, and thus effectively increase the network throughput.

Consider the scenario in Fig. 1, where node u has a data packet for destination node v and a cooperative relay node c is within the transmission range of both u and v and can overhear the transmission from u to v. Assuming a set of feasible transmission rates $r_k, k = 1, \ldots, K$ in the network, existing cooperative protocols such as rDCF [22] and CoopMAC [13] aim to replace a slower link with a faster two-hop "detour" so that total transmission time can be reduced. On the other hand, in protocols such as COBRA-MAC [1] and CMAC [18], the relay helps with retransmitting on behalf of the source *opportunistically*, only when the original transmission is unsuccessful. For example, suppose that the successful delivery ratio of a packet of size L from u to the receiver and relay nodes is as shown in Table 1 (assume that the link quality between c to v is 100% perfect at all available rates). Without cooperation, the expected transmission time of a packet (including retransmissions by u) is $\frac{L}{0.7 \cdot 5.5Mbps} \approx 0.26L\mu s$ if the rate $r_1 = 5.5Mbps$ is used, or

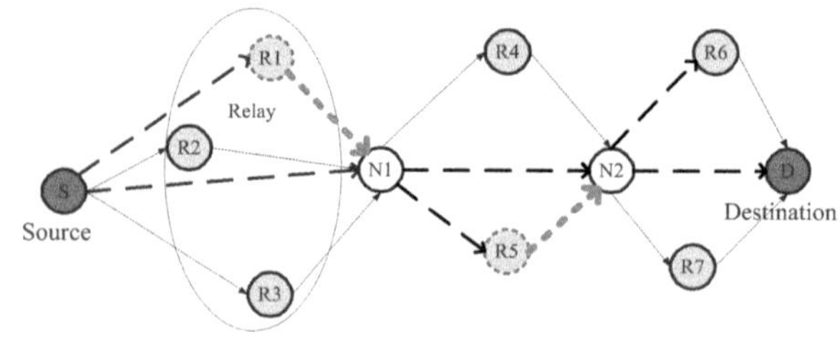

Figure 2: Multi-hop cooperative routing.

$\frac{L}{0.2 \cdot 11Mbps} \approx 0.45L\mu s$ for the rate $r_2 = 11Mbps$. Thus, the lower rate is preferred; in other words, the shorter transmission time does not compensate the significantly lower success rate of the higher rate. On the other hand, with the cooperation of c, node u only needs to keep retransmitting until the packet is received either at v or at c; in the latter case, the final retransmission will then be undertaken by c (at the higher rate of r_2, since the link is assumed to be perfect anyway). Thus, the expected packet transmission times, for rates r_1 and r_2 (used initially by node u) respectively, are

$$\frac{\mathrm{L}}{[1-(1-0.95)(1-0.7)] \cdot 5.5\mathrm{Mbps}} + \frac{0.95(1-0.7)}{1-(1-0.95)(1-0.7)} \cdot \frac{\mathrm{L}}{11\mathrm{Mbps}} \approx 0.21\mathrm{L}\mu\mathrm{s}$$

and

$$\frac{\mathrm{L}}{[1-(1-0.8)(1-0.2)] \cdot 11\mathrm{Mbps}} + \frac{0.8(1-0.2)}{1-(1-0.8)(1-0.2)} \cdot \frac{\mathrm{L}}{11\mathrm{Mbps}} \approx 0.18\mathrm{L}\mu\mathrm{s}.$$

Accordingly, in the presence of the cooperative relay, it is preferable for u to use the higher rate: even though it leads to a very low success probability on the direct link, this is compensated by the overhearing probability of the relay (which decreases only marginally at the higher rate) and the shorter packet transmission time. This example shows that the traditional approach to rate selection, based on the statistics of the direct link alone, is inadequate in the presence of cooperation, and that the questions of rate and relay selection should be addressed *jointly*. To that end, we design a metric (ORETT) that reflects the expected transmission time on a link, accounting both for the MAC-layer cooperative retransmission and the ability to use multiple rates.

3. SYSTEM MODEL

We consider a multi-hop wireless network represented by a directed graph $G = (V, E)$, where V is the set of nodes and E is the set of directed links. A directed link $\langle u, v\rangle$ from node u to node v exists if node v is within the transmission range of node u; more specifically, we require the packet delivery probability of link $\langle u, v\rangle$ to be non-zero for at least one transmission rate in the set of feasible rates $\{r_k\}, k = 1, \ldots, K$. The set of nodes $\{v : \langle u, v\rangle \in E\}$ forms the one-hop neighborhood N_u of node u. Each node is equipped with one radio. All radios are tuned to a common channel with fixed transmission power for all transmission rates.

We now explain the assumed cooperative retransmission scheme using a simple topology depicted in Fig. 2. Suppose that source S has a (network-layer) data packet for destination node D, which is to be forwarded along the route $S \to N1 \to N2 \to D$. For concreteness, we describe the cooperation protocol assumptions using the first-hop link (from S to N1); cooperation along subsequent links is similar. Assume that a set of neighbor nodes, $R = \{R1, R2, R3\}$, are inside the shared transmission area of S and N1 and can overhear and cooperatively retransmit any ongoing transmissions; furthermore, assume that node R1 is identified to be the best relay node in this set (the relay node selection process is described in detail in the next section). In this case, the cooperative scheme works as follows.

If the first transmission attempt from node S to N1 is successful, the transmission is completed. In that case relay node R1 plays no role. Otherwise, if the transmission from S is overheard by relay R1 but fails to be received by N1, then R1 will take over the role of retransmitting the lost packet. Specifically, relay R1 will send an ACK packet to the packet originator S, and thereafter, R1 follows the standard ARQ mechanism to retransmit the packet to node N1. Finally, as long as neither node N1 nor R1 receives the packet, it is retransmitted by node S. For further details on our cooperative retransmission approach, in particular the protocol actions after ACK losses and recovery from possible duplicate packets, please refer to our previous work [17].

We point out that the use of an explicit ACK by node R1, confirming that it takes responsibility for the packet from that point, differs from the approach of COBRA-MAC; this allows the relay to retransmit the packet several times if necessary, whereas in COBRA-MAC the relay is limited to one attempt before the retransmission must revert to the original source. As a result, the ORETT metric (introduced in the next section) can attain a lower value than the similar metric CETT based on COBRA-MAC [19], even when multiple rates are not used.

4. THE ORETT CROSS-LAYER ROUTING METRIC

For a multi-rate wireless mesh network, an optimum rate as well as best relay node needs to be selected so that total transmission time is minimized. This section discusses the rate and best relay node selection procedure in the computation of the ORETT metric. The computation of ORETT for a one-hop sender-receiver pair uses the packet delivery ratio (PDR) in the links among all the nodes involved, which is a function of the transmission rate used. More specifically, for a given set of three nodes u, v and c as in Fig. 1, we define $\mathrm{ORETT}_{u,r,v}^{(c)}$ as expected transmission time (including retransmission time) required for one-hop transmission of a data packet from node u to node v with node c as the cooperating node at rate r. Let $\mathrm{ORETT}_{u,v}^{(c)}$ as minimal ORETT for the one-hop transmission from node u to node v with the help of the node c across all available transmission rates. $\mathrm{ETT}_{u,r,v}$ is defined as expected transmission time (including retransmissions) required for the one-hop transmission from node u to node v at a rate r. If PDR of link $(u-v)$ is defined as $p_{u,r,v}$ at transmission rate r, the ETT of packet L is given by [6]

$$\mathrm{ETT}_{u,r,v} = \frac{\mathrm{L}}{r} \times \frac{1}{p_{u,r,v}}. \tag{1}$$

In computing ORETT, we assume that the sender keeps transmitting the packet until it receives an ACK from either the receiver or the relay. Similarly, we assume that, after the cooperating node receives the packet, it continues to retransmit the packet until it is acknowledged by the receiver. For simplicity, we henceforth assume that an ACK from either the receiver or the cooperating node is received at the sender with probability one. This is due to the fact that the loss rate of ACK packets (14 bytes in the 802.11 MAC) is typically very low in wireless networks, compared to that of data packets [17]; however, we point out that the ORETT metric can be extended to explicitly account for a positive ACK loss probability in a straightforward manner.

Assuming that the links between u, v and c (and the corresponding successful transmission probabilities $p_{u,r,v}$, $p_{u,r,c}$ and $p_{c,r,v}$) are statistically independent, the one-hop transmission of a data packet from node u to node v with node c as the cooperating node has three possible outcomes:

- With probability $(1-p_{u,r,v})(1-p_{u,r,c})$, neither node v nor node c receives the packet. In that case, the expected future retransmission time is again $\mathrm{ORETT}_{u,r,v}^{(c)}$.
- With probability $(1-p_{u,r,v})p_{u,r,c}$, the transmission from node u is overheard by node c, but node v fails to receive the packet. Since node c will take over the role of retransmitting the packet to node v, the expected future retransmission time in this state is $\mathrm{ETT}_{c,v}$.
- With probability $p_{u,r,v}$, node v receives the packet, and hence no additional retransmissions are required.

Let us define the transmission time of data packet between links (u-v), (u-c), and (c-v) are $\mathrm{T}_{u,r,v}$, $\mathrm{T}_{u,r,c}$, and $\mathrm{T}_{c,r,v}$, respectively. Summing over the above outcomes, $\mathrm{ORETT}_{u,r,v}^{(c)}$ can be derived as

$$\begin{aligned}\mathrm{ORETT}_{u,r,v}^{(c)} = &(1-p_{u,r,v})(1-p_{u,r,c})(\mathrm{T}_{u,r,v} + \mathrm{ORETT}_{u,r,v}^{(c)})\\ &+ (1-p_{u,r,v})p_{u,r,c}(\mathrm{T}_{u,r,v} + \mathrm{ETT}_{c,v})\\ &+ p_{u,r,v}\mathrm{T}_{u,r,v}\ .\end{aligned} \tag{2}$$

Rearranging (2), we have

$$\mathrm{ORETT}_{u,r,v}^{(c)} = \frac{\mathrm{T}_{u,r,v} + (1-p_{u,r,v})p_{u,r,c}\mathrm{ETT}_{c,v}}{p_{u,r,v} + p_{u,r,c} - p_{u,r,v}p_{u,r,c}} \tag{3}$$

where

$$\mathrm{ETT}_{c,v} = \min_{r}\ \mathrm{T}_{c,r,v} \times \frac{1}{p_{c,r,v}}\ . \tag{4}$$

The optimal transmission rate for $(u-v)$ link in presence of relay c is found by exhaustive search among all possible rates so as to minimize the value of ORETT:

$$r^* = \arg\min_{r}\ \mathrm{ORETT}_{u,r,v}^{(c)} \tag{5}$$

Finally, if more than one potential relay node exists in the vicinity of u and v, the above process is repeated for each candidate, and the relay node c^* (and corresponding transmission rate r^*) is chosen to be the one that minimizes the ORETT value overall. Let $\mathrm{ORETT}_{u,v}$ denote the minimally required ORETT for the one-hop transmission from node u to node v with the help of the best cooperating node:

$$\mathrm{ORETT}_{u,v} = \mathrm{ORETT}_{u,r^*,v}^{(c^*)}. \tag{6}$$

Similarly, the optimal rate for the link ($c^* - v$) can be derived by exhaustive search among all possible rates so as to minimize the value of ETT

$$r_c^* = \arg\min_r \ \mathrm{ETT}_{c^*,r,v} \ . \tag{7}$$

In a situation where no suitable relay exists between node u and node v, the transmission rate

$$r_u^* = \arg\min_r \ \frac{\mathrm{L}}{r} \times \frac{1}{p_{u,r,v}}. \tag{8}$$

and corresponding cost

$$\mathrm{ETT}_{u,v} = \min_r \ \frac{\mathrm{L}}{r} \times \frac{1}{p_{u,r,v}} \tag{9}$$

are used for direct communication between node u and node v instead, and the ORETT metric is defined to be equal to ETT for consistency.

THEOREM 1. $\mathrm{ORETT}_{u,v} < \mathrm{ETT}_{uv}$ if ETT_{cv} is less than ETT_{uv}.

PROOF. To satisfy the above theorem, the difference between Eq. 3 and Eq. 1 must be less than zero. Comparing (1) and (3) and after some algebraic manipulation, we have $\mathrm{ECTT}_{u,r,v}^{(c)} - \mathrm{ETT}_{u,r,v}$ as follows:

$$\frac{(p_{u,r,v}\mathrm{T}_{c,r,v} - p_{c,r,v}\mathrm{T}_{u,r,v})(1-p_{u,r,v})p_{u,r,c}}{p_{u,r,v}(1-p_{u,r,v})p_{u,r,c}p_{c,r,v} + p_{u,r,v}^2 p_{c,r,v}} \ .$$

Clearly, $(p_{u,r,v}\mathrm{T}_{c,r,v} - p_{c,r,v}\mathrm{T}_{u,r,v}) < 0$ when $\mathrm{ETT}_{c,v}$ is less than $\mathrm{ETT}_{u,v}$. This means, as long as $p_{u,r,c} > 0$, using c for cooperative retransmission reduces the expected transmission time. □

5. CROSS-LAYER ROUTING WITH ORETT

The objective of cooperative routing based on the ORETT metric in multi-hop WMNs is to find the optimal path for a unicast session that minimizes the sum of ORETT required for delivering data packet from a source to destination. Minimizing the per-packet transmission time generally reduces the radio resources usage, and results in lower interference in the network and eventually higher network throughput. Below we consider the adaptations required at the lower layers to enable ORETT-based routing, specifically, transmission rate adaptation at the physical layer and the changes required to the distributed coordination (DCF) among the nodes at the MAC layer. With regard to the routing protocol at the network layer, we use DSR [9] for evaluation purposes in this work (for a discussion of some minor adaptations in DSR to support cooperative retransmissions, see [17]).

In our cooperative retransmission policy, each node periodically sends a broadcast (probe) packet at all 802.11b bit-rates. Every node keeps track of the fraction of probe packets it receives from each of its neighbors at each bit rate. Based on the predicted link quality values gathered from probing, each node calculates the optimum transmit rate as discussed in section 4 before sending packets to the physical layer. We do not follow cooperative communication schemes such as [13,23] where the transmit bit-rate selection is based on observing the signal-to-noise ratio (SNR) of the preceding RTS at the receiver. The latter approach to rate selection is more suitable when nodes are mobile, so that immediate channel quality is preferable over long-term measurements where the relationship between the SNR and error rate is predictable [2]; on the other hand, ORETT is designed for wireless mesh networks with stationary nodes, thus rate selection based on recent packet loss history is better-suited to this context. In our implementation, one probe packet of size 512 bytes is sent once per second on average. More specifically, we consider three 802.11b rates (2, 5.5 and 11 Mbps) to evaluate the performance of our ORETT based cooperative transmission policy. We have used 802.11b in our simulations for simplicity; however, our scheme is applicable to 802.11a/g/n standards as well. In each three-second interval, our implementation sends a probe at each of the three rates at an independently chosen random point. Finally, each receiving node expired its record of received probes after 60 seconds.

The Δ-MAC protocol introduced in our previous work [17] uses a mechanism similar to that of RTS/CTS handshaking in basic IEEE 802.11 DCF to alleviate collisions, with a minor modification to include the cooperative relay node as well as the sender and receiver. In this paper, we employ the same Δ-MAC protocol for evaluation purposes of the ORETT metric, extending it only to include the selection of the best rate as well as the best relay as described in section 4.

6. EVALUATION

6.1 Simulation Configuration

We use QualNet [16] to simulate a WMN of 50 stationary mesh nodes that are randomly distributed in an area of 1600 m × 1600 m. Each node has one interface with fixed transmission power. All experiments use the two-ray propagation path loss model, with free space path loss exponent of 2 for near sight and plane earth path loss exponent of 4 for far sight. The physical layer uses PHY802.11b and restricts transmission rates to 2, 5.5, or 11 Mbps. A number of source-destination pairs (varying between 1 and 9 randomly selected pairs) simultaneously transmit constant bit rate (CBR) traffic at 50 packets per second, with a data payload size of 512 bytes. In addition, background multicast flows in the form of multicast CBR (MCBR) sessions are randomly generated, running ODMRP [12] as the multicast routing protocol. This allows us to check the impact of ORETT-based routing in a more challenging situation where multicast background traffic increases the chance of packet collision. In our simulation setup, we use 3 multicast sources and 5 group members, with each source generating MCBR traffic at a rate of 1 packet per second, with a data payload size of 128 bytes. The total simulation time is 550 seconds, allowing the first 120 seconds for warm-up to acquire link PDR's for initial routing set-up and remove transients. In each of the scenarios below, we perform the experiment for 25 independent runs and show the average as well as the standard deviation bars in the graphs. For the estimation of link PDR's required for the metric calculations, we apply the probing technique of [5].

We focus on three important performance measures: (1) End-to-end packet delivery ratio for the unicast flows, (2) Total ORETT, i.e. the sum of all MAC-layer transmission times (including retransmissions) to deliver a data packet in a unicast flow, and (3) Cooperation ratio, defined as the proportion of cooperative relay retransmissions out of the total number of packets transmitted in the network.

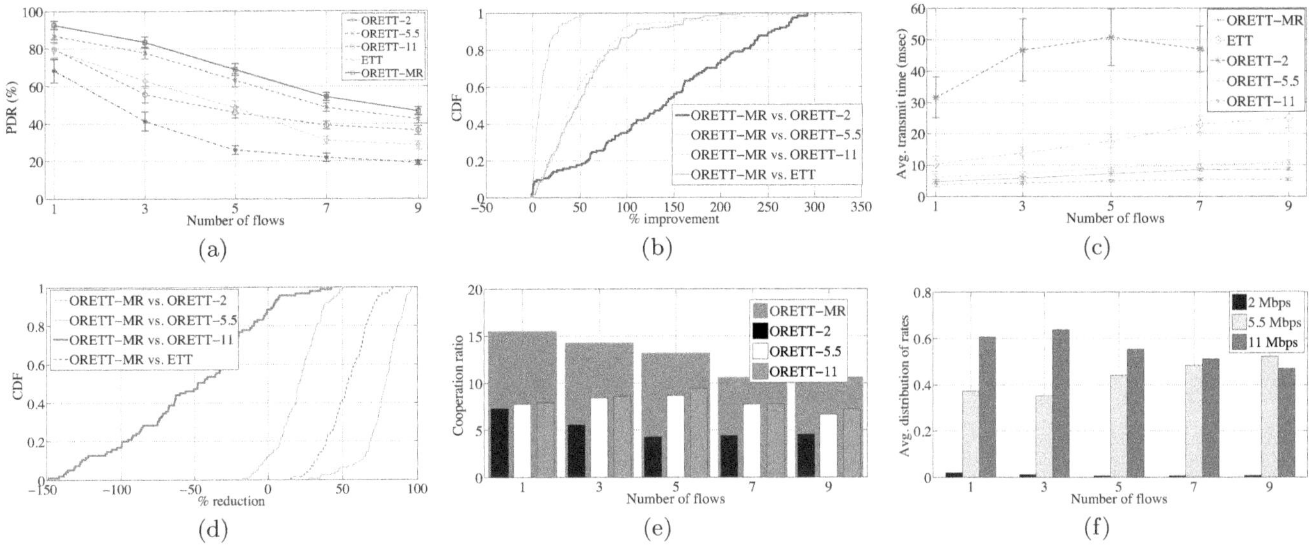

Figure 3: Performance comparison (a) -(b) Line graph display the avg. result of end-to-end PDR. CDF shows the PDR improvement(%) of ORETT-MR over other metrics. (%) of improvement = $(\frac{PDR_{ORETT-MR}}{PDR_{other\ metric}}-1)\times 100\%$. (c) - (d) Line plot display total transmit time per packet. CDF shows (%) of reduction in terms of packet transmit time of ORETT-MR over other metrics. Percentage of reduction = $(1-\frac{TxTIME_{ORETT-MR}}{TxTIME_{other\ metric}})\times 100\%$.(e) Percentage of packets transmitted by relay nodes, and (f) Transmit rate distribution in ORETT-MR based routing.

6.2 Simulation Results and Discussion

Our simulations consider the following five scenarios: DSR in conjunction with ETT, DSR in conjunction with ORETT for three single rates (ORETT-2 at 2 Mbps, ORETT-5.5 at 5.5 Mbps, and ORETT-11 at 11 Mbps), and DSR in conjunction with ORETT for a multi-rate network (ORETT-MR). Figure 3(a) shows the average end-to-end PDR for the five different setups. Almost in all simulation runs, ORETT-MR based routing presents higher PDR compared to ETT. For example, ORETT-MR presents 34% higher PDR than ETT when there are three simultaneous flows in the network. Without considering the rate diversity, single-rate cooperative routing shows lower PDR than ORETT-MR. With five simultaneous flows, ORETT-MR achieves 102%, 9%, and 51% higher PDR than routing at ORETT-2, ORETT-5.5, ORETT-11, respectively. In Figure 3(b), the cumulative distribution function (CDF) for the improvement (in percentage terms) of ORETT-MR over ETT and single-rate routing is presented. To calculate the CDF, we have considered 125 different PDR measurements for each routing metric. As the figure shows, the median improvement of PDR is 47% compared to ETT, and the improvement of ORETT-MR over single-rate routing is similarly significant.

Generally, the paths between nodes become longer (in terms of hop count) at higher rates due to possible loss of connectivity on some links. As a result, a larger proportion of packets are lost due to retransmit limits being reached at ORETT-11. Conversely, under clear channel conditions, using a lower rate leads to higher end-to-end PDR. However, under heavy load and channel errors, the longer channel occupancy times lead to increased incidence of collisions and other failures, and the actual PDR can become much lower; indeed ORETT-2 shows the worst performance as expected. For ETT, which does not utilize cooperative retransmission policy, the performance in terms of PDR is clearly different from ORETT-MR. Overall, multi-rate ORETT based routing outperforms single-rate and ETT based routing, because the path selection accounts for the advantages of rate diversity and results in paths with a higher reliability.

The result in Figure 3(c) shows the average end-to-end packet transmit time at different traffic configurations. In comparison to ETT, ORETT-MR consumes 51% less transmission time. It consistently outperforms single rate cooperative routing metrics regardless of network load. For example, ORETT-MR reduces transmission time by 79.15% and 20.30% when compared with ORETT-2 and ORETT-5.5 respectively in the case of three simultaneous flows. Although ORETT-11 requires less packet transmission time, the actual end-to-end PDR is much lower compared to ORETT-MR. In Figure 3(d), a CDF is shown over the 125 different measurements. The median transmission time is reduced by 53% with ORETT-MR compared to ETT. For up to 60% of simulation runs, ORETT-MR reduces transmission time by more than 15% over routing with ORETT-5.5. Interestingly, in some simulation runs, ORETT-5.5 requires less transmission time than ORETT-MR; indeed, because of low link PDR at higher rates, in some instances ORETT-MR prefers 2 Mbps links with a higher transmission delay. However, the overall reduction of transmission time for ORETT-MR compared to ETT clearly shows the effectiveness of cooperative retransmission policy. The results confirm that, by exploiting the rate diversity, ORETT consumes less transmission time than routing at single-rate or ETT.

Figure 3(e) shows how often relay nodes cooperate to forward data packets. ORETT-MR shows higher cooperation ratio irrespective of traffic load in the network; in other words, relay nodes are much more likely to be selected when multiple transmission rates are considered. This effect is clearly evident in Figure 3(e) where a large fraction of packets has been forwarded by relay nodes when ORETT-MR is used. Note that the cooperation ratio for ORETT-MR varies from 11% to 16% even when the number of flows is increased from one to nine. Figure 3(f) shows how optimal rate selection by ORETT-MR makes use of all the available rates, explaining why ORETT-MR always performs better than single rate routing. The rate selection mostly oscillates between 5.5 Mbps and 11 Mbps; however, in absence of alternate high speed links, a few node pairs selected 2 Mbps as an optimal rate. For example, we have 55% of node pairs using 11 Mbps, 44% using 5.5 Mbps, and 1% using 2 Mbps as the optimal rate when five simultaneous flows are present in the network.

7. CONCLUSION

In this paper, we have introduced ORETT as a new high throughput routing metric for multi-hop multi-rate WMNs. ORETT effectively utilizes wireless broadcast advantages for MAC-layer relay aided retransmission mechanism on an opportunistic basis, and includes an adaptive transmission rate control to minimize the expected transmission time in the presence of such cooperative retransmission. Our simulation experiments have confirmed that routing in multi-rate WMNs using the ORETT metric achieves a significant reduction of packet latency and higher end-to-end packet delivery ratio than either traditional ETT or single-rate cooperative routing.

8. ACKNOWLEDGMENTS

This work was supported in part by ARC Discovery Grant DP1096359. We thank Jun Guo for his input.

9. REFERENCES

[1] N. Agarwal, D. ChanneGowda, L. Kannan, M. Tacca, and A. Fumagalli. IEEE 802.11b cooperative protocols: A performance study. In *Proc. IFIP NETWORKING '07*, pages 415–426, GA, USA, 2007.

[2] D. Aguayo, J. Bicket, and R. Morris. SrcRR: A high throughput routing protocol for 802.11 mesh networks (draft). `http://pdos.csail.mit.edu/rtm/srcrr-draft.pdf`, 2005.

[3] J. Bicket, D. Aguayo, S. Biswas, and R. Morris. Architecture and evaluation of an unplanned 802.11b mesh network. In *Proc. ACM MobiCom '05*, pages 31–42, Cologne, Germany, Aug. 2005.

[4] A. Bletsas, A. Khisti, D. Reed, and A. Lippman. A simple cooperative diversity method based on network path selection. *Selected Areas in Communications, IEEE Journal on*, 24(3):659–672, 2006.

[5] D. S. J. De Couto, D. Aguayo, J. Bicket, and R. Morris. A high-throughput path metric for multi-hop wireless routing. In *Proc. ACM MobiCom '03*, pages 134–146, CA, USA, 2003.

[6] R. Draves, J. Padhye, and B. Zill. Routing in multi-radio, multi-hop wireless mesh networks. In *Proc. ACM MobiCom '04*, pages 114–128, PA, USA, 2004.

[7] G. Holland, N. Vaidya, and P. Bahl. A rate-adaptive MAC protocol for multi-hop wireless networks. In *Proc. ACM MobiCom '01*, pages 236–251, Rome, Italy, Jul. 2001.

[8] IEEE Computer Society. Wireless LAN medium access control (MAC) and physical layer (PHY) specifications. IEEE 802.11 Standard, 2007.

[9] D. Johnson, Y. Hu, and D. Maltz. The dynamic source routing protocol (DSR) for mobile ad hoc networks for IPv4. RFC 4728 (Experimental), Feb. 2007.

[10] A. Kamerman and L. Monteban. WaveLAN®-ii: a high-performance wireless LAN for the unlicensed band. *Bell Labs technical journal*, 2(3):118–133, 1997.

[11] M. Lacage, M. Manshaei, and T. Turletti. IEEE 802.11 rate adaptation: A practical approach. In *Proc. ACM MSWiM '04*, pages 126–134, Venice, Italy, Oct. 2004.

[12] S. J. Lee, M. Gerla, and C. C. Chiang. On-demand multicast routing protocol. In *Proc. IEEE WCNC '99*, pages 1298–1302, LA, USA, Sep. 1999.

[13] P. Liu, Z. Tao, S. Narayanan, T. Korakis, and S. S. Panwar. CoopMAC: a cooperative MAC for wireless LANs. *Selected Areas in Communications, IEEE Journal on*, 25(2):340–354, Feb. 2007.

[14] V. Nikolyenko and L. Libman. Coop80211: Implementation and evaluation of a SoftMAC-based Linux kernel module for cooperative retransmission. In *Proc. IEEE WCNC '11*, pages 239–244, Cancun, Mexico, Mar. 2011.

[15] B. Sadeghi, V. Kanodia, A. Sabharwal, and E. Knightly. OAR: an opportunistic auto-rate media access protocol for ad hoc networks. *Wireless Networks*, 11(1-2):39–53, 2005.

[16] Scalable Network Technologies. `http://www.scalable-networks.com/`.

[17] B. Sen, J. Guo, X. Zhao, and S. Jha. ECTX: A high-throughput path metric for multi-hop wireless routing exploiting MAC-layer cooperative retransmission. In *Proc. IEEE WoWMoM '12*, San Francisco, CA, USA, Jun. 2012.

[18] S. Shankar, C. Chou, and M. Ghosh. Cooperative communication MAC (CMAC) – a new MAC protocol for next generation wireless LANs. In *Proc. IWCMC '05*, Maui, HI, USA, Jun. 2005.

[19] K. Vijayasankar, L. Kannan, M. Tacca, and A. Fumagalli. CETT: A cross layer routing metric for cooperative wireless ad hoc networks. In *Proc. IEEE/ACM MASCOTS '09*, London, United Kingdom, 2009.

[20] C. Yu, K. G. Shin, and L. Song. Link-layer salvaging for making routing progress in mobile ad hoc networks. In *Proc. ACM MobiHoc '05*, pages 242–253, IL, USA, May 2005.

[21] S. Zheng, P. Kong, and C. Tham. A novel routing metric for multi-hop cooperative wireless networks. In *Proc. IEEE PIMRC '09*, pages 2911–2915, Tokyo, Japan, Sep. 2009.

[22] H. Zhu and G. Cao. *r*DCF: a relay-enabled medium access control protocol for wireless ad hoc networks. *Mobile Computing, IEEE Transactions on*, 5(9):1201–1214, Sep. 2006.

[23] S. Zou, B. Li, H. Wu, Q. Zhang, W. Zhu, and S. Cheng. A relay-aided media access (RAMA) protocol in multirate wireless networks. *Vehicular Technology, IEEE Transactions on*, 55(5):1657–1667, Sep. 2006.

Energy-Efficient Three-Phase Task Scheduling Heuristic for Supporting Distributed Applications in Cyber-Physical Systems

Wei Li[1], Flávia C. Delicato[2,1], Paulo F. Pires[2,1], Albert Y. Zomaya[1]

1. The Centre for Distributed and High Performance Computing, School of Information Technologies, The University of Sydney, Australia

2. Department of Computer Science, Federal University of Rio De Janeiro, Brazil

liwei@it.usyd.edu.au, fdelicato@gmail.com, paulo.f.pires@gmail.com, albert.zomaya@sydney.edu.au

ABSTRACT
Cyber physical systems (CPS) have recently emerged as a promising approach to improve the synergy between physical and virtual worlds. CPS applications can be built on top of Wireless Sensor Networks (WSNs) exploiting the physical information collected by them to bridge real and cyber realms. In this paper, we investigated the emerging problem of how to schedule multiple applications onto cyber-physical systems encompassed of multiple WSNs, while meeting demands of these applications. Our goal is to find out an optimal scheduling scheme for each application so that: (i) it minimizes the overall energy consumption, (ii) it meets the application's deadline, (iii) it provides the required data accuracy to applications, and (iv) it balances the workload of the system. To achieve such goals, we developed a polynomial-time three phase task scheduling heuristic, named HTPTS. Experimental results show that the time and energy performance of HTPTS are close to the time and energy of the benchmark in most of the cases, while load balance is satisfied.

Categories and Subject Descriptors
C.2.2 [Computer-Communication Networks]: Network Protocols—Wireless communication; **C.2.4 [Computer-Communication Networks]**: Distributed Systems

Keywords
Wireless sensor networks, task scheduling, energy efficiency, multi-objective optimization, cyber-physical system

1. INTRODUCTION
Wireless Sensor Networks (WSNs) are a unique paradigm of distributed systems that present several specific features and impose new challenges on researchers. The first generation of WSNs adopted an application-specific design, with the primary goal of achieving energy efficiency to prolong the system lifetime. In recent years, a new trend of designing WSN infrastructures that encompass a multitude of heterogeneous sensor networks has emerged as a cyber-physical system [1] for performing multiple applications. The main benefits with such new design approach are to fully utilize hardware resources on the one hand, and to enhance system flexibility by dynamically allocating sensor nodes to meet different applications demands on the other hand. Real world examples [2] include building automation, Internet of Things (IoT), and integrated environmental monitoring. These applications normally utilize a number of sensor nodes across the cyber-physical system during their execution. Due to the inherent high distribution, dynamic execution context and heterogeneity of the sensor nodes in the cyber-physical systems, performing such applications successfully is non-trivial. One major challenge researchers are currently facing is how to effectively allocate available resources to contending applications subject to the resource constraints of sensor nodes, while meeting the different application demands.

MSWiM'12, October 21–25, 2012, Paphos, Cyprus.

From the application point of view, the WSN application consists of a number of functional tasks (sensing, data processing and communication), which represent the required action(s) performed by a sensor node upon occurrence of an event of interest for the application. In order to enable collaborative processing across a number of WSNs within the system, how to schedule tasks in a systematical way, including the assignment of tasks to sensor nodes, and determining their execution and communication sequence is the most important issue to address. For purposes of task scheduling, a WSN application is represented by a task graph, which is often simply referred to as a directed acyclic graph (DAG) [3]. In contrast to application allocation in the first generation of WSNs, in which the generated scheme was only concerned with the energy efficiency, the new generation design must not only distribute the tasks of the applications based on the principle of energy efficiency, but also needs to take the different (and sometimes conflicting) application's requirements into account. The application requirements are generally represented by QoS parameters, which are specified from two perspectives: the application's and network's. In this paper, we use data accuracy, application deadline (both application's QoS parameters) and system lifetime (network's QoS) as the selected parameters for generating the task scheduling scheme since they are strongly associated with the entire system performance.

To address the aforementioned challenge, we study the problem of how to schedule multiple applications onto cyber-physical systems while meeting each application's demands. Our goal is to find out an optimal scheduling scheme for an application that: (i) minimizes the overall energy consumption, (ii) meets the application's deadline, (iii) provides the satisfied data accuracy to applications, and (iv) balances the workload of the system. We propose a high level heterogeneous WSNs based three-phase task scheduling (HTPTS) scheme to achieve such goal, which operates with the support from low level protocols. In the first phase, a deadline distribution algorithm is employed to decompose the overall application's deadline into sub-deadlines and assign them to the corresponding tasks in order to guarantee the completion of the application in time. Second, an adaptive task graph partitioning algorithm is developed to distribute tasks to the selected WSNs according to the overall goal of minimizing the energy consumption and balancing the workload of WSNs while meeting different application QoS requirements. Finally, a multi-objective scheduling algorithm generates the task scheduling scheme in terms of energy consumption, latency, data accuracy and load balancing among sensor nodes in the selected WSNs.

The research in this paper makes the following contributions:

1. developing a three-phase task scheduling heuristic for distributing tasks of multiple applications, even simultaneously arrived, onto the cyber-physical system so as to successfully perform these applications,

2. considering multiple objectives (and sometimes conflicting) in the task scheduling decision process for each application, including data accuracy, application deadline, energy consumption and load balance among WSNs, and providing user-defined parameters to adjust the weight of objectives for each application execution.

The rest of this paper is organized as follows. A review of related works is provided in Section 2. Afterwards, we introduce the models used in our work and formally define the research problem in Section 3. In Section 4, the details of our proposed algorithm is presented and discussed. The simulation results and analysis are given in Section 5. Finally, we conclude our study in Section 6.

2. RELATED WORK

To schedule tasks onto multiple sensor nodes has been attracted many researchers recently, since it is considered as a promising way to fully exploit the physical network infrastructure improving the utilization of the limited resources of cyber-physical systems. The existing solutions can be classified, according to the network topology, into two categories: task scheduling on a single WSN, and task scheduling on multiple WSNs.

For task scheduling on a single WSN, the problem has been well studied. Yu and Prasanna [4] developed an energy-balanced task allocation (EBTA) algorithm to meet the deadline of a real-time application running on homogeneous sensor nodes connected via multiple wireless channels. However, they did not consider the broadcasting nature of wireless communication in their model and the multiple wireless channel technique is not widely used in the real world sensor nodes. Xie and Qin [5] presented a task scheduling algorithm called BEATA to solve the energy-delay dilemma that exists in heterogeneous WSN. BEATA aims at minimizing the energy consumption while confining schedule lengths through task allocation. The aforementioned techniques are concentrated on task scheduling in one WSN with single hop topology, but in practical applications, sensor nodes are normally randomly deployed in an area of interest and form an irregular topology. For solving this problem, Yuan et al. [6] proposed an algorithm called MTMS to handle the task scheduling issue in WSN with multi-hop topology. However, their algorithm only considers one-hop communication collision and the hidden communication collision may not be properly handled in the data communication process between sensor nodes.

Since all the aforementioned works assume that the application is only performed inside a single WSN, they are not able to explore the potential cooperation among multiple WSNs of a cyber-physical system. In order to better utilize the resource of the cyber-physical system and extend its lifetime, Li et al. [7] proposed a heuristic called TPTS for enabling the task scheduling over multiple WSNs. TPTS is designed to find out a scheduling scheme that minimizes the overall energy consumption and balances the workload of the system while meeting the application deadline. However, TPTS considers that all sensor nodes contain the same sensing units, which means each WSN within the system is homogeneous in terms of its capability. Moreover, TPTS considers that each application is processed by all WSNs within the system. HTPTS augments TPTS addressing these two constraints, thus providing a more general solution to the problem of task scheduling on multiple WSNs.

3. MODELS AND PROBLEM DEFINITION

This section presents the models used to define the cyber-physical system as well as the problem statement.

3.1 System Model

The cyber-physical system considered in this work encompasses three core components, namely, a central node, a set of sink nodes and a set of heterogeneous sensor nodes. The system architecture is shown in the Figure 1.

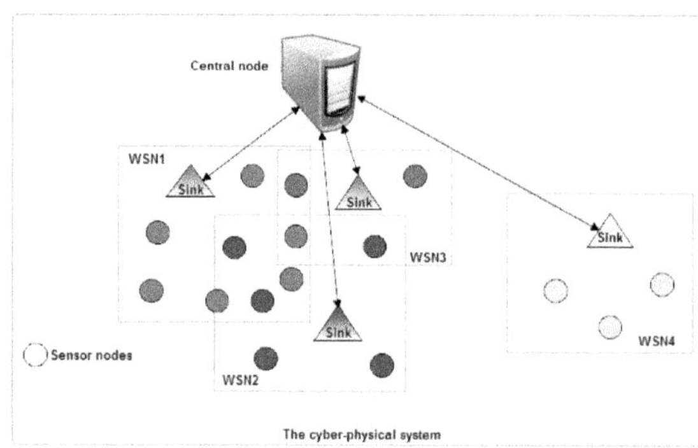

Figure 1. The cyber-physical system architecture

The central node acts as a facade to the entire system for handling the dynamic arrival of different applications, decomposing an application into tasks, determining of the specific WSNs required for processing the application, deciding how to dispatch the tasks to the selected WSNs and gather the latest information about the status and execution context from all WSNs.

Each WSN provides various sensing functions to the applications in its coverage area, depending on the composition of sensor nodes it contains. Any given WSN is modeled as an undirected graph $G = (V, E)$, where $V = (v_1, v_2, ..., v_n)$ represents the set of sensor nodes and $E = (e_1, e_2, ..., e_m)$ represents the set of all possible communication links among the sensor nodes. Each WSN has a single static sink node which has two responsibilities:

it enables communication between the sensor nodes and external networks, and maintains the most updated WSN information. The information includes, for each sensor node, its capabilities, current status (residual energy and operation mode) and geographical location. Such updated information is acquired by messages sent from sensor nodes whenever required by sink nodes (normally at the time of a successful completion of the allocated tasks of an application). According to the gathered information, sink nodes cooperate with sensor nodes and with the central node to assign and execute the applications.

In the WSNs, sensor nodes have three general functional units: communication, sensing and data processing. We assume that each functional unit has two states, *active* and *idle*. Each unit can only stay in one of two states. All the sensor nodes are mainly responsible for performing the allocated tasks, and none of them is equipped with dynamic voltage scaling (DVS) for energy saving. Sensor nodes are considered to be heterogeneous in terms of the supported sensing functions, since a sensor node can be equipped with one or more sensing devices. However, the sensing range of sensor nodes is considered as the same value. We also assume that the sensor nodes are homogeneous for any given computation task, which means all the sensor nodes are identical in their processing speed. Similarly, all the sensor nodes have the same communication range with diameter $diam(v) \leq k$, where k is the maximum communication range that a sensor node can reach. Furthermore, all the communication links are assumed to work at the same frequency, and each of them is not allowed to transmit and receive data simultaneously. In the multi-hop WSNs, simultaneous communications could occur and thus potential interference needs to be addressed. The graph-based protocol model [8] is used to determine whether primary and secondary interferences [8] occurs between the scheduled communication tasks.

To examine the sensing availability of a given sensor node, we introduce a judging function that determines whether a sensor node can provide the required sensing function at the specified area. The judging function is shown in equation (1),

$$A(s,x,y,i)=\begin{cases}1, & \text{if a sensor node is available}\\ 0, & \text{if a sensor node is unavaiable}\end{cases} \qquad (1)$$

where s denotes the sensing function that an application requires, x and y are the location information of the monitoring event and i is the index of the sensor node.

In this paper, we consider the possibility that the sensing range of any given WSN overlaps other WSNs. This means more than one sensor node can provide the same sensing function in the specific target area. However, each node will provide a different value of data accuracy. The data accuracy of sensor nodes is normally affected by a number of factors [9], such as: nominal precision of the sensing device, the environmental noise of its measurements, the number of neighboring nodes and proximity to the target area. In this paper, we only take the environmental noise of a sensor measurement (SM_i) into account since it affects the quality of data collection the most. The parameter SM_i is mainly affected by physical characteristics of the location, such as: electric signals, local temperature and humidity. The parameter SM_i is a normalized value that depends on the actual level of environmental noise N_i, where N_i ranges from 0 to 100. Real values of environmental noise should be detected by the sensor node upon the reading; the values adopted in our work are relative values, utilized for purposes of comparison, so that sensor nodes which can provide less noisy measurements take precedence in the selection process. The measurement of environmental noise is then given as equation (2).

$$SM_i = 1 - N_i / 100 \qquad (2)$$

3.2 Application Model

We model a WSN application as a Directed Acyclic Graph (DAG) [3] to indicate the precedence between tasks. A DAG $T=(V,E)$ consists of a set of vertices V representing a number of non-preemptive tasks and a set of directed and weighted edges E representing the tasks dependence and the communication load among tasks. The computation cost of a task is denoted by the number of CPU clock cycles to execute the task. The time for performing a task on the sensor node i can be determined by $t_i^{comp} = N(i)/CPU(i)$ where $N(i)$ is the computation cost of a task and $CPU(i)$ is the CPU speed of sensor node i. For any given edge E_{ij}, it represents the dependence between two vertices V_i and V_j which V_i is indicated as the parent task of V_j and V_j is indicated as the child task of V_i. Consequently, V_j cannot start executing until it receives all the data from its parent task. To determine the communication time between two tasks, the following equation is used, $t_i^{comm} = data(i,j)/bandwidth(i,j)$ where data(i,j) is the total data needs to be transferred from task i to task j and *bandwidth(i,j)* is the bandwidth in the communicating link between sensor nodes which were assigned with task i and task j. In DAG, a task without parent tasks is called entry-task and a task without child tasks is called exit-task. We stipulate that the entry tasks of a DAG, are sensing tasks, meaning the tasks that represent the different types of sensing functions provided by sensor nodes. Similarly to the computation tasks, the measurement of the cost of sensing tasks also depends on time, but we stipulate that the related interesting events must occur within the considered time period, and the required data can be successfully collected before the end of such period once the sensing task begins.

3.3 Energy Model

The energy consumption of a sensor node is a function of the node state and the time during which the sensor node remains in that state. The state of a sensor node is described by the combined states of all individual functional units (communication, sensing and data processing). If the states of a sensor node are determined, it consumes a fixed energy rate in the following time period until the state is changed. When a functional unit stays in the active state, its energy consumption is much greater than when it stays in the inactive state. Therefore, we do not include the energy consumption when the functional units stay in their inactive state. For these three functional units, the major energy consumer is the communication unit. Regarding the other two units, data processing unit mainly spends energy for performing computational functions like data aggregation, computing of routing tables, and management, etc, while the sensing unit consumes energy for a variety of sensing devices and ADC converters. Overall, the energy consumption of a sensor node during time t is calculated by $E(t) = E_c + E_p + E_s$, where E_c represents the energy consumption of communication unit, E_s represents the energy consumption of sensing unit and E_p represents the energy consumption of data processing unit. As assumed in the system model, we consider multi-hop network topology in each WSN and the communication path between any two nodes follows the principle of shortest path routing. Please note that, the detail of routing protocol is out of the scope of this paper.

The energy consumption of transmitting *l*-bit data over distance d is defined as $E_{tx}(l,d)$:

$$E_{tx}(l,d) = E_{elec} \times l + \varepsilon_{amp} \times l \times d^2 \quad (3)$$

$$E_{rx}(l) = E_{elec} \times l \quad (4)$$

where E_{elec} and ε_{amp} are hardware-related parameters, which represents radio energy dissipation and transmission amplifier energy dissipation respectively. The above equations (3) and (4) are known as first order radio model [10], which is a well known and generally accepted formula for estimating energy consumption of communication in the field of low-energy radios (e.g. sensor), during the last decade. The energy consumption of the data processing unit in executing N clock cycles with CPU speed f as follows:

$$E_{comp}(V_{dd}, f) = NCV_{dd}^2 + V_{dd}(I_0 e^{\frac{V_{dd}}{nV_T}})(\frac{N}{f}) \quad (5)$$

$$f \approx K(V_{dd} - c) \quad (6)$$

where V_T denotes the thermal voltage and C, I_0, n, K and c are processor-dependent parameters. The energy consumption of the sensing unit is calculated by a linear equation:

$$E_{s_i} = ER_{s_i} \times t_{s_i} \quad (7)$$

where ER_{s_i} represents the energy consumed by an active sensing device i in the unit of time and t_{s_i} represents the time period of a sensing devices i remains in the active state. It is important to note that the equations for calculating energy consumption of a sensor node that we have introduced in this section only consider the energy expenditure directly associated with tasks execution. The other possible energy consumptions for a sensor node, e.g. idle listening, state transition are not explicitly taken into consideration. The main reason for making such assumption is due to HTPTS focuses on how to save energy by allocating tasks onto sensor nodes in the execution period of the application subject to certain given constraints so as to extend the system lifetime. The generated scheduling scheme decides the duty cycle and sleep cycle of sensor nodes in each application execution period, where sensor nodes switch to the sleep mode for reducing energy consumption when no tasks are scheduled on them. The interesting readers should refer to [7] for more detailed discussion.

3.4 Problem Statement

Let M represent the mapping function of a task assignment which means task V_{ij} from application j is assigned to sensor i. After the task assignment, the communication link between nodes is considered whenever correlated tasks are not allocated to the same sensor node. Let's suppose that $E_{comp}^M(V_{ij})$ represents the computation energy consumption of task V_{ij}, $E_{comm}^M(V_{ij})$ represents the communication energy consumption and $E_{sen}^M(V_{ij})$ represents the sensing energy consumption. The communication energy consumption includes transmitting the intermediate result of V_{ij} to its successor task(s), and receiving the data from the predecessor task(s). Thus, the related energy is $E_{comm}^M(V_{ij}) = E_{tx}(l,d) + E_{rx}(l)$. The goal of our algorithm is to find out a scheme that prolongs the system lifetime by managing the energy consumption (including reducing and balancing energy consumption of performing applications within the system) subject to meet the data accuracy of sensing tasks and the time constraint of the applications. System lifetime can be seen as an indicator for the maximum utility that a sensor network is able to provide. According to [11], there are different definitions of system lifetime for WSNs. In this paper, we adopt the definition of the time of the first sensor node has its energy completed depleted, since the system may be still functional after the first sensor node dies, the quality of service is no longer to be 100% guaranteed as before. The problem statement is given below:

Find a scheme M^0:

$$arg\ min\ energy(M) \quad (8)$$

To minimize:

$$E_{total}^M = \sum_{i=1}^{n} E_{comp}^M(v_{ij}) + \sum_{i=1}^{n} E_{comm}^M(v_{ij}) + \sum_{i=1}^{n} E_{sen}^M(v_{ij}) \quad (9)$$

Subject to:

$$\psi_{A_i} = f(E_{A_i}, D_{A_i}) \text{ and } Length(M) \leq ADL \quad (10)$$

where E_{total}^M and *Length(M)* are respectively the overall energy consumption and the schedule length of M, ψ_{A_i} represents the overall requirement of application A_i and it is a cost function of energy consumption E_{A_i} and data accuracy D_{A_i} of the sensing tasks belonging to the application A_i, and ADL is the deadline of the application A_i. In general, the task allocation and scheduling issue has been proven as an NP-complete problem [3]. As a consequence, heuristic algorithms are employed to solve the problem in polynomial time.

4. SCHEDULING HEURISTIC DESIGN

The HTPTS heuristic includes three major phases, (i) the deadline distribution algorithm, (ii) the energy-load tunable graph partitioning algorithm and (iii) the multi-objective contention aware list scheduling algorithm. The first two algorithms run on the central node, which is referred to as *global scheduler* and their goal is to generate sub-task graphs with relative deadlines for the selected WSNs. In this process, the algorithm takes not only the remaining energy of each WSN into consideration, but also the sensing functions and the sensing quality of each WSN. After these two steps, the task graph is partitioned into several parts and distributed to the sink node of the selected WSNs. Afterwards, the multi-objective contention aware list scheduling algorithm continues the processing on the sink nodes of the selected WSNs, which are referred to as *local schedulers* in our paper, and its goal is to generate a scheduling scheme in those selected WSNs in order to perform the allocated tasks. After this step, the sink node assigns each task of the allocated sub-task graph to the proper sensor node to meet the application QoS metrics, data accuracy and deadline, with the possible minimum energy consumption as well as balancing the load among the sensor nodes within the WSN.

4.1 The Deadline Distribution Algorithm

We consider a time-constrained application, which is restricted to commence at a specified time and to be finished within a given time span called the *end-to-end deadline*. In our case, each application is performed on at least one WSN. A good strategy to guarantee the application can be finished in time is to distribute the end-to-end deadline over tasks before the application is partitioned into sub-task graphs and allocated to different WSNs. The global scheduling problem is therefore converted to a local scheduling problem to which each local scheduler finds out its best scheme. If each local scheduler can guarantee its own sub-deadline, the entire application execution will be guaranteed to complete within the end-to-end deadline. the individual task deadline can also provide more information to the local scheduler so that it can make fine tuned and more efficient decisions at the scheduling phase.

In this paper, we adopted and modified the deadline distribution algorithm proposed in [12] to distribute the end-to-end deadline to each task prior to the task-processor assignment phase. The main idea of this deadline distribution algorithm is to put all the tasks (sensing, computation and communication) into a set $\prod$ and find out a critical path Φ for the tasks in the set $\prod$ at each loop until the set is empty. The pseudo code of the deadline distribution algorithm is given in Figure 2.

```
Input: Task graph, end-to-end deadline
Output: Task graph with sub-deadlines
Put all tasks and edges of task graph into a set ∏
While the size of ∏ > 0 do
  Find a critical path Φ in ∏ that minimizes metric R_nonsens
  Compute the task sub-deadline based on equation 9
  For each task t_i in Φ do
    For each parent task t_p of t_i do
      deadline[t_p] = ready time[t_i];
    End for
    For each child task t_c of t_i do
      ready time[t_c] = deadline[t_i];
    End for
  End for
  Remove all tasks in Φ from ∏
End while
```

Figure 2. The pseudo code of deadline distribution algorithm

In the deadline distribution algorithm, several definitions of the metric R are discussed in [12] for finding the critical path, including (i) allocate the slack to the sum of actual execution times of all tasks in a path, (ii) allocate the slack to the proportion of task execution time, and (iii) allocate the slack to the sum of estimated execution time of all tasks in a path. In this paper, we adopt the (i) approach with a small change, which excludes the sensing tasks from the slack distribution. This is because sensing devices have no existing mature technique to save energy without degrading the sensing quality. Therefore, allocating additional time to sensing tasks has no contribution to energy saving. According to this reasoning, we propose a new approach, which only distributes the available path slack to the sum of computation and communication tasks execution time in the critical path. The ratio of slack that each non-sensing task in the critical path can be allocated is computed as below:

$$R_{nonsens} = (D_{\phi} - \sum\nolimits_{t_i \in \phi} c_i) / \sum\nolimits_{tnonsens_j \in \phi} c_j \qquad (11)$$

where D_{ϕ} denotes the end-to-end deadline of a critical path, $\sum c_{t_i \in \phi_i}$ denotes the total cost of all tasks in a critical path, and $\sum_{tnonsens_j \in \phi} c_j$ denotes the total non-sensing tasks cost in a critical path. Therefore, the deadline of a non-sensing task can be computed by $\text{ReadyTime}(t_j) + c_j(1 + R_{nonsens})$, which means the available time slack is assigned proportionally to all non-sensing tasks execution time. This approach gives more time for the computation and communication tasks to ensure they can meet their deadlines when they have a late start due to waiting for the availability of allocated sensor or the communication contention taking place in the system.

4.2 The Energy-Load Tunable Graph Partitioning Algorithm

The distribution of tasks among WSNs is a key factor to balance the workload in the system so as to prolong the system lifetime. This step attempts to find a solution to divide the task graph into *k* disjoint partitions and distribute them over WSNs to balance the workload of computation tasks by considering the residual energy of these selected WSNs, while minimizing communication cost between WSNs to reduce total energy consumption of performing an application, shorten the schedule length and provide the satisfied data accuracy to the applications (more precisely, sensing tasks). This process is the well-known graph partitioning problem. Since the complexity of graph partitioning problem is normally NP-complete [3], a heuristic method has been developed to obtain sub-optimal result in reasonable time. The pseudo code of the graph partitioning algorithm is given in Figure 3.

The graph partitioning algorithm starts with a locality-aware sensing function discovery. When a WSN application arrives, the global scheduler will analyze the composition of the application. The information to generate primary task-sensor assignment is extracted from the analysis process, including the number of tasks, the structure of task graph, the requested sensing function of each task, and the geographical location of interest. For the sensing tasks, all possible sensor nodes capable of providing the sensing function that the task requires are identified by equation (1) by the local scheduler. Due to hardware limitation of sensor nodes, each of them is constrained by its sensing range and only the physical phenomenon occurring within this range can be detected. As a result, sensor nodes are filtered by considering the location of available sensor nodes, their sensing ranges and the location of requested task. The resulting filtered sensor nodes are put into a candidate sensor list.

To determine the best sensor node to process a specific task, we derive an equation to support flexible tradeoff between data accuracy and energy consumption, and quantitatively select the best candidate, which has the greatest value from the candidate sensor list based on the given tradeoff. Since the QoS parameters, data accuracy and energy consumption, represent different types of performance metrics, the normalization technique is employed to generate the following equation:

$$\alpha \frac{D_{s_i}}{D_{best}} + \beta \frac{E_{residual} - E_{s_i}}{E_{residual}} \quad (12)$$

where α and β are user-defined coefficients for the normalized data accuracy and energy consumption, respectively.

```
Input: The original task graph; system topology; application QoS
requirements; user controlled parameters; tolerance values,
Output: The optimal partitioned graph
//Determine how many WSNs are used for processing an application
While count(sensing tasks)>0 do
  For a given task, check the service ability of sensor nodes by using equation (1)
  Create candidate sensors list
End While
//Best candidate selection
For each candidate sensors list
   Identify the best sensor-task assignment based on the application requirements considering the residual energy of sensor by using equation (12).
   Put all the result in a list Qr in an increasing order.
End For
For (i=0; i<| Qr |, i++)
     Estimate whether the selected WSNs can finish the allocated tasks before deadline in order
     IF True
       Break;
     Else
       Continue;
     End IF
End For
//Calculate maximum workload by input tolerance
While(num of vertex > num of the selected WSNs) do
  /*Step 1: Using local heaviest edge first merge method to merge vertex*/
  While (the heaviest weight coarsen vertex< preset maximum workload) do
    Find out all entry-tasks in the task graph
    Find out the maximum edge connected to each entry-task
    Select the most heaviest edge among entry-tasks
    If merge objects contains totally one entry-point then
      Merge non-entry task vertex to entry task vertex
      Remove the edge from list
      Consider the coarse node as new entry task
    Else
      Remove the edge from list
    End If
  End while
  /*Step 2: Trade-off based vertices merge*/
  While(unprocessed edge size > 0) do
    Select the lightest weight edge from the unprocessed edge list
    For each coarse entry-task do
      Evaluate the tradeoff using equation (14)
      Merge the vertex pair with the best tradeoff
    End for
  End while
  /*Step 3: Narrow down */
  Narrow down to the number of clusters by merging vertices using equation (14)
End while
Return the partitioned graph
```

Figure 3. The pseudo code of Energy-Load tunable graph partitioning algorithm

And these two coefficients have a relationship β=1-α. For a given sensing task, D_{best} represents the best data accuracy that can be provided by the sensors in the candidate sensors list and $E_{residual}$ represents the residual energy of the WSN whose the selected sensor belongs to. D_{s_i} denotes the data accuracy value for any sensor node in the candidate sensor list and its range from 0 to D_{best}. E_{s_i} denotes the energy consumption value for any sensor node in the candidate sensor list and its range from E_{best} to ∞. Since the E_{best} is generated by a greedy search method, the possibility of the same sensor node is frequently selected can be high, thus the energy of such sensor can be overused and the system lifetime will be shorten. Hence, the best sensor node for energy-conservation may not be the best choice for the purposes of system lifetime extension. To alleviate this issue, as shown in the second term of equation (12), we employ a concise look-ahead mechanism. The look-ahead mechanism allows the algorithm to select a sensor node that is not the minimum energy consumer, but the one which contributes most to the system lifetime extension considering a specific sensing task and system status at the time. After this time, the sensor-task assignment with the minimum energy consumption is selected and committed. Being processed by the normalization technique, both normalized value of data accuracy and energy consumption range from 0 to 1. Eventually, the algorithm selects the sensor node which has the largest value calculated from equation (12) as the best candidate to perform the given sensing task.

As discussed earlier, equation (12) is only applied for the sensing tasks allocation among WSNs, without considering the remaining computation and communication tasks allocation between sensor nodes. In the cyber-physical system, the central node does not have direct control over all sensor nodes. Instead, it only distributes tasks to the selected sink nodes with appropriate resource and application QoS requirements. Moreover, multiple applications are likely to be executed within the system, but the computation and communication tasks are performed serially in sensor nodes. If the processor or the communication channel of the selected sensor node is busy, those unprocessed tasks have to wait until the related units become available for their processing. Therefore, it is very important to estimate the waiting time on targeted WSNs in order to meet the deadline of DAG partitions if they are running on those WSNs.

The wait time on targeted WSNs (also known as queue wait time prediction) has attracted the attention of many researchers in last decade. However, the works in this area did not show significant impact on scheduling strategy design until the successful deployment of QBETS (Queue Bounds Estimation from Time Series) [13] on more than a dozen clustered computers, offering two types of online queue delay predictions for individual tasks:

1. Given the task characteristics, QBETS is able to predict a statistical upper bound on how long the task is possible to spend waiting in the queue before its execution;
2. Given the task characteristics and a sub-deadline of the task, QBETS is able to calculate the probability that the task begins execution by the deadline.

Before we can actually measure the DAG partition waiting time on a selected WSN, the size of that DAG partition is a necessary condition to perform the estimation. However, the precise DAG partition size can only be known after the entire graph partitioning algorithm is performed. Thus, a circular dependency between the number of WSNs and the estimated waiting time occurs. One possible idea to break this circular dependency is to generate the DAG partitions close to the ratio of remaining energy of the selected WSNs. According to this idea, an approximation is made to obtain the size of DAG partitions based on the remaining energy of the selected WSNs. Please note that, the size of DAG partitions only takes the cost of computation tasks into account. The estimated processing time of a given DAG partition k is measured by the smallest earliest finish time (EFT) and the equation is given as below:

$$EFT(k) = exec(k, q, p) + comm(k, W(q)) + QBETS(W(q), q, p) \quad (13)$$

The first item *exec(k,q,p)* denotes the estimated execution time of task partition k running on the selected WSN. Based on the Amdahls' law, $exec(k,q,p) = \frac{w_{k,q}}{\sqrt{q}}$, where $w_{k,q}$ is the execution time for DAG partition k running on single processor in queue q. The second item *comm(k, W(q))* denotes the communication time for intermediate result exchange between WSNs. The third item *QBETS(W(q), q, p)* denotes the estimated queue waiting time for a task partition if it is scheduled to queue q on WSN W with request for p processors. This estimation process continues until the satisfied number of WSNs is determined. In other words, all the DAG partitions can be completed before their deadlines on the selected WSNs. After this step, the WSNs used for processing the application are finally determined. In the following, the actual task partitioning algorithm is performed.

The actual task partitioning algorithm is composed of three steps, namely, (i) local heaviest edge first merge, (ii) trade-off based vertices merge and (iii) narrow down. The rationale behind the partitioning algorithm is to group tasks into same partitioning by collapsing vertices and edges of the task graph until the number of task partitions is equals to the number of selected WSNs. This process is known as coarsening phase [3], and the Heavy Edge First Merge (HEFM) [3] is the most popular algorithm for coarsening due to its simplicity and relative good performance. For a given task graph, HEFM selects the most heaviest weight edge from the task graph and merges the vertices connected to that edge. The algorithm runs iteratively, until the expected number of clusters is reached. Based on the residual energy of the selected WSNs and the given tolerance value ranging from 0 to 1, the maximum workload of a selected WSN is computed. The tolerance value is a parameter to set how much load imbalance can be tolerated between the selected WSNs. The larger value of tolerance indicates the less importance of load balancing. Therefore, the maximum workload is used as an upper bound to prevent the heaviest edge first merge algorithm to generate a coarse vertex with very large weight. In step 1, for each entry task, we select the heaviest weight edge connected to it and put this edge into a list. Then, we select the heaviest weight edge from the list. If the vertices connected to the selected edge contain only one entry-task after the merger, the action is actually performed and the selected edge is removed from the list; otherwise, this edge is simply removed from the list without merge. The procedure continues until the maximum workload constraint for clusters no longer holds; then either the algorithm moves into step 2 or, if no more vertices can be merged, the algorithm terminates. In step 2, all the remaining unprocessed edges are put into a new list, and the lightest weight edge is selected from this list. For all the coarse vertices containing an entry-task, we evaluate the quality of a potential solution by merging the vertex connected to the lightest weight edge with every coarse vertex. The solution providing the least value is selected and the real merge is processed. In this step, we intend to select the lightest weight edge from the unprocessed edge list for reducing the inter-cluster communication cost when we process the vertex merging. Let us assume that after the step 1 processing, several remaining vertices are still waiting for further process to be merged with m coarse vertices. The merge decision in the step 2 is determined by the equation (14).

$$P = \mu \frac{C_1^0}{C_1} + (1-\mu)\frac{C_2^0}{C_2} \quad (14)$$

where C_1 denotes the minimum sum of inter-WSNs communication cost between the designated numbers of partitions, C_2 denotes the minimum sum of load imbalance between the selected WSNs, and the value of C_2 is generated by the best-fit rule [14], C_1^0 and C_2^0 are the inter-WSNs communication cost and load imbalance of a partitioning solution generated by our proposed graph partitioning algorithm, respectively, and user-defined parameter μ is used to tune the importance of each objective. The interesting readers is referred to [7] for more detailed discussion.

The communication cost is a major source of energy consumption in WSN, as well as the inter-WSNs communication cost consumes more energy than intra-WSN communication cost [7]. Consequently, select the lightest weight edge aims mainly at reducing the possible inter-WSNs communication cost. As mentioned before, the comparable standard C_1 for the minimum inter-WSNs communication is generated by using HEFM algorithm. The comparable standard C_2 for achieving the best load balance is modeled as a multibin packing problem, and the best-fit rule [14] is used to obtain the result. If a task graph contains the same number of entry-tasks as the number of selected WSNs in the system, the algorithm is completed and terminates at this point, otherwise, the algorithm reaches step 3. Until this step, the number of the coarse vertices is still greater than the number of selected WSNs. The purpose of this step is to keep merging the remaining coarse vertices subject to the stated objectives. Based on the equation (14), the algorithm will evaluate every possible combination of remaining vertices, find out the best solution and combine the selected two vertices from the remaining vertices into one. The procedure is continued until the remaining number of vertices is eventually equal to the number of selected WSNs.

4.3 The Multi-Objective Contention Aware List Scheduling Algorithm

As mentioned before, the sink node plays the role of local scheduler on each individual WSNs. The sink node receives tasks dispatched by the central node and handles these tasks as directly submitted by the users. After that, it generates an appropriate scheduling scheme to meet the sub-deadline of the allocated tasks in each WSN, while managing the energy consumption. In our case, multiple applications may arrive and execute in the system simultaneously. The new arrival tasks may influence the scheduling of uncompleted tasks from the previous arrived applications. To deal with such issue and make sure the tasks can be finished in time, the insertion technique [3] is used in the multi-objective contention aware list scheduling algorithm. The pseudo code for the multi-objective contention aware list scheduling algorithm is given in Figure 4.

Input: Partitioned task graph, WSN topology
Output: Schedule scheme

```
//Sort tasks according to task sub-deadline and precedence constraints, and put them into queue Q
For each task n_i ∈ Q do
    For each s_j ∈ S in the WSN do
        Compute finish time t_f of n_i on sensor s_j with the insertion technique.
    End for
    Remove the sensors which violate sub-deadline from S
    For the sensors which NOT violate sub-deadline from S do
        Compute the overall energy consumption by equation 9
    End for
End for
Select s_j from the list that offers the lowest energy consumption and meets sub-deadline
Assign n_i to s_j
Update the schedule on sensor s_j
Compute the computation and communication energy consumption on sensor s_j
```

Figure 4. The pseudo code of multi-objective contention aware list scheduling algorithm

In the traditional scheduling scheme, tasks execution sequence is set to FCFS (first comes first serves), which means new tasks can start processing only when all the previously arrived tasks are completed. As discussed in section 4.1, each computation task is associated with additional time for its processing, however the processor is not endowed with DVS technique, which means no processing speed variation occurs during the task execution. Under certain circumstances, a task can be scheduled earlier, if there is a time period between two tasks already scheduled on a given sensor node S_i, where S_i runs idle. In other words, the algorithm (Figure 4, line 4) returns more possible candidate sensor nodes for complete processing task n_j before its deadline, compared to the priority scheme used in FCFS. An example is given in the Figure 5 (a), in the given processor, an idle time period between task *a* and *b* is found to process the task *c* before its deadline without delaying the start of task *b*. Therefore this sensor node is selected as an available processor for executing task *c*, and it will be further evaluated by the remaining steps of the extended multi-objective contention aware list scheduling algorithm to determine whether it is the best candidate to process task c. Furthermore, as shown in Figure 5 (b), even no idle time period exists in between task *a* and *b*, the insertion technique is still able to find out a time period for processing task *c* in the given processor. This is because each computation task is associated with additional time for its processing in the extended deadline distribution algorithm. In this case, the earliest finish time for task *a* and the latest start time for task *b* can be easily found out, if the time period between these two values is enough for processing task *c*, then this sensor node can be seen as an available selection, otherwise, this sensor node is discarded.

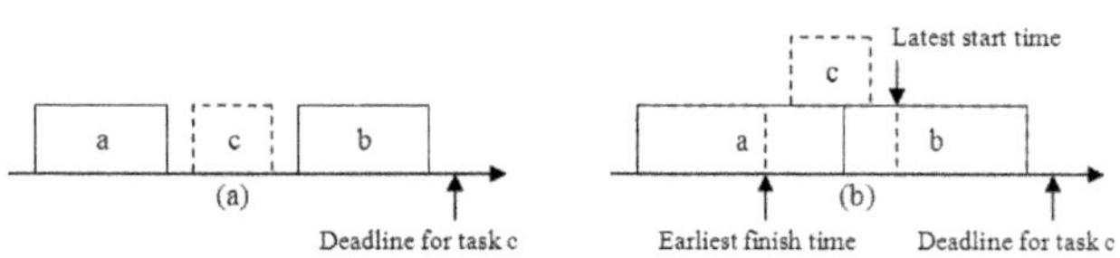

Figure 5. A partial schedule example for the insertion technique, (a) insert the task c in the existing idle period between task a and b, (b) insert the task c in the found idle period between task a and b.

Eventually, the algorithm determines the allocation list as well as the execution sequence, and then each task will be distributed to the optimal sensor node for its processing. After the processing, the energy consumption of each sensor node will be calculated and the updated information of sensor nodes will be collected by the sink nodes later on.

5. PERFORMANCE EVALUATION

In this section, the performance of the proposed HTPTS algorithm is evaluated in a simulated environment similar to the one presented in [7], using the original TPTS algorithm [7] with the similar design objectives and application scenarios as the benchmark to reveal the cyber-physical system performance difference by running these two algorithms. In addition, two theoretic benchmarks are also introduced for the performance comparison; the length of critical path is used as *lower bound* and the value hops × critical path is used as *upper bound*. In order to achieve a fair comparison, we study the system performance according to the following metrics: schedule length, system lifetime, energy consumption and load balancing. However, it is important to notice that the data accuracy requirements for the sensing tasks are always guaranteed in the HTPTS algorithm. Due to the page limitation, full simulation results can not be presented, but we introduce some selective tests in this paper to show the performance of these two algorithms.

The following assumptions are necessary to run the simulations. Firstly of all, we assume that all computation tasks can be performed on any sensor nodes. Secondly, each individual task (sensing task or computation task) once assigned to a sensor node, it needs to be finished inside that WSN. No task is allowed to execute across multiple WSNs. Thirdly, the overlapping of WSNs is set to light, which is 10%. Each sensor node is equipped with 1 to 3 kinds of sensing devices to provide the relative functions. Finally, we assume the queue wait time prediction capability is defined by system design. The task graphs used in the simulations are generated from TGFF [15], which is a well-known pseudo-random task graph generator widely adopted in simulation studies. In the following, all the presented data is averaged over at least 300 times, consequently, a 95% confidence interval with a 5% (or better) precision is achieved.

5.1 Effect of Task Graph

To evaluate the effect of task graphs, simulations are run on a cyber-physical system with 2 WSNs. There are 5 sensor nodes in each WSN to construct a multiple-hop network. The task graphs are generated based on three common parameters: the number of entry-tasks, the in-degree of a task and the out-degree of a task. In order to ensure the original TPTS algorithm working properly, the minimum number of entry-tasks is set to not less than the number of WSNs. All the entry tasks are assumed to be sensing tasks. The in-degree of a task and the out-degree of a task are both bounded to 3. To investigate the system performance when different applications are performed in the system, different sizes of DAGs are randomly generated. In our simulation, four sets of DAGs with 10, 40, 70 and 100 tasks are generated. In the simulations, the CCR is set to 1, which means the ratio of the total cost of communication tasks to computation tasks is 1.

As shown in Figure 6 (a), the application makespan is proportional to the number of tasks. Since the limitation of the number of sensor nodes in each WSN, the application makespan is significantly increased after the number of tasks reaches 40. With the limited sensor nodes, the computation tasks may not be processed as soon as they arrive and the delay is cumulated.

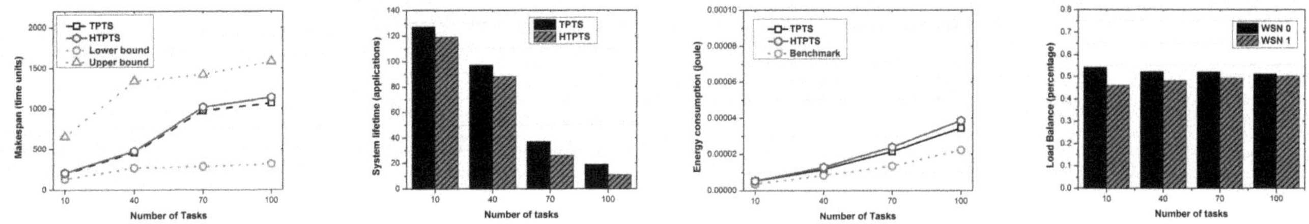

(a) (b) (c) (d)

Figure 6. Effect of the number of tasks. (a) makespan; (b) system lifetime; (c) energy consumption; (d) load balance.

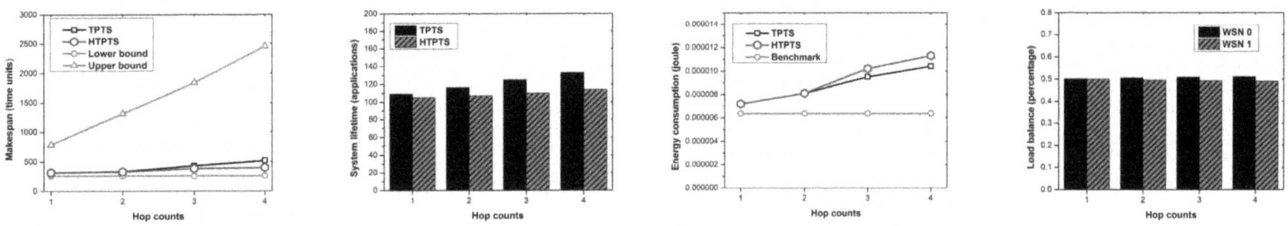

(a) (b) (c) (d)

Figure 7. Effect of internal change of the WSNs. (a) makespan; (b) system lifetime; (c) energy consumption; (d) load balance.

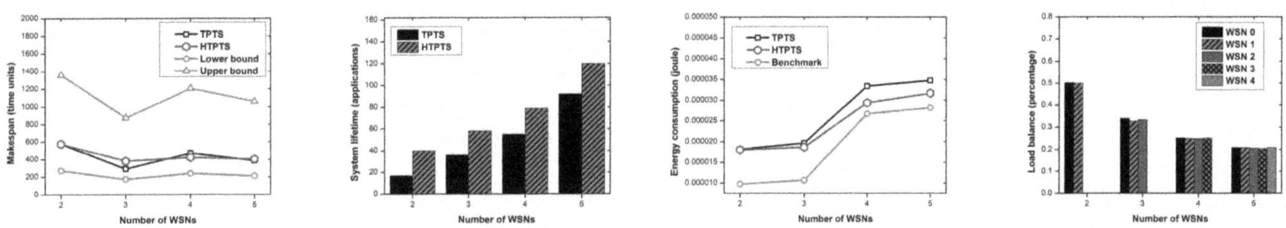

(a) (b) (c) (d)

Figure 8. Effect of changing the number of WSNs. (a) makespan; (b) system lifetime; (c) energy consumption; (d) load balance.

Thus, the curve of makespan moves toward the upper bound. Although the heterogeneous sensor nodes in WSNs limits the entry tasks distribution, the schedule length of HTPTS is still close to the schedule length of TPTS at all times. This is because, on the one hand, the sensing tasks start being processed immediately since the collected data can be shared between them, on the other hand, the insertion technique fully utilizes the gap between tasks to shorten makespan of each application, especially when multiple applications are simultaneously running in the system. The system energy consumption per application is also affected by the increment of the number of tasks. More tasks a DAG contains, more computation and communication energy the system will consume. The HTPTS consumes more energy than TPTS, especially when the number of tasks of an application is increased. The extra energy consumption is mainly caused by the increase of communication between sensor nodes when HTPTS tries to distribute tasks of the same DAG partition to more available sensor nodes for meeting the application deadline. As a result, the system lifetime is in inverse proportion to the number of tasks. Compared with the HTPTS, the TPTS has longer system lifetime due to it proactively adjusts to the system changes and it has less restriction from the sensing tasks. However, the difference between TPTS and HTPTS is not obvious since even the entry tasks need to be allocated to some specific sensor nodes but the equation (12) has effect to balance the workload between WSNs. Figure 6 (d) shows that different degrees of load imbalance between WSNs occur in HTPTS. Within the system lifetime, there is only a small amount (1-10%) of load imbalance (difference in the energy consumption) between two WSNs.

5.2 Effect of WSN size

In this section, the performance impact of changing WSN size is evaluated. The impact can originate from two sources, one is the internal WSN change, e.g. topology change and the other one is the external WSN change, e.g. the total number of WSNs. We will investigate how HTPTS performs when these changes occur.

First, we study the impact of internal change of a WSN on the aspects of topology change and the increase of number of sensor nodes. The simulations are tested with a randomly generated 30 tasks DAG scheduled on 1 to 4-hop random WSNs, respectively. The total number of sensor nodes in each WSN is calculated by 4×hops. In the simulations, the CCR is set to 1. In Figure 7(a), the makespan slightly increases when the hop count increases. Although more sensor nodes provide the possibility of parallel tasks execution, the makespan for both algorithms are increased due to the extra intra-WSN communication time spent on exchanging data with other sensor nodes caused by the increase of hop counts. Moreover, the inter-WSNs communication is also affected since the communication between sensor nodes and sink node is intra-WSN communication as well. Compared to the TPTS, the makespan of HTPTS is less affected by the increment of hops. Unlike homogeneous WSNs, the sensing tasks can not be intensively allocated to the sensor nodes with more residual energy to fully exploit the capabilities of all sensor nodes inside a WSN. Instead, the sensing tasks can only be allocated to the sensor nodes which are able to provide the relative services within their sensing ranges with the appropriate data accuracy. For a given computation task, HTPTS always tends to find the close neighbor nodes to process its child tasks in order to meet their deadlines. As a result, the conjoint tasks within the same DAG partition are unlikely to be allocated to the sensor nodes with long distance. The sensor nodes far away from the sink node have chances to process tasks unless some sensing tasks are required to collect data from the areas covered by them or their adjacent neighbor nodes are allocated computation tasks. For these reasons, the makespan of HTPTS is less than TPTS. The same reasons are also affected the energy consumption of both algorithms for processing an application and the result is shown in Figure 7 (c). From Figure 7 (b), the increment of sensor nodes does not result in the corresponding scale of system lifetime extension. This indicates us that the system lifetime is not only simply affected by the number of sensor nodes, but also by the WSN topology. The sensor nodes close to the sink node are more responsible for transmitting data to the other sensor nodes or sink node. According to our system lifetime definition, when the first sensor node dies the system dies, therefore it is hard for HTPTS to extend the system lifetime under the multiple applications scenario within the heterogeneous WSNs. However, HTPTS traces the system change and dynamically partitions the DAG according to the residual cluster energy, which properly balances the load between WSNs regardless of the variation of the size of the WSNs.

Second, we study how the number of WSNs change will affect HTPTS performs in the system. A randomly generated DAGs having 50 tasks are used for scheduling on multi-WSNs (from 2 to 5). The number of sensor nodes in each WSN is 5. In the simulations, the CCR is set to 1. From Figure 8(a) and 8(c), we can observe the HTPTS algorithm is not significantly affected by the change of number of WSNs. The results are still close to the benchmark and HTPTS performs better than TPTS when the number of WSNs reaches 4. The better performance of HTPTS is caused by the applications running in the system may not need to be distributed and performed over all WSNs in the system, some inter-WSNs communications performed in TPTS are reduced in HTPTS. The system lifetime is still prolonged in proportion to the number of WSNs. Compared to the TPTS, the system lifetime of HTPTS is averagely 30% longer. No apparent load imbalance is shown with the number of WSNs changes.

6. CONCLUSION

In this paper, we addressed the problem of scheduling multiple applications onto a cyber-physical system with multiple WSNs, where each WSN has a multi-hop network topology composed of a number of heterogeneous sensor nodes. The design objective of the proposed scheduling heuristic, HTPTS, is to schedule the tasks by considering energy consumption, load balance and data accuracy so as to extend the system lifetime, while completing the application before its deadline. HTPTS provides several parameters to allow users tune the performance of the algorithm based on different requirements and application scenarios. The experimental results show the time and energy performance of HTPTS are close to the time and energy of the benchmark in most of the cases, while load balance is satisfied.

In the future, we plan to implement and evaluate our approach in real systems. We are especially interested in integrating our solution with a service-oriented middleware, which can coordinate the resource consumption and adopt the various application demands in an automated and user-transparent manner.

7. ACKNOWLEDGMENTS

The work of Professors Delicato and Pires were partially supported by the Brazilian National Council for Scientific and Technological Development (CNPq), under grants numbers 311363/2011-3, 470586/2011-7, and 311515/2009-6. Professor Zomaya's work was supported by an Australian Research Council Grant number DP1097111.

8. REFERENCES

[1] Marwedel, P., Embedded System Design: Embedded Systems Foundations of Cyber-Physical Systems. 2010: Springer Verlag.

[2] Atzori, L., A. Iera, and G. Morabito, The Internet of Things: A survey. Computer Networks, 2010. 54(15): p. 2787-2805.

[3] Zomaya, A.Y., Parallel and distributed computing handbook. 1995: McGraw-Hill Professional.

[4] Yu, Y. and V.K. Prasanna, Energy-Balanced Task Allocation for Collaborative Processing in Wireless Sensor Networks. Mobile Networks and Applications, 2005. 10(1): p. 115-131.

[5] Xie, T. and X. Qin, An Energy-Delay Tunable Task Allocation Strategy for Collaborative Applications in Networked Embedded Systems. Computers, IEEE Transactions on, 2008. 57(3): p. 329-343.

[6] Yuan, T. and E. Eylem, Cross-Layer Collaborative In-Network Processing in Multihop Wireless Sensor Networks. Mobile Computing, IEEE Transactions on, 2007. 6(3): p. 297-310.

[7] Li, W., F.C. Delicato, and A.Y. Zomaya, Adaptive energy-efficient scheduling for hierarchical wireless sensor networks. ACM Trans. Sen. Netw., 2013. 9(4).

[8] Cardieri, P., Modeling Interference in Wireless Ad Hoc Networks. Communications Surveys & Tutorials, IEEE, 2010. 12(4): p. 551-572.

[9] Delicato, F., et al., An efficient heuristic for selecting active nodes in wireless sensor networks. Computer Networks, 2006. 50(18): p. 3701-3720.

[10] Heinzelman, W.R., A. Chandrakasan, and H. Balakrishnan. Energy-efficient communication protocol for wireless microsensor networks. in System Sciences, 2000. Proceedings of the 33rd Annual Hawaii International Conference on. 2000.

[11] Dietrich, I. and F. Dressler, On the lifetime of wireless sensor networks. ACM Trans. Sen. Netw., 2009. 5(1): p. 1-39.

[12] Jonsson, J. and K.G. Shin. Deadline assignment in distributed hard real-time systems with relaxed locality constraints. in Distributed Computing Systems, 1997., Proceedings of the 17th International Conference on. 1997.

[13] Nurmi, D., J. Brevik, and R. Wolski, QBETS: Queue Bounds Estimation from Time Series Job Scheduling Strategies for Parallel Processing, E. Frachtenberg and U. Schwiegelshohn, Editors. 2008, Springer Berlin / Heidelberg. p. 76-101.

[14] Lemaire, P., G. Finke, and N. Brauner, The Best-Fit Rule for Multibin Packing: An Extension of Graham's List Algorithms

[15] Multidisciplinary Scheduling: Theory and Applications, G. Kendall, et al., Editors. 2005, Springer US. p. 269-286.

[16] Dick, R.P., D.L. Rhodes, and W. Wolf, TGFF: task graphs for free, in Proceedings of the 6th international workshop on Hardware/software codesign. 1998, IEEE Computer Society: Seattle, Washington, United States. p. 97-101.

Dynamic TXOP HCCA Reclaiming Scheduler with Transmission Time Estimation for IEEE 802.11e Real-Time Networks

Gabriele Cecchetti*
g.cecchetti@sssup.it

Anna Lina Ruscelli*
a.ruscelli@sssup.it

Antonia Mastropaolo*
antonella.mastropaolo@gmail.com

Giuseppe Lipari*†
giuseppe.lipari@lsv.ens–cachan.fr

*ReTiS Lab. - TeCIP Institute
Scuola Superiore S. Anna
Via G. Moruzzi 1, 56124 Pisa
Italy

†Laboratoire Spécification et Vérification
École Normal Supérieure Cachan
61, Avenue du Président Wilson, 94235 Cachan
France

ABSTRACT

IEEE 802.11e HCCA reference scheduler guarantees Quality of Service only for Constant Bit Rate traffic streams, whereas its assignment of scheduling parameters (transmission time TXOP and polling period) is too rigid to serve Variable Bit Rate (VBR) traffic.

This paper presents a new scheduling algorithm, *Dynamic TXOP HCCA* (DTH). Its scheduling scheme, integrated with the centralized scheduler, uses both a statistical estimation of needed transmission duration and a bandwidth reclaiming mechanism with the aim of improving the resource management and providing an instantaneous dynamic *Transmission Opportunity* ($TXOP$), tailored to multimedia applications with variable bit rate. Performance evaluation through simulation, confirmed by the scheduling analysis, shows that DTH is suitable to reduce the transmission queues length. This positively impacts on the delay and on packets drop rate experienced by VBR traffic streams.

Categories and Subject Descriptors

C.2.1 [**Computer-Communication Networks**]: Network Architecture and Design—*Wireless Communication*

General Terms

Algorithms, Performance

Keywords

Quality of Service, real-time scheduling algorithms, wireless LAN

1. INTRODUCTION

The wide spreading of multimedia applications over *Wireless Local Area Networks* (WLAN) is a challenge for the resource management in order to provide the required *Quality of Service* (QoS). Multimedia streaming, *Voice Over IP* (VoIP), *High Definition TV* (HDTV), mobile applications etc. require a service differentiation based on tailored specification of QoS parameters like end-to-end delay, jitter, packet loss etc.

IEEE 802.11e amendment [19] to IEEE 802.11b [18] adds *Hybrid Coordination Channel Access* (HCCA) MAC protocol, a parameterized polling mechanism. Each *Station with QoS* (QSTA) negotiates the service that better meets the requirements of its *Traffic Streams* (TSs) with *QoS Access Point* (QAP).

The service differentiation of IEEE 802.11e *reference* scheduler, whose guidelines are drawn by the standard, is performing for *Constant Bit Rate* (CBR) TSs, but not for *Variable Bit Rate* (VBR) TSs, as shown by numerous studies [10,16,27,33]. Indeed, the reference scheduler suggests computing the main protocol parameters, during the admission control: a fixed *Transmission Opportunity* $TXOP_{AC,i}$ for each $QSTA_i$ and a same *Service Interval SI*, (the polling period), for all stations. Hence, it provides a sufficient but not optimal service, set around mean values streams parameters, but this is not satisfying for VBR traffic, where bit data rate can vary with respect to its mean value, and where traffic periods can be different.

Traffic variations require a flexible resource management, in order to better follow the traffic profile. For instance, in the case of highly variable bit rate, when data rate falls below its average value, the assigned $TXOP_{AC}$ is not exhausted. Instead, when data rate goes up and overcomes its mean value, the accorded transmission time is not sufficient to dispatch the incoming streams and to absorb bursts of traffic. Consequently, transmission queues length increases, jeopardizing the assured QoS and real-time network performance, expressed in terms of end-to-end delay, is worse. Also the packets drop rate, due to expiration of packets validity time, rises. A very simple solution is allocating more resources than needed in normal conditions. However, overprovisioning reduces the number of QSTAs admitted to transmit, and wastes resources when there is no high data rate. Also dynamically modify, at each polling, the $TXOP$ computation can yet jeopardizes the admission control feasibility test. Instead, a method suitable for providing a dynamic $TXOP$, without changes in the accorded resources threshold, is a bandwidth reclaiming scheme [6,9,24,28,30].

In this paper, a new scheduling algorithm *Dynamic TXOP HCCA* (DTH) is presented. DTH integrates a bandwidth reclaiming mechanism into a centralized HCCA scheduler, improving the transmission capacity management and providing further resources to VBR TSs, when needed. It is able to recover the unused portion of transmission time and assign that to the next polled station, which can need a longer transmission interval, and it prevents losing the spare time. Hence, DTH computes an instantaneous dynamic *TXOP*, tailored to the considered traffic type. This computation is further refined introducing an estimation of the previously used transmission time of each QSTA, with the aim of adapting the currently assigned *TXOP* to the effective resource needs. Moreover, since DTH takes action only at the polling time for computing the current *TXOP*, it does not modify the centralized scheduling policy. Finally, since it simply recovers unused time, it does not overcome the maximum threshold of the accorded resources, set during the admission control.

Analytical study and performance evaluation through simulation will show that DTH is able to reduce the transmission queues length and delay, improving real-time performance and following traffic variations.

This paper is organized as follows: in Section 2 a brief summary of most meaningful HCCA schedulers with QoS is listed, in Section 3 the DTH scheduling algorithm is illustrated, whereas in Section 5 performance evaluation through simulation is shown. Finally Section 6 concludes the work.

2. RELATED WORKS

The IEEE 802.11e HCCA *reference* scheduler [19] suggests how to compute the main protocol parameters: polling period *Service Interval* (SI), and transmission duration *Transmission Opportunity* (TXOP). They are calculated as fixed values during the admission control phase and are recomputed only if a new *Traffic Stream* (TS) is admitted to transmit.

Numerous studies about the reference scheduler, [10,16,27,33], shown that it is suitable to serve only CBR traffic, whereas it is unable to efficiently manage VBR TSs, since all different TSs are polled with the same period and are served with the same computation time. Hence, many alternative scheduling algorithms have been proposed to improve the HCCA QoS provisioning [12,14,26, 33], but the support of the real-time requirements has been investigated only by few studies [4]. In the following, at the best of our knowledge, we summarize some significant HCCA real-time schedulers.

Fair HCF (FHCF) [1] computes variable *TXOPs* by means of a mathematical model that distinguishes between packet queuing delay, influenced by the variations in packet size and data rate, and waiting time delay until the polling time.

Feedback Based Dynamic Scheduler (FBDS) [3] uses a closed loop feedback control to limit the maximum delay. The queue length estimation corrected by the queue length information sent by each QSTA, is used to dynamically calculate *TXOPs*; *SI* is fixed.

Most of the works based on IEEE 802.11e *HCCA-EDCA Mixed Mode* (HEMM) are focused on the optimization of the HCCA-EDCA duration ratio to efficiently manage multimedia traffic over both these functions [22,23,31].

Instead *Overboost* [29] uses EDCA resource to improve the HCCA traffic scheduling. This local node scheduler moves the traffic exceeding the assigned HCCA *TXOP* from the HCCA queue to the higher priority Access Category EDCA queue, collaborating with the centralized HCCA scheduler that continues to manage admission control, scheduling parameters, and polling list.

Scheduling Estimated Transmission Time - Earliest Due Date (SETT-EDD) [15] uses a token bucket scheme to vary *TXOPs* according to the station requirements. Each *SI* is computed taking into account its traffic profile, while *Earliest Deadline First* (EDF) [25] determines the polling order.

Fan [11] uses a timer-based scheduler to compute the transmissions deadlines as the smallest between the downlink and uplink ones, and EDF schedules traffic streams.

Real-Time HCCA (RTH) scheduler [8] manages *TXOP* as critical section by using *Stack Resource Policy* (SRP) algorithm [2]. An EDF-based algorithm schedules TS transmissions.

Adaptive Resource Reservation Over WLANs (ARROW) [32] dynamically computes each *TXOP* by taking into account the actual different buffered TSs data at the beginning of the polling. Moreover, Maximum SI is bounded in order to ensure no deadline miss and delay requirements. Finally *Earliest Due Date* (EDD) [21] manages the QSTAs polling list.

Application-Aware Adaptive HCCA Scheduler [20] distinguishes uplink and downlink schedulers, while EDF [25] defines the polling order. The uplink scheduler assigns each QSTA a minimum and a maximum *SI*, adapted to application and network conditions and to the buffered traffic.

Wireless Capacity-Based Scheduler (WCBS) [5] dynamically updates QSTA polling list using EDF algorithm and uses static and dynamic parameters to adapt the transmissions scheduling to the TSs characteristics.

All these mentioned schedulers try their best to compute variable TXOP and/or SI in order to improve the VBR traffic management, but they do not recover assigned but not used resources, that could be present in the case of highly variable TSs.

The bandwidth reclaiming mechanism proposed by Larcheri [24] uses two different *max-min fairness* algorithms to reduce the delay produced by HCCA in the case of VBR traffic by recovering spare resources through a proportional controller.

Its non linearities are overcame by Palopoli [28] with an optimal controller, based on Model Predictive Control [13].

Moreover Lo Cigno applies [9] both cited proportional controller as optimal controller algorithms to multi-class traffic with different priorities and to heterogeneous traffic.

Unused Time Shifting Scheduler [30] is based on a greedy reclaiming scheme that recovers the unspent time from previous polled stations and assigns that to the next polled one in order to provide further resources to serve bursty traffic.

3. DYNAMIC TXOP HCCA SCHEDULER

Traffic dynamicity of VBR applications suggests an adaptive computation of the *TXOP* protocol parameter.

Modifying *TXOP* by finding additional resources, without upgrading the assigned resources threshold but recovering the unused ones, can be obtained knowing the effectively used transmission time at each polling and foreseeing the needed one, and evaluating the presence of resources that can be recovered.

As far as transmission time assignment is concerned, sending duration is dependent of the traffic profile, in particular of the bit data rate. Thus, if the traffic profile is unknown, as in the case of TSs transmission, analyzing the trend of used transmission time is functional to compute the suitable $TXOP'$. Moreover IEEE 802.11e HCCA helps to find further resources, without implying a global assignment recomputation. Indeed, QAP takes control of the medium when a station has exhausted its transmission time, or when it has no more data to send and its *TXOP* is not finished. In this last situation the remaining portion of transmission opportunity is no more used and can be reclaimed.

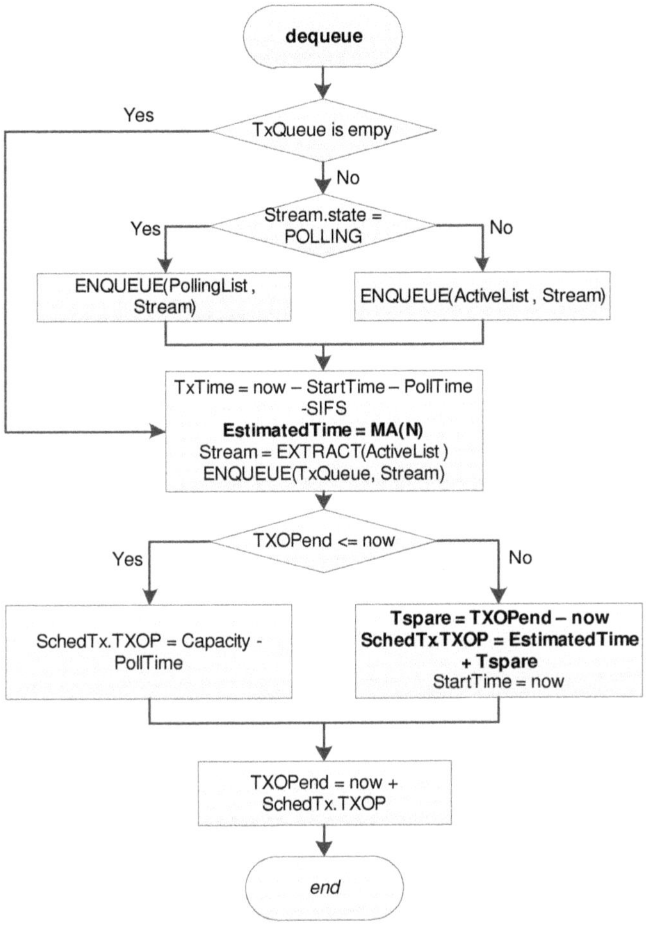

Figure 1: DTH dequeue flow diagram

These are the basic ideas behind Dynamic TXOP HCCA algorithm. DTH aims at integrating the centralized scheduler with an additional mechanism, exclusively focused on the computation of an instantaneous $TXOP_i^j$ appropriate to the scheduled TS_i. DTH does not deal with general protocol parameters computation, polling list sorting, polling of stations, access medium management. These tasks are still performed by the QAP scheduling algorithm, as it will be demonstrated in Section 4.

In particular, DTH is composed by two different mechanisms, shown in bold style in Fig. 1 that summarizes the function `dequeue` used to extract a TS from the polling queue and assign the polling time and the transmission time. The first, which deals with the estimation of effectively used transmission time, aims at refining the current $TXOP$ allocation. It tries to foresee the transmission time that will be effectively used during the next polling, studying the previously used transmission intervals. The second mechanism is focused on finding the spare transmission time.

3.1 Statistical estimation of needed transmission time

The estimation of the transmission time expected to be effectively used during the next polling is inspired by the concept of *time series*. This is used in statistics, signal processing, econometrics and mathematical finance, to study a stochastic phenomenon by monitoring its temporal behavior. A time series is composed by a set of temporally ordered random variables, with the aim at describing its temporal variations. This theoretical tool is perfectly suitable to describe the trend of transmission duration and, consequently, to foresee the length of the next sending interval. Thus, the effectively used transmission time Δ of QSTA$_j$ is described by the time series:

$$\Delta = \{t_{used_{j,t}}, t \in T\},$$

where the time t is the state variable, t_{used_t} is the observation of Δ at time t, and T is the discrete *index set*, that is referred to the time periods, i.e., the sampling periods. In our case they are coincident with traffic periods, since we assume as sample each transmission of the considered QSTA during subsequent *Controlled Access Phases* (CAPs). In *time series forecasting Moving Average (MA)* is suitable to obtain an evaluation of the trend component of an investigated phenomenon from the more recent samples y_i. MA of order $m_1 + m_2$ at time t is expressed as follows:

$$MA(t) = \sum_{i=-m_1}^{m_2} \theta_i \cdot y_{t+i}$$

where m_1 is the number of periods before t, m_2 after t, and θ_i is the weight of the element i. In the case of DTH all samples have the same importance, thus $\theta_i = \frac{1}{m_1+m_2}, \forall i$. Moreover, since all samples of a QSTA$_j$ have the same period, SI_j, the *Simple Moving Average* (SMA) is used. Its computation is coincident with that of the arithmetic average, except that it uses a *mobile window* to collect the samples. Thus, $t_{est_j}(t)$, the estimated transmission time of a station j at time t, before the $(N+1)^{th}$ polling, is computed through the SMA of the effectively used transmission time during the previous N polling:

$$t_{est_j}(t) = \frac{\sum_{i=1}^{N} t_{used_{j,i}}}{N},$$

where $t_{used_{j,i}}$ is the transmission time effectively used during the ith polling time by QSTA$_j$, see Figure 2. In this computation the index set T, and, then, the length of observations interval, the sampling window, are chosen dependently from the traffic type and period, in order to have a sufficient number of samples. For instance, if we assume that T is composed by 250 elements, in the case of a video streaming TS with a period of 40 s, the sampling window is long 10 s. Then, shifting the window along the time axis is possible to have an updated estimation of the transmission time used during the last 250 polling.

This efficient approach can be improved through the integration of a mechanism to recover the unspent transmission time. Moreover this method helps to diagnose the inversion in the data rate trend.

3.2 Resources recovery computation

The time remaining from the previous transmission can provide the additional resources needed when data rate exceeds its average value, without overprovisioning.

The unused time to recover, T_{spare}, is computed with the same mechanism of UTSS, as follows:

$$T_{spare} = t_{end} - t_{stop},$$

where $t_{end} = t_p + TXOP$ is the ending transmission time when $TXOP$ is completely exhausted (t_p is the polling time), and t_{stop} is the actual finishing transmission time.

However the statistical evaluation of the current transmission time makes the DTH $TXOP$ computation completely different. Indeed, in UTSS, no estimation mechanism was adopted and T_{spare}

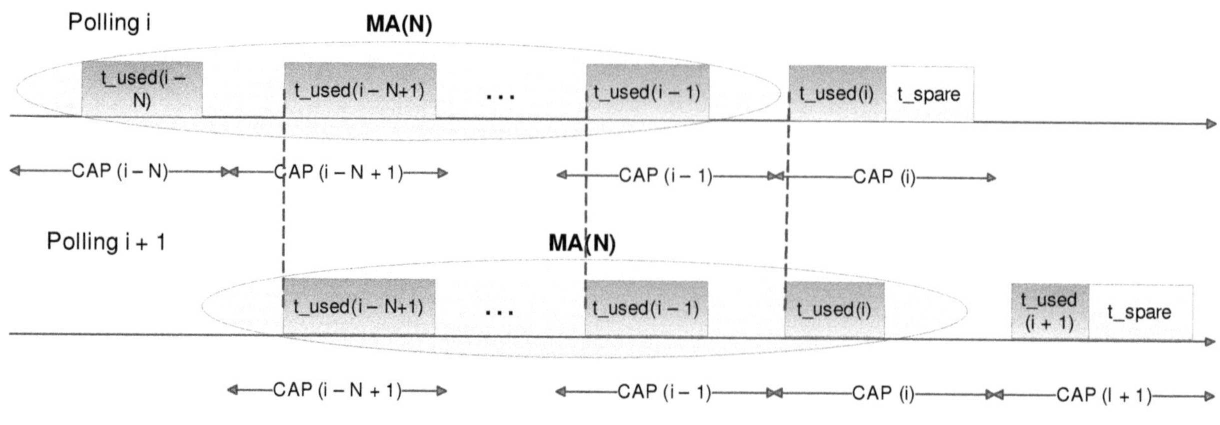

Figure 2: Transmission time estimation mechanism.

was simply shifted to the next polled station, without verifying if needed or not. This means that it was likely to assign the recovered time to station that did not use that and, consequently, the same T_{spare} had to be taken twice. Instead DTH tries to avoid unnecessary assignments. It makes the dynamic $TXOP$ computation more precise with respect to the traffic trend.

3.3 Transmission Opportunity computation

Finally, $TXOP_i'$ for the next polling of $QSTA_i$ is computed as follows:

$$TXOP_i' = \begin{cases} TXOP_{AC(i)} & \text{if } T_{spare} \equiv 0 \\ t_{est(i)} + T_{spare} & \text{if } T_{spare} > 0 \end{cases} \quad (1)$$

4. DTH SCHEDULER ANALYSIS

In this section an analytical study of DTH is presented. It aims at investigating its properties and the impact of its integration with a centralized scheduling algorithm on the real-time properties of the global scheduler. Hence, first of all, it is shown that the assigned $TXOP'$ is tailored to follow traffic variability. Then DTH statistical features are analyzed. As far as real-time properties are concerned the effect of the bandwidth reclaiming scheme and of the transmission time estimation on the admission control feasibility test and on the respect of deadlines constraints is studied.

PROPOSITION 1. *Setting $TXOP_i' = TXOP_{AC,i}$ when $T_{spare} = 0$ is suitable to avoid saturation effect when data rate begins growing and only t_{est} is considered and, then, it is suitable to efficiently manage variations of VBR streams data rate.*

PROOF. In order to highlight the scheduler behavior without the resource integration mechanism, assume $T_{spare} = 0$. Fig. 3 illustrates the centralized scheduler + DTH behavior, first, when data rate R decreases and, then, when increases.

At i^{th} polling the assigned $TXOP_i'$ is:

$$TXOP_i' = t_{est(i)} = \frac{\sum_{j=i-N}^{i-1} t_{used(j)}}{N}.$$

At $(i+1)^{th}$ polling $TXOP'$ is:

$$TXOP_{i+1}' = t_{est(i+1)} = \frac{\sum_{j=i-N+1}^{i} t_{used(j)}}{N}.$$

If R is decreasing, $TXOP_{i+1}' < TXOP_i'$ since MA is based on samples from $i-N+1$ to $i+1$, included $t_{used(i)} < TXOP_i'$, thus the average decreases. The same behavior happens for $(i+2)^{th}$ polling:

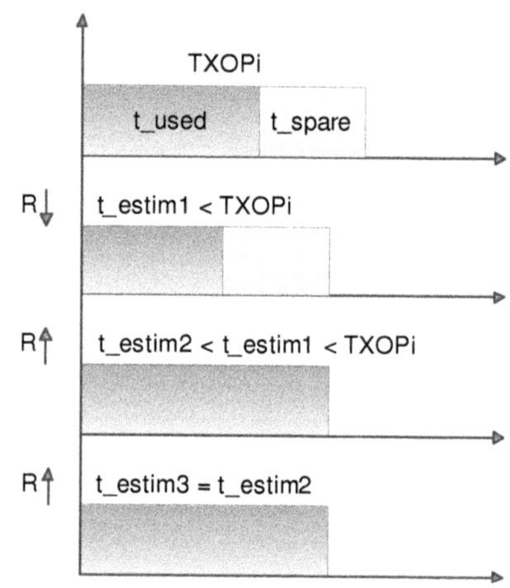

Figure 3: Example of saturation.

correctly the scheduler reduces the assigned transmission time in order to follow R, which is decreasing. But if at polling $(i+3)^{th}$ R changes its trend and begins increasing, $t_{used(i+3)} \equiv TXOP_{i+3}'$. If R continues to grow and we continue to assign $TXOP_i' \equiv t_{used(i)}$, i.e. we continue to assign $TXOP'$ considering only the estimation of previously used transmission time, at next polling

$$TXOP_{i+4}' = t_{est(i+4)} = \frac{\sum_{j=i-N+3}^{i+3} t_{used(j)}}{N} > TXOP_{i+3}',$$

and $t_{used,(i+4)} \equiv TXOP_{i+4}'$ since now all assigned resources are necessary. If at polling $(i+5)^{th}$, assuming R is yet increasing,

$$TXOP_{i+5}' = t_{est(i+5)} = \frac{\sum_{j=i-N+4}^{i+4} t_{used(j)}}{N} \equiv TXOP_{i+4}'.$$

from this point onwards, as shown, if R is increasing, the mechanism is saturated since, at each polling, QSTA uses the whole accorded transmission time but, if $T_{spare} = 0$, there is no chance to upgrade $TXOP'$. Hence, the choice of assigning $TXOP_{AC,i}$ when $T_{spare} = 0$ is a reasonable tradeoff in order to follow traffic variability and, in particular, changes of R trend. Indeed, even if this solution is superabundant when R decreases, when data rate increases and random recovered resources are not present, it provides the maximum allowed capacity. This is a prudential method since the algorithm is primarily focused on preserving real-time performances as well as saving resources. Furthermore it is possible to highlight that DTH shows a certain inertia, which enables to ignore traffic spikes and to follow the general traffic trend. □

THEOREM 1. *The admission control feasibility test is respected when the DTH mechanism is integrated with the centralized scheduler.*

PROOF. Assume to consider the admission control feasibility test suggested by the HCCA reference scheduler. It is expressed as follows:

$$\sum_{i=1}^{M} TXOP_i = T_{CAP} = T_H - T_{CP} \qquad (2)$$

where T_{CAP} is the CAP length, T_H is the duration of the hyperperiod and T_{CP} is that of *Contention Period*, and M is the number of stations composing the *Basic Service Set* related to the considered QAP and admitted to transmit. Without loss of generality, assuming to allow T_{spare} propagation between the last polled station during a CAP and the first one at the subsequent CAP, when DTH is used, at j^{th} CAP the admission control expression becomes:

$$\sum_{i=1}^{M} TXOP_i = \sum_{i=1}^{M} \{t_{est(i)} + T_{spare(i)} - T_{spare(i+1)}\} = \sum_{i=1}^{M} t_{est(i)}$$

where $T_{spare,i}$ is the recovered time assigned to QSTA$_i$, and $T_{spare,(i+1)}$ is the time unused by station i^{th} and shifted to station $(i+1)^{th}$. The expression holds since T_{spare} is simply propagated along the QSTAs chain. Moreover, since $t_{est,i}$ is function of used transmission time during the previous N CAPs,

$$\begin{aligned}
\sum_{i=1}^{M} t_{est(i)} &= \sum_{i=1}^{M} MA_i(t) \\
&= \sum_{i=1}^{M} \{ \sum_{l=j-N}^{j-1} t_{used(i,l)} \} \\
&= \sum_{l=j-N}^{j-1} \frac{t_{est(1,l)} + T_{spare(1,i)} - T_{spare(2,i)}}{N} + \\
&\quad \sum_{l=j-N}^{j-1} \frac{t_{est(2,l)} + T_{spare(2,i)} - T_{spare(3,i)}}{N} + \ldots + \\
&\quad \sum_{l=j-N}^{j-1} \frac{t_{est(M,l)} + T_{spare(M,i)} - T_{spare(1,i)}}{N} \\
&\leq T_{CAP} = T_H - T_{CP}
\end{aligned}$$

since during each one of the N polling considered by MA T_{spare} is propagated between the polled QSTAs. □

THEOREM 2. *The global scheduler, composed by the centralized algorithm and DTH, continues meeting deadlines.*

PROOF. Independently from the type of T_{spare} propagation, if or not enabled between consecutive CAPs, DTH does not modify the centralized scheduling policy. In the case of $T_{spare} > 0$ it simply shifts that to the next polled station. This means that this station receives an upgrade of resources and is polled in advance of the same $T_{spare} > 0$, with respect to the polling time without DTH. Since the portion of added capacity is added polling in advance of the same time interval the station, then it is possible to deduct that deadlines are respected. □

In order to investigate the impact of DTH on delay and packets drop rate, the behavior of queues length is analyzed.

THEOREM 3. *The introduction of DTH is suitable to reduce transmission queues length.*

PROOF. In order to highlight the effect of an increasing data rate R on the transmission queues length, the traffic arriving from the previous transmission ending the ratio η between the incoming traffic during two subsequent polling is considered. Without loss of generality we assume no residual traffic from previous polling, thus queue length is due to the incoming streams. If at polling i^{th} $R \equiv R_{mean}$, and at polling $(i+1)^{th}$ $R > R_{mean}$, in worst condition, at polling time t_i η_i is equal to:

$$\eta_i = \frac{(t_i - t_{end(i-1)})(R_{max} - R_{mean})}{(t_i - t_{end(i-1)})R_{mean}} > 1,$$

where the traffic arriving from the previous transmission ending time $t_{end(i-1)}$ and the current time t_{now} are considered. R_{max} is the max value of data rate reached during the considered interval. Thus, transmission queue length increases. However, when $T_{spare} > 0$, part of this further traffic can be dispatched using this added capacity. This introduces a reduction of transmission queue length with respect to no use of DTH. Indeed, since the delay experience by packets to send is due both to transmission duration that to waiting time into the transmission queues, DTH positively affects the delay experienced by TSs, as it will be shown in Section 5. □

5. EXPERIMENTAL RESULTS

This section presents the results of DTH performance evaluation obtained through simulations.

We first describe the simulation settings and the used traffic model. Then we discuss the analysis results about admission control, polling interval, and experienced null rate in order to investigate the effect of DTH on the centralized scheduling algorithm. Then transmission queues length and mean access delay are analyzed. The obtained results are compared with reference and WCBS schedulers. Indeed, being WCBS an EDF-based scheduler, it does not show good delay performances [4], thus it is a candidate suitable to highlight the DTH effects on real-time performances.

5.1 Simulation settings, network scenario and traffic model

The simulations has been carried out using Network Simulator 2 (*ns-2*) [17], with the HCCA extension described in [7].

The Physical layer is specified in IEEE 802.11g standard, which uses OFDM (*Orthogonal Frequency-Division Multiplexing*) as mandatory modulation scheme; its parameters are listed in Table 1.

MAC level fragmentation, multirate support, RTS/CTS protection mechanism are disabled and we assume that all nodes can directly communicate with each other, without the hidden node problem.

Parameters	*Value*	*Parameters*	*Value*
SIFS (μs)	10	PLCP header (b)	24
DIFS (μs)	28	Preamble (b)	72
PIFS (μs)	19	Data Rate ($Mbit/s$)	54
Slot Time (μs)	9	Basic Rate ($Mbit/s$)	1

Table 1: MAC/PHY simulation parameters.

The presented results have been obtained simulating independent replications of 700 s with a warm-up time of 100 s until the 95% confidence interval is reached for each measure. Confidence intervals are not drawn whenever negligible.

The chosen scenario is composed by one QAP and five QSTAs. Each station transmits *one* uplink TS, which is received by QAP. QSTA named *VP* sends VoIP traffic encoded with G.729A codec, whose parameters are shown in Table 2.

Parameter	*Value*	*Parameter*	*Value*
Frame size *(B)*	10	Payload size *(B)*	20
Frame per packet	2	IP/UDP/RTP	
Period *(s)*	0.02	Header size *(B)*	40
Data rate *(kb/s)*	24	SDU size *(B)*	60

Table 2: G.729A VoIP traffic stream parameters.

QSTA named *VC* transmits a pre-encoded video conference session. The remaining QSTAs, from *VS1* to *VS3*, stream pre-encoded high quality MPEG4 videos. The traffic specifications of these four stations are listed in Table 3.

Parameter	VC	VS1	VS2	VS3
Mean frame size (B)	3800	3800	2900	2900
Maximum frame size (B)	11386	11386	22239	15251
Period (s)	0.040	0.040	0.040	0.040
Mean data rate (kb/s)	770	770	580	580
Maximum data rate(kb/s)	3300	3300	4400	3100

Table 3: Video streaming and video conference parameters.

5.2 Performance evaluation

The efficiency of HCCA schedulers can be evaluated looking at the number of the admitted TSs, *null rate* and *polling interval*.

Since DTH uses the same admission control test of WCBS thus the number of admitted TSs is the same, and it is higher than the one produced by the reference scheduler, due to improved efficiency in resource management.

Null rate is the sending rate of `CF-Null` packets as answer to a polling when the station has no data to transmit; it is due to an uncorrect polling time and can increase the network overhead.

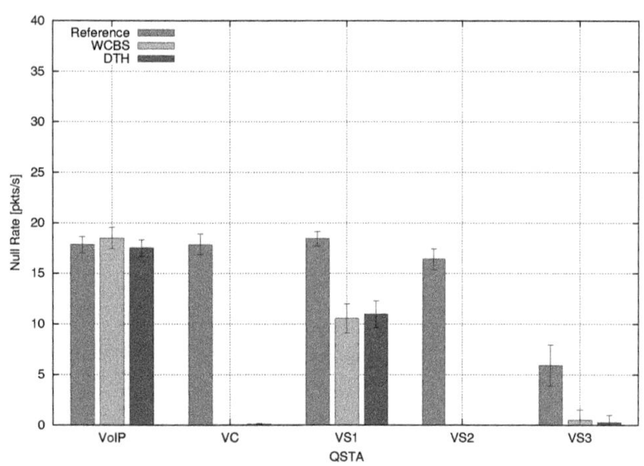

Figure 4: Null rate.

Fig. 4 shows as reference, WCBS, and DTH schedulers have similar results about null rate only for VoIP traffic. When VBR traffic is transmitted, DTH and WCBS obtain lower values since they assign *SI* and $TXOP$ considering the QSTAs requirements.

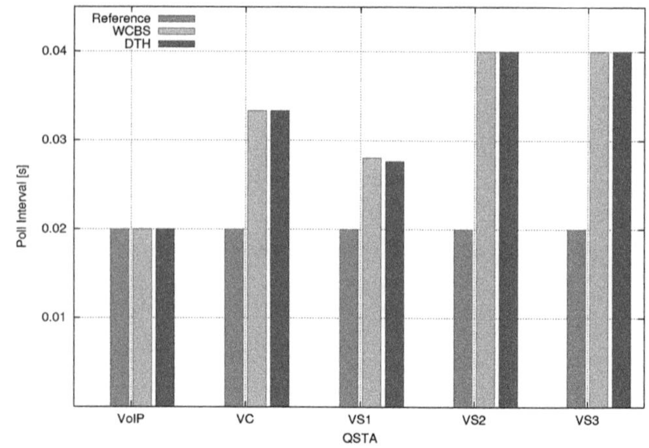

Figure 5: Polling interval.

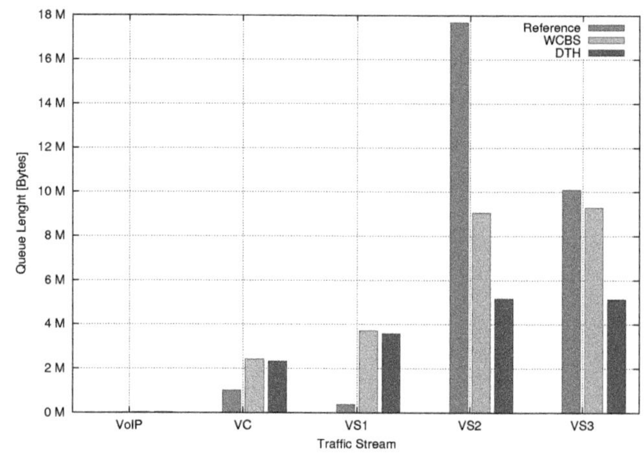

Figure 6: Traffic Stream queue length: 99th percentile.

In Fig. 5 the *polling interval* analysis is illustrated. As stated by the standard, reference scheduler polls all TSs with the same period. Instead DTH and WCBS are able to efficiently assign differentiated polling periods depending on traffic type. DTH provides values similar to those obtained by WCBS: the start time of each Contention Free Period is the same for both these algorithms, and, even if in DTH $TXOP$ can vary over the time, the following shift of polling time for TSs belonging to the same CAP is almost negligible.

Fig. 6 shows the 99th percentile of queue length for each TS. Since DTH assigns $TXOP$ taking into account the average duration of recent transmissions and adding the further resources recovered through T_{spare}, it is able to transmit more packets, when need, taking advantage of this increased capacity, as deduced in Section 4. In this way the queues where packets wait to be transmitted are generally shorter. This improvement is particularly evident when a TS with an highly VBR, like VS2, is transmitted (in this case the DTH percentile is just 27% w.r.t reference and 55% of WBCS values). Fig. 7 focus the analysis on a single QSTA, VS3, in order to deeply study the effect of DTH on transmission queue length. In this case performance evaluation confirms that DTH is suitable to reduce queue length. In particular it limits HCCA queue at 5 MB for about 90%, and, in general, it has shorter queues than WCBS and reference schedulers for every percentage of the distribution.

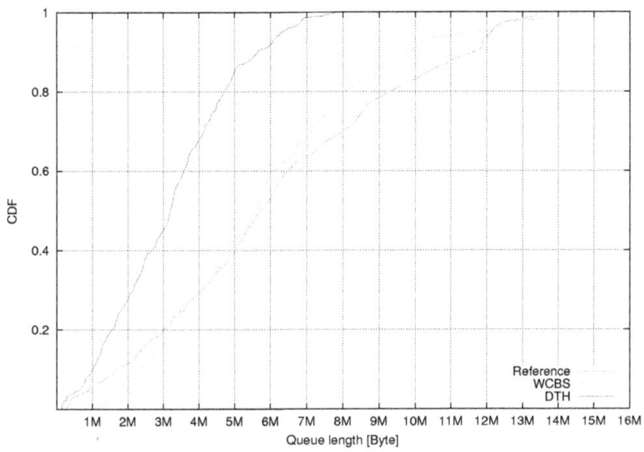

Figure 7: Queue length CDF for VS3 stream.

Throughput has been measured too but there are not significative differences among the considered schedulers.

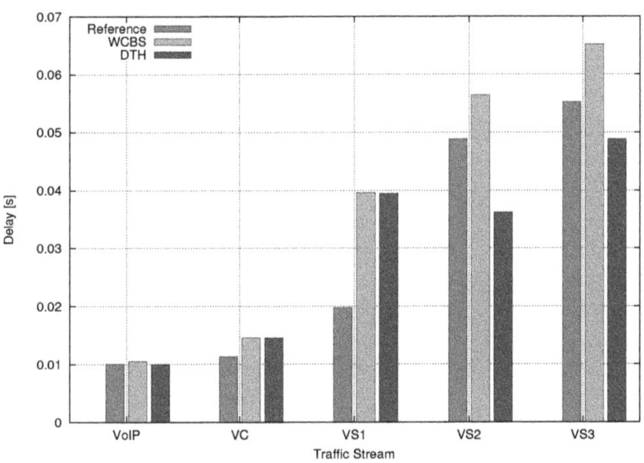

Figure 8: Mean Access Delay.

Mean access delay is illustrated in Fig. 8: here generally DTH performs as WCBS scheduler, except with high VBR TS, like VS3, where it is suitable to reduce access delay. This is due to the fact that it is able to adapt *TXOP* to TS needs, providing more resources when needed.

Further investigating its behavior with VS3 by considering *Cumulative Distributed Function* (CDF) of the access delay, Fig. 9 confirms that the DTH performs better respect to WCBS and reference.

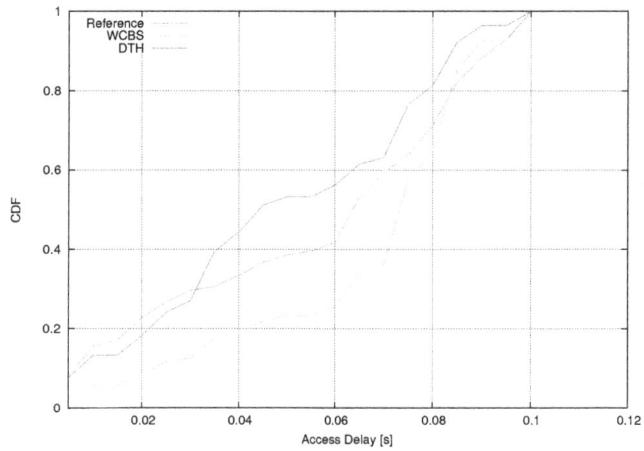

Figure 9: CDF of VS3 Access Delay.

Finally Fig. 10 shows the number of packets per second discarded at the QAP due to an experimented delay longer than the maximum allowed delay bound. DTH has the best performance compared to the other schedulers, and this matches the results obtained about queues length, and access delay.

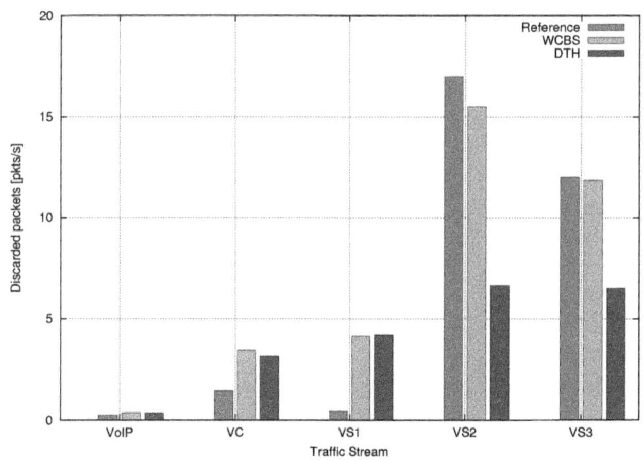

Figure 10: Discarded packets

6. CONCLUSIONS

In this paper a new HCCA scheduling algorithm, *Dynamic TXOP HCCA* (DTH), has been presented. It is suitable to cooperate with the centralized scheduler in order to provide a more efficient mechanism to recover unused, but assigned, transmission time. Furthermore this recovery mechanism refines the computation of the current *TXOP* considering the effective used transmission time of the current station during previous polling. It is computed using time series forecasting based on the Moving Average of the previous sending durations. This solution allows to assign a capacity tailored to the affective needs of the station.

Performance evaluation through simulation corroborated by the analytical study, confirms that DTH does not impact on the policy of the centralized scheduler but it is suitable to improve its performance in terms of transmission queues length, delay and number of dropped packets and absorb and follow data rate variations.

In the future works the use of the Moving Average to estimate the needed transmission time will be deeper investigated in order to find a setting of the mobile sampling windows more tailored to the different type of VBR traffic.

Acknowledgements

The research leading to these results has received funding from the European Community's Seventh Framework Programme under grant agreement n. 246556, "RBUCE-UP".

7. REFERENCES

[1] P. Ansel, Q. Ni, and T. Turletti. FHCF: A simple and efficient scheduling scheme for IEEE 802.11e wireless LAN. *Mobile Networks and Applications*, 11(3):391–403, June 2006.

[2] T. P. Baker. Stack-based scheduling for real-time processes. *Real-Time Syst.*, 3(1):67–99, 1991.

[3] G. Boggia, P. Camarda, L. G. Grieco, and S. Mascolo. Feedback-based control for providing real-time services with the IEEE 802.11e MAC. *IEEE/ACM Trans. Net.*, 2(15):323–333, Apr. 2007.

[4] G. Cecchetti and A. L. Ruscelli. Real-Time support for HCCA function in IEEE 802.11e networks: a performance evaluation. *Wiley Security and Communication Networks*, 4(3):299–315, 2011. Special Issue on Security for QoS Assured Wireless Networks.

[5] G. Cecchetti, A. L. Ruscelli, and F. Checconi. W-CBS: a scheduling algorithm for supporting QoS in IEEE 802.11e. In *Proceedings of IEEE/ACM QSHINE*, pages 1–7, Vancouver, British Columbia, Canada, 2007.

[6] G. Cecchetti, A. L. Ruscelli, A. Mastropaolo, and G. Lipari. Providing variable TXOP for IEEE 802.11e HCCA real-time networks. In *Proceedings of IEEE WCNC*, Paris, France, Apr. 2012.

[7] C. Cicconetti, L. Lenzini, E. Mingozzi, and G. Stea. A software architecture for simulating IEEE 802.11e HCCA. In *Proceedings of IPS MoMe*, pages 97–104, Warsaw, Poland, Mar. 2005.

[8] C. Cicconetti, L. Lenzini, E. Mingozzi, and G. Stea. Design and performance analysis of the Real-Time HCCA scheduler for IEEE 802.11e WLANs. *Computer Networks*, 51(9):2311–2325, 2007.

[9] R. L. Cigno, L. Palopoli, and A. Colombo. Analysis of different scheduling strategies in 802.11e networks with multi-class traffic. In *Proceedings of IEEE LCN*, Dublin, Ireland, Oct. 2007.

[10] J. Cowling and S. Selvakennedy. A detailed investigation of the IEEE 802.11e HCF reference scheduler for VBR traffic. In *Proceedings of 13th ICC*, pages 453–459, Chicago, US, Oct. 2004.

[11] Y. Fan and C. Huang. Real-Time traffic sheduling algorithm in WLAN. In *Proceedings of 4GMF*, San Diego, CA, 2005.

[12] H. Fattah and C. Leung. An overview of scheduling algorithms in wireless multimedia networks. *IEEE Wireless*, 9(5):76–83, Oct. 2002.

[13] C. Garcia, D. Prett, and M. Morari. Model predictive control: theory and practice: a survey. *Automatica*, 25(3):335–348, 1989.

[14] A. Grilo, M. Macedo, and M. Nunes. A service discipline for support of IP QoS in IEEE 802.11 networks. In *Proceedings of PWC*, Laapenranta, Finland, Aug. 2001.

[15] A. Grilo, M. Macedo, and M. Nunes. A scheduling algorithm for QoS support in IEEE 802.11e networks. *IEEE Wireless Comm.*, 10(3):36–43, June 2003.

[16] A. Grilo and M. Nunes. Performance evaluation of IEEE 802.11e. In *Proceedings of 13th IEEE PIMRC*, volume 1, pages 511–517, Lisboa, Portugal, Sept. 2002.

[17] http://www.isi.edu/nsnam/ns/. *Network Simulator 2*, 1995.

[18] IEEE802.11b. Wireless LAN medium access control (MAC) and physical layer (PHY) specification: Higher-speed physical layer extension in the 2.4 GHz band, 1999.

[19] IEEE802.11e. 802.11e-2005 IEEE standard for information technology telecommunications and information exchange between systems local and metropolitan area networks specific requirements part 11: Wireless LAN medium access control (MAC) and physical layer (PHY) specifications: Amendment 8: Medium access control (MAC) quality of service enhancements, 2005.

[20] I. Inanc, F. Keceli, and E. Ayanoglu. An adaptive multimedia QoS scheduler for IEEE 802.11e wireless LANs. In *Proceedings of IEEE ICC*, Istanbul, Turkey, June 2006.

[21] J. R. Jackson. Scheduling a production line to minimize maximum tardiness. Research Report 43, University of California, Los Angeles, CA, 1955. Management Science Research Project.

[22] C. Kuan and K. Dimyati. Utilization model for HCCA/EDCA mixed mode in IEEE 802.11e. *ETRI Journal*, 29(6):829–831, Dec. 2007.

[23] W. K. Lai, C. Shien, and C. Jiang. Adaptation of HCCA/EDCA ratio in IEEE 802.11 for improved system performance. *Int. J. on Innovative Computing, Inf. and Contr.*, 5(11(B)):4177Ű–4188, Nov. 2009.

[24] P. Larcheri and R. L. Cigno. Scheduling in 802.11e: Open-loop or closed-loop? In *Proceedings of IFIP WONS*, Les Ménuires, France, Jan. 2006.

[25] C. L. Liu and J. W. Layland. Scheduling algorithms for multiprogramming in a hard-real-time environment. *J. ACM*, 20:46–61, Jan. 1973.

[26] S. Lu, V. Bharghavan, and R. Srikant. Fair scheduling in wireless packet networks. *IEEE/ACM Trans. Net.*, 7(4):473–489, Aug. 1999.

[27] S. Mangold, S. Choi, P. May, O. Klein, G. Hiertz, and L. Stibor. IEEE 802.11e wireless LAN for quality of service. In *Proceedings of European Wireless*, volume 1, pages 32–39, Florence, Italy, Feb. 2002. invited paper.

[28] L. Palopoli, R. L. Cigno, and A. Colombo. Control and optimization of HCCA 802.11e access scheduling. In *Proceedings of IEEE CDC*, New Orleans, LA, Dec. 2007.

[29] A. L. Ruscelli, G. Cecchetti, A. Alifano, and G. Lipari. Enhancement of QoS support of HCCA schedulers using EDCA function in IEEE 802.11e networks. *Ad Hoc Networks*, 10(2):147–161, 2012. Special Issue on Recent advances in analysis and deployment of IEEE 802.11e and IEEE 802.11p protocol families.

[30] A. L. Ruscelli, G. Cecchetti, A. Mastropaolo, and G. Lipari. A greedy reclaiming scheduler for IEEE 802.11e HCCA networks. In *Proceedings of ACM MSWIM*, Miami, Florida, Nov. 2011.

[31] V. A. Siris and C. Courcoubetis. Resource control for the EDCA and HCCA mechanisms in IEEE 802.11e networks. In *Proceedings of WiOpt*, pages 22–27, Boston, Massachusetts, 2006.

[32] D. Skyrianoglou, N. Passas, and A. K. Salkintzis. ARROW: An efficient traffic scheduling algorithm for IEEE 802.11e HCCA. *IEEE Transaction on Wireless Communications*, 5(12):3558–3567, Dec. 2006.

[33] S.-L. Tsao. Extending Earliest Due Date scheduling algorithms for wireless networks with location-dependent errors. In *Proceedings of 52nd IEEE VTC-Fall*, volume 1, pages 223–228, Boston, MA, Sept. 2000.

On Bounding the Number of Mobiles Sharing a Slot in a Point-to-Multipoint Network

Phani Krishna P
Indian Institute of Technology Madras, India
p.phanikrishna@gmail.com

Siva Ram Murthy C
Indian Institute of Technology Madras, India
murthy@iitm.ac.in

ABSTRACT

Multiple antenna techniques are widely being recognized as front-runners in improving the performance of a wireless network. Sharing a slot between two Mobile Stations (MSs) is defined in a Point-to-Multipoint (PMP) network, independent of the number of antennas at each MS. However, sharing a slot among more than two MSs in a PMP network is not studied in the literature. In this paper, we study if more than two MSs can share a slot for transmission. We find that there exists few realistic scenarios where more than two MSs can share a slot, thus improving the throughput of the network. We observe that the number of MSs sharing a slot is dependent not only on the number of antennas at each MS but also the multiple antenna technique utilized for transmission. When each MS has a constant rate requirement of 2 bps/Hz, we found that 3 MSs can share a slot when each MS uses Spatial Multiplexing techniques, while 4 MSs can share a slot when each MS uses Spatial Diversity techniques for transmission.

Categories and Subject Descriptors

C.2.1 [**Network Architecture and Design**]: Wireless communication

General Terms

Performance

Keywords

WiMAX, MIMO, Signal to Interference Ratio

1. INTRODUCTION

We consider a Point-to-Multipoint (PMP) network, in which several Mobile Stations (MSs) are directly connected to a Base Station (BS). Orthogonal Frequency Division Multiple Access (OFDMA) technology is considered at the physical layer, so that the multi-path interference can be minimized. In each frame, an MS requests the BS, the number of slots required in the next frame. BS, in response, allocates the available OFDM slots to each of the MSs depending on their individual requirements. An OFDM slot may be allocated to different MSs based on several parameters such as geographical location, channel quality and quality of service requirements of each MS. The problem of choosing MSs that share an OFDM slot is studied by the research community under the name 'pairing' and the technique is termed as Collaborative Spatial Multiplexing (C-SM).

In IEEE 802.16 Networks [1], two MSs with 1 or 2 antennas, can share an OFDM slot for transmission. However, sharing a slot with more than two MSs is not studied for a PMP network by considering the interference limitations. Most of the studies [12, 8, 10] are based on a decentralized network, wherein several user (transmitter-receiver) pairs share the same slot for transmission.

This setup can be mimicked in a PMP network by placing all the receivers of a decentralized network at a centralized location. However, the centralized entity needs sufficient number of antennas to support transmission for all MSs sharing a slot in the network. In such scenario, the signals of different MSs sharing same slot interfere with each other at the receiving antennas. Furthermore, as the number of MSs that share a slot increases, the complexity to decode signals increases at the BS (receiver). Using Interference Alignment (IA), signals are aligned along different vectors at antennas of each MS (thus minimizing the interference).

Interference in a PMP network is caused due to two reasons: a) Inter Carrier Interference (ICI) and b) Multiple Input Multiple Output (MIMO) Interference. Interference contributing factors such as Carrier and Doppler Frequency Offsets, and Synchronization caused due to the relative mobility between transmitter and receiver is accounted as part of ICI. This is highly dependent on the *PHY* layer technology used, and has a wide range of solutions [14, 9] particularly when OFDM technique is used. However, most of the solutions use the concept of redundancy to eliminate ICI. They transmit same data symbol along a group of adjacent subcarriers [14]. The size of this group containing redundant data is analytically derived to be two [9], when linear combining techniques are utilized at the receiver. In [9], ICI is shown to be a non-Gaussian random process that has higher impact on the network performance.

When several MSs with several antennas share a slot for transmission, interference occurring in the network is accounted as part of MIMO interference. It is the responsibility of the BS to determine which MSs share a slot and which MSs utilize multiple antennas for transmission, such

MSWiM'12, October 21–25, 2012, Paphos, Cyprus.

that the interference can be mitigated. The BS provides the beam vectors along which data is transmitted at each antenna of the MS.

While it has been a working assumption that two MSs can share a slot for transmission in a WiMAX network [1], there has been no theoretical or analytical study to find the maximum number of MSs that can share a slot. We take initial step towards the solution by finding out the maximum MSs that can share a slot, by considering Signal to Interference Ratio (SIR) as a parameter to verify if the data is successfully received. The problem that we consider in this paper can be formally stated as '*What is the maximum number of MSs that can share an OFDM slot in a PMP network given that each MS is equipped with same number of antennas and each MS has equal rate requirements in a realistic scenario?*' This study also points us to probe into several questions such as: Does this number depend on the *PHY* layer parameters of each MS participating in the share? Does this number vary with the rate requirements of each MS? Does this number vary as the number of antennas at each MS varies? Can we determine the lower bound on number of antennas required at the BS?

We organize the remainder of the paper as follows: In Section 2, keeping the problem statement in view, we review relevant work in the literature that closely addresses the problem. We provide the parameters required for studying the problem and the system model considered for evaluation in Section 3. We analyze the numerical results obtained in Section 4. Finally, we conclude the paper in Section 5 providing pointers towards work for the future.

2. RELATED WORK

IA is a technique in which several user pairs with single antenna transmit and receive data successfully by aligning their data along pre-determined beam vectors [3]. Though IA is proposed for a decentralized network wherein K user-pairs with single antenna share a slot, the technique can be applied in a PMP network where the BS has antennas equal to the number of MSs sharing a slot (wherein each MS is equipped with one antenna). The number of times throughput of the network is improved was shown to be $K/2$ using IA technique. Beam vectors are assumed to be computable and communicated to each MS as and when they are determinable, using the algorithms proposed in [11].

The IA technique is effectively used [6] and experimentally shown to be efficient in designing a protocol for utilizing same slots in an 802.11 network (this work considers the MSs endowed with multiple antennas up to 'K'). The important result being, 'In a distributed network where a transmit-receive pair has equal number of antennas, a pair having more number of antennas than the spatial streams that are being transmitted currently, can utilize the same slot for transmission'. The implications of this result in a PMP network are studied in this paper.

Peters and Heath-Jr. [10] study multiple antenna interference channels and propose a greedy algorithm to partition available users into several groups such that users in a group use IA to transmit data in a single slot, while users across groups use TDMA mechanism to transmit data across different slots. They model users (transmit-receiver pairs) in the network as a connected graph, and partition the graph such that users across groups transmit mutually orthogonally, while users inside a group use IA for transmission.

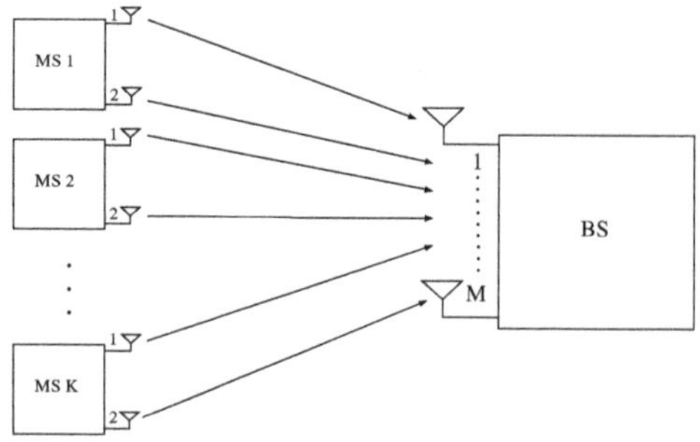

Figure 1: System Model.

The users are partitioned based on the position information of each transmitter and receiver in the network. However, their solution assumes that the user pairs are decentralized in the network, which is not the case in a PMP network under consideration.

When each MS is equipped with only one antenna, receiver filters that mitigate the ICI are designed in [4]. The receiver complexity was proven to be reduced from $O(N^3)$ to $O(Nlog_2N)$ using these receiver filters. The complexity of these filters is dependent on the Fast Fourier Transform (FFT) size (N) used in the OFDMA technique. In a typical WiMAX network, the FFT size is constant depending on the bandwidth allocated at the BS [1]. A set of linear filters (MMSE/ML/ZF receivers) that minimize the interference are assumed to be available at the BS so that all the streams are received without errors.

3. SYSTEM MODEL

For a signal to be interference free, the value of SIR should be above a threshold β, where β is dependent on the bit transmission rate. The SIR for a signal can be defined as

$$SIR = \frac{P(R)}{P(I)} \tag{1}$$

where $P(R)$ is the power of the required signal at the receiver and $P(I)$ is the power of interfered signal at the receiver. β can be derived from popular Shannon's theorem as follows: $C = B\ log_2(1 + SIR) \Rightarrow SIR = 2^{(C/B)} - 1$, where C is achieved capacity of the channel given B bandwidth. Similar expression is also used in [12] as $2^R - 1$, where R is the required rate of transmission in bps/Hz.

We use SIR to measure power of the received signal at the BS. Also, average SIR of the signal at each antenna of the BS is measured in one OFDM slot and plotted to determine effective interference for each signal.

Let 'K' MSs share the same OFDM slot for transmitting to the BS. Each MS is equipped with 2 antennas and the BS is equipped with $M(\geq 2K)$ antennas. An antenna at each MS transmits the data that is aligned along a vector, beamed to a specific antenna at the BS. In other words, alignment vectors specific to each antenna of the MS are transmitted by the BS for each frame. The overhead of transmitting these beam vectors in each frame is presumed to be negligible when compared to the achievable data rates. The data received at antenna i of the BS, similar to Equation 6 in [7], can be written as

$$y_i(s) = \sum_{j=1}^{2K} d_{ij}^{-\rho} W_j H_{ij} x_j(s) W_i^H + n_{ij} \quad (2)$$

where W_j is the beam-formed (weighted) vector along which data is transmitted,
W_i^H is the inverse of the beam-formed matrix used at the BS,
$x_j(s)$ is the data transmitted across antenna j on subcarrier s,
H_{ij} is the channel coefficient between the transmitted antenna j and received antenna i,
n_{ij} is the Additive White Gaussian Noise associated with the channel,
d_{ij} is the distance between antenna i and antenna j,
and ρ is the path loss exponent of the channel. The ICI in a channel using OFDM is mainly due to adjacent subcarriers [7]. Also, assuming the channel to be frequency flat fading, Equation 2 can be modified as follows:

$$y_i(s) = \sum_{j=1}^{2K} d_{ij}^{-\rho} W_j H_{ij} x_j(s) W_i^H + \sum_{j=1}^{2K} d_{ij}^{-\rho} W_j H_{ij} x_j(s-1) W_i^H + \sum_{j=1}^{2K} d_{ij}^{-\rho} W_j H_{ij} x_j(s+1) W_i^H + n_{ij} \quad (3)$$

Equation 3 assumes that ICI occurs every time in the network. The probability of occurrence of ICI in a network [13] is shown to be a random number uniformly generated in the range of $[0, 0.1]$.

The SIR of ith stream at antenna i on a OFDM slot using subcarrier s can be determined as

$$SIR = \frac{\left|W_i W_i^H H_{ii}(s)\right|^2}{\sum_{l=s} \sum_{\substack{j=1 \\ j \neq i}}^{2K} \left|W_j W_i^H H_{ij}(l)\right|^2 + \sum_{\substack{l=s-1 \\ l=s+1}} \sum_{j=1}^{2K} \left|W_j W_i^H H_{ij}(l)\right|^2} \quad (4)$$

where $H_{ij}(l)$ is the channel coefficient for the lth subcarrier between transmit antenna j and receiving antenna i ($\because d_{ij}$, ρ are constant for all MSs in the system).

From the concept of IA, it is obvious that

$$W_j W_i^H = \left\{ \begin{array}{ll} 1 & \text{if } i = j \\ 0 & \text{otherwise} \end{array} \right\}$$

Hence, $W_j W_i^H$ is always binary (i.e., 0 or 1). However, due to the factors such as imperfections in measured channel conditions and highly varying nature of the channel the IA vectors may not remain fixed even for 20 ms (same as time duration of a typical super-frame in WiMAX Networks [1]) time frame, $W_l \times W_m^H$ will not be zero all the time and is considered to have minute errors in evaluations.

Different MIMO techniques can be achieved by varying the transmitted data $x_j(l)$ across different antennas with alignment vectors W_j. In Spatial Multiplexing technique, two antennas of an MS transmit different data (X_1 and X_2) along IA vectors W_1 and W_2, as shown in Fig. 2. Antennas at the BS receive data aligned along vectors W_1^H and W_2^H. The signals aligned along remaining vectors is considered as interference. In Spatial Diversity technique, two antennas of an MS transmit variants of the same data (X_1 and $\tilde{X}_1$) along IA vectors W_1 and W_1, as shown in Fig. 3. Two antennas at the BS receive data along same IA decoding vector W_1^H. We consider both multiplexing and diversity techniques for determining the maximum number of MSs that can share a slot for transmission. A thorough study on the consequences of using different MIMO techniques is provided in the next section.

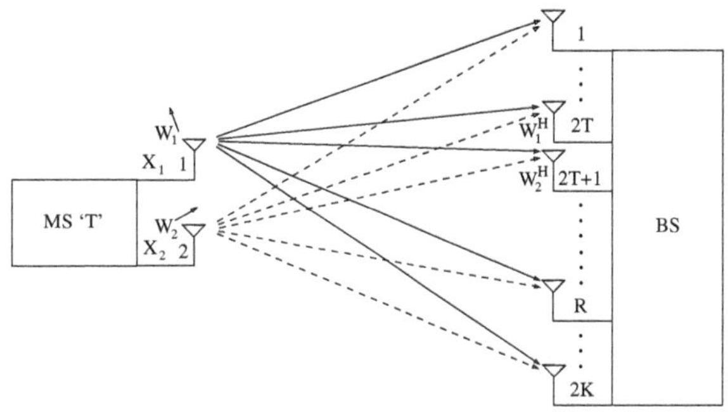

Figure 2: Spatial Multiplexing Technique using IA Vectors for Transmission.

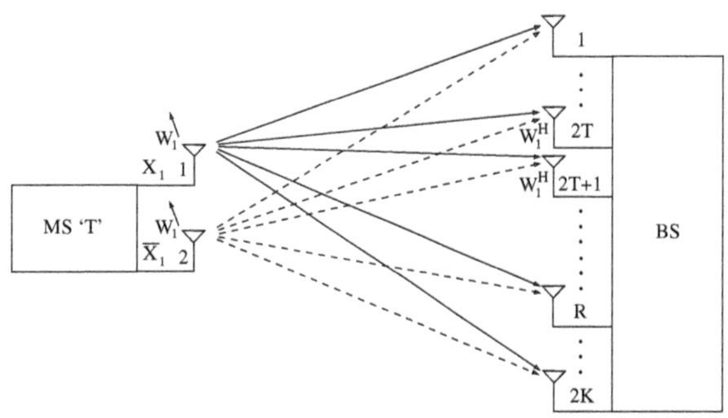

Figure 3: Spatial Diversity Technique using IA Vectors for Transmission.

For the numerical results, initially we place all MSs adjacent to each other as shown in Fig. 1. Since each MS shares approximately the same channel to the BS, the SIR estimate remains same for several MSs (Note: Distance remains the same for all MSs). Later SIR value of each MS is made to be independent of other MSs, representing a more realistic scenario.

4. EXPERIMENTAL DETAILS

We consider only one slot is being used for transmission by all MSs in the network and increase the number of MSs in the network from 1 to 20. The authors in [5] propose a transmitter centric algorithm after realizing that the uplink interference in a multi-user system is highly fluctuant. Since each MS receives different SIR values for each transmission, we calculate SIR of required signal at each antenna of the BS, and plot the average SIR across all antennas observed at the BS. This average SIR is used to determine the achievable transmission rate (in bps/Hz) for each MS in the network.

Table 1: **Average SIR determined at different antennas of the receiver when each MS exhibits different channel quality.**

Number of MSs Sharing a Slot	No ICI One Antenna	Full ICI One Antenna	Two antennas No ICI Diversity	Two antennas Full ICI Diversity	Two antennas No ICI Multiplexing	Two antennas Full ICI Multiplexing
1	Infinity +/- NaN	4.21 +/- 0.01	Infinity +/- NaN	Infinity +/- NaN	65.64 +/- 0.54	53.88 +/- 0.1
2	59.71 +/- 0.51	3.29 +/- 0.01	19.54 +/- 0.04	10.68 +/- 0.02	5.31 +/- 0.00	8.57 +/- 0.0
3	12.58 +/- 0.40	3.22 +/- 0.03	7.20 +/- 0.00	4.77 +/- 0.03	3.17 +/- 0.01	3.10 +/- 0.0
4	7.06 +/- 0.03	2.67 +/- 0.01	4.32 +/- 0.02	3.02 +/- 0.00	2.30 +/- 0.00	1.63 +/- 0.0
5	5.06 +/- 0.01	2.32 +/- 0.03	2.75 +/- 0.01	2.20 +/- 0.01	1.82 +/- 0.01	1.32 +/- 0.0
6	3.75 +/- 0.01	1.95 +/- 0.00	2.58 +/- 0.00	1.69 +/- 0.01	1.58 +/- 0.00	1.47 +/- 0.0
7	3.07 +/- 0.02	1.74 +/- 0.01	1.81 +/- 0.00	1.55 +/- 0.00	1.46 +/- 0.00	1.13 +/- 0.0
8	2.59 +/- 0.02	1.50 +/- 0.00	1.57 +/- 0.01	1.44 +/- 0.00	1.35 +/- 0.01	0.98 +/- 0.0
9	2.26 +/- 0.01	1.39 +/- 0.00	1.39 +/- 0.00	1.34 +/- 0.00	1.09 +/- 0.00	0.79 +/- 0.00
10	1.99 +/- 0.00	1.24 +/- 0.01	1.31 +/- 0.00	1.11 +/- 0.00	0.97 +/- 0.00	0.80 +/- 0.0
11	1.75 +/- 0.00	1.14 +/- 0.00	1.22 +/- 0.00	1.10 +/- 0.00	0.84 +/- 0.00	0.66 +/- 0.0
12	1.61 +/- 0.00	1.05 +/- 0.00	1.20 +/- 0.00	1.00 +/- 0.00	0.72 +/- 0.00	0.59 +/- 0.00
13	1.47 +/- 0.00	0.99 +/- 0.00	1.11 +/- 0.00	0.89 +/- 0.00	0.66 +/- 0.00	0.57 +/- 0.00
14	1.36 +/- 0.00	0.93 +/- 0.00	0.99 +/- 0.00	0.79 +/- 0.00	0.64 +/- 0.00	0.56 +/- 0.0
15	1.26 +/- 0.00	0.86 +/- 0.00	0.89 +/- 0.00	0.77 +/- 0.00	0.62 +/- 0.00	0.51 +/- 0.0
16	1.17 +/- 0.00	0.82 +/- 0.00	0.89 +/- 0.00	0.76 +/- 0.00	0.54 +/- 0.00	0.45 +/- 0.00
17	1.09 +/- 0.00	0.77 +/- 0.00	0.85 +/- 0.00	0.69 +/- 0.00	0.52 +/- 0.00	0.43 +/- 0.0
18	1.02 +/- 0.00	0.73 +/- 0.00	0.80 +/- 0.00	0.65 +/- 0.00	0.49 +/- 0.00	0.42 +/- 0.0
19	0.97 +/- 0.00	0.70 +/- 0.00	0.74 +/- 0.00	0.63 +/- 0.00	0.47 +/- 0.00	0.39 +/- 0.0

For the sake of numerical calculations, we assume the channel coefficients to be uniformly distributed Gaussian random variables with zero mean and unit variance while noise is circularly symmetric and uniformly distributed with zero mean and 0.1 variance.

Initially, we compute the maximum number of MSs that can share a slot when channel conditions are identical for each MS. Later, we compute the same with channel conditions being completely independent for each MS. Also, a Selection Combining receiver is used at the BS when Diversity techniques are employed.

Considering the total capacity of the network to be constant, the number of MSs sharing a slot is maximized when the requirement of each MS is minimized. We consider 2 bps/Hz as a norm in the analysis for the following reason: It is highly visible in WiMAX networks, even when channel quality is as low as $0-5$ dB for an MS. In a WiMAX network, the rate can scale to higher values as the channel quality increases for each MS (Details in Appendix).

4.1 Numerical Results

When the number of MSs that share a slot is 1, the SIR value is maximum for all the four studied scenarios: Each MS with 1 (2) antennas and ICI can (not) be mitigated completely. As the number of MSs sharing a slot increases, the interference in the network also will increase, and a reduction in SIR value can be observed. The maximum MSs that can share a slot is limited to that number, where the average SIR falls below the threshold $\beta = 3$ (pertaining to R=2 bps/Hz). Also, note that β is calculated from Section 3. The values shown in Table 1 are generated after performing 10,000 iterations. Since β is dependent on rate requirement of each MS, number of MSs that share a slot depends on rate requirement of each participating MS.

When an MS with one antenna transmits only one stream of data, SIR value would be maximum as there is no interference in the network. Table 1 provides the results for the scenario where the channel quality of each MS is independent of other MSs. Also, unless mentioned, the results are analyzed for a constant rate requirement of 2 bps/Hz and $\beta = 3$. We also observed that with different rate requirements, the number of MSs that can share a slot varies. This is represented for different rate requirements $(1,2,3$ bps/Hz) in Table 1 using Plum, blue, and ForestGreen colors to denote the maximum number of MSs sharing each slot.

Though multiplexing techniques multi-fold the achievable transmit rates depending on number of antennas used for transmission, prior work showed that using multiplexing techniques improves performance of the network only in high SIR regions [1]. Thus, it can be misleading to consider that every MS uses multiplexing techniques for transmission. We observed that diversity techniques attain more SIR values compared to that of multiplexing techniques reiterating this notion. Hence, it becomes noteworthy that in a real world scenario, the number of MSs that share a slot also depends on the MIMO technique employed by each MS sharing the slot.

We observed that the number of MSs that can share a slot is 7 when ICI is eliminated and 3 when ICI is not mitigated in the network. Hence, the number of MSs sharing a slot also depends on the mitigating ICI in the network. Also, it can be observed from Table 1 that the SIR values remain almost same when the number of MSs sharing a slot is above 8, irrespective of the number of antennas at each MS and ICI mitigation in the network. This confirms that the upper bound for the number of MSs that can share a slot is bounded to 8 if we plan to exploit MIMO techniques available at each MS.

We can also observe that when 2 MSs with one antenna share an OFDM slot for transmission, the SIR value remains close to the threshold β, conforming to the C-SM specifications in WiMAX standards [1]. However, we observed that more MSs can share a slot when each MS is equipped with multiple antennas. A maximum of 7 MSs can share an OFDM slot for a constant rate requirement of 2 bps/Hz in the best possible settings. Consequently, a BS must need at least 14 antennas to receive data from each MS participating in the share. Thus, we provide a lower bound on the number of antennas required at the BS in a PMP network, such that rate requirements of each MS is satisfied. However, the number MSs sharing a slot is completely dependent on channel characteristics, rate requirements, and number of antennas at each MS participating in the share.

4.2 Discussion

The above study determines the maximum number of MSs that can share a slot. It can be misleading that the throughput of the network increases as the number of MSs sharing a slot increases. This is because of the requirement of IA vectors to be available at each MS before transmission. This calculation of IA vectors is done by the centralized entity (BS), and the BS transmits the IA vectors to each MS participating in the share. Though algorithms to calculate IA vectors are proposed in the literature for different channels, the time and space complexities of those algorithms are very high. Thus, the challenge shifts from *being able to share a slot among many MSs* to *providing on-line algorithms that can schedule many MSs* in a specific network.

5. CONCLUSION

In this paper, we studied the effect of increasing the number of MSs that share a slot in a PMP network using simulations. For a constant rate requirement of 2 bps/Hz, we observed that more than 2 MSs can share a slot when each MS is equipped with multiple antennas, which is in stark contrast to the current wireless network standards [1], that allows only 2 MSs to share a slot for transmission. The number of MSs sharing the slot varies for different rate requirements at each MS and different MIMO techniques applied at each participating MS.

A thorough analysis of the presented system to determine the maximum number of MSs that can share a slot is currently being investigated. The analysis is expected to be robust of any receiver combining techniques when diversity techniques are used for transmission. Studying the performance of a network when number of antennas at each MS is not constant across the network is left for the future work.

6. ACKNOWLEDGEMENTS

The first author would like to thank the timely support of Giridhari while preparing the manuscript. This work was supported by the Department of Science and Technology, New Delhi, India.

7. REFERENCES

[1] IEEE Standard for Local and Metropolitan Area Networks Part 16: Air Interface for Broadband Wireless Access Systems. *IEEE Std 802.16-2009 (Revision of IEEE Std 802.16-2004)*, pages 1–2082, May 2009.

[2] K. Balachandran, D. Calin, N. Gopalakrishnan, J. H. Kang, A. Kogiantis, S. Li, L. Ozarow, S. Ramakrishna, A. N. Rudrapatna, and R. Sun. Design and performance analysis of collaborative spatial multiplexing for IEEE 802.16e-based systems. *Bell Labs Technical Journal*, 13(4):97–117, Feb. 2009.

[3] V. Cadambe and S. Jafar. Interference alignment and degrees of freedom of the k -user interference channel. *IEEE Transactions on Information Theory*, 54(8):3425–3441, Aug. 2008.

[4] C.-Y. Hsu and W.-R. Wu. ICI mitigation for OFDMA uplink systems. In *Proceedings of IEEE International Symposium on Personal, Indoor and Mobile Radio Communications*, pages 1–5, 2008.

[5] Y. Li, G. Zhu, and S. Zhang. Transmitter centric scheduling in multi-cell MIMO uplink system. In *Proceedings of the IEEE International Conference on Communications*, pages 1–6, 2010.

[6] K. C.-J. Lin, S. Gollakota, and D. Katabi. Random access heterogeneous MIMO networks. In *Proceedings of the ACM SIGCOMM Conference*, pages 146–157, 2011.

[7] C. O. Martin Garcia. Intercarrier interference in OFDM: A general model for transmissions in mobile environments with imperfect synchronization. *EURASIP Journal on Wireless Communications and Networking*, 2009(1):101–111.

[8] P. Mohapatra, K. Nissar, and C. Murthy. Interference alignment algorithms for the K user constant MIMO interference channel. *IEEE Transactions on Signal Processing,*, 59(11):5499–5508, Nov. 2011.

[9] K. W. Park and Y. S. Cho. An MIMO-OFDM technique for high-speed mobile channels. *IEEE Communications Letters*, 9(7):604–606, July 2005.

[10] W. S. Peters and W. R. Heath-Jr. User partitioning for less overhead in MIMO interference channels. *CoRR*, abs/1007.0512, 2010.

[11] M. Shen, A. Host-Madsen, and J. Vidal. An improved interference alignment scheme for frequency selective channels. In *Proceedings of the IEEE International Symposium on Information Theory*, pages 559–563, 2008.

[12] R. Vaze. How many transmit antennas to use in a MIMO interference channel. *CoRR*, abs/1105.6170, 2011.

[13] T. Yucek and H. Arslan. Carrier frequency offset compensation with successive cancellation in uplink OFDMA systems. *IEEE Transactions on Wireless Communications,*, 6(7):3546–3551, Oct. 2007.

[14] Y. Zhao and S.-G. Haggman. Intercarrier interference self-cancellation scheme for OFDM mobile communication systems. *IEEE Transactions on Communications*, 49(7):1185–1191, July 2001.

APPENDIX

Here, we provide a reality check on the of assumptions made in this paper, and argue for their validity in a real world scenario.

A. RATE REQUIREMENT ASSUMPTION

A bit rate requirement of 2 bps/Hz is considered for the numerical analysis in this paper due to the following reasons:

- These rates are highly visible in a WiMAX network even in the SNR range of $0 - 5$dB [2].

Table 2: Typical CBR requirements for VoIP traffic

	Rate (in bps)	Bandwidth	Rate (in bps/Hz)
CBR	256 Kbps	10 MHz	0.02621
CBR	384 Kbps	10 MHz	0.03932
CBR	512 Kbps	10 MHz	0.05242
CBR	1024 Kbps	10 MHz	0.10485

- The rate requirements of a typical MS in a WiMAX network (that uses 10 MHz bandwidth) are tabulated in Table 2. Clearly, 2 bps/Hz rate requirement is significantly higher than rate requirements in a typical network.

B. ICI REDUCTION

The existing ICI elimination/reduction techniques include: frequency domain equalization, time domain windowing, peak to average power ratio techniques, and self cancellation techniques. In frequency domain equalization techniques, the pattern at which ICI occurs for each OFDM symbol is calculated and estimated for the remaining OFDM symbols. These are used to reduce the fading distortions in the channel. In time domain windowing technique, linear distortions are eliminated by using adaptive filters. In peak to average power ratio reduction technique, the input data block is partitioned into disjoint sub-blocks, and several sub-blocks are combined to minimize peaks. In self cancellation technique, one data symbol is mapped on a group of subcarriers for transmission. However, self cancellation techniques reduce the effective bandwidth of the network by half, thus reducing the network throughput. Since the filters are heuristic in nature, they do not completely eliminate ICI in the network. ICI occurs with probability [0,0.1] in the network [13].

Joint Frequency Assignment and Optimal Association of Stations to Access Points in IEEE 802.11 WLAN

Babul P. Tewari
Department of Computer Science & Engineering
NSHM Knowledge Campus
Durgapur Group of Institutions
Durgapur 713 212, West Bengal, India
tewari.babul@gmail.com

Sasthi C. Ghosh
Advanced Computing and Microelectronics Unit
Indian Statistical Institute
203 B. T. Road
Kolkata-700 108, West Bengal, India
sasthi@isical.ac.in

ABSTRACT

Frequency channels are allocated to the access-points (APs) subject to an acceptable co-channel interference level in an IEEE 802.11 WLAN. Due to limited number of the non-overlapping frequency channels, all APs in a given area may not be activated simultaneously. Therefore which subset of the APs to select for activation is a major concern. Given a set of selected APs, often a station (STA) can potentially associate with more than one AP. The association of an STA to an inappropriate AP will not only lead to degraded service for the concern STA but also may pull down the throughput of the other STAs associated with that AP. Thus finding the optimal association between STAs and APs is another important concern. We argue that treating these two issues in succession may lead to suboptimal solutions, whereas, ignoring the co-channel interference may result over estimation of the throughput. In this paper, an integrated model based on integer programming and an efficient greedy algorithm are proposed that address both aspects simultaneously and maximizes the overall throughput while taking care of load balancing across different APs. Computational results show that indeed the integrated approach is superior to both two-step approach and methods that ignore the interference.

Categories and Subject Descriptors

C.2.1 [**Computer-Communication Networks**]: Network Architecture and Design—*Wireless Communication*

General Terms

Experimentation, Algorithms, Performance

Keywords

Frequency assignment, access points, optimal association, throughput maximization, wireless LAN, integer programming.

1. INTRODUCTION

An IEEE 802.11 wireless LAN (WLAN) consists of a set of APs connected to a wired backbone and a set of STAs. An STA is able to connect to the Internet via an AP selected from the set of APs. For an infrastructure based WLAN an STA can send data frames via an AP only when it is associated with that AP. This requires an STA to AP association that is used by the Distribution System (DS) to accomplish its distribution service. When many APs are deployed in a region of IEEE 802.11 WLAN network, it is possible that an STA may be in the range of several APs. In such situation, a well-known problem is which AP an STA selects to associate with. According to the conventional association policy an STA selects an AP from which it receives the strongest signal strength by Received Signal Strength Indicator (RSSI) measurement. This traditional RSSI strategy might lead to poor performance when a large number of STAs get strong signal from a common AP. Then their association with that AP may result a degraded throughput for each of them and hence may pull down the overall throughput due to non-uniform load distribution across different APs. Thus finding an appropriate association between STAs and APs is an important design concern.

A frequency channel is required to be allocated to an AP in order to serve its associated STAs. In IEEE 802.11 wireless WLAN, frequency channels are allocated to the APs subject to an acceptable co-channel interference level. If two closely located APs are allocated the same channel they start interacting and communication quality degrades. In this type of network the number of available frequency channels is very limited. For example, in IEEE 802.11b (operates in the unlicensed ISM band at 2.4 GHz) though there are 14 channels available, due to inter-channel interference, only 3 of them (1, 6 and 11) can be considered to be non-overlapping [10]. Due to this limited number of non-overlapping frequency channels if the number of APs in a given area increases in large scale then frequency assignment becomes almost impossible for increased interference level. Thus all the APs in a given area may not be activated to associate with the STAs. Therefore which subset of the APs to select for activation together with their particular frequency assignment is another major design concern.

In this type of networking situations, two major planning tasks are to decide, given a set of STAs, a set of APs and a set of frequency channels, which of the APs from the available APs are to be selected for activation by allocating frequency channels such that it satisfies the co-channel interference constraints and how to associate the selected APs with the set of STAs so that overall throughput of the association is maximized.

In this paper, we propose a joint approach to select a subset of APs realizing the frequency assignment and to determine an optimal association between STAs and the selected APs. This integrated approach is very essential as treating these two issues in succession may lead to suboptimal solutions. On the other hand, if proper frequency planning is assumed ignoring the co-channel interference, as has been done in [1, 2], the overall throughput achieved may either be impractical or difficult to obtain. It may also lead to an impractical association. To address these issues and to sum-up, the key contributions of our work are:

- An integrated model based on integer programming (IP) is proposed that addresses both the problems simultaneously.
- An efficient greedy algorithm is also developed for solving the large scale instances of the joint problem. The performance of the proposed greedy is compared against the formulated IP by considering various simulation setup. It has been shown that the proposed greedy achieves almost uniform load distribution across the selected APs.
- We have measured the level of over estimation of the throughput that may arise when the co-channel interference is ignored, as has been done in [1].
- We have demonstrated through a motivational example that only a suboptimal solution may be achieved if these two problems are considered in succession.

The rest of the paper is organized as follows. Section 2 presents the related works in the areas of STA-AP association and frequency assignment problems in WLAN. Section 3 provides the networking model, the details of throughput calculations, the joint problem formulation and a motivational example. Section 4 presents the proposed integrated model based on integer programming (IP). Section 5 describes the proposed greedy algorithm to solve the joint problem. In Section 6 we discuss the performance of the proposed greedy and compare it to that of the IP formulation and earlier works. Section 7 concludes the paper.

2. RELATED WORKS

In IEEE 802.11 network when several APs are available the AP selection of an STA heavily influences the performance of its own and others. Intensive studies were done for a feasible association between STAs and APs based on different metrics and different objectives.

Kumar et al. [1] addresses an optimal association policy between STAs and APs with the objective of maximizing the throughput of the association. They, however, ignores the co-channel interference by assuming the proper frequency planning of the APs. Many authors advocate the issue of load balancing in their work in order to optimize the performance. It has been shown in [2, 5, 4] that the traditional association based on the Received Signal Strength Indicator (RSSI) measurement may lead to poor user experience due to non-uniform load distribution across different APs. To address this issue, some approaches have been proposed which are essentially AP access control mechanisms [4, 5]. Bejerano et al. [2] proposed an association model considering both load balance and fairness. It also includes a complicated central AP association mechanism. An oppose to the traditional RSSI, in [3] a load balancing heuristic is proposed that helps users to select the AP to associate with on the basis of the number of users already associated with that AP and the mean RSSI. Vasudevan et al. in [6] addresses an AP selection mechanism where the bandwidth that an STA is likely to receive if it is associated with that AP has been considered as the selection criterion. Each user estimates the potential bandwidth of neighbor APs by passive measurement and selects the AP with the highest potential bandwidth to connect to. In [17] an AP association strategy is proposed to maximize the minimal throughput for all STAs. An association policy depending on link rate and realized throughput is addressed in [18] for a multi channel multiple AP network. In [8] the AP selection is modeled as many to one matching problem in a bipartite graph in order to maximize the total load among all the APs. In [9] the association metric considers the throughput degradation when an STA joins an AP. The metric is a weighted sum of the throughput provided to the new STA and the average transmission time of STAs that are already associated with the concerned AP.

Some authors consider the dynamic scenarios where users come and go. In [7] a heuristic association strategy for a 2-AP topology with the flexibility of STAs to join and leave the network dynamically is addressed. Before an STA associates with an AP it calculates the related throughput and associates with the AP where it can achieve the maximum throughput.

Many authors address the issue of frequency assignment in their work. In [15] the author addresses an association mechanism for IEEE 802.11h specification where a station estimates the channel conditions before it associates with the AP. Leung et al. [11, 12] addresses the frequency assignment problem and proposed heuristic approaches for IEEE 802.11 wireless network considering traffic load at the MAC layer. Park et al. [13] addresses a heuristic frequency assignment that allocates a collection of frequency channels to a set of transmitters using the smallest span of frequency band with satisfactory communication quality. The author addresses the equivalent graph coloring problem for the same.

Bosio et al. [10] addressed a two step problem in order to optimize the access efficiency of wireless medium. In the first step, finding the location of APs is considered and taking the solution of it they found the frequency assignment to the APs considering a fixed STA-AP association to achieve the objective. Andreas et al. [14] proposed an integrated model for optimum AP placement and channel assignment in order to maximize the net throughput that an STA can expect.

In our work instead of considering the AP placement problem, we have assumed that the AP placement is fixed and given. We select an appropriate subset of APs from the given set of APs such that there exists a feasible frequency assignment of the selected APs satisfying the necessary interference criterion along with finding the optimal association between STAs and the selected APs. Here, the objective is to maximize the overall throughput achieved by all the STAs while taking care of the load balancing across the selected APs. To the best of our knowledge this problem has not been dealt before in a combined way.

3. JOINT FREQUENCY ASSIGNMENT AND OPTIMAL ASSOCIATION PROBLEM

In this section, we discuss the networking model, the details of throughput calculations, the joint problem formulation and a motivational example.

3.1 The network model

We consider an infrastructure based WLAN, where n number of APs are connected to a wired infrastructure. There is m number of STAs which access this network through these APs. An AP establishes an IEEE 802.11 cell and coordinates all communications that take place in that cell. We have assumed that the radio ranges of all the APs are equal and known. An area is covered by several APs where a single STA can be associated with at most one AP but a single AP may serve multiple STAs. The set of m STAs is represented by $S_{STA} = \{1, 2, 3, \ldots, m\}$ and indexed by $1 \leq i \leq m$. Similarly the set of n APs is represented by $S_{AP} = \{1, 2, 3, \ldots, n\}$ and indexed by $1 \leq j \leq n$. The data rate obtained by STA i if it is associated with AP j is denoted as w_{ij}, where $w_{ij} \in \mathcal{C}$, a finite data rate set. For example, $\mathcal{C} = \{1, 2, 5.5, 11\}$ Mbps for IEEE 802.11b. We assume that w_{ij}'s are known and given by an $m \times n$ data rate matrix W. The set of k frequencies is represented by $F = \{1, 2, 3, \ldots, k\}$ and indexed by $1 \leq f \leq k$ which will be used to activate the set of APs of the given area. A single AP will be allocated a single frequency channel to serve all its associated STAs. The APs with common interfering range requires the allocation of different channel for satisfying the co-channel interference. Two APs are within the common interfering range of each other if their intermediate distances are less than certain threshold. If STA i is associated with AP j then the throughput obtained by it and the overall throughput obtained by all the STAs are calculated as described in the following section.

3.2 Throughput calculation

The throughput calculation used here is similar to the *random polling access method* as described in [1]. Suppose there are m_j STAs $i_1, i_2, \ldots, i_{m_j}$ associated with AP j and these are associated with rate $w_{i_p j}$ where $1 \leq p \leq m_j$. Then Kumar et al. [1], [16] shown by analysis of the IEEE 802.11 Distributed Coordinate Function (DCF) that the aggregate uplink throughput obtained by all the STAs served by AP j is of the form:

$$\frac{1}{\frac{1}{m_j}\sum_{p=1}^{m_j}\frac{1}{w_{i_p j}} + h_{m_j}} \tag{1}$$

Where h_{m_j} is an overhead term that depends on m_j, the IEEE 802.11 DCF parameters and the packet length. If overheads are not taken into account Equation (1) reduces to the simpler formula:

$$\frac{m_j}{\sum_{p=1}^{m_j}\frac{1}{w_{i_p j}}} \tag{2}$$

In random polling access method it is assumed that the individual throughput obtained by all the STAs served by a common AP is equal. Let q_j be the individual throughput of all the STAs served by AP j. Then q_j is given by:

$$q_j = \frac{1}{\sum_{p=1}^{m_j}\frac{1}{w_{i_p j}}} \tag{3}$$

It is observed by the above formula that one STA associated at a low rate can pull down the throughput seen by other STAs, even though these others may be associated at high rates. The overall throughput obtained by all the STAs is given by $\sum_{j=1}^{n} m_j \ln(q_j)$. So the overall throughput of a particular association is nothing but the the sum total of the logarithmic of the throughput obtained by each STA present in the association. In all our simulation we have used the random polling access method for throughput calculation. The other methods such as *max-min fair rates* and *proportionally fair access* as described in [1] are equally applicable.

3.3 The joint problem

As discussed earlier, an STA can be in the range of several APs so there can be many possible associations between STAs and APs. The optimal association problem is to find out an association between the APs and the STAs such that the overall throughput is maximized. Since the number of non-overlapping frequency channels are limited so all the APs in a given area may not be activated. So we need to select a subset of APs such that there exists a frequency assignment to them satisfying the co-channel interference. Therefore the joint frequency assignment and optimal association problem deals with selecting an appropriate subset of APs by allocating the available non-overlapping frequency channels and find out an association between the selected APs and the STAs with the objective of maximizing the overall throughput simultaneously. Note that neither our objective is to maximize the frequency reuse capability in order to activate more number of APs nor do we want to ignore the co-channel interference constraints for finding out an optimal association to maximize the overall throughput. We are interested in selecting those APs by allocating the available non-overlapping frequency channels, avoiding co-channel interference so that we can find out an association between the STAs and the selected APs that maximizes the overall throughput.

3.4 A motivational example

In this subsection we evaluate the need for an integrated approach of frequency assignment and finding an optimal association between STAs and APs by revealing the weaknesses of a two step solution of the problem and a solution that ignores the interference. To demonstrate our idea we consider an IEEE 802.11 infrastructure network where there are 4 STAs that needs to be served and 4 APs are available to cover the region. All the STAs are allowed to freely choose an AP to associate with. The AP placements are such that it generates the interference graph as shown in Fig. 1. In the interference graph, an edge means the corresponding APs are within the interfering range of each other. The data rate matrix between STAs and APs are given in Table 1. In this example we have assumed that there is only *one frequency channel* available.

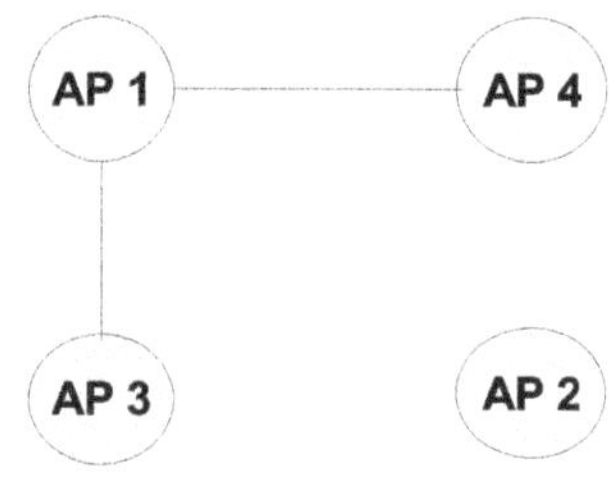

Figure 1: The AP placement for the motivational example

We now consider three cases with the objective of assigning frequencies to a set of APs and to find out an association between the selected APs and the STAs that maximizes the overall throughput.

Table 1: Data rate matrix for motivational example

	$AP1$	$AP2$	$AP3$	$AP4$
$STA1$	11	5.5	2	2
$STA2$	11	5.5	2	2
$STA3$	2	5.5	2	2
$STA4$	2	5.5	2	2

If we want to solve the problem in a two step process then at the first step in order to maximize the number of APs that can be activated, we assign the frequency to $AP2$, $AP3$ and $AP4$ ensuring co-channel interference constraints. It can be seen from Fig. 1 that $AP2$, $AP3$ and $AP4$ are non-interfering to each other and so the same frequency can be assigned to all of them. At the second step we try to find out an optimal association between these selected APs and the STAs. If we consider all possible associations between these APs and the STAs we see that the maximum overall throughput (as computed using the throughput calculation method described in Subection 3.2) is found to be 3.4095 with the association $STA1 \rightarrow AP2, STA2 \rightarrow AP2, STA3 \rightarrow AP3, STA4 \rightarrow AP4$.

Instead of solving the two problems in succession let us now consider our combined approach. We are interested in activating only those APs by allocating appropriate frequencies to them satisfying the co-channel interference as well as results an association that maximizes the overall throughput. Keeping view to this fact we activate $AP1$ and $AP2$ ensuring the co-channel interference and find the optimal association as $STA1 \rightarrow AP1$, $STA2 \rightarrow AP1$, $STA3 \rightarrow AP2$, $STA4 \rightarrow AP2$ resulting an overall throughput of 5.4728 which is substantially greater than the previous one. It can be seen from Fig. 1 that $AP1$ and $AP2$ are non-interfering to each other and hence only one frequency is sufficient to activate both of them.

If we relax the limitation of available frequency channel and assume a proper frequency planning ignoring the co-channel interference then any subset of APs can be activated. Such assumption leads to an optimal throughput of 5.8072 with the selection of $AP1$, $AP2$ and $AP3$. The resulted optimal association is found to be $STA1 \rightarrow AP1, STA2 \rightarrow AP1, STA3 \rightarrow AP2, STA4 \rightarrow AP3$. It is evident that the obtained throughput is much greater than both the previous two. But it is important to note that $AP1$, $AP2$ and $AP3$ can not be activated simultaneously with only one frequency. This is because $AP1$ and $AP3$ are interfering to each other as can be seen from Fig. 1 and hence we need at least two frequencies to activate them simultaneously.

Thus we observe that the optimal association achieved by the two step solution gives only a sub-optimal throughput. The joint approach results an achievable optimal throughput. When we remove the interference constraints we may get better result but that overestimates the throughput as the selected APs may not be possible to activate simultaneously with limited number of frequencies.

4. INTEGER PROGRAMMING FORMULATION OF THE JOINT PROBLEM

Our focus is on selecting an appropriate subset of APs together with a feasible frequency assignment to them satisfying the co-channel interference and finding an optimal association with a view to maximizing the overall throughput. The interferences between the APs are represented as the $n \times n$ interference matrix $A = (a_{jj'})$ where $a_{jj'} = 1$ if AP j and AP j' are interfering to each other and 0, if they are non-interfering.

For $i \in S_{STA}$, $j \in S_{AP}$, and $f \in F$ we define the following binary variables:

$$x_{ij} = \begin{cases} 1 & \text{if STA } i \text{ is associated with AP } j \\ 0 & \text{otherwise.} \end{cases}$$

$$y_j = \begin{cases} 1 & \text{if AP } j \text{ is selected from } S_{AP} \\ 0 & \text{otherwise.} \end{cases}$$

$$z_{jf} = \begin{cases} 1 & \text{if AP } j \text{ is assigned with frequency } f \\ 0 & \text{otherwise.} \end{cases}$$

Now the joint problem can be represented by the following integer programming where the objective function is nonlinear but all the constraints are linear.

$$Maximize \sum_{j \in S_{AP}} m_j \ln(\frac{1}{Q_j}) = - \sum_{j \in S_{AP}} m_j \ln(Q_j)$$

subject to the following constraints:

$$\sum_{j \in S_{AP}} x_{ij} \leq 1 \; \forall i \in S_{STA} \quad (4)$$

$$x_{ij} \leq y_j \; \forall i \in S_{STA}, j \in S_{AP} \quad (5)$$

$$\sum_{f \in F} z_{jf} = y_j \; \forall j \in S_{AP} \quad (6)$$

$$z_{jf} + z_{j'f} \leq 1 + \infty(1 - a_{jj'}) \; \forall j, j' \in S_{AP}, f \in F \quad (7)$$

$$m_j = \sum_{i \in S_{STA}} x_{ij} \; \forall j \in S_{AP} \quad (8)$$

$$Q_j = \sum_{i \in S_{STA}} \frac{1}{w_{ij}} x_{ij} \; \forall j \in S_{AP} \quad (9)$$

$$\sum_{j \in S_{AP}} w_{ij} x_{ij} \geq D^{ths} \sum_{j \in S_{AP}} x_{ij} \; \forall i \in S_{STA} \quad (10)$$

where m_j and Q_j are integer and real variables respectively, to store the intermediate results. The ∞ represents a big number.

Constraint (4) ensures that each STA can be associated with at most one AP. Constraint (5) ensures that an AP must be chosen to be selected before any STA is associated with it. Constraint (6) ensures that a selected AP must be assigned a frequency from the set of available frequencies. Constraint (7) ensures that APs within interfering range of each other (adjacent APs) cannot be assigned the same frequency. Constraint (8) calculates the number of STAs associated to a particular AP. Constraint (9) calculates the value of Q_j where $q_j = \frac{1}{Q_j}$ is the individual throughput achieved by each STA served by AP j, as defined in Section 3.2. Constraint (10) ensures that STA i can be associated to AP j if and only if it receives a data rate from it which is greater that some threshold D^{ths} value. The objective function calculates the overall throughput obtained by all the STAs as defined in Section 3.2. The number of binary decision variables and the constraints in this formulation are $(n + nm + nk)$ and $(2m + 3n + nm + kn^2)$, respectively. Here m, n and k are the cardinality of S_{STA}, S_{AP}, and F respectively.

5. THE PROPOSED GREEDY ALGORITHM FOR THE JOINT PROBLEM

In this section, we propose a greedy algorithm for solving the joint frequency assignment and optimal association problem. Initially all APs are *not activated* and all STAs are *not associated.* The STAs are sorted based on their maximum data rate and the count on number of APs from which it receives this maximum data rate. The STAs are then considered for association one by one in that sorted order. The first STA in that order gets associated with the AP that provides maximum throughput among all others and the corresponding AP is activated by assigning the least available frequency to it. Next for associating an STA to an AP, it first computes the maximum overall throughput that can be achieved if the STA is associated with one from the set of already activated APs. Followed by, it computes the maximum overall throughput that can be achieved if the STA is associated with one from the non-activated APs where a frequency can be assigned satisfying the co-channel interference with the already activated APs. Depending on these two values the STA selects the AP to associate with and activate the selected AP if it is not already been activated by assigning a frequency to it. The set of activated APs are then updated accordingly and the obtained throughput and the association are retained. In effect the STAs get associated with the APs from either already activated or not yet activated set that leads to the maximum improvement in the overall throughput. The procedure continues until all the STAs are associated with and finally return the set of activated APs, the optimal association and the maximum throughput. The throughput obtained by an STA from an AP and the overall throughput obtained by all the STAs have been computed as described in Section 3.2. More specifically, the proposed greedy can be described as follows:

Input: S_{AP}, S_{STA}, F, data rate matrix and interference matrix.

Step 1: Initialization

Let V be the set of already activated APs and X be the set of already associated STAs. Initially all APs are *not activated* and all STAs are *not associated* and thus $V = \emptyset$ and $X = \emptyset$.

Step 2: Sort the STAs

Consider an STA and compute the data rates it receive from all the APs. Save the maximum data rate it receives and count the number of APs from which it receives this maximum data rate. After computing these values for all STAs, sort the STAs according to descending order of the maximum data rates they receive. Rearrange the STAs receiving the same maximum data rate according to the ascending order of the count on number of APs from which it gets this maximum data rate. Let STA 1, STA 2, ..., STA m be this sorted ordering.

Step 3: Finding the association between STAs and APs

For association purpose, consider the STAs in the order as obtained in Step 2 above. Find AP j from S_{AP} which provides the maximum throughput to STA 1. Designate this throughput as the overall throughput $optsum$. Activate AP j with frequency $f = 1$, associate STA 1 with AP j. Update $V = V \cup AP\ j$ and $X = X \cup STA\ 1$.

For $STA\ i$, $i = 2, 3, \ldots m$ do the following:

Find AP j from V such that if STA i is associated with AP j, the overall throughput obtained by all the STAs in $X \cup STA\ i$ is maximized. Let us denote this resultant overall throughput as $optsum_1$. Find AP j^* from $S_{AP} \setminus V$ such that if STA i is associated with AP j^*, the overall throughput obtained by all the STAs in $X \cup STA\ i$ is maximized and there exist a frequency in F that can be assigned to AP j^* by satisfying the co-channel interference with the APs in V. Let us denote this resultant overall throughput as $optsum_2$. Set $optsum$ as the maximum of $optsum_1$ and $optsum_2$. If $optsum_1$ and $optsum_2$ are equal take $optsum$ as $optsum_1$. If $optsum = optsum_1$, associate $STA\ i$ with AP j and update $X = X \cup STA\ i$. Otherwise, if $optsum = optsum_2$ assign the *least frequency* f in F that satisfies the co-channel interference with the APs in V and associate $STA\ i$ with AP j^*. Update $V = V \cup AP\ j^*$ and $X = X \cup STA\ i$.

Output: $optsum$, V, and the optimal association.

6. EXPERIMENTAL RESULTS

In this section we evaluate the performance of our proposed greedy algorithm for solving joint frequency assignment and optimal association problem. We compare our results with the result obtained by formulated Integer Programming (IP) as well as with the formulation reported in [1]. We provide different experimental results considering different networking parameters.

Table 2: Data rate matrix for greedy vs IP

	$AP1$	$AP2$	$AP3$	$AP4$
$STA1$	11	11	2	11
$STA2$	11	11	2	2
$STA3$	2	5.5	2	5.5
$STA4$	2	2	2	11

6.1 Simulation set up

We have considered an infrastructure based WLAN in which the data rate obtained by an STA depends on the distance of it from the serving AP. Depending on the positions of STAs and APs we define that the data rate receives by the STAs within 50 meters from the AP is 11 Mbps, 5.5 Mbps between 50 and 80 meters, 2 Mbps between 80 and 120 meters and when the distance is between 120 and 150 meters the data rate is 1 Mbps. These values are common with that provided by IEEE 802.11 vendors. This implies that the AP coverage range is 150 meters. The STAs that are beyond the distance of 150 meters from an AP will be considered as out of the coverage range of the concern AP and will get a 0 data rate from that AP. If an STA receives 0 data rate from all the APs then that STA will not be associated to any AP. In our simulation, 25 APs are uniformly placed in a 300×300 $meters^2$ area. The placement of APs are shown in Fig. 2. We consider that two APs are within the interference range of each other if their distance is within certain threshold. We generate different interference matrices by considering this threshold value as 100 meters, 125 meters and 150 meters to reflect different level of interferences. We have considered 100, 125 and 150 STAs and placed them uniformly in a 300×300 $meters^2$ area to represent different user load distribution. The number of frequencies considered varies from 1 to 11 for dealing with the co-channel interference. In our experimentation, we consider that the APs within the interfering range of each other require different frequency channels for being activated simultaneously.

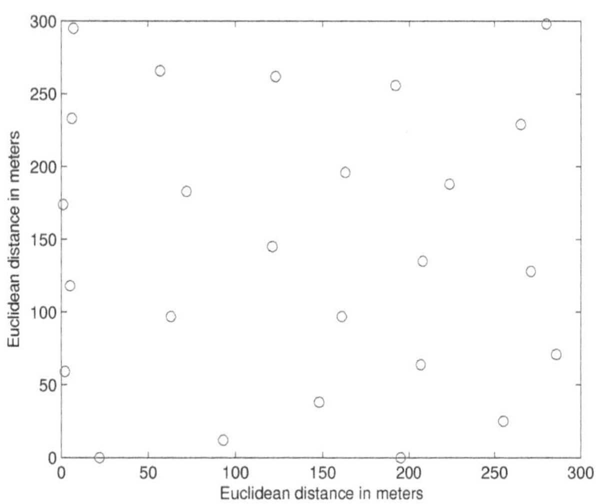

Figure 2: The positions of access points

6.2 Simulation results

We first compare the performance of our proposed greedy with the formulated IP for a small instance where the IP can produce optimal result. Then the performance of the greedy is evaluated for larger instances with more number of APs and STAs with different load distribution and different number of available frequencies that may exist in a practical situation.

6.2.1 *IP vs greedy*

To compare the performance of our proposed greedy with the formulated integer programming we chose a small example of 4 APs and 4 STAs where the data rate matrix is as shown in Table 2. We consider that all APs are within the interference range of each other. We vary the number of frequencies from 1 to 4. It is evident that to activate all 4 APs we need at least 4 frequencies. With 1, 2, 3 frequencies at most 1, 2 and 3 APs can activated, respectively. This allows us to understand the ability of finding an optimal association in conjunction with frequency assignment satisfying the co-channel interference. The Table 3 shows the resultant throughput, selected APs and the optimal association obtained by both the IP and the greedy for different number of frequencies varying from 1 to 4. The IP was solved using the optimization package LINGO from LINDO system [19]. We observe that our proposed greedy approach gives surprisingly equal performance when the number of frequency is 1, 2 and 4 whereas it gives nearly equal result when the number of frequency is 3.

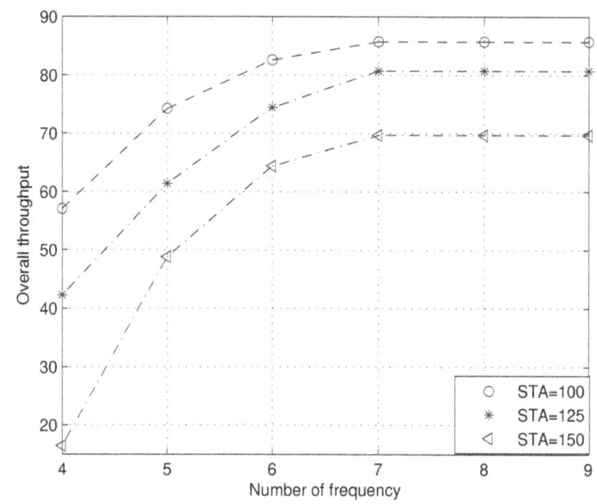

Figure 3: Frequency vs throughput for interference range 125 meters

6.2.2 *Performance evaluation of the greedy for larger instances*

As shown in Fig. 3, we study how the overall throughput varies with number of frequencies for different number of STAs (100, 125 and 150) representing different load distribution when the interference range has been fixed to 125 meters. We plot the number of frequencies on X-axis and the obtained overall throughput on Y-axis. Here, the throughput value represents the average value obtained by considering 10 different random placement of the specified number of $STAs$. We observe that the throughput is increased with the increase in number of frequencies and becomes saturated when the number of frequencies are 7. It justifies the fact that with 7 frequencies all the required number of APs are activated for maximizing the throughput and hence increasing the number of frequencies further does not change the overall throughput. For a fixed number of frequencies, with lesser number of STAs more throughput is obtained, as expected.

Next as shown in Fig. 4, we conduct an experiment to see how the overall throughput varies with the number of frequencies for a fixed number of STAs (100) with varying interference ranges (100, 125 and 150 meters). Here, the throughput value represents the average value obtained by

Table 3: Optimal throughput and associations for frequencies $F = \{1, 2, 3, 4\}$

F	*IP*			*Greedy*		
	Throughput	*Selected APs*	*OptimalAssociation* $(STA \rightarrow AP)$	*Throughput*	*Selected APs*	*OptimalAssociation* $(STA \rightarrow AP)$
1	0.5864	4	$1 \rightarrow 4, 2 \rightarrow 4, 3 \rightarrow 4, 4 \rightarrow 4$	0.5864	4	$1 \rightarrow 4, 2 \rightarrow 4, 3 \rightarrow 4, 4 \rightarrow 4$
2	6.0081	1, 4	$1 \rightarrow 1, 2 \rightarrow 1, 3 \rightarrow 4, 4 \rightarrow 4$	6.0081	1, 4	$1 \rightarrow 1, 2 \rightarrow 1, 3 \rightarrow 4, 4 \rightarrow 4$
3	7.5121	1, 2, 4	$1 \rightarrow 4, 2 \rightarrow 1, 3 \rightarrow 2, 4 \rightarrow 4$	7.3943	1, 2, 4	$1 \rightarrow 2, 2 \rightarrow 1, 3 \rightarrow 2, 4 \rightarrow 4$
4	7.8868	1, 2, 3, 4	$1 \rightarrow 1, 2 \rightarrow 2, 3 \rightarrow 3, 4 \rightarrow 4$	7.8868	1, 2, 3, 4	$1 \rightarrow 2, 2 \rightarrow 1, 3 \rightarrow 3, 4 \rightarrow 4$

considering 10 different random placement of 100 STAs. We observe that the throughput is increased with the increase in number of frequencies and becomes saturated when the number of frequencies are 7, 7 and 9 for the cases where interference ranges are 100, 125 and 150 meters, respectively. This is because the increased interference range results more interferences and thus requires more number of frequencies to activate the desired subset of APs for maximizing the throughput. We observe in the diagram that for a fixed number frequencies the overall throughput is less when the range is large, as expected.

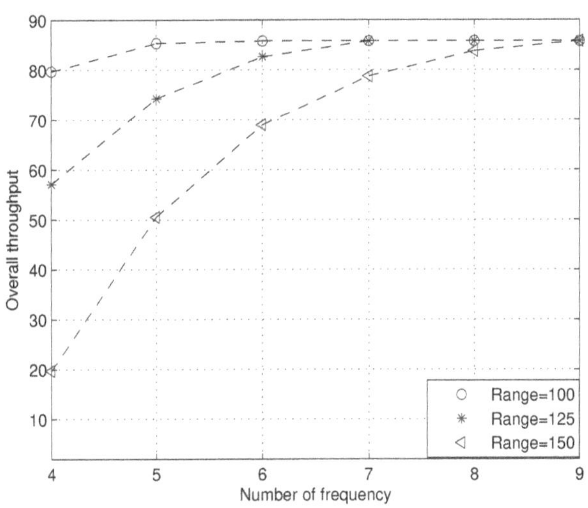

Figure 4: Frequency vs throughput for STA 100

We conduct an experiment to count the number of activated APs with the increase of number of frequency channels. For a fixed number of STAs (125) we have considered a varying interference ranges (100, 125 and 150 meters) to show the effect co-channel interference. Fig. 5 shows that the number of activated APs is increased with the increase in number of frequencies and becomes saturated when the number of frequencies are 5, 7 and 9 for the interference ranges of 100, 125 and 150 meters, respectively. When we increase interference ranges it leads to more interferences so more frequencies are required to activate the desired subset of APs that maximizes the throughput. This justifies our argument that the constraint of available number of frequency channels must play a role while activating the available APs which in turn has an effect on the overall throughput.

We also conduct a simulation to show how the load across

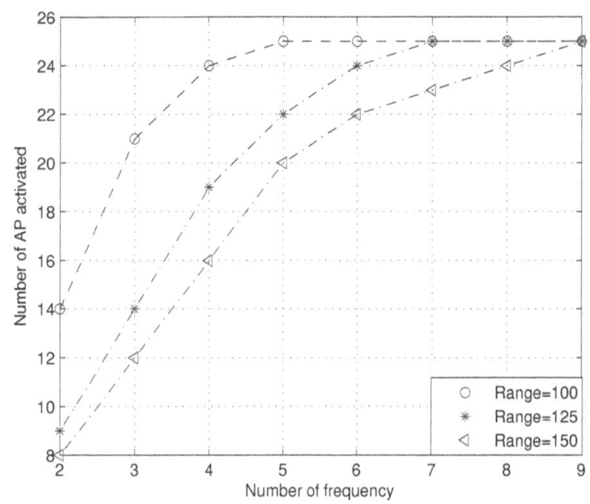

Figure 5: Frequency vs AP for STA 125

different APs are balanced by our proposed greedy algorithm. We have taken 9 frequencies so that all 25 APs become activated for each of the cases where number of STAs are 100, 125 and 150. We show the number of STAs served by each AP for all the three cases in Fig. 6. It shows that for all three cases the load distribution among the APs are almost uniform. It implies that our proposed greedy approach takes care of load balancing while finding an association with a view to maximize the overall throughput.

7. CONCLUSION

In this paper, an integrated model based on integer programming is proposed to solve the joint optimal association and frequency assignment problem. The need for such an integrated approach is established by revealing the weaknesses of a two step solution of the problem and a solution that ignores the interference exists between the APs. To solve the problem in large scale a greedy algorithm has also been developed. Through extensive simulation we have shown that for small scale the greedy algorithm provides comparable results to that of optimal values given by integer programming formulation. We have also simulated our proposed algorithm considering various networking load, different level of interferences, and different number of available frequencies and the result shows that our algorithm provides almost uniform load distribution across different APs and obtained the maximum achievable throughput. As a final note it has

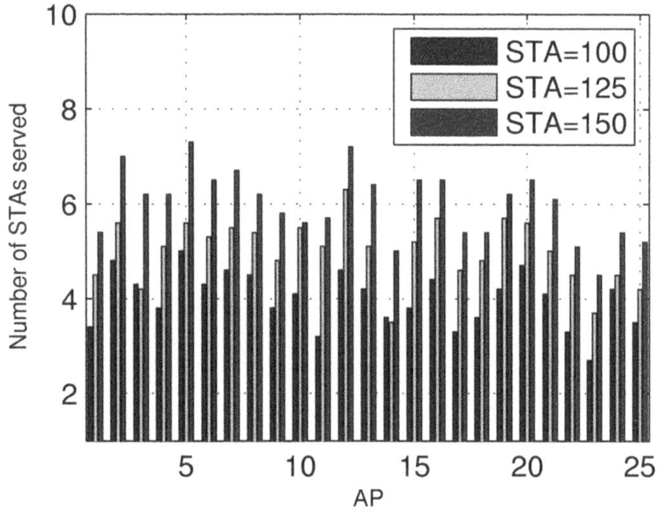

Figure 6: AP vs number of STAs served

been shown that the integrated approach is superior to both two-step approach and the methods that ignore the interference.

8. REFERENCES

[1] A. Kumar, and V. Kumar. Optimal association of stations and APs in IEEE 802.11 WLAN. *National Communications Conference (NCC)*, January 2005.

[2] Y. Bejrtano, S-J. Han, and L. Li. Fairness and load balancing in wireless LANs using association control. *In Proceedings of the 10th Annual International Conference on Mobile computing and Networking (ACM MobiCom)*, Philadelphia, Pennsylvania, USA, pages 315–329, Sept-Oct 2004.

[3] I. Papanikos, and M. Logothetis. A study on dynamic load balance for IEEE 802.11b wireless LAN. *In Proceedings of the 8th International Conference on Advances in Communications and Control (COMCON)*, Rethymna, Crete/Greece, 2001.

[4] G. Judd, and P. Steenkiste. Fixing 801.11 access point selection. *In Poster in Proceedings of ACM MobiCom 2002.*

[5] A. Balachandran, P. Bahl, and G. Voelker. Hot-spot congestion relief and service guarantees in public-area wireless networks. *SIGCOMM Computer Communication Review*, Vol. 32(1), 2002.

[6] S. Vasudevan, K. Papagiannaki, C. Diot, J. Kurose, and D. Towsley. Facilitating access point selection in IEEE 802.11 wireless networks. *In Proceedings of the 5th ACM SIGCOMM Conference on Internet Measurement (IMC'05)*, pages 293-298, 2005.

[7] G. S. Kasbekar, P. Nuggehalli, and J. Kuri. Online client-AP association in WLANs. *In Proceedings of the Workshop on Resource Allocation in Wireless Networks (RAWNET)*, Boston, Massachusetts, 03-06 April 03-04, 2006.

[8] Mingming Lu, and Jie Wu. Localized access point association in wireless LANs with bounded approximation ratio. *In Proceedings of IEEE 17th International Conference on Computer Communications and Networks (ICCCN)*, Vol. 1-2, pages 656-663, 2008.

[9] M. Abusubaih, J. Gross, S. Wiethoelter, and A. Wolisz. On access point selection in IEEE 802.11 wireless local area networks. *In proceedings of the 31st IEEE Conference on Local Computer Networks*, 2006.

[10] S. Bosio, and Di Yuan. Modeling and solving AP location and frequency assignment for maximizing access efficiency in Wi-Fi networks. *CoRR abs/1101.5027:* 2011.

[11] K. K. Leung, and Byoung-Jo "J" Kim. Frequency assignment for IEEE 802.11 wireless networks. *In proceedings of the IEEE 58th Vehicular Technology Conference (VTC-Fall)*, Murray Hill, NJ, USA, pages 1422–1426, 2003.

[12] K. K. Leung, Byoung-Jo "J" Kim. Frequency assignment for multi-cell IEEE 802.11 wireless networks. *In proceedings of the IEEE 58th Vehicular Technology Conference (VTC-Fall)*, 2003.

[13] T. Park and, Chae Y. Lee. Application of graph coloring algorithm to the frequency assignment problem. *Journal of the Operational Research Society of Japan*, Vol-39, No-2, June 1996.

[14] A. Eisenblatter, H.-F. Geerdes, and I. Siomina. Integrated access point placement and channel assignment for wireless LANs in an indoor office environment. *In Proceedings of the IEEE International Symposium on a World of Wireless, Mobile and Multimedia Networks (WoWMoM 2007)*, pages 1-10, 2007.

[15] T. Korakis, O. Ercetin, S. Krishnamurthy, L. Tassiulas, and S. Tripathi. Link quality based association mechanism in IEEE 802.11h compliant wireless LANs. *In Proceedings of the Workshop on Resource Allocation in Wireless Networks (RAWNET)*, Trentino, Italy, 2005.

[16] A.Kumar, E. Altman, D. Miorandi, and M. Goyal. New insights from a fixed point analysis of single cell IEEE 802.11 wireless LANs. *In Proceedings of the INFOCOM 2005*, Miami, March 13-17, pages. 1550-1561, 2005.

[17] F. Xu, C. C. Tan, Q. Li, G. Yan, and Jie Wu, Designing a practical access point association protocol. *In Proceedings of the INFOCOM 2010*, San Diego, USA, March 15-19, 2010.

[18] T. Bonald, A. Ibrahim, and J. Roberts. The impact of association on the capacity of WLANs. *In Proceedings of the 7th International Symposium on Modeling and Optimization in Mobile, Ad Hoc, and Wireless Networks*, 2009.

[19] LINGO User's Guide, LINDO Systems, Inc., Chicago, IL, email:info@lindo.com, 2004.

Traffic-Aware Channel Width Adaptation in Long-Distance 802.11 Mesh Networks

Sofia Pediaditaki*
Intel Research Center
Intel Labs Barcelona – UPC

Mahesh K. Marina and Daniel Tyrode
School of Informatics
University of Edinburgh

ABSTRACT

We consider the traffic adaptive channel allocation problem in long-distance 802.11 mesh networks. Our approach is to exploit the capability provided by 802.11 hardware to use different channel widths and assign channel widths to links based on their relative traffic volume. We show that this traffic-aware channel width assignment problem is NP-complete and propose a polynomial time, greedy channel allocation algorithm that guarantees valid channel allocations for each node. Evaluation of the proposed algorithm via simulations of real network topologies shows that it consistently outperforms the current approach of fixed width allocation due to its ability to adapt to spatio-temporal variations in traffic demands.

Categories and Subject Descriptors

C.2.1 [**Computer-Communication Networks**]: Network Architecture and Design—*Wireless communication*

General Terms

Algorithms, Design, Performance

Keywords

Long-distance 802.11 mesh networks, Channel allocation, Traffic awareness, Channel width assignment, Adaptive spectrum management, Resource allocation

1. INTRODUCTION

The remarkable success of WiFi (based on the IEEE 802.11 standard) has led to its use in originally unintended scenarios. Long-distance 802.11 mesh network scenario, the focus of this paper, is one of those that is making a huge impact in the real world in helping bringing low cost Internet access to rural areas and developing regions (e.g., [1, 2]) by enabling affected communities and new Wireless Internet Service Providers (WISPs). The main impediment for provisioning broadband access in these regions is the deployment cost to serve either low density scattered communities or populations with limited incomes. It will take years to fully penetrate into these underserved regions offering limited profitability with broadband access technologies prevalent in urban areas (fiber, DSL, cable, 3G/4G) because of their high infrastructure costs. The long-distance 802.11 mesh scenario is also of interest as part of larger and scalable mesh networks in urban settings along the lines of the architecture considered in [3]. It also fits well with the needs of WISPs that span both rural and urban areas.

*This work was done when this author was at the University of Edinburgh.

MSWiM'12, October 21–25, 2012, Paphos, Cyprus.

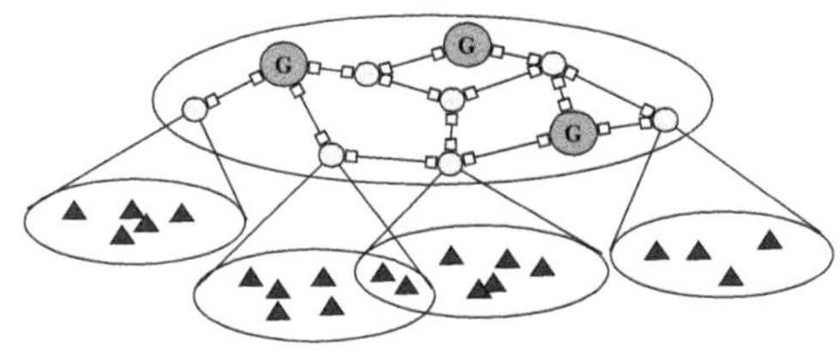

Figure 1: Multi-tier 802.11 based mesh network model.

We consider a tiered 802.11 based mesh network model shown in Fig. 1 applicable to rural and urban settings mentioned above. Two tiers are shown of which *our focus is on the topmost "directional backhaul" tier.* Nodes in the top tier could be separated potentially by long distances in the order of several Kms, hence their interconnection into a network is achieved with the use of a pair of high-gain directional antennas per link. As such this tier can be seen as a point-to-point wireless network. Some of these nodes in the top tier called gateway nodes connect to the wired Internet infrastructure (e.g., nodes labeled 'G' in Fig. 1). While some nodes in the top tier only have the router role to forward data between other top tier nodes, several nodes additionally provide connectivity to the lower tier subnets using point-to-multipoint wireless links as illustrated. Therefore the latter set of nodes can be seen as traffic aggregation points in the directional backhaul tier. Each of the subnets in the lower tier could in turn be an omnidirectional mesh network with each node representing a rooftop mesh access point (in a village or urban neighborhood). One can imagine an additional tier (not shown in the figure) connecting devices inside homes to a rooftop access point.

The directional backhaul tier in the network model is a specific type of multi-radio multi-channel mesh network. Each node in the backhaul has as many radio interfaces as the number of incident links (each connected to a directional antenna), and each of these links are assigned a different channel to avoid *side-lobe interference* that occurs with commonly used high-gain directional antennas [4] — non-negligible side-lobe energy from directional transmission on a link appears as interference to reception on other co-incident links, so such interference needs to be avoided. Moreover, for long-distance communication, besides directional antennas, higher radio transmit power may also be needed. Therefore, such long-distance point-to-point wireless communication is restricted by spectrum regulatory bodies to a few specified frequency bands with relatively higher transmit limits. The 5.8GHz frequency band is one such band and is available in most regions of the world. Consequently, the total amount of spectrum available for the directional backhaul tier is limited (e.g., 100MHz in the 5.8GHz band as opposed to more than 500MHz available for indoor wireless LAN operating on the 5GHz unlicensed bands).

Since the directional backhaul tier serves as an intermediate data transport network between the wired Internet and large number of client devices in the lower tiers, limited available spectrum needs to be managed judiciously and adaptively in response to varying traffic demands. Long-distance mesh deployments in practice tend to skirt around this important issue for lack of a suitable adaptive channel allocation framework. In fact, it is common to assign identically sized but possibly different channels to network links at deployment time and have them remain unchanged (e.g., [5]) or even use only a single channel for the whole network (e.g., [4]).

In this paper, we aim to fill this void by viewing *channel width* as a knob to enable traffic-aware channel allocation in long-distance 802.11 based mesh networks, i.e., focusing on the directional backhaul tier in Fig. 1. The intuition behind our approach is as follows: since the total amount of available spectrum is limited, adapting to spatio-temporal variations in traffic demand can be achieved by allocating "wider" channels to links with higher demand by taking spectrum away from links with less demand. In other words, greater capacity is assigned to heavily utilized links, thereby benefiting the flows passing through them. Following the work of Chandra et al. [6] who first demonstrated experimentally the throughput, range and energy efficiency benefits of channel width adaptation in an isolated 802.11 link scenario, other research efforts have since highlighted the value of channel width adjustment in 802.11 wireless LANs [7, 8]. To the best of our knowledge, we are not aware of work that considers channel width adaptation in the context of long-distance mesh networks.

Our contributions are as follows: (1) We present a graph theoretic formulation of the traffic-aware channel width assignment problem in long-distance mesh networks and show that it is NP-complete (Section 3). (2) We develop a polynomial time algorithm for assigning channel widths to links based on their relative traffic volume; the algorithm ensures that every node gets a valid channel allocation (Section 4). (3) Our simulation based evaluation of the proposed algorithm using real network topologies shows that it delivers substantial improvements in performance (40-70% throughput improvements) from adapting the channel allocation in response to variations in spatio-temporal traffic demands (Section 5).

We start by looking at related work in a bit more detail in the following section.

2. RELATED WORK

Related work falls into three categories as discussed below.

2.1 Channel Allocation and MAC Design for Long-Distance Mesh Networks

Most of the work in this space focuses on TDMA-based MAC protocols as an alternative to 802.11. While part of the motivation behind these protocols is the detrimental impact of high propagation delays on 802.11 performance for *very* long distance wireless links in the order of 100Kms, the rest has to do with the so-called "side-lobe interference" issue [4]. The latter refers to the interference among incident (directional) links at a node using the *same* channel, especially when one or more of them are transmitting and other links are receiving. This type of interference occurs with commonly used high gain directional antennas having non-negligible side lobes in their radiation pattern. The 2P protocol [4] is the first alternative design in the literature to address the above problems using a TDMA based approach that requires each node to alternate between transmitting (on all incident links) and receiving (again, on all incident links). Several subsequently proposed channel allocation protocols assume 2P as the underlying MAC protocol (e.g., [9, 10]). However, the 2P protocol works only if the network topology is a bipartite graph. This limitation has been addressed in a later proposal called JazzyMAC [11]. By their very nature, both 2P and JazzyMAC need inter-nodal time synchronization, an additional requirement.

Our work on channel allocation in this paper instead assumes standard 802.11 MAC, based on the following two observations: (1) Real world long-distance wireless links are typically in the order of several Kms to few tens of Kms for which 802.11 MAC gives acceptable performance through suitable adjustment of the built-in ACK timeout. This is experimentally shown in [12] and is also confirmed by our experience deploying and monitoring the Tegola network in rural Scotland [13] for the past three years with links in the range of 2-20Km. (2) The use of multiple channels alleviates the side-lobe interference problem as different links at a node can be assigned to different non-interfering channels.

The recent channel assignment work of Dutta et al. [14] is similar in spirit to ours in that it also assumes standard 802.11 MAC, thereby leveraging readily available commodity 802.11 hardware. However, that work focuses on *directed* edge coloring, requiring a channel for each *directed* link, potentially resulting in inefficient spectrum utilization with limited spectrum; it also increases the cost and deployment complexity — to support directed edge coloring, each node requires two directional antennas (and radios) per link and those antennas need to be carefully separated to manage side lobe interference.

Crucially, to the best of our knowledge, the existing channel allocation literature on long-distance mesh networks does not consider channel width adaptation, which is the key aspect of our work.

2.2 Channel Width Adaptation in 802.11 Networks

Chandra et al. [6] were the first to examine the impact of channel width adaptation on throughput, range and power consumption and obtain experimental evidence of the potential benefits of adjusting the width of a channel in 802.11 networks. Their focus, however, is on the simplest case, i.e., single link, for which they propose a channel width adaptation algorithm called SampleWidth. Most of the subsequent work on channel width adaptation in 802.11 networks has focussed on the wireless LAN (WLAN) scenario. Moscibroda et al. [7] consider channel width adaptation for achieving load balancing in multi-AP WLANs as an alternative approach to transmit power and client association controls. They show that the dynamic width assignment problem is NP-hard and propose heuristics that are evaluated using simulations. Yuan et al. [15] take a game-theoretic approach to address the same problem and propose a decentralized learning-based algorithm for achieving optimal allocation. Recently, Rayanchu et al. [8] find that realizing the benefits of variable channel widths in multi-AP 802.11 based WLANs requires reliable characterization of interference (conflicts) and they propose a mechanism to efficiently model such conflicts.

Note that adapting these WLAN channel width adaptation proposals to our multihop wireless network context is not straightforward because of the fundamental differences between the two scenarios. For instance, we have the requirement to maintain network connectivity *wirelessly*, whereas access points (APs) in a wireless LAN are interconnected via a wired backhaul network.

In [16], the authors show that splitting the spectrum into variable width channels among mutually interfering transmitters and keeping them active simultaneously is more effective than having them share a larger, fixed-width channel because of SNR increase at smaller channel widths for a given transmit power. A similar argument can be used in our scenario for determining the feasible combinations of channel widths and bit-rates (modulation and coding schemes) for an installed long-distance link with fixed transmit power and antenna gains. We also note that Orthogonal Frequency Division Multiple Access (OFDMA) used in WiMAX [17], a fine-grained approach operating at the subcarrier level for spectrum sharing among multiple users, is related to channel width adaptation in 802.11 based networks.

2.3 Channel Allocation in Multi-Radio Multi-Channel Wireless Mesh Networks

Much work has been done on channel allocation in multi-radio multi-channel wireless mesh networks using omnidirectional antennas (e.g., [18, 19, 20]). Though the network model we consider is a special type of a multi-radio multi-channel wireless mesh network, our interference model is different from mesh networks based on omnidirectional antennas; more importantly, we have an added decision variable — channel width — that is not present in the above works. Recently a few papers looked into this aspect (e.g., [21, 22]), focusing on the omnidirectional mesh network scenario and mainly employing mathematical optimization methods (e.g., mixed integer linear programming). More recently, Wu et al. [23] consider the adaptive width channel allocation in multi-radio wireless mesh networks from a game theoretic perspective but their work is associated with unrealistic assumptions such as all nodes lying within a single collision domain, and each node participating in only one communication session and that too over a single hop.

In DMesh [24], the authors propose a wireless mesh network architecture based on directional antennas with the goal of exploiting spatial reuse benefit of directional communication to go along with the benefit of using multiple channels. We, on the other hand, are more interested in the range benefit offered by directional antennas.

3. MODEL AND PROBLEM FORMULATION

As stated at the outset, we consider a multi-tier 802.11 mesh network scenario as shown in Fig. 1 and our focus in this paper is on the topmost directional backhaul tier, especially keeping the rural wireless Internet access use case in mind. We model the network topology of the directional backhaul tier as an undirected graph, $\mathcal{T} = (\mathcal{N}, \mathcal{L})$. Each node n at the backhaul tier is equipped with $K_n (K_n \geq 1)$ 802.11 wireless interface cards, each attached to a directional antenna forming an end of a long-distance point-to-point wireless link with a neighboring node. We use the notation $n_p, 1 \leq p \leq K_n$ to refer to the pth interface at n. Note that K_n is equal to the number of point-to-point wireless links incident at node n.

We use (n, m) to denote the logical point-to-point link between two nodes n and m. And we use (n_p, m_q) to denote the actual physical bidirectional point-to-point link between the pth interface of node n and the qth interface of neighbor m. When we need to refer to the direction of a link, we will use the notation $n_p \rightarrow m_q$ to refer to the direction from n to m. Note that, in practice, each link in a long-distance 802.11 mesh network is determined at deployment time by pointing a pair of directional antennas located at two mast sites towards each other. These links remain fixed when the network goes into operational stage; our focus in this paper is solely on adapting the "channel" used by a link based on the traffic volume measured over it.

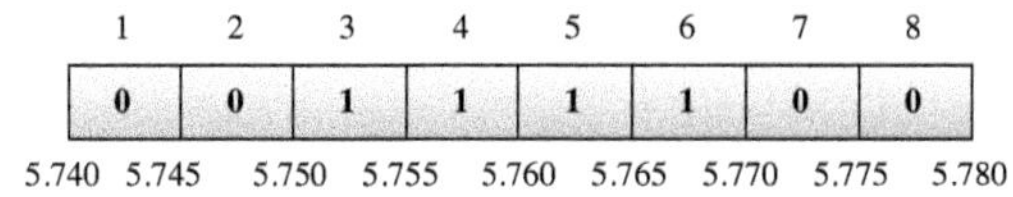

Figure 2: An example to illustrate *Spectrum Allocation Map (SAM) of a node*. In this example, the total spectrum available is assumed to be 40MHz, which results in 8 5MHz wide blocks. Each 5MHz wide spectrum block is shown using a box with a block identifier above the box and start frequency of the block underneath. Based on SAM of the node (i.e., the values shown inside boxes), blocks 3-6 are used while the rest are free. Assuming the node has only one link and all the used blocks correspond to the channel allocated to that link, the *block assignment (channel) for the link* is: $ch = < f_s, w > = < 5.75GHz, 20MHz >$ with center frequency, $f_c = 5.76GHz$.

We use $\mathcal{F}$ to denote the total available spectrum. In the case of widely used 5.8GHz band, $\mathcal{F}$ = 100MHz, ranging from 5.725GHz to 5.85GHz with small guard bands on either end[1]. This band can accommodate 5 20MHz channels (the default channel width in 802.11a) — channel numbers: 149, 153, 157, 161 and 165. The same band can be used to accommodate up to 2 40MHz channels. We assume the set of available channel widths to be 5, 10, 20, 40 based on what is currently supported by commodity 802.11 hardware.

A channel in our context is defined by the tuple $ch =< f_c, w >$, where f_c represents the center frequency and w the width of the channel taking one of the 4 values just mentioned — frequencies of ch range from $f_c - w/2$ to $f_c + w/2$. For example, 802.11a channel number 149 corresponds to a 20MHz channel centered at 5.745GHz. Given $\mathcal{F}$, the range of frequencies that fall within the spectrum and available channel widths, several channels can be realized by choosing various center frequencies and widths. We can alternatively look at ch using its start frequency, $f_s = f_c - w/2$ and width w; in this case, ch ranges from f_s to $f_s + w$. We use the notation $ch_{(n_p,m_q)}$ to refer to the channel assigned to the link (n_p, m_q). Note that $ch_{n_p \rightarrow m_q} = ch_{m_q \rightarrow n_p}$.

For convenience, we view the given spectrum as a sequence of atomic 5MHz wide *blocks*. For example, when the available spectrum is 100MHz, we have 20 5MHz wide blocks. If S denotes the number of blocks[2] (20 in the example), then we assume that $S \geq 2 * \Delta(\mathcal{T}) - 1$, where $\Delta(\mathcal{T})$ is the maximum node degree in $\mathcal{T}$. This is quite a reasonable assumption since a node in the backhaul directional tier typically has at most around a handful of incident point-to-point wireless links. The relevance of this assumption will become clear later on in Section 4.

We now define a *spectrum allocation map for each node* n: SAM_n that represents the spectrum usage of the interfaces at n. Specifically, this map is a sequence of bits associated with the blocks, where a bit is set to 1 if the corresponding block is occupied by some interface of the node. Otherwise, it is 0. Fig. 2 shows

[1] 10MHz at the lower end and 15MHz on the upper end.

[2] Henceforth, we just use the term 'block' as a shorthand for '5MHz wide block'.

an example. Using the notion of blocks, the channel assigned to a link (n_p, m_q), $ch_{(n_p,m_q)}$, can be seen as contiguous and identical set of blocks in $\mathtt{SAM}_n$ and $\mathtt{SAM}_m$; we refer to such assignment of contiguous set of blocks to a link (n_p, m_q) as the *block assignment for the link*, denoted by $\mathtt{BA}_{n_p,m_q}$ (see Fig. 2 for an example).

Having described what a channel means in our model, we now introduce the three key constraints in our channel allocation problem.

Side-lobe interference constraint:
$BA_{n_p,m_q} \cap BA_{n_r,l_s} = \emptyset; n, m, l \in \mathcal{N}, \exists(n_p, m_q), \exists(n_r, l_s),$ $p \neq r, m \neq l$. **(C1)**

This constraint essentially requires that any two incident links at a node are assigned different *non-overlapping* channels[3].

Minimum channel constraint:
$ch_{n_p,m_q} = < f_c, w >$ s. t. $w \geq 5; \exists(n_p, m_q)$. **(C2)**

This constraint requires that each link is assigned at least a 5MHz wide channel (i.e., a block). This is to make sure that network topology always remains intact with each link having a usable channel, which can be used for exchanging at least control traffic (e.g., routing messages).

Total spectrum constraint:
$0 < \sum_{m,\exists(n_p,m_q)} ch_{n_p,m_q} \leq \mathcal{F}$. **(C3)**

This constraint makes sure that channels allocated to incident links at a node do not exceed the total spectrum available.

Before going to the objective function, we need to model flow (traffic) on a link and the link capacity based on the channel allocated to it. Let f_{n_p,m_q} denote the total traffic flow passing through a link (n_p, m_q): f_{n_p,m_q}[4] $= f_{n_p \to m_q} + f_{m_q \to n_p}$. The capacity of a link (n_p, m_q) denoted by C_{n_p,m_q} is dependent on the channel it is allocated and link characteristics (link distance, etc.). Channel width and best bit-rate (modulation and coding scheme) supported by the link are inter-dependent, and together determine the raw link capacity, C_{n_p,m_q}[5].

Objective function: our objective can be stated as *minimizing the maximum excess link load across all links in the network* subject to constraints **C1**-**C3** at each node, where excess link load of a link (n_p, m_q) is defined as $max(f_{n_p,m_q} - \delta * C_{n_p,m_q}, 0)$. Here δ models the fraction of raw link capacity effectively available at network layer after discounting overhead related to link and physical layers (e.g., headers, inter-frame spaces) as well as routing control traffic overhead. For example, if 20MHz channel width can support 54Mbps physical layer bit-rate, then effective achievable capacity above the link layer is at most 30Mbps dependent on frame length. Assuming some portion of it (say 10%) is consumed by control traffic, the maximum effective capacity is 27Mbps, leading to a δ value of 0.5.

The intuition behind using this objective function is to evenly distribute the available spectrum resource among links based on their traffic demands so that capacity assigned to a link matches its load as closely as possible.

We refer to the decision problem equivalent of the above optimization problem as *Channel Width Assignment for MinMax Excess Link Load*, which can be stated as follows: Is there a channel width assignment such that the maximum excess link load $\leq$ B (where B is a non-negative integer)?

THEOREM 1. *The* Channel Width Assignment for MinMax Excess Link Load *decision problem as stated above is NP-complete.*

PROOF. A channel width assignment can be verified in polynomial time, thus the problem is clearly NP.

The rest of the proof is by restriction. We show that the above channel width assignment problem contains the minimum edge coloring problem (also called the minimum chromatic index) - a known NP-complete problem - as a special case [29].

Specifically, we show that a specific instance of the problem at hand is identical to the minimum edge coloring problem. For this instance, the following constraints hold: (i) only one channel width is allowed (5MHz), (ii) the total amount of spectrum available = maximum node degree * 5MHz, (iii) $f_{n_p,m_q} << \delta * C_{n_p,m_q}$, for all links (n_p, m_q) when using a 5MHz wide channel, and (iv) the graph is chosen to be identical for both problems and bound for number of colors (chromatic index) for the edge coloring is set to the maximum node degree.

The above specified instance of our problem is identical to the minimum edge coloring problem, which completes the proof. □

Note that using the above stated objective function and per-node constraints **C1**-**C3**, we can formulate this problem as a mixed integer linear program for obtaining a lower bound on the optimum.

4. CHANNEL WIDTH ASSIGNMENT ALGORITHM

In this section, we describe a polynomial time greedy heuristic for traffic-aware channel width assignment in long distance 802.11 mesh networks.

In our approach, channel allocation is performed independently from routing but influences it and vice versa. Moreover, channel reallocation is done at a relatively slower timescale than routing. This is because channel width changes are more disruptive as each such change at least requires endpoints of a link to reconfigure (and even restart) their corresponding wireless interfaces. Exact frequency of channel width adaptation is a tradeoff between responsiveness to traffic dynamics and keeping network disruption and overhead low; we will discuss this issue further in Section 6. Keeping routing and channel allocation independent has the advantage that any routing protocol can be used on top of the channel width adaptation algorithm. While a traffic adaptive routing protocol such as [30] would be an ideal companion for the channel allocation algorithm, some network deployments may not plan for enough redundancy in the

[3]Here we make the simplifying assumption that side lobe interference can be avoided if co-incident links are assigned channels with frequency ranges that do not overlap. We can extend it to incorporate a sophisticated non-binary model of interference between any two co-incident links based on the work by Angelakis et al. [25] that takes into account antenna radiation patterns, inter-antenna distance, separation between center frequencies of channels assigned to the two links and channel widths. We elaborate more on the latter in Section 6.

[4]A lightweight sampling method to continually estimate the total traffic flow on a link based on MRTG and SNMP is described in [26]; Fig. 7 is generated from data obtained using this method in the Tegola network [13]

[5]Prior work [27, 28] has shown that typical rural long-distance wireless links, the particular focus of our work, experience negligible link quality variations. In such cases, we only need to consider interaction between channel width and bit-rate to estimate link capacity based on [6] for a given link quality (that can be obtained via measurement) . As a corollary, bit-rate would also be stable in the fixed width case.

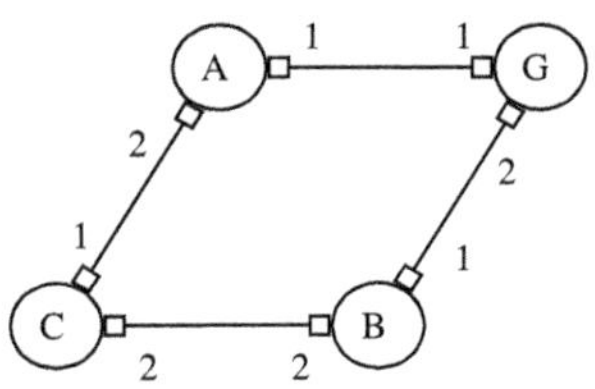

Figure 3: An example network.

topology for cost reasons, making the complexity of using a traffic adaptive routing protocol questionable; in such cases, a single path routing protocol can be used with the burden of traffic adaptivity shifted largely to the channel allocation algorithm. For our simulation based evaluations, we have implemented P-STARA [30] and use it as the default routing protocol.

Before describing the algorithm, we introduce the concepts of *feasible* and *valid* channel allocations. A feasible channel allocation for a node n is an allocation that respects constraints **C1-C3** (see Section 3). As an example, consider node A in the example network shown in Fig. 3 and suppose that the total amount of available spectrum is 40MHz. Also suppose that interface 1 (link to G) at A has a higher priority than interface 2 at A (link to C) — setting of node and link priorities is discussed shortly. Under these assumptions, feasible channel width combinations for node A are shown in Fig. 4 in the order of decreasing preference[6]. Not all of these combinations are valid depending on when node A gets to do its channel allocation with respect to its neighboring nodes. To see this, consider the network in Fig. 3 again. Suppose that node G allocates a 20MHz channel to its interface numbered 1, thus coloring[7] link to A and corresponding interface at A. This coloring decision by G immediately reduces the possible width combinations to the set shown in Fig. 5. Moreover, depending on where channel for link (G_1, A_1) lies in the spectrum, the possible width combinations may reduce even further. If the $\texttt{SAM}_A$ after link (G_1, A_1) is allocated 20MHz channel looks like in Fig. 2, then the combination (1) in Fig. 5 is no longer possible. In the worst case, for this example, depending on the width and center frequency used by node G for link (G_1, A_1), A may not have a possible width combination that ensures a minimum channel for its remaining interface (numbered 2, link to C). Thus, node G while coloring link (G_1, A_1) should make sure that A has at least a block for its remaining uncolored interfaces. A valid channel allocation then is a channel allocation that is not only feasible from a node's perspective but also is consistent with channel allocations at other nodes.

Based on the above, we seek a channel allocation that results in a valid channel width combination at every node in the network. Before going to the actual algorithm, we need to introduce two more concepts: *guard block assignments (gBAs)* and *guard spectrum allocation maps (gSAMs)*. Recall from constraint **C2** in Section 3 that we require each link in the network to be assigned at least a

	Interface 1	Interface 2
(1)	20	20
(2)	20	10
(3)	20	5
(4)	10	20
(5)	10	10
(6)	10	5
(7)	5	20
(8)	5	10
(9)	5	5

Figure 4: Feasible channel width combinations for node A in Fig. 3 under the assumption that A_1 has a higher priority than A_2.

	Interface 1	Interface 2
(1)	20	20
(2)	20	10
(3)	20	5

Figure 5: Valid channel width combinations for node A in Fig. 3 after its interface 1 is colored by node G with a 20MHz channel.

minimum channel (block). We ensure that this constraint is met at all times (i.e., prior to, during and after completion of channel allocation algorithm execution) via gBAs and gSAMs. A valid block is identified for each link at the initialization stage of the algorithm as described below in step 1. The block so identified for a link is referred to as the gBA for that link. The gSAM of a node is a collective representation of gBAs of all its incident links. Like SAMs, gSAMs are also bit-vectors. After every link gets a gBA, it is straightforward to determine node gSAMs. For example, consider the network shown in Fig. 3 and suppose that the total amount of available spectrum is 40MHz with the block identifiers as in Fig. 2. A possible (though not optimal) gBA assignment for links G-A, A-C, C-B and B-G is blocks 1, 2, 3, 2 respectively. For that assignment, $\texttt{gSAM}_A$ is $< 1, 1, 0, 0, 0, 0, 0, 0 >$, $\texttt{gSAM}_B$ is $< 0, 1, 1, 0, 0, 0, 0, 0 >$ and so forth. The gBA assignments can be seen as proactively *reserving* a minimum sized channel for each link that results in a valid channel allocation to start with. Later on when the algorithm finds a larger width channel that is commensurate with the load on a link and does not compromise validity of the channel allocation, the gBA for that link is *released* in exchange for a valid block assignment corresponding to the larger width channel; gSAMs and SAMs of the end nodes of that link are also accordingly updated then.

Our channel width assignment algorithm consists of the following sequence of steps:

1. *Initialize gBAs and gSAMs:* We do this by applying an edge coloring heuristic on the network, viewing each individual block in the given spectrum as a potential color. Even if this coloring is done greedily we are guaranteed to have a proper edge coloring given our assumption in Section 3 about the relationship between maximum degree of the network $(\Delta(\mathcal{T}))$ and the total number of spectrum blocks (S). See [31]. Our assumption and choice of the heuristic are driven by the fact that greedy edge coloring can be easily implemented in a distributed manner [32].

2. *Assign node and link priorities based on traffic load:* The rest of the algorithm is also greedy, driven by priorities assigned to nodes and links. Node priorities determine the order in which nodes allocate channels to their interfaces (incident

[6]Relative preference among channel width combinations at a node can be determined, for example, by computing a function (e.g., product, sum) of *link utilizations* with each combination and then ranking the combinations based on their function values. Each link's utilization is computed by taking the ratio of its load to capacity, the latter based on channel width corresponding to the link in the chosen width combination. We use product of link utilizations as the default method in this paper.

[7]Henceforth we use the terms of coloring and channel allocation interchangeably.

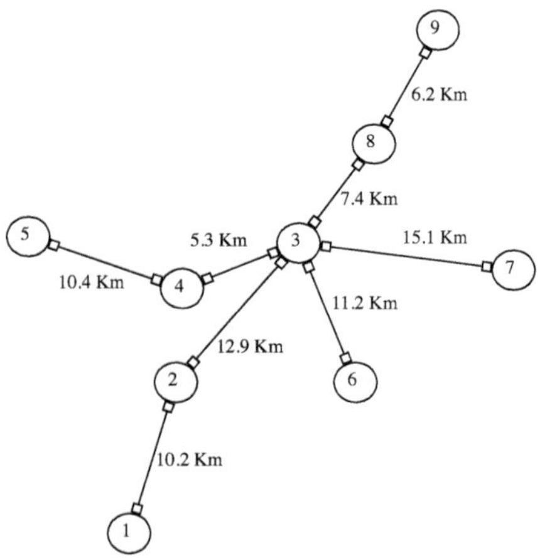

Figure 6: Aravind telemedicine network topology [2].

links) — higher a node's priority sooner its turn for channel allocation. Specifically, the priority of a node n, P_n is determined as follows:

$$P_n = \sum_{m,\exists(n_p,m_q)} f_{n_p,m_q} \tag{1}$$

The above equation favors nodes with larger traffic volume when assigning priorities since the goal of the algorithm is to adapt based on traffic demands. Priority of a link (n, m) is set to the average of the priorities of end nodes, i.e., $P_{n_p,m_q} = avg\{P_n, P_m\}$.

3. *Steps taken when a node x is the next highest priority node to be processed:*

   (a) Find the *list* of feasible width combinations at node x taking into account priorities of incident links at x and order the combinations based on their relative preference as described earlier in this section (see Fig. 4 and corresponding text).

   (b) Prune the list from the previous step (3.a) to retain only potentially valid width combinations. Depending on the priority of node x relative to its neighbors, some of its incident links may already be colored by higher priority neighbors. In such a case, some of the combinations from (3.a) may not be valid causing their removal from the list. Fig. 5 illustrates this step.

   (c) Find a valid channel width combination from the remaining list from step (3.b) considering combinations in the order of preference and stopping when a valid combination is found. First check for validity of a combination by verifying if the widths in the combination for uncolored incident links (x, y) at node x can be satisfied based on the current `SAMs` of x and all such neighbors y while not violating constraints **C1** and **C3**. If the first check is successful then the combination is checked for violation of `gBAs` for uncolored incident links (y, z) z $\neq$ x of neighbors y. If both these checks are successful, then search for a valid combination is successful. At that point, uncolored links of x are colored based on the combination found and block assignments for those links and `SAMs` of x and affected neighbors y are updated. Moreover, `gBAs` for the newly colored links of x are released and `gSAMs` of end nodes accordingly updated. Note that this step will result in a valid combination being found because the combination corresponding to `gBA` assignment for the uncolored links of x prior to this step is always among the combinations searched during this step. Before finishing this sub-step, node x tries to move the existing `gBAs` to increase flexibility of remaining channel width assignments.

At the end of the execution of the above algorithm, we are guaranteed a valid channel allocation because validity of the channel allocation is maintained at each step of the algorithm. The fact that the algorithm in fact terminates and runs in polynomial time is also evident from the above description. At termination, `gBAs` for all links are released and `gSAMs` of all nodes become null vectors. Theoretically characterizing the approximation guarantee of this algorithm is an issue for future work. We discuss the practical aspects and distributed operation later in Section 6.

5. EVALUATION

In this section, we evaluate the effectiveness of our channel width assignment algorithm described in the previous section using simulations to understand its ability to adapt to spatio-temporal variations in traffic demands. Our goal is to study the benefit of adapting channel width across a diverse set of scenarios in terms of effective and fair allocation of the limited spectrum resource. Since we are not aware of channel width assignment algorithms for long-distance mesh networks, we conduct this study in comparison with a variant of [14] that does (the more common) undirected edge coloring. Within this benchmark, we consider several alternatives each based on a different fixed size channel width, starting from 5MHz (minimum sized channels in current 802.11 systems). We focus on rural wireless access network scenarios because they are a compelling real-world use case for long-distance 802.11 mesh networks.

We use the QualNet simulator that has a built-in detailed model for standard 802.11 CSMA/CA MAC protcol. We have added the variable channel width functionality to QualNet and validated it against the results reported in [6]. We use the Traffic-Gen application in the simulator for flexible realization of different traffic patterns with variable session durations and traffic loads. We use 1KB packets throughout. In all our experiments, we set the total available spectrum to 100MHz to match with the commonly used 5.8GHz band. For the bit-rate (modulation and coding scheme), we use 6Mbps unless mentioned otherwise. For the adaptive channel width case, the channel width assignment algorithm is executed periodically. Unless otherwise specified, we use a simulation length of 25 minutes with traffic flows starting after one minute.

For our evaluations, we use three real long-distance wireless network topologies:

1. Aravind telemedicine network in Southern India [2] consisting of 9 backhaul wireless nodes. Fig. 6 shows the topology of this network — node 3 is the hospital situated in a town and nodes 1, 5, 6, 7 and 9 are remote vision centers.

2. Connected Communities (ConCom) network [33] is a relatively large broadband wireless access network covering the Western Isles of Scotland with a population around 26,000 spread across 11 islands and span of over 200Km. This network consists of 34 backhaul sites interconnected by point-

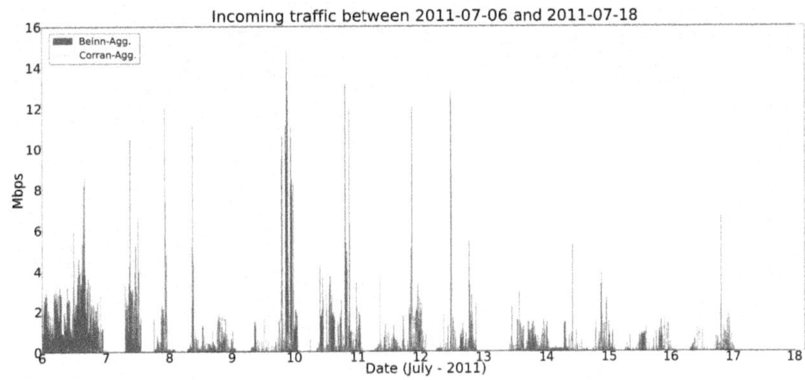

Figure 7: The trace of aggregated download traffic seen at two most used backhaul wireless sites in the Tegola network [13] over a two week period.

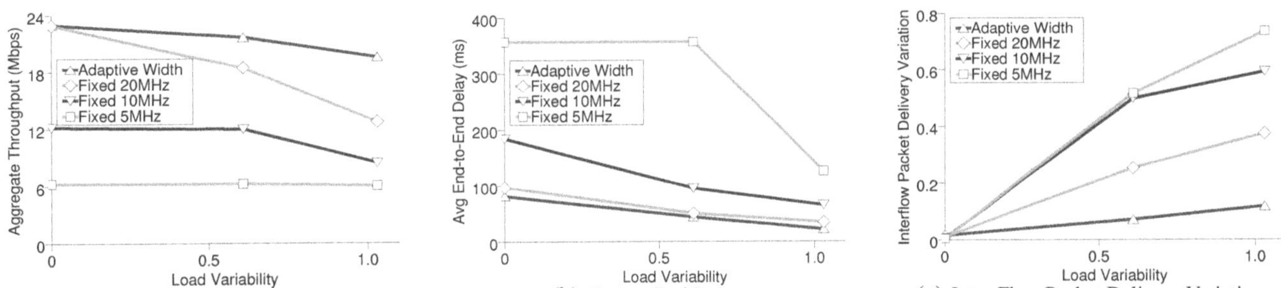

(a) Throughput (b) End-to-End Delay (c) Inter-Flow Packet Delivery Variation

Figure 8: Performance impact of spatial variation in traffic demands (variable load across flows) for the Aravind network topology.

to-point wireless links with widely different link lengths. It provides connectivity to public buildings (e.g., schools, community centers) as well as residential users.

3. Tegola consisting of 5 backhaul wireless nodes is a network we have deployed in rural Scotland [13]. Though originally intended as a research testbed, it currently also serves as a community wireless network connecting real users to the Internet. We have used this topology for our preliminary evaluations, but the results reported in this paper consider the other two relatively larger network topologies.

Before going to our simulation experiments, we first consider *a key aspect that motivates the need for adaptive spectrum management mechanisms, i.e., the spatio-temporal variability in traffic seen by different backhaul nodes in a long-distance wireless mesh network.* We illustrate the existence of such variability in real deployments in Fig. 7 with a 2-week trace of traffic at two Tegola network backhaul sites Beinn (shown in red) and Corran (shown in green) with most number of associated users (16 and 6, respectively). This trace corresponds to aggregated download traffic seen at each site; upload traffic (not shown) also exhibits similar behavior though as expected lower in volume compared to download. While variability in traffic across these sites and over time is visually evident, we also quantified this diversity by computing the pearson's correlation coefficient between the traffic at the two sites. This comes out to be 0.12, which indicates little or no correlation. In the following, we examine the performance of our channel width adaptation algorithm against such spatio-temporal traffic variations.

5.1 Spatial Traffic Variation

In the first experiment, we study the impact of spatial variation in traffic load. For this we consider the Aravind topology with 5 traffic flows from each of the edge nodes (1, 5, 6, 7, 9) to the middle node (3). We uniformly vary the traffic load across the flows while keeping the mean traffic load constant around 5Mbps. Fig. 8 shows the results for aggregate throughput, average end-to-end delay and packet delivery variation across flows as a function of load variability. Load variability is essentially the coefficient of variation[8] calculated using loads of individual flows. Packet delivery variation across flows shown in Fig. 8(c) is also a coefficient of variation but calculated using the packet delivery ratios of individual flows. For the Aravind network topology (as well as the ConCom topology), all links cannot be assigned 40MHz channels given the limited total amount of spectrum (100MHz), so only 20MHz, 10MHz and 5MHz fixed width allocations are shown as alternatives to adaptive channel width. As expected, the opportunity for adapting channel width is marginal when the traffic is uniform, while significant gains are achieved as load becomes more variable, resulting in throughput improvement around 53%. Improvements in variation of packet delivery across flows (Fig. 8(c)) at high load variability are even more remarkable as all the fixed width cases fail to support flows with high load. The big drop in delay for 5MHz fixed width case (Fig. 8(b)) can be explained by the fact that most of the high delay packets corresponding to high load flows are dropped. This is also reflected in the large variation of packet delivery across flows for this width in Fig.8(c).

[8]The ratio of standard deviation to the mean.

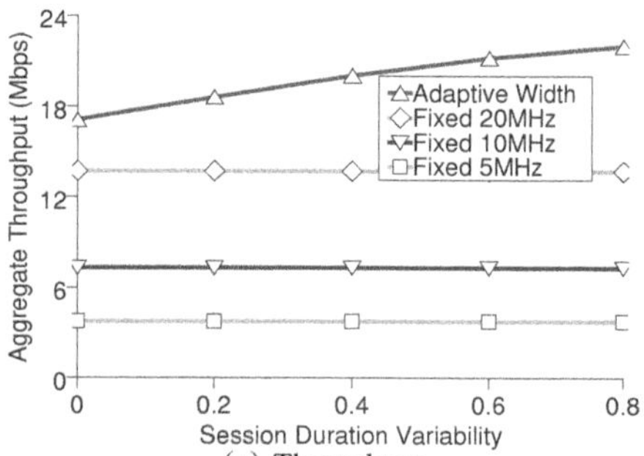

(a) Throughput

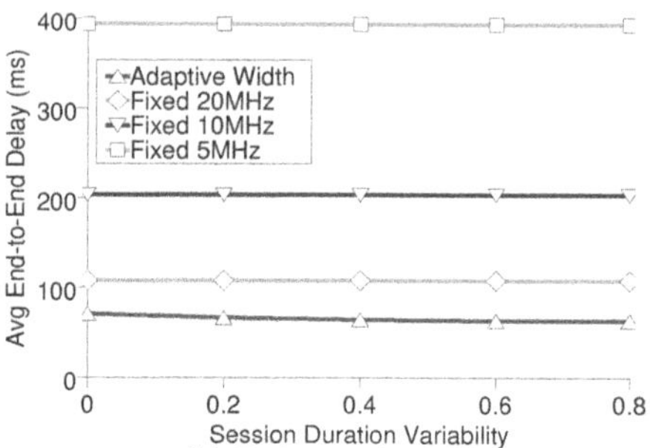

(b) End-to-End Delay

Figure 9: Performance impact of temporal variation in traffic demands (variable session durations across flows) for the Aravind network topology.

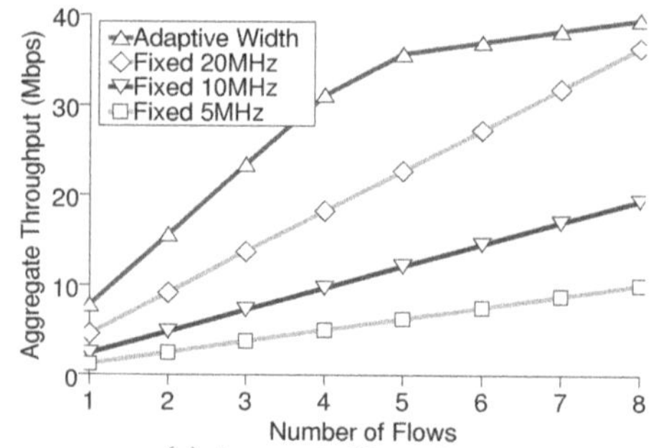

(a) Aggregate Throughput

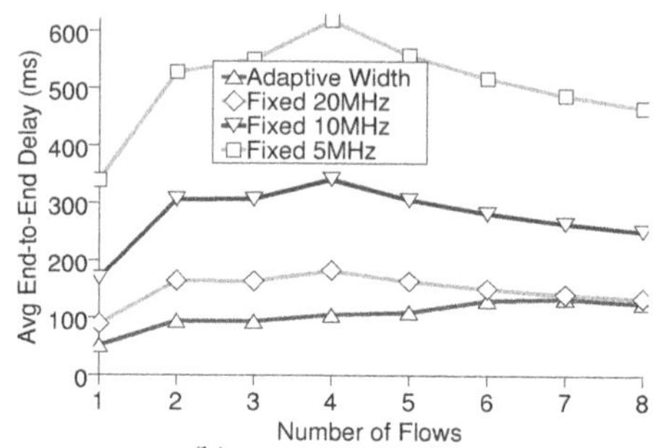

(b) End-to-End Delay

Figure 10: Performance impact from using adaptive channel width for the larger ConCom network with increasing number of flows.

5.2 Temporal Traffic Variation

We now consider the impact of temporal variation in traffic demands, again using the Aravind network topology. For this experiment, we consider three flows (5-3, 6-3 and 7-3), each having the same load around 10Mbps but varying in session durations. Fig. 9 shows the throughput and delay results as a function of session duration variability, computed as the coefficient of variation using session durations of individual flows. Mean session duration was kept constant at 25 minutes for all the data points. Like in the previous spatial variation experiment, we find that adaptive width offers greater gains as temporal variability increases (up to 45% improvement in throughput) by adaptively reallocating spectrum as flows come and go.

5.3 Larger Network Scenario

We now examine the benefit of channel width adaptation in a larger network using the ConCom topology. For the traffic pattern, we consider a common use-case for the ConCom network, i.e., tele-commuting/tele-education. Sites connecting office buildings and schools act as traffic sources (8 in number) while traffic destinations for the flows are randomly distributed from among the remaining sites. We keep the load of each flow constant at around 10Mbps and increase the number of flows. Results are shown in Fig. 10. We observe that increasing the number of flows has the effect of making the traffic pattern more uniform, limiting the benefit of adaptive width. When the traffic pattern is non-uniform and less constrained by the amount of available spectrum (left half of the figure), adaptive width results in throughput improvement over 70% compared to the best fixed width alternative. The drop in delay from midway (especially for fixed width cases) seen in Fig. 10(b) is a result of packets with large delays getting dropped as contention increases and queues build up.

5.4 Effect of Bit-Rate

We also study the interaction between bit-rates and channel widths and their net effect on performance using the Aravind topology and considering two other bit-rates for the spatial variation experiment: 12Mbps and 24Mbps. The choice of these rates is based on the fact that 24Mbps is the maximum bit-rate that can be supported by link 3-7 for 40MHz. While the results shown in Fig. 11 qualitatively are similar to those in Fig. 8, latter corresponding to 6Mbps bit-rate, improvements from using adaptive channel width differ because the opportunity provided by use of increased widths somewhat reduces with increased bit-rates as also shown experimentally in Fig. 2 of [6].

6. DISCUSSION

For small to medium scale scenarios like the ones considered in this paper, channel width adaptation can be carried out in a centralized fashion at a gateway node. Gateway in such an implementation acts as a channel allocation server with each node periodically reporting measured link level traffic volume information to the server. Every adaptation interval, the channel allocation server uses that information and recomputes the new channel allocations for each link. The server then communicates them back to the network one node at a time, waiting for confirmation from the node that it completed channel reconfiguration locally through coordination with its neighboring nodes. This approach to implementing traffic-aware channel allocation is practical as demonstrated earlier for the omnidirectional mesh scenario by Ramachandran et al. [18].

Concerning the length of the adaptation interval itself, it depends on the traffic dynamics as well as network overhead for channel reallocation. Due to the aggregation of traffic from individual users at the backhaul nodes, the variability of traffic seen at a backhaul node over time is slower in the order of several minutes; this ob-

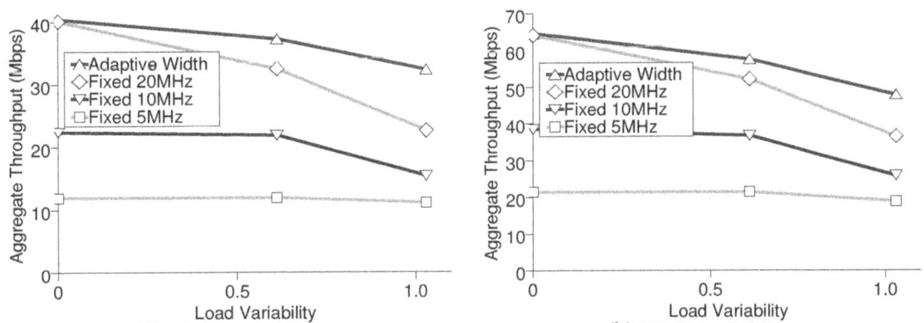

(a) Bit-Rate=12Mbps (b) Bit-Rate=24Mbps

Figure 11: Effect of inter-dependence between channel width and bit-rate on performance for the case of spatial traffic variation with Aravind network topology.

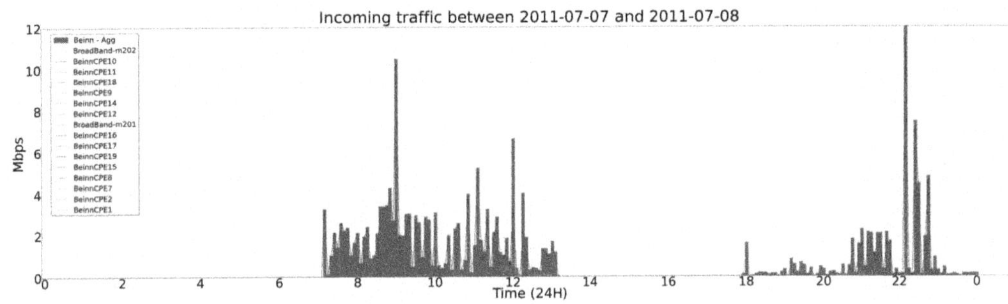

Figure 12: The trace of aggregated and individual user download traffic seen at Beinn backhaul site in the Tegola network [13] on a typical week day. Aggregated traffic is shown in red.

servation is also confirmed by our traffic traces from the Tegola network (Fig. 12). Also in rural wireless networks such as Tegola, aggregate traffic patterns are reasonably predictable with highs and lows around the same time each day. This could be exploited to schedule global channel width adaptations a priori.

For larger networks like [5], distributed implementation is required for scalability reasons. Here we outline the distributed implementation of our proposed channel width assignment algorithm below. Every instantiation of the algorithm consists of three phases corresponding to the three high-level steps described in Section 4: (1) initial distributed edge coloring for guard block assignments; (2) localized node priority assignment based on locally exchanging load information and using node IDs for breaking ties; (3) committing a node's coloring decision after finding a valid width combination — this phase requires the use of a distributed mutual exclusion mechanism along the lines of [19] to limit the number of nodes in a local neighborhood that can concurrently update their channel allocation. We leave the detailed specification of this distributed algorithm and its prototype implementation for future work.

The channel width assignment algorithm can be extended to consider inter-channel separation as a means to minimise adjacent channel interference as follows: We define the minimum required separation distance between any two incident channels ch1 and ch2 as a number of 5MHz blocks (i.e., $(f_{ch2} - w_{ch2}/2) - (f_{ch1} + w_{ch1}/2) \geq 5 * d$, where $f_{ch2} > f_{ch1}$ and d is the separation distance). We refer to these blocks as the *padding blocks*. Each link is then allocated the required channel for transmission as well as the *padding blocks*, which are chosen to the right of the *transmission channel* (except when the transmission channel lies at the upper end of the available spectrum, in which case no padding channel is necessary). Although this requires a larger number of total available blocks to ensure valid channel allocation, one block channel separation between co-incident links may be sufficient. This, however, should be experimentally validated by considering other factors, such as antenna separation.

Finally, the effect of frequency-dependant attenuation, which increases as the carrier frequency increases can be considered in the proposed algorithm as follows. As attenuation increases, the signal-to-noise plus interference ratio, which determines whether a packet is correctly decoded at the receiver, decreases. This causes links operating at higher frequencies to use lower data rates, which restricts the available capacity. To capture this, we consider the effect of frequency on the raw capacity in our algorithm by measuring the effective data rate at each width using the lowest frequency in the available spectrum and use a coefficient to scale down capacity as the frequency increases. Moreover, the feasible width combinations at a node are ordered based on their relative preference assuming the center frequency of each width is the lowest possible frequency (step (3.a) of our algorithm in Section 4). Then, as each combination is checked for its validity (step (3.c)), its preference is scaled down based on the part of the spectrum each width actually occupies. At the end, a set of valid width combinations is formed among which the node chooses the one with the highest preference.

7. CONCLUSIONS

In this paper, we have studied the traffic-aware channel allocation in long-distance 802.11 mesh networks. We leverage the flexibility of using variable channel widths to adapt the channel allocation in response to spatio-temporal variations in traffic demands. We show that the traffic-aware channel width assignment problem is NP-complete by establishing a relationship with the well known edge coloring problem. Our proposed polynomial time greedy heuristic algorithm results in a valid channel allocation for every node. Our simulation based evaluation of the algorithm using real network topologies shows that it substantially improves network performance (e.g., up to around 70% throughput improvement) relative to the existing fixed width allocation approach. As part of future work, we plan to develop a prototype implementation of the algorithm and conduct experimental evaluation over our Tegola network. Theoretical performance characterization of the proposed algorithm is another issue for future work. We also intend to come up with a detailed design of distributed channel width assignment protocol for long-distance 802.11 mesh networks.

8. REFERENCES

[1] Wireless Networking in the Developing World. http://wndw.net/index.html, Dec 2007. Second Edition.

[2] S. Surana, R. Patra, S. Nedevschi, M. Ramos, L. Subramanian, Y. Ben-david, and E. Brewer. Beyond Pilots: Keeping Rural Wireless Networks Alive. In *Proc. USENIX NSDI*, 2008.

[3] W. Tu, C. J. Sreenan, C. T. Chou, A. Misra, and S. Jha. Resource-Aware Video Multicasting via Access Gateways in Wireless Mesh Networks. In *Proc. IEEE ICNP*, 2008.

[4] B. Raman and K. Chebrolu. Design and Evaluation of a new MAC Protocol for Long-Distance 802.11 Mesh Networks. In *Proc. ACM MobiCom*, 2005.

[5] NGI SpA. http://www.ngi.it/.

[6] R. Chandra, R. Mahajan, T. Moscibroda, R. Raghavendra, and P. Bahl. A Case for Adapting Channel Width in Wireless Networks. In *Proc. ACM SIGCOMM*, 2008.

[7] T. Moscibroda, R. Chandra, Y. Wu, S. Sengupta, P. Bahl, and Y. Yuan. Load-Aware Spectrum Distribution in Wireless LANs. In *Proc. IEEE ICNP*, 2008.

[8] S. Rayanchu, V. Shrivastava, S. Banerjee, and R. Chandra. FLUID: Improving Throughputs in Enterprise Wireless LANs through Flexible Channelization. In *Proc. ACM MobiCom*, 2011.

[9] B. Raman. Channel Allocation in 802.11-based Mesh Networks. In *Proc. IEEE INFOCOM*, 2008.

[10] P. Dutta, S. Jaiswal, and R. Rastogi. Routing and Channel Allocation in Rural Wireless Mesh Networks. In *Proc. IEEE INFOCOM*, 2007.

[11] S. Nedevschi, R. K. Patra, S. Surana, S. Ratnasamy, L. Subramanian, and E. Brewer. An Adaptive, High Performance MAC for Long-Distance Multihop Wireless Networks. In *Proc. ACM MobiCom*, 2008.

[12] R. Patra, S. Nedevschi, S. Surana, a. Sheth, L. Subramanian, and E. Brewer. WiLDNet: Design and Implementation of High PerformanceWiFi Based Long Distance Networks. In *Proc. USENIX NSDI*, 2007.

[13] G. Bernardi, P. Buneman, and M. K. Marina. Tegola Tiered Mesh Network Testbed in Rural Scotland. In *Proc. ACM MobiCom Workshop on Wireless Networks and Systems for Developing Regions (WiNS-DR'08)*, Sep 2008.

[14] P. Dutta, S. Jaiswal, D. Panigrahi, and R. Rastogi. A New Channel Assignment Mechanism for Rural Wireless Mesh Networks. In *Proc. IEEE INFOCOM Mini-Conference*, 2008.

[15] W. Yuan, W. Liu, and W. Cheng. Capacity Maximization for Variable-Width WLANs: A Game-Theoretic Approach. In *Proc. IEEE ICC*, 2010.

[16] R. Gummadi, R. Patra, H. Balakrishnan, and E. Brewer. Interference Avoidance and Control. In *Proc. ACM HotNets*, 2008.

[17] J. G. Andrews, A. Ghosh, and R. Muhamed. *Fundamentals of WiMAX*. Prentice Hall, 2007.

[18] K. Ramachandran, E. Belding, K. Almeroth, and M. Buddhikot. Interference-Aware Channel Assignment in Multi-Radio Wireless Mesh Networks. In *Proc. IEEE INFOCOM*, 2006.

[19] B. J. Ko, V. Misra, J. Padhye, and D. Rubenstein. Distributed Channel Assignment in Multi-Radio 802.11 Mesh Networks. In *Proc. IEEE WCNC*, 2007.

[20] S. Pediaditaki, P. Arrieta, and M. K. Marina. A Learning-based Approach for Distributed Multi-Radio Channel Allocation in Wireless Mesh Networks. In *Proc. IEEE ICNP*, 2009.

[21] M. F. Uddin, H. Alazemi, and C. Assi. Joint Routing, Scheduling and Variable-Width Channel allocation for Multi-hop WMNs. In *Proc. IEEE ICC*, 2010.

[22] L. Li, C. Zhang, and Y. Li. QoS-Aware On-Demand Channel Width Adaptation Protocols for Multi-Radio Ad-hoc Networks. In *Proc. IEEE WCNC*, 2009.

[23] F. Wu, N. Singh, N. Vaidya, and G. Chen. On Adaptive-Width Channel Allocation in Non-Cooperative, Multi-Radio Wireless Networks. In *Proc. IEEE INFOCOM*, 2011.

[24] S. M. Das, H. Pucha, D. Koutsonikolas, Y. C. Hu, and D. Peroulis. DMesh: Incorporating Practical Directional Antennas in Multichannel Wireless Mesh Networks. In *IEEE JSAC*, 2006.

[25] V. Angelakis, N. Kossifidis, S. Papadakis, V. Siris, and A. Traganitis. The Effect of Using Directional Antennas on Adjacent Channel Interference in 802.11a: Modeling and Experience With an Outdoors Testbed. In *Proc. IEEE WiNMee Workshop*, 2008.

[26] D. Tyrode. On the Predictability of Traffic and Energy Resource Availability for Self-Powered Wireless Network Routers. Master's thesis, The University of Edinburgh, Aug 2011.

[27] A. Sheth, S. Nedevschi, R. Patra, S. Surana, and E. Brewer. Packet Loss Characterization in WiFi-based Long Distance Networks. In *Proc. IEEE INFOCOM*, 2007.

[28] D. Gokhale, S. Sen, K. Chebrolu, and B. Raman. On the Feasibility of the Link Abstraction in (Rural) Mesh Networks. In *Proc. IEEE INFOCOM*, 2008.

[29] M. R. Garey and D. S. Johnson. *Computers and Intractability: A Guide to the Theory of NP-Completeness*. W. H. Freeman, 1979.

[30] V. Raghunathan and P. R. Kumar. Wardrop Routing in Wireless Networks. In *IEEE Transactions on Mobile Computing*, 2009.

[31] D. B. West. *Introduction to Graph Theory*, chapter 7. Prentice Hall, 2000.

[32] M. Kodialam and T. Nandagopal. Characterizing Achievable Rates in Multi-hop Wireless Networks: The Joint Routing and Scheduling Problem. In *Proc. ACM MobiCom*, 2003.

[33] Connected Communities Network. http://www.connectedcommunities.co.uk/.

Efficient Conditional-Probability Link Modeling Capturing Temporal Variations in Body Area Networks

David B. Smith*
Networks Research Group
NICTA†
Canberra, ACT, Australia
David.Smith@nicta.com.au

Athanassios Boulis, and
Yuriy Tselishchev‡
Networks Research Group
NICTA†
Sydney, NSW, Australia
Athanassios.Boulis@nicta.com.au,
Yuri.Tselishchev@nicta.com.au

ABSTRACT

This paper presents conditional probabilistic modeling suitable to characterize the temporal variation of links in wireless body area networks (BAN); according to short, medium and long term fading characteristics. The approach captures first and second order statistics appropriately; and using conditional probabilities, predicts what signal power levels can be expected from 10 ms ahead to many seconds into the future in the context of typical channel coherence times. Hundreds of hours of link measurements, in "Everyday" mixed activity BAN, are used to generate these conditional models. We show that such modeling has an important effect to higher level simulation results (e.g., packets received at the application layer) for different simulation scenarios. Moreover we show that the model is computationally efficient as its introduction adds only 8% of simulation time on average. It is also shown that short and medium-term conditional modeling can vary considerably from long-term modeling, particularly given lower instantaneous path loss.

Categories and Subject Descriptors

C.4 [**Performance of Systems**]: Modeling Techniques; C.2.1 [**Network Architecture and Design**]: Wireless communication

Keywords

Body area networks, channel modeling, conditional probability, time-varying channels

*D. Smith holds an adjunct appointment with the Australian National University.

†NICTA is funded by the Australian Government as represented by the Department of Broadband, Communications and the Digital Economy and the Australian Research Council through the ICT Centre of Excellence program.

‡Y. Tselishchev is also with the University of Sydney.

MSWiM'12, October 21–25, 2012, Paphos, Cyprus.

1. INTRODUCTION

Wireless Body Area Networks (BANs) represent the next generation of personal area networks. A BAN typically consists of a wireless communicating network of sensors (and/or actuators) placed on the human body. There are various application of BANs [11], including health-care and consumer fitness. The recently approved IEEE 802.15.6 BAN Standard [2] underlines the need for reliability at low-power for BAN [19]. In this context, variation of path loss, i.e., channel gain, with time, although not rapid has shown to be significant in many scenarios, e.g., [5, 6, 14, 18]. Thus it is important to understand the impact of the time varying channel on the performance of IEEE 802.15.6 compliant network implementations. Thus appropriate modeling, to event-driven network level simulators, of temporal variation of channel gain is required; as such simulators are the main evaluation tools for communication protocols in BANs.

Previously long-term fading statistics (large-scale and small-scale fading) have been derived from measurements to model the time-varying BAN channel, with appropriate distributions fitted to these statistics for BAN link data, either for individual BAN links, e.g. [5,6] or combined multi-link BAN data, e.g., [15]. Typically second order statistics measures of temporal variation such as average fade duration, autocorrelation, Doppler spread and level crossing rates have been derived in conjunction with first-order statistics capturing distributions of channel gains, e.g., [6, 14, 15].

Here, we attempt to answer the following: given a current signal sample, and particular channel gain, *what channel gains can be anticipated according to accurate probabilistic modeling for a signal sample over short, medium and long terms with respect to coherence time?* Such specific conditional probabilistic modeling, as far as we are aware, has not been applied to describe the time-varying BAN channel across particular BAN links, specifically for use in event-driven network simulation. We argue that such conditional modeling is far more useful than traditional long-term modeling, with an associated second-order statistic, for event-driven simulation. Such simulation can demand short-term or long-term wireless channel state information, e.g., after 10 ms or 1 s respectively. And thus conditional probability distributions are best suited to capturing these varying demands, given a current signal sample and channel gain.

Here we use a very large set of narrowband empirical BAN channel-gain data collected using small wearable radios at 2.36 GHz carrier frequency, with ten adult human subjects engaged in "everyday" activity, for periods of at least two hours. This large set of dynamic data for the BAN channel facilitates good overall long-term statistical characterization, which can be found in [14]. Instead of long-term statistics, here we are particularly motivated to use our empirical data to provide short and medium-term models capturing temporal variation, which can be implemented with computational efficiency in appropriate network-level simulation tools such as the open-source Castalia sensor network simulator [3]. These models are conditional probabilistic models. According to this modeling applied to network-level BAN simulation, here we show the following:

1. According to a coherence time of around 400 ms for the everyday BAN channel, we propose very short-term, medium term, and medium-to-long-term conditional probabilistic models for BAN. Given a current channel gain, compared with a typical channel gain; we accurately anticipate a channel gain 10 ms (short-term), 250 ms (medium-term) and 1 s (medium-to-long-term) in to the future. These three prediction time intervals mean temporal variations of channel gain can be accurately anticipated with respect to the rates of variation in channel gain in everyday BAN; and *importantly they are enough to characterize across small-to-long intervals for all time.*

2. Short and medium-term estimates of path loss, given a current path loss that is less than a median path loss for everyday BAN, after 10 ms and 250 ms respectively, are often best described by a Normal distribution. Such models provide a marked difference from most long-term BAN statistical modeling that has been reported, where time-varying path loss is often lognormal, and sometimes gamma or Weibull distributed, e.g., [14].

3. By a method called "inverse transform sampling" from the inverse cumulative distribution functions, cdfs, of the time-varying BAN channel, we derive models whereby any anticipated channel gain can efficiently, and simply, be chosen from a uniformly randomly chosen sample. Tails of distributions, representing very high and very low conditional channel gains, are also efficiently represented.

4. We have applied this temporal BAN model in the Castalia simulator [3], and using IEEE 802.15.6 MAC [2] and IEEE 802.15.4 MAC [1] implementations we show that: (i) there is a significant impact in higher-layer simulation results in terms of correctly received packets; and (ii) computational overhead of the model is small in comparison to static modeling.

2. EXPERIMENTAL SETUP AND BAN CHANNEL COHERENCE TIME

2.1 Experimental Setup

The experimental set-up was for on-body area communications encompassing everyday activity of long periods using multiple small body-mounted wearable radios (as Tx and Rx). The activity is predominantly that of an office-worker over several hours in an indoor office, including some activity at home, and also jogging in an outdoor suburban environment. The measurements were made at 2360 MHz, one of the carrier frequencies in the IEEE 802.15.6 BAN standard [2].

8 adult test subjects wore the small radios at different times, with some subjects having multiple measurements at different times. One measurement dataset used here, with measurements over 9 hours, is open-access [10]. A full description of the radios can be found in [9]. Subjects wore a varying number of these radios, with some operating as Rx, some as Tx and Rx, and some as Tx (the total amount of Tx and Rx varied from 3 to 20). The complete map of all possible Tx/Rx positions are given in Table 1. Individual subject measurements generally lasted for a period of two or more hours. The digital RSSI (Received Signal Strength Indicator) log made upon successful packet detection by the radios was used to measure channel gain. With two Tx, the RSSI for any Tx/Rx link was logged every 10 ms, whereas with one Tx, the RSSI for any Tx/Rx link was logged every 5 ms. A total of 140 Tx/Rx link measurements were made over all test subjects.

2.2 Everyday BAN coherence time

In order to provide accurate short and medium-term conditional modeling for time-varying BAN, of key importance is the coherence time of the BAN channel. Coherence time is traditionally derived from the typical time taken for the autocorrelation of any BAN link to drop from 1 to 0.7. The autocorrelation is calculated according to

$$\rho(\tau) = \frac{\frac{1}{M}\sum_{m=1}^{M-\tau}\{x(m)-m_x\}\{x(m+\tau)-m_x\}^*}{\sqrt{\sum_{m=1}^{M-\tau}\{x(m)-m_x\}^2}\sqrt{\sum_{m=1}^{M-\tau}\{x(m+\tau)-m_x\}^2}}, \quad (1)$$

, where m_x is the mean channel gain with respect to any particular link, with τ being a time-lag samples index. We calculate the autocorrelation ρ for a 2.5 s period (sampling at 100 Hz, 2500 samples τ; sampling at 200 Hz, 5000 samples τ). The autocorrelation, averaged across measurements for all 140 links, is shown below in Figure 1. The coherence time, accordingly, is 400 ms.

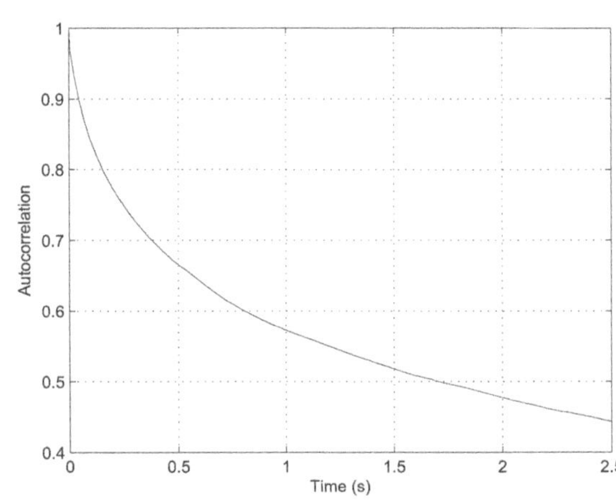

Figure 1: Autocorrelation across all measurements.

RECEIVER (Rx) LOCATIONS	TRANSMITTER (Tx) LOCATIONS									
	RIGHT HIP					LEFT CHEST				
FRONT	RA	LA	RK	LK	RW	LW	RE	LE	LH	HEAD
BACK	RA	LA	RK	LK	RW	LW	RE	LE	LH	RH

Table 1: Channel Sounders placements; R,L-Right,Left. A:Ankle,K:Knee,E:Elbow,H:Hip,W:Wrist

3. SHORT-TO-LONG TERM CONDITIONAL PROBABILISTIC MODELS

3.1 Conditional Probabilities

Across all measurements we note that the median channel gain is -72.2 dB. It has previously been argued that the median provides a clearer indication of the most typical channel gain across any BAN measurement data set [14]. Given the large range in channel gain in most BAN link measurement data sets, the mean channel gain is significantly skewed to larger channel gains in the particular measurement set, rather than the typical channel gain represented by the median, which represents the mid-point between the upper-half and lower-half of channel gain magnitudes, either across any link measurement data set or across all data sets. Hence our conditional probability distributions that we derive are based on given signal levels *with respect to the median channel gain.* We also note here that we do not differentiate between different Tx and Rx locations, but provide probabilities for agglomerate links. The validity of this approach using agglomerated data is rigorously justified in [17].

In order to be most appropriate to typical channel gain profiles from our everyday BAN measurements, we derive conditional probabilities at suitable time intervals with respect to signal levels that are: a) any values less than -25 dB, b) from -25 dB to 15 dB in one-dB steps (where we round to the nearest dB value), and c) above 15 dB with respect to the median channel gain[1]. Thus initial channel gain samples are placed in 43 groups, for specifying conditional probabilities. According to the general BAN channel coherence time of 400 ms we are interested in providing conditional probabilities at short-term (i.e., small fraction of coherence time), medium term (i.e., close to the coherence time) and medium-to-long term (i.e, some multiple of the coherence time). Thus we choose 10 ms, 250 ms and 1000 ms intervals, *Est.interval* respectively, appropriate to specifying our conditional probabilities. Note that our conditional probability distributions are dependent/conditioned both on current channel gain (quantised in one of the 43 channel gain groups) and the 3 *Est.Intervals*. For example, for short-term 10 ms interval with an initial channel gain rounded to -10 dB with respect to median, what is the conditional probabilistic model for the channel gain 10 ms ahead?

We derive conditional probabilities according to choosing between six standard models commonly used to describe small-scale fading; i.e., Rayleigh, normal, lognormal, gamma, Weibull and Nakagami-m distribution models; where we use the negative log-likelihood criterion over maximum-likelihood (ML) parameter estimates to choose the best model. Accordingly given our 43 groups of channel gain samples, and our 3 prediction intervals, we derive 129 distributions from our hundreds of hours of everyday BAN measurements (hundreds of millions of samples). These distributions, in general, provide a good fit to relevant data samples. The models are summarized in Table 2. Please note where the median-removed channel gain range $Pow.Range_{mdr}$ is specified as $[x, y]$, this implies $y - x + 1$ initial (rounded) channel gain samples with 1 dB spacing between integer x and integer y. We note that in no case was a conditional model described by a Weibull or a Rayleigh distribution, and in only 2 cases are they described by a Nakagami-m distribution. 42 of the conditional distributions are normal-distributed, 59 are lognormal-distributed and 26 are gamma distributed. We do not give parameters here, as due to the inverse transform sampling derived in the next section, simulation can be performed without knowledge of these parameters (in fact even the underlying distribution models).

$Pow.Range_{mdr}$ (in dB)	*BestDist*	*Est.Interval*
<-25;[-25,-14];-12;>15	Lognormal	10 ms
-13;[-11,-9]	Gamma	10 ms
[-8,15]	Normal	10 ms
<-25;[-25,-6]	Lognormal	250 ms
[-5,3];>15	Gamma	250 ms
[4,15]	Normal	250 ms
<-25;[-25,-4]	Lognormal	1 s
[-3,7];16	Gamma	1 s
[9,10]	Nakagami-m	1 s
8;[11,15]	Normal	1 s

Table 2: Best Conditional Probability Distributions, *BestDist*, with respect to initial median removed channel gain ranges, $Pow.Range_{mdr}$, at *Est.Interval* time in the future

We note that most long-term fading characteristics, not conditional distributions, derived in the literature are not normal-distributed, e.g., [4, 7, 8, 12–14]; they are often lognormal distributed, and may also be gamma or Weibull distributed [14]. Whereas for the conditional probabilities here, predicted channel gains short and medium-term at 10 ms and 250 ms ahead where the initial channel gains are higher, i.e., lower instantaneous path loss, the predicted channel gains are often normal-distributed. Moreover, approximately $\frac{1}{3}$ of the conditional distributions are normal distributions, this is a significant finding providing a contrast to non-conditional fading statistics previously derived for the on-body BAN radio channel, e.g., [14]. This also demonstrates that the proposed conditional probabilities are useful, and provide more accuracy than those derived for overall models; as an overall model may not represent well the short and medium term characteristics of the BAN channel, as these are often better represented by a different model.

3.2 Inverse Transform Sampling

According to the conditional probabilistic temporal mod-

[1]It should be noted that in our measurements the majority of channel gain samples were in the range from 25 dB below the typical median channel gain to 15 dB above this median.

els, and associated ML distribution parameter estimates, we apply a method known as *inverse transform sampling* for simple computationally efficient network-level simulation and implementation[2]. For each of the 129 conditional distributions and parameter estimates, we use the inverse cdf, known as the quantile distribution, to efficiently draw samples from these distributions in a one-step process (and approximately 2% of the time a 2-step process to capture distribution tails.) The integer drawn selects a specific quantile, and the content of the quantile is the returned distribution value. In 2% of the cases the content of the quantile points us to perform another uniform integer drawn from a smaller set of quantiles. More specifically, for each of the 129 conditional distributions, we create 109 quantiles, with the middle 99 quantiles evenly spaced, and the bottom and top 5 quantiles more finely resolved used to capture the lower and upper distribution tails respectively. The quantile of a random variable, generated from a particular distribution, specifies for a given probability the value which the random variable will be at, or below, with that probability. The complete quantiles are from [0.0001,0.0081] in 0.002 steps, [0.01,0.99] in 0.01 steps, and [0.9919,0.9999] in 0.02 steps.

In order to avoid overweighting the distribution tails, we separate the quantiles into two levels for inverse transform sampling. Hence we use a uniform random integer sampling m between 1 and 101 for the first level; if the random sample m is from 2 to 100 we choose the $m-1$ th quantile in [0.01,0.99]. If the random sample $m=1$ or $m=101$, we go to the second level and we use a random integer sampling n between 1 and 5; if $m=1$ we choose the n th quantile in [0.0001,0.0081] and if $m=101$ we choose the n th quantile in [0.9919,0.9999].

This is best illustrated by an example. For our $Est.interval =$ 10 ms, assume the initial channel gain sample is rounded to 15 dB below the median channel gain, i.e. $Pow_{mdr} = -15$ dB. From our conditional modeling the estimated channel gain is best described by a lognormal distribution with parameter estimates log-mean $\mu = -1.73$ and log-standard deviation $\sigma = 0.31$. From this distribution, prior to any simulation, we generate our 109 quantiles according to the inverse cdf of this lognormal distribution for each of these quantiles. The quantiles are generated in terms of channel gain magnitudes and then converted to dB-values with respect to the median. Then when it comes to simulation, assume from our uniform random sampling $m = 8$, hence we choose the $m-1=7$ th quantile from [0.01,0.99], which specifies the value that the random variable will be at with an approximate probability of 0.07; and this quantile is a value of -18.3 dB with respect to the median, which is the determined signal level at an interval of 10 ms in the future.

Thus we have generated $109 \times 3 \times 43 = 14,061$ quantiles for our complete conditional probabilistic temporal modeling for anticipated signal levels. Here we store these values in our BAN simulation tool, e.g. our NICTA Castalia simulator [3]. $14,061$ quantiles, generated with only 5 significant digit precision, represent a very small relative memory requirement for modern-day computation. Text files, with these $14,061$ quantiles can be found in [16].

Importantly, although we only have conditional probabilities for 3 prediction intervals, $Est.interval$, it is readily apparent that we can iteratively apply our inverse transform sampling to derive predicted channel gains for any multiple of 10 ms into the future; e.g. if we want to anticipate a channel gain 50 ms ahead, we round our current channel gain to nearest dB values, and then apply 5 iterations of our inverse transform sampling with respect to $Est.interval = 10$ ms, where at each multiple of 10 ms we adjust our choice of inverse transform predicted sample according to our last predicted sample.

The choice of 3 particular short, medium and medium-to-long term $Est.interval$ at 10 ms, 250 ms and 1 s is done with respect to coherence time as stated. *Our empirical testing has confirmed that more than 3 intervals provides no real extra accuracy, and increases memory requirements. Using less than these 3 intervals, is less accurate however.* E.g., 50 iterations of conditional channel gains at 10 ms to obtain a channel gain 500 ms into the future is, in general, less accurate than 2 iterations of 250 ms (as well as being computationally more expensive). Similarly 40 iterations at 250 ms interval to obtain a channel gain 10 s into the future is less accurate than 10 iterations of 1 s interval (as well as being computationally more expensive). However, due to the coherence time being significantly less than 1 s; there is no gain in having any $Est.Interval$ longer than 1 s.

4. NETWORK LEVEL SIMULATION USING CONDITIONAL TEMPORAL MODELS

In order to show the importance of the temporal modeling, we use the Castalia sensor network simulator [3] and investigate the effect of the presence or absence of temporal modeling on the application level results. The basic simulation setup comprises 8 nodes, i.e., 7 on-body sensors connecting to a hub centrally located on the body. The 7 sensors send data to the hub at a constant rate. We vary this rate from 10 to 50 packets per second. We record the application packets received in the simulation, and use this as a performance metric. For both the temporal models presence and its absence (i.e., static model) we assume a typical path loss, i.e., the median, of 72 dB. Hence this is fixed for all time for the static model. Configuration files for the simulation can be found in [3]. It is important to note here that we use both the IEEE 802.15.6 MAC [2] and for the IEEE 802.15.4 MAC [1] and make sure that the parameter settings are comparable. More specifically, we choose the same superframe duration, same slot size, same number of slots assigned to contention-based access and same number of slots assigned to TDMA access as follows:

- Each superframe has 16 slots, where each slot is 5ms long. Hence, superframes are 80ms long.
- Each node requests 2 slots for contention-free access upon connecting to the hub.
- Once all 7 nodes are connected, there remains only 2 contention based slots, which come immediately after the beacon. The 14 contention-free slots follow the contention based slots.
- Hence in general, each node has 10ms of contention access and 10ms of contention-free access to the hub.

[2]We note that using these models, rather than directly using data samples allows finer modeling precision with respect to the sampling and allows us to better represent distribution tails.

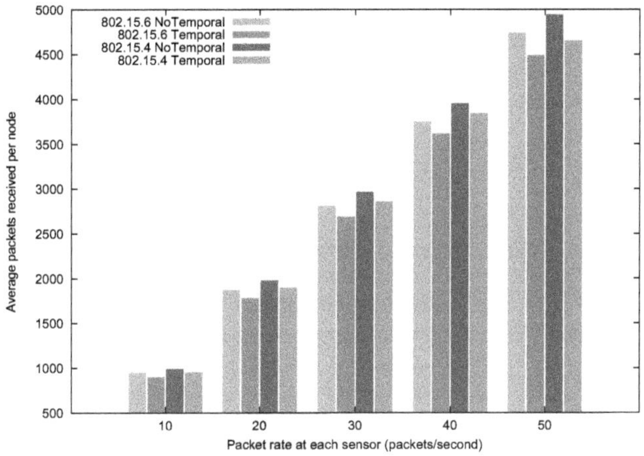

Figure 2: Average packets received in terms of packet transmission rate per node (sensor), with and without conditional temporal model, in the context of IEEE 802.15.4 MAC [1]and IEEE 802.15.6 MAC [2]. Median path loss of 72 dB is assumed.

- Both the 802.15.4 and the 802.15.6 protocol can only fit 2 contention-free transmissions in a single slot, or 4 transmissions in 2 consecutive slots. In contention-based slots the number of contention-based transmissions varies, but generally 802.15.4 contends quicker, thus can fit more.
- Thus at 50 packets/s for any given link there are 4 packets sent from the sensor in any 80 ms superframe.
- We choose a simulation time of 100 s for each sensor-to-hub link packet rate, with a 1 s delay before packets are generated. Hence at 10 packets/s for each link 990 packets are generated for each link in the simulation.

In Figure 2, we show a bar graph of the number of packets received for each link packet rate from the simulation results; where we have averaged the number of packets received over the 7 sensor-to-hub links (i.e., node to hub). We note that the number of packets correctly received at the application layer when the conditional temporal model is applied is approximately 7% less than when simply applying a static, "no-temporal", model for all link packet rates for the same MAC protocol; where for the case of the static model nearly all packets are correctly received when applying the IEEE 802.15.4 MAC protocol. The 4 to 5% superior performance of IEEE 802.15.4 over IEEE 802.15.6, with and without the temporal model is mainly due to the quicker contention of IEEE 802.15.4. To illustrate this we disable contention-based transmissions, and only allow contention-free transmissions.

In Figure 3 we show the results with only contention-free transmission, where for the static model the number of packets received for IEEE 802.15.4 are almost identical to those for IEEE 802.15.6, similarly for the temporal model. Importantly, up to a packet transmission rate of 40 packets/s there is very little degradation (1 to 2%, in packets received), applying the temporal model when compared to applying the static model. This is due to the packet retransmission scheme the MAC protocols employ, which is able to recover most packets. However, the importance of applying

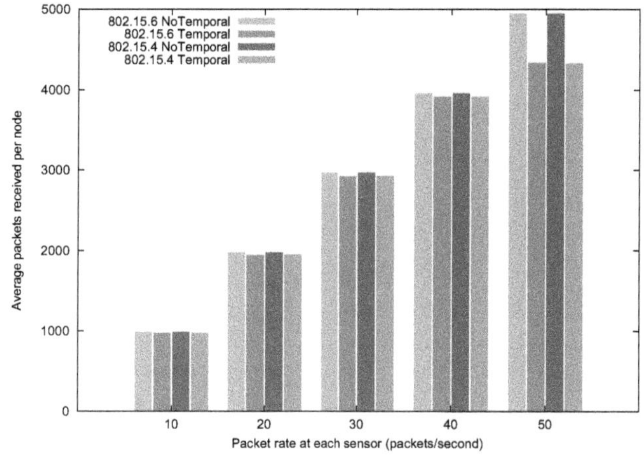

Figure 3: Average packets received in terms of packet transmission rate per node (sensor), with and without conditional temporal model, in the context of IEEE 802.15.4 MAC and IEEE 802.15.6 MAC where there is no contention access period. Median path loss of 72 dB is assumed.

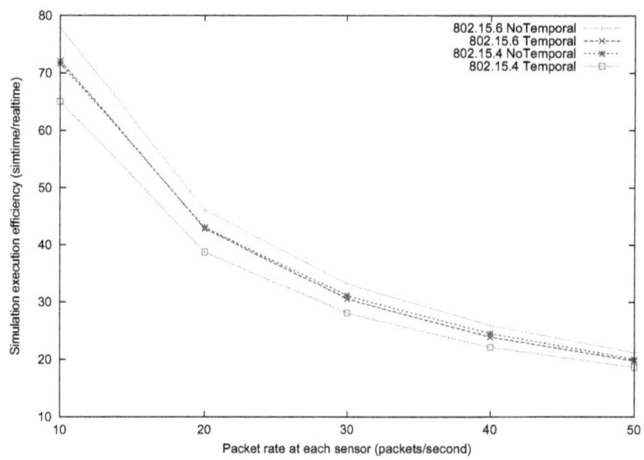

Figure 4: Run-time efficiency for IEEE 802.15.4 and IEEE 802.15.6 implementation, comparing using the static, "no temporal", model and the temporal model run on the Castalia sensor network simulator.

the temporal model is emphasized for the fastest packet rate (at which the network is close to saturation and there are no spare slots to accommodate retransmissions), where the degradation in the number of packets received is approximately 12% for both IEEE 802.15.4 and IEEE 802.15.6.

Another important aspect of applying the conditional probabilistic temporal models is their run-time efficiency, as they use the applied inverse transform sampling to improve simulation run-time efficiency. The smaller the extra computation time, with the conditional temporal model as opposed to the static, "no temporal" model, the better. Once again, both contention-free and contention-based access periods are assumed for both IEEE 802.15.4 and 802.15.6 MAC protocols. Figure 4 shows there is only approximately 8% degradation in run-time efficiency (8% extra time for computations), when comparing applying the temporal model to the static model for both cases of IEEE 802.15.4 and IEEE 802.15.6 implementations. There is also an approximate 8% improvement of IEEE 802.15.6 BAN MAC implementation over IEEE 802.15.4 BAN MAC implementation.

5. CONCLUDING REMARKS

A novel conditional probabilistic temporal model appropriate to the "everyday" on-body wireless body area networks (BAN) channel has been presented, which has been derived from extensive empirical data. The model captures short, medium and long-term fading characteristics effectively; and can be used in simulation with excellent computational efficiency by the method of inverse transform sampling presented. Conditional probabilistic temporal models are appropriately derived with respect to an approximate overall BAN channel coherence time of 400 ms. Importantly, short and medium-term fading models based on predicting channel gains at time intervals of 10 ms and 250 ms respectively, conditioned on particular signal levels with respect to a typical (i.e., median) path loss, can show significant differences from longer-term fading models; as shown by the normal distribution often found when the current channel gain is higher.

Network-level simulation implementation highlights the need for the temporal model, where assuming a static model is unrealistic and overestimates the number of packets received for realistic everyday on-body BAN by about 7%. Hence good system design needs such a temporal model as presented here, where there is only an 8% overhead in runtime efficiency, in network-level simulation. Further previous fading characteristics presented in the literature do not properly capture short to medium term variations needed in practical link-simulations, as these can vary significantly from long-term characteristics; further highlighting that conditional probabilistic models for BAN, of the type presented here suitable for event-based network simulations, are needed.

6. REFERENCES

[1] IEEE standard for local and metropolitan area networks–part 15.4: Low-rate wireless personal area networks (LR-WPANs). *IEEE Std 802.15.4-2011 (Revision of IEEE Std 802.15.4-2006)*, pages 1 –314, Sep. 2011.

[2] IEEE standard for local and metropolitan area networks part 15.6: Wireless body area networks. *IEEE Std 802.15.6-2012*, pages 1 –271, Feb. 2012.

[3] A. Boulis. Castalia wireless sensor network simulator. online `http://castalia.npc.nicta.com.au/`, May 2012.

[4] S. L. Cotton and W. G. Scanlon. A statistical analysis of indoor multipath fading for a narrowband wireless body area network. In *IEEE Personal, Indoor and Mobile Radio Communications Symposium, PIMRC 2006*, pages 1–5, Helsinki, Finland, Sept. 2006.

[5] S. L. Cotton and W. G. Scanlon. Higher order statistics for lognormal small-scale fading in mobile radio channels. *IEEE Antennas and Wireless Propagation Letters*, 6:540–543, 2007.

[6] R. D'Errico and L. Ouvry. Time-variant BAN channel characterization. In *Personal, Indoor and Mobile Radio Communications, 2009 IEEE 20th International Symposium on*, pages 3000 –3004, Sep. 2009.

[7] R. D'Errico, R. Rosini, and M. Maman. A performance evaluation of cooperative schemes for on-body area networks based on measured time-variant channels. In *Communications (ICC), 2011 IEEE International Conference on*, pages 1 –5, Kyoto, Japan, June 2011.

[8] A. Fort, C. Desset, P. deDoncker, P. Wambacq, and L. van Biesen. An ultra-wideband body area propagation channel model-from statistics to implementation. *IEEE Transactions on Microwave Theory and Techniques*, 54(4):1820–1826, Apr. 2006.

[9] L. Hanlen et al. Open-source testbed for body area networks: 200 sample/sec, 12 hrs continuous measurement. In *Personal Indoor and Mobile Radio Communication, 2010 IEEE 20th International Symposium on*, pages 1–5, Istanbul,Turkey, Sep. 2010.

[10] L. W. Hanlen and D. Rodda. Public dataset for BAN radio channel models. online `www.nicta.com.au/research/projects/human_performance_improvement/researchoutcomes/wireless`, March 2010.

[11] D. Lewis. 802.15.6 call for applications - response summary ID: 802.15-08-0407-05. IEEE submission, July 2008.

[12] W. G. Scanlon and S. L. Cotton. Understand on-body fading channel at 2.45 GHz using measurements based on user state and environment. In *Loughborough Antennas and Propagation Conference*, pages 10–13, Loughborough, UK, March 2008.

[13] D. Smith et al. Statistical characterization of the dynamic narrowband body area channel. *Applied Sciences on Biomedical and Communication Technologies, 2008. ISABEL '08. First International Symposium on*, pages 1–5, Oct. 2008.

[14] D. Smith, L. Hanlen, J. Zhang, D. Miniutti, D. Rodda, and B. Gilbert. First- and second-order statistical characterizations of the dynamic body area propagation channel of various bandwidths. *Annals of Telecommunications*, 66:187–203, 2011. 10.1007/s12243-010-0233-8.

[15] D. Smith, J. Zhang, L. Hanlen, D. Miniutti, D. Rodda, and B. Gilbert. Temporal correlation of dynamic on-body area radio channel. *Electronics Letters*, 45(24):1212 –1213, Nov. 2009.

[16] D. B. Smith. Conditional probabilistic everyday BAN models from inverse transform sampling, text files. online can be found in `/Simulations/Parameters/WirelessChannel/BANmodels/` directory in Castalia download from `http://castalia.npc.nicta.com.au`, February 2010.

[17] D. B. Smith, L. W. Hanlen, and T. A. Lamahewa. A new look at the body area network channel model. In *European Conference on Antennas and Propagation*, pages 2987–2991, Italy, Rome, Apr. 2011.

[18] K. Yazdandoost and K. Sayrafian-Pour. TG6 channel model ID: 802.15-08-0780-12-0006. IEEE submission, Nov. 2010.

[19] B. Zhen, M. Patel, S. Lee, and E. Won. Body Area Network (BAN) Technical Requirements. July 2008. IEEE802.15.6 technical contribution, document ID:15-08-0037-01-0006-ieee-802-15-6-technical-requirements-document-v-4-0.

Characterisation and Application of Idle Period Durations in IEEE 802.11 DCF-based Multihop Wireless Networks

Van Nam Nguyen
Université Claude Bernard Lyon 1 - IFI
IRD, UMI 209 UMMISCO, MSI Team
LIP (UMR ENS Lyon - INRIA - CNRS - UCBL)
Lyon, France
nguyen.van.nam@ens-lyon.fr

Isabelle Guérin Lassous
Université Claude Bernard Lyon 1
LIP (UMR ENS Lyon - INRIA - CNRS - UCBL)
Lyon, France
isabelle.guerin-lassous@ens-lyon.fr

Victor Moraru
Institut de la Francophonie pour l'Informatique (IFI)
IRD, UMI 209 UMMISCO, MSI Team
Hanoi, Vietnam
victor.moraru@auf.org

Tahiry Razafindralambo
IRCICA/LIFL
Univ. Lille 1, CNRS UMR 8022, INRIA LNE/ POPS
tahiry.razafindralambo@inria.fr

ABSTRACT

Multihop wireless networks are used to provide internet connectivity to the users and the level of performance and quality expected by these users are increasing. In order to meet these performance and quality requirements, wireless communications should be enhanced. Previous works from the literature show that the performance and quality provided by an IEEE 802.11-based multihop wireless network are far from optimal and that there exist different ways to increase the efficiency and the quality of service of such a network. Some studies show that using the medium state as a parameter to tune the behaviour of an IEEE 802.11-based multihop network is an appropriate way to proceed. A station in a IEEE 802.11-based multihop wireless network senses the medium either busy or idle. The durations of idle periods and busy periods and their distributions have a clear impact on the network and nodes performance. The understanding of the relationship between these indicators, namely idle and busy periods, the network topology and the traffic, would give new insights to enhance the performance and quality of multihop wireless networks. Due to its multihop and distributed nature, the characterisation of idle period durations is difficult in such a network. This paper explores the characterisation of idle period distribution by proposing a new analytical model and provides an application of this characterisation with the design of an adaptive backoff algorithm based on idle periods.

MSWiM'12, October 21–25, 2012, Paphos, Cyprus.

Categories and Subject Descriptors

C.2.1 [**Computer-Communication Networks**]: Network Architecture and Design—*Wireless Communication*

General Terms

Measurement, Performance

Keywords

Idle Periods, Multihop wireless networks, IEEE 802.11, Quality of service

1. INTRODUCTION

A station in a IEEE 802.11-based multihop wireless network senses the medium either busy or idle. The durations of idle periods and busy periods have a clear impact on the network and nodes performance. The characterisation of idle period durations is an important issue in wireless and mobile networks. For example, in cognitive radio networks, a secondary node must be able to adapt its behavior based on observations of the radio medium and on its experience in the past. The cognitive process is composed of the analysis of conditions observed on the network, of the decision-making and of the learning based on its experience. Under these conditions, the observation and the analysis of idle period durations and their statistical parameters are important for the proper functioning of this approach [9]. It is also possible to rely on observed idle periods to estimate the number of active nodes in the network, when all nodes are in the same carrier sense area [2]. Observation of idle period durations are also used in QoS solutions, like in [10] that describes an available bandwidth estimation between two nodes based on the nodes' average idle time durations. It is also the authors point of view that it should be possible to improve efficiency and fairness of MAC solutions for multihop wireless networks thanks to a knowledge of idle period durations.

As far as we know, very few works have attempted to characterise idle period durations in multihop wireless networks. In [11], the authors characterise interarrival times (of TCP flows) on any node via an analytical model. The considered networks are WLANs with and without infrastructure. They show that interarrival times have a multimodal distribution. The proposed model could be used to infer idle time durations but the work assumes that there is no hidden node. In [3], the authors seek to characterize idle periods to estimate the number of nodes in competition on a CSMA/CA network. This work is valid for single-hop networks with nodes that have always a packet to send and that always use the same backoff stage (the initial one). In [9], the authors characterize idle period durations in 802.11e single-hop networks (based on EDCA) with Poisson traffic arriving at each node. They show that, when the traffic intensity is non negigible, the distribution of idle period durations can be approximated by a Gamma distribution. All these works are mainly focused on single-hop networks and do not consider multihop networks whose features impact differently idle periods.

In studies of multihop wireless networks, assumptions on idle periods are often considered. For instance, Garetto *et al.* [5] model the wireless channel as a stochastic process in which the durations of busy and idle periods of nodes are exponentially distributed. Kolar *et al.* [7] model the backoff and transmission periods as an *Alternating Renewal Process* in which backoff time and packet size are exponentially distributed. In [10], the authors assume that idle periods, for which the total duration is known (by measurement), are uniformly distributed in a given period of observation. As far as we know, there is no evidence that these assumptions are appropriate.

In this paper, we study the distribution of idle periods in multihop wireless networks based on IEEE 802.11. As opposed to the other existing works, we consider multihop networks with potential hidden stations and under non-negligible traffic intensity. In this work, we first study the distribution of idle periods in a basic multihop configuration. Then, we propose an analytical model to characterise idle period durations in multihop wireless networks. Finally, we use the previous analysis to adress one problem in multihop wireless networks, namely the design of a fair MAC solution.

This article is organized as follows: in Section 2, we summarize the main principles of 802.11 DCF and we give the main assumptions of our study. In Section 3, we study, by simulation, the distribution of idle periods in a basic multihop wireless scenario. Then, in Section 4, we propose an analytical model for the distribution of idle periods for a node in an arbitrary multihop toplogy. The model is validated with simulations in Section 5. In Section 6, we use the study of the idle period distribution, carried out in the previous sections, to propose a simple and fair MAC protocol for multihop wireless networks.

2. 802.11 DCF PRINCIPLES AND ASSUMPTIONS

As we consider 802.11-based multihop wireless networks, we first describe the main principles of the DCF mode of 802.11 that impact idle periods in the network. We then discuss the main assumptions we make for our study.

2.1 DCF main principles

IEEE 802.11 DCF [8] is widely used as the MAC layer of multihop wireless networks due to its distributed nature. In the DCF (Distributed Coordination Function) mechanism, a node that wants to transmit a packet has to wait for a fixed period DIFS and a random period, called *backoff*, randomly chosen in a contention window. During these periods, the medium must be idle. During the backoff counting process, if the medium is detected as busy, the backoff counter is frozen and will be resumed as soon as the medium returns to the idle state. As soon as the backoff counter reaches zero and if the channel is idle, the node starts the transmission of its packet. In 802.11 DCF, a unicast packet is considered to be successfully transmitted if the sender receives an acknowledgement from the receiver after the fixed period SIFS. Otherwise, the sender considers that a collision has occurred. The packet is then retransmitted by using the same process but with a contention window size that is doubled. After each successive collision, the contention window size is doubled (the contention window is also called the backoff stage and the stage number depends on the number of successive undergone collisions). The number of packet retransmissions is limited to a maximal value defined by the standard. The value of the backoff stage is set to its initial value if the transmission succeeds or if the packet is dropped after a maximal number of retransmissions without success.

2.2 Assumptions

In this article, we assume that the RTS/CTS mechanism is disabled. We also assume that the packet size, denoted by T hereafter, is fixed and expressed in number of slots, that the used modulation is fixed and that the physical layer is perfect (*i.e.* there is no packet loss due to physical error). We do not consider SIFS time between a packet and its MAC acknowlegment like an idle period since it is part of an entire 802.11 transmission. In our study, when a backoff will be considered, we will assume that the DIFS time is included since a backoff is always associated to a DIFS.

Finally, we consider that an *idle period* perceived by a node is the time during which no signal is sensed on the wireless channel with a power higher than the carrier sense threshold.[1]

3. FIRST STUDY: A BASIC SCENARIO

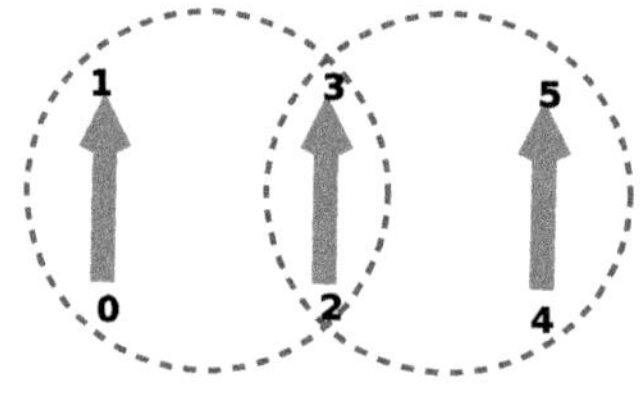

Figure 1: Flow In the Middle (FIM) scenario

In this section, we study, by simulation, the distribution of idle periods in the basic Flow in the Middle (FIM) scenario (Fig. 1) [4]. The scenario consists of six nodes and three flows. The nodes $(0, 1, 2, 3)$ are in the same sensing range,

[1]Note that other mechanisms exist to determine whether the medium is idle; this is called the Clear Channel Assessment process.

Propagation model	2-Ray Ground
SIFS	10 μs
DIFS	50 μs
Backoff time slot	20 μs
Transmision range	200 m
Carrier sensing range	250 m([2])
Physical rate	2 Mb/s
Flows	TCP
Packet size	1 kbyte
Retransmision limit	7
Contention window size (min, max)	32, 1024
Routing Protocol	ASR([3])
Simulation time	100 s

Table 1: Nodes configuration with Ns2.33

like the nodes $(2, 3, 4, 5)$, while the nodes $(0, 1)$ are hidden to the nodes $(4, 5)$ (and vice versa). In the simulations carried out with Ns2.33, all nodes are configured according to the parameters given in Table 1.

The distribution of idle periods of nodes are depicted in Figs. 2 and 3. In Fig. 2, the interior figure is a focus on the part of the distribution around 5000 μs. From these figures, we find that the nodes $(0, 1, 4, 5)$ have identical bi-modal distributions and that Nodes 2 and 3 have also the same mono-modal distribution.

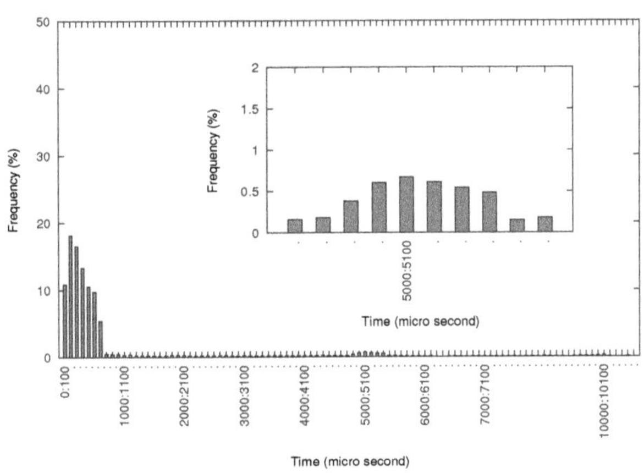

Figure 2: Distribution of idle period durations for the nodes $(0, 1, 4, 5)$ in the FIM scenario

We identify three main kinds of idle periods in this scenario:

- *Entire backoff time (denoted as B):* In IEEE 802.11 DCF, a random backoff time is always associated to a packet transmission. If the backoff counter is not interrupted, then, for each backoff stage, the backoff time is uniformly distributed in the associated contention window. In the tested FIM scenario, collisions are only possible if the backoff counters of two (or more) sources in the same sensing range reach zero at the same time. This happens very rarely. Therefore, the sources mainly draw their backoff in the minimum contention window. We notice that the first peak of each distribution[4] is mostly included in the interval $[0 : 620\mu s]$ which corresponds to the minimum contention window size. Thus, this peak represents backoff time and idle periods corresponding to entire backoff times are included in this first peak. However, the first peak is not uniform, which means that the nodes' backoff counter is sometimes interrupted.

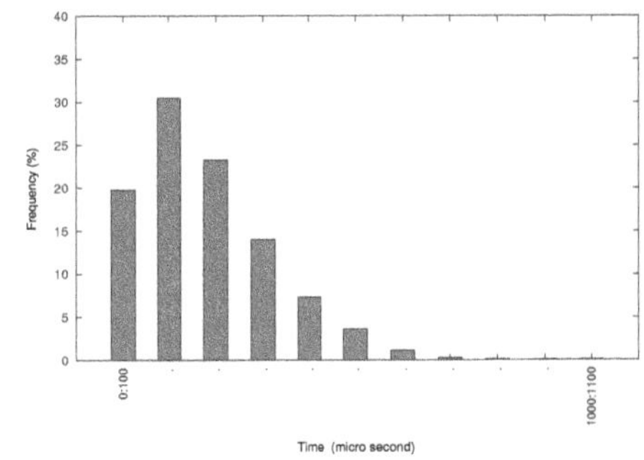

Figure 3: Distribution of idle period durations for the nodes $(2, 3)$ in the FIM scenario

- *Interrupted backoff time (denoted as $\breve{B}$)*: When the backoff counter of a node is interrupted, the backoff time is separated into several smaller periods. These idle periods are also included in the first peak of the distribution of the six nodes. Depending on their neighbors' activity in terms of packets to send, the nodes have different frequencies of backoff interruptions. For instance, Node 2 is impacted by the packet transmissions of five nodes $(0, 1, 3, 4, 5)$, while Node 0 is only impacted by the packet transmissions of three nodes $(1, 2, 3)$. Therefore, the frequency of backoff interruptions of Node 2 is greater than the one of Node 0, which explains that Node 2 has smaller idle periods in its distribution.

- *Backoff and interruption time (denoted as $\widehat{B}$)*: In the tested FIM scenario with TCP flows, nodes have sometimes no packet to send at the MAC level while neighbors in the same sensing range have packets to transmit. However, these nodes sense the medium idle during a period larger than a backoff. This fact can be explained by the transmissions of nodes that are hidden to nodes that have no packet to send and that freeze the backoff counter of the neighbor nodes. For instance, in the FIM scenario, when the backoff counter of Node 2 is interrupted by a packet transmission of Node 4, the backoff of Node 2 is frozen during a packet transmission. During this period, Node 0, that has no packet to send, considers that the channel is idle during a period larger than a backoff. Such a phenomenon is depicted on Figure 4. The second peak of the idle period distribution for the nodes $(0, 1, 4, 5)$ corresponds to this kind of idle period. As with TCP

[2]This range may be low compared to reality, but with this value, we easily simulate hidden nodes.

[3]To ensure that no routing control packet is sent and competes with data packets, we use a static routing protocol (called ASR). It simplifies the analysis but not the shape of the distribution.

[4]that is the single peak for Nodes 2 and 3.

flows, the probability that the packets interarrival time on the MAC level of a node is greater than zero is non-negligible, it explains why the idle periods due to backoff and interruptions of hidden nodes exist in this scenario even if their frequency is small.

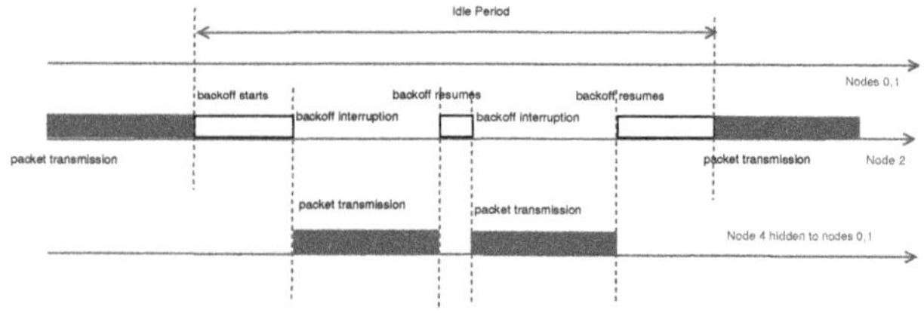

Figure 4: Idle period due to backoff and interruptions from hidden nodes

From the previous observations, we see that three kinds of idle period impact the idle period distribution of nodes in the FIM scenario: the entire backoff times, the interrupted backoff times and the backoff and interruption times from hidden nodes. The entire backoff time depends on the backoff stage in which the backoff is drawn. The backoff stage depends on the collision probability. Therefore, the collision probability impacts the width of the peaks of the distribution. The interrupted backoff depends on the activity of the neighbor nodes in the same sensing range. Therefore, the packet transmission rate of neighbor nodes impacts the height of the peaks of the distribution. The backoff and interruption time depends on the number of hidden nodes and their activity. Therefore, the number of hidden nodes impact the number of peaks in the distribution and the packet transmission rate of hidden nodes impact the width of the distribution.

4. ANALYTICAL MODEL

In this section, we propose an analytical model, based on the three kinds of idle periods identified previously, in order to compute the distribution of idle periods of a node in an arbitrary multihop topology.

4.1 Model assumptions

In this paper, we assume that the traffic pattern transmitted by the nodes in the multihop wireless network is stochastic with a non-negligible traffic intensity. With this assumption, the probability that a node has packets to send is non-negligible and the probability that the wireless channel be saturated is also non-negligible. We also assume that the hidden stations (as defined in Section 4.2) of a same node have independent activities.

4.2 System Model

We model the wireless channel of a given node in a multihop wireless network as follows. Given a node A, the wireless channel of A is impacted by the nodes which are in the same sensing range, like B, C and D (see Fig. 5; the dotted circle represents the carrier sensing area of A). The wireless channel of A is also impacted by hidden nodes of A which are out of its sensing range (like E, F and G in Fig. 5), but which impact the behavior of nodes in the carrier sensing area of A. Note that, here, the term hidden nodes has a large meaning. Indeed, it corresponds to nodes that can not be sensed by Node A but these nodes' activity may provoke collisions on packets transmitted in the carrier sensing area of A or may prevent a node in the carrier sensing area of A from transmitting.

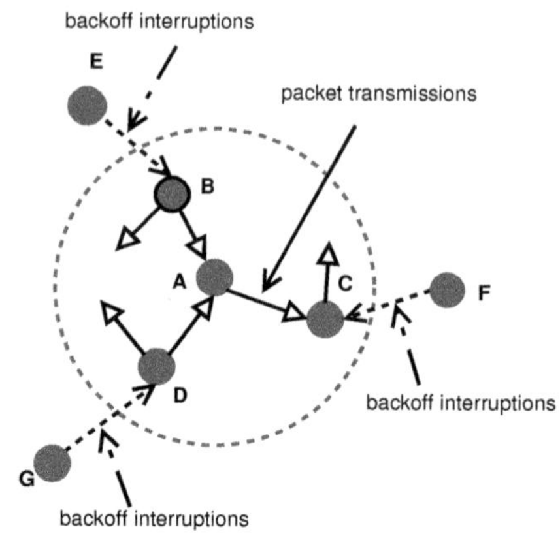

Figure 5: System model for the wireless channel of a node

These impacts are represented by the following parameters:

- *Packet transmission rate* (λ)*:* This is the average packet output rate at the MAC level from all the nodes which are in the sensing range of A (including A).
- *Hidden transmission rate* (λ_h)*:* This is the average packet output rate at the MAC level from hidden nodes of A. These hidden nodes are in the sensing range of active neighbors of A.
- *Collision probability* (p)*:* This is the average collision probability for a packet transmitted in the carrier sensing area of A.

4.3 Distribution of entire backoff times

First, we determine the idle periods of Node A due to entire (not interrupted) backoff drawn by A or by nodes in the carrier sensing area of A. Given a contention window, the drawn backoff is uniformly distributed in this window. Once a packet has collided, it will be retransmitted with a contention window whose size is doubled. The backoff time depends therefore on the number of retransmissions.

Given the collision probability p, the probability mass function for the number of retransmissions per packet [1] is:

$$Pr(n = i) = \begin{cases} p^i * (1-p), & 0 \le i < M \\ p^M, & i = M \end{cases} \tag{1}$$

where M is the retransmission limit.

Let's define N such that $2^N * CW_{min}$ is the maximal contention window size (CW_{max}). For a value t (expressed in number of slots), we denote $m = \left\lfloor log_2(\frac{t}{CW_{min}}) \right\rfloor$. The probability that the backoff is equal to t can then be expressed as follows:

- *If* $N < M-1$:

$$P(B=t) = \begin{cases} \sum_{i=m}^{N} Pr(n=i) * \frac{1}{2^i * CW_{min}}, & 0 < m \le N \\ \sum_{i=m}^{N} Pr(n=i) * \frac{1}{2^i * CW_{min}} + \frac{m-N}{2^N * CW_{min}}, & \\ N < m \le M-1 & \\ 0, otherwise & \end{cases} \quad (2)$$

- *If* $M-1 \le N$:

$$P(B=t) = \begin{cases} \sum_{i=m}^{M-1} Pr(n=i) * \frac{1}{2^i * CW_{min}}, \\ 0 \le m \le M-1 \\ 0, \ otherwise \end{cases} \quad (3)$$

4.4 Distribution of interrupted backoff times

In this section, we determine the idle periods of Node A that correspond to backoff interrupted by the transmissions of A or by nodes in the carrier sensing area of A. In [9], the authors show that the interrupted backoff times (including also entire backoff times) follow a Gamma distribution. The probability that the interrupted backoff time is equal to t (expressed in number of slots) can be expressed as follows:

$$Pr(\breve{B}=t) = \frac{\beta^\alpha}{\Gamma(\alpha)} t^{\alpha-1} e^{-\beta t} \quad (4)$$

Given the mean $E(\breve{B})$ and variance $Var(\breve{B})$, the parameters α and β can be approximated as follows:

$$\beta = \frac{E(\breve{B})}{Var(\breve{B})} \quad (5)$$

$$\alpha = E(\breve{B})\beta \quad (6)$$

Note that, the interrupted backoff times are caused by the packet transmission rate (λ) and the sizes of the associated contention windows to transmit these packets (CW). Both $E(\breve{B})$ and $Var(\breve{B})$ can therefore be calculated from λ and CW: $E(\breve{B})$ can be estimated as $\frac{1}{\lambda}$ and $Var(\breve{B})$ can then be derived from $E(\breve{B})$ and CW by using the formulas in [6].

4.5 Distribution of backoff and interruption times

In this part, we determine the idle periods of Node A larger than a backoff and that are due to an empty queue at the MAC level for Node A and transmissions of nodes hidden to A that prevent nodes in the carrier sensing area of A from transmitting. Since we assume that hidden nodes of a same node have independent activities, then we consider that the packet arrival from these hidden nodes follows a Poisson distribution. If there are k arrivals during an entire backoff time B, then the interruption time due to packet transmissions can be expressed as follows:

$$S^k = k * T \quad (7)$$

where T is the packet transmission time. The values of S^k are discrete and multiple of T.

The probability that S^k is equal to t (expressed in number of slots) can be estimated as follows:

$$Pr(S^k = t) = (\lambda_h * T)^k * \frac{e^{-\lambda_h * T}}{k!} * Pr(k*T = t) \quad (8)$$

The idle periods due to backoff and interruptions from hidden nodes can be expressed as follows:

$$\widehat{B} = B + S^k \quad (9)$$

Given the distributions of B and S^k, we can derive the distribution $\widehat{B}$ by the convolution approach.

Since the difference between two consecutive values of B is much smaller than the difference between two consecutive values of S^k, the distribution of $\widehat{B}$ is multimodal. Moreover, the difference between two consecutive peaks is equal to a packet transmission time T.

4.6 General distribution of idle periods

Finally, we determine the general distribution for the idle periods of Node A. Given the output rate λ in the wireless channel of Node A and the packet transmission time T, the offered load of the wireless channel of Node A is $\rho = \lambda * T$. As a packet transmission is always associated to a backoff time, the probability that the idle periods of Node A corresponds to entire and interrupted backoff times is then ρ. The probability of other kinds of idle periods is then $1-\rho$. Therefore, we estimate the distribution of idle periods (denoted IP) of Node A as follows:

$$Pr(IP=t) = \rho * Pr(\breve{B}=t) + (1-\rho) * Pr(\widehat{B}=t) \quad (10)$$

$(1-\rho) * Pr(\widehat{B}=t)$ corresponds to the probability that nodes in the carrier sensing range of A be blocked by transmissions of hidden nodes of A and that A has no packet to send.

4.7 Discussion

The distribution of idle periods given in Equation 10 reflects what we have analyzed with the FIM scenario (Section 3). The contention window size determines the width of the distributions of $\breve{B}$ and B and impacts the width of the distribution $\widehat{B}$. Since there is only one part in the distribution of $\breve{B}$, the number of peaks of the general idle period distribution is the number of peaks in the distribution of $\widehat{B}$ plus one. These minor peaks are due to hidden nodes.

Since we consider networks under non-negligible traffic intensity, ρ is rather high. The height of the distribution of $\breve{B}$ will determine the height of the general idle period distribution.

5. MODEL VALIDATION

In this section, we validate our proposed model by simulations with Ns2.33.

5.1 Topologies

We generate two random topologies on an area of $(1000m, 1000m)$. The first one (called 20/47 topology) consists of 20 nodes and 47 single-hop TCP flows. The second one (called 100/176 topology) includes 100 nodes and 176 single-hop TCP flows. The nodes are now configured with more realistic communication range ($250m$) and sensing range ($550m$) (all the other parameters of Ns2.33 are identical to the ones given in Table 1).

Then, we randomly select two nodes in each topology and we compare, for each node, the idle period distribution obtained by simulation and the one computed with our analytical model. Note that, with our model, we use simulation results to compute λ, λ_h, p, $E(\breve{B})$ and $Var(\breve{B})$ required for the computation of $Pr(\breve{B} = t)$, rather than general formulas.

5.2 Simulation Results

The distributions of idle periods of Nodes 0 $(39m, 546m)$ and 5 $(570m, 525m)$ of the 20/47 topology are given in Figs. 6 and 7. The interior figures show the different parts of the distribution other than the first peak.

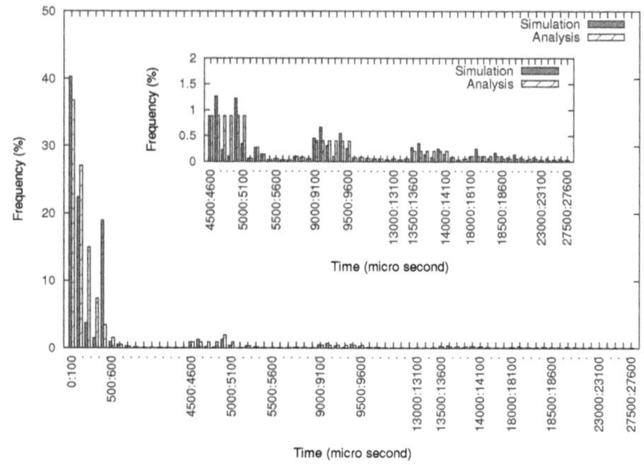

Figure 6: Distribution of idle periods of Node 0 in the 20/47 topology

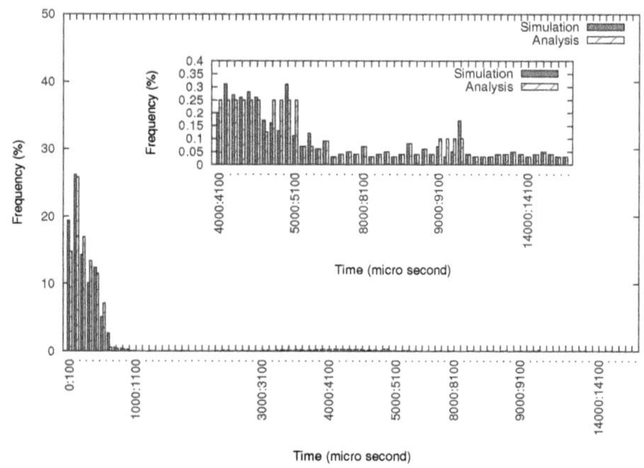

Figure 7: Distribution of idle periods of Node 5 in 20/47 topology

First, we see that the results computed from our proposed model are close to the ones obtained by simulation. From these two figures, we find that the two nodes have the first part of their distribution in the range ([0 : 600] μs approximately). It corresponds to backoff times (interrupted or not). This part represents more than 90% of the idle periods. The same kind of result is observed on the other nodes of the network.

However, these distributions have other parts that are smaller (in terms of frequency) than the first part. When we plot them in another scale, we clearly see that the distributions are multimodal. The peaks of these distributions are multiple of $4500\mu s$ which is the transmission time for a packet of 1 kbyte transmitted at 2 Mb/s. These parts correspond to idle periods due to backoff and interruptions of hidden nodes. There are five peaks in the distribution of Node 0 (at $4500\mu s$, $9000\mu s$, $13500\mu s$, $18000\mu s$, $22500\mu s$), while there are only three peaks in the distribution of Node 5 (at $4500\mu s$, $9000\mu s$, $13500\mu s$).

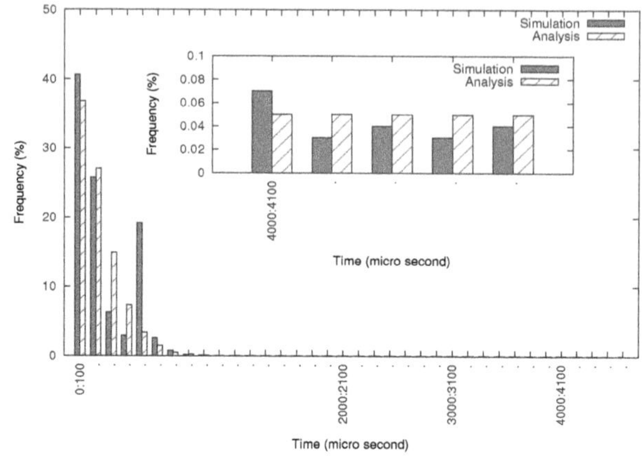

Figure 8: Distribution of idle periods of Node 61 in the 100/176 topology

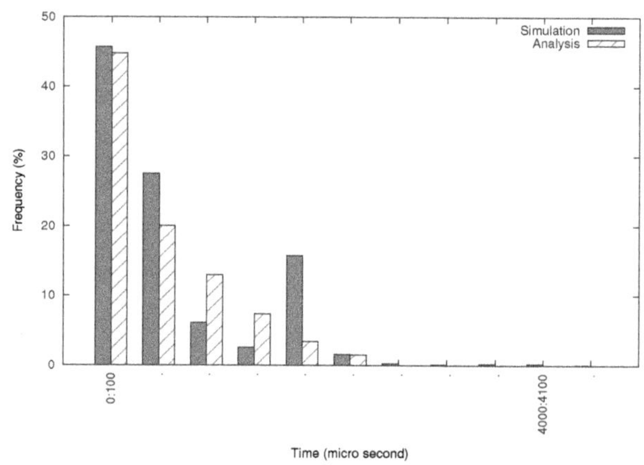

Figure 9: Distribution of idle periods of Node 50 in 100/176 topology

Simulation results for the idle period distributions of Nodes 61 $(64m, 756m)$ and 50 $(180m, 726m)$ of the 100/76 topology are given in Figs. 8 and 9. In the distribution of Node 50, there is only one peak that corresponds to backoff interruptions, while there are two parts in the distribution of Node 61. As the 100/176 topology is very loaded, the frequency of idle periods due to backoff and interruptions of hidden nodes is smaller than with the 20/47 topology.

These differents results, observed on the selected nodes, but also on the other nodes, show that the first parts of the distributions are relatively similar between the nodes, while the other parts may be very different from one node to another. These parts are correlated to the number of hidden nodes and the way these hidden nodes are located in the topology (relative to the neighbors and to the other hidden nodes). This can be very different from node to node and from topology to topology.

6. APPLICATION: IDLE PERIOD-BASED ADAPTIVE BACKOFF SCHEME (IABS)

In this section, we discuss a possible use of idle periods and their distribution in multihop wireless networks under non-negligible traffic intensity. We propose a simple backoff algorithm, called idle period-based adaptive backoff scheme (iABS), to solve the unfairness phenomenon that arises in multihop wireless networks. This scheme uses the backoff and interruption times from hidden nodes observed on minor peaks of the idle period distribution and the neighbors' activity.

We define CX, the contention index of a node as:

$$CX = \left\lfloor \frac{\widehat{B}_{sum}}{L_{arrival}} \right\rfloor \tag{11}$$

where $\widehat{B}_{sum}$ is the total backoff and interruption times (expressed in the number of slots) of the node during an observation period (and extracted from the secondary peaks in the idle period distribution) and $L_{arrival}$ is the total data packet arrivals on the radio medium from neighboring nodes. CX is close to zero if the hidden nodes' activity is much less important than the neighbors' activity. CX is large if the hidden nodes' activity is much more important than the neighbors' activity.

iABS operates as follows: upon reception of an acknowledgement, a node modifies its minimum contention window CW_{min} for the new packet to send as follows:

$$CW_{min}^{new} = CW_{min}^{original} + \omega * CX^{current} \tag{12}$$

where ω is a scaling factor which is constant for each topology, $CW_{min}^{original}$ is the defaut value of CW_{min} and $CX^{current}$ is the contention index of the node during the current observation period. Note that, at the end of an observation period, we set CW_{min} to its default value $CW_{min}^{original}$. It ensures that CW_{min} does not increase to a large value through time.

With this algorithm, the backoff time of a node is increased according to its contention index. If its contention index is high, it means that hidden nodes have much activity than its neighbors. Therefore, the minimal contention window size is increased in order to reduce the channel access probability of the node and to increase the one of its neighbors. If its contention index is low, it means that its neighbors have much activity. Therefore, the minimal contention window size is decreased in order to increase the channel access probability of the node and to decrease the one of its neighbors. The goal of the adjustment of the minimal contention window size is to make the contention indexes of the nodes in competition in the wireless channel access converge to a same value.

We have evaluated iABS with Ns2.33. The communication and sensing ranges of the nodes are $250m$ and $550m$, respectively (the other parameters are given in Table 1). Fig. 10 shows the results obtained on the FIM scenario with and without iABS. We see that, with iABS, the flow rates converge quickly to the same rate and that the unfairness issue raised by this topology is solved.

We have also tested iABS on three random topologies which are generated in an area of $(1000m, 1000m)$. The first topology consists of 10 nodes and 14 TCP one-hop flows (and is denoted 10/14 topology). The second one consists of 30 nodes and 37 TCP one-hop flows, while the third topology

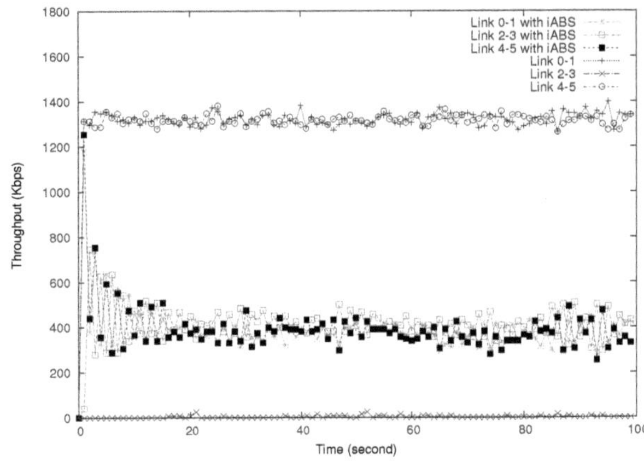

Figure 10: iABS for the FIM scenario

includes 50 nodes and 97 TCP one-hop flows. In the 50/97 topology, we randomly select three TCP flows: Flow 0 (from Node 0 ($811m$, $446m$) to Node 24 ($828m$, $489m$)), Flow 50 (from Node 3 ($390m$, $377m$) to Node 49 ($349m$, $396m$)) and Flow 90 (from Node 49 to Node 9 ($158m$, $426m$)). Fig. 11 shows the rate of these three flows without and with the use of iABS. We see that the use of iABS provides fairer results than 802.11 DCF. This observation is also confirmed by the results provided in Fig. 12 that shows the Jain fairness index for the three tested topologies. An index close to 0 indicates an unfair situation, while an index close to 1 means that flows obtained similar throughputs. From this figure, we see that iABS reduces the unfairness issue on the three topologies, specifically on the 30/37 and 50/97 topologies.

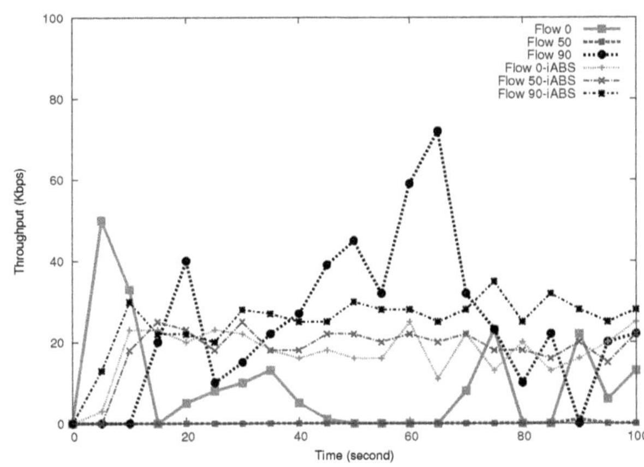

Figure 11: The rates of flows with the use of iABS in the 50/97 topology

On these preliminary tests, we can conclude that iABS provides very efficient results in terms of fairness without message overhead since iABS uses only the idle period distribution (and more precisely the secondary peaks induced by hidden nodes) and the neighbors' activity.

7. CONCLUSION

In this paper, we propose a novel analytical model for the distribution of idle periods of nodes in relatively saturated 802.11-based multihop wireless networks. Most of

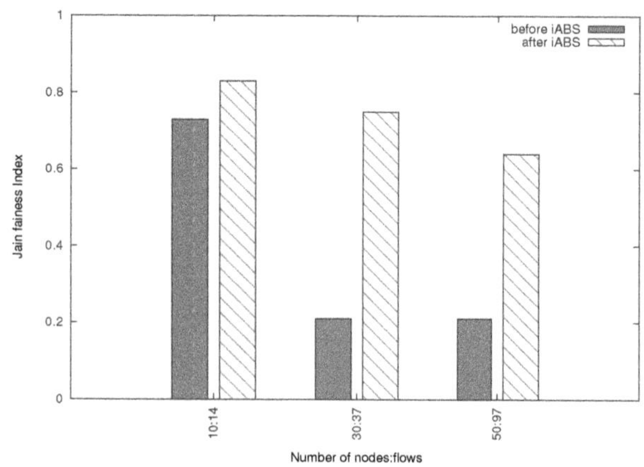

Figure 12: Jain fairness index obtained on the three topologies with and without iABS

propoposed models in literature mainly consider common ilde periods of the nodes such as the backoff times. Our model takes into account longer idle periods, called backoff and interruption times and induced by hidden nodes. This kind of idle periods can be very different from one node to another. The distribution of idle periods, provided by our model, is therefore characterized by the number of peaks rather than the frequency of idle periods. The validation in random and large topologies shows that our analytical model matches well the simulation results in Ns2.33.

Based on the analytical model and its analysis, we also propose a simple and efficient adaptive backoff algorithm to solve the contention unfairness in these networks. This algorithm relies on local information and has no communication overhead. It enables an equalization not only of the neighbors' activities but also those of hidden nodes. The Jain fairness index obtained from simulations in Ns2.33 in random and large topologies proves the efficiency and the simplicity of the algorithm.

The proposed analytical model is extensible for different packet sizes and for unsaturated wireless networks. The different packet sizes may impact the distance between the peaks. Moreover, they also imply changes into the packet collisions and therefore the width of the peaks in the distribution. In unsaturared networks, there may exist other kinds of idle periods such as the residual interarrival times. If the interarrival times are continuous then there is no more new peak in the distribution. It may be more difficult to distinguish the peaks. However, the general form of the distribution remains unchanged. It is also possible to take into account errors from the physical layer and/or the different modulations by adjusting the dynamic paramerers used in our model.

In future work, we plan to use the propeties of the idle period distribution characterized by the proposed analytical model in some solutions, like, for instance, the available bandwidth estimation. For instance, neighbor nodes could exchange the number of peaks of their distribution, rather than the whole distribution parameters, and each node could try to infer the multihop topology (like the number of hidden nodes) in which it is located based on this information. Such exchanges are easy to do compared to exchanges of the whole distribution and reduce the communication overhead. We also plan to compare the performance of our proposed solution iABS with existing algorithms.

Acknowledgments

This work is partially funded by the French National Research Agency (ANR) under the project ANR VERSO RESCUE (ANR-10-VERS-003)

8. REFERENCES

[1] G. Bianchi. Performance analysis of the ieee 802.11 distributed coordination function. *Selected Areas in Communications, IEEE Journal on*, 18(3):535–547, 2000.

[2] R. Bowden and A. Coyle. Idle times analysis in a csma/ca network. In *Proceedings of the 15th IEEE International Conference on Networks, ICON 2007, 19-21 November 2007, Adelaide, Australia*, pages 443–448. IEEE, 2007.

[3] R. Bowden and A. Coyle. Idle times analysis in a csma/ca network. In *ICON'07*, pages 443–448, 2007.

[4] C. Chaudet, I. Guérin Lassous, E. Thierry, and B. Gaujal. Study of the impact of asymmetry and carrier sense mechanism in IEEE 802.11 multi-hops networks through a basic case. In *ACM Workshop on Performance Evaluation of Wireless Ad Hoc, Sensor, and Ubiquitous Networks (PE-WASUN'04),*, Venice, Italy, Oct. 2004.

[5] M. Garetto, T. Salonidis, and E. W. Knightly. Modeling per-flow throughput and capturing starvation in csma multi-hop wireless networks. In *In Proc. of IEEE Infocom*, 2006.

[6] S. Hozo, B. Djulbegovic, and I. Hozo. Estimating the mean and variance from the median, range, and the size of a sample. *BMC Medical Research Methodology*, 5(1):13+, 2005.

[7] V. Kolar, K. Bharath, N. B. Abu-Ghazaleh, and J. Riihijarvi. Contention in multi-hop wireless networks: model and fairness analysis. In *Proceedings of the 12th ACM international conference on Modeling, analysis and simulation of wireless and mobile systems*, MSWiM '09, pages 21–29, New York, NY, USA, 2009. ACM.

[8] L.S.Committee. *ANSI/IEEE Std 802.11:Wireless LAN Medium Access Control(MAC) and Physical Layer (PHY) Specifications.* IEEE Computer Society, 1999.

[9] J. V. Misic and V. B. Misic. Characterization of idle periods in ieee 802.11e networks. In *WCNC*, pages 1004–1009. IEEE, 2011.

[10] C. Sarr, C. Chaudet, G. Chelius, and I. Guérin Lassous. Bandwidth Estimation for IEEE 802.11-Based Ad Hoc Networks. *IEEE Transactions on Mobile Computing*, 7(10):1228–1241, 2008.

[11] O. Tickoo and B. Sikdar. On the impact of ieee 802.11 mac on traffic characteristics. *IEEE Journal on Selected Areas in Communications*, 21(2):189–203, 2003.

Modelling and Analysis of the Split Phase Multi-Channel MAC Approach

Abdelaziz El Fatni
University of Toulouse, INSA
135 Avenue de Rangueil
31400 Toulouse, France
abdelaziz.elfatni@insa-toulouse.fr

ABSTRACT

The throughput of wireless networks can be improved significantly by using multiple channels communications in MAC protocols. However, due to the multi-channel environment, a novel issue must be addressed in MAC protocols which is called the channel assignment. Such mechanism decides which channel can be used for data transmission. There are many variations for the channel assignment which have been defined and studied in the literature. In this paper we propose a formal modelling and analysis of the multi-channel MAC protocols based on the Split Phase Approach (SPA) using Stochastic Timed Petri Net (STPN). We analyze qualitatively and quantitatively this approach. Firstly, we verify the properties inherent to the typical characteristics of the split phase approach. Secondly, we analyze quantitatively the benefit of this approach in terms of the network throughput with respect to a single channel MAC approach.

Categories and Subject Descriptors

B.4.4 [**Performance Analysis and Design Aids**]: *Formal models, Verification*; C.2.5 [**Computer communication Networks**]: *Local and Wide-Area Networks– Access schemes*; C.4 [**Performances of Systems**]: *Modeling techniques.*

General Terms

Algorithms, Design, Performance, Verification.

Keywords

Wireless networks, Multi-channel MAC protocols, Formal Modelling, Petri Net, Verification, Quantitative analysis.

1. INTRODUCTION

Using multiple channels in MAC protocols can significantly improve the network throughput compared with single channel MAC protocols since multiple transmissions can take place simultaneously without interfering. However, due to the multi-channel environment, a novel issue must be addressed in MAC protocols which is called the channel assignment mechanism in order to coordinate the channels access for data transmission.

Several multi-channel MAC protocols have been proposed which differ in how they assign channels to nodes for data transfer. Mo et al [11] classified multi-channel MAC protocols according to the channel assignment into two categories: single rendezvous and parallel rendezvous. In the first case, the exchange of control information occurs on a single channel, even though the rendezvous channel may change over the time. Three types of multi-channel MAC protocols are distinguished in this category: common hopping [3, 4], dedicated control channel [5, 6], and split phase [7, 8, 9, 10]. In the second category, multiple MAC entities can use different channels in parallel to exchange control information and make agreements simultaneously on distinct channels. Examples of this category include parallel hopping approach [1, 2].

The channel assignment mechanism plays a fundamental role in the efficiency of multi-channel MAC protocols. In fact, this mechanism should provide a better way to allocate the wireless channel resources to nodes. However, the nature distributed of the channel allocation (no central control), and the characteristics of wireless channels (air propagation time, collision detection) make the design of this mechanism more sophisticated. In the literature, most studies of these protocols have been evaluated mainly by simulation [1-10], or by approximate analytical approach [11, 12]. Although these studies work well in general, they do not allow to represent explicitly the different aspects of the channel assignment mechanism and also do not show how this mechanism works in terms of synchronization, concurrent and parallel events between nodes (concurrent channel reservations, accesses to shared resources (channels) in a distributed way, etc).

Another alternative to study these protocols is to use formal models and analyzing techniques which are able to do high level descriptions of systems and allow to perform both verification of system's properties and performance evaluation. In this paper we propose a formal STPN model and analysis of the multi-channel MAC protocols based on the split phase approach which requires one radio. We verify the properties inherent to the characteristics of the split phase approach and we analyze quantitatively the benefit in terms of the network throughput with respect to a single channel MAC approach. The paper is organized as follows. In the second section we present the context of the study. The third section describes the formal modelling of the split phase approach. The fourth and fifth sections present respectively the qualitative and the quantitative analysis. The sixth section concludes the paper.

2. CONTEXT OF STUDY

2.1 Split Phase Approach (SPA)

In this approach [7-10], the time is divided into cycles composed of two phases: the first one is a control phase; the second one is a data phase. During the control phase, all nodes tune to the default channel (default channel is one of the available channels which is considered as the control channel during this phase), where they must listen in order to maintain a channel usage list, and attempt to negotiate a data channel to be used during the data phase. In fact, when a node has a data packet to send it contends for the control channel to exchange control packets based on the channel usage list with the intended receiver. If a common channel is free the sender and the receiver wait for the start of the data phase, and switch to the agreed channel to transmit data. Note that when the number of the agreements allowed during the control phase is not limited by the number of the channels, multiple pairs of nodes can reserve the same channel during the same data phase. Consequently, during this phase, the nodes may either compete following the four way handshake (RTS/CTS/DATA/ACK) [8] as in IEEE 802.11 [15], or may use a scheduled access [7, 9]. In this approach, the fundamental property is the channel load balancing which is essential to improve the total network throughput. To achieve this goal, the channel assignment procedure must attempt to balance the channel load as much as possible. Figure 1 represents a scenario which illustrates the split phase approach where the data transmissions during the data phase are scheduled according to the channel load balancing policy. We consider three non-overlapping channels (Ch_1, Ch_2, and Ch_3) where Ch_1 is the control channel, and eight nodes (A,B,C,D,E,F,G,H) within the communication range of each other. We suppose that A, C, E and G have one data packet to send respectively to B, D, F and H, and the length of data packets (ld) is defined such that: $ld(E) > ld(A) > ld(C)$. During the control phase, we still suppose that the order of reservation is A, C, E, G and that A, C, E have reserved respectively Ch_1, Ch_2, Ch_3. Consequently, by considering the availability degree, G reserves Ch_2 which has the best availability. In the data phase, A, C and E start their transmission on respectively Ch_1, Ch_2, and Ch_3. When C ends to transmit, G initiates the data transfer in channel Ch_2.

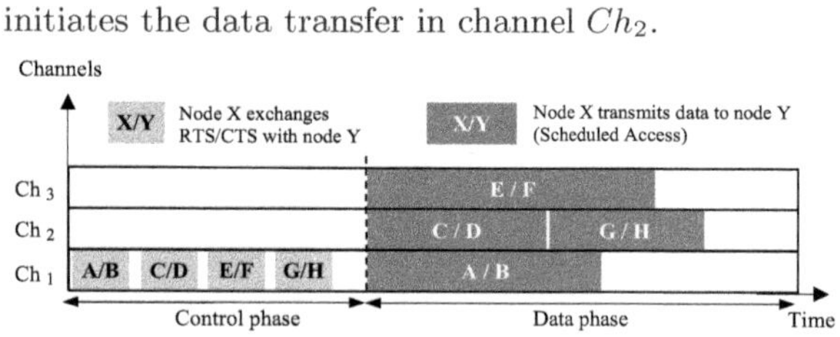

Figure 1: Illustration of the split phase approach

2.2 Formal models and analysis

In recent years, there is a vast literature devoted to formal analysis for single channel wireless communication MAC protocols in terms of verification techniques for analyzing correctness and / or quantitative performance. However, there are few formal studies for multi-channel MAC protocols. To the best of our knowledge, there is no prior work defining a formal specification with both verification and performance activities to study the split phase multi-channel MAC protocols. In what follows, we cite two works relevant to our study.

In [13], we proposed a formal representation and verification of the common hopping multi-channel MAC approach using STPN model. In this approach the channel allocation is independent of the channel state (no negotiation process is needed between nodes to select channels) and the data transfer takes place immediately after a successful control exchange, i.e., no data transfer policy is needed. In this work, we do not address performance issues. Recently, the formal model SRN (Stochastic Reward Net) has been used for the modelling of the bidirectional multi-channel MAC protocol [14] concerning two channel selection policies: a random channel selection and a fast channel-first selection. This work focuses specifically on performance evaluation, and the channel assignment is represented by an abstract view which does not reflect the underlying characteristics of this mechanism. Our goal in this paper is precisely to combine both qualitative analysis and quantitative performances.

2.3 STPN model

An STPN model [17-19] is an extension of the Petri Net model where we associate to each transition a firing time and a Probability Density Function (PDF). Many PDFs can be considered. In this paper, we consider two types of PDFs: (i) an exponential distribution; (ii) a deterministic distribution with two kinds of time memory: enabling memory and age memory [18]. When the firing time is equal to 0, the transition is called an immediate transition and it is fired as soon as it is enabled. Priorities (natural numbers) are associated to immediate transitions in order to have a deterministic behaviour when these transitions are in competition. So, an immediate transition is enabled if no higher priority transition is enabled. The STPN dynamic behaviour is represented by the RSG (Randomized State Graph) which has a qualitative component (transition T_i) and quantitative components (the firing probability and the firing date of the transition T_i). The analysis are done from the RSG. Here, we consider two types of analysis: (i) The qualitative analysis are done by considering only the qualitative components. We can verify the general properties (reachability, liveness, deadlock ...) and the specific properties in terms of relations between events of the protocols that we want to observe (observable events) by using a projection method based on the observational equivalence relation [20]. (ii) The quantitative analysis is done by considering the quantitative components in order to evaluate performance parameters such as the throughput.

3. FORMAL MODELLING OF THE SPLIT PHASE APPROACH

3.1 Assumptions and methodology

Split phase protocols developed in [7, 8, 9] differ in how to handle the channel access during the data phase (contention / scheduling) and also according to the considered wireless context (single-hop or multi-hop). In order to realize a common specification of this approach and also to keep the model as clear as possible, we made a common assumption: A3 (below). The rest of the assumptions are as follow:

- A1: n non-overlapping channels are available numbered as follow: $1, 2, ..., n$; and assumed to be errors free.
- A2: The length of the control phase (lcp) and the length of the data phase (ldp) are considered constant. All nodes maintain synchronized clocks.
- A3: We consider a single domain collision, and the data phase is organized into equal time slots where the length of a time slot is equal to the length of the data packet (ld).
- A4: For simplicity, we consider three time slots for the data phase ($ldp = 3ld$). Hence, we define three degrees of availability for a channel used in the data phase: High (H) i.e. not yet reserved; Medium (M) i.e. the first slot is reserved; Low (L) i.e. the second slot is reserved. When Low, the channel can be still reserved once (third slot). Then it becomes Non available (N).
- A5: Each node maintains a channel availability list where each entry comprises the channel number and its state (its degree of availability): ($<Ch_i$, $S_i>$), where $i \in \{1, ..., n\}$ and $S_i \in \{H, M, L, N\}$.

This approach comprises three main components (figure 2): the clock, the channel assignment process (preselection and reservation), and the data transfer on the data channel reserved.
1. Clock: its represents the cycle structuring in terms of the duration of the control phase and the data phase. There is one clock per node.
2. Channel assignment (preselection and reservation): it concerns each node having a data packet to send (request) and it consists of two phases: (i) The node preselects an available channel based on the channel information available locally; (ii) And then it can undertake the reservation phase which will be done through a dialogue on the control channel with the receiver. Such dialogue must not exceed the end of the control phase. Therefore, before starting, the node must check the two following conditions:
Cond1: there is at least one available channel ; i.e.:

$$\forall i, \exists S_i : S_i \in \{H, M, L\} \tag{1}$$

Cond2: the remaining time from the arrival instant of a request until the end of the control phase is at least equal to the minimum duration of the control exchange (noted lc). Let us define t_{cp}, $t_{arrival}$ and $t_{ex-limit}$ (exchange limit) three instants which represent respectively the start of a cycle, the request arrival of a data packet, and the last instant from which the node can still initiate a dialogue: $t_{ex-limit} = t_{cp} + lcp - lc$ (see figure 2). The arrival instant must be so that:

$$t_{cp} \leq t_{arrival} \leq t_{ex-limit} \tag{2}$$

If the two conditions (1) and (2) are satisfied, the node starts the channel assignment process (preselection and reservation phases). Otherwise, the node must wait until the next cycle.

2.a) *Preselection:* in order to achieve a better channel load balancing with respect to the assumption A4, the node preselects the channel with highest availability (H has higher availability than M which has a higher availability than L). When two or more channels are in the same state, the preselection is done in ascending order of the channel numbers in order to have a deterministic choice.
2.b) *Reservation:* the node initiates the RTS/CTS (Request To Send / Clear To Send) exchange in the control channel in order to reserve the data channel preselected. This phase is slightly different to the classical RTS/CTS exchange defined in IEEE 802.11 [15], since this dialogue contains a supplementary information about the channel will be used for data transfer. There are many important mechanisms to model such as: the collision detection (caused by simultaneous request), the cancellation of the preselection, the backoff mechanism and the successful reservation of a data channel.
3. Data transfer: with respect to the assumptions A3 and A4, we consider that the channel access is based on the FIFO scheduling policy which depends on the order of the reservation requests and the channel availability degree.

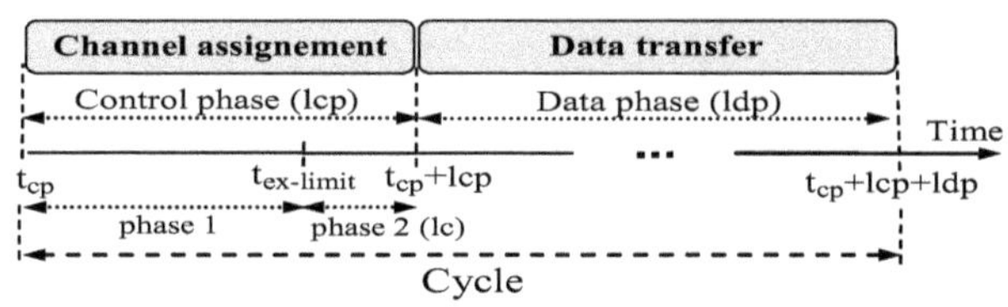

Figure 2: The main components of the split phase approach

3.2 A detailed model

For graph clarity reasons, we adopt the notation presented in the figure 3. The detailed model of the MAC entity of each node includes four parts represented on the figures 4, 5, 6 and 7. Table 1 presents the priorities of immediate transitions. The model of each node is connected with the models of the other nodes by means of the global places "Channels: Ch_i" and "Channel States: CS_i where $i \in \{1, ...n\}$".

3.2.1 *Clock, request arrival and the start of the channel assignment / the data transfer model*

The model comprises four submodels (SM): SM41, SM42 SM43 and SM44 which model respectively the clock, the request arrival, the start of the channel assignment and the start of the data transfer.
Submodel SM41. it describes the alternate time of the control and the data phases during a cycle from the place *Control-phase* to the place *Data-phase* and inversely. The transitions *Tcphase1*, *Tcphase2* and the place *Ex-limit* (Exchange limit) model the two periods of the control phase. Note that the deterministic transition *Tcphase1* has the age memory property in order to keep constant the *Tcphase1* duration when a request arrives in the interval $]tcp, tex - limit]$ (see figure 2). The transition *Tdphase* models the duration of the data phase.
Submodels SM42, SM43 and SM44. The transition *Tarrival* (exponential distribution) models the arrival of data request with rate λ. For simplicity, we suppose that we have a buffer with one packet per cycle. As soon as the transition *Tarrival* is fired, we have a request (place *Request* marked), and if the MAC entity is ready (place *Ready* marked), we can start the first step of the channel assign-

ment, i.e., preselection of a data channel (place *preselection* becomes marked). This operation is enabled only if the two conditions (1) and (2) are satisfied as stated in the previous section. The condition (1) is modeled by a guard function associated to the transition *Tstart-CA* such as : $(\#CS_1+....+\#CS_n) < 3n$ (availability condition), where n is the total number of channels and number 3 means three availability degrees. The condition (2) is modeled by the marking of the place *Control-phase* (*#Control-phase*=1). The transition *Tstart-DT* models the fact that the request has been successfully reserved and the node waits for the data phase period to start the data transfer (place *Transfer* marked).

3.2.2 Channel assignment : preselection model

The model comprises three submodels (SM): SM51, SM52 and SM53 which correspond respectively to the preselection of a channel in one of the three states H, M, *and* L. This operation is guided by the marking of the global places CS_i (Channel State) with $i \in \{1, ...n\}$, which can be in one of the following states: H (CS_i is empty), M (CS_i contains 1 token), L (CS_i contains 2 tokens), and N (CS_i contains 3 tokens); where each token means a reservation. Hence, as soon as the place *Preselection* is marked, the node preselects the channel with highest availability modeled by the firing of one of the following transitions *Tps-H*$_i$ (state H), *Tps-M*$_i$ (state M), and *Tps-L*$_i$ (state L). In order to have a deterministic choice, these transitions are mutually exclusive due to their different priority levels (see table1) so that the channel with the lower number and the best availability has a higher priority to be preselected. When the preselection is done, the node starts the reservation phase (a token is deposited in the place *Start-res*). Note that we also record the channel number preselected in the place *Ps-ch* and its state in the place *State* (the availability degree). The first operation serves to reserve the channel number preselected while the second operation serves for the scheduling of data packets between nodes.

3.2.3 Channel assignment : reservation model

The model includes three interconnected submodels: SM61, SM62, and SM63.

Submodel SM61. It represents the RTS/CTS exchange on the control channel Ch_1 for the reservation of the preselected channel. According to the state of the control channel, we have two cases: (i) If the control channel Ch_1 is busy ($\#Ch_1 \neq 0$), the exchange is deferred (transition *Tdefer* is fired) which induces the cancellation of the preselected channel. In fact, the firing of *Tdefer* flushes all tokens from places *Ps-ch* and *States* (reset their markings to 0), and the token enters in the submodel SM63; i.e, the backoff mechanism (place *BO*). At the end of the backoff time, we can try a new preselection (place *Request* marked) according to the conditions (1) and (2). (ii) If the control channel Ch_1 is free ($\#Ch_1 = 0$), the access is done (transition *Taccess* is fired) and then we go into the vulnerability period modeled by the place *Vperiod* and the transition *Tvperiod* (*CCATime + RxTxTurnaroundTime + AirPropagationTime*). At the end of the this period, the transmission becomes perceptible by the other nodes. At this point there are two possibilities: failed or successful exchange. A failed exchange corresponds to a collision between simultaneous reservation requests from two or more nodes during the vulnerability period. Because of the collision detection is not possible during the transmission phase, we model this case by an artifact model inspired from [16] as follows: if two or more tokens are in the global place at the same time, a collision occurs causing the firing of the transition *Tcoll* and puts a token in the place *Coll* which serves as memory of the collision detection. When the sender ends to transmit the RTS frame, it waits for the CTS timeout (place *Wait-CTS*). According to the token in place *Coll* or not, one of the timed transitions *Ttmout* (failure) or *Tsucc* (success) is fired.

Submodels SM62 and SM63. The success guarantees the channel reservation (submodel SM62). Depending on the preselected channel number, one of the transitions $Tres_i$ is fired moving one token in places CS_i and *#Ps-ch tokens* (channel number) in place *Rec (Record)* which induces the firing of one of the transitions *Trecord-L, Trecord-M, Trecord-H* in order to memorize the channel state and the channel number. The transitions $Tres_i$ (respectively the transitions *Trecord-L*/M/H) are mutually exclusive due to their priority levels: $Tres_n$ has higher priority ($Trecord-L$ has higher priority). In the case of a failure (submodel SM63), we have a behavior like when we have a postponement based on the Backoff mechanism modeled by the timed transition T_{BO} with age memory property in order to capture correctly the frozen backoff time. Note that the place *Rtx* represents the retransmission count which is permitted (value R is set to 6 [15]) if we have successive collisions. When the value R is reached, the transition *Tdrop* is fired (request is dropped) and we reinitialize the data request production (places $Buffer$ and *Ready*). The backoff procedure can be stopped when the control phase is finished. The transition *Tpostp* (request is postponed) models this situation and it is associated to a guard function: (*#Data-phase=1* $\wedge$ *#BO =1*).

3.2.4 Data transfer model

The start of the data transfer on a channel Ch_i, modeled by the transition *Tstart-DT*, occurs when the reservation of a data channel has been done (place *Req-Reserved* is marked), and the control phase is finished (place *Data-phase* marked). As stated above, the data transfer is based on the scheduling policy according to the availability degrees represented now by the name of the places *H*, *M*, and *L*. Thus, a data packet for which the channel has been reserved with quality H will be transmitted before the one with quality M, which itself will be transmitted before the one with quality L. So, we have three cases: (i) if $\#H = i$ (state H), the transition TH_i is fired as soon as the place *Transfer* is marked (the place $Data-tx_i$ inducing the start of the data transfer in channel Ch_i); (ii) if $\#M = i$ (state M), the transition TM_i will be fired after the firing of the transition TH_i which represents the start of the data transfer which had the reservation H by another node; (iii) if $\#L = i$ (state L), we have to wait two firings (by two other nodes): TH_i and TM_i before to fire the transition TL_i followed by the firing of the timed transition *Tdata-tx*$_i$. The transitions TH_i, TM_i and TL_i are associated to a guard function: $\#ch_i$=0. At the end of the transfer, we reinitialize the condition of the possibility of a new transfer need (places *Buffer* and *Ready*).

Remark: the detailed model can be easily extended to include: multiple requests in the same cycle per node, several channel states (> 3), and the channel switching time.

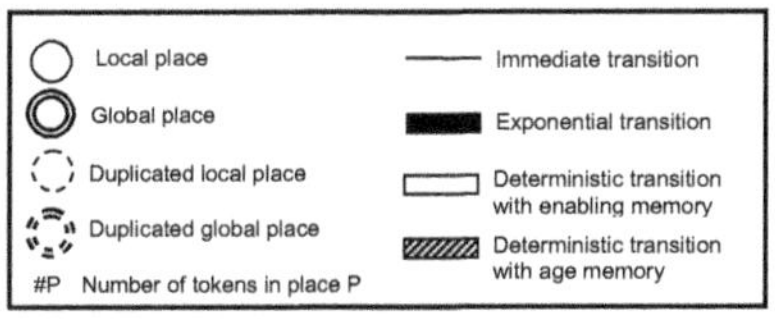

Figure 3: Notation

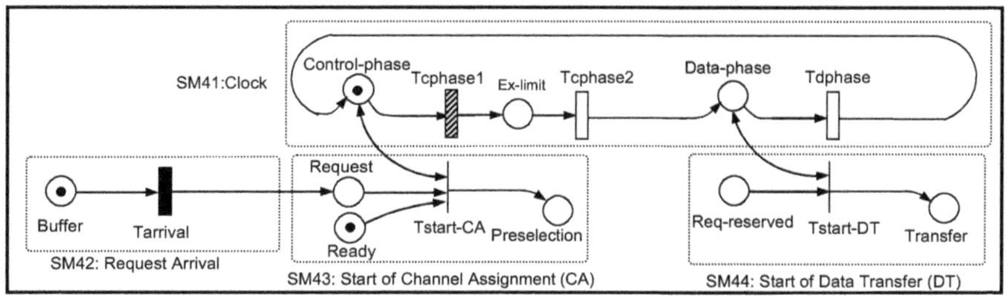

Figure 4: Clock, request arrival, and the start of the Channel Assignment (CA) / the Data Transfer (DT) model

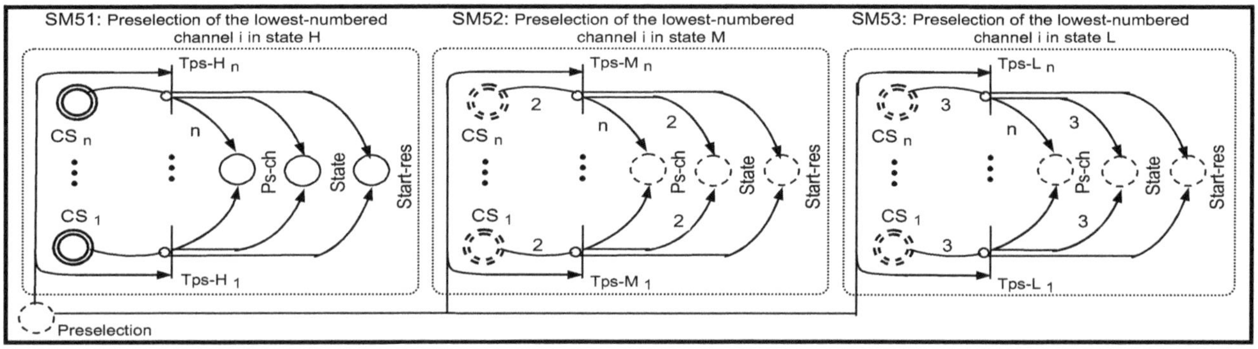

Figure 5: Channel assignment: preselection model

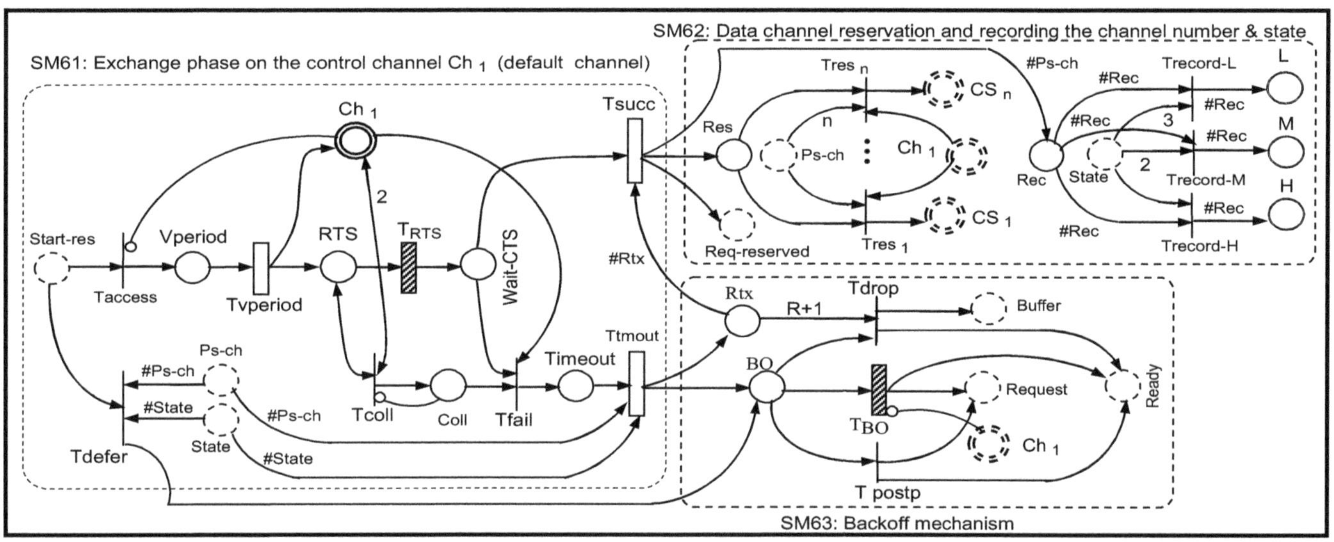

Figure 6: Channel assignment: reservation model

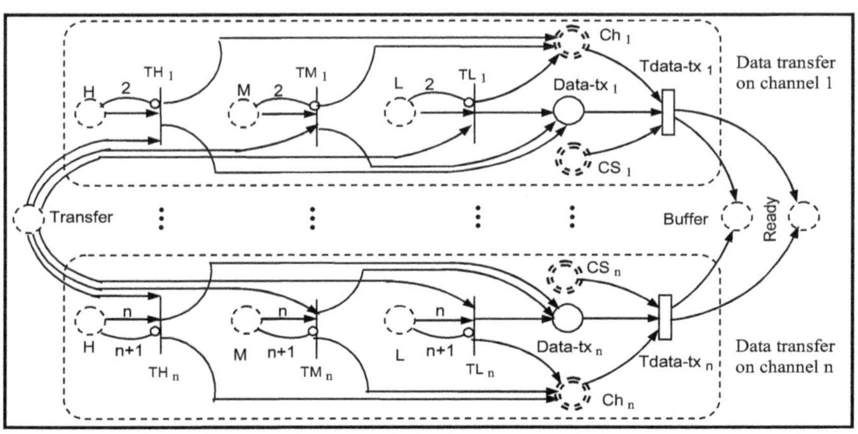

Figure 7: Data Transfer model

Immediate transitions	Priorities
$Tps-H_i$	$(3n+1)-i$
$Tps-M_i$	$(2n+1)-i$
$Tps-L_i$	$(n+1)-i$
$Tres_i$	i
$Taccess$	3
$Tdefer$	2
$Tdrop$	2
$DefaultPriority$	1

Table 1: Priorities of immediate transitions

3.2.5 *Specification of the time distributions*

Let us consider Δ = *aAirPropagationTime*, τ = *RxTx-TurnaroundTime*, γ =*CCATime*, l_{RTS} and l_{CTS} are respectively the duration of RTS and CTS. We assume that the length of data (*ld*) includes the length of ACK (Acknowledgment) and the channel switching time. Let *lc* denote the minimum duration of the exchange where $lc = DIFS + l_{RTS} + SIFS + l_{CTS} + 2\Delta$.

- $Tcphase1$: $lcp - lc$
- $Tcphase2$: lc
- $Tdphase$: $3ld$
- $Tvperiod$: $\gamma + \tau + \Delta$
- $Tdata - tx_i$: ld
- T_{RTS} : $DIFS + l_{RTS} - \gamma - \tau$
- T_{succ} : $SIFS + l_{CTS} + \Delta$
- $Ttmout$: $CTS\_timeout$
- T_{BO} : $bc * aSlotTime$

Where *bc* (backoff counter) is an integer uniformly distributed over $[0, CW]$ where CW is the contention window, the durations SIFS and DIFS represent respectively the Short InterFrame Space and the Distributed InterFrame Space defined in IEEE 802.11 [15].

3.3 A compact model

The model presented in the previous section allows making a detailed representation of the different mechanisms of the split phase approach. Such model is pertinent for the qualitative analysis of the relevant aspects based on the observation of a subset of events which are representative of the desired behavior of this approach (see next section). However, they are not easy to use for performance evaluation accounting for the combinatorial explosion of the space state when we have a great number of nodes and channels (several transitions between nodes are enabled concurrently). So we present now a more scalable model "compact model" derived from the detailed model by folding all the detailed models of nodes (the methodology is similar to [16]). The resulting model, presented in the figure 8, is parametric with respect to the number of nodes k, and the number of channel states. The meaning of most places and transitions are the same. In what follows, we explain the main differences with the detailed model.

The compact model comprises six submodels: SM81, SM82, SM83, SM84, SM85, and SM86.

Submodel SM81 (Clock). It is obviously identical to the detailed model.

Submodel SM82 (Arrival). It represents the arrival model in all nodes in the network, i.e., the traffic is superposed (one data packet by node). It is modeled by the place *Nodes* and the transition *Tarrival.* Note that the request arrivals can be considered to try a preselection (transition *Tstart-CA*) only when we have less than $n * s$ reservations which have been done (availability condition). The place *Nb-res* (Number of reservation) is equivalent to the all previous places CS_i where $i \in \{1, ...n\}$ and $s = 3$, i.e., *H, M, L*.

Submodel SM83 (Preselection). The order of the channels preselection is preserved based on the increasing numbers of the channels $(Ch_1, ..., Ch_n, Ch_1, ...)$ by means of the place *Ch-nbr* (Channel number to be preselected). The first firing of the transition *Tps-ch* induces the preselection of the channel Ch_1 (place *Ch-nbr* contains one token initially) and the start of the exchange phase (place *Start-res* becomes marked). It is important to notice that a channel can be preselected several times as long as it has not yet been reserved. When the reservation of the channel Ch_1 is done with success (submodel SM84 (transition *Tsucc*) and submodel SM85 (transition $Tres_i$)), the place *Ch-nbr* is incremented which means the next channel should be preselected (i.e., Ch_2) and so on. Note that we make loops in the preselection model exactly s times at the maximum expressed by the firing of the transition *Tloop* which reinitializes the number of channels when the total number of channels n is reached. In this model, we do not represent the preselection by considering explicitly the availability qualities (*H, M, L*) but the fact of the loops is a way to represent them implicitly. At the end of the control phase, we reinitialize the place *Ch-nbr* by the firing of the transition *Treinitialize* which is associated to a guard function so that: $\#Data - phase = 1 \wedge \#Ch - nbr > 1$.

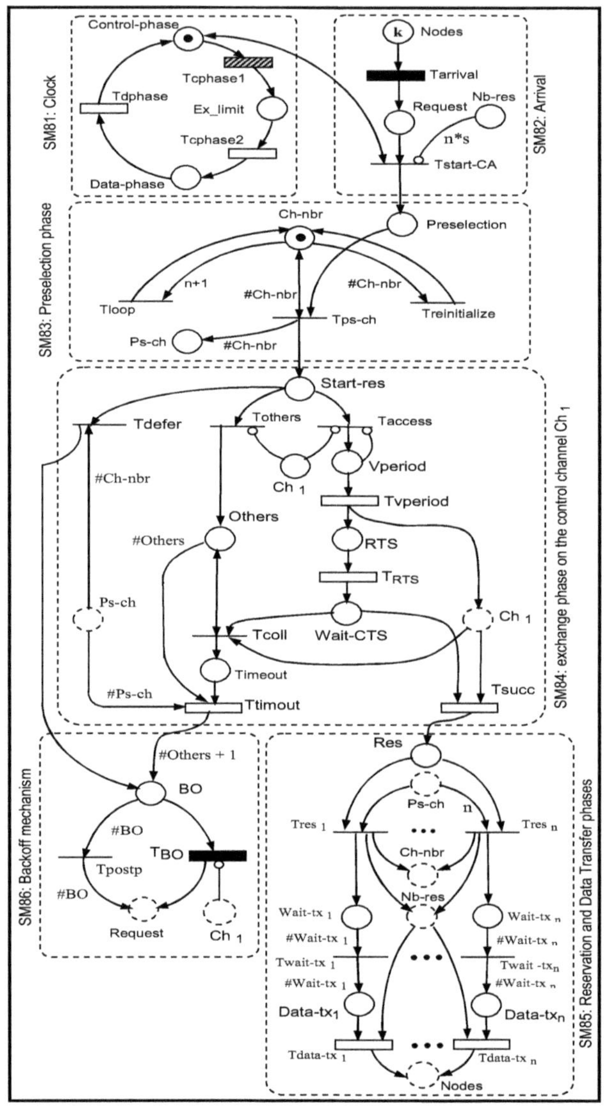

Figure 8: Compact model (all immediate transitions are mutually exclusive)

Submodel SM84 (Exchange). It is slightly different to the previous submodel SM61 (figure 6). When a node preselects a channel, it starts the exchange phase and enters to the vulnerability period (place *Vperiod*). If any other node arrives in this period, it preselects the same channel (as long as the marking of the place *Ch-nbr* is not changed) and initiates an exchange (place *Others* becomes marked, due to the priorities: $Priority(Taccess) > Priority(Tothers) > Priority(Tdefer)$. Hence, a collision occurs. The number of reservation attempts done in the vulnerability period for the same channel is represented by the marking of both places *Vperiod* and *Others*. Therefore, we have a failure (transitions *Tcoll* and *Ttmout*) if both places *Vperiod* and *Others* are marked: ($\#Vperiod = 1 \wedge \#Others \geq 1$). Otherwise, we have a success (transition *Tsucc*). A P-invariant analysis are used to ensure the behavior correctness, in both cases success or collision, we have always the equation satisfied:

$$\#Ch - nbr * (\#Others + \#Vperiod) = \#Ps - ch \quad (3)$$

Submodel SM85 (Reservation and data transfer). it represents the following operations: (i) The reservation of each channel with the consequent modification of the channel number to be preselected (place *Ch-nbr*). (ii) The increase of the number of reservations (place *Nb-res*). (iii) The data transfer on each channel when the control phase is finished (place *Data-phase* is marked). Concerning the data transfer, as the compact model is global, the FIFO scheduling is natural and then there is no need for scheduling policy. Note that the transitions $Twait - tx_i$ (waiting for the data phase) are associated to a guard function such that: $\#Data - phase = 1 \wedge \#Wait - tx_i \geq 1$.

Submodels SM86 (backoff). It represents the cancellation of the preselection after the postponement or a failure in the exchange phase which is followed by the backoff mechanism. In this submodel two modifications were made: (i) The backoff is simplified and it is approximated by an exponential transition with infinite server semantics. Its rate is equal to the reciprocal of the mean delay of the first backoff. (ii) The retransmission count is not considered. Note that when the exchange is deferred, the transition *Tdefer* (SM84) flushes only *#Ch-nbr* tokens from the place *Ps-ch* which correspond to the last preselection.

4. VERIFICATION

We verify specific properties of the split phase approach which are very important from the point of views of a designer, and are representative of the desired behaviour of this approach. The analysis is done from the RSG of the detailed model by considering only the qualitative labels and using the projection method on the events that we want to observe on the basis of the observational equivalence relation [20].

4.1 Observational relation and quotient automaton

Informally, two states are observationally equivalent if the observed behaviour of the system being modeled from these states are similar. Two states are equivalent with respect to an observable event e (transition) at the order k if for any e-experimentation from one state, we have a e-experimentation from the other state which is equivalent at the order $k-1$. Let us consider S the set of the states of the RSG graph, E its set of labels (events) associated to the transitions and E' is the subset of the observable events (transitions). The observational equivalence relation (noted $\equiv$) is defined as a decreasing sequence of equivalence relation by recurrence [17, 20] such that: (i) $\forall m,\, p \in S,\; m \equiv_0 p$; (ii) $m \equiv_{k+1} p\, iff\; m \equiv_k p\; and\, \forall e \in E' \cup \{\varepsilon\}$:

$$\begin{cases}(m \stackrel{e}{\Rightarrow} u)\ implies\ (\exists q \in S\ : p \stackrel{e}{\Rightarrow} q\ et\ u \equiv_k p)\\(p \stackrel{e}{\Rightarrow} q)\ implies\ (\exists u \subset S\ : m \stackrel{e}{\Rightarrow} u\ et\ q \equiv_k u)\end{cases}$$

Figure 9 shows an example of quotient automaton obtention, from a simple RSG graph with only the qualitative labels where $e1$ and $e4$ are the observable events (this example show how are the automatons derived from the specification).

4.2 Verification of specific properties of the split phase approach

There are three specific properties that we want to verify:

- Property 1: Load balancing between the channels, i.e., the channel assignment (preselection and reservation) is done according to the best availability degree.
- Property 2: Reservation consistency, i.e., reservation in mutual exclusion and all local modifications (i.e., in one node) are observed in the same manner by the other nodes (global knowledge).
- Property 3: Free collision data transfer, i.e., the data transfer is scheduled in a free collision way.

In order to make this verification, we consider a system with three non-overlapping channels (Ch_1,Ch_2 and Ch_3); and two nodes N1 and N2 within the communication range of each other. We analyze the four scenarios presented in the table 2 where we have only one arrival of a data request in N1 and in N2 per cycle. The three first scenarios consider the system start-up, i.e., with initial conditions where none channel is used (all channels have the availability H). The fourth scenario considers a system in which the channels have different initial conditions in terms of availability degree: M for Ch_1 and Ch_2 ; H for Ch_3. The initial conditions are represented by the initial markings of the global places CS_1, CS_2 and CS_3. The differences between the three first scenarios results from the differences in the instants of the arrivals of the data request in the MAC entities of N1 and N2:

- Scenario 1: these instants are such that the request of the node N1 is made first, and when the request of the node N2 is done, the request of the node N1 has already induced the reservation of a channel.
- Scenario 2: these instants are such that the request of the node N1 is made first, and when the request of the node N2 is done, the process inherent to the request in the node N1 is still in the reservation process (exchange of RTS/CTS in the control channel).
- Scenario 3: the requests in the nodes N1 and N2 are done at the same instant.
- Scenario 4: we consider the differences in the occurrences of the data requests in the nodes N1 and N2 like in the scenario 1.

For this analysis, we consider the following parameters: $lcp = 9.00ms$ ($lcp = 10 * \overline{lc}$ where $\overline{lc}$ corresponds to the estimation of the time needed for a successful exchange: $\overline{lc} = 900\mu s$), $ldp = 3ld$, the bit rate $B = 2Mbps$, $ld = 512bytes/B$ and $l_{vperiod} = 0.02ms$ (vulnerability period). In this section, we focus on qualitative aspects and then we consider a deterministic arrival of data packet request for both N1 and

N2 per cycle (we modify the arrival condition such as the request arrival is done in the control phase). We want, in particular, to study the influence of this duration and the durations of the different nodes. The observable events of N1 and N2 are represented by the following transitions: (i) $Tarrival$ (arrival of data packet); (ii) $Tps-H_i$, $Tps-M_i$ and $Tps-L_i$ (preselection of a data channel Ch_i); (iii) $Tdefer$, $Tfail$, T_{BO}, $Tsucc$ and $Tres_i$ (control exchange and reservation of data channel on the control channel Ch_1); and (iv) $Tdata-tx_i$ (data transmission on channel Ch_i).

Scenarios	Initial marking	Arrival instant of N1 and N2
Scenario1	$\#CS_1 = 0$ $\#CS_2 = 0$ $\#CS_3 = 0$	$t_{arrival}(N1) = 0, 1ms$ $t_{arrival}(N2) = 1, 1ms$
Scenario2		$t_{arrival}(N1) = 0, 1ms$ $t_{arrival}(N2) = 0, 15ms$
Scenario3		$t_{arrival}(N1) = 0, 1ms$ $t_{arrival}(N2) = 0, 1ms$
Scenario4	$\#CS_1 = 1$,$\#CS_2 = 1$, $\#CS_3 = 0$	$t_{arrival}(N1) = 2, 1ms$ $t_{arrival}(N2) = 3, 1ms$

Table 2: Initial marking and request arrival instants

We get the automatons which are represented in the figures 10, 11, 12 and 13 (note that previously we have checked the boundedness and the deadlock free properties). Each automaton is composed of set of states classes which are observationally equivalent with respect to the observable events. All these automatons show the correctness of the three properties 1, 2 and 3.

In the automaton 1 of the figure 10, we observe the reservation of Ch_2 by N2 with availability H after the reservation of the Ch_1 by N1 with availability H. The data transfer phase (sequences between classes C_9, C_{10}, C_{11}, and C_1) shows the parallelism in the two channels (interleaving of the data packet transfers).

In the automaton 2 of the figure 11, we observe, with respect to the automaton 1 that, as the arrival of the request in N2 occurs just after the sending of the reservation of N1, N2 sees that the control channel is busy and defers its reservation. After the backoff (sequence C_9, C_{10}), consequent of the deferring (sequence C_5 and C_6), N2 makes the preselection of Ch_2 (it shows that N2 has the knowledge that Ch_1 is already reserved).

In the automaton 3 of the figure 12, we observe, with respect to the two previous automatons, the consequence of the simultaneous arrivals of the data requests in N1 and N2. The sequence between the classes C_1 and C_9 shows the interleaving of the local actions both in N1 and N2 (arrivals and preselection of Ch_1 for both N1 and N2). Then, we have of course the collision in the control channel (sequences between classes C_9 and C_{12}). As we have considered the case where the end of the backoff for N2 is a lot less than the end of the backoff for N1, we can see that N2 reserves Ch_1 and N1 reserves Ch_2. At the start of the data phase (from the class C_{20}) , N1 and N2 transmit data respectively in Ch_2 and Ch_1 simultaneously.

In the automaton 4 of the figure 13, we visualize the influence of the initial conditions on the preselection and the reservation process (choice by N1 of Ch_3 which has the best availability, which, after the reservation, gives the same availability for all the channels; then N2 will preselect Ch_1 with quality M i.e., the first lowest-numbered channel with the best availability). Furthermore, the sequence: C_9, C_{10}, C_1 shows the precedence between the data transfer on Ch_3 by N1, and on Ch_1 by N2. In fact, the data transfer of N1 (sequence C_9, C_{10}) is done before N2 because the channel Ch_1 are used by one communication (from the initial conditions, the first communication has a quality H). These actions illustrate the FIFO policy scheduling.

5. QUANTITATIVE ANALYSIS

The goal of this section is to study the benefit of the multichannel communications of the Split Phase Approach (SPA) in terms of the network throughput with respect to the single channel MAC protocol IEEE 802.11 DCF (Distributed Coordination Function) [15]. Therefore, we consider two STPN compact models:

- (i) Model1: the compact model of the SPA described in the figure 8.
- (ii) Model2: the compact model of the IEEE 802.11 DCF. It comprises three submodels SM82, SM84, and SM86 of the figure 8 by considering a single channel (without the clock, the preselection/reservation and the transfer models). In this model, the firing time of the transition $Tsucc$ comprises the duration of the exchange RTS/CTS and the duration of a data length.

We get the average throughput system η as follows. Let us consider η_i the average throughput of the transition $Tdata-tx_i$ (the firing frequency of $Tdata-tx_i$ on channel i) per unit of time normalized to bit rate. Let A_i denote the subset of all states S_j in which the transition $Tdata-tx_i$ is enabled, and π_j is the steady state probability of S_j. Let P_i, θ_i are respectively the firing probability and the firing rate of $Tdata-tx_i$, and n the number of channels. The average throughput system η is defined as:

$$\eta = \sum_{i=1}^{n} \eta_i \quad where\ \eta_i = \sum_{j:S_j \in A_i} \pi_j \frac{P_i}{\theta_i} \tag{4}$$

For this purpose, we use TimeNET software package [21] which enables to obtain easily η. For this analysis, we consider the packet length is $ld = 512bytes/B$, the control phase duration $lcp = 9.00ms$ (average duration), the data phase duration $ldp = 3ld$, the bit rate for each channel is $B = 2Mbps$, and the number of nodes $k = 10$. Let VL denote the virtual load defined as follow: $VL = (\frac{K \times ld}{B})\ \lambda$, where λ is the packet arrival rate.The parameters that we vary are: the virtual load VL (we vary λ to vary VL), the number of channels n, and also the length of the control phase. Figure 14 and 15 show the results respectively for 2 and 3 channels, and figure 16 shows the results respectively for three durations of the control phase: $lcp1$ ($= lcp/2$), $lcp2$ ($= lcp$), and $lcp3$ ($= 2lcp$).

In figure 14, we observe that the throughput is similar for both IEEE 802.11 and SPA when the virtual load is low. When the load increases, SPA outperforms IEEE 802.11. When the virtual load reaches 1, IEEE 802.11 achieves its higher throughput, while SPA continues to increase. SPA achieves 1,53 times the throughput than IEEE 802.11. In figure 15, we see that when the number of channels increases, SPA outperforms significantly IEEE 802.11, and

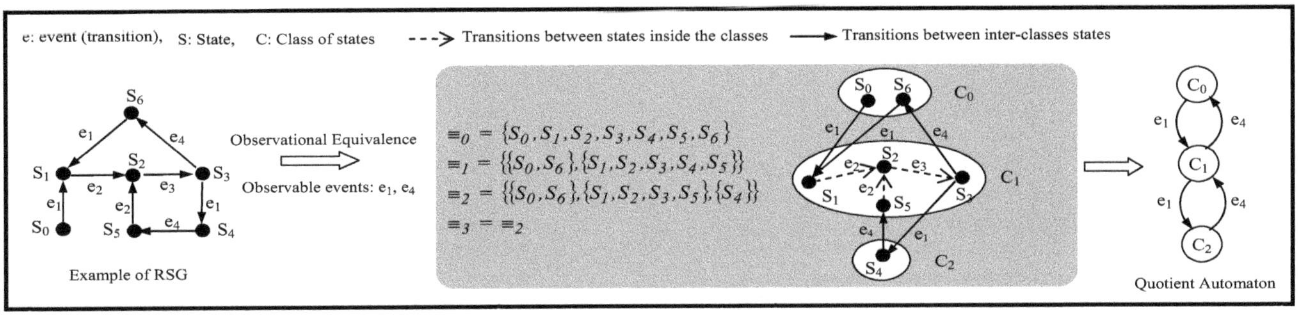

Figure 9: Quotient automaton obtention

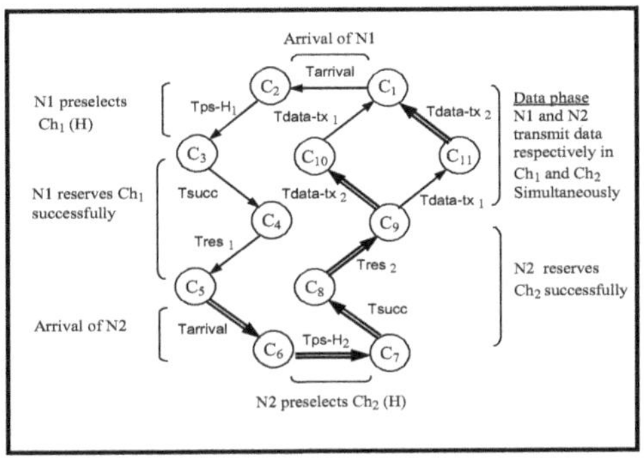

Figure 10: Scenario 1

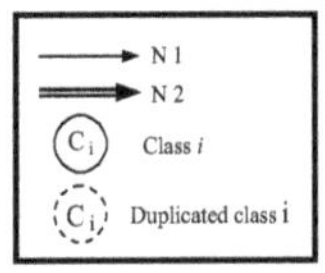

Legend of scenarios

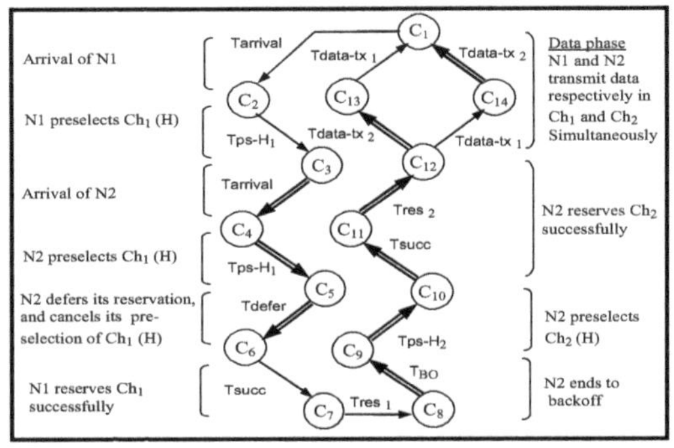

Figure 11: Scenario 2

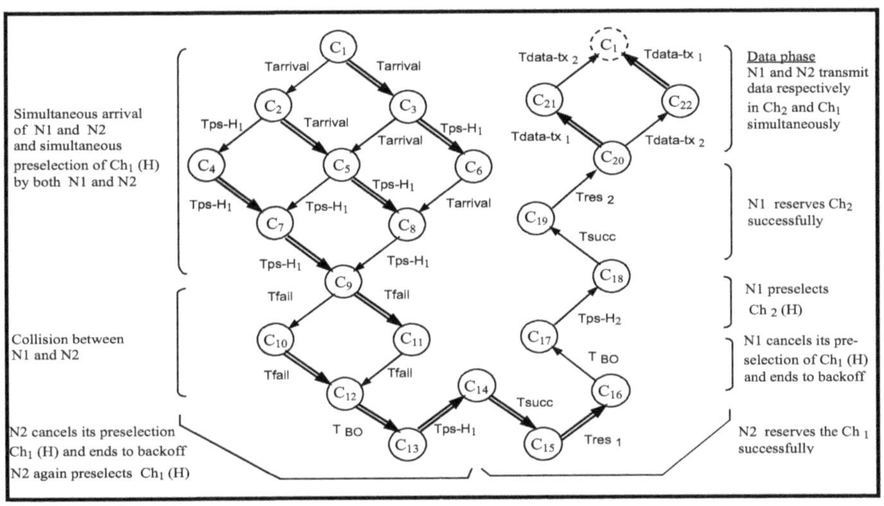

Figure 12: Scenario 3

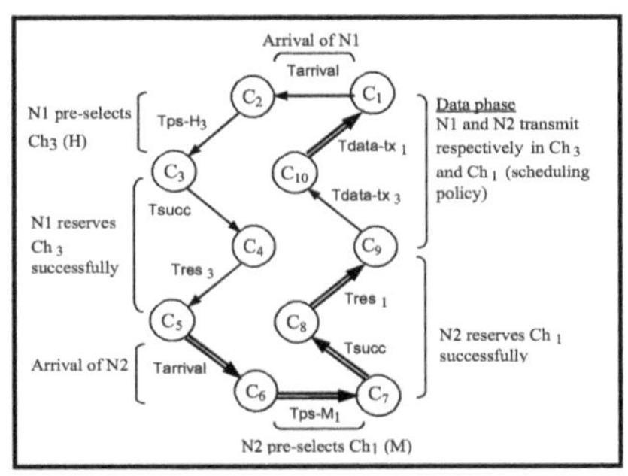

Figure 13: Scenario 4

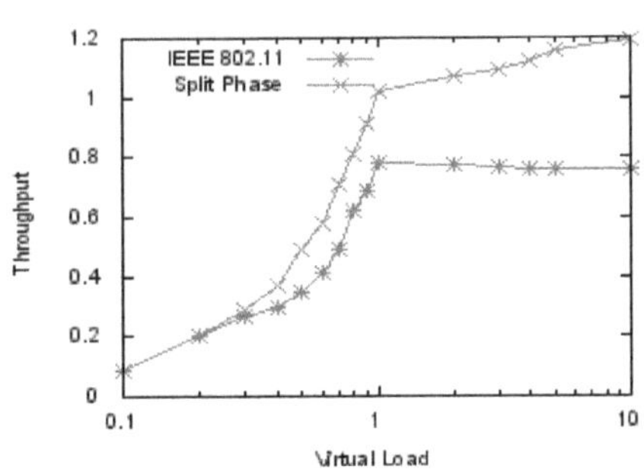

Figure 14: Throughput vs. VL (2 channels)

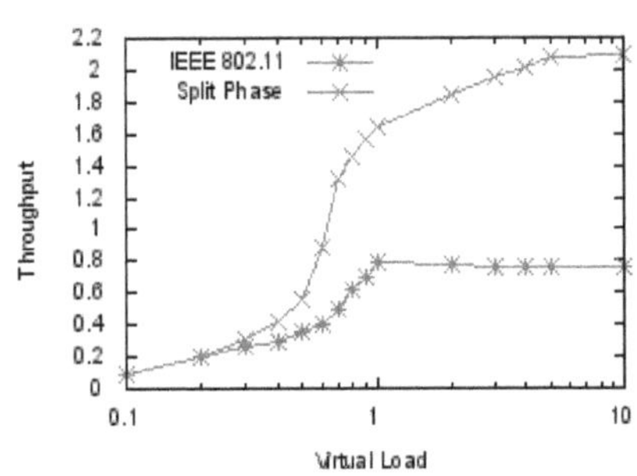

Figure 15: Throughput vs. VL (3 channels)

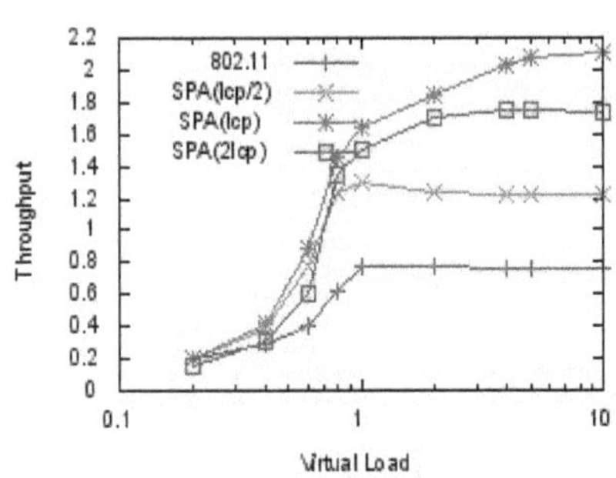

Figure 16: Throughput vs. VL / Impact of the control phase duration (3 channels)

achieves 2,75 times better than IEEE 802.11. Therefore, consequently to the use of multiple channels, SPA improves the network throughput compared to the IEEE 802.11 DCF. However, the performance of the SPA does not reach n times (the improvement factor is equal to the number of channels) the performance of IEEE 802.11, although the free contention communication during the data phase (scheduled access). We can explain this situation because of the exchange overhead in the control phase when the load increases. This later can lead to waste the bandwidth due to the contention, the collisions and the backoff.

In figure 16, we observe when the length of the control phase is smaller ($4.5ms$) or larger ($18ms$) compared to the average length of lcp ($9ms$) for 10 nodes and 3 channels, the throughput is less important. This is because when the length is small, few negotiations took place leading to a partial use of the bandwidth during the data phase. Inversely, when the length is large, the bandwidth during the control phase is used partially. Therefore, the choice of the control phase duration has a significant impact on the performance and should be made according to the traffic load.

6. CONCLUSION

In this paper, we proposed a formal specification and analysis of the split phase multi-channel MAC approach. Firstly, we proposed a detailed model which allows to specify the different aspects of the channel assignment mechanism. Consequently, we can make a qualitative analysis of important properties, from the user point of view based on the equivalence observational relation, which are very representative of the desired behavior of the split phase such as the reservation consistency, the channel load balancing, the collision detection and the backoff mechanism. Secondly, to overcome the space state explosion, an analytically tractable model "compact" derived from the detailed model is formulated which still captures the most relevant aspects of the system. The quantitative analysis conduced from this later shown that the split phase approach outperforms significantly the single channel MAC IEEE 802.11 DCF in terms of the network throughput.

For future work, we shall extend the quantitative analysis to cover more sophisticated scenarios. We plan also to make a similar study based on the parallel hopping multi-channel MAC approach [1, 2] in order to compare the main multi-channel MAC approaches.

7. REFERENCES

[1] P. Bahl, R. Chandra, and J. Dunagan, "SSCH: Slotted Seeded Channel Hopping for Capacity Improvement in IEEE 802.11 Ad-Hoc Wireless Networks," *In Proc. ACM MobiCom'04, pp. 216-230, USA* , Oct. 2004.

[2] J. Mo, H.-S. So, and J. Walrand, "McMAC: A Parallel Rendezvous Multi-Channel MAC protocol," *In Proc. IEEE WCNC'07, pp. 334-339, Hong Kong* , Oct. 2007.

[3] A. Tzamaloukas and J. Garcia-Luna-Aceves, "Channel-Hopping Multiple Access," *In Proc. IEEE ICC'00, pp. 415-419, New Orleans* , June 2000.

[4] A. Tzamaloukas and J. Garcia-Luna-Aceves, "Channel-Hopping Multiple Access with Packet Trains for Ad Hoc Networks," *In Proc. IEEE MoMuC'00, 6 pages, Tokyo* , Oct. 2000.

[5] S.-L. Wu, C.-Y. Lin, Y.-C. Tseng, and J.-P. Sheu, "A new Multi-Channel MAC Protocol with On- Demand Channel Assignment for Multi-hop Mobile Ad Hoc Networks," *In Proc. ISPAN'00, pp. 232-237, Dallas, USA* , Dec. 2000.

[6] M. Seo, Y. Kim, and J. Ma, "Multi-channel MAC protocol for multi-hop wireless networks: Handling multi-channel hidden node problem using snooping," *In Proc. IEEE MILCOM'08. pp. 1-7* , Nov. 2008.

[7] J. Chen, S. Sheu, and C. Yang, "A new multichannel access protocol for IEEE 802.11 ad hoc wireless LANs," *In Proc. PIMRC'03, pp. 2291 - 2296* , 2003.

[8] J. So and N. Vaidya, "Multi-Channel MAC: Handling Multi-Channel Hidden Terminals Using a Single Tranceiver," *In Proc. ACM MobiHoc'04, pp.222-233, Tokyo* , May 2004.

[9] J. Zhang, G. Zhou, C. Huang, S. H. Son, and J. A. Stankovic, "TMMAC: An Energy Efficient Multi-Channel MAC Protocol for Ad Hoc Networks," *In Proc. IEEE ICC'07, pp. 3554-3561* , 2007.

[10] R. Maheeshwari, H. Gupta, and S. R. Das, "Multichannel MAC Protocols for Wireless Networks," *In Proc. IEEE SECON'06, pp. 393-401, Boston, VA, USA* , Sep. 2006.

[11] J. Mo, H.-S. So, and J. Walrand, "Comparison of Multi-channel MAC Protocols," *IEEE Transactions On Mobile Computing, Vol. 7, No. 1, pp. 50-65* , Jan. 2008.

[12] L. Le, "Performance Analysis of Multi-channel MAC Protocols in Multi-hop Ad-hoc Networks," *IEEE Globecom, pp. 1-6* , Dec. 2010.

[13] A. El Fatni, G. Juanole and T. Val, "Formal Modelling and Verification of the Common Hopping Multi-Channel MAC Protocols," *In Proc. ACM MSWIM'10, pp. 311-318, Turkey* , Oct. 2010.

[14] V. Gupta, M. Gong, S. Dharmaraja, and C. Williamson, "Analytical modeling of bidirectional multi-channel IEEE 802.11 MAC protocols," *Int. Journal of Communication Systems, Vol. 24, pp. 647-665* , 2011.

[15] IEEE 802.11 Working Group, "802.11 IEEE Standard, Wireless LAN Medium Access Control (MAC) and Physical Layer (PHY) Specifications," 2007.

[16] A. Heindl and R. Germain, "Performance Evaluation IEEE 802.11 Wireless LANs with Stochastic Petri Nets," *Journal Performance Evaluation, Vol. 44, pp. 139-164* , April 2001.

[17] G. Juanole and L. Gallon, "Concept of quantified abstract quotient automaton and its advantage," *In Proc. FORTE/PSTV'97, Osaka, Japan* , Nov. 1997.

[18] A. Marsan, G. Balbo, G. Conte, S. Donatelli, and G. Franceschinis, "Modelling with Generalised Stochastic Petri Nets," *John Wiley and Sons* , 1996.

[19] C. Lindemann, "Performance Modelling with Deterministic and Stochastic Petri Nets," *John Wiley and Sons* , 1998.

[20] R. Milner, "Communication and Concurrency," *International Series in Computer Science, Prentice Hall* , 1989.

[21] http://www.tu-ilmenau.de/TimeNET, "TimeNET software package 4.0," April 2007.

Towards Trustworthy Simulation of Wireless MAC/PHY Layers: A Comparison Framework

Laura Marie Feeney
Swedish Institute of Computer Science
lmfeeney@sics.se

ABSTRACT

Simulation is an essential tool for evaluating the performance of wireless protocols and applications because of its cost effectiveness and its high level of control over the simulated system. As the research community envisions increasingly complex applications in areas like mobile social networks and wireless sensor networks, the need for trustworthy simulation of core communication protocols increases.

This paper presents a comparative evaluation of several IEEE 802.11 and IEEE 802.15.4 simulators. The paper defines a set of simple experiments that are intended to highlight handling of backoff and contention.

A wide variation in results is observed. Although this is partially explained by differences in wireless communication models, the experiments also reinforce existing concerns about the reliability of simulation results. A long term goal is that this methodology will encourage the adoption of common test and evaluation scenarios. This would lead to better understanding of simulation behavior and help to improve the quality of – and confidence in – simulation results.

Categories and Subject Descriptors

I.6 [**Computing Methodologies**]: Simulation and Modeling

Keywords

simulation, wireless, MAC/PHY, OMNeT++

1. INTRODUCTION

Testbed experiments are often used in evaluating wireless protocols and applications and are generally viewed as the gold standard for performance evaluation. However, testbed results reflect the irregularities of a specific radio environment and the effects of external interferences, as well as complex cross-layer interactions in the communicating devices. Although testbeds provide "ground truth", this lack of control can make it extremely difficult to reason about observed results.

By contrast, simulation provides complete visibility and control of the environment in which the simulated system operates. This allows the user to explore a wide range of scenarios, to create structured experiments, and to reason carefully about system behavior. But it is not possible to take advantage of these capabilities, unless the underlying simulations are accurate and well understood.

This is especially important in areas like mobile social networks and wireless sensor networks because their low power device-to-device communication architecture results in very complex interactions between wireless devices. Moreover, these networks are very application-oriented and their design and performance can depend heavily on external processes like human interaction or natural phenomena. As a result, there is increasing participation from research communities whose expertise lies outside wireless communication. A researcher studying information dissemination in mobile social networks or distributed data processing algorithms for monitoring air quality may not have – and should not need – expertise in low-level details of wireless simulation.

There is therefore a growing demand for tools that provide trustworthy components for simulation of core wireless communication protocols. However, there is currently no consensus of high confidence in wireless simulation results, due to concerns about software quality and validation. The community also lacks a high-level systematic understanding of the behavior of various wireless simulation environments. Attempts to validate simulators with testbed experiments have generally proven somewhat unsatisfying, largely due to limitations in parameterizing simulations to reflect real radio hardware and environments.

This paper presents a comparative evaluation of several IEEE 802.11 and IEEE 802.15.4 simulators. The paper defines a set of simple, well-designed experiments that highlight handling of backoff and contention in the presence of hidden terminals. The issue of parameterization is carefully addressed: Comparable scenarios are defined and normalized in terms of nominal transmit range, which encapsulates the variation in wireless communication models.

The results demonstrate significant variation in behavior. Some of these reflect legitimate differences between simulators, while others reveal issues that need further investigation. It is hoped that the availability of simple, well-defined, and revealing test and evaluation scenarios will improve quality and confidence in simulation results. Specifically, this paper:

MSWiM'12, October 21–25, 2012, Paphos, Cyprus.

- Characterizes the structure and parameterization of wireless network simulations, with particular attention to testbed vs simulation experiments.
- Proposes a small set of well-designed scenario-based experiments that demonstrate the behavior of key components of the simulation. Nominal transmit range is used as the basis of comparable simulations.
- Compares five OMNeT++-based simulators supporting IEEE 802.11 and IEEE 802.15.4 wireless standards and observes some significant variation. (Note that the scenarios themselves are not OMNeT++ specific.)
- Concludes that diversity of simulators is generally positive, not least because it enables this kind of cross-comparison. The long-term goal is that simulation quality is improved by the wider adoption of common test and evaluation scenarios, leading to increased confidence in simulation results.

2. RELATED WORK

This section briefly presents the rather sparse body of existing work comparing simulation results and proposing common environments. The discussion focuses on work comparing simulation and testbed results and its implications for comparing simulation results directly. These ideas are important for understanding the problem of creating sound comparison experiments in both areas.

Comparing simulations There is a large survey literature (e.g. [12]) comparing the features and usability of various wireless simulation environments. But there have been very few large-scale comparisons of the results obtained from different simulators

In 2002, paper [4] reported significantly different results when simulating a simple IEEE 802.11 broadcast flooding scenario in ns-2, OPNET and Glomosim. An unpublished technical report [6] from 2008 compares ns-2 (versions 2.26, 2.29 and 2.31) and OMNeT++'s inet and mobility-fw. Like our work, this work emphasizes simple hidden terminal scenarios. However, it attempts to create comparable scenarios by adjusting radio sensitivity parameters, which seems potentially problematic because these are fixed hardware properties. The results suggest that inet is more permissive than mobility-fw in hidden terminal scenarios. This is the opposite of the result seen in our study of more recent software.

Comparisons with testbeds It is often suggested that simulations should be validated by experimental measurements. Unfortunately, there have been few satisfying results comparing simulation and testbed results.

It is hard to parameterize simulation models for a specific communication environment and hardware. Visible effects, such as packet reception rate or link quality indication reflect the combined effect of many channel and hardware parameters. When a single simulation parameter, such as path-loss, is tuned to match testbed results, there is risk of over-fitting.

It is also sometimes possible to find a parameterization that gives a fairly close match in high-level statistics such as throughput or packet reception ratio (PRR), while detailed statistics such as latency or jitter remain highly inconsistent. This inconsistency makes it difficult to use results to reason about low-level internal behavior. Moreover, even the best work does not seem to provide robust conclusions: Does a parameterization obtained in one environment give similar results when tested in another, similar environment?

One way to try minimize the minimize the complexity of the wireless environment and the associated parameterization problems is to use a very restricted testbed scenario. This approach is sometimes helpful, but in other cases it reduces the usefulness of the results, without significantly reducing complexity.

Comparisons that use specialized hardware to essentially eliminate all wireless effects and focus on the MAC protocol implementation often show very good results. For example, paper [3] compares a four node 802.11g network with an OMNeT++ simulator (based on inet + cocorada). The nodes were in an anechoic chamber and separated by only 0.5-1.5m, giving near perfect connectivity. The simulation results suggested that the IEEE 802.11g protocol simulation itself was quite accurate. Testbed and simulation results for throughput were nearly identical and in 99% of cases, even inter-packet times were in agreement. Deviation in the other 1% of measurements suggest some occasional phenomena were not accounted for. Obviously, this approach has limited relevance for simulating wireless communication, but it is very valuable for investigating the accuracy of protocol models and inconsistencies in vendor implementations.

Other experiments use fully connected networks with short distances and and unobstructed radio environments (e.g. sports fields). This eliminates the effects of hidden terminals and unstable links. It also reduces the relevance of the results, because these are essential factors in the performance of wireless networks and (equivalently) a large proportion of simulation code is not exercised in such a test.

Paper [1] compares ns-2 and two OMNeT++ simulators, Castalia and the MAC simulator (an older framework of which MiXiM is a descendant). The experiment used a fully connected network of 12 Tmote-Sky (IEEE 802.15.4) nodes, deployed outdoors in a 5m area minimizing reflections and interference. Unlike [10], the channel parameters (path-loss exponent $= 2.4$ and $\sigma = 4$) are defined *a priori* (radio hardware parameters were not discussed). The authors claim that for ns-2, the shadowing component has no effect on results, while for Castalia log-normal shadowing significantly underestimates the network performance. Overall, the simplest communication models for both ns-2 and Castalia give results within 10-30% of testbed results.

Paper [10] compares ns-2 (ns-2.29) simulation of IEEE 802.11 with a multi-hop indoor network of 16 nodes. The simulation was parameterized by simulating the network for various values of path-loss and selecting the value that gave the best overlap between the set of connected node pairs in the simulated and testbed topologies. It is not clear whether the (rather large) chosen path-loss value of 5.7 reflects the actual path-loss or whether it is acting more like an arbitrary tuning parameter. For video traffic, good agreement was seen for packet delivery ratios, but latencies varied by much as -68% and +125%, depending on hopcount.

Paper [14] compares ns-2 (ns-2.34) and Qualnet to results obtained in indoor and outdoor testbeds using a variety of IEEE 802.11 mesh network hardware. There is little discussion of parameterization, but neither simulator was able to model the behavior of even simple UDP traffic, especially with respect to environmental effects. Later experiments suggested ns-2 path-loss parameters of 2.4 (indoor) and 2.5 (outdoor), but these values were not used in further com-

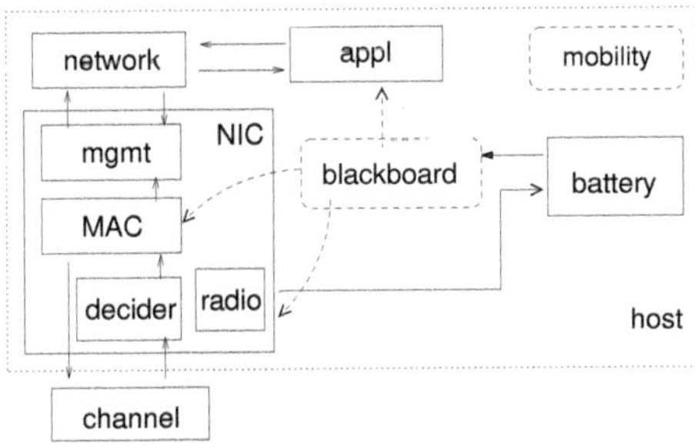

Figure 1: Example of a simplified wireless host.

simulator	version	IEEE 802.11	IEEE 802.15.4	cites
inet	20110225	b, AP	no	133
MiXiM	2.1 (pre-release)	b, DCF	csma	218
inetmanet	151cd4	a/b/g/e, AP	PAN only	80
Castalia	3.2	no	PAN only	200
mobility-fw	2.0p3 (extended)	b, DCF	csma	137

Table 1: The simulators used in this paper are all available via www.omnetpp.org. The citation counts are *very rough* estimates based on Google Scholar.

parison. The authors also note significant effects related to bitrate adaptation, dual antennas and flow interference.

Another way to avoid parameterization issues is to use testbed-specific empirical data. Paper [9] uses a detailed Tossim simulation of a 22-node multi-hop sensor network (mica2-like nodes). Each link in the simulated network was parameterized using signal strength measurements for the corresponding link in the testbed. When using an SNR-based communication model and empirical signal strength measurements, there was very good agreement in packet delivery ratio for several sensor network protocols. But more detailed metrics, such as latency, showed less agreement.

Similarly, paper [5] compares the OMNeT++ MAC simulator with a fully connected four-node network of Tmote-Sky nodes (IEEE 802.15.4). Each link was parameterized with frame reception probabilities based on testbed measurements. This technique gave good agreement for throughput, but the testbed used for parameterization and for testing seem to have been identical, making the results unclear.

Shared evaluation scenarios One contribution of this paper is the idea that well-designed common test and evaluation scenarios will allow developers to detect problems and may increase the user community's confidence in simulation results. The idea of creating a common basis for comparing simulation environments is also explored in [16]. However, this work focuses on configuration and mobility modeling and does not present comparative simulation results.

3. SIMULATORS AND TOOLS

OMNeT++[15] is a public source, highly modular, general purpose discrete event simulation environment. An OMNeT++ simulation is a defined as a network (directed graph) of modules that exchange messages and event notifications. By itself, OMNeT++ does not provide any domain-specific functionality: It "merely" provides the discrete event scheduling and message passing engine, the network and message description languages, and support for configuring and managing simulations, gathering and analyzing statistics, and visualization.

The community has developed OMNeT++-based simulators for a wide range of application areas. OMNeT++'s modular structure often makes it feasible to integrate different simulators, making it potentially very well-suited for the kind of multidisciplinary wireless network research described earlier. The modular structure of the MiXiM wireless host is shown in Figure 1.

The five OMNeT++-based wireless network simulators used in this work are described below. Although the OMNeT++ architecture naturally leads to some similarity of structure, they are largely independent in their design and implementation. This combination of independent simulators and a common management framework is ideal for our experiments. However, the approach and scenarios described here are not specific to OMNeT++.

The simulators used in this experiment are inet[13], an inet fork called inetmanet[13], the mobility-framework [7] and its successor MiXiM [11, 17], and Castalia [2], described in more detail below and in Table 1. The simulators are all open source and freely available for (at least) academic use and have been widely cited in the literature.

Inet is one of the largest and most important OMNeT++ network simulation frameworks and its development team includes some OMNeT++ core developers. Inet mainly provides support for (wired) Internet protocols, but it also includes IEEE 802.11b in both ad hoc and AP modes.

Inetmanet is an independent fork of inet that focuses on wireless networks and MANET protocols and includes models ported from a wide variety of sources. It includes support for IEEE 802.15.4 PANs (but not unslotted CSMA), as well as IEEE 802.11a/e/g and IEEE 802.16. Its IEEE 802.11b implementation shares code with inet, so only inet results are reported.

MiXiM is the successor to several earlier OMNeT++ based simulators, including mobility-fw (below). It focuses on radio signal modeling, supporting not only IEEE 802.11b (ad hoc mode only) and IEEE 802.14.5 (unslotted CSMA only), but also UWB and multi-channel models.

Mobility-fw is the oldest and generally fastest and simplest of the simulators and is based on an earlier version of OMNeT++. The simulators above all have at least a little bit of mobility-fw in their ancestry, so it is included, despite its age. It includes support for IEEE 802.11b and a separate package, ucsma, is a generic unslotted CSMA model that can be configured to model IEEE 802.15.4.

Castalia has the least in common with the other simulators. It is intended for simulating sensor networks, especially body area networks, and emphasizes detailed modeling of the wireless channel and radio, including specialized proprietary noise mapping functionality (not studied here). Castalia provides support for IEEE 802.15.4 PANs (but not unslotted CSMA) and IEEE 802.15.6 BANs, as well as some wireless sensor network MAC's.

ApplTestTool Implementing comparisons in this paper were made much easier by the OMNeT++ ApplTestTool[8] test framework. A core module of the ApplTestTool is the EnsembleApplLayer, which is a configurable traffic generator and statistics collector. Since this module is an application layer, it is relatively easy to port to each of the OMNeT++ simulation frameworks. For this work, Ensem-

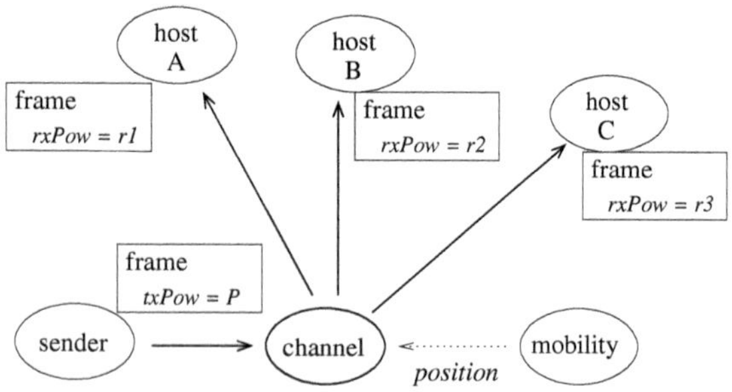

Figure 2: Example of channel propagation.

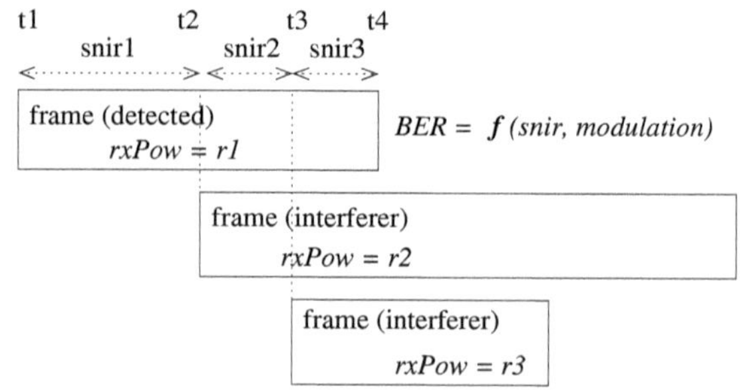

Figure 3: Example of frame reception.

bleApplLayer was ported to run directly above the NIC/MAC modules in each simulator. This prevented the simulators' very different addressing and ARP implementations from affecting the experiment results.

4. SIMULATING WIRELESS NETWORKS

This section outlines in general terms how wireless networks are modeled, with a focus on OMNeT++-based simulators and their parameterization. For our purposes, this functionality can be divided into three parts: The first models the transmission and propagation of RF signals and the second models their detection and decoding. The third models sending and receiving of data and control frames according to MAC and higher layer management protocols.

Transmission For each outgoing transmission, the propagation model determines the resulting RF energy at each receiver. This computation depends on the antenna (ignored here, since all of the simulators assume omni-directional antennas with unit gain) and channel models.

Figure 2 shows the basic structure of most OMNeT++-based wireless simulators: When a host has determined that it is allowed to access the channel, a host (sub-)module sends a message (i.e. an event) to the global channel control module. This "airframe" message includes the transmit power, which the channel control uses to compute the received signal strength at each of the other hosts.

Transmit power is an example of a parameter that is usually controlled externally by higher layer protocol or application. For both IEEE 802.11 and IEEE 802.15.4, limits on transmit power are defined by the standard and available settings are given in various hardware specifications. Since there is no natural default transmit power, experiments are explicitly configured with consistent values.

Propagation All of the simulators in this paper use analytic, rather than empirical, channel models. These are based on the usual well-known models. Nevertheless, both the range of models and their instantiation vary considerably between simulators. Mobility-fw provides only a simple path-loss model, while MiXiM uses complex data structures supporting signal mapping in time, frequency, and space, though this functionality is not explored fully here. MiXiM, Castalia and inet all provide a log-normal shadowing model, but with non-trivial differences in implementation. This variation complicates the problem of creating "comparable" simulation configurations.

Exponential path loss, with signal strength proportional to $d^{-\alpha}$, is the most significant effect. Because this parameter is common to all models, it is explicitly configured. (MiXiM's breakpoint path-loss model, with two path-loss parameters, is not studied here. But its existence highlights the issue of creating parameter-identical comparisons.)

Where possible, this is combined with normal variation with mean μ and standard deviation σ. In general, the default values for these parameters vary between simulators, as there is no agreement on a "typical" environment, so these are also explicitly matched among models that use them. Further parameters are defined as needed, generally following each simulator's defaults. MiXiM's log-normal model takes a time scale parameter, while Castalia's takes a parameter for bi-directional variance.

The channel module sends a copy of the "airframe" message to the appropriate sub-module of each host, annotated with the received signal strength. A host's frame reception model therefore has information about the signal strength of all ongoing transmissions.

All of the simulators do filtering to avoid the cost of computing interferences that are extremely unlikely to affect detection or frame reception. Filtering is intended to have negligible impact on simulation accuracy and is meaningful only in large multi-hop networks, so the default parameters for each simulator's filtering mechanism are used.

Position information is usually managed by the mobility module, but details of mobility modeling and interaction with the channel control module vary considerably between simulators. Only static networks are considered here.

Detection and frame reception The PHY model must determine whether (or which) frame arriving at the receiver is successfully decoded. The model must also determine whether the channel is observed to be clear or busy; this is needed for MAC layers that perform clear channel assessment (CCA). An example of a model in which signal strength is constant, but interference varies over time is shown in Figure 3.

Differences in detection and frame reception models have a large impact on simulation results and the simulators' reception models are even more varied than their transmission and propagation models. This includes not only detection and capture models, but also the detail with which SNIR is modeled over the duration of a frame and the computation that determines whether a frame is correctly received.

The simplest model, mobility-fw, uses a simple SNIR threshold. The first arriving frame is accepted if the minimum SNIR over the duration of the packet is greater than some threshold. Inet uses a similar model. For this model, key parameters are the radio sensitivity and this threshold. As these values are hardware specific, each simulator's default values are used. For IEEE 802.14.5, inetmanet uses a sensitivity parameter of -85dBm (the minimum allowed by the standard) and mobility-fw uses -95dBm (typical of commercial hardware). The threshold value for inetmanet is 4dB, while the mobility-fw uses 5dB (the standard describes 5-6dB as an achievable value for low cost hardware).

Castalia and MiXiM use more complex calculations that model the bit (or symbol) error rate as function of the SNIR and the frame modulation. The two vary somewhat in their level of detail modeling of time-variation of interference and modulation over the duration of the frame, but the underlying computations are well defined. CCA is determined by radio sensitivity, and MiXiM's model also applies start of frame detection (SFD). (The inetmanet reception model can include a BER calculation, but this does not seem to be fully implemented.)

A successfully decoded frame is passed up to the MAC layer and possibly higher layers. Alternatively, the MAC may be notified that the channel has become busy or idle.

This discussion again highlights the complexity surrounding "consistent" parameterization of different simulators: Different models take different parameters or use different semantics for similar parameters. Even models that take the same parameters may define different, but entirely plausible defaults (e.g. standard defined limits vs typical hardware).

MAC protocols Simulating communication over a wireless channel is a challenging problem in finding suitable abstractions for representing complex physical phenomena. By contrast, modeling the handing of data frames in the MAC layer is more straightforward.

The simulation is intended to mirror the operation of well-defined protocols like IEEE 802.15.4 or IEEE 802.11, which are specified by standards documents. Default values for parameters such as header lengths and backoff and retransmission limits are defined in standards documents.

It is reasonable to expect simulators' default configuration should match these defaults. Among the simulators in this work, only inet uses non-default MAC parameters (a smaller initial contention window and large retry limit for IEEE 802.11). These values were not modified for this study.

The main differences between simulators arise from the detail with which NIC internal operations such as processing and radio turnaround are modeled. These timing values parameters are generally defined in the standard or in hardware specifications. Each simulator's defaults are assumed to represent typical hardware and are not modified.

Because protocols are well defined, accurate protocol simulation is largely dependent on good software engineering. However there may still be issues where the specification is ambiguous (e.g. interframe space after failed channel access in IEEE 802.15.4). Clarity is also important for behavior that is defined as implementation dependent (e.g. IEEE 802.11 bitrate adaptation). Similar variation is observed – for similar reasons – among hardware from different vendors.

5. NOMINAL TRANSMIT RANGE

This section describes the radio characterization experiments that are used to determine the nominal transmit range for each configuration. This transmit range is used to create scenarios that are normalized across simulators.

By now it should be clear that it is not possible to create identical parameterizations for these simulators, due to their very different models of wireless propagation and frame reception. Nominal transmit range is intended to encapsulate the combined effects of the various propagation and frame reception models and parameters. Topologies that are defined in terms of transmit range have similar effective node densities, despite large differences in absolute range. However, this cannot eliminate the inherent differences in propagation and reception models and the topologies may have different interference and other characteristics.

The radio characterization experiment measures the packet reception rate (PRR) sending 1000 frames from a sender to a receiver at each of several distances. The nominal transmit range is defined as the largest distance that gives at least 92% PRR in the absence of interfering transmissions. For IEEE 802.11, the payload is 1024-bytes and for IEEE 802.15.4, it is 100-bytes.

The IEEE 802.11 results are shown in Table 4. The transmit ranges for MiXiM and mobility-fw are in agreement, which is unsurprising given their shared history. The values are large compared to realistic networks, however. The range for both inet and inet with log-normal shadowing configurations are smaller, reflecting the higher SNIR threshold required to successfully receive the frame. Including log-normal shadowing with $\sigma = 10$ increases the nominal range. This is due to the gradual decrease in the PRR, which first drops below 99% at distances similar to the much sharper transition in the simple path loss model.

The IEEE 802.15.4 results are shown in Table 5. Castalia and inetmanet simulations use IEEE 802.15.4 PAN in direct transfer mode, while MiXiM and mobility-fw use unslotted CSMA. However, PAN operation do not significantly affect the radio characterization: The test transmissions are simply slightly delayed to allow association to complete. The transmit ranges, especially for MiXiM and the mobility-fw, seem rather large, but there does not seem to be any clear determining factor.

	MiXiM	mobility fw	inet-manet	Castalia
shadowing	$\mu = 0, \sigma = 0$	no	no	$\mu = 0$, $\sigma = 4$ bi-$\sigma = 1.0$
sensitivity	-100dBm	-95 dBm	-85 dBm	-95dBm
threshold	n/a	5dB	4dB	n/a
BER model	yes	no	no	yes
PAN	no	no	yes	yes

Table 2: IEEE 802.15.4 simulation parameters.

	MiXiM	inet	mobility fw
shadowing	$\mu = 0, \sigma = 10$ t = 1s	$\mu = 0, \sigma = 10$	no
sensitivity	-119 dBm	-85 dBm	-119 dBm
threshold	0.1259 (-9dB)	4dB	-9dB
cwMin	31	7 (data)	31
cwMax	1023	1023	1023
retry (short)	4	7	4
retry (long)	7	7	7

Table 3: IEEE 802.11b simulation parameters.

6. SCENARIO EXPERIMENTS

In this section, we present results from three scenarios. The very stylized sync-star scenario is used to examine contention and backoff behavior and is the core of comparison

	MiXiM ± 2.5m	MiXiM w/ shadowing μ=0, σ=10 ± 2.5m	mobility fw ± 2.5m	inet ± 1m	inet w/ shadowing μ=0, σ=10 ± 1m
2 mW					
α=3.0	287.5	418.9	287.5	40.0	62.0
4.0	67.5	92.0	72.5	16.0	22.0
10 mW					
α=3.0	492.5	716.0	492.5	67.0	105.0
4.0	102.5	138.0	102.5	24.0	33.0
50 mW					
α=3.0	837.5	1224.0	847.5	117.0	179.0
4.0	157.5	207.0	157.5	36.0	50.0

Table 4: Nominal transmit ranges IEEE 802.11b @ 2Mbps: 92% frame reception rate with 1024 byte payload.

	MiXiM ± 1m	mobility-fw ± .5m	inet-manet ± .5m	Castalia ± .5m
-25dBm				
α = 2.0	81.5	31.5	9.5	2.5
2.5	33.0	15.5	6.5	2.5
3.0	18.0	9.5	4.5	1.5
-15dBm				
α = 2.0	255.0	99.5	31.5	8.5
2.5	85.0	39.5	15.5	5.5
3.0	41.0	21.5	9.5	4.5
-10dBm				
α = 2.0	453.0	176.5	55.5	15.5
2.5	135.0	62.5	25.5	9.5
3.0	60.0	31.5	14.5	6.5
-5dB m				
α = 2.0	805.0	314.5	99.5	28.5
2.5	212.0	99.5	39.5	14.5
3.0	87.0	46.5	21.5	9.5
0dBm				
α = 2.0	> 1K	559.5	176.6	49.5
2.5	260.0	157.5	62.5	22.5
3.0	103.0	67.5	31.5	13.5

Table 5: Nominal transmit ranges IEEE 802.15.4: 92% frame reception rate over 100 byte payload.

framework. The slightly more realistic WSN-like scenario represents traffic to a central gateway and the AP-like scenario represents traffic to and from an access point.

In many cases, there is considerable variation between simulation results and some speculation is made as to the sources of these differences. But it should be emphasized that the goal of this work is *not* to perform five-way cross-simulator debugging, but rather to demonstrate the value of the comparison scenarios to highlight behavior worth further investigation.

6.1 Sync-star

As the name suggests, this is a star topology, with nodes sending synchronized, periodic transmissions to the central receiver. This scenario is intended to highlight contention and backoff behavior by ensuring that all senders are equidistant from the receiver and simultaneously contending for the channel. The period is such that the contentions are essentially independent experiments; each frame has either succeeded or exhausted its retry opportunities in each period.

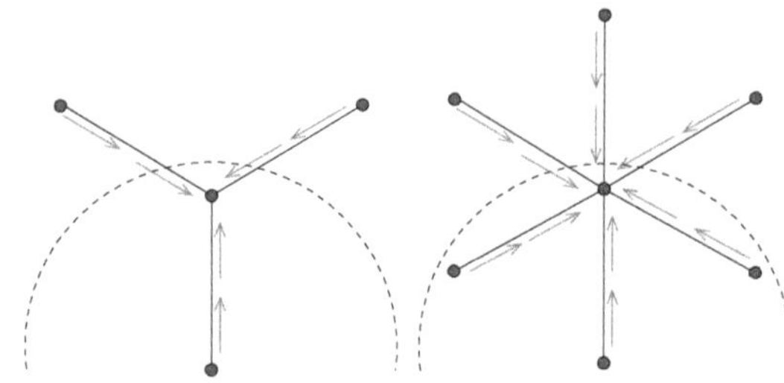

Figure 4: Hidden terminals in the sync-star scenario. The dotted line shows the range. The communication model determines which nodes are able to detect each other's transmissions.

Varying the radius and density of the star controls the existence of hidden terminals (Fig. 4). In these experiments, the radius is either 0.2 × nominal transmit range, giving a fully connected network with high signal strengths, or 0.9 × nominal transmit range, so that nodes on opposite sides of the star are hidden terminals.

The experiment measures the total number of packets received, the packet reception rate and the latency, for networks with varying numbers of senders. The latency is the per-packet application latency. Because this scenario explicitly avoids queuing, the latency reflects channel access time for each frame, plus any processing delays modeled by the simulator. (Note that the 95% CI is for the mean latency across multiple simulation runs, not the variation in latency within a simulation.)

IEEE 802.15.4 The IEEE 802.15.4 results are shown in Figure 5. Each sender generates 250 packets with 100 bytes payload and the experiment is repeated up to 10 times.

In the fully connected network, mobility-fw and MiXiM show peak throughput with around seven to nine senders. Mobility-fw, which has a simple SNIR threshold packet reception model, gives a higher packet delivery and lower latency than MiXiM and supports more senders. The reason for this is unclear, since all frames are received by all nodes with high signal strength, the effect of differences in interference and reception models should be small.

The opposite is true in the hidden terminal case, where MiXiM's more sensitive radio model presumably allows senders to detect competing transmissions more accurately. For both MiXiM and Castalia, their probabilistic reception models also seem to lead to more variation between simulation runs.

Unsurprisingly, the beacon-enabled PAN's modeled by inetmanet and Castalia support significantly more traffic than either of the unslotted CSMA networks, albeit with high latency (unreasonably so in the case of inetmanet). In the hidden terminal case, the Castalia PAN continues to have the expected high throughput, but inetmanet PAN performs worse than the CSMA networks. This behavior almost certainly reflects some significant issue in the inetmanet implementation. (Inetmanet's IEEE 802.15.4 simulation is documented as preliminary work.)

Another interesting feature is the throughput-PRR anomaly in the hidden terminal case: When three senders are distributed on opposite sides of the star, they cannot detect

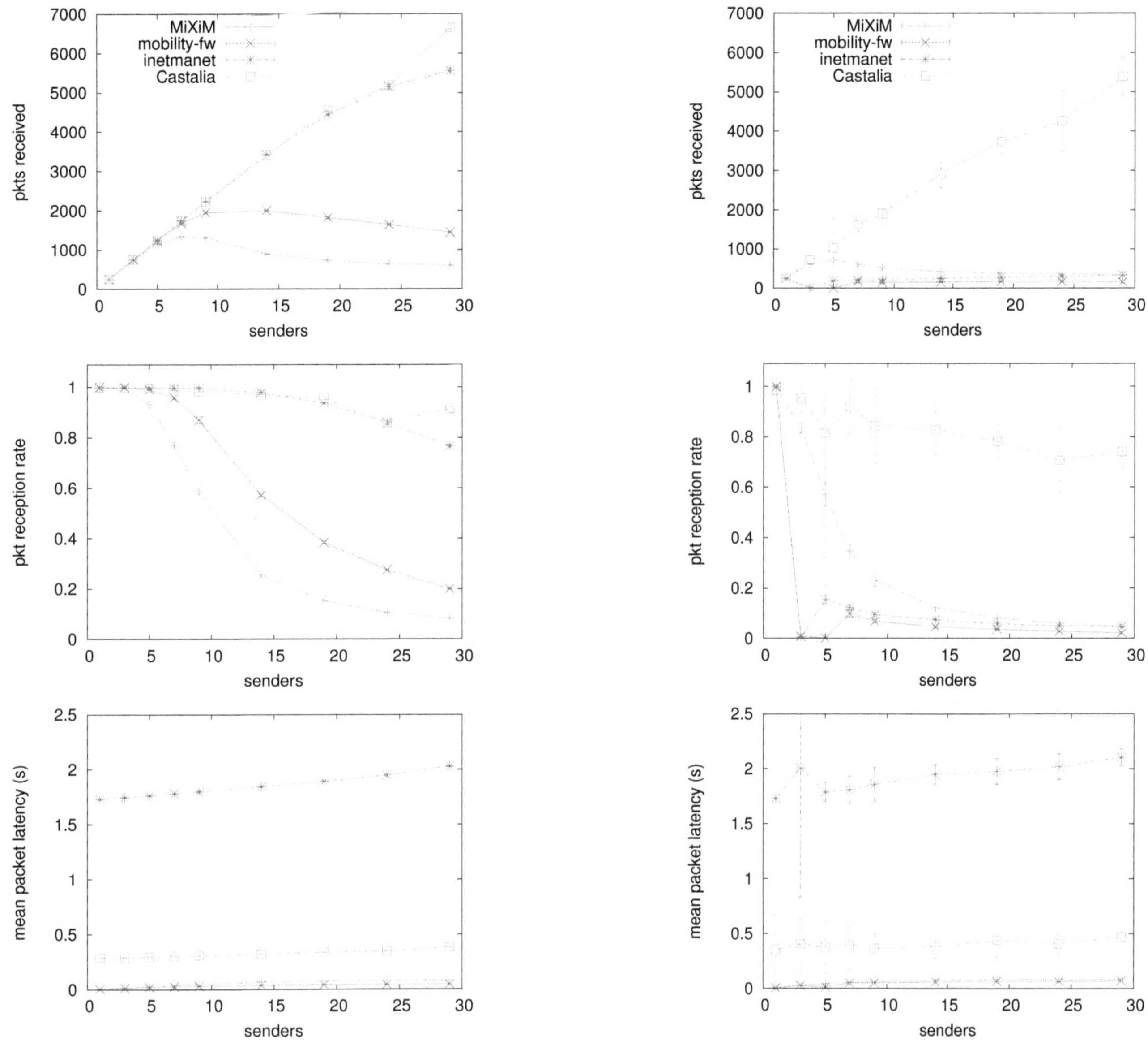

Fully connected (radius = 0.2 × nominal range). Hidden terminals (radius = 0.9 × nominal range).

Figure 5: Sync-star: Star topology with synchronized senders (IEEE 802.15.4). Path loss α = 2.5; shadowing μ=0, σ=4; txPower = -10dBm. The error bars give the 95% CI over 10 simulation runs.

each other's transmissions, so they consistently, incorrectly assess the channel as clear, resulting in collisions at each transmission attempt. As the number of nodes grows, node become closer to their left- and right-hand neighbors and are again able to defer to some contending transmissions (Fig. 4). The effect is seen in both mobility-fw and inetmanet, though more strongly in the former, where it persists even with five senders. Even Castalia shows a small effect; there is considerably more variability in the PRR and latency for small numbers of senders. By contrast, MiXiM's sensitive radio model seems to allow it detect to competing senders, even in sparse networks.

IEEE 802.11 The IEEE 802.11 experiment is shown in Figure 6. Each sender generates 100 packets with 1024-byte payload, transmitting one packet every 10s.

For the fully connected case, all the simulators agree on nearly 100% PRR in all cases, though both inet variants show more rapidly increasing latencies than either MiXiM or mobility-fw. For both MiXiM and inet, the variant that includes shadowing shows larger latencies than its simple-path loss counterpart.

The hidden terminal case is more interesting. Mobility-fw shows threshold behavior, with nearly perfect PRR up to seven senders, and about 10% PRR thereafter. Inet with simple path loss has significant loss with three senders and fails to deliver any packets with more than five senders, despite having a similar frame reception model. This result almost certainly signifies some issue in inet, given its much more plausible behavior under log normal shadowing.

The other three simulators have more generally consis-

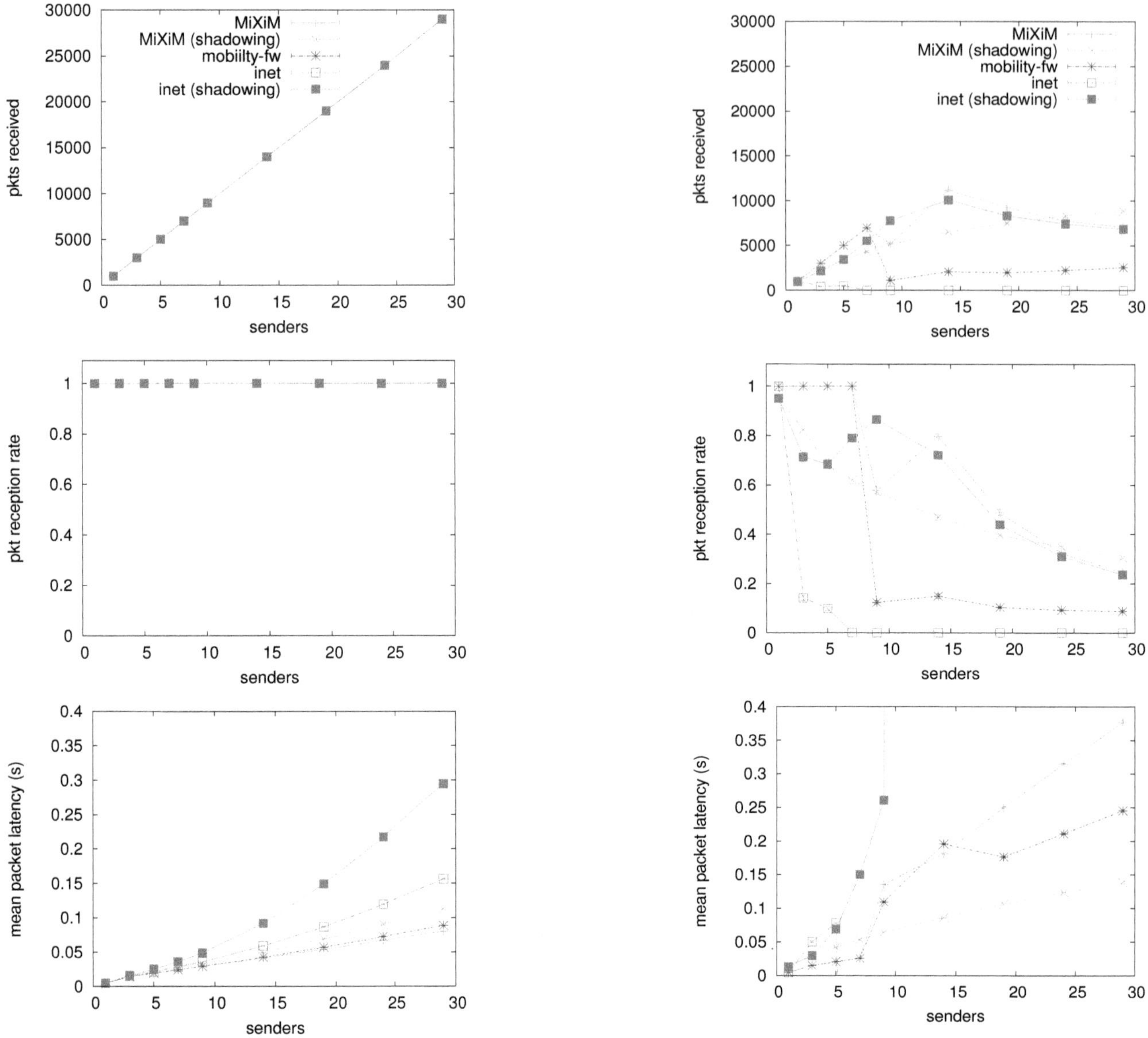

Fully connected (radius = 0.2 × nominal range).

Hidden terminals (radius = 0.9 × nominal range).

Figure 6: Sync-star: star topology with synchronized senders (IEEE 802.11b). Path loss α = 3.5; shadowing μ=1, σ = 10; txPower = 50mW. The (very small) error bars give the 95% CI over 10 simulation runs.

tent behavior. MiXiM follows mobility-fw in having near 100% PRR up to seven senders, but its PRR then declines more gradually, though the anomaly at nine senders remains mysterious as it does not seem to be related to the topological anomaly described above. When log-normal shadowing is added, MiXiM's results are completely smooth. Inet with log-normal shadowing seems to show a trace of the anomaly seen in the IEEE 802.15.4 data. At higher loads, all three have similar throughput, but inet's latency becomes extremely high. This seems to be associated with a very high number of duplicate packets (i.e. lost ACKs) and is not seen in the MiXiM data.

6.2 WSN-like traffic

The second scenario, called "WSN-like", is a simple representation of single-hop sensor network, with randomly placed senders generating data and forwarding it to a sink. Compared to sync-star, this scenario introduces differences in received signal strength and more complex contention timing scenarios. Only results for IEEE 802.15.4 are reported, further results are available online.

A designated node (sink) is placed "near" the center of the rectangular field. Specifically, its x,y coordinates are uniformly randomly distributed within ±20% of the center. The other nodes are uniformly and randomly distributed over the field, which is dimensioned such that all nodes are within nominal transmit range of the sink. Consequently, the network is mostly connected: the field is a square whose edge is approximately 1.2 × nominal range. In these experiments, senders generate frames with exponentially dis-

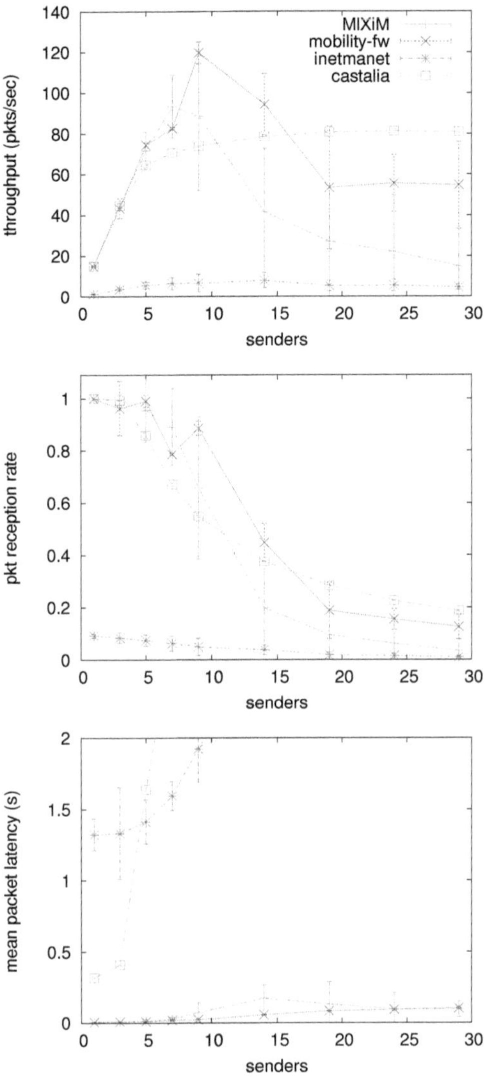

Figure 7: IEEE 802.15.4 WSN-like scenario: random topology, exponentially distributed uplink traffic. Path loss α=3.0; shadowing μ=0, σ=4; txPower = -5dBm. Error bars show 95% CI over 10 runs.

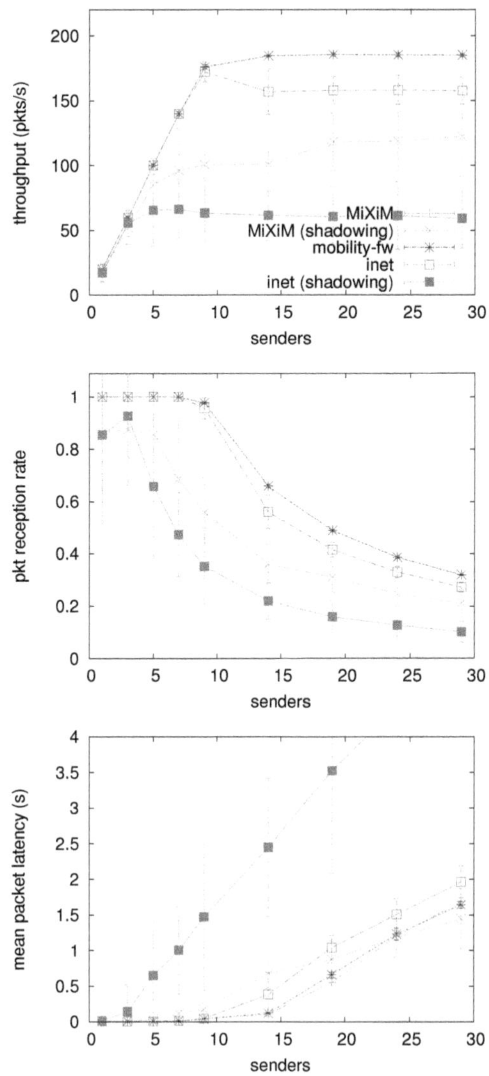

Figure 8: IEEE 802.11 AP-like scenario: random topology, exponentially distributed uplink and downlink traffic. Path loss α = 3.5; shadowing μ=0, σ = 10; txPower = 50mW. Error bars show 95% CI over 10 runs.

tributed arrivals at rate λ = 15 and transmit them to the sink. Each experiment is 300 s long and is repeated up to 10 times. The results are shown in Figure 7. The metrics are as in the sync-star experiment, though here latency most definitely reflects queuing as well as channel access delay.

MiXiM and mobility-fw both show a fairly plausible peak application throughout of about 80-95 kbps on the 250kbps channel. As in the fully connected sync-star case, mobility-fw has slightly higher throughput. MiXiM's higher variation between simulations may reflect increased sensitivity to topology: With mobility-fw's simple SNIR threshold radio model, nodes are likely to be either be hidden terminals, or not. MiXiM's radio model gives probabilistic losses in intermediate cases.

For the Castalia PAN network, the peak throughput is slightly lower than in the unslotted CSMA case, but the PAN scheduling allows the network to maintain that throughput at high load. This is very much the expected behavior.

The outlier is inetmanet, which shows poor performance in all cases, contrasting with its plausible (though very high latency) throughput in the sync-star scenario. Again, this model is documented as preliminary and the result clearly reflects this (some simulations did not run to completion).

6.3 AP-like traffic

The third scenario, called "AP-like", is similar to the WSN-like scenario. The topology is defined the same way, but with both uplink and downlink traffic to the designated node (AP). This traffic is roughly similar to an IEEE 802.11 AP although it does not reflect the flow control or timing associated with TCP or multimedia traffic. IEEE 802.11 results are presented, further results are available online.

Each sender generates 1024-byte packets, with exponentially distributed arrivals with rate λ = 10 and transmits them to the AP. The AP also generates traffic, with intensity $\lambda_{AP} = \lambda \times numSenders$, and the destination of each frame being randomly chosen from among the senders. Each experiment is 300 s long and is repeated up to 10 times. For this scenario, IEEE 802.11 RTS/CTS is used, otherwise parameters are as in the sync-star experiment.

The results are shown in Figure 8. Three simulators show similar throughput and packet reception rate. The AP-like scenario is a mostly connected network, so it is not surprising that it largely reflects the good agreement between simulators seen in the sync-star case. The network capacity is a quite plausible 1.4Mbps for the mobility-fw and MiXiM, inet is slightly lower. The main differences between simulators are small variations in latency, which may reflect differences in queuing as much as channel access.

The MiXiM and inet models that include log-normal shadowing both show much lower throughput than their simple path-loss counterparts. This result is especially confusing given that both of these simulators showed relatively higher throughput at higher loads in the hidden terminal sync-star scenario. (It is also noted that some MiXiM simulations with log-normal shadowing did not run to completion.)

7. CONCLUSION

The research community envisions increasingly complex applications in mobile social networks and wireless sensor networks. There is therefore a growing need for trustworthy simulation components for core protocols elements such as IEEE 802.11 and IEEE 802.15.4.

This work contributes to that goal by introducing a scenario framework for comparing results obtained in different simulation environments. Because it is not possible to create equivalent parameterizations of simulators with very different propagation and frame reception models, the framework defines nominal transmit range as an (imperfect) metric that encapsulates channel parameters and frame reception model.

Normalized test and evaluation scenarios are then defined in terms of nominal transmit range. Experimental results for five different OMNeT++-based simulations of IEEE 802.11 and IEEE 802.15.4 are presented. The differences that they uncovered demonstrate their effectiveness in comparing simulation results and highlighting behavior worth further investigation, even when the cause of the anomaly remained unclear. This approach allows us to leverage the diversity of models and implementations to identify problems, while still respecting valid differences between simulators.

Clearly, a result showing good agreement among simulators (or even the identification of a single outlier) would have been more reassuring. However, a long-term goal is that the framework described here (which is not specific to OMNeT++) will lead to the adoption of common test and evaluation scenarios and increase the quality of and confidence in simulation results. Most of the code and further output are available at `http://www.sics.se/nets`.

8. REFERENCES

[1] L. Bergamini, C. Crociani, A. Vitaletti, and M. Nati. Validation of WSN simulators through a comparison with a real testbed. In *Proc. of the 7th ACM Workshop on Performance Evaluation of Wireless Ad Hoc, Sensor, and Ubiquitous Networks*, PE-WASUN, 2010.

[2] A. Boulis. Castalia: Revealing pitfalls in designing distributed algorithms in WSN. In *Proc. of the 5th Int'l Conf. on Embedded Networked Sensor Systems*, SenSys, 2007.

[3] M. Bredel and M. Bergner. On the accuracy of IEEE 802.11g wireless LAN simulations using OMNeT++. In *Proc. of the 2nd Int'l Conf. on Simulation Tools and Techniques*, Simutools, 2009.

[4] D. Cavin, Y. Sasson, and A. Schiper. On the accuracy of MANET simulators. In *Proc. of the 2nd ACM Int'l Workshop on Principles of Mobile Computing*, POMC, 2002.

[5] U. M. Colesanti, C. Crociani, and A. Vitaletti. On the accuracy of OMNeT++ in the wireless sensor networks domain: simulation vs. testbed. In *Proc. of the 4th ACM Workshop on Performance Evaluation of Wireless Ad Hoc, Sensor, and Ubiquitous Networks*, PE-WASUN, 2007.

[6] P. Di, Y. Houri, K. Kutzner, and T. Fuhrmann. Towards comparable network simulations. Interner Bericht 2008-9, Dept. of Computer Science, Universität Karlsruhe (TH), Aug 2008.

[7] W. Drytkiewicz, S. Sroka, V. Handziski, A. Köpke, and H. Karl. A mobility framework for OMNeT++. TU-Berlin Tech. Report, 2003.

[8] L. M. Feeney. Poster abstract: Two simple tools for testing wireless communication modules in OMNeT++. In *Proc. of the 4th Int'l Workshop on OMNeT++*, 2011.

[9] G. P. Halkes and K. G. Langendoen. Experimental evaluation of simulation abstractions for wireless sensor network mac protocols. *EURASIP J. Wirel. Commun. Netw.*, 2010:24:1–24:2, 2010.

[10] S. Ivanov, A. Herms, and G. Lukas. Experimental validation of the ns-2 wireless model using simulation, emulation, and real network. In *Proc. of the 4th Workshop on Mobile Ad-Hoc Networks*, WMAN, 2007.

[11] A. Köpke, M. Swigulski, K. Wessel, D. Willkomm, P. K. Haneveld, T. Parker, O. Visser, H. S. Lichte, and S. Valentin. Simulating wireless and mobile networks in OMNeT++: The MiXiM vision. In *Proc. of the 1st Int'l Workshop on OMNeT++*, 2008.

[12] J. Lessmann, P. Janacik, L. Lachev, and D. Orfanus. Comparative study of wireless network simulators. In *Proc. of the 7th Int'l Conf. on Networking*, ICN, 2008.

[13] The OMNeT++ homepage. `http://www.omnetpp.org`.

[14] K. Tan, D. Wu, A. Chan, and P. Mohapatra. Comparing simulation tools and experimental testbeds for wireless mesh networks. *Pervasive and Mobile Computing*, 7(4):434–448, 2011.

[15] A. Varga. OMNeT++. In K. Wehrle, M. Günes, and J. Gross, editors, *Modeling and Tools for Network Simulation.* Springer-Verlag, 2010.

[16] T. Voigt, J. Eriksson, F. Österlind, R. Sauter, N. Aschenbruck, P. J. Marrón, V. Reynolds, L. Shu, O. Visser, A. Koubaa, and A. Köpke. Towards comparable simulations of cooperating objects and wireless sensor networks. In *Proc. of the 4th Int'l Conf. on Performance Evaluation Methodologies and Tools*, VALUETOOLS, 2009.

[17] K. Wessel, M. Swigulski, A. Köpke, and D. Willkomm. Mixim – The physical layer: An architecture overview. In *Proc. of the 2nd Int'l Workshop on OMNeT++*, 2009.

Measuring Channel Occupancy for 802.11 Wireless LAN in the 2.4 GHz ISM Band

Till Wollenberg
Universität Rostock
Rostock, Germany
till.wollenberg@uni-rostock.de

Sebastian Bader
Universität Rostock
Rostock, Germany
sebastian.bader@uni-rostock.de

Andreas Ahrens
Hochschule Wismar
Wismar, Germany
andreas.ahrens@hs-wismar.de

ABSTRACT

In this paper we focus on measuring channel occupancy in the 2.4 GHz ISM band in a way that transmissions from Wireless LAN stations and other spectrum users can be distinguished. Most current 802.11 hardware does not offer methods to capture channel occupancy caused by non-802.11 transmissions accurately. We therefore combine a commodity 802.11 Wireless LAN card with an inexpensive 2.4 GHz RF transceiver IC which we use for spectrum measurements. We build a probabilistic model which allows to calculate the overall channel occupancy and to determine the fractions of 802.11 and non-802.11 activity. Our model covers adjacent channel interference, detects both narrow-band and wide-band interferers, and adapts to different signal strengths. We evaluate our approach in a test bed with controlled interference, estimate the achievable precision, and identify open problems.

Categories and Subject Descriptors

C.2.1 [**Computer-Communication Networks**]: Network Architecture and Design—Wireless Communication

Keywords

802.11, Channel Occupancy, Spectrum, Measurement, Model

1. INTRODUCTION

In 2012, 15 years after its first standardization, IEEE 802.11 Wireless LAN is omnipresent. In densely populated regions, the 2.4 GHz ISM band can often be found to be crowded. Particularly in this band, various other spectrum users make the congestion situation worse [8, 3, 6]. Despite Wireless LAN, the band is also used by Bluetooth, ZigBee, wireless input devices, cordless phones, audio and video transmission systems, microwave ovens and suchlike.

Determining the performance of a Wireless LAN link is useful to estimate the achievable network quality in terms of throughput, latency, or packet loss. This information can be used for example to proactively select the optimal compression ratio for multimedia streaming, to refrain from offering services to the user which are not usable under the current network quality, to select an optimal Wireless LAN channel in network planning, or to identify causes for low overall network performance.

The quality of a wireless link is determined by the packet error rate (PER) and the channel occupancy [5, 2]. The PER for each supported 802.11 data rate can be predicted from the channel state information [4]. Since the wireless channel is a shared medium only one station can transmit at any given time. The channel occupancy[1] describes the probability to find the channel idle when a station wants to transmit. Carrier sense and collision avoidance techniques are used to ensure that all stations within reception range get their fair share of channel time. However, these techniques only work for decodable 802.11 frames. Transmissions from other spectrum users, from 802.11 stations on neighboring channels, and from weak stations cannot be decoded and are considered as interference. They result in a reduction of free channel time and can cause collisions when they occur during the transmission of a frame. In order to estimate the channel occupancy properly, we have to distinguish between channel occupancy caused by (decodable) 802.11 transmissions on the one hand and channel occupancy caused by interference on the other hand. In the latter case we cannot expect any cooperative behavior.

In this work, we aim for a model-driven approach to estimate the average channel occupancy rate in time frames of the order of 30 seconds. For deciding whether the channel is idle or busy at a given moment, our goal is to emulate the behavior of an average Wireless LAN network adapter.

2. BACKGROUND AND CONCEPT

For assessing whether the channel is idle or busy the IEEE 802.11-2007 standard defines both physical and virtual carrier sense (CS) functions. The physical CS is done through the clear channel assessment (CCA) function as part of the physical layer (PHY). For OFDM PHY, the 802.11 standard defines an energy detection scheme with fixed thresholds. In order to prevent false detects of a busy channel caused by spurious noise, a permanent high noise floor, or transmissions on neighboring channels, hardware vendors may have decided to implement more advanced sensing algorithms. However, details about these methods are usually not pub-

[1] also referred to as *channel busy time*, *channel load*, or *airtime utilization*

MSWiM'12, October 21–25, 2012, Paphos, Cyprus.

lished. Notable exceptions are U.S. Patents No. 7,349,503 and No. 7,522,669 by Atheros Qualcomm which give an impression of the methods used in practice.

Due to the heterogeneity of Wireless LAN chipsets we cannot assume any details about the CCA algorithm used by a specific station. For our model we therefore decided not to focus on a specific behavior and reducing the error for that special case. Instead, we decided to keep the model more general and accept a greater error in specific cases in order to obtain a good accuracy across different chipsets.

In the context of our work we distinguish between the three terms CO_{WLAN}, $CO_{\text{interference}}$, and CO_{total}. CO_{WLAN} is the channel occupancy caused by 802.11 transmissions while $CO_{\text{interference}}$ is the channel occupancy caused by "interference", i.e. by other spectrum users or undecodable 802.11 transmissions. We call the overall channel occupancy CO_{total}. We estimate CO_{WLAN} from successfully decoded 802.11 frames. Since strong interference likely corrupts a transmitted frame and consequently prevents a successful decoding of that frame, we approximate CO_{total} as follows:

$$CO_{\text{total}} \approx CO_{\text{WLAN}} + CO_{\text{interference}} \tag{1}$$

Since we cannot measure $CO_{\text{interference}}$ directly, our approach is based on measuring CO_{total} and CO_{WLAN} and deriving $CO_{\text{interference}}$ from that.

2.1 Determining CO_{WLAN}

For determining the average occupancy for a particular channel or set of channels, we configure the wireless adapter to decode and deliver all correctly received 802.11 frames to our measurement application. This operating mode is usually referred to as "monitor mode". We get a list of all received frames with the length, data rate, preamble length, and channel for each frame. From this data we can compute the duration of each transmission. We now sum the durations of all frames transmitted on a particular channel, divide the result by the total time spent on this channel and hereby get the fraction of busy time for this channel.

A channel hopping scheme can be used to survey more than one channel. In this case, frames transmitted on all but the current channel are normally not received and therefore not included in the computation of the channel occupancy. However, we assume that for dwell times being short in relation to the overall survey time the error is acceptable. For our model we consider 13 distinct channels in the 2.4 GHz band. Since the channels overlap, frames sent on one particular channel cause the neighboring channels to be sensed busy with some probability as well. We take this into account by distinguishing between the events of "frame transmission on channel i" and "channel i is busy".

Below, we use the following notation. $P(S_i)$ is the probability of a transmission on channel i, whereas $P(B_i)$ is the probability that channel i is sensed busy. $P(B_i|S_j)$ is the conditional probability that channel i is sensed busy when there is a transmission on channel j.

The fraction of busy time for each channel which we computed above is identical to $P(S_i)$ and therefore known. Furthermore, $P(B_i|S_j)$ is determined by the effect of adjacent channel interference (ACI) to the CCA procedure. We use the values from Table 1 which we determined in our experimental setup described in § 3. Eventually, we want to compute $P(B_i)$ for every "interesting" channel i. To simplify the following steps we assume that the events S_i and S_j are

Channel separation	Probability	
	DSSS,low duty	OFDM,high duty
0	100 %	100 %
1	99.3 %	99.1 %
2	99.0 %	98.6 %
3	98.3 %	88.6 %
4	20.0 %	26.4 %

Table 1: Average probability of a busy channel $P(B_i|S_j)$ for transmissions on neighboring channels.

independent for $i \neq j$ and discuss this decision towards the end of this section. Now we define $S = \{S_1, \ldots, S_{13}\}$ and $T \in \mathcal{P}(S)$ whereas $\mathcal{P}(S)$ is the power-set of S.

$P(T)$ is the probability for the case that all $s \in T$ are true while all $s \notin T$ are false. Since we assume all S_i to be independent, we can calculate $P(T)$ as follows:

$$P(T) := \prod_{t \in T} P(t) * \prod_{t \notin T} (1 - P(t)) \tag{2}$$

We can now calculate the probability to find channel i busy given a set T of channels on which a transmission occurs:

$$P(B_i|T) = \sum_{S_j \in T} P(B_i|S_j) * P(S_j) \tag{3}$$

Now let TT be the set of all T. Since TT contains all possible elementary events the law of total probability applies:

$$P(B_i) = \sum_{T \in TT} P(B_i|T) * P(T) \tag{4}$$

We have now calculated $P(B_i)$, the probability to find channel i busy based on our observations of transmitted frames. $P(B_i)$ is equivalent to CO_{WLAN} for channel i.

2.2 Determining CO_{total}

We found that almost no current Wireless LAN chipset exposes functions for spectrum measurement. Notable exceptions are chipsets from the AR9200 family manufactured by Atheros Qualcomm, which offer some severely limited spectrum measurement functions [8]. To measure channel occupancy, usually dedicated measurement hardware is used [7, 9]. These devices usually imply high cost and power consumption. We aim towards a solution that can be integrated with a broad range of wireless devices and is efficient in terms of size, cost, and power consumption. Therefore, we decided to use a separate IC for spectrum measurement.

We use the consumer-grade AirView USB spectrum analyzer manufactured by Ubiquiti Networks. This device is based on a low-cost 2.4 GHz transceiver IC by Texas Instruments. The IC is programmed to sweep over the 2399–2485 MHz frequency range in steps of 500 kHz. For each step, herein referred to as subchannel, the IC reports a power level in dBm with a resolution of 1 dBm. The sweep time and the number of channels are fixed. We get ≈ 3.7 reports per second with measurements of all 173 subchannels. This corresponds to a dwell time of ≈ 1.56 ms per subchannel.

We want to determine the fraction of busy time of a Wireless LAN channel j in a time frame of n sweeps. We therefore first apply a threshold function $\text{thr}(x)$ to every subchannel measurement $SC_i(t)$ to classify idle ($x < \gamma$) and busy ($x \geq \gamma$) samples. We then compute the average occupancy

Case	CO_{total}	CO_{WLAN}	RX_{CLR}
A	0.050 (0.001)	0.031 (0.001)	0.040 (0.001)
B	0.856 (0.012)	0.812 (0.011)	0.858 (0.001)
C	0.079 (0.012)	0.056 (0.001)	0.063 (0.005)

Table 2: Average channel occupancy (standard deviation) for three test cases. A: DSSS, frame duration 3.2 ms, frame rate 10/s. B: OFDM, 268 µs, 3045/s. C: OFDM, 536 µs/40 µs (alternating), 197/s.

Ch.	A		B	
sep.	CO_{total}	RX_{CLR}	CO_{total}	RX_{CLR}
1	0.052	0.0397	0.880	0.8524
2	0.041	0.0396	0.845	0.8475
3	0.008	0.0393	0.804	0.7615
4	0.000	0.0080	0.000	0.2272

Table 3: Average CO_{total} for transmissions on adjacent channels in comparison to RX_{CLR}. Test cases A and B according to description of Table 2.

$\overline{SC_i}$ for all subchannels i in a 16.5 MHz span (33 subchannels) around the subchannel j that corresponds to the center frequency of the observed Wireless LAN channel.

$$\text{thr}(x) = \begin{cases} 0, & \text{if } x < \gamma \\ 1, & \text{if } x \geq \gamma \end{cases}$$

$$\overline{SC_i} = \frac{1}{n} \sum_{t=1}^{n} \text{thr}(SC_i(t)) \tag{5}$$

$$CO_{\text{total}}(j) = \max(\overline{SC_i}), i = j - 16 \,..\, j + 16$$

We tried different functions for combining the subchannel occupancies SC_i and found the maximum to work best for both wide-band signals (> 10 MHz) as well as narrow-band signals (≤ 1 MHz). An important problem is the long channel dwell time of 1.56 ms compared to the duration of 802.11 frames which we found usually to be below 1 ms for OFDM transmissions. This may create a round-up effect resulting in a slight overestimation of the channel occupancy. The slow dwell time also results in an undersampling for short transmissions which can in part be compensated for by averaging over a time window. In our experiments we found a window size of $n = 50 \,..\, 100$ (13 .. 27 s) to be a useful trade-off between accuracy and time resolution.

The computation of CO_{total} is strongly dependent on the choice of γ. Instead of using a fixed threshold, we found that γ needs to be adjusted dynamically to reflect different CCA implementations and signal level constellations. We propose to adjust γ to minimize the difference between CO_{WLAN} and CO_{total} in time frames with very high CO_{WLAN}. In such time frames it is plausible to assume that there is almost no interference and we can therefore assume $CO_{\text{interference}}$ to be negligible. If there was significant interference, the sending station would sense the channel to be busy and/or the receiving station would not be able to decode the frames due to collisions and consequently CO_{WLAN} would be low.

3. EXPERIMENTAL EVALUATION

We built a small test bed to evaluate the general fitness of our model and the achievable accuracy. The test bed consists of two Wireless LAN access points (AP) based on Atheros 5212 Wireless LAN chipsets. We configure one AP as sender and set the other one to "monitor mode" and use it as receiver. Our testbed also contains one AirView USB spectrum analyzer, a digital storage oscilloscope (Agilent DSO90804A) and a Bluetooth module which we use as narrow-band interferer. All components except the Bluetooth module are connected with shielded RF cables and RF combiners. We use RF attenuators to compensate for the missing free-space path loss.

We first generate bursts at different signal levels and with different durations to evaluate the measurement accuracy of the AirView spectrum analyzer. Our experiments reveal that the readings from the AirView spectrum analyzer show an average deviation of -0.9 dBm from the actual signal level across durations of 400 µs .. 2 ms. We consider this accuracy to be entirely sufficient for our purposes.

We then conduct three experiments to evaluate our model in three different scenarios. First, we set sender and receiver to the same channel and send 802.11 data frames at three different data rates and intervals. We adjust γ to minimize the difference between CO_{total} and CO_{WLAN} in a high-duty case and compare the accuracy for low-duty cases. We use a 30 s window for computing CO_{total} and a 10 s window for CO_{WLAN}. We compare both numbers with the channel busy time according to the RX_{CLR} profile counter [6, 2, 1] from the receiver's Wireless LAN adapter. This counter is incremented at a rate of 44 MHz every time the adapter's receiver logic senses the channel busy. In Table 2 we give the average and standard deviation of all three variables for all three cases (A, B, C). Comparing CO_{total} with RX_{CLR} we see a minor overestimation ($\leq$ 1pp). Comparing CO_{WLAN} with RX_{CLR} we see a slight underestimation which we consider acceptable particularly since it only occurs under very high load. All computations show a very low variance.

Second, we set sender and receiver 1 .. 4 channels apart and repeat the measurement for cases A and B. In Table 3 we summarize the computed CO_{total} and show the RX_{CLR} values for comparison. The computations show a low variance (not displayed in table). We also generate $P(B_i|S_j)$ shown in Table 1 from this experiment. Comparing CO_{total} to RX_{CLR} we see a good accuracy for channel separations of 1 and 2 but find a lower accuracy for separations ≥ 3.

Third, we apply our model in a scenario with both 802.11 transmissions and an active interferer. We set sender and receiver to the same channel. We use a radio test mode of the Bluetooth module to send Bluetooth frames with a duration of ≈ 400 µs and different intervals of 625 µs, 1000 µs, 2000 µs, and 4000 µs on a fixed frequency within the Wireless LAN channel. We compute the average values of CO_{total}, CO_{WLAN}, and RX_{CLR} over the course of the experiment using a sliding window as in the first experiment and repeat the experiment for all three configurations of data rate, frame length and interval (A, B, C). In Figure 1 we show the results of one experiment (case B) as an example. Over all three configurations we find CO_{total} to be accurate for low and medium overall channel occupancy. For overall channel occupancies $\geq 40\,\%$ we see overestimations of up to 16 pp. CO_{WLAN} is accurate for low interference cases but drops with increasing interference.

4. DISCUSSION

Regarding CO_{WLAN} our model delivers a very good accuracy for low interference, but unveils problems in situations with high interference, when a significant amount of

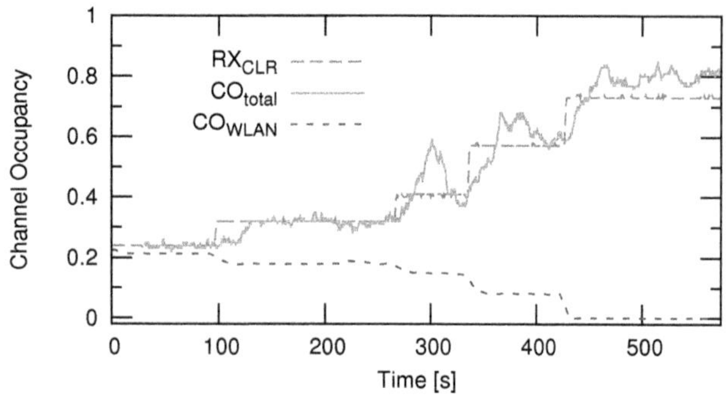

Figure 1: Computed values of CO_{total}, CO_{WLAN}, and RX_{CLR} over the course of one experiment (OFDM frames, frame duration 268 µs, frame rate 900/s, interferer sequentially set to 0 %, 10 %, 21 %, 42 %, and 63 % duty cycle).

802.11 frames are lost because of collisions. The corrupt frames do not show up in the computation of CO_{WLAN} and are consequently counted as interference. Our model could be extended to detect such situations on Wireless LAN adapters which provide statistics about corrupt frames (incorrect frame check sequence, loss of synchronization). The computation of CO_{WLAN} is straight-forward and implementable on a broad range of Wireless LAN chipsets. Since most drivers currently implement the 802.11 protocol stack on the host system and therefore need access to all 802.11 frames, the "monitor mode" is a common feature. However, our model needs to be extended to reflect the interdependency of transmissions on neighboring channels properly. Additionally, our model does not cover the interframe spaces and back-off periods in which there is no transmission, while these periods are considered as "busy" by the Wireless LAN adapter. Nevertheless, if we took the virtual CS into account, the resulting channel occupancy could not be compared to CO_{total} anymore.

The computation of CO_{total} is quite precise considering the consumer-grade performance of the measurement device. We see a slight overestimation of CO_{total} and assume this to be caused by round-up effects due to the long subchannel dwell time. Additionally, the undersampling seems to generate aliasing effects. Also, the accuracy gets lower for channel separations ≥ 3. Some of the effects may be compensated for by applying more processing to the raw samples. It may also be possible to distinguish different interferers already from the AirView samples similar to AirShark [8] and to process them separately. Finally, when combining measurements from both the Wireless LAN adapter and an external IC the different antennas and RF paths need to be considered. We designed our model with low-gain omnidirectional antennas in mind and assume similar conditions for both RF paths. Due to the limited time resolution of our measurement devices, the overall system is not well-suited to accurately describe short, bursty channel usage. Nevertheless, the system is able to sufficiently determine the distribution of channel usage from long-term measurements.

5. CONCLUSION

The presented model provides a good approximation of the channel occupancy in many situations. In the evaluation, we found the error of the estimated total channel occupancy to be very low for low to medium interference and borderline acceptable in high interference situations. The estimation of channel occupancy caused by Wireless LAN is accurate for low interference situations but deviates with increasing interference. We identified 802.11 frames which get destroyed by collisions and counted as interference as the reason for that. In the current work state, the model can already be used for coarse-grained estimations and is implementable on a broad range of Wireless LAN adapters. The spectrum analyzer IC allows a cost-effective and energy-efficient spectrum sensing for the entire 2.4 GHz ISM band which is independent from the Wireless LAN adapter. However, our model needs to be extended to cover channel interdependency and to improve the accuracy in situations with high interference. Also, an extended evaluation with multiple senders and varying signal levels is desirable.

Acknowledgements

We thank Uwe Starke for his help during the experiments. Till Wollenberg and Sebastian Bader are supported by the DFG (German Research Foundation) through the research training group "MuSAMA" (GRK 1424).

6. REFERENCES

[1] M. Christine, M. Krishnan, S. Ng, E. Haghani, and A. Zakhor. Local estimation of collision probabilities in 802.11 wlans: An experimental study. In *2011 IEEE Wireless Communications and Networking Conference (WCNC)*, pages 43–48. IEEE, 2011.

[2] P. Dely, A. Kassler, and D. Sivchenko. Theoretical and experimental analysis of the channel busy fraction in ieee 802.11. In *Future Network and Mobile Summit*, pages 1–9. IEEE, 2010.

[3] R. Gummadi, D. Wetherall, B. Greenstein, and S. Seshan. Understanding and mitigating the impact of rf interference on 802.11 networks. In *ACM SIGCOMM CCR*, volume 37, pages 385–396. ACM, 2007.

[4] D. Halperin, W. Hu, A. Sheth, and D. Wetherall. Predictable 802.11 packet delivery from wireless channel measurements. In *ACM SIGCOMM CCR*, volume 40, pages 159–170. ACM, 2010.

[5] K. Jamieson, B. Hull, A. Miu, and H. Balakrishnan. Understanding the real-world performance of carrier sense. In *Proc. of the 2005 ACM SIGCOMM workshop on Experimental approaches to wireless network design and analysis*, pages 52–57. ACM, 2005.

[6] K. Lakshminarayanan, S. Seshan, and P. Steenkiste. Understanding 802.11 performance in heterogeneous environments. In *Proc. of the 2nd ACM SIGCOMM workshop on Home networks*, pages 43–48. ACM, 2011.

[7] M. Oularbi, A. Aissa-El-Bey, and S. Houcke. Physical layer IEEE 802.11 channel occupancy rate estimation. In *5th Int. Symposium on I/V Communications and Mobile Network (ISVC)*, pages 1–4. IEEE, 2010.

[8] S. Rayanchu, A. Patro, and S. Banerjee. Airshark: Detecting Non-WiFi RF Devices using Commodity WiFi Hardware. In *Proc. of the ACM SIGCOMM Internet Measurement Conference (IMC)*, 2011.

[9] M. Wellens, J. Riihijärvi, and P. Mähönen. Empirical time and frequency domain models of spectrum use. *Physical Communication*, 2(1):10–32, 2009.

Efficient Energy Management in Wireless Rechargeable Sensor Networks*

Constantinos Marios Angelopoulos
Computer Technology Institute
& Press "Diophantus" (CTI)
and University of Patras,
Greece
aggeloko@ceid.upatras.gr

Sotiris Nikoletseas
Computer Technology Institute
& Press "Diophantus" (CTI)
and University of Patras,
Greece
nikole@cti.gr

Theofanis P. Raptis
Computer Technology Institute
& Press "Diophantus" (CTI)
and University of Patras,
Greece
traptis@ceid.upatras.gr

Christoforos Raptopoulos
Computer Technology Institute
& Press "Diophantus" (CTI),
Greece
raptopox@ceid.upatras.gr

Filippos Vasilakis
University of Patras, Greece
vasilakis@ceid.upatras.gr

ABSTRACT

Through recent technology advances in the field of wireless energy transmission, Wireless Rechargeable Sensor Networks (WRSN) have emerged. In this new paradigm for WSNs a mobile entity called Mobile Charger (MC) traverses the network and replenishes the dissipated energy of sensors. In this work we first provide a formal definition of the charging dispatch decision problem and prove its computational hardness. We then investigate how to optimize the trade-offs of several critical aspects of the charging process such as a) the trajectory of the charger, b) the different charging policies and c) the impact of the ratio of the energy the MC may deliver to the sensors over the total available energy in the network. In the light of these optimizations, we then study the impact of the charging process to the network lifetime for three characteristic underlying routing protocols; a greedy protocol, a clustering protocol and an energy balancing protocol. Finally, we propose a Mobile Charging Protocol that locally adapts the circular trajectory of the MC to the energy dissipation rate of each sub-region of the network. We compare this protocol against several MC trajectories for all three routing families by a detailed experimental evaluation. The derived findings demonstrate significant performance gains, both with respect to the no charger case as well as the different charging alternatives; in particular, the performance improvements include the network lifetime, as well as connectivity, coverage and energy balance properties.

*This work was partially supported by EU/FIRE HOBNET project - STREP ICT-257466

Categories and Subject Descriptors

C.2.1 [**Network Architecture and Design**]: Distributed Networks, Network Communications, Wireless Communication

General Terms

Algorithms, Experimentation, Performance

Keywords

Sensor Networks, Energy Efficiency, Mobility, Distributed Algorithms, Wireless Recharging

1. INTRODUCTION AND CONTRIBUTION

Recent advances in the fields of *wireless energy transmission* and batteries material offer new possibilities for managing the available energy in WSNs. In the first field, the technology of highly-efficient wireless energy transmission was proposed for efficient, non-radiative energy transmission over mid-range. The work in [19] and [17] has shown that through strongly coupled magnetic resonances, the efficiency of transferring 60 watts of power over a distance in excess of 2 meters is as high as 40%. Industry research also demonstrated that it is possible to improve transferring 60 watts of power over a distance of up to two to three feet with efficiency of 75% [8]. At present, commercial products utilizing wireless energy transmission have been available on the market such as that in [14].

In the second field, there is the new battery material for ultra-fast charging. Ultra-fast charging was recently realized in LiFePO4 by creating a fast ion-conducting surface phase through controlled off-stoichiometry [11]. It inherits and combines the advantages of both conventional Li-ion batteries and super-capacitors, that bring high energy density and can be charged as high as $400C^1$. This way, the time to fully charge a battery is shortened to few seconds.

[1] C is determined by the nominal capacity of the battery. For a battery with the capacity of $1000mAh$, $C = 1000mA$.

These technologies lead the way towards *a new paradigm* for wireless sensor networks; the Wireless Rechargeable Sensor Networks (WRSNs). WRSNs consist of sensor motes that may be either stationary or mobile, as well as few mobile motes with high energy supplies. The latter, by using wireless energy transmission technologies are capable of fast recharging sensor motes. This way, the highly constrained resource of energy can be managed in great detail and more efficiently. Another important aspect is the fact that energy management in WRSN can be performed passively from the perspective of sensor motes and without the computational and communicational overhead introduced by complex energy management algorithms. Finally, WRSNs allow energy management to be studied and designed independently of the underlying routing protocol used for data propagation.

The Problem. Let a Wireless Rechargeable Sensor Network consisting of a set of stationary sensor motes and a special mobile node called *Mobile Charger*. The sensor motes are deployed uniformly at random over a network area and propagate data to a *Sink* using a routing protocol P. The Mobile Charger has finite energy supplies, that are significantly greater than those of a single sensor mote, and is capable of recharging the sensors. The problem we study is identifying best possible configuration of the Mobile Charger in order to improve energy efficiency and to prolong the lifetime of the network. Such mobile charger configuration alternatives include the charging policy followed (i.e. whether to recharge motes fully or partially), the ratio of energy available to the charger over the energy initially deployed at the sensor motes, as well as the optimization of the network traversal by the charger e.g. finding which are the best trajectories the charger should follow. Although the problem might have some similarities with related research topics, such as data collection via mobile Sinks in sensor networks or sensor actor networks, there are special features in wireless recharging and the particular trade-offs necessitate the optimization of the new aforementioned crucial parameters.

Our Contribution. While considerable research efforts have been invested into energy-efficient scheduling of the mobile charger, proposed solutions in the literature so far require a global (or at best, a semi-global) knowledge of the state of the network. On the contrary, the solutions proposed in this paper is fully distributed and adaptive, and rely solely on local information. Furthermore, our proposed algorithm for the MC can be used in combination with any underlying routing protocol and adapts of the distribution of sensors in the network area.

More specifically, we investigate alternative strategies for efficient recharging in WRSNs via a Mobile Charger. First, we identify several critical aspects of the charging procedure and the corresponding trade-offs. We provide a formal definition of the charging problem and prove that it is computationally hard. We then try to optimize each trade-off and, lastly, we investigate the impact of the charging process over three characteristic families of routing protocols. The trade-offs we have identified include: a) how the total available energy of the network should be split between sensor motes and the MC b) given that the energy the MC may deliver to the motes is finite, whether each sensor will be fully or partially charged (and to what extent) and c) what is the trajectory the MC should follow in order to charge the sensor motes. The implementation of the alternative strategies and their detailed simulation evaluation demonstrates significant performance gains with respect to several metrics such as network lifetime, connectivity and coverage, as well as energy balance properties.

2. RELATED WORK AND COMPARISON

The application of wireless energy transfer is a rather new technology and has not been analysed extensively in sensor networks. However, the idea behind WRSNs was first studied by Estrin et al. in [15]. The authors study the feasibility of extending the lifetime of a wireless sensor network by exploiting mobility. In particular, small percentage of network nodes which are autonomously mobile, search for new energy, recharge their batteries and deliver it to the rest of the network. In that paper the authors define the energy properties that a sensor network should have in order to be self-contained; that is, continue work forever. In [10] the authors analyse the problem of optimal scheduling in WRSN of both charger's path and sensor's sleeping policy for stochastic event capture. The maximization of the quality of monitoring is also studied. The work in [2] describes applications suited for wireless recharging sensor networks and analyse the advantages over classic sensor networks but also their limitations.

In [18] the authors consider a sensor network in which a mobile entity is employed which (in contrast to our approach) serves also as a data collector and as an energy transporter that charges the static sensors on its migration tour. They provide a two-step approach: in the first step the mobile entity selects the maximum number of anchor points such that the sensors located in these anchor points hold the least energy and meanwhile the tour length is no more than a threshold. In the second step they formulate a utility maximization problem on a flow-level network model in order to determine how to gather data from sensors. However this algorithm requires global information, thus making it not very practical in even medium-sized sensor networks. In [12] the authors analyse again the possibility of practical and efficient joint routing and charging schemes. They propose a sensor network in which both a mobile charger and a base station appear. Each sensor sends data hop-by-hop to the Sink periodically using the Collection Tree Protocol (CTP). Also, measurements of other local properties such as energy level, consumption rate, etc. are piggybacked along with data and reported to the Sink. Then, the base station, according to sensors information, schedules future charging activities and commands the mobile charger through long range radio to execute the schedules. Authors show that the network lifetime is prolonged by the mobile charger which mostly moves in energy-minimum paths. However, each sensor has to send more data to the Sink and the charger has to know the location of each sensor a priori. Authors in [16] consider the scenario of a mobile charging vehicle periodically travelling inside the sensor network and charging the battery of each sensor node wirelessly. The necessary and sufficient conditions are introduced and the problem is studied as an optimization problem, with the objective of maximizing the ratio of the wireless charging vehicle's vacation time over the cycle time. However, in their model authors use global knowledge.

Overall, in all above methodologies the knowledge of the model is much stronger than ours allowing for off-line and/or centralized optimizations under high levels of network information. Also, in several of these approaches the charging problem is coupled together with routing, while in our method the charging policy implicitly adapts to any underlying routing policy.

Regarding the three families of routing protocols we use to investigate the impact of our methods, we refer to [7] for clustering, [3] for routing and [9, 4] for energy balanced data propagation.

3. THE MODEL

3.1 Deployment and energy model

Our model features three types of devices: static sensors, a mobile charger (MC) and one static Sink. We assume that there are N sensors uniformly distributed at random in a circular area of radius R. The communication range of sensors (we denote it by r) varies according to requirements of the underlying routing protocol. The network density is $\rho = \frac{N}{\pi R^2}$. The Sink S lies at the center of the circular area. In our model we assume that the MC does not perform any data gathering process. For simplicity we assume that all sensors have the same data generation rate of λ packets per unit time. Also we assume that E_{total} is the total available energy in the network. Initially, $E_{total} = E_{sensors} + E_{MC}^{init}$ where $E_{sensors}$ is the amount of energy shared among the sensor nodes and E_{MC}^{init} is the total amount of energy that the MC may deliver to the network by recharging sensors. At any given time the energy left to the MC for sensor charging is denoted as E_{MC}^{curr}. The maximum amount of energy that a single sensor may store is denoted as E_{sensor}^{max} and is the initial energy given to each sensor, i.e. $E_{sensor}^{max} = \frac{E_{sensors}}{N}$.

For transmitting and receiving a message, we assume that the radio module dissipates an amount of energy proportional to the message size. To transmit a k-bit message, the radio expends $E_T(k) = \epsilon_{trans} \cdot k$ and to receive a k-bit message, the radio expends $E_R(k) = \epsilon_{recv} \cdot k$ where ϵ_{trans} and ϵ_{recv} are constants that depend on the radio module and the transmission range of the sensors. As usual, the power needed to transmit a message at distance d is roughly d^α where $2 \leq \alpha \leq 6$ a constant; for simplicity we take $\alpha = 2$.

3.2 Charging Model

In our model the charging is performed point-to-point, i.e. only one sensor may be charged at a time from the MC by approaching it at a very close distance so that the charging process has maximum efficiency. The time that elapses while the MC moves from one sensor to another is considered to be very small when compared to the charging time; still the trajectory followed (and particularly its length) is of interest to us, since it may capture diverse cost aspects. We assume that the charging time is equal for every sensor and independent of its battery status.

4. THE CHARGER DISPATCH DECISION PROBLEM

Below we give a formal definition for the decision version of the problem we consider.

DEFINITION 1 (CDDP). *Suppose that we are given a set S of sensors each one capable to store E units of energy and for each sensor $s \in S$ a list L_s of pairs $(t_s^j, e_s^j), j \geq 1$ in which t_s^j corresponds to the time that the j-th message of s was generated and e_s^j the energy that the sensor used to transmit it. We are also given an $|S| \times |S|$ matrix D, where $D_{i,j}$ is the distance[2] between sensors i and j and a mobile charger M which can charge a sensor in one time unit to its initial energy. The charger dispatch decision problem (CDDP) is to determine whether there is a feasible schedule for M to visit the sensors so that no message is lost due to insufficient energy.*

Notice that we neglect to include energy needed by sensors in order to receive messages. Moreover, the messages a sensor s might receive by other nodes are included in each L_s. Thus we suppose that these messages are generated by the sensor itself. This allows the consideration of different routing policies in a unified manner.

We show that the general version of the CDDP is $\mathcal{NP}$-complete.

THEOREM 1. *CDDP is $\mathcal{NP}$-complete.*

PROOF. We first note that, given a certain walk W of the charger visiting sensors in S, we can verify whether this walk is sufficient so that no message is lost, i.e. no message x is generated on a sensor s such that x is the j-th message of s and s has less than e_s^j available energy at time t_s^j. In particular, this can be done in $O(T \cdot |W|)$ time, where T is the total number of events generated in the network. Therefore CDDP $\in \mathcal{NP}$.

For the hardness part we use the Geometric Travelling Salesman Problem (G-TSP in short, see [5], page 212). Let $P \subseteq \mathbb{Z} \times \mathbb{Z}$ and $B \in \mathbb{N}$ be the input of G-TSP. We now transform this into an input for CDDP as follows: We use a set S of $|S| = |P|$ sensors and set $D_{i,j}$ equal to the Euclidean distance between the i-th and j-th point in P. Furthermore, for each sensor $s \in S$ we define its event list to be $L_s = \{(0, E), (\frac{B}{v}, 1)\}$, where v is the charger's speed. That is, two events happen in each sensor s, namely one at time 0 depleting all the energy available in s and one at time $\frac{B}{v}$ requiring energy 1. Notice that a solution to this instance of the CDDP problem would provide an answer to G-TSP, which means that G-TSP $\leq_m$ CDDP. This completes the proof. □

5. TRADE-OFFS OF THE CHARGING PROCESS

Theorem 1 states that the problem is hard to solve, even with global knowledge. The heuristics that we present encounter an even harder problem, since the knowledge is limited to a local level. In order to properly understand and investigate the charging process we first analyse several of its aspects and specify their inherent trade-offs.

5.1 Energy percentage available to charger

In order to be fair in our evaluation, we assume that the total available energy to the network E_{total} is finite and same in all cases. This way, we will be able to investigate whether the energy efficiency is increased (and to what extent) with and without the introduction of the MC and the charging process to the network. This particular trade-off consists in how much energy (in terms of the total energy available) should the MC be initially equipped with. On

[2]Notice that if we assume that the charger moves with constant speed v then the time needed to travel between i and j is $\frac{D_{i,j}}{v}$.

the other hand, more energy to the MC leads to better on-line management of energy in the network. However, since $E_{total} = E_{sensors} + E_{MC}^{init}$, more energy to the MC also means that the sensor motes will initially be only partially charged. Therefore, it will be more likely that they run out of energy before the MC charges them leading to possible network disconnection and low coverage of the network area. In order to determine the optimal energy amount available to charger as a percentage of E_{total}, we conduct experiments for various ratios between $E_{sensors}$ and E_{MC}^{init}.

A key conclusion is that a rather modest percentage of energy at the MC is a wise strategy (see section 7.2 and Fig. 2b), with a 20% percentage value being the best choice.

5.2 Full versus partial charging

Each time the MC visits a mote a straightforward strategy would be to fully charge that mote. This way the MC would maximize the time interval of revisiting the mote before it runs out of energy. However, as the network operates, energy is dissipated in motes due to data propagation and in the MC due to the recharging process. Therefore, the MC will have increasingly less energy to distribute to more and more motes.

Another approach is to judiciously spread the precious available energy to as many motes as possible in order to extent the network lifetime. Following this rationale, the amount of energy the MC delivers to a mote i is proportional to the residual charging energy of the MC. More formally, MC charges a mote until its energy becomes

$$e_i \approx \frac{E_{MC}^{curr}}{E_{MC}^{init}} \cdot E_{sensor}^{max} \quad (1)$$

In order to determine the best strategy we conduct detailed experiments comparing the full charging strategy against our adaptive, partial charging strategy.

Our basic result is that partial recharging is more efficient than the full case (see section 7.1 and Fig. 2a).

6. TRAVERSAL STRATEGIES OF THE MOBILE CHARGER

6.1 Global knowledge traversal strategy

The global-knowledge charger we study is an on-line solution that in each round minimizes the product of each sensor's energy with its distance from the current position of the charger. More specifically, in each moving step the global charger minimizes the following product:

$$P = min\left\{\left(1+\frac{E_{curr}}{E_{init}}\right)\cdot\left(1+\frac{dist_{curr}}{2R}\right)\right\} \quad (2)$$

where E_{curr}, E_{init} and $dist_{curr}$ are respectively the current energy, initial energy and distance of each sensor, with the minimum taken over all sensors in the network (or at least a large part of it). Since this strategy requires a global knowledge of the state of the network, it is expected to outperform all other strategies that use only local information, thus somehow representing a performance bound. However, it would not be suitable for real life networks as it introduces great communication overhead (i.e. every mote has to propagate its residual energy to the MC) and does not scale well with network size.

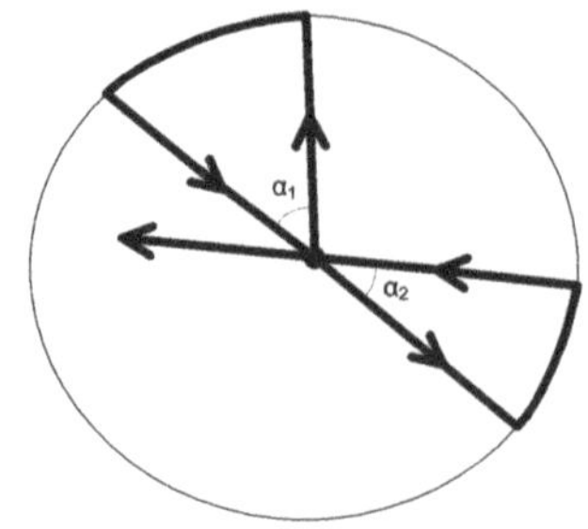

Figure 1: Diameter traversal strategy

6.2 Spiral traversal strategy

Starting from the Sink, the MC traverses a path which forms a set of concentric circles, centered around the Sink but with increasing radii. Thus it forms a spiral until it reaches the boundaries of the network area. Then the MC follows the same path towards the opposite direction. The advantages of this movement is that due to its space filling attributes, the MC covers the whole network and almost every node is charged, until the energy of MC is totally depleted. On the other hand, this movement is not adaptive, i.e. it does not take into account differences to the energy depletion rates of each subregion of the network caused by the underlying routing protocols, such as bottleneck areas.

6.3 Diameter traversal strategy

Starting from the Sink, located at the center of the circular network area, the MC arbitrarily chooses a direction and moves towards the perimeter of the network moving along the corresponding diameter. Once it has reached the edge of the network area, the MC traverses the perimeter of the network area for an angle of d degrees, i.e. it moves over $\frac{d}{2\pi}2\pi R = dR$ length along the perimeter. Then, it starts moving again along the new corresponding diameter until it reaches the opposing edge of the network area. This procedure is repeated until there is no energy left in MC for charging. In order for the areas that the MC charges not to overlap, in every step d is chosen uniformly at random from the interval of $(0, 90]$ degrees. By following this trajectory pattern the MC manages to charge sensors that lie both close to and far from the Sink. However, although simple, it is also a non-adaptive strategy as well. An instance of the diameter traversal strategy is depicted in Fig. 1

6.4 Random walk traversal strategy

In a simple, blind random walk, each next move of the MC is random and stochastically independent of the previous ones. Furthermore, given that the current node is i, the probability of moving to any neighbouring node j is $p_{i,j} = \frac{1}{deg(i)}$. This method is very robust, since it probabilistically guarantees that eventually all network regions and nodes will be visited, but it may become inefficient on networks with bottleneck areas, or under routing protocols with special nodes (like clustering protocols) or nodes that serve a lot of traffic, as in the case of multi-hop data propagation.

6.5 Our adaptive circular traversal strategy

Given the symmetric geometry, uniform density and uniform data generation rate of the network, we propose that the MC follows a circular trajectory around the Sink. The radius of the trajectory varies and adapts to the energy depletion rates of each subregion of the network. Starting from the Sink, the mobile charger traverses a path which forms a set of concentric circles, centered around the Sink with varying (increasing or decreasing) radii. In particular, the MC charges the sensors inside the ring which contains the corresponding trajectory. The width d of each ring is pre-specified and constant. For example, if we denote by S the set of the sensors to be charged at the k-th ring, then $|S| = \frac{N}{R^2}((kd)^2 - (k-1)^2 d^2) = \frac{N}{R^2}(2kd - d^2)$.

Let e_i denote the current residual energy of node i. While at a given distance from the Sink, the MC records the mean value of the energy of the sensors lying on the corresponding circular trajectory; we denote this value by $\overline{E}_{current}$. Accordingly, the MC keeps record of the mean value of the energy of the sensors lying on its previous circular trajectory; we denote this value by $\overline{E}_{previous}$. Based on these two values, the MC tries to optimize its trajectory in terms of recharging the nodes that deplete their energy faster. The algorithm is shown below:

while $E_{MC}^{curr} > 0$ **do**
  $E_{tmp} = 0$
  for every $i \in S$ **do**
    $E_{tmp} += e_i^c$
    Charge until $e_i \approx \frac{E_{MC}^{curr}}{E_{MC}^{init}} \cdot E_{sensor}^{max}$
  end for
  $\overline{E}_{current} = \frac{E_{tmp}}{|S|} = \frac{\sum e_i}{|S|}$
  if $\overline{E}_{current} \approx \overline{E}_{init}$ **then**
    if $\overline{E}_{previous} \geq \overline{E}_{current}$ **then**
      Keep direction
    else
      Change direction
    end if
  end if
end while

The MC commences traversing the network from the first center of the network area by setting $R_{MC} = 1$, i.e. it will first visit the nodes that lie one hop away from the Sink. Once all corresponding sensors are charged, i.e. the nodes the MC encounters have $e_i \simeq E_{sensor}^{max}$, the MC increases its radius R_{MC} by visiting nodes that are two hops away from the Sink. By comparing the values of $\overline{E}_{current}$ and $\overline{E}_{previous}$ the MC is able to figure out whether it moves towards areas that are stressed by the routing protocol or not. More specifically, if $\overline{E}_{current} < \overline{E}_{previous}$, then the MC assumes it is moving towards a stressed area of the network. Otherwise, if $\overline{E}_{current} > \overline{E}_{previous}$, then it assumes it is moving away from the stressed areas and therefore should change direction.

7. EXPERIMENTAL EVALUATION

Our simulation environment for conducting the experiments is Matlab 7.11.0. The Sink is placed at the center (0,0) of the circular deployment area. For statistical smoothness, we apply several times the deployment of nodes in the network and repeat each experiment 100 times. For each experiment we simulate large numbers of data propagations and the average value is taken. The statistical analysis of the findings (the median, lower and upper quartiles, outliers of the samples) demonstrate very high concentration around the mean, so in the following figures we only depict average values.

Since the N sensors are uniformly distributed in the circular area $\mathcal{A}$ of radius R, we apply a well known connectivity threshold, in order to maximize the probability that the produced random instances are connected. More strictly, since $\mathcal{A} \subset \mathbb{R}^2$, an instance of the *random geometric graphs model* $\mathcal{G}(\mathcal{X}_N; r)$ is constructed as follows: select N points $\mathcal{X}_N$ uniformly at random in $\mathcal{A}$. The set $V = \mathcal{X}_N$ is the set of vertices of the graph and we connect two vertices if their euclidean distance is at most r. In [6, 13] it is shown that the connectivity threshold for $\mathcal{G}(\mathcal{X}_N; r)$ is $r_c = \sqrt{\frac{\ln N}{\pi N}}$. In this paper we consider random instances of $\mathcal{G}(\mathcal{X}_N; r)$ of varying density, by selecting $r = \sqrt{\frac{c \ln N}{\pi N}}$, for different values of $c > 1$, which guarantees that the produced random instance is connected with high probability. Throughout the experiments we fix the parameter c to higher values for the Greedy routing protocol because it is more prone to early disconnections.

Since the network is dense enough, we assume that each transmission costs r^2 in terms of energy, where r is the transmission range of a sensor node. Since the sensors are uniformly distributed and the event generation is also uniform, the average number of hops needed for an event to travel from the source node to the Sink is $\frac{R}{2} \cdot \frac{1}{r}$. Consequently, the average energy spent for the routing of an event a source node to the Sink is $r^2 \frac{R}{2r} = \frac{rR}{2}$, and for the overall energy spent in the network we have $E_{total} = \frac{\mu r R}{2}$, where μ is the number of the events. We provide the network with $E_{total} = \frac{h \mu r R}{4}$, for $h > 1$, in order to prevent some nodes from dying during the experiments. The value of E_{total} is the same across different simulation scenarios.

We focus on the following performance metrics: a) **alive nodes over time**, that is the number of nodes with enough residual energy to operate, during the progress of the experiment, b) **network energy map**, which is a depiction of the whole network in terms of spatial aspects, after the generation of a number of events, c) **connectivity over time**, in terms of the nodes' average degree during the progress of the experiment and d) **coverage ageing**, that is the average coverage number (number of sensors having the point in their range) of 1000 randomly selected points in the network over time. At first we investigate the above mentioned charging trade-offs and traversals by using the alive nodes over time as our lead metric. Then, in the light of these results, we apply the best possible configurations on the MC's parameters and conduct a detailed experimental analysis of the MC's impact of the charging process to each of the underlying routing protocols. Due to lack of space, we graphically present the experiments with the E_i protocol as underlying routing protocol and we omit the Greedy and LEACH cases, since the results are similar. A more detailed presentation of the results can be found at the full paper [1].

7.1 Full versus partial charging

The first trade-off to be investigated is at which level should the sensor motes be charged by the MC. We compare two possible strategies: a) full charging of the nodes and b) partial charging of the nodes using our approach mentioned at section 5.2. At this stage we arbitrarily set the initial energy of the MC to be 40% of E_{total}. This ratio will be investigated individually and optimized later on. The results on the life evolution of the network are shown in Fig. 2a. Our partial charging strategy outperforms the full charging strategy, after a specific number of generated

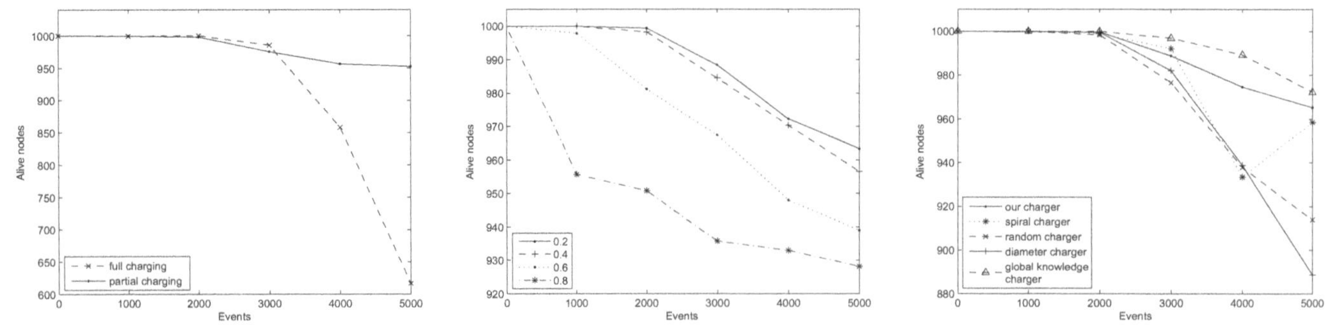

(a) Full and partial charging, MC provided with 40% of E_{total} (b) Various MC initial energy percentages, partial charging selected (c) Various MC traversals, partial charging selected, 20% of E_{total} to the MC selected

Figure 2: Alive nodes over time for E_i Protocol

events. This behaviour is the outcome of the fact that the MC is spending more energy in sensor charging when following the full charging strategy. Thus, the MC's energy is consumed earlier, resulting to a sharp increase in the node death rate.

7.2 Energy available to the charger

The next trade-off to be investigated is the optimal amount of total energy given to the charger with respect to the total energy of the network E_{total}. We conducted a comparison among several percentages of initial energy given to the charger. More specifically, we investigate the cases of 20%, 40%, 60%, and 80% of the total energy to be given to the MC, using our partial charging strategy, since it is found to be more efficient. The results are shown in Fig. 2b.

It is clear that providing the MC with more than 40% of the total energy is negatively affecting the life evolution of the network, for all three routing protocols. This result comes from two facts. First, as $E_{total} = E_{sensors} + E_{MC}^{init}$, the more is the energy of the MC and the less is the initial energy of the nodes, resulting in a faster node death rate. A second corollary effect is that if the MC is provided with a high amount of energy, then it can not fast enough to distribute the whole amount of energy to the network, resulting in high residual energy at the MC by the time the network becomes disconnected.

7.3 Traversals comparison

Following the clarification of the previous questions, which resulted in an optimized configuration of our charger (20% initial MC energy, partial charging strategy), comes an experimental comparison of the aforementioned possible traversal strategies of the MC.

More specifically, Fig. 2c shows the impact on the life evolution of the network for each of the proposed strategies, with E_i as underlying routing protocol. The global knowledge charger outperforms, as expected, all the other chargers, until its energy is depleted. The spiral charger has a rather unstable impact on the number of alive nodes over time. Our charger's performance is very close to the global knowledge charger in all three cases.

The *distance travelled* by each charger during the experiment is given in Table 1. The random walk and diameter chargers are not only inefficient at managing the network energy but they also cover the greatest distance. Our charger travels significantly less distance than the spiral and the global knowledge charger.

	Greedy	E_i	LEACH
global knowledge	48974	64330	57458
spiral	64167	64167	64167
random walk	181135	222462	222447
diameter	152252	172734	173169
our charger	42412	37856	38641

Table 1: Distance travelled by chargers

7.4 Overall improvements on the routing protocols

After the line of experiments conducted, we are able to identify various parameters of the MC in order to achieve high performance. Following the previous results, the MC will be provided with 20% of the network's total energy, using our partial charging strategy and our adaptive traversal. We study the effect of our charger on the three routing protocols, on the number of alive nodes over time, the energy balance of the network, the average connectivity over time and the coverage ageing.

The overall death rate (in terms of *alive nodes* over time) of the network is vastly reduced, as shown in Fig. 3a.

Figs. 4,5,6 depict the energy map of the network over time, for the three routing protocols. More specifically, we present graphically the spatial evolution of energy dissipation in the network after 5.000 event generations. Nodes with low residual energy are depicted with dark colours. In contrast, nodes with high residual energy are depicted with bright colours. The *energy balance* property is crucial for the networks lifetime, since early disconnections are avoided and nodes tend to die in a uniform manner.

For all three protocols, we observe that the use of our charger improves the energy balance property. For the Greedy protocol, the stressed nodes closer to the Sink are used very frequently, thus their energy is depleted quickly. Using the charger, not only do the inner nodes not die that fast, but the network's energy in total is more balanced. For the E_i and LEACH protocols, in which more distant nodes tend to be overused, the energy dissipation is higher at the distant form the Sink nodes. The use of charger has a similar effect with the Greedy case, balancing the network by charging more distant from the Sink nodes.

Connectivity is critical for sensor networks, as information collected needs to be sent to remote control centers. This is only possible if there is a path from each node to that center. The connectivity of a sensor network is usually studied by considering a graph associated with that network. As far as connectivity is concerned, the average node degree over time is depicted in Fig. 3b. We observe that the degree number degradation follows a pattern identical to the node death rate in Fig. 3a.

Point *coverage* problem is regarding how to ensure that each of some selected points in the network are covered by enough sensors. Coverage is an important aspect in sensor networks (e.g. localization, data accuracy etc.). A point that is covered by k sensors is called k-covered. The coverage ageing of 1000 randomly selected points in the network is shown in Fig. 7. We examine how many points are $< 2, 2, 3, > 3$ covered for 5000 generated events. We observe that during the experiment without charger, the number of < 2-covered points is increasing and the number of > 3-covered points is decreasing. The use of the charger is improving the coverage ageing of the selected points. More specifically, we observe that the absolute difference of the number of < 2-covered points and > 3-covered points, between different time instances of the experiment, is not increasing quickly, compared to the experiment without the charger.

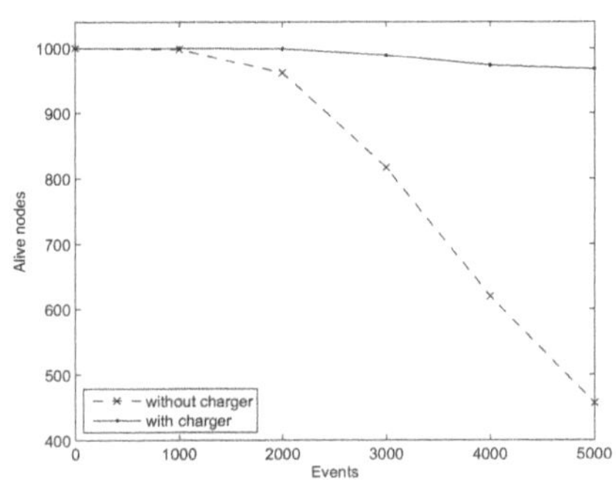

(a) Alive nodes over time

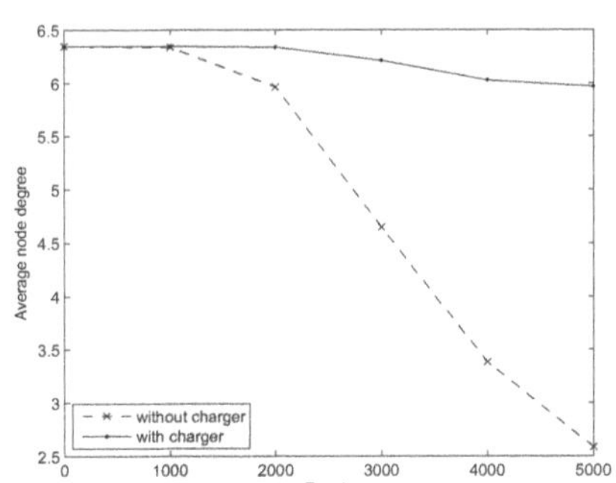

(b) Average node degree over time

Figure 3: E_i with and without charger

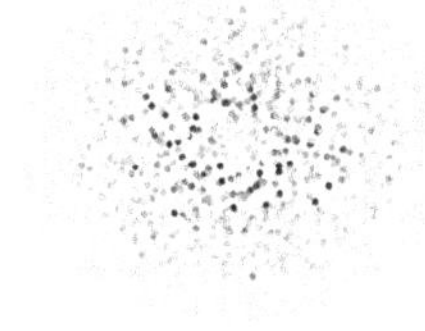

(a) Without charger

(b) With charger

Figure 4: Energy map for Greedy protocol.

(a) Without charger

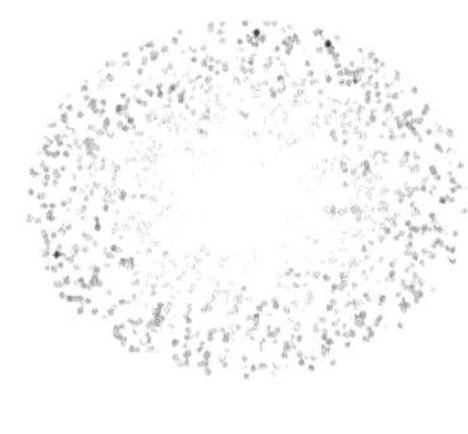

(b) With charger

Figure 5: Energy map for E_i protocol.

(a) Without charger

(b) With charger

Figure 6: Energy map for LEACH protocol.

8. CONCLUSIONS AND FUTURE WORK

We investigate diverse energy efficiency aspects of the recent paradigm of Wireless Rechargeable Sensor Networks (WRSNs), in which a mobile charger traverses a WSN and transfers energy to sensor motes in a wireless manner. The ability to add energy to the network during its evolution probably strengthens the power of the model; however a careful investigation of alternative configurations of the charging process is necessary to fully exploit the potential benefits and identify the extent of possible improvements.

In this work, we first provide a formal definition of the wireless recharging problem and prove that it is computationally hard. We then investigate three key aspects of recharging: a) whether motes should be recharged fully or partially (and to what extent) b) what is the best split of total available energy between the motes and the mobile charger c) what is the trajectory the charger should follow.

For each of these issues we investigate (via detailed simulations) several alternatives and identify best options in each case. Overall, we show that the best strategy is partial recharge of the motes, by a charger with a moderate fraction of the total available energy, which follows a circular trajectory around the Sink of a radius which locally adapts

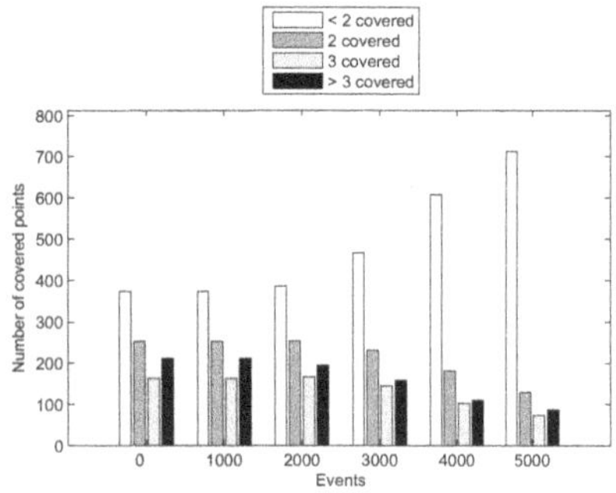

(a) Without charger

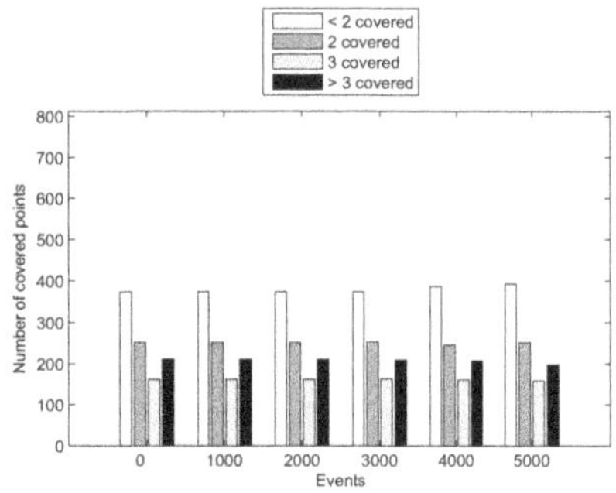

(b) With charger

Figure 7: Coverage ageing for E_i protocol.

to spatial variations of energy in the network. In particular, our simulation findings suggest significant performance gains with respect to various metrics such as network lifetime, connectivity and coverage as well as energy balance.

Future research directions include the design and rigorous analysis of efficient approximation algorithms with provable performance guarantees. In this respect, we note that the reduction used in Theorem 1 for proving the computational hardness of the problem is not preserving the approximation ratios (i.e. known efficient algorithms for the geometric TSP problem may not be very efficient for the charging problem), so this task seems quite challenging.

We also plan to study charging policies in non-uniform network deployments; this may actually lead to new algorithms for the mobile charger policy. Another research direction is the modeling, analysis and simulation of other proposed charging models, such as the simultaneous charging of more than one sensors. Success ratio will also be considered as an additional metric, since the impact of the charger on the network's lifetime guarantees improvements on the aspect of lost messages.

9. REFERENCES

[1] C. M. Angelopoulos, S. Nikoletseas, T. P. Raptis, C. Raptopoulos, and F. Vasilakis. Efficient energy management in wireless rechargeable sensor networks, full paper: http://students.ceid.upatras.gr/~traptis/mswim12.pdf.

[2] M. Buettner, B. Greenstein, D. Wetherall, and J. R. Smith. Revisiting smart dust with rfid sensor networks. 2008.

[3] I. Chatzigiannakis, S. Nikoletseas, and P. Spirakis. Smart dust protocols for local detection and propagation. In *POMC*, 2002.

[4] C. Efthymiou, S. Nikoletseas, and J. Rolim. Energy balanced data propagation in wireless sensor networks. *Wireless Networks*, 2006.

[5] M. R. Garey and D. S. Johnson. *Computers and intractability*. W. H. Freeman and Company, 1979.

[6] P. Gupta and P. R. Kumar. Critical power for asymptotic connectivity in wireless networks. *Stochastic Analysis, Control, Optimization and Applications*, 1998.

[7] W. R. Heinzelman, A. Chandrakasan, and H. Balakrishnan. Energy-efficient communication protocol for wireless microsensor networks. In *HICSS*, 2000.

[8] Intel. Wireless resonant energy link (wrel) demo, http://software.intel.com/en-us/videos/wireless-resonant-energy-link-wrel-demo/.

[9] A. Jarry, P. Leone, O. Powell, and J. Rolim. An optimal data propagation algorithm for maximizing the lifespan of sensor networks. In *DCOSS*, 2006.

[10] F. Jiang, S. He, P. Cheng, and J. Chen. On optimal scheduling in wireless rechargeable sensor networks for stochastic event capture. In *MASS*, 2011.

[11] B. Kang and G. Ceder. Battery materials for ultrafast charging and discharging. In *Nature 458, pp.190-193*, 2009.

[12] Z. Li, Y. Peng, W. Zhang, and D. Qiao. J-roc: A joint routing and charging scheme to prolong sensor network lifetime. In *ICNP*, 2011.

[13] M. Penrose. *Random Geometric Graphs*. Oxford University Press, 2003.

[14] Powercast. http://www.powercastco.com/.

[15] M. Rahimi, H. Shah, G. S. Sukhatme, J. Heideman, and D. Estrin. Studying the feasibility of energy harvesting in a mobile sensor network. In *Robotics and Automation*, 2003.

[16] Y. Shi, L. Xie, Y. T. Hou, and H. D. Sherali. On renewable sensor networks with wireless energy transfer. In *INFOCOM*, 2011.

[17] W. Wang, V. Srinivasan, and K.-C. Chua. Extending the lifetime of wireless sensor networks through mobile relays. In *IEEE/ACM Trans. Netw.*, 2008.

[18] M. Zhao, J. Li, and Y. Yang. Joint mobile energy replenishment and data gathering in wireless rechargeable sensor networks. In *Teletraffic Congress (ITC)*, 2011.

[19] C. Zhu, K. Liu, C. Yu, R. Ma, and H. Cheng. Simulation and experimental analysis on wireless energy transfer based on magnetic resonances. In *Vehicle Power and Propulsion Conference*, 2008.

Transmission Range Adaptation Based Energy Efficient Neighborhood Discovery

Jovan Radak
Inria Lille - Nord Europe, France
jovan.radak@inria.fr

Nathalie Mitton
Inria Lille - Nord Europe, France
nathalie.mitton@inria.fr

ABSTRACT

Neighborhood discovery - detection of the devices within communication range using hello messages - is a fundamental mechanism in wireless sensor networks (WSN) that enables usage of many different topology control and routing algorithms. In this work, we combine parameters of the neighborhood discovery (sending frequency of hello messages and changes in the neighborhood tables) and transmission range of the nodes. We present two algorithms that adapt transmission range of the sensors in a mobile WSN by still adapting frequency of hello messages in order to save energy and get accurate neighborhood tables. The first solution is based on the knowledge of turnover - change in the number of neighbors in consecutive iterations of the neighborhood discovery - used in conjunction with an adaptation of the message frequency and the transmission range, minimizing overall transmission cost of hello messages. The second solution is based on the computation of optimal range knowing the nodes' speed. Both algorithms are based on theoretical analysis. Results show that they are energy efficient and outperform solutions of the literature by maintaining high accuracy.

Categories and Subject Descriptors

C.2.1 [**Computer-Communications Networks**]: Network Architecture and Design—*wireless communication, sensor networks*

General Terms

Algorithms

Keywords

Wireless sensor networks, neighborhood discovery, transmission range, hello message

1. INTRODUCTION

Wireless sensor networks (WSN) can be defined as networks of small spatially distributed devices, called sensor nodes, which are working cooperatively - exchanging messages wirelessly - on the same application. Mobility of the sensor nodes is an emerging issue and nowadays it becomes more and more under the attention of scientific community. Included in the WSN, mobility arises new questions such as optimization of energy consumption, connectivity of the WSN, routing in mobile networks and a lot more. Nodes in WSN can have mobility included inherently, as it is case with nodes which are attached to animals with the purpose to track animals' habits and their natural habitat[1]. In this case mobility pattern is very often hard to estimate and protocols built upon this issue must take into account possible losses of connectivity or delays in the transmission of messages. Other possibility would include mobile agents (robots) which may have been given mobile pattern to follow to improve some of the parameters in the WSN [9]. In this case, overall energy efficiency of transmission can be significantly augmented using controlled mobility based smart placement of nodes related to the routing path. Due to the specific nature of mobile networks, some of the characteristic mechanisms used in static WSNs need to be redefined and adapted to the specific types of node mobility.

Neighborhood discovery is one of the most important protocols in WSN. The mechanism behind this protocol is rather simple, it includes periodical sending of a specific type of messages, called *hello messages* (also known as *beacon* messages) and gathering data from the received hello messages. Hello messages contain the data of the sender id, unique identification number for the node in the WSN – usually MAC address in practical applications. Each node, usually, acquires data from all hello messages that it has received and organizes them into the *neighborhood table* which can be further used for some kind of topology control [11] or proactive routing [6]. The main challenges, in the practical usage of the neighborhood discovery mechanism are accuracy of the neighborhood table, size of hello messages, as well as their content and the frequency at which they are being sent in order not to spent network resource uselessly. Indeed if the hello frequency is too low, nodes may not be detected by their neighbors, leading to deprecated neighborhood tables, and protocol failures are likely to occur. But if the frequency is too high, neighborhood tables are up to date, but then energy and bandwidth are wasted to the detriment of data traffic. Similarly, nodes can adapt their range. With the longer range, more neighbors are going to be detected with less changes in neighborhood but with higher energy expenditure. It is shown [12] that varying these parameters with other parameters of the protocol stack in WSN (MAC layer, clustering) can have significant impact on the performance of the nodes and of the whole network. In the presence of mobility, we must observe that each node in successive periods of time Δt receives hello messages from different nodes. Also, hello messages that the node sends are going to be received by different sets of nodes. This means that neighborhood table changes due to the mobility of the nodes. The question that arises is: how can we use a neighborhood

[1]http://wasp-project.org

discovery mechanism in the presence of the mobility and which values for parameters are the best in these cases?

Our work, presented in this paper, relies on the Turnover based Adaptive hello Protocol (TAP) [5]. TAP exploits the idea of the existence of an optimal hello messages frequency. TAP derives from it a number of changes that appear in the neighborhood table, called *turnover*. TAP computes the optimal value for the turnover in the mobile WSNs and sets the basis for the future works which can include prediction of the mobility or some kind of topology control in the mobile WSNs. Coming from the fact that overall energy efficiency can be minimized when manipulating both power of transmission and the frequency of hello messages, we present two algorithms for the adaptation of the frequency of hello messages in parallel with the adaptation of the power of transmission of each node. Both algorithms rely on theoretical analysis. They are periodically executed by each node (independently of the other nodes) in the mobile network, upon gathering the data from the hello messages when node receives them. To the traditional neighborhood table we have added also history table in which we are keeping successive neighborhood tables with the idea to track the changes between the neighborhood tables and according to those changes properly changes the factors of neighborhood discovery and transmission range of each node. Results show that our solutions maintain same accuracy as TAP protocol but saving more energy. To the best of our knowledge the algorithms that we are presenting are the first that handle both frequency of the hello messages and the transmission range of the nodes when nodes are not aware of their position.

Section 2 gives review of the literature related to our work. Section 3 provides theoretical preliminaries and analysis that gave us starting point to build our algorithms. Section 4 presents our algorithms. Simulation results are detailed in Section 5. Section 6 gives conclusion and directions of future work.

2. RELATED WORK

Hello protocol was first introduced in OSPF [10] and works as follows. Each node sends hello messages at a fixed frequency, noted f_{HELLO} to allow other nodes detect its presence. At the same time node collects data from hello messages received from other nodes *i.e.* if node u receives hello message from the node v it registers node v in its neighborhood table, similarly if node v receives hello message from node u he will put it in his neighborhood table.

Neighborhood discovery mechanism as described in here assumes that f_{HELLO} is fixed. In mobile networks this assumption in not adequate since it is more natural to assume that the nodes which are moving faster are also changing their neighbors faster thus they need to update neighborhood table more often but also since they are losing (and gaining) their neighbors more often it is logical that they send hello messages at a higher rate.

As stated earlier, a low hello frequency leads to obsolete tables while a high one may saturate bandwidth to the detriment of data traffic. The trade-off depends on characteristics (*e.g.* density, speed) that may evolve over time, leading to the need for an adaptive protocol. Nevertheless, only a few studies have tried to tackle this problem. In [4], a simple adaptive protocol is proposed, in which nodes are monitoring their neighborhoods and evaluating two values: time link failure (TLF) and time without change (TWC). Moreover, they periodically send hello at a frequency f_{low} again. If a node notices that the measured TWC is greater than a given threshold, it switches to the "high dynamics rate" and sends hello messages at a frequency f_{high}. On the contrary, if the estimated TLF is smaller than a threshold, it goes back to the "low dynamics rate" and sends hello at a frequency f_{low}. In this solution, finding the correct thresholds is not obvious since the thresholds may also need to evolve over time.

In [3], authors propose three protocols in order to achieve the best trade-off. The first one is called *Adaptive hello protocol*: each node simply sends a hello each time it has gone through X meters. Though straightforward, this scheme assumes that nodes are aware of their speed and their location. The second protocol is called *Reactive hello protocol*. It is based on the idea that nodes should build their neighborhood table only when needed. Thus, when a node sends a data packet, it first sends a hello message and waits during a time t for an answer. If no answer is received, then it repeats the same behavior up to X times. Upon reception of a hello message, nodes trigger a timeout before answering, to avoid collisions. While this scheme minimizes the quantity of hello messages, it introduces a high latency before sending data packets, and should not be used in networks with high mobility. The third protocol proposed in [3] is called *event-based hello protocol*. Nodes perform the classic periodic hello protocol, but if they do not receive any message and do not need to send data packets during a given time period, then they stop sending beacon messages until reception of a hello message. The main drawback of this protocol is that some nodes may never be detected by mobile nodes.

Turnover based adaptive hello protocol (TAP) [5] is the first algorithm which adapts frequency of hello messages without using any additional hardware. In this paper authors propose to exploit the *turnover* value which can be calculated considering the differences of two neighborhood tables at fixed moments in time. The first neighborhood table is the current one and the other is the one obtained and saved at the previous moment. This protocol thus only requires periodical save of current neighborhood table in so called *history* table which will be then used for the calculation of the turnover.

Knowledge of the relative speed of the nodes can also lead into the detection and estimation of the type of the mobility. In [8] authors are presenting Autoregressive hello protocol (ARH) and evaluate its performance alongside TAP. Different approach is presented in [7] where the authors are relying on the signal strength descriptors which are embedded in wireless radio (again without using any additional hardware) to detect mobility of the nodes in the closed space (room). While giving excellent results in detection and prediction of mobility these works do not tackle the possibility of changing the transmission range of the nodes.

Distinction is made between initial and continuous neighborhood discovery in [1]. Initial neighborhood discovery is applied when the sensor is unaware of its immediate environment, while the continuous neighborhood discovery is performed when the node is already aware of the neighborhood. Initial neighborhood discovery needs to be done by each sensor separately while continuous discovery can be applied as joint task of the segment of the neighborhood, and not each sensor. This joint task allows single node to spend more time in sleep mode and relies to the part of its neighborhood which is currently in active state. In this way they can lower energy consumption with high probability that new node is going to be properly discovered. The algorithm guarantees that new node is going to be discovered in given time slot with requested probability.

In [14], an optimal hello frequency which depends on the relative speed between objects is described. The idea is that a node which strides more than a given distance in the communication area of another node has to be detected by the latter. If the two nodes, with transmission range R, move with a speed S, the optimal frequency f_{opt} is equal to:

$$f_{opt} = \frac{2S}{aR} \quad (1)$$

where $a < 1$ is constant that determines threshold distance – the part of the transmission range R used in computation.

3. THEORETICAL ANALYSIS

3.1 Model and notations

Wireless networks are represented by a graph $G = (V, E)$ in which V is the set of nodes and E is the set of edges such that $E \subseteq V^2$. We suppose that nodes are randomly deployed using a Poisson Point Process [13], with node positions which are independent and $\lambda > 0$, where λ represents the mean number of nodes per surface unit. Each node u has transmission range R_u such that $0 < R_u < R_{max}$. For the pair of nodes u and v, the Euclidean distance between them is defined either as $|uv|$ or as d.

In this paper, since nodes can adapt their range, they do not have necessarily the same transmission range. According to this property, we differentiate two types of neighbors: ***bilateral neighbors*** – two nodes u and v with property $|uv| < min(R_u, R_v)$ and ***unilateral neighbors*** – node u is unilateral neighbor of node v if $min(R_u, R_v) < |uv| < max(R_u, R_v)$

Every node moves at constant speed S in random direction. Position of node u at time t_0 is given as u_0 (respectively node v at t_0 has position v_0) and at time t_1 position is u_1 (respectively v_1). Distance covered by a node during time Δt is given as $\Delta d = S \times \Delta t$. Angles α and β are given as $\angle \overrightarrow{u_1, v_1}, \overrightarrow{u_1, u_0}$ and $\angle \overrightarrow{u_1, v_1}, \overrightarrow{v_1, v_0}$ respectively, and represent the directions from which nodes u and v come. In the worst case, node direction is random and thus α and β are uniformly distributed in $[-\pi, \pi]$.

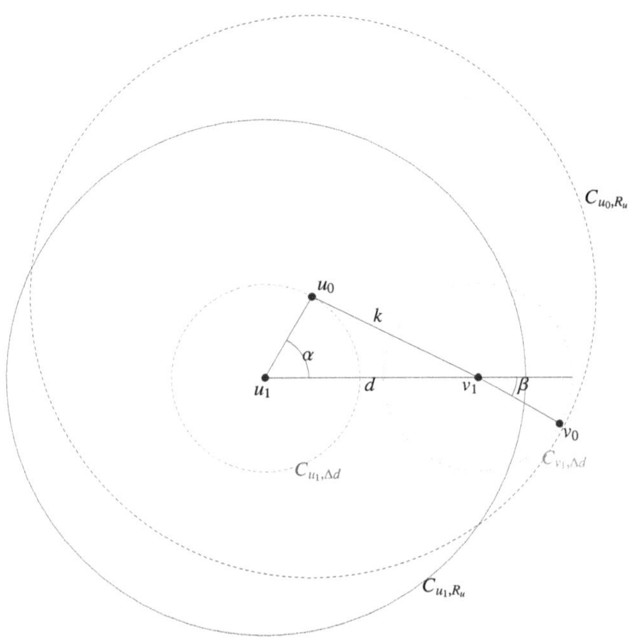

Figure 1: Global view. Circle C_{u,R_u} is centered at the position of the node u with radius R_u. In this case, node v, with radius R_v is a new neighbor if and only if v_0 does not lie in the area delimited by C_{u_0,R_u} and if v_1 lies in the area delimited by C_{u_1,R_u}. The blue dashed circle $C_{u_1,\Delta d}$ and the red dotted circle $C_{v_1,\Delta d}$ represent the possible positions of the u_0 and v_0.

3.2 Theoretical analysis on probable number of new neighbors

In our analysis, we are interested in the mean number of **new** neighbors that node u is going to meet during time interval Δt. We focus on a typical node u. Let $N_{bi}(u)_{\Delta t}$ be the number of new bilateral neighbors of node u and $N_{uni}(u)_{\Delta t}$ the number of new unilateral neighbors that node u detects during period Δt. Let v be a node at distance d ($d < R_{max}$) from node u at time $t_1 = t_0 + \Delta t$. We note: P^{bi} – the probability that node v is new bilateral neighbor of node u and P^{uni} – the probability that node v is new unilateral neighbor of node u. Both probabilities depends on the distance between nodes, d and on transmission ranges of nodes R_u and R_v. From these functions we can determine the average values of new neighbors $N_{bi}(u)_{\Delta t}$ and $N_{uni}(u)_{\Delta t}$:

$$E[N_{bi}] = \int_{R_u=0}^{R_{max}} \int_{R_v=0}^{R_{max}} \int_{d=0}^{R_u} P(R_u) \cdot P(R_v) \cdot \lambda \pi d \cdot P^{bi} \, dd R_u dR_v \tag{2}$$

$$E[N_{uni}] = \int_{R_u=0}^{R_{max}} \int_{R_v=0}^{R_{max}} \int_{d=0}^{R_u} P(R_u) \cdot P(R_v) \cdot \lambda \pi d \cdot P^{uni} \, dd R_u dR_v \tag{3}$$

where $P(R_u)$ is the probability that u has radius R_u and $P(R_v)$ is theprobability that v has radius R_v. Fig. 1 illustrates our model when $R_u < R_v$. Circle C_{u,R_u} is centered at node u with radius R_u. In this case, node v, with radius R_v is a new neighbor if and only if v_0 does not lie in the area delimited by C_{u_0,R_u} and if v_1 lies in the area delimited by C_{u_1,R_u}. The blue dashed circle $C_{u_1,\Delta d}$ and the red dotted circle $C_{v_1,\Delta d}$ represent the possible positions of u_0 and v_0.

3.2.1 *Computing the number of new bilateral neighbors* $E[N_{bi}]$

We are interested in the probability that at time t_0, u and v were either not neighbors ($|u_0 v_0| > max(R_u, R_v)$) or only unilateral neighbors ($max(R_u, R_v) > |u_0 v_0| > min(R_u, R_v)$). We are interested in probability P^{bi} that given R_u, R_v and d, $|u_0 v_0| > min(R_u, R_v)$ knowing that $|u_1 v_1| \leq min(R_u, R_v)$.

We make the distinction between two cases:

Case 1: $\mathbf{R_u \leq R_v}$ We note $P^{bi}_{R_u \leq R_v}$ the probability that v is a new bilateral neighbor of u if $R_u \leq R_v$; for this case $P(R_v) = \frac{1}{R_{max}}$ and $P(R_u \mid R_u < R_v) = \frac{R_v}{R_{max}}$,

Case 2: $\mathbf{R_u > R_v}$ We note $P^{bi}_{R_u > R_v}$ the probability that v is a new bilateral neighbor of u if $R_u > R_v$; for this case $P(R_v) = \frac{1}{R_{max}}$ and $P(R_u \mid R_u \leq R_v) = \frac{R_{max} - R_v}{R_{max}}$.

From these two cases we have:

$$\begin{aligned} E[N_{bi}] &= \int_{R_u=0}^{R_{max}} \int_{R_v=0}^{R_{max}} \int_{d=0}^{R_u} P(R_u) \cdot P(R_v) \cdot \lambda \pi d \cdot P^{bi} \, dd R_u dR_v = \\ &= \int_{R_v=0}^{R_{max}} \int_{R_u=0}^{R_v} \int_{d=0}^{R_u} \frac{R_v}{R_{max}} \cdot \frac{1}{R_{max}} \cdot \lambda \pi d \cdot P^{bi}_{R_u \leq R_v} \, dd(dR_u) dR_v + \\ &+ \int_{R_v=0}^{R_{max}} \int_{R_u=R_v}^{R_{max}} \int_{d=0}^{R_u} \frac{R_{max} - R_v}{R_{max}} \cdot \frac{1}{R_{max}} \cdot \lambda \pi d \cdot P^{bi}_{R_u > R_v} \, dd(dR_u) dR_v \end{aligned} \tag{4}$$

Case 1: $\mathbf{R_u \leq R_v}$ We first note that if $d \geq R_u$, $P^{bi}_{R_u \leq R_v}(d) = 0$ since v is not a neighbor of u at a time t_1. Next, we find that there exists a value $d_{min} < R_{max}$ such that, if $d < d_{min}$, nodes u and v were already neighbors at time t_0 regardless of α and β *i.e.* v cannot be a new neighbor of u. Justification and computation details are given in Appendix A (Eq. 19). We have that $d_{min} = max(0, R_u - 2\Delta d)$ which leads to:

$$P^{bi} = \begin{cases} \int_{-\pi}^{\pi} P^{bi}_{R_u \leq R_v}(d, R_u, R_v) d\alpha & \text{if} \quad d_{min} < d < R_u \\ 0 & \text{otherwise} \end{cases} \tag{5}$$

where $P^{bi}_{R_u \le R_v}(d, R_u, R_v)$ is the probability that node v at distance d with radius R_v is a *new* bilateral neighbor of node u with radius R_u coming from direction α, assuming $d_{min} < d < R_u$.

Now, we can compute $P^{bi}_{R_u \le R_v}(d, R_u, R_v)$. For this purpose, we can notice that, for a given value of $d_{min} < d < R_u$, there exists a value α_{min} such that, for $\alpha < \alpha_{min}$, node v was a bilateral neighbor of node u at t_0 regardless of its direction β, thus it cannot be a *new* neighbor. This is illustrated on Fig. 5(b). Indeed, for $\alpha < \alpha_{min}$, the whole circle $C_{v_1,\Delta d}$ lies inside Circle C_{u_0,R_u} thus, for any direction that node v may have had, it was already in the bilateral neighborhood of node u. As a result we get:

$$P^{bi} = 0 \qquad \text{if} \qquad \alpha < \alpha_{min} \tag{6}$$

where $\alpha_{min} = \arccos(\frac{d^2 + 2R_u\Delta d - R_u^2}{2d\Delta d})$. Computation of α_{min} is given in Appendix A.2 Eq. 23.

For any α, such that $\alpha > \alpha_{min}$, computing $P^{bi}_{R_u \le R_v}(d, R_u, R_v)$ is equal to the computing of the probability that node v comes from the dotted blue angular sector on Fig. 5(c). Node v is a new neighbor of node u if and only if β is such that $|u_0v_1| > R_u$ *i.e.* such that v is outside of C_{u_0,R_u} at t_0. In this case node v is a new neighbor of u if and only if $\beta^- < \beta < \beta^+$, where β^- and β^+ are the angles of the intersection points between C_{u_0,R_u} and $C_{v_1,\Delta d}$, as illustrated on Fig. 5(c). As a result we have:

$$P^{bi}_{R_u \le R_v}(d, R_u, R_v) = \int_{\beta^-}^{\beta^+} \frac{d\beta}{2\pi} = \frac{\beta^+ - \beta^-}{2\pi} = \frac{1}{\pi}\arccos(\frac{R_u^2 - \Delta d^2 - k^2}{2d\Delta d})$$

if $\alpha > \alpha_{min}$ where $k = |u_0v_1| = \sqrt{\Delta d^2 + d^2 - 2d\Delta d\cos\alpha}$. The details of computation of $\beta^+ - \beta^-$ and k are given in Appendix A.3 Eq. 22. Thus, when $d > d_{min}$:

$$P^{bi}_{R_u \le R_v}(d, R_u, R_v) = \begin{cases} \frac{1}{\pi}\arccos(\frac{R_u^2 - \Delta d^2 - k^2}{2d\Delta d}) & \text{if} \quad \alpha > \alpha_{min} \\ 0 & \text{otherwise} \end{cases} \tag{7}$$

From Eq. 7, we derive:

$$P^{bi}_{R_u \le R_v}(d, R_u, R_v) = 2\int_0^{\pi} \frac{P^{bi}_{R_u \le R_v}(d, R_u, R_v, \alpha)}{\pi} d\alpha$$

$$= \begin{cases} \frac{2}{\pi^2}\int_{\alpha_{min}}^{\pi} \arccos(\frac{R_u^2 - \Delta d^2 - k^2}{2d\Delta d})d\alpha & \text{if} \quad max(0, R_u - 2\Delta d) \le d \le R_u \\ 0 & \text{otherwise} \end{cases}$$

Case 2: $\mathbf{R_u > R_v}$ Computing of $P^{bi}_{R_u > R_v}(d, R_u, R_v)$ is similar to the computing of the $P^{bi}_{R_u \le R_v}(d, R_u, R_v)$. The differences are in the values of $d_{min}, \beta^+, \beta^-$ and α_{min} as shown in Appendix.

$$P^{bi}_{R_u > R_v}(d, R_u, R_v) = \begin{cases} \frac{2}{\pi^2}\int_{\alpha_{min}}^{\pi} \arccos(\frac{R_v^2 - \Delta d^2 - k^2}{2k\Delta d})d\alpha \\ \text{if } \max(0, R_u - 2\Delta d) \le d \le R_u \wedge R_v - 2\Delta d < R_u \\ 0 \quad \text{otherwise} \end{cases}$$

3.2.2 *Computing the number of new unilateral neighbors $E[N_{uni}]$*

Value that is interesting for our analysis is the probability P^{uni} that at time t_0, when u and v are not neighbors with given R_u, R_v and d. We distinguish two cases: $R_u \le R_v$, which we note $P^{uni}_{R_u \le R_v}$ and $R_u > R_v$, noted $P^{uni}_{R_u > R_v}$

From these two cases we have:

$$E[N_{bi}] = \int_{R_u=0}^{R}\int_{R_v=0}^{R}\int_{d=0}^{R_u} P(R_u)\cdot P(R_v)\cdot \lambda\pi d\cdot P^{uni}\, dddR_udR_v =$$

$$= \int_{R_u=0}^{R_v}\int_{R_v=0}^{R}\int_{d=0}^{R_u} \frac{R_v}{R_{max}}\cdot\frac{1}{R_{max}}\cdot\lambda\pi d\cdot P^{uni}_{R_u > R_v}\, dddR_udR_v +$$

$$+ \int_{R_u=R_v}^{R}\int_{R_v=0}^{R}\int_{d=0}^{R_u} \frac{R_{max} - R_v}{R_{max}}\cdot\frac{1}{R_{max}}\cdot\lambda\pi d\cdot P^{uni}_{R_u > R_v}\, dddR_udR_v \tag{8}$$

Case 1: $\mathbf{R_u \le R_v}$ In such a case, probability for node v to be a new unilateral neighbor of node u is null. Indeed, since $R_u \le R_v$, this is indeed u which is a bilateral neighbor of node v and not the opposite, thus we have:

$$P^{uni}_{R_u > R_v}(d, R_u, R_v) = 0 \tag{9}$$

Case 2: $\mathbf{R_u > R_v}$ We compute the probability $P^{uni}_{R_u > R_v}$ for node v to be a new unilateral neighbor of node u knowing R_u, R_v and d. This study is similar to the one given for the bilateral neighbors with the differences in values for $d_{min}, \beta^+, \beta^-$ and α_{min} as shown in Appendix.

$$P^{uni}_{R_u > R_v}(d, R_u, R_v) = \begin{cases} \frac{2}{\pi^2}\int_{\arccos(\frac{d^2 + 2R_u\Delta d - R_u^2}{2d\Delta d})}^{\pi} \arccos(\frac{R_u^2 - \Delta d^2 - k^2}{2k\Delta d})d\alpha \\ \text{if } max(R_v, R_u - 2\Delta d) \le d \le R_u \\ 0 \quad \text{otherwise} \end{cases} \tag{10}$$

3.3 Theoretical analysis on minimization of energy cost

In this section, we analyze jointly the optimal frequency and transmission range in order to define the function that will be used in our algorithms. Indeed, the energy spent in period of time Δt by a node u can be expressed as the number of messages sent by u in Δt multiplied by the cost of a message. The number of messages sent by u during Δt is $\Delta t \cdot f_u(R_u, t)$ where $f_u(R_u, t)$ is the hello frequency of node u and R_u is the range of node u at time t. We assume that the cost of a message follows energetic model given in [2] with $E(R_u) = R_u(t)^A + C$, where A is a real constant (> 1) that represents the signal attenuation and C is the overhead due to signal processing. We assume that Δt is such that u does not change its range nor its frequency during this period of time, so $R_u(t) = R$. Thus, energy spent by node u with transmission range R during Δt is:

$$cost_{\Delta t}(R) = \Delta t\cdot f(R)\times\left(R^A + C\right) \tag{11}$$

Based on this equation and combining it to Eq. 1, we obtain a system with two equations and two unknowns: R and f that we compute in Section 4.

Partial derivative is applied in order to get the minimum of the cost function when it is equal to 0:

$$\frac{\partial cost(R)}{\partial R} = \Delta t A f(R)\times R^{A-1} + \Delta t\frac{\partial f(R)}{\partial R}\times(R^A + C) = 0 \tag{12}$$

which gives us

$$\frac{\partial f(R)}{\partial R} = -Af(R)\frac{R^{A-1}}{R^A + C} \tag{13}$$

This differential equation gives as a solution:

$$f(R) = \frac{C_1}{R^A + C} \tag{14}$$

where C_1 is a constant defined by initial conditions for the differential equation. For parameters used later on in our simulations, *e.g.* $A = 4$, $C = 10^8$ given in [2] gives $C_1 = 2 \times 10^8$. It becomes:

$$f(R) = \frac{200000000}{R^4 + 100000000} \tag{15}$$

Optimum function $f(R)$ for energy consumption is depicted on Fig. 2. Note that, as expected, the hello frequency decreases when the range increases.

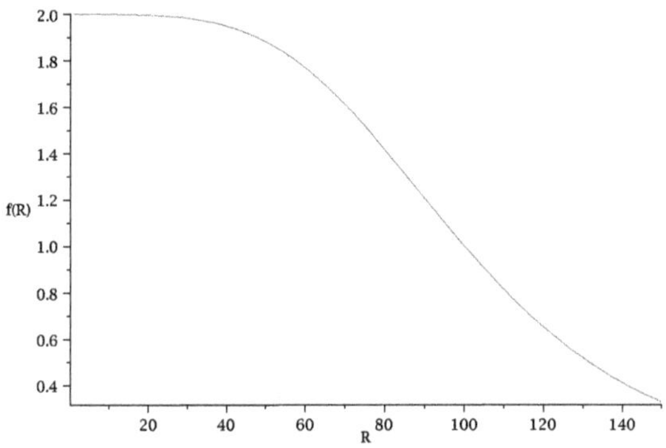

Figure 2: Solution of differential equation obtained for optimal energy cost. Frequency f[Hz] as a function of transmission range R[m]

4. ALGORITHMS FOR ADAPTATION OF FREQUENCY AND RANGE

In this section we present our algorithms used to adapt both frequency of hello messages and node transmission range simultaneously. The first algorithm is based on [5], it relies on the calculation of turnover. The turnover of [5] has been redefined to capture both unilateral and bilateral neighbors (that appear because of different transmission ranges of each node) along with optimal frequency of hello messages [14]. This turnover helps in determining the hello frequency. Algorithm is then declined in two variants to define the transmission range. Second algorithm is designed combining results for optimal frequency and dependency between f and transmission range to adapt both frequency and transmission range without knowledge of turnover.

Both algorithms are based on the existence of an optimal frequency of hello messages [14] given by Eq. 1.

4.1 Using the turnover

To be able to define this algorithm, the first step is to calculate turnover r, in a real scenario. The number of new neighbors (theoretically calculated in previous section) is translated into the calculation of number the new neighbors with given weights.

Current turnover r, is calculated by each node independently according to their own track of the changes between the types of the new neighbors, using equation:

$$r = \frac{x + 2y}{x_t + y_t} \cdot \frac{T_{HELLO}}{\Delta t} \tag{16}$$

where Δt is the time passed between updates of two tables that we are comparing, T_{HELLO} is the period of hello messages, $x_t + y_t$ presents total number of neighbors in neighborhood table and it is the sum of all unilateral and bilateral neighbors. Total number of new unilateral neighbors x found in given period of time Δt is multiplied by constant 1, while total number of new bilateral neighbors y found in given period of time Δt is multiplied by constant 2. These constants are defined so that they reflect importance of the specific type of the new neighbor. For new unilateral neighbors we use smaller constant because we consider those neighbors weaker and we want to give them less importance in calculation of turnover. Bilateral neighbors are considered stronger and they are multiplied by a bigger constant.

First solution is detailed in Algorithm 1 *Turnover based Power Transmission Adjustment*. Algorithm 1 is executed by each node independently only based in the observation of its neighborhood. Algorithm 1 is TAP-fashion algorithm which adapts the hello frequency dynamically based on changes on node neighborhood. It aims to reach an optimal turnover previously computed thanks to computing of new neighbors provided in Section 3.2. Starting points for the algorithms are neighborhood table and *history table*, and all other values are calculated using these values, including the turnover. Neighborhood table is the standard neighborhood table, as explained in Section 2. History table is a table in which we are preserving the values which we have obtained for a given time moment, number of new bilateral and unilateral neighbors and the changes between the type of the neighbors which allows calculating the turnover at that moment.

Adjustment of f is calculated through the period between two hello messages T_{HELLO}, where $f = \frac{1}{T_{HELLO}}$:

$$T_{HELLO} = \begin{cases} T_{HELLO} + \frac{T_{HELLO}}{4} \cdot g(r) & \text{if } r \le r_{opt} \\ T_{HELLO} - \frac{T_{HELLO}}{4} \cdot g(r) & \text{otherwise} \end{cases} \tag{17}$$

Function $g(r)$ is retrieved using turnover:

$$g(r) = \begin{cases} (\frac{r - r_{opt}}{r_{opt}})^2 & \text{if } r < 2 \cdot r_{opt}, \\ 1 & \text{otherwise} \end{cases} \tag{18}$$

Algorithm 1 Turnover based Power Transmission Adjustment

1: **while** 1 **do**
2:    **CalculateCurrentTurnover** $r = \frac{x+2y}{x_t+y_t} \cdot \frac{T_{HELLO}}{\Delta t}$
3:    **if** $r \le r_{opt}$ **then**
4:       *Lower* f with Eq. 17
5:    **else if** $r > r_{opt}$ **then**
6:       *Augment* f with Eq. 17
7:    **end if**
8:    *Adapt Power of Transmission*
9: **end while**

Both variants differ in the call to *Adapt Power of Transmission* function in Line 8 of Algorithm 1. First variant uses Eq. 1 while second variant uses the analysis given in Section 3.3. As a result, we have $R = \sqrt[A]{\frac{C_1}{f} - C}$ as a function for transmission range.

4.2 Minimizing the energy consumption

This second algorithm is based on the idea that there exists a joint way to optimize the transmission range and the hello frequency. Replacing the optimal frequency f (given in Eq. 1) in Eq. 14, we retrieve equation $R^A - \frac{aRC_1}{2S} + C = 0$ that we need to solve to get transmission range R for different speeds S and constants A and C.

Algorithm 2 Cost Based Transmission Power Adjustment

To get R, solve $R^A - \frac{aRC_1}{2S} + C = 0$ based on A values.
$f = \frac{2S}{aR}$
Return (f,R)

5. SIMULATION RESULTS

We obtained our results using WSNET[2] simulator. We use a model in which the range of transmission can be adapted according to the values that we retrieve using proposed algorithms. In our simulation setup 100 simulation nodes are randomly deployed in a rectangular area of $1000 \times 1000m$. Mobility model used for the nodes uniform linear velocity with perfect deflection *i.e.* billiard model of motion, although, not realistic this mobility model can for example approximate behavior of the animals in natural habitat. Moreover, this kind of mobility matches with presented theoretical analysis and helps us to evaluate simulation results in terms of the results obtained theoretically.

In our simulations we have tracked important parameters for neighborhood discovery and their dependence with respect to the given speed of the nodes. In the following graphs we present values for the accuracy of the neighborhood tables, range of transmission, period of hello messages, state of the nodes batteries and observed turnover as the functions of the nodes' speed. In these graphs Algorithm 2 is called *NoTAP* since it does not use turnover in calculations, Algorithm 1 with usage of optimal frequency is called *Fopt*, TAP algorithm [5] is referred as TAP while Algorithm 1 with minimized cost is called *Cost*. We present results retrieved from the simulations with comments on the performance of the Fopt, NoTAP and TAP algorithm. Simulation results for TAP algorithm are obtained using the turnover as defined in Eq. 16, using information on both unilateral and bilateral neighbors. Separate discussion will be given on the results retrieved from the Cost algorithm for adaptation of the range using minimized energy consumption because results obtained using this algorithm are unrealistic (too good) and deserve additional explanation. Each figure depicts an average of 500 runs and 95% confidence intervals.

Accuracy of the neighborhood tables (shown on Fig. 3) is one of the most important values to follow and it presents ratio of the real neighbors against the neighbors found in the neighborhood table. Accuracy is calculated periodically checking the state of neighborhood table and comparing if the nodes listed as the neighbors are still in the neighborhood of the node that is being checked. All algorithms are showing tendency of decreasing accuracy with increase of the speed which is logical since the faster nodes spend less time in physical neighborhood of their neighboring node. Accuracy of the neighborhood tables are higher than 90% even for the highest speed of nodes.

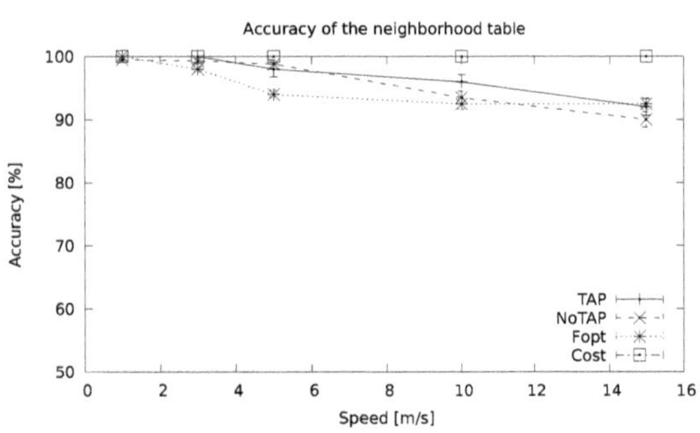

Figure 3: Accuracy of the neighborhood table as function of speed

Figure 4(a) shows transmission range adaptation as the function of speed of the nodes. We can notice that transmission range for the TAP algorithm is held on the same level, using the same approach (unit disk graph) as in [5]. In the case for Fopt and NoTAP algorithm ranges are increasing as the speed is increasing with the difference that NoTAP algorithm increases range linearly. This behavior of these two algorithms is expected because when nodes are moving faster then they are also changing their neighbors faster. Then in order to maintain the number of new neighbors both of these two algorithms are increasing the range. We have to note also that Fopt increases the range more than NoTAP since its range is not bounded to single value for a given speed as it is the case for NoTAP. On the other side smaller increase of range of NoTAP algorithm is compensated (as given in Eq.1) with the increase of the period of hello messages (Fig. 4(b)). Fopt is keeping the period on almost constant value because the adaptation to the higher speed is done with an increase of the range. TAP is decreasing hello period with the increase of speed because its range is fixed and the only way to adapt its behavior when nodes are moving faster is to lower the period (hence increase the frequency of hello messages). From Fig. 4(c) we can see that all algorithms better balance energy than TAP which keeps transmission range on the same level thus it has worse results than others. We can also observe that energy loss increases as the speed of the nodes increases, this is due to the adaptation of algorithms and their attempt to balance the values of turnover, frequency and range compensated in higher energy loss.

In this case we have to point out that we used linear discharge model for the battery, decreasing certain amount energy from the battery multiplied by range of transmission each time when we transmit packet, constant value for each received packet and loss of energy in idle mode represented by constant value multiplied by the time spent in idle mode (the time between receptions or between reception and transmit and vice versa). This representation of battery is not the best one since it overstates the impact of transmission range which is in real case smaller and is given with the increase of power of transmission.

The results for turnover, Fig. 4(d), are more straightforward, we can see that for the algorithms that take turnover (Fopt and TAP) into account there is an increase of turnover with the increase of speed while NoTAP changes turnover independently on the speed. Fopt has slightly bigger change of turnover as a result of adaptation both frequency and range.

5.1 Comment on the Cost algorithm

Cost algorithm shows almost perfect results when taking into account state of the battery and accuracy of the neighborhood table, but looking into dependency of the turnover and range gave us the clue what is the reason for this. This algorithm tries to minimize the energy, and it does it well, but at the cost of keeping the range of transmission on the lowest value thus gaining new neighbors occasionally and with big accuracy. Also since the transmission range is minimal it also preserves battery in the best way. What we can conclude from this is that the model that we imposed for this algorithm is too ideal and does not take into account the nuances this adaptation might have. Minimal transmission range also has another consequence, keeping transmission range on such a low value is ensuring low power consumption but at the same time threatens the correct neighborhood discovery since each node with such a small transmission range detects small number of neighbors and keeps its neighborhood table almost empty.

6. CONCLUSION

In this paper we present neighborhood discovery algorithms that take into account both frequency of hello messages and the transmission range in order to save energy. We are using results obtained by the theoretical analysis on the neighborhood discovery to calculate

[2]http://wsnet.gforge.inria.fr/

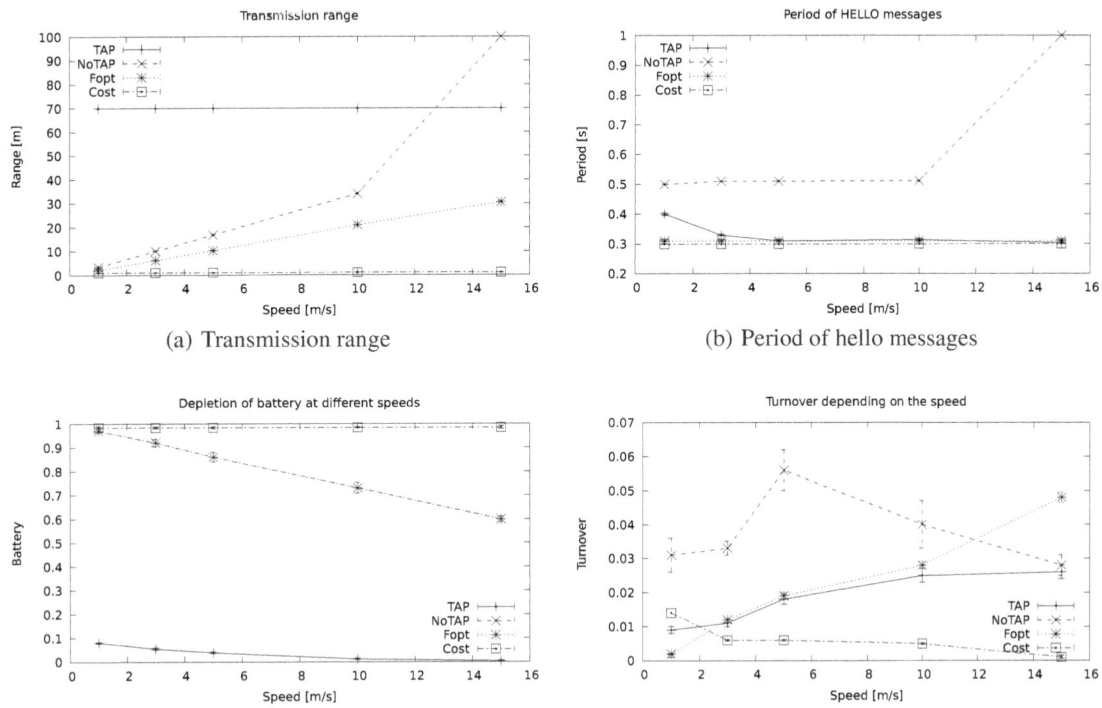

(a) Transmission range

(b) Period of hello messages

(c) Battery level

(d) Turnover

Figure 4: Different parameters of the WSN as the function of nodes speed

turnover and its impact on the overall energy consumption. These results are combined with optimal frequency of hello messages for mobile nodes and to conceive two algorithms which are adapting both frequency of hello messages and transmission range. Simulations have proven that our algorithms outperform TAP (which only adapts frequency with fixed range) in terms of energy consumption while keeping the accuracy of the neighborhood table.

Future work will include further improvement of proposed algorithms. We will take into account the total number of neighbors and received messages thus ensuring both energy efficient and realistic neighborhood discovery (limiting the transmission range to a higher values). Also, as the real sensor hardware transmission range cannot be directly manipulated so one of the ideas is to make the connection between transmission range and power of transmission. Using this connection our algorithms can be run on the real wireless sensor networks.

7. ACKNOWLEDGMENTS

This work was partially supported by CPER NPdC/FEDER Campus Intelligence Ambiante and the ANR F-Lab project.

APPENDIX

A. COMPUTATION OF THE LIMITS FOR THE INTEGRALS

A.1 Computing d_{min}

We need to compute the mean number of new neighbors node u detects for a Δt time period. For it, we suppose a node v at a distance d from node u at time $t_1 = t_0 + \Delta t$, such that $0 \le d \le R$.

Case 1 – *new bilateral neighbors* In this case v is a new bilateral neighbor of node u which means that $|u_0 v_0| > min(R_u, R_v)$ and $|u_1 v_1| < min(R_u, R_v)$. Value d_{min} is such that, if $d < d_{min}$, nodes u and v are neighbors at time t_0 whatever the directions α and β. As illustrated by the Fig. 5(a), the longest distance l between u_0 and v_0 at time t_0 is when both node directions are opposite (for example when $\alpha = 0$ and $\beta = \pi$). We have $l = 2\Delta d + d$, where $\Delta d = \Delta t \times S$. For v to be a new neighbor of node u, we need $l > min(R_u, R_v)$, which leads to:

$$
\begin{aligned}
& 2\Delta d + d > min(R_u, R_v) \\
\Leftrightarrow\ & d_{min} = 2\Delta d - min(R_u, R_v)
\end{aligned}
\tag{19}
$$

Since, depending of the node speed, we may have $2\Delta d < min(R_u, R_v)$, we finally get:

$$
d_{min} = max(0, 2\Delta d - min(R_u, R_v)) \tag{20}
$$

Case 2 – *new unilateral neighbors* By definition, for v to be a new unilateral neighbor of node u, d is such that $d > R_v$. Then, similarly to the previous case, d_{min} is such that if $d > d_{min}$, nodes u and v were already neighbors at the time t_0, whatever the values α and β.

Since, depending on the node speed, we may have $2\Delta d < min(R_u, R_v)$, we finally get:

$$
d_{min} = max(R_v, 2\Delta d - min(R_u, R_v)) \tag{21}
$$

A.2 Computing α_{min}

In this section, we are interested in computing the value α_{min} such that, given $d > d_{min}$ and α, if $\alpha < \alpha_{min}$ then for any β (no matter the value which it takes) v was neighbor of u at time t_0.

For it, we introduce k such that $k = |u_0 v_1|$, illustrated in Fig 5(b).

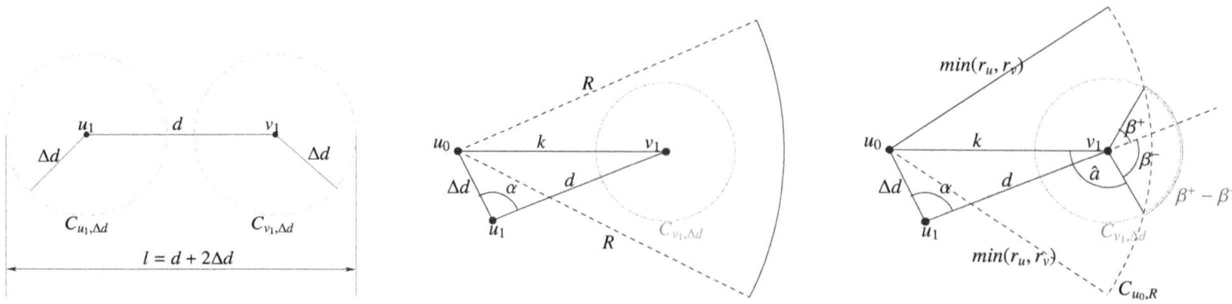

(a) Calculating d_{min} (b) Calculating α_{min} (c) Calculating $\beta^+ - \beta^-$

Figure 5: Computation of the limits for the integrals

According to Pythagore's theorem, we can compute the value of k:

$$\Delta d^2 - (\Delta d \cos(\pi - \alpha))^2 = \quad k^2 - (d + \Delta d \cos(\pi - \alpha))$$
$$\Leftrightarrow k = \quad \sqrt{\Delta d^2 + d^2 - 2d\Delta d \cos\alpha} \qquad (22)$$

Case 1 – *new bilateral neighbors* For v to be new bilateral neighbor of node u we need that $|u_0v_0| > min(R_u, R_v)$ for any β *i.e.* that the disks delimited by $C_{v_1,\Delta d}$ and $C_{u_0,min(R_u,R_v)}$ overlap:

$$k - \Delta d > min(R_u, R_v)$$
$$\Leftrightarrow k^2 > (min(R_u, R_v))^2 + \Delta d^2 + 2\Delta d \cdot min(R_u, R_v)$$
$$\Leftrightarrow d^2 - 2d\Delta d \cos\alpha > (min(R_u, R_v))^2 + 2\Delta d \cdot min(R_u, R_v)$$
$$\Leftrightarrow \alpha_{min} = \arccos(\frac{d^2 - (min(R_u, R_v))^2 - 2\Delta d \cdot min(R_u, R_v)}{2d\Delta d}) \qquad (23)$$

Case 2 – *new unilateral neighbors* For v to be a new unilateral neighbor of node u we need that $|u_0v_0| > max(R_u, R_v)$ for any β *i.e.* that the disks delimited by $C_{v_1,\Delta d}$ and $C_{u_0,max(R_u,R_v)}$ overlap:

$$k - \Delta d > max(R_u, R_v)$$
$$\Leftrightarrow k^2 > (max(R_u, R_v))^2 + \Delta d^2 + 2\Delta d \cdot max(R_u, R_v)$$
$$\Leftrightarrow d^2 - 2d\Delta d \cos\alpha > (max(R_u, R_v))^2 + 2\Delta d \cdot max(R_u, R_v)$$
$$\Leftrightarrow \alpha_{max} = \arccos(\frac{d^2 - (max(R_u, R_v))^2 - 2\Delta d \cdot max(R_u, R_v)}{2d\Delta d}) \qquad (24)$$

A.3 Computing $\beta^+ - \beta^-$

For computing of the $(\beta^+ - \beta^-)$ we use notations from the Fig 5(c). We can notice that:

$$2\pi = (\beta^+ - \beta^-) + 2\hat{a} \qquad (25)$$

Case 1 – *new bilateral neighbors* In this case we consider

$$\hat{a} = \arccos(\frac{\Delta d^2 + k^2 - min(R_u, R_v)}{2k\Delta d}) \qquad (26)$$

We deduce that:

$$(\beta^+ - \beta^-) = 2\arccos(\frac{min(R_u, R_v) - \Delta d^2 - k^2}{2k\Delta d}) \qquad (27)$$

Case 2 – *new unilateral neighbors* In this case we consider

$$\hat{a} = \arccos(\frac{\Delta d^2 + k^2 - max(R_u, R_v)}{2k\Delta d}) \qquad (28)$$

We deduce that:

$$(\beta^+ - \beta^-) = 2\arccos(\frac{max(R_u, R_v) - \Delta d^2 - k^2}{2k\Delta d}) \qquad (29)$$

B. REFERENCES

[1] R. Cohen and B. Kapchits. Continuous neighbor discovery in asynchronous sensor networks. *Networking, IEEE/ACM Transactions on*, 19(1):69 –79, feb. 2011.

[2] E. Fleury and D. Simplot-Ryl, editors. *Réseaux de capteurs*. Hermes Science - Lavoisier, 2009.

[3] V. Giruka and M. Singhal. Hello protocols for ad hoc networks : Overhead and accuracy trade-offs. In *IEEE International Symposium on a World of Wireless Mobile and Multimedia Networks (WOWMOM)*, 2005.

[4] C. Gomez, A. Cuevas, and J. Paradells. A two-state adaptive mechanism for link connectivity maintenance in AODV. In *REALMAN*, 2006.

[5] F. Ingelrest, N. Mitton, and D. Simplot-Ryl. A turnover based adaptive hello protocol for mobile ad hoc and sensor networks. In *MASCOTS*, pages 9–14, Istanbul, Turkey, Oct. 2007.

[6] P. Jacquet, P. Muhlethaler, T. Clausen, A. Laouiti, A. Qayyum, and L. Viennot. Optimized link state routing protocol for ad hoc networks. In *Proceedings IEEE INMIC 2001.*, pages 62 – 68, 2001.

[7] K. Kleisouris, B. Firner, R. Howard, Y. Zhang, and R. P. Martin. Detecting intra-room mobility with signal strength descriptors. In *Proceedings ACM MobiHoc*, pages 71–80, New York, NY, USA, 2010.

[8] X. Li, N. Mitton, and D. Simplot-Ryl. Mobility prediction based neighborhood discovery in mobile ad hoc networks. In *Networking*, pages 241–253, 2011.

[9] V. Loscrì, E. Natalizio, and C. Costanzo. Simulations of the impact of controlled mobility for routing protocols. *EURASIP J. Wireless Comm. and Networking*, 2010(1), 2010.

[10] J. Moy. OSPF - Open Shortest Path First, Mar. 1994. RFC 1583.

[11] J. Radak, N. Mitton, and D. Simplot-Ryl. Using Battery Level as Metric for Graph Planarization. In *10th International Conference on Ad Hoc Networks and Wireless (AdHocNow)*, pages 58–71, Padderborn, Germany, July 2011.

[12] T. Razafindralambo and N. Mitton. Analysis of the impact of hello protocol parameters over a wireless network self-organization. In *PE-WASUN*, pages 46–53, Chania, Greece, Oct. 2007. ACM.

[13] D. Stoyan, S. Kendall, and J. Mecke. *Stochastic geometry and its applications, second edition*. John Wiley & Sons, 1995.

[14] A. Troël. *Prise en compte de la mobilité dans les interactions de proximité entre terminaux à profils hétérogènes*. PhD thesis, Universite de Rennes, France, 2004. In French.

Characterization of 3G Control-Plane Signaling Overhead from a Data-Plane Perspective

Li Qian† Edmond W. W. Chan† Patrick P. C. Lee‡ Cheng He†
†Noah's Ark Lab, Huawei Technologies, Shenzhen, China
‡Dept of Computer Science & Engineering, The Chinese University of Hong Kong, Hong Kong
{qianli,edmond.chan,hecheng}@huawei.com, pclee@cse.cuhk.edu.hk

ABSTRACT

In 3G networks, when user applications of mobile subscribers send or receive data-plane traffic, control-plane signaling messages will be triggered to initiate or release radio resources for the data transfer. Such signaling messages can increase the processing and transmission overheads of the 3G cellular network infrastructure, and this in turns degrades the performance experience of mobile subscribers. Thus, understanding the signaling overhead of a 3G network becomes critical. In this paper, we conduct the first comprehensive measurement study of the signaling overhead of a city-wide 3G operational network in China. Our main contributions are two-fold. First, based on real cellular traces collected from both data and control planes, we validate that by simply monitoring data-plane packets, we can accurately profile the control-plane signaling overhead due to the initiations of radio resources for data transfer. Second, using data-plane signaling profiling, we characterize the signaling overhead due to common transport protocols and network applications. Our measurement methodology and results presented in this paper would be useful for network operators to better understand how data-plane traffic patterns influence the control-plane signaling overhead of a 3G network.

Categories and Subject Descriptors

C.2.1 [**Computer-Communication Networks**]: Network Architecture and Design—*Wireless communication*; C.4 [**Performance of Systems**]: Measurement techniques

Keywords

Signaling overhead, 3G networks

1. INTRODUCTION

Third generation (3G) cellular networks have been widely deployed worldwide. With the emergence of smartphones and mobile applications, the performance of operational cellular networks has been challenged. One potential challenge is the control-plane signaling overhead. According to existing 3G standards, signaling messages will be triggered when the user applications of mobile subscribers initiate or release radio resources for data transfer [2]. Different types of user applications (e.g., bulk-data transfer versus interactive applications) have different data transfer characteristics, thereby having different impacts on triggering the signaling messages. Since signaling messages consume radio resources, excessive signaling messages can increase both the processing and transmission overheads within the cellular infrastructure, and hence degrades performance experience of mobile subscribers. From the network operators' perspectives, it is critical to characterize the control-plane signaling overhead of a 3G network due to different types of data-plane traffic so as to devise effective strategies for radio resource provisioning and network planning.

In this paper, we conduct an in-depth measurement study of the signaling overhead of a city-wide 3G operational cellular network in China. At the core of the 3G network, we collect and analyze a 24-hour span of IP data packets and radio resource control (RRC) logs, which correspond to data-plane and control-plane traffic traces, respectively. Our goal is to provide a comprehensive view of how data-plane traffic is correlated with the control-plane signaling overhead of a currently deployed 3G network, from the network operators' perspectives. To our knowledge, this is the first measurement study that quantifies the signaling overhead of a commercially deployed 3G network.

Our contributions are two-fold. We start with considering a data-plane signaling profiling approach that characterizes the 3G signaling overhead by monitoring only raw IP packets in the data plane, without looking into the control-plane information. Our methodology is built on [12, 14], and infers the state transitions for radio resource allocation triggered by data-plane packets. Using our collected cellular traces, we validate that by simply monitoring data-plane packets, we can accurately profile the state transitions due to RRC connection setups, which account for a large proportion of signaling messages. Thus, data-plane signaling profiling simplifies the complexities of collecting and correlating both data-plane and control-plane traces, while still providing highly accurate signaling profiling results.

We further use the data-plane signaling profiling approach to analyze the signaling overhead due to different characteristics of the transport and application layers in the data plane. In each protocol layer, we identify the types of messages that trigger the most signaling messages. In general, we observe that nearly 90% of all state transitions (which trigger signaling messages) are due to small-size data packets with payload less than 200 bytes. We also identify several signaling-prone applications that we observe.

The remaining of the paper proceeds as follows. In Section 2, we review the design of a UMTS network and its radio resource control mechanism. In Section 3, we present and validate our data-plane signaling profiling approach. In Section 4, we report our analysis results on how the signaling load is influenced by the transport-

MSWiM'12, October 21–25, 2012, Paphos, Cyprus.

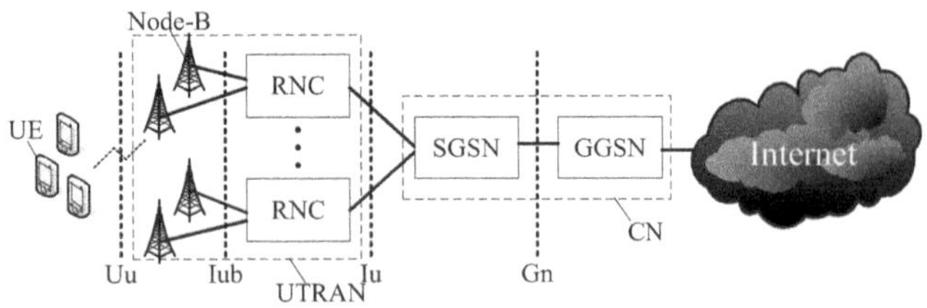

Figure 1: The UMTS network architecture.

layer and application-layer behaviors in the data plane. Section 5 reviews related work. Finally, Section 6 concludes the paper.

2. BACKGROUND

2.1 The UMTS network

Universal Mobile Telecommunication System (UMTS) [1] is one of the most popular 3G mobile communication technologies deployed nowadays. It comprises three interacting domains: Core Network (CN), UMTS Terrestrial Radio Access Network (UTRAN), and User Equipment (UE). The CN mainly provides switching, routing, and transit for user traffic and is further divided into circuit switched (CS) and packet switched (PS) domains. In this paper, we focus on the PS domain, which manages IP packet transmissions. Figure 1 shows a typical UMTS network. The UTRAN comprises the base stations (Node-Bs) for serving multiple UEs and the Radio Network Controllers (RNCs). The CN comprises the Serving GPRS Support Node (SGSN) and the Gateway GPRS Support Node (GGSN), each of which communicates with several RNCs for traffic exchange. Moreover, UMTS defines several interfaces, including Gn (between GGSN and SGSN), Iu (between SGSN and RNC), Iub (between RNC and Node-B), and Uu (between Node-B and UE).

2.2 RRC state machine

The Radio Resource Control (RRC) protocol belongs to the control plane of UMTS for system-wide control of communication resources and services. It controls the radio bearer resources for data-plane traffic of each UE. In the protocol, each UE is associated with a *state machine*, which is typically composed of three states: IDLE, CELL_DCH (or DCH), and CELL_FACH (or FACH). Each state corresponds to different levels of radio resource allocation. Initially, when there is no data traffic, a UE is at IDLE and no RRC connection has been established. After an RRC connection is established for data-plane transmission, the UE either operates at DCH for high-speed data transmission through dedicated transport resources, or at FACH for low-speed data transmission in the shared channel to save both radio resources and the UE's energy consumption. Note that in each state transition, the UE and its associated RNC exchange signaling messages through the Iub and Uu interfaces.

Figure 2 depicts the RRC state machine for a commercial UMTS network considered in this paper. When the UE is at IDLE and has uplink (UL)/downlink (DL) data to send/receive, it is always allocated a dedicated transport channel after an RRC connection is set up (i.e., IDLE→DCH). The subsequent state transitions depend on two inactivity timers denoted by T_{FACH} and T_{IDLE}, both of which can be configured by network operators, as well as the service type, which can be specified by the UE in its RRC connection request. For best-effort services (e.g., Web Browsing and Instant Messaging), the RNC resets T_{FACH} to 5s when any UL/DL data traffic is observed, or demotes the state to FACH (i.e., DCH→FACH) when T_{FACH} expires. At FACH, the RNC promotes the state to DCH (i.e., FACH→DCH) again when the Radio Link Control (RLC) buffer size from either the UE (for UL packets) or the RNC (for DL packets) is greater than a threshold of 1024 bytes. However, for real-time services (VoIP and Streaming), DCH is always maintained for the UE to fully utilize radio resources.

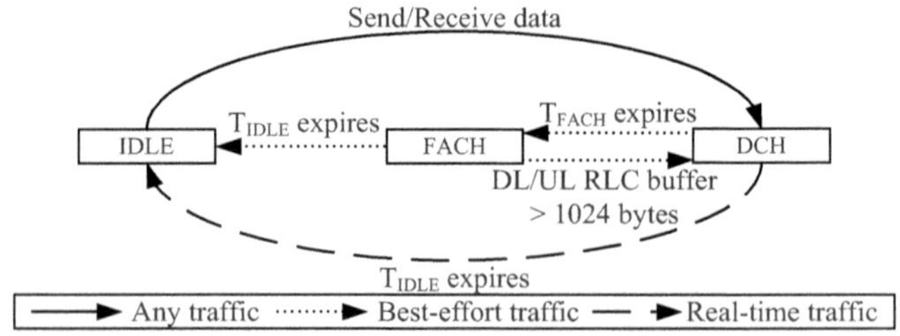

Figure 2: The RRC state machine for the UMTS network considered in this paper.

Table 1: Numbers of signaling messages generated per transition by each type of RRC state transitions.

RRC state transition	No. of signaling messages
IDLE→DCH	23
DCH→FACH	4
FACH→DCH	10
DCH→IDLE	8
FACH→IDLE	6

To release the RRC connection, the RNC also maintains a T_{IDLE} timer for the UE regardless of the service type. At DCH or FACH, T_{IDLE} is reset to 18s whenever UL/DL data traffic is observed. When T_{IDLE} expires, the RNC demotes the state to IDLE from either DCH (i.e., DCH→IDLE) or FACH (i.e., FACH→IDLE) .

We notice two distinct features from the two RRC state machines reported in [12, 14] compared with ours. First, the transitions from the two state machines do not depend on the service type, thus trading the real-time service performance for improved channel resource utilization among different services. Second, the DCH→IDLE transition never occurs in their state transitions, but instead the RRC connection always switches to FACH before reaching IDLE.

For each state transition, the associated RNC exchanges a few round-trips of signaling messages in the control plane with the UE, SGSN, and Node-B to manage the usage of radio resources allocated for the UE. Table 1 shows the number of signaling messages generated per transition by each type of the RRC state transitions as shown in Figure 2, whose numbers are obtained through internal conversations with our product line engineers. We see that the state promotions (i.e., IDLE→DCH and FACH→DCH) trigger more signaling messages than the state demotions (i.e., DCH→FACH, DCH→IDLE, and FACH→IDLE). Specifically, the IDLE→DCH transition generates the most number of signaling messages. Note that frequent RRC state transitions introduce massive signaling messages, which can overload the processing capacity of the RNC and exhausting radio resources [9].

3. METHODOLOGY VALIDATION

As described in Section 2.2, the state transitions, and hence the signaling messages, are triggered by the data transmissions (either uplink or downlink) for a UE. To understand the root causes of the signaling load of a network, one may need to correlate the control-plane signaling messages and the data-plane packets. Such correlation requires that both traces of signaling messages and data packets be available. In this section, we consider a signaling profiling strategy that uses only data-plane packets to infer the signaling overhead of a network. Using real traffic traces, we validate that data-plane profiling, even without control-plane information, can achieve high inference accuracy in practice. This simplifies the complexities of collecting and correlating the data-plane and control-plane traces.

3.1 Data-Plane Signaling Profiling Overview

We consider a data-plane signaling profiling approach, in which we infer the signaling overhead due to RRC state transitions based on raw IP packet traces captured from the Iu/Gn interfaces. Our profiling approach is built on [12, 14]. Each raw IP packet contains a private IP address that corresponds to a UE. We extract from the traces each *UE session*, which consists of all packets having the same private IP address and packet inter-arrival times less than 60 seconds. We assume that *each private IP address corresponds to a unique UE*, based on the observation that a private IP address is less likely reused by different UEs within a short time period.

For each UE session, we infer a sequence of RRC states based on the inter-arrival times of adjacent IP packets and the known RRC state machine shown in Figure 2. Also, the service type is typically specified in the RRC connection request and is not available in data packets (see Section 2.2). To infer the service type (i.e., real-time or best-effort services) for each IP packet, we use a commercial Deep Packet Inspection (DPI) tool to classify each packet into one of the 14 application types: Database, Email, File Access, Game, Instant Messaging (IM), Network Admin, Network Storage, Peer-to-peer (P2P), Remote Connectivity, Stock, Streaming, Tunneling, Voice over IP (VoIP), and Web Browsing. After identifying the service type, we initiate the required inactivity timers and determine the corresponding state transitions (see Section 2.2). Based on each state transition, we then determine the corresponding number of signaling messages stated in Table 1.

We point out that using data-plane packet traces to infer radio resource usage is not entirely new. Qian *et al.* employ raw IP packet traces collected from two UMTS networks to infer the state machine configurations and energy consumption in each state [12, 14], and propose a client-based profiling tool [13] to perform a similar analysis from the traces collected within a UE. We elaborate the differences of our work and the related studies [12, 13, 14] in Section 5.

3.2 Dataset

It is important that our data-plane signaling profiling approach provides high inference accuracy in practice. In this subsection, we describe the real traffic traces that we use for validating our profiling approach.

In this paper, we analyze 24-hour IP packet traces collected from a commercial UMTS network from a city in China on 1 December 2010. The raw IP packets with full payload are captured (without sampling) from the Iu interface (i.e., the RNC-SGSN interface) and stored in the PCAP format. The network has a total of 16 RNCs deployed in the whole city, yet our analysis mainly focuses on a particular RNC deployed in the urban region. Thus, after the extraction of useful data, we conduct our analysis on 306M IP packets with 682K UE sessions.

During the period, we also collected Performance Call History Record (PCHR) log files from the particular RNC we consider. A PCHR describes the control-plane information about RRC connection setups and releases from a specific RNC for both PS and CS domains. Note that we only use PCHR logs for our validation, while the profiling approach itself does not involve PCHR logs.

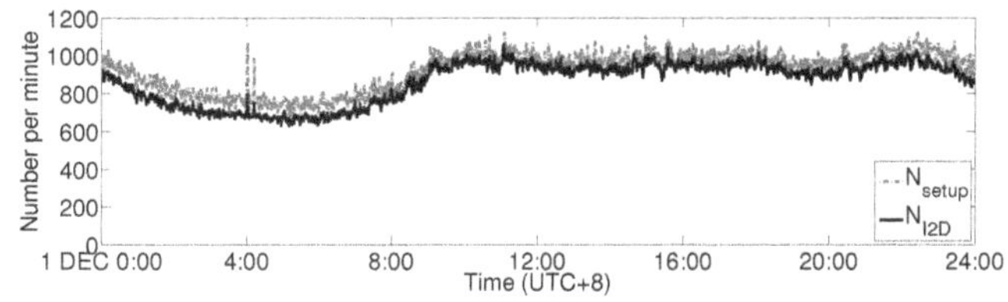

Figure 3: 24-hour time series for the number (per minute) of RRC connection setups (N_{setup}) and the inferred number of IDLE→DCH transitions (N_{I2D}) on 1 December 2010.

3.3 Validation

We now validate the accuracy of our data-plane signaling profiling approach using PCHR logs. We extract all RRC connection setup records in the PS domain from the PCHR log files of the RNC that we consider. Since a single RRC connection setup is triggered by the IDLE→DCH transition of a UE, we can use the setup records to obtain the actual number of IDLE→DCH transitions.

Figure 3 shows the time series of the number of IDLE→DCH transitions (denoted by N_{I2D}) inferred from the data-plane signaling profiling approach and the number of RNC connection setups (denoted by N_{setup}) obtained from the PCHR logs on 1 December 2010. The x-axis is presented in the local UTC+8 China time. Each data point is measured every minute. The figure reveals strong correlation between the inferred number of IDLE→DCH transitions and the actual number of connection setups. Using the profiling approach, we can identify the diurnal pattern of the RRC connection setups. The number of IDLE→DCH transitions drops to below 700 per minute during the midnight, and consistently stays above 800 per minute during the office hour and the evening.

Figure 4 depicts the CCDF of the relative difference between N_{I2D} and N_{setup} using our one-day dataset. The relative difference is computed as $(N_{I2D} - N_{setup})/N_{setup}$. As shown in the figure, the number of RRC connection setups and the inferred number of IDLE→DCH transitions are *very similar*, where more than 80% of the estimates have the relative differences less than 10%. We observe that our profiling approach tends to underestimate the number of IDLE→DCH transitions. The missing IDLE→DCH transitions are likely due to an early RRC connection release by a UE before the 18-second idle time. In this case, the successor of the UE's IP address can immediately transmit a packet through a new RRC connection. Therefore, our signaling profiling approach will miss one IDLE→DCH transition if the inter-arrival time between the last packet and the first packet in the two adjacent RRC connections is less than the dormancy time. Note that the inference accuracy can be improved by using International Mobile Subscriber Identification Number (IMSI) information to distinguish separate RRC connections.

4. SIGNALING PROFILING ANALYSIS

We apply our signaling profiling approach to study the transport-layer and application-layer characteristics for the IP packets after state transitions and hence the signaling overhead on the UMTS network that we consider. We also analyze the root causes of our profiling results, and discuss the possible strategies that can mitigate the signaling load. Our measurement methodology and results provide insights into the impact of data-plane traffic on the signaling load of a commercial UMTS network, and we expect that such insights are of interest to network operators for better network planning.

Our analysis is applied to the data-plane traces (i.e., raw IP pack-

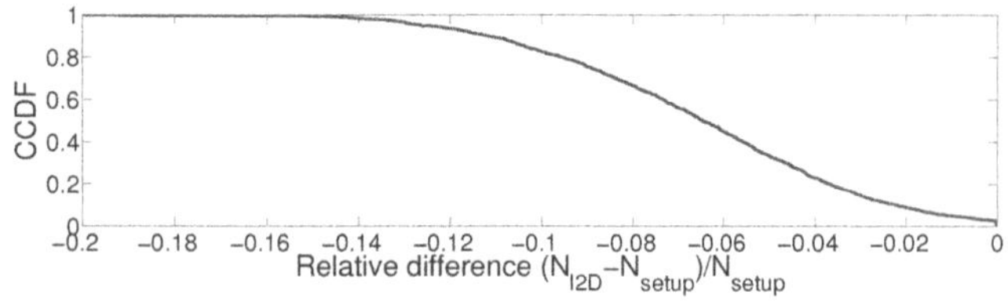

Figure 4: CCDF of the relative difference between the numbers of the RRC connection setups and the inferred IDLE→DCH transitions.

Table 2: Percentage of signaling messages contributed by the state transitions in our dataset

RRC state transition	Percentage of signaling messages
IDLE→DCH	42.84%
DCH→FACH	14.12%
FACH→DCH	25.03%
DCH→IDLE	11.86%
FACH→IDLE	6.15%

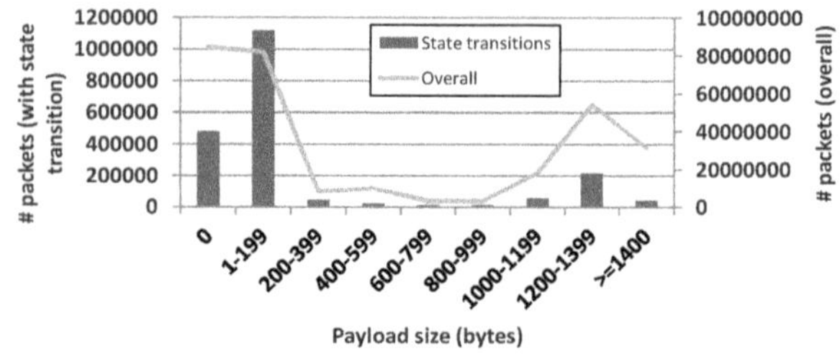

Figure 5: Distributions of payload sizes of overall traffic and the packets inducing state transitions.

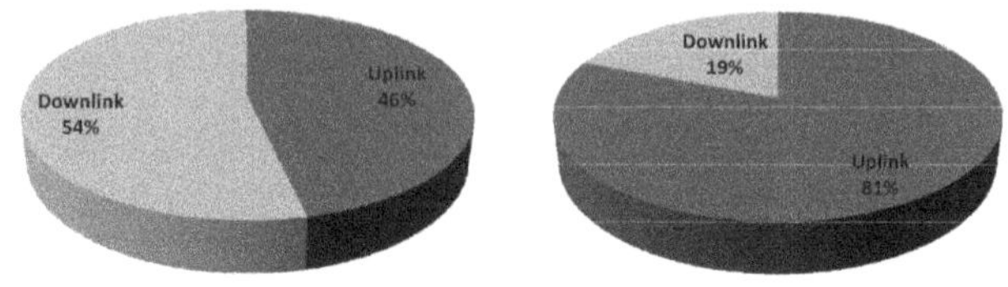

(a) Overall traffic. (b) Only traffic triggering state transitions.

Figure 6: Distributions of uplink and downlink packet numbers.

ets) gleaned from a particular RNC in our one-day dataset. We collect the *first* IP packets observed right after one of the three state transitions: IDLE→DCH (or I→D in short), FACH→DCH (F→D), and DCH→FACH (D→F). As discussed in Section 2.2, the state promotions I→D and F→D are triggered due to packet transmissions. The state demotion D→F, on the other hand, can be regarded as the result of the delayed packet transmission. By considering the first packets after D→F, we can also characterize how frequently a particular type of application initiates the data transmission in a (low-speed) shared channel. Therefore, analyzing the first packet right after each such transition allows us to infer the root causes of the signaling overheads induced by these state transitions.

On the other hand, we ignore the state demotions DCH→IDLE and FACH→IDLE because they are resulted from inactivity timer expiries and not associated with any packet transmission.

Table 2 shows the percentage of signaling messages contributed by each type of state transitions in our dataset, obtained by counting the number of occurrences of each type of state transitions in the dataset (based on our data-plane signaling profiling) and the number of signaling messages introduced per transition (based on Table 1). We note that the DCH→IDLE and FACH→IDLE transitions altogether contribute only 18% of the total number of signaling messages in the dataset, so their influences on the overall analysis are expected to be insignificant.

We analyze the transport-layer protocol and network application type for each of the collected IP packets. The analysis only focuses on the TCP and UDP packets, both of which contribute the majority of the data traffic. We extract the transport-layer information, such as source/destination ports, TCP flags, and payload sizes from their transport-layer headers. For the application-layer analysis, we use a commercial DPI tool to classify each TCP/UDP packet into one of the 14 application types (see Section 3.2). In addition, we classify whether the packet is observed from the uplink (i.e., from UE to remote destination) or downlink. For an uplink (or downlink) packet, its source (or destination) IP address from the GTPU header is equal to the RNC's IP address. Thus, we can use this information to distinguish between uplink and downlink packets.

We address the limitations of our analysis. First, our dataset is collected in December 2010. Given the emergence of mobile Internet access, the scale and pattern of our dataset may not entirely reflect the real usage of today's 3G networks. Second, although we manually validate some of the results of our DPI tool, we cannot comprehensively verify its correctness as its source code is not available. Third, our dataset, while representing one-day traffic of a city-wide network, may be less representative for some networks in national scales. Nevertheless, our analysis is built on the existing 3G UMTS standard, and we expect that the same analysis methodology is applicable for many 3G operational networks. Our analysis results can provide guidelines for network planning and future analysis.

4.1 State Transition Behavior: An Overview

We first provide a general overview of how the state transition behavior is influenced by the packet size and the directions of packet transmissions.

Effect of TCP/UDP payload size. Figure 5 plots the TCP/UDP payload size distributions of overall traffic and the packets inducing state transitions. We observe that 88.6% of the state transitions are induced by small TCP/UDP packets with payload size less than 200 bytes, which altogether contribute 56.4% of the total number of TCP/UDP packets observed from the RNC. Moreover, the packets with zero payload size contribute 23.9% of the state transitions, and all of them are due to *TCP control messages* (e.g., pure ACKs, SYNs, RSTs, FINs) that do not carry any payload.

Uplink versus downlink packets. Figure 6(a) plots the distributions of the total numbers of uplink and downlink TCP/UDP packets observed, and Figure 6(b) shows the distributions for only the packets that induce state transitions. As shown in Figure 6(a), the numbers of uplink and downlink packets are fairly close, such that the downlink packets contribute 8% more than the uplink packets. However, Figure 6(b) reveals that more than 80% of the state transitions are triggered by the uplink packets.

We also plot the breakdowns of the TCP/UDP packet-induced state transitions and their corresponding numbers of signaling messages, as shown in Figures 7(a)–7(b). We observe from Figure 7(a) that I→D contributes the most state transitions for both uplink and downlink directions (i.e., 59.1% and 57.3%, respectively), whereas

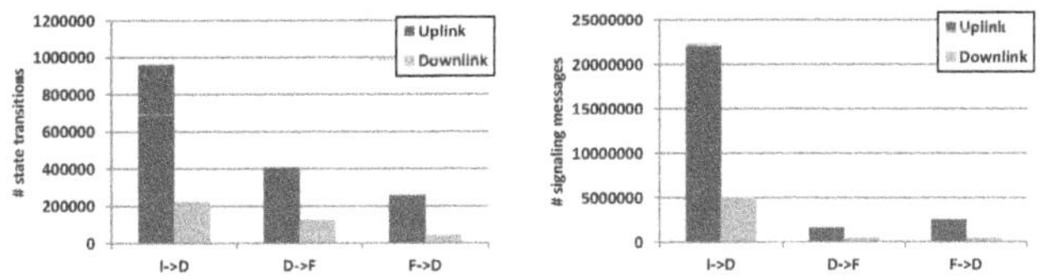

(a) Number of state transitions. (b) Number of signaling messages.

Figure 7: State transition breakdowns of uplink and downlink traffic.

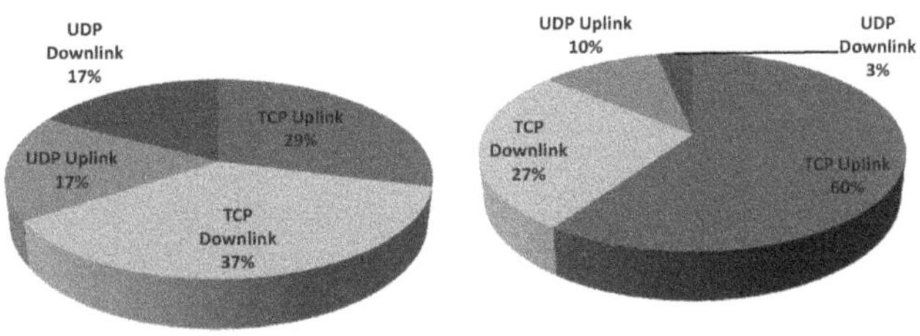

(a) Overall traffic. (b) Only traffic triggering state transitions.

Figure 8: Distributions of the TCP and UDP traffic from uplink and downlink directions.

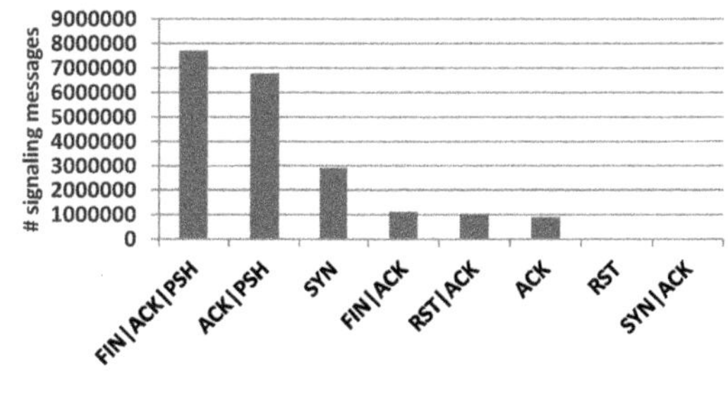

(a) Uplink traffic.

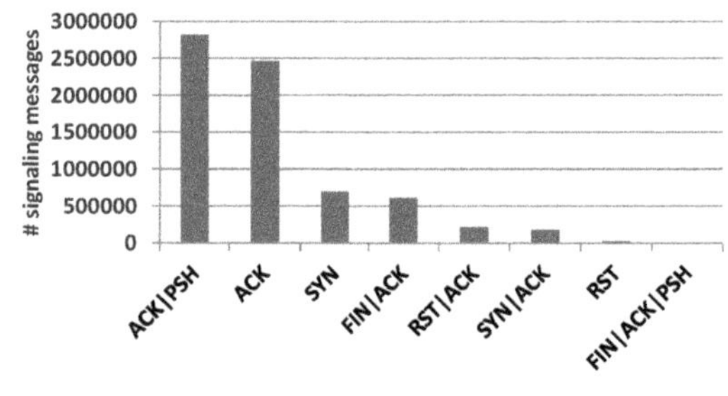

(b) Downlink traffic.

Figure 9: Distribution of signaling messages triggered by different types of TCP packets.

F→D contributes the least (i.e., 15.8% and 10.9%, respectively). Since each I→D triggers at least twice more signaling messages than other state transitions (see Table 1), Figure 7(b) shows that the contributions of signaling messages of I→D are even more dominant, with 84.1% and 84.8%) in the uplink and downlink directions, respectively. The number of signaling messages due to F→D in the uplink direction is higher than the number due to D→F, mainly because each F→D generates four more signaling messages than each D→F.

Summary of observations. Most signaling messages are triggered by small-size packets (with no more than 200 bytes), including TCP control messages that have zero payload size. Also, most signaling messages triggered by uplink packets initiated by UEs. In the following discussion, we report our transport-layer analysis and application-layer analysis to explain the uneven distributions of the signaling messages observed in the uplink and downlink directions. We also characterize the state transitions induced by the TCP control messages.

4.2 Transport-layer Analysis

In this subsection, we look into the transport layer and analyze how TCP and UDP packets each contribute to the signaling load. In particular, we focus on signaling load due to different types of TCP packets.

TCP versus UDP. Figure 8(a) shows the distributions of the numbers of all TCP and UDP packets, and Figure 8(b) shows the distributions for only the packets that induce state transitions. Figure 8(a) shows that TCP packets contribute a higher proportion than UDP packets (66% versus 34%). We observe more downlink TCP packets (37%) than uplink TCP packets (29%). We can also observe an even distribution of the uplink/downlink UDP traffic (17% each). On the other hand, TCP traffic is more likely to trigger state transitions than UDP traffic. Figure 8(b) reveals that a total of 87% of the packets that trigger state transitions are due to TCP.

TCP flag analysis. Since most state transitions are induced by TCP packets, we investigate different types of TCP packets to identify the root cause of generating state transitions. We classify the TCP packets by their flags set.

Figure 9 shows the number of signaling messages generated by different types of uplink and downlink TCP packets. Both Figures 9(a) and 9(b) show only the top eight types of TCP packets (in descending order) that trigger the most signaling messages. Figure 9(a) shows that among all uplink packets, a significant proportion of signaling messages are due to the FIN/ACK/PSH packets (37.9%) and ACK/PSH packets (33.2%). Figure 9(b) shows that among all downlink packets, ACK/PSH packets generate the most signaling messages (40.2%). Since the signaling messages are mostly triggered by uplink packets, the downlink ACK/PSH packets contribute only 10.3% of the overall signaling messages. Another important observation is that uplink packets carrying control flags SYN, FIN, or RST contribute a significant proportion (46.5%) of signaling messages, while the packets from downlink only introduce a small proportion (6.3%) of messages.

We study how each type of TCP packets (classified by TCP flags and direction) triggers different types of state transitions. Here, we only consider the top eight types of uplink/downlink TCP packets that trigger the most signaling messages. Figure 10 shows the proportions of different types of state transitions for the top eight types of TCP packets (arranged in descending order). For each of the top-four TCP packets FIN/ACK/PSH (UL), ACK/PSH (UL), SYN (UL), and ACK/PSH (DL), more than 50% of the state transitions are due to I→D. Their I→D transitions altogether contribute 45.7% of the state transitions (corresponding to 64.1% of the signaling messages) among all TCP packets that trigger state transitions. Note that almost all (at least 99%) FIN/ACK/PSH (UL), ACK/PSH (UL), and ACK/PSH (DL) packets contain TCP payload, while all of the SYN (UL) packets are pure control messages with zero payload size.

Recall from Figure 9 that RST packets trigger relatively fewer frequent state transitions compared to other types of TCP packets. It is expected because in normal situation a TCP sender immedi-

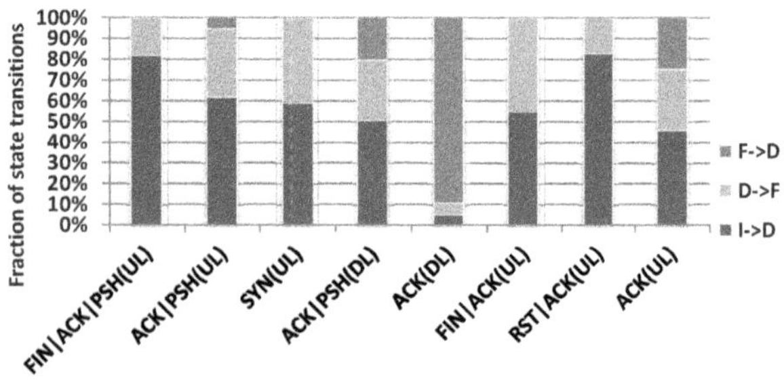

Figure 10: Breakdown of the state transitions triggered by top eight TCP packets (in descending order) that trigger the largest number of signaling messages.

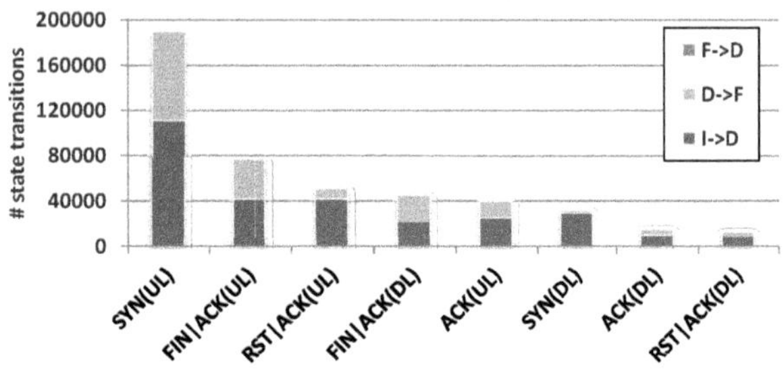

Figure 11: Breakdown of the state transitions triggered by top eight TCP packets without TCP payload (in descending order) that trigger the largest number of signaling messages.

ately dispatches a pure TCP RST (without payload) when it detects incorrect information from an incoming TCP packets [11]. Therefore, TCP RST packets should less likely trigger I→D transitions after an idle period. However, Figure 10 shows that 82.7% of the state transitions induced by RST/ACK (UL) are due to the I→D. One plausible explanation is that some network applications exploit TCP RST messages (instead of TCP FIN messages) to abort TCP connections.

We also note that downlink packets trigger the majority (86.9%) of the F→D transitions. 99.7% of the transitions are due to ACK (DL) and ACK/PSH (DL) packets, and all of them are pure data packets (i.e., ACK with payload). Uplink packets, on the other hand, are more likely to encounter the D→F transitions (76.9%).

TCP control messages. As shown in Figure 5, a substantial proportion (23.9%) of state transitions are due to the TCP control messages with zero payload size. Figure 11 plots the state transition breakdowns for the top TCP control packets that trigger the most signaling messages. The top three of them are SYN, FIN/ACK, and RST/ACK, all of which are in the uplink directions. Due to small packet sizes, it is unlikely for the control packets to induce the F→D transition.

Summary of observations. We observe that most signaling messages are triggered by TCP packets. Many of them are uplink packets initiated by UEs and also carry flags SYN, FIN, or RST, thus accounting for the 64.3% of the I→D transitions inferred from the uplink traffic. We also note that TCP control messages (with null payload) trigger a significant number of I→D transitions.

4.3 Application-layer Analysis

In this subsection, we study how different types of applications (classified by our commercial DPI tool as stated in Section 3.2) contribute to the signaling load. We also seek to differentiate between the signaling-prone and signaling-averse applications.

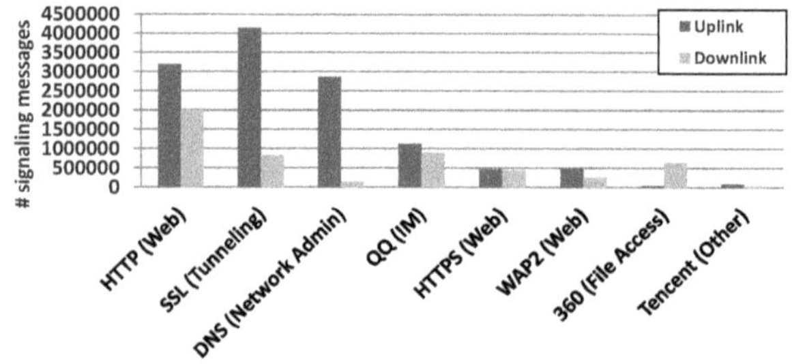

Figure 12: Signaling message distribution of the top eight network applications inducing the largest number of signaling messages (in descending order).

Signaling loads triggered by applications. Figure 12 shows the top eight network applications that trigger the most signaling messages (arranged in descending order). Except for 360 (File Access), all applications generate more signaling messages in uplink packets. In the uplink direction, SSL introduces the most signaling messages (20.3%); in the downlink direction, HTTP introduces the most signaling messages (28.5%).

Figure 13 provides the state transition breakdowns for the top eight applications. We observe that in the uplink direction, most state transitions are I→D. In particular, for SSL, DNS, and QQ, the I→D account for 50% of state transitions. In the downlink direction, DNS's and 360's downlink packets are prone to triggering I→D transitions (with at least 50%), while HTTP, SSL, HTTPS, and WAP2 are prone to triggering F→D transitions (67.7%, 85%, 48.1%, and 62.1%, respectively). These applications introduce significant numbers of ACK or ACK/PSH packets in the downlink, thus accounting for the majority of F→D transitions (see Section 4.2). All the top eight applications are best-effort services; at least 20% of the transitions observed from each such application are due to D→F.

Signaling-prone and signaling-averse applications. In the above application-layer analysis, the contribution of signaling messages of each application depends on the amount of traffic being generated. We now attempt to identify how the uplink and downlink traffic patterns of an application affect the signaling load, independent of the amount of traffic being generated.

Here, we look into the *signaling density* of each application. Consider the uplink traffic of an application. Let $N_{packets}$ be the total number of uplink packets and N_{trans} be the total number of induced state transitions. We then compute the signaling density, defined by $N_{trans}/N_{packets}$, for each type of network applications whose $N_{packets}$ is greater than 0.1% of the overall uplink traffic observed from the RNC. This avoids having a very large ratio due to the very small number of packets observed. Thus, an application is *signaling-prone* if $N_{trans}/N_{packets}$ is high, or *signaling-averse* if $N_{trans}/N_{packets}$ is low. We also repeat the above procedures with the downlink traffic to identify signaling-prone and signaling-averse applications.

Figures 14 and 15 report the signaling density and the fraction of traffic for the top signaling-prone and signaling-averse applications determined by their uplink and downlink traffic, respectively. As shown in Figure 14, several applications (e.g., SSL and QQ) are signaling-prone for both uplink and downlink directions, while some of them, especially network administration applications like SSDP, induce signaling overheads only from a particular direction of traffic. On the other hand, Figure 15 shows that bulk-transfer applications such as streaming, peer-to-peer, and file access applica-

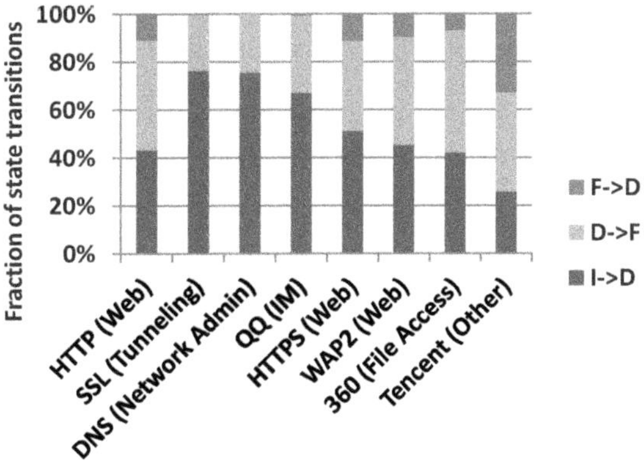

(a) Uplink traffic.

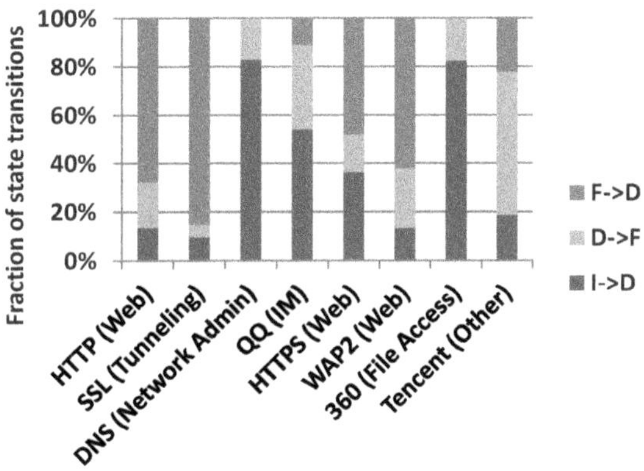

(b) Downlink traffic.

Figure 13: Breakdown of the state transitions triggered by uplink and downlink traffic for the top eight network applications.

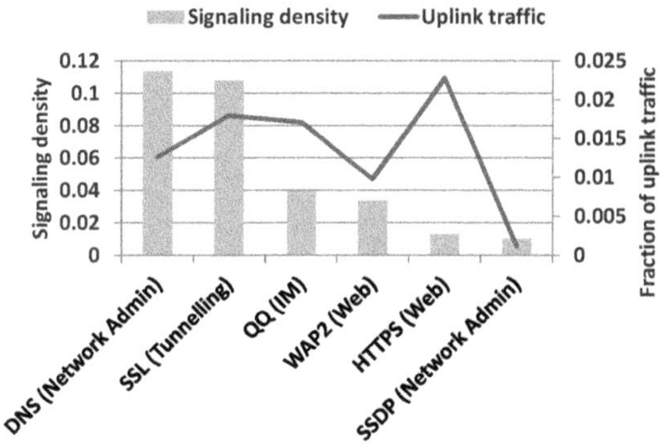

(a) Uplink traffic.

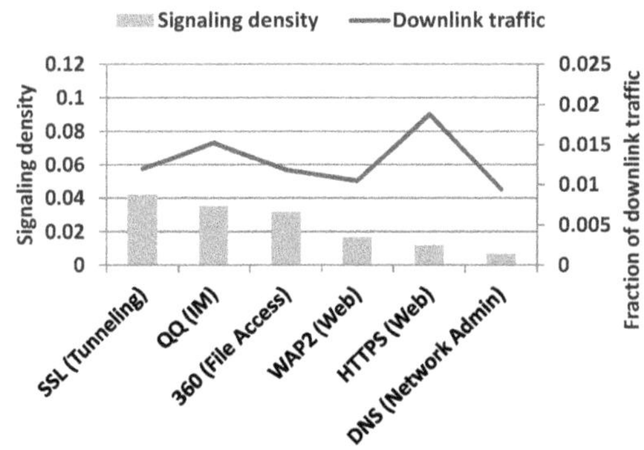

(b) Downlink traffic.

Figure 14: Signaling density and traffic distributions of the top signaling-prone network applications in the uplink and downlink directions.

tions are less likely to introduce signaling messages in both uplink and downlink directions.

Summary of observations. We observe that the signaling-prone applications that generate the most signaling messages are all interactive applications, including Web, Tunneling, Network Admin, and IM. Such interactive applications generate messages with a large inter-packet time, so they generate signaling messages if the inter-packet time is larger than the idle period. The protocol messages of these applications are generally of small size, so this explains why most signaling messages are triggered by small-size packets (see Section 4.1).

4.4 Discussion

In this section, we analyze data traffic traces of a 3G operational network, and identify different types of transport-layer and application-layer data traffic that trigger most signaling messages. We observe that most signaling messages are attributed to the I$\rightarrow$D transitions, i.e., when the RRC connection is activated from the idle state. To mitigate the signaling load (from a network operator's perspective), one possible solution is to keep the RRC connection of a mobile active for a longer period of time, such as by polling the mobile or increasing the inactivity timeout value. The trade-off is that it increases the energy consumption of the mobile [12, 14]. Another possible solution is to optimize the signaling mechanism, although it may also contradict the existing 3G standard. In future work, we explore strategies that can mitigate the signaling load, based on the transport-layer and application-layer traffic patterns.

5. RELATED WORK

There have been various measurement studies on cellular data networks. Existing studies mainly analyze the *data-plane* performance by looking into the cellular traces at the network core. Examples of analysis include: similarities with IP wireline traffic [15], performance of TCP flows [8], user browsing behavior [7], spatial/temporal traffic variations of mobile subscribers and network resource usage [10], as well as traffic patterns of smartphone applications [16].

Recent measurement studies on cellular data networks also look into the *control-plane* performance, mainly from the security perspective. Lee *et al.* [9] show that low-volume data traffic with specific patterns can overload the signaling plane of a Radio Network Controller (RNC) in a 3G network. Zhao *et al.* [17] consider a similar type of signaling attack, but focus on the IP Multimedia Subsystem (IMS) services deployed in a 3G network. Both studies [9, 17] focus on the signaling load on a 3G network similar to ours. However, their studies are based on simulations, while ours is based on real traces of a 3G operational network.

The closest related works to ours are [12, 14], which infer the radio resource control (RRC) state machine using data packets collected at the core of a 3G network. Qian *et al.* [12] uses the RRC state machine to characterize the energy consumption of a mobile in each RRC state. Their follow-up work [14] proposes a fast-dormancy-based solution to proactively notify the 3G network of the release of the radio resources so as to achieve better power saving. On the other hand, our work has several key differences from [12, 14]. First, we focus on the signaling load of a cellular network and hence we are more concerned with the state transitions, each

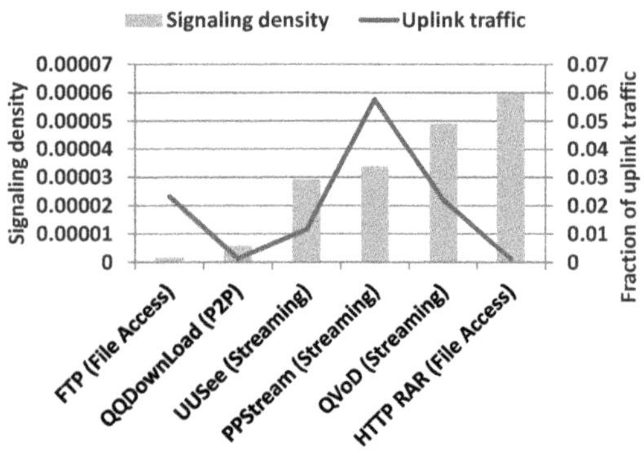

(a) Uplink traffic.

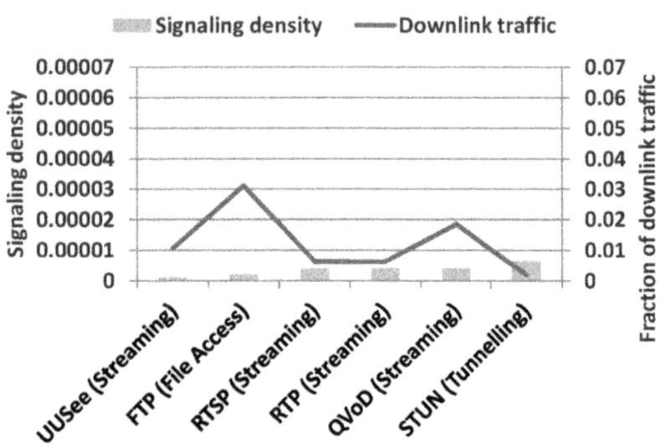

(b) Downlink traffic.

Figure 15: Signaling density and traffic distributions of the top signaling-averse network applications in the uplink and downlink directions.

of which triggers a different load of signaling messages. On the other hand, [12, 14] focus on the energy consumption of a mobile handset, which is mainly attributed to the times spent on non-idle states (i.e., FACH and DCH). Second, we use a more comprehensive dataset which includes all IP-level packets with full payload, while [12, 14] focus on TCP traffic only. We also use deep packet inspection (DPI) to identify the signaling loads of different types of applications. Finally, and most notably, we use the control-plane RRC logs to validate the accuracy of our signaling inference, while the validation is not considered in [12, 14].

Besides using cellular traces, some studies [3, 4, 6, 13] collect traces directly within mobile handsets. In particular, [13] uses the traces of a mobile handset to infer the RRC state machine. Our work conducts measurements from a network operator's perspective, such that we have full access to the traffic traces at the core.

Our prior work [5] conducts a measurement study of the data/control-plane performance of different types of mobile terminals in a 3G network. This work complements our prior work by quantifying the signaling overhead of a 3G network and studying how the signaling load is influenced by the characteristics of the transport and application layers.

6. CONCLUSIONS

To our knowledge, we provide the first measurement study of the impact of raw data packets (in the data plane) on the signaling load (in the control plane) of a commercial city-wide 3G network in China, based on real traffic traces collected at the network core. We consider a data-plane signaling profiling approach that uses only raw IP packet traces to infer the signaling load, and most notably, we validate with real traces that such a profiling approach achieves high inference accuracy. We proceed to use the data-plane signaling profiling approach to analyze the interactions between raw data packets and the signaling load. In particular, we identify the types of data packets, in both transport and application layers, that contribute most to the overall signaling load.

7. ACKNOWLEDGMENTS

The work of Patrick Lee is supported in part by grant GRF CUHK 413711 from the Research Grant Council of Hong Kong.

8. REFERENCES

[1] 3GPP. UMTS. http://www.3gpp.org/article/umts.

[2] 3GPP. UTRAN Functions, Examples on Signaling Procedures (Release 1999), Jun 2002. TR 25.931 v3.7.0.

[3] H. Falaki, D. Lymberopoulos, R. Mahajan, S. Kandula, and D. Estrin. A First Look at Traffic on Smartphones. In *Proc. of ACM IMC*, Nov 2010.

[4] H. Falaki, R. Mahajan, S. Kandula, D. Lymberopoulos, R. Govindan, and D. Estrin. Diversity in Smartphone Usage. In *Proc. of ACM MobiSys*, Jun 2010.

[5] X. He, P. P. C. Lee, L. Pan, C. He, and J. C. S. Lui. A Panoramic View of 3G Data/Control-Plane Traffic: Mobile Device Perspective. In *Proc. of IFIP/TC6 Networking*, 2012.

[6] J. Huang, Q. Xu, B. Tiwana, Z. M. Mao, M. Zhang, and P. Bahl. Anatomizing Application Performance Differences on Smartphones. In *Proc. of ACM MobiSys*, 2010.

[7] R. Keralapura, A. Nucci, Z.-L. Zhang, and L. Gao. Profiling Users in a 3G Network Using Hourglass Co-Clustering. In *Proc. of ACM MobiCom*, 2010.

[8] J. Kilpi and P. Lassila. Micro- and macroscopic analysis of RTT variability in GPRS and UMTS networks. In *Proc. of NETWORKING*, 2006.

[9] P. P. C. Lee, T. Bu, and T. Woo. On the Detection of Signaling DoS attacks on 3G/WiMax Wireless Networks. *Computer Networks*, 53(15):2601–2616, Oct 2009.

[10] U. Paul, A. P. Subramanian, M. M. Buddhikot, and S. R. Das. Understanding Traffic Dynamics in Cellular Data Networks. In *Proc. of IEEE INFOCOM*, 2011.

[11] J. Postel. Transmission Control Protocol, Sep 1981. RFC 793.

[12] F. Qian, Z. Wang, A. Gerber, Z. Mao, S. Sen, and O. Spatscheck. Characterizing Radio Resource Allocation for 3G Networks. In *Proc. of ACM/USENIX IMC*, 2010.

[13] F. Qian, Z. Wang, A. Gerber, Z. Mao, S. Sen, and O. Spatscheck. Profiling Resource Usage for Mobile Applications: a Cross-Layer Approach. In *Proc. of ACM/USENIX MobiSys*, 2011.

[14] F. Qian, Z. Wang, A. Gerber, Z. M. Mao, S. Sen, and O. Spatscheck. TOP: Tail Optimization Protocol for Cellular Radio Resource Allocation. In *Proc. of IEEE ICNP*, 2010.

[15] J. Ridoux, A. Nucci, and D. Veitch. Seeing the Difference in IP Traffic: Wireless versus Wireline. In *Proc. of IEEE INFOCOM*, 2006.

[16] Q. Xu, J. Erman, A. Gerber, Z. M. Mao, J. Pang, and S. Venkataraman. Identifying Diverse Usage Behaviors of Smartphone Apps. In *Proc. of ACM IMC*, Nov 2011.

[17] B. Zhao, C. Chi, W. Gao, S. Zhu, and G. Cao. A Chain Reaction DoS Attack on 3G Networks: Analysis and Defenses. In *Proc. of IEEE INFOCOM*, 2009.

Combining Local and Global Profiles for Mobility Prediction in LTE Femtocells*

Dominique Barth
Universite de Versailles
45, Av. des Etats-Unis
78000 Versailles, France
barth@prism.uvsq.fr

Samir Bellahsene
Universite de Versailles
45, Av. des Etats-Unis
78000 Versailles, France
sabe@prism.uvsq.fr

Leïla Kloul
Universite de Versailles
45, Av. des Etats-Unis
78000 Versailles, France
kle@prism.uvsq.fr

ABSTRACT

We propose in this paper a mobility prediction model based on the notions of local and global mobile-user profiles. The local profiles are associated with a mobile user and correspond to its frequent and similar movements, whereas the global profiles match with the frequent and similar movements of the majority of users in the covered area. We consider the LTE network architecture with possible deployment of femtocells. The prediction model combines two complementary algorithms: the global profiles-based algorithm and the local profiles-based one. The former is implemented in the enhanced Node B and the home enhanced Node B and the latter works at the user terminal level. An algorithmic approach is used to identify such local and global profiles from real cellular network datasets and we show how to use them for an efficient mobility prediction. Simulation results show that our approach is significantly efficient in predicting both random and regular movements.

Categories and Subject Descriptors

C.2.1 [**Network Architecture and Design**]: Wireless communication; C.4 [**Performance of Systems**]: Metrics—*Modeling techniques, performance attributes*

General Terms

Algorithms, Performances

Keywords

Mobile Networks, Mobility Prediction, Quality of Service, Markov chains, Graph theory, Femtocells

1. INTRODUCTION

The success of the third generation (3G) mobile networks led to the explosion of broadband services and the emergence of new mobile applications with stringent delay constraints. Thus, telecommunication actors are facing the needs of increasing networks capacity and providing the quality of service required by these new mobile applications. To satisfy these needs, the International Telecommunication Union (ITU) has fixed a set of recommendations for standardization bodies and kicked off the competition to establish the fourth generation (4G) of mobile networks. Wimax and Long Term Evolution (LTE) each one developed respectively by the Forum Wimax and the 3rd Generation Partnership Project (3GPP) are the two technologies recently accepted as 4G technologies.

The enhancements that LTE brings are due to several characteristics such as the use of new methods for channel sharing, advanced antennas and enhanced system architecture. Moreover, to meet the quality of service requirements, provide more efficient coverage in indoor and outdoor configurations, the so-called Femto Base Stations are proposed for LTE technology. The Femto Base Station is characterized by a low cost plug-and-play installation and a low power.

However, the massive deployment of femtocells on the one hand and the need to reduce the size of cells to increase the network capacity on the other hand generate difficulties in managing mobility of users. In fact, the handover procedure which consists in handing off a call to a new cell when the mobile user crosses the current-cell boundaries and moves to an adjacent cell while the call is in progress may constitute a serious problem to guarantee services continuity. The handover is initiated if the RSRP (Reference Signal Received Power) of the adjacent cell is greater than the RSRP of the current cell. Once the handover procedure is initiated, if the new cell does not have enough channels to support the handoff, or if the session with the new base station is not established yet, the call is dropped. As the coverage density in mobile networks is often important, in particular in urban areas, the time needed by any mobile terminal to scan all its neighbouring cells and classify those offering the best RSRP can be long enough to become a critical issue for service continuity. In LTE, a 6ms measurement gap allows measuring up to 3 neighbouring cells [24, 26, 6]. The periodicity of this measurement gap is generally 40ms or 120ms. Thus, if the neighbourhood of the cell serving a mobile terminal consists of 8 cells, at least 98ms will be required to scan all the cells in the neighbourhood. Consequently, in critical situation, the handover in LTE can introduce an interruption time, which can last more than 100 ms. Thus, for multimedia applications that require service interruption time lower

*This work is supported by the European Celtic project Homesnet.

MSWiM 2012, October 21–25, 2012, Paphos, Cyprus.
Copyright 2012 ACM 978-1-4503-1628-6/12/10 ...$10.00.

than 150ms, the number of cells to scan, before a handover is initiated, may play a decisive role in service continuity.

Therefore, in the context of HomesNet, a European Celtic project which supports the work presented in this paper, we are focusing on methods that provide efficient handover and ensure continuity of services, like VoIP. Mobility prediction, if well performed, may constitute the solution to limit the number of cells to scan during handover and to solve the problem of mobility issues in the case of LTE macrocells and femtocells deployments. It anticipates the preparation of the handover in the predicted cell(s) or femtocell(s), reduces call dropping probability and limits the blocking probability of new calls.

In this paper, we propose a mobility prediction model in which we take advantage of two types of regularities. The first type of regularity is related to the specific moves that a user performs frequently. The second type of regularity is related to repetitive movements that are shared by a large number of users. In the first case, we talk about the user local profiles and the resulting prediction model is based on learning the past movements of the user. In the second case, we talk about global profiles and the prediction model is based on statistics of all users and on a topological data (main streets, subway lines,...). The objective of the proposed approach is to combine the notion of local and global profiles in order to compute, and thus predict, the probability of the next access point for each user in the network, and this regardless of the randomness degree characterizing the user mobility.

We consider here the notion of user profile we have investigated in [4] and which is defined as a set of closed sequences of cells the user has the tendency to regularly cross in the same way. Similarly, we will also define global profiles as a set of cell sequences often followed by a large number of users.

Structure of the paper: Section 2 is dedicated to the related work. In Section 3, we present the mobility management in LTE architecture. The notion of mobile user profiles is detailed in Section 4. We describe our prediction model based on mobile user profiles in Section 5. A case study is given in Section 6. In Section 7, we present and discuss the simulation results obtained. Finally concluding remarks and future works are given in the Section 8.

2. RELATED WORK

Many studies were carried out on the mobile user behaviour. Based on six months tracking of 100,000 mobile phone users, González et al. have proved in [20] that humans follow simple reproducible patterns. The human trajectories show a high degree of temporal and spatial regularity, each individual being characterised by a time-independent characteristic travel distance and a significant probability to return to a few highly frequented locations. To detect mobile user patterns trajectories, clustering was proposed in [21]. Rinzivillo et al. suppose that each cluster corresponds to one typical trip, like going from home to work. Similarity is also defined between a cluster and routes, like the peripherals, the inwards or the outwards routes.

In [22], Giannotti et al. assess a new multidisciplinary research area on mobility, data mining, and privacy. For the authors, which relate concepts of movement data and knowledge discovery from wireless network and next-generation mobile technologies, a global prediction model built on user patterns can be made to predict next location. In this context, mobility prediction models based on user patterns were proposed in [14, 8, 11]. In these models, a server is supposed to contain all the users paths. The prediction process consists then in comparing the current user path with its stored paths in the database. In [15], additional data such as the geometry of cells, the received signal measures and the mobile velocity, are considered. The proposed models are interesting because the complete itinerary is predicted, not only the next cell. However, they are limited because all mobile users data are saved in a single server. Given that the current mobile networks are very dense and the number of subscribers is important, the access delays to the mobility data can be very important. Moreover, the comparison process of the paths is time and resource consuming and this can affect the performances of the network. Finally, privacy-wise, most of mobile users do not agree to let the network be aware of their mobility history.

Mobility history was also combined with received radio signal measurements [16, 15] or cell shape and position updates [9, 10, 3] to build mobility prediction models. These approaches face the dense deployment of cells in the current mobile networks. The sizes of cells are reduced and generally the coverage beam does not respect any geometric shape. With fast movements of mobile users and the important density of buildings in urban areas it is difficult to continuously make the appropriate measurements on signals received from the serving and all the neighbouring base stations.

Markov chains are widely used as modelling technique for mobility prediction [18, 19, 12, 17, 3, 5, 6]. The fundamental input parameter used to build the Markov model is the mobility history. In [19], the authors explore the characteristics of wireless infrastructure using campus-wide Wi-Fi real data. The authors propose two prediction models which are based on the first order of Markov chains and the second order of Markov chains, respectively. Their results show that the second order based model performs better predictions than the first order based one. Similar results are presented in [18] where different location predictors were evaluated. Moreover, the authors prove that an approach using a second order with a possible fallback to the first order of Markov chains provides better predictions than all the others location predictors. This study was also based on a Wi-Fi data, collected from 6000 users. In [5, 6], the second order-based location predictor with simple fallback was enhanced by proposing an approach that uses, according to the available information, both second order and first order of Markov chains and the visit frequency to a cell. In [4], Markov chains were combined with a notion of local profiles to predict the next cell of a user. This notion supposes that a user follows in its movements similar paths. The proposed model provides significant results, however, when a user is newly connected to the network, the profiles cannot be defined and this affect the prediction model.

Probabilistic modelling methods are used in [13, 2, 8]. These methods make use of the mobility history and the radio signal strength measures to predict a group of the most probable cells for a mobile user. Once the prediction is made, resources are reserved in all the cells belonging to this group. In general, this type of approaches increases the blocking probability of the new calls and works only with regular movements.

3. MOBILITY MANAGEMENT IN LTE ARCHITECTURE

E-UTRAN (Evolved Universal Terrestrial Radio Access Network) is the access network of LTE and consists of evolved Node Bs (eNBs) and Home evolved Node Bs (HeNBs) providing user plane (Radio Link Control, Medium Access Control, Physical Layer, ...) and control plane (Ressource Control) protocol termination for User Equipment (UE) [25].

Both types of base stations, eNB (macrocell) and HeNB (femtocell), host the same functions and are connected, using S1 interface, to the Mobility Management Entity (MME) and to the Serving Gateway (S-GW), two entities in the Evolved Packet Core (EPC). HeNBs can either be directly connected to MME/S-GW or through a Home eNB Gateway (HeNB GW), an additional gateway deployed in E-UTRAN architecture (see Figure 1).

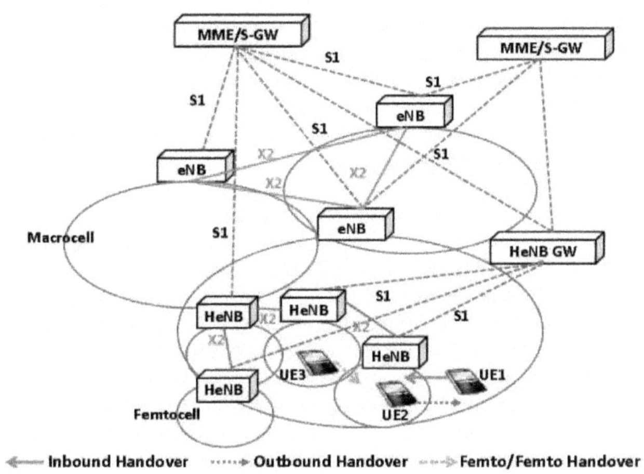

Figure 1: Handover scenarios in LTE Femtocells

The deployment of femtocells in LTE involves three different handover scenarios: femtocell to femtocell handover, macrocell to femtocell handover and femtocell to macrocell handover. The first mobility scenario involves UEs moving between covered areas of two neighbouring HeNBs (see Figure 1). This type of mobility occurs in, for example, business environments where femtocells are deployed. In the second scenario, also called inbound mobility, the UE leaves the current base station (eNB) and enters the air interface of the home base station (HeNB). In the last scenario, the UE leaves the air interface of the HeNB for the eNB coverage. This mobility type is known as the outbound mobility.

In order to achieve an efficient mobility management, in LTE architecture, additional functionalities are assigned to UEs, eNBs and HeNBs, the objective being a decentralization of the mobility management. Thus, the handover decision which is based on the UEs measurement reports is made by the eNBs without the involvement of the MME. As specified in the LTE standard, the functions supported by a HeNB are the same as those supported by an eNB. Therefore, the HeNB is also able to make handover decisions. Moreover, in order to speed up the handover procedure, a new interface (X2) between eNBs is used (see Figure 1). This interface is used for direct packet data forwarding and control messages exchanging. It allows reducing the number of control messages exchanged between eNBs and EPC. Therefore, it represents, compared to the legacy 3GPP technologies, an important enhancement as it allows fast transfers.

Mobility management in LTE HeNBs is facing several technical challenges, including the small size of the femtocells (average 10m), their massive deployment in the future, the radio frequency interferences and the speed of the user equipment. Indeed, currently the standards of LTE specify that X2-based handover between femtocells is allowed only if no access control at the MME is needed, that is when the handover is between HeNBs having the same Closed Subscriber Group Identity (CSG ID) or when the target HeNB offers open access. Otherwise, when the access control is needed at the MME, the direct interface X2 between HeNBs cannot be used. The handover management strategy rely solely on S1 interface and this in the three mobility scenarios (inbound, outbound and femtocell to femtocell). Unfortunately, interface S1 is based on the internet subscriber link. Clearly, connection to an operator's network via internet introduces uncertain delays and hence unreliability into the access architecture. The handovers occurring in this context cannot be transparent to mobile users and assuring continuity of service in this case becomes a challenging task, in particular for multimedia applications.

Thus and in order to avoid dropping the call of a mobile user when this one crosses a cell boundaries and moves to an adjacent cell while its call is in progress, a solution which allows overcoming the delays problem of S1 interface, the long scanning time of neighbouring cells and the small size of cells is expected. Mobility prediction, if well performed, may constitute a solution for such a problem. It anticipates the preparation of a handover in the predicted cell, and by doing so, allows dealing with the latency introduced by S1 interface. Such an approach may lead to the reduction of the call dropping rate and the limitation of the blocking rate of new calls.

4. MOBILE USER PROFILES

In the mobility prediction approach developed in [4], the notion of mobile user profile has been introduced in the context of GSM networks. This notion is based on the assumption that the mobile user has recurrent behaviours which do not necessarily translate into a unique path (sequence of crossed cells), but as a set of *sufficiently* closed paths.

The approach we propose in this paper is not only based on this notion, to which we refer here as the *local profile* of the mobile user, but also on the notion of *global profile*. The objective of this new notion is to catch the tendancy of several mobile users to behave similarly in a geographical covered area. In the scope of our work, we are interested in the sequence of all visited cells (macrocells, femtocells) by the mobile users during a fixed time period.

In order to detail the notion of both local and global profiles, we need to introduce first some notations and definitions.

Covering graph: let $G = (V, E)$ be a covering graph of the network such that, for $1 \leq i, j, \leq |V|$, each vertex $v_i \in V$ is a cell and each edge $[v_i, v_j] \in E$ indicates that cells v_i and v_j overlap.

The movements of a mobile user u are modelled as a sequence $v_1^u, v_2^u, \ldots, v_N^u$ of adjacent cells such that $[v_i^u, v_{i+1}^u] \in E$, for all $i = 1, \ldots, N-1$, N being the size of the sequence. Moreover, with each cell sequence, we associate a time sequence $t_1^u, t_2^u, \ldots, t_N^u$ where t_i^u is the sojourn time of the mobile user in cell v_i^u. Consequently, we denote by L the *trace* recording the movements of the target mobile user during a time period T such that $L = v_1^u v_2^u \ldots v_N^u$ and $T = \sum_{i=1}^{i=N} t_i^u$. The duration T has to be long enough (days/weeks/months) to allow the identification of repetitive behaviours of the mobile user.

Potential Stationary Vertex (PSV): a PSV is defined as a vertex v_i^u in L such that t_i^u is greater or equal to a chosen threshold T_{dest}. The PSV corresponds to a cell where the mobile user spends a lot of time, for example, cells covering home, office, etc. Because of ping pong handovers, identifying PSVs by analysing a mobility trace may not be straightforward. Indeed, as we focus on urban areas where the density of cell coverage is important, several adjacent cells may be, at the same time, candidate to serve a mobile user. The signal levels of these cells can be close enough to observe situations where a mobile user switches several times between these cells, and this in very short times [5, 4, 6]. Such situations may occur even if the mobile user is not moving or moves just few meters. Thus, the identification process of the PSVs in trace L consists in identifying first those which may be hidden because of the ping pong phenomenon. Then all the cells (vertices) involved in each ping pong situation are *aggregated* in a single cell (vertex) (see [4, 5] for more details about the ping pong phenomenon).

Path: a mobile user (or user equipement) path in L is defined as a sub-sequence of L starting from a PSV v_i^u (the source node) and ending at the following PSV v_j^u (the destination node) in L, with $1 \leq i < j \leq N$. The details of the identification process of the paths in a mobile user trace can be found in [4].

4.1 Local profiles

This notion [4] is based on the hypothesis that a mobile user has the tendency to regularly behave in the same way. Thus, the objective of a local profile is to catch recurrent behaviours like going to the office or leaving it every day of the week at the same hours and using, more likely, the same paths. As the behaviour of the mobile user is not unique, each mobile user can be characterized by several local profiles.

Thus, let F^u be a set of local profiles of mobile user u. A local profile $f^u \in F^u$ is a set of similar paths which are pairwise closed and such that $f^u \subset \mathcal{P}^u$, $\mathcal{P}^u$ being the set of all detected paths of mobile user u. Two paths in $\mathcal{P}^u$ are similar if they follow the same pattern, that is:

- they start at the same source node and end at the same destination node,
- they have, at least, Z_{min} identical nodes (cells) which appear in the same positions in both paths,
- if both paths have, in the same position, two different nodes, these ones should be connected in the covering graph G.

The last condition is related to the ping pong phenomenon and possible small changes in the user's habits. In these cases, a pattern occurring several times, during the time period of L, may slightly change from one occurrence to another in L. Indeed, due to small changes in the user's habits, the paths in local profile f^u may not completely match because of the presence of different nodes. Thus, if two different nodes are in the same position in two paths, we require these nodes to be connected in the covering graph G, or have to contain some identical nodes if they are aggregated nodes. Consider, for example, the two paths in Figure 2. These paths have only five nodes in common. However, as the non shared nodes, like v_{41}, are connected by the covering graph (v_{41} is connected to v_2, v_3 and v_4 in the other path), the paths are considered as similar.

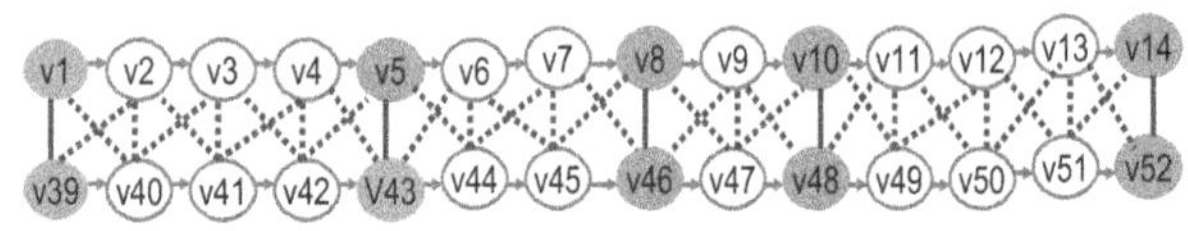

Figure 2: Two similar paths

4.2 Global profiles

In this paper, we introduce the notion of global profile which translates to a set of frequently followed paths by a certain *percentage* of mobile users in the network (main streets, subway lines, ...). We suppose that the network is able to collect anonymous lists of followed paths during a certain period of time. Anonymity allows using paths for building the global profiles, and this without any information about the identity of the mobile users following them.

We define the set H of global profiles as follows. Consider $\mathcal{P}$, the set of all the paths collected by the network management. Each path $p \in \mathcal{P}$ is associated with a traffic volume $t(p)$ corresponding to the number of mobile users following this path. We consider the *similarity graph* $G_1 = (\mathcal{P}, E_1)$ in which a pair of paths $[p, p']$ in $\mathcal{P}$ is an edge in E_1 if these two paths are similar, considering the same similarity criteria as the ones used for local profiles. Let $\mathcal{C}$ be the set of maximal cliques in G_1 (obtained using an exhaustive evaluation) of size, say γ, at least. By definition, a clique is a set of nodes pairwise adjacent. We then define each clique $C_i \in \mathcal{C}$, $i = 1, \ldots, |\mathcal{C}| = W$, as a global profile $h_i \in H$, $i = 1, \ldots, W$, associated with D_i mobile users, $D_i = \sum_{p \in C_i} t(p)$. The value of γ, which is based on the percentage of the mobile users following the paths in the global profile h_i, and thus on D_i, is experimentally chosen.

The detection and the construction of the global profiles follow the same procedure as local profiles. Note however that local profiles are not used for the global profiles detection because there is a possibility that a path belongs to a global profile without belonging to any local profile.

5. MOBILITY PREDICTION MODEL

In [4], we have proposed a mobility prediction model, in the context of GSM networks, which is based solely on the

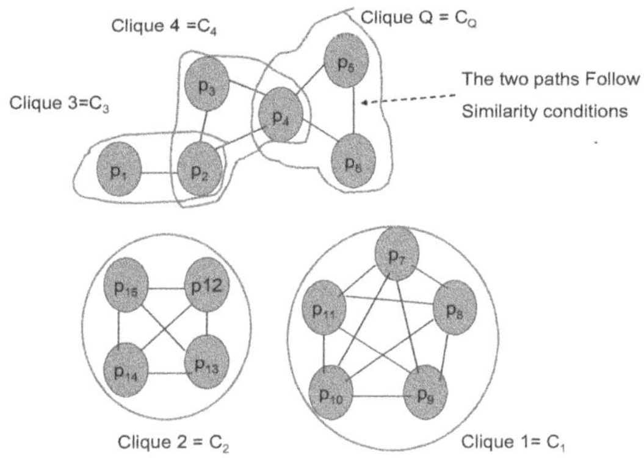

Figure 3: Global profiles detection

notion of local profiles. The model provides significant results, except when the network does not have enough information about the mobile user (a new mobile user, new paths, ...). In this case, the accuracy level of the prediction model degrades because the system cannot build the set of local profiles from a poor mobility trace L.

To enable seamless handovers in LTE femtocells architecture, we propose a Local and Global profiles-based Prediction Model (LGPM) which combines two complementary algorithms: a Local Profiles-based Prediction Algorithm (LPPA) and a Global Profiles-based one (GPPA).

In the new prediction model, the priority is given to the specific behaviour of the mobile user, that is the information provided by its local profiles. In this case, the mobility prediction is made by LPPA using these information. However, if the user's behaviour is unknown to the network, the prediction model assumes that the mobile user has a high probability to behave in the same way as some of the mobile users in the network. In this case, the mobility prediction is made by GPPA, based on the information provided by the global profiles known by the network.

Taking into account the enhancements made in terms of mobility management within E-UTRAN and the ability of any eNB or HeNB to make handover decisions without involving the core network, we propose to implement LPPA in the user equipments and GPPA within eNBs and HeNBs.

Our implementation scheme is also motivated by the fact that local profiles must remain the propriety of the mobile users and the data confidentiality of each user has to be guaranteed. Furthermore, we suppose that each eNB or HeNB is aware of the global profiles and is able to update them. These updates can be made by importing the set of all global profiles and the corresponding statistical information from the core network. These information are unique and are those used by all HeNBs and eNBs in the covered area. Note that, like in [4], we assume the update procedure to be performed during off-peak periods. However, at the user equipment level, such procedure should record the sequence of all crossed cells or femtocells, during a time period T. This duration time should be large enough (days/weeks/months) to obtain significant data and allow further identification and extraction of the local profiles.

5.1 Local Profiles-based Algorithm (LPPA)

We apply the algorithm described in [4] for mobility prediction in the context of LTE architecture. This algorithm consists in identifying the next cell(s) to be likely visited by the user, using the list of local profiles or the mobility trace if no profiles are available. Because of the aggregated nodes, there may be more than one predicted cell at once. At each prediction, the LPPA in the user equipment sends the list of predicted cell(s) to the serving eNB (or HeNB in the case of femtocell), which prepares the handover of the mobile user to the base station(s) of the predicted cell(s).

In order to identify the next cell(s) to be likely visited by a mobile user u, the LPPA identifies, in each current node (cell or femtocells) of the user, the next local profile $f_i^u \in F^u$, $1 \leq i \leq K$, to be likely followed by the mobile user. For that, let $v_c \in V$ and $F_c^u \in F^u$ be the current cell and the set of all local profiles of the mobile user containing v_c, respectively. If $v_{c-1} \in V$ is the previous node visited by the user, the probability P_f for the mobile user to follow a local profile, $f_i^u \in F_c^u$ is the following:

$$P_f = P(f_i^u|v_{c-1}) = \frac{S_{f_i^u|v_{c-1}}}{\sum_{k=1}^{K_c} S_{f_k^u|v_{c-1}}} \tag{1}$$

where K_c is the number of local profiles containing the current node v_c, that is $K_c = |F_c^u|$, and $S_{f_i^u|v_{c-1}}$ (respectively $S_{f_k^u|v_{c-1}}$) is the number of times profile f_i^u (respectively f_k^u) has been followed by the mobile user, given the previous visited node, $v_{c-1} \in f_i^u$.

The success of such an approach depends on the contents of the user local profiles list. If this list is empty or does not provide enough information, the algorithm will be unable to perform predictions. An empty list means that no local profile has been followed (yet) by the mobile user and the current node v_c is not contained in any local profile. Consequently, probability P_f is equal to zero.

In both cases (empty or poor profile list), we propose to exploit the repeated sequences in the mobility trace L. The movements of the mobile user are modelled using a continuous-time Markov process whose discrete states are the nodes of the graph G (cells or femtocells of the network) [5, 6]. To perform the predictions, a second order Markov based predictors are used. Thus the probability of the next node to be likely visited by a mobile user depends not only on its current node but also on the previously visited one. Then for each mobile user, the transition probability from the current node v_c to each node in its neighbourhood (adjacent cells of v_c) is computed.

For that, let $\Gamma(v_c)$ be the neighbourhood of node v_c and let (M, Υ, t) be the tuple associated with v_c such that:

- for each pair of nodes v_i and v_j in $\Gamma(v_c)$, $M(v_i v_c, v_j)$ is the number of transitions of the mobile user from cell v_c to cell v_j in the past, knowing that each time such a transition occurred the mobile user was previously in cell v_i.
- $\Upsilon(v_i, v_c)$ is the number of transitions of the user from cell v_i to cell v_c.
- $t(v_c)$ indicates the average residence time of a mobile user in cell v_c.

Let $X = v_{c-1}v_c$ be the sequence, in L, of the previously visited node and the current node of the user. Assuming that $Y = v_c v_{c+1}$ is the sequence of the current node and the future node to be likely visited, the estimated transition probability P_{e_1} from node v_c to v_{c+1} is given by:

$$P_{e_1} = P(X_{c+1} = Y/X_c = X) = \frac{M(X, v_{c+1})}{\Upsilon(v_{c-1}, v_c)} \qquad (2)$$

However, if the previous node v_{c-1} appears for the fist time in the mobility trace of the mobile user, $M(X, v_{c+1})$ is equal to zero for all neighbouring cells of v_c. In this situation, we fall back on the first order Markov chain, that is, only sequences formed by two consecutive visited cells (v_c, v_{c+1}) are used. Thus, the estimated probability P_{e_2} to go to the next node v_{c+1} is:

$$P_{e_2} = P(X_{c+1} = Y/X_c = X) = \frac{\Upsilon(v_c, v_{c+1})}{Z(v_c)} \qquad (3)$$

where $Z(v_c)$ is the number of times cell v_c appears in the mobility history of the mobile user. Whenever the current cell v_c appears for the first time in the mobility trace, P_{e_2} is equal to zero for all the neighbouring nodes. In this case, we propose to use the visit frequency to each node from its neighbouring nodes. For that, let $H(v_c) = \sum_{j=1}^{A} Z(v_j)$ where A is the total number of adjacent cells to v_c. The probability value of the next move is then given by:

$$P_{e_3} = \frac{Z(v_{c+1})}{H(v_c)} \qquad (4)$$

The overall prediction process using the proposed approach is very simple. Each time a new cell is visited, the algorithm checks first if the list of user profiles provides enough information to predict the next profile, and thus the next node(s) to be likely visited by the mobile. If it is the case, the probability for the mobile user to move to an adjacent cell is given by Equation 1. Otherwise, this probability takes the general form: $P_e = P$, where P depends on the contents of the trace. If the trace provides enough information about the previously visited cell, the second order Markov based-predictor is used and $P = P_{e_1}$ (Equation 2). Otherwise, the first order Markov based-predictor is used and $P = P_{e_2}$ (Equation 3). However, as the first order cannot be used if $\Upsilon(C_i, C_{i+1}) = 0$ the algorithm uses the ultimate information that can be found in the mobility trace, that is, the visit frequency to a cell. In this case, $P = P_{e_3}$ (Equation 4).

Note that, implementing the LPPA at the user equipment level allows the network to avoid any scalability constraints when modelling a larger number of users. Each mobile terminal processes itself its associated mobility data to extract the list of profiles and the information needed to perform the predictions.

5.2 Global Profiles-based Algorithm (GPPA)

Like LPPA, the objective of this algorithm is to identify, at each current cell (node), the more likely cell(s) to be visited by a mobile user in its near future. For that, GPPA exploits two important information: the set of global profiles H and the percentage of mobile users following each profile $h_i \in H$, $i = 1, \ldots, W$.

The algorithm consists then in computing, at each current cell (node) of mobile user u, the probability that this mobile user follows each global profile $h_i \in H$ crossing the current cell (node). The global profile with the highest probability is then predicted. Since a profile is a sequence of cells, predicting the next global profile allows predicting the next cell the mobile user will visit. The predicted cell is the one following the current cell in the cell sequence (global profile).

Let $v_c \in V$ and $H_c^u \in H$ be the current cell and the set of all global profiles of mobile user u containing v_c, respectively. The probability Q_g that mobile user u follows profile $h_i^u \in H_c^u$, $1 \leq i \leq W$, is defined as follows:

$$Q_g = P(h_i^u | v_{c-1}) = \frac{S_{h_i^u | v_{c-1}}}{\sum_{k=1}^{W_c} S_{h_k^u | v_{c-1}}} \qquad (5)$$

where W_c is the number of global profiles containing the current node v_c, that is $W_c = |H_c^u|$, and $S_{h_i^u | v_{c-1}}$ (respectively $S_{h_k^u | v_{c-1}}$) is the percentage of mobile users following global profile h_i^u (respectively h_k^u) in the current node v_c, given the previous visited node, $v_{c-1} \in f_i^u$.

5.3 The complete prediction model

When a mobile user enters a new cell $v_c \in V$, the LPPA, within the user terminal, tries to make prediction exactly as described in Section 5.1. The LPPA computes the probability, for this mobile user, to follow each local profile $f_i^u \in F_c^u$, after leaving v_c. The local profile with the highest probability is predicted by the user equipment which informs its serving eNB or HeNB about the prediction. If the list of local profiles is not available or not sufficient to perform the prediction, the LPPA uses the Markov chains or the visit frequency to predict the next cell. However, if LPPA fails to make the prediction, the user equipment informs immediately its serving eNB or HeNB. Consequently, the GPPA in the serving base station computes the probability to follow each global profile $h_i \in H_c$. Based on the computed probabilities, the serving eNB or HeNB predicts the next cell as described in the previous section and anticipates the handover preparation.

Our approach makes no distinction between a macrocell and a femtocell. Each visited macrocell or femtocell is considered by our model as a visited cell. Consequently, no changes are needed in the prediction algorithms when predicting a macrocell or a femtocell. Our prediction model is also able to make predictions in the case of overlapping deployment of macrocells and femtocells. The prediction is made immediately when a user enters a cell. Thus, enough time left to the network to prepare the handover to the predicted cell. Consequently, small size of cells or speed of users do not affect the efficiency of our prediction model.

6. CASE STUDY

In order to assess the performances of our mobility prediction model (LGPM), we use the same field data as in [4]. The cellular network data have been gathered from 10 participants within the coverage area of Houston, a major US urban area [7].

The data collection lasted between three and six weeks, depending on the participants. Seven of the ten participants are subscribers of USA T-Mobile operator. The other three participants use US Cingular Wireless.

For the purpose of our study, and in order to be able to build the set H of global profiles, we need to consider the data of mobile users which share the same cells. Thus, we consider the data of the seven subscribers of T-Mobile operator, the largest participant group.

The measures made, every 60 *seconds*, on the mobile terminals allowed to record information such as the current date (day, hour, minute, second), the network availability, cell tower identity, signal strength, and the channel of the currently-associated GSM cell and those of up to 6 other visible cells from the same operator.

Note that, as the commercialization of LTE started recently, the mobility data of LTE terminals are not yet available. For this reason, we simply use the mobility data of GSM terminals. This does not affect the obtained results. The data required by our prediction model are related to the sequence of crossed cells, not to the characteristics of the architecture.

Before assessing our prediction model, a pre-processing step of these data is required. From the data of the seven subscribers, we build the global covering graph G that represents the network coverage of T-Mobile operator. Then, for each mobile participant u, we extract its mobility trace L and its set of followed paths $\mathcal{P}^u$. Finally, the set of local profiles of the mobile user is extracted from its corresponding set $\mathcal{P}^u$.

As results, we obtained a global covering graph with 1284 nodes. The ratio between the number of visited nodes and the total number of nodes in the graph is 60%. The maximum degree of G is greater than 40 nodes. In addition, the average degree which is defined as the average value of the nodes degree of G is about 31 nodes. This value reflects the average number of neighbouring cells in the network and implies that T-Mobile network coverage in Houston is dense.

Seven mobility traces are obtained. Each trace is associated with one mobile participant. The longest trace consists of 7286 nodes and the smallest one consists of 1629 nodes. The results are shown in Table 1 where the number of extracted local profiles are given too. More details about how the mobility traces and the local profiles were obtained are given in [4]. Results interpretation and the values of used parameters are also detailed in [4].

User	$\|L\|$	Z_{min}	Number of local profiles
U_2	1629	3	6
U_3	5294	3	17
U_4	7286	4	13
U_6	4467	3	6
U_7	4496	4	40
U_8	5496	3	5
U_{10}	2690	3	18

Table 1: Results of field data analysis

Now, in order to obtain the set H of global profiles, we use the global profiles detection method described in Section 4 on the sets of paths of the seven participants. The total number of extracted paths is 336. The value of both parameters Z_{min} and γ are set to 3. This means that a clique C_k, $k = 1, \ldots, W$ must have a size $|C_k| \geq \gamma = 3$. Consequently, we obtain a total of 28 global profiles.

7. SIMULATION RESULTS

In order to assess the efficiency of our mobility prediction approach, we investigate the accuracy of the predictions using the following equation:

$$Accuracy = \frac{Number\ of\ successful\ predictions}{Total\ number\ of\ predictions}$$

The evaluation of our prediction model is performed by following two phases: learning phase and assessment phase. During the learning phase, for each mobile participant, we extract its local profiles from the first half of its mobility trace. Then, using the data in this first half and in the complete traces of the remaining six participants, we extract the list of global profiles. During the assessment phase, we use the second part of the mobility trace of the target user as a scenario that reproduces real-time situation. Thus, for each participant, we obtain a different list of global profiles, as we consider that the data in the second half of the mobility trace of the target user are unknown for the network when starting the prediction process.

In the following, we first compare and analyse the prediction results obtained with LPPA and GPPA. The second part of the performance analysis consists in assessing the efficiency of the complete mobility prediction approach (LGPM).

7.1 Evaluation of LPPA and GPPA

Consider users U_4, U_7 and U_2. The field data of mobile participant U_4 were collected over 36 *days* leading to a mobility trace L of 7286 nodes (see Table 1). Using our approach, we divide this mobility trace in two parts of 3414 nodes and 3872 nodes, respectively. The first part corresponds to the data collected during the first 18 *days* and the second one to those collected during the following 18 *days*. By considering only the first part of the trace, we obtain 9 local profiles for this user. In addition, 22 global profiles are obtained by using this first part of the trace and the complete traces of the remaining six participants (U_2, U_3, U_6, U_7, U_8, U_{10}).

For user U_2, we obtain only 4 local profiles, while the number of global profiles obtained by considering additionally the traces of the corresponding remaining six participants (U_3, U_4, U_6, U_7, U_8, U_{10}) is equal to 17. In the case of participant U_7, we detected 8 local profiles and 19 global profiles. The complete traces of mobile participants U_2 and U_7 are collected over 32 *days* and 29 *days*, respectively.

Note that if we compare the total number of global profiles (28) extracted from all the traces to the ones obtained for users U_2, U_4 or U_7, we can conclude that the data of each of these participants contribute significantly in building the global profiles.

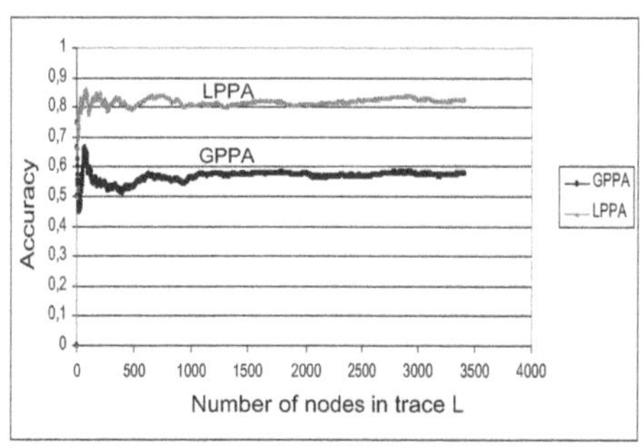

Figure 4: Comparing GPPA and LPPA for U_4

Figure 4 provides the prediction accuracy of LPPA and GPPA when applied on the second half of user U_4 data. It shows that LPPA is more efficient than GPPA in predicting the next moves of this user. That means that, U_4 has a recurrent behaviour and, during the learning phase, LPPA learned a lot about its moves. Thus only few new paths appeared in the second half of its trace. Moreover, considering the general behaviour of the other users in the covered area is not very informative about the habits of U_4 as he/she has different habits. Thus, for U_4, local information (local profiles and mobility trace) are, generally, sufficient to perform good predictions (80% and 85%).

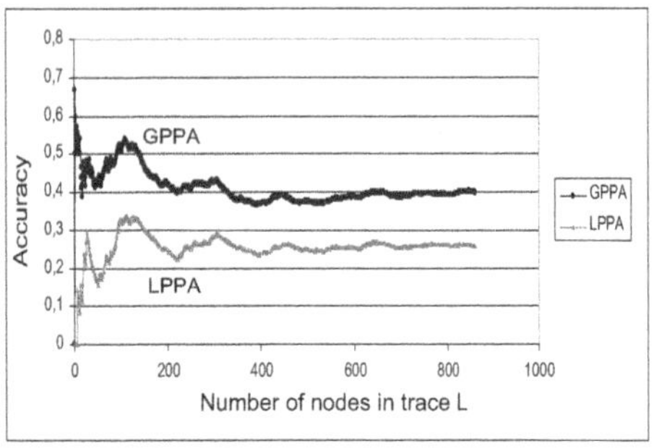

Figure 5: Comparing GPPA and LPPA for U_2

When the behaviour of a mobile user, like U_2, is not well known, the accuracy level of the predictions made by LPPA is not good (Figure 5). The number of local profiles obtained for this user is only 4 which means that not enough data about its behaviour are provided in the first half of its trace. Note that U_2 has the smallest trace (Table 1). Consequently the prediction accuracy level of LPPA does not exceed 33%. This reflects also a certain randomness degree in the behaviour of U_2 in the second half of its mobility trace. The use of global profiles provides better results as the accuracy level is maintained between 38% and 66%. That means that U_2 is quite similar to the other users in its behaviour.

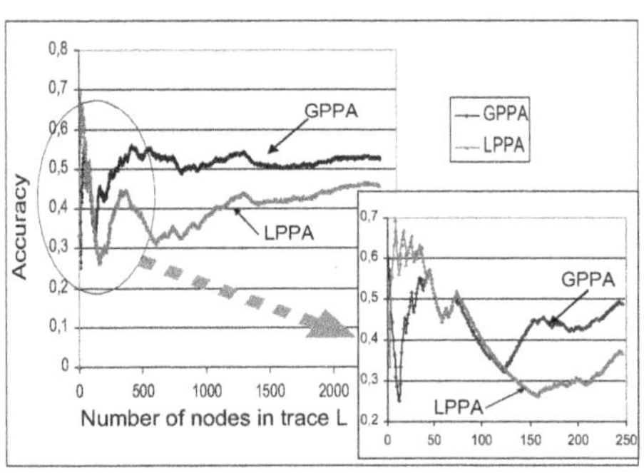

Figure 6: Comparing GPPA and LPPA for U_7

Figure 6 confirms the benefits of using global profiles when the mobile user (U_7) does not follow its local profiles, that is when new paths are met. At the beginning of the trace ($10 \leq |L| \leq 80$), LPPA makes better predictions than GPPA. In this case, the accuracy of LPPA is between 0.5 and 0.7. The accuracy levels of the predictions made by LPPA and GPPA are nearly the same for $80 \leq |L| \leq 130$ and decrease quickly from 0.5 to 0.32. Both GPPA and LPPA find difficulties to perform good predictions in this interval because of the lack of information in the mobility trace. Then, when $|L| \geq 130$, GPPA is more efficient as the prediction accuracy averages 0.5. User U_7 has the tendency to follow the global profiles not its local ones.

7.2 Combining LPPA and GPPA

Consider again the results obtained for U_4 (Figure 7). We have seen that LPPA gives much better results (between 80% and 90%) than GPPA. In this case, local profiles and mobility trace are sufficient to perform good predictions. As, in the global model LGPM, predictions are performed using first local information if these are available it is then clear that the results obtained for U_4 when using LGPM are the same as those of LPPA (Figure 7). LGPM does not have to rely on the global profiles.

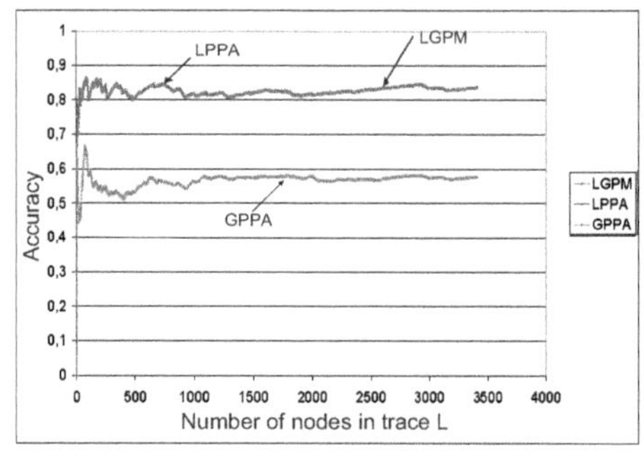

Figure 7: Accuracy of LGPM for U_4

A similar scenario is depicted in Figure 8 where the results obtained for user U_3 are provided. For information, the data of this participant were collected during 33 *days*, leading to a mobility trace of 5294 nodes. The first 16 *days* provide a mobility trace of 2805 nodes. The number of local and global profiles for this user are 6 and 7, respectively. The comparison of LPPA and GPPA in this case shows that, except when $50 \leq |L| \leq 250$, LPPA provides slightly better results than GPPA. Thus user U_3 has the tendency to follow its local profiles.

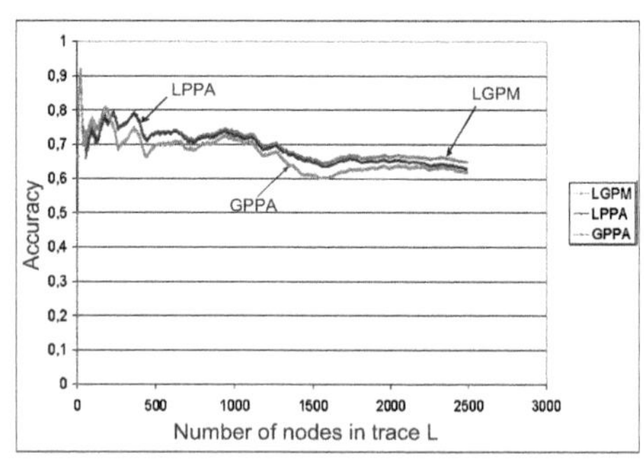

Figure 8: Accuracy of LGPM for U_3

Figure 8 shows also that the global model (LGPM) leads to better predictions, except when $50 \leq |L| \leq 250$. The use of global profiles, in addition to local ones, improves slightly the prediction accuracy level. Even if the improvement is small, it is significant to handover anticipation. The particularity of interval between $|L| = 50$ and $|L| = 250$ is discussed latter.

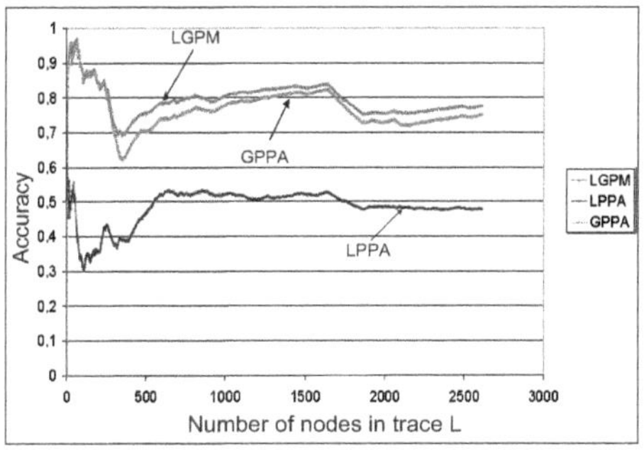

Figure 9: Accuracy of LGPM for U_8

The results of users U_8, U_6, and U_7 show a significant contribution of global information in enhancing the prediction accuracy reached by LGPM. For example, the data of user U_8 were collected over 26 *days* leading to a complete mobility trace of 5496 nodes. No local profiles are extracted from the first 13 days whereas the number of obtained global profiles is 9. Figure 9 shows GPPA performs much better predictions than LPPA. It also show that LGPM, which combines local and global information, performs better predictions than GPPA. This is due to the fact that, in addition to local profiles and mobility trace, global profiles help LGPM to estimate accurately the next destination of the user which seems to often move along the same paths as those followed by most of corresponding remaining users ($U_2, U_3, U_4, U_6, U_7, U_10$). Initially, without the use of global profiles, the level of prediction is between 0.48 and 0.52 (LPPA). After the use of global information, this level reaches nearly 0.78 (LGPM).

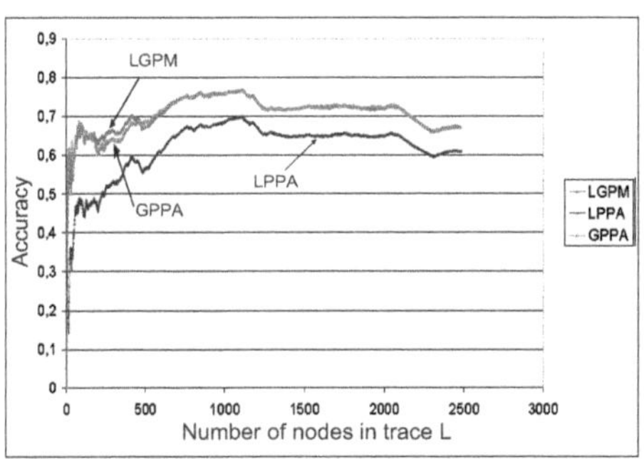

Figure 10: Accuracy of LGPM for U_6

Similarly, global profiles improve the accuracy level of the predictions made in the case of user U_6 (see Figure 10). The number of obtained local profiles for this user is only 1 and the number of extracted global profiles is equal to 11. Generally, the results obtained for user U_6 are similar to those of user U_8. The only difference is that the accuracy level of the predictions made using local information is much better for user U_6. Moreover, the results obtained with LGPM and GPPA are almost the same, except when $130 \leq |L| \leq 500$, where the general model provides slightly better predictions.

Generally, what happens in the case of this user is that LGPM makes predictions based on local information. However, as only one local profile is detected, the prediction is often limited to the use of Markov chains or visit frequency of cells. Each time local information is not sufficient to perform predictions, LGPM uses global profiles which succeeds to improve prediction level by 10%.

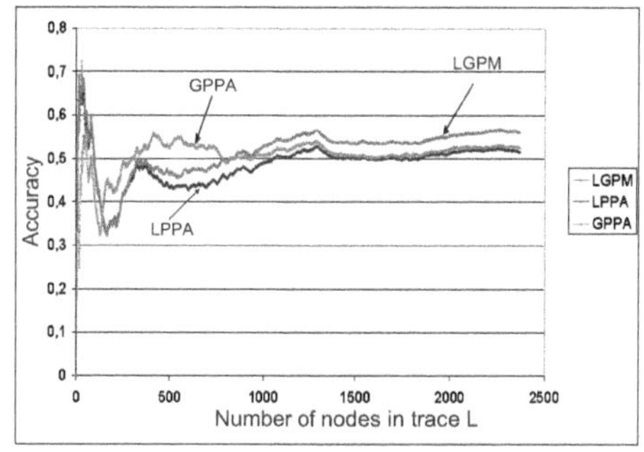

Figure 11: Accuracy of LGPM for U_7

The analysis of the results we obtain for users U_7 and U_2 allows us to see a particular situation where GPPA performs better predictions than LGPM. This is also the same for user U_3 when $50 \leq |L| \leq 250$. Logically, LGPM must always offer better predictions than the two sub-algorithms. However, by following the working process of LGPM, we rapidly see that sometimes our logic cannot be met. To perform predictions, LGPM uses first the local informations (local profiles and mobility trace) to estimate the probabilities of the next moves. If these information are lacking, the global profiles are used. Thus, when the local information is not available, the use of global profiles enhance the prediction level given by the LGPM. However, when both local and global information are available, if local information are used, LPPA makes less successful predictions than GPPA. Therefore, LGPM will be less efficient than GPPA.

For example, in the case of user U_7, Figure 11 shows that, when $130 \leq |L| \leq 800$, GPPA performs better predictions than LGPM. For this user, 8 local profiles and 19 global profiles are extracted. In the rest of the mobility trace L, LGPM is more efficient than LPPA and GPPA.

8. CONCLUSION

In this paper, we have proposed an approach for mobility prediction based on the notion of user profiles, Markovian processes and visit frequency of cells. In addition to the local profiles, we defined a new type of user profile: the global profile. The proposed prediction model (LGPM) combines two algorithms: LPPA and GPPA. According to available information, LPPA makes prediction based on the local profiles, Markov chains or visit frequencies to the cells and GPPA makes prediction based on the list of global profiles. In order to assess the performance of our prediction model, we used data sets from real networks. The analysis of these data shows two important characteristics of users movements: any user in the network follows in its movements similar patterns and the majority of users have the tendency to follow frequent paths. The obtained results show that 80% of the predictions made by our prediction model can be successful and this in a network of high coverage density (average degree of a graph ≈ 30 nodes). The results prove that the use of global profiles in addition to the local profiles and mobility trace enhances the performance of our prediction model and solves the problem of new users in the network. Thus, LGPM can be a good alternative for managing mobility in LTE networks with dense deployment of femtocells. In addition to the mobility issues, in the future, we are interested in the use of our prediction model in LTE networks to develop some applications, like traffic offloading and resource management.

9. REFERENCES

[1] Celtic telecommunication solution. HomesNet Project web site. [Online] Available: http://www.celtic-initiative.org/Projects/HOMESNET/abstract.asp

[2] A. Aljadhai, T. Znati. Predictive mobility support for QoS provisioning in mobile wireless environments. IEEE Journal on Selected Areas in Communications, 19(10), October 2001.

[3] S. Bellahsene, L. Kloul, D. Barth. A hierarchical prediction model for two nodes-based IP mobile networks. Proceedings of the 12th ACM MSWIM, pp 173-180. Tenerife, Canary Islands, Spain, 2009.

[4] D. Barth, S. Bellahsene, L. Kloul. Mobility Prediction Using Mobile User Profiles. Proceedings of the 19th IEEE MASCOTS, pp 286-294. Singapore, 2011.

[5] S. Bellahsene, and L. Kloul. A New Markov-Based Mobility Prediction Algorithm for Mobile Networks. 7th European Performance Engineering Workshop (EPEW), Bertinoro, Italy, 2010.

[6] S. Bellahsene, L. Kloul, P. Reinecke, and K. Wolter. Case Study: Mobile Networks. Book chapter in Resilience Assessment and Evaluation of Computing Systems. ISBN 978-3-642-29031-2. June, 2012.

[7] A. Rahmati, L. Zhong. CRAWDAD data set rice/context (v. 2007-05-23). [Online] Available: http://crawdad.cs.dartmouth.edu/rice/context. May, 2007.

[8] J. Chan, S. Zhou, and A. Seneviratne. A QoS adaptive mobility prediction scheme for wireless networks. In Proceedings of IEEE GLOBECOM, Sydney, November 1998.

[9] R. Chellappa, A. Jennigs, and N. Shenoy. The sectorized mobility prediction algorithm for wireless networks. In Proceeding of the International Conference on Information and Communication Technologies (ICT), April 2003.

[10] R. Chellappa, A. Jennigs, and N. Shenoy. A review on current work in mobility prediction for wireless networks. In Proceedings of the 3rd Asian International Mobile Computing Conference, Kasetsart University, 2004.

[11] F. Erbas, J. Steuer, D.Eggeiseker, K. Kyamakya, K. Jobmann. A regular path recognition method and prediction of user movements in wireless networks. In Proceedings of Vehicular Technology Conference (VTC), Atlantic City, October 2001.

[12] M. H. Sun and D. M. Blough. Mobility prediction using future knowledge. In Proceedings of the 10th ACM Symposium on Modelling, analysis, and simulation of wireless and mobile systems, Chania, Crete Island, Greece, 2007.

[13] D. Levine, I. F. Akyildiz, and M. Naghshineh. The shadow cluster concept for resource allocation and call admission in ATM-based wireless networks. In Proceedings of the ACM Conferences on Mobile Computing and Networking, Berkeley, November 1995.

[14] G. Liu and G. Maguire, Jr. A class of mobile motion prediction algorithms for wireless mobile computing and communication. Mobile Networks and Applications, 1(2):113–121, 1996.

[15] T. Liu, P. Bahl, and I. Chlamtac. Mobility modelling, location tracking, and trajectory prediction in wireless ATM networks. IEEE Journal on Selected Areas in Communications, 1998.

[16] R. Sankar and N. Savkoor. A combined prediction system for handoffs in overlaid networks. In Proceeding of IEEE International Conference on Communications (ICC), Vancouver, June 1999.

[17] W.-S. Soh and H. S. Kim. A predictive bandwidth reservation scheme using mobile positioning and road topology information. In Proceedings IEEE/ACM Transactions on Networking, San Francisco, October 2006.

[18] L. Song, D. Kotz, R. Jain, and X. He. Evaluating location predictors with extensive wi-fi mobility data. In Twenty-third Annual Joint Conference of the IEEE Computer and Communications Societies (INFOCOM), March 2004.

[19] F. Chinchilla, M. Lindsey, and M. Papadopouli. Analysis of wireless information locality and association patterns in a campus. In Twenty third Annual Joint Conference of the IEEE INFOCOM, March 2004.

[20] M. C. Gonzàlez, C. A. Hidalgo, A-L. Barabàsi. Understanding individual human mobility patterns. Nature, International weekly journal of science, 453, 779-782, June 2008.

[21] S. Rinzivillo, D. Pedreschi, M. Nanni, F. Giannotti, N. Andrienko, Andrienko. Visually driven analysis of movement data by progressive clustering. Information Visualization 7, 3, 225-239, June 2008.

[22] F. Giannotti , D. Pedreschi. Mobility, Data Mining and Privacy, Geographic Knowledge Discovery. ISBN: 978-3-540-75176-2, 2008.

[23] GeoPKDD: Geographic Privacy-aware Knowledge Discovery and Delivery, European project website. [Online] Available: http://www.geopkdd.eu/

[24] J. Kurjenniemi, T. Henttonen. Effect of measurement bandwidth to the accuracy of inter-frequency RSRP measurements in LTE. Proceedings of the 19th International Symposium Personal, Indoor and Mobile Radio Communications (PIMRC), Cannes, France, September 2008.

[25] 3GPP TS 36.300, "Evolved Universal Terrestrial Radio Access (E-UTRA) and Evolved Universal Terrestrial Radio Access Network (E-UTRAN); Overall description; Stage 2", Release 10, v10.5.0, September 2011.

[26] 3GPP TS 36.331, "LTE and Evolved Universal Terrestrial Radio Access (E-UTRA); Radio Resource Control (RRC); Protocol specification". Release 9, v9.4.0, September 2010.

A Reactive Solution with a Redundancy-based Error Correction Mechanism for Video Dissemination over Vehicular Ad Hoc Networks*

Cristiano Rezende
PARADISE Research Laboratory
SITE - University of Ottawa
Ottawa - Canada
creze019@site.uottawa.ca

Richard W. Pazzi
Faculty of Business and Information Technology
University of Ontario Institute of Technology - Canada
richard.pazzi@uoit.ca

Azzedine Boukerche
PARADISE Research Laboratory
SITE - University of Ottawa
Ottawa - Canada
boukerch@site.uottawa.ca

ABSTRACT

Vehicular Ad Hoc Networks (VANETs) are challenging environments for video dissemination since stringent requirements have to be satisfied. We have designed a protocol that is reactive to the dynamic topology of VANETs, adaptive to the non-uniformity of vehicles density throughout a city or a highway and that deliver packets in time for the playback of video frames. The Reactive, Density-Aware and Timely Dissemination protocol (REACT-DIS) was able to achieve low end-to-end latency with an acceptable number of transmissions. In order to meet delivery ratio requirements, we have used Network Coding to implement an additional redundancy control aimed at handling packet loss. By this manner, lower rates of frame loss were achieved which fulfill delivery ratios requirements for video-streaming while maintaining suitable levels of end-to-end latency and overhead.

Categories and Subject Descriptors

H.4 [**Information Systems Applications**]: Miscellaneous; D.2.8 [**Software Engineering**]: Metrics—*complexity measures, performance measures*

General Terms

Network

Keywords

VANETs, Video dissemination

*This research is partially sponsored by grants from NSERC DIVA Research Network, the Canada Research Chair Program, the Ontario Distinguished Researcher Award, the Ontario Research Funds (ORF), and the Ontario Centres of Excellence (OCE).

MSWiM'12, October 21–25, 2012, Paphos, Cyprus.

1. INTRODUCTION

We have witnessed in these last years an accelerating evolution in vehicular technology. A lot has been done in terms of equipments aimed at improving security and there are increasing investments in the development of solutions that improve comfort and provide on-board entertainment options. Vehicular Ad Hoc Networks (VANETs) are networks formed by the connection of these modern vehicles. Although this process may be assisted by roadside infrastructure, it is assumed that a vehicle-to-vehicle exchange of information is frequently required.

For many VANETs applications, video streaming capabilities are either crucially necessary or greatly valuable for a high level of user satisfaction. Computational devices and display screens available inside vehicles do not suffer from the same constraints imposed to MANETs' usual devices (e.g. PDAs, cellphones). Therefore, they are capable of reproducing high quality videos which increase the pressure on Quality-of-Service (QoS) expected from VANETs.

In this work, we focus on the task of disseminating video content in a VANET. We consider dissemination as the process of broadcasting data to all nodes within a distance to the source node. All our assumptions can be easily adapted to a geocast premiss where data is broadcast to all nodes within a defined region (not necessarily around the source node). Video dissemination can be used for a variety of services where interest in video content is shared among vehicles in the same region. Examples of services are the broadcast of local traffic to incoming vehicles, a video of nearby crashes to drivers, the notification to users of local events that are currently happening or commercial services that are provided around users' vicinity.

The Reactive, Density-Aware and Timely Dissemination Protocol (REACT-DIS) uses a receiver-based approach to react to the frequent changes on VANETs' topology. It is also density-aware as it takes different decisions in dense or sparse regions. Besides that, it solves the large delays observed in receiver-based solutions by changing the perspective on the selection of forwarding nodes from being per packet but rather for a period of time. Through exhaustive simulations, we have observed that REACT-DIS meets latency challenges on video-streaming but it does not achieve the necessary delivery ratios at high data rates.

There are several error-correction mechanisms for packet loss in networks. We have decided to use additional redun-

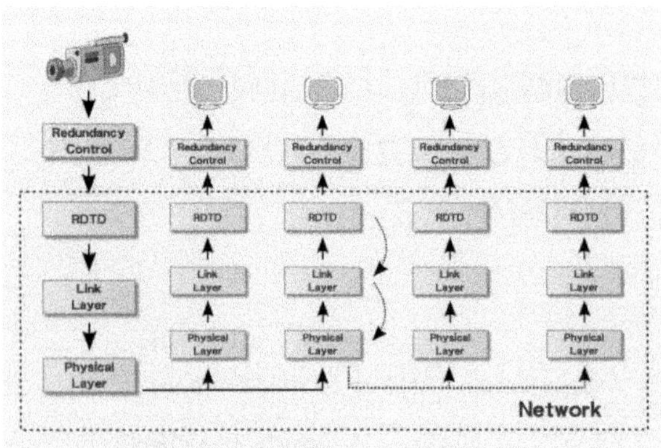

Figure 1: System Architecture

dancy for the purpose of achieving higher delivery ratios. The use of redundancy requires no synchronization nor interaction between source and destination nodes. In this work, we show how we could use redundancy with REACT-DIS to fulfill video-streaming requirements in terms of delivery ratio.

Figure 1 shows the architecture we envision. The redundancy mechanism that we have implemented and evaluated works separately from the network infrastructure. The dissemination of messages is controlled by REACT-DIS and any relay node retransmits received packets without any interaction with the application nor the redundancy control (as exemplified by the second receiving node in the figure). Our dissemination protocol interacts with the link layer for the local broadcast of packets and, ideally, only a percentage of receiving nodes forward packets further.

The most challenging point in video streaming over Vehicular Networks is how to achieve high rates of delivery ratio. Higher rates of delivery ratio are not easily achieved due to VANET's topology dynamism, the wireless nature of communications and the frequent occurrence of disconnected platoons. End-to-end delay is also a challenging issue in the provision of video streaming dissemination support in VANETs. Although some solutions might provide a low delay average, its variation could negatively impact the ability of receivers to reproduce incoming videos. Jitter is the metric that represents delay variation.

In [14], a study is conducted to determine the network requirements in video transmission. Although this values may vary depending on the video coding technique used, they are a reliable guideline to define minimum requirements. The authors have categorized video transmission into video-streaming and interactive-video, tn this work, the focus is on the first set of applications - Video-streaming. The work has concluded that for video-streaming the packet loss must not exceed 5% while delay must be limited to 5 seconds, jitter does not impose a significant constraint in video-streaming.

In the best of our knowledge, we are the first to propose a complete and feasible solution for video dissemination over VANETs that have taken in consideration video-streaming requirements and evaluate the impact of different data rates in the protocol's performance. We show in this work a solution that is extremely reactive to topology changes while delivering frames with a low end-to-end delay even for nodes located far from the source node. Furthermore, our solution considers the diversity in terms of vehicle density as it adapts in order to assure connectivity through sparse regions and to avoid unnecessary transmissions in dense spots.

We discuss some of the existing work in data and video dissemination in the following section. Section 3 describes in detail our REACT-DIS protocol, its evaluation and an analysis of its performance. After that, we discuss in Section 4 the impact of redundancy in the delivery ratio achieved by REACT-DIS. Our final remarks are presented in Section 5.

2. RELATED WORK

There are some works on data dissemination in VANETs in the literature but they have a substantially different perspective. In [7], the authors discuss dissemination techniques for the exchange of messages containing either traffic conditions or vehicles current status information (e.g. speed, direction), thus, focusing in a scenario where information shared is not as large (in terms of bytes) as videos. Besides that, every node in the network is not only an interested party but also a potential source of data.

In [17], the authors propose a data pouring scheme based on available roadside infrastructure that take in consideration vehicles movement to avoid an excessive number of transmissions. Furthermore, the authors suggest placing this roadside infrastructure strategically on intersections in order to optimize coverage of each transmission. However, this solution assumes that applications are reasonably delay tolerant which it is not suitable to describe video-streaming services.

Receiver-based solutions have been proposed for data dissemination [8, 2]. These works take into consideration overheard retransmissions of packets for each node's decision to broadcast further or drop received messages. The time a node waits to decide to retransmit is biased towards favoring farther nodes. In [8], the waiting time is chosen randomly from 0 to a maximum waiting time that decreases exponentially with the distance from the previous node. Additionally, both solutions requires nodes to keep track of overheard retransmissions of the same received packets, while in [8] this is done by comparing relay nodes that are closer to or farther from the previous node, in [2], the comparison is done based on inter-arrival time of duplicate packets. Although these two mechanisms depend on the number of overheard transmissions, they do not directly reflect the non-uniformity of VANETs' density. Our approach in this work implements better the greedy heuristic of choosing farther nodes to rebroadcast received packets and it is better suitable to VANETs' density distribution.

The applicability of Network Coding in VANETs has already been studied [5, 10, 16]. In [5], Network Coding is used for content distribution over VANETs and the authors present an efficiency analysis of Network Coding in VANETs based on computational costs and both works show the impact of network coding on delivery ratio by measuring the time necessary to distribute stored content within a VANET. However, QoS requirements on content distribution are much less rigorous than those on video-streaming.

There are some works that use network coding for video dissemination over VANETs [10, 16]. Park et al. [10] suggest the use of network coding in video dissemination by forcing relay nodes to wait the reception of whole decodable blocks so they can forward newly encoded packets (with a new set of random coefficients) and by this manner improving bandwidth consumption as unique packets are sent by different nodes. However, this approach incurs into prohibitively delays after a few hops of communication. Yang et al. [16] use neighboring cooperation in a way that relay nodes can detect nodes that have not received enough packets to decode blocks. When this is detected, relay nodes encode new packets from the same blocks and broadcast them. Encoding by intermediary nodes is one of the advantages of network coding, however, its use in video-streaming leads to an excessive delay as new encoded packets can only be created after relay nodes receive enough packets to already decode a whole block. In this work, we refrain from encoding new packets at intermediary nodes but rather limit these nodes to forward packets encoded at the source node.

3. REACT-DIS PROTOCOL

The contribution of this work is twofold in providing an effective video dissemination protocol and in studying the impact of the use of redundancy to improve delivery ratio. For the first purpose, we have designed a Reactive, Density-aware and Timely Dissemination protocol (REACT-DIS). REACT-DIS is implemented to fulfill (or at least get close to it) the aforementioned requirements of a successful video-streaming dissemination service. For this reason, we have used three premises to guide the development of our solution: reactiveness, density-aware and timely.

The first principle is that any protocol for VANETs must be reactive and respond to constant topology changes (e.g. links breakage and links creation). There are several solutions to MANETs that already take in consideration the necessity to be reactive. However, many of these solutions still rely on some sort of neighboring table, AODV and DSR are some examples. These kind of solutions are not reactive enough to VANETs' dynamism.

A more suitable paradigm is the receiver-based approach. Through this perspective, the selection of relaying or forwarding nodes is performed at the receiver's side instead of at senders. The most common way by which this is performed is that any forwarding node simply broadcasts the message and any node within range triggers a mechanism to decide if it is going to broadcast the packet further. This mechanism consists of scheduling itself to make the decision of forwarding in t time in the future. The value of t is chosen from a range $[\tau, \Gamma]$ based on a heuristic that determines nodes' suitability as forwarding nodes (lower values of t to better candidates). Before t expires, all scheduled nodes observe their own channels keeping track if the same received message is forwarded by any other node. When t expires, nodes take into consideration the overheard broadcasts to decide if they should forward further the received message or not.

The majority of existing receiver-based solutions use geographically greedy solutions. In these solutions, t is inversely proportional to the distance between the receiving and forwarding nodes. REACT-DIS chooses t as a simple linear transposition of the distance between the nodes divided by the node communication range[1], which can be defined by the following equation (where $\delta(n_s, n_r)$ is the distance between sender and receiver nodes):

$$t = \tau + \left[\left(1 - \frac{\delta(n_s, n_r)}{R}\right) \times (\Gamma - \tau)\right] \qquad (1)$$

The distribution of vehicles throughout a city or a highway varies significantly. In highways, we observe that vehicles with similar mobility patterns (i.e. speed and direction) form connected clusters that may be disconnected from other clusters though. Besides that, constructions, traffic or other particular events create spots of high density while other regions become sparsely populated. Non uniform density is even more evident in urban environments. In a macro perspective, traffic seasonality, popular neighborhoods (either for commercial, entertainment or residential reasons) or common routes tend to attract concentration of nodes from some other areas. Furthermore, lights, intersections or stop signs cause density fluctuation to be even more dynamic.

The main problem with density non-uniformity is that solutions usually try to be optimized for the trade-off of overhead and effectiveness and this become complicate with such wide density variety. For example, routing protocols focus on achieving high delivery ratios with the least amount of transmissions. In order to be able to successfully deliver packets from sources to destinations, a routing protocol has to be able to relay packets through low density regions but solutions for these lead probably to a unnecessary highly overhead over denser regions.

The density-aware aspect of REACT-DIS comes into play when nodes have to decide if they are going to forward further received packets after the time t expires. The decision of forwarding is based on the number of overheard retransmissions of duplicated packets. Each node keeps track of how many duplicates messages are received during the time slot of t seconds. Then, a node forward the packet with a probability ρ inversely exponentially proportional to the number of overheard copies c. This probability follows equation 2.

$$\rho = \frac{1}{\Upsilon^c} \qquad (2)$$

By this manner, the relay of a packet depends on nodes' suitability to be a forwarding node (based on its distance to the node in the last hop) and also on the density of its vicinity. Regions with a high density of vehicles are going to have only nodes with a low value of t forwarding packets while low density regions are going to permit that less suitable candidates relay further the message. This is the expected result which dynamically balances the cost of transmissions with the need of high reachability.

The problem with receiver-based solutions for video dissemination is that the waiting time t at each hop leads to an excessive end-to-end delay. In order to have a timely solution, REACT-DIS tackles this issue through the perspective that in each hop there is a competition to determine which nodes are suitable to be forwarding nodes. As many successive packets are transmitted in video dissemination and repeated competitions are unnecessary, in REACT-DIS, whenever a node *wins* as one of the suitable forwarding nodes, we consider that these nodes continue to be the best options to relay packets for a predefined amount of time Φ. Therefore, within this time window, nodes are selected as relay nodes and any packet received is immediately forwarded. This solution does not impact the ability to react to link breaks, if a relay node is no longer within reach of the previous relay

[1]In this work, I consider the communication range to be know by receiving nodes. This value can be easily estimated based on previous received messages or included in each packet's header.

node (or the source node), other eligible neighboring nodes have higher chances of forwarding the packet and to consider themselves as new relay nodes.

REACT-DIS is a non-deterministic solution as nodes are subjected to probability tests in order to decide if they are going to relay further received packets. When a node decides to forward a packet it continues forwarding packets that are received subsequently. During this time, an excessive number of nodes might consider themselves suitable to relay packets. For this reason, we have also adopted the density-aware principle for the forwarding of packets by relay nodes. This is done by using a probability ϱ to decide if specific packets are forwarded. This probability is given by the following function:

$$\varrho = \begin{cases} 1.0 & \text{if } c \leq k \\ 1.0 - [(c-k) \times \Lambda] & \text{if } c > k \end{cases} \quad (3)$$

In this work, we have chosen k based on how many relay nodes we would expect to have per hop. We understand that if more than four nodes are relay nodes within the same broadcasting zone of another relay node, its transmission is likely not crucial. For this reason, we have chosen $k = 4$. Although the choice of k is not empirically checked, experiments are performed varying the value of Λ and its impact on REACT-DIS' performance is analyzed.

In the next section, we describe in more details the peculiarities of REACT-DIS and its implementation.

3.1 Implementation

In order to evaluate our proposed protocol's effectiveness and efficiency in VANETs, we have implemented REACT-DIS using the Network Simulator (ns2)[9]. We have used the Two-Ray Ground propagation model, IEEE 802.11 for the link layer and a radio range of 300 meters. The packet payload size is of 1,000 bytes. This configuration is used throughout this work if not otherwise specified.

Before we continue describing the simulated scenario, we want to point out an important mechanism that is often omitted in other receiver-based works. As forwarding nodes are chosen locally by receiving node themselves, simultaneous transmissions[2] happen often which leads to many collisions. The most common and effective solution to this issue is to add a delay that is randomly chosen between 0 and θ. The value of θ is subject to a trade-off between low end-to-end delay and collisions avoidance. We have modelled a scenario where n nodes are placed uniformly in a region where all nodes are mutually reachable with a one-hop communication. We have varied θ and analyzed the delivery ratio of packets triggered to be broadcast at the exactly same time by all nodes (with additional random delays within the mentioned range). We have analyzed this under four different data rates (80kbps, 400kbps, 800kbps and 1600kbps). We have observed that $\theta = 10ms$ is the most suitable value for up to 10 nodes communicating at the same time (the results are not shown due to lack of space).

Bai et. al. [1] describe a mobility model for the city environment named Manhattan Model based on streets placed as in a grid. In this model, each street has two lanes (one for each direction) and vehicles move at a random speed within a predefined range. When a vehicle reaches an intersection it randomly chooses which direction to pursue. Vehicles have a 50% chance of continuing straight and 25% chance of turning left and the remaining 25% probability of turning right.

[2]Transmissions are considered simultaneous if any two or more transmissions are performed by different mutually reachable nodes with a time difference between them smaller than the one necessary for all of them to hear each other.

Table 1: Urban Mobility Model Parameters

Parameter	Value
Number of Vehicles	1,000
Area	3,000m x 3,000m
Block Length	100 m
Speed Range	1 to 30 m/s

In order to emulate mobility, we use the Urban Mobility Model (UMM) [13] which extends Bai's Manhattan Model taking into consideration traffic conditions. In real life, the speed of vehicles in a city is not only subject to speed limits but greatly by traffic conditions as well. For this reason, UMM uses a reducing factor that is proportional to the number of vehicles in the street segment the vehicle is moving to. There is also a maximum capacity regarding the number of vehicles in a segment at any given time. This capacity is calculated based on an estimate of vehicles' and segments' length. Therefore, the speed of a vehicle in UMM changes every time it passes through an intersection. This speed is initially randomly chosen between speed limits and then multiplied by the reducing factor that is based on traffic density.

Traffic plays an important role in the movement behavior of the network since it affects not only the movement speed but the density around the network. It is believed that, with traffic awareness, UMM represents movement aspects of real life scenarios with higher fidelity . Since UMM takes traffic into consideration, the distribution of nodes through the network area is not uniform. Regions with higher density have vehicles moving at lower speed which increases even further the local density. It is crucial for a mobility model for VANETs to represent density non-uniformity as it can significantly impact the performance of protocols that are focused only in either preventing overhead in dense regions or maintain connectivity over sparse areas.

Table 1 summarizes the values of the main parameters for the generation of the 32 distinct instances of the Urban Mobility Model. It is important to point out that, although the speed is initially randomly chosen with the defined range, vehicles may move slower than the lower limit since the final speed is subject to traffic conditions. The reducing factor is a function on the current number of vehicles in a segment and it is only used if there are more than 5 vehicles in such segment. The video source is an extra static node in the network. We have placed it in a corner of the network area instead of having it in the center of the simulated area. There is no particular reason in the network to believe that the performance in different quarters around a disseminating node in the middle would not be symmetrical. Besides that, a source node placed in a corner forces the network to create longer paths to reach farther distances.

We have considered that all nodes in our simulations are interested in the disseminated content. The majority of services we envision would be composed of videos that are of the interests of nodes 4,300 meters (the farthest possible distance) or less apart from the source node. The use of our solution for dissemination over a smaller region would only require an additional field on packets headers that delimits the dissemination range.

As mentioned in the previous section, the time t that nodes wait to try to forward incoming packets is chosen within $[\tau, \Gamma]$ based on their distance to the node that has lastly broadcast the packet to them. The values assumed by τ and Γ are greatly influenced by the value of θ. In order to prevent new nodes from broadcasting packets in the place of previously selected ones, we have $\tau = \theta = 10ms$ so any relay node forwards further the received packet before any new candidate. We have varied Γ and analyzed its impact on the performance of REACT-DIS, the results are shown in the next section.

The process of buffering packets is necessary as the transmission data rate is usually large enough so that nodes receive multiple packets before their waiting time t expires. If any packet is received before t expires, it is stored in this buffer. For this reason, whenever t expires and the node has to calculate its probability ρ of forwarding buffered packets, the attenuator c (number of duplicates) is divided by the number of packets in the buffer. The value used is the largest previous integer to this division. Therefore, with s as the number of packets in the buffer, a more precise description of the equation used to calculate ρ is:

$$\rho = \frac{1}{\Upsilon^{\lfloor c/s \rfloor}} \tag{4}$$

Once a node decides to forward buffered packets, it does following the same data rate as the packets were sent by the source node. If new packets are received before all buffered packets are broadcast, the new incoming packets are added to the end of the buffer.

Through preliminary experiments, we have observed that as many packets are sent, nodes with very low probabilities ρ of forwarding eventually become relay nodes and they keep forwarding packets for a long period of time at regions with already enough relay nodes. In order to avoid this, ρ is rounded to zero if it is less than 1%. Besides that, we have also noticed that the initial process of determining the first relay nodes have a great impact in the dissemination of the first packets. We have decided to broadcast a *null* message one second before starting to disseminate the video content. This additional delay is definitely acceptable and it prevents the loss of many initial packets.

Although it is important for relay nodes to be subjected to a forwarding probability based on density, these probabilities should not be too prohibitively to the dissemination of packets. For this reason, we have defined a lower bound for ϱ which is of 40% in this work.

In the next section, we define the metrics used for the evaluation of REACT-DIS, we discuss the observed results and we summarize our findings.

3.2 Evaluation

REACT-DIS' performance is evaluated through simulations in the Network Simulator (ns2), as mentioned. Each plotted result is an average of 15 to 20 runs, each with a different instance of the same mobility model and different seeds for the generation of random numbers. Confidence intervals are also plotted and they are calculated based on Student's t-distribution with a confidence level of 95%. All

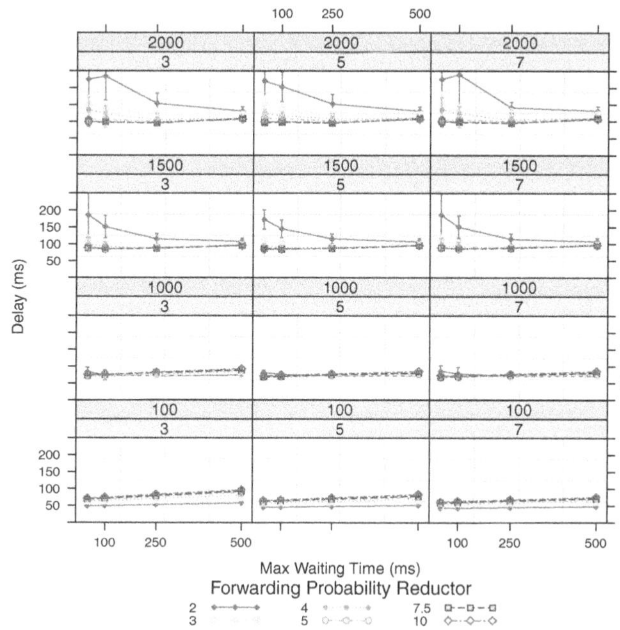

Figure 2: Delay in Video Dissemination - columns vary in Φ and rows in data rate (in kbps).

graphs and statistical measurements are made with the use of the free software environment for statistical computing and graphics called R [12]. We have observed that these considerations were enough to achieve trustworthy results as the calculated confidence intervals are short and compared results can be statistically differentiated.

In this work, we have used EvalVid *–A Video Quality Evaluation Tool-set* [4] –in order to get results relevant to video streaming. The video transmitted in our simulations is well-known and widely available online (akiyo_cif). It is a MPEG video with resolution of 360x486 composed of 300 frames that could be fitted into 353 packets.

REACT-DIS performance is greatly influenced by the configuration of its main four parameters. These parameters are the maximum waiting time Γ, the forwarding probability reducing factor Υ, the time a node remains as a relay node Φ and the relay node forwarding probability reducing factor Λ. These four parameters determine REACT-DIS' configuration quadruple $(\Gamma, \Upsilon, \Phi, \Lambda)$.

EvalVid provide tools to evaluate the transmission of video with metrics such as frame loss, frame delay, jitter, PSNR, MIV and MOS. In order to do so, it requires to convert the simulation output which consists of a summary of packets sent by the source and one list of the received packets per node. After the simulation, these lists are used to assemble what would be the received videos. Then, these videos are compared to the original reference video. This process is extremely time consuming and has to be performed 1,000 times per simulation (one for each receiving node).

The tunning of REACT-DIS' parameters is done through the evaluation of 600 unique combinations of its configuration. The evaluation of all these results using EvalVid's video's metrics is unfeasible. Consequently, for this part we have decided to measure the network metrics of delivery ratio, packet delay and number of packets sent.

A crucial distinction on the dissemination of video con-

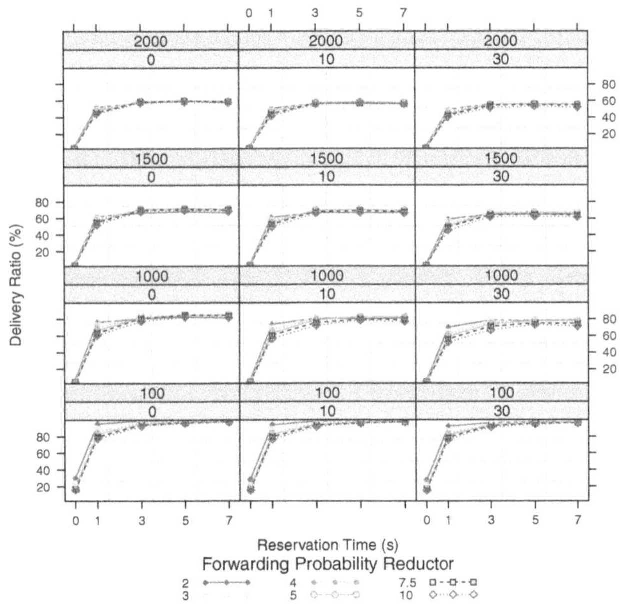

Figure 3: Delivery Ratio in Video Dissemination - columns vary in Λ and rows in data rate (in kbps).

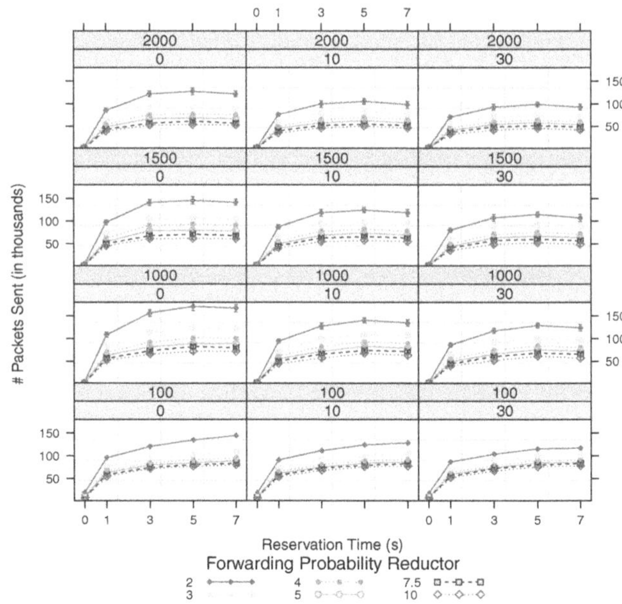

Figure 4: Number of Packets Sent in Video Dissemination - columns vary in Λ and rows in data rate (in kbps).

tent rather than other type of data is high data rate of transmission required for higher levels of video quality. Accordingly, we evaluate our protocol through four different data rates which support different levels of quality. We have used 100kbps (kilobits per second), 1,000kbps, 1,500kbps and 2,000kbps.

Our first observation is that REACT-DIS' approach in extending the selection of relay nodes for more than a single packet but instead to a time period has shown to be extremely successful in its attempt to be suitable to the dissemination of video content. Whenever we consider a reservation time greater than zero ($\Phi > 0$), end-to-end delay requirement is fulfilled (Figure 2). And when this mechanism is used, the impact of the maximum waiting time Γ is minimum. Based on this and the observed results, we have fixed $\Gamma = 250ms$ for the remainder of this work as this improves significantly the clarity of plotted results.

Figure 3 shows the results for a deeper investigation on REACT-DIS' ability to successfully delivery packets to receivers. The first surprising observation is that both forwarding probability reducing factors (Υ and Λ) had little impact in the achieved delivery ratio (a slight decrease is observed with the highest value of these parameters). The time Φ that a node considers itself as a relay node, however, influences greatly REACT-DIS' performance. The highest delivery are reached by values of Φ greater or equals to 3 seconds. Furthermore, we can see that delivery ratios high enough to fulfill video-streaming requirements are not reached with data rates of 1,000kbps or higher. At 1,000kbps, the delivery ratio is already around 80% and it decreases even further for higher data rates with a level around 60% at 2,000kbps. The second part of this work (Section 4), discusses further the issue of reaching high delivery ratios. The evaluation so far consisted in measuring the benefits of REACT-DIS. In order to estimate its feasibility, it is fundamental to evaluate also its cost. The number of video packets transmitted is the measurement of cost in this work. We show the results for this metric in Figure 4. As mentioned before, higher values of Υ and Λ do not lead to lower delivery ratios, however, in this figure, we see the efficiency of REACT-DIS as the same higher values reduce dramatically the number of packets sent.

The trade-off between benefit and cost is clearer seen in Figure 5. We have used a metric that divides delivery ratio (already normalized) by a normalized version of packets sent. The later is the division of the number of packets sent by the maximum possible number of packets that can be sent (the number of nodes in the network multiplied by the number of unique packets sent by the source node). Therefore, higher values of this metric reflects a better use of the network's resources. REACT-DIS' efficiency decreases with higher periods of time that nodes remain as relay nodes. However, this is necessary to fulfill the minimum requirements of delivery ratio. The impact of increasing probability reducing factors is significant and leads to more efficient REACT-DIS.

We have selected four configurations that better fulfill video-streaming requirements and better exploit the benefit and cost trade-off. These configurations are the four combinations with Υ equals to 5.0 or 10.0 and Φ equals to 3 or 5 seconds.

3.3 Conclusions

Our Reactive, Density-Aware and Timely Dissemination Protocol (REACT-DIS) was described in detail and evaluated in this section. This extensive evaluation allowed us to properly tune REACT-DIS' parameters. We have also measured REACT-DIS' performance under different data rates in order to observe its behavior for a variation of video quality demands.

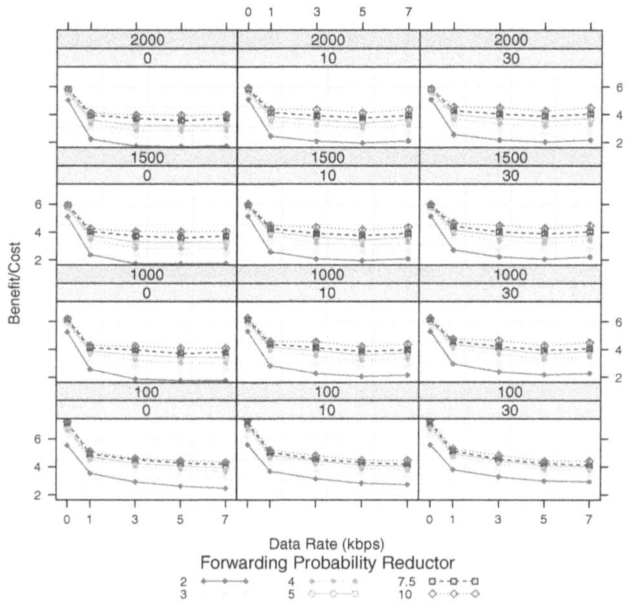

Figure 5: Benefit vs Cost in Video Dissemination - columns vary in Λ and rows in data rate (in kbps).

We have observed that REACT-DIS could only fulfill delivery ratio requirements for video streaming at very low data rates (100kbps). However, reasonable levels of delivery ratio were achieved for data rates up to 2,000kbps while maintaining end-to-end delay under one second. The resources consumption by REACT-DIS (in terms of packets sent) is not insignificant but it is definitely feasibly sustained in a VANET.

In the next Section, we are going to discuss and evaluate the use of redundancy with REACT-DIS to achieve higher delivery ratios that could conform with video-streaming requirements.

4. REDUNDANCY IN VIDEO DISSEMINATION

The most common approach to deal with packet loss is through the use of acknowledgments (acks). Acks can be either to inform the successful reception of a packet (positive acknowledgments) or to warn about the lost of one (negative acknowledgments). Both of these solutions are used in well-known network protocols such as TCP and CTS/RTS.

The use of acknowledgments in video-streaming is known to be highly discouraged, specially in network with dynamic topology such as VANETs. The delay in perceiving that packets were lost, the transmission of the acknowledgment back to the source to then trigger the re-transmission of lost packets makes it problematic to assemble video frames with these packets in time for real-time playback. Besides that, the use of acknowledgment in dissemination is prohibitively since many nodes are involved in the process as receivers.

Redundancy is an error correction mechanism that requires no synchronization between sources and receivers. It consists on making source nodes to broadcast information with a higher entropy than the one necessary to retrieve the original information. In other words, it means that sending redundant data, receivers can assemble the whole original data upon the reception of only part of all information sent by the source. By this manner, with no need of interaction between receivers and sources, it is possible to handle packet loss and achieve higher delivery ratios.

There are several ways of increasing the entropy of outgoing messages. The most simple approach is to simply broadcast copies of packets to be sent, however, it allows only large increasing steps on number of packets sent (100%,200% and so forth) and this solution is strongly prone to failure depending on the packet loss pattern. Another error correction mechanism based on additional redundancy is a coding solution where the source transmitted extra packets formed by `xor` operations between combinations of original packets, thus, lost packets may be recovered when these extra packets are received. One main disadvantage of this approach is that encoded packets vary of importance based on how many original packets are encoded into them. The reception of packets resulted from encoding of many original packets is worthless if encoded packets from few original packets are not received as well.

In this work, we use Network Coding which is a well-known and more advanced technique. Network Coding [6] is an information theory technique where segments of original information is merged together through a linear combination of them into data with the same size as each segment. By this manner, when nodes receive these coded packets, a linear system is built and, if enough combinations are received, this system can be solved to obtain the original information. Although this technique has been created to optimize the distribution of information in a multicast scenario, it can be directly adapted to add redundancy in a network in order to handle packet loss.

Video content is coded with network coding by first dividing the original data into blocks. These blocks are then subdivided into η segments of same sizes. Segments are then linear combined using κ (with $\kappa \geq \eta$) sets of η coefficients to create κ unique frames (to be included in the transmitted packets). Each one of these κ unique sets have to be linear independent of all the other sets. Therefore, nodes that receive η of κ packets sent can decoded the original data broadcast.

The use of Network Coding at intermediary nodes would require that these nodes wait the reception of a full block in order to be able to transmit newly encoded ones. This would increase substantially the efficiency on the use of the shared medium, however, it would lead to prohibitively end-to-end delays for nodes located further from the source node. Nevertheless, the shared medium is better used through Network Coding than other coding techniques (such as Reed Solomon [15]) as the overhearing of some packets might be useful to nodes who have not received such packet before due to collisions. This property of Network Coding is particularly useful in dissemination when compared to its use in unicast.

The increases in redundancy with network coding is still discrete but it is more fragmented than the one with simple copies. Another advantage of using network coding is that all packets sent have the same role in the decoding process at receivers. The reception of any η packets from the κ broadcast by the source (considering they are all linearly independent) allows receivers to decode the η original segments.

In the next section, we describe in detail the use of network coding with REACT-DIS.

4.1 Network Coding in REACT-DIS

The choice of the coefficients for the linear combination of the original data in network coding is a fundamental aspect of its implementation. In this work, this process is done through randomly selecting these coefficients. This is known as Random Network Coding [3] and with a proper selection of the space where coefficients are chosen from, the probability of having sets of coefficient that are co-linear is insignificant. Although this approach requires that η coefficients are included in each coded packet's header, we consider that the overall gain exceeds this cost.

The implementation of network coding in the transmission of packets through a network requires that multiplication operations involving the coefficients and the original data have results that can be represented with the same number of bytes. Finite field arithmetics (or Galois field) provide the necessary mathematical background for this purpose. Therefore, all multiplications into the coded data, the divisions required for the solution of the linear systems at receivers and the choice of coefficients follows the premisses of Galois field arithmetics.

We have used the Fast Galois Field Arithmetic Library in C/C++ [11] for the implementation of galois field operations. Coefficients and all operations are in the $G(2^{32})$ field which is sufficient to guarantee the unlikelihood of co-linear vectors of random coefficients. For all combinations of η and κ, the computation cost is irrelevant and it does not lead to any significant additional delay.

Each data segment of each block of the original video is analog to the payload added in each network packet. In this work, we have reserved 1,000 bytes in each packet for this payload. Therefore, the block size is the 1,000 bytes times the number η of segments per block. The percentage of redundancy added by source nodes is the ratio between the number of encoded packets sent by block κ and the number of segments η minus 100%.

In the next section, we describe the evaluation of REACT-DIS with Network Coding (REACT-DIS-NC) and we show the observed results for different combinations of η and κ.

4.2 Evaluation

The evaluation of the efficiency and effectiveness of the use of Random Network Coding with REACT-DIS follows the same configuration setup as the one described in Section 3. The statistical methodology is also the same as what was used before in this work. The topology and the position of the source node is the same as well and the same video is used. EvalVid is used for measuring video related metrics in the following simulations.

The use of Network Coding affects how EvalVid can be used to measure our protocol's performance. This happens because when a coded packet is sent it contains information from all packets in the same block. For this reason, we have decided that the time used to determine when a original packet was sent is when the first packet of its block is sent.

For this part, the network packet delivery ratio and delay are metrics that do not reflect the experienced service by users. The packets that are sent contain encoded information and its use depends on receivers' capability of decoding their contents. Consequently, the use of EvalVid tool-set is crucial to the proper evaluation of the impact of using network coding.

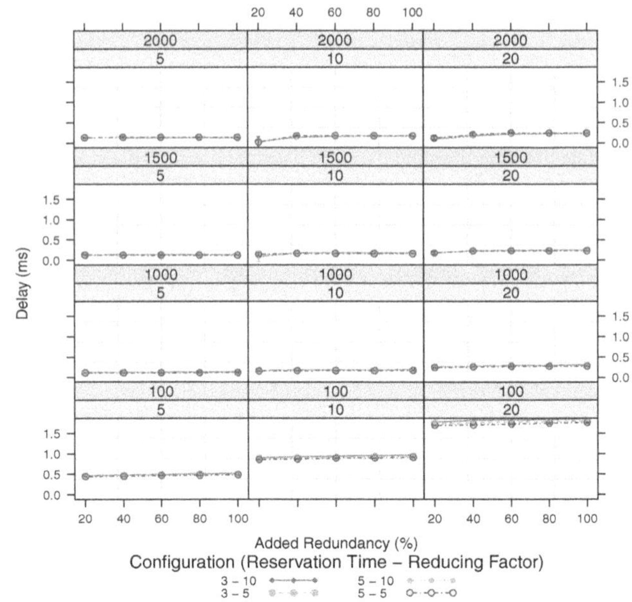

Figure 6: Delay in Video Dissemination with Network Coding

The delivery ratio shown in the following figures is the value resulted from subtracting the overall frame loss percentage from 100%. Delay is calculated between the time the first packet from a block that has data from a frame to the time this frame could be assembled at receivers. The resource consumption continues to be measured as the number of packets sent (here, the number of encoded packets).

Network Coding affects the transmission of video in the sense that the network data rate does not resemble the data rate by which decoded information is obtained by receivers. In all figures in this section, data rate shown is the network data rate. This is mentioned in all the following discussions but the reader must be aware that this has to be taken in consideration. For example, a network data rate of 2,000kbps with 100% of additional redundancy leads, in the best case, to a 1,000kbps data rate of decoded information at receivers.

In Figure 6, we show the delay in the reception of decoded frames at receivers. The first observation is that through low data rates, the end-to-end delay is strongly influenced by the number κ of packets sent per block of original data. This happens because at these low data rates, the time elapsed until the reception of at least η packets is significant and it overshadows the delay of the dissemination of packets through the network. However, REACT-DIS' performance without network coding for data rates of 100kbps already fulfills all requirements for video-streaming. Therefore, if the video broadcast requires this low data rates, it is recommended that network coding is simply not used. For all the other levels of data rate, delay requirements are satisfied.

The success of nodes in receiving decoded information is shown in Figure 7. It is clear that increasing the percentage

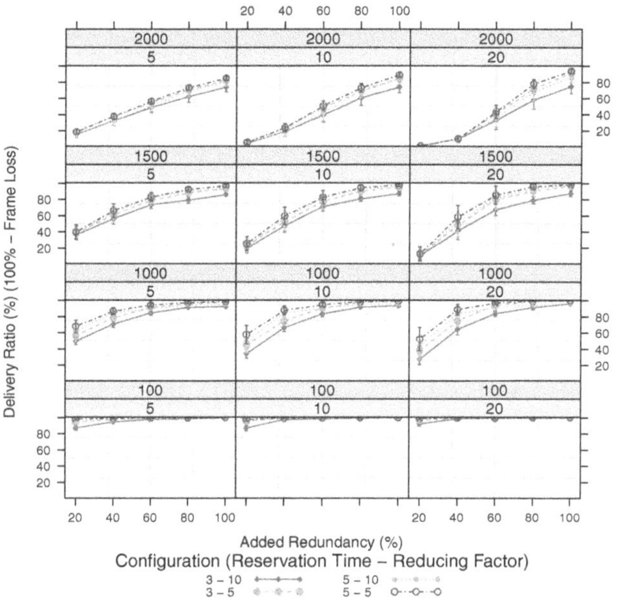

Figure 7: Delivery Ratio in Video Dissemination with Network Coding

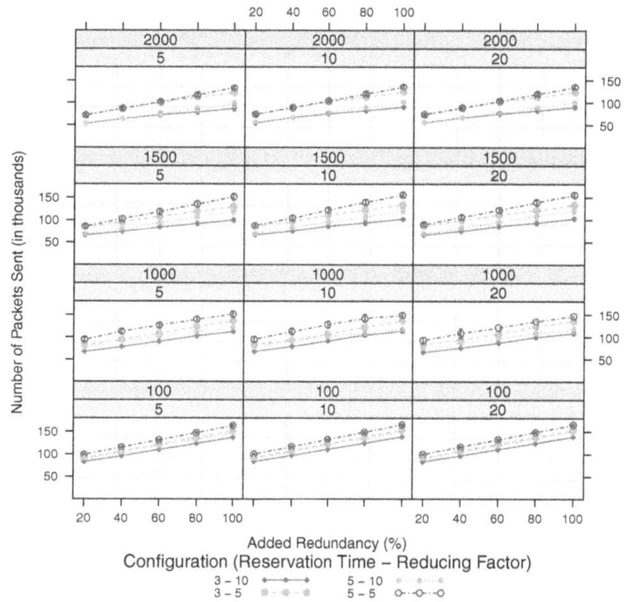

Figure 8: Number of Packets Sent in Video Dissemination with Network Coding

of additional redundancy leads to higher delivery ratios[3]. For higher data rates, with few additional redundancy, the delivery ratio is worse than that without network coding. The reason for this is that the network packet delivery ratios reached by REACT-DIS is too low and many of the received packets can not be decoded into useful information. However, high delivery ratios that can comply with video-streaming requirements are achieved with higher percentages of additional redundancy.

In a quantitative analysis of the figures, we see that with a network data rate of 500kbps with 40% of additional redundancy, delivery ratio requirements are met. This represents a user perspective of video content receiving data rate of 357kbps. At 1,000kbps, this level of delivery ratio is reached with 60% of redundancy (625kbps of video data). With 80% of additional redundancy, a network data rate of 1,500kbps leads to a feasible reception of video at 833kbps. At a network data rate of 2,000kbps, 100% of additional redundancy is necessary for a satisfying level of delivery ratio at receivers which leads to a video reception rate of 1,000kbps.

Figure 8 shows the overhead caused by forcing additional redundancy from the source node. The straightforward expected outcome is a linear increase in the number of packets sent as the percentage of redundancy increases. The mark of 150,000 packets sent is the same performance as if 42% of all nodes participate in the network participate in the broadcast of each non-encoded packet. An interesting observation is that although η itself had no impact in the number of packets sent, higher values of this parameter resulted in higher delivery ratios. Therefore, as long as data rate is not too low, the use of larger values for η (such as 10 and 20) is recommended.

[3] As mentioned before, here delivery ratio is calculated as 100% minus frame loss (in %).

4.3 Conclusion

Redundancy is an error correction mechanism that does not require any synchronization or interaction between receiving and source nodes. It is the ideal solution for a network with such highly dynamic topology as VANETs' topology.

In this Section, we have showed how Network Coding can be used as an effective way of using additional redundancy to increase delivery ratio at higher data rates to levels that satisfy video-streaming requirements (high delivery ratios with low end-to-end latency).

It is important to notice that the use of redundancy decreases the experienced receiving video data rate when compared to the used network data rate. However, higher data rates at users perspective with the necessary level of delivery ratio are achieved with network coding that could not be by the dissemination of raw data with REACT-DIS.

5. FINAL REMARKS

The Reactive, Density-Aware and Timely Dissemination Protocol (REACT-DIS) is described in detail and thoroughly evaluated in this work. This extensive evaluation allowed us to properly tune REACT-DIS' parameters and observe different levels of performance by unique configurations. Therefore, different configurations can be use depending on different Quality of Service (QoS) policies. We have also measured REACT-DIS' performance under different data rates in order to observe its behavior for a variation of video quality demands.

We have observed that REACT-DIS achieved delivery ratios higher than 60% with an end-to-end latency no longer than 200ms. Although delay requirements were fulfilled, REACT-DIS only fulfilled delivery ratio requirements for video streaming at very low data rates (100kbps). The resources consumption by REACT-DIS (in terms of packets sent) is not insignificant but it is definitely feasibly sustained in a VANET. For this reason, our next step in this work was to study error correction approaches that could improve the delivery ratios achieved.

Redundancy is a error correction mechanism that does not require any synchronization or interaction between source and receiving nodes. It is the ideal solution for a network with such highly dynamic topology. We have showed how Network Coding can be used as an effective way of using additional redundancy to increase delivery ratio at higher data rates to levels that satisfy video-streaming requirements sustaining a small end-to-end delay.

It is important to notice that the use of redundancy decreases the experienced receiving video data rate when compared to the used network data rate. Nevertheless, REACT-DIS with Network Coding could reach delivery ratios that satisfy video-streaming requirements for video data rates as high as 1,000kbps. At this rate, videos with reasonable quality can be displayed. Besides that, our simulations were limited to a network with bandwidth of 2.0mbps, thus, if radio interfaces support higher bandwidths, REACT-DIS would be able to support videos at higher quality.

REACT-DIS with Network Coding fulfills video-streaming requirements while preventing the overhead caused by excessive transmissions. This is due to the error correction characteristics of Network Coding, the reactiveness of receiver-based solutions, a change of perspective for the choice of relay nodes from each transmission to a period of time and because of the density-aware aspect of it that prevents an excessive number of transmissions while preserving connectivity.

In the future, we plan to investigate the suitability of other redundancy approaches and other coding techniques. Besides that, we want to design and evaluate a solution for video-unicast that follows the same premises discussed in this work and used for designing REACT-DIS.

6. REFERENCES

[1] F. Bai, N. Sadagopan, and A. Helmy. Important: a framework to systematically analyze the impact of mobility on performance of routing protocols for adhoc networks. *Twenty-Second Annual Joint Conference of the IEEE Computer and Comm. Societies.*, 2:825–835, 2003.

[2] M. Bakhouya, J. Gaber, and P. Lorenz. An adaptive approach for information dissemination in vehicular ad hoc networks. *J. Netw. Comput. Appl.*, 34:1971–1978, November 2011.

[3] T. Ho, M. Medard, R. Koetter, D. Karger, M. Effros, J. Shi, and B. Leong. A random linear network coding approach to multicast. *Information Theory, IEEE Transactions on*, 52(10):4413 –4430, oct. 2006.

[4] J. Klaue, B. Rathke, and A. Wolisz. Evalvid – a framework for video transmission and quality evaluation. In P. Kemper and W. Sanders, editors, *Computer Performance Evaluation. Modelling Techniques and Tools*, volume 2794 of *Lecture Notes in Computer Science*, pages 255–272. Springer Berlin / Heidelberg, 2003.

[5] S.-H. Lee, U. Lee, K.-W. Lee, and M. Gerla. Content distribution in vanets using network coding: The effect of disk i/o and processing o/h. In *Sensor, Mesh and Ad Hoc Communications and Networks, 2008. SECON '08. 5th Annual IEEE Communications Society Conference on*, pages 117 –125, june 2008.

[6] S.-Y. Li, R. Yeung, and N. Cai. Linear network coding. *Information Theory, IEEE Transactions on*, 49(2):371 –381, feb. 2003.

[7] T. Nadeem, P. Shankar, and L. Iftode. A comparative study of data dissemination models for vanets. *Mobile and Ubiquitous Systems - Workshops, 2006. 3rd Annual Int. Conference on*, pages 1 –10, jul. 2006.

[8] M. Nekovee and B. Bogason. Reliable and effcient information dissemination in intermittently connected vehicular adhoc networks. In *Vehicular Technology Conference, 2007. VTC2007-Spring. IEEE 65th*, pages 2486 –2490, april 2007.

[9] ns 2. The network simulator. http://www.isi.edu/nsnam/ns.

[10] J.-S. Park, U. Lee, S. Y. Oh, M. Gerla, and D. S. Lun. Emergency related video streaming in vanet using network coding. In *VANET '06: Proceedings of the 3rd international workshop on Vehicular ad hoc networks*, pages 102–103, New York, NY, USA, 2006. ACM.

[11] J. S. Plank and Y. Ding. Note: Correction to the 1997 tutorial on Reed-Solomon coding. *Software – Practice & Experience*, 35(2):189–194, February 2005.

[12] R Development Core Team. *R: A Language and Environment for Statistical Computing.* R Foundation for Statistical Computing, Vienna, Austria, 2011. ISBN 3-900051-07-0.

[13] C. Rezende, A. Boukerche, R. W. Pazzi, B. P. Rocha, and A. A. Loureiro. The impact of mobility on mobile ad hoc networks through the perspective of complex networks. *Journal of Parallel and Distributed Computing*, 71(9):1189 – 1200, 2011. Special Issue on Advancement of Research in Wireless Access and Mobile Systems.

[14] T. Szigeti and C. Hattingh. *End-to-End QoS Network Design: Quality of Service in LANs, WANs, and VPNs (Networking Technology).* Cisco Press, 2004.

[15] S. Wicker and V. Bhargava. *Reed-Solomon Codes and Their Applications.* John Wiley & Sons, 1999.

[16] Z. Yang, M. Li, and W. Lou. Codeplay: Live multimedia streaming in vanets using symbol-level network coding. In *Network Protocols (ICNP), 2010 18th IEEE International Conference on*, pages 223 –232, oct. 2010.

[17] J. Zhao, Y. Zhang, and G. Cao. Data pouring and buffering on the road: A new data dissemination paradigm for vehicular <emphasis emphasistype="italic">Ad hoc</emphasis> networks. *IEEE Transactions on Vehicular Technology*, 56(6):3266–3277, Nov. 2007.

An Efficient QoS Based Gateway Selection Algorithm for VANET to LTE Advanced Hybrid Cellular Network

Ghayet el mouna Zhioua[†‡], Houda Labiod[†], Nabil Tabbane[‡], Sami Tabbane[‡]
[†]InfRes - Telecom ParisTech, 46 rue Barrault 75634 Paris Cedex 13 - FRANCE
[‡]MEDIATRON - Sup'Com, City of Communication Technologies 2083, Ariana - TUNISIA
{zhioua, houda.labiod}@telecom-paristech.fr
{zhioua, nabil.tabbane, sami.tabbane}@supcom.rnu.tn

ABSTRACT

In this paper, we address the gateway selection issue in a hybrid system composed of a VANET network and a 4G LTE Advanced infrastructure. The VANET network is structured via a clustering scheme. Our algorithm is based on a multi-criteria and QoS related attributes approach used to make a decision of the appropriate gateway that might be used to connect a source vehicle to the LTE Advanced infrastructure. Considered criteria for making the decision include load, Received Signal Strength and Link Connectivity Duration of the cluster head and other candidate gateways. Simulation results demonstrate that our algorithm makes efficient decisions for electing the appropriate gateway with the best VANET and infrastructure features.

Categories and Subject Descriptors

C.2.1 [**COMPUTER-COMMUNICATION NETWORKS**]: Wireless communication

Keywords

VANET, Cluster, LTE-A Infrastructure, Gateway, QoS classes

1. INTRODUCTION

Actually, Vehicular Ad hoc Networks (VANETs) attract not only research community but also the automotive industry. In fact, according to the ETSI 102 638 technical report, they estimate that by 2027 almost 100% of the vehicles will be equipped with On Board Units (OBUs). OBUs are devices that provide communications among neighboring vehicles, i.e Vehicle to Vehicle (V2V) or between vehicles and nearby fixed equipments, i.e. Vehicle to Infrastructure (V2I) communications [1]. On the other side, LTE Advanced cellular network is a well designed system that offers data rate that could reach 1Gb/s and presents a latency of about 5ms which meets the IMT Advanced project requirements for 4G cellular systems [2]. In this paper, we propose a cooperative traffic transmission algorithm in a joint LTE Advanced-VANET hybrid network architecture. The focus of this paper is proposing an algorithm that selects the appropriate gateway that will connect the source vehicle to the LTE Advanced infrastructure under the scope of V2I communications. Many authors [3, 4, 5] propose algorithms for clustering and Cluster Head (CH) selection where, in most cases and in spite of its role of initiating the communication and controlling the flow of signaling messages within the cluster, the CH is set as the gateway to the infrastructure for source vehicles of its cluster. Some references [6, 7] that do not not consider clustering, elects the gateway after information broadcast and gathering from all nodes leading to high overhead. Obviously, centralizing the connectivity to the infrastructure for nodes belonging to one cluster appears as a good approach. The key advantage is to decrease the cellular network resources consumption, by multiplexing distinct source node flows into one gateway that handles to send them to the infrastructure. However, this scheme generates serious drawbacks such as inducing CH overload and increasing delays which is intolerable for delay sensitive services. Moreover, the cluster head might not be the optimal gateway to the infrastructure as almost all algorithms consider only VANET features for CH election and ignore infrastructure features. Therefore, it is essential to propose a method that considers traffic priority and sensitivity and on the other side take aware of VANET and infrastructure features. In this paper, we first present our system model. In the third section, we describe the proposed gateway selection algorithm. Finally, simulation results and conclusions are presented in the last sections.

2. SYSTEM MODELING

The system model (see fig.1) is based on a hybrid network architecture that consists of two systems: an LTE Advanced infrastructure and a VANET network. The architecture comprises IEEE 802.11p based VANET vehicles and an LTE Advanced eNodeB connected via S1 interface to core network, i.e. Evolved Packet Core (EPC). We assume that all vehicles are equipped with OBUs that contain two communication interfaces: the first one is an IEEE 802.11p wireless interface and the second is an LTE Advanced interface. Each vehicle uses its VANET interface to communicate with its neighboring vehicles. We made the assumption that the VANET network is already clustered and the cluster head is elected. Clustering

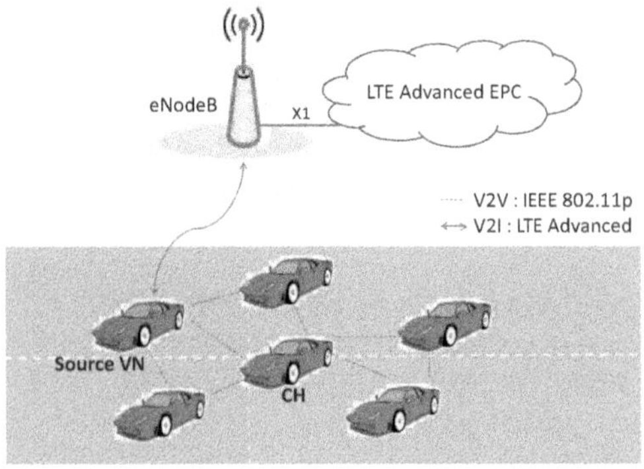

Figure 1: System Model

and cluster head selection techniques are out of the scope of this paper. The CH selection scheme we adopt is described in [3].

The goal of this work is to make the decision for a source vehicle upon the gateway to select that will play a proxy role to the LTE Advanced infrastructure. There are three possible alternatives for gateway selection: the gateway could be the cluster head of the cluster to which the source vehicle belongs to, another Gateway candidate (GwC) vehicle which belong to the same cluster of the source vehicle or there may be also no gateway between the source and the infrastructure where the source vehicle will be directly attached to the eNodeB. Therefore, our goal is make the decision upon the appropriate gateway for a source node (see section 3).

3. GATEWAY SELECTION ALGORITHM IN CLUSTERED VANET

We design an adaptive multi-criteria and attributes decentralized gateway selection algorithm and we propose to implement it in vehicles' OBUs. Our QoS based Gateway Selection (QGwS) algorithm assumes input parameters for making final decision upon the gateway. On one side, QGwS algorithm considers QoS requirements of the source vehicle and, on the other side, CH and infrastructure features.

3.1 Input parameters

Input parameters are divided into two categories: attribute and criteria.

3.1.1 Attributes:

In this paper, we consider only one attribute which is the type of traffic of the flow to be transmitted from the source vehicle to the infrastructure. Three classes of traffics are considered: Class 1, Class 2 and Class 3. As the source vehicle is a VANET node, generated traffic types are governed by IEEE 802.11p standard. On the other side, the infrastructure is an LTE Advanced cellular network. Thus, to achieve systems' compatibility, a mapping between IEEE 802.11p access categories (ACs) and LTE Advanced priority classes is done. The mapping is performed based of QoS constraints (packets average delay budget and packets loss) of IEEE 802.11p and 3GPP Release 10 standards. Traffic mapping is presented in Table 1. We define only three QoS category classes because for

QoS Classes	Class 1 (Voice)	Class 2 (Streaming)	Class 3 (Data)
IEEE 802.11p Designation	Alerts / Voice AC	Video AC	Best Effort and Background AC
LTE Advanced Designation	Conversational services and Control Signaling	Streaming and Interactive services	Background services
Average Delay	<100ms	< 150ms	< 300ms

Table 1: Traffic mapping

first deployments, a majority of 4G operators will likely start with three basic service classes containing voice, and best-effort data. Moreover, for sake of simplicity, in this paper, we will call Class 1 traffic as Voice traffic, Class 2 as Streaming traffic and Class 3 as Data traffic.

3.1.2 Criteria

Criteria are input parameters gathered by the source vehicle. QGwS algorithm considers the following criteria:

- ***Best Received Signal Strength (RSS) of source vehicle***: As source mobile vehicle could detect several LTE Advanced base stations, the best RSS is the best eNodeB received power level detected by the vehicle.
- ***Load of the Base Station:*** it is the resources occupation ratio of the selected eNodeB
- ***Best Received Signal Strength (RSS) of CH***: is the eNodeB signal power level received by the cluster head of the cluster to which the source vehicle belongs.
- ***Load of the CH:*** it is the occupation ratio of the buffering queue of the cluster head
- ***CH Link Connectivity Duration (LCD):*** this criterion is used to reflect the stability of the link and its lifetime between two vehicles, in this case it is the source vehicle and the CH. LCD is computed using this formula, inspired from [3]:

$$LCD = \frac{\sqrt{(\alpha^2+\gamma^2)R^2-(\alpha\delta-\beta\gamma)^2}-(\alpha\beta+\gamma\delta)}{\alpha^2+\gamma^2} \quad (1)$$

where,
$\alpha = v_i cos\theta_i - v_j cos\theta_j$
$\beta = x_i - x_j$
$\gamma = v_i sin\theta_i - v_j sin\theta_j$
$\delta = y_i - y_j$

(x_i, y_i), respectively (x_j, y_j), is the Cartesian coordinates of two neighboring vehicles i and j and that have an inclination of θ_i, respectively θ_j ($0<\theta_i, \theta_j<2\Pi$) with respect to the x-axis and moving at v_i, respectively v_j, speed. R is the IEEE 802.11p wireless transmission range. We suppose that the CH and source vehicle are adjacent, otherwise LCD is the minimum of LCD_i (1<i<n), where n is the number of hops between the source and the CH.

- ***Other GwC RSS, Load, LCD:*** there could be another gateway candidate (GwC) in the cluster for the source vehicle with better RSS, load and LCD as compared to the CH. Each Ordinary Vehicle (OV) that experience high RSS ($>RSS_{th}$), low buffering queue load ($<L_{th}$) and high LCD ($>LCD_{th}$) becomes a GwC and sends a response to the source vehicle GATEWAY SOLICITATION message, see later. The latter will elect and keep track of the best GwC features.

3.2 Proposed algorithm

In this section, we present in details the QGwS algorithm.

3.2.1 Decision phase

In the first phase, source vehicle measures RSS of detected eNodeBs, and chooses the BS with the highest RSS. Then, it checks its load. If the eNodeB load is acceptable, i.e. $<L_{th}$ (see Table2), and the traffic type is Voice, the source vehicle will be attached directly to the eNodeB without any gateway to the infrastructure. This decision is taken because Class 1 traffics are very delay sensitive and not resources' greedy. However, with the Class 2 & 3 traffics, in spite of RSS level and eNodeB load, source vehicle will check the CH features.

Cluster Head Criteria Check.

With Class 2 and Class 3 traffics, source vehicle verifies CH criteria. If the CH metrics values, i.e. RSS, load and source vehicle-to-CH LCD, are acceptable (as compared to the thresholds of Table2), the decision over the gateway is immediately taken, without solicitate other gateway candidates. Therefore, for this source vehicle, the gateway to the infrastructure is the CH.

Gateway Candidates Solicitation.

If, at least, one level of the three criteria of the CH is not acceptable (see thresholds of Table2), source vehicle broadcasts a GATEWAY SOLICITATION REQuest message through its cluster to solicitate OVs. Each OV with high RSS ($>RSS_{th}$), low load ($<L_{th}$) **and** high LCD ($>LCD_{th}$) becomes a gateway candidate (GwC) and respond with a GATEWAY SOLICITATION RESponse message, containing its metrics features (RSS, load, velocity and position (used to compute LCD)). Afterward, source vehicle collects information about all GwCs and computes corresponding weights using an additive metric of the three criteria. It selects then the GwC with the highest weight as the gateway to the infrastructure, see later. If source vehicle don't receive GATEWAY SOLICITATION RESponse messages, it repeats measurement from step 3.2.1.

Gateway Selection.

Source vehicle computes and affects a weight to each GwC. Weights computing are performed using (2), which is a simple additive metric of the three aforementioned criteria.

QoS Classes	Class 1	Class 2	Class 3
RSS_{th}	-120dBm	-120dBm	-120dBm
L_{th}	90%	85%	95%
LCD_{th}	30 s	60 s	60 s

Table 2: QGwS algorithm parameters

$$W_i = \frac{RSS_i}{RSS_{th}} + \frac{LCD_i}{LCD_{th}} + \frac{L_{th}}{L_i} \tag{2}$$

where,
W_i is the weight of gateway candidate i (GwC_i), RSS_i is the RSS measured by GwC_i, LCD_i is the source vehicle-to-GwC_i LCD, computed by source vehicle, L_i is the load of GwC_i; RSS_{th}, LCD_{th} and L_{th} are defined in Table2. Afterward, source vehicle keeps track of the best GwC by electing the gateway to the infrastructure as the GwC with the highest weight.

This algorithm is the baseline for making the decision over the gateway. As it can be noticed, a source vehicle checks in priority the CH state (RSS, load and LCD). If the latter ensures efficient proxy capabilities, it will be selected as the gateway and source vehicle will not send the GATEWAY SOLICITATION REQuest message. Otherwise, source vehicle will solicitate other GwCs. Thanks to this step, gateway efficient decision is taken without leading to useless extra overhead due to gateway solicitation requests and responses broadcast, see [8] for more algorithm details.

3.2.2 Maintenance phase

After making the decision over the gateway, if there is no gateway between source vehicle and the infrastructure, the link maintenance phase is handled by periodic Channel Quality Indicator (CQI) reports. However, if a gateway is elected, source vehicle initiates its timer and starts transmitting until $Timer = LCD_{GW} - \varepsilon$. Afterward, if there is more packets to be transmitted, source vehicle will check its CH and/or GwC RSS, load and LCD to re-elect the appropriate gateway for the remaining packets.

4. PERFORMANCES EVALUATION

Through simulations, we compare our QGwS algorithm with classical gateway selection scheme, based on choosing the CH as a gateway to the infrastructure. Simulations were made using Network Simulator NS2.33. We implement an IEEE 802.11p package in order to enable VANET communications among high-speed vehicles. Vehicles mobility is generated with VanetMobiSim [9]. In the simulation scenario, the cluster is composed of N_v vehicles, where the CH is elected using [3]. The source node belongs to the cluster and uses the QGwS or the deterministic scheme, called here Default Gateway selection scheme (DGwS), for gateway selection. With QGwS algorithm, source vehicle takes the decision over the gateway based on computed criteria. Simulation scenarios have been repeated several times, where on each scenario, source vehicle velocity is varied. We investigate upon Class 3 traffic services and LCD and RSS metrics. In our future works, further simulations will be handled for remaining traffics.

Figure 2 presents the obtained packets average delay for data traffic (Class 3) while varying the IEEE 802.11p wireless transmission range of the QGwS algorithm as compared to deterministic approach that selects the CH as gateway. The figure shows better results of QGwS algorithm as compared to deterministic approach. This is due to the fact that the selected gateway has better features than the CH. In fact, Cluster traffic is centralized in the CH, which leads having high load in the CH, which generates delay increase. Thus

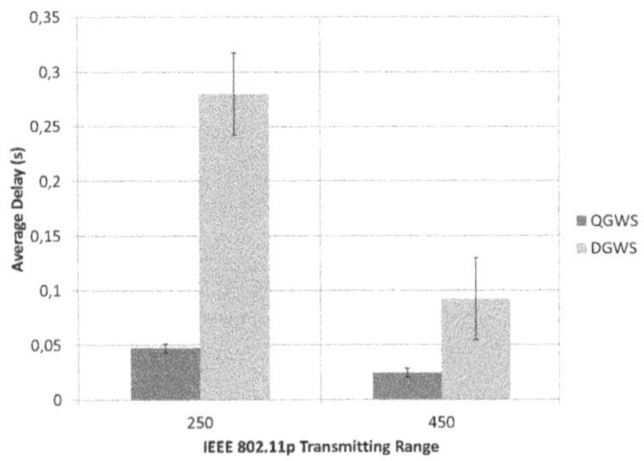

Figure 2: Average Delay Vs IEEE 802.11p wireless transmitting range (Class 3 Traffic)

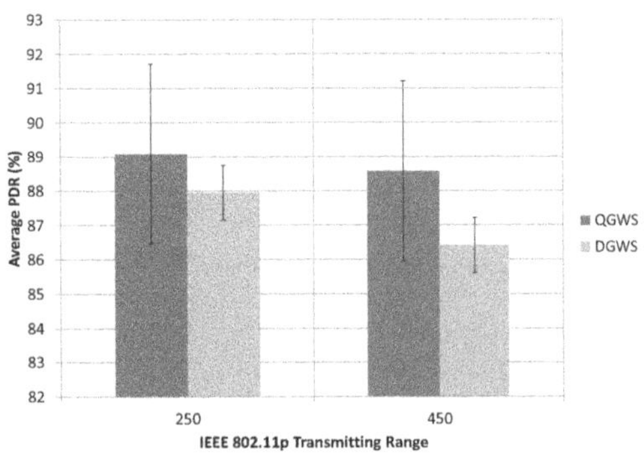

Figure 3: Average PDR Vs IEEE 802.11p transmitting range (Class 3 Traffic)

selecting appropriate gateway depending on traffic priority is primordial in that situation.

Results shown in figure 3 enlighten higher packet delivery ratio (PDR) for the QGwS algorithm. In fact, our scheme has selected a more appropriate gateway to the infrastructure that presents lower load than the CH in the deterministic scheme leading to less packets dropping. Moreover, as the IEEE 802.11p transmitting range increases, the PDR decreases. It is obvious as more vehicles will attempt to access to the same medium leading to more collisions and increasing number of packets buffered in the queues waiting to access to the medium.

In figure 4, we plot the metrics improvement while comparing our algorithm to the DGwS scheme. Two metrics are considered in this paper: RSS and LCD. I_{RSS} and I_{LCD} are computed based metrics ratio of QGwS and DGwS algorithms. According the figure, there is an improvement of more than 30% in terms of RSS and 80% in terms of LCD.

5. CONCLUSIONS

This paper presented a new method for gateway selection from a clustered VANET network to the LTE Advanced infrastructure. In addition to VANET and LTE Advanced networks features, the originality of our algorithm is to take into consideration the QoS class of the traffic to be transmitted to the infrastructure to elect the adequate gateway. This new approach has been compared to the standard deterministic approach that uses the CH as the default gateway and simulation results show that our algorithm performs better results. We are currently extending this work by evaluating its performances for other traffic classes and while varying clustering and CH election algorithms.

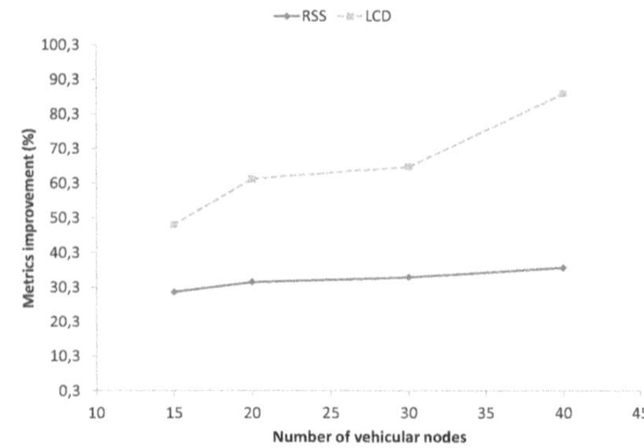

Figure 4: RSS/ LCD improvement

6. REFERENCES

[1] D. Camara, C. Bonnet, and F. Filali, "Propagation of public safety warning messages a delay tolerant network approach," *in Proc. IEEE WCNC*, pp. 1–6, 2010.

[2] "Itu-r m.2134," *http://www.itu.int.*

[3] A. Benslimane, T. Taleb, and R. Sivaraj, "Dynamic clustering-based adaptive mobile gateway management in integrated vanet-3g heterogeneous wireless networks," *in Proc. IEEE JSAC*, vol. vol. 29, No. 3, pp. 1–8, 2011.

[4] E. Dror, C. Avin, and Z. Lotker, "Fast randomized algorithm for hierarchical clustering in vehicular ad-hoc networks," *IFIP Med-Hoc-Net 2011*, pp. 1–8, 2011.

[5] N. Maslekar, M. Boussedjra, J. Mouzna, and H. Labiod, "Direction based clustering algorithm for data dissemination in vehicular networks," *in Proc. IEEE VNC*, pp. 1–6, 2009.

[6] R. Manoharan, S. Rajarajan, S. Sashtinathan, and K. Sriram, "A novel multi-hop b3g architecture for adaptive gateway management in heterogeneous wireless networks," *in Proc. IEEE WiMob*, pp. 447 – 452, 2009.

[7] F. Setiawan, S. Bouk, and I. Sasase, "An optimum multiple metrics gateway selection mechanism in manet and infrastructured networks integration," *in Proc. IEEE WCNC*, pp. 2229 – 2234, 2008.

[8] G. e. m. Zhioua, H. Labiod, N. Tabbane, and S. Tabbane, "A multi-metric qos-balancing scheme for gateway selection in a clustered hybrid vanet network," *in Proc. IEEE WiMob*, 2012.

[9] "Vanetmobisim project," *Website: http://vanet.eurecom.fr/.*

Enabling Mobile Distributed Social Networking on Smartphones

Kanchana Thilakarathna, Henrik Petander
NICTA & UNSW, Australia
[kanchana.thilakarathna; henrik.petander] @nicta.com.au

Julián Mestre
USYD, Australia
mestre @it.usyd.edu.au

Aruna Seneviratne
NICTA & UNSW, Australia
aruna.seneviratne @nicta.com.au

ABSTRACT

Distributed social networking services show promise to solve data ownership and privacy problems associated with centralised approaches. Smartphones could be used for hosting and sharing users data in a distributed manner, if the associated high communication costs and battery usage issues of the distributed systems could be mitigated. We propose a novel mechanism for reducing these costs to a level comparable with centralised systems by using a connectivity aware replication strategy. To this end, we develop an algorithm based on a combination of bipartite b-matching and a greedy heuristics for grouping devices into tribes among intended content consumers. The tribes replicate content and serve it using low-cost network connections by exploiting time elasticity of user generated content sharing. The performance is evaluated using three real world trace data sets. The results show that a persistent low-cost network availability can be achieved with an average of two replicas per content. Additionally, a content creator can reduce 3G traffic by up to 43% and device energy use by up to 41% on average compared to content sharing in non-mobile-optimised distributed social networking approaches. Moreover, the results show that the proposed mechanism can provide the benefits of a distributed content sharing system for monetary and energy costs comparable to those of a centralised server based system.

Categories and Subject Descriptors

C.2.1 [**Computer-Communication Networks**]: Network Architecture and Design—*Store and forward networks*

Keywords

User Generated Content Sharing, Cellular Data Traffic Offloading, Mobile Social Networking

1. INTRODUCTION

The increasing capabilities of mobile devices are driving rapid growth in the use of these devices in social networking applications/services. Majority of current social network services, such as Facebook, Twitter, YouTube and Google+, use the mobile device as a thin client with all the data associated with at the applications being centrally hosted on the servers of the service provider. This centralised approach has several benefits, for both users and the service provider. For the service provider the benefits of the centralised architecture are the ease of management and the ability for application/service provider to easily mine user data. For the user the advantages are the ability to access the service and the data from any device and perception that the data is safe. The primary disadvantage for the service provider is the cost associated with storage and distribution of the data. However, the benefits that can be obtained by mining the user data outweighs all the disadvantages of hosting the application/service centrally. The disadvantages for the user are twofold. Firstly, perceived advantage of the data being safe comes at the cost of the users loosing the ownership of their data. In some cases, they may not even be able to export it out of the service, thus locking them in to a single service. Secondly, as the data is mined by the application/service provider, the users loose their privacy. In contrast to the service provider, there are no over arching benefits for the user that will outweigh these disadvantages.

This has lead to the development of distributed decentralised social networking architectures, which are being proposed to be used for the next-generation of social networking services [9, 5, 16, 13]. The distributed architectures provide users with control of their data by enabling an individual user or a community of users to host their data, thus overcoming the disadvantages associated with loosing control and leakage of privacy due to mining of their data by the service provider. In general, this requires the users need to provide "always on" devices which will host the data or rent resources from a hosting service. The hosting service can be provided by the service provider themselves or a third party. If the hosting service is provided by the service provider, the users may still loose control or and more importantly the providers may mine the data. If provided by a third party, inconvenience and the costs of hosting will limits its appeal and usability.

Today most users have devices such as smartphones and tablets which have significant computing power, storage capacities and are connected to a network. Its is likely that these devices in the future will have even more computing power, and storage. Although there have been numerous proposals for peer-to-peer services which potentially utilise these capabilities, they have not been harnessed to date for providing distributed social network services. This is primarily due to the communications costs and battery usage. For example, with mobile data plans with capped limits, if Safebook [5] and MyZone [13] were to be provided using mobile device resources, it would be prohibitively expensive as they replicate content on a number of devices without consideration of communications costs or power usage. To this end, we showed that if it is possible to exploit the time elasticity of social networking content and the fact that the user will have access to high speed low-cost network such as WiFi networks, having spare capacity at some time periods of the day, it is possible to provide social networking services for sharing user generated content [19]. The proposed MobiTribe architecture provides user the necessary control of their data, and also protect their privacy.

This paper provides a deeper analysis of the MobiTribe architecture described in [19] to show that it is possible to provide a persistent content availability through low-cost networks by replicating the content only twice. We show that peer-to-peer storage schemes such as MobiTribe, can provide the same functionality as centralised architectures with the same or lower distribution costs and energy usage and it is always cost efficient than content sharing via non-mobile-optimised distributed social networking. The contributions of this paper are:

- Show that the content replication problem in distributed peer-to-peer architectures such as MobiTribe is *NP-Hard* and then present a new algorithm based on a combination of a bipartite b-matching and a greedy heuristic, which minimises content replication and maximises content availability whilst ensuring fairness;
- Models for determining the communication cost and the device energy consumption; and
- Using real data traces, we show the viability of the proposed algorithm, namely that it is possible to reduce the network load created by content sharing without compromising the availability of the content. For the particular tracks, MobiTribe content creator can save up to 43% of 3G bandwidth and up to 41% of device energy compared to non-mobile-optimised distributed social networking architecture.

The remainder of the paper is organised as follows: The next section presents related work. This is followed by an overview of the MobiTribe system architecture in Section 3. In Section 4, we formally define the content replication problem in peer-to-peer architectures such as MobiTribe and Section 5 presents our device grouping algorithm. The evaluation of the device grouping algorithm is presented in Section 6. Finally, Section 7 concludes the paper.

2. RELATED WORK

Diaspora [9] can be considered as the only widely used distributed social networking architecture. It is fully funded by user donations and estimated to have over 1.5 million users in March 2012 indicating a high demand for privacy-aware social networking. In order to host data, Diaspora users need to set up their own server for hosting their content. PrPl [17] is another proposed service which enables a user to run the service on a home server, or rent a vendor to

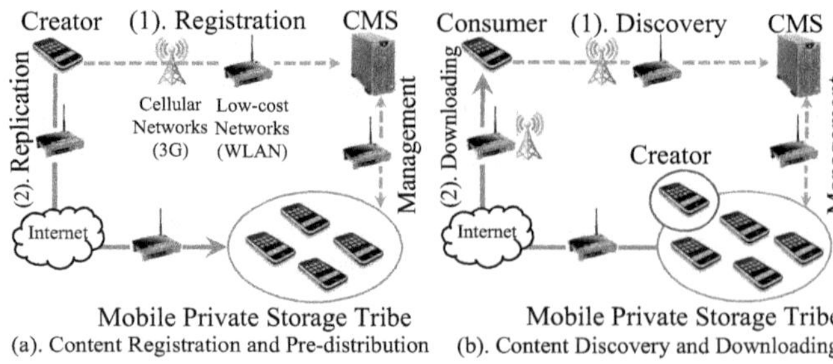

Figure 1: MobiTribe System Overview

host own data. If a mobile device is used to host the content as proposed in these services, either the availability has to be compromised or incur increased communications costs.

Safebook [5] is based on the concept of decentralisation and collaboration among friends and creates a secure social network. Similar to MobiTribe, friends are assumed to be cooperative and friends devices are used for replication to increase the availability. MyZone [13] also deploys user profile replicas on the devices of trusted friends to increase the availability of the profile. However, either system does not explicitly consider mobile devices, and the availability is dependent on the number of replicas. Hence, similar to Diaspora, if used with mobile devices, it would result in increased communication costs and energy consumption of devices. Contrail [18] is a cloud-based architecture to enable distributed social networking on smartphones which is conceptually similar to MobiTribe. The cloud relays content between users and stores content for offline recipients to increase the reliability. Though, it is not completely centralised, the content is stored in a third-party location, whereas MobiTribe propose to store content only among intended consumers.

Tribler [16] is a peer-to-peer file sharing system, which takes advantage of social relationships. Peers replicate their contextual information related to interests to enable local searching capability for relevant content. Also, peers are clustered into social groups based on similar interests. Again although there are similarities to MobiTribe, Tribler does not consider methods to increase the availability of the content or minimise communication costs.

In ActiveCast [12], the content is pre-fetched and cached in different locations including mobile phone via low-cost connectivity. An operator managed memory partition inside the mobile phone is used to cache content. ActiveCast does not address uploading or hosting of user generated content. Han et al. [8] have proposed a target set selection for propagation of the content with opportunistic communication. This mechanism will only work well for popular content, whereas popularity distribution of user generated content has a long tail [4]. Moreover, the focus of all this work is dissemination of data from a "centralised" data store and as a result does not address the privacy and censorship issues which MobiTribe addresses.

3. THE MOBITRIBE ARCHITECTURE

The primary objectives of MobiTribe are to provide users control of their data and to protect their privacy, without negatively impacting the usability and cost. The proposed MobiTribe architecture is shown in Figure 1. A *Creator* distributes the content to be shared to a group of devices, *Mobile Private Storage Tribe*, which would be selected as described in Section 3.2. The pre-distribution of content is carried out as described in Section 3.1. A *Consumer* would then access the shared content through the mobile private storage tribe. The architecture relies on a centralised entity *(Content Management Server - CMS)* to track the addresses and current connectivity of devices hosting content and to mange the peer-to-peer content distribution.

3.1 Sharing and accessing of content

A creator could share content via a social networking application similar to usual social networking applications/services. Then, the underneath MobiTribe platform initiates the content sharing process in two steps: *1). Registration:* The content would be registered with the CMS immediately after content creation. The registration can be done through any available network as shown in Figure 1a. Then, a link to the content, pointing to the CMS, is posted to the users who are supposed to view the content. *2). Replication:* Devices that take part in MobiTribe are grouped into tribes, as described below in Section 3.2. The content to be shared is replicated on the devices of the tribe, thus improving its availability to all users. The replication of the content is done when the creator and the devices in the storage tribe have access to low-cost networks. The low-cost network connections can be spare cellular network capacity, WLAN access or direct connectivity. The CMS would coordinate the content replication process acting as a tracker. After this pre-distribution process has been completed, the content will have very high availability over spare network capacity as a result of the device grouping strategy described in Section 4. The strategy ensures that the devices in the tribe have complementary low-cost network access patterns, thus making it highly probable that one of them can upload the content over a low-cost network connection. The pre-distribution is not instantaneous, and time required will depend on the network connectivity available to the creator and devices in the tribe. If during this time another user wanted to access the content, the content creator has the option of delivering the content over the least congested data connection available to the creator or make it appear as unavailable until the pre-distribution process has been completed.

Figure 1b shows the content access mechanism in MobiTribe. A consumer could initiate the content downloading process by clicking on a shared notification appearing in the social networking application. The notification feed contains the link to the CMS and trigger the content downloading process. *1). Discovery:* When the CMS receives a query for content, it directs requests to the devices in the mobile private storage tribe that the device should connect to get the requested content. The choice of the devices is based on the network connectivity and the load of available devices in the tribe. *2). Downloading:* A content consumer has the option of downloading the content over whatever network connection available or could be scheduled to download through low-cost networks. In contrast, the upload part of the path, i.e. from tribe to fixed network edge, would always be over spare network capacity. In addition, during the download and after it, the consumer would also act as a seeder in BitTorrent to share the clip temporarily helping the system to scale with demand.

To perform the content sharing between the content creator, devices in the tribe and content consumers, a peer-to-peer protocol, such as BitTorrent, could be used with the help of the CMS acting as a tracker to balance the traffic load and increase the scalability and data rates of transfers. We consider the details of the peer-to-peer mechanisms to be outside the scope of this paper due to lack of space.

3.2 Making a tribe for replicating content

A mobile device, if used as a content store, cannot provide access to the content continuously for three primary reasons, namely the communications costs, battery usage, and possibly not being "on" all the time. However, if devices which have complementary connectivity, battery power and being "on" can be found, and if the content can be replicated on these devices, it will be possible to provide similar availability to content that is hosted in a centralised server. Moreover, through this process, if the traffic can be routed through a low-cost and/or less congested network, it will result in significant benefits to both the service providers and the users. This can be achieved by having a dedicated encrypted storage partition in the devices in the tribe, which is controlled by the service and the content to be only accessible through an interface provided by the service. The down sides of the content replication are the overheads of replication such as the additional use of cellular bandwidth, energy and storage capacity of the mobile devices. Therefore, device grouping is crucial for the viability and also efficiency and performance of MobiTribe.

The device grouping is performed by the CMS, as it has a global view of the system state in terms of both mobile devices and network infrastructure. CMS carries out the grouping periodically or in response to a significant degradation to the availability of a tribe's content. To perform device grouping, all devices need to update the CMS with their context information, namely the low-cost network availability, spare storage capacity of the device and the remaining battery life of the device periodically. This context information is used to create tribes as described in Section 5. Although storing context information on a centralised server may result in some loss of privacy, this is insignificant in comparison to exposing all user information.

4. DEVICE GROUPING PROBLEM

The primary objective of the device grouping is to locate a tribe of devices for content replication such that the tribe maximises the availability of a content via low-cost networks. To enable that at least one device in the tribe needs to be always available at a low-cost connection with sufficient battery resources. Hence, a device could only be used to host content at a given time instance, if it has low-cost connectivity, sufficient battery and spare storage capacity. To this end, we name the behavioural pattern of capability to host content on the device as "Device Availability" pattern. The device availability patterns can be generated based on the device context information such as available network types, the battery level, AC power availability and the amount of spare storage capacity on the mobile device. The communication cost can be further optimised by analysing network infrastructure context information available at the central management entity such as connected cell ids, network load and relative importance of the traffic.

In our previous work [19], we showed that the history of device availability patterns can be effectively used to infer

Table 1: Contingency Table

		Device 2		
		p	q	r
Device 1	p:available	n_{pp}	n_{pq}	n_{pr}
	q:unavailable	n_{qp}	n_{qq}	n_{qr}
	r:unknown	n_{rp}	n_{rq}	n_{rr}

future availability patterns of a device and groups of devices. Therefore, we aim to group devices into tribes based on history of device availability patterns of devices. Suppose that there are two devices having device availability patterns for M time instances. The first step to determine the potential for pairing is to derive a *Contingency Table*, shown in Table 1, that can be considered as the summary of behaviour of the availability patterns. The table contains elements n_{ij} indicating the number of time instances that one device in state i and the other in state j, where i, j stand for the availability states of the each device, available: p, unavailable: q and unknown: r.

If at least one device in the tribe has availability, the content can be considered as available, as it enables the delivery of the content to the consumer. Hence, the probability of availability of a tribe can be defined as $P_a = \frac{n_{pp}+n_{qp}+n_{pq}+n_{pr}+n_{rp}}{M}$. In addition, pre-distribution of content is performed only when both end devices are connected to low-cost networks. To this end, we define the device grouping metric $P_a^{W_o}$, a modified probability of availability;

$$P_a^{W_o} = \frac{(n_{pp} + n_{qp} + n_{pq} + n_{pr} + n_{rp}) + W_o \times n_{pp}}{M + W_o \times n_{pp}}$$

The overlapping weight, W_o is used to control the impact of overlapping low-cost availability during the group selection, which is evaluated in Section 6. If $P_a^{W_o}$ is higher than a certain threshold - the "Threshold of Pairing", $P_{threshold}$, we consider that the tribe satisfies the availability constraint. However, we cannot implicitly select tribes based on threshold of pairing because replication consumes scarce resources in a mobile environment such as network bandwidth, storage and energy of devices. To this end, we consider two additional constraints in device grouping to manage the resource limitations in mobile networks.

Minimum Replication: It is obvious that the availability of the content increases with the level of replication. However, a high level of replication will disrupt the system performance by creating additional traffic flows and overheads, wasting battery life and spare storage of mobile devices. Hence, the device grouping should be able to mange the trade-off between replication and availability. Here after, the minimum replication is confined by the "Limit of Replication", i.e. the maximum allowed size of a tribe.

Fairness: The system should not place an extra burden on the resources of devices with good low-cost network availability. For instance, if such a device is selected by many users to host data, it would drain energy and storage of that device. Therefore, the replication mechanism should make sure to replicate content fairly among the devices. In the remainder of the paper, the fairness of the system is quantified by the "Limit of Hosting", i.e. the maximum number of tribes that a single device can belong to.

Each device in the system should be able to locate a tribe of devices while satisfying the threshold of pairing, limit of replication and limit of hosting. After considering all these constraints when a device is able to find its replication tribe, we say that the device is *covered*. Thus, the content replication problem becomes *"how to find a maximum cover while ensuring minimum replication and fairness constraints"*.

PROBLEM DEFINITION

A hypergraph is a pair (V, E), where V is a ground set of elements and E is a collection of subsets of V. The rank of a hypergraph $H = (V, E)$ is defined as $\Delta_H = \max_{e \in E} |E|$; we will drop the subscript H when the hypergraph is clear from the context. For a subset C of E and a vertex $u \in V$, we denote the *degree of* u *in* C by $\deg_C(u) = |\{e \in C : u \in e\}|$.

In the DEVICE GROUPING (DG) problem we are given a hypergraph $H = (V, E)$ and a capacity function $c : V \to \mathbb{Z}^+$. The objective is to find a subset $C \subseteq E$ that maximises the number of vertices spanned by C, namely;

$$\begin{array}{ll} \text{Maximise} & |\{u \in V : \deg_C(u) \geq 1\}| \\ \text{subject to} & \deg_C(u) \leq c(u) \text{ for all } u \in V \end{array}$$

The relation between DG and our application is as follows. The vertex set V represents the set of devices. The hyperedge set E is the collection of subsets of V satisfying the availability constraint, $P_{threshold}$. The parameter Δ is the maximum allowed group size, i.e. the limit of replication. Finally, the capacity function c is used to upper bound the maximum number of tribes that a single device can belong to, i.e. the limit of hosting. The objective is to perform a grouping that covers (spans) as many devices as possible. Our first result is to show that DC is computationally hard even for hypergraph of rank 3.

THEOREM 1. DG *is NP-Hard even when* $\Delta = 3$ *and* $c(u) = 1$ *for all* $u \in V$.

PROOF. Let (V, H) be a hypergraph. A matching is a subset C of E, where the sets in C are pairwise disjoint. The matching is perfect if every vertex is covered by some edge in C; namely, $\cup_{e \in C} e = V$. When $\Delta = 3$, testing if such a matching exists is known as the 3D-matching problem and is one of Karp's original 21 NP-Hard problems [10].

The hardness of DG follows immediately since the case specified in the theorem statement is the problem of finding the largest subset $C \subseteq E$ such that sets in C are pairwise disjoint. □

5. DEVICE GROUPING ALGORITHM

5.1 A tractable special case of DG

Even though the problem is hard when $\Delta \geq 3$, we can show that the problem is solvable in polynomial time when $\Delta = 2$, i.e. when (V, E) is a graph. This shows that the hardness proof in the previous section cannot be pushed to an even simpler version of the problem.

First, we reduce DG to the MINIMUM WEIGHT DEGREE CONSTRAINED SUBGRAPH (MWDCS) problem for $\Delta = 2$. Let, a graph (V, E), a weight function $w : E \to R^+$, and lower and upper degree constraints $L(u) \leq U(u)$ for each $u \in V$ as inputs to MWDCS. The objective is to find a minimum weight subset $F \subseteq E$ such that $L(u) \leq \deg_F(u) \leq U(u)$ for all $u \in V$. Gabow [6] gave an $O(U(V)|E| \log |V|)$ time algorithm for this problem, where $U(V) = \sum_{v \in V} U(v)$.

Figure 2: An example of how F is pruned in the proof of Theorem 3. First edges outside the cycle are pruned, and then edges inside the cycle are pruned.

THEOREM 2. DG *can be solved in* $O(c(V)|E|\log|V|)$ *time when* $\Delta = 2$.

PROOF. Given an instance (V, E, c) of DG, we construct an instance of WMDCS (V', E', U, L, w). The graph (V', E') is obtained by taking (V, E) and adding an edge (u, u') for each $u \in V$; in other words, $V' = \{u, u' : u \in V\}$ and $E' = E \cup \{(u, u') : u \in V\}$. We set $w(e) = 0$ for all $e \in E$ and $w(e) = 1$ for all $e \in E' \setminus E$. Finally we set $L(v) = 1$ and $U(v) = c(v)$ for all $v \in V$, and $L(v') = 0$ and $U(v') = 1$ for all $v \in V$. It is trivial to check that a solution for the DG problem covering X vertices induces a solution for the degree constrained problem with weight $|V| - X$, and vice versa. Therefore, a minimum weight solution for the degree constrained problem corresponds to a maximum coverage solution for the DG problem. □

While this settles the complexity of DG when $\Delta = 2$, this result does not constitute a satisfactory solution for our problem because Gabow's algorithm [6] involves complex data structures and there are no implementations of it available. Instead, we show that in most cases our problem can be reduced to a single bipartite b-matching computation for which highly tuned implementations are available (the problem reduces in a straightforward way to maximum flow). An instance of the b-matching problem is specified by a bipartite graph (A, B, E) and a function $b : A \cup B \to \mathbb{Z}^+$. The objective is to find a maximum cardinality subset M of E such that $\deg_M(u) \leq b(u)$ for all $u \in A \cup B$.

THEOREM 3. *When* $\Delta = 2$ *and* $c(u) > 1$ *for all* $u \in V$, *we can reduce* DG *to a single bipartite b-matching computation in linear time.*

PROOF. Let (V, E, c) be our DG instance. We construct a bipartite graph (A, B, E) where $A = \{v : v \in V\}$ and $B = \{v' : v \in V\}$; for each $(u, v) \in E$ we include two edges (u, v') and (v, u') in E. In addition, we set $c(u) = 1$ and $c(u') = b(u)$ for each $u \in V$.

We claim that any solution to the DG instance induces a solution to the b-matching problem with the same value, and vice versa. To prove the forward direction, let $C \subseteq E$ be a solution to DG. For each vertex u covered by C, we choose an arbitrary edge $(u, v) \in C$ incident on it, and add (u, v') to the matching. It is trivial to show that the cardinality of the matching equals the number of devices covered by C and that the matching is feasible. The other direction takes more effort.

Let M be some b-matching. We construct a directed graph (V, F) such that if $(u, v') \in M$, we add the directed edge (u, v) to F. The meaning that we want to attach to edges is that device u is *being covered* by v. Our ultimate goal is to find a solution for DG that covers the same number of devices as M. We use $\deg^+_F(u)$ to denote the number of edges in F going out of u, and $\deg^-_F(u)$, the number of edges coming into u. Notice that for every vertex $u \in V$, we have $\deg^+_F(u) \leq 1$ and $\deg^-_F(u) \leq c(u)$; this follows immediately from b-matching constraints. This means that in the underlying undirected graph induced by F, each connected component will be made up of one-trees (a tree plus one edge, or equivalently, a graph with a single cycle).

It is tempting to think that we could disregard the directions of the edges in F and use this as our DG solution. This, however, will not work because the combined in and out degree of a vertex u can be $c(u) + 1$. Ultimately, we will ignore the direction of edges in F, but first we need to prune some unnecessary edges. We say a node is a *leaf* in F if $\deg^-_F(u) = 0$. We repeatedly perform the following operation: If there is a node u such that $\deg^+_F(u) = 1$ and $\deg^-_F(u) \geq 1$ and its in-neighbourhood is made up of leaves, then delete from F the edge going out of u. Once we reach a point where we cannot delete any more edges, we must be left with a collection of disjoint stars and directed cycles with some additional edges pointing to vertices in the cycles as in the second graph in Figure 2. If a vertex u in cycle has an incoming edge from outside the cycle, then we remove from F its outgoing edge as in the third graph in Figure 2.

At this point it is not difficult to show that if a vertex u still has its out-going edge, it must be the case that u is a leaf or it was a cycle node with a single incoming edge from other vertex in a cycle. Notice that in either case $\deg^-_F(u) + \deg^+_F(u) \leq 2$. For all other vertices with no out-going edges we have $\deg^-_F(u) + \deg^+_F \leq c(u)$. Therefore, if we just disregard the directions of edges in F, the solution is now feasible for DG because the $c(u) > 1$ for all $u \in V$ by the assumption made in the theorem statement.

The only thing left to show is that every vertex covered by the b-matching is also covered by the DG solution. Notice that every time we removed an edge (u, v) from F we always had another edge (x, u) with $\deg^-_F(x) = 0$; thus, we are guaranteed that (x, u) will survive further pruning and remain in F until the end. Therefore, vertex u will be covered by the (undirected) edge (x, v) in our DG solution. □

5.2 A greedy approximation algorithm for general instances

For general hypergraphs (V, E), it is not possible to solve the problem in polynomial time since it is NP-Hard. However, we will be able to approximate the objective efficiently. This section shows that the following greedy heuristic has an approximation ratio bounded by Δ, the size of the largest set in E. The algorithm works in iterations by building a solution C. In each iteration, it finds an edge e that can safely added to C without violating the capacity constraints so as to maximise the number of newly covered elements.

Algorithm 1 GREEDY(V, E, b)

```
C ← D ← ∅
while ∃e ∈ E : deg_{C+e}(u) ≤ c(u) for all u ∈ e do
    Let e an edge in E maximising |e \ D|
    such that deg_{C+e}(u) ≤ c(u) for all u ∈ e
    C ← C + e
    D ← D ∪ e                {vertices in e are now covered}
return C
```

We note in passing that the set systems defined by the feasible solutions of the DG was shown to be Δ-extendible by Mestre [14] and the maximization objective is submodular. For such a problem, Chekuri et al. [3] showed that greedy is a $\Delta + 1$ approximation. However, for our special objective function, a slightly better approximation is possible.

THEOREM 4. *Let (V, E, b) be an instance of* DG, *then* GREEDY *is a Δ-approximation algorithm.*

PROOF. Let O be an optimal solution. The idea is to assign devices covered by O to devices covered by C so no device covered by C is assigned more than Δ devices covered by O. First though, we will assign sets in O to sets in C.

Suppose that GREEDY picks a set e in a given iteration. For each $u \in e$ we find some set f_u in O, if any, such that $u \in f_u$; we assign f_u to e and we remove f_u from O. We claim that at the end of the algorithm all sets in O must be assigned. Indeed, let f be some set in O. If $\deg_C(u) = c(u)$ for some $u \in f$, then clearly f must have been assigned. Otherwise, $\deg_C(u) < c(u)$ for all $u \in f$, but this means that GREEDY left the while-loop prematurely because f can be safely added to C, a contradiction.

Because of our greedy choice, when adding e to C the number of newly covered devices cannot be larger if we had added f_u to C. More specifically, $|e \setminus D| \geq |f_u \setminus D|$ for the set D just before adding e to C. Thus, we can distribute the devices covered by $\{f_u\}$ to devices covered by e so that each newly-covered device in e receives at most Δ devices from $\{f_u\}$. Therefore, the overall number of devices covered by O is no more than Δ times the overall number of devices covered by C. $\square$

5.3 Device Grouping in practice

DEVICE GROUPING will be carried out at a central entity. Each device periodically updates the central entity with its recent device availability patterns. From the device availability patterns, an undirected graph among all pairs of devices will be generated such that there exists an edge between two devices u and v if $P_a^{W_o} > P_{threshold}$. We use this undirected graph to solve DG for the case of $\Delta = 2$, i.e. a maximum tribe size of two. Next, we construct a bipartite graph to implement the single bipartite b-matching reduction as in Theorem 3. The maximum flow computation is used to find b-matching by introducing two extra nodes s and t. All incoming edges to node t from a device u have the capacity equal to $c(u)$, i.e. the limit of hosting. All the other edges have the capacity of one. Thus, the limit of hosting ensures that the node u only takes part in a maximum of $c(u)$ tribes. Once we compute the maximum flow from s to t, we prune extra edges from the resulted directed graph as illustrated in Figure 2. Finally, we ignore the direction of the edges and consider it as the solution of DG problem for the case of $\Delta = 2$. As the next step, if there are ungrouped devices, the GREEDY approximation algorithm proved in Theorem 4 is used to find tribes for the remaining devices.

6. PERFORMANCE EVALUATION

We evaluate the effectiveness of the DEVICE GROUPING algorithm in the context of device grouping using WiFi connectivity patterns of three data sets. Effectiveness of DEVICE GROUPING can be quantified by the low-cost network availability of a selected tribe, required replication for a content and the maximum number of tribes that a single device belongs to. Since MobiTribe depends on user behaviour, an analysis on real mobile users connectivity patterns is the best method to perform a comprehensive and realistic evaluation. To this end, we were able to find three data sets which are applicable to our particular problem, but with limited context information. Therefore, in this evaluation, the context information considered for low-cost network selection is limited to WiFi connectivity patterns of the devices.

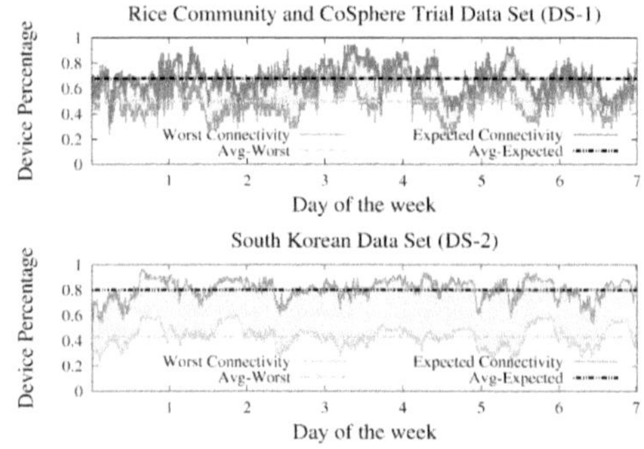

Figure 3: Percentage of devices which have WLAN connectivity

6.1 Dynamics of Data Sets

The Rice Community [1] trace data set has WiFi connectivity information with a one minute granularity from 14 users in Houston for 6 weeks in Feb. 2007. The CoSphere trial [1] contains WiFi connectivity patterns of 12 users for 6 weeks in Mar. 2007. We refer to these two sets as DS1. Further, we were able to use WiFi connectivity patterns of 97 iPhone users from a recent study conducted by Lee et al. [11] in South Korea in Feb. 2010 which spanned over 18 days with a 3 minute granularity. We refer to this data set as DS2. However, neither data set has consistent connectivity patterns, i.e. there are time instances for which the exact state of WiFi connectivity is unknown. In cases where there is only one time stamp gap between data points, we filled it to match the surrounding state if the surrounding states are the same. Otherwise, the gap was left as "unknown". However, we had to remove traces of 18 devices due to a high portion of unknown states. If a device has WiFi connectivity, we consider that the device is available via a low-cost network and vice versa. If we consider all "unknown" states as "unavailable", that gives the worst possible low-cost network availability for a particular device. On the other hand, we can remove the "unknown" states from the evaluation, which then gives the realistic low-cost network availability of the devices. Hence, these two cases are used to obtain bounds for the evaluation metrics as the worst and the expected cases where necessary.

The portion of devices having WiFi connectivity at a given time is shown in Figure 3 for both expected and worst cases. The worst case average WiFi availability for DS1 data set is 50%, i.e. each device has on the average WiFi available for 50% of the time, and for DS2 43%. The difference between expected and worst case is higher in DS2 suggesting that it has more unknown states compared to DS1. Altogether, the data sets contain 1470 days of connectivity patterns from 105

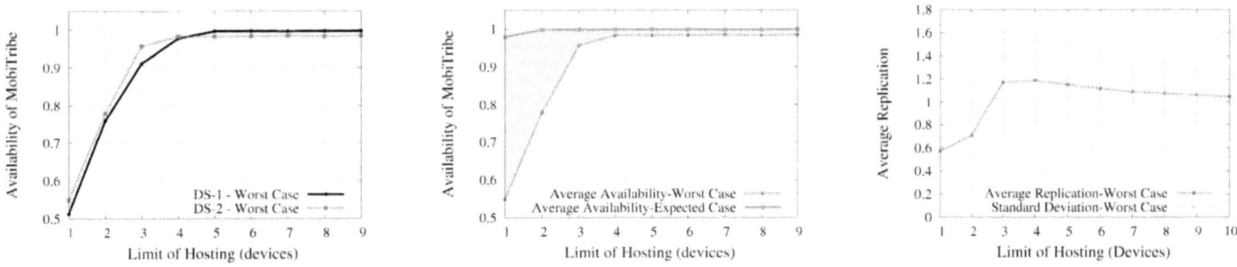

(a) For DS1, $P_a = 0.99, W_o = 0.5$. For DS2, $P_a = 0.95, W_o = 0.5$

(b) Expected and worst case availability of DS2

(c) Expected and worst case replication of DS2

Figure 4: Availability and replication of MobiTribe vs limit of hosting

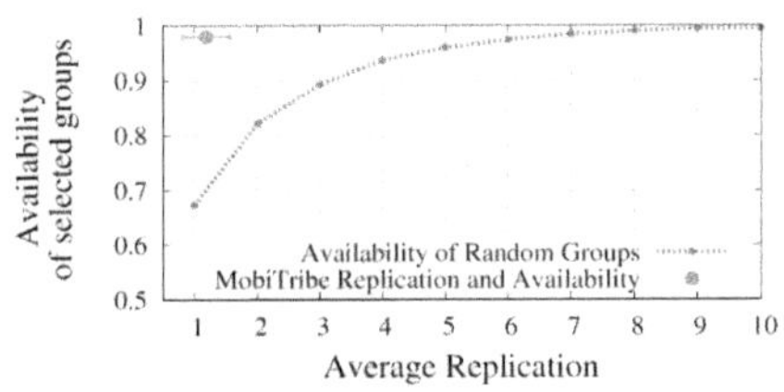

Figure 5: Availability of randomly selected groups

users in three different cites at three different times. Thus, the combination of DS1 and DS2 contains a wide range of WiFi connectivity patterns of real mobile users which can be considered as large enough and diverse enough for the evaluation of the proposed DEVICE GROUPING algorithm.

6.2 Availability, Replication and Fairness

Availability in MobiTribe is the probability of low-cost network availability (P_a) of content hosted in a selected tribe. We performed DEVICE GROUPING over a seven day training period[1], calculated P_a for all selected tribes and averaged it to obtain the availability of MobiTribe for the entire system. This procedure was repeated over the entire trace data period to obtain average system evaluation metrics. It was observed that the fairness constraint described in Section 4, limit of hosting, i.e. the maximum number of tribes that a single device belongs to, affects the efficiency and performance of grouping. The main reason for this is that the devices with high availability would be preferred by many others to host content. Hence, we measured the availability of MobiTribe against the limit of hosting.

Figure 4a shows the availability variation of DS1 and DS2 with respect to limit of hosting. Note that both data sets give similar availability behaviour irrespective of their different connectivity characteristics. This indicates that the DEVICE GROUPING algorithm have the potential to adjust to different types of mobile environments. Figure 4b depicts that in the expected case, we can achieve nearly 100% availability irrespective of the limit of hosting. In the worst case, limit of hosting has to be more than 4 for the best performance. Hence, the availability of tribes will fall on to the shaded area in Figure 4b. In the remainder of this paper, only DS2 is used since it has a larger device pool and the worst case analysis is used in every case to benchmark the performance of MobiTribe with alternative architectures.

Replication of the system is the average number of replicas per content in addition to the original content (R_a). When the limit of hosting is lower, some devices may not be able to find a tribe, which lowers the average replication as shown in Figure 4c. When the limit of hosting is higher, devices are allowed to select a single device with high availability instead of several devices with low availability, which again lowers the average replication. The maximum average replication required to satisfy the availability requirement of $P_a = 0.95$ is $R_a = 1.19$ with standard deviation of 0.39. That is, in this particular social network, each user only needs to group with one or two other users to increase the low-cost network availability of its own content.

This analysis shows that a very high low-cost network availability of content (nearly 100%), can be achieved with a very low number of replicas (approximately 2), if the devices are carefully grouped into tribes. Although the result is subjective to these trace data sets, we argue that the DEVICE GROUPING algorithm has the capability to group devices appropriately when there is such diversity in the WiFi connectivity patterns. This minimises the overhead of replication in terms of network bandwidth, storage and battery consumption. Earlier proposed systems, such as Safebook [5] or MyZone [13], do not select devices for replication based on the connectivity patterns. Instead, they use random replication, which requires a higher number of replicas than MobiTribe to reach a similar level of availability. We performed a random group selection in order to compare the performance with MobiTribe strategy as shown in Figure 5. Random selection requires 7 replicas to reach the same availability level that is provided by MobiTribe with 1.19 replicas. Hence, the random replication would not be suitable for mobile environment, since it consumes more scarce resources in mobile networks. For content sharing through distributed architectures such as Diaspora [9] which do not replicate the content, the availability would depend on whether the user's device has low-cost network availability at the time when another user wants to access the content. In the case of such systems if the device has no network access or is switched off, the content will not be accessible at all.

6.3 Delay Tolerance

After a device has registered its content with the CMS, it has to wait until one of the devices in the tribe has overlapping low-cost network availability to pre-distribute the content, namely the pre-distribution delay. As described in

[1]The effect of training period selection is analysed in [19] with a heuristic device grouping algorithm.

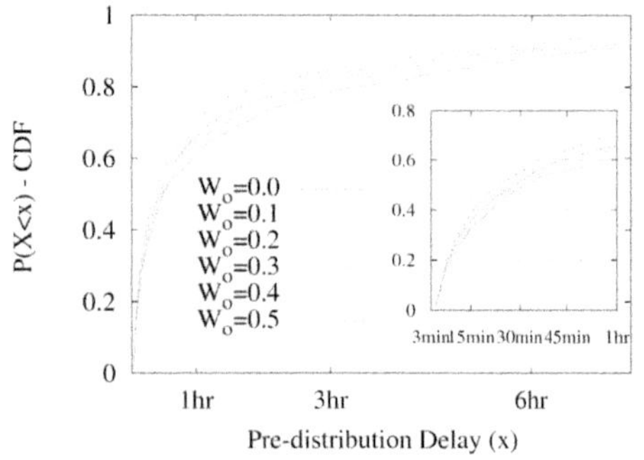

Figure 6: The CDF of the pre-distribution delay

Table 2: Network Usage Models for MobiTribe. The probability of successful pre-distribution (M_a), the average replication per content (R_a), the probability of availability of a selected tribe (P_a), the number of downloads (N), and the probability of WLAN availability of a device (W_a), size of the content (F_s)

MobiTribe WLAN Usage Model	
Uplink	$[M_aR_a + (1 - M_a)W_aN + M_aP_aN]\,F_s$
Downlink	$[M_aR_a + W_aN]\,F_s$
MobiTribe 3G Usage Model	
Uplink	$[M_a(1 - P_a) + (1 - M_a)(1 - W_a)]\,NF_s$
Downlink	$(1 - W_a)NF_s$

Section 3.1, if the content is accessed before this time, it will be delivered from the originating device over the available lowest cost network. Hence, this pre-distribution delay is critical for the performance of MobiTribe. We calculated the pre-distribution delay of selected tribes repeatedly for seven days. The CDF of the delay for different overlapping factors are shown in Figure 6. As expected, the higher the overlapping factor W_o the lower the delay. However, the effect of W_o seems not very significant compared to the added complexity to DEVICE GROUPING. The probability of the delay being less than 1 hour is over 0.6 irrespective of W_o. There is 90% probability for the pre-distribution delay being less than 6 hours. In [20], it has been observed that 55% of Flicker content is uploaded after a lag of more than one day. Hence, we believe that the results of more than 6 hours delay tolerance would be practically significant.

6.4 Time Complexity of Device Grouping

We have noticed that we can cover the majority of devices by using bipartite b-matching algorithm for all three data sets. Hence, the time complexity of the DEVICE GROUPING is dominated by the bipartite b-matching algorithm. Since bipartite b-matching is directly reduced to the maximum flow, the complexity is bounded by the maximum flow computation. There are many fine tuned algorithms available for maximum flow computation in the literature [7]. In this evaluation, there were 2905 sets which satisfied the $P_{threshold} = 0.95$ for 84 devices in DS2. As per our notation in Section 4, $|V| = 84$ and $|H| = 2905$. In general, if a system has n devices, there can be a maximum of $\frac{n(n-1)}{2}$ number of pairs or edges in the hypergraph (V, H) for the case of $\Delta = 2$. Then, the bipartite flow network (A, B, E) consists of $|A \cup B| = 2|V| + 2 = 2n + 2 \approx 2n$ vertices and $|E| = 2(\frac{n(n-1)}{2})$ edges. Hence, if we use Goldberg's algorithm [7], the upper bound for the bipartitie b-matching would be $O\left(|A \cup B||E| \log \frac{|A \cup B|^2}{|E|}\right) \approx O(n^3)$ in general. This can be considered as a practical time complexity for our initial NP-Hard DEVICE GROUPING problem.

6.5 3G Bandwidth Saving Comparison

Cellular network capacity is typically scarce and minimizing its usage is one of the key goals of MobiTribe. To evaluate its efficiency in this regard, we benchmark the performance of MobiTribe with a centralised server architecture, which is currently the common method of content sharing in social networks such as Facebook. Further, we compare the performance of MobiTribe with a simple Diaspora [9] like non-mobile-optimised distributed architecture.

MobiTribe: The network usage models for MobiTribe is presented in Table 2. From previous analysis, we observed that $P_a = 0.98$, $R_a = 1.19$ and $W_a = 0.43$ for DS2 in the worst case. Since M_a is directly related to the delay tolerance as in Figure 6, we use the delay to represent M_a. We used a simple assumption that all downloads occur exactly after the delay and all consumers are selfish, i.e. the consumers will not share the downloaded content with others.

Centralised server: The content will be uploaded to a centralised server at the time of sharing and all consumers access the content through the server. From the same set of parameters used in the MobiTribe model, we can derive the 3G bandwidth usage of the centralised server system as $C^{3g}_{server} = (1 + N)(1 - W_a)F_s$.

Mobile server: Diaspora [9] allows users to host data on their own servers. The content will not be uploaded to a centralised server at the time of sharing. Instead, the content consumers access the shared content via creator's own server. Then, the 3G bandwidth consumption model would be $M^{3g}_{server} = 2(1 - W_a)NF_s$.

In all architectures, we consider that on-the-spot data offloading via WLAN is possible since majority of the modern smartphones are equipped with WiFi prioritization over cellular network. The high replication overheads of the non-mobile-optimised replicating systems such as SafeBook [5] and MyZone [13], would almost certainly lead to higher bandwidth consumption. This is due to the large number of replicas that would be needed to provide consistent availability of the content via low-cost networks when using a replication algorithm that is not aware of network availability patterns. Therefore, we do not compare MobiTribe against those architectures in this comparison.

Figure 7a shows that the normalised 3G bandwidth usage of MobiTribe decreases with content access delay because a higher delay allows more time to pre-distribute the content to the tribe via low-cost network connections. Note that we considered worst case content popularity model in terms of bandwidth usage, i.e. everyone in the group accessing all generated content. Even in the worst case situation, MobiTribe performs better than the mobile server system irrespective of the delay tolerance. When the delay is larger than 6 hours, i.e. $M_a \simeq 1$, MobiTribe 3G bandwidth consumption is almost equal to the centralised server system, whereas the mobile server system consumes twice as much bandwidth. To this end, in this particular trace environment, MobiTribe justifies its key design goal of cellular bandwidth saving, and stands out as a mobile opti-

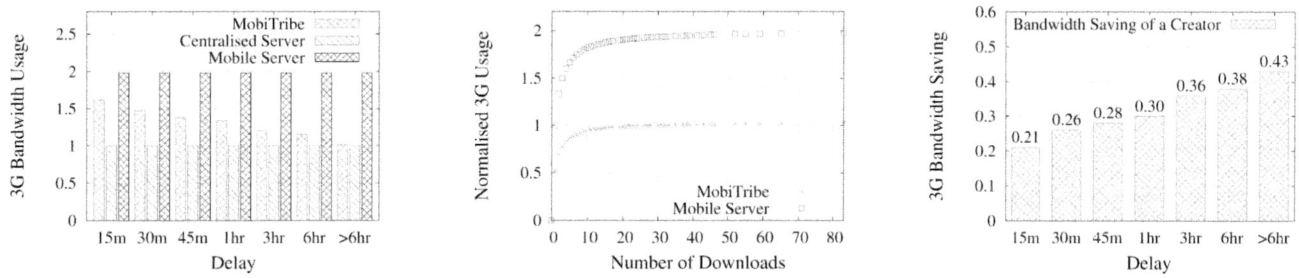

(a) 3G bandwidth usage normalised by centralised server architecture. N=83-worst case content popularity

(b) Normalised 3G bandwidth usage against number of downloads. $N \sim$Pareto-(80-20 Rule). $M_a \simeq 1$

(c) 3G bandwidth saving of a content creator. $N \sim$Pareto-(80-20 Rule)

Figure 7: Comparison of 3G bandwidth usage of MobiTribe with other alternative architectures

mised distributed content sharing platform for online social networking from its other contemporary architectures.

However, user generated content popularity has a "long-tail" distribution as observed in the literature [4], i.e. the majority of the content is accessed by a very few consumers. Figure 7b shows normalised 3G bandwidth usage with a long-tail content popularity model, where N is distributed as a Pareto variable. MobiTribe consumes less bandwidth than the centralised server system in the majority of the instances, whereas the mobile server system performs better only when $N = 0$. To this end, MobiTribe consumes even less 3G bandwidth than a centralised server system in real situations by providing monetary cost benefits for the user.

Usually, in distributed content sharing the content is hosted by the creator or a group of users including the creator, which could possibly lead the content creators to consume more bandwidth than others. Therefore, we compare the 3G bandwidth consumption of content creators in MobiTribe and mobile server architectures. Figure 7c shows that a MobiTribe content creator can achieve 21%-43% of bandwidth saving compared to a Diaspora like content creator. Since more than 6 hours access delay could be a realistic assumption to be made [20], savings would be more than 43%. This would incentivise mobile users to use MobiTribe for distributed social networking, who would otherwise have been reluctant to do so because of the high monetary cost involved in the distributed architectures.

In all three architectures, 3G downlink usage $(1-W_a)NF_s$ is similar because it only depends on network availability of the consumers at the time of downloading. In future work, we intend to extend MobiTribe to replicate content where it is likely to be consumed. However, this approach requires modelling of user consumption and was left out of the scope of this paper. In addition, we are planning on extending MobiTribe to pre-distribute content created out side the MobiTribe system. The algorithm presented in this paper could be directly applied to this extension as well. This would allow us to further reduce the network utilization by a significant amount from downlink bandwidth usage.

6.6 Energy Saving Comparison

We analyse MobiTribe energy consumption to compare it with alternative architectures. The energy consumption of WLAN, E_w, is significantly lower than of the 3G networks, E_{3g}, due to the heavy traffic load in cellular networks leading to lower speeds and therefore to longer durations of data transfers as observed by [2, 15]. The experimental study conducted by Petander [15] shows that E_{3g}/E_w can be varied up to 100 based on the network status at the time of transfer. Hence, a comprehensive energy evaluation would only be possible by an experimental implementation of MobiTribe. At this stage, we evaluate the energy consumption of MobiTribe and other alternatives with respect to the energy ratio E_{3g}/E_w and network usage models described in Section 6.5. We consider that the energy ratio is calculated after considering all the factors which affect energy consumption such as WLAN association energy, signal strength and data rate.

Figure 8a shows the energy consumption of each system, normalised by the energy consumption of a centralised server system for the worst case content popularity model. A mobile server system always consumes approximately 2 times more energy than a centralised server system, whereas MobiTribe energy consumption exponentially decays with E_{3g}/E_w and consumes almost the same energy as the centralised server system for higher ratios, even with the worst case content popularity model. The high overheads of the non-mobile-optimised replicating systems, such as SafeBook [5] and MyZone [13], would lead to higher energy consumption. To this end, MobiTribe does not consume extra energy, unlike other distributed systems in general.

Similar to 3G bandwidth comparison, we evaluated the two systems against realistic content popularity distribution. Figure 8b shows that MobiTribe consumes less energy when the number of downloads is smaller, even for $E_{3g}/E_w = 10$. Although, MobiTirbe does not consume any 3G bandwidth when $N = 0$, the creator still uploads the created content to the tribe via low-cost networks. Thus, energy consumption of MobiTribe is 0.4 times the centralised server system when $N = 0$, while mobile server based creators consume zero energy at the same time. However, even with this extra energy consumption associated with pre-distribution, a MobiTribe content creator will be able to gain 18%-41% energy saving over a mobile server creator as shown in Figure 8c. To make it more realistic, the average saving is obtained by varying E_{3g}/E_w from 1 to 100 and N as Pareto distribution.

The energy consumption of a MobiTribe creator does not improve with the number of downloads since we used the simple assumption of selfish consumers. However, in reality the performance would keep on improving with the number of downloads spanning a longer period of time because there can be others who wish to share the content. Moreover, the proposed extension with predictive pre-fetching would reduce the energy consumption of consumers.

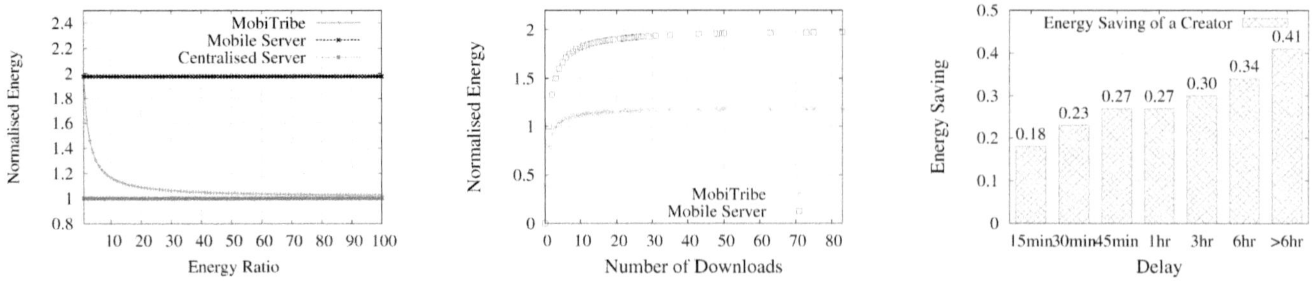

(a) Energy consumption normalised by centralised server architecture. N=83-worst case content popularity

(b) Normalised energy consumption against number of downloads. $N \sim$Pareto, $M_a \simeq 1$, $E_{3g}/E_w = 10$

(c) Energy saving of a content creator. $N \sim$Pareto-(80-20 Rule) and $E_{3g}/E_w = (1:100)$

Figure 8: Comparison of energy consumption of MobiTribe with other alternative architectures

7. CONCLUSION

Data ownership and privacy problems inherent in centralised approaches could be solved by using distributed social networking services. Smart phones could contribute to these services by hosting and sharing distributed social network data, if the associated high communication costs and battery usage were mitigated. We propose a novel mechanism to address these challenges by taking advantage of network availability patterns among groups of users to provide continuous availability of content over low-cost network connections. We show that the content replication problem in distributed peer-to-peer architectures such as MobiTribe is NP-Hard, but in case of our test data sets, it can be reduced to a P-hard problem. A new algorithm based on a combination of a bipartite b-matching and a greedy heuristic is developed, which minimises content replication and maximises content availability whilst ensuring fairness. The performance of the algorithm was evaluated using three empirical data sets and the results show that a persistent low-cost network availability of content can be achieved with only two replicas. In addition, the analysis of the system showed that a content creator can save up to 43% of 3G bandwidth and up to 41% of device energy on average compared to non-mobile-optimised distributed social networking approaches.

8. REFERENCES

[1] Crawdad:. In *http://crawdad.cs.dartmouth.edu.*

[2] N. Balasubramanian, A. Balasubramanian, and A. Venkataramani. Energy consumption in mobile phones: a measurement study and implications for network applications. In *Proc. of the 9th ACM SIGCOMM IMC '09*, pages 280–293, Chicago, 2009.

[3] G. Calinescu, C. Chekuri, M. Pál, and J. Vondrák. Maximizing a monotone submodular function subject to a matroid constraint. *SIAM J. Comput*, 2009.

[4] M. Cha, H. Kwak, P. Rodriguez, Y. Ahn, and S. Moon. Analyzing the video popularity characteristics of large-scale user generated content systems. *IEEE/ACM Transactions on Networking (TON)*, 17(5):1357–1370, 2009.

[5] L. Cutillo, R. Molva, and T. Strufe. Safebook: A privacy-preserving online social network leveraging on real-life trust. *Communications Magazine, IEEE*, 47(12):94–101, 2009.

[6] H. N. Gabow. An efficient reduction technique for degree-constrained subgraph and bidirected network flow problems. In *Proc. of the 15th ACM Symposium on Theory of Computing*, pages 448–456, 1983.

[7] A. V. Goldberg and R. E. Tarjan. A new approach to the maximum-flow problem. *J. ACM*, 35:921–940, 1988.

[8] B. Han, P. Hui, V. Kumar, M. Marathe, J. Shao, and A. Srinivasan. Mobile data offloading through opportunistic communications and social participation. *Mobile Computing, IEEE Transactions on*, (99):1–1, 2011.

[9] http://joindiaspora.org.

[10] R. M. Karp. Reducibility among combinatorial problems. In *Complexity of Computer Computations*, pages 85–103. Plenum Press, 1972.

[11] K. Lee, I. Rhee, J. Lee, S. Chong, and Y. Yi. Mobile data offloading: how much can wifi deliver? In *Proc. of the Co-NEXT '10*, pages 1–12, Philadelphia, 2010.

[12] P. Lungaro, Z. Segall, and J. Zander. Activecast-a network and user aware mobile content delivery system. In *Proc. of ICUFN*, pages 309–313, June 2010.

[13] A. Mahdian, J. Black, R. Han, and S. Mishra. Myzone: A next-generation online social network. In *Tech Report: Dpt. of Computer Science, Uni. of Colorado, Boulder*, 2011.

[14] J. Mestre. Greedy in approximation algorithms. In *Proceedings of the 14th Annual European Symposium on Algorithms*, pages 528–539, 2006.

[15] H. Petander. Energy-aware network selection using traffic estimation. In *Proc. of the 1st ACM workshop MICNET '09*, pages 55–60, Beijing, 2009.

[16] J. Pouwelse, P. Garbacki, J. Wang, A. Bakker, J. Yang, A. Iosup, D. Epema, M. Reinders, M. Van Steen, and H. Sips. Tribler: a social-based peer-to-peer system. *Concurrency and Computation: Practice and Experience*, 20(2):127–138, 2008.

[17] S. Seong, J. Seo, M. Nasielski, D. Sengupta, S. Hangal, S. Teh, R. Chu, B. Dodson, and M. Lam. Prpl: a decentralized social networking infrastructure. In *Proc. of the ACM Workshop on MCS: Social Networks and Beyond*, page 8. ACM, 2010.

[18] P. Stuedi, I. Mohomed, M. Balakrishnan, Z. Mao, V. Ramasubramanian, D. Terry, and T. Wobber. Contrail: Enabling decentralized social networks on smartphones. *Middleware 2011*, pages 41–60, 2011.

[19] K. Thilakarathna, H. Petander, and A. Seneviratne. Performance of content replication in mobitribe:a distributed architecture for mobile ugc sharing. In *Proc. of IEEE LCN*, pages 558–566, oct 2011.

[20] I. Trestian, S. Ranjan, A. Kuzmanovic, and A. Nucci. Taming user-generated-content in mobile networks via drop zones. In *INFOCOM'11*, pages 2840–2848, 2011.

SAGA: Socially- And Geography-Aware Mobility Modeling Framework

Bruno A. A. Nunes
Department of Computer Engineering
University of California, Santa Cruz, CA, USA
bnunes@soe.ucsc.edu

Katia Obraczka
Department of Computer Engineering
University of California, Santa Cruz, CA, USA
katia@soe.ucsc.edu

Abel Rodriguez
Department of Applied Math
University of California, Santa Cruz, CA, USA
abel@soe.ucsc.edu

ABSTRACT

In this paper, we introduce a user mobility modeling framework that accounts for both the users' social structure as well as the geographic diversity of the region of interest. SAGA, or Socially– and Geography-Aware mobility model, captures social features through the use of *communities* which cluster users with similar features such as average time in a cell, average speed, and pause time. SAGA accounts for geographic diversity by considering that different communities exhibit different interests for different locales; therefore, different communities are attracted to certain physical locations with different intensities. Besides introducing SAGA, the contributions of this work include: a model calibration approach based on formal statistical procedures to extract social structures and geographical diversity from real traces and set SAGA's parameters; and validation of SAGA by applying it to real mobility traces. Our experimental results show that, when compared to existing mobility regimes such as Random-Waypoint and Preferential-Attachment based mobility, SAGA is able to preserve the desired non-uniform node spatial density present in real user mobility, creating and maintaining clusters and accounting for differential node popularity and transitivity.

Categories and Subject Descriptors

I.6 [**SIMULATION AND MODELING**]: Model Validation and Analysis

Keywords

Socially Inspired Mobility Models, Realistic Mobility, Node Density Distribution

1. INTRODUCTION

Mobility models are an indispensable tool in the design and evaluation of wireless networks and their protocols. Therefore, defining realistic mobility models is critical when simulating applications and protocols for mobile networked environments. This is the case of infrastructure-based networks (e.g., wireless LANs or WLANs), but even more so in the case of infrastructure-less networks, a.k.a., wireless, self-organizing networks (WSONs), which include wireless mobile ad-hoc networks (MANETs), wireless sensor networks (WSNs), and disruption-tolerant networks (DTNs). In these networks, unlike their infrastructure-based counterparts where only the end user node is mobile, every node may have unlimited mobility. Indeed, most network simulators include mobility generators, which, following a pre-specified mobility regime, determine the position of network nodes over time during simulation runs. These synthetic mobility generators have been extensively used in the study of wireless networks [1].

An example of a widely used synthetic mobility model is the Random Waypoint (RWP) mobility regime. However, RWP mobility has been known to exhibit some undesirable properties. For example, Bettstetter et al. [2] show that random mobility leads to "homogeneous" steady-state spatial node distribution. This means that if there were any spatial density heterogeneity, or node clustering, in the initial spatial distribution of the nodes, random mobility does not preserve it.

To address the problem of unrealistic mobility regimes, a number of models that are based on real traces and mobility patterns have been proposed; notable examples include [3, 4, 5, 6, 7]. Additionally, several efforts have focused on the "scale-free" properties observed in many real networks such as the Internet, the Web, and some social networks, to name a few. In particular, the seminal work of Barabási and Albert [8] proposes a model that generates scale-free networks, i.e., networks whose node degrees follow a power law distribution. One key concept underpinning the Barabási-Albert model is the *preferential attachment* principle which states that "the more connected a node is, the more likely it is to receive new links". Several recently proposed mobility models (e.g., [9, 10, 11, 12, 13, 14]) try to mimic real human mobility by following the preferential attachment principle. This is done by defining *attraction points*, whose probabilities of attracting other nodes increase as more nodes congregate around them.

In our previous work, we have shown that using the preferential attachment principle to model human mobility leads to undesirable long-term behavior [15]. More specifically, preferential attachment based mobility regimes do not preserve the original spatial node density distribution and lead to steady-state behavior similar to random mobility as exemplified by the Random Waypoint model. Instead, real

MSWiM'12, October 21–25, 2012, Paphos, Cyprus.

human mobility exhibits persistent density heterogeneity as shown in [15].

More recently, alternate approaches to developing realistic mobility regimes rely on social interactions between nodes to drive the selection of movement direction and destination [16, 17, 18, 19, 20, 21, 13]. These models usually construct a social network either using outside information unrelated to the mobility trace, or on the basis of geographical collocation. As will become clear in the remainder of this paper, we take a completely different approach to modeling human mobility; more specifically, our proposed framework accounts for the social structure in the data by identifying user *communities*. We thus refer to our models as "socially-inspired". We then use of the preferential attachment principles, "attached" to the built community structure to model the interaction of nodes within– and between communities. By doing so, SAGA addresses the undesirable behavior exhibited by mobility models that rely solely on pure preferential attachment.

The mobility modeling framework proposed here is also "geography-aware" as it accounts for the heterogeneity of the different locales in the region of interest. These two features allow our model to account for features observed in real life networks such as differential popularity, *transitivity* (i.e., two individuals that have a tie to a third one are more likely to be tied together [22]) and *clustering* (i.e., the tendency of individuals to form communities beyond what can be explained by transitivity [23]). Moreover, to calibrate our model, we apply formal statistical learning techniques to extract, from real mobility traces, both the nodes' social structure, i.e., the communities to which they belong, as well as the region's geographical diversity.

Overall, the contributions of our work are four-fold: (1) we introduce a novel mobility modeling framework that is both socially-inspired and geography-aware, (2) we propose an efficient method to extract social structures and geographical diversity from real traces to calibrate and set the model's parameters based on formal statistical procedures, (3) we validate our work by applying it to real mobility scenarios and comparing our results against real measurements, and (4) we show that our model is able to preserve the desired heterogeneous node spatial density, creating and maintaining clusters and accounting for differential node popularity and transitivity. Thus, a by-product of our work is to solve the problem exhibited by mobility models that rely solely on pure preferential attachment.

The remainder of this paper is organized as follows. We present our mobility modeling framework in detail in Section 2. In Section 3 the real GPS mobility traces used to calibrate and validate our approach are presented. Section 4 describes how we calibrate our model parameters while in Section 5 presents the validation of our work towards modeling real human mobility. Section 6 places our work in perspective by presenting related work in mobility modeling and characterization. Finally, Section 7 concludes the paper.

2. SOCIALLY- AND GEOGRAPHY-AWARE (SAGA) MOBILITY MODEL

Our model, SAGA for Socially- And Geography-Aware mobility model, captures both social interactions among individuals, as well as the geographical diversity of the region of interest. At the social level, our model captures *transitivity* in the relationship among subjects and could easily be adapted to capture *homophily* by attributes, i.e., the tendency of individuals with similar or completely opposite features to have a higher probability of being linked [24]. As described in detail below, SAGA captures these features through the use of *communities* or *clusters*. SAGA also accounts for geographical diversity: it considers that different node communities do not have the same interests and thus are "attracted" to certain locales more than others. For example, in a University campus, students are attracted to the library, classrooms, or cafeteria in a different way than faculty or staff are. Similarly, it is often the case that students and faculty spend time together in the same physical location (e.g., classrooms). However, it is not as common for staff members and students to be collocated the same way as faculty and students tend to be.

We now proceed to provide a more detailed description of SAGA. We use a *waypoint*-based mobility regime in which: (1) a node chooses its next destination following some given probability distribution, then (2) moves to that destination at a randomly chosen speed uniformly distributed between $[V_{min}, V_{max}]$, (3) pauses for some time, and (4) repeats from step (1).

In our model, the area of interest is divided in equally sized squares, or *cells*. Every mobile node independently chooses as their next destination a cell $i \in \{1, \ldots, I\}$, where I is the total number of cells in the simulation area according to $p_{i,k,t}$, the probability that destination i is chosen by a node belonging to community $k \in \{1, \ldots, K\}$ at time t, where K is the total number of communities in the system. Here $p_{i,k,t}$ is constructed on the basis of the intensity $\mu_{i,k,t} \in \mathbb{R}$, where $\mu_{i,k,t} > 0$, which represents the attractiveness of destination i to community k at time t. In other words, $\mu_{i,k,t}$ captures geographic diversity in terms of the different levels of interests each community has in particular regions at different points in time, along with the preferences of individuals to being in the same locations with members of the same or other communities.

We employ a set of community indicators $\xi_1, \ldots, \xi_n, \ldots, \xi_N$ for each user $n \in \{1, \ldots, N\}$ at all points in time, where N is the total number of users in the system. Community indicator ξ_n indicates which community user n belongs to. Also, $\Pr(\xi_n = k) = w_k$ is the probability that user n belongs to community k, where $\sum_{k=1}^{K} w_k = 1$ and K is the total number of communities that exist in the user population. In other words, w_k defines the portion of the population that belongs to community k. We can now write $p_{i,k,t}$, the probability that a node n, belonging to community k, chooses cell i as its next destination at time t as:

$$p_{i,k,t} = \frac{\mu_{i,k,t}}{\sum_{j=1}^{I} \mu_{j,k,t}} \quad \text{for all} \quad i, j \in \{1, \ldots, I\}. \tag{1}$$

where the intensity $\mu_{i,k,t}$ of destination i towards nodes belonging to community k at time t is defined as:

$$\mu_{i,k,t} = \exp\left\{ c_{i,k} + \sum_{l=1}^{K} \theta_{k,l} \log(1 + a_{i,l,t-1}) \right\} \tag{2}$$

In Equation 2, $c_{i,k} \in \mathbb{R}$ is a constant that reflects the *intrinsic attraction* that cell i exerts on members of community k, independent of the number of other nodes present in the cell. $\theta_{k,l} \in \mathbb{R}$ expresses the *affinity* between nodes in community k and l. Positive values $\theta_{k,l}$ imply that nodes belonging to community k are attracted to nodes belonging to community l (and vice-versa, since we make $\theta_{k,l} = \theta_{l,k}$). On the other hand, negative values imply that nodes belonging to community k are repelled by those in community l. Both parameters $c_{i,k}$ and $\theta_{k,l}$ can be extracted from real live traces using the statistical methods we present in Section 4.

The quantity $a_{i,l,t-1}$ is the total number of nodes in cell i that belong to community l at time $t-1$ and can be written as:

$$a_{i,l,t-1} = \sum_{m=1}^{N} \mathbb{1}_{\{m \in \mathbb{N}_{i,t-1}\}}, \tag{3}$$

where $\mathbb{N}_{i,t-1}$ is the set of nodes currently present or going towards a location inside cell i at time $t-1$ and $\mathbb{1}_{\{m \in \mathbb{N}_{i,t}\}}$ is an indicator function equal to 1 if node m belongs to the set $\mathbb{N}_{i,t}$ ($m \in \mathbb{N}_{i,t}$) or 0 otherwise.

Note that we express $\mu_{i,k,t}$ as an exponential in (2) rather than the more traditional linear structure for the intensity. We do it for two reasons. Firstly, the exponential transformation automatically ensures that intensities are positives without any need to impose constrains in the space of parameters, and allows for both assortative and dissortative patterns of behavior. Secondly, the representation agrees with the structure of a multivariate logistic regression model (see Section 4.2), which simplifies the estimation of parameters from the data. Table 1 summarizes SAGA's parameters.

Param.	Description
N,I, K	Numb. of nodes, cells and communities respectively.
$a_{i,l,t-1}$	Numb. of nodes in cell i belonging to community l at time $t-1$.
$\theta_{k,l}$	Affinity between nodes from communities k and l.
$c_{i,k}$	Intrinsic attraction of node n towards cell i.
$\mathbb{N}_{i,t-1}$	Set of nodes present or going to cell i at time $t-1$.
$p_{i,k,t}$	Prob. that node n, from community k, chooses cell i at time t.
$\mu_{i,k,t}$	Intensity of nodes from community k towards cell i at time t.

Table 1: Summary of SAGA parameters: The methodology used to extract these parameters from real traces is described in Sections 4.1 and 4.2

Collecting the affinity parameters into a symmetric affinity matrix $\Theta = [\theta_{k,l}]$, as illustrated by our case study in Table 4, facilitates identifying different association patterns, such as: (1) total assortative behavior, when nodes only interact with other nodes from the same community, (2) total dissortative behavior, where nodes only interact with other nodes from different communities, or (3) an arbitrary association pattern.

3. REAL TRACES

Table 2 presents a summary of the GPS traces we use in our analysis. *Quinta*, refers to the "Quinta da Boa Vista Park" trace, first presented in [25]. It is a GPS trace collected at a park in the city of Rio de Janeiro, Brazil. The park has many trees, lakes, caves, and trails. It houses the National Museum of Natural History and the city Zoo. The *Statefair* trace [26] corresponds to daily GPS track logs collected at the NC State Fair held in North Carolina, USA.

Given that the traces of some of the nodes presented some discontinuity, we selected sections of the raw traces where no discontinuity occurred. We selected only nodes where we could find a continuous sequence of GPS fixes that were as long as 900 and 8000 seconds for the Quinta and Statefair traces, respectively. These were the total durations of the traces.

The information in table 2 reflects the speed and pause time measured from the traces and averaged over all nodes, at every sample, during the duration of the trace. The last line in the table show the sizes of the communities identified in the traces, after applying our community identification method. This method is described in the following section.

Parameter	Quinta	Statefair
Avg. Speed ($\pm\sigma$)(m/s)	1.27($\pm$0.16)	0.58($\pm$0.12)
Avg. Pause ($\pm\sigma$)(sec)	4.61($\pm$1.60)	7.20($\pm$6.18)
Area (meters x meters)	840x840	1260x1260
Duration (sec)	900	8000
Sampling frequency (sec)	1	30
# nodes	98	19
# communities (size in nodes)	5 (29,12,17,27,13)	2 (10,9)

Table 2: Summary of the GPS traces we use to calibrate and validate SAGA.

4. MODEL CALIBRATION

One of the contributions of the SAGA mobility modeling framework is its calibration from real mobility traces using formal, yet simple, statistical learning techniques. In this section we describe our model calibration approach in detail.

4.1 Community Identification

Communities of nodes, which are represented by the indicators $\xi_1, \xi_2, \ldots, \xi_N$, are identified using the *complete-linkage agglomerative hierarchical clustering* technique (see for example [27]). Each node is characterized through a vector of features that includes *average time spent* in each cell, *average node speed*, and *pause time*. We are then able to build an observation matrix $M_{N\times(I+2)}$ with the mobility features extracted from the trace. This matrix has N rows (one row per node) and $I+2$ columns (one column per cell, plus the last two columns for the average speed and pause time measured).

Since we measure dissimilarities among nodes using Euclidean distances, we work with the *logit* transformation of the average time spent on each cell. In addition, to deal with issues associated with zero counts for some cells, we add a single pseudo count to each cell before applying any transformation. The number of clusters is selected using the *Bayesian Information Criteria* (BIC), as discussed in [28]. The BIC associated with a vector of clustering indicators $\xi = (\xi_1, \xi_2, \ldots, \xi_N)$ is computed as

$$BIC(\xi) = loglik(\xi) - \frac{1}{2}K\log(N)$$

where $loglik(\xi)$ is the logarithm of the likelihood associated with a Gaussian finite mixture model where observations are clustered according to ξ. K represents the number of distinct communities (or clusters) in ξ, and N is the total number of nodes. BIC is computed for different models and number K of clusters. Finally, the number K of communities associated with the largest BIC is selected. All computations were carried out using the package `mclust` in the statistical software `R` [29]; sample code for our model calibration approach is available at [30].

4.2 Extracting Model Parameters from Traces

Once communities are identified using the approach described in the previous section, a multivariate logit model [31] is used to estimate the geographical preference parameters $c_{i,k}$ and the affinity matrix $\Theta = [\theta_{k,l}]$. More specifically, let $y_{n,t} = i$ if node n is in cell i at time t. A multivariate logit model assumes that

$$\log\left\{\frac{\Pr(y_{n,t}=i)}{\Pr(y_{n,t}=1)}\right\} = c_{i,\xi_n} + \sum_{l=1}^{K} \theta_{\xi_n,l}\log(1+a_{i,l,t-1})$$

or, equivalently,

$$\Pr(y_{n,t}=i) \propto \exp\left\{c_{i,\xi_n} + \sum_{l=1}^{K} \theta_{\xi_n,l}\log(1+a_{i,l,t-1})\right\},$$

where, for identifiability, and without any loss of generality, we set $c_{1,k} = 0$. Hence, the logit model agrees with the structure of the intensity in (2).

Computations were carried out using the package `mlogit` in the statistical software **R**. This package uses a Fisher scoring algorithm, which is a variant of the Newton method, to find the maximum likelihood estimates for the parameters of the model along with their standard errors. As before, sample code is available at [30]. To reduce the computational burden associated with fitting these models, we exploit the conditional independence assumption underlying this model and subsample some of the traces. This subsampling has minimal effect on the results as discussed in more detail in Section 5.

We constructed our model using the logarithm of the number of nodes on each cell to measure preferential attachment. This is at odds with the choices made by previous authors, who have preferred the use of linear terms. Our choice was motivated by a formal model comparison among the class of models based on the Bayesian information criteria (BIC). More generally, it should be possible to estimate the functional form of the preferential attachment using nonparametric techniques. This avenue will be explored elsewhere.

Selection of cell sizes is another important issue for our model. In our examples we have selected the sizes in a mostly ad hoc manner. However, since our model implicitly assumes that the mobility model follows a non-homogeneous Poisson process, the problem of selecting the cell sizes is similar to the problem of selecting bandwidths in kernel density estimate problem, which can be used through the use of generalized cross-validation procedures.

5. VALIDATION AND EXPERIMENTAL RESULTS

In this section we validate SAGA in two case studies corresponding to the traces described in the previous section, namely the Quinta and Statefair traces. We show how SAGA manages to capture the community structure and geography diversity of the traces. As baseline for our comparison, we also use two other mobility regimes, namely Random Waypoint (RWP) [1] and Natural [11]. We start by briefly describing the RWP and Natural mobility; then we present the setup we used to generate mobility traces under these regimes using the Scengen [32] simulation tool.

5.1 Baseline Mobility Regimes

5.1.1 RWP Mobility

Random Waypoint (RWP) mobility has been widely used in the study of multi-hop ad-hoc wireless networks (MANETs). Under RWP mobility, mobile nodes are initially placed in the area being simulated according to a given distribution (typically, a uniform distribution). Each node remains in its position for a given period of time, called *pause time* P uniformly chosen in the interval $[0, P_{max}]$, where P_{max} is a pre-specified parameter. After this period, the mobile node chooses a new destination uniformly distributed in the simulation area, and a velocity, also uniformly distributed in the interval $[v_{min}, v_{max}]$, where both v_{min} and v_{max} are pre-specified parameters. Once the destination is reached, the node pauses again and chooses another destination and velocity, as described above.

5.1.2 The Natural Mobility Regime

We also compare our results against mobility regimes that follow the *preferential attachment* principle. As representative of this family of mobility regimes, we use the Natural mobility model, or simply *Natural* [11].

As discussed in Section 6, *Natural* is based on attraction points, where the attractiveness of each point is proportional to the *attractor*'s popularity given by the number of nodes at or going towards it and inversely proportional to the distance to it. Thus, the probability $\Pi(a_i)$ that a node z_k chooses an attractor a_i among all possible attractors is proportional to the portion of the total attractiveness it carries: $\Pi(a_i) = \frac{\mathcal{A}_{a_i,z_k}}{\sum_j \mathcal{A}_{a_j,z_k}}$. The attractiveness of an attractor is then defined as:

$$\mathcal{A}_{a_i,z_k} = \frac{(1+\sum_{z_j\in\mathbb{Z}, z_j\neq z_k} B(a_i,z_k))}{\sqrt{(X_{a_i}-X_{z_k})^2+(Y_{a_i}-Y_{z_k})^2}} \qquad (4)$$

where $B(a_i, z_k)$ is an indicator variable, with $B = 1$ if the individual z_k is going toward or staying at attractor a_i and 0 otherwise, and X and Y are the coordinates of a node and an attractor. In our implementation, we divided the simulation area in equally sized squares, or cells, and consider each cell to be an attraction point. The coordinates (X_{a_i}, Y_{a_i}) mark the center of the i-th attraction point. Once the new destination is known, the node travels towards it with a velocity uniformly distributed in the interval $[v_{min}, v_{max}]$. A pause time is randomly selected once arriving at the destination before choosing another destination and beginning the process again.

5.2 Case Study: Quinta

We capture the social structure of the subjects present in the Quinta trace by applying the community identification method described in Section 4.1. As a result, we identify 5 distinct communities of sizes 13, 27, 17, 12, and 29 nodes. Figure 5.2 shows the average intensity over time associated with each of these communities as well as the average intensity associated with the whole population. In addition, Table 3 shows the average speed and average pause time for each community, along with 95% confidence intervals. Note that, for the most part, communities are well separated by the predilection of their members for different geographical regions. The exception are communities 3 and 4, whose members seem to prefer very similar locations. However, it is very clear that members of these two communities have different speeds and pause times.

Due to the high computational burden associated with fitting multivariate logit models when both the number of cells and the number of observations is large, we computed

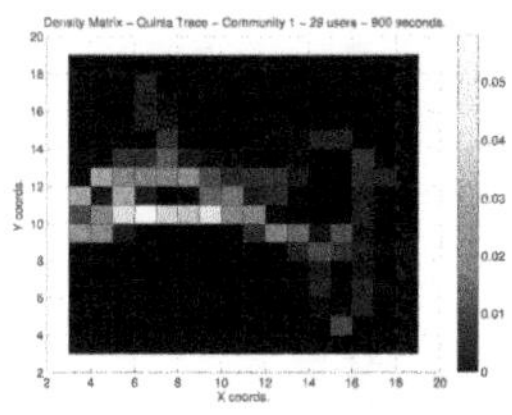

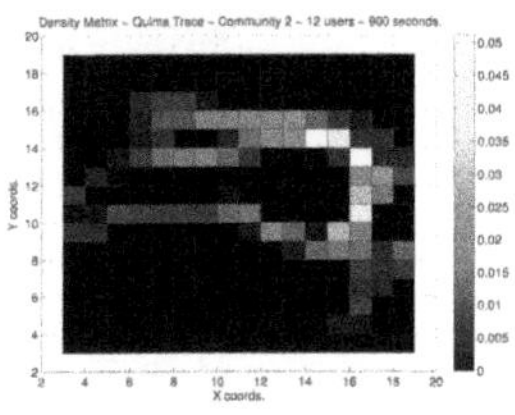

(a) Community 1 (29 nodes) (b) Community 2 (12 nodes)

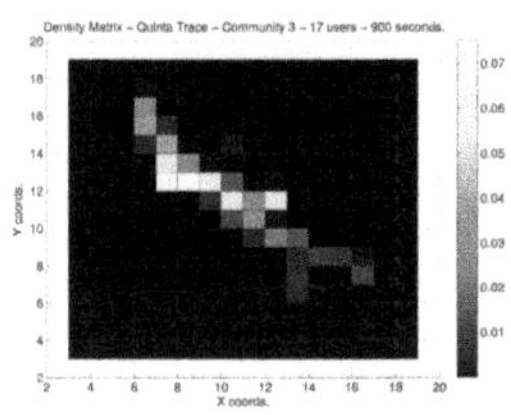

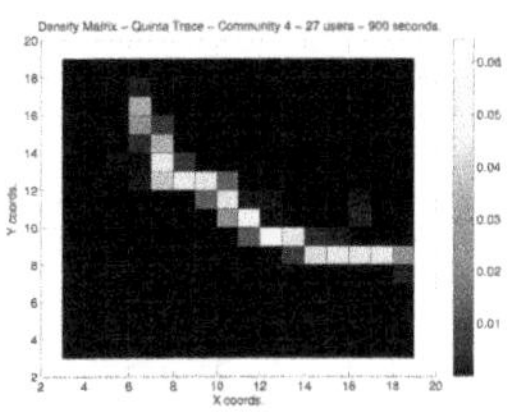

(c) Community 3 (17 nodes) (d) Community 4 (27 nodes)

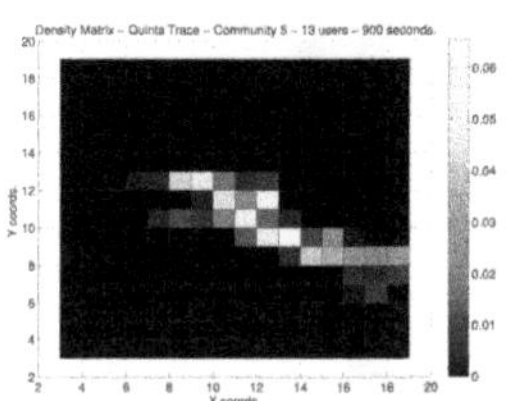

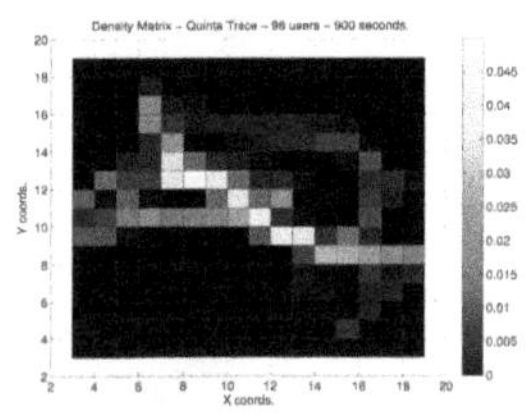

(e) Community 5 (13 nodes) (f) All Nodes (98 nodes)

Figure 1: Average intensity over time for the user communities in the Quinta trace (from 1(a) to 1(e)) and overall intensity for all nodes from all communities (1(f)).

Community	Speed (m/s)	Pause Time (sec)
1	1.124 (1.117, 1.132)	3.316 (2.303, 4.330)
2	1.453 (1.444, 1.462)	2.365 (1.356, 3.375)
3	1.028 (1.017, 1.038)	5.951 (1.974, 9.928)
4	1.139 (1.132, 1.146)	2.036 (1.432, 2.640)
5	1.103 (1.092, 1.114)	4.320 (3.092, 5.548)

Table 3: Speed and pause times for user communities in the Quinta trace.

our estimates based on a subsample of the original 900 seconds available for inference. This is well justified in this case because of the Markovian structure of the model. To assess the robustness of the procedure we consider various subsample rates (every other second, every five seconds, and every ten seconds). The point estimators were very similar in every case; all the results reported below are based on a subsample of 450 data points generated by retaining every other second of the original trace.

Table 4 presents the estimates of the affinity parameters generated by the multivariate logit model, along with 95% confidence intervals. Note that there is evidence of both assortative and dissortative communities. Also, the confidence intervals suggests that a few affinity parameters are indistinguishable from 0 (e.g., $\theta_{2,3}$), but the majority are clearly statistically significant.

As previously pointed out, we use the *spatial density* of mobile nodes to characterize and describe their mobility. In other words, rather than focusing on individual nodes, we are interested in characterizing how node mobility affects the behavior of the network as a whole. In the case of spatial node density, it directly affects fundamental network properties, such as connectivity and capacity, and thus has considerable impact on core network functions, e.g., medium access and routing. Spatial node density can be defined as the number of nodes located in a given unit area. In order to study node's spatial density behavior, we define the node density distribution as the percentage of cells containing more than n nodes.

Θ	1	2	3	4	5
1	-0.138 (-0.258,-0.012)	0.299 (0.239,0.363)	-0.135 (-0.21,-0.071)	0.369 (0.307,0.495)	-0.072 (-0.123,-0.008)
2		-0.411 (-0.479,-0.341)	-0.004 (-0.061,0.039)	0.425 (0.357,0.497)	0.151 (0.110,0.205)
3			0.183 (0.113,0.269)	0.494 (0.417,0.561)	0.339 (0.284,0.374)
4				0.512 (0.405,0.657)	0.452 (0.393,0.518)
5					-0.077 (-0.133,-0.022)

Table 4: Affinity parameters associated with the five communities identified by our clustering algorithm. Since the matrix is symmetric, only its upper triangular part is presented.

Figure 2 shows the spatial node density distribution for the Quinta trace along with density distributions associated with SAGA, RWP, and Natural mobility regimes. The curve labeled "*Initial*" shows the trace's density distribution at the beginning of the trace collection interval, while the curve labeled "*Quinta*" shows the distribution at the end of the trace collection interval. The same initial distribution collected from the trace and shown in this figure was used in all the simulations to feed the initial node's placement, leveling the initial conditions in the simulations for all the studied mobility regimes. Data points for the RWP, Natural, and SAGA mobility regimes were averaged over 10 runs.

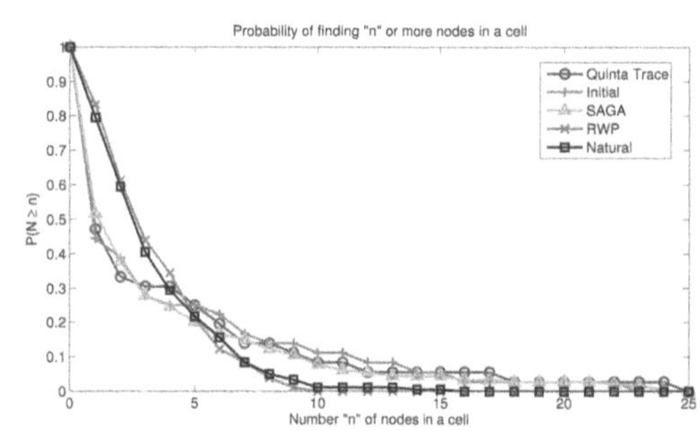

Figure 2: Initial and final spatial node density distribution for the Quinta trace compared with the node density distribution of RWP, Natural, and SAGA.

Note that SAGA's density distribution follows closely the distribution of the real trace. In the case of RWP, the majority of cells present similar number of nodes, and no cells contain significantly higher concentration of nodes. This is also the behavior observed for Natural which can be explained as follows: in the beginning of the simulation the initial distribution weighs on the computation of the attractiveness of the cells, but as the simulation proceeds more and more nodes may choose, with small, but greater than zero probability, cells where there are no other nodes present. Over time, this behavior spreads the nodes around different cells, leveling the attractiveness of cells in general, approaching uniformity and RWP's density distribution. SAGA, on the other hand, is able to maintain the original spatial node density.

Next we consider the average intensities over time, which are presented in Figure 3. These plots denote the average percentage of time, nodes choose to be in each cell. More specifically, Figure 3(a) shows the average intensity for Quinta and demonstrates that nodes have a strong preference for some areas of the park, which is partially driven by its geographic characteristics (e.g., the presence of a lake in the top right corner of the park). Note that SAGA (Figure 3(b) is capable of preserving this observed spatial preferences, while both RWP (Figure 3(c)) and Natural (Figure 3(d)) tend to converge towards a homogeneous distribution over the whole region, which is clearly suboptimal in this context.

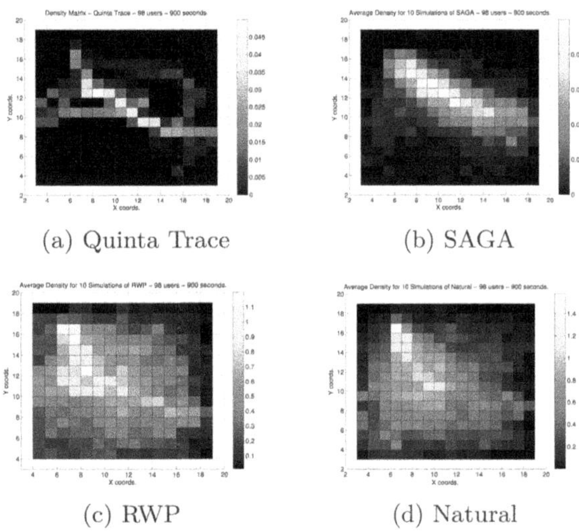

(a) Quinta Trace (b) SAGA

(c) RWP (d) Natural

Figure 3: Density over time for the Quinta scenario.

Another way to validate SAGA is through a *collocation matrix*, which represents the average amount of time that any pair of nodes spend in the same cell. Figure 4 presents the collocation matrices for the Quinta trace, as well as RWP, Natural, and SAGA. Note that Quinta's collocation matrix exhibits a pattern that matches the community structure we identified in Section 4.1 quite well. More specifically, note that members of communities 3 and 4 tend to have high collocation. However, members of communities 1, 2, and 5 have fairly low collocation with other members of their same community. In this case, these patterns agree with the structure of the affinity matrix, but this is not necessarily the case in general (for example, see our analysis of the Statefair trace in the following subsection). Our model replicates this structure fairly well, while RWP and Natural (which treat all nodes identically) exhibit uniform collocation matrices.

5.3 Case study: Statefair

Similarly to the Quinta trace, we use the method described in Section 4.1 to capture the social structure of the subjects present in the Statefair trace. In this trace, we identify 2 distinct communities of sizes 9 and 10 nodes. Figure 5.3 shows the average intensity over time associated with each of these communities as well as the average intensity associated with the whole user population. In addition, Table 5 shows the average speed and average pause time for each community, along with 95% confidence intervals. Like in the Quinta trace, we manage to clearly differentiate one community from the other and show how members of each community favor different geographical regions. In this case study, not only their preferred location are very distinct for each community, but also their speeds and pause times.

The Statefair trace is longer but was sampled at a lower frequency than Quinta. Hence, in this scenario, we were able to use the totality of the measurements (267 samples)

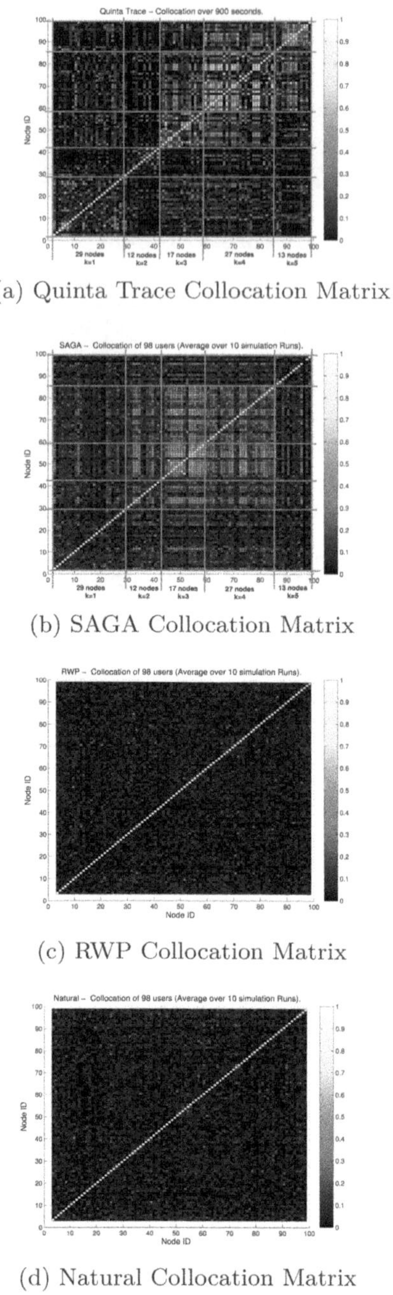

(a) Quinta Trace Collocation Matrix

(b) SAGA Collocation Matrix

(c) RWP Collocation Matrix

(d) Natural Collocation Matrix

Figure 4: Collocation matrices for the Quinta scenario.

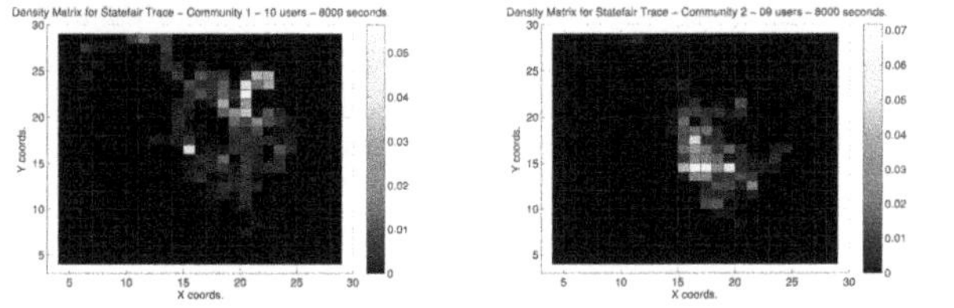

(a) Community 1 (9 nodes) (b) Community 2 (10 nodes)

Figure 5: Average intensity over time for the user communities in the Statefair trace (5(a) and 5(b)) and overall intensity for all nodes from all communities can be observed in Figure 6.

Community	Speed (m/s)	Pause Time (sec)
1	0.606 (0.596, 0.616)	4.663 (1.259, 8.066)
2	0.435 (0.426, 0.444)	8.574 (0.077, 17.071)

Table 5: Speed and pause times for user communities in the statefair trace.

in the trace. and there was no need for subsampling due to the lower sampling frequency in this trace (30 seconds).

Table 6 presents the estimates of the affinity parameters generated by the multivariate logit model, along with 95% confidence intervals. Different from our first case study, in this scenario there is only evidence of assortative behavior between the two communities.

Figure 6 shows the spatial node density distribution metric, as defined above, for the Statefair trace. This figure also shows the density distributions associated with SAGA, RWP, and Natural mobility. Once again, we show that the clustering behavior observed in real live traces is preserved by SAGA. This is in contrast with the preferential attachment based mobility regime, which, by accounting only for the number of nodes in a given location and disregarding social interactions and geographical preferences, leads to undesirable homogeneous node spatial density in steady state. Such steady-state behavior approaches the behavior observed in random mobility regimes as exemplified by the RWP. Table 7 summarizes the results for density shown in Figures 2 and 6, showing the mean square error for final density curves, between the GPS traces and the mobility regimes, for the Quinta and Statefair case studies.

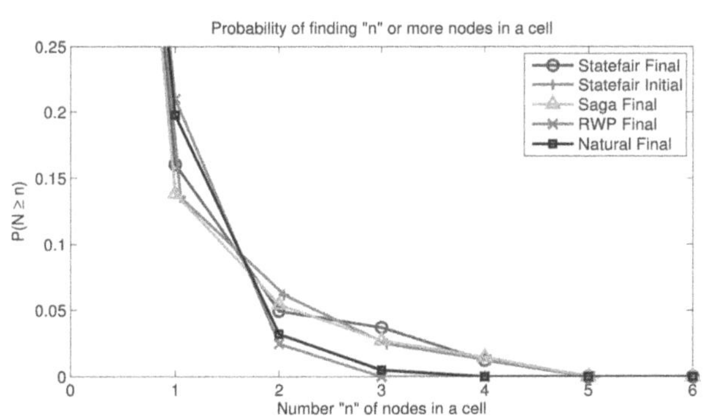

Figure 6: Initial and final spatial node density distribution for the Statefair trace compared with the node density distribution of RWP, Natural, and SAGA.

We now consider the average intensities over time, which are presented in Figure 7. As also observed in the Quinta scenario, the average intensity measured for the Statefair scenario, as shown in Figure 7(a), presents some concentration of nodes in certain areas. Figure 7(b) shows that SAGA is capable of preserving the spatial preferences observed in the real trace, while both RWP (Figure 7(c)) and Natural (Figure 7(d)) tend to degrade towards a homogeneous distribution.

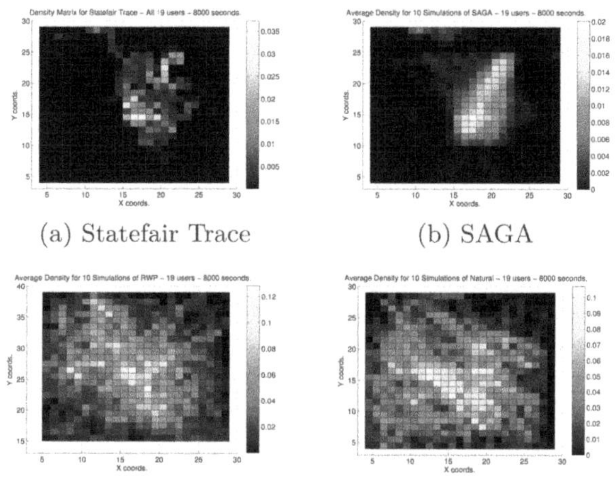

(a) Statefair Trace (b) SAGA

(c) RWP (d) Natural

Figure 7: Statefair mean density over time.

Θ	1	2
1	0.386 (0.223, 0.551)	0.716 (0.634, 0.799)
2		0.907 (0.811, 1.004)

Table 6: Affinity parameters associated with the two communities identified by our clustering algorithm.

	SAGA	Natural	RWP
Quinta	0.0179	0.0802	0.0913
Statefair	0.0093	0.0316	0.0378

Table 7: Mean square error for final density curves, between the GPS traces and the mobility regimes, for the Quinta and Statefair case studies.

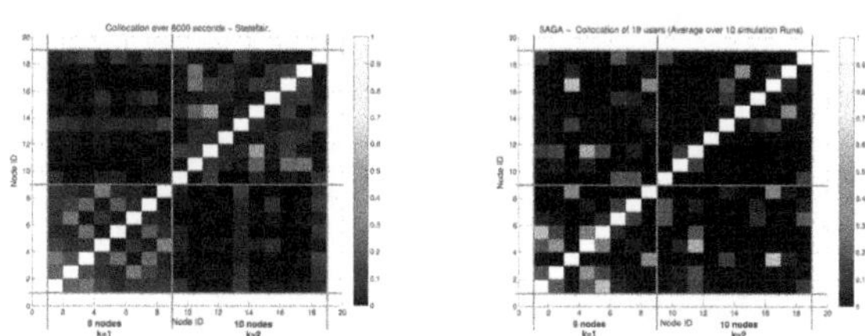

(a) Statefair (b) SAGA

Figure 8: Collocation matrices for the Statefair scenario.

Figure 8 presents the collocation matrices for the Statefair trace, as well as SAGA. In the case of RWP and Natural, the collocation matrices for the Statefair scenario look very similar to those presented in our first case study, where it is not possible to see any collocation of pairs of nodes for more than 10% of the time. For that reason, and due to space constrains, we do not show these plots. The collocation matrices for the trace and SAGA exhibit a pattern that matches the community structure we identified in Section 4.1 quite well. More specifically, note that members of communities 1 and 2 tend to have high colocation, among nodes in their own community, but we also observe nodes from different communities with a higher degree of collocation. This is consistent with the fact that these are highly assortative communities, as described by their affinity matrices (Table 6).

6. RELATED WORK

Motivated by the growing need to evaluate mobile networks and their protocols under realistic conditions, there has been considerable work in trying to depart from random– towards more realistic mobility models. In this section, we do not attempt to present an exhaustive survey of the state-of-the-art in mobility modeling but, rather, highlight efforts that are more closely related to our work on socially-inspired mobility modeling.

As pointed out in Section 1, an important milestone on the subject is the work by Barabási and Albert [8] which demonstrated that many real large networks are scale free.The authors discuss the mechanism responsible for the emergence of scale-free networks and argue that understanding this problem will require a shift from modeling network topology to modeling "network assembly and evolution". To this end, they define the Barabási-Albert model based on *growth* and *preferential attachment*. Growth refers to the fact that the number of nodes in the network increases over time, where a new node is placed with m edges connecting it to other m nodes. Preferential attachment means that a node will choose to connect to another node i with probability $\Pi(k_i) = \frac{k_i}{\sum_j k_j}$ based on the degree k_i of node i and any node j connected to node i.

Several mobility models have been inspired by the Barabási-Albert preferential attachment principle. For example, in [11], a model based on preferential attachment has been proposed, where the choice of going towards an attraction region is weighted proportionally to the region's popularity (i.e., the number of other nodes that chose it) and inversely proportional to the distance to it. The work proposed in [14] is another example of a model that follows Barabási-Albert's growth and preferential attachment principles. The authors even show a figure where they present their initial (after growth) and steady-state spatial distribution. It is possible to see how clusters dissipate and fade away over time. The same concept is also used in [12] where nodes are also driven by identified social interactions.

More recent approaches try to account for social interaction between nodes, and those are the most closely related to our research. Some examples of mobility modeling for social networks include the work in [33] and [13]. In these models, node movement is influenced by the strength of social ties and the choice of an attraction point is based on the history of visits of other nodes to that location.

The model proposed in [21] defines a set of locations called anchors; dictated by an anchor association function, nodes are associated with specific anchors at specific times during the simulation. Also, nodes are associated with each other by a "social acquaintance" function, where concepts of "attraction" and "repulsion" are also defined.

Two basic behaviors, namely *socialize* and *isolate* are defined in [18]. The strength of social interactions are represented by an *acquaintance matrix*. This is then used as input to define the attraction and repulsion forces acting on each node, which define the node's direction while moving, as well as its acceleration and velocity. The work in [17] also computes attraction and repulsion forces between nodes which define a social interaction matrix. Here, in addition to the social forces between nodes, they also take into account the forces exerted by locations (attraction points).

In [20], the notion of communities is used and incorporate assortative/dissortative behavior given by an *Attractor Matrix*. This matrix describes the attributes of each node, such as gender, age, and field of work; these attributes are matched through an *Attribute Matching Matrix*. Nodes then move through communities looking for communities where they find other nodes that are a better match to their attributes. Another mobility model based on the formation of communities is introduced in [16]. It builds a k-clique overlapping community structure which satisfies the common statistical features observed from real social networks. Functions that describe the attraction of a node to a group and a location are defined and these graph connections change over time based on the defined overlapping graph.

In [34], authors propose a model where each node is randomly assigned to one or more home location areas on the simulation area. A simple 2-state Markov Chain governs the transitions in, out, and between home locations. Nodes perform random waypoint trips in and out home locations, with a probability $1-p$ of roaming and a probability p of staying or returning to their home location.

In [35] authors propose a model where a transition probability (weighted by a Caveman graph [36]) to a given location does not depend only on the number of nodes currently at that location but on the total weight of nodes assigned to the destination as their home location, independently of their current location. Social ties in this case depend on the set of home locations attributed to each node and no explicitly social structure can be defined, other than the one defined by the Caveman graph.

The work proposed in [37] uses the measured Levy flight property of human displacements, given a distribution of the travelled distance between two waypoints follows a Pareto law.

Most existing social mobility regimes, such as the one described here, construct a social network using outside information unrelated to the mobility itself, such as a "social trace", i.e., an a priori description of the relationship between nodes, their attributes, etc. SAGA, on the other hand, accounts for the social structure inherent in the mobility data itself, through the identification of user communities. We then use of the preferential attachment principles, "attached" to the built community structure to model the interaction of nodes within– and between communities. By doing so, SAGA addresses the undesirable behavior exhibited by mobility models that rely solely on pure preferential attachment. SAGA also considers the heterogeneity of the different locales in the region of interest. Thus, through the coupling of social– and geography-awareness, SAGA accounts for features observed in real life networks such as differential popularity, transitivity, and clustering, while the existing mobility models account for homophily by attributes. Moreover, our modeling framework applies formal statistical learning techniques to extract, from real mobility traces, both the nodes' social structure as well as the region's geographical diversity for model calibration.

We argue that being able to calibrate the model in order to improve realism and approach mobility metrics to the ones measured from real live traces is important. Our approach allows that, also presenting a flexible framework where synthetic social structures and cell attractiveness can also be defined if needed. Moreover, while most existing mobility models are able to reflect contact duration distributions similar to real life measurements, very few effort focused on the actual size of the clusters and *which* specific nodes are part of a given cluster [38]. Our results show by the density distribution curves and collocation matrix, that not only SAGA is able to reflect realistic cluster sizes, it also uses the affinity matrix to account for which node spends time in each cluster.

7. CONCLUSION

In this work, we introduced a novel mobility model that accounts for both the social structure exhibited by the user population as well as the geographic diversity of the region of interest. We call our model SAGA for Socially– And Geography-Aware mobility model. SAGA captures social features through the use of *communities* which group users with similar features such as average time in a cell, average speed, and pause time. Acknowledging the importance of preserving– and being able to control node density over time, SAGA also accounts for geographic diversity by considering that different communities exhibit different interests for different locales. Besides introducing SAGA, the contributions of this work include: a model calibration approach based on formal statistical procedures to extract social structures and geographical diversity from real traces and set SAGA's parameters. We validate our work by comparing SAGA with two real mobility traces and two other mobility regimes, namely Random Waypoint and Natural, the latter representing preferential-attachment based mobility models. We show how to extract social structures and the geographical diversity exhibited by real traces. Moreover, our experimental results demonstrate that, when compared to existing mobility regimes, SAGA is able to preserve the desired heterogeneous node spatial density present in real user mobility, creating and maintaining clusters and accounting for differential node popularity and transitivity.

Acknowledgements

Financial support was granted by the CAPES Foundation, Ministry of Education of Brazil, Caixa Postal 250, Brasilia - DF 70040-020 Brazil. This work was partially supported by NSF grant CCF-091694 and a US Army-ARO MURI grant.

References

[1] T. Camp, J. Boleng, and V. Davies, "A survey of mobility models for ad hoc network research," *Wireless Communications and Mobile Computing*, vol. 2, pp. 483–502, 2002.

[2] C. Bettstetter, M. Gyarmati, and U. Schilcher, "An inhomogeneous spatial node distribution and its stochastic properties," in *Proceedings of the 10th ACM Symposium on Modeling, analysis, and simulation of wireless and mobile systems*, MSWiM '07, 2007.

[3] M. Musolesi and C. Mascolo, "Mobility models for systems evaluation," in *Middleware for Network Eccentric and Mobile Applications*, pp. 43–62, Springer, 2009.

[4] A. Balachandran, G. M. Voelker, P. Bahl, and P. V. Rangan, "Characterizing user behavior and network performance in a public wireless lan," in *ACM SIGMETRICS 2002.*

[5] M. Balazinska and P. Castro, "Characterizing mobility and network usage in a corporate wireless local-area network," in *MobiSys 2003*, pp. 303–316.

[6] D. Tang and M. Baker, "Analysis of a local-area wireless network," in *ACM MOBICOM 2000.*

[7] T. Henderson, D. Kotz, and I. Abyzov, "The changing usage of a mature campus-wide wireless network," in *ACM MOBICOM 2004.*

[8] R. Albert and A.-L. Barabási, "Statistical mechanics of complex networks," *Rev. Mod. Phys.*, vol. 74, no. 1, pp. 47–97, 2002.

[9] A. Jardosh, E. Belding-Royer, K. Almeroth, and S. Suri, "Real-world environment models for mobile network evaluation," *Selected Areas in Communications, IEEE Journal on*, vol. 23, no. 3, pp. 622 – 632, 2005.

[10] M. Kim and D. Kotz, "Extracting a mobility model from real user traces," in *IEEE INFOCOM*, 2006.

[11] V. Borrel, M. D. de Amorim, and S. Fdida, "On natural mobility models," in *WAC*, 2005.

[12] A. Mei and J. Stefa, "Swim: A simple model to generate small mobile worlds," in *IEEE INFOCOM 2009.*

[13] M. Musolesi and C. Mascolo, "Designing mobility models based on social network theory," *ACM SIGMOBILE Mobile Computing and Communication Review*, vol. 11, pp. 59–70, 2007.

[14] S. Lim, C. Yu, and C. Das, "Clustered mobility model for scale-free wireless networks," in *Local Computer Networks, Proceedings 2006 31st IEEE Conference on*, 2006.

[15] B. A. Nunes and K. Obraczka, "On the invariance of spatial node density for realistic mobility modeling," in *Mobile Adhoc and Sensor Systems (MASS), 2011 IEEE 8th International Conference on*, pp. 322 –331, oct. 2011.

[16] S. Yang, X. Yang, C. Zhang, and E. Spyrou, "Using social network theory for modeling human mobility," *Network, IEEE*, vol. 24, pp. 6 –13, september-october 2010.

[17] C. Zhao and M. Sichitiu, "N-body: Social based mobility model for wireless ad hoc network research," in *Sensor Mesh and Ad Hoc Communications and Networks (SECON), 2010 7th Annual IEEE Communications Society Conference on*, pp. 1 –9, june 2010.

[18] V. Borrel, F. Legendre, M. Dias de Amorim, and S. Fdida, "Simps: Using sociology for personal mobility," *Networking, IEEE/ACM Transactions on*, vol. 17, pp. 831 –842, june 2009.

[19] L. Harfouche, S. Boumerdassi, and E. Renault, "Towards a social mobility model," in *Personal, Indoor and Mobile Radio Communications, 2009 IEEE 20th International Symposium on*, pp. 2876 –2880, sept. 2009.

[20] J. Wang, J. Yuan, X. Shan, Z. Feng, J. Geng, and I. You, "Samob: A social attributes based mobility model for ad hoc networks," in *Innovative Mobile and Internet Services in Ubiquitous Computing (IMIS), 2011 Fifth International Conference on*, pp. 444 –449, 30 2011-july 2 2011.

[21] D. Fischer, K. Herrmann, and K. Rothermel, "Gesomo: A general social mobility model for delay tolerant networks," in *Mobile Adhoc and Sensor Systems (MASS), 2010 IEEE 7th International Conference on*, pp. 99 –108, nov. 2010.

[22] H. White, S. Boorman, and R. Breiger, "Social structure from multiple networks: I. blockmodels of roles and positions," *American Journal of Sociology*, vol. 81, no. 4, pp. 730–80, 1976.

[23] S. Wasserman and K. Faust, *Social Network Analysis: Methods and Applications*. No. 8, Cambridge University Press, 1994.

[24] M. McPherson, L. Smith-Lovin, and J. M. Cook, "Birds of a feather: Homophily in social networks," *Annual Review of Sociology*, vol. 27, no. 1, pp. 415–444, 2001.

[25] T. Azevedo, R. Bezerra, C. Campos, and L. de Moraes, "An analysis of human mobility using real traces," in *IEEE WCNC 2009*.

[26] I. Rhee, M. Shin, S. Hong, K. Lee, and S. Kim, "CRAWDAD http://crawdad.cs.dartmouth.edu/ ncsu/mobilitymodels/GPS/KAIST v.2009-07-23," July 2009.

[27] B. S. Everitt, S. Landau, M. Leese, and D. Stahl, "Cluster analysis," *Wiley. Fifth Edition*, 2011.

[28] C. Fraley and A. E. R. (2002), "Model-based clustering, discriminant analysis, and density estimation," *Journal of the American Statistical Association*, 2002.

[29] "R: A language and environment for statistical computing."

[30] A. Rodriguez and B. A. A. Nunes and K. Obraczka, "http://inrg.cse.ucsc.edu/inrgwiki/projects," 2012.

[31] K. Train, "Discrete choice modelling, with simulations," *Cambridge University Press*, 2004.

[32] The Scenario Generator, "http://isis.poly.edu/ qiming/scengen/index.html."

[33] L. Harfouche, S. Boumerdassi, and E. Renault, "Weighted social manhattan: Modeling and performance analysis of a mobility model," in *Personal Indoor and Mobile Radio Communications (PIMRC), 2010 IEEE 21st International Symposium on*, 2010.

[34] W. J. Hsu, T. Spyropoulos, K. Psounis, and A. Helmy, "Modeling time-variant user mobility in wireless mobile networks," *IEEE Infocom*, 2007.

[35] C. Boldrini and A. Passarella, "Hcmm: Modelling spatial and temporal properties of human mobility driven by users social relationships," *Elsevier Computer Commununication*, 2010.

[36] M. E. J. Newman, "The structure and function of complex networks," *March*, 2003.

[37] I. Rhee, M. Shin, S. Hong, K. Lee, and S. Chong, "On the levy-walk nature of human mobility," *IEEE Infocom*, 2008.

[38] T. Hossmann, T. Spyropoulos, and F. Legendre, "Putting contacts into context: mobility modeling beyond inter-contact times," in *Proceedings of the Twelfth ACM International Symposium on Mobile Ad Hoc Networking and Computing*, MobiHoc '11, 2011.

Understanding and Modeling the Small-World Phenomenon in Dynamic Networks

Anh-Dung Nguyen
ISAE-University of Toulouse
LAAS-CNRS
Toulouse, France
anh-dung.nguyen@isae.fr

Patrick Senac
ISAE-University of Toulouse
LAAS-CNRS
Toulouse, France
patrick.senac@isae.fr

Michel Diaz
LAAS-CNRS
Toulouse, France
michel.diaz@laas.fr

ABSTRACT

The small-world phenomenon first introduced in the context of static graphs consists of graphs with high clustering coefficient and low shortest path length. This is an intrinsic property of many real complex static networks. Recent research has shown that this structure is also observable in dynamic networks but how it emerges remains an open problem. In this paper, we propose a model capable of capturing the small-world behavior observed in various real traces. We then study information diffusion in such small-world networks. Analytical and simulation results with epidemic model show that the small-world structure increases dramatically the information spreading speed in dynamic networks.

Categories and Subject Descriptors

G.2.2 [**Graph Theory**]: Network problems; C.2.1 [**Network Architecture and Design**]: Store and forward networks

General Terms

Theory, Measurement

Keywords

Small-world, dynamic networks, information diffusion

1. INTRODUCTION

In his famous experiment Milgram showed that the human acquaintance network has a diameter of the order of six, so leading to the small-world qualification. Watts and Strogatz introduced later a model of small-world phenomenon for static graphs [10]. They proposed a random rewiring procedure for interpolating between a regular ring lattice and a random graph. Between these two extrema, the graph exhibits an exponential decay of the average shortest path length contrasted to a slow decay of the average clustering coefficient. Interestingly, numerous real static networks exhibit such properties.

From a communication perspective, it was shown that a small-world network behaves as random network in terms of information diffusion. However, the great majority of studies on small-world networks properties and behaviors focused on static graphs and ignore the dynamics of real mobile networks. For example, in a static graph an epidemic cannot break out if the initial infected node is in a disconnected component of the network; conversely in a mobile network, nodes movements can ensure a property of temporal connectivity of the underlying dynamic graph. Moreover, an epidemic can take off or die out depending not only on the network structure and the initial carrier, but also on the time when the disease begins to spread. These aspects cannot be captured by a static small-world model.

In this paper, we address this issue by introducing a model for dynamic small-world networks. In particular, this model aims to capture the small-world phenomenon that emerges from delay/disruption tolerant networks. Starting from a highly clustered dynamic network where nodes tend to stay all the time in their preferential zones, we introduce a dynamic rewiring procedure which consist in progressively increasing the ratio of nomadic nodes. This allows to cover a whole scope of dynamic networks from highly clustered ones to fully random ones.

In order to formalize the dynamic small-world phenomenon, we extend to dynamics networks the notions of dynamic clustering coefficient and shortest dynamic path. By studying the evolution of these metrics from our rewiring process, we exhibit the emergence of a class of dynamic small-world networks with high dynamic clustering coefficients and low shortest dynamic path length. This simulation-based work is then extended by an analytical study of information spreading in dynamic small-world networks.

The rest of the paper is structured as follows. Section 2 introduces previous works that inspired our contribution. In Section 3, after having formally defined the notions of dynamic clustering coefficient and shortest dynamic path length, we analyze various real traces and show the intrinsic properties of dynamic networks that induce the small-world phenomenon. We introduce then a dynamic small-world model which captures these properties in Section 4. In Section 5, from simulations results and analytical formulation we study information diffusion performances in dynamic

MSWiM'12, October 21–25, 2012, Paphos, Cyprus.

small-world networks. Finally, Section 6 will conclude the paper and discuss about future works.

2. RELATED WORKS

In [10], Watts and Strogatz introduce a model of small-world phenomenon in static graphs. From a regular ring lattice, they rewire randomly edges in this graph with a probability varying from 0 (i.e. leading to a regular network) to 1 (i.e. leading a random graph). During this process they observed an abrupt decrease of the average shortest path length, leading to short path of the same order of magnitude as observed in random graphs, while the clustering coefficient is still of the same order of magnitude than the one of a regular graph. This features suggested the emergence of the small-world phenomenon. The authors also demonstrate that this graph structure allows information to diffuse as fast as in a random graph.

Kleinberg [3] extended the model to 2-d lattices and introduced a new rewiring process. This time, the edges are not uniformly rewired but follow a power law $\frac{1}{d^{\alpha}}$ where d is the distance on the lattice from the starting node of the edge and α is the parameter of the model. Newman [5] proposed another definition of the clustering coefficient which has a simple interpretation and is easier to process. The authors argue that the definition in [10] favors vertices with low degree and introduces a bias towards networks with a significant number of such vertices.

Although research pays much attention to the small-world phenomenon in static networks, whether through modeling or analytical analysis, the small-world phenomenon in dynamic networks (i.e. networks with time varying edges and nodes) is still not well understood. This is partly due to the lack of models and metrics for dynamic graphs. Recently, J.Tang et al. [8] defined several metrics for time varying graphs, including temporal path length, temporal clustering coefficient and temporal efficiency. They showed that these metrics are useful to capture temporal characteristics of dynamic networks that cannot be captured by traditional static graph metrics. The definition of dynamic path, introduced in this paper, is close to their definition (we had the dual metric of the number of hops). We introduce a new definition of dynamic clustering coefficient which, as shown in the next section, captures more accurately the dynamics of networks.

In [9], the authors highlight the existence of the small-world behavior in real traces. Using the definition of temporal correlation introduced in [2], they show that real dynamic networks have a high temporal correlation and low temporal shortest path, suggesting a dynamic small-world structure. In this paper, we simply and consistently extend to dynamic networks the initial small-world metrics (i.e. shortest path length and clustering coefficient) defined in [10]. Finally, the analytical study of information diffusion in dynamic small-world introduced in this paper is inspired from the method applied for static networks described by Newman in [6].

3. SMALL-WORLD PHENOMENON IN DYNAMIC NETWORKS

In this section, we formalize the notion of small-world phenomenon in dynamic networks by introducing two metrics used for qualifying such phenomenon: shortest dynamic path length and dynamic clustering coefficient. And then by analyzing extensively real network traces, we show fundamental characteristics which are at the origin of the dynamic small-world phenomenon.

3.1 Dynamic Small-world Metrics

3.1.1 Shortest Dynamic Path Length

Basically, the shortest path problem in static graphs consists in finding a path such that the sum of the weights of its constituent links is minimized. From a graph theory point of view a dynamic networks can be described by a temporal graph, that is a temporal sequence of graphs that describe the discrete evolution of the network according to nodes and links creation and destruction events. A path in a dynamic network can be seen as an ordered set of temporal links that allow a message to be transferred using the store-move-forward paradigm between two nodes. Formally, let l_{ij}^t be a link between node i and node j at instant t. A dynamic path from node u to node v from time t_0 to time t is described by a time ordered set $p_{uv}(t_0,t) = \{l_{ui}^{t_0}, l_{ij}^{t_1}, \ldots, l_{wv}^{t}\}$ where $t_{k+1} > t_k$. We consider two metrics of dynamic paths:

- Delay : the sum of the inter-contact times between consecutive links which constitutes the path.
- Number of hops : the number of temporal links which constitutes the path.

This leads to the following definition of shortest dynamic path length.

Definition 1. The shortest dynamic path is the path giving the minimum amount of delay[1]. If there are several paths giving the same delay, then we select the one giving the minimum number of hops. Formally, the shortest dynamic path length between i and j from time t_0 is

$$\mathcal{L}_{ij}^{t_0} = \inf\{t - t_0 | \exists p_{ij}(t_0,t)\} \quad . \tag{1}$$

The shortest dynamic path length of a network of N nodes from time t_0 is the average of the shortest dynamic path lengths of all pairs of nodes in the network

$$\mathcal{L}^{t_0} = \frac{\sum_{ij} \mathcal{L}_{ij}^{t_0}}{N(N-1)} \quad . \tag{2}$$

To find the shortest dynamic path length of all pairs of nodes, we propose an algorithm leveraging on the following interesting property of adjacency matrix in static graphs.

The adjacency matrix A is defined as the matrix in which the element $(A)_{i,j} \in \{0,1\}$ at row i and column j denotes the existence of a link between node i and node j. If we process the power n of such matrix, then its element $(A^n)_{i,j}$ gives the number of paths of length n between i and j. Indeed, for example, when $n = 2$, $(A^2)_{i,j} = \sum_k (A)_{i,k} \times (A)_{k,j}$ sums all the possibilities to go from i to j through an intermediate node k. We extend this property to dynamic networks. Let $A_t, t = 0, 1, \ldots, n$ be the adjacency matrix of a dynamic network at time t. The matrix C_t obtained as follows

$$C_t = \mathbf{A}_t \vee \mathbf{A}_t^2 \vee \ldots \vee \mathbf{A}_t^n \quad , \tag{3}$$

where $\mathbf{A}_t^i$ denotes the binary version of the matrix A_t^i (i.e. the element $(\mathbf{A}_t)_{i,j}$ equals to 1 if $(A_t)_{i,j} > 0$ and 0 otherwise)

[1] we can also have another definition for minimizing the number of hops. In this work, we focus only on delay constrained path.

has its elements $(C_t)_{i,j}$ which indicate if there is a direct or indirect link (up to n hops) between i and j at time t. Indeed, $(C_t)_{i,j}$ is the logical sum of all possibilities to have a direct or indirect link (up to n hops) between i and j at time t. In consequence, the product

$$D_t = C_0 C_1 \dots C_t \tag{4}$$

results in a matrix in which the element $(D_t)_{i,j}$ specifies, when not null, that there is dynamic path of delay t between i and j. Therefore the shortest dynamic path length from node i to node j is given by the smallest value of t such as $(D_t)_{i,j}$ equals to 1.

It is straight-forward to demonstrate that if a node k belongs to a shortest path between node i and node j then the i to k sub-path gives the shortest path between i and k. Therefore the shortest path between two nodes i and j can be easily backwardly reconstructed. Note that at time t two nodes can be connected to each other via a multiple hops link. To find the complete spatio-temporal path with all the intermediate nodes, we simply apply a breath first search each time we find a spatial multiple hops link. In practice, as it's unlikely to have a large number of nodes connected to each other at a given moment, we can optimize the algorithm by limiting the number of iterations n in Equation 3 to an upper bound of the network diameter. Finally, this algorithm is more efficient (time complexity $O(n^3)$) than the depth first search approach as proposed in [8] (time complexity $O(n^4)$).

3.1.2 *Dynamic Clustering Coefficient*

As defined in [10], the clustering coefficient measures the cliquishness of a typical friend circle. The clustering coefficient of a node is calculated as the fraction of actual existing links between its neighbors and the number of possible links between them. The clustering coefficient of a network is calculated by averaging the clustering coefficients of all the nodes in the network. In [5], Newman defines the clustering coefficient in term of transitivity. The connection between nodes u, v, w is said to be transitive if u connected to v and v connected to w implies that u is connected to w. The clustering coefficient of a network is then calculated as the fraction of the number of closed path of length two over the number of path of length two, where a path of length two is said to be closed if it is a transitive path. This definition is simple to interpret and easy to calculate. Considering that, the initial definition gives more weight to nodes with low degree and introduces a bias towards graph composed of several of these nodes, in this work, we favor Newman's definition and extend it to dynamic networks.

In [8], Tang et al. first introduced a generalization of Watts and Strogatz's definition for temporal graph. The temporal clustering coefficient of a node during a time interval t is the fraction of existing contacts [2] (multiple contacts count once) between the neighbors of the node over the number of possible contacts between them during t. The clustering coefficient of the network is the average of the clustering coefficients of all the nodes. This is simply the application of Watts and Strogatz's definition on a time snapshot of the network. While this definition was shown to better capture temporal characteristics of time varying graph, it depends strongly on the larger of the chosen time interval. If this interval tends to infinity, the temporal clustering coefficient tends to 1 as all the nodes meet each other with a high probability. On the other hand, if the interval tends to 0, then conversely the temporal clustering coefficient tends to 0. Figure 1 illustrates the influence of the time window size on measures of the clustering coefficient on a real mobility trace.

[2] A contact is a intermittent link between two nodes

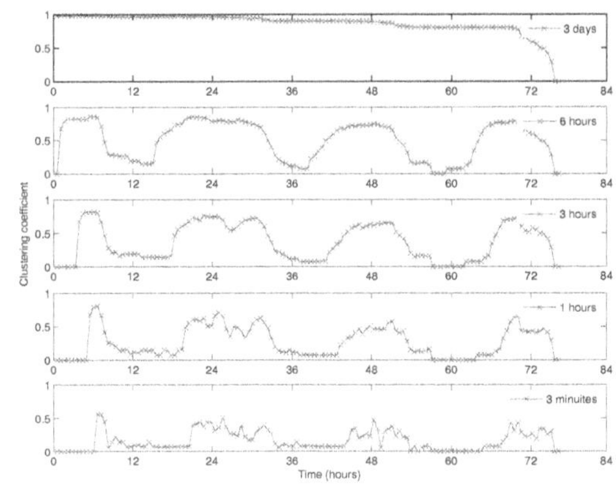

Figure 1: Time window size effect on measure of temporal clustering coefficient (as defined in [11])

Trace	Intel	Cambridge	Infocom05	Infocom06
Number of nodes	9	12	41	78
Duration (days)	6	6	4	4
Granularity (seconds)	120	120	120	120
Connectivity	Bluetooth	Bluetooth	Bluetooth	Bluetooth
Environment	Office	Office	Conference	Conference

Table 1: Data sets of real traces

In order to avoid this temporal bias, we propose a new definition of dynamic clustering coefficient which captures the dynamics of the degree of transitivity and is independent of the measuring time interval.

Definition 2. A dynamic path from node i to node j is transitive if there exists a node k and time t_1, t_2, t_3 so that i is connected to k at t_1, k is connected to j at t_2, i is connected to j at t_3 and $t_1 \le t_2 \le t_3$. The dynamic clustering coefficient of node i from time t_0 is measured by the inverse of the time $t - t_0$ where t is the first instant from time t_0 when the transitive path from i to j is formed, that is $\mathcal{C}_i^{t_0} = \frac{1}{t-t_0}$. The dynamic clustering coefficient of a network of N nodes is then calculated by averaging over the dynamic clustering coefficient of all the nodes from time t_0

$$\mathcal{C} = \frac{1}{N} \sum_i \mathcal{C}_i \ . \tag{5}$$

3.2 Network Traces Analysis

In this section, we analyze extensively real human connectivity traces to understand how small-world behavior emerges in dynamic networks. We first apply the above definitions to highlight the existence of dynamic small-world phenomenon on these traces. For that, we use the data sets from the Haggle project [1, 4] which consists of the recording of opportunistic bluetooth contacts between users in conference or office environments. The settings of these data sets are summarized in Table 1.

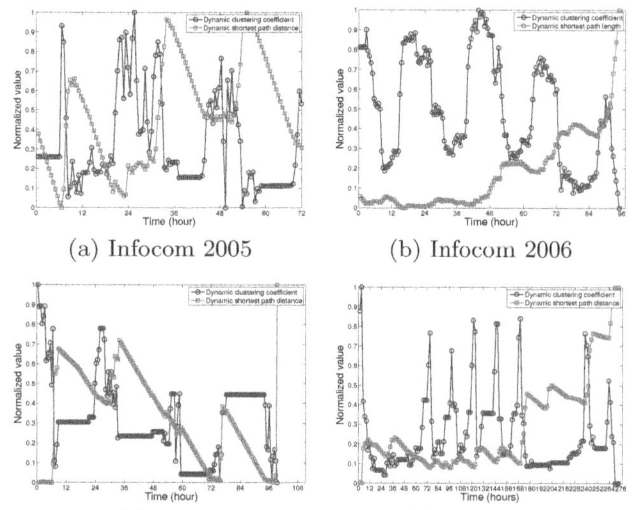

(a) Infocom 2005 (b) Infocom 2006

(c) Intel (d) Cambridge

Figure 2: Small world phenomenon observed in real mobility traces

For each trace, we measure the shortest dynamic path length and the dynamic clustering coefficient every 2000 seconds to see the evolution of these metrics over time. The obtained values are then normalized and plotted on Figure 2. We can observe a periodic pattern of both metrics with a typical period of 24 hours and a phase change every 12 hours. This can be easily explained by the fact that human daily activity is periodic with nigh/day phases. Indeed, people are more nomadic during day while they are mostly sedentary at night. Besides, it is interesting to note that the dynamic clustering coefficient and the shortest dynamic path length evolve in opposite phase. Despite of a slight diversity in different traces due to differences in nodes number and density, traces durations, etc, during the dynamic phase (i.e., day, e.g., the period around 24 hours in Infocom2005 trace), these networks always exhibit high dynamic clustering coefficients and low shortest dynamic path lengths, suggesting the existence of a dynamic small-world phenomenon.

To explain the emergence of this phenomenon, let us focus on and analyze the structure of these networks during the dynamic phases. In the Watts and Strogatz model, the small-world phenomenon emerges when shortcut edges are randomly added to a regular graph. These shortcuts allow the average shortest path length to be reduced significantly while conserving network nodes's cliquishness. We argue that in dynamic networks, mobile nodes are implicitly at the origin of these shortcuts. Indeed, it is known that people spend their daily life among different social communities at specific locations at different times (e.g. colleagues at office in the morning, family at home in the evening). A community can be disconnected from the others in space and/or in time. Besides, some people are more "mobile" than other, they have contacts in many communities and move often between these communities or areas. These "nomadic" nodes contribute to reducing significantly the shortest dynamic path length from nodes in a disconnected component to the rest of network and hence contribute to the emergence of the dynamic small-world phenomenon.

To identify the spatio-temporal shortcuts in dynamic net-

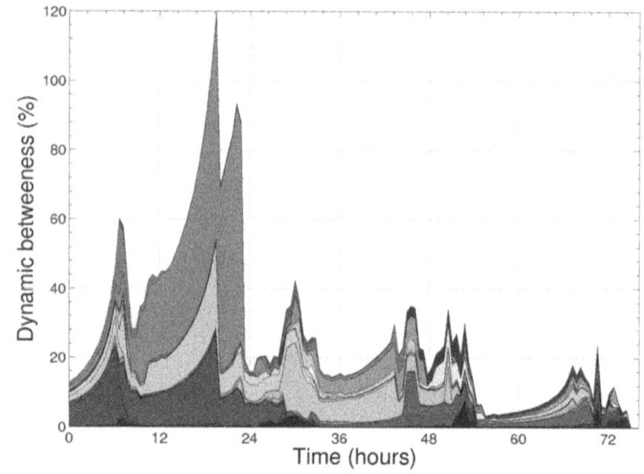

Figure 3: Evolution of dynamic betweeness centrality of nodes in Infocom05 trace

works, we introduce a metric that measures the influence a node has on the characteristic dynamic path length of the network. The nodes with highest influence are the ones whose removal from the network increases the most the average shortest dynamic path length of the network. To identify these nodes one may adapt to dynamic networks the notion of betweeness centrality already introduced for static networks. In the context of dynamic networks, we call it *dynamic betweeness centrality.* Consider a node i, first of all we measure the average of the shortest dynamic path length between all pairs of nodes s, t except paths from and to i. Then, we remove i from the network and perform the same measure. The dynamic betweeness centrality of i is defined as the ratio between these two measures. Formally, that is

$$x_i = \frac{\sum_{st} \mathcal{L}'_{st}}{\sum_{st} \mathcal{L}_{st}} , \tag{6}$$

where $\mathcal{L}'_{st}$ and $\mathcal{L}_{st}$ are respectively the shortest dynamic path lengths from s to t after and before removing i.

We measure the dynamic betweeness of all nodes at different times on the traces. Figure 3 shows the temporal evolution of the dynamic betweeness for the Infocom05 trace. On the figure, the different areas represent the measured metric (in %) of different nodes. We can observe that the dynamic betweeness of a nodes changes over time and that there are some nodes with very high influence whose removal would dramatically increase the dynamic characteristic path length of the network. Indeed, at the time of the highest pick, the most influential node decreases of almost 70% the average shortest dynamic path length of the network. We can also see the abrupt decrease of the average shortest dynamic path length at time 24 hours on Figure 2(a). The time average of the dynamic betweeness centrality of each network node allows one to rank the influence of each node during a time window. For instance in Figure 3 the highest value for the daytime period corresponds to node with ID 34.

Now let us look at the highest pick to see what happened in the network at that moment. We measure the impact of node 34 on the shortest dynamic paths from and to each other node (respectively denoted out/in shortest dynamic path on Figure 4). Figure 4 shows that node 34 has very high influence on node 31. We then apply our all-pairs-shortest-dynamic-path algorithm to find all the shortest dy-

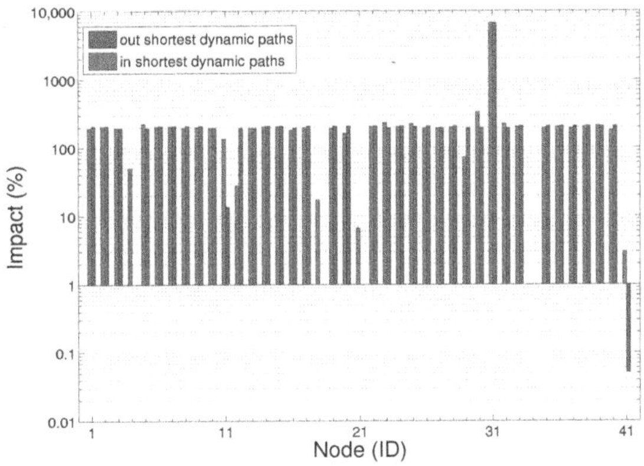

Figure 4: Impact of node 34 on the shortest dynamic paths from and to the other nodes at $t = 22$h (Infocom05 trace)

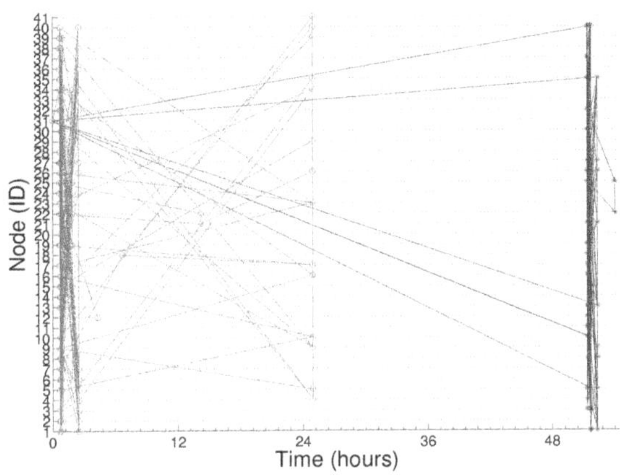

Figure 5: Shortest dynamic paths from node 31 at $t = 22$h in Infocom05 trace, before (in red) and after (in blue) removal of node 34

namic paths from node 31 to others before and after removing node 34. Figure 5 shows that node 34 is crucial for node 31 to efficiently communicate with other nodes at that instant. Therefore node 34 plays the role of a spatio-temporal shortcut between node 31 and the other nodes. This result can be observed for any node with a high dynamic betweeness. Indeed, if the so disconnected network part is a giant component, then the impact of the removal of a dynamic shortcut can dramatically increase the characteristic dynamic path length of the network.

4. DYNAMIC SMALL-WORLD MODELING

The previous section showed that the dynamic work small world phenomenon is intrinsic to a great diversity of dynamic network traces. Therefore, the definition of a model that abstracts this phenomenon in order to study the impact of its emergence on dynamic networks communications performances is a significant issue. In this section, we introduce a model for dynamic small-worlds and show its modeling and expressive power. As evoked in the introduction, this model aims to capture the dynamic small-world phenomenon resulting from human mobility behavior through human characteristics described in the previous section. That is:

- People move between clusters or communities
- Some people are nomad while the others are more sedentary. These nomadic nodes move from cluster to cluster and hence contribute to reducing the network diameter.

4.1 The model

Consider a mobile network in which each node belongs to a preferential attachment zone or clusters. We introduce two types of nodes: sedentary nodes which move only inside their preferential zone[3] and nomadic nodes which can move between zones. Nodes moving in two different zones cannot be connected and therefore nodes in different zones can communicate only through the movement of nomadic nodes. Initially, nomadic nodes are uniformly distributed over all clusters. The fraction of nomadic nodes is p which is a parameter of the model.

Obviously, when p equals to 0, there is no possible communication between nodes in different zones and hence the network is totally partitioned in disconnected zones. On the contrary, when p equals to 1, all the nodes are nomadic and hence there are dynamic paths between clusters and in consequence the network is highly connected. In consequence, the *rewiring process* consist in varying p from 0 to 1. We are interested in the properties of networks which are formed between these two extrema.

As we have demonstrated in a previous work [7], human mobility can be modeled with STEPS—a simple model implementing preferential attachment and attractor via a power law. We apply this mobility model to the movement of nomadic nodes. The node chooses the zone to move in with a probability inversely proportional to the distance between his preferential zone and the chosen zone. That is

$$P(Z = z_i) = \frac{\lambda}{(1 + d_{z_0 z_i})^\alpha}, \qquad (7)$$

where z_0 is the preferential cluster of the node, λ is constant of normalization and the power law exponent α a parameter of STEPS model.

4.2 Simulation Results

We simulate in MATLAB a mobile network of 1000 mobile nodes. The simulated network area of size 200×200 m^2 divided into 20×20 zones as shown in Figure 6 gives a density of 25000 nodes/km^2 which corresponds to the population density of a large city like Paris. We distribute uniformly nodes over zones and set the radio range to 10 m which corresponds to bluetooth technology. While varying the fraction of nomadic nodes from 0 to 1, we measure the dynamic clustering coefficient and shortest dynamic path length of resulting networks.

Figure 7 plots the resulting average over 10 simulations with the corresponding 95% confidence intervals. We observe that the shortest dynamic path length drops rapidly as soon as we introduce a small percentage of nomadic nodes (i.e. less than 10 %) into the network whilst the clustering coefficient remains very high. This result obtained from a simple model of evolution of a dynamic network demonstrates, identically to the simple model used to exhibit the

[3]following a random mobility model such as Random Waypoint

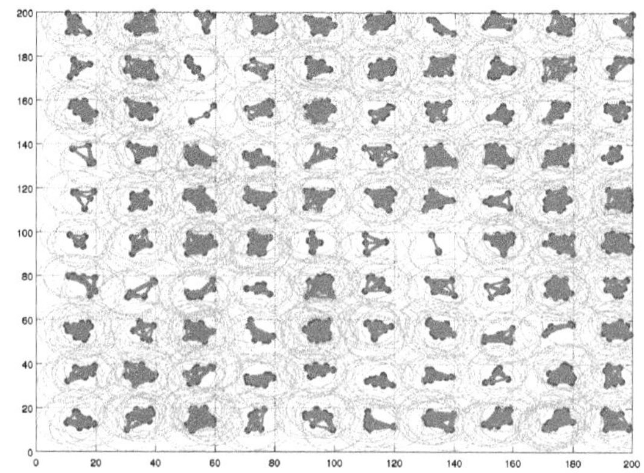

Figure 6: Small-world network configuration

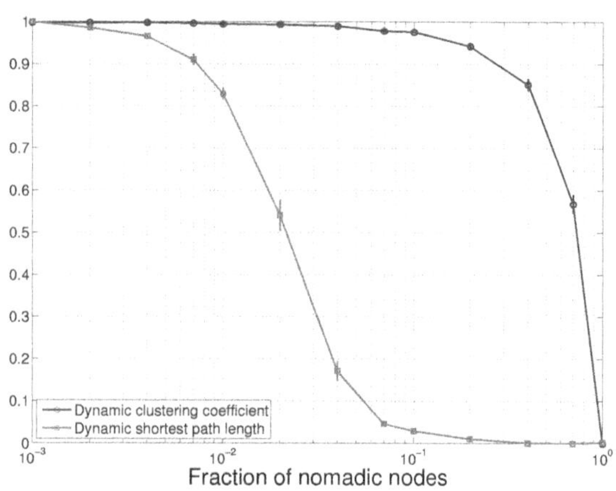

Figure 7: Small-world phenomenon in dynamic networks

emergence of the small world phenomenon in static networks [10], how the small-world phenomenon emerges from dynamic networks.

5. INFORMATION DIFFUSION IN DYNAMIC SMALL-WORLD NETWORKS

Following the understanding and modeling of the small world phenomenon in dynamic networks, this section aims to study the effect of the dynamic small-world phenomenon on information diffusion performances. We present analytical and simulation results obtained with the Susceptible Infected (SI) epidemic model. We apply SI model on a simplified version of the dynamic small-world model used in the previous section in which we make abstraction of nodes displacement inside sites. Nodes movements are limited to jumping from one site to another. Simulation results for the complete version of the model are also provided. Let us consider a square lattice of N sites with initially 1 node in each site, node i is associated to site number i. Assume that when 2 nodes are in the same site, they are connected. At each time t, mobile nodes jump to another site while sedentary nodes stay where they are. Initially, we suppose 1 infected node and 100% of other nodes susceptible. We formulate the dynamics of the infection of this network through a differential equation as follows.

Consider a node i. Let $x_i(t)$ be the random variable which denotes the probability of node i being infected at time t. Let us express the probability that i becomes infected between time t and $t+dt$. A static node i can be infected by a mobile node m only. This happens with the probability δ_{mi} that node m jumps to site i. The probability that the infection (i.e. information) is transmitted during the interval dt is βdt where β is the transmission rate—a standard parameter of the SI epidemic model. Summing over all the mobile nodes and then multiplying by the probability $1-x_i$ that i is not infected at time t , we obtain the differential equation

$$\frac{dx_i}{dt} = \beta(1-x_i)\sum_{m}\delta_{mi}x_m \ . \tag{8}$$

Conversely, for a mobile node, i, to be infected, it must be in contact with a mobile or static node already infected. If a node i jumps to a site j where resides an static node, this happens with probability $x_j + \sum_{m\neq i}\delta(mi)x_m$ and hence

$$\frac{dx_i}{dt} = \beta(1-x_i)(x_j + \sum_{m\neq i}\delta_{mj}x_m) \ . \tag{9}$$

If the destination site j is associated to a mobile node then this probability is $\sum_{m\neq i}\delta_{mj}x_m$ and hence

$$\frac{dx_i}{dt} = \beta(1-x_i)\sum_{m\neq i}\delta_{mj}x_m \ . \tag{10}$$

Combining all these cases, we obtain the following matrix differential equation describing the dynamics of the system

$$\begin{aligned}\frac{d\mathbf{x}}{dt} =& \beta(1-\mathbf{x})\circ\langle(1-m)\circ[\mathbf{D}(m\circ\mathbf{x})]+ \\ &+ m\circ\{\mathbf{D}\left[(1-m)\circ\mathbf{x}+\mathbf{D}(m\circ\mathbf{x})\right]\}\rangle \ ,\end{aligned} \tag{11}$$

where $\mathbf{x} = [x_1 \ldots x_N]'$ is the random vector of nodes infection probability and m is the binary vector in which $m_i = 1$ for a mobile node and $m_i = 0$ for a static node. The matrix

$$\mathbf{D} = \begin{bmatrix}\delta_{11} & \ldots & \delta_{1N}\\ \vdots & \ddots & \vdots \\ \delta_{N1} & \ldots & \delta_{NN}\end{bmatrix} = \lambda\begin{bmatrix}1 & \ldots & \frac{1}{(1+d_{1N})^\alpha}\\ \vdots & \ddots & \vdots \\ \frac{1}{(1+d_{N1})^\alpha} & \ldots & 1\end{bmatrix}$$

is the stochastic matrix describing the stationary state of the system.

To have an approximated solution of Equation 11, we integrate it by numerical method. The results are shown in Figure 8 which plots the time evolution of the fraction of infected nodes for a 10000 nodes network. The four curves correspond a fraction of mobile nodes p_m equals to $0.1\%, 1\%, 10\%, 100\%$ respectively. The coefficient α for mobile node is equal to 0 and the transmission rate β is equal to 1. It is interesting to see that for a low fraction of mobile nodes, epidemic spreads very slowly because this fraction is still below the percolation threshold of the network, while with only 10% (i.e. as soon as the small-world structure emerges) the epidemic breaks out rapidly. We also ran simulations with a network of 100 nodes, α equals to 0, β equals to 1 and different values for the fraction of mobile nodes. The average over 100 simulations is then compared with the corresponding analytical results. Figure 9 shows

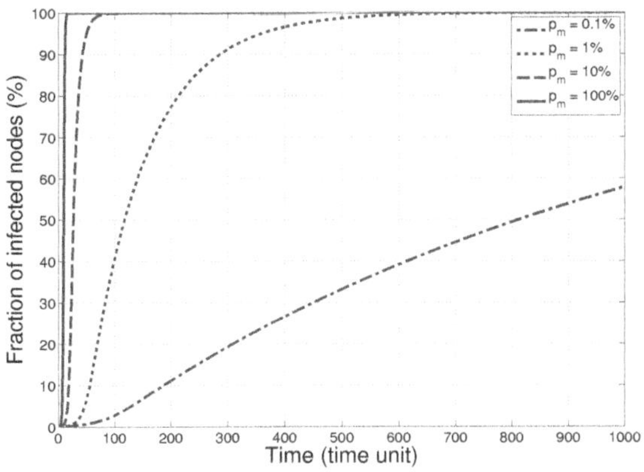

Figure 8: Evolution of fraction of infected nodes over time (theory)

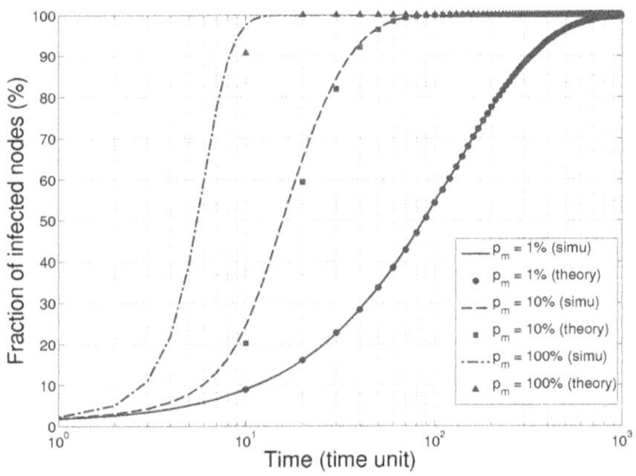

Figure 9: Comparison of analytical and simulation results

the strong accuracy and compliance of our analytical model with the simulation results. Finally, we run simulation with the complete version of the small-world model. The result is the average over 10 simulations of a 100 nodes network with the same configurations. Figure 10 shows a similar result compared to the first case. The effect of displacement time between zones only spreads the curves so that the epidemic take much longer time to take off.

6. CONCLUSION

In this paper, we first proposed a model that explains the formation, evolution, behavior and performances of dynamic small-world networks. In this model, from a clusterized network composed of nodes with localized moves, we perform a rewiring process which consists in introducing nomadic nodes into the network. These nodes play the role of bridges between disconnected clusters and reduce significantly the dynamic shortest paths between nodes in the network. From the study of real mobility traces we showed that our model expresses salient properties observed in real dynamic networks. Finally, based on this model and from a joint use of analytical analysis and simulations, we have studied information diffusion in dynamic small-world networks, for the SI epidemic model. We showed that the emergence of a dynamic small-world structure in dynamic networks is accompanied by a sudden improvement of information diffusion performances in the network. Our on-going work aims to address this issue for more complex epidemic models such as SIR or SIRS. A study of the percolation threshold and percolation process in dynamic small-world networks is also an significant problem to be considered. Although in this paper, one contributing factor was found, further unique properties of dynamic networks might be contributing as well and hence should also be considered. For our best knowledge, this is the first model of small-world phenomenon in dynamic networks. This model could be integrated in existing dynamic network simulators (e.g. theONE) to evaluate protocol performances (at network or transport level) in such type of network.

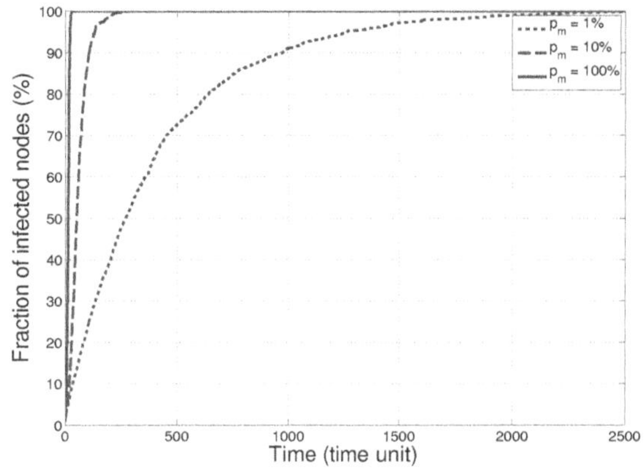

Figure 10: Evolution of fraction of infected nodes over time with the complete version of the model (simulation result)

7. REFERENCES

[1] Augustin Chaintreau, Pan Hui, Jon Crowcroft, Christophe Diot, Richard Gass, and James Scott. Impact of human mobility on opportunistic forwarding algorithms. *IEEE Transactions on Mobile Computing*, 6(6):606–620, June 2007.

[2] Aaron Clauset and Nathan Eagle. Persistence and periodicity in a dynamic proximity network. In *DIMACS/DyDAn Workshop on Computational Methods for Dynamic Interaction Networks*, 2007.

[3] JM Kleinberg. Navigation in a small world. *Nature*, 406(6798):845, August 2000.

[4] Jérémie Leguay, Anders Lindgren, James Scott, Timur Friedman, and Jon Crowcroft. Opportunistic content distribution in an urban setting. In *Proceedings of the 2006 SIGCOMM workshop on Challenged networks - CHANTS '06*, pages 205–212, New York, New York, USA, September 2006. ACM Press.

[5] M. E. J. Newman. The Structure and Function of Complex Networks. *SIAM Review*, 45(2):167, November 2003.

[6] MEJ Newman. *Networks: an introduction.* Oxford University Press, 2010.

[7] A.D. Nguyen, Patrick Sénac, Victor Ramiro, and Michel Diaz. STEPS-an approach for human mobility modeling. In *NETWORKING 2011*, pages 254–265. Springer, 2011.

[8] John Tang, Mirco Musolesi, Cecilia Mascolo, and Vito Latora. Temporal distance metrics for social network analysis. In *Proceedings of the 2nd ACM workshop on Online social networks - WOSN '09*, page 31, New York, New York, USA, August 2009. ACM Press.

[9] John Tang, Mirco Musolesi, Cecilia Mascolo, and Vito Latora. Characterising temporal distance and reachability in mobile and online social networks. *ACM SIGCOMM Computer Communication Review*, 40(1):118–124, January 2010.

[10] D.J. Watts and S.H. Strogatz. Collective dynamics of 'small-world'networks. *Nature*, 393(6684):440–442, June 1998.

Adaptive Neighbor Discovery for Mobile and Low Power Wireless Sensor Networks

Dimitrios Amaxilatis
Computer Technology Institute & Press "Diophantus"
Computer Engineering and Informatics Department, University of Patras
Patras, Greece
amaxilat@cti.gr

Georgios Oikonomou
Computer Engineering and Informatics Department, University of Patras
Patras, Greece
oikonomou@ceid.upatras.gr

Ioannis Chatzigiannakis
Computer Technology Institute & Press "Diophantus"
Computer Engineering and Informatics Department, University of Patras
Patras, Greece
ichatz@cti.gr

ABSTRACT

Wireless Sensor Networks are by nature highly dynamic and communication between sensors is completely ad hoc, especially when mobile devices are part of the setup. Numerous protocols and applications proposed for such networks operate on the assumption that knowledge of the neighborhood is a priori available to all nodes. As a result, WSN deployments need to use or implement from scratch a neighborhood discovery mechanism. In this work we present a new protocol based on adaptive periodic beacon exchanges. We totally avoid continuous beaconing by adjusting the rate of broadcasts using the concept of consistency over the understanding of neighborhood that nearby devices share. We propose, implement and evaluate our adaptive neighborhood discovery protocol over our experimental testbed and using large scale simulations. Our results indicate that the new protocol operates more efficiently than existing reference implementations while it provides valid information to applications that use it. Extensive performance evaluation indicates that it successfully reduces generated network traffic by 90% and increases network lifetime by 20% compared to existing mechanisms that rely on continuous beaconing.

Categories and Subject Descriptors

C.2.3 [**Network Operations**]: Network monitoring; C.2.4 [**Distributed Systems**]: Distributed applications

General Terms

Design, Experimentation, Performance Evaluation

Keywords

Wireless Sensor Networks, Low Power Networks, Mobile Networks, Neighborhood Discovery, Adaptive Protocols, Protocol Design, Experimental Evaluation

1. INTRODUCTION

Over recent years, Wireless Sensor Networks are proposed as a valuable tool towards the realization of Future and Smart Cities. Many recent academic and industrial studies provided the knowledge to make large scale WSN installations reliable. In many cases WSNs are characterized as low power and lossy networks. Devices operate solely on small power cells with no or minimal ability to recharge while communication with nearby nodes is neither guaranteed nor cheap in terms of power requirements. As a result, already known and studied protocols for Computer Networks are not always suitable for WSNs. Although communication failures can be dealt with methods like retransmission, energy concerns rise and impose the need for a different way of dealing with such problems. Concepts like Fault Tolerance and Sleep Scheduling were introduced in the theoretical models to design robust and self maintained algorithms.

One of the most important issues faced by algorithms running on WSNs is the erratic behavior of the communication channel. It is quite common for sudden changes in the quality of the communication medium or nodes with poor communication channel (i.e., due to local interference, or node positioning), to greatly affect the stability and performance of an entire network running an otherwise stable software module. It is therefore critical to provide Neighborhood Discovery mechanisms for establishing local connectivity awareness with the ability to characterize the stability of the neighboring nodes so that the higher applications can adjust their operation to frequent and/or significant topology changes. The need for such a mechanism is common in many protocols (i.e., the routing protocol [18] and the clustering protocols [3, 4, 17]).

A number of different methodologies are proposed in the relevant bibliography. We here focus on some very recent results. In [22] a leader based neighborhood discover protocol is presented. Initially, a node is selected as leader that starts the discovery process by broadcasting a special request message. Nodes respond to the leader with short feedback messages. When the leader has identified all its neighbors, the leadership passes to the next node until all nodes have discovered their neighbors. [23] proposes a probabilistic scheme for solving the neighborhood discovery problem. Each neighbor is characterized in 2 states (Good or Bad) and nodes become neighbors or are removed from the neighborhood based on a specific probability function. In [14], a polling based methodology is adopted for neighborhood discovery as an integrated part of the MAC layer for WSNs. Also in [13] an adaptive beaconing protocol that detects only the size of the neighborhood is presented.

Our work is inspired by Trickle [16], a protocol that comes to solve the problem of consensus in Distributed Systems using the concept of "polite gossip". Devices, in need of a consistent shared piece of information, notify the rest of the network of their own version. Incoming versions are merged with the local one and a new common version is generated. The "polite" part of the solution comes up when information exchanged requires no merging. Then the information exchange rate is relaxed, thus less traffic is generated and nodes save their much needed power. If, for any reason, an inconsistent version is received, messaging rates rise to maximum to make sure that all devices get notified immediately. In scenarios where no changes happen in the network (i.e., during the night in an office building monitoring scenario) message rates can be reduced and energy requirements can be lowered, while during highly unstable situations (i.e., during working hours) higher message rates can be used to achieve the required results.

In this work we present an adaptive fully decentralized neighborhood discovery mechanism that is based on the periodic broadcasting of beacon messages. In contrast to [23], we do not use simple HELLO messages; our beacons include MAC information of all the neighbors of each node thus allowing nodes to establish a common understanding of how the local neighborhood is perceived by nearby nodes. Moreover, we totally avoid a fixed time interval of periodically broadcasting the message updates; instead our protocol proactively increases the period when nodes reach a consensus in the spirit of Trickle. We further improve the performance of the protocol by introducing mechanisms to adapt to the size and communication quality of the network without any prior knowledge of the topology. Our experimental findings provide strong evidence that we manage to reduce the overhead imposed on the communication medium when operated in low dynamics networks. Furthermore, we propose a series of improvements over the initial protocol to make it suitable for topologies with mobile nodes. Our protocol manages to limit regular message exchanges to mobile nodes thus minimizing the energy consumption of static nodes resulting. Our experimental evidence indicate that the adaptive mechanism successfully reduces message transmissions and minimizes the energy consumption. Duty cycling can further reduce power needs and thus lead to longer operation of the WSN.

Our implementation is part of the Wiselib WSN code library [7], that allows OS-independent implementations for various WSN platforms such as Contiki, TinyOS and iSense. To reduce the complexity and code size of the protocols running on each sensor, we follow the approach of [19]. We implement our protocol as an external service that other protocols can use. The Unifying Link Abstraction layer lies between the physical Link Layer and the other protocols and applications executed on the sensor node. Its purpose is to abstract the link layer's details from protocols' send/receive functions and transmit node's messages efficiently. Higher-layer protocols rely on the ULA to deliver messages to all nodes in the neighborhood and ULA can then decide on the optimal transmission rate of messages or broadcasting power required based on predefined thresholds and options. ULA also notifies other protocols of neighborhood changes.

We conduct extensive simulations and real device experiments as we consider both to be essential parts of the development process of robust and efficient algorithms. For the simulations we scale up to 1000 nodes while in our experiments with real devices we work in testbeds of up to 35 nodes, and prove that on average our protocol achieves neighborhood discovery using 90% less messages and thus allows battery powered devices to consume less power and operate for 20% more time on the same power.

In Section 2 we present a representative periodic beaconing mechanism with fixed transmission periods. The algorithm is implemented as part of ULA, along with various extensions to improve its suitability for low power and Lossy networks. We use this algorithm as a reference to evaluate the performance of our algorithm that is presented in details in Section 3. We showcase the results of the evaluation we performed on large scale simulations and real hardware platforms in Section 5. Finally the paper is concluded in Section 6 where we also discuss future research directions.

2. FIXED NEIGHBOR DISCOVERY

Beaconing is currently the most effective methodology for neighborhood discovery. Many protocols for Ad hoc Mobile Networks and WSNs use it to identify neighbors either as an embedded sub-protocol or as an external service provided by another protocol. ULA attempts to define the characteristics of a neighborhood discovery protocol for WSN and indicates that each implementation should provide a registration mechanism for applications to exchange data on top of periodic beacons (using piggybacking) and a Notification mechanism for higher applications to receive changes in the neighborhood.

With the term beaconing [3] we refer to the exchange of special messages containing information not only for the presence of the node itself but also for its view of the neighborhood. Each beacon contains at least the full list of neighbors available at the moment which in general are the nodes whose beacons are regularly received and comply with a set of given requirements. Also these special messages we use, unlike the HELLO messages in [23], contain indicators to estimate the state of the communication channel and classify neighbors. Based on the information contained (more details in 4.1) nodes can detect bad neighbors and assess link symmetry.

In this work we use the protocol presented in [3] (which we call the Fixed-ND protocol) as a reference protocol for evaluating our solution by conducting a comparative experimental study. The selection of [3] is because the protocol is already implemented in Wiselib (see Sec.4), is thoroughly tested in the context of project WISEBED [11] and used in other experiments (e.g., [8],[10]). The protocol uses beaconing on fixed intervals to detect changes in the topology. When no beacons are received from a particular node after a predefined timeout period, the node is dropped from the neighborhood. The beacons are also used to assess the quality and symmetry (whether they are Unidirectional or Bidirectional) of the links. In a Unidirectional link only one of the nodes can receive the beacons of the second while Bidirectional nodes can have two way communication.

In more details, for each received beacon message, Fixed-ND examines a number of indicator values which define the quality of the communication channel with it's neighboring node. Then, based on these indicators, it decides if a node should be considered as a "stable" neighboring node. More specifically, if the Fixed-ND receives at least 3 consecutive beacons from a single neighbor whose indicator values are above certain thresholds then the neighbor is marked "stable". The stable neighboring nodes are attached to each broadcast packet and when a node receives a beacon, it checks whether its address is listed in the received message. In this way, Fixed-ND is able to recognize whether the communication with the neighboring nodes is bi-directional or not. If Fixed-ND stops receiving beacons from a neighbor, or if the beacons received do not match the defined indicators, it will remove the node from the list of neighbors after a predefined amount of time.

The indicators used for examining the quality of the communication channels are (a) the LQI of the received beacons, (b) the RSSI of the received beacons, (c) the link associativity [21] (the time, in beacons, that nodes are associated, i.e., they retain a connection), (d) the node stability [6] (the average link associativity for all neighboring nodes) and (e) the reverse node stability [10] (defined as a result of experimentation, indicating the average numbers of consecutive beacons that successfully delivered on a destination).

The goal is to filter out the neighbors that have poor communication channel quality. The nodes will consider the broadcaster as a neighboring node only if it receives a number of consecutive beacon messages with RSSI/LQI above a certain threshold. In order to avoid occasional short channel degradation to negatively impact the above strategy, we allow a node to miss a number of beacons within a given period of time before removing it (called the *timeout period*). However we set a second RSSI/LQI threshold; message beacons with RSSI/LQI below this threshold are dropped (and therefore count towards the number of beacons that are missed).

Fixed-ND can parameterize its mode of operation based on the requirements of the higher-level applications. Depending on the requests, it fine tunes (i) the periodicity of examining the condition of the nearby devices, (ii) the frequency of generating notifications for the changes in the nearby topology, (iii) the indicators that will be used to characterize the quality of the communication channel, and (iv) the contents of the beacon message.

3. ADAPTIVE NEIGHBOR DISCOVERY

We now present our adaptive neighbor discovery algorithm which we call Adaptive-ND. Our goal is to avoid constant periodic beacon exchanges, instead we wish to adapt the beacon broadcast period to achieve low energy consumption when the network is relatively stable. The beacon interval is controlled by each node independently based on its local perception on the consistency of the network, i.e., the changes to the local neighborhood detected by the node. We use a basic rule: as long as a node does not detect any changes on its neighborhood (i.e., *it is consistent*) the beaconing interval period is doubled until a *maximum beaconing interval period* is reached. Whenever a changed or a new neighboring node is detected (or lost), the beaconing interval is reduced to a predefined *minimum beaconing interval period.*

In more detail, Adaptive-ND nodes transmit a maximum of one message Beacon during each interval period. During this period, the node examines all beacons received by the neighboring nodes; a beacon that has not been sent from a known neighbor (e.g., a new node) is said to be *consistent* and if the node only receives consistent beacons it is characterized as *consistent.* If an *inconsistent* message beacon is received, i.e., a beacon that originates from an unknown node, the node becomes *inconsistent.* Therefore, consistency is defined by the merging incoming beacons with the local neighborhood status.

When an interval expires, nodes decide on the next interval's duration and the resulting message rate. If the node is consistent, the next interval will be of double duration and the message rate will be cut down to half. If on the contrary, the node is not consistent, the interval goes down to a minimum duration and message rate rises to the maximum. The minimum and maximum beaconing intervals are properly selected by the network designer in order to achieve fast convergence and low message rates during consistent periods (in the Sec. 5 we conduct experiments on real devices and large scale simulations to understand the effect of the minimum and maximum beaconing intervals on the performance of our protocol). Whenever a node decides on a new beaconing interval, a time moment t is selected and marked as the transmission event time. At the time of t, the node will check the number of consistent beacons received and decide whether to send his own beacon or not. When the node is inconsistent, a message is sent to notify others of its current state. The time t is a random time during the interval to ensure low congestion at the moment of transmission.

3.1 Control of the Adaptation Process

We control the adaptation process of Adaptive-ND to the unique characteristics of different WSNs by setting the following three protocol parameters:

I_{min} – the minimum time interval

I_{max} – the maximum time interval

k – the number of consistent messages per interval required for a node to become consistent

I_{min} **and** I_{max} control the beaconing rate of our protocol. Inconsistent nodes transmit a beacon every I_{min}. To achieve better response time to changes in the neighborhood we need to set I_{min} to a very low value but not too low as this would cause flooding and would degrade the communication channels. I_{max} affects only consistent nodes and represents how relaxed beaconing can become when consistency is achieved. The more we relax beaconing the higher response times to changes become.

The value for k is used to decide the state of a node and in our scenario it depends completely on the size of the neighborhood of each individual setup. k is compared during the transmission events to an internal counter c of the consistent messages received. If enough consistent messages are received ($c >= k$) there is no need to send a beacon as we assume that the network is stable. If on the other hand, less than k consistent messages are received ($c < k$), we consider that our knowledge of the neighborhood may be outdated or incomplete and therefore a new Beacon is transmitted. The value of c is set to 0 after each new Interval to count new beacons during it. As a result, when we have a dense network, higher values for k are required to keep track of the neighborhood. When we have a sparse setup with a wrong value for k, the consistency requirements may never be met.

3.2 Characterizing Node Consistency

From the above it is clear how the value of k is crucial to the overall performance of our protocol. Choosing a certain value may cause false inconsistencies or drop events resulting to unnecessary high message exchange rates that cancel out the advantages of the adaptive operation. We propose and evaluate two different versions of Adaptive-ND that are influenced differently by this parameter. Experiments presented in Sec. 5 clearly highlight the advantages and disadvantages of each approach.

FixK maintains a constant value for k during the lifetime of the network. A node expects a neighborhood of size close to k and does so uniformly during its whole operation. This design choice is more straight forward and suitable for evenly populated networks where we can decide prior to installation what is the size of the expected neighborhood.

AverageK chooses different values for k on each node of the network after every transmission event. Each node adjusts its k value according to the number of its consistency during past Intervals. The value of k is re-calculated using Alg. 1: The advantage of this solution is that each node

Algorithm 1 Calculating new value for k

$interval_{counter} + +$
$consistent\_mess_{total} += consistent\_mess_{received}$
$k_{current} \leftarrow ceil(consistent\_mess_{total}/interval_{counter})$
$k_{new} \leftarrow k_{prev} * w + k_{current} * (1 - w)$
$consistent\_mess_{received} \leftarrow 0$

uses an estimate of the expected consistent messages and the decision for the next transmission event is more accurate on a local scale. We expect this approach to be more efficient in terms of congestion and power in dynamic networks where the neighborhoods are periodically changing, or networks with unevenly distributed sensors. To avoid sharp changes in the value of k, the new value is actually a linear combination of both the old and the new value with adjustable weights.

3.3 Assessing Link Quality Indicators

Link Quality Indicators (LQI) provided by the MAC layer for each received incoming beacon are used to better evaluate the neighboring nodes that are either too far away or with limited transmission capabilities. The goal is to filter out the neighbors that have poor communication channel quality. Using such indicators has proven to be a reliable solution when trying to define the exact neighborhood of a node [3]. The nodes will consider the broadcaster as a neighboring node only if it receives a number of consecutive beacon messages with LQI above a certain threshold. In order to avoid occasional short channel degradation to negatively impact the above strategy, the algorithm allows a node to miss a number of beacons within a given period of time before removing it (called the timeout period). A second LQI threshold is also set; message beacons with LQI below this threshold are dropped (and therefore count towards the number of beacons that are missed).

During the first steps of our hardware evaluation we realized that certain nodes on the outskirts of the neighborhoods were constantly causing instability mainly because of high message loss rate. We decided to address that problem by introducing a mechanism to identify and exclude these weak connections based on the LQI value of each incoming message. We use 2 different LQI thresholds, one strict and one more relaxed to characterize incoming beacons as presented in Alg. 2. The first is used to accept any node in the neighborhood and the second more relaxed threshold is required for the rest of the time, as the node has proven its reliability.

Internally, 2 LQI values are kept for each individual Neighbor. The LQI value of the last incoming message, and a value provided by the Neighbor which is the LQI value of the last outgoing message. These values can also set a node to an inconsistent state. If a node receives a beacon where his reported LQI value is greatly different than the previous one, he considers himself inconsistent.

Algorithm 2 Processing Beacons

```
if node is in neighbor_list then
    if lqi ≥ threshold_min then
        neighbor_list.update(node)
    else
        drop beacon
    end if
else
    if lqi ≥ threshold_max then
        neighbor_list.add(node)
    end if
end if
```

3.4 Dealing with Low Power and Lossy Links

Link failures are not unusual events in WSNs even for neighbors whose LQI are above certain thresholds. Experimental evidence indicates that message transmissions, even from nearby nodes, will not be received all the time. To avoid removing from our neighborhood nodes that have proven their consistency but for some reason occasionally lose a message or two, we define a timeout period for which a link is considered valid even if a message beacon is not received.

Previous Experiments [3, 10, 23] have shown that up to two beacons may be lost (especially on dense networks) so as a good practice we set the timeout close to the time period equivalent to 3 consecutive beacons. However, this may turn out to be non-trivial as beacon broadcasting periods are not predefined and change during the execution of the protocol as part of the adaptation process. Note that using a higher timeout period will cause slow responses to nodes leaving the neighborhood, while lower timeouts will cause events that do not correspond to physical events. As an extra feature to ensure that beaconing rates do not fall too low and reduce the validity of the reported neighborhood we add an override transmission event to ensure that beacons will be transmitted to refresh the information available on the nodes.

Taking into consideration the relaxed beaconing of Adaptive-ND defining the timeout period is not as straightforward as for Fixed-ND. Timeouts should not only be bounded by I_{max} as nodes can choose not to transmit a message during every interval. Therefore, to avoid falsely deleted neighbors in our implementation, we consider that the timeout should not also be less than $\frac{5}{2}\ I_{max}$. This value is computed as the more relaxed interval between two successfully received consecutive beacons while 2 communication failures can happen. $Timeout = 2T_{lost}^{max} + T_{succesful}^{max} = 2I_{max} + \frac{I_{max}}{2} = \frac{5I_{max}}{2}$
Of course the timeout selection depends completely on the

special network characteristics but the above threshold provided a reliable behaviour during all our experiments.

4. IMPLEMENTATION DETAILS

We implement the Adaptive-ND algorithm using Wiselib [7], a code library for WSNs, that allows OS-independent implementations for various WSN platforms. The Fixed-ND algorithm is already provided by the library [3]. Algorithms are implemented on a subset C++ using templates and callbacks, but not virtual inheritance or exceptions that are not available on older platforms. Algorithm implementations can be recompiled for several platforms and firmwares, without any code changes. Wiselib currently supports platforms like iSense, Contiki, TelosB and the WSN simulator Shawn[12] for testing and debugging, with plans to support mobile devices like Android and iPhone. The ability to run the same application on a simulator (Shawn) and with no changes evaluate the operation of our implementation on real hardware is a huge advantage that really speeds up development.

4.1 Message Beacons

The beacons exchanged by Adaptive-ND and Fixed-ND follow the ULA format. The beacon header consists of a $1Byte$ Message id (required for all Wiselib Applications) and 2 counters for the number on neighbors and application payloads contained in the message. The list of neighbors is an array of $2Byte$ neighbor MAC addresses in the plain implementation while when using LQI values we add after each address the respective LQI value as an unsigned $8bit$ integer. Finally after the information about the Neighborhood we add, if they exist, the payloads of the applications registered on our algorithm, as triples of Application Id, Size and Data. Due to hardware limitations from the Radio of Wiselib platforms, a single beacon cannot contain an unlimited number of neighbors or piggybacked payloads of very big size. In Wiselib, packets are limited to $116Bytes$ and as a result, it may include a maximum of 56 neighbors on each beacon when we do not use LQI thresholds or 37 neighbors with LQI. If we need to operate on a larger neighborhood we can use the Wiselib Fragmenting Radio and transmit beacons larger than a single message reach a neighbor of up to 255 nodes which is the limit of the unsigned integer counter used for the neighbors counter in the message header.

Working with Wiselib also gives us the freedom to work on an abstraction from the actual MAC layer implementation and helps move with minimal effort to 6LoWPAN [15] and IPv6 addressing. In Wiselib data types like Node Id or platform specific characteristics like maximum packet length are defined during compilation target platform and applications are developed in a generic format disconnected from the actual deployment platform (i.e., using `make isense` or `make tinyos-telosb` we compile the same application for iSense or TelosB devices). As a result, the transition to 6LoWPAN, when this becomes available in Wiselib, will simply be a matter of compiling the application to the new target. 6LoWPAN offers packets of $1280Bytes$ and $8Byte$ addresses, making it possible to store up to 150 unique neighbors in each beacon.

4.2 Piggybacking

It is common among many existing algorithms to periodically exchange data as new events are generated. Normally this is done by broadcasting messages containing the updated information. This can often lead to the generation of excess messages waste of precious memory space when more that one algorithm is running on the same device and needlessly duplicate the previous mechanism. Fixed-ND provides a piggybacking technique that can be significantly reduce the rate of broadcasting messages, by allowing higher-layer applications to take advantage of its own broadcasting mechanism. The piggybacking technique exposes an interface that allows other applications to attach their data on the beacon messages of the Fixed-ND, and get notified by a callback mechanism when new data are received. Fixed-ND will gather all the data from the different applications and will add them to the payload of the next beacon message, and upon receiving the beacon it will unpack the data and deliver them to the relevant receivers. For example, a Clustering protocol can use this technique so that the nodes also broadcast their role (cluster head, cluster member).

5. PERFORMANCE EVALUATION

To evaluate the performance of Adaptive-ND we conduct a series of large scale simulations on Shawn and real world experiments in Testbeds of up to 35 iSense[1] nodes. We comparatively evaluate the performance of our protocol with the Fixed-ND protocol which we consider to be the reference. We consider as pointed in [5, 20] that an extensive analysis of any WSN application based on simulations before deployment on real hardware is crucial and can provide a beneficial initial feedback on fully managed conditions and no outside interference. The metrics we concentrate on during experiments and simulations as well as the experimental setup are described in the rest of this section.

Messages: With the term *messages* we describe the Beacons each algorithm transmits over the 802.15.4 wireless network. Message exchanges are one of the most costly operations in WSNs and with Adaptive-ND we limit the amount of messages needed to identify and maintain neighbors.

Events: All changes a neighborhood discovery protocol reports to any registered applications and/or higher-level protocol. In general we report 4 types of events. The NewNeighbor (`NB`) events, occur when a node detects for the first time a new neighbor and represent uni-directional links. The NewBidi (`NBB`) events occur when both 2 nodes consider each other as neighbors and form a bi-directional link.

The NeighborLost (`NBL`) events when nodes demote a bi-directional to a uni-directional neighbor and finally the NeighborDrop (`NBD`) events when there is no communication with a neighbor and nodes remove him from their lists.

Lifetime: This term is used only in the device experiments where nodes operate using a limited power cell. *Lifetime* of a node is the time until the device runs out of power and is not part of the neighborhood any more.

Stability: A neighborhood is stable when all nodes and neighbors are correctly identified. We expect that in a stable neighborhood a very limited number of changes that result in new *events* occur.

Density and Diameter: *Density* of a WSN is the number of existing nodes divided by the area covered by the network. The more nodes we place in a specific area, the bigger neighborhoods get, and the more messages nodes need to process during the execution of the neighborhood discovery mechanism. *Diameter* is the longest shortest path between any two nodes of the WSN.

Simulation and Experimental goals: The goal of the Shawn simulations is to examine the scalability and performance of Adaptive-ND in controlled environments (topologies) of up to 1000 nodes. Each topology describes different a WSN based on two characteristics critical for the operation of WSN algorithms, density and diameter. Also we performed a number of comparison experiments between the previous ND protocol available in Wiselib and Adaptive-ND.

The experiments on real devices were divided into 3 stages. First we tested the functionality and performance of our implementation with repeated 20 minute experiments, to eliminate any error and confirm the correctness of our implementation on real hardware. Also we used these experiments to understand how the different choices in I_{min}, I_{max} and k affect Adaptive-ND. Secondly, we experimented with battery powered devices and compared the lifetime of a network operating on the two protocols and proved that the Adaptive-ND approach would consume less power even without any duty cycling. Finally we experimented with mobile nodes to identify how Adaptive-ND adapts movement and the behavior in terms of messages, events and stability.

5.1 Large-Scale Simulations

For the simulations presented below we used topologies with fixed network density of 9 neighbors and increasing diameter, ranging from 1 to 112 hops The results presented in the rest of this section were achieved using the AverageK version of Adaptive-ND. In the simulations, we do not use any LQI threshold as the LQI values in Shawn are constant based on the distance between nodes and do not impact on the results. Time is calculated in synchronous simulation rounds of Shawn where Incoming messages are processed, internal timers and operations are executed and then all existing outgoing messages are sent.

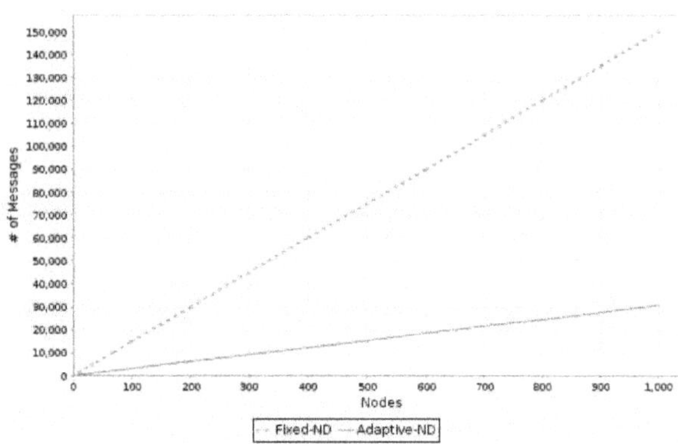

Figure 1: Transmitted Messages from Fixed-ND and Adaptive-ND in a Reliable Network of different sizes ($n \in [10, 1000]$)

The first set of experiments acts as a comparison of Adaptive-ND and Fixed-ND algorithms and a proof of the correctness of the Adaptive-ND implementation. We run multiple experiments in the topologies described above ranging from 10 to 1000 nodes and for the duration of 300 rounds. Adaptive-ND and Fixed-ND have the same timeout period of 20 rounds and the beaconing period of Fixed-ND is set to be the same as the I_{min} of Adaptive-ND, 2 rounds. We note that both protocols report the same neighborhoods and generate the same *events* in all experiments. This is an indicator for the correctness of the implementation since in all simulations no communication or device failures occur. Therefore it was easy to compare the results of each protocol. However, as we present in Figure 1 there is a great difference on the amount of traffic generated by the two algorithms. As there are no failures, nodes quickly identify their neighborhoods and then reduce the rate of the transmission events. We calculate that Adaptive-ND only needs an average of 10% of the *messages* exchanged by Fixed-ND in all topologies to achieve the same result.

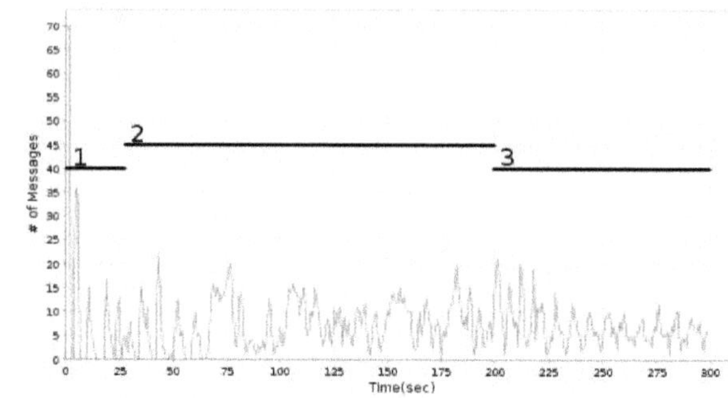

Figure 2: Transmitted Messages from Adaptive-ND in a Lossy Network (with a 50% message loss rate) of 70 nodes, Period-1: Initialization, Period-2: Lossy, Period-3: Reliable

In Figure 2, we showcase the behaviour of Adaptive-ND in a 70 node topology with lossy links (50% of the message exchanges fail) for the duration of 300 rounds. First we have the initialization phase (first 60 rounds of the simulation), when all devices are in an inconsistent state and high beaconing rates are required to quickly identify neighbors. After all neighbors are correctly identified and reported, all nodes are consistent and the beaconing rates go down to a minimum. Then we introduce a 50% message failure chance until round 200 and as we can see more messages are transmitted over time as nodes lose their consistency and start beaconing to identify the new state of the neighborhood. After round 200 we remove the message failures and observe how nodes become consistent once more and intervals between beacons become bigger. This experiment proves how Adaptive-ND adapts to the quality of the links to maintain communication with their neighbors.

The behavior of our protocol during the lossy period (see previous experiment) led us to the next experiment. We compare Fixed-ND and Adaptive-ND on the same topologies while communication failures occur. We expect that as we use same timeout periods in all simulations, Fixed-ND will have an advantage since to maintain a connection a node needs to receive a single beacon every 20 rounds. The probability for all the messages to fail during a 20 round period is $P^{\text{Fixed-ND}}_{drop\ all} = 0.5^{10}$. Since Adaptive-ND on a consistent node attempts to send a beacon every $I_{max} = 8$ rounds and the timeout period is 20 rounds there may be only 2 attempts to send new beacon during the timeout period. The probability for both messages to fail is $P^{\text{Adaptive-ND}}_{drop\ all} = 0.5^2 = 0.25 >> P^{\text{Fixed-ND}}_{drop\ all}$. This results to Figure 3(a) where Fixed-ND generates far less events. We can see that difference also in Figure 3, since Adaptive-ND attempts to broadcast 57% less

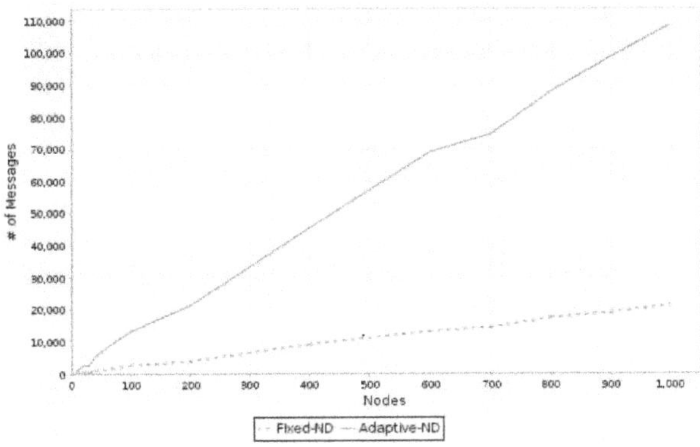

(a) Events Generated

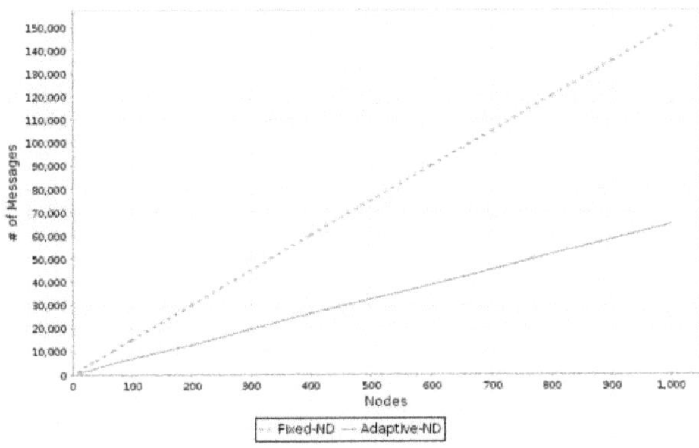

(b) Transmitted Messages

Figure 3: Fixed-ND and Adaptive-ND in an Lossy Network (50% message loss) of different sizes ($n \in [10, 100]$)

messages than Fixed-ND. Even in networks with low quality communication channels a loss rate of 50% is an extreme situation but if this is the case we can modify the Adaptive-ND I parameters to transmit messages more frequently and achieve an Fixed-ND like behaviour. The design choice to minimize transmission of messages revealed the trade-off between beaconing intervals and stability.

5.2 Real Device Experiments

We run our real device experiments on Wisebed[9, 11], a large-scale heterogeneous WSN research testbed installed in 9 geographically disparate European Universities and Research Institutes. We carefully selected 15 iSense devices from the CTI Testbed (node positions are available in Figure 4) and 35 iSense of the UZL Testbed that are very close to the topologies examined in [3]. We did this to produce comparable results using the WISEBED flexible testbed services. iSense are equipped with an IEEE 802.15.4 compliant radio, a $16MHz$ $32-bit$ RISC controller, $96kbytes$ of memory, and a highly accurate clock.

We report that the measurements collected from the execution of Fixed-ND are very close to those reported in [3, 10, 8]. For this reason we are confident that the protocols studied in [3, 10, 8] can improve their performance by switching Fixed-ND with Adaptive-ND.

We set the parameters I_{min} & I_{max} of Adaptive-ND to be 2 and 8 seconds respectively close to the parameters used

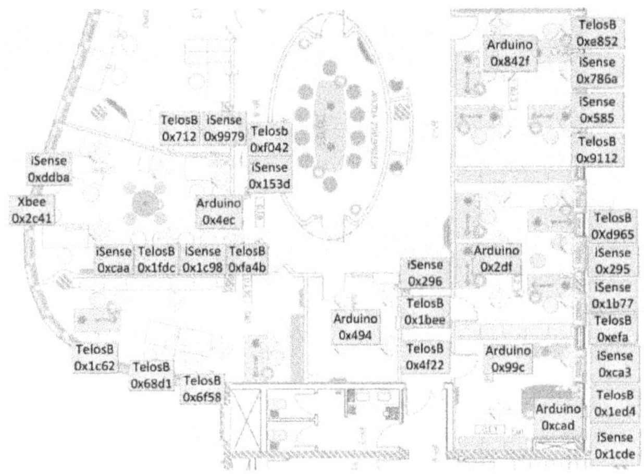

Figure 4: Map of the CTI WSN testbed

in the simulations. For the the FixK implementation we choose $k = 2$. For all experiments the timeout period is 20 seconds and the duration of all experiments is 20 minutes. All experiments were performed during the night when the noise in the testbed environment was minimal.

5.2.1 *Operation of* Adaptive-ND *and* Fixed-ND

In Figure 5, we can see how *messages* are transmitted over time by Fixed-ND (dashed line) and Adaptive-ND (solid line). We see that for Fixed-ND a new beacon is broadcasted every 2 seconds while Adaptive-ND sends less messages and over variable intervals. As each node chooses to send at random intervals, fewer transmission events collide resulting to less congestion and interference between beacons, avoiding degradation of the communication quality. This contributes to successful delivery of more messages and less conflicts also with applications that messages over the network. We note that for the duration of all experiments nodes reported the same neighbors and so other parameters of the mentioned above like *stability and events* were the same proving the correctness of the operation of Adaptive-ND. The results were

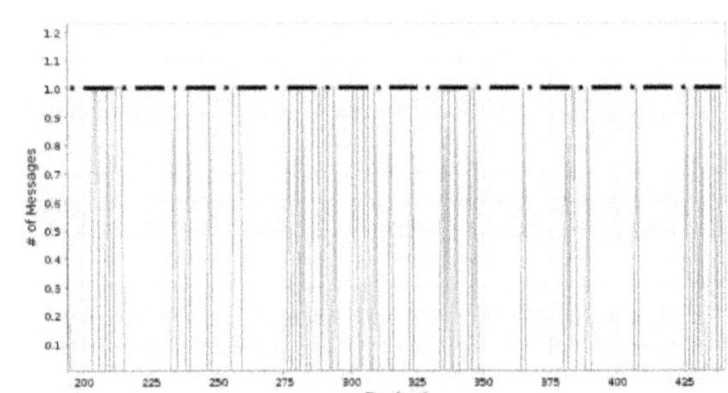

Figure 5: Transmitted Messages over time for Fixed-ND and Adaptive-ND for a specific node

very similar to the respective topologies that were used during simulations. The performance of the protocols in each case is very similar.

5.2.2 *FixK vs AverageK*

Here we experiment in order to better understand the differences between FixK and AverageK versions of Adaptive-

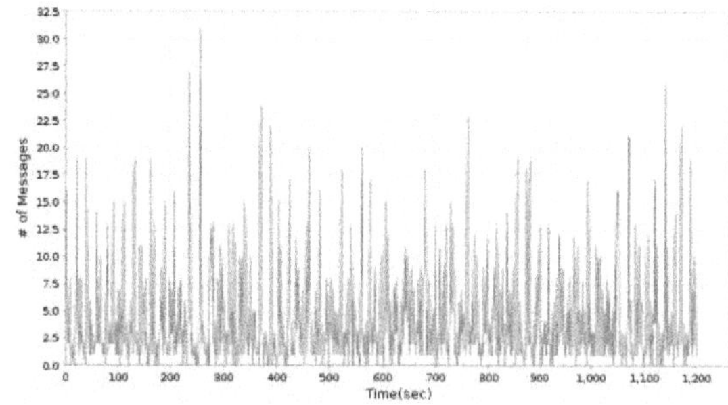

(a) FixK with $k = 2$

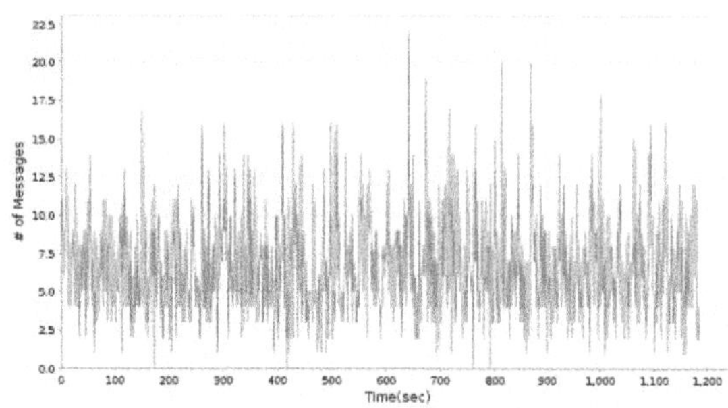

(b) AverageK

Figure 6: Transmitted Messages in a Real Network of 35 nodes

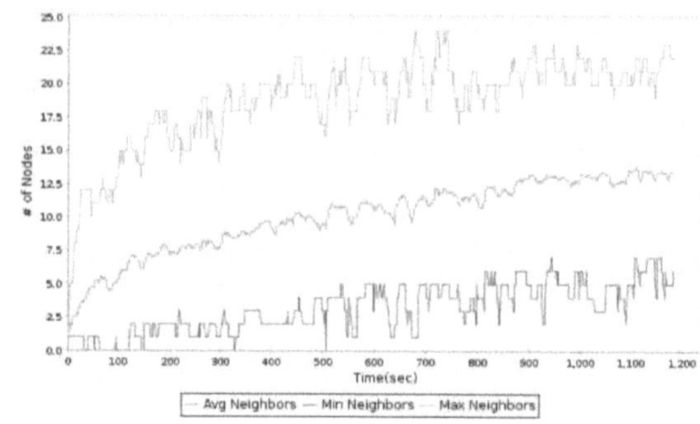

(a) Adaptive-ND AverageK, without LQI threshold

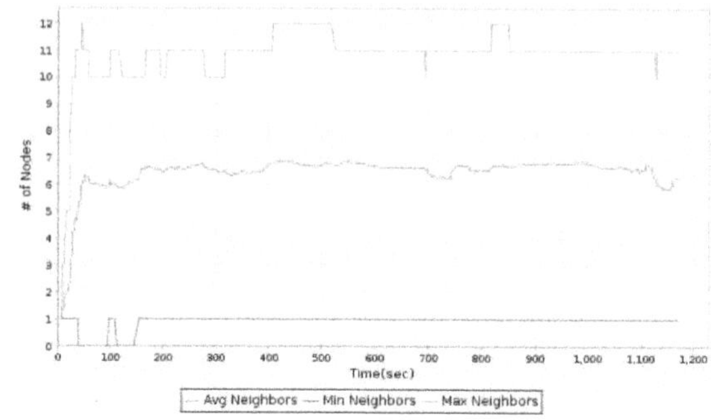

(b) Adaptive-ND AverageK, LQI threshold 180

Figure 7: Neighborhood Statistics

ND. Figure 6(a) depicts the *messages* transmitted by FixK. With FixK, less messages are transmitted over time as neighborhoods are larger than 3 nodes and so the consistency requirement is met even when nodes lose some of the messages. On the other hand, the AverageK version has a slightly higher message rate as it adapts to the size of the reported neighborhood. We note that the Adaptive-ND version in both experiments reported the same neighborhoods confirming the correctness of the implementations. The specific characteristics of a WSN define the version of Adaptive-ND we should choose to achieve our main goal and minimize network traffic during stable periods. When the nodes are uniformly distributed over the network without prior knowledge of the deployed topology we can use FixK with the network density as the k value. AverageK is a more adaptive implementation and operates reliably on any given network as each device uses a different k value based on its own neighborhood with a small overhead on the number of messages.

5.2.3 *Using LQI for refined neighborhood*

Next we examined the effect of using LQI thresholds on the operation of Adaptive-ND. For this experiment we present the results of experiments with the AverageK implementation with an LQI threshold of 180. We present the neighborhood statistics in Figure 7. Using LQI to ignore bad neighbors provides much more stable links for the whole duration of the experiment, eliminating devices on the outskirts of the neighborhood that have high rate of communication failures and allows the nodes to reach a consistent state easily. Not using LQI thresholds allows more distant nodes in the neighborhood, and results in less stability as message loss rates are higher over high distances. Without LQI thresholds in real networks, it is very difficult to reach a consistent state and even with LQI we need to adapt the LQI value to the special needs of the WSN.

5.2.4 *Mobility Experiments*

In order to evaluate the performance of the two protocols in case where nodes are mobile, we defined a path within our network deployment. A member of our team carried a sensor node while moving on the predefined path with different walking speeds. The path we followed as well as the positions of the sensors are depicted in Figure 8. To better evaluate the operation of our algorithm with mobile devices we performed slow (1 step every 10 seconds) and fast (1 step every 5 seconds) walk experiments.

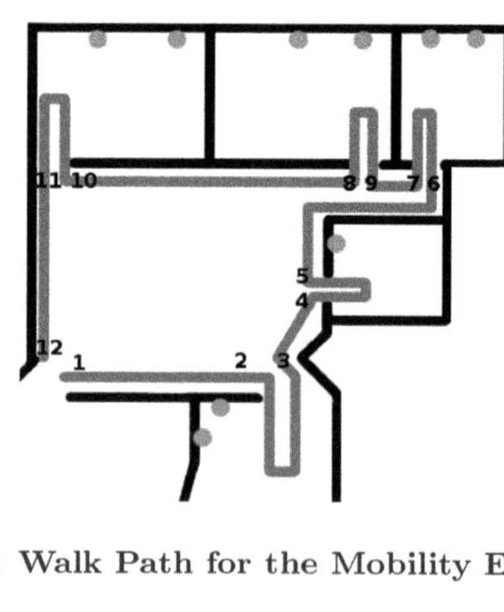

Figure 8: Walk Path for the Mobility Experiment

In Figure 9 we show that the mobile node transmits over 80% more *messages* during the walk while all devices transmit more messages than they would without movement.

This is because mobile nodes are always expected to be inconsistent and transmit regularly, while other nodes operate on longer intervals for the largest part of the experiment. Comparison of the *events* reported by the nodes simply confirms that all nodes identified the mobile node in their neighborhood an the mobile node identified the changes around him. At this point we need to mention that the mobile node was detected to the neighborhood of fixed nodes faster (on average after 8 seconds) than it did detect them (on average 13 seconds) since its neighborhood was always changing and its beaconing rate was higher. Also results for Fixed-ND do not provide any notable results as message rate is constant and higher than Adaptive-ND in all nodes.

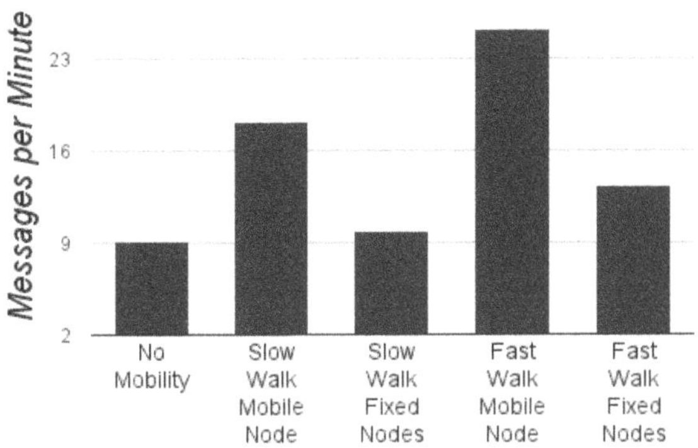

Figure 9: Messages Transmitted per Minute with no Mobile devices, by Mobile and Fixed Nodes during Slow and Fast Walks

5.2.5 *Network Lifetime experiments*

Results from previous experiments indicate that Adaptive-ND achieves the design goal of minimizing message exchanges and while maintaining the neighborhood as the previous approach of Fixed-ND. Our next step is to evaluate the performance of Adaptive-ND on battery powered devices to find out how it affects the *lifetime* of a WSN in comparison to Fixed-ND. We used 6 devices powered with 2 AA Rechargeable Batteries each and them placed in a single neighborhood around our office rooms (see Figure 10(a)) and a gateway device connected and powered via USB that collected information from all the devices and could provide information about the operation of the battery powered devices. We run consecutive experiments where we used the same 2 AA Batteries for each device, fully recharged for every experiment and confirmed the Voltage of all batteries was the same before each experiment. As we can see from Figure 10(b) nodes in a network running Fixed-ND had an average lifetime of 1332 minutes while both Adaptive-ND implementations achieved a longer lifetime of 1612 minutes for the FixK version and 1406 for the AverageK. We see that in the FixK implementation we achieved on average 21% longer lifetime. We have to mention here that during these experiments all algorithms used no duty cycling and devices operated continuously during the whole experiment. Also taking into account that computations consume far less power than message exchanges the longer lifetime achieved by Adaptive-ND is almost completely due to minimizing beacon exchanges. We believe that introduction of duty cycling will further increase the operation of Adaptive-ND over Fixed-ND, and will

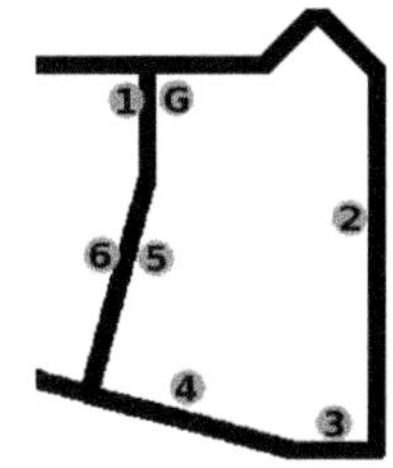

(a) Topology

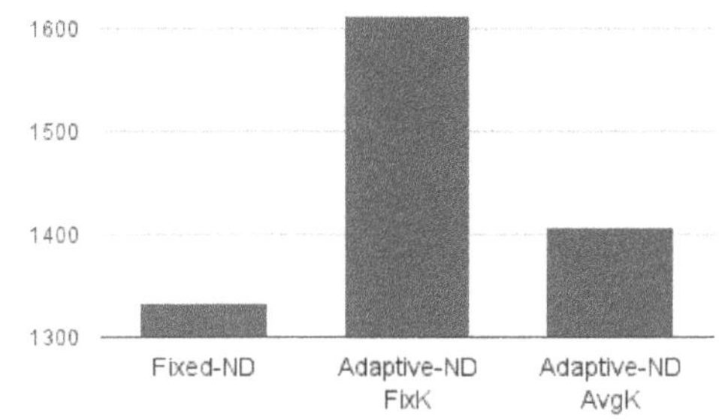

(b) Average lifetime per Algorithm

Figure 10: Lifetime Experiment: Setup and Results

result in even longer lifetimes but due to space and time limitations we could not perform.

6. CONCLUSIONS AND FUTURE WORK

In this work we presented a Neighborhood Discovery protocol that uses Adaptive Beaconing to detect and maintain Links between devices using a more efficient methodology than HELLO message exchanges. We did an extensive evaluation of the implemented protocol both in large scale simulations and real world experiments to prove its correctness and efficiency. The results of all the experiments showed that the new protocol achieves in most cases a reduced beaconing rate by 90% with almost no negative effects on the identified neighborhood. The only scenario where fixed interval beaconing outperformed our algorithm was the heavy message loss scenario, where message exchanges failed with a probability of more than 50%. Still, even in such scenarios Adaptive-ND suitably adapts and operates with a similar behavior to Fixed-ND in order to achieve the same stability and effectiveness. Also we showcased how our protocol is suitable for networks that include mobile nodes and more efficient on low power devices where we achieved 20% longer network lifetimes just by minimizing the total number of transmitted messages. Finally, we provided a second version of our protocol that adapts well to WSNs with different characteristics without any a priori knowledge for them and compared it to the original version to show how it suits to networks that are not uniformly populated.

One of our next steps will be the introduction of duty cycling on the individual nodes to investigate how devices would respond to situations where not the whole neighborhood is available at any given time. Our algorithm will allow protocols running on top to operate without knowledge or information on the real state of the neighborhood (alive or sleeping nodes) but provide an abstraction of an always online system. We anticipate that duty cycling together with the decreased message exchange rates will be a huge advantage for our implementation as energy requirements will be even lower and result to even longer network lifetimes, lower power consumption and less network traffic thus an increased quality in the communication channels.

Another direction is to investigate how the operation of Adaptive-ND affects protocols (i.e., routing[10], clustering[3, 10],tracking[8]) available on Wiselib[7]. These results will be a helpful feedback to better understand how our algorithm impacts on above protocols and how we will be able to further its operation and the coupling with those protocols.

7. ACKNOWLEDGMENTS

This work is partially supported by the European Union under contract numbers ICT-258885 (SPITFIRE [2]).

8. REFERENCES

[1] Coalesenses isense sensors. http://www.coalesenses.com/.

[2] Semantic service provisioning for the internet of things using future internet research by experimentation, (SPITFIRE). http://spitfire-project.eu.

[3] D. Amaxilatis, I. Chatzigiannakis, S. Dolev, C. Koninis, A. Pyrgelis, and P. Spirakis. Adaptive Hierarchical Network Structures for Wireless Sensor Networks. In *Int. Conference on Ad Hoc Networks*, AdHocNets '11.

[4] A. D. Amis, R. Prakash, T. H. Vuong, D. T. Huynh, T. H. P, V. Dung, and T. Huynh. Max-min d-cluster formation in wireless ad hoc networks. In *IEEE Information Communications Marketplace*, INFOCOMM '00, pages 32–41.

[5] D. J. Anthony, W. P. Bennett, M. C. Vuran, M. B. Dwyer, S. Elbaum, and F. Chavez-Ramirez. Simulating and testing mobile wireless sensor networks. In *Int. Conference on Modeling, analysis, and simulation of wireless and mobile systems*, MSWIM '10, pages 49–58.

[6] A. Bamis, A. Boukerche, I. Chatzigiannakis, and S. Nikoletseas. A mobility aware protocol synthesis for efficient routing in ad hoc mobile networks. *Computer Networks*, 52(1):130 – 154, 2008. MSWIM'06.

[7] T. Baumgartner, I. Chatzigiannakis, S. Fekete, C. Koninis, A. Kröller, and A. Pyrgelis. Wiselib: A generic algorithm library for heterogeneous sensor networks. *Computer Engineering*, page 16, 2011.

[8] K. Chantzis, C. Koninis, I. Chatzigiannakis, and J. Rolim. Design and evaluation of a lightweight tracking algorithm using wireless sensor networks. In *Int. Conference on Mobile Ad-hoc and Sensor Systems*, MASS '11.

[9] I. Chatzigiannakis, C. Koninis, G. Mylonas, S. Fischer, and D. Pfisterer. WISEBED: an open large-scale wireless sensor network testbed. In *Int. Conference on Sensor Networks Applications, Experimentation and Logistics*, 2009.

[10] I. Chatzigiannakis and P. Spirakis, editors. *Distributed Self-organized Societies of Tiny Artefacts: Design & Implementation.* FET Pervasive Adaptation Research, FRONTS Project, 2011.

[11] G. Coulson, B. Porter, I. Chatzigiannakis, C. Koninis, S. Fischer, D. Pfisterer, D. Bimschas, T. Braun, P. Hurni, M. Anwander, G. Wagenknecht, S. Fekete, A. KrÃűller, and T. Baumgartner. Flexible experimentation in wireless sensor networks. *Communications of the ACM*, 55(1):82–90, 2012.

[12] S. P. Fekete, A. Kröller, S. Fischer, and D. Pfisterer. Shawn: The fast, highly customizable sensor network simulator. In *Int. Conference on Networked Sensing Systems*, INSS 2007.

[13] V. Iyer, A. Pruteanu, and S. Dulman. Netdetect: Neighborhood discovery in wireless networks using adaptive beacons. In *5th Int. Conference Self-Adaptive and Self-Organizing Systems*, SASO '11, pages 31–40.

[14] G. Jakllari, W. Luo, and S. V. Krishnamurthy. An integrated neighbor discovery and mac protocol for ad hoc networks using directional antennas. In *Int Symposium on a World of Wireless Mobile and Multimedia Networks*, WoWMoM '05, pages 11–21.

[15] N. Kushalnagar, G. Montenegro, and C. P. Schumacher. IPv6 over low-power wireless personal area networks (6LoWPANs): Overview, assumptions, problem statement, and goals. RFC 4919, RFC Editor, 2007.

[16] P. Levis, N. Patel, D. Culler, and S. Shenker. Trickle: a self-regulating algorithm for code propagation and maintenance in wireless sensor networks. In *Symposium on Networked Systems Design and Implementation*, NSDI '04, pages 15–28.

[17] P. Nie, Z. Jin, and Y. Gong. Mires++: a reliable, energy-aware clustering algorithm for wireless sensor networks. In *Int. Conference on Modeling, analysis, and simulation of wireless and mobile systems*, MSWIM '10, pages 178–186.

[18] B. Pavković, F. Theoleyre, and A. Duda. Multipath opportunistic rpl routing over ieee 802.15.4. In *Int. Conference on Modeling, analysis and simulation of wireless and mobile systems*, MSWiM '11, pages 179–186.

[19] J. Polastre, J. Hui, P. Levis, J. Zhao, D. Culler, S. Shenker, and I. Stoica. A unifying link abstraction for wireless sensor networks. In *Int. Conference on Embedded networked sensor systems*, SenSys '05, pages 76–89.

[20] R. Steinmetz. Combining simulations, models and testbeds: possibilities and pitfalls. In *Int. Conference on Modeling, analysis and simulation of wireless and mobile systems*, MSWiM '09, pages 1–1.

[21] C. K. Toh. *Wireless ATM and Ad-hoc Networks: Protocols and Architectures.* Kluwer Academic Publishers, 1997.

[22] L. Wang, Y. Li, Z. Xu, and S. V. Krishnamurthy. A novel neighbor discovery protocol for ultraviolet wireless networks. In *Int. Conference on Modeling, analysis and simulation of wireless and mobile systems*, MSWiM '11, pages 135–142.

[23] Z. R. Zaidi, M. Portmann, and W. L. Tan. Analysis of link break detection using hello messages. In *Int. Conference on Modeling, analysis and simulation of wireless and mobile systems*, MSWiM '11, pages 143–150.

Routing Through Holes in Wireless Sensor Networks

Ahmed Mostefaoui
FEMTO-St Institute DISC dep.
University of Franche Comte
Belfort, 90000, France
ahmed.mostefaoui@univ-fcomte.fr

Mahmoud Melkemi
LMIA laboratory
University of Haute Alsace
Mulhouse 68093, France
mahmoud.melkemi@uha.fr

Azzedine Boukerche
PARADISE Research Lab,
University of Ottawa
Ottawa, Canada
boukerch@site.uottawa.ca

ABSTRACT

Besides of its high efficiency, in particular its scalability, geographic greedy routing is of a memoryless nature (i.e., nodes do not maintain any routing tables) which makes it well suitable for large-scale sensor networks with limited resources. Nevertheless, the occurrence of local minimum phenomenon, especially in networks with many holes, limits considerably the applicability of such routing protocol. The local minimum situations occur when the packet could not be delivered by the current node based only on distance to the destination. This problem has been tackled in previous research works to guarantee packets delivery by routing around the boundaries of the hole but at an excessive consumption of control overheads. In this paper, we propose and study a novel approach that is of memoryless nature and performs better that the state-of-art approaches in terms of guarantying packet delivery and deriving efficient routing paths. We provide in this paper proofs of its correctness (guarantying packet delivery) whereas showing, through simulations, its performance effectiveness in terms of reducing routing paths.

Categories and Subject Descriptors

C.2.2 [**Computer communication networks**]: Routing protocols

Keywords

wireless sensor networks, geographic greedy routing, local minimum problem, hole traversal, localized algorithms

1. INTRODUCTION

Last years have witnessed a tremendous technological progress that makes the deployment of large scale sensor networks a reality in many applications ranging from healthcare applications to military applications [2, 7]. In such networks, sensor nodes, provided with sensing, computing and wireless communication capabilities, are often deployed in a random fashion, to monitor a region of interest. Due to their limited resources, mainly energy resource, designing efficient localized multihop routing for wireless sensor network, where nodes are mostly static, becomes an important research challenge [1]. Indeed, the main objective behind is to let packets originated from source node to reach their destination with the lowest overhead possible.

To this end, greedy forwarding strategy [11], as simple, efficient and scalable approach seems fulfilling this requirement. In this approach, every node is supposed to be aware of its geographical location, by means of a location service [12] or derived from the localization phase [3] which is initiated just after deployment. Hence, a source node, knowing the location of the destination node, sends the packet to its 1-hop neighbor which is the closest node to the destination among all neighbors. Knowing the destination location from the received message, every node in the path repeats this process until the message reaches the destination node. It has been shown that geographic routing protocols are not only efficient in the average case; they are also worst-case optimal localized routing protocols with the hop-counts of the resulting paths bounded by $O(d^2)$, where d is the distance between the source and the destination [11].

This simple process does not induce any additional communication overhead since it requires that every node knows only its 1-hop neighbors. However, it suffers from a serious drawback called *local minimum phenomenon*. This "case" occurs when the packet could not be delivered to the next hop as the current node is the closest one to the destination. Figure 1 shows the local minimum phenomenon, where node N_{local}, receiving the packet from node N_4 (originated from N_s) could not deliver it in the direction of N_d.

Many related routing algorithms were developed in the literature to solve or at least to reduce the local minimum problem. These works fall in general into two categories: (a) graph based approaches and (b) non-graph-based approaches. In graph based approaches, as GPSR [8], Compass Routing II [9], GOAFR [11], GOAFR+ [10], etc. recovery from local minimum is performed by routing along the boundaries of holes, according to a local planar graph. The *right-hand rule* is then adopted which states that when arriving to a node n_x from node n_y, the next link traversed is the next one sequentially counterclockwise hinged at n_x from link $E(n_x, n_y)$. This rule requires however that all the links are non-crossing. For this reason, planarization techniques such as The Gabriel Graph (GG) [5], the Relative Neighborhood Graph (RNG) [15], etc. are usually used to derive planar graph from the Unit Disk Graph (UDG) of

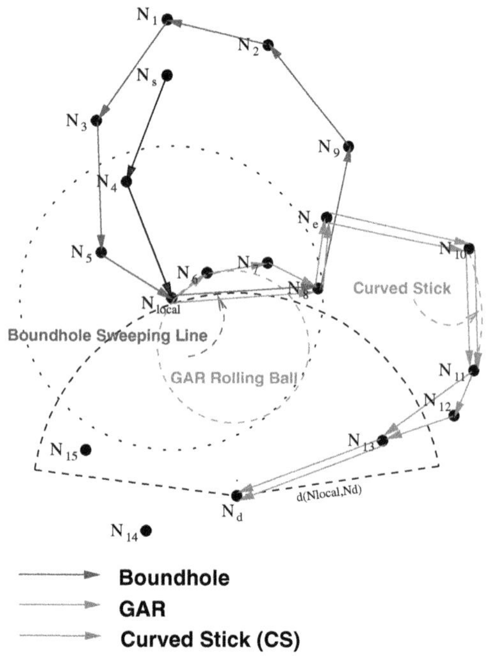

Figure 1: Local minimum phenomenon in greedy forwarding routing.

the underlying network. A planar graph represents the same connectivity as the original network with non-crossing links. In non-graph approaches, the basic idea behind is to first localize nodes on the holes boundaries and then to derive a detour path so to avoid routing in the direction of holes. In [6], Jia et al. presented HAIR (Hole Avoiding In advance Routing) protocol to bypass holes in advance. HAIR operates as follows: during the first step, nodes recognize themselves as nodes facing holes (i.e., local minimum nodes). They then, during the second step, inform their neighbors to mark them as hole nodes. Routing is then performed on non-hole nodes when possible. This process is then repeated leading thus larger holes perimeters. Taking advantage of earlier knowledge about holes positions, HAIR achieves shorter routing paths, and thus reduces energy consumption. However it suffers from a serious drawback: great energy depletion of nodes along the detour paths. Moreover the holes become larger requiring new detour paths and so on. GEAR (Geographic and Energy Aware Routing) protocol [16] also works in two phases, in phase 1 energy aware next hop selection is performed when routing a packet toward the a region of interest, in phase 2 flooding or recursive geographical forwarding is used to disseminate the packet inside the region.

Though these approaches resolve the local minimum problem, they require however to locally store the entire network graph or parts of it for graph based approaches leading thus to poor scalability. Besides their success to find a routing path, the non-graph-based approaches derive however either long routing paths, as the case of GAR (Greedy Anti-Void Routing) [13] protocol, or make use of expensive techniques in term of energy consumption (local flooding for instance) to meet the memoryless requirement, as it is the case of BOUNDHOLE [4] approach. The behavior of these two most efficient protocols is illustrated into Figure 1. It has to be noted that Boundhole approach, in spite of its efficiency, presents a great risk of falling into looping. This situation, also reported by the authors, comes from the fact that Boundhole could not, by construction, handle *intersection situations* i.e., N_e could communicate with N_{10} while neither N_8 nor N_9 could communicate with N_{10}. To overcome this situations, also called false boundaries detection problem, GAR protocol makes use of a rolling ball of R/2 of radius that is hinged at the current node. The next node hit by the rolling ball is selected as the next hop. By doing so, GAR ensures that packets could be delivered and do not fall into looping situations as it is the case for Boundhole. However, GAR generates longuer paths and visites unnecessary nodes as illustrated into Figure 1.

Our contribution in this paper is to propose a novel approach, named ***Curved Stick*** (CS) that prevents, by construction, the false boundaries detection problem observed previously within Boundhole approach, while deriving more efficient routing paths than GAR protocol. We outline the basic idea behind our approach in Figure 1. Unlike Boundhole approach, we used a curved stick instead a sweeping line to select the next hop. For instance node N_{local} will select node N_8 as its next hop avoiding hence the selection of unnecessary nodes as it is the case within GAR protocol. Similarly, node N_8 will select node N_e as its next hope rather selecting N_9 as it was the case in Boundhole approach. By doing so, CS ensures that it will not fall into looping (avoiding false boundaries detection problem) while it highly limits the number of unnecessary visited nodes. We provide in this paper proofs of the correctness of our proposed approach and showed through simulation its effectiveness compared to the state-of-the-art approaches.

The rest of the paper is organized as follows: Section 2 provides details about the used network model and outlines the problem of interest. The proposed "Curved Stick (CS)" approach is detailed in Section 3. Section 4 presents proofs of the correctness of CS. In Section 5, we presents the results of simulations we have conducted in order to evaluate the proposed approach and compare its performances to the state-of-art approaches. Finally, Section 6 concludes this paper.

2. PROBLEM STATEMENT

We consider a geographic wireless sensors network in which all nodes are homogeneous and static with R as their communication range. The network could then be represented by a Unit Disc Graph (UDG) $G(N, E)$, where $N = \{N_1, N_2, \ldots, N_n\}$ is a finite set of sensor nodes and $E = \{E_{i,j} \mid d(N_i, N_j) \leq R\}$ is a finite set of links. Note that $d(N_i, N_j)$ stands for the euclidean distance between nodes N_i and N_j. The locations (x, y) of the set N are known (each node is aware of its location by means of a positioning system like GPS or as the result of a localization process [3]). Nodes are also aware of their neighboring nodes with their corresponding locations. We further assume that the location of destination node is known by the source node. Thus intermediates nodes can only know the location of the destination node by receiving the packet from source node.

Under these assumptions, when two nodes want to communicate (asking for a routing path), greedy forwarding approach could be performed by each intermediate node based only on local information until the packet reaches the destination node or falls into local minimum as mentioned before. When the local minimum situation is met i.e., the packet is on the boundary of a hole, the Curved Stick approach is used to get the packet out of the hole by traversing holes boundaries nodes. Hence, the main issue is how to design a boundaries traversal algorithm that ensures true boundaries detection while at the same time derives efficient communicating paths in terms of number of hops?

Before diving in details of our proposal, let's first introduce some definitions and clearly state the false boundaries detection problem..

Definition 1. *(Hole): we define a hole as a cyclic sequence of nodes:* $H = \{N_i, N_{i+1}, ..., N_{i+j}, N_i\}$ *where it does not exist a link* $E_{k,l}$ *such that* N_k *is inside the hole and* N_l *is outside the hole or* $N_l \in H$.

For instance, in our running example, the sequence $H_1 = \{N_{local}, N_6, N_7, N_8, N_e, N_9, N_2, N_s, N_4, N_{local}\}$ is considered as a hole, whereas the sequence $H_2 = \{N_{local}, N_6, N_7, N_8, N_9, N_2, N_s, N_4, N_{local}\}$ is not, since there exists a link $(E_{e,10})$ between n_e located inside the polygon $H2$ and N_{10} which is located outside. We also can note from the above definition that the corresponding sequence of links $E = \{E_{i,i+1}, E_{i+1,i+2}, ..., E_{i+j-1,i+j}, E_{i+j,i}\}$ delimits the hole boundary.

Definition 2. *(Boundary Node): we define a node as a Boundary Node (BN) if it is located on the boundaries of the network or of a hole inside the network.*

For instance in our running example, in addition to nodes belonging to H_1, cited previously, $N_1, N_3, \ldots$ are considered as boundary nodes.

Now let's introduce the communication intersection situation which is behind the false boundary detection problem in Boundhole approach.

Problem 1. *(Communication Intersection Problem) Communication intersection situations occur when there exists a communication between two boundary nodes* BN_a *and* BN_b *that is "crossed" by a communication* $E_{c,d}$ *where* N_d *is not visible neither to* BN_a *nor to* BN_b.

Figure 2 illustrates the false boundary detection problem. Our objective here is to develop an approach that handles **by construction** the intersection problem while guarantying at the same time packet delivery and generating efficient routing paths.

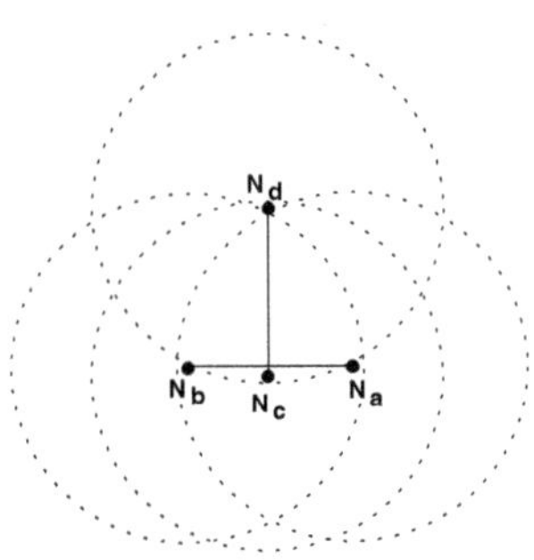

Figure 2: Communication Intersection Problem.

3. CURVED STICK (CS) APPROACH

The main idea behind our proposed approach is to well characterize nodes responsible of intersection situations when traversing hole/network boundaries so that these nodes will be directly selected by the previous hop node. Of course, the distributed nature of the proposed approach must remains i.e., no extra-information than the one mentioned before is needed. It means that each node knows only its immediate neighbors and their locations and based on that it will select the next hop. To this end, we have not make use of a sweeping line which generates, as mentioned before, intersection situation but we used a curved stick that selects nodes behind intersection situation as shown in Figure 1 (node N_e is selected instead of node N_9).

3.1 Curved Stick form

The question that follows immediately is: what is the optimum form of this curved stick in order to select only nodes responsible of intersection situations and no more on one hand and on the other hand to be sure that it will not "miss" intersection cases? In others terms, how to prevent to select nodes that could not be responsible of intersection situations, as it is the case for GAR protocol (generating longer routing paths) and ensures that all intersection situations are handled contrary to Boundhole approach?

We begin by introducing the following definition.

Definition 3. *(Potential-Exit-Gate node): We say a node* N_c *is a potential-exit-gate (PEG) in the direction* $\angle N_a N_c N_b$ *if there exists a location* $X \in \mathbb{R}^2$ *into* N_c'*s transmission range and outside the transmission ranges of* N_a *and* N_b *which are boundary nodes.*

For instance in Figure 3, node N_c is considered as a potential-Exit-Gate. The locations X are represented by the dashed region in Figure 3.

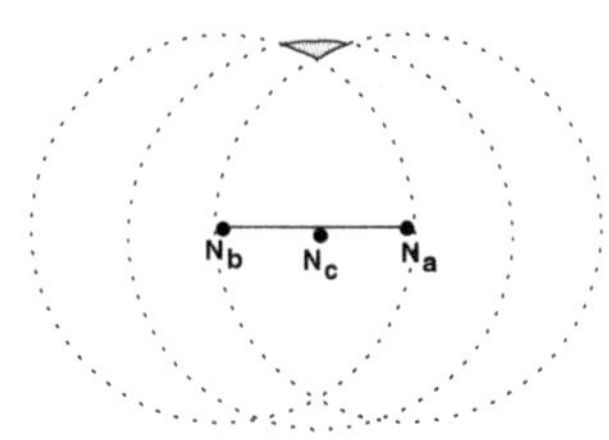

Figure 3: N_c **is a potential-exit-gate PEG.**

Hence our objective is to characterize PEG nodes and select them when performing next hop selection, without missing any one, otherwise intersection problem occurs.

Later on, we will use notation of Figure 4. Consider that θ_0 is the angle greater than $\pi/2$ formed by a point located in the left side of the line oriented from N_a to N_b (i.e., $\angle N_a N_0 N_b$), and verifying:

$$\sin\theta_0 = \frac{d(N_a, N_b)}{2R}. \quad (1)$$

Property 1. *(Angle characterization of PEG node): Let* N_c *be a node in the left side of the line oriented from* N_a *to* N_b*. The node* N_c *is a PEG if and only if* $\angle N_a N_c N_b > \theta_0$.

Proof. Let us refer to Figure 4. Consider a point N_0 such that the discs of radii R and centered at N_a, N_b and N_0 intersect at the location X. Figure 4 shows that:

$$d(X, N_b) = d(X, N_a) = d(X, N_0) = R \quad (2)$$

Hence, $\gamma = (\pi - 2\alpha) + (\pi - 2\beta) = 2\pi - 2(\alpha + \beta) = 2\pi - 2\theta_0$. Then $\sin \frac{\gamma}{2} = \sin \theta_0$, where $\alpha = \angle X N_b N_0 = \angle X N_0 N_b$, $\beta = \angle X N_c N_a = \angle X N_a N_c$ and $\gamma = \angle N_a X N_b$. On the other hand, from Equation 2, we have:

$$\sin \frac{\gamma}{2} = \frac{d(N_a, N_b)/2}{R} = \sin \theta_0 \quad (3)$$

Now, Let us prove the first implication. If N_c is a PEG node then the communication range of N_c contains the point X, therefore N_c is necessarily into the disc centered at X (the dotted blue disc), thus $\angle N_a N_0 N_b > \angle N_a N_c N_b = \theta_0$. Conversely, if the angle $\angle N_a N_c N_b > \theta_0$ then N_c is into the disc centered at X, this means that $d(X, N_c) < R$. Consequently, the communication range of N_c contains the point X, hence N_c is PEG. □

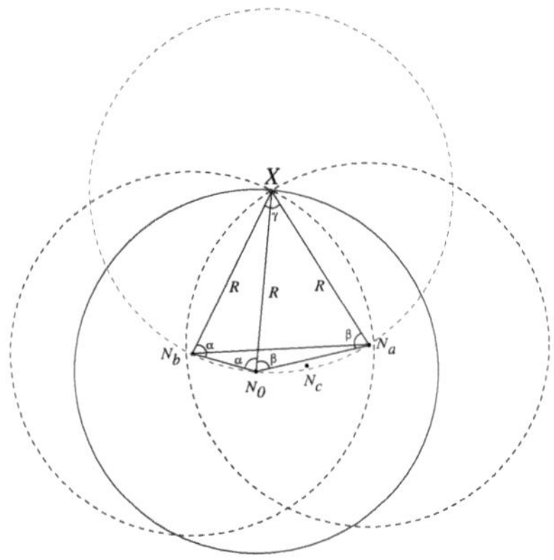

Figure 4: Characterization of a Potential-Exit-Gate node N_c.

This property could be used by the current node (which has to select its next hop) as a criterion to reduce the number of candidates for the next hop. In other terms, we are interested only in nodes that are PEG nodes. This is not the case for instance for GAR approach in which the use of a rolling ball of $R/2$ of radius is not relevant since it considers non PEG nodes. This important drawback of GAR approach is illustrated in Figure 5 where the rolling ball passes through many nodes that are not PEG nodes leading therefore to unnecessary longer routing paths. The derived path using GAR approach is $N_a, N_1, N_2, \ldots, N_b$ whereas the use of property 1 ensures that none of these nodes is PEG and then avoids selecting them.

Property 2. *(Region characterization of PEG node): All PEG nodes in the direction $\angle N_a N_c N_b$ are located in the region delimited by the segment $]N_a, N_b]$ and the arc of circle of radius R passing through nodes N_a and N_b.*

Proof. By definition, all PEG are located under the segment $]N_a, N_b]$, otherwise there is no communication intersection. On the other hand, as shown in Figure 4, from the critical case, Equation 2 holds, i.e., nodes N_a, N_b and N_c are equidistant to location X and this distance is equal to the

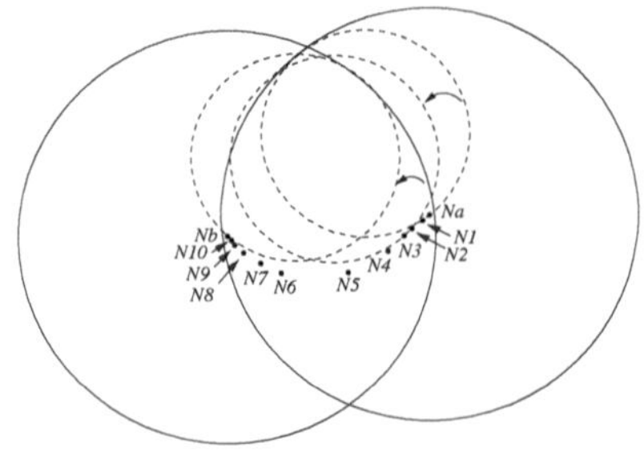

Figure 5: The dotted circles are example of GAR-rolling ball where $Path = \{N_a, N_1, \ldots, N_{10}, N_b\}$ is the GAR generated path

communication range R. Consequently, all nodes that have their locations inside the arc of circle that passes through nodes N_a and N_b and having a radius of R are potential-exit-gate nodes □

The interest of Property 2 comes from the fact that it precisely defines the location region of all PEG nodes with regard to the current hop node. Therefore, it is sufficient to the latter to just *"sweep"* this region for looking to a PEG node to select it. If this region is empty, this means that there is no node that could rise intersection situation, however if there is at least one PEG, the sweeping process will certainly select it. The important issue becomes then how could the current hop node sweep **exactly** this region?

Our answer is to use a **Curved Stick which has the form of an arc with radius of R and hinged at the current node** e.g., node N_a as illustrated in Figure 6. Here we ensure that only this region is visited and more importantly that if there is a PEG node, it is selected first avoiding hence the intersection situations.

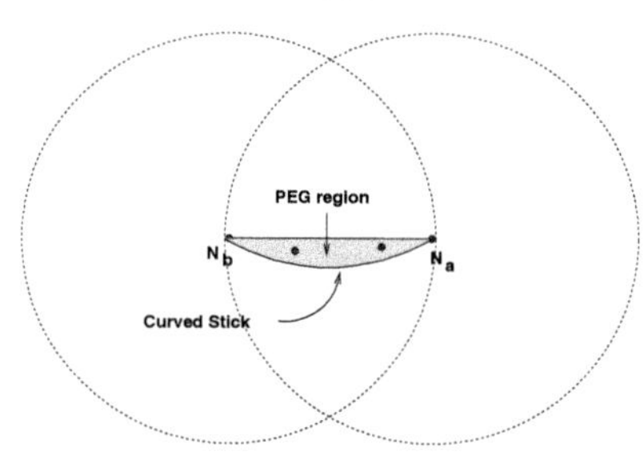

Figure 6: The dashed region is the potential-exit-gates locations.

Theorem 1.
The Curved Stick approach resolves the false boundary detection problem.

Proof. let's use the notation of Figure 2. Assume that exists a node N_c which is a Potential-Exit-Gate node in the direction $\angle N_a N_c N_b$ and this node has not been selected by the curved stick. This means that the next hop of node N_a is N_b. From the description of the curved stick approach, the region delimited by the segment $]N_a, N_b]$ and the arc of the circle of radius R which has its center in the direction $\angle N_a N_c N_b$ and passing through N_a and N_b is **empty** of nodes. Consequently N_c is located outside this region. This means that the set of locations $X \in \mathbb{R}^2$ where $X = \{(x, y) \mid d(X, N_c) \leq R \text{ and } d(X, N_a) > R \text{ and } d(X, N_b) > R\}$ is empty, which contradicts the assumption that a PEG N_c exists□

3.2 Curved Stick Routing Algorithm

Having proved that the proposed curved stick (CS) approach resolves the false boundary detection problem, we provide below details on the proposed CS routing algorithm. Three phases characterize our proposal: (a) engaging phase, (b) curved stick boundary traversal phase and (c) termination phase.

3.2.1 Engaging phase

Similarly to other approaches, The CS algorithm is of a greedy nature: the message between two communicating nodes is geographically forwarded hop by hop until it reaches the destination node or falls into local minimum situation. When the message gets stuck i.e., it could not be delivered, the current hop node, called *the initiator node* (N_{init}), starts applying the curved stick boundary traversal rule.

3.2.2 Curved Stick Boundary Traversal rule

To apply the CS boundary traversal rule, each visited node needs the location of the initiator as an input from the previous hop node. It first computes the two distances e.g., its own distance to destination and the one from the initiator to destination. If it is closer to destination than the initiator, this means that the message gets out of the local minimum. In this case, the current node starts applying greedy forwarding policy. Otherwise, the message is still on the boundary of a hole and the CS rule has to be applied. To this end, the current node determines its *Starting Point (SP)* (see below), from which the curved stick is swept until a node is hit. The latter is then selected as the next hop and notified by the current node.

Definition 4. *(Starting Point): we define the Starting Point of node N_{i+1}, noted SP_{i+1} as the intersection point between the two circles of range R, centered at N_i and N_{i+1} and following the direction $\angle N_{i-1}N_iN_{i+1}$ where N_{i-1} is the previous CS hop of node N_i and N_{i+1} is the next CS hop of N_i.*

The Starting Point is used to prevent generating links intersection, as illustrated in Figure 7, and then guaranteeing progress at each step. However, for the initiator node, that has not a CS previous hop (the message comes from a greedy forwarding previous hop), its corresponding Starting Point, noted ISP_{init} is considered as the intersection point between the circle of range R, centered at this node and the segment between N_{init} and the destination (see Figure 7). Note that we make difference between SP_{init} which corresponds to the starting point of node N_{init} when the message comes from a previous CS hop (node N_{cs} in Figure 7) and ISP_{init} when the message comes from a previous greedy forwarding hop (node N_g in Figure 7). This difference is essential for the termination of the algorithm as explained below.

3.2.3 Termination phase

For the termination of the algorithm, there are only two possible cases:

1. In the first case, the message is received by the destination. This means that there is at least one routing path between the source node and the destination node.

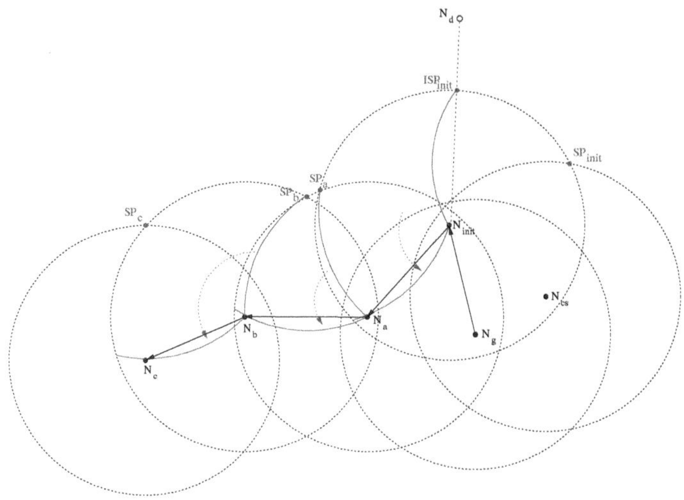

Figure 7: CS Rule

2. In the second case, the message travels the hole boundary using CS rule and returns back to the initiator node. In this case, the initiator sweeps the curved stick from its SP_{init} to its ISP_{init}. If a node is hit, this node is notified as the next hop and then the CS process continues, otherwise the initiator node could not apply CS rule and deduces that it does not exists any route from the source node to the destination node (see the demonstration in the next section). In this case, the initiator node informs the source node that the message could not be delivered.

Algorihm 1 summarizes the proposed routing algorithm.

4. PROOF OF CORRECTNESS

In this section the correctness of CS approach is proven. In order to guarantee packet delivery, we firstly prove that CS approach generates a finite sequence of links which means that the algorithm terminates. From that, we secondly prove that CS is able to find at least one path connecting the source node to the destination node whereas such paths exist. Oppositely if it does not exist any path between the two communicating nodes, CS is able to detect it and to notify the source node.

Property 3. *(Links Intersection): Two links generated by CS scheme do not intersect.*

Property 4. *(Consecutive Angles)*
The angle between two consecutive links in the CS sequence $\angle N_{i-2}N_{i-1}N_i$ is greater than or equal to $\pi/3$.

Proofs. Because of the limited space, we do not provide in this paper details about the proofs of these two properties. Nevertheless, we present succinctly the demonstration ideas. For the first property, we can proof it by contradiction i.e., we assume that two consecutive links intersect and we then show that this leads to a contradiction. In the second property, we have to consider two cases :(a) when the next hop is visible to the previous one and when it is not. In the two cases, we can show that the angle in greater than or equal to $\pi/3$.

Algorithm 1 The CS routing algorithm.

```
Require: Receive message M from the previous hop
if (M.Initiator.ID == null) then
   // greedy forwarding policy is used
   select the next hop based on greedy forwarding ;
   if (Local minimum occurs) then
      set M.initiator.ID to MyID
      select the next hop based on CS rule
   end if
else
   if M.Initiator.ID != MyID then
      compute: d(N_current, N_dest) and d(N_init, N_dest)
      if d(N_current, N_dest) < d(N_init, N_dest) then
         // the message is out of the hole
         set M.initiator.ID to null
         select the next hop based on greedy forwarding
      else
         select the next hope based on CS rule
      end if
   else
      // the message traveled the hole boundary and comes
      back to the initiator
      sweep the curved stick from SP_init to ISP_init
      if (a node is hit then
         select the next hop based on CS rule
      else
         inform the source node that the message could not be
         delivered
      end if
   end if
end if
send the message
```

Theorem 2.
The Curved Stick scheme terminates and generates a finite sequence of links.

Proof. The CS algorithm terminates if any node will not appear in the sequence infinitely many times. In fact, each node can only appear in the sequence at most 6 times. Consider the sequence $N_0N_1 \cdots N_iN_{i+1} \cdots N_jN_{j+1}$ $(i < j)$. Assume that $N_i = N_{j+1}$, that is the message gets back to N_i which it has been visited. CS scheme ensures that the angle $\angle N_{i-1}N_iN_{i+1}$ is smaller than the angle $\angle N_{i-1}N_iN_j$ as illustrated in Figure 8 (the grey area swept by the curved stick is empty of nodes). Also the node N_{j+2} does not belong to the grey area, consequently the angle $\angle N_{i-1}N_iN_{i+1}$ does not overlap with the angle $\angle N_jN_iN_{j+2}$, two illutrations of the possible locations of N_{j+2} are shown by Figure 8, one on the right side and the other on the left side of the line directed from N_j to N_i. Since the two angles $\angle N_{i-1}N_iN_{i+1}$ and $\angle N_jN_iN_{j+2}$, do not overlap and are at are at least $\pi/3$, as shown in Property 4. So N_i can appear at most 6 times in the sequence.□

Theorem 3.
Given a source node N_s and a destination node N_d, if it exists at least one path between N_s and N_d then CS algorithm guarantees packet delivery.

Proof. Because of the greedy nature of CS algorithm, if it does not exist any local minimum between N_s and N_d, then the packet is guaranteed to be delivered. Now, consider the case where the Greedy Forwarding algorithm meets a stuck node N_{local} which satisfies $d(N_{local}, N_d) < d(N_k, N_d)$ for all $N_k \in \mathcal{V}(N_{local})$ where $\mathcal{V}(N_{local})$ is the set of N_{local} neighbors. Then node N_{local} starts curved stick boundary

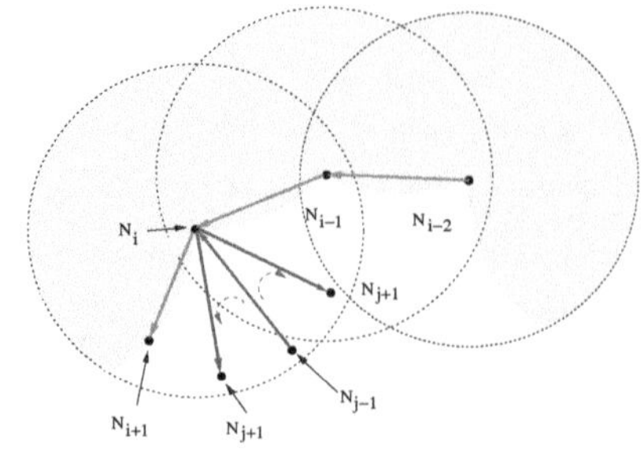

Figure 8: Related to the proof of Theorem 2

traversal rule. Assume that the generated sequence of links is: $N_{local} \ldots N_{end}$. As proven before, Theorem 1 ensures that this is a finite sequence. Now there are only two cases to consider:

- **Case 1:** $N_{local} \neq N_{end}$. By construction, this means that $d(N_{local}, N_d) > d(N_{end}, N_d)$. Hence the greedy forwarding could be used again to reach N_d
- **Case 2:** $N_{local} = N_{end}$. In this case, the curved stick draws a closed curve which divide the plane into two simply connected regions Z and its complementary $\bar{Z}$ (see Figure 9). The region Z contains N_{local} and $\bar{Z}$ contains node N_d. In addition, all the nodes connected to N_{local} are into Z. Consequently Z contains N_s. Also, since the area swept by the curved stick (the grey area in Figure 9) is empty, then there is no connection between nodes belonging to the region Z and those located outside Z. Therefore there is no path connecting N_s to N_d.□

5. PERFORMANCE EVALUATION

In order to evaluate the effectiveness of our proposed routing approach, we have conducted several series of simulations using NS-3 [14] network simulator. We have implemented and compared the three routing approaches presented previously, namely: Boundhole approach [4], GAR approach [13] and our proposed approach. The main metric used for comparison is the lengths of generated paths between communicating pairs of nodes (i.e., hop counts). We have used the same simulation settings as those reported in GAR approach and presented in Table 1. The simulation scenarios were constructed as fellow: we firstly randomly placed artificial holes in the network area so to emulate practical obstacles or dead-zones (nodes run out of energy). The height of those holes was varied in order to increase the occurrence of minimum local situations. Once the holes placed, we generated networks with randomly placed nodes within the network area except holes i.e., no node is placed into the holes. All nodes are considered with the same communication range. In our scenarios, we have varied this parameter in order to simulate dense networks when the communication range is large and sparse network when the communication range is low. It has to be noted that small holes other than the artificial ones might occur in regions where node density is low.

For each scenario, we generated 30 networks to repeat the simulation. For each simulation, we selected source nodes as

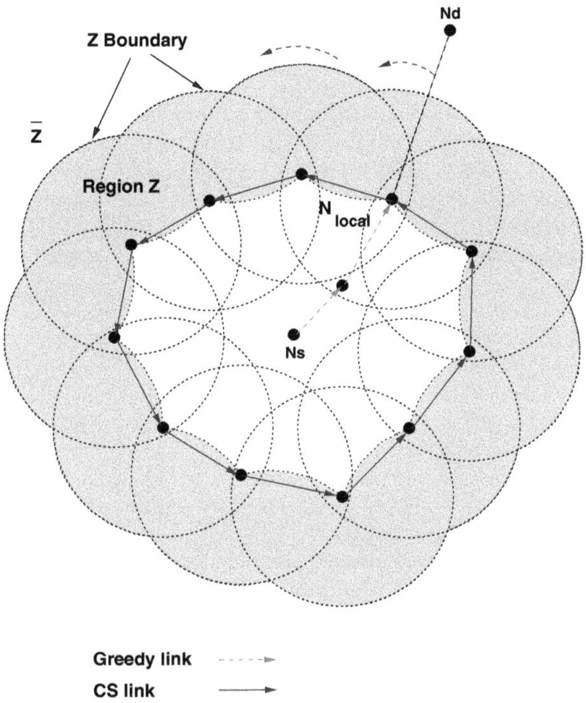

Figure 9: CS approach sweeps the bounded region Z when there is no path connecting N_s to N_d

Parameter	value(s)
Network Area (m^2)	1000 x 800
Number of nodes	500 nodes
Transmission range (m)	50, 100 and 150
Number of holes	3
Holes width (m)	300
Holes height (m)	100, 200, 300, 400 and 500
Source limit (m)	100
Destination limit (m)	900

Table 1: Simulations parameters

nodes that have their x coordinate below source limit (e.g., 100 m) and destination nodes as nodes that have their x coordinate beyond destination limit (e.g., 900m). By doing so, we located the communicating pairs at the left and the right boundaries and then increase the risk of raising local minimum situations. We then generated all communication pair composed of those source nodes and destination nodes. From those communicating pairs, we have selected only those that fall into local minimum situations, since all the approaches are of greedy nature. It means that on communication pair that do not raise local minimum, the three algorithms gave the same routing paths.

Figure 10 plots the number of local minimum communicating pairs with regard to the height of holes. As expected, the number of local minimum pairs increases with the increase of the holes heights. Similarly, the number of local minimum cases is greater in sparse networks (e.g., communication range equal to 50m) than in dense networks (e.g., communication range equal to 150 m for instance). This is due to the fact that in dense networks each node has many neighbors and consequently many routing possibilities. As illustrated in the figure, we also note that in dense networks, small artificial holes (height equal to 100m) do not generate any local minimum situations or very few (18 situations over 1978 communicating pairs when the communication range is equal to 100m).

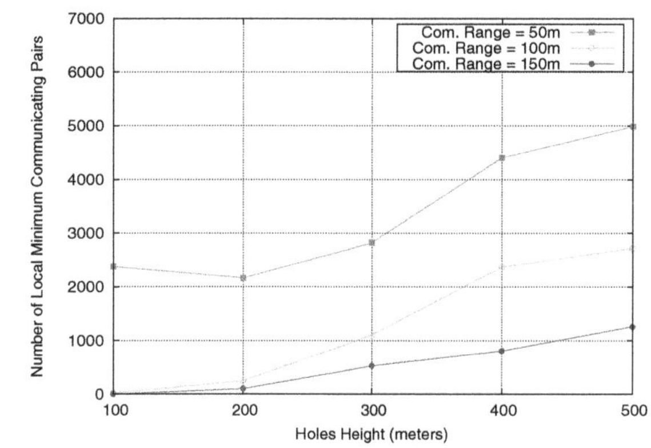

Figure 10: Local minimum number evolution.

In Figures 11, we plotted the average routing path lengths generated by the three approaches in function of holes heights when the communication range is set to 50m. As expected and confirmed in the figure, generated routing paths lengths increase with the increase of the heights of holes and this remains true for the three compared approaches. We also noted that GAR and CS approaches outperform Boundhole approach because we have observed that the latter failed into *intersection situations*, reported previously, more frequently in sparse networks. Hence this situation leads Boundhole approach to generate longer routing paths. Similarly, CS approach experienced better performance, compared to GAR approach, because the latter visits unnecessary nodes as explained in previous sections. Consequently, we have observed a mean improvement of about 3 hops.

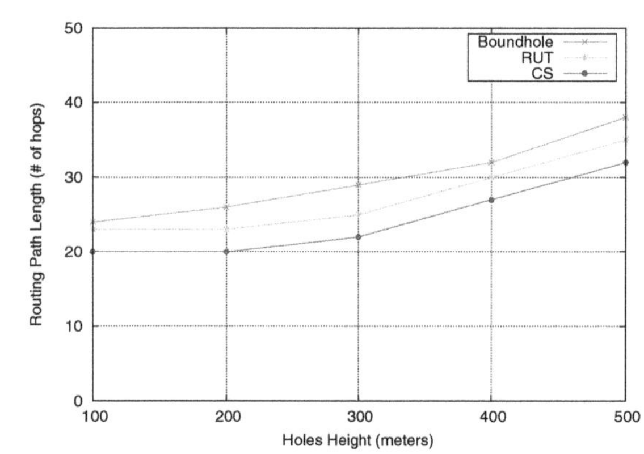

Figure 11: Routing performance for a communication range equal to 50m.

We have conducted same simulation series, by using two other values of the communication range, 100m and 150m, to better capture different network densities. The results are plotted into Figures 12 and 13, respectively. It has to be noted that the lengths of the generated routing paths are, to a certain extent, "inversely proportional" to the communication range. In fact, greater the communication range is, smaller the lengths of the routing paths are.

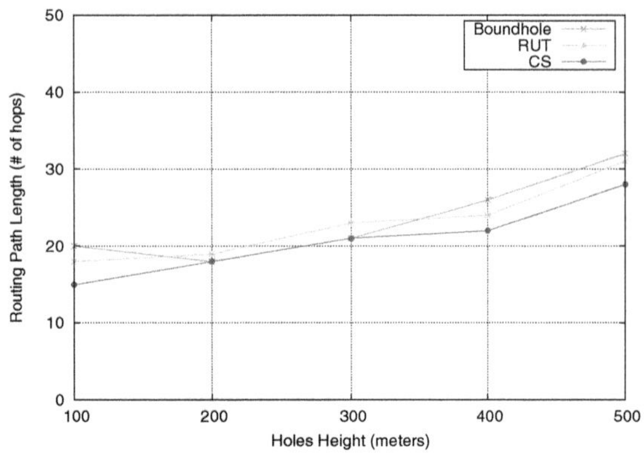

Figure 12: Routing performance for a communication range equal to 100m.

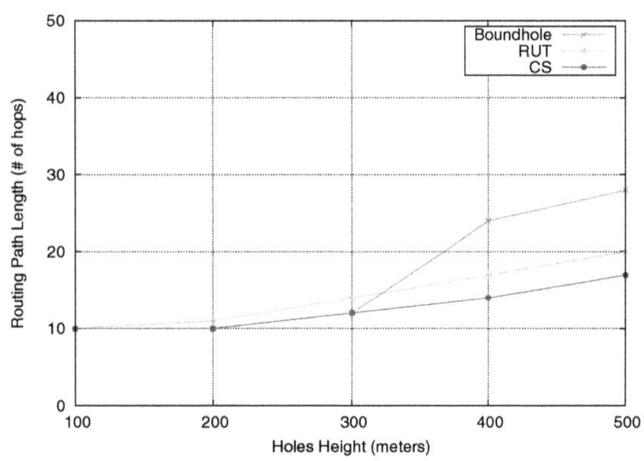

Figure 13: Routing performance for a communication range equal to 150m.

Again, we have noticed that CS approach outperforms GAR approach. The mean improvement ranges from 1 to 3 hops, depending on the holes heights. We observed that important improvement was reached when the networks is more constrained i.e., larges holes, as reported by the figures.

We also observed that in some cases, Boundhole approach outperfoms GAR approach. We then conducted a detailed analysis of the obtained results in the 60 simulations for these scenarios and we have found that the intersection situations, responsible of Boundhole bad performing, are less frequent in dense networks than in sparse networks. That is why figures showed a roughly comparable performance between CS approach and Boundhole approach when the holes height is medium. The situation changes when the holes are important as reported in the figures.

In summary, CS outperforms the other approaches in deriving more shorter routing paths while guarantying at the same packets delivery. This performance is noticeable, as seen from the results, in more constrained networks, which makes CS approach well suitable for monitoring area with many holes.

6. CONCLUSION

In this paper, we presented a novel geographic routing approach that efficiently resolves the local minimum problem compared to the state-of-the-art approaches. Our approach is localized and memoryless i.e., no extra-information rather than the localization of nodes is needed and no information is maintained by nodes (contrary to graph-based approaches for instance), which makes it scalable and well suitable for large distributed wireless sensor networks. We also provided formal proofs of its correctness (free of looping and data delivery guaranteed). The simulation evaluation and comparison that we have conducted showed clearly its effectiveness in terms of shorter path derivation in comparison to the state-of-art approaches. We have noted that the improvement is more important in constrained networks i.e., networks with many holes.

In near future, we plan the extension of our work to 3D networks where geographic greedy forwarding remains very efficient (hop count bounded by d^3, where d is the distance between the source node and the destination node) and still suffers from the local minimum problem.

7. REFERENCES

[1] K. Akkaya and M. Younis. A survey of routing protocols for wireless sensor networks. *Elsevier Ad Hoc Network Journal*, 3(3):325–349, 2005.

[2] I. Akyildiz, Y. S. W. Su, and E. Cayirci. A survey of wireless sensor networks. *IEEE Communications magazine*, 40(8), 2002.

[3] J. Bahi, A. Makhoul, and A. Mostefaoui. Localization and coverage for high density sensor networks. *Computer Communications*, 31(4):770–781, 2008.

[4] Q. Fang, J. Gao, , and L. Guibas. Locating and bypassing routing holes in sensor networks. In *Proc. IEEE INFOCOM 04*, pages 2458–2468, 2004.

[5] K. Gabriel and R. Sokal. A new statistical approach to geographic variation analysis. *Systematic Zoology*, 18(3):259–278, 1969.

[6] W. Jia, T. Wang, G. Wang, and M. Guo. Hole avoiding in advance routing in wireless sensor networks. In *Wireless Communications and Networking Conference*, pages 3519–3523, 2007.

[7] H. Karl and A. Willig. *Protocols and Architectures for Wireless Sensor Networks.* Wiley, 2005.

[8] B. Karp and H. Kung. Gpsr: Greedy perimeter stateless routing for wireless networks. In *ACM MobiCom*, pages 243–254, 2000.

[9] E. Kranakis, H. Singh, and J. Urrutia. Compass routing on geometric networks. In *Canadian Conf. Computational Geometry (CCCG Š99)*, pages 51–54, 1999.

[10] F. Kuhn, R. Wattenhofer, Y. Zhang, and A. Zollinger. Geometric ad-hoc routing: Of theory and practice. In *ACM Symp. Principles of Distributed Computing (PODC Š03)*, pages 63–72, 2003.

[11] F. Kuhn, R. Wattenhofer, and A. Zollinger. Worst-case optimal and average-case efficient geometric ad-hoc routing. In *ACM MobiCom*, pages 267–278, 2003.

[12] J. Li, J. Jannotti, D. DeCouto, D. Karger, and R. Morris. A scalable location service for geographic ad-hoc routing. In *6th Annu. ACM/IEEE International Conference on Mobile Computeing and Networking*, pages 120–130, 2000.

[13] W.-J. Liu and K.-T. Feng. Greedy routing with anti-void traversal for wireless sensor networks. *IEEE Transactions on Mobile Computing*, 8(7):910–922, 2009.

[14] NS3. http://www.nsnam.org.

[15] G. Toussaint. The relative neighborhood graph of a finite planar set. *Pattern Recognition*, 12(4):261–268, 1980.

[16] Y. Yu, D. Estrin, and R. Govindan. Geographical and energy-aware routing: A recursive data dissemination protocol for wireless sensor networks. Technical report, UCLA Computer Science Department Technical Report, UCLA-CSD TR-01-0023, 2001.

S-NC: Structure-Free Network Coding-Aware Routing in Wireless Sensor Networks

Ehsan Enayati-Noabadi
Department of Computer Engineering
Sharif University of Technology
enayati@ce.sharif.edu

Hamed Yousefi
Department of Computer Engineering
Sharif University of Technology
hyousefi@ce.sharif.edu

Ali Movaghar
Department of Computer Engineering
Sharif University of Technology
movaghar@sharif.edu

ABSTRACT

Network Coding (NC) is an emerging technique that helps wireless networks to have higher throughput and better energy efficiency. In this paper, a novel network coding-aware routing (S-NC) which uses NC to improve reliability and robustness in unreliable WSNs is proposed. It is based on a structure-free routing algorithm as the underlying routing engine which employs neighbor queue length to increase coding opportunity; while exploiting a spatial coding method to control the amount of redundancy. To the best of our knowledge, this is the first study on the topic of network coding-aware routing in WSNs. Simulation results in TOSSIM demonstrate that S-NC significantly improves the performance in terms of reliability, end-to-end delay, and energy consumption in the network.

Categories and Subject Descriptors

C.2.1 [**Computer-Communication Networks**]: Network Architecture and Design—*Wireless communication*

General Terms

Algorithms, Performance

Keywords

Wireless Sensor Networks, Network Coding, Structure-free Routing

1. INTRODUCTION

Wireless Sensor Networks (WSNs) consist of a large number of sensor nodes that cooperate to monitor environmental conditions. They lend themselves to a wide range of applications. Some of them involve working in harsh environments where there are lots of noises and obstacles which can lead to frequent packet losses and node failures, hence it is important for the data to survive. To preserve reliability of the entire network, a node might have to send some redundant data. However, the mechanism for adding this redundancy to the network plays a great role in controlling the network's efficiency, especially in terms of nodes' energy, memory, and CPU usages.

It was shown in multiple researches that Network Coding has a significant effect on the reliability of wireless networks while keeping the amount of redundancy controllable [7][8]. Unlike simple relaying of packets, NC lets intermediate nodes mix multiple packets intelligently to convey more information in each transmission on the medium. As an example, consider the network in Figure 1 in which Node A wants to send Packet p1 to Node B and Node B wants to send Packet p2 to Node A, where Node R is a relay node and can hear both Nodes A and B. If we use the traditional forwarding mechanism, then Node R should once send p1 to B and once p2 to A, while by using network coding, Node R can send p1+p2 to both of them; in this way, node A can extract its intended packet, p2, by calculating p1+(p1+p2) and node B can, similarly, get p1 by calculating p2+(p1+p2). NC can improve the network throughput, energy consumption, and reliability. It was shown that making routing algorithms aware of network coding results in further improvement in terms of throughput, reliability, and end-to-end delay [6][9][12][13].

There are many previous studies on network coding-aware routing in wireless networks [2][3][6][9][12][13], but none of them are suitable for WSN needs, limitations, and traffic patterns. A queue-based metric named CRM in [9] can compare the possibility of coding on different routes. However, it may lead to the congestion; also, the metric calculation is not energy and memory efficient. A theoretical framework for coding-aware routing is proposed in [12] to show that this approach can reduce the number of transmissions and improve throughput; however, their solution is centralized and difficult to be implemented in WSNs. C2AR [6] is a Coding and Capacity-Aware Routing which combines the advantages of load-balancing and network coding. However, in this routing algorithm a node must have a good knowledge about its neighbors and also keep information of all the flows passed by it in its memory and, in turn, is completely inefficient for memory-limited sensor motes. Moreover, [13] proposed a coding-aware algorithm based on opportunistic routing to take advantage of the broadcast characteristic of a wireless channel for data delivery for wireless mesh networks. However, all the coding-aware routings discussed above have

MSWiM'12, October 21–25, 2012, Paphos, Cyprus.

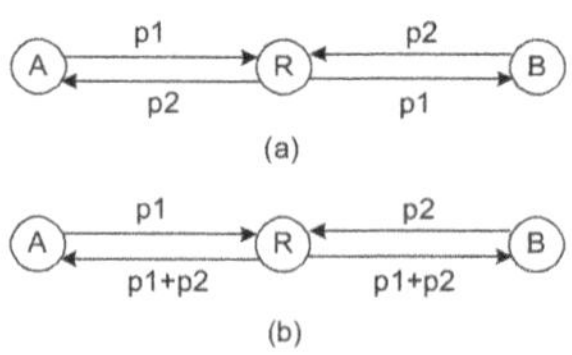

Figure 1: (a) No Coding (b) Network Coding

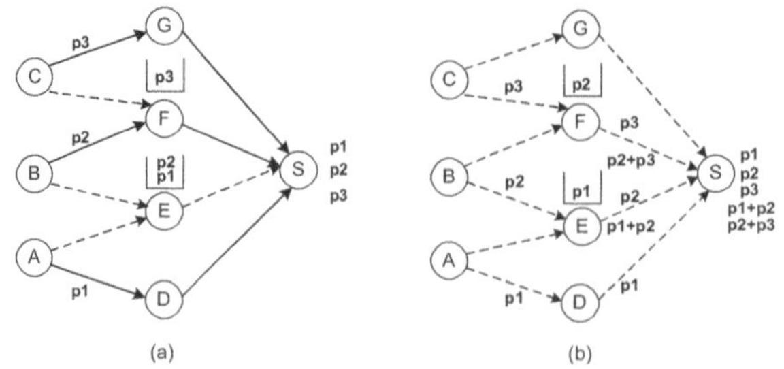

Figure 2: (a) Spatial coding on a structure-based routing (b) Spatial coding on a structure-free routing

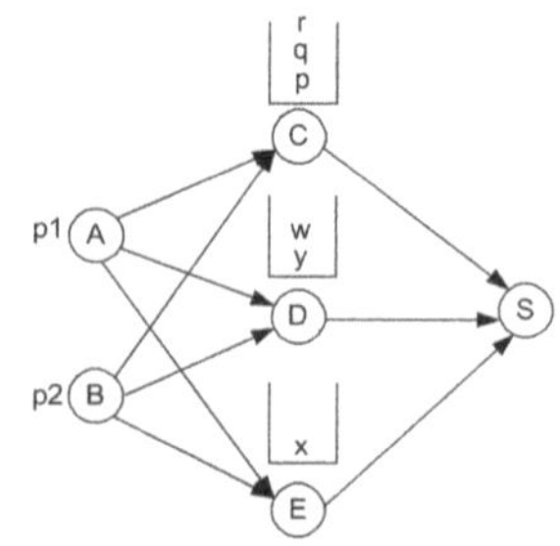

Figure 3: Queue length effect

a common problem. They assume there are flows in the network in different directions and they code the flows in opposite directions, but in common WSN applications where all the sensors try to send their data to a single sink this kind of traffic pattern does not exist. In other words, the major communication mode in WSNs is many-to-one flows. Recently, some studies [2], [3] and [8] specifically applied NC to WSNs while considering the many-to-one flow directions. A network coding-based approach for wireless mesh and sensor networks was designed by in [3] to protect many-to-one flows. Network coding is used to tolerate multiple packet losses in [2]. Moreover, relative indexing was introduced to reduce coding coefficients overhead. Keller et al. [8] suggested SenseCode that exploits overhearing to allow intermediate nodes to send overheard packets along with their own packets. A concept named spatial coding was presented in their study. In spatial coding a packet may enter the network "coded" or "uncoded". They showed that with spatial coding it is possible to control the amount of redundancy and have an acceptable reliability. In addition, they demonstrated that network coding makes the network robust against node failures and efficient in energy consumption. They used CTP [5] as their routing algorithm which is a structured routing scheme. However, the network coding-based algorithms proposed for WSNs do not consider coding opportunities in the route decision phase. Hence, they might not be able to achieve the full benefits of NC. From another point of view, the prior works have focused on tree-based structured approaches which are not robust against network dynamics especially node failures. In contrast, the structure-free approaches perform dynamic routing using local information, so they do not spend extra energy to build a structure.

In this paper, a network coding-aware algorithm which employs a structure-free routing scheme as the underlying routing engine as well as a spatial coding method for data transmission is proposed. Moreover, it takes queue length into account in the route decision phase as a metric for coding awareness to increase the chance of decoding at the sink. Our approach is based on anycasting at the MAC layer to determine the next-hop node for each transmission on-the-fly. To the best of our knowledge, this is the first study on the topic of network coding-aware routing in WSNs. We show that the combination of network coding and structure-free routing makes significant improvements in terms of reliability and energy consumption of WSNs.

The rest of this paper is organized as follows. In Section 2, we explore motivation of this work by two examples. Section 3, explains our proposed algorithm in detail. The simulation results are presented and discussed in Section 4. Finally, Section 5 concludes our work.

2. MOTIVATION

Consider a WSN which uses spatial coding to send/forward packets. Spatial coding was defined in [8] as a way to add controllable redundancy to the network. For example, to send X packets in the network with double redundancy, one might send each packet twice, or 2X coded packets, or by spatial coding, X packets coded and X packets "uncoded". It was shown in [8] that spatial coding has a better performance in terms of reliability in comparison with the other two methods where reliability is defined as the number of packets received and decoded at the sink divided by the number of packets sent by sources. Hence, each intermediate node which receives a new packet will send that once untouched and once coded with other packets (received or overheard ones) in its queue. Here, two examples are given to illustrate: (1) structure-free routing exploiting spatial coding has better reliability than the structured one and (2) queue length is a good metric for the network coding opportunities; therefore, it improves the reliability gain of spatial coding.

In this section, we just provide the examples and the details of our routing algorithm will be discussed in Section 3. Assume that S-NC chooses a neighbor as the next hop if it has more packets in its queue. In addition, spatial coding is used to send packets, so once a packet has been coded with other packets it will be deleted from the queue. For (1), consider Figure 2 where Nodes A, B and, C are sources, D, E, F, and G are relay nodes, and S is the sink. Source nodes send Packets p1, p2, and p3, respectively. Packet names on a link means the packet has been sent through that and boxes above nodes show their overheard packet queue. Figure 2-a shows a structured routing where solid lines point to pre-chosen parents based on the routing algorithm and dashed lines are "overhear-possible" routes. As it is shown just Packets p1, p2 and p3 may reach the sink and in the presence of packet loss, the sink may never receive a particular packet. Thus, it cannot tolerate any packet loss. In Figure 2-b a structure-free routing is used, instead. In this example, without loss of generality, we assume that Node A will select Node D as its next hop, so Node E will overhear p1. When Node B decides to send Packet p2, there are two choices: E and F. It will choose E because it has a packet in its queue while F has nothing there. The same scenario happens when Node C wants to send Packet p3. As it can be seen, the sink may receive 5 packets and that is because of using coding opportunities in the network. As a result, in the presence of packet loss, reliability of the second approach is better than the first one. It is worth mentioning that if sending order changes, the sink might receive four packets but even in this case the second approach excels the first one.

While the different flows have the chance to spatially converge at the same point, in the structured routing algorithms, they may take partially (even completely) different routes to the sink. Therefore, finding a way to join the routes somewhere in the path can lead to the creation of packets with more redundancy and help the sink to decode the original ones more easily. We used structure-free routing to achieve this opportunity.

Before demonstrating Example 2, we intend to define a new metric named "decoding gain". Decoding gain is produced by dividing the number of real packets in the network to the number of individual packets (coded or not) that hold them inside. For example, if Packets p1 and p2+p3 are sent, decoding gain is 1.5 because of transmission of 3 real packets into 2 individual ones. In the presence of noise in the network, the higher value of decoding gain will lead to faster decoding at the sink and improve the network reliability. To show (2) is correct, consider the scenario in Figure 3 in which Nodes A and B are sources, C, D, and E are relays, and S is the sink. The network is working for a while and the boxes above relays show their queue at the moment. A structured routing algorithm like CTP may choose any of the paths between sources and the sink based on its metric(s), but our structure-free routing (S-NC) only chooses one path. Table 1 shows all the possible route selection choices along with their decoding gains.

As it is shown in Table 1, the probability of selecting the path with the highest decoding gain in the structured routing scheme is 1/6 while S-NC always chooses that path. One may argue here why decoding gain is an important metric. Our answer is, in the presence of high level of noise in the system and when we have no feedback from the sink on the required packets, it is better to add more redundancy to the network. Decoding gain is a metric that shows which routes can add the highest redundancy with the lowest number of transmissions. For example, if we choose relays C and D as the next hops, then by sending only four packets in the network we are adding a redundancy equal to sending nine packets and give the sink a higher chance of decoding the packets in its queue. Of course, it is clear that this method might not lead to the optimal path from a source to the sink. In fact, this is not our goal because by considering the limitations of WSNs we just want to make optimal decisions in a hop-by-hop manner not for a total path.

3. S-NC

Structure-free routing was proposed to improve aggregation efficiency in WSNs [4]. Our approach is based on anycasting at the MAC layer to determine the next-hop node for each transmission on-the-fly. In other words, S-NC makes a dynamic forwarding decision by providing the freedom for the MAC layer to decide among a set of nodes (rather than a single next-hop). Therefore, it can achieve a higher coding

Table 1: Comparison Between Structure-based and Structure-free Routing with Respect to Decoding Gain

Routing Scheme	Selected Relays	Received packets at the sink	Decoding Gain
Structure-based (CTP)	Both E	p1, p2, p1+x	1.5
	Both D	p1, p2, p1+y+w	1.6
	Both C	p1, p2, p1+p+q+r	2
	C,D	p1, p2, p1+p+q+r, p2+y+w	2.25
	D,E	p1, p2, p1+y+w, p2+x	1.75
	C,E	p1, p2, p1+p+q+r, p2+x	2
Structure-free (S-NC)	C,D	p1, p2, p1+p+q+r, p2+y+w	2.25

opportunity by exploiting the information about the existence and number of data in neighbor nodes; also it can balance energy consumption in the network. Our dynamic routing scheme uses only the local information of a node to make the routing decisions; therefore, it is simple and scalable.

In anycasting, each node broadcasts an RTS packet and waits to receive a CTS one from one neighbor before it starts sending. These packets are employed to determine the next hop in S-NC. When a node receives the RTS packet, it starts a timer and when the timer expires, it sends a CTS packet back to the RTS sender. Actually, the node which is the most beneficial for coding obtains a higher priority and sends CTS first.

Since the selection of the next hop is done a posteriori, no topological knowledge nor routing tables are needed at each node, but the position information is enough. Nodes have to know their location and also the sink's location. Furthermore, nodes have to know their neighbors' location. Neighbors' location can be either contained in the RTS packet so the receivers can learn the sender's location, or exchanged between neighbors during network deployment [4].

Next hop relay nodes are prioritized based on queue length, residual energy, and distance to the sink. It potentially causes the receivers to randomly delay the CTS transmission to avoid CTS collision. Choosing a shorter random delay before sending the CTS is a sign of higher priority. Moreover, nodes will cancel their CTS transmission if they overhear any CTS during the prioritized delay, so multiple CTS transmissions can be prevented because of the interference between the neighbors of the sender. After this, the node will act as a source or relay and send its packets to the next hop while exploiting spatial coding.

- A node acts as a source if it receives a request packet broadcast by the sink through the network in each round. In this case, it will send one raw response packet and R-1 coded packets which are a combination of the response and other packets in its queue. R is a factor which determines the amount of redundancy we would like to add to the network. The number of packets coded together to form a single packet is limited to 6, because bigger numbers either lead to slow decoding or linearly dependence of packets.
- A node acts as a relay if it receives a packet from another neighbor and wants to forward it. If the packet is raw, the node will send it as it is, but if it is coded the node will code it again with other packets in its queue.

The neighbors are assigned three classes of priorities in responding to RTS as follows:

- Class A: The receiver is closer to the sink than the sender, its queue length is not zero, and its residual energy is bigger than half of the initial energy.
- Class B: The receiver is closer to the sink, its queue length is not zero, but the residual energy is below half of initial energy.
- Class C: The receiver is closer to the sink, but its queue length is zero.

Being closer to the sink exists in all our priorities. In this way, we avoid being trapped in loops. In addition, we consider residual energy in the neighbor selection to spread energy consumption among network nodes. Hence, the network will have a longer lifetime. To avoid a collision between these three classes of neighbors three different slots are reserved in responder nodes for sending CTS packets. This policy is implemented as the priority of Class A is over Class B, and Class B is over Class C. Moreover, to avoid a collision between nodes in the same class, they select a mini-slot for CTS transmission. As in [4], the main goal actually is to stagger the starting time of CTS transmissions based on the priorities using the slots and mini-slots. Note that the actual transmission time of the CTS could be larger than the mini-slot or slot time. However, considering the assumption of interference between neighbors, we expect only the first CTS transmission to succeed since the resulting interference will suppress the transmissions of the others. Among the forwarding nodes in the same class, a higher priority (earlier mini-slot) is given to the node which has a larger queue length in A or B, or to the nodes which are closer to the sink in C.

Structure-free routing mitigates memory usages because it obviates the need for memorizing any kind of information about the neighbors or the network. Hence, it removes all the periodic messages that were sent to update routing tables. Moreover, if there are node failures in the network, it can survive because a failed node would not respond to any RTS packets, so it cannot be chosen as the next hop. Besides, the main advantage of structure-free routing which motivates us to use it is the fact that a node can be informed of its neighbors status exactly at the sending moment, while other methods which use periodic beacons may have obsolete data at that time.

4. PERFORMANCE EVALUATION

In this section, simulation results are presented. S-NC is implemented in TinyOS2.1 [11] platform and simulated using TOSSIM [10]. In SenseCode [8], the authors have compared their proposed method with CTP [5], as the best existing collection protocol in terms of reliability and energy efficiency, to show how it achieves higher reliability (typically 15 - 20%) by means of network coding, while consuming a comparable amount of network resources. Thus, our results are only compared with SenseCode as the best, to our knowledge, network coding-based algorithm designed for WSNs. The goal of our simulation is to evaluate effectiveness of S-NC, as a structure-free coding-aware routing protocol, in terms of reliability, end-to-end delay, and energy consumption in the network with changing topology.

To simulate changing network topology, we used the same method as SenseCode. A two-state Markov Chain decides whether a node is connected to the network or not. The model uses two parameters: Mean Time Between Failures (MTBF) which indicates the average time a node is in a connected state, and Mean Time To Repair (MTTR) which indicates the average length of time a node is in a disconnected state. Every point in the figures in this section is an average over a large number of experiments (ranging from 10 to 50). The simulation scenarios run on a 5×5 grid network and end as soon as the sink receives a total of 3000 packets. SenseCode executed with redundancy (R) of 2 and sampling period (round) of 60 seconds. The sampling period is the period in which the sink broadcasts the request message.

Figure 4 depicts reliability as a function of MTBF. As we mentioned before, to compute the reliability, the number of received and decoded packets at the sink is divided by the total number of packets sent from the sources. Thus, a higher reliability means a higher amount of decoding happening at the sink. In this experiment, we set the environment noise so that it does not make any packet loss, MTTR is set to 10 seconds, and we vary MTBF to evaluate the effect of node failures in WSNs. The less the MTBF value, the more nodes go to the disconnected state. We observe that S-NC shows an average of 30% improvement in terms of reliability compared to SenseCode. The first reason is that it detects better routes exactly at the moment of transmission using RTS/CTS packets. If a node has gone to the disconnected state, then it has no chance of being chosen as the next hop for any transmission. Therefore, S-NC selects alive neighbors automatically while SenseCode or other structure-based routing algorithms insist on packet retransmission to a pre-chosen next hop until they make sure that the node has failed. The second reason is the decoding gain defined in Section 2. S-NC selects routes with a higher decoding gain (shown in Figure 7) and this results in more decoding at the sink. As it can be seen, the difference between algorithms is more evident when the node failure occurs more frequently.

In Figure 5, the average end-to-end delay tolerated by all the packets in the network is plotted as a function of MTBF. End-to-end delay is defined as the duration between the transmission of a packet at the source and its reception at the sink. As expected, when MTBF increases, more nodes are alive in the network. This increases the number of routes to the sink, so all packets can attend to it more quickly. Our simulation demonstrates that S-NC tolerates an average of 40% less end-to-end delay than SenseCode. The reason would be the fact that S-NC adapts more quickly to the disconnection of nodes and does not insist on sending packets to pre-chosen neighbors which might be in a disconnected state. Moreover, because of higher decoding gains in S-NC, the sink can decode packets quite quickly while in SenseCode, the sink experiences more delay before it gathers all the required packets to decode a coded packet. The results shown in Figure 5 would be more interesting if we consider energy consumption, too. Figure 6 illustrates the average energy consumption over all network nodes. In wireless motes, transmitting and receiving are the main sources of energy consumption. Therefore, to measure energy consumption, the length of time the motes' radio was ON for sending/receiving is computed and multiplied by the power consumed by MicaZ motes at TX/RX mode [1]. It can be

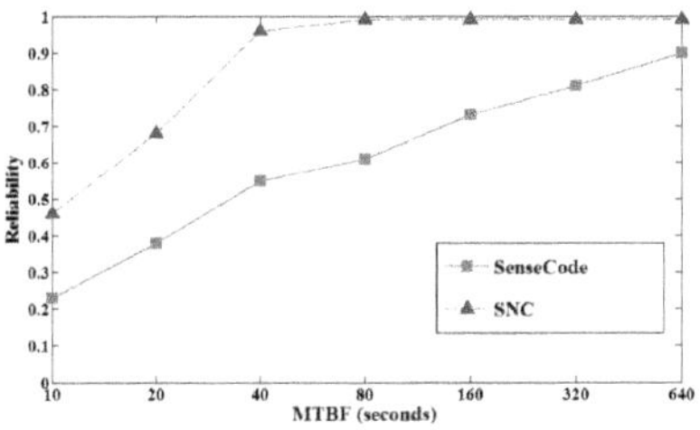

Figure 4: Reliability vs. MTBF

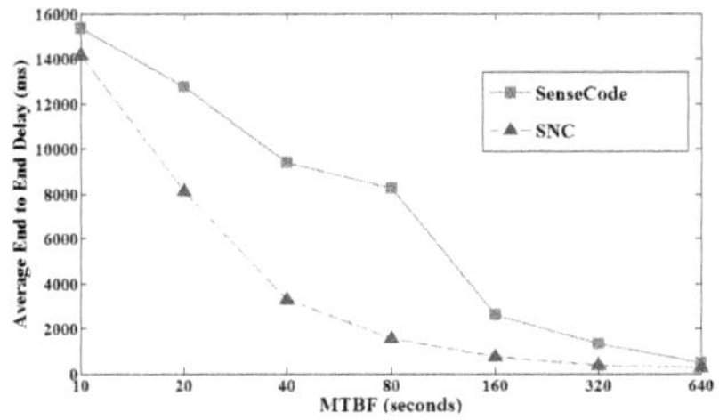

Figure 5: End-to-end delay vs. MTBF

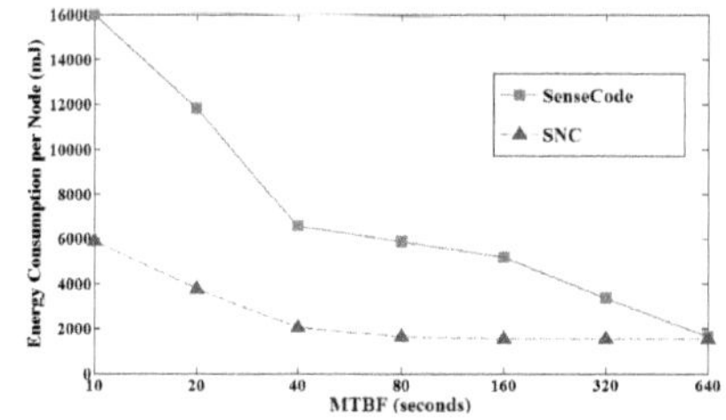

Figure 6: Energy consumption vs. MTBF

observed that S-NC consumes an average of 40% less energy than SenseCode. The difference is more evident at higher frequency of node failures. The reason is that S-NC does its best to spread transmission among different network nodes. So, the average amount of energy consumption among all nodes will become less. In addition, higher reliability of S-NC means that fewer packets are sent to the network, which leads to lower energy consumption.

In Figure 7 decoding gain is plotted as a function of MTBF. As it is shown S-NC always has better coding gain than SenseCode, so it can do better decoding at the sink.

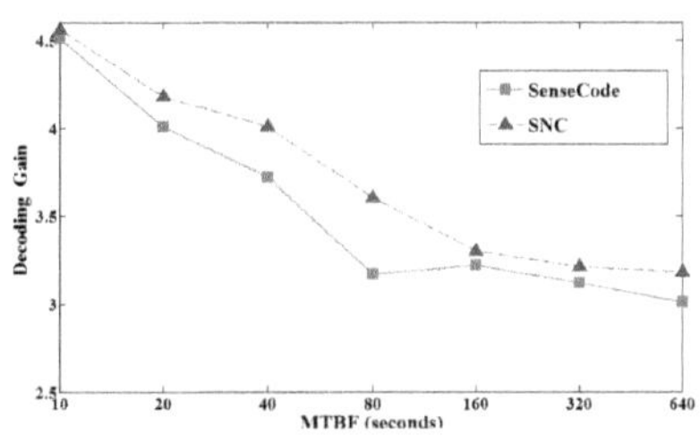

Figure 7: Decoding gain vs. MTBF

5. CONCLUSION

In this paper, we explored the combination of structure-free routing and network coding. This work attempted to design a coding-aware routing protocol that fits in with all WSN limitations like energy and memory as well as the many-to-one traffic pattern. The simulation results demonstrated that the spatial coding on a structure-free routing algorithm has better reliability than a structured routing algorithm, especially in the presence of high level of node failure. Moreover, S-NC can balance energy consumption all over the network by spreading the traffic while it does not impose any extra end-to-end delay to the packets during the transmission.

6. ACKNOWLEDGEMENT

The authors would like to thank Lorenzo Keller and Christina Fragouli who provided them with the implementation of SenseCode generously and took the time to answer their questions kindly.

7. REFERENCES

[1] TI CC2420 RF Transciever Datasheet, http://www.ti.com/lit/gpn/cc2420.

[2] O. Al-Kofahi and A. E. Kamal. Scalable redundancy for sensors-to-sink communication. In *Proceedings of IEEE GLOBECOM'08*, pages 459–464, 2008.

[3] O. Al-Kofahi and A. E. Kamal. Network coding-based protection of many-to-one wireless flows. *IEEE Journal on Selected Areas in Communications*, 27:797–813, 2009.

[4] K. Fan, S. Liu, and P. Sinha. Structure-free data aggregation in sensor networks. *IEEE Transaction on Mobile Computing*, 6:929–942, 2007.

[5] O. Gnawali, R. Fonseca, K. Jamieson, D. Moss, and P. Levis. Collection tree protocol. In *Proceedinsgs of ACM SenSys'09*, pages 1–14, 2009.

[6] M. Jhang, S. Lin, and W. Liao. C2AR: Coding and capacity aware routing for wireless ad hoc networks. In *Proceedings of IEEE ICC'10*, pages 1–5, 2010.

[7] S. Katti, H. Rahul, W. Hu, D. Katabi, M. Medard, and J. Crowcroft. XORs in the air: Practical wireless network coding. In *Proceedings of ACM SIGCOMM'06*, pages 243–254, 2006.

[8] L. Keller, E. Atsan, K. Argyraki, and C. Fragouli. SenseCode: Network coding for reliable sensor networks. In *EPFL Technical Report*, July 2010.

[9] J. Le, J. C. S. Lui, and D. Chiu. DCAR: Distributed coding-aware routing in wireless networks. *IEEE Transaction on Mobile Computing*, 9:596–608, 2010.

[10] P. Levis, N. Lee, M. Welsh, and D. E. Culler. TOSSIM: Accurate and scalable simulation of entire TinyOS applications. In *Proceedings of ACM SenSys'03*, pages 126–137, 2003.

[11] P. Levis, S. Madden, D. Gay, J. Polastre, R. Szewczyk, A. Woo, E. A. Brewer, and D. E. Culler. The emergence of networking abstractions and techniques in TinyOS. In *Proceedings of USENIX NSDI'04*, pages 1–14, 2004.

[12] S. Sengupta, S. Rayanchu, and S. Banerjee. An analysis of wireless network coding for unicast sessions: The case for coding-aware routing. In *Proceedings of IEEE INFOCOM'07*, pages 1028–1036, 2007.

[13] Y. Yan, B. Zhang, J. Zheng, and J. Ma. CORE: A coding-aware opportunistic routing mechanism for wireless mesh networks. *IEEE Wireless Communications Magazine*, 17:96–103, 2010.

Distributed Opportunistic and Diffused Coding with Multiple Decoders in Wireless Mesh Networks

Thuong Van Vu
University of Pierre and Marie Curie
Laboratory of Computer Science of Paris 6
4 place Jussieu, 75005 Paris, France
thuong-van.vu@lip6.fr

Thi Mai Trang Nguyen
University of Pierre and Marie Curie
Laboratory of Computer Science of Paris 6
4 place Jussieu, 75005 Paris, France
thi-mai-trang.nguyen@lip6.fr

Guy Pujolle
University of Pierre and Marie Curie
Laboratory of Computer Science of Paris 6
4 place Jussieu, 75005 Paris, France
guy.pujolle@lip6.fr

ABSTRACT

In wireless mesh networks for unicast traffic, opportunistic network coding are specifically introduced for improved network utilization. Some recent practical implementations like COPE, BEND, DCAR and DODE have shown the promising results over the original 802.11 conventional forwarding mechanism. For better performance, the aim of all mentioned network coding implementations is trying to find more and more coding chances in network topologies. However, they restricted the finding within a simple rule "a pair of coder and decoder". For every coded packet (the combination of natives packets of traffic flows) sent on the traffic flow, we always have one coder (which creates the coded packet) and one decoder (which retrieves the desired packet for the destination). The trivial rule limited coding chances much. In this paper, we are loosing this noose: we decouple the coding and decoding functions from strictly a pair of coder and decoder. For one coder, we can have multiple decoders on the path (up to the number of other traffic flows involved in the coding). With this, more coding chances are found, thus, improving the network performance. We extend our proposed DODE with this new idea, called *Distributed Opportunistic and Diffused Coding with Multiple Decoders* - DODEX. We implement DODEX system in NS-2. The simulation results show that DODEX can outperform its previous introducing systems.

Categories and Subject Descriptors

C.2.1 [**Network Architecture and Design**]: Wireless communication

Keywords

network coding; opportunistic coding; forwarding

1. INTRODUCTION

Since its first introduction in [1], network coding has gained a significant attention from the research communities in the need of improving the way of communication in computer networks. Network coding is a technique which allows the nodes to combine several *native* packets into one *coded* packet for transmission (i.e., coding packets) instead of simply forwarding packets one by one. The network then can save the number of transmissions to reduce the data transfer time and increase the throughput. Based on that, *opportunistic network coding* (ONC) is firstly introduced in COPE [3], the practical network coding system for the unicast traffic. Afterwards, more extended works are presented like BEND [4], DCAR [5] and DODE [6]. Providing a particular coding condition, all these network coding systems can ensure after the coded packets are opportunistically created and broadcasted in the air by a coder (i.e., node intersecting many flows), there will be **one** decoder (i.e., node retrieving the desired packet for the destination) on the traffic flow decoding the coded packet successfully. The trivial rule limited coding chances much. Our previous work DODE has combined the features from COPE (the opportunistic listening), BEND (the diffused gain), and DCAR (the generalized coding condition) to create a better network coding architecture. DODE also exploited the routing metric SPENM (*Shortest Paths of Enriched Neighbors Metric*) which helps to choose the path with the most coding chances. DODE can indeed grasp more coding chance and the obtained throughput is significantly improved.

In this paper, we extend DODE to resolve the restricting rule: we decouple the coding and decoding functions from strictly a pair of a coder and a decoder. After one coder creating the coded packets, multiple decoders can be the next on the transmission path to retrieve the desired packet for the destination. More coding chances can be found, leading to an improved network performance. Our contribution are two-fold:

- We redesign the coding condition in DODEX for each coded packet at the coder, which allows multiple decoders to be on the traffic flow instead of only one in order to find more coding chance and improve network utilization.
- We implement the DODEX system in NS-2 and compare the result with other implementation 802.11, COPE, BEND, DCAR and DODE. The simulation results show that DODEX can outperform these systems.

The remainder of this paper is organized as follows. In sec-

MSWiM'12, October 21–25, 2012, Paphos, Cyprus.

tion 2, we describe the related work which inspires our proposal. Section 3 details the design of DODEX. Simulation results are presented and analysed in section 4. Finally, we give the conclusion in section 5.

2. RELATED WORK

2.1 Previous Opportunistic Coding systems

By *opportunistically listening* to other transmission, wireless nodes in COPE [3] can use overheard packets to perform ONC. In Fig 1(b), node $0, 2, 3, 4$ wish to send a packet p_0, p_2, p_3, p_4 to its opposite node $2, 0, 4, 3$ respectively, via the forwarder 1. The dashed lines imply related nodes (e.g., 0 and 1) are 1-hop neighbors to each other and the arrows indicate the traffic flows. Node 1 codes four native packets to broadcast ($p_0 \oplus p_2 \oplus p_3 \oplus p_4$) and receivers can extract the desired packets without failure. For examples, node 0 can get p_2 successfully because p_3, p_4 are overheard by transmissions of $(3 \to 1), (4 \to 1))$, and p_0 is its own packet. To perform opportunistic coding, COPE designs its own particular coding condition. Each pair of n native packets will be checked with the coding condition if they can be coded together before nodes can process n-packet coding (the combination of n native packets). By this way, the coders ensure their coded packets are received and decoded successfully. the authors of COPE designed the *two next-hop coding pattern*: two flows intersecting at a node c are codable if the next hop of c on a flow is the previous forwarder of c on the other flow (Fig 1(a)) **or** its neighbor (Fig 1(b)), and vice versa.

BEND argued a problem presented by the nature of ONC: concentrating traffic via a node for coding can lead to packet collisions and drops. BEND suggested the diffused gain, the neighbors of the node intersecting flows (like 3, 4 of coder *I* for flows $(0 \to 1 \to 2)$ and $(2 \to 1 \to 0)$ in Fig 1(b)) can actively overhear the transmission and broadcast the coded packets. Instead of only one coder, we can have advantages of *multiple coders* sharing the coding process. From that point of view, BEND keeps the two next-hop coding pattern from COPE, but it increases the number of coders on the flows, which alleviates the packet collisions and drops at the coders, and improves the network performance. DCAR takes another approach for network enhancement: DCAR breaks the two next hop coding pattern provided by COPE. The authors of DCAR creates the general coding condition which allows the coder and decoders not to necessarily be neighbors but any nodes which can intersect flows (coders) and overhear the transmission properly (decoders), respectively. For examples, in Fig 1(c), we can see 3 as coder and 4, 7 as decoders while COPE, and BEND cannot detect that.

3. DISTRIBUTED OPPORTUNISTIC AND DIFFUSED CODING WITH MULTIPLE DECODERS - DODEX

3.1 DODE

In this part, we introduce DODE, which our proposition in this paper is heavily based on. DODE is a network coding architecture that works at MAC layer but also uses information from upper layers (the packet queues and the routing protocol) to do the coding, decoding processes. In summary, DODE contains several features: the opportunistic listening and encoding from COPE, the diffused gain by neighborhood from BEND and the generalized coding condition from DCAR. The coding condition of DODE can be seen as a combination of two features: the diffused gain from BEND and the general coding condition from DCAR. At first, DODE can increase the coding chances by adding more coders (via the diffused gain). Next, the general coding condition is applied, helping DODE to discover the coding chances over the whole traffic paths, better than the two next hop coding pattern from BEND or COPE. Moreover, as already explained in *Related work*, the diffused gain can alleviate the concentrating traffic problem as DCAR inherits it from COPE. Particularly, DODE gains a better performance by taking the advantages form the previous and resolving their problems. Besides, DODE also provide a new routing metric called *SPENM* (*Shortest Path with Enriched Neighbors Metric*) over the routing protocol DSDV [2] to cope with the new coding condition, helping to find routing paths with enhanced coding chances. Due to the lack of space, more technical explanation can be found in our previous work [6].

3.2 Coding chance improvement

As mentioned in *Related work*, each opportunistic coding architecture designs its own coding condition to gain more chances, thus, improving the network performance. In DODEX, coders can combine many native packets for transmission and multiple decoders next on the flow will recover the desired packet. Consider Fig 3(a), we can see three flows $F_1(0 \to 3 \to 4 \to 5)$, $F_2(2 \to 3 \to 6)$ and $F_3(8 \to 7 \to 3 \to 1)$ can be coded together but on the flow F_1 there are two decoders 4 (for leaving out packet p_2) and 5 (removing p_3 to get the desired p_1). With previous implementation, only 2-packet coding can be found instead of 3-packet coding like our proposition DODEX. The coded packets are not maintained *"untouched"* from the coder to the destination but eventually *"peeled out"* the unnecessary packets until only the intended packet can reach to destination. Therefore, the coding condition can be rewritten as followings:

Generalized coding conditions in DODEX
for each pair of n flows F_i and F_j intersecting at node c $c \in ((F_i \cap F_j) \cup (N(nexthop(c, F_i)) \cap N(nexthop(c, F_j)) \cap N(prevhop(c, F_i)) \cap N(prevhop(c, F_j))))$
1. There exists a node $d_{c,i,j} \in (D(c, F_i) \cap (U(c, F_j) \cup N(U(c, F_j)))$.
2. There exists a node $d_{c,j,i} \in (D(c, F_j) \cap (U(c, F_i) \cup N(U(c, F_i)))$.

where:

- F_i denotes the ith flow in the network, $i \in \{1..n\}$
- $nexthop(c, F)/prevhop(c, F)$ denotes the nexthop/previous hop of node c on flow F
- $D(c, F)/U(c, F)$ denotes the downstream/upstream nodes of flow F.
- $N(s_i)$ denotes the set of all neighbors nodes of node s_i
- $d_{c,i,j}$ denotes the decoder of c on flow F_i to remove the packet of flow F_j

3.3 Detailed design

To perform the coding condition above, DODEX has to collect the information *"who sends what"* on the transmission path of each flow and detect the coding chances based on that. Moreover, DODEX needs to store, forward native packets, create and forward coded packets when coding

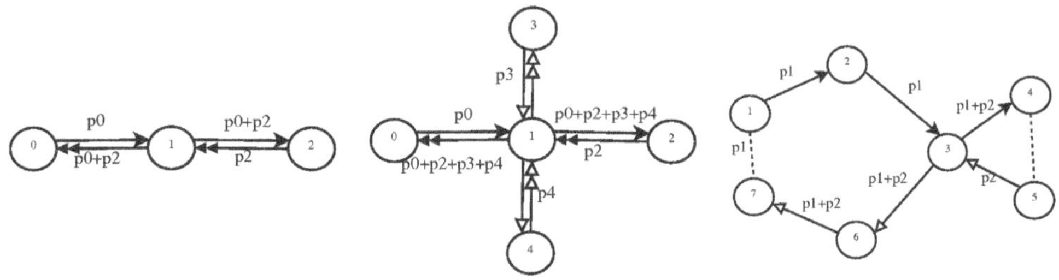

(a) (b) (c)

Figure 1: Opportunistic network coding scenarios

chances exist. Particularly, Each nodes also change their behaviors to also intercept the coded traffic. and have to collect the neighbor list and the source routing list to check the coding condition above, the queuing system to manipulate packets.

3.3.1 Neighbor list and source routing list

The routing protocol DSDV is modified to cope with carrying more information to detect coding chances. Besides original information of DSDV, each route entry in the update message will also carry the list of nodes constituting the route and the neighbor list of each node along the route. The neighbor list is obtained by collecting the senders' addresses of DSDV messages. Before broadcasting out the DSDV messages, the sender adds its address to the current list of routing path in the entry. After receiving the routing updates from neighbors, each node will process the routing update messages as the original DSDV routing protocol and also cache the information about the neighbors and detailed routing path. With this, all nodes in the network gradually acquire enough information to perform the correct coding.

3.3.2 Queuing system

DODEX uses four different packet queues: Q_{native} for the native packets, Q_ovrhrd for storing the overheard packets, $Q_{codable}$ for the linked lists of packet that can be combined together to create coded packets, and Q_{coded} for coded packets that need forwarding. Figure 2 shows the illustration of queues.

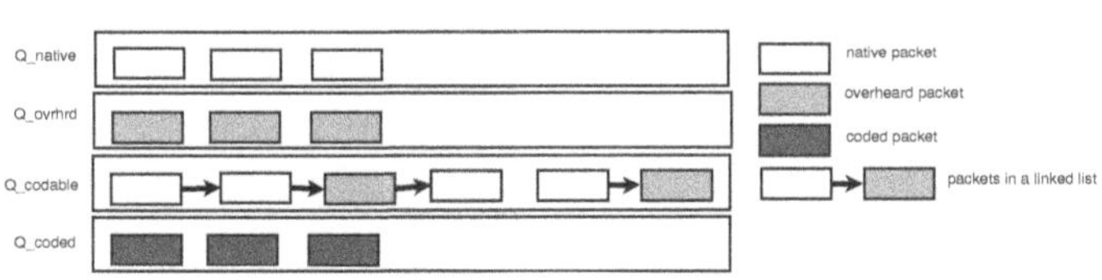

Figure 2: Queuing system

3.3.3 Node behavior

Each node in DODEX has to perform native packet interception, coded packet interception and packet transmission which is described in Section A.

4. SIMULATION AND RESULTS

We use NS-2 as the simulator to compare the performances of DODEX with the previous architectures: IEEE 802.11, COPE, BEND, DCAR and the non-extended DODE. We use two topologies as illustrated in Fig 3(a) and 3(b). The first topology (Fig 3(a)) is provided for test scenario of maximum 3-packet coding with light traffic. The second topology (Fig 3(b)) is used for test scenario of maximum 4-packet coding with stress traffic causing high packet collisions and drops.

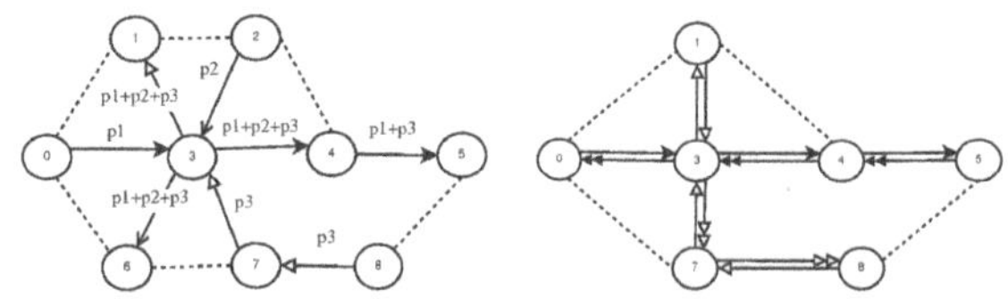

(a) Test topology 1 for 3-packet coding (b) Test topology 2 for 4-packet coding

Figure 3: Test scenarios for exposing 3 and 4-packet coding in DODEX

Each topology is created in a flat area of $1000m \times 1000m$. The data traffic in the network are all CBR (Constant Bit Rate) flows sent over UDP (User Datagram Protocol) using 1000-byte datagrams with an arrival interval of 0.01s and traffic generation duration at source of 150s. The performances is then evaluated by two performance metrics, the throughput and the number of coded packets. We vary the traffic flows in test scenarios as shown in table *Flows in Test scenarios*. Afterwards, the test scenarios will be executed with each traffic-flow variety for all implementations and the result is collected with a 95% confidence interval.

Table 1: Flows in Test scenarios

3 flows varied in test scenario 1	$F_1(0 \to 3 \to 4 \to 5)$, $F_2(2 \to 3 \to 6)$, $F_3(8 \to 7 \to 3 \to 1)$
6 flows varied in test scenario 1	$F_1(0 \to 3 \to 4 \to 5)$, $F_2(2 \to 3 \to 6)$, $F_3(8 \to 7 \to 3 \to 1)$, $F_4(5 \to 4 \to 3 \to 0)$, $F_5(6 \to 3 \to 2)$, $F_6(1 \to 3 \to 7 \to 8)$
7 flows varied in test scenario 2	$F_1(6 \to 2 \to 3 \to 1)$, $F_2(1 \to 3 \to 2 \to 6)$, $F_3(0 \to 3 \to 4 \to 5)$, $F_4(5 \to 4 \to 3 \to 0)$, $F_5(5 \to 6)$, $F_6(1 \to 4)$, $F_7(0 \to 2)$
10 flows varied in test scenario 2	$F_1(6 \to 2 \to 3 \to 1)$, $F_2(1 \to 3 \to 2 \to 6)$, $F_3(0 \to 3 \to 4 \to 5)$, $F_4(5 \to 4 \to 3 \to 0)$, $F_5(5 \to 6)$, $F_6(1 \to 4)$, $F_7(0 \to 2)$, $F_8(5 \to 6 \to 2)$, $F_9(2 \to 6 \to 5)$, $F_{10}(0 \to 2 \to 6)$

As shown in Fig 4(a) and 4(b), with the light traffic, DODEX outperforms previous implementations significantly. In the case of 3 flows, only DODEX discovered the 3-packet coding, and the throughtput gain over the previous is quite promising (32% over our old DODE and 30%-45% over the others). First, COPE and BEND only detects the 2-packet coding between F_2 and F_1. However, coder 3 also serves as a forwarder for F_3, increasing packet collisions and drops at node 3 due to concentrating traffic via the coder. Second, DODE and DCAR are also able to detect the 2 packet-coding between F_1 and F_3 thanks to the general coding condition but the same problem happened too because the dif-

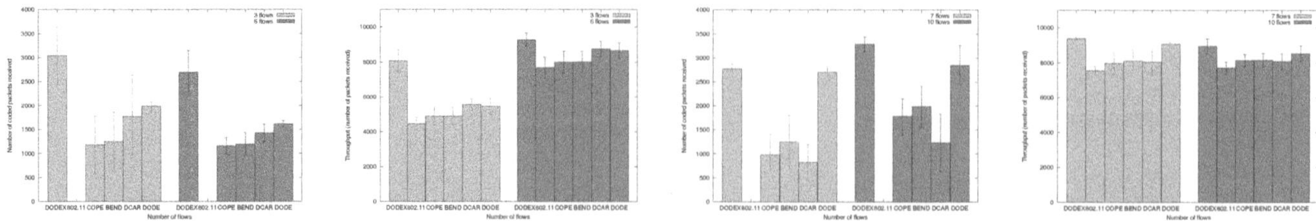

(a) Coded packets result for Test scenario 1 (b) Throughput result for Test scenario 1 (c) Coded packets result for Test scenario 2 (d) Throughput result for Test scenario 2

Figure 4: Test scenario results

fused gain does not help BEND, DODE or even DODEX much (only 3 as the coder). With 3-packet coding, DODEX allows more packets delivered to destination, result in higher throughput. For the next case (6 flows), because there are more coding chances (F_4, F_5 and F_6 are in reverse directions with F_1, F_2 and F_3, respectively), the previous works can compensate the throughput gain and the number of coded packets by creating the coded packets from these flows with opposite ways. Nevertheless, DODEX still maintains higher throughput because DODEX can detect not only all coding chances like the previous but also the 3-packet coding with multiple decoders.

In the second topology exposing the coding 3 or 4 packets in a single transmission, the performances of all previous and DODEX are presented in Figs 4(c) and 4(d). We would like to check if under heavy traffic, DODEX still keeps the high throughput and coded packets. By applying 4-packet coding with multiple decoders, DODEX transfers more data even in the interference of the non-codable flows. Compared to DODEX, in case of 10 flows (4-6 of them are non-codable), all previous implementations are losing throughput because it takes more transmissions for 2-packet coding. Due to high interference from the non-codable flows, the throughputs gained from all architectures are reduced, both from competition of accessing MAC layer to send coded packets or the packet collisions and drops. DODEX can alleviate the problem by discovering more coding chances via coding with multiple decoders, draining the packets from the queue at the forwarder faster than the others (4-packet coding compared to 2-packet coding), thus, giving DODEX a better performance over the formers (6% over DODE, and 10%-20% over the others as Figs 4(c) and 4(d) shows)

5. CONCLUSION

In this paper, we have proposed a nouvel and practical network coding architecture for unicast traffic in wireless mesh network. We extend the result gained from DODE to explore a new way of intercepting the coding chance in opportunistic network coding, which helps to improve the network performance. We allow many decoders on the flow to share the decoding process instead of only one like the previous systems. More coding chances are discovered, and help to achieve the better network utilization.

6. ACKNOWLEDGMENTS

We appreciate the support and suggestion from Professor Nadia Boukhatem from Computer Science & Network Department, TELECOM ParisTech to accomplish this work.

7. REFERENCES

[1] R. Ahlswede, N. Cai, S.-Y. R. Li, and R. W. Yeung, "Network Information Flow", IEEE Transactions on Information Theory, IT-46, pp. 1204-1216, 2000.

[2] C. E. Perkins, P. Bhagwat, "Highly Dynamic Destination-Sequenced Distance-Vector Routing (DSDV) for Mobile Computers", SIGCOMM 94 -8/94 London England UK, 1994.

[3] S. Katti, H. Rahul, W. H. D. Katabi, M. Medard, J. Crowcroft, "XORs in The Air: Practical Wireless Network Coding", SIGCOMM 06, September 11âĂŞ15, 2006, Pisa, Italy.

[4] J. Zhang, Y. P. Chen, I. Marsic, "MAC-layer proactive mixing for network coding in multi-hop wireless networks", Computer Networks 54 (2010) 196âĂŞ207.

[5] J. Le, J. C.S. Lui, D. M. Chiu, "DCAR: Distributed Coding-Aware Routing in Wireless Networks", IEEE Transaction on Mobile Computing, Vol. 9, No. 4, April 2010.

[6] T. V. Vu, T. M. T. Nguyen, G. Pujolle, "Distributed Opportunistic and Diffused Coding in Multi-hop Wireless Networks", IEEE ICC 2012 Coconet4, June, 2012.

APPENDIX

A. NODE BEHAVIOR

Algorithm 1 Native packet interception

$p_i \leftarrow$ native packet
if $\exists$ linked list $L \subset Q_{codable}$ / p_i is codable with $\forall p_j, p_j \in L$ **then**
  Add p_i to the list
else if there exists $\exists$ $p_j \in Q_{native}$ / p_i is codable with p_j **then**
  Remove p_i, p_j from Q_{native}, create the linked list with p_i, p_j and add the list to $Q_{codable}$
else if there exists $\exists$ $p_j \in Q_{ovrhrd}$ / p_i is codable with p_j **then**
  Remove p_i, p_j from Q_{ovhrd}, create the linked list with p_i, p_j and add the list to $Q_{codable}$
else if p_i is the intended packet **then**
  Add p_i to Q_{native}
else
  Add p_i to Q_{ovhrd}
end if

Algorithm 2 Coded packet interception

$p_i \leftarrow$ coded packet
if p'_i is an intended packet for the node **then**
  Remove as many as possible the non-intended native packets in p'_i
  if only one desired packet remained **then**
    Process *Algorithm 1 Native packet interception*
  else if there exists coded packet $p"_i$ **and** there still exists decoders on path to destination **then**
    Add $p"_i$ to Q_{coded}
  end if
end if

Algorithm 3 Packet transmission

Access the medium
if $Q_{codable} \neq \emptyset$ **then**
  Remove the linked list from $Q_{codable}$
  Create the coded packet from the first linked list
  Send the packet
else if $Q_{coded} \neq \emptyset$ **then**
  Remove the first coded packet from Q_{coded}
  Send the packet
else if $Q_{native} \neq \emptyset$ **then**
  Remove the first native packet from Q_{native}
  Send the packet
end if

Author Index

www.ingramcontent.com/pod-product-compliance
Ingram Content Group UK Ltd.
Pitfield, Milton Keynes, MK11 3LW, UK
UKHW051511290926
15471UKWH00008B/1550

9 781450 316286